STUDENT'S SOLUTIONS MANUAL

DANIEL S. MILLER

Niagara County Community College

PRECALCULUS

FOURTH EDITION

Robert Blitzer

Miami Dade College

Prentice Hall
is an imprint of

The author and publisher of this book have used their best efforts in preparing this book. These efforts include the development, research, and testing of the theories and programs to determine their effectiveness. The author and publisher make no warranty of any kind, expressed or implied, with regard to these programs or the documentation contained in this book. The author and publisher shall not be liable in any event for incidental or consequential damages in connection with, or arising out of, the furnishing, performance, or use of these programs.

Reproduced by Pearson Prentice Hall from electronic files supplied by the author.

Copyright © 2010 Pearson Education, Inc.
Publishing as Pearson Prentice Hall, Upper Saddle River, NJ 07458.

ISBN-13: 978-0-321-57532-6
ISBN-10: 0-321-57532-6

1 2 3 4 5 6 BRR 12 11 10 09

Prentice Hall
is an imprint of

www.pearsonhighered.com

TABLE OF CONTENTS for STUDENT SOLUTIONS

PRECALCULUS 4E

Chapter P	Prerequisites: Fundamental Concepts of Algebra	1
Chapter 1	Functions and Graphs	83
Chapter 2	Polynomial and Rational Functions	165
Chapter 3	Exponential and Logarithmic Functions	267
Chapter 4	Trigonometric Functions	320
Chapter 5	Analytic Trigonometry	436
Chapter 6	Additional Topics in Trigonometry	510
Chapter 7	Systems of Equations and Inequalities	620
Chapter 8	Matrices and Determinants	709
Chapter 9	Conic Sections and Analytic Geometry	782
Chapter 10	Sequences, Induction, and Probability	860
Chapter 11	Introduction to Calculus	925

Chapter P
Prerequisites: Fundamental Concepts of Algebra

Section P.1

Check Point Exercises

1.
$$8+6(x-3)^2 = 8+6(13-3)^2$$
$$= 8+6(10)^2$$
$$= 8+6(100)$$
$$= 8+600$$
$$= 608$$

2. Since 2010 is 10 years after 2000, substitute 10 for x.
$$T = 17x^2 + 261x + 3257$$
$$= 17(10)^2 + 261(10) + 3257$$
$$= 7567$$
If trends continue, the tuition and fees will be $7567

3. The elements common to {3, 4, 5, 6, 7} and {3, 7, 8, 9} are 3 and 7.
$$\{3,4,5,6,7\} \cap \{3,7,8,9\} = \{3,7\}$$

4. The union is the set containing all the elements of either set.
$$\{3,4,5,6,7\} \cup \{3,7,8,9\} = \{3,4,5,6,7,8,9\}$$

5. $\left\{-9,\ -1.3,\ 0,\ 0.\overline{3},\ \dfrac{\pi}{2},\ \sqrt{9},\ \sqrt{10}\right\}$

   a. Natural numbers: $\sqrt{9}$ because $\sqrt{9}=3$

   b. Whole numbers: $0,\ \sqrt{9}$

   c. Integers: $-9,\ 0,\ \sqrt{9}$

   d. Rational numbers: $-9,\ -1.3,\ 0,\ 0.\overline{3},\ \sqrt{9}$

   e. Irrational numbers: $\dfrac{\pi}{2},\ \sqrt{10}$

   f. Real numbers:
$$\left\{-9,\ -1.3,\ 0,\ 0.\overline{3},\ \frac{\pi}{2},\ \sqrt{9},\ \sqrt{10}\right\}$$

6. **a.** $\left|1-\sqrt{2}\right|$

Because $\sqrt{2} \approx 1.4,$ the number inside the absolute value bars is negative. The absolute value of x when $x < 0$ is $-x.$ Thus,
$$\left|1-\sqrt{2}\right| = -\left(1-\sqrt{2}\right) = \sqrt{2}-1$$

   b. $\left|\pi-3\right|$

Because $\pi \approx 3.14,$ the number inside the absolute value bars is positive. The absolute value of a positive number is the number itself. Thus,
$$\left|\pi-3\right| = \pi-3.$$

   c. $\dfrac{|x|}{x}$

Because $x > 0,\ \ |x| = x.$

Thus, $\dfrac{|x|}{x} = \dfrac{x}{x} = 1$

7. $\left|-4-(5)\right| = \left|-9\right| = 9$
The distance between -4 and 5 is 9.

8.
$$7(4x^2+3x)+2(5x^2+x)$$
$$= 7(4x^2+3x)+2(5x^2+x)$$
$$= 28x^2+21x+10x^2+2x$$
$$= 38x^2+23x$$

9.
$$6+4[7-(x-2)]$$
$$= 6+4[7-x+2)]$$
$$= 6+4[9-x]$$
$$= 6+36-4x$$
$$= 42-4x$$

Exercise Set P.1

1. $7+5(10) = 7+50 = 57$

3. $6(3)-8 = 18-8 = 10$

5. $8^2+3(8) = 64+24 = 88$

7. $7^2 - 6(7) + 3 = 49 - 42 + 3 = 7 + 3 = 10$

9. $4 + 5(9-7)^3 = 4 + 5(2)^3$
$\qquad = 4 + 5(8) = 4 + 40 = 44$

11. $8^2 - 3(8-2) = 64 - 3(6)$
$\qquad = 64 - 18 = 46$

13. $\dfrac{5(x+2)}{2x-14} = \dfrac{5(10+2)}{2(10)-14}$
$\qquad = \dfrac{5(12)}{6}$
$\qquad = 5 \cdot 2$
$\qquad = 10$

15. $\dfrac{2x+3y}{x+1}; x = -2, y = 4$
$\qquad = \dfrac{2(-2)+3(4)}{-2+1} = \dfrac{-4+12}{-1} = \dfrac{8}{-1} = -8$

17. $C = \dfrac{5}{9}(50-32) = \dfrac{5}{9}(18) = 10$
$\qquad 10^{\circ}$C is equivalent to 50°F.

19. $h = 4 + 60t - 16t^2 = 4 + 60(2) - 16(2)^2$
$\qquad = 4 + 120 - 16(4) = 4 + 120 - 64$
$\qquad = 124 - 64 = 60$
Two seconds after it is kicked, the ball's height is 60 feet.

21. $\{1,2,3,4\} \cap \{2,4,5\} = \{2,4\}$

23. $\{s,e,t\} \cap \{t,e,s\} = \{s,e,t\}$

25. $\{1,3,5,7\} \cap \{2,4,6,8,10\} = \{\ \}$
The empty set is also denoted by $\varnothing$.

27. $\{a,b,c,d\} \cap \varnothing = \varnothing$

29. $\{1,2,3,4\} \cup \{2,4,5\} = \{1,2,3,4,5\}$

31. $\{1,3,5,7\} \cup \{2,4,6,8,10\}$
$\qquad = \{1,2,3,4,5,6,7,8,10\}$

33. $\{a,e,i,o,u\} \cup \varnothing = \{a,e,i,o,u\}$

35. **a.** $\sqrt{100}$

b. $0, \sqrt{100}$

c. $-9, 0, \sqrt{100}$

d. $-9, -\dfrac{4}{5}, 0, 0.25, 9.2, \sqrt{100}$

e. $\sqrt{3}$

f. $-9, -\dfrac{4}{5}, 0, 0.25, \sqrt{3}, 9.2, \sqrt{100}$

37. **a.** $\sqrt{64}$

b. $0, \sqrt{64}$

c. $-11, 0, \sqrt{64}$

d. $-11, -\dfrac{5}{6}, 0, 0.75, \sqrt{64}$

e. $\sqrt{5}, \pi$

f. $-11, -\dfrac{5}{6}, 0, 0.75, \sqrt{5}, \pi, \sqrt{64}$

39. 0

41. Answers may vary. An example is 2.

43. true; -13 is to the left of -2 on the number line.

45. true; 4 is to the right of -7 on the number line.

47. true; $-\pi = -\pi$

49. true; 0 is to the right of -6 on the number line.

51. $|300| = 300$

53. $|12 - \pi| = 12 - \pi$

55. $\left|\sqrt{2} - 5\right| = 5 - \sqrt{2}$

57. $\dfrac{-3}{|-3|} = \dfrac{-3}{3} = -1$

59. $\left||-3| - |-7|\right| = |3 - 7| = |-4| = 4$

61. $|x + y| = |2 + (-5)| = |-3| = 3$

63. $|x| + |y| = |2| + |-5| = 2 + 5 = 7$

65. $\dfrac{y}{|y|} = \dfrac{-5}{|-5|} = \dfrac{-5}{5} = -1$

67. The distance is $|2-17| = |-15| = 15$.

69. The distance is $|-2-5| = |-7| = 7$.

71. The distance is $|-19-(-4)| = |-19+4| = |-15| = 15$.

73. The distance is
$|-3.6-(-1.4)| = |-3.6+1.4| = |-2.2| = 2.2$.

75. $6+(-4) = (-4)+6$;
commutative property of addition

77. $6+(2+7) = (6+2)+7$;
associative property of addition

79. $(2+3)+(4+5) = (4+5)+(2+3)$;
commutative property of addition

81. $2(-8+6) = -16+12$;
distributive property of multiplication over addition

83. $\dfrac{1}{x+3}(x+3) = 1; x \neq -3$,

inverse property of multiplication

85. $5(3x+4)-4 = 5 \cdot 3x + 5 \cdot 4 - 4$
$\qquad\qquad = 15x+20-4$
$\qquad\qquad = 15x+16$

87. $5(3x-2)+12x = 5 \cdot 3x - 5 \cdot 2 + 12x$
$\qquad\qquad\quad = 15x-10+12x$
$\qquad\qquad\quad = 27x-10$

89. $7(3y-5)+2(4y+3)$
$= 7 \cdot 3y - 7 \cdot 5 + 2 \cdot 4y + 2 \cdot 3$
$= 21y-35+8y+6$
$= 29y-29$

91. $5(3y-2)-(7y+2) = 15y-10-7y-2$
$\qquad\qquad\qquad\quad = 8y-12$

93. $7-4[3-(4y-5)] = 7-4[3-4y+5]$
$\qquad\qquad\qquad = 7-4[8-4y]$
$\qquad\qquad\qquad = 7-32+16y$
$\qquad\qquad\qquad = 16y-25$

95. $18x^2+4-\left[6(x^2-2)+5\right]$
$= 18x^2+4-\left[6x^2-12+5\right]$
$= 18x^2+4-\left[6x^2-7\right]$
$= 18x^2+4-6x^2+7$
$= 18x^2-6x^2+4+7$
$= (18-6)x^2+11 = 12x^2+11$

97. $-(-14x) = 14x$

99. $-(2x-3y-6) = -2x+3y+6$

101. $\dfrac{1}{3}(3x)+[(4y)+(-4y)] = x+0$
$\qquad\qquad\qquad\qquad\qquad = x$

103. $|-6| \;\square\; |-3|$

$\qquad 6 \;\square\; 3$

$\qquad 6 > 3$

Since $6 > 3$, $|-6| > |-3|$.

105. $\left|\dfrac{3}{5}\right| \;\square\; |-0.6|$

$|0.6| \;\square\; |-0.6|$

$\quad 0.6 \;\square\; 0.6$

$\quad 0.6 = 0.6$

Since $0.6 = 0.6$, $\left|\dfrac{3}{5}\right| = |-0.6|$.

107. $\dfrac{30}{40}-\dfrac{3}{4} \;\square\; \dfrac{14}{15} \cdot \dfrac{15}{14}$

$\dfrac{30}{40}-\dfrac{30}{40} \;\square\; \dfrac{\cancel{14}}{\cancel{15}} \cdot \dfrac{\cancel{15}}{\cancel{14}}$

$\qquad\quad 0 \;\square\; 1$

$\qquad\quad 0 < 1$

Since $0 < 1$, $\dfrac{30}{40}-\dfrac{3}{4} < \dfrac{14}{15} \cdot \dfrac{15}{14}$.

109. $\dfrac{8}{13} \div \dfrac{8}{13} \;\square\; |-1|$

$\dfrac{8}{13} \cdot \dfrac{13}{8} \;\square\; 1$

$\qquad\quad 1 \;\square\; 1$

$\qquad\quad 1 = 1$

Since $1 = 1$, $\dfrac{8}{13} \div \dfrac{8}{13} = |-1|$.

111. $8^2 - 16 \div 2^2 \cdot 4 - 3 = 64 - 16 \div 4 \cdot 4 - 3$
$$= 64 - 4 \cdot 4 - 3$$
$$= 64 - 16 - 3$$
$$= 48 - 3$$
$$= 45$$

113. $\dfrac{5 \cdot 2 - 3^2}{[3^2 - (-2)]^2} = \dfrac{5 \cdot 2 - 9}{[9 - (-2)]^2}$
$$= \dfrac{10 - 9}{[9 + 2]^2}$$
$$= \dfrac{10 - 9}{11^2}$$
$$= \dfrac{1}{121}$$

115. $8 - 3[-2(2 - 5) - 4(8 - 6)] = 8 - 3[-2(-3) - 4(2)]$
$$= 8 - 3[6 - 8]$$
$$= 8 - 3[-2]$$
$$= 8 + 6$$
$$= 14$$

117. $\dfrac{2(-2) - 4(-3)}{5 - 8} = \dfrac{-4 + 12}{-3}$
$$= \dfrac{8}{-3}$$
$$= -\dfrac{8}{3}$$

119. $\dfrac{(5 - 6)^2 - 2|3 - 7|}{89 - 3 \cdot 5^2} = \dfrac{(-1)^2 - 2|-4|}{89 - 3 \cdot 25}$
$$= \dfrac{1 - 2(4)}{89 - 75}$$
$$= \dfrac{1 - 8}{14}$$
$$= \dfrac{-7}{14}$$
$$= -\dfrac{1}{2}$$

121. $x - (x + 4) = x - x - 4 = -4$

123. $6(-5x) = -30x$

125. $5x - 2x = 3x$

127. $8x - (3x + 6) = 8x - 3x - 6 = 5x - 6$

129. **a.** $H = \dfrac{7}{10}(220 - a)$
$$H = \dfrac{7}{10}(220 - 20)$$
$$= \dfrac{7}{10}(200)$$
$$= 140$$
The lower limit of the heart rate for a 20-year-old with this exercise goal is 140 beats per minute.

b. $H = \dfrac{4}{5}(220 - a)$
$$H = \dfrac{4}{5}(220 - 20)$$
$$= \dfrac{4}{5}(200)$$
$$= 160$$
The upper limit of the heart rate for a 20-year-old with this exercise goal is 160 beats per minute.

131. **a.** $T = 15{,}395 + 988x - 2x^2$
$$= 15{,}395 + 988(7) - 2(7)^2$$
$$= 22{,}213$$
The formula estimates the cost to have been $22,213 in 2007.

b. This underestimates the value in the graph by $5.

c. $T = 15{,}395 + 988x - 2x^2$
$$= 15{,}395 + 988(10) - 2(10)^2$$
$$= 25{,}075$$
The formula projects the cost to be $25,075 in 2010.

133. **a.** $0.05x + 0.12(10{,}000 - x)$
$$= 0.05x + 1200 - 0.12x$$
$$= 1200 - 0.07x$$

b. $1200 - 0.07x = 1200 - 0.07(6000)$
$$= \$780$$

135. – 143. Answers may vary.

145. does not make sense; Explanations will vary. Sample explanation: Models do not always accurately predict future values.

147. makes sense

149. false; Changes to make the statement true will vary. A sample change is: Some rational numbers are not integers.

151. true

153. false; Changes to make the statement true will vary. A sample change is: The term x has a coefficient of 1.

155. false; Changes to make the statement true will vary. A sample change is: $-x - x = -2x$.

157. $\sqrt{2} \approx 1.4$

$1.4 < 1.5$

$\sqrt{2} < 1.5$

159. $-\dfrac{3.14}{2} = -1.57$

$-\dfrac{\pi}{2} \approx -1.571$

$-1.57 > -1.571$

$-\dfrac{3.14}{2} > -\dfrac{\pi}{2}$

160. a. $b^4 \cdot b^3 = (b \cdot b \cdot b \cdot b)(b \cdot b \cdot b) = b^7$

b. $b^5 \cdot b^5 = (b \cdot b \cdot b \cdot b \cdot b)(b \cdot b \cdot b \cdot b \cdot b) = b^{10}$

c. add the exponents

161. a. $\dfrac{b^7}{b^3} = \dfrac{b \cdot b \cdot b \cdot b \cdot b \cdot b \cdot b}{b \cdot b \cdot b} = b^4$

b. $\dfrac{b^8}{b^2} = \dfrac{b \cdot b \cdot b \cdot b \cdot b \cdot b \cdot b \cdot b}{b \cdot b} = b^6$

c. subtract the exponents

162. $6.2 \times 10^3 = 6.2 \times 10 \times 10 \times 10 = 6200$

It moves the decimal point 3 places to the right.

Section P.2

Check Point Exercises

1. a. $\left(2x^3 y^6\right)^4 = (2)^4 \left(x^3\right)^4 \left(y^6\right)^4 = 16x^{12} y^{24}$

b. $\left(-6x^2 y^5\right)\left(3xy^3\right) = (-6) \cdot 3 \cdot x^2 \cdot x \cdot y^5 \cdot y^3$

$\qquad\qquad\qquad = -18x^3 y^8$

c. $\dfrac{100x^{12} y^2}{20x^{16} y^{-4}} = \left(\dfrac{100}{20}\right)\left(\dfrac{x^{12}}{x^{16}}\right)\left(\dfrac{y^2}{y^{-4}}\right)$

$\qquad\qquad = 5x^{12-16} y^{2-(-4)}$

$\qquad\qquad = 5x^{-4} y^6$

$\qquad\qquad = \dfrac{5y^6}{x^4}$

d. $\left(\dfrac{5x}{y^4}\right)^{-2} = \dfrac{(5)^{-2} (x)^{-2}}{\left(y^4\right)^{-2}}$

$\qquad\qquad = \dfrac{(5)^{-2} (x)^{-2}}{\left(y^4\right)^{-2}}$

$\qquad\qquad = \dfrac{5^{-2} x^{-2}}{y^{-8}}$

$\qquad\qquad = \dfrac{y^8}{5^2 x^2}$

$\qquad\qquad = \dfrac{y^8}{25x^2}$

2. a. $-2.6 \times 10^9 = -2,600,000,000$

b. $3.017 \times 10^{-6} = 0.000003017$

3. a. $5,210,000,000 = 5.21 \times 10^9$

b. $-0.00000006893 = -6.893 \times 10^{-8}$

4. $410 \times 10^7 = \left(4.1 \times 10^2\right) \times 10^7$

$\qquad\qquad = 4.1 \times \left(10^2 \times 10^7\right)$

$\qquad\qquad = 4.1 \times 10^9$

5. **a.** $\left(7.1\times10^5\right)\left(5\times10^{-7}\right)$

$= 7.1\cdot5\times10^5\cdot10^{-7}$

$= 35.5\times10^{-2}$

$= \left(3.55\times10^1\right)\times10^{-2}$

$= 3.55\times\left(10^1\times10^{-2}\right)$

$= 3.55\times10^{-1}$

b. $\dfrac{1.2\times10^6}{3\times10^{-3}} = \dfrac{1.2}{3}\cdot\dfrac{10^6}{10^{-3}}$

$= 0.4\times10^{6-(-3)}$

$= 0.4\times10^9$

$= 4\times10^8$

6. $\dfrac{13\times10^9}{5.1\times10^6} = \dfrac{13}{5.1}\cdot\dfrac{10^9}{10^6}$

$\approx 2.5\cdot10^3$

≈ 2500

The average Pell grant was $2500 in 2006.

7. $S = (1.76\times10^5)[(1.44\times10^{-2})-r^2]$

$= (1.76\times10^5)[(1.44\times10^{-2})-0^2]$

$= 2534.4$

The speed of the blood at the central axis of the artery is 2534.4 centimeters per second.

Exercise Set P.2

1. $5^2\cdot2 = (5\cdot5)\cdot2 = 25\cdot2 = 50$

3. $(-2)^6 = (-2)(-2)(-2)(-2)(-2)(-2) = 64$

5. $-2^6 = -2\cdot2\cdot2\cdot2\cdot2\cdot2 = -64$

7. $(-3)^0 = 1$

9. $-3^0 = -1$

11. $4^{-3} = \dfrac{1}{4^3} = \dfrac{1}{4\cdot4\cdot4} = \dfrac{1}{64}$

13. $2^2\cdot2^3 = 2^{2+3} = 2^5 = 2\cdot2\cdot2\cdot2\cdot2 = 32$

15. $(2^2)^3 = 2^{2\cdot3} = 2^6 = 2\cdot2\cdot2\cdot2\cdot2\cdot2 = 64$

17. $\dfrac{2^8}{2^4} = 2^{8-4} = 2^4 = 2\cdot2\cdot2\cdot2 = 16$

19. $3^{-3}\cdot3 = 3^{-3+1} = 3^{-2} = \dfrac{1}{3^2} = \dfrac{1}{3\cdot3} = \dfrac{1}{9}$

21. $\dfrac{2^3}{2^7} = 2^{3-7} = 2^{-4} = \dfrac{1}{2^4} = \dfrac{1}{2\cdot2\cdot2\cdot2} = \dfrac{1}{16}$

23. $x^{-2}y = \dfrac{1}{x^2}\cdot y = \dfrac{y}{x^2}$

25. $x^0y^5 = 1\cdot y^5 = y^5$

27. $x^3\cdot x^7 = x^{3+7} = x^{10}$

29. $x^{-5}\cdot x^{10} = x^{-5+10} = x^5$

31. $(x^3)^7 = x^{3\cdot7} = x^{21}$

33. $(x^{-5})^3 = x^{-5\cdot3} = x^{-15} = \dfrac{1}{x^{15}}$

35. $\dfrac{x^{14}}{x^7} = x^{14-7} = x^7$

37. $\dfrac{x^{14}}{x^{-7}} = x^{14-(-7)} = x^{14+7} = x^{21}$

39. $(8x^3)^2 = 8^2(x^3)^2 = 8^2x^{3\cdot2} = 64x^6$

41. $\left(-\dfrac{4}{x}\right)^3 = \dfrac{(-4)^3}{x^3} = -\dfrac{64}{x^3}$

43. $(-3x^2y^5)^2 = (-3)^2(x^2)^2\cdot(y^5)^2$

$= 9x^{2\cdot2}y^{5\cdot2}$

$= 9x^4y^{10}$

45. $(3x^4)(2x^7) = 3\cdot2x^4\cdot x^7 = 6x^{4+7} = 6x^{11}$

47. $(-9x^3y)(-2x^6y^4) = (-9)(-2)x^3x^6yy^4$

$= 18x^{3+6}y^{1+4}$

$= 18x^9y^5$

49. $\dfrac{8x^{20}}{2x^4} = \left(\dfrac{8}{2}\right)\left(\dfrac{x^{20}}{x^4}\right) = 4x^{20-4} = 4x^{16}$

51. $\dfrac{25a^{13}\cdot b^4}{-5a^2\cdot b^3} = \left(\dfrac{25}{-5}\right)\left(\dfrac{a^{13}}{a^2}\right)\left(\dfrac{b^4}{b^3}\right)$

$= -5a^{13-2}b^{4-3}$

$= -5a^{11}b$

53. $\dfrac{14b^7}{7b^{14}} = \left(\dfrac{14}{7}\right)\left(\dfrac{b^7}{b^{14}}\right) = 2 \cdot b^{7-14} = 2b^{-7} = \dfrac{2}{b^7}$

55. $(4x^3)^{-2} = (4^{-2})(x^3)^{-2}$

$\qquad\qquad = 4^{-2}x^{-6}$

$\qquad\qquad = \dfrac{1}{4^2 x^6}$

$\qquad\qquad = \dfrac{1}{16x^6}$

57. $\dfrac{24x^3 \cdot y^5}{32x^7 y^{-9}} = \dfrac{3}{4} x^{3-7} y^{5-(-9)}$

$\qquad\qquad = \dfrac{3}{4} x^{-4} y^{14}$

$\qquad\qquad = \dfrac{3y^{14}}{4x^4}$

59. $\left(\dfrac{5x^3}{y}\right)^{-2} = \dfrac{5^{-2}x^{-6}}{y^{-2}} = \dfrac{y^2}{25x^6}$

61. $\left(\dfrac{-15a^4 b^2}{5a^{10} b^{-3}}\right)^3 = \left(\dfrac{-3b^{2-(-3)}}{a^{10-4}}\right)^3$

$\qquad\qquad\qquad = \left(\dfrac{-3b^5}{a^6}\right)^3$

$\qquad\qquad\qquad = \dfrac{-27b^{15}}{a^{18}}$

63. $\left(\dfrac{3a^{-5}b^2}{12a^3 b^{-4}}\right)^0 = 1$

65. $3.8 \times 10^2 = 380$

67. $6 \times 10^{-4} = 0.0006$

69. $-7.16 \times 10^6 = -7,160,000$

71. $7.9 \times 10^{-1} = 0.79$

73. $-4.15 \times 10^{-3} = -0.00415$

75. $-6.00001 \times 10^{10} = -60,000,100,000$

77. $32,000 = 3.2 \times 10^4$

79. $638,000,000,000,000,000$

$\qquad = 6.38 \times 10^{17}$

81. $-5716 = -5.716 \times 10^3$

83. $0.0027 = 2.7 \times 10^{-3}$

85. $-0.00000000504 = -5.04 \times 10^{-9}$

87. $(3 \times 10^4)(2.1 \times 10^3) = (3 \times 2.1)(10^4 \times 10^3)$

$\qquad\qquad\qquad = 6.3 \times 10^{4+3} = 6.3 \times 10^7$

89. $(1.6 \times 10^{15})(4 \times 10^{-11}) = (1.6 \times 4)(10^{15} \times 10^{-11})$

$\qquad\qquad\qquad\qquad = 6.4 \times 10^{15+(-11)}$

$\qquad\qquad\qquad\qquad = 6.4 \times 10^4$

91. $(6.1 \times 10^{-8})(2 \times 10^{-4}) = (6.1 \times 2)(10^{-8} \times 10^{-4})$

$\qquad\qquad\qquad\qquad = 12.2 \times 10^{-8+(-4)}$

$\qquad\qquad\qquad\qquad = 12.2 \times 10^{-12}$

$\qquad\qquad\qquad\qquad = 1.22 \times 10^{-11}$

93. $(4.3 \times 10^8)(6.2 \times 10^4)$

$\qquad = (4.3 \times 6.2)(10^8 \times 10^4)$

$\qquad = 26.66 \times 10^{8+4}$

$\qquad = 26.66 \times 10^{12}$

$\qquad = 2.666 \times 10^{13} \approx 2.67 \times 10^{13}$

95. $\dfrac{8.4 \times 10^8}{4 \times 10^5} = \dfrac{8.4}{4} \times \dfrac{10^8}{10^5}$

$\qquad\qquad = 2.1 \times 10^{8-5} = 2.1 \times 10^3$

97. $\dfrac{3.6 \times 10^4}{9 \times 10^{-2}} = \dfrac{3.6}{9} \times \dfrac{10^4}{10^{-2}}$

$\qquad\qquad = 0.4 \times 10^{4-(-2)}$

$\qquad\qquad = 0.4 \times 10^6 = 4 \times 10^5$

99. $\dfrac{4.8 \times 10^{-2}}{2.4 \times 10^6} = \dfrac{4.8}{2.4} \times \dfrac{10^{-2}}{10^6}$

$\qquad\qquad = 2 \times 10^{-2-6} = 2 \times 10^{-8}$

101. $\dfrac{2.4 \times 10^{-2}}{4.8 \times 10^{-6}} = \dfrac{2.4}{4.8} \times \dfrac{10^{-2}}{10^{-6}}$

$\qquad\qquad = 0.5 \times 10^{-2-(-6)}$

$\qquad\qquad = 0.5 \times 10^4 = 5 \times 10^3$

103. $\dfrac{480,000,000,000}{0.00012} = \dfrac{4.8\times10^{11}}{1.2\times10^{-4}}$

$\qquad\qquad = \dfrac{4.8}{1.2}\times\dfrac{10^{11}}{10^{-4}}$

$\qquad\qquad = 4\times10^{11-(-4)}$

$\qquad\qquad = 4\times10^{15}$

105. $\dfrac{0.00072\times0.003}{0.00024}$

$= \dfrac{\left(7.2\times10^{-4}\right)\left(3\times10^{-3}\right)}{2.4\times10^{-4}}$

$= \dfrac{7.2\times3}{2.4}\times\dfrac{10^{-4}\cdot10^{-3}}{10^{-4}} = 9\times10^{-3}$

107. $\dfrac{\left(x^{-2}y\right)^{-3}}{\left(x^{2}y^{-1}\right)^{3}} = \dfrac{x^{6}y^{-3}}{x^{6}y^{-3}}$

$\qquad\qquad = x^{6-6}y^{-3-(-3)} = x^{0}y^{0} = 1$

109. $\left(2x^{-3}yz^{-6}\right)\left(2x\right)^{-5} = 2x^{-3}yz^{-6}\cdot2^{-5}x^{.-5}$

$= 2^{-4}x^{-8}yz^{-6} = \dfrac{y}{2^{4}x^{8}z^{6}} = \dfrac{y}{16x^{8}z^{6}}$

111. $\left(\dfrac{x^{3}y^{4}z^{5}}{x^{-3}y^{-4}z^{-5}}\right)^{-2} = \left(x^{6}y^{8}z^{10}\right)^{-2}$

$\qquad\qquad = x^{-12}y^{-16}z^{-20} = \dfrac{1}{x^{12}y^{16}z^{20}}$

113. $\dfrac{\left(2^{-1}x^{-2}y^{-1}\right)^{-2}\left(2x^{-4}y^{3}\right)^{-2}\left(16x^{-3}y^{3}\right)^{0}}{\left(2x^{-3}y^{-5}\right)^{2}}$

$= \dfrac{\left(2^{2}x^{2}y^{2}\right)\left(2^{-2}x^{8}y^{-6}\right)(1)}{\left(2^{2}x^{-6}y^{-10}\right)}$

$= \dfrac{x^{18}y^{6}}{4}$

115. **a.** 2.52×10^{12}

b. 3×10^{8}

c. $\dfrac{2.52\times10^{12}}{3\times10^{8}} = \dfrac{2.52}{3}\times\dfrac{10^{12}}{10^{8}}$

$\qquad\qquad = 0.84\times10^{4}$

$\qquad\qquad = 8400$

$\$8400$ per American

117. $1450\times10^{9}\cdot6.60 = 1.45\times10^{12}\cdot6.6$

$\qquad\qquad = 1.45\cdot6.6\times10^{12}$

$\qquad\qquad = 9.57\times10^{12}$

Box-office receipts were $\$9.57\times10^{12}$ in 2006.

119. $5.3\times10^{-23}\cdot20,000 = 5.3\times10^{-23}\cdot2\times10^{4}$

$\qquad\qquad = 5.3\cdot2\times10^{-23}\cdot10^{4}$

$\qquad\qquad = 10.6\times10^{-19}$

$\qquad\qquad = 1.06\times10^{1}\cdot10^{-19}$

$\qquad\qquad = 1.06\times10^{-18}$

The mass is 1.06×10^{-18} gram.

121. $3.2\times10^{7}\cdot127 = 3.2\times10^{7}\cdot1.27\times10^{2}$

$\qquad\qquad = 3.2\cdot1.27\times10^{7}\cdot10^{2}$

$\qquad\qquad = 4.064\times10^{9}$

Americans eat 4.064×10^{9} chickens per year.

123. – 129. Answers may vary.

131. does not make sense; Explanations will vary.
Sample explanation: $36(x^{3})^{9} = 36x^{27}$ not $36x^{12}$.

133. does not make sense; Explanations will vary.
Sample explanation: 4.6×10^{12} represents over 4 trillion. The entire world population is measured in billions (10^{9}).

135. false; Changes to make the statement true will vary.
A sample change is: $4^{-2} > 4^{-3}$.

137. false; Changes to make the statement true will vary.
A sample change is: $(-2)^{4} \neq 2^{-4}$ because $16 \neq \dfrac{1}{16}$.

139. false; Changes to make the statement true will vary.
A sample change is: $534.7 \neq 5347$.

141. false; Changes to make the statement true will vary.
A sample change is:
$(7\times10^{5}) + (2\times10^{-3}) = 700,000.002$.

143. The doctor has gathered:
$2^{-1} + 2^{-2} = \dfrac{1}{2} + \dfrac{1}{2^{2}} = \dfrac{2}{4} + \dfrac{1}{4} = \dfrac{3}{4}$

So, $1 - \dfrac{3}{4} = \dfrac{1}{4}$ is remaining.

145. $\dfrac{70 \text{ bts}}{\text{min}} \cdot \dfrac{60 \text{ min}}{\text{hr}} \cdot \dfrac{24 \text{ hrs}}{\text{day}} \cdot \dfrac{365 \text{ days}}{\text{yr}} \cdot 80 \text{ yrs}$

$= 70 \cdot 60 \cdot 24 \cdot 365 \cdot 80 \text{ beats}$

$= 2943360000 \text{ beats}$

$= 2.94336 \times 10^9 \text{ beats}$

$\approx 2.94 \times 10^9 \text{ beats}$

The heartbeats approximately 2.94×10^9 times over a lifetime of 80 years

147. a. $\sqrt{16} \cdot \sqrt{4} = 4 \cdot 2 = 8$

    b. $\sqrt{16 \cdot 4} = \sqrt{64} = 8$

    c. $\sqrt{16} \cdot \sqrt{4} = \sqrt{16 \cdot 4}$

148. a. $\sqrt{300} \approx 17.32$

    b. $10\sqrt{3} \approx 17.32$

    c. $\sqrt{300} = 10\sqrt{3}$

149. a. $21x + 10x = 31x$

    b. $21\sqrt{2} + 10\sqrt{2} = 31\sqrt{2}$

Section P.3

Check Point Exercises

1. a. $\sqrt{81} = 9$

    b. $-\sqrt{9} = -3$

    c. $\sqrt{\dfrac{1}{25}} = \dfrac{1}{5}$

    d. $\sqrt{36 + 64} = \sqrt{100} = 10$

    e. $\sqrt{36} + \sqrt{64} = 6 + 8 = 14$

2. a. $\sqrt{75} = \sqrt{25 \cdot 3} = \sqrt{25}\sqrt{3} = 5\sqrt{3}$

    b. $\sqrt{5x} \cdot \sqrt{10x} = \sqrt{5x \cdot 10x}$

$= \sqrt{50x^2}$

$= \sqrt{25 \cdot 2x^2}$

$= \sqrt{25x^2} \cdot \sqrt{2}$

$= 5x\sqrt{2}$

3. a. $\sqrt{\dfrac{25}{16}} = \dfrac{\sqrt{25}}{\sqrt{16}} = \dfrac{5}{4}$

    b. $\dfrac{\sqrt{150x^3}}{\sqrt{2x}} = \sqrt{\dfrac{150x^3}{2x}}$

$= \sqrt{75x^2}$

$= \sqrt{25x^2} \cdot \sqrt{3}$

$= 5x\sqrt{3}$

4. a. $8\sqrt{13} + 9\sqrt{13} = (8+9)\sqrt{3}$

$= 17\sqrt{13}$

    b. $\sqrt{17x} - 20\sqrt{17x}$

$= 1\sqrt{17x} - 20\sqrt{17x}$

$= (1 - 20)\sqrt{17x}$

$= -19\sqrt{17x}$

5. a. $5\sqrt{27} + \sqrt{12}$

$= 5\sqrt{9 \cdot 3} + \sqrt{4 \cdot 3}$

$= 5 \cdot 3\sqrt{3} + 2\sqrt{3}$

$= 15\sqrt{3} + 2\sqrt{3}$

$= (15 + 2)\sqrt{3}$

$= 17\sqrt{3}$

    b. $6\sqrt{18x} - 4\sqrt{8x}$

$= 6\sqrt{9 \cdot 2x} - 4\sqrt{4 \cdot 2x}$

$= 6 \cdot 3\sqrt{2x} - 4 \cdot 2\sqrt{2x}$

$= 18\sqrt{2x} - 8\sqrt{2x}$

$= (18 - 8)\sqrt{2x}$

$= 10\sqrt{2x}$

6. a. If we multiply numerator and denominator by $\sqrt{3}$, the denominator becomes $\sqrt{3} \cdot \sqrt{3} = \sqrt{9} = 3$. Therefore, multiply by 1, choosing $\dfrac{\sqrt{3}}{\sqrt{3}}$ for 1.

$$\frac{5}{\sqrt{3}} = \frac{5}{\sqrt{3}} \cdot \frac{\sqrt{3}}{\sqrt{3}} = \frac{5\sqrt{3}}{\sqrt{9}} = \frac{5\sqrt{3}}{3}$$

b. The *smallest* number that will produce a perfect square in the denominator of $\dfrac{6}{\sqrt{12}}$ is $\sqrt{3}$ because $\sqrt{12} \cdot \sqrt{3} = \sqrt{36} = 6$. So multiply by 1, choosing $\dfrac{\sqrt{3}}{\sqrt{3}}$ for 1.

$$\frac{6}{\sqrt{12}} = \frac{6}{\sqrt{12}} \cdot \frac{\sqrt{3}}{\sqrt{3}} = \frac{6\sqrt{3}}{\sqrt{36}} = \frac{6\sqrt{3}}{6} = \sqrt{3}$$

7. Multiply by $\dfrac{4-\sqrt{5}}{4-\sqrt{5}}$.

$$\frac{8}{4+\sqrt{5}} = \frac{8}{4+\sqrt{5}} \cdot \frac{4-\sqrt{5}}{4-\sqrt{5}}$$

$$= \frac{8(4-\sqrt{5})}{4^2-(\sqrt{5})^2}$$

$$= \frac{8(4-\sqrt{5})}{16-5}$$

$$= \frac{8(4-\sqrt{5})}{11} \text{ or } \frac{32-8\sqrt{5}}{11}$$

8. a. $\sqrt[3]{40} = \sqrt[3]{8 \cdot 5} = \sqrt[3]{8} \cdot \sqrt[3]{5} = 2\sqrt[3]{5}$

b. $\sqrt[5]{8} \cdot \sqrt[5]{8} = \sqrt[5]{64} = \sqrt[5]{32} \cdot \sqrt[5]{2} = 2\sqrt[5]{2}$

c. $\sqrt[3]{\dfrac{125}{27}} = \dfrac{\sqrt[3]{125}}{\sqrt[3]{27}} = \dfrac{5}{3}$

9. $3\sqrt[3]{81} - 4\sqrt[3]{3}$

$= 3\sqrt[3]{27 \cdot 3} - 4\sqrt[3]{3}$

$= 3 \cdot 3\sqrt[3]{3} - 4\sqrt[3]{3}$

$= 9\sqrt[3]{3} - 4\sqrt[3]{3}$

$= (9-4)\sqrt[3]{3}$

$= 5\sqrt[3]{3}$

10. a. $25^{\frac{1}{2}} = \sqrt{25} = 5$

b. $8^{\frac{1}{3}} = \sqrt[3]{8} = 2$

c. $-81^{\frac{1}{4}} = -\sqrt[4]{81} = -3$

d. $(-8)^{\frac{1}{3}} = \sqrt[3]{-8} = -2$

e. $27^{-\frac{1}{3}} = \dfrac{1}{27^{\frac{1}{3}}} = \dfrac{1}{\sqrt[3]{27}} = \dfrac{1}{3}$

11. a. $27^{\frac{4}{3}} = \left(\sqrt[3]{27}\right)^4 = (3)^4 = 81$

b. $4^{\frac{3}{2}} = \left(\sqrt[2]{4}\right)^3 = (2)^3 = 8$

c. $32^{-\frac{2}{5}} = \dfrac{1}{32^{\frac{2}{5}}} = \dfrac{1}{\left(\sqrt[5]{32}\right)^2} = \dfrac{1}{2^2} = \dfrac{1}{4}$

12. a. $\left(2x^{4/3}\right)\left(5x^{8/3}\right)$

$= 2 \cdot 5 x^{4/3} \cdot x^{8/3}$

$= 10 x^{(4/3)+(8/3)}$

$= 10 x^{12/3}$

$= 10 x^4$

b. $\dfrac{20x^4}{5x^{3/2}} = \left(\dfrac{20}{5}\right)\left(\dfrac{x^4}{x^{3/2}}\right)$

$= 4x^{4-(3/2)}$

$= 4x^{(8/2)-(3/2)}$

$= 4x^{5/2}$

13. $\sqrt[6]{x^3} = x^{3/6} = x^{1/2} = \sqrt{x}$

Exercise Set P.3

1. $\sqrt{36} = \sqrt{6^2} = 6$

3. $-\sqrt{36} = -\sqrt{6^2} = -6$

5. $\sqrt{-36}$, The square root of a negative number is not real.

7. $\sqrt{25-16} = \sqrt{9} = 3$

9. $\sqrt{25} - \sqrt{16} = 5 - 4 = 1$

11. $\sqrt{(-13)^2} = \sqrt{169} = 13$

13. $\sqrt{50} = \sqrt{25 \cdot 2} = \sqrt{25}\sqrt{2} = 5\sqrt{2}$

15. $\sqrt{45x^2} = \sqrt{9x^2 \cdot 5}$
$= \sqrt{9x^2}\sqrt{5}$
$= \sqrt{9}\sqrt{x^2}\sqrt{5}$
$= 3|x|\sqrt{5}$

17. $\sqrt{2x} \cdot \sqrt{6x} = \sqrt{2x \cdot 6x}$
$= \sqrt{12x^2}$
$= \sqrt{4x^2} \cdot \sqrt{3}$
$= 2x\sqrt{3}$

19. $\sqrt{x^3} = \sqrt{x^2} \cdot \sqrt{x} = x\sqrt{x}$

21. $\sqrt{2x^2} \cdot \sqrt{6x} = \sqrt{2x^2 \cdot 6x}$
$= \sqrt{12x^3}$
$= \sqrt{4x^2} \cdot \sqrt{3x}$
$= 2x\sqrt{3x}$

23. $\sqrt{\dfrac{1}{81}} = \dfrac{\sqrt{1}}{\sqrt{81}} = \dfrac{1}{9}$

25. $\sqrt{\dfrac{49}{16}} = \dfrac{\sqrt{49}}{\sqrt{16}} = \dfrac{7}{4}$

27. $\dfrac{\sqrt{48x^3}}{\sqrt{3x}} = \sqrt{\dfrac{48x^3}{3x}} = \sqrt{16x^2} = 4x$

29. $\dfrac{\sqrt{150x^4}}{\sqrt{3x}} = \sqrt{\dfrac{150x^4}{3x}}$
$= \sqrt{50x^3}$
$= \sqrt{25x^2} \cdot \sqrt{2x}$
$= 5x\sqrt{2x}$

31. $\dfrac{\sqrt{200x^3}}{\sqrt{10x^{-1}}} = \sqrt{\dfrac{200x^3}{10x^{-1}}}$
$= \sqrt{20x^{3-(-1)}}$
$= \sqrt{20x^4}$
$= \sqrt{4 \cdot 5x^4}$
$= 2x^2\sqrt{5}$

33. $7\sqrt{3} + 6\sqrt{3} = (7 + 6)\sqrt{3} = 13\sqrt{3}$

35. $6\sqrt{17x} - 8\sqrt{17x} = (6 - 8)\sqrt{17x} = -2\sqrt{17x}$

37. $\sqrt{8} + 3\sqrt{2} = \sqrt{4 \cdot 2} + 3\sqrt{2}$
$= 2\sqrt{2} + 3\sqrt{2}$
$= (2 + 3)\sqrt{2}$
$= 5\sqrt{2}$

39. $\sqrt{50x} - \sqrt{8x} = \sqrt{25 \cdot 2x} - \sqrt{4 \cdot 2x}$
$= 5\sqrt{2x} - 2\sqrt{2x}$
$= (5 - 2)\sqrt{2x}$
$= 3\sqrt{2x}$

41. $3\sqrt{18} + 5\sqrt{50} = 3\sqrt{9 \cdot 2} + 5\sqrt{25 \cdot 2}$
$= 3 \cdot 3\sqrt{2} + 5 \cdot 5\sqrt{2}$
$= 9\sqrt{2} + 25\sqrt{2}$
$= (9 + 25)\sqrt{2}$
$= 34\sqrt{2}$

43. $3\sqrt{8} - \sqrt{32} + 3\sqrt{72} - \sqrt{75}$
$= 3\sqrt{4 \cdot 2} - \sqrt{16 \cdot 2} + 3\sqrt{36 \cdot 2} - \sqrt{25 \cdot 3}$
$= 3 \cdot 2\sqrt{2} - 4\sqrt{2} + 3 \cdot 6\sqrt{2} - 5\sqrt{3}$
$= 6\sqrt{2} - 4\sqrt{2} + 18\sqrt{2} - 5\sqrt{3}$
$= 20\sqrt{2} - 5\sqrt{3}$

45. $\dfrac{1}{\sqrt{7}} = \dfrac{1}{\sqrt{7}} \cdot \dfrac{\sqrt{7}}{\sqrt{7}} = \dfrac{\sqrt{7}}{7}$

47. $\dfrac{\sqrt{2}}{\sqrt{5}} = \dfrac{\sqrt{2}}{\sqrt{5}} \cdot \dfrac{\sqrt{5}}{\sqrt{5}} = \dfrac{\sqrt{10}}{5}$

49.
$$\frac{13}{3+\sqrt{11}} = \frac{13}{3+\sqrt{11}} \cdot \frac{3-\sqrt{11}}{3-\sqrt{11}}$$
$$= \frac{13(3-\sqrt{11})}{3^2 - (\sqrt{11})^2}$$
$$= \frac{13(3-\sqrt{11})}{9-11}$$
$$= \frac{13(3-\sqrt{11})}{-2}$$

51.
$$\frac{7}{\sqrt{5}-2} = \frac{7}{\sqrt{5}-2} \cdot \frac{\sqrt{5}+2}{\sqrt{5}+2}$$
$$= \frac{7(\sqrt{5}+2)}{(\sqrt{5})^2 - 2^2}$$
$$= \frac{7(\sqrt{5}+2)}{5-4}$$
$$= 7(\sqrt{5}+2)$$

53.
$$\frac{6}{\sqrt{5}+\sqrt{3}} = \frac{6}{\sqrt{5}+\sqrt{3}} \cdot \frac{\sqrt{5}-\sqrt{3}}{\sqrt{5}-\sqrt{3}}$$
$$= \frac{6(\sqrt{5}-\sqrt{3})}{(\sqrt{5})^2 - (\sqrt{3})^2}$$
$$= \frac{6(\sqrt{5}-\sqrt{3})}{5-3}$$
$$= \frac{6(\sqrt{5}-\sqrt{3})}{2}$$
$$= 3(\sqrt{5}-\sqrt{3})$$

55. $\sqrt[3]{125} = \sqrt[3]{5^3} = 5$

57. $\sqrt[3]{-8} = \sqrt[3]{(-2)^3} = -2$

59. $\sqrt[4]{-16}$ is not a real number.

61. $\sqrt[4]{(-3)^4} = |-3| = 3$

63. $\sqrt[5]{(-3)^5} = -3$

65. $\sqrt[5]{-\frac{1}{32}} = \sqrt[5]{-\frac{1}{2^5}} = -\frac{1}{2}$

67. $\sqrt[3]{32} = \sqrt[3]{8 \cdot 4} = \sqrt[3]{8}\sqrt[3]{4} = 2 \cdot \sqrt[3]{4}$

69. $\sqrt[3]{x^4} = \sqrt[3]{x^3 \cdot x} = x \cdot \sqrt[3]{x}$

71. $\sqrt[3]{9} \cdot \sqrt[3]{6} = \sqrt[3]{54} = \sqrt[3]{27 \cdot 2} = \sqrt[3]{27}\sqrt[3]{2} = 3\sqrt[3]{2}$

73. $\dfrac{\sqrt[5]{64x^6}}{\sqrt[5]{2x}} = \sqrt[5]{\dfrac{64x^6}{2x}} = \sqrt[5]{32x^5} = 2x$

75. $4\sqrt[5]{2} + 3\sqrt[5]{2} = 7\sqrt[5]{2}$

77.
$$5\sqrt[3]{16} + \sqrt[3]{54} = 5\sqrt[3]{8 \cdot 2} + \sqrt[3]{27 \cdot 2}$$
$$= 5 \cdot 2\sqrt[3]{2} + 3\sqrt[3]{2}$$
$$= 10\sqrt[3]{2} + 3\sqrt[3]{2}$$
$$= 13\sqrt[3]{2}$$

79.
$$\sqrt[3]{54xy^3} - y\sqrt[3]{128x}$$
$$= \sqrt[3]{27 \cdot 2xy^3} - y\sqrt[3]{64 \cdot 2x}$$
$$= 3y\sqrt[3]{2x} - 4y\sqrt[3]{2x}$$
$$= -y\sqrt[3]{2x}$$

81. $\sqrt{2} + \sqrt[3]{8} = \sqrt{2} + 2$

83. $36^{1/2} = \sqrt{36} = 6$

85. $8^{1/3} = \sqrt[3]{8} = 2$

87. $125^{2/3} = \left(\sqrt[3]{125}\right)^2 = 5^2 = 25$

89. $32^{-4/5} = \dfrac{1}{32^{4/5}} = \dfrac{1}{2^4} = \dfrac{1}{16}$

91.
$$\left(7x^{1/3}\right)\left(2x^{1/4}\right) = 7 \cdot 2x^{1/3} \cdot x^{1/4}$$
$$= 14 \cdot x^{1/3+1/4}$$
$$= 14x^{7/12}$$

93.
$$\frac{20x^{1/2}}{5x^{1/4}} = \left(\frac{20}{5}\right)\left(\frac{x^{1/2}}{x^{1/4}}\right)$$
$$= 4 \cdot x^{1/2-1/4}$$
$$= 4x^{1/4}$$

95. $\left(x^{2/3}\right)^3 = x^{2/3 \cdot 3} = x^2$

97. $(25x^4y^6)^{1/2} = 25^{1/2}x^{4 \cdot 1/2}y^{6 \cdot 1/2} = 5x^2|y|^3$

99. $\dfrac{\left(3y^{\frac{1}{4}}\right)^3}{y^{\frac{1}{12}}} = \dfrac{27y^{\frac{3}{4}}}{y^{\frac{1}{12}}} = 27y^{\frac{3}{4}-\frac{1}{12}}$

$= 27y^{\frac{8}{12}} = 27y^{\frac{2}{3}}$

101. $\sqrt[4]{5^2} = 5^{2/4} = 5^{1/2} = \sqrt{5}$

103. $\sqrt[3]{x^6} = x^{6/3} = x^2$

105. $\sqrt[6]{x^4} = \sqrt[6/2]{x^{4/2}} = \sqrt[3]{x^2}$

107. $\sqrt[9]{x^6 y^3} = x^{\frac{6}{9}} y^{\frac{3}{9}} = x^{\frac{2}{3}} y^{\frac{1}{3}} = \sqrt[3]{x^2 y}$

109. $\sqrt[3]{\sqrt[4]{16} + \sqrt{625}} = \sqrt[3]{2 + 25} = \sqrt[3]{27} = 3$

111. $\left(49x^{-2}y^4\right)^{-1/2}\left(xy^{1/2}\right)$

$= \left(49\right)^{-1/2}\left(x^{-2}\right)^{-1/2}\left(y^4\right)^{-1/2}\left(xy^{1/2}\right)$

$= \dfrac{1}{49^{1/2}}x^{(-2)(-1/2)}y^{(4)(-1/2)}\left(xy^{1/2}\right)$

$= \dfrac{1}{7}x^1 y^{-2} \cdot xy^{1/2} = \dfrac{1}{7}x^{1+1}y^{-2+(1/2)}$

$= \dfrac{1}{7}x^2 y^{-3/2} = \dfrac{x^2}{7y^{3/2}}$

113. $\left(\dfrac{x^{-5/4}y^{1/3}}{x^{-3/4}}\right)^{-6} = \left(x^{(-5/4)-(-3/4)}y^{1/3}\right)^{-6}$

$= \left(x^{-2/4}y^{1/3}\right)^{-6} = x^{(-2/4)(-6)}y^{(1/3)(-6)}$

$= x^3 y^{-2} = \dfrac{x^3}{y^2}$

115. **a.** In 2004, we have $x = 5$.

$y = 20.8\sqrt{5} + 21 \approx 67.5$

According to the model, 67.5% of email was spam in 2004.
This underestimates the actual value shown in the bar graph by 0.5%.

b. In 2011, we have $x = 12$.

$y = 20.8\sqrt{12} + 21 \approx 93.1$

According to the model, 93.1% of email will be spam in 2011.
This overestimates the value given in the bar graph by 21.1%.

117. $\dfrac{2}{\sqrt{5}-1} \cdot \dfrac{\sqrt{5}+1}{\sqrt{5}+1} = \dfrac{2(\sqrt{5}+1)}{5-1}$

$= \dfrac{2(\sqrt{5}+1)}{4}$

$= \dfrac{\sqrt{5}+1}{2}$

≈ 1.62

About 1.62 to 1.

119. Perimeter:

$P = 2l + 2w$

$= 2 \cdot \sqrt{125} + 2 \cdot 2\sqrt{20}$

$= 2 \cdot \sqrt{25 \cdot 5} + 4\sqrt{4 \cdot 5}$

$= 2 \cdot 5\sqrt{5} + 4 \cdot 2\sqrt{5}$

$= 10\sqrt{5} + 8\sqrt{5}$

$= 18\sqrt{5}$ feet

Area:

$A = lw$

$= \sqrt{125} \cdot 2\sqrt{20}$

$= 2\sqrt{125 \cdot 20}$

$= 2\sqrt{2500}$

$= 2 \cdot 50$

$= 100$ square feet

121. – 127. Answers may vary.

129. does not make sense; Explanations will vary.
Sample explanation: The denominator is rationalized correctly.

131. does not make sense; Explanations will vary.
Sample explanation: $2\sqrt{20} + 4\sqrt{75}$ simplifies to $4\sqrt{5} + 20\sqrt{3}$ and thus the radical terms are not common.

133. false; Changes to make the statement true will vary.
A sample change is: $7^{\frac{1}{2}} \cdot 7^{\frac{1}{2}} = 7^1 = 7$.

135. false; Changes to make the statement true will vary.
The cube root of –8 is the real number –2.

137. $\left(5 + \sqrt{\boxed{3}}\right)\left(5 - \sqrt{\boxed{3}}\right) = 22$

$25 - \boxed{3} = 22$

$\boxed{3} = 3$

139. $\sqrt{13+\sqrt{2}+\dfrac{7}{3+\sqrt{2}}}$

$$= \sqrt{13+\sqrt{2}+\frac{7}{3+\sqrt{2}}\cdot\frac{3-\sqrt{2}}{3-\sqrt{2}}}$$

$$= \sqrt{13+\sqrt{2}+\frac{21-7\sqrt{2}}{9-2}}$$

$$= \sqrt{13+\sqrt{2}+\frac{21-7\sqrt{2}}{7}}$$

$$= \sqrt{13+\sqrt{2}+3-\sqrt{2}}$$

$$= \sqrt{16}$$

$$= 4$$

141. a. $2^{\frac{5}{2}}\cdot 2^{\frac{3}{4}} \div 2^{\frac{1}{4}} = \dfrac{2^{\frac{5}{2}}\cdot 2^{\frac{3}{4}}}{2^{\frac{1}{4}}} = 2^{\frac{5}{2}+\frac{3}{4}-\frac{1}{4}} = 2^3 = 8$

Her son is 8 years old.

b. Son's portion:

$$\frac{8^{-\frac{4}{3}}+2^{-2}}{16^{-\frac{3}{4}}+2^{-1}} = \frac{\dfrac{1}{\left(\sqrt[3]{8}\right)^4}+\dfrac{1}{2^2}}{\dfrac{1}{\left(\sqrt[4]{16}\right)^3}+\dfrac{1}{2}}$$

$$= \frac{\dfrac{1}{2^4}+\dfrac{1}{4}}{\dfrac{1}{2^3}+\dfrac{1}{2}}$$

$$= \frac{\dfrac{1}{16}+\dfrac{1}{4}}{\dfrac{1}{8}+\dfrac{1}{2}}$$

$$= \frac{\dfrac{5}{16}}{\dfrac{5}{8}}$$

$$= \frac{8}{16}$$

$$= \frac{1}{2}$$

Mom's portion:

$$\frac{1}{2}\left(1-\frac{1}{2}\right) = \frac{1}{2}\left(\frac{1}{2}\right) = \frac{1}{4}$$

142. $(2x^3y^2)(5x^4y^7) = 10x^7y^9$

143. $2x^4(8x^4 + 3x) = 2x^4(8x^4) + 2x^4(3x)$
$$= 16x^8 + 6x^5$$

144. $2x(x^2 + 4x + 5) + 3(x^2 + 4x + 5)$
$$= 2x^3 + 8x^2 + 10x + 3x^2 + 12x + 15$$
$$= 2x^3 + 8x^2 + 3x^2 + 10x + 12x + 15$$
$$= 2x^3 + 11x^2 + 22x + 15$$

Section P.4

Check Point Exercises

1. **a.** $(-17x^3 + 4x^2 - 11x - 5) + (16x^3 - 3x^2 + 3x - 15)$
$$= (-17x^3 + 16x^3) + (4x^2 - 3x^2) + (-11x + 3x) + (-5 - 15)$$
$$= -x^3 + x^2 - 8x - 20$$

    b. $(13x^2 - 9x^2 - 7x + 1) - (-7x^3 + 2x^2 - 5x + 9)$
$$= (13x^3 - 9x^2 - 7x + 1) + (7x^3 - 2x^2 + 5x - 9)$$
$$= (13x^3 + 7x^3) + (-9x^2 - 2x^2) + (-7x + 5x) + (1 - 9)$$
$$= 20x^3 - 11x^2 - 2x - 8$$

2. $(5x - 2)(3x^2 - 5x + 4)$
$$= 5x(3x^2 - 5x + 4) - 2(3x^2 - 5x + 4)$$
$$= 5x \cdot 3x^2 - 5x \cdot 5x + 5x \cdot 4 - 2 \cdot 3x^2 + 2 \cdot 5x - 2 \cdot 4$$
$$= 15x^3 - 25x^2 + 20x - 6x^2 + 10x - 8$$
$$= 15x^3 - 31x^2 + 30x - 8$$

3. $(7x - 5)(4x - 3) = 7x \cdot 4x + 7x(-3) + (-5)4x + (-5)(-3)$
$$= 28x^2 - 21x - 20x + 15$$
$$= 28x^2 - 41x + 15$$

4. **a.** $(7x - 6y)(3x - y) = (7x)(3x) + (7x)(-y) + (-6y)(3x) + (-6y)(-y)$
$$= 21x^2 - 7xy - 18xy + 6y^2$$
$$= 21x^2 - 25xy + 6y^2$$

    b. $(2x + 4y)^2 = (2x)^2 + 2(2x)(4y) + (4y)^2$
$$= 4x^2 + 16xy + 16y^2$$

5. **a.** $(3x+2+5y)(3x+2-5y) = (3x+2)^2 - (5y)^2$

$$= 9x^2 + 12x + 4 - 25y^2$$

$$= 9x^2 + 12x - 25y^2 + 4$$

b. $(2x+y+3)^2 = (2x+y)^2 + 2(2x+y)(3) + 3^2$

$$= 4x^2 + 4xy + y^2 + 12x + 6y + 9$$

$$= 4x^2 + 4xy + 12x + y^2 + 6y + 9$$

Exercise Set P.4

1. yes; $2x + 3x^2 - 5 = 3x^2 + 2x - 5$

3. no; The form of a polynomial involves addition and subtraction, not division.

5. $3x^2$ has degree 2
$-5x$ has degree 1
4 has degree 0
$3x^2 - 5x + 4$ has degree 2.

7. x^2 has degree 2
$-4x^3$ has degree 3
$9x$ has degree 1
$-12x^4$ has degree 4
63 has degree 0
$x^2 - 4x^3 + 9x - 12x^4 + 63$ has degree 4.

9. $(-6x^3 + 5x^2 - 8x + 9) + (17x^3 + 2x^2 - 4x - 13) = (-6x^3 + 17x^3) + (5x^2 + 2x^2) + (-8x - 4x) + (9 - 13)$

$$= 11x^3 + 7x^2 - 12x - 4$$

The degree is 3.

11. $(17x^3 - 5x^2 + 4x - 3) - (5x^3 - 9x^2 - 8x + 11) = (17x^3 - 5x^2 + 4x - 3) + (-5x^3 + 9x^2 + 8x - 11)$

$$= (17x^3 - 5x^3) + (-5x^2 + 9x^2) + (4x + 8x) + (-3 - 11)$$

$$= 12x^3 + 4x^2 + 12x - 14$$

The degree is 3.

13. $(5x^2 - 7x - 8) + (2x^2 - 3x + 7) - (x^2 - 4x - 3) = (5x^2 - 7x - 8) + (2x^2 - 3x + 7) + (-x^2 + 4x + 3)$

$$= (5x^2 + 2x^2 - x^2) + (-7x - 3x + 4x) + (-8 + 7 + 3)$$

$$= 6x^2 - 6x + 2$$

The degree is 2.

15. $(x+1)(x^2 - x + 1) = x(x^2) - x \cdot x + x \cdot 1 + 1(x^2) - 1 \cdot x + 1 \cdot 1$

$$= x^3 - x^2 + x + x^2 - x + 1$$

$$= x^3 + 1$$

17. $(2x-3)(x^2 - 3x + 5) = (2x)(x^2) + (2x)(-3x) + (2x)(5) + (-3)(x^2) + (-3)(-3x) + (-3)(5)$

$$= 2x^3 - 6x^2 + 10x - 3x^2 + 9x - 15$$

$$= 2x^3 - 9x^2 + 19x - 15$$

19. $(x+7)(x+3) = x^2 + 3x + 7x + 21 = x^2 + 10x + 21$

21. $(x-5)(x+3) = x^2 + 3x - 5x - 15 = x^2 - 2x - 15$

23. $(3x+5)(2x+1) = (3x)(2x) + 3x(1) + 5(2x) + 5 = 6x^2 + 3x + 10x + 5 = 6x^2 + 13x + 5$

25. $(2x-3)(5x+3) = (2x)(5x) + (2x)(3) + (-3)(5x) + (-3)(3) = 10x^2 + 6x - 15x - 9 = 10x^2 - 9x - 9$

27. $(5x^2-4)(3x^2-7) = (5x^2)(3x^2) + (5x^2)(-7) + (-4)(3x^2) + (-4)(-7) = 15x^4 - 35x^2 - 12x^2 + 28 = 15x^4 - 47x^2 + 28$

29. $\left(8x^3+3\right)\left(x^2-5\right) = \left(8x^3\right)\left(x^2\right) + \left(8x^3\right)(-5) + (3)\left(x^2\right) + (3)(-5) = 8x^5 - 40x^3 + 3x^2 - 15$

31. $(x+3)(x-3) = x^2 - 3^2 = x^2 - 9$

33. $(3x+2)(3x-2) = (3x)^2 - 2^2 = 9x^2 - 4$

35. $(5-7x)(5+7x) = 5^2 - (7x)^2 = 25 - 49x^2$

37. $(4x^2+5x)(4x^2-5x) = (4x^2)^2 - (5x)^2 = 16x^4 - 25x^2$

39. $\left(1-y^5\right)\left(1+y^5\right) = (1)^2 - \left(y^5\right)^2 = 1 - y^{10}$

41. $(x+2)^2 = x^2 + 2 \cdot x \cdot 2 + 2^2 = x^2 + 4x + 4$

43. $(2x+3)^2 = (2x)^2 + 2(2x)(3) + 3^2 = 4x^2 + 12x + 9$

45. $(x-3)^2 = x^2 - 2 \cdot x \cdot 3 + 3^2 = x^2 - 6x + 9$

47. $(4x^2-1)^2 = (4x^2)^2 - 2(4x^2)(1) + 1^2 = 16x^4 - 8x^2 + 1$

49. $(7-2x)^2 = 7^2 - 2(7)(2x) + (2x)^2 = 49 - 28x + 4x^2 = 4x^2 - 28x + 49$

51. $(x+1)^3 = x^3 + 3 \cdot x^2 \cdot 1 + 3x \cdot 1^2 + 1^3 = x^3 + 3x^2 + 3x + 1$

53. $(2x+3)^3 = (2x)^3 + 3 \cdot (2x)^2 \cdot 3 + 3(2x) \cdot 3^2 + 3^3 = 8x^3 + 36x^2 + 54x + 27$

55. $(x-3)^3 = x^3 - 3 \cdot x^3 \cdot 3 + 3 \cdot x \cdot 3^2 - 3^3 = x^3 - 9x^2 + 27x - 27$

57. $(3x-4)^3 = (3x)^3 - 3(3x)^2 \cdot 4 + 3(3x) \cdot 4^2 - 4^3 = 27x^3 - 108x^2 + 144x - 64$

59. $(x+5y)(7x+3y) = x(7x) + x(3y) + (5y)(7x) + (5y)(3y)$
$$= 7x^2 + 3xy + 35xy + 15y^2$$
$$= 7x^2 + 38xy + 15y^2$$

61. $(x-3y)(2x+7y) = x(2x) + x(7y) + (-3y)(2x) + (-3y)(7y)$
$$= 2x^2 + 7xy - 6xy - 21y^2$$
$$= 2x^2 + xy - 21y^2$$

63. $(3xy-1)(5xy+2) = (3xy)(5xy)+(3xy)(2)+(-1)(5xy)+(-1)(2)$
$$= 15x^2y^2+6xy-5xy-2$$
$$= 15x^2y^2+xy-2$$

65. $(7x+5y)^2 = (7x)^2+2(7x)(5y)+(5y)^2 = 49x^2+70xy+25y^2$

67. $(x^2y^2-3)^2 = (x^2y^2)^2-2(x^2y^2)(3)+3^2 = x^4y^4-6x^2y^2+9$

69. $(x-y)(x^2+xy+y^2) = x(x^2)+x(xy)+x(y^2)+(-y)(x^2)+(-y)(xy)+(-y)(y^2)$
$$= x^3+x^2y+xy^2-x^2y-xy^2-y^3$$
$$= x^3-y^3$$

71. $(3x+5y)(3x-5y) = (3x)^2-(5y)^2 = 9x^2-25y^2$

73. $(x+y+3)(x+y-3) = (x+y)^2-3^2 = x^2+2xy+y^2-9$

75. $(3x+7-5y)(3x+7+5y) = (3x+7)^2-(5y)^2 = 9x^2+42x+49-25y^2$

77. $[5y-(2x+3)][5y+(2x+3)] = (5y)^2-(2x+3)^2 = 25y^2-(4x^2+12x+9) = 25y^2-4x^2-12x-9$

79. $(x+y+1)^2 = (x+y)^2+2(x+y)+1 = x^2+2xy+y^2+2x+2y+1$

81. $(2x+y+1)^2 = (2x+y)^2+2(2x+y)+1 = 4x^2+4xy+y^2+4x+2y+1$

83. $(3x+4y)^2-(3x-4y)^2 = \left[(3x)^2+2(3x)(4y)+(4y)^2\right]-\left[(3x)^2-2(3x)(4y)+(4y)^2\right]$
$$= \left(9x^2+24xy+16y^2\right)-\left(9x^2-24xy+16y^2\right)$$
$$= 9x^2+24xy+16y^2-9x^2+24xy-16y^2$$
$$= 48xy$$

85. $\left(5x-7\right)\left(3x-2\right)-\left(4x-5\right)\left(6x-1\right)$
$$= \left[15x^2-10x-21x+14\right]-\left[24x^2-4x-30x+5\right]$$
$$= \left(15x^2-31x+14\right)-\left(24x^2-34x+5\right)$$
$$= 15x^2-31x+14-24x^2+34x-5$$
$$= -9x^2+3x+9$$

87. $\left(2x+5\right)\left(2x-5\right)\left(4x^2+25\right)$
$$= \left[(2x)^2-5^2\right]\left(4x^2+25\right)$$
$$= \left(4x^2-25\right)\left(4x^2+25\right)$$
$$= \left(4x^2\right)^2-(25)^2$$
$$= 16x^4-625$$

89. $\dfrac{(2x-7)^5}{(2x-7)^3} = (2x-7)^{5-3}$

$$= (2x-7)^2$$

$$= (2x)^2 - 2(2x)(7) + (7)^2$$

$$= 4x^2 - 28x + 49$$

91. a. $M = 177x^2 + 288x + 7075$

$M = 177(16)^2 + 288(16) + 7075 = 56,995$

The model estimates the median annual income for a man with 16 years of education to be \$56,995.
The model underestimates the actual value of \$57,220 shown in the bar graph by \$225.

b. $M - W = (-18x^3 + 923x^2 - 9603x + 48,446) - (17x^3 - 450x^2 + 6392x - 14,764)$

$M - W = -18x^3 + 923x^2 - 9603x + 48,446 - 17x^3 + 450x^2 - 6392x + 14,764$

$M - W = -18x^3 - 17x^3 + 923x^2 + 450x^2 - 9603x - 6392x + 48,446 + 14,764$

$M - W = -35x^3 + 1373x^2 - 15,995x + 63,210$

c. $M - W = -35x^3 + 1373x^2 - 15,995x + 63,210$

$M - W = -35(14)^3 + 1373(14)^2 - 15,995(14) + 63,210 = 12,348$

The difference in the median income between men and women with 14 years experience is \$12,348.

d. $44,404 - 33,481 = 10,923$

The actual difference displayed in the graph in the median income between men and women with 14 years experience is \$10,923.
The model overestimates this difference by $\$12,348 - \$10,923 = \$1425$.

93. $x(8-2x)(10-2x) = x(80 - 36x + 4x^2)$

$$= 80x - 36x^2 + 4x^3$$

$$= 4x^3 - 36x^2 + 80x$$

95. $(x+9)(x+3) - (x+5)(x+1)$

$= x^2 + 12x + 27 - (x^2 + 6x + 5)$

$= x^2 + 12x + 27 - x^2 - 6x - 5$

$= 6x + 22$

97. – 101. Answers may vary.

103. makes sense

105. makes sense

107. $(x+3)(x-1) + ((x+3) - x)(x - (x-1))$

$= (x+3)(x-1) + 3(x - x + 1)$

$= x^2 - x + 3x - 3 + 3$

$= x^2 + 2x$

109. $(x+5)(2x+1)(x+2)-3 \cdot x(x+5)$

$= (2x^2+11x+5)(x+2)-3x^2-15x$

$= 2x^3+15x^2+27x+10-3x^2-15x$

$= 2x^3+12x^2+12x+10$

111. $(x+3)(x+\boxed{4})=x^2+7x+12$

112. $(x-\boxed{2})(x-12)=x^2-14x+24$

113. $(4x+1)(2x-\boxed{3})=8x^2-10x-3$

Section P.5

Check Point Exercises

1. **a.** $10x^3-4x^2$

$= 2x^2(5x)-2x^2(2)$

$= 2x^2(5x-2)$

b. $2x(x-7)+3(x-7)$

$= (x-7)(2x+3)$

2. $x^3+5x^2-2x-10$

$= (x^3+5x^2)-(2x+10)$

$= x^2(x+5)-2(x+5)$

$= (x+5)(x^2-2)$

3. Find two numbers whose product is 40 and whose sum is 13. The required integers are 8 and 5. Thus,
$x^2+13x+40=(x+5)(x+8)$ or $(x+8)(x+5)$

4. Find two numbers whose product is -14 and whose sum is -5. The required integers are -7 and 2. Thus,
$x^2-5x-14=(x-7)(x+2)$ or $(x+2)(x-7)$.

5. Find two First terms whose product is $6x^2$.
$6x^2+19x-7=(6x \quad)(x \quad)$
$6x^2+19x-7=(3x \quad)(2x \quad)$

Find two Last terms whose product is -7.
The possible factors are $1(-7)$ and $-1(7)$.

Try various combinations of these factors to find the factorization in which the sum of the Outside and Inside products is $19x$.

Possible Factors of $6x^2 + 19x - 7$	Sum of Outside and Inside Products (Should Equal $19x$)
$(6x+1)(x-7)$	$-42x + x = -41x$
$(6x-7)(x+1)$	$6x - 7x = -x$
$(6x-1)(x+7)$	$42x - x = 41x$
$(6x+7)(x-1)$	$-6x + 7x = x$
$(3x+1)(2x-7)$	$-21x + 2x = -19x$
$(3x-7)(2x+1)$	$3x - 14x = -11x$
$(3x-1)(2x+7)$	$21x - 2x = 19x$
$(3x+7)(2x-1)$	$-3x + 14x = 11x$

Thus, $6x^2 + 19x - 7 = (3x-1)(2x+7)$ or $(2x+7)(3x-1)$.

6. Find two First terms whose product is $3x^2$.
$3x^2 - 13xy + 4y^2 = (3x \quad)(x \quad)$

Find two Last terms whose product is $4y^2$.
The possible factors are $(2y)(2y)$, $(-2y)(-2y)$, $(4y)(y)$, and $(-4y)(-y)$.

Try various combinations of these factors to find the factorization in which the sum of the Outside and Inside products is $-13xy$.

$3x^2 - 13xy + y^2 = (3x - y)(x - 4y)$ or $(x - 4y)(3x - y)$.

7. Express each term as the square of some monomial. Then use the formula for factoring $A^2 - B^2$.
   a. $x^2 - 81 = x^2 - 9^2 = (x+9)(x-9)$

   b. $36x^2 - 25 = (6x)^2 - 5^2 = (6x+5)(6x-5)$

8. Express $81x^4 - 16$ as the difference of two squares and use the formula for factoring $A^2 - B^2$.
$81x^4 - 16 = (9x^2)^2 - 4^2 = (9x^2 + 4)(9x^2 - 4)$

   The factor $9x^2 - 4$ is the difference of two squares and can be factored. Express $9x^2 - 4$ as the difference of two squares and again use the formula for factoring $A^2 - B^2$.
$(9x^2 + 4)(9x^2 - 4) = (9x^2 + 4)\left[(3x)^2 - 2^2\right] = (9x^2 + 4)(3x+2)(3x-2)$

   Thus, factored completely,
$81x^4 - 16 = (9x^2 + 4)(3x+2)(3x-2)$.

9. **a.** $x^2 + 14x + 49 = x^2 + 2 \cdot x \cdot 7 + 7^2 = (x+7)^2$

b. Since $16x^2 = (4x)^2$ and $49 = 7^2$, check to see if the middle term can be expressed as twice the product of $4x$ and 7. Since $2 \cdot 4x \cdot 7 = 56x$, $16x^2 - 56x + 49$ is a perfect square trinomial. Thus,

$$16x^2 - 56x + 49 = (4x)^2 - 2 \cdot 4x \cdot 7 + 7^2$$
$$= (4x-7)^2$$

10. **a.** $x^3 + 1 = x^3 + 1^3$
$$= (x+1)(x^2 - x \cdot 1 + 1^2)$$
$$= (x+1)(x^2 - x + 1)$$

b. $125x^3 - 8 = (5x)^3 - 2^3$
$$= (5x-2)\left[(5x)^2 + (5x)(2) + 2^2\right]$$
$$= (5x-2)(25x^2 + 10x + 4)$$

11. Factor out the greatest common factor.
$$3x^3 - 30x^2 + 75x = 3x\left(x^2 - 10x + 25\right)$$
Factor the perfect square trinomial.
$$3x\left(x^2 - 10x + 25\right) = 3x(x-5)^2$$

12. Reorder to write as a difference of squares.
$x^2 - 36a^2 + 20x + 100$
$= x^2 + 20x + 100 - 36a^2$
$= \left(x^2 + 20x + 100\right) - 36a^2$
$= (x+10)^2 - 36a^2$
$= (x+10+6a)(x+10-6a)$

13. $x(x-1)^{-\frac{1}{2}} + (x-1)^{\frac{1}{2}}$
$= (x-1)^{-\frac{1}{2}}\left[x + (x-1)^{\frac{1}{2}-\left(-\frac{1}{2}\right)}\right]$
$= (x-1)^{-\frac{1}{2}}\left[x + (x-1)\right]$
$= (x-1)^{-\frac{1}{2}}(2x-1)$
$= \dfrac{(2x-1)}{(x-1)^{\frac{1}{2}}}$

Exercise Set P.5

1. $18x + 27 = 9 \cdot 2x + 9 \cdot 3 = 9(2x+3)$

3. $3x^2 + 6x = 3x \cdot x + 3x \cdot 2 = 3x(x+2)$

5. $9x^4 - 18x^3 + 27x^2$
$= 9x^2(x^2) + 9x^2(-2x) + 9x^2(3)$
$= 9x^2(x^2 - 2x + 3)$

7. $x(x+5) + 3(x+5) = (x+5)(x+3)$

9. $x^2(x-3) + 12(x-3) = (x-3)(x^2 + 12)$

11. $x^3 - 2x^2 + 5x - 10 = x^2(x-2) + 5(x-2)$
$$= (x^2 + 5)(x-2)$$

13. $x^3 - x^2 + 2x - 2 = x^2(x-1) + 2(x-1)$
$$= (x-1)(x^2 + 2)$$

15. $3x^3 - 2x^2 - 6x + 4 = x^2(3x-2) - 2(3x-2)$
$$= (3x-2)(x^2 - 2)$$

17. $x^2 + 5x + 6 = (x+2)(x+3)$

19. $x^2 - 2x - 15 = (x-5)(x+3)$

21. $x^2 - 8x + 15 = (x-5)(x-3)$

23. $3x^2 - x - 2 = (3x+2)(x-1)$

25. $3x^2 - 25x - 28 = (3x-28)(x+1)$

27. $6x^2 - 11x + 4 = (2x-1)(3x-4)$

29. $4x^2 + 16x + 15 = (2x+3)(2x+5)$

31. $9x^2 - 9x + 2 = (3x-1)(3x-2)$

33. $20x^2 + 27x - 8 = (5x+8)(4x-1)$

35. $2x^2 + 3xy + y^2 = (2x+y)(x+y)$

37. $6x^2 - 5xy - 6y^2 = (3x + 2y)(2x - 3y)$

39. $x^2 - 100 = x^2 - 10^2 = (x + 10)(x - 10)$

41. $36x^2 - 49 = (6x)^2 - 7^2 = (6x + 7)(6x - 7)$

43. $9x^2 - 25y^2 = (3x)^2 - (5y)^2$
$$= (3x + 5y)(3x - 5y)$$

45. $x^4 - 16 = (x^2)^2 - 4^2$
$$= (x^2 + 4)(x^2 - 4)$$
$$= (x^2 + 4)(x + 2)(x - 2)$$

47. $16x^4 - 81 = (4x^2)^2 - 9^2$
$$= (4x^2 + 9)(4x^2 - 9)$$
$$= (4x^2 + 9)[(2x)^2 - 3^2]$$
$$= (4x^2 + 9)(2x + 3)(2x - 3)$$

49. $x^2 + 2x + 1 = x^2 + 2 \cdot x \cdot 1 + 1^2 = (x + 1)^2$

51. $x^2 - 14x + 49 = x^2 - 2 \cdot x \cdot 7 + 7^2$
$$= (x - 7)^2$$

53. $4x^2 + 4x + 1 = (2x)^2 + 2 \cdot 2x \cdot 1 + 1^2$
$$= (2x + 1)^2$$

55. $9x^2 - 6x + 1 = (3x)^2 - 2 \cdot 3x \cdot 1 + 1^2$
$$= (3x - 1)^2$$

57. $x^3 + 27 = x^3 + 3^3$
$$= (x + 3)(x^2 - x \cdot 3 + 3^2)$$
$$= (x + 3)(x^2 - 3x + 9)$$

59. $x^3 - 64 = x^3 - 4^3$
$$= (x - 4)(x^2 + x \cdot 4 + 4^2)$$
$$= (x - 4)(x^2 + 4x + 16)$$

61. $8x^3 - 1 = (2x)^3 - 1^3$
$$= (2x - 1)[(2x)^2 + (2x)(1) + 1^2]$$
$$= (2x - 1)(4x^2 + 2x + 1)$$

63. $64x^3 + 27 = (4x)^3 + 3^3$
$$= (4x + 3)[(4x)^2 - (4x)(3) + 3^2]$$
$$= (4x + 3)(16x^2 - 12x + 9)$$

65. $3x^3 - 3x = 3x(x^2 - 1) = 3x(x + 1)(x - 1)$

67. $4x^2 - 4x - 24 = 4(x^2 - x - 6)$
$$= 4(x + 2)(x - 3)$$

69. $2x^4 - 162 = 2(x^4 - 81)$
$$= 2[(x^2)^2 - 9^2]$$
$$= 2(x^2 + 9)(x^2 - 9)$$
$$= 2(x^2 + 9)(x^2 - 3^2)$$
$$= 2(x^2 + 9)(x + 3)(x - 3)$$

71. $x^3 + 2x^2 - 9x - 18 = (x^3 + 2x^2) - (9x + 18)$
$$= x^2(x + 2) - 9(x + 2)$$
$$= (x^2 - 9)(x + 2)$$
$$= (x^2 - 3^2)(x + 2)$$
$$= (x - 3)(x + 3)(x + 2)$$

73. $2x^2 - 2x - 112 = 2(x^2 - x - 56) = 2(x - 8)(x + 7)$

75. $x^3 - 4x = x(x^2 - 4)$
$$= x(x^2 - 2^2)$$
$$= x(x - 2)(x + 2)$$

77. $x^2 + 64$ is prime.

79. $x^3 + 2x^2 - 4x - 8 = (x^3 + 2x^2) + (-4x - 8)$

$= x^2(x+2) - 4(x+2) = (x^2 - 4)(x+2) = (x^2 - 2^2)(x+2) = (x-2)(x+2)(x+2) = (x-2)(x+2)^2$

81. $y^5 - 81y$

$= y(y^4 - 81) = y[(y^2)^2 - 9^2] = y(y^2 + 9)(y^2 - 9) = y(y^2 + 9)(y^2 - 3^2) = y(y^2 + 9)(y+3)(y-3)$

83. $20y^4 - 45y^2 = 5y^2(4y^2 - 9) = 5y^2[(2y)^2 - 3^2] = 5y^2(2y+3)(2y-3)$

85. $x^2 - 12x + 36 - 49y^2$

$= (x^2 - 12x + 36) - 49y^2 = (x-6)^2 - 49y^2 = (x-6+7y)(x-6-7y)$

87. $9b^2x - 16y - 16x + 9b^2y$

$= (9b^2x + 9b^2y) + (-16x - 16y) = 9b^2(x+y) - 16(x+y) = (x+y)(9b^2 - 16) = (x+y)(3b+4)(3b-4)$

89. $x^2y - 16y + 32 - 2x^2$

$= (x^2y - 16y) + (-2x^2 + 32) = y(x^2 - 16) - 2(x^2 - 16) = (x^2 - 16)(y-2) = (x+4)(x-4)(y-2)$

91. $2x^3 - 8a^2x + 24x^2 + 72x$

$= 2x(x^2 - 4a^2 + 12x + 36) = 2x[(x^2 + 12x + 36) - 4a^2] = 2x[(x+6)^2 - 4a^2] = 2x(x+6-2a)(x+6+2a)$

93. $x^{\frac{3}{2}} - x^{\frac{1}{2}} = x^{\frac{1}{2}}\left(x^{\frac{3}{2}-\frac{1}{2}}\right) - 1 = x^{\frac{1}{2}}(x-1)$

95. $4x^{-\frac{2}{3}} + 8x^{\frac{1}{3}} = 4x^{-\frac{2}{3}}\left(1 + 2x^{\frac{1}{3}-\left(-\frac{2}{3}\right)}\right) = 4x^{-\frac{2}{3}}(1+2x) = \dfrac{4(1+2x)}{x^{\frac{2}{3}}}$

97. $(x+3)^{\frac{1}{2}} - (x+3)^{\frac{3}{2}} = (x+3)^{\frac{1}{2}}\left[1 - (x+3)^{\frac{3}{2}-\frac{1}{2}}\right] = (x+3)^{\frac{1}{2}}[1-(x+3)] = (x+3)^{\frac{1}{2}}(-x-2) = -(x+3)^{\frac{1}{2}}(x+2)$

99. $(x+5)^{-\frac{1}{2}} - (x+5)^{-\frac{3}{2}} = (x+5)^{-\frac{3}{2}}\left[(x+5)^{-\frac{1}{2}-\left(-\frac{3}{2}\right)} - 1\right] = (x+5)^{-\frac{3}{2}}[(x+5)-1] = (x+5)^{-\frac{3}{2}}(x+4) = \dfrac{x+4}{(x+5)^{\frac{3}{2}}}$

101. $(4x-1)^{\frac{1}{2}} - \dfrac{1}{3}(4x-1)^{\frac{3}{2}}$

$= (4x-1)^{\frac{1}{2}}\left[1 - \dfrac{1}{3}(4x-1)^{\frac{3}{2}-\frac{1}{2}}\right] = (4x-1)^{\frac{1}{2}}\left[1 - \dfrac{1}{3}(4x-1)\right] = (4x-1)^{\frac{1}{2}}\left[1 - \dfrac{4}{3}x + \dfrac{1}{3}\right]$

$= (4x-1)^{\frac{1}{2}}\left(\dfrac{4}{3} - \dfrac{4}{3}x\right) = (4x-1)^{\frac{1}{2}}\dfrac{4}{3}(1-x) = \dfrac{-4(4x-1)^{\frac{1}{2}}(x-1)}{3}$

103. $10x^2(x+1) - 7x(x+1) - 6(x+1) = (x+1)(10x^2 - 7x - 6) = (x+1)(5x-6)(2x+1)$

105. $6x^4 + 35x^2 - 6 = (x^2 + 6)(6x^2 - 1)$

107. $y^7 + y = y\left(y^6 + 1\right) = y\left[\left(y^2\right)^3 + 1^3\right] = y\left(y^2 + 1\right)\left(y^4 - y^2 + 1\right)$

109. $x^4 - 5x^2y^2 + 4y^4 = \left(x^2 - 4y^2\right)\left(x^2 - y^2\right) = (x + 2y)(x - 2y)(x + y)(x - y)$

111. $(x - y)^4 - 4(x - y)^2$

$= (x - y)^2\left((x - y)^2 - 4\right) = (x - y)^2\left((x - y) + 2\right)\left((x - y) - 2\right) = (x - y)^2(x - y + 2)(x - y - 2)$

113. $2x^2 - 7xy^2 + 3y^4 = \left(2x - y^2\right)\left(x - 3y^2\right)$

115. a. $(x - 0.4x) - 0.4(x - 0.4x) = (x - 0.4x)(1 - 0.4) = (0.6x)(0.6) = 0.36x$

    b. No, the computer is selling at 36% of its original price.

117. a. $(3x)^2 - 4 \cdot 2^2 = 9x^2 - 16$

    b. $9x^2 - 16 = (3x + 4)(3x - 4)$

119. a. $x(x + y) - y(x + y)$

    b. $x(x + y) - y(x + y) = (x + y)(x - y)$

121. $V_{shaded} = V_{outside} - V_{inside}$

$= a \cdot a \cdot 4a - b \cdot b \cdot 4a$

$= 4a^3 - 4ab^2$

$= 4a\left(a^2 - b^2\right)$

$= 4a(a + b)(a - b)$

123. – 129. Answers may vary.

131. makes sense

133. makes sense

135. true

137. false; Changes to make the statement true will vary. A sample change is: $x^3 - 64 = (x - 4)(x + 4x + 16)$

139. $-x^2 - 4x + 5 = -1\left(x^2 + 4x - 5\right) = -1(x + 5)(x - 1) = -(x + 5)(x - 1)$

141. $(x - 5)^{-\frac{1}{2}}(x + 5)^{-\frac{1}{2}} - (x + 5)^{\frac{1}{2}}(x - 5)^{-\frac{3}{2}} = (x - 5)^{-\frac{3}{2}}(x + 5)^{-\frac{1}{2}}\left[(x - 5)^{-\frac{1}{2} - \left(-\frac{3}{2}\right)} - (x + 5)^{\frac{1}{2} - \left(-\frac{1}{2}\right)}\right]$

$= (x - 5)^{-\frac{3}{2}}(x + 5)^{-\frac{1}{2}}\left[(x - 5) - (x + 5)\right]$

$= (x - 5)^{-\frac{3}{2}}(x + 5)^{-\frac{1}{2}}(-10) = \dfrac{-10}{(x - 5)^{\frac{3}{2}}(x + 5)^{\frac{1}{2}}}$

143. $b = 0, 3, 4,$ or $-c(c + 4),$ where $c > 0$ is an integer.

144. $\dfrac{x^2 + 6x + 5}{x^2 - 25} = \dfrac{(x+5)(x+1)}{(x+5)(x-5)} = \dfrac{x+1}{x-5}$

145. $\dfrac{5}{4} \cdot \dfrac{8}{15} = \dfrac{5}{4} \cdot \dfrac{4 \cdot 2}{5 \cdot 3} = \dfrac{1}{1} \cdot \dfrac{2}{3} = \dfrac{2}{3}$

146. $\dfrac{1}{2} + \dfrac{2}{3} = \dfrac{3}{6} + \dfrac{4}{6} = \dfrac{7}{6}$

Section P.6

Check Point Exercises

1. **a.** The denominator would equal zero if $x = -5,$ so -5 must be excluded from the domain.

   b. $x^2 - 36 = (x+6)(x-6)$
   The denominator would equal zero if $x = -6$ or $x = 6,$ so -6 and 6 must both must be excluded from the domain.

2. **a.** $\dfrac{x^3 + 3x^2}{x+3} = \dfrac{x^2(x+3)}{x+3}$ Because the denominator is $x + 3,$ $x \neq -3$

   $\phantom{\dfrac{x^3 + 3x^2}{x+3}} = \dfrac{x^2(x+3)}{x+3}$

   $\phantom{\dfrac{x^3 + 3x^2}{x+3}} = x^2, \; x \neq -3$

   b. $\dfrac{x^2 - 1}{x^2 + 2x + 1} = \dfrac{(x-1)(x+1)}{(x+1)(x+1)}$ Because the denominator is

   $\phantom{\dfrac{x^2 - 1}{x^2 + 2x + 1}} = \dfrac{x-1}{x+1}, x \neq -1$ $(x+1)(x+1), x \neq -1$

3. $\dfrac{x+3}{x^2 - 4} \cdot \dfrac{x^2 - x - 6}{x^2 + 6x + 9}$

   $= \dfrac{x+3}{(x+2)(x-2)} \cdot \dfrac{(x-3)(x+2)}{(x+3)(x+3)}$ Because the denominator has factors of $x + 2,$ $x - 2,$ and $x + 3,$ $x \neq -2,$ $x \neq 2,$ and $x \neq -3.$

   $= \dfrac{x+3}{(x+2)(x-2)} \cdot \dfrac{(x-3)(x+2)}{(x+3)(x+3)}$

   $= \dfrac{x-3}{(x-2)(x+3)}, \; x \neq -2, x \neq 2, x \neq -3$

4. $\dfrac{x^2-2x+1}{x^3+x} \div \dfrac{x^2+x-2}{3x^2+3}$

$= \dfrac{x^2-2x+1}{x^3+x} \cdot \dfrac{3x^2+3}{x^2+x-2}$

$= \dfrac{(x-1)(x-1)}{x(x^2+1)} \cdot \dfrac{3(x^2+1)}{(x+2)(x-1)}$

$= \dfrac{3(x-1)}{x(x+2)}, \; x \neq 0, \; x \neq -2, \; x \neq 1$

5. $\dfrac{x}{x+1} - \dfrac{3x+2}{x+1} = \dfrac{x-3x-2}{x+1}$

$= \dfrac{-2x-2}{x+1}$

$= \dfrac{-2(x+1)}{x+1}$

$= -2, x \neq -1$

6. $\dfrac{3}{x+1} + \dfrac{5}{x-1}$

$= \dfrac{3x(x-1)+5(x+1)}{(x+1)(x-1)}$

$= \dfrac{3x-3+5x+5}{(x+1)(x-1)}$

$= \dfrac{8x+2}{(x+1)(x-1)}$

$= \dfrac{2(4x+1)}{(x+1)(x-1)}$

$= \dfrac{2(4x+1)}{(x+1)(x-1)}, \; x \neq -1 \text{ and } x \neq 1.$

7. Factor each denominator completely.
$x^2-6x+9 = (x-3)^2$
$x^2-9 = (x+3)(x-3)$

List the factors of the first denominator.
$x-3, \; x-3$

Add any unlisted factors from the second denominator.
$x-3, \; x-3, \; x+3$

The least common denominator is the product of all factors in the final list.
$(x-3)(x-3)(x+3)$ or $(x-3)^2(x+3)$ is the least common denominator.

8. Find the least common denominator.

$x^2 - 10x + 25 = (x-5)^2$

$2x - 10 = 2(x-5)$

The least common denominator is $2(x-5)(x-5)$.

Write all rational expressions in terms of the least common denominator.

$$\frac{x}{x^2 - 10x + 25} - \frac{x-4}{2x-10}$$

$$= \frac{x}{(x-5)(x-5)} - \frac{x-4}{2(x-5)}$$

$$= \frac{2x}{2(x-5)(x-5)} - \frac{(x-4)(x-5)}{2(x--5)(x-5)}$$

Add numerators, putting this sum over the least common denominator.

$$= \frac{2x - (x-4)(x-5)}{2(x-5)(x-5)}$$

$$= \frac{2x - (x^2 - 5x - 4x + 20)}{2(x-5)(x-5)}$$

$$= \frac{2x - x^2 + 5x + 4x - 20}{2(x-5)(x-5)}$$

$$= \frac{2x - x^2 + 5x + 4x - 20}{2(x-5)(x-5)}$$

$$= \frac{-x^2 + 11x - 20}{2(x-5)(x-5)}$$

$$= \frac{-x^2 + 11x - 20}{2(x-5)^2}, \ x \neq 5$$

9.

$$\frac{\dfrac{1}{x} - \dfrac{3}{2}}{\dfrac{1}{x} + \dfrac{3}{4}} = \frac{\dfrac{2}{2x} - \dfrac{3x}{2x}}{\dfrac{4}{4x} + \dfrac{3x}{4x}}, \ x \neq 0$$

$$= \frac{\dfrac{2-3x}{2x}}{\dfrac{4+3x}{4x}}, \ x \neq \frac{-4}{3}$$

$$= \frac{2-3x}{2x} \div \frac{4+3x}{4x}$$

$$= \frac{2-3x}{2x} \cdot \frac{4x}{4+3x}$$

$$= \frac{2-3x}{4+3x} \cdot \frac{4}{2}$$

$$= \frac{2-3x}{4+3x} \cdot \frac{2}{1}$$

$$= \frac{2(2-3x)}{4+3x}, \ x \neq 0 \ and \ x \neq \frac{-4}{3}$$

10. Multiply each of the three terms, $\dfrac{1}{x+7}$, $\dfrac{1}{x}$, and 7 by the least common denominator of $x(x+7)$.

$$\dfrac{\dfrac{1}{x+7}-\dfrac{1}{x}}{7} = \dfrac{x(x+7)\left(\dfrac{1}{x+7}\right)-x(x+7)\left(\dfrac{1}{x}\right)}{7x(x+7)}$$

$$= \dfrac{x-(x+7)}{7x(x+7)}$$

$$= \dfrac{-7}{7x(x+7)}$$

$$= -\dfrac{1}{x(x+7)}, \; x \neq 0, \; x \neq -7$$

11. $\dfrac{\sqrt{x}+\dfrac{1}{\sqrt{x}}}{x} = \dfrac{\sqrt{x}+\dfrac{1}{\sqrt{x}}}{x} \cdot \dfrac{\sqrt{x}}{\sqrt{x}} = \dfrac{x+1}{x^{3/2}}$

12. $\dfrac{\sqrt{x+3}-\sqrt{x}}{3}$

$$= \dfrac{\sqrt{x+3}-\sqrt{x}}{3} \cdot \dfrac{\sqrt{x+3}+\sqrt{x}}{\sqrt{x+3}\cdot\sqrt{x}}$$

$$= \dfrac{\left(\sqrt{x+3}\right)^2-(\sqrt{x})^2}{3\left(\sqrt{x+3}+\sqrt{x}\right)}$$

$$= \dfrac{x+3-x}{3\left(\sqrt{x+3}+\sqrt{x}\right)}$$

$$= \dfrac{1}{\sqrt{x+3}+\sqrt{x}}$$

Exercise Set P.6

1. $\dfrac{7}{x-3}, \; x \neq 3$

3. $\dfrac{x+5}{x^2-25} = \dfrac{x+5}{(x+5)(x-5)}, \; x \neq 5, -5$

5. $\dfrac{x-1}{x^2+11x+10} = \dfrac{x-1}{(x+1)(x+10)}, \; x \neq -1, -10$

7. $\dfrac{3x-9}{x^2-6x+9} = \dfrac{3(x-3)}{(x-3)(x-3)}$

$$= \dfrac{3}{x-3}, \; x \neq 3$$

9. $\dfrac{x^2-12x+36}{4x-24}=\dfrac{(x-6)(x-6)}{4(x-6)}=\dfrac{x-6}{4}$.

$x\neq 6$

11. $\dfrac{y^2+7y-18}{y^2-3y+2}=\dfrac{(y+9)(y-2)}{(y-2)(y-1)}=\dfrac{y+9}{y-1}$,

$y\neq 1,2$

13. $\dfrac{x^2+12x+36}{x^2-36}=\dfrac{(x+6)^2}{(x+6)(x-6)}=\dfrac{x+6}{x-6}$,

$x\neq 6,-6$

15. $\dfrac{x-2}{3x+9}\cdot\dfrac{2x+6}{2x-4}=\dfrac{x-2}{3(x+3)}\cdot\dfrac{2(x+3)}{2(x-2)}$

$\qquad\qquad =\dfrac{2}{6}=\dfrac{1}{3}, x\neq 2,-3$

17. $\dfrac{x^2-9}{x^2}\cdot\dfrac{x^2-3x}{x^2+x-12}$

$=\dfrac{(x-3)(x+3)}{x^2}\cdot\dfrac{x(x-3)}{(x+4)(x-3)}$

$=\dfrac{(x-3)(x+3)}{x(x+4)}, x\neq 0,-4,3$

19. $\dfrac{x^2-5x+6}{x^2-2x-3}\cdot\dfrac{x^2-1}{x^2-4}$

$=\dfrac{(x-3)(x-2)}{(x-3)(x+1)}\cdot\dfrac{(x+1)(x-1)}{(x-2)(x+2)}$

$=\dfrac{x-1}{x+2}, x\neq -2,-1,2,3$

21. $\dfrac{x^3-8}{x^2-4}\cdot\dfrac{x+2}{3x}=\dfrac{(x-2)(x^2+2x+4)}{(x-2)(x+2)}\cdot\dfrac{x+2}{3x}$

$=\dfrac{x^2+2x+4}{3x}, x\neq -2,0,2$

23. $\dfrac{x+1}{3}\div\dfrac{3x+3}{7}=\dfrac{x+1}{3}\div\dfrac{3(x+1)}{7}$

$=\dfrac{x+1}{3}\cdot\dfrac{7}{3(x+1)}$

$=\dfrac{7}{9}, x\neq -1$

25. $\dfrac{x^2-4}{x}\div\dfrac{x+2}{x-2}=\dfrac{(x-2)(x+2)}{x}\cdot\dfrac{x-2}{x+2}$

$\qquad\qquad =\dfrac{(x-2)^2}{x}; x\neq 0,-2,2$

27. $\dfrac{4x^2+10}{x-3}\div\dfrac{6x^2+15}{x^2-9}$

$=\dfrac{2(2x^2+5)}{x-3}\div\dfrac{3(2x^2+5)}{(x-3)(x+3)}$

$=\dfrac{2(2x^2+5)}{x-3}\cdot\dfrac{(x-3)(x+3)}{3(2x^2+5)}$

$=\dfrac{2(x+3)}{3}, x\neq 3,-3$

29. $\dfrac{x^2-25}{2x-2}\div\dfrac{x^2+10x+25}{x^2+4x-5}$

$=\dfrac{(x-5)(x+5)}{2(x-1)}\div\dfrac{(x+5)^2}{(x+5)(x-1)}$

$=\dfrac{(x-5)(x+5)}{2(x-1)}\cdot\dfrac{(x+5)(x-1)}{(x+5)^2}$

$=\dfrac{x-5}{2}, x\neq 1,-5$

31. $\dfrac{x^2+x-12}{x^2+x-30}\cdot\dfrac{x^2+5x+6}{x^2-2x-3}\div\dfrac{x+3}{x^2+7x+6}$

$=\dfrac{(x+4)(x-3)}{(x+6)(x-5)}\cdot\dfrac{(x+2)(x+3)}{(x+1)(x-3)}\cdot\dfrac{(x+6)(x+1)}{x+3}$

$=\dfrac{(x+4)(x+2)}{x-5}$

$x\neq -6,-3,-1,3,5$

33. $\dfrac{4x+1}{6x+5}+\dfrac{8x+9}{6x+5}=\dfrac{4x+1+8x+9}{6x+5}$

$=\dfrac{12x+10}{6x+5}$

$=\dfrac{2(6x+5)}{6x+5}=2, x\neq -\dfrac{5}{6}$

41. $\dfrac{3}{x+4}+\dfrac{6}{x+5}=\dfrac{3(x+5)+6(x+4)}{(x+4)(x+5)}$

$=\dfrac{3x+15+6x+24}{(x+4)(x+5)}$

$=\dfrac{9x+39}{(x+4)(x+5)}, x\neq -4,-5$

43. $\dfrac{3}{x+1}-\dfrac{3}{x}=\dfrac{3x-3(x+1)}{x(x+1)}$

$=\dfrac{3x-3x-3}{x(x+1)}=-\dfrac{3}{x(x+1)},\ x\neq -1,0$

45. $\dfrac{2x}{x+2}+\dfrac{x+2}{x-2}=\dfrac{2x(x-2)+(x+2)(x+2)}{(x+2)(x-2)}$

$=\dfrac{2x^2-4x+x^2+4x+4}{(x+2)(x-2)}$

$=\dfrac{3x^2+4}{(x+2)(x-2)},\ x\neq -2,2$

47. $\dfrac{x+5}{x-5}+\dfrac{x-5}{x+5}$

$=\dfrac{(x+5)(x+5)+(x-5)(x-5)}{(x-5)(x+5)}$

$=\dfrac{x^2+10x+25+x^2-10x+25}{(x-5)(x+5)}$

$=\dfrac{2x^2+50}{(x-5)(x+5)},\ x\neq -5,5$

49. $\dfrac{3}{2x+4}+\dfrac{2}{3x+6}=\dfrac{3}{2(x+2)}+\dfrac{2}{3(x+2)}$

$=\dfrac{9}{6(x+2)}+\dfrac{4}{6(x+2)}$

$=\dfrac{9+4}{6(x+2)}$

$=\dfrac{13}{6(x+2)}$

$x\neq -2$

51. $\dfrac{4}{x^2+6x+9}+\dfrac{4}{x+3}=\dfrac{4}{(x+3)^2}+\dfrac{4}{x+3}$

$=\dfrac{4+4(x+3)}{(x+3)^2}=\dfrac{4+4x+12}{(x+3)^2}=\dfrac{4x+16}{(x+3)^2},$

$x\neq -3$

53. $\dfrac{3x}{x^2+3x-10}-\dfrac{2x}{x^2+x-6}$

$=\dfrac{3x}{(x+5)(x-2)}-\dfrac{2x}{(x+3)(x-2)}$

$=\dfrac{3x(x+3)-2x(x+5)}{(x+5)(x-2)(x+3)}$

$=\dfrac{3x^2+9x-2x^2-10x}{(x+5)(x-2)(x+3)}$

$=\dfrac{x^2-x}{(x+5)(x-2)(x+3)},\ x\neq -5,2,-3$

55. $\dfrac{x+3}{x^2-1}-\dfrac{x+2}{x-1}$

$=\dfrac{x+3}{(x+1)(x-1)}-\dfrac{x+2}{x-1}$

$=\dfrac{x+3}{(x+1)(x-1)}-\dfrac{(x+1)(x+2)}{(x+1)(x-1)}$

$=\dfrac{x+3}{(x+1)(x-1)}-\dfrac{x^2+3x+2}{(x+1)(x-1)}$

$=\dfrac{x+3-x^2-3x-2}{(x+1)(x-1)}$

$=\dfrac{-x^2-2x+1}{(x+1)(x-1)}$

$x\neq 1,-1$

57. $\dfrac{4x^2+x-6}{x^2+3x+2}-\dfrac{3x}{x+1}+\dfrac{5}{x+2}$

$=\dfrac{4x^2+x-6}{(x+1)(x+2)}+\dfrac{-3x}{x+1}+\dfrac{5}{x+2}$

$=\dfrac{4x^2+x-5}{(x+1)(x+2)}+\dfrac{-3x(x+2)}{(x+1)(x+2)}+\dfrac{5(x+1)}{(x+1)(x+2)}$

$=\dfrac{4x^2+x-6-3x^2-6x+5x+5}{(x+1)(x+2)}$

$=\dfrac{x^2-1}{(x+1)(x+2)}$

$=\dfrac{(x-1)(x+1)}{(x+1)(x+2)}$

$=\dfrac{x-1}{x+2};x\neq -2,-1$

59. $\dfrac{\frac{x}{3}-1}{x-3}=\dfrac{3\left[\frac{x}{3}-1\right]}{3\left[x-3\right]}=\dfrac{x-3}{3(x-3)}=\dfrac{1}{3},\ \ x\neq 3$

61. $\dfrac{1+\frac{1}{x}}{3-\frac{1}{x}}=\dfrac{x\left[1+\frac{1}{x}\right]}{x\left[3-\frac{1}{x}\right]}=\dfrac{x+1}{3x-1},\ x\neq 0,\ \frac{1}{3}$

63. $\dfrac{\frac{1}{x}+\frac{1}{y}}{x+y}=\dfrac{xy\left[\frac{1}{x}+\frac{1}{y}\right]}{xy\left[x+y\right]}=\dfrac{y+x}{xy(x+y)}=\dfrac{1}{xy},$

$x\neq 0,\ y\neq 0,\ x\neq -y$

65. $\dfrac{x-\frac{x}{x+3}}{x+2}=\dfrac{(x+3)\left[x-\frac{x}{x+3}\right]}{(x+3)(x+2)}=\dfrac{x(x+3)-x}{(x+3)(x+2)}$

$=\dfrac{x^2+3x-x}{(x+3)(x+2)}=\dfrac{x^2+2x}{(x+3)(x+2)}$

$=\dfrac{x(x+2)}{(x+3)(x+2)}=\dfrac{x}{x+3},\ x\neq -2,\ -3$

67. $\dfrac{\frac{3}{x-2}-\frac{4}{x+2}}{\frac{7}{x^2-4}}=\dfrac{\frac{3}{x-2}-\frac{4}{x+2}}{\frac{7}{(x-2)(x+2)}}$

$=\dfrac{\left[\frac{3}{x-2}-\frac{4}{x+2}\right](x-2)(x+2)}{\left[\frac{7}{(x-2)(x+2)}\right](x-2)(x+2)}$

$=\dfrac{3(x+2)-4(x-2)}{7}$

$=\dfrac{3x+6-4x+8}{7}=\dfrac{-x+14}{7}$

$=-\dfrac{x-14}{7}\ \ x\neq -2,\ 2$

69. $\dfrac{\frac{1}{x+1}}{\frac{1}{x^2-2x-3}+\frac{1}{x-3}}=\dfrac{\frac{1}{x+1}}{\frac{1}{(x+1)(x-3)}+\frac{1}{x-3}}$

$=\dfrac{\dfrac{(x+1)(x-3)}{x+1}}{\dfrac{(x+1)(x-3)}{(x+1)(x-3)}+\dfrac{(x+1)(x-3)}{x-3}}$

$=\dfrac{x-3}{1+x+1}$

$=\dfrac{x-3}{x+2}\ \ x\neq -2,-1,3$

71. $\dfrac{\frac{1}{(x+h)^2}-\frac{1}{x^2}}{h}=\dfrac{\frac{x^2(x+h)^2}{(x+h)^2}-\frac{x^2(x+h)^2}{x^2}}{hx^2(x+h)^2}$

$=\dfrac{x^2-(x+h)^2}{hx^2(x+h)^2}$

$=\dfrac{x^2-(x^2+2hx+h^2)}{hx^2(x+h)^2}$

$=\dfrac{x^2-x^2-2hx-h^2}{hx^2(x+h)^2}$

$=\dfrac{-2hx-h^2}{hx^2(x+h)^2}$

$=\dfrac{-h(2x+h)}{hx^2(x+h)^2}$

$=-\dfrac{(2x+h)}{x^2(x+h)^2}$

73. $\dfrac{\sqrt{x}-\frac{1}{3\sqrt{x}}}{\sqrt{x}}=\dfrac{\left(\sqrt{x}-\frac{1}{3\sqrt{x}}\right)(3\sqrt{x})}{\sqrt{x}(3\sqrt{x})}$

$=\dfrac{3x-1}{3x}$

$=1-\dfrac{1}{3x},\ x>0$

75.

$$\frac{\frac{x^2}{\sqrt{x^2+2}}-\sqrt{x^2+2}}{x^2}$$

$$=\frac{\left(\frac{x^2}{\sqrt{x^2+2}}-\sqrt{x^2+2}\right)\sqrt{x^2+2}}{x^2\sqrt{x^2+2}}$$

$$=\frac{x^2-(x^2+2)}{x^2\sqrt{x^2+2}}$$

$$=-\frac{2}{x^2\sqrt{x^2+2}}$$

77. $\dfrac{\dfrac{1}{\sqrt{x+h}}-\dfrac{1}{\sqrt{x}}}{h}=\dfrac{\left(\dfrac{1}{\sqrt{x+h}}-\dfrac{1}{\sqrt{x}}\right)\sqrt{x+h}\sqrt{x}}{h\sqrt{x+h}\sqrt{x}}$

$$=\frac{\sqrt{x}-\sqrt{x+h}}{h\sqrt{x(x+h)}},\ h\neq 0$$

79. $\dfrac{\sqrt{x+5}-\sqrt{x}}{5}=\dfrac{\sqrt{x+5}-\sqrt{x}}{5}\cdot\dfrac{\sqrt{x+5}+\sqrt{x}}{\sqrt{x+5}+\sqrt{x}}$

$$=\frac{(\sqrt{x+5})^2-(\sqrt{x})^2}{5(\sqrt{x+5}+\sqrt{x})}$$

$$=\frac{x+5-x}{5(\sqrt{x+5}+\sqrt{x})}$$

$$=\frac{1}{\sqrt{x+5}+\sqrt{x}}$$

81. $\dfrac{\sqrt{x}+\sqrt{y}}{x^2-y^2}=\dfrac{\sqrt{x}+\sqrt{y}}{x^2-y^2}\cdot\dfrac{\sqrt{x}-\sqrt{y}}{\sqrt{x}-\sqrt{y}}$

$$=\frac{(\sqrt{x})^2-(\sqrt{y})^2}{(x+y)(x-y)(\sqrt{x}-\sqrt{y})}$$

$$=\frac{x-y}{(x+y)(x-y)(\sqrt{x}-\sqrt{y})}$$

$$=\frac{1}{(x+y)(\sqrt{x}-\sqrt{y})}$$

83. $\left(\dfrac{2x+3}{x+1}\cdot\dfrac{x^2+4x-5}{2x^2+x-3}\right)-\dfrac{2}{x+2}=\left(\dfrac{(2x+3)}{x+1}\cdot\dfrac{(x+5)(x-1)}{(2x+3)(x-1)}\right)-\dfrac{2}{x+2}=\dfrac{x+5}{x+1}-\dfrac{2}{x+2}$

$$=\frac{(x+5)(x+2)}{(x+1)(x+2)}-\frac{2(x+1)}{(x+1)(x+2)}=\frac{(x+5)(x+2)-2(x+1)}{(x+1)(x+2)}=\frac{x^2+2x+5x+10-2x-2}{(x+1)(x+2)}=\frac{x^2+5x+8}{(x+1)(x+2)}$$

85. $\left(2-\dfrac{6}{x+1}\right)\left(1+\dfrac{3}{x-2}\right)=\left(\dfrac{2(x+1)}{(x+1)}-\dfrac{6}{(x+1)}\right)\left(\dfrac{(x-2)}{(x-2)}+\dfrac{3}{(x-2)}\right)$

$$=\left(\dfrac{2x+2-6}{x+1}\right)\left(\dfrac{x-2+3}{x-2}\right)=\left(\dfrac{2x-4}{x+1}\right)\left(\dfrac{x+1}{x-2}\right)=\dfrac{2\cancel{(x-2)}\ \cancel{(x+1)}}{\cancel{(x+1)}\ \cancel{(x-2)}}=2$$

87. $\dfrac{y^{-1}-(y+5)^{-1}}{5}=\dfrac{\dfrac{1}{y}-\dfrac{1}{y+5}}{5}$

$LCD=y(y+5)$

$$\dfrac{\dfrac{1}{y}-\dfrac{1}{y+5}}{5}=\dfrac{y(y+5)\left(\dfrac{1}{y}-\dfrac{1}{y+5}\right)}{y(y+5)(5)}=\dfrac{y+5-y}{5y(y+5)}=\dfrac{5}{5y(y+5)}=\dfrac{1}{y(y+5)}$$

89. $\left(\dfrac{1}{a^3-b^3}\cdot\dfrac{ac+ad-bc-bd}{1}\right)-\dfrac{c-d}{a^2+ab+b^2}=\left(\dfrac{1}{(a-b)(a^2+ab+b^2)}\cdot\dfrac{a(c+d)-b(c+d)}{1}\right)-\dfrac{c-d}{a^2+ab+b^2}$

$$=\left(\dfrac{1}{\cancel{(a-b)}(a^2+ab+b^2)}\cdot\dfrac{(c+d)\cancel{(a-b)}}{1}\right)-\dfrac{c-d}{a^2+ab+b^2}=\dfrac{c+d}{a^2+ab+b^2}-\dfrac{c-d}{a^2+bd+b^2}$$

$$=\dfrac{c+d-c+d}{a^2+ab+b^2}=\dfrac{2d}{a^2+ab+b^2}$$

91. a. $\dfrac{130x}{100-x}$ is equal to

1. $\dfrac{130\cdot40}{100-40}=\dfrac{130\cdot40}{60}=86.67$,

    when $x=40$

2. $\dfrac{130\cdot80}{100-80}=\dfrac{130\cdot80}{20}=520$,

    when $x=80$

3. $\dfrac{130\cdot90}{100-90}=\dfrac{130\cdot90}{10}=1170$,

    when $x=90$

It costs \$86,670,000 to inoculate 40% of the population against this strain of flu, and \$520,000,000 to inoculate 80% of the population, and \$1,170,000,000 to inoculate 90% of the population.

b. For $x=100$, the function is not defined.

c. As x approaches 100, the value of the function increases rapidly. So it costs an astronomical amount of money to inoculate almost all of the people, and it is impossible to inoculate 100% of the population.

93.   **a.**   Substitute 4 for x in the model.

$$W = -66x^2 + 526x + 1030$$

$$W = -66(4)^2 + 526(4) + 1030$$

$$W = 2078$$

According to the model, women between the ages of 19 and 30 with this lifestyle need 2078 calories per day. This underestimates the actual value shown in the bar graph by 22 calories.

   b.   Substitute 4 for x in the model.

$$M = -120x^2 + 998x + 590$$

$$M = -120(4)^2 + 998(4) + 590$$

$$M = 2662$$

According to the model, men between the ages of 19 and 30 with this lifestyle need 2662 calories per day. This underestimates the actual value shown in the bar graph by 38 calories.

   c. $\dfrac{W}{M} = \dfrac{-66x^2 + 526x + 1030}{-120x^2 + 998x + 590}$

$$= \dfrac{2(-33x^2 + 263x + 515)}{2(-60x^2 + 499x + 295)}$$

$$= \dfrac{-33x^2 + 263x + 515}{-60x^2 + 499x + 295}$$

95. $P = 2L + 2W$

$$= 2\left(\dfrac{x}{x+3}\right) + 2\left(\dfrac{x}{x-4}\right)$$

$$= \dfrac{2x}{x+3} + \dfrac{2x}{x+4}$$

$$= \dfrac{2x(x+4)}{(x+3)(x+4)} + \dfrac{2x(x+3)}{(x+3)(x+4)}$$

$$= \dfrac{2x^2 + 8x + 2x^2 + 6x}{(x+3)(x+4)}$$

$$= \dfrac{4x^2 + 14x}{(x+3)(x+4)}$$

97. –107.   Answers may vary.

109. does not make sense; Explanations will vary. Sample explanation: $\dfrac{3x-3}{4x(x-1)} = \dfrac{3(1)-3}{4(1)(1-1)} = \dfrac{0}{0}$ which is undefined.

111. does not make sense; Explanations will vary. Sample explanation: The first step is to invert the second fraction.

113. false; Changes to make the statement true will vary. A sample change is: $\dfrac{x^2-25}{x-5} = \dfrac{(x+5)(x-5)}{x-5} = x+5$

115. true

117.
$$\frac{1}{x^n-1} - \frac{1}{x^n+1} - \frac{1}{x^{2n}-1} = \frac{x^n+1}{x^{2n}-1} - \frac{x^n-1}{x^{2n}-1} - \frac{1}{x^{2n}-1}$$
$$= \frac{x^n+1-x^n+1-1}{x^{2n}-1}$$
$$= \frac{1}{x^{2n}-1}$$

119.
$$(x-y)^{-1} + (x-y)^{-2} = \frac{1}{(x-y)} + \frac{1}{(x-y)^2} = \frac{(x-y)}{(x-y)(x-y)} + \frac{1}{(x-y)^2} = \frac{x-y+1}{(x-y)^2}$$

121.
$$2(x-3)-17 = 13-3(x+2)$$
$$2(6-3)-17 = 13-3(6+2)$$
$$2(3)-17 = 13-3(8)$$
$$6-17 = 13-24$$
$$-11 = -11, \text{ true}$$

122.
$$12\left(\frac{x+2}{4} - \frac{x-1}{3}\right) = 12\left(\frac{x+2}{4}\right) - 12\left(\frac{x-1}{3}\right)$$
$$= 3(x+2) - 4(x-1)$$
$$= 3x+6-4x+4$$
$$= -x+10$$

123.
$$\frac{-b-\sqrt{b^2-4ac}}{2a} = \frac{-(9)-\sqrt{(9)^2-4(2)(-5)}}{2(2)}$$
$$= \frac{-9-\sqrt{81+40}}{4}$$
$$= \frac{-9-\sqrt{121}}{4}$$
$$= \frac{-9-11}{4}$$
$$= -5$$

121.
$$2(x-3)-17=13-3(x+2)$$
$$2(6-3)-17=13-3(6+2)$$
$$2(3)-17=13-3(8)$$
$$6-17=13-24$$
$$-11=-11, \text{ true}$$

122.
$$12\left(\frac{x+2}{4}-\frac{x-1}{3}\right)=12\left(\frac{x+2}{4}\right)-12\left(\frac{x-1}{3}\right)$$
$$=3(x+2)-4(x-1)$$
$$=3x+6-4x+4$$
$$=-x+10$$

123.
$$\frac{-b-\sqrt{b^2-4ac}}{2a}=\frac{-(9)-\sqrt{(9)^2-4(2)(-5)}}{2(2)}$$
$$=\frac{-9-\sqrt{81+40}}{4}$$
$$=\frac{-9-\sqrt{121}}{4}$$
$$=\frac{-9-11}{4}$$
$$=-5$$

Section P.7

Check Point Exercises

1.
$$4(2x+1)-29=3(2x-5)$$
$$8x+4-29=6x-15$$
$$8x-25=6x-15$$
$$8x-25-6x=6x-15-6x$$
$$2x-25=-15$$
$$2x-25+25=-15+25$$
$$2x=10$$
$$\frac{2x}{2}=\frac{10}{2}$$
$$x=5$$

Check:
$$4(2x+1)-29=3(2x-5)$$
$$4[2(5)+1]-29=3[2(5)-5]$$
$$4[10+1]-29=3[10-5]$$
$$4[11]-29=3[5]$$
$$44-29=15$$
$$15=15 \text{ true}$$

The solution set is {5}.

2.
$$\frac{x-3}{4}=\frac{5}{14}-\frac{x+5}{7}$$
$$28\cdot\frac{x-3}{4}=28\left(\frac{5}{14}-\frac{x+5}{7}\right)$$
$$7(x-3)=2(5)-4(x+5)$$
$$7x-21=10-4x-20$$
$$7x-21=-4x-10$$
$$7x+4x=-10+21$$
$$11x=11$$
$$\frac{11x}{11}=\frac{11}{11}$$
$$x=1$$

Check:
$$\frac{x-3}{4}=\frac{5}{14}-\frac{x+5}{7}$$
$$\frac{1-3}{4}=\frac{5}{14}-\frac{1+5}{7}$$
$$\frac{-2}{4}=\frac{5}{14}-\frac{6}{7}$$
$$-\frac{1}{2}=-\frac{1}{2}$$

The solution set is {1}.

3.
$$\frac{6}{x+3}-\frac{5}{x-2}=\frac{-20}{x^2+x-6}$$
$$\frac{6}{x+3}-\frac{5}{x-2}=\frac{-20}{(x+3)(x-2)}$$
$$\frac{6(x+3)(x-2)}{x+3}-\frac{5(x+3)(x-2)}{x-2}=\frac{-20(x+3)(x-2)}{(x+3)(x-2)}$$
$$6(x-2)-5(x+3)=-20$$
$$6x-12-5x-15=-20$$
$$x-27=-20$$
$$x=7$$

The solution set is $\{7\}$.

4.
$$\frac{1}{x+2} = \frac{4}{x^2-4} - \frac{1}{x-2}$$

$$\frac{1}{x+2} = \frac{4}{(x+2)(x-2)} - \frac{1}{x-2}$$

$$\frac{1(x+2)(x-2)}{x+2} = \frac{4(x+2)(x-2)}{(x+2)(x-2)} - \frac{1(x+2)(x-2)}{x-2}$$

$$x-2 = 4-(x+2)$$

$$x-2 = 4-x-2$$

$$x-2 = 2-x$$

$$2x = 4$$

$$x = 2$$

2 must be rejected. The solution set is $\{\ \}$.

5.
$$\frac{1}{p} + \frac{1}{q} = \frac{1}{f}$$

$$\frac{1pqf}{p} + \frac{1pqf}{q} = \frac{1pqf}{f}$$

$$qf + pf = pq$$

$$qf - pq = -pf$$

$$q(f-p) = -pf$$

$$\frac{q(f-p)}{f-p} = \frac{-pf}{f-p}$$

$$q = \frac{pf}{p-f}$$

6.
$$4|1-2x| - 20 = 0$$

$$4|1-2x| = 20$$

$$|1-2x| = 5$$

$$1-2x = 5 \quad \text{or} \quad 1-2x = -5$$

$$-2x = 4 \qquad\qquad -2x = -6$$

$$x = -2 \qquad\qquad x = 3$$

The solution set is $\{-2, 3\}$.

7. **a.**
$$3x^2 - 9x = 0$$

$$3x(x-3) = 0$$

$$3x = 0 \quad \text{or} \quad x-3 = 0$$

$$x = 0 \qquad\qquad x = 3$$

The solution set is $\{0, 3\}$.

b.
$$2x^2 + x = 1$$

$$2x^2 + x - 1 = 0$$

$$(2x-1)(x+1) = 0$$

$$2x-1 = 0 \quad \text{or} \quad x+1 = 0$$

$$2x = 1 \qquad\qquad x = -1$$

$$x = \frac{1}{2}$$

The solution set is $\left\{\frac{1}{2}, -1\right\}$.

8. **a.**
$$3x^2 - 21 = 0$$

$$3x^2 = 21$$

$$\frac{3x^2}{3} = \frac{21}{3}$$

$$x^2 = 7$$

$$x = \pm\sqrt{7}$$

The solution set is $\left\{-\sqrt{7}, \sqrt{7}\right\}$.

b.
$$(x+5)^2 = 11$$

$$x+5 = \pm\sqrt{11}$$

$$x = -5 \pm \sqrt{11}$$

The solution set is $\left\{-5+\sqrt{11}, -5-\sqrt{11}\right\}$.

9.
$$x^2 + 4x - 1 = 0$$

$$x^2 + 4x = 1$$

$$x^2 + 4x + 4 = 1 + 4$$

$$(x+2)^2 = 5$$

$$x+2 = \pm\sqrt{5}$$

$$x = -2 \pm \sqrt{5}$$

10. $2x^2 + 2x - 1 = 0$

$a = 2, b = 2, c = -1$

$x = \dfrac{-b \pm \sqrt{b^2 - 4ac}}{2a}$

$= \dfrac{-2 \pm \sqrt{2^2 - 4(2)(-1)}}{2(2)}$

$= \dfrac{-2 \pm \sqrt{4 + 8}}{4}$

$= \dfrac{-2 \pm \sqrt{12}}{4}$

$= \dfrac{-2 \pm 2\sqrt{3}}{4}$

$= \dfrac{2(-1 \pm \sqrt{3})}{4}$

$= \dfrac{-1 \pm \sqrt{3}}{2}$

The solution set is $\left\{ \dfrac{-1+\sqrt{3}}{2}, \dfrac{-1-\sqrt{3}}{2} \right\}$.

11. $3x^2 - 2x + 5 = 0$

$a = 3, \ b = -2, \ c = 5$

$b^2 - 4ac = (-2)^2 - 4 \cdot 3 \cdot 5 = 4 - 60 = -56$

The discriminant is –56. The equation has two complex imaginary solutions.

12. $\sqrt{x+3} + 3 = x$

$\sqrt{x+3} = x - 3$

$\left(\sqrt{x+3}\right)^2 = (x-3)^2$

$x + 3 = x^2 - 6x + 9$

$0 = x^2 - 7x + 6$

$0 = (x-6)(x-1)$

$x - 6 = 0 \quad$ or $\quad x - 1 = 0$

$x = 6 \qquad\qquad x = 1$

1 does not check and must be rejected.

The solution set is $\{6\}$.

Exercise Set P.7

1. $7x - 5 = 72$

$7x = 77$

$x = 11$

Check:

$7x - 5 = 72$

$7(11) - 5 = 72$

$77 - 5 = 72$

$72 = 72$

The solution set is $\{11\}$.

3. $11x - (6x - 5) = 40$

$11x - 6x + 5 = 40$

$5x + 5 = 40$

$5x = 35$

$x = 7$

The solution set is $\{7\}$.

Check:

$11x - (6x - 5) = 40$

$11(7) - [6(7) - 5] = 40$

$77 - (42 - 5) = 40$

$77 - (37) = 40$

$40 = 40$

5. $2x - 7 = 6 + x$

$x - 7 = 6$

$x = 13$

The solution set is $\{13\}$.

Check:

$2(13) - 7 = 6 + 13$

$26 - 7 = 19$

$19 = 19$

7. $7x + 4 = x + 16$

$6x + 4 = 16$

$6x = 12$

$x = 2$

The solution set is $\{2\}$.

Check:

$7(2) + 4 = 2 + 16$

$14 + 4 = 18$

$18 = 18$

9. $3(x-2)+7 = 2(x+5)$
$3x-6+7 = 2x+10$
$3x+1 = 2x+10$
$x+1 = 10$
$x = 9$
The solution set is $\{9\}$.

Check:
$3(9-2)+7 = 2(9+5)$
$3(7)+7 = 2(14)$
$21+7 = 28$
$28 = 28$

11. $\dfrac{x+3}{6} = \dfrac{3}{8} + \dfrac{x-5}{4}$

$24\left[\dfrac{x+3}{6} = \dfrac{3}{8} + \dfrac{x-5}{4}\right]$

$4x+12 = 9+6x-30$
$4x-6x = -21-12$
$-2x = -33$
$x = \dfrac{33}{2}$

The solution set is $\left\{\dfrac{33}{2}\right\}$.

13. $\dfrac{x}{4} = 2 + \dfrac{x-3}{3}$

$12\left[\dfrac{x}{4} = 2 + \dfrac{x-3}{3}\right]$

$3x = 24 + 4x - 12$
$3x-4x = 12$
$-x = 12$
$x = -12$
The solution set is $\{-12\}$.

15. $\dfrac{x+1}{3} = 5 - \dfrac{x+2}{7}$

$21\left[\dfrac{x+1}{3} = 5 - \dfrac{x+2}{7}\right]$

$7x+7 = 105 - 3x - 6$
$7x+3x = 99 - 7$
$10x = 92$
$x = \dfrac{92}{10}$
$x = \dfrac{46}{5}$

The solution set is $\left\{\dfrac{46}{5}\right\}$.

17. a. $\dfrac{1}{x-1} + 5 = \dfrac{11}{x-1} \quad (x \neq 1)$

b. $\dfrac{1}{x-1} + 5 = \dfrac{11}{x-1}$
$1 + 5(x-1) = 11$
$1 + 5x - 5 = 11$
$5x - 4 = 11$
$5x = 15$
$x = 3$
The solution set is $\{3\}$.

19. a. $\dfrac{8x}{x+1} = 4 - \dfrac{8}{x+1} \quad (x \neq -1)$

b. $\dfrac{8x}{x+1} = 4 - \dfrac{8}{x+1}$
$8x = 4(x+1) - 8$
$8x = 4x + 4 - 8$
$4x = -4$
$x = -1 \Rightarrow$ no solution
The solution set is the empty set, $\varnothing$.

21. a. $\dfrac{3}{2x-2} + \dfrac{1}{2} = \dfrac{2}{x-1} \quad (x \neq 1)$

b. $\dfrac{3}{2x-2} + \dfrac{1}{2} = \dfrac{2}{x-1}$
$\dfrac{3}{2(x-1)} + \dfrac{1}{2} = \dfrac{2}{x-1}$
$3 + 1(x-1) = 4$
$3 + x - 1 = 4$
$x = 2$
The solution set is $\{2\}$.

23. a. $\dfrac{2}{x+1} - \dfrac{1}{x-1} = \dfrac{2x}{x^2-1} \quad (x \neq 1, x \neq -1)$

b. $\dfrac{2}{x+1} - \dfrac{1}{x-1} = \dfrac{2x}{x^2-1}$
$\dfrac{2}{x+1} - \dfrac{1}{x-1} = \dfrac{2x}{(x+1)(x-1)}$
$2(x-1) - 1(x+1) = 2x$
$2x - 2 - x - 1 = 2x$
$-x = 3$
$x = -3$
The solution set is $\{-3\}$.

25. a. $\dfrac{1}{x-4} - \dfrac{5}{x+2} = \dfrac{6}{(x-4)(x+2)} ; (x \neq -2, 4)$

b. $\dfrac{1}{x-4} - \dfrac{5}{x+2} = \dfrac{6}{x^2 - 2x - 8}$

$\dfrac{1}{x-4} - \dfrac{5}{x+2} = \dfrac{6}{(x-4)(x+2)}$

$(x \neq 4, x \neq -2)$

$1(x+2) - 5(x-4) = 6$

$x + 2 - 5x + 20 = 6$

$-4x = -16$

$x = 4$

The solution set is the empty set, $\varnothing$.

27. $I = Prt$

$P = \dfrac{I}{rt};$

interest

29. $T = D + pm$

$T - D = pm$

$\dfrac{T - D}{m} = \dfrac{pm}{m}$

$\dfrac{T - D}{m} = p$

total of payment

31. $A = \dfrac{1}{2} h(a + b)$

$2A = h(a + b)$

$\dfrac{2A}{h} = a + b$

$\dfrac{2A}{h} - b = a$

area of trapezoid

33. $S = P + Prt$

$S - P = Prt$

$\dfrac{S - P}{Pt} = r;$

interest

35. $B = \dfrac{F}{S - V}$

$B(S - V) = F$

$S - V = \dfrac{F}{B}$

$S = \dfrac{F}{B} + V$

37. $IR + Ir = E$

$I(R + r) = E$

$I = \dfrac{E}{R + r}$

electric current

39. $\dfrac{1}{p} + \dfrac{1}{q} = \dfrac{1}{f}$

$qf + pf = pq$

$f(q + p) = pq$

$f = \dfrac{pq}{p + q}$

thin lens equation

41. $f = \dfrac{f_1 f_2}{f_1 + f_2}$

$f(f_1 + f_2) = f_1 f_2$

$ff_1 + ff_2 = f_1 f_2$

$ff_1 - f_1 f_2 = -ff_2$

$f_1(f - f_2) = -ff_2$

$\dfrac{f_1(f - f_2)}{f - f_2} = \dfrac{-ff_2}{f - f_2}$

$f_1 = \dfrac{ff_2}{f_2 - f}$

focal length

43. $|x - 2| = 7$

$x - 2 = 7 \quad x - 2 = -7$

$x = 9 \qquad x = -5$

The solution set is $\{9, -5\}$.

45. $|2x - 1| = 5$

$2x - 1 = 5 \quad 2x - 1 = -5$

$2x = 6 \qquad 2x = -4$

$x = 3 \qquad x = -2$

The solution set is $\{3, -2\}$.

47. $2|3x - 2| = 14$

$|3x - 2| = 7$

$3x - 2 = 7 \quad 3x - 2 = -7$

$3x = 9 \qquad 3x = -5$

$x = 3 \qquad x = -5/3$

The solution set is $\{3, -5/3\}$

49. $2\left|4-\dfrac{5}{2}x\right|+6=18$

$\qquad 2\left|4-\dfrac{5}{2}x\right|=12$

$\qquad \left|4-\dfrac{5}{2}x\right|=6$

$4-\dfrac{5}{2}x=6 \qquad$ or $\qquad 4-\dfrac{5}{2}x=-6$

$-\dfrac{5}{2}x=2 \qquad\qquad -\dfrac{5}{2}x=-10$

$\qquad x=-\dfrac{4}{5} \qquad\qquad\qquad x=4$

The solution set is $\left\{-\dfrac{4}{5},4\right\}$.

51. $|x+1|+5=3$

$\qquad |x+1|=-2$

No solution

The solution set is $\{\ \}$.

53. $|2x-1|+3=3$

$\qquad |2x-1|=0$

$\qquad 2x-1=0$

$\qquad 2x=1$

$\qquad x=1/2$

The solution set is $\left\{\dfrac{1}{2}\right\}$.

55. $x^2-3x-10=0$

$(x-5)(x+2)=0$

$x-5=0 \quad$ or $\quad x+2=0$

$\qquad x=5 \qquad\qquad x=-2$

The solution set is $\{-2,5\}$.

57. $\qquad x^2=8x-15$

$x^2-8x+15=0$

$(x-3)(x-5)=0$

$x-3=0 \quad$ or $\quad x-5=0$

$\qquad x=3 \qquad\qquad x=5$

The solution set is $\{3,5\}$.

59. $\qquad 5x^2=20x$

$5x^2-20x=0$

$5x(x-4)=0$

$5x=0 \quad$ or $\quad x-4=0$

$\quad x=0 \qquad\qquad x=4$

The solution set is $\{0,4\}$.

61. $\quad 3x^2=27$

$\qquad x^2=9$

$\qquad \sqrt{x^2}=\pm\sqrt{9}$

$\qquad x=\pm 3$

The solution set is $\{\pm 3\}$.

63. $\quad 5x^2+1=51$

$\qquad 5x^2=50$

$\qquad x^2=10$

$\qquad \sqrt{x^2}=\pm\sqrt{10}$

$\qquad x=\pm\sqrt{10}$

The solution set is $\left\{\pm\sqrt{10}\right\}$.

65. $\quad 3(x-4)^2=15$

$\qquad (x-4)^2=5$

$\qquad \sqrt{(x-4)^2}=\pm\sqrt{5}$

$\qquad x-4=\pm\sqrt{5}$

$\qquad x=4\pm\sqrt{5}$

The solution set is $\left\{4\pm\sqrt{5}\right\}$.

67. $\qquad x^2+6x=7$

$x^2+6x+9=7+9$

$\qquad (x+3)^2=16$

$\qquad x+3=\pm 4$

$\qquad x=-3\pm 4$

The solution set is $\{-7,1\}$.

69. $\qquad x^2-2x=2$

$x^2-2x+1=2+1$

$\qquad (x-1)^2=3$

$\qquad x-1=\pm\sqrt{3}$

$\qquad x=1\pm\sqrt{3}$

The solution set is $\left\{1+\sqrt{3},1-\sqrt{3}\right\}$.

71. $x^2 - 6x - 11 = 0$

$x^2 - 6x = 11$

$x^2 - 6x + 9 = 11 + 9$

$(x - 3)^2 = 20$

$x - 3 = \pm\sqrt{20}$

$x = 3 \pm 2\sqrt{5}$

The solution set is $\left\{3 + 2\sqrt{5}, 3 - 2\sqrt{5}\right\}$.

73. $x^2 + 4x + 1 = 0$

$x^2 + 4x = -1$

$x^2 + 4x + 4 = -1 + 4$

$(x + 2)^2 = 3$

$x + 2 = \pm\sqrt{3}$

$x = -2 \pm \sqrt{3}$

The solution set is $\left\{-2 + \sqrt{3}, -2 - \sqrt{3}\right\}$.

75. $x^2 + 8x + 15 = 0$

$x = \dfrac{-8 \pm \sqrt{8^2 - 4(1)(15)}}{2(1)}$

$x = \dfrac{-8 \pm \sqrt{64 - 60}}{2}$

$x = \dfrac{-8 \pm \sqrt{4}}{2}$

$x = \dfrac{-8 \pm 2}{2}$

The solution set is $\{-5, -3\}$.

77. $x^2 + 5x + 3 = 0$

$x = \dfrac{-5 \pm \sqrt{5^2 - 4(1)(3)}}{2(1)}$

$x = \dfrac{-5 \pm \sqrt{25 - 12}}{2}$

$x = \dfrac{-5 \pm \sqrt{13}}{2}$

The solution set is $\left\{\dfrac{-5 + \sqrt{13}}{2}, \dfrac{-5 - \sqrt{13}}{2}\right\}$.

79. $3x^2 - 3x - 4 = 0$

$x = \dfrac{3 \pm \sqrt{(-3)^2 - 4(3)(-4)}}{2(3)}$

$x = \dfrac{3 \pm \sqrt{9 + 48}}{6}$

$x = \dfrac{3 \pm \sqrt{57}}{6}$

The solution set is $\left\{\dfrac{3 + \sqrt{57}}{6}, \dfrac{3 - \sqrt{57}}{6}\right\}$

81. $4x^2 = 2x + 7$

$4x^2 - 2x - 7 = 0$

$x = \dfrac{2 \pm \sqrt{(-2)^2 - 4(4)(-7)}}{2(4)}$

$x = \dfrac{2 \pm \sqrt{4 + 112}}{8}$

$x = \dfrac{2 \pm \sqrt{116}}{8}$

$x = \dfrac{2 \pm 2\sqrt{29}}{8}$

$x = \dfrac{1 \pm \sqrt{29}}{4}$

The solution set is $\left\{\dfrac{1 + \sqrt{29}}{4}, \dfrac{1 - \sqrt{29}}{4}\right\}$.

83. $x^2 - 4x - 5 = 0$

$(-4)^2 - 4(1)(-5)$

$= 16 + 20$

$= 36$; 2 unequal real solutions

85. $2x^2 - 11x + 3 = 0$

$(-11)^2 - 4(2)(3)$

$= 121 - 24$

$= 97$; 2 unequal real solutions

87. $x^2 = 2x - 1$

$x^2 - 2x + 1 = 0$

$(-2)^2 - 4(1)(1)$

$= 4 - 4$

$= 0$; 1 real solution

89. $x^2 - 3x - 7 = 0$

$(-3)^2 - 4(1)(-7)$
$= 9 + 28$
$= 37; \text{ 2 unequal real solutions}$

91. $\quad\quad 2x^2 - x = 1$

$2x^2 - x - 1 = 0$
$(2x+1)(x-1) = 0$
$2x+1 = 0 \text{ or } x - 1 = 0$
$2x = -1$
$x = -\dfrac{1}{2} \text{ or } x = 1$

The solution set is $\left\{ -\dfrac{1}{2}, 1 \right\}$.

93. $\quad\quad 5x^2 + 2 = 11x$

$5x^2 - 11x + 2 = 0$
$(5x-1)(x-2) = 0$
$5x - 1 = 0 \text{ or } x - 2 = 0$
$5x = 1$
$x = \dfrac{1}{5} \text{ or } x = 2$

The solution set is $\left\{ \dfrac{1}{5}, 2 \right\}$.

95. $3x^2 = 60$

$x^2 = 20$
$x = \pm\sqrt{20}$
$x = \pm 2\sqrt{5}$

The solution set is $\left\{ -2\sqrt{5}, 2\sqrt{5} \right\}$.

97. $\quad\quad x^2 - 2x = 1$

$x^2 - 2x + 1 = 1 + 1$
$(x-1)^2 = 2$
$x - 1 = \pm\sqrt{2}$
$x = 1 \pm \sqrt{2}$

The solution set is $\left\{ 1+\sqrt{2}, 1-\sqrt{2} \right\}$.

99. $\quad\quad (2x+3)(x+4) = 1$

$2x^2 + 8x + 3x + 12 = 1$
$2x^2 + 11x + 12 = 1$
$2x^2 + 11x + 11 = 0$
$x = \dfrac{-11 \pm \sqrt{11^2 - 4(2)(11)}}{2(2)}$
$x = \dfrac{-11 \pm \sqrt{121 - 88}}{4}$
$x = \dfrac{-11 \pm \sqrt{33}}{4}$

The solution set is $\left\{ \dfrac{-11+\sqrt{33}}{4}, \dfrac{-11-\sqrt{33}}{4} \right\}$.

101. $(3x-4)^2 = 16$

$3x - 4 = \pm\sqrt{16}$
$3x - 4 = \pm 4$
$3x = 4 \pm 4$
$3x = 8 \text{ or } 3x = 0$
$x = \dfrac{8}{3} \text{ or } x = 0$

The solution set is $\left\{ 0, \dfrac{8}{3} \right\}$.

103. $3x^2 - 12x + 12 = 0$

$x^2 - 4x + 4 = 0$
$(x-2)(x-2) = 0$
$x - 2 = 0$
$x = 2$

The solution set is $\{2\}$.

105. $4x^2 - 16 = 0$

$4x^2 = 16$
$x^2 = 4$
$x = \pm 2$

The solution set is $\{-2, 2\}$.

107.
$$x^2 = 4x - 2$$
$$x^2 - 4x + 2 = 0$$
$$x = \frac{-b \pm \sqrt{b^2 - 4ac}}{2a}$$
$$x = \frac{-(-4) \pm \sqrt{(-4)^2 - 4(1)(2)}}{2(1)}$$
$$x = \frac{4 \pm \sqrt{8}}{2}$$
$$x = 2 \pm \sqrt{2}$$
The solution set is $\left\{ 2 \pm \sqrt{2} \right\}$.

109. $2x^2 - 7x = 0$
$$x(2x - 7) = 0$$
$$x = 0 \text{ or } 2x - 7 = 0$$
$$2x = 7$$
$$x = 0 \text{ or } x = \frac{7}{2}$$
The solution set is $\left\{ 0, \frac{7}{2} \right\}$.

111. $\dfrac{1}{x} + \dfrac{1}{x+2} = \dfrac{1}{3}; x \neq 0, -2$
$$3x + 6 + 3x = x^2 + 2x$$
$$0 = x^2 - 4x - 6$$
$$x = \frac{-(-4) \pm \sqrt{(-4)^2 - 4(1)(-6)}}{2(1)}$$
$$x = \frac{4 \pm \sqrt{16 + 24}}{2}$$
$$x = \frac{4 \pm \sqrt{40}}{2}$$
$$x = \frac{4 \pm 2\sqrt{10}}{2}$$
$$x = 2 \pm \sqrt{10}$$
The solution set is $\{ 2 + \sqrt{10},\ 2 - \sqrt{10} \}$.

113.
$$\frac{2x}{x-3} + \frac{6}{x+3} = \frac{-28}{x^2 - 9}; x \neq 3, -3$$
$$2x(x+3) + 6(x-3) = -28$$
$$2x^2 + 6x + 6x - 18 = -28$$
$$2x^2 + 12x + 10 = 0$$
$$x^2 + 6x + 5 = 0$$
$$(x+1)(x+5) = 0$$
The solution set is $\{ -5,\ -1 \}$.

115.
$$\sqrt{3x + 18} = x$$
$$3x + 18 = x^2$$
$$x^2 - 3x - 18 = 0$$
$$(x+3)(x-6) = 0$$
$$x + 3 = 0 \quad x - 6 = 0$$
$$x = -3 \qquad x = 6$$
$$\sqrt{3(-3) + 18} = -3 \quad \sqrt{3(6) + 18} = 6$$
$$\sqrt{-9 + 18} = -3 \qquad \sqrt{18 + 18} = 6$$
$$\sqrt{9} = -3 \text{ False} \quad \sqrt{36} = 6$$
The solution set is $\{6\}$.

117.
$$\sqrt{x+3} = x - 3$$
$$x + 3 = x^2 - 6x + 9$$
$$x^2 - 7x + 6 = 0$$
$$(x-1)(x-6) = 0$$
$$x - 1 = 0 \quad x - 6 = 0$$
$$x = 1 \qquad x = 6$$
$$\sqrt{1+3} = 1 - 3 \qquad \sqrt{6+3} = 6 - 3$$
$$\sqrt{4} = -2 \quad \text{False} \quad \sqrt{9} = 3$$
The solution set is $\{6\}$.

119.
$$\sqrt{2x + 13} = x + 7$$
$$2x + 13 = (x+7)^2$$
$$2x + 13 = x^2 + 14x + 49$$
$$x^2 + 12x + 36 = 0$$
$$(x+6)^2 = 0$$
$$x + 6 = 0$$
$$x = -6$$
$$\sqrt{2(-6) + 13} = -6 + 7$$
$$\sqrt{-12 + 13} = 1$$
$$\sqrt{1} = 1$$
The solution set is $\{-6\}$.

121.

$$x - \sqrt{2x+5} = 5$$
$$x - 5 = \sqrt{2x+5}$$
$$(x-5)^2 = 2x+5$$
$$x^2 - 10x + 25 = 2x + 5$$
$$x^2 - 12x + 20 = 0$$
$$(x-2)(x-10) = 0$$
$$x - 2 = 0 \quad x - 10 = 0$$
$$x = 2 \qquad x = 10$$
$$2 - \sqrt{2(2)+5} = 5 \quad 10 - \sqrt{2(10)+5} = 5$$
$$2 - \sqrt{9} = 5 \qquad 10 - \sqrt{25} = 5$$
$$2 - 3 = 5 \quad \text{False} \quad 10 - 5 = 5$$

The solution set is $\{10\}$.

123. $\sqrt{2x+19} - 8 = x$

$$\sqrt{2x+19} = x + 8$$
$$\left(\sqrt{2x+19}\right)^2 = (x+8)^2$$
$$2x + 19 = x^2 + 16x + 64$$
$$0 = x^2 + 14x + 45$$
$$0 = (x+9)(x+5)$$
$$x + 9 = 0 \quad \text{or} \quad x + 5 = 0$$
$$x = -9 \qquad x = -5$$

−9 does not check and must be rejected.
The solution set is $\{-5\}$.

125.

$$25 - [2 + 5y - 3(y+2)] = -3(2y-5) - [5(y-1) - 3y + 3]$$
$$25 - [2 + 5y - 3y - 6] = -6y + 15 - [5y - 5 - 3y + 3]$$
$$25 - [2y - 4] = -6y + 15 - [2y - 2]$$
$$25 - 2y + 4 = -6y + 15 - 2y + 2$$
$$-2y + 29 = -8y + 17$$
$$6y = -12$$
$$y = -2$$

The solution set is $\{-2\}$.

127. $7 - 7x = (3x+2)(x-1)$

$$7 - 7x = 3x^2 - x - 2$$
$$0 = 3x^2 + 6x - 9$$
$$0 = x^2 + 2x - 3$$
$$0 = (x+3)(x-1)$$
$$x + 3 = 0 \quad \text{or} \quad x - 1 = 0$$
$$x = -3 \qquad x = 1$$

The solution set is $\left\{-3, 1\right\}$.

129. $\left|x^2 + 2x - 36\right| = 12$

$$x^2 + 2x - 36 = 12 \qquad x^2 + 2x - 36 = -12$$

$$x^2 + 2x - 48 = 0 \quad \text{or} \quad x^2 + 2x - 24 = 0$$

$$(x+8)(x-6) = 0 \qquad (x+6)(x-4) = 0$$

Setting each of the factors above equal to zero gives
$x = -8, \quad x = 6, \quad x = -6, \quad \text{and} \quad x = 4.$

The solution set is $\{-8, -6, 4, 6\}$.

131.

$$\frac{1}{x^2 - 3x + 2} = \frac{1}{x+2} + \frac{5}{x^2 - 4}$$

$$\frac{1}{(x-1)(x-2)} = \frac{1}{x+2} + \frac{5}{(x+2)(x-2)}$$

Multiply both sides of the equation by the least common denominator, $(x-1)(x-2)(x+2)$. This results in the following:

$$x + 2 = (x-1)(x-2) + 5(x-1)$$

$$x + 2 = x^2 - 2x - x + 2 + 5x - 5$$

$$x + 2 = x^2 + 2x - 3$$

$$0 = x^2 + x - 5$$

Apply the quadratic formula:
$a = 1 \quad b = 1 \quad c = -5.$

$$x = \frac{-1 \pm \sqrt{1^2 - 4(1)(-5)}}{2(1)} = \frac{-1 \pm \sqrt{1 - (-20)}}{2}$$

$$= \frac{-1 \pm \sqrt{21}}{2}$$

The solution set is $\left\{ \dfrac{-1 \pm \sqrt{21}}{2} \right\}$.

133. $\sqrt{x+8} - \sqrt{x-4} = 2$

$$\sqrt{x+8} = \sqrt{x-4} + 2$$

$$x + 8 = \left(\sqrt{x-4} + 2\right)^2$$

$$x + 8 = x - 4 + 4\sqrt{x-4} + 4$$

$$x + 8 = x + 4\sqrt{x-4}$$

$$8 = 4\sqrt{x-4}$$

$$2 = \sqrt{x-4}$$

$$4 = x - 4$$

$$x = 8$$

$$\sqrt{8+8} - \sqrt{8-4} = 2$$

$$\sqrt{16} - \sqrt{4} = 2$$

$$4 - 2 = 2$$

The solution set is $\{8\}$.

135. Values that make the denominator zero must be excluded.

$$2x^2 + 4x - 9 = 0$$

$$x = \frac{-b \pm \sqrt{b^2 - 4ac}}{2a}$$

$$x = \frac{-(4) \pm \sqrt{(4)^2 - 4(2)(-9)}}{2(2)}$$

$$x = \frac{-4 \pm \sqrt{88}}{4}$$

$$x = \frac{-4 \pm 2\sqrt{22}}{4}$$

$$x = \frac{-2 \pm \sqrt{22}}{2}$$

137. $\dfrac{W}{2} - 3H = 53$

$$\frac{W}{2} - 3(6) = 53$$

$$\frac{W}{2} - 18 = 53$$

$$\frac{W}{2} - 18 + 18 = 53 + 18$$

$$\frac{W}{2} = 71$$

$$2 \cdot \frac{W}{2} = 2 \cdot 71$$

$$W = 142$$

According to the formula, the healthy weight of a person of height 5'6" is 142 pounds. This is 13 pounds below the upper end of the range shown in the bar graph.

139.

$$C = \frac{x + 0.1(500)}{x + 500}$$

$$0.28 = \frac{x + 0.1(500)}{x + 500}$$

$$0.28(x + 500) = x + 0.1(500)$$

$$0.28x + 140 = x + 50$$

$$-0.72x = -90$$

$$\frac{-0.72x}{-0.72} = \frac{-90}{-0.72}$$

$$x = 125$$

125 liters of pure peroxide must be added.

141. $f(x) = 0.013x^2 - 1.19x + 28.24$

$3 = 0.013x^2 - 1.19x + 28.24$

$0 = 0.013x^2 - 1.19x + 25.24$

Apply the quadratic formula:

$a = 0.013 \quad b = -1.19 \quad c = 25.24$

$x = \dfrac{-(-1.19) \pm \sqrt{(-1.19)^2 - 4(0.013)(25.24)}}{2(0.013)}$

$\quad = \dfrac{1.19 \pm \sqrt{1.4161 - 1.31248}}{0.026}$

$\quad = \dfrac{1.19 \pm \sqrt{0.10362}}{0.026}$

$\quad \approx \dfrac{1.19 \pm 0.32190}{0.026}$

$\quad \approx 58.15 \text{ or } 33.39$

The solutions are approximately 33.39 and 58.15. Thus, 33 year olds and 58 year olds are expected to be in 3 fatal crashes per 100 million miles driven. The function models the actual data well.

143. $M = 0.7\sqrt{x} + 12.5$

$15.1 = 0.7\sqrt{x} + 12.5$

$2.6 = 0.7\sqrt{x}$

$\dfrac{2.6}{0.7} = \sqrt{x}$

$\left(\dfrac{2.6}{0.7}\right)^2 = \left(\sqrt{x}\right)^2$

$14 \approx x$

There will be 15.1 cluttered minutes 14 years after 1996, or 2010.

145. – 157. Answers may vary.

159. does not make sense; Explanations will vary. Sample explanation: Substitute $n = 6$ into the equation to find P.

161. does not make sense; Explanations will vary. Sample explanation: The factoring method would be quicker.

163. false; Changes to make the statement true will vary. A sample change is: $(2x - 3)^2 = 25$

$\sqrt{(2x-3)^2} = \pm\sqrt{25}$

$2x - 3 = \pm 5$

165. true

167. $\dfrac{7x + 4}{b} + 13 = x$

$\dfrac{7(-6) + 4}{b} + 13 = -6$

$\dfrac{-38}{b} = -19$

$-19b = -38$

$b = 2$

169. $V = C - \dfrac{C - S}{L}N$

$VL = CL - (C - S)N$

$VL = CL - CN + SN$

$CN - CL = NS - LV$

$C(N - L) = NS - LV$

$\dfrac{C(N - L)}{N - L} = \dfrac{NS - LV}{N - L}$

$C = \dfrac{NS - LV}{N - L} \text{ or } \dfrac{LV - NS}{L - N}$

171. $x + 150$

172. $20 + 0.05x$

173. $4x + 400$

Section P.8

Check Point Exercises

1. Let x = the average salary for women
    Let $x + 14{,}037$ = the average salary for men
    $$x + (x + 14{,}037) = 130{,}015$$
    $$x + x + 14{,}037 = 130{,}015$$
    $$2x + 14{,}037 = 130{,}015$$
    $$2x = 115{,}978$$
    $$x = 57{,}989$$
    $$x + 14{,}037 = 72{,}026$$
    In 2007 the average teaching salary for women was
    \$57,989 and the average salary for men was \$72,026.

2. Let x = the number of years since 1969.
    $$88 - 1.1x = 33$$
    $$-1.1x = 33 - 88$$
    $$-1.1x = -55$$
    $$x = \frac{-55}{-1.1}$$
    $$x = 50$$
    33% of female freshmen will respond this way 50
    years after 1969, or 2019.

3. Let x = the computer's price before the reduction.
    $$x - 0.30x = 840$$
    $$0.70x = 840$$
    $$x = \frac{840}{0.70}$$
    $$x = 1200$$
    Before the reduction the computer's price was \$1200.

4. Let x = the width of the court.
    Let $x + 44$ = the length of the court.
    $$2l + 2w = P$$
    $$2(x + 44) + 2x = 288$$
    $$2x + 88 + 2x = 288$$
    $$4x + 88 = 288$$
    $$4x = 200$$
    $$x = \frac{200}{4}$$
    $$x = 50$$
    $$x + 44 = 94$$
    The dimensions of the court are 50 by 94.

5. $$(16 + 2x)(12 + 2x) = 320$$
    $$192 + 56x + 4x^2 = 320$$
    $$4x^2 + 56x - 128 = 0$$
    $$x^2 + 14x - 32 = 0$$
    $$(x + 16)(x - 2) = 0$$
    $$x + 16 = 0 \quad \text{or} \quad x - 2 = 0$$
    $$x = -16 \qquad\qquad x = 2$$
    -16 must be rejected.
    The path must be 2 feet wide.

6. $$a^2 + b^2 = c^2$$
    $$a^2 + (50)^2 = (130)^2$$
    $$a^2 + 2500 = 16{,}900$$
    $$a^2 = 14{,}400$$
    $$a = \pm 120$$
    -120 must be rejected.
    The tower is 120 yards tall.

7.

The original amount of money per person.

reduction per winner

The new amount of money per person.

$$\overbrace{\frac{5,000,000}{x}} - \overbrace{375,000} = \overbrace{\frac{5,000,000}{x+3}}$$

$$x(x+3)\left(\frac{5,000,000}{x} - 375,000\right) = x(x+3)\frac{5,000,000}{x+3}$$

$$5,000,000(x+3) - 375,000x(x+3) = 5,000,000x$$

$$5,000,000x + 15,000,000 - 375,000x^2 - 1,125,000x = 5,000,000x$$

$$-375,000x^2 - 1,125,000x + 15,000,000 = 0$$

$$x^2 + 3x - 40 = 0$$

$$(x+8)(x-5) = 0$$

$$x+8 = 0 \quad \text{or} \quad x-5 = 0$$

$$x = -8 \qquad\qquad x = 5$$

-8 must be rejected. There were 5 people in the original group.

Exercise Set P.8

1. Let $x =$ the time spent listening to radio.
Let $x + 581 =$ the time spent watching TV.

$$x + (x+581) = 2529$$
$$x + x + 581 = 2529$$
$$2x + 581 - 581 = 2529 - 581$$
$$2x = 1948$$
$$x = 974$$
$$x + 581 = 1555$$

Americans spent 974 hours listening to radio and 1555 hours watching TV.

3. Let $x =$ the average salary for carpenters.
Let $2x - 7740 =$ the average salary for computer programmers.

$$x + (2x - 7740) = 99,000$$
$$x + 2x - 7740 = 99,000$$
$$3x - 7740 = 99,000$$
$$3x - 7740 + 7740 = 99,000 + 7740$$
$$3x = 106,740$$
$$x = 35,580$$
$$2x - 7740 = 63,420$$

The average salary for carpenters is \$35,580 and the average salary for computer programmers is \$63,420.

5. Let $x =$ the number of years since 1983.

$$43 + 1.5x = 100$$
$$1.5x = 100 - 43$$
$$1.5x = 57$$
$$x = \frac{57}{1.5}$$
$$x = 38$$

All American adults will approve 38 years after 1983, or 2021.

7. **a.** $y = 24,000 - 3000x$

b.
$$y = 24,000 - 3000x$$
$$9000 = 24,000 - 3000x$$
$$9000 - 24,000 = -3000x$$
$$-15,000 = -3000x$$
$$x = \frac{-15,000}{-3000}$$
$$x = 5$$

The car's value will drop to $9000 after 5 years.

9. Let x = the number of years after 2005
$$13,300 + 1000x = 26,800 - 500x$$
$$1500x = 13,500$$
$$\frac{1500x}{1500} = \frac{13,500}{1500}$$
$$x = 9$$

The two colleges will have the same enrollment about 9 years after 2005, or 2014.
$$13,300 + 1000(9) = 22,300$$
and
$$26,800 - 500(9) = 22,300$$

The college's enrollments will be 22,300 at that time.

11. Let x = the cost of the television set.
$$x - 0.20x = 336$$
$$0.80x = 336$$
$$x = 420$$

The television set's price is $420.

13. Let x = the nightly cost
$$x + 0.08x = 162$$
$$1.08x = 162$$
$$x = 150$$

The nightly cost is $150.

15. Let c = the dealer's cost
$$584 = c + 0.25c$$
$$584 = 1.25c$$
$$467.20 = c$$

The dealer's cost is $467.20.

17. Let w = the width of the field
Let $2w$ = the length of the field
$$P = 2(\text{length}) + 2(\text{width})$$
$$300 = 2(2w) + 2(w)$$
$$300 = 4w + 2w$$
$$300 = 6w$$
$$50 = w$$

If $w = 50$, then $2w = 100$. Thus, the dimensions are 50 yards by 100 yards.

19. Let w = the width of the field
Let $2w + 6$ = the length of the field
$$228 = 6w + 12$$
$$216 = 6w$$
$$36 = w$$

If $w = 36$, then $2w + 6 = 2(36) + 6 = 78$. Thus, the dimensions are 36 feet by 78 feet.

21. Let x = the width of the frame.
Total length: $16 + 2x$
Total width: $12 + 2x$
$$P = 2(\text{length}) + 2(\text{width})$$
$$72 = 2(16 + 2x) + 2(12 + 2x)$$
$$72 = 32 + 4x + 24 + 4x$$
$$72 = 8x + 56$$
$$16 = 8x$$
$$2 = x$$

The width of the frame is 2 inches.

23. Let w = the width
Let $w + 3$ = the length
$$\text{Area} = lw$$
$$54 = (w + 3)w$$
$$54 = w^2 + 3w$$
$$0 = w^2 + 3w - 54$$
$$0 = (w + 9)(w - 6)$$
$$w + 9 = 0 \qquad w - 6 = 0$$
$$w = -9 \qquad w = 6$$

Disregard –9 because we can't have a negative length measurement. The width is 6 feet and the length is $6 + 3 = 9$ feet.

25. Let x = the length of the side of the original square
Let $x + 3$ = the length of the side of the new, larger square

$$(x+3)^2 = 64$$

$$x^2 + 6x + 9 = 64$$

$$x^2 + 6x - 55 = 0$$

$$(x+11)(x-5) = 0$$

Apply the zero product principle.

$$x + 11 = 0 \qquad x - 5 = 0$$

$$x = -11 \qquad x = 5$$

The solution set is $\{-11, 5\}$. Disregard -11 because we can't have a negative length measurement. This means that x, the length of the side of the original square, is 5 inches.

27. Let x = the width of the path

$$(20 + 2x)(10 + 2x) = 600$$

$$200 + 40x + 20x + 4x^2 = 600$$

$$200 + 60x + 4x^2 = 600$$

$$4x^2 + 60x + 200 = 600$$

$$4x^2 + 60x - 400 = 0$$

$$4(x^2 + 15x - 100) = 0$$

$$4(x + 20)(x - 5) = 0$$

Apply the zero product principle.

$$4(x + 20) = 0 \qquad x - 5 = 0$$

$$x + 20 = 0 \qquad x = 5$$

$$x = -20$$

The solution set is $\{-20, 5\}$. Disregard -20 because we can't have a negative width measurement. The width of the path is 5 meters.

29.

$$(20 + 2x)(30 + 2x) - (20)(30) = 336$$

$$600 + 100x + 4x^2 - 600 = 336$$

$$4x^2 + 100x - 336 = 0$$

$$x^2 + 25x - 84 = 0$$

$$(x - 3)(x + 28) = 0$$

$$x - 3 = 0 \quad \text{or} \quad x + 28 = 0$$

$$x = 3 \qquad x = -28$$

-28 must be rejected.
The width of the path is 3 feet

31.

$$a^2 + b^2 = c^2$$

$$a^2 + 15^2 = 20^2$$

$$a^2 + 225 = 400$$

$$a^2 = 175$$

$$a = \pm\sqrt{175}$$

$$a \approx \pm 13.2$$

-13.2 must be rejected.
The ladder reaches 13.2 feet up the house.

33.

$$a^2 + b^2 = c^2$$

$$5^2 + x^2 = (x+1)^2$$

$$x^2 + 25 = x^2 + 2x + 1$$

$$25 = 2x + 1$$

$$24 = 2x$$

$$x = 12$$

$$x + 1 = 13$$

The wire is 13 feet long.

35. Let x be the width.

$$a^2 + b^2 = c^2$$

$$x^2 + (2x)^2 = 64^2$$

$$x^2 + 4x^2 = 4096$$

$$5x^2 = 4096$$

$$x^2 = \frac{4096}{5}$$

$$x = \pm\sqrt{\frac{4096}{5}}$$

$$x \approx 28.62 \text{ feet}$$

$$2x \approx 57.24 \text{ feet}$$

The distance along the length and width is about $28.62 + 57.24$, or about 85.9 feet. A person could save $85.9 - 64$, or about 21.9 feet.

37.

The original amount of money per person.

reduction per winner

The new amount of money per person.

$$\frac{20,000,000}{x} - 500,000 = \frac{20,000,000}{x+2}$$

$$x(x+2)\left(\frac{20,000,000}{x} - 500,000\right) = x(x+2)\frac{20,000,000}{x+2}$$

$$20,000,000(x+2) - 500,000x(x+2) = 20,000,000x$$

$$20,000,000x + 40,000,000 - 500,000x^2 - 1,000,000x = 20,000,000x$$

$$40,000,000 - 500,000x^2 - 1,000,000x = 0$$

$$x^2 + 2x - 80 = 0$$

$$(x+10)(x-8) = 0$$

$$x+10 = 0 \quad \text{or} \quad x-8 = 0$$
$$x = -10 \qquad\qquad x = 8$$

−10 must be rejected. There were 8 people in the original group.

39. Let x be the car's average velocity.

car's time traveled

bus's time traveled

$$\frac{300}{x} = \frac{180}{x-20}$$

$$300(x-20) = 180x$$

$$300x - 6000 = 180x$$

$$120x = 6000$$

$$x = 50$$

$$x - 20 = 30$$

The average velocity of the car is 50 miles per hour. The average velocity of the bus is 30 miles per hour.

41. Let x be the average velocity on the return trip.

$$\frac{5}{x+9} + \frac{5}{x} = \frac{7}{6}$$

$$6x(x+9)\left(\frac{5}{x+9} + \frac{5}{x}\right) = 6x(x+9)\frac{7}{6}$$

$$30x + 30(x+9) = 7x(x+9)$$

$$30x + 30x + 270 = 7x^2 + 63x$$

$$0 = 7x^2 + 3x - 270$$

$$0 = (x-6)(7x+45)$$

$$x-6 = 0 \quad \text{or} \quad 7x+45 = 0$$
$$x = 6 \qquad\qquad x = -\frac{45}{7}$$

$-\frac{45}{7}$ must be rejected. The average velocity on the return trip is 6 miles per hour.

43. Let x = number of hours
$35x$ = labor cost
$35x + 63 = 448$
$\qquad 35x = 385$
$\qquad\quad x = 11$
It took 11 hours.

45. Let x = inches over 5 feet

$$100 + 5x = 135$$
$$5x = 35$$
$$x = 7$$

A height of 5 feet 7 inches corresponds to 135 pounds.

47. Let x be the number of consecutive hits.

$$\frac{35 + x}{140 + x} = 0.30$$
$$35 + x = 0.30(140 + x)$$
$$35 + x = 42 + 0.30x$$
$$350 + 10x = 420 + 3x$$
$$7x = 70$$
$$x = 10$$

You must get 10 consecutive hits to increase your batting average to 0.30.

49. – 51. Answers may vary.

53. does not make sense; Explanations will vary. Sample explanation: Though mathematical models can often provide excellent estimates about future attitudes, they cannot guaranty perfect precision.

55. does not make sense; Explanations will vary. Sample explanation: The correct equation is $x - 0.35x = 780$.

57. Let x be the length of one leg.

$$a^2 + b^2 = c^2$$
$$x^2 + (x+1)^2 = \left[12 - x - (x+1)\right]^2$$
$$x^2 + x^2 + 2x + 1 = \left[12 - x - x - 1\right]^2$$
$$2x^2 + 2x + 1 = (11 - 2x)^2$$
$$2x^2 + 2x + 1 = 121 - 44x + 4x^2$$
$$0 = 2x^2 - 46x + 120$$
$$0 = x^2 - 23x + 60$$
$$0 = (x - 3)(x - 20)$$
$$x - 3 = 0 \quad \text{or} \quad x - 20 = 0$$
$$x = 3 \qquad\qquad x = 20$$
$$x + 1 = 4$$
$$12 - (3 + 4) = 5$$

20 must be rejected, as it is greater than the perimeter.
The lengths of the sides are 3, 4, and 5.

59. Let x = woman's age

$$3x = \text{Coburn's age}$$
$$3x + 20 = 2(x + 20)$$
$$3x + 20 = 2x + 40$$
$$x + 20 = 40$$
$$x = 20$$

Coburn is 60 years old the woman is 20 years old.

61. Let x = mother's amount

$2x$ = boy's amount

$\dfrac{x}{2}$ = girl's amount

$$x + 2x + \frac{x}{2} = 14,000$$
$$\frac{7}{2}x = 14,000$$
$$x = \$4,000$$

The mother received $4000, the boy received $8000, and the girl received $2000.

63. Answers may vary.

64.
$$3 - 2x \le 11$$
$$3 - 2(-1) \le 11$$
$$3 + 2 \le 11$$
$$5 \le 11, \text{ true}$$

-1 is a solution.

65.
$$-2x - 4 = x + 5$$
$$-2x - x = 5 + 4$$
$$-3x = 9$$
$$x = \frac{9}{-3}$$
$$x = -3$$

The solution set is $\{-3\}$.

66.
$$\frac{x+3}{4} = \frac{x-2}{3} + \frac{1}{4}$$
$$12\left(\frac{x+3}{4}\right) = 12\left(\frac{x-2}{3} + \frac{1}{4}\right)$$
$$3(x + 3) = 4(x - 2) + 3$$
$$3x + 9 = 4x - 8 + 3$$
$$3x + 9 = 4x - 5$$
$$3x - 4x = -5 - 9$$
$$-x = -14$$
$$x = 14$$

The solution set is $\{14\}$.

Section P.9

Check Point Exercises

1.   a. $[-2,\ 5) = \left\{x\middle|-2 \le x < 5\right\}$

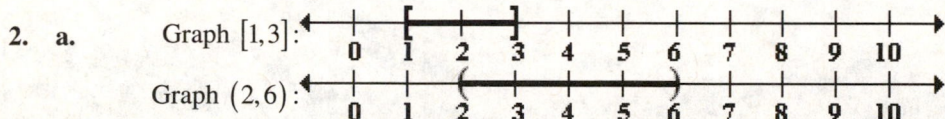

   b. $[1,\ 3.5] = \left\{x\middle|1 \le x \le 3.5\right\}$

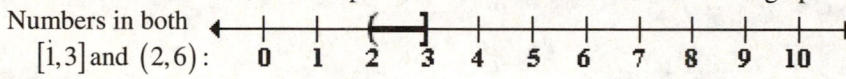

   c. $[-\infty,\ -1) = \left\{x\middle|x < -1\right\}$

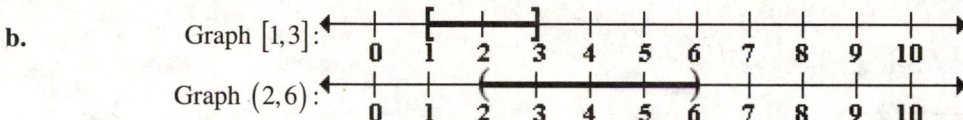

2.   a.   Graph $[1,3]$:

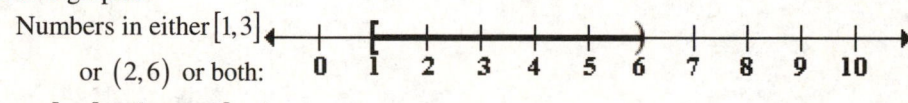

Graph $(2,6)$:

To find the intersection, take the portion of the number line that the two graphs have in common.

Numbers in both
$[1,3]$ and $(2,6)$:

Thus, $[1,3] \cap (2,6) = (2,3]$.

   b.   Graph $[1,3]$:

Graph $(2,6)$:

To find the union, take the portion of the number line representing the total collection of numbers in the two graphs.

Numbers in either $[1,3]$
or $(2,6)$ or both:

Thus, $[1,3] \cup (2,6) = [1,6)$.

3. $2 - 3x \le 5$

$-3x \le 3$

$x \ge -1$

The solution set is $\left\{x\middle|x \ge -1\right\}$ or $[-1, \infty)$.

4. $3x + 1 > 7x - 15$

$-4x > -16$

$\dfrac{-4x}{-4} < \dfrac{-16}{-4}$

$x < 4$

The solution set is $\left\{x\middle|x < 4\right\}$ or $(-\infty, 4]$.

5. $\dfrac{x-4}{2} \geq \dfrac{x-2}{3} + \dfrac{5}{6}$

$6\left(\dfrac{x-4}{2}\right) \geq 6\left(\dfrac{x-2}{3} + \dfrac{5}{6}\right)$

$3(x-4) \geq 2(x-2) + 5$

$3x - 12 \geq 2x - 4 + 5$

$3x - 12 \geq 2x + 1$

$3x - 2x \geq 1 + 12$

$x \geq 13$

The solution set is $\{x | x \geq 13\}$ or $[13, \infty)$.

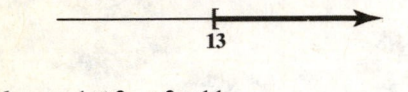

6. $1 \leq 2x + 3 < 11$

$-2 \leq 2x < 8$

$-1 \leq x < 4$

The solution set is $\{x | -1 \leq x < 4\}$ or $[-1, 4)$.

7. $|x - 2| < 5$

$-5 < x - 2 < 5$

$-3 < x < 7$

The solution set is $\{x | -3 < x < 7\}$ or $(-3, 7)$.

8. $-3|5x - 2| + 20 \geq -19$

$-3|5x - 2| \geq -39$

$\dfrac{-3|5x - 2|}{-3} \leq \dfrac{-39}{-3}$

$|5x - 2| \leq 13$

$-13 \leq 5x - 2 \leq 13$

$-11 \leq 5x \leq 15$

$\dfrac{-11}{5} \leq \dfrac{5x}{5} \leq \dfrac{15}{5}$

$-\dfrac{11}{5} \leq x \leq 3$

The solution set is $\left\{x \left| -\dfrac{11}{5} \leq x \leq 3\right.\right\}$ or $\left[-\dfrac{11}{5}, 3\right]$.

9. $18 < |6 - 3x|$

$$6 - 3x < -18 \quad \text{or} \quad 6 - 3x > 18$$
$$-3x < -24 \qquad\qquad -3x > 12$$
$$\frac{-3x}{-3} > \frac{-24}{-3} \qquad\qquad \frac{-3x}{-3} < \frac{12}{-3}$$
$$x > 8 \qquad\qquad x < -4$$

The solution set is $\{x | x < -4 \text{ or } x > 8\}$

or $(-\infty, -4) \cup (8, \infty)$.

10. Let x = the number of miles driven in a week.

$260 < 80 + 0.25x$

$180 < 0.25x$

$720 < x$

Driving more than 720 miles in a week makes Basic the better deal.

Exercise Set P.9

1. $1 < x \le 6$

3. $-5 \le x < 2$

5. $-3 \le x \le 1$

7. $x > 2$

9. $x \ge -3$

11. $x < 3$

13. $x < 5.5$

15. Graph $(-3, 0)$:

Graph $[-1, 2]$:

To find the intersection, take the portion of the number line that the two graphs have in common.

Numbers in both $(-3, 0)$ and $[-1, 2]$:

Thus, $(-3, 0) \cap [-1, 2] = [-1, 0)$.

17.
Graph $(-3, 0)$:
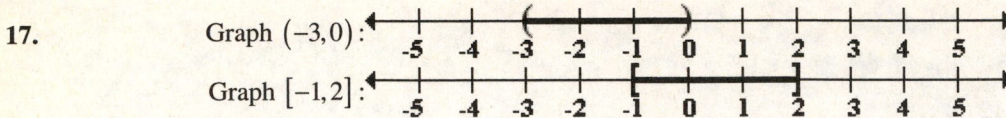

Graph $[-1, 2]$:
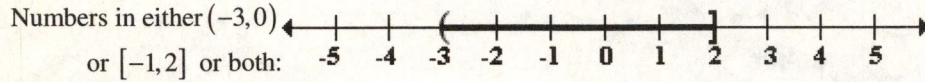

To find the union, take the portion of the number line representing the total collection of numbers in the two graphs.

Numbers in either $(-3, 0)$ or $[-1, 2]$ or both:
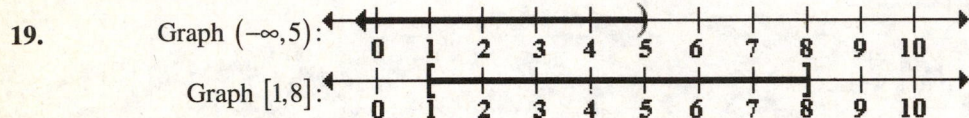

Thus, $(-3, 0) \cup [-1, 2] = (-3, 2]$.

19.
Graph $(-\infty, 5)$:
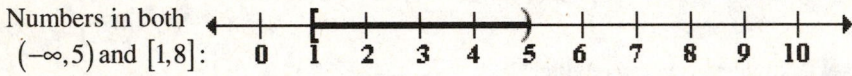

Graph $[1, 8]$:

To find the intersection, take the portion of the number line that the two graphs have in common.

Numbers in both $(-\infty, 5)$ and $[1, 8]$:
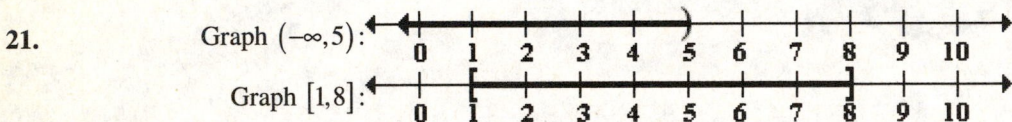

Thus, $(-\infty, 5) \cap [1, 8] = [1, 5)$.

21.
Graph $(-\infty, 5)$:
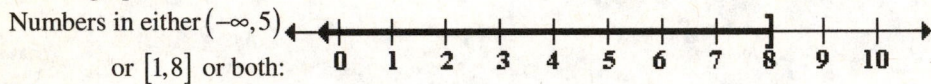

Graph $[1, 8]$:

To find the union, take the portion of the number line representing the total collection of numbers in the two graphs.

Numbers in either $(-\infty, 5)$ or $[1, 8]$ or both:
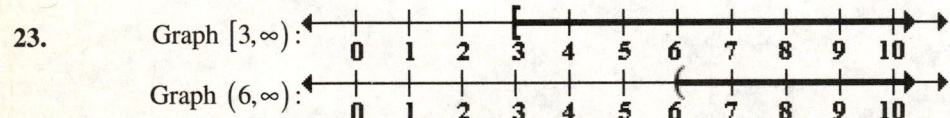

Thus, $(-\infty, 5) \cup [1, 8] = (-\infty, 8]$.

23.
Graph $[3, \infty)$:
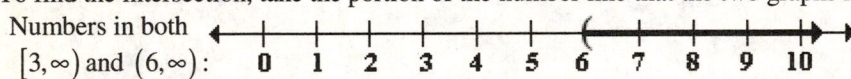

Graph $(6, \infty)$:

To find the intersection, take the portion of the number line that the two graphs have in common.

Numbers in both $[3, \infty)$ and $(6, \infty)$:
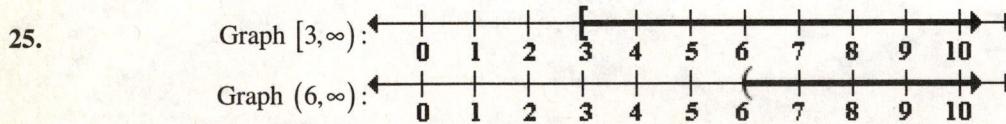

Thus, $[3, \infty) \cap (6, \infty) = (6, \infty)$.

25.
Graph $[3, \infty)$:
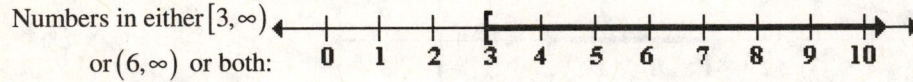

Graph $(6, \infty)$:

To find the union, take the portion of the number line representing the total collection of numbers in the two graphs.

Numbers in either $[3, \infty)$ or $(6, \infty)$ or both:

Thus, $[3, \infty) \cup (6, \infty) = [3, \infty)$.

27. $5x + 11 < 26$

$5x < 15$

$x < 3$

The solution set is $\{x \mid x < 3\}$, or $(-\infty, 3)$.

29. $3x - 7 \geq 13$

$3x \geq 20$

$x \geq \dfrac{20}{3}$

The solution set is $\left\{ x \mid x > \dfrac{20}{3} \right\}$, or $\left[\dfrac{20}{3}, \infty \right)$.

31. $-9x \geq 36$

$x \leq -4$

The solution set is $\{x \mid x \leq -4\}$, or $(-\infty, -4]$.

33. $8x - 11 \leq 3x - 13$

$8x - 3x \leq -13 + 11$

$5x \leq -2$

$x \leq -\dfrac{2}{5}$

The solution set is $\left\{ x \mid x \leq -\dfrac{2}{5} \right\}$, or $\left(-\infty, -\dfrac{2}{5} \right]$.

35. $4(x + 1) + 2 \geq 3x + 6$

$4x + 4 + 2 \geq 3x + 6$

$4x + 6 \geq 3x + 6$

$4x - 3x \geq 6 - 6$

$x \geq 0$

The solution set is $\{x \mid x > 0\}$, or $[0, \infty)$.

37. $2x - 11 < -3(x + 2)$

$2x - 11 < -3x - 6$

$5x < 5$

$x < 1$

The solution set is $\{x \mid x < 1\}$, or $(-\infty, 1)$.

39. $1-(x+3) \geq 4-2x$

$1-x-3 \geq 4-2x$

$-x-2 \geq 4-2x$

$x \geq 6$

The solution set is $\{x \mid x \geq 6\}$, or $[6, \infty)$.

41. $\dfrac{x}{4} - \dfrac{3}{2} \leq \dfrac{x}{2} + 1$

$\dfrac{4x}{4} - \dfrac{4 \cdot 3}{2} \leq \dfrac{4 \cdot x}{2} + 4 \cdot 1$

$x - 6 \leq 2x + 4$

$-x \leq 10$

$x \geq -10$

The solution set is $\{x \mid x \geq -10\}$, or $[-10, \infty)$.

43. $1 - \dfrac{x}{2} > 4$

$-\dfrac{x}{2} > 3$

$x < -6$

The solution set is $\{x \mid x, -6\}$, or $(-\infty, -6)$.

45. $\dfrac{x-4}{6} \geq \dfrac{x-2}{9} + \dfrac{5}{18}$

$3(x-4) \geq 2(x-2) + 5$

$3x-12 \geq 2x-4+5$

$x \geq 13$

The solution set is $\{x \mid x \geq 13\}$, or $[13, \infty)$.

47. $3[3(x+5)+8x+7] + 5[3(x-6)-2(3x-5)] < 2(4x+3)$

$3[3x+15+8x+7] + 5[3x-18-6x+10] < 8x+6$

$3[11x+22] + 5[-3x-8] < 8x+6$

$33x+66-15x-40 < 8x+6$

$18x+26 < 8x+6$

$10x < -20$

$x < -2$

The solution set is $\{x \mid x < -2\}$ or $[-\infty, -2)$.

49. $6 < x + 3 < 8$
$6 - 3 < x + 3 - 3 < 8 - 3$
$3 < x < 5$
The solution set is $\{x \mid 3 < x < 5\}$, or (3, 5).

51. $-3 \le x - 2 < 1$
$-1 \le x < 3$
The solution set is $\{x \mid -1 \le x < 3\}$, or [–1, 3).

53. $-11 < 2x - 1 \le -5$
$-10 < 2x \le -4$
$-5 < x \le -2$
The solution set is $\{x \mid -5 < x \le -2\}$, or
(–5, –2].

55. $-3 \le \dfrac{2}{3}x - 5 < -1$

$2 \le \dfrac{2}{3}x < 4$

$3 \le x < 6$
The solution set is $\{x \mid 3 \le x < 6\}$, or [3, 6).

57. $|x| < 3$
$-3 < x < 3$
The solution set is $\{x \mid -3 < x < 3\}$, or (–3, 3).

59. $|x - 1| \le 2$
$-2 \le x - 1 \le 2$
$-1 \le x \le 3$
The solution set is $\{x \mid -1 \le x \le 3\}$, or [–1, 3].

61. $|2x - 6| < 8$
$-8 < 2x - 6 < 8$
$-2 < 2x < 14$
$-1 < x < 7$
The solution set is $\{x \mid -1 < x < 7\}$, or (–1, 7).

63. $|2(x - 1) + 4| \le 8$
$-8 \le 2(x - 1) + 4 \le 8$
$-8 \le 2x - 2 + 4 \le 8$
$-8 \le 2x + 2 \le 8$
$-10 \le 2x \le 6$
$-5 \le x \le 3$
The solution set is $\{x \mid -5 \le x \le 3\}$, or
[–5, 3].

65. $\left| \dfrac{2y + 6}{3} \right| < 2$

$-2 < \dfrac{2y + 6}{3} < 2$

$-6 < 2y + 6 < 6$
$-12 < 2y < 0$
$-6 < y < 0$
The solution set is $\{x \mid -6 < y < 0\}$, or (–6, 0).

67. $|x| > 3$
$x > 3$ or $x < -3$
The solution set is $\{x \mid x > 3 \text{ or } x < -3\}$, that is,
$(-\infty, -3)$ or $(3, \infty)$.

69. $|x - 1| \ge 2$
$x - 1 \ge 2$ or $x - 1 \le -2$
$x \ge 3$ $\qquad$ $x \le -1$
The solution set is $\{x \mid x \le -1 \text{ or } x \ge 3\}$, that is,
$(-\infty, -1]$ or $[3, \infty)$.

71. $|3x - 8| > 7$
$3x - 8 > 7$ or $3x - 8 < -7$
$3x > 15$ $\qquad$ $3x < 1$

$x > 5$ $\qquad$ $x < \dfrac{1}{3}$

The solution set is $\left\{ x \mid x < \dfrac{1}{3} \text{ or } x > 5 \right\}$, that is,

$\left(-\infty, \dfrac{1}{3} \right)$ or $(5, \infty)$.

73. $\left| \dfrac{2x + 2}{4} \right| \ge 2$

$\dfrac{2x + 2}{4} \ge 2$ or $\dfrac{2x + 2}{4} \le -2$

$2x + 2 \ge 8$ $\qquad$ $2x + 2 \le -8$
$2x \ge 6$ $\qquad$ $2x \le -10$
$x \ge 3$ $\qquad$ $x \le -5$
The solution set is $\{x \mid x \le -5 \text{ or } x \ge 3\}$, that is,
$(-\infty, -5]$ or $[3, \infty)$.

75. $\left|3-\dfrac{2}{3}x\right|>5$

$3-\dfrac{2}{3}x>5$ or $3-\dfrac{2}{3}x<-5$

$-\dfrac{2}{3}x>2$ $-\dfrac{2}{3}x<-8$

$x<-3$ $x>12$

The solution set is $\{x\,|\,x<-3 \text{ or } x>12\}$, that is,

$(-\infty,-3)$ or $(12,\infty)$.

77. $3|x-1|+2\geq 8$
$3|x-1|\geq 6$
$|x-1|\geq 2$

$x-1\geq 2$ or $x-1\leq -2$

$x\geq 3$ $x\leq -1$

The solution set is $\{x\,|\,x\leq 1 \text{ or } x\geq 3\}$, that is,

$(-\infty,-1]$ or $[3,\infty)$.

79. $-2|x-4|\geq -4$

$\dfrac{-2|x-4|}{-2}\leq\dfrac{-4}{-2}$

$|x-4|\leq 2$

$-2\leq x-4\leq 2$

$2\leq x\leq 6$

The solution set is $\{x\,|\,2\leq x\leq 6\}$.

81. $-4|1-x|<-16$

$\dfrac{-4|1-x|}{-4}>\dfrac{-16}{-4}$

$|1-x|>4$

$1-x>4$ $1-x<-4$

$-x>3$ or $-x<-5$

$x<-3$ $x>5$

The solution set is $\{x\,|\,x<-3 \text{ or } x>5\}$.

83. $3\leq|2x-1|$

$2x-1\geq 3$ $2x-1\leq -3$

$2x\geq 4$ or $2x\leq -2$

$x\geq 2$ $x\leq -1$

The solution set is $\{x\,|\,x\leq -1 \text{ or } x\geq 2\}$.

85. $5>|4-x|$ is equivalent to $|4-x|<5$.

$-5<4-x<5$

$-9<-x<1$

$\dfrac{-9}{-1}>\dfrac{-x}{-1}>\dfrac{1}{-1}$

$9>x>-1$

$-1<x<9$

The solution set is $\{x\,|-1<x<9\}$.

87. $1<|2-3x|$ is equivalent to $|2-3x|>1$.

$2-3x>1$ $2-3x<-1$

$-3x>-1$ $-3x<-3$

$\dfrac{-3x}{-3}<\dfrac{-1}{-3}$ or $\dfrac{-3x}{-3}>\dfrac{-3}{-3}$

$x<\dfrac{1}{3}$ $x>1$

The solution set is $\left\{x\,\bigg|\,x<\dfrac{1}{3} \text{ or } x>1\right\}$.

89. $12<\left|-2x+\dfrac{6}{7}\right|+\dfrac{3}{7}$

$\dfrac{81}{7}<\left|-2x+\dfrac{6}{7}\right|$

$-2x+\dfrac{6}{7}>\dfrac{81}{7}$ or $-2x+\dfrac{6}{7}<-\dfrac{81}{7}$

$-2x>\dfrac{75}{7}$ $-2x<-\dfrac{87}{7}$

$x<-\dfrac{75}{14}$ $x>\dfrac{87}{14}$

The solution set is $\left\{x\,\bigg|\,x<-\dfrac{75}{14} \text{ or } x>\dfrac{87}{14}\right\}$, that is,

$\left(-\infty,-\dfrac{75}{14}\right)$ or $\left(\dfrac{87}{14},\infty\right)$.

91. $4+\left|3-\dfrac{x}{3}\right|\geq 9$

$\left|3-\dfrac{x}{3}\right|\geq 5$

$3-\dfrac{x}{3}\geq 5$ or $3-\dfrac{x}{3}\leq -5$

$-\dfrac{x}{3}\geq 2$ $-\dfrac{x}{3}\leq -8$

$x\leq -6$ $x\geq 24$

The solution set is $\{x\,|\,x\leq -6 \text{ or } x\geq 24\}$, that is,

$(-\infty,-6]$ or $[24,\infty)$.

93.
$$y \geq 4$$
$$1 - (x+3) + 2x \geq 4$$
$$1 - x - 3 + 2x \geq 4$$
$$x - 2 \geq 4$$
$$x \geq 6$$
The solution set is $[6, \infty)$.

95.
$$y \leq 4$$
$$7 - \left| \frac{x}{2} + 2 \right| \leq 4$$
$$-\left| \frac{x}{2} + 2 \right| \leq -3$$
$$\left| \frac{x}{2} + 2 \right| \geq 3$$

$$\frac{x}{2} + 2 \geq 3 \quad \text{or} \quad \frac{x}{2} + 2 \leq -3$$
$$x + 4 \geq 6 \qquad x + 4 \leq -6$$
$$x \geq 2 \qquad x \leq -10$$
The solution set is $(-\infty, -10] \cup [2, \infty)$.

97. Let x be the number.
$$|4 - 3x| \geq 5 \quad \text{or} \quad |3x - 4| \geq 5$$

$$3x - 4 \geq 5 \qquad 3x - 4 \leq -5$$
$$3x \geq 9 \quad \text{or} \quad 3x \leq -1$$
$$x \geq 3 \qquad x \leq -\frac{1}{3}$$

The solution set is $\left\{ x \mid x \leq -\frac{1}{3} \text{ or } x \geq 3 \right\}$ or

$\left(-\infty, -\frac{1}{3} \right] \cup [3, \infty)$.

99. $(0, 4)$

101. passion $\leq$ intimacy or intimacy $\geq$ passion

103. passion<commitment or commitment > passion

105. 9, after 3 years

107. $3.1x + 25.8 > 63$

$3.1x > 37.2$

$x > 12$

Since x is the number of years after 1994, we calculate 1994+12=2006. 63% of voters will use electronic systems after 2006.

109. $28 \leq 20 + 0.40(x - 60) \leq 40$
$$28 \leq 20 + 0.40x - 24 \leq 40$$
$$28 \leq 0.40x - 4 \leq 40$$
$$32 \leq 0.40x \leq 44$$
$$80 \leq x \leq 110$$
Between 80 and 110 ten minutes, inclusive.

111.
$$\left| \frac{h - 50}{5} \right| \geq 1.645$$

$$\frac{h - 50}{5} \geq 1.645 \quad \text{or} \quad \frac{h - 50}{5} \leq -1.645$$
$$h - 50 \geq 8.225 \qquad h - 50 \leq -8.225$$
$$h \geq 58.225 \qquad h \leq 41.775$$
The number of outcomes would be 59 or more, or 41 or less.

113. $15 + 0.08x < 3 + .12x$
$$12 < 0.04x$$
$$300 < x$$
Plan A is a better deal when driving more than 300 miles a month.

115. $2 + 0.08x < 8 + 0.05x$
$$0.03x < 6$$
$$x < 200$$
The credit union is a better deal when writing less than 200 checks.

117. $3000 + 3x < 5.5x$
$$3000 < 2.5x$$
$$1200 < x$$
More then 1200 packets of stationary need to be sold each week to make a profit.

119. $245 + 95x \leq 3000$
$$95x \leq 2755$$
$$x \leq 29$$
29 bags or less can be lifted safely.

121. a.
$$\frac{86+88+x}{3} \geq 90$$
$$\frac{174+x}{3} \geq 90$$
$$174+x \geq 270$$
$$x \geq 96$$
You must get at least a 96.

b.
$$\frac{86+88+x}{3} < 80$$
$$\frac{174+x}{3} < 80$$
$$174+x < 240$$
$$x < 66$$
This will happen if you get a grade less than 66.

123. Let x = the number of times the bridge is crossed per three month period
The cost with the 3-month pass is
$C_3 = 7.50 + 0.50x$.
The cost with the 6-month pass is $C_6 = 30$.

Because we need to buy two 3-month passes per 6-month pass, we multiply the cost with the 3-month pass by 2.
$$2(7.50 + 0.50x) < 30$$
$$15 + x < 30$$
$$x < 15$$
We also must consider the cost without purchasing a pass. We need this cost to be less than the cost with a 3-month pass.
$$3x > 7.50 + 0.50x$$
$$2.50x > 7.50$$
$$x > 3$$
The 3-month pass is the best deal when making more than 3 but less than 15 crossings per 3-month period.

125. – 131. Answers may vary.

133. makes sense

135. makes sense

137. false; Changes to make the statement true will vary.
A sample change is: $(-\infty,3) \cup (-\infty,-2) = (-\infty,3)$

139. true

141. a. $|x-4| < 3$

b. $|x-4| \geq 3$

143. $y = 4-x$

x	$y = 4-x$
–3	$4-(-3)=7$
–2	$4-(-2)=6$
–1	$4-(-1)=5$
0	$4-(0)=4$
1	$4-(1)=3$
2	$4-(2)=2$
3	$4-(3)=1$

144. $y = 4-x^2$

x	$y = 4-x^2$
–3	$4-(-3)^2=-5$
–2	$4-(-2)^2=0$
–1	$4-(-1)^2=3$
0	$4-(0)^2=4$
1	$4-(1)^2=3$
2	$4-(2)^2=0$
3	$4-(3)^2=-5$

145. $y = |x+1|$

x	$y = |x+1|$
–4	$|-4+1|=3$
–3	$|-3+1|=2$
–2	$|-2+1|=1$
–1	$|-1+1|=0$
0	$|0+1|=1$
1	$|1+1|=2$
2	$|2+1|=3$

Chapter P Review Exercises

1. $3+6(x-2)^3 = 3+6(4-2)^3$
    $= 3+6(2)^3$
    $= 3+6(8)$
    $= 3+48$
    $= 51$

2. $x^2-5(x-y) = 6^2-5(6-2)$
    $= 36-5(4)$
    $= 36-20$
    $= 16$

3. $S = 0.015x^2+x+10$
    $S = 0.015(60)^2+(60)+10$
    $= 0.015(3600)+60+10$
    $= 54+60+10$
    $= 124$

4. $A=\{a,b,c\}\quad B=\{a,c,d,e\}$
    $\{a,b,c\}\cap\{a,c,d,e\}=\{a,c\}$

5. $A=\{a,b,c\}\quad B=\{a,c,d,e\}$
    $\{a,b,c\}\cup\{a,c,d,e\}=\{a,b,c,d,e\}$

6. $A=\{a,b,c\}\quad C=\{a,d,f,g\}$
    $\{a,b,c\}\cup\{a,d,f,g\}=\{a,b,c,d,f,g\}$

7. $A=\{a,b,c\}\quad C=\{a,d,f,g\}$
    $\{a,d,f,g\}\cap\{a,b,c\}=\{a\}$

8. a. $\sqrt{81}$

    b. $0,\sqrt{81}$

    c. $-17,0,\sqrt{81}$

    d. $-17,-\dfrac{9}{13},0,0.75,\sqrt{81}$

    e. $\sqrt{2},\pi$

    f. $-17,-\dfrac{9}{13},0,0.75,\sqrt{2},\pi,\sqrt{81}$

9. $|-103|=103$

10. $\left|\sqrt{2}-1\right|=\sqrt{2}-1$

11. $\left|3-\sqrt{17}\right|=\sqrt{17}-3$ since $\sqrt{17}$ is greater than 3.

12. $|4-(-17)|=|4+17|=|21|=21$

13. $3+17=17+3$;
    commutative property of addition.

14. $(6\cdot3)\cdot9=6\cdot(3\cdot9)$;
    associative property of multiplication.

15. $\sqrt{3}(\sqrt{5}+\sqrt{3})=\sqrt{15}+3$;
    distributive property of multiplication over addition.

16. $(6\cdot9)\cdot2=2\cdot(6\cdot9)$;
    commutative property of multiplication.

17. $\sqrt{3}(\sqrt{5}+\sqrt{3})=(\sqrt{5}+\sqrt{3})\sqrt{3}$;
    commutative property of multiplication.

18. $(3\cdot7)+(4\cdot7)=(4\cdot7)+(3\cdot7)$;
    commutative property of addition.

19. $5(2x-3)+7x=10x-15+7x=17x-15$

20. $\dfrac{1}{5}(5x)+[(3y)+(-3y)]-(-x)=x+[0]+x=2x$

21. $3(4y-5)-(7y+2)=12y-15-7y-2=5y-17$

22. $8-2[3-(5x-1)]=8-2[3-5x+1]$
    $= 8-2[4-5x]$
    $= 8-8+10x$
    $= 10x$

23. $P=-0.05x^2+3.6x-15$
    $P=-0.05(21)^2+3.6(21)-15$
    $= 38.55$
    38.55% of 21 year olds have been tested.
    This overestimates the percent displayed by the bar graph by 3.55%.

24. $(-3)^3(-2)^2=(-27)\cdot(4)=-108$

25. $2^{-4} + 4^{-1} = \dfrac{1}{2^4} + \dfrac{1}{4}$

$\qquad = \dfrac{1}{16} + \dfrac{1}{4}$

$\qquad = \dfrac{1}{16} + \dfrac{4}{16}$

$\qquad = \dfrac{5}{16}$

26. $5^{-3} \cdot 5 = 5^{-3} 5^1 = 5^{-3+1} = 5^{-2} = \dfrac{1}{5^2} = \dfrac{1}{25}$

27. $\dfrac{3^3}{3^6} = 3^{3-6} = 3^{-3} = \dfrac{1}{3^3} = \dfrac{1}{27}$

28. $(-2x^4 y^3)^3 = (-2)^3 (x^4)^3 (y^3)^3$

$\qquad = (-2)^3 x^{4\cdot3} y^{3\cdot3}$

$\qquad = -8x^{12} y^9$

29. $(-5x^3 y^2)(-2x^{-11} y^{-2})$

$\qquad = (-5)(-2)x^3 x^{-11} y^2 y^{-2}$

$\qquad = 10 \cdot x^{3-11} y^{2-2}$

$\qquad = 10x^{-8} y^0$

$\qquad = \dfrac{10}{x^8}$

30. $(2x^3)^{-4} = (2)^{-4} (x^3)^{-4}$

$\qquad = 2^{-4} x^{-12}$

$\qquad = \dfrac{1}{2^4 x^{12}}$

$\qquad = \dfrac{1}{16x^{12}}$

31. $\dfrac{7x^5 y^6}{28x^{15} y^{-2}} = \left(\dfrac{7}{28}\right)(x^{5-15})(y^{6-(-2)})$

$\qquad = \dfrac{1}{4} x^{-10} y^8$

$\qquad = \dfrac{y^8}{4x^{10}}$

32. $3.74 \times 10^4 = 37,400$

33. $7.45 \times 10^{-5} = 0.0000745$

34. $3,590,000 = 3.59 \times 10^6$

35. $0.00725 = 7.25 \times 10^{-3}$

36. $(3\times10^3)(1.3\times10^2) = (3\times1.3)\times(10^3\times10^2)$

$\qquad = 3.9\times10^5$

$\qquad = 390,000$

37. $\dfrac{6.9\times10^3}{3\times10^5} = \left(\dfrac{6.9}{3}\right)\times10^{3-5}$

$\qquad = 2.3\times10^{-2}$

$\qquad = 0.023$

38.   **a.** $257\times10^9 = 2.57\times10^2 \cdot 10^9 = 2.57\times10^{11}$

      b. $175\times10^6 = 1.75\times10^2 \cdot 10^6 = 1.75\times10^8$

39. $\dfrac{2.57\times10^{11}}{1.75\times10^8} = \dfrac{2.57}{1.75} \cdot \dfrac{10^{11}}{10^8} \approx 1.469\times10^3 = 1469$

The average tax return cost $1469.

40. $\sqrt{300} = \sqrt{100\cdot3} = \sqrt{100} \cdot \sqrt{3} = 10\sqrt{3}$

41. $\sqrt{12x^2} = \sqrt{4x^2 \cdot 3} = \sqrt{4x^2} \cdot \sqrt{3} = 2|x|\sqrt{3}$

42. $\sqrt{10x} \cdot \sqrt{2x} = \sqrt{20x^2}$

$\qquad = \sqrt{4x^2} \cdot \sqrt{5}$

$\qquad = 2x\sqrt{5}$

43. $\sqrt{r^3} = \sqrt{r^2} \cdot \sqrt{r} = r\sqrt{r}$

44. $\sqrt{\dfrac{121}{4}} = \dfrac{\sqrt{121}}{\sqrt{4}} = \dfrac{11}{2}$

45. $\dfrac{\sqrt{96x^3}}{\sqrt{2x}} = \sqrt{\dfrac{96x^3}{2x}}$

$\qquad = \sqrt{48x^2}$

$\qquad = \sqrt{16x^2} \cdot \sqrt{3}$

$\qquad = 4x\sqrt{3}$

46. $7\sqrt{5} + 13\sqrt{5} = (7+13)\sqrt{5} = 20\sqrt{5}$

47. $2\sqrt{50} + 3\sqrt{8} = 2\sqrt{25\cdot2} + 3\sqrt{4\cdot2}$

$\qquad = 2\cdot5\sqrt{2} + 3\cdot2\sqrt{2}$

$\qquad = 10\sqrt{2} + 6\sqrt{2}$

$\qquad = 16\sqrt{2}$

48. $4\sqrt{72} - 2\sqrt{48} = 4\sqrt{36 \cdot 2} - 2\sqrt{16 \cdot 3}$
$= 4 \cdot 6\sqrt{2} - 2 \cdot 4\sqrt{3}$
$= 24\sqrt{2} - 8\sqrt{3}$

49. $\dfrac{30}{\sqrt{5}} = \dfrac{30}{\sqrt{5}} \cdot \dfrac{\sqrt{5}}{\sqrt{5}} = \dfrac{30\sqrt{5}}{5} = 6\sqrt{5}$

50. $\dfrac{\sqrt{2}}{\sqrt{3}} = \dfrac{\sqrt{2}}{\sqrt{3}} \cdot \dfrac{\sqrt{3}}{\sqrt{3}} = \dfrac{\sqrt{6}}{3}$

51. $\dfrac{5}{6+\sqrt{3}} = \dfrac{5}{6+\sqrt{3}} \cdot \dfrac{6-\sqrt{3}}{6-\sqrt{3}}$
$= \dfrac{5(6-\sqrt{3})}{36-3}$
$= \dfrac{5(6-\sqrt{3})}{33}$

52.
$\dfrac{14}{\sqrt{7}-\sqrt{5}} = \dfrac{14}{\sqrt{7}-\sqrt{5}} \cdot \dfrac{\sqrt{7}+\sqrt{5}}{\sqrt{7}+\sqrt{5}}$
$= \dfrac{14(\sqrt{7}+\sqrt{5})}{7-5}$
$= \dfrac{14(\sqrt{7}+\sqrt{5})}{2}$
$= 7(\sqrt{7}+\sqrt{5})$

53. $\sqrt[3]{125} = 5$

54. $\sqrt[5]{-32} = -2$

55. $\sqrt[4]{-125}$ is not a real number.

56. $\sqrt[4]{(-5)^4} = \sqrt[4]{625} = \sqrt[4]{5^4} = 5$

57. $\sqrt[3]{81} = \sqrt[3]{27 \cdot 3} = \sqrt[3]{27} \cdot \sqrt[3]{3} = 3\sqrt[3]{3}$

58. $\sqrt[3]{y^5} = \sqrt[3]{y^3 y^2} = y\sqrt[3]{y^2}$

59. $\sqrt[4]{8} \cdot \sqrt[4]{10} = \sqrt[4]{80} = \sqrt[4]{16 \cdot 5} = \sqrt[4]{16} \cdot \sqrt[4]{5} = 2\sqrt[4]{5}$

60. $4\sqrt[3]{16} + 5\sqrt[3]{2} = 4\sqrt[3]{8 \cdot 2} + 5\sqrt[3]{2}$
$= 4 \cdot 2\sqrt[3]{2} + 5\sqrt[3]{2}$
$= 8\sqrt[3]{2} + 5\sqrt[3]{2}$
$= 13\sqrt[3]{2}$

61. $\dfrac{\sqrt[4]{32x^5}}{\sqrt[4]{16x}} = \sqrt[4]{\dfrac{32x^5}{16x}} = \sqrt[4]{2x^4} = x\sqrt[4]{2}$

62. $16^{1/2} = \sqrt{16} = 4$

63. $25^{-1/2} = \dfrac{1}{25^{1/2}} = \dfrac{1}{\sqrt{25}} = \dfrac{1}{5}$

64. $125^{1/3} = \sqrt[3]{125} = 5$

65. $27^{-1/3} = \dfrac{1}{27^{1/3}} = \dfrac{1}{\sqrt[3]{27}} = \dfrac{1}{3}$

66. $64^{2/3} = (\sqrt[3]{64})^2 = 4^2 = 16$

67. $27^{-4/3} = \dfrac{1}{27^{4/3}} = \dfrac{1}{(\sqrt[3]{27})^4} = \dfrac{1}{3^4} = \dfrac{1}{81}$

68. $(5x^{2/3})(4x^{1/4}) = 5 \cdot 4x^{2/3 + 1/4} = 20x^{11/12}$

69. $\dfrac{15x^{3/4}}{5x^{1/2}} = \left(\dfrac{15}{5}\right)x^{3/4 - 1/2} = 3x^{1/4}$

70. $(125 \cdot x^6)^{2/3} = (\sqrt[3]{125x^6})^2$
$= (5x^2)^2$
$= 25x^4$

71. $\sqrt[6]{y^3} = (y^3)^{1/6} = y^{3 \cdot 1/6} = y^{1/2} = \sqrt{y}$

72. $(-6x^3 + 7x^2 - 9x + 3) + (14x^3 + 3x^2 - 11x - 7) = (-6x^3 + 14x^3) + (7x^2 + 3x^2) + (-9x - 11x) + (3 - 7)$

$$= 8x^3 + 10x^2 - 20x - 4$$

The degree is 3.

73. $(13x^4 - 8x^3 + 2x^2) - (5x^4 - 3x^3 + 2x^2 - 6) = (13x^4 - 8x^3 + 2x^2) + (-5x^4 + 3x^3 - 2x^2 + 6)$

$$= (13x^4 - 5x^4) + (-8x^3 + 3x^3) + (2x^2 - 2x^2) + 6$$

$$= 8x^4 - 5x^3 + 6$$

The degree is 4.

74. $(3x - 2)(4x^2 + 3x - 5) = (3x)(4x^2) + (3x)(3x) + (3x)(-5) + (-2)(4x^2) + (-2)(3x) + (-2)(-5)$

$$= 12x^3 + 9x^2 - 15x - 8x^2 - 6x + 10$$

$$= 12x^3 + x^2 - 21x + 10$$

75. $(3x - 5)(2x + 1) = (3x)(2x) + (3x)(1) + (-5)(2x) + (-5)(1)$

$$= 6x^2 + 3x - 10x - 5$$

$$= 6x^2 - 7x - 5$$

76. $(4x + 5)(4x - 5) = (4x^2) - 5^2 = 16x^2 - 25$

77. $(2x + 5)^2 = (2x)^2 + 2(2x) \cdot 5 + 5^2 = 4x^2 + 20x + 25$

78. $(3x - 4)^2 = (3x)^2 - 2(3x) \cdot 4 + (-4)^2 = 9x^2 - 24x + 16$

79. $(2x + 1)^3 = (2x)^3 + 3(2x)^2(1) + 3(2x)(1)^2 + 1^3 = 8x^3 + 12x^2 + 6x + 1$

80. $(5x - 2)^3 = (5x)^3 - 3(5x)^2(2) + 3(5x)(2)^2 - 2^3 = 125x^3 - 150x^2 + 60x - 8$

81. $(x + 7y)(3x - 5y) = x(3x) + (x)(-5y) + (7y)(3x) + (7y)(-5y)$

$$= 3x^2 - 5xy + 21xy - 35y^2$$

$$= 3x^2 + 16xy - 35y^2$$

82. $(3x - 5y)^2 = (3x)^2 - 2(3x)(5y) + (-5y)^2$

$$= 9x^2 - 30xy + 25y^2$$

83. $(3x^2 + 2y)^2 = (3x^2)^2 + 2(3x^2)(2y) + (2y)^2$

$$= 9x^4 + 12x^2y + 4y^2$$

84. $(7x + 4y)(7x - 4y) = (7x)^2 - (4y)^2$

$$= 49x^2 - 16y^2$$

85. $(a - b)(a^2 + ab + b^2)$

$= a(a^2) + a(ab) + a(b^2) + (-b)(a^2)$

$+ (-b)(ab) + (-b)(b^2)$

$= a^3 + a^2b + ab^2 - a^2b - ab^2 - b^3$

$= a^3 - b^3$

86. $[5y-(2x+1)][5y+(2x+1)]$

$\qquad = (5y)^2 - (2x+1)^2$

$\qquad = 25y^2 - (4x^2+4x+1)$

$\qquad = 25y^2 - 4x^2 - 4x - 1$

87. $(x+2y+4)^2$

$\qquad = (x+2y+4)(x+2y+4)$

$\qquad = x(x+2y+4) + 2y(x+2y+4) + 4(x+2y+4)$

$\qquad = x^2 + 2xy + 4x + 2xy + 4y^2 + 8y + 4x + 8y + 16$

$\qquad = x^2 + 4xy + 4y^2 + 8x + 16y + 16$

88. $15x^3 + 3x^2 = 3x^2 \cdot 5x + 3x^2 \cdot 1$

$\qquad = 3x^2(5x+1)$

89. $x^2 - 11x + 28 = (x-4)(x-7)$

90. $15x^2 - x - 2 = (3x+1)(5x-2)$

91. $64 - x^2 = 8^2 - x^2 = (8-x)(8+x)$

92. $x^2 + 16$ is prime.

93. $3x^4 - 9x^3 - 30x^2 = 3x^2(x^2 - 3x - 10)$

$\qquad = 3x^2(x-5)(x+2)$

94. $20x^7 - 36x^3 = 4x^3(5x^4 - 9)$

95. $x^3 - 3x^2 - 9x + 27 = x^2(x-3) - 9(x-3)$

$\qquad = (x^2 - 9)(x-3)$

$\qquad = (x+3)(x-3)(x-3)$

$\qquad = (x+3)(x-3)^2$

96. $16x^2 - 40x + 25 = (4x-5)(4x-5)$

$\qquad = (4x-5)^2$

97. $x^4 - 16 = (x^2)^2 - 4^2$

$\qquad = (x^2+4)(x^2-4)$

$\qquad = (x^2+4)(x+2)(x-2)$

98. $y^3 - 8 = y^3 - 2^3 = (y-2)(y^2 + 2y + 4)$

99. $x^3 + 64 = x^3 + 4^3 = (x+4)(x^2 - 4x + 16)$

100. $3x^4 - 12x^2 = 3x^2(x^2 - 4)$

$\qquad = 3x^2(x-2)(x+2)$

101. $27x^3 - 125 = (3x)^3 - 5^3$

$\qquad = (3x-5)[(3x)^2 + (3x)(5) + 5^2]$

$\qquad = (3x-5)(9x^2 + 15x + 25)$

102. $x^5 - x = x(x^4 - 1)$

$\qquad = x(x^2-1)(x^2+1)$

$\qquad = x(x-1)(x+1)(x^2+1)$

103. $x^3 + 5x^2 - 2x - 10 = x^2(x+5) - 2(x+5)$

$\qquad = (x^2 - 2)(x+5)$

104. $x^2 + 18x + 81 - y^2 = (x^2 + 18x + 81) - y^2$

$\qquad = (x+9)^2 - y^2$

$\qquad = (x+9-y)(x+9+y)$

105. $16x^{-\frac{3}{4}} + 32x^{\frac{1}{4}} = 16x^{-\frac{3}{4}}\left(1 + 2x^{\frac{1}{4} - \left(-\frac{3}{4}\right)}\right)$

$\qquad = 16x^{-\frac{3}{4}}(1+2x)$

$\qquad = \dfrac{(1+2x)}{16x^{\frac{3}{4}}}$

106. $(x^2-4)(x^2+3)^{\frac{1}{2}} - (x^2-4)^2(x^2+3)^{\frac{3}{2}}$

$\qquad = (x^2-4)(x^2+3)^{\frac{1}{2}}\left[1 - (x^2-4)(x^2+3)\right]$

$\qquad = (x-2)(x+2)(x^2+3)^{\frac{1}{2}}\left[1 - (x-2)(x+2)(x^2+3)\right]$

$\qquad = (x-2)(x+2)(x^2+3)^{\frac{1}{2}}(-x^4 + x^2 + 13)$

107. $12x^{-\frac{1}{2}} + 6x^{-\frac{3}{2}} = 6x^{-\frac{3}{2}}(2x+1) = \dfrac{6(2x+1)}{x^{\frac{3}{2}}}$

108. $\dfrac{x^3 + 2x^2}{x+2} = \dfrac{x^2(x+2)}{x+2} = x^2, \, x \neq -2$

109. $\dfrac{x^2 + 3x - 18}{x^2 - 36} = \dfrac{(x+6)(x-3)}{(x+6)(x-6)} = \dfrac{x-3}{x-6}$,

$\qquad x \neq -6, 6$

110. $\dfrac{x^2 + 2x}{x^2 + 4x + 4} = \dfrac{x(x+2)}{(x+2)^2} = \dfrac{x}{x+2}$,

$\qquad x \neq -2$

111. $\dfrac{x^2+6x+9}{x^2-4} \cdot \dfrac{x+3}{x-2} = \dfrac{(x+3)^2}{(x-2)(x+2)} \cdot \dfrac{x+3}{x-2}$

$\qquad\qquad = \dfrac{(x+3)^3}{(x-2)^2(x+2)},$

$x \neq 2, -2$

112. $\dfrac{6x+2}{x^2-1} \div \dfrac{3x^2+x}{x-1}$

$= \dfrac{2(3x+1)}{(x-1)(x+1)} \div \dfrac{x(3x+1)}{x-1}$

$= \dfrac{2(3x+1)}{(x-1)(x+1)} \cdot \dfrac{x-1}{x(3x+1)}$

$= \dfrac{2}{x(x+1)},$

$x \neq 0, 1, -1, -\dfrac{1}{3}$

113. $\dfrac{x^2-5x-24}{x^2-x-12} \div \dfrac{x^2-10x+16}{x^2+x-6}$

$= \dfrac{(x-8)(x+3)}{(x-4)(x+3)} \div \dfrac{(x-2)(x-8)}{(x+3)(x-2)}$

$= \dfrac{x-8}{x-4} \cdot \dfrac{x+3}{x-8}$

$= \dfrac{x+3}{x-4},$

$x \neq -3, 4, 2, 8$

114. $\dfrac{2x-7}{x^2-9} - \dfrac{x-10}{x^2-9} = \dfrac{2x-7-(x-10)}{x^2-9}$

$\qquad\qquad = \dfrac{x+3}{(x+3)(x-3)}$

$\qquad\qquad = \dfrac{1}{x-3},$

$x \neq 3, -3$

115. $\dfrac{3x}{x+2} + \dfrac{x}{x-2} = \dfrac{3x}{x+2} \cdot \dfrac{x-2}{x-2} + \dfrac{x}{x-2} \cdot \dfrac{x+2}{x+2}$

$= \dfrac{3x^2-6x+x^2+2x}{(x+2)(x-2)}$

$= \dfrac{4x^2-4x}{(x+2)(x-2)}$

$= \dfrac{4x(x-1)}{(x+2)(x-2)},$

$x \neq 2, -2$

116. $\dfrac{x}{x^2-9} + \dfrac{x-1}{x^2-5x+6}$

$= \dfrac{x}{(x-3)(x+3)} + \dfrac{x-1}{(x-2)(x-3)}$

$= \dfrac{x}{(x-3)(x+3)} \cdot \dfrac{x-2}{x-2} + \dfrac{x-1}{(x-2)(x-3)} \cdot \dfrac{x+3}{x+3}$

$= \dfrac{x(x-2)+(x-1)(x+3)}{(x-3)(x+3)(x-2)}$

$= \dfrac{x^2-2x+x^2+2x-3}{(x-3)(x+3)(x-2)}$

$= \dfrac{2x^2-3}{(x-3)(x+3)(x-2)}$

$x \neq 3, -3, 2$

117. $\dfrac{4x-1}{2x^2+5x-3} - \dfrac{x+3}{6x^2+x-2}$

$= \dfrac{4x-1}{(2x-1)(x+3)} - \dfrac{x+3}{(2x-1)(3x+2)}$

$= \dfrac{4x-1}{(2x-1)(x+3)} \cdot \dfrac{3x+2}{3x+2}$

$\quad - \dfrac{x+3}{(2x-1)(3x+2)} \cdot \dfrac{x+3}{x+3}$

$= \dfrac{12x^2+8x-3x-2-x^2-6x-9}{(2x-1)(x+3)(3x+2)}$

$= \dfrac{11x^2-x-11}{(2x-1)(x+3)(3x+2)},$

$x \neq \dfrac{1}{2}, -3, -\dfrac{2}{3}$

118. $\dfrac{\dfrac{1}{x}-\dfrac{1}{2}}{\dfrac{1}{3}-\dfrac{x}{6}} = \dfrac{\dfrac{1}{x}-\dfrac{1}{2}}{\dfrac{1}{3}-\dfrac{x}{6}} \cdot \dfrac{6x}{6x}$

$\qquad = \dfrac{6-3x}{2x-x^2}$

$\qquad = \dfrac{-3(x-2)}{-x(x-2)}$

$\qquad = \dfrac{3}{x},$

$x \neq 0, 2$

119. $\dfrac{3+\dfrac{12}{x}}{1-\dfrac{16}{x^2}} = \dfrac{3+\dfrac{12}{x}}{1-\dfrac{16}{x^2}} \cdot \dfrac{x^2}{x^2}$

$\qquad = \dfrac{3x^2+12x}{x^2-16}$

$\qquad = \dfrac{3x(x+4)}{(x+4)(x-4)}$

$\qquad = \dfrac{3x}{x-4},$

$\qquad x \neq 0,\, 4,\, -4$

120. $\dfrac{3-\dfrac{1}{x+3}}{3+\dfrac{1}{x+3}} = \dfrac{3-\dfrac{1}{x+3}}{3+\dfrac{1}{x+3}} \cdot \dfrac{x+3}{x+3}$

$\qquad = \dfrac{3(x+3)-1}{3(x+3)+1}$

$\qquad = \dfrac{3x+9-1}{3x+9+1}$

$\qquad = \dfrac{3x+8}{3x+10},$

$\qquad x \neq -3,\, -\dfrac{10}{3}$

121. $\dfrac{\sqrt{25-x^2}+\dfrac{x^2}{\sqrt{25-x^2}}}{25-x^2}$

$\qquad = \dfrac{\left(\sqrt{25-x^2}+\dfrac{x^2}{\sqrt{25-x^2}}\right)\sqrt{25-2x^2}}{(25-x^2)\sqrt{25-x^2}}$

$\qquad = \dfrac{25-x^2+x^2}{(25-x^2)\sqrt{25-x^2}}$

$\qquad = \dfrac{25}{\sqrt{(25-x^2)^3}}$

$\qquad = \dfrac{25}{\sqrt{(25-x^2)^3}} \cdot \dfrac{\sqrt{25-x^2}}{\sqrt{25-x^2}}$

$\qquad = \dfrac{25\sqrt{25-x^2}}{(25-x^2)}$

$\qquad = \dfrac{25\sqrt{25-x^2}}{(5-x)^2(5+x)^2}$

122. $1-2(6-x) = 3x+2$

$\qquad 1-12+2x = 3x+2$

$\qquad -11-x = 2$

$\qquad -x = 13$

$\qquad x = -13$

The solution set is $\{-13\}$.

This is a conditional equation.

123. $2(x-4)+3(x+5) = 2x-2$

$\qquad 2x-8+3x+15 = 2x-2$

$\qquad 5x+7 = 2x-2$

$\qquad 3x = -9$

$\qquad x = -3$

The solution set is $\{-3\}$.

This is a conditional equation.

124. $2x-4(5x+1) = 3x+17$

$\qquad 2x-20x-4 = 3x+17$

$\qquad -18x-4 = 3x+17$

$\qquad -21x = 21$

$\qquad x = -1$

The solution set is $\{-1\}$.

This is a conditional equation.

125. $x \neq 1,\, x \neq -1$

$\qquad \dfrac{1}{x-1} - \dfrac{1}{x+1} = \dfrac{2}{x^2-1}$

$\qquad \dfrac{1}{x-1} - \dfrac{1}{x+1} = \dfrac{2}{(x+1)(x-1)}$

$\qquad x+1-(x-1) = 2$

$\qquad x+1-x+1 = 2$

$\qquad 2 = 2$

The solution set is all real numbers except 1 and -1.

126. $x \neq -2,\, x \neq 4$

$\qquad \dfrac{4}{x+2} + \dfrac{2}{x-4} = \dfrac{30}{(x+2)(x-4)}$

$\qquad 4(x-4)+2(x+2) = 30$

$\qquad 4x-16+2x+4 = 30$

$\qquad 6x-12 = 30$

$\qquad 6x = 42$

$\qquad x = 7$

The solution set is $\{7\}$.

127. $-4|2x+1|+12=0$

$$-4|2x+1|=-12$$

$$|2x+1|=3$$

$2x+1=3$ or $2x+1=-3$

$2x=2$ $2x=-4$

$x=1$ $x=-2$

The solution set is $\{-2,1\}$.

128. $2x^2-11x+5=0$

$(2x-1)(x-5)=0$

$2x-1=0$ $x-5=0$

$x=\dfrac{1}{2}$ or $x=5$

The solution set is $\left\{\dfrac{1}{2},5\right\}$.

129. $(3x+5)(x-3)=5$

$3x^2+5x-9x-15=5$

$3x^2-4x-20=0$

$x=\dfrac{4\pm\sqrt{(-4)^2-4(3)(-20)}}{2(3)}$

$x=\dfrac{4\pm\sqrt{16+240}}{6}$

$x=\dfrac{4\pm\sqrt{256}}{6}$

$x=\dfrac{4\pm16}{6}$

$x=\dfrac{20}{6},\dfrac{-12}{6}$

$x=\dfrac{10}{3},-2$

The solution set is $\left\{-2,\dfrac{10}{3}\right\}$.

130. $3x^2-7x+1=0$

$x=\dfrac{7\pm\sqrt{(-7)^2-4(3)(1)}}{2(3)}$

$x=\dfrac{7\pm\sqrt{49-12}}{6}$

$x=\dfrac{7\pm\sqrt{37}}{6}$

The solution set is $\left\{\dfrac{7+\sqrt{37}}{6},\dfrac{7-\sqrt{37}}{6}\right\}$.

131. $x^2-9=0$

$x^2=9$

$x=\pm3$

The solution set is $\{-3,3\}$.

132. $(x-3)^2-24=0$

$(x-3)^2=24$

$\sqrt{(x-3)^2}=\pm\sqrt{24}$

$x-3=\pm2\sqrt{6}$

$x=3\pm2\sqrt{6}$

133. $\dfrac{2x}{x^2+6x+8}=\dfrac{x}{x+4}-\dfrac{2}{x+2}$

$\dfrac{2x}{(x+4)(x+2)}=\dfrac{x}{x+4}-\dfrac{2}{x+2}$

$\dfrac{2x(x+4)(x+2)}{(x+4)(x+2)}=(x+4)(x+2)\left(\dfrac{x}{x+4}-\dfrac{2}{x+2}\right)$

$2x=x(x+2)-2(x+4)$

$2x=x^2+2x-2x-8$

$0=x^2-2x-8$

$0=(x+2)(x-4)$

$x+2=0$ or $x-4=0$

$x=-2$ $x=4$

-2 must be rejected. The solution set is $\{4\}$.

134. $\sqrt{8-2x}-x=0$

$\sqrt{8-2x}=x$

$\left(\sqrt{8-2x}\right)^2=x^2$

$8-2x=x^2$

$0=x^2+2x-8$

$0=(x+4)(x-2)$

$x+4=0$ or $x-2=0$

$x=-4$ $x=2$

-4 must be rejected. The solution set is $\{2\}$.

135. $\sqrt{2x-3}+x=3$

$$\sqrt{2x-3}=3-x$$
$$2x-3=9-6x+x^2$$
$$x^2-8x+12=0$$
$$x^2-8x=-12$$
$$x^2-8x+16=-12+16$$
$$(x-4)^2=4$$
$$x-4=\pm2$$
$$x=4+2$$
$$x=6,\,2$$

The solution set is {2}.

136. $vt+gt^2=s$

$$gt^2=s-vt$$
$$\frac{gt^2}{t^2}=\frac{s-vt}{t^2}$$
$$g=\frac{s-vt}{t^2}$$

137. $\qquad T=\dfrac{A-P}{Pr}$

$$Pr(T)=Pr\frac{A-P}{Pr}$$
$$PrT=A-P$$
$$PrT+P=A$$
$$P(rT+1)=A$$
$$P=\frac{A}{1+rT}$$

138. $\qquad x^2=2x-19$

$$x^2-2x+19=0$$
$$b^2-4ac=(-2)^2-4(1)(19)=-72$$

$-72<0,$ thus the equation has no real solutions.

139. $9x^2-30x+25=0$

$$b^2-4ac=(-30)^2-4(9)(25)=0$$

$b^2-4ac=0,$ thus the equation has one repeated real solution.

140. Let $x=$ millions of barrels of oil consumed each day by Japan.

Let $x+0.8=$ millions of barrels of oil consumed each day by China.

Let $x+15=$ millions of barrels of oil consumed each day by the United States.

$$x+(x+0.8)+(x+15)=32.3$$
$$x+x+0.8+x+15=32.3$$
$$3x+15.8=32.3$$
$$3x=16.5$$
$$x=5.5$$
$$x+0.8=6.3$$
$$x+15=20.5$$

The daily oil consumption of the United States, China, and Japan is 20.5 million barrels, 6.3 million barrels, and 5.5 million barrels, respectfully.

141. Let $x=$ the number of years after 2000.

$$17.5+0.4x=25.1$$
$$0.4x=7.6$$
$$x=19$$

The percentage of people in the U.S. that will speak a language other than English at home will reach 25.1% 19 years after 2000, or 2019.

142. Let $x=$ the original price of the phone

$$48=x-0.20x$$
$$48=0.80x$$
$$60=x$$

The original price is $60.

143. Let $x=$ the amount sold to earn $800 in one week

$$800=300+0.05x$$
$$500=0.05x$$
$$10,000=x$$

Sales must be $10,000 in one week to earn $800.

144. Let $w=$ the width of the playing field,

Let $3w-6=$ the length of the playing field

$$P=2(\text{length})+2(\text{width})$$
$$340=2(3w-6)+2w$$
$$340=6w-12+2w$$
$$340=8w-12$$
$$352=8w$$
$$44=w$$

The dimensions are 44 yards by 126 yards.

145. b. Check some points to determine that
$y_1 = 14,100 + 1500x$ and
$y_2 = 41,700 - 800x$. Since
$y_1 = y_2 = 32,100$ when $x = 12$, the two
colleges will have the same enrollment in the
year $2007 + 12 = 2019$. That year the
enrollments will be 32,100 students.

146. $A = lw$

$15 = l(2l - 7)$

$15 = 2l^2 - 7l$

$0 = 2l^2 - 7l - 15$

$0 = (2l + 3)(l - 5)$

$l = 5$

$2l - 7 = 3$

The length is 5 yards, the width is 3 yards.

147. Let x = height of building
$2x$ = shadow height

$x^2 + (2x)^2 = 300^2$

$x^2 + 4x^2 = 90,000$

$5x^2 = 90,000$

$x^2 = 18,000$

$x \approx \pm 134.164$

Discard negative height.
The building is approximately 134
meters high.

148. $(10 + 2x)(16 + 2x) = 280$

$160 + 52x + 4x^2 = 280$

$4x^2 + 52x - 120 = 0$

$x^2 + 13x - 30 = 0$

$(x + 15)(x - 2) = 0$

$x + 15 = 0$ or $x - 2 = 0$

$x = -15$ $x = 2$

−15 must be rejected. The width of the frame is 2
inches.

149.
$$\frac{1500}{x} + 100 = \frac{1500}{x - 4}$$

$$x(x-4)\left(\frac{1500}{x} + 100\right) = x(x-4)\frac{1500}{x-4}$$

$$1500(x-4) + 100x(x-4) = 1500x$$

$$1500x - 6000 + 100x^2 - 400x = 1500x$$

$$15x - 60 + x^2 - 4x = 15x$$

$$x^2 - 4x - 60 = 0$$

$$(x+6)(x-10) = 0$$

$x + 6 = 0$ or $x - 10 = 0$

$x = -6$ $x = 10$

−6 must be rejected. There were originally 10 people.

150. $\{x | -3 \le x < 5\}$

151. $\{x | x > -2\}$

152. $\{x | x \le 0\}$

153. Graph $(-2,1]$: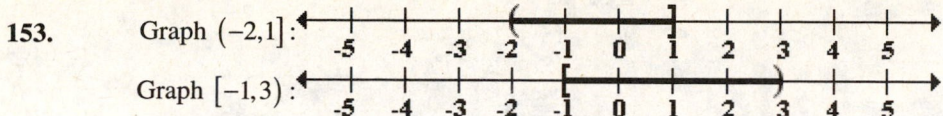

Graph $[-1,3)$: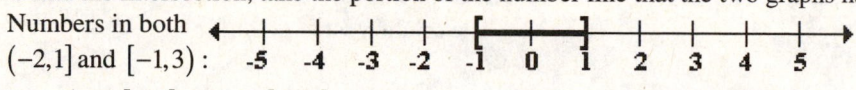

To find the intersection, take the portion of the number line that the two graphs have in common.

Numbers in both
$(-2,1]$ and $[-1,3)$: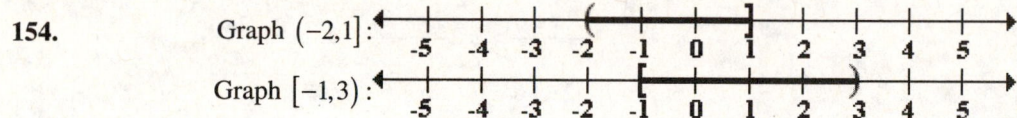

Thus, $(-2,1] \cap [-1,3) = [-1,1]$.

154. Graph $(-2,1]$: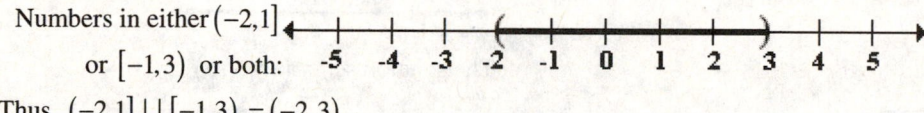

Graph $[-1,3)$:

To find the union, take the portion of the number line representing the total collection of numbers in the two graphs.

Numbers in either $(-2,1]$
or $[-1,3)$ or both:

Thus, $(-2,1] \cup [-1,3) = (-2,3)$.

155. Graph $[1,3)$: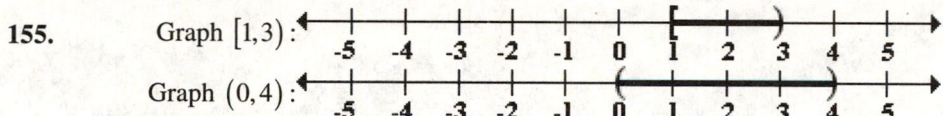

Graph $(0,4)$:

To find the intersection, take the portion of the number line that the two graphs have in common.

Numbers in both
$[1,3)$ and $(0,4)$: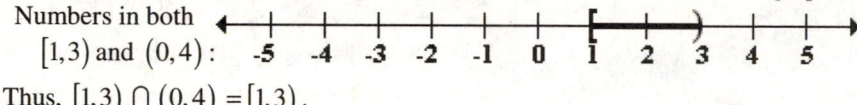

Thus, $[1,3) \cap (0,4) = [1,3)$.

156. Graph $[1,3)$: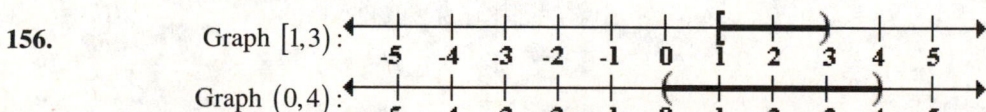

Graph $(0,4)$:

To find the union, take the portion of the number line representing the total collection of numbers in the two graphs.

Numbers in either $[1,3)$
or $(0,4)$ or both: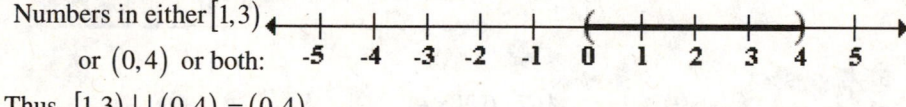

Thus, $[1,3) \cup (0,4) = (0,4)$.

157. $-6x + 3 \le 15$

$-6x \le 12$

$x \ge 2$

The solution set is $[-2,\infty)$.

158. $6x - 9 \geq -4x - 3$

$10x \geq 6$

$x \geq \dfrac{3}{5}$

The solution set is $\left[\dfrac{3}{5}, \infty\right)$.

159. $\dfrac{x}{3} - \dfrac{3}{4} - 1 > \dfrac{x}{2}$

$12\left(\dfrac{x}{3} - \dfrac{3}{4} - 1\right) > 12\left(\dfrac{x}{2}\right)$

$4x - 9 - 12 > 6x$

$-21 > 2x$

$-\dfrac{21}{2} > x$

The solution set is $\left(-\infty, -\dfrac{21}{2}\right)$.

160. $6x + 5 > -2(x - 3) - 25$

$6x + 5 > -2x + 6 - 25$

$8x + 5 > -19$

$8x > -24$

$x > -3$

The solution set is $(-3, \infty)$.

161. $3(2x - 1) - 2(x - 4) \geq 7 + 2(3 + 4x)$

$6x - 3 - 2x + 8 \geq 7 + 6 + 8x$

$4x + 5 \geq 8x + 13$

$-4x \geq 8$

$x \leq -2$

The solution set is $[-\infty, -2)$.

162. $7 < 2x + 3 \leq 9$

$4 < 2x \leq 6$

$2 < x \leq 3$

$(2, 3]$

The solution set is $[2, 3)$.

163. $|2x + 3| \leq 15$

$-15 \leq 2x + 3 \leq 15$

$-18 \leq 2x \leq 12$

$-9 \leq x \leq 6$

The solution set is $[-9, 6]$.

164. $\left|\dfrac{2x + 6}{3}\right| > 2$

$\dfrac{2x + 6}{3} > 2 \qquad \dfrac{2x + 6}{3} < -2$

$2x + 6 > 6 \qquad 2x + 6 < -6$

$2x > 0 \qquad 2x < -12$

$x > 0 \qquad x < -6$

The solution set is $(-\infty, -6)$ or $(0, \infty)$.

165. $|2x + 5| - 7 \geq -6$

$|2x + 5| \geq 1$

$2x + 5 \geq 1$ or $2x + 5 \leq -1$

$2x \geq -4 \qquad 2x \leq -6$

$x \geq -2 \quad$ or $\quad x \leq -3$

The solution set is $(-\infty, -3]$ or $[-2, \infty)$.

166. $-4|x + 2| + 5 \leq -7$

$-4|x + 2| \leq -12$

$|x + 2| \geq 3$

$x + 2 \geq 3 \qquad x + 2 \leq -3$

$\qquad$ or

$x \geq 1 \qquad x \leq -5$

The solution set is $(-\infty, -5] \cup [1, \infty)$.

167. $0.20x + 24 \leq 40$

$0.20x \leq 16$

$\dfrac{0.20x}{0.20} \leq \dfrac{16}{0.20}$

$x \leq 80$

A customer can drive no more than 80 miles.

168. $80 \le \dfrac{95+79+91+86+x}{5} < 90$

$400 \le 95+79+91+86+x < 450$

$400 \le 351+x < 450$

$49 \le x < 99$

A grade of at least 49% but less than 99% will result in a B.

Chapter P Test

1. $5(2x^2-6x)-(4x^2-3x)=10x^2-30x-4x^2+3x$

$=6x^2-27x$

2. $7+2[3(x+1)-2(3x-1)]$

$=7+2[3x+3-6x+2]$

$=7+2[-3x+5]$

$=7-6x+10$

$=-6x+17$

3. $\{1,2,5\}\cap\{5,a\}=\{5\}$

4. $\{1,2,5\}\cup\{5,a\}=\{1,2,5,a\}$

5. $\dfrac{30x^3y^4}{6x^9y^{-4}}=5x^{3-9}y^{4-(-4)}=5x^{-6}y^8=\dfrac{5y^8}{x^6}$

6. $\sqrt{6r}\cdot\sqrt{3r}=\sqrt{18r^2}=\sqrt{9r^2}\cdot\sqrt{2}=3r\sqrt{2}$

7. $4\sqrt{50}-3\sqrt{18}=4\sqrt{25\cdot2}-3\sqrt{9\cdot2}$

$=4\cdot5\sqrt{2}-3\cdot3\sqrt{2}$

$=20\sqrt{2}-9\sqrt{2}$

$=11\sqrt{2}$

8. $\dfrac{3}{5+\sqrt{2}}=\dfrac{3}{5+\sqrt{2}}\cdot\dfrac{5-\sqrt{2}}{5-\sqrt{2}}$

$=\dfrac{3(5-\sqrt{2})}{25-2}$

$=\dfrac{3(5-\sqrt{2})}{23}$

9. $\sqrt[3]{16x^4}=\sqrt[3]{8x^3\cdot2x}$

$=\sqrt[3]{8x^3}\cdot\sqrt[3]{2x}$

$=2x\sqrt[3]{2x}$

10. $\dfrac{x^2+2x-3}{x^2-3x+2}=\dfrac{(x+3)(x-1)}{(x-2)(x-1)}=\dfrac{x+3}{x-2}$,

$x\ne2,1$

11. $\dfrac{5\times10^{-6}}{20\times10^{-8}}=\dfrac{5}{20}\cdot\dfrac{10^{-6}}{10^{-8}}=0.25\times10^2=2.5\times10^1$

12. $(2x-5)(x^2-4x+3)$

$=2x^3-8x^2+6x-5x^2+20x-15$

$=2x^3-13x^2+26x-15$

13. $(5x+3y)^2=(5x)^2+2(5x)(3y)+(3y)^2$

$=25x^2+30xy+9y^2$

14. $\dfrac{2x+8}{x-3}\div\dfrac{x^2+5x+4}{x^2-9}$

$=\dfrac{2(x+4)}{x-3}\div\dfrac{(x+1)(x+4)}{(x-3)(x+3)}$

$=\dfrac{2(x+4)}{x-3}\cdot\dfrac{(x-3)(x+3)}{(x+1)(x+4)}$

$=\dfrac{2(x+3)}{x+1}$,

$x\ne3,-1,-4,-3$

15. $\dfrac{x}{x+3}+\dfrac{5}{x-3}$

$=\dfrac{x}{x+3}\cdot\dfrac{x-3}{x-3}+\dfrac{5}{x-3}\cdot\dfrac{x+3}{x+3}$

$=\dfrac{x(x-3)+5(x+3)}{(x+3)(x-3)}$

$=\dfrac{x^2-3x+5x+15}{(x+3)(x-3)}$

$=\dfrac{x^2+2x+15}{(x+3)(x-3)}$, $x\ne3,-3$

16.
$$\frac{2x+3}{x^2-7x+12}-\frac{2}{x-3}$$

$$=\frac{2x+3}{(x-3)(x-4)}-\frac{2}{x-3}$$

$$=\frac{2x+3}{(x-3)(x-4)}-\frac{2}{x-3}\cdot\frac{x-4}{x-4}$$

$$=\frac{2x+3-2(x-4)}{(x-3)(x-4)}$$

$$=\frac{2x+3-2(x-4)}{(x-3)(x-4)}$$

$$=\frac{2x+3-2x+8}{(x-3)(x-4)}$$

$$=\frac{11}{(x-3)(x-4)},$$

$$x\neq 3,4$$

17.
$$\frac{1-\frac{x}{x+2}}{1+\frac{1}{x}}=\frac{\left(1-\frac{x}{x+2}\right)(x+2)x}{\left(1+\frac{1}{x}\right)(x+2)x}$$

$$=\frac{x(x+2)-x^2}{x(x+2)+(x+2)}$$

$$=\frac{x^2+2x-x^2}{(x+1)(x+2)}$$

$$=\frac{2x}{x^2+3x+2},x\neq 0$$

18.
$$\frac{2x\sqrt{x^2+5}-\frac{2x^3}{\sqrt{x^2+5}}}{x^2+5}$$

$$=\frac{\left(2x\sqrt{x^2+5}-\frac{2x^3}{\sqrt{x^2+5}}\right)\sqrt{x^2+5}}{(x^2+5)\sqrt{x^2+5}}$$

$$=\frac{2x(x^2+5)-2x^3}{(x^2+5)\sqrt{x^2+5}}$$

$$=\frac{2x^3+10x-2x^3}{(x^2+5)\sqrt{x^2+5}}$$

$$=\frac{10x}{\sqrt{(x^2+5)^3}}$$

19. $x^2-9x+18=(x-3)(x-6)$

20. $x^3+2x^2+3x+6=x^2(x+2)+3(x+2)$
$$=(x^2+3)(x+2)$$

21. $25x^2-9=(5x)^2-3^2=(5x-3)(5x+3)$

22. $36x^2-84x+49=(6x)^2-2(6x)\cdot 7+7^2$
$$=(6x-7)^2$$

23. $y^3-125=y^3-5^3=(y-5)(y^2+5y+25)$

24. $(x^2+10x+25)-9y^2$
$$=(x+5)^2-9y^2=(x+5-3y)(x+5+3y)$$

25.
$$x(x+3)^{-\frac{3}{5}}+(x+3)^{\frac{2}{5}}=(x+3)^{-\frac{3}{5}}x\ (x\quad 3)$$

$$=(x+3)^{-\frac{3}{5}}(2x+3)=\frac{2x+3}{(x+3)^{\frac{3}{5}}}$$

26. $-7,-\frac{4}{5},0,0.25,\sqrt{4},\frac{22}{7}$ are rational numbers.

27. $3(2+5)=3(5+2)$;
commutative property of addition

28. $6(7+4)=6\cdot 7+6\cdot 4$
distributive property of multiplication over addition

29. $0.00076=7.6\times 10^{-4}$

30. $27^{-\frac{5}{3}}=\frac{1}{27^{\frac{5}{3}}}=\frac{1}{\left(\sqrt[3]{27}\right)^5}=\frac{1}{(3)^5}=\frac{1}{243}$

31. $2\left(6.6\times 10^9\right)=13.2\times 10^9=1.32\times 10^{10}$

32. a. 2003 is 14 years after 1989.
$M=-0.28n+47$
$M=-0.28(14)+47$
$\quad=43.08$
In 2003, 43.08% of bachelor's degrees were awarded to men. This overestimates the actual percent shown by the bar graph by 0.08%.

b. $R=\frac{M}{W}=\frac{-0.28n+47}{0.28n+53}$

c. $R=\frac{-0.28n+47}{0.28n+53}$
$R=\frac{-0.28(25)+47}{0.28(25)+53}$
$\quad=\frac{2}{3}$

Three women will receive bachelor's degrees for every two men. This describes the projections exactly.

33. $7(x-2) = 4(x+1) - 21$

$7x - 14 = 4x + 4 - 21$

$7x - 14 = 4x - 17$

$3x = -3$

$x = -1$

The solution set is $\{-1\}$.

34. $\dfrac{2x-3}{4} = \dfrac{x-4}{2} - \dfrac{x+1}{4}$

$2x - 3 = 2(x-4) - (x+1)$

$2x - 3 = 2x - 8 - x - 1$

$2x - 3 = x - 9$

$x = -6$

The solution set is $\{-6\}$.

35. $\dfrac{2}{x-3} - \dfrac{4}{x+3} = \dfrac{8}{(x-3)(x+3)}$

$2(x+3) - 4(x-3) = 8$

$2x + 6 - 4x + 12 = 8$

$-2x + 18 = 8$

$-2x = -10$

$x = 5$

The solution set is $\{5\}$.

36. $2x^2 - 3x - 2 = 0$

$(2x + 1)(x - 2) = 0$

$2x + 1 = 0 \quad \text{or} \quad x - 2 = 0$

$x = -\dfrac{1}{2} \quad \text{or} \quad x = 2$

The solution set is $\left\{-\dfrac{1}{2}, 2\right\}$.

37. $(3x-1)^2 = 75$

$3x - 1 = \pm\sqrt{75}$

$3x = 1 \pm 5\sqrt{3}$

$x = \dfrac{1 \pm 5\sqrt{3}}{3}$

The solution set is $\left\{\dfrac{1 - 5\sqrt{3}}{3}, \dfrac{1 + 5\sqrt{3}}{3}\right\}$.

38. $x(x-2) = 4$

$x^2 - 2x - 4 = 0$

$x = \dfrac{-b \pm \sqrt{b^2 - 4ac}}{2a}$

$x = \dfrac{2 \pm \sqrt{(-2)^2 - 4(1)(-4)}}{2}$

$x = \dfrac{2 \pm 2\sqrt{5}}{2}$

$x = 1 \pm \sqrt{5}$

The solution set is $\left\{1 - \sqrt{5}, 1 + \sqrt{5}\right\}$.

39. $\sqrt{x-3} + 5 = x$

$\sqrt{x-3} = x - 5$

$x - 3 = x^2 - 10x + 25$

$x^2 - 11x + 28 = 0$

$x = \dfrac{11 \pm \sqrt{11^2 - 4(1)(28)}}{2(1)}$

$x = \dfrac{11 \pm \sqrt{121 - 112}}{2}$

$x = \dfrac{11 \pm \sqrt{9}}{2}$

$x = \dfrac{11 \pm 3}{2}$

$x = 7 \quad \text{or} \quad x = 4$

4 does not check and must be rejected.

The solution set is $\{7\}$.

40. $\sqrt{8-2x} - x = 0$

$\sqrt{8-2x} = x$

$\left(\sqrt{8-2x}\right)^2 = (x)^2$

$8 - 2x = x^2$

$0 = x^2 + 2x - 8$

$0 = (x+4)(x-2)$

$x + 4 = 0 \quad \text{or} \quad x - 2 = 0$

$x = -4 \qquad\qquad x = 2$

−4 does not check and must be rejected.

The solution set is $\{2\}$.

41. $\left|\dfrac{2}{3}x - 6\right| = 2$

$\dfrac{2}{3}x - 6 = 2 \qquad \dfrac{2}{3}x - 6 = -2$

$\dfrac{2}{3}x = 8 \qquad\quad \dfrac{2}{3}x = 4$

$x = 12 \qquad\qquad x = 6$

The solution set is $\{6, 12\}$.

42. $-3|4x - 7| + 15 = 0$

$-3|4x - 7| = -15$

$|4x - 7| = 5$

$\begin{array}{ll} 4x - 7 = 5 & 4x - 7 = -5 \\ 4x = 12 & \text{or} \quad 4x = 2 \\ x = 3 & x = \dfrac{1}{2} \end{array}$

The solution set is $\left\{\dfrac{1}{2}, 3\right\}$

43. $\dfrac{2x}{x^2 + 6x + 8} + \dfrac{2}{x + 2} = \dfrac{x}{x + 4}$

$\dfrac{2x}{(x+4)(x+2)} + \dfrac{2}{x+2} = \dfrac{x}{x+4}$

$\dfrac{2x(x+4)(x+2)}{(x+4)(x+2)} + \dfrac{2(x+4)(x+2)}{x+2} = \dfrac{x(x+4)(x+2)}{x+4}$

$2x + 2(x+4) = x(x+2)$

$2x + 2x + 8 = x^2 + 2x$

$2x + 8 = x^2$

$0 = x^2 - 2x - 8$

$0 = (x-4)(x+2)$

$\begin{array}{ll} x - 4 = 0 & \text{or} \quad x + 2 = 0 \\ x = 4 & x = -2 \ (\text{rejected}) \end{array}$

The solution set is $\{4\}$.

44. $3(x + 4) \geq 5x - 12$

$3x + 12 \geq 5x - 12$

$-2x \geq -24$

$x \leq 12$

The solution set is $(-\infty, 12]$.

45. $\dfrac{x}{6} + \dfrac{1}{8} \leq \dfrac{x}{2} - \dfrac{3}{4}$

$4x + 3 \leq 12x - 18$

$-8x \leq -21$

$x \geq \dfrac{21}{8}$

The solution set is $\left[\dfrac{21}{8}, \infty\right)$.

46. $-3 \leq \dfrac{2x + 5}{3} < 6$

$-9 \leq 2x + 5 < 18$

$-14 \leq 2x < 13$

$-7 \leq x < \dfrac{13}{2}$

The solution set is $\left[-7, \dfrac{13}{2}\right)$.

47. $|3x + 2| \geq 3$

$\begin{array}{ll} 3x + 2 \geq 3 & \text{or} \quad 3x + 2 \leq -3 \\ 3x \geq 1 & 3x \leq -5 \\ x \geq \dfrac{1}{3} & x \leq -\dfrac{5}{3} \end{array}$

The solution set is $\left(-\infty, -\dfrac{5}{3}\right] \cup \left[\dfrac{1}{3}, \infty\right)$.

48. $V = \dfrac{1}{3}lwh$

$3V = lwh$

$\dfrac{3V}{lw} = \dfrac{lwh}{lw}$

$\dfrac{3V}{lw} = h$

$h = \dfrac{3V}{lw}$

49.
$$y - y_1 = m(x - x_1)$$
$$y - y_1 = mx - mx_1$$
$$-mx = y_1 - mx_1 - y$$
$$\frac{-mx}{-m} = \frac{y_1 - mx_1 - y}{-m}$$
$$x = \frac{y - y_1}{m} + x_1$$

50.
$$R = \frac{as}{a+s}$$
$$R(a+s) = as$$
$$Ra + Rs = as$$
$$Ra - as = -Rs$$
$$a(R - s) = -Rs$$
$$\frac{a(R - s)}{R - s} = \frac{-Rs}{R - s}$$
$$a = \frac{Rs}{s - R}$$

51.
$$43x + 575 = 1177$$
$$43x = 602$$
$$x = 14$$
The system's income will be $1177 billion 14 years after 2004, or 2018.

52.
$$B = 0.07x^2 + 47.4x + 500$$
$$1177 = 0.07x^2 + 47.4x + 500$$
$$0 = 0.07x^2 + 47.4x - 677$$
$$0 = 0.07x^2 + 47.4x - 677$$
$$x = \frac{-b \pm \sqrt{b^2 - 4ac}}{2a}$$
$$x = \frac{-(47.4) \pm \sqrt{(47.4)^2 - 4(0.07)(-677)}}{2(0.07)}$$
$$x \approx 14, \quad x \approx -691 \text{ (rejected)}$$
The system's income will be $1177 billion 14 years after 2004, or 2018.

53. The formulas model the data quite well.

54. Let $x =$ the number drive-in theaters.
Let $x + 16 =$ the number movie theaters.
Let $x + 64 =$ the number video rental stores.
$$(x) + (x + 16) + (x + 64) = 83$$
$$x + x + 16 + x + 64 = 83$$
$$3x + 80 = 83$$
$$3x = 3$$
$$x = 1$$
$$x + 16 = 17$$
$$x + 64 = 65$$
For every one million U.S. residents, there is 1 drive-in theater, 17 movie theaters, and 65 video rental stores.

55.
$$29,700 + 150x = 5000 + 1100x$$
$$24700 = 950x$$
$$26 = x$$
In 26 years, the cost will be $33,600.

56.
$$l = 2w + 4$$
$$A = lw$$
$$48 = (2w + 4)w$$
$$48 = 2w^2 + 4w$$
$$0 = 2w^2 + 4w - 48$$
$$0 = w^2 + 2w - 24$$
$$0 = (w + 6)(w - 4)$$

$$w + 6 = 0 \qquad w - 4 = 0$$
$$w = -6 \qquad w = 4$$
$$2w + 4 = 2(4) + 4 = 12$$
width is 4 feet, length is 12 feet

57. $24^2 + x^2 = 26^2$

$576 + x^2 = 676$

$x^2 = 100$

$x = \pm 10$

The wire should be attached 10 feet up the pole.

58. Let x = the original selling price

$20 = x - 0.60x$

$20 = 0.40x$

$50 = x$

The original price is $50.

59.

$$\frac{600,000}{x} - 6000 = \frac{600,000}{x+5}$$

$$x(x+5)\left(\frac{600,000}{x} - 6000\right) = x(x+5)\frac{600,000}{x+5}$$

$$600,000(x+5) - 6000x(x+5) = 600,000x$$

$$600,000x + 3,000,000 - 6000x^2 - 30,000x = 600,000x$$

$$-6000x^2 - 30,000x + 3,000,000 = 0$$

$$x^2 + 5x - 500 = 0$$

$$(x+25)(x-20) = 0$$

$x + 25 = 0 \quad$ or $\quad x - 20 = 0$

$\quad x = -25 \qquad\qquad x = 20$

−25 must be rejected. There were originally 20 people.

60. Let x = the number of local calls

The monthly cost using Plan A is $C_A = 25$.

The monthly cost using Plan B is $C_B = 13 + 0.06x$.

For Plan A to be better deal, it must cost less than Plan B.

$C_A < C_B$

$25 < 13 + 0.06x$

$12 < 0.06x$

$200 < x$

$x > 200$

Plan A is a better deal when more than 200 local calls are made per month.

Chapter 1
Functions and Graphs

Section 1.1

Check Point Exercises

1.

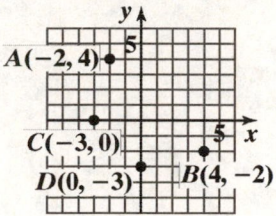

2.

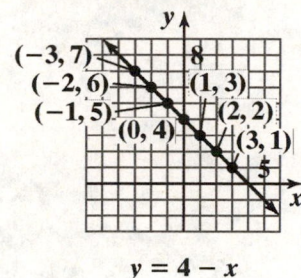

$$y = 4 - x$$

$x = -3, y = 7$

$x = -2, y = 6$

$x = -1, y = 5$

$x = 0, y = 4$

$x = 1, y = 3$

$x = 2, y = 2$

$x = 3, y = 1$

3.

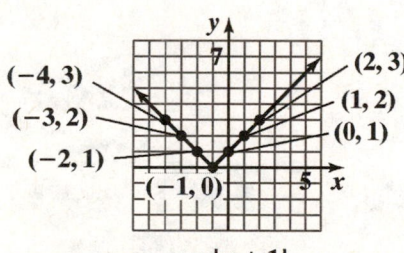

$$y = |x + 1|$$

$x = -4, y = 3$

$x = -3, y = 2$

$x = -2, y = 1$

$x = -1, y = 0$

$x = 0, y = 1$

$x = 1, y = 2$

$x = 2, y = 3$

4. The meaning of a
$[-100,100,50]$ by $[-100,100,10]$
viewing rectangle is as follows:

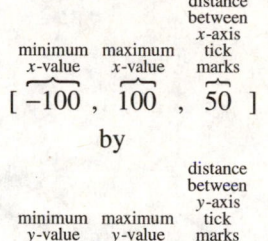

$$\underbrace{[\ -100}_{\substack{\text{minimum} \\ x\text{-value}}}, \underbrace{100}_{\substack{\text{maximum} \\ x\text{-value}}}, \underbrace{50\ ]}_{\substack{\text{distance} \\ \text{between} \\ x\text{-axis} \\ \text{tick} \\ \text{marks}}}$$

by

$$\underbrace{[\ -100}_{\substack{\text{minimum} \\ y\text{-value}}}, \underbrace{100}_{\substack{\text{maximum} \\ y\text{-value}}}, \underbrace{10\ ]}_{\substack{\text{distance} \\ \text{between} \\ y\text{-axis} \\ \text{tick} \\ \text{marks}}}$$

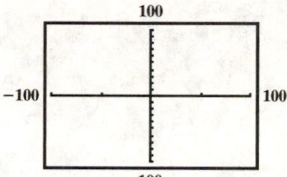

5. a. The graph crosses the x-axis at $(-3, 0)$.
Thus, the x-intercept is -3.
The graph crosses the y-axis at $(0, 5)$.
Thus, the y-intercept is 5.

b. The graph does not cross the x-axis.
Thus, there is no x-intercept.
The graph crosses the y-axis at $(0, 4)$.
Thus, the y-intercept is 4.

c. The graph crosses the x- and y-axes at
the origin $(0, 0)$.
Thus, the x-intercept is 0 and the
y-intercept is 0.

6. a. $d = 4n + 5$

$d = 4(15) + 5 = 65$

65% of marriages end in divorce after 15 years
when the wife is under 18 at the time of
marriage.

b. According to the line graph, 60% of marriages
end in divorce after 15 years when the wife is
under 18 at the time of marriage.

c. The mathematical model overestimates the
actual percentage shown in the graph by 5%.

Functions and Graphs

Exercise Set 1.1

1.

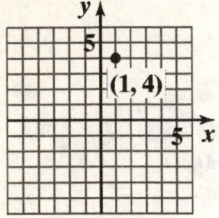

3.

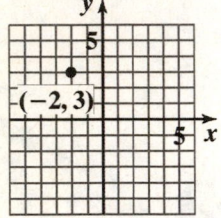

5.

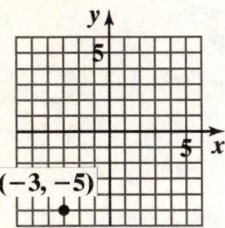

7.

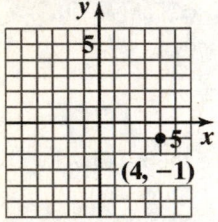

9.

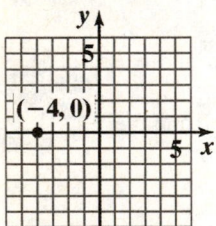

11.

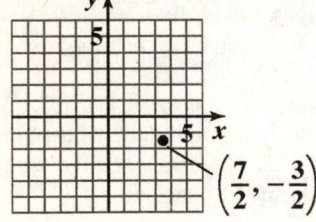

13.

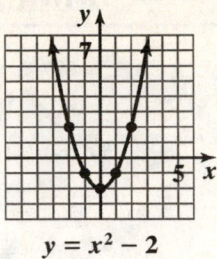

$y = x^2 - 2$

$x = -3, y = 7$

$x = -2, y = 2$

$x = -1, y = -1$

$x = 0, y = -2$

$x = 1, y = -1$

$x = 2, y = 2$

$x = 3, y = 7$

15.

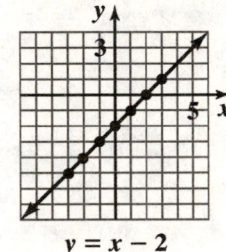

$y = x - 2$

$x = -3, y = -5$

$x = -2, y = -4$

$x = -1, y = -3$

$x = 0, y = -2$

$x = 1, y = -1$

$x = 2, y = 0$

$x = 3, y = 1$

17.

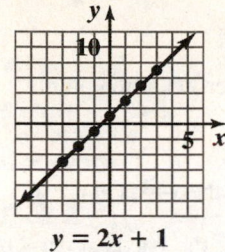

$$y = 2x + 1$$

$x = -3, y = -5$

$x = -2, y = -3$

$x = -1, y = -1$

$x = 0, y = 1$

$x = 1, y = 3$

$x = 2, y = 5$

$x = 3, y = 7$

21.

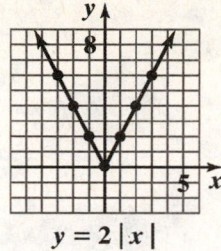

$$y = 2\,|x|$$

$x = -3, y = 6$

$x = -2, y = 4$

$x = -1, y = 2$

$x = 0, y = 0$

$x = 1, y = 2$

$x = 2, y = 4$

$x = 3, y = 6$

19.

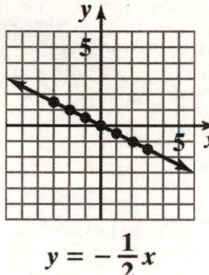

$$y = -\frac{1}{2}x$$

$x = -3, y = \dfrac{3}{2}$

$x = -2, y = 1$

$x = -1, y = \dfrac{1}{2}$

$x = 0, y = 0$

$x = 1, y = -\dfrac{1}{2}$

$x = 2, y = -1$

$x = 3, y = -\dfrac{3}{2}$

23.

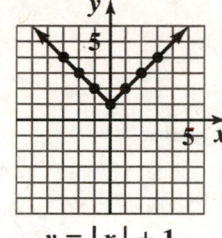

$$y = |x| + 1$$

$x = -3, y = 4$

$x = -2, y = 3$

$x = -1, y = 2$

$x = 0, y = 1$

$x = 1, y = 2$

$x = 2, y = 3$

$x = 3, y = 4$

25.

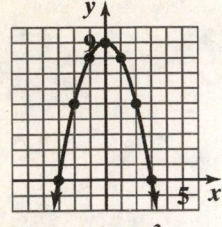

$$y = 9 - x^2$$

$x = -3, y = 0$

$x = -2, y = 5$

$x = -1, y = 8$

$x = 0, y = 9$

$x = 1, y = 8$

$x = 2, y = 5$

$x = 3, y = 0$

27.

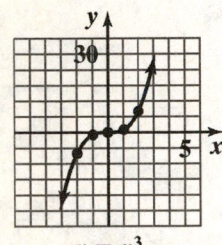

$$y = x^3$$

$x = -3, y = -27$

$x = -2, y = -8$

$x = -1, y = 1$

$x = 0, y = 0$

$x = 1, y = 1$

$x = 2, y = 8$

$x = 3, y = 27$

29. (c) x-axis tick marks $-5, -4, -3, -2, -1, 0, 1, 2, 3, 4,$ 5; y-axis tick marks are the same.

31. (b); x-axis tick marks $-20, -10, 0, 10, 20, 30, 40, 50,$ $60, 70, 80$; y-axis tick marks $-30, -20, -10, 0, 10, 20,$ $30, 40, 50, 60, 70$

33. The equation that corresponds to Y_2 in the table

is (c), $y_2 = 2 - x$. We can tell because all of

the points $(-3, 5)$, $(-2, 4)$, $(-1, 3)$, $(0, 2)$,

$(1, 1)$, $(2, 0)$, and $(3, -1)$ are on the line

$y = 2 - x$, but all are not on any of the others.

35. No. It passes through the point $(0, 2)$.

37. $(2, 0)$

39. The graphs of Y_1 and Y_2 intersect at the points

$(-2, 4)$ and $(1, 1)$.

41. **a.** 2; The graph intersects the x-axis at $(2, 0)$.

   b. -4; The graph intersects the y-axis at $(0, -4)$.

43. **a.** 1, -2; The graph intersects the x-axis at $(1, 0)$ and $(-2, 0)$.

   b. 2; The graph intersects the y-axis at $(0, 2)$.

45. **a.** -1; The graph intersects the x-axis at $(-1, 0)$.

   b. none; The graph does not intersect the y-axis.

47.

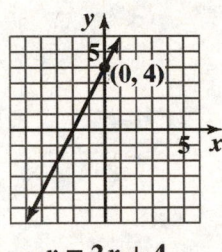

$$y = 2x + 4$$

49.

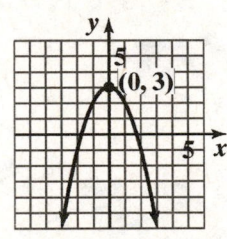

$$y = 3 - x^2$$

51.

x	(x, y)
-3	$(-3, 5)$
-2	$(-2, 5)$
-1	$(-1, 5)$
0	$(0, 5)$
1	$(1, 5)$
2	$(2, 5)$
3	$(3, 5)$

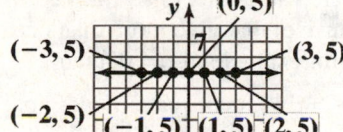

$$y = 5$$

53.

x	(x, y)
-2	$\left(-2, -\dfrac{1}{2}\right)$
-1	$(-1, -1)$
$-\dfrac{1}{2}$	$\left(-\dfrac{1}{2}, -2\right)$
$-\dfrac{1}{3}$	$\left(-\dfrac{1}{3}, -3\right)$
$\dfrac{1}{3}$	$\left(\dfrac{1}{3}, 3\right)$
$\dfrac{1}{2}$	$\left(\dfrac{1}{2}, 2\right)$
1	$(1, 1)$
2	$\left(2, \dfrac{1}{2}\right)$

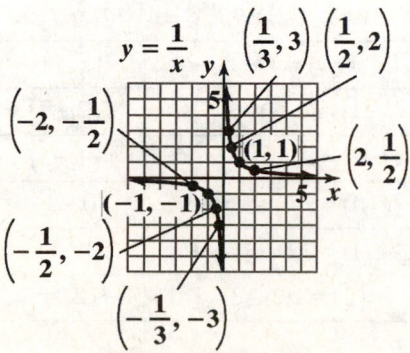

55.

a. According to the line graph, 20% of seniors used marijuana in 2005.

b. 2005 is 25 years after 1980.
$$M = -0.4n + 28$$
$$M = -0.4(25) + 28 = 18$$
According to formula, 18% of seniors used marijuana in 2005. This underestimates the value in the graph by 2%.

c. According to the line graph, about 45% of seniors used alcohol in 2006.

d. 2006 is 26 years after 1980.
$$A = -n + 70$$
$$A = -(26) + 70 = 44$$
According to formula, 44% of seniors used alcohol in 2006. This underestimates the value in the graph.

e. The minimum for marijuana was reached in 1990.
According to the line graph, about 14% of seniors used marijuana in 1990.

57. At age 8, women have the least number of awakenings, averaging about 1 awakening per night.

59. The difference between the number of awakenings for 25-year-old men and women is about 1.9.

61. – 65. Answers may vary.

67. makes sense

69. does not make sense; Explanations will vary. Sample explanation: These three points are not collinear.

71. false; Changes to make the statement true will vary. A sample change is: The product of the coordinates of a point in quadrant III is also positive.

73. true

75. (a)

77. (b)

79. (b)

81. (c)

83. Set 1 has each x-coordinate paired with only one y-coordinate.

84.

x	$y = 2x$	(x, y)
-2	$y = 2(-2) = -4$	$(-2, -4)$
-1	$y = 2(-1) + 4 = 2$	$(-1, -2)$
0	$y = 2(0) = 0$	$(0, 0)$
1	$y = 2(1) = 2$	$(1, 2)$
2	$y = 2(2) = 4$	$(2, 4)$

x	$y = 2x + 4$	(x, y)
-2	$y = 2(-2) + 4 = 0$	$(-2, 0)$
-1	$y = 2(-1) + 4 = 2$	$(-1, 2)$
0	$y = 2(0) + 4 = 4$	$(0, 4)$
1	$y = 2(1) + 4 = 6$	$(1, 6)$
2	$y = 2(2) + 4 = 8$	$(2, 8)$

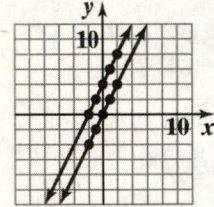

85. **a.** When the x-coordinate is 2, the y-coordinate is 3.

    b. When the y-coordinate is 4, the x-coordinates are -3 and 3.

    c. The x-coordinates are all real numbers.

    d. The y-coordinates are all real numbers greater than or equal to 1.

Section 1.2

Check Point Exercises

1. The domain is the set of all first components: {0, 10, 20, 30, 36}. The range is the set of all second components: {9.1, 6.7, 10.7, 13.2, 17.4}.

2. **a.** The relation is not a function since the two ordered pairs (5, 6) and (5, 8) have the same first component but different second components.

    b. The relation is a function since no two ordered pairs have the same first component and different second components.

3. **a.** $2x + y = 6$

        $y = -2x + 6$

    For each value of x, there is one and only one value for y, so the equation defines y as a function of x.

    b. $x^2 + y^2 = 1$

        $y^2 = 1 - x^2$

        $y = \pm\sqrt{1 - x^2}$

    Since there are values of x (all values between -1 and 1 exclusive) that give more than one value for y (for example, if $x = 0$, then $y = \pm\sqrt{1 - 0^2} = \pm 1$), the equation does not define y as a function of x.

4. **a.** $f(-5) = (-5)^2 - 2(-5) + 7$

             $= 25 - (-10) + 7$

             $= 42$

    b. $f(x + 4) = (x + 4)^2 - 2(x + 4) + 7$

               $= x^2 + 8x + 16 - 2x - 8 + 7$

               $= x^2 + 6x + 15$

    c. $f(-x) = (-x)^2 - 2(-x) + 7$

              $= x^2 - (-2x) + 7$

              $= x^2 + 2x + 7$

5.

x	$f(x) = 2x$	(x, y)
-2	-4	$(-2, -4)$
-1	-2	$(-1, -2)$
0	0	$(0, 0)$
1	2	$(1, 2)$
2	4	$(2, 4)$

x	$g(x) = 2x - 3$	(x, y)
-2	$g(-2) = 2(-2) - 3 = -7$	$(-2, -7)$
-1	$g(-1) = 2(-1) - 3 = -5$	$(-1, -5)$
0	$g(0) = 2(0) - 3 = -3$	$(0, -3)$
1	$g(1) = 2(1) - 3 = -1$	$(1, -1)$
2	$g(2) = 2(2) - 3 = 1$	$(2, 1)$

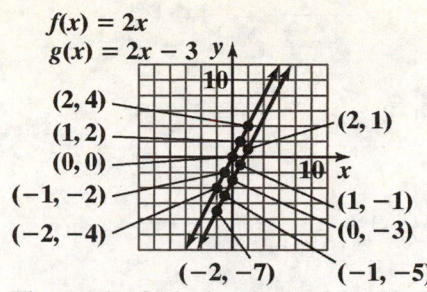

$f(x) = 2x$
$g(x) = 2x - 3$

The graph of g is the graph of f shifted down 3 units.

6.   The graph (c) fails the vertical line test and is therefore not a function.
     y is a function of x for the graphs in (a) and (b).

7.   **a.** $f(5) = 400$

     b. $x = 9$, $f(9) = 100$

     c.   The minimum T cell count in the asymptomatic stage is approximately 425.

8.   **a.**   domain: $\{x | -2 \le x \le 1\}$ or $[-2,1]$.
             range: $\{y | 0 \le y \le 3\}$ or $[0,3]$.

     b.   domain: $\{x | -2 < x \le 1\}$ or $(-2,1]$.
             range: $\{y | -1 \le y < 2\}$ or $[-1,2)$.

     c.   domain: $\{x | -3 \le x < 0\}$ or $[-3,0)$.
             range: $\{y | y = -3, -2, -1\}$.

Exercise Set 1.2

1.   The relation is a function since no two ordered pairs have the same first component and different second components. The domain is {1, 3, 5} and the range is {2, 4, 5}.

3.   The relation is not a function since the two ordered pairs (3, 4) and (3, 5) have the same first component but different second components (the same could be said for the ordered pairs (4, 4) and (4, 5)). The domain is {3, 4} and the range is {4, 5}.

5.   The relation is a function because no two ordered pairs have the same first component and different second components The domain is {3, 4, 5, 7} and the range is {−2, 1, 9}.

7.   The relation is a function since there are no same first components with different second components. The domain is {−3, −2, −1, 0} and the range is {−3, −2, −1, 0}.

9.   The relation is not a function since there are ordered pairs with the same first component and different second components. The domain is {1} and the range is {4, 5, 6}.

11. $x + y = 16$
     $y = 16 - x$
     Since only one value of y can be obtained for each value of x, y is a function of x.

13. $x^2 + y = 16$
     $y = 16 - x^2$
     Since only one value of y can be obtained for each value of x, y is a function of x.

15. $x^2 + y^2 = 16$
     $y^2 = 16 - x^2$
     $y = \pm\sqrt{16 - x^2}$
     If $x = 0$, $y = \pm 4$.
     Since two values, $y = 4$ and $y = -4$, can be obtained for one value of x, y is not a function of x.

17. $x = y^2$
     $y = \pm\sqrt{x}$
     If $x = 1$, $y = \pm 1$.
     Since two values, $y = 1$ and $y = -1$, can be obtained for $x = 1$, y is not a function of x.

19. $y = \sqrt{x + 4}$
     Since only one value of y can be obtained for each value of x, y is a function of x.

21. $x + y^3 = 8$
     $y^3 = 8 - x$
     $y = \sqrt[3]{8 - x}$
     Since only one value of y can be obtained for each value of x, y is a function of x.

23. $xy + 2y = 1$
     $y(x + 2) = 1$
     $y = \dfrac{1}{x + 2}$
     Since only one value of y can be obtained for each value of x, y is a function of x.

25. $|x| - y = 2$

$-y = -|x| + 2$

$y = |x| - 2$

Since only one value of y can be obtained for each value of x, y is a function of x.

27. a. $f(6) = 4(6) + 5 = 29$

b. $f(x + 1) = 4(x + 1) + 5 = 4x + 9$

c. $f(-x) = 4(-x) + 5 = -4x + 5$

29. a. $g(-1) = (-1)^2 + 2(-1) + 3$

$= 1 - 2 + 3$

$= 2$

b. $g(x + 5) = (x + 5)^2 + 2(x + 5) + 3$

$= x^2 + 10x + 25 + 2x + 10 + 3$

$= x^2 + 12x + 38$

c. $g(-x) = (-x)^2 + 2(-x) + 3$

$= x^2 - 2x + 3$

31. a. $h(2) = 2^4 - 2^2 + 1$

$= 16 - 4 + 1$

$= 13$

b. $h(-1) = (-1)^4 - (-1)^2 + 1$

$= 1 - 1 + 1$

$= 1$

c. $h(-x) = (-x)^4 - (-x)^2 + 1 = x^4 - x^2 + 1$

d. $h(3a) = (3a)^4 - (3a)^2 + 1$

$= 81a^4 - 9a^2 + 1$

33. a. $f(-6) = \sqrt{-6 + 6} + 3 = \sqrt{0} + 3 = 3$

b. $f(10) = \sqrt{10 + 6} + 3$

$= \sqrt{16} + 3$

$= 4 + 3$

$= 7$

c. $f(x - 6) = \sqrt{x - 6 + 6} + 3 = \sqrt{x} + 3$

35. a. $f(2) = \dfrac{4(2)^2 - 1}{2^2} = \dfrac{15}{4}$

b. $f(-2) = \dfrac{4(-2)^2 - 1}{(-2)^2} = \dfrac{15}{4}$

c. $f(-x) = \dfrac{4(-x)^2 - 1}{(-x)^2} = \dfrac{4x^2 - 1}{x^2}$

37. a. $f(6) = \dfrac{6}{|6|} = 1$

b. $f(-6) = \dfrac{-6}{|-6|} = \dfrac{-6}{6} = -1$

c. $f(r^2) = \dfrac{r^2}{|r^2|} = \dfrac{r^2}{r^2} = 1$

39.

x	$f(x) = x$	(x, y)
-2	$f(-2) = -2$	$(-2, -2)$
-1	$f(-1) = -1$	$(-1, -1)$
0	$f(0) = 0$	$(0, 0)$
1	$f(1) = 1$	$(1, 1)$
2	$f(2) = 2$	$(2, 2)$

x	$g(x) = x + 3$	(x, y)
-2	$g(-2) = -2 + 3 = 1$	$(-2, 1)$
-1	$g(-1) = -1 + 3 = 2$	$(-1, 2)$
0	$g(0) = 0 + 3 = 3$	$(0, 3)$
1	$g(1) = 1 + 3 = 4$	$(1, 4)$
2	$g(2) = 2 + 3 = 5$	$(2, 5)$

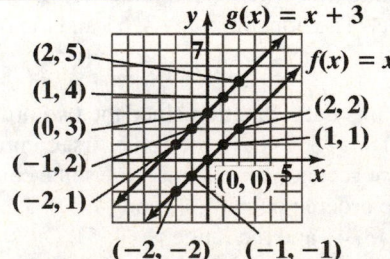

The graph of g is the graph of f shifted up 3 units.

41.

x	$f(x) = -2x$	(x, y)
–2	$f(-2) = -2(-2) = 4$	$(-2, 4)$
–1	$f(-1) = -2(-1) = 2$	$(-1, 2)$
0	$f(0) = -2(0) = 0$	$(0, 0)$
1	$f(1) = -2(1) = -2$	$(1, -2)$
2	$f(2) = -2(2) = -4$	$(2, -4)$

x	$g(x) = -2x - 1$	(x, y)
–2	$g(-2) = -2(-2) - 1 = 3$	$(-2, 3)$
–1	$g(-1) = -2(-1) - 1 = 1$	$(-1, 1)$
0	$g(0) = -2(0) - 1 = -1$	$(0, -1)$
1	$g(1) = -2(1) - 1 = -3$	$(1, -3)$
2	$g(2) = -2(2) - 1 = -5$	$(2, -5)$

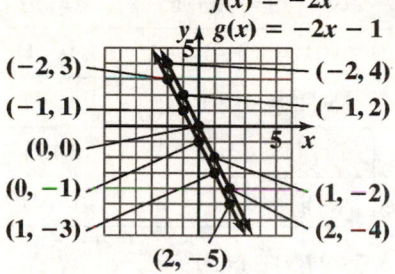

The graph of g is the graph of f shifted down 1 unit.

43.

x	$f(x) = x^2$	(x, y)
–2	$f(-2) = (-2)^2 = 4$	$(-2, 4)$
–1	$f(-1) = (-1)^2 = 1$	$(-1, 1)$
0	$f(0) = (0)^2 = 0$	$(0, 0)$
1	$f(1) = (1)^2 = 1$	$(1, 1)$
2	$f(2) = (2)^2 = 4$	$(2, 4)$

x	$g(x) = x^2 + 1$	(x, y)
–2	$g(-2) = (-2)^2 + 1 = 5$	$(-2, 5)$
–1	$g(-1) = (-1)^2 + 1 = 2$	$(-1, 2)$
0	$g(0) = (0)^2 + 1 = 1$	$(0, 1)$
1	$g(1) = (1)^2 + 1 = 2$	$(1, 2)$
2	$g(2) = (2)^2 + 1 = 5$	$(2, 5)$

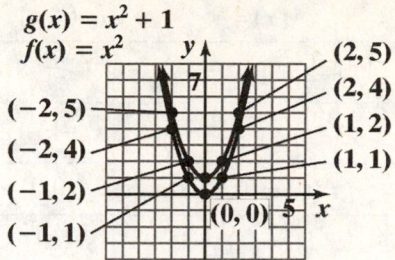

The graph of g is the graph of f shifted up 1 unit.

45.

x	$f(x) = |x|$	(x, y)
–2	$f(-2) = |-2| = 2$	$(-2, 2)$
–1	$f(-1) = |-1| = 1$	$(-1, 1)$
0	$f(0) = |0| = 0$	$(0, 0)$
1	$f(1) = |1| = 1$	$(1, 1)$
2	$f(2) = |2| = 2$	$(2, 2)$

x	$g(x) = |x| - 2$	(x, y)
–2	$g(-2) = |-2| - 2 = 0$	$(-2, 0)$
–1	$g(-1) = |-1| - 2 = -1$	$(-1, -1)$
0	$g(0) = |0| - 2 = -2$	$(0, -2)$
1	$g(1) = |1| - 2 = -1$	$(1, -1)$
2	$g(2) = |2| - 2 = 0$	$(2, 0)$

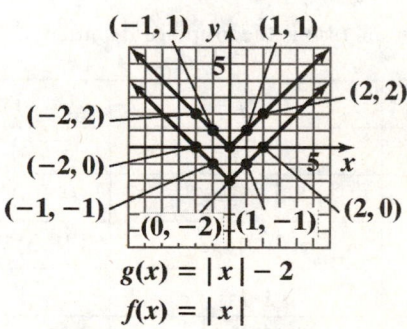

The graph of g is the graph of f shifted down 2 units.

47.

x	$f(x)=x^3$	(x,y)
-2	$f(-2)=(-2)^3=-8$	$(-2,-8)$
-1	$f(-1)=(-1)^3=-1$	$(-1,-1)$
0	$f(0)=(0)^3=0$	$(0,0)$
1	$f(1)=(1)^3=1$	$(1,1)$
2	$f(2)=(2)^3=8$	$(2,8)$

x	$g(x)=x^3+2$	(x,y)
-2	$g(-2)=(-2)^3+2=-6$	$(-2,-6)$
-1	$g(-1)=(-1)^3+2=1$	$(-1,1)$
0	$g(0)=(0)^3+2=2$	$(0,2)$
1	$g(1)=(1)^3+2=3$	$(1,3)$
2	$g(2)=(2)^3+2=10$	$(2,10)$

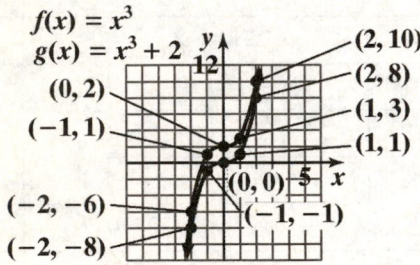

The graph of g is the graph of f shifted up 2 units.

49.

x	$f(x)=3$	(x,y)
-2	$f(-2)=3$	$(-2,3)$
-1	$f(-1)=3$	$(-1,3)$
0	$f(0)=3$	$(0,3)$
1	$f(1)=3$	$(1,3)$
2	$f(2)=3$	$(2,3)$

x	$g(x)=5$	(x,y)
-2	$g(-2)=5$	$(-2,5)$
-1	$g(-1)=5$	$(-1,5)$
0	$g(0)=5$	$(0,5)$
1	$g(1)=5$	$(1,5)$
2	$g(2)=5$	$(2,5)$

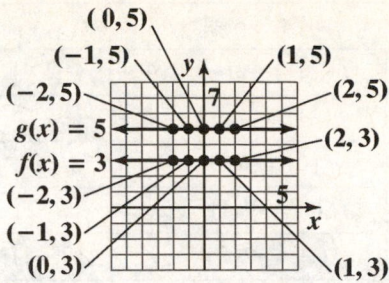

The graph of g is the graph of f shifted up 2 units.

51.

x	$f(x)=\sqrt{x}$	(x,y)
0	$f(0)=\sqrt{0}=0$	$(0,0)$
1	$f(1)=\sqrt{1}=1$	$(1,1)$
4	$f(4)=\sqrt{4}=2$	$(4,2)$
9	$f(9)=\sqrt{9}=3$	$(9,3)$

x	$g(x)=\sqrt{x}-1$	(x,y)
0	$g(0)=\sqrt{0}-1=-1$	$(0,-1)$
1	$g(1)=\sqrt{1}-1=0$	$(1,0)$
4	$g(4)=\sqrt{4}-1=1$	$(4,1)$
9	$g(9)=\sqrt{9}-1=2$	$(9,2)$

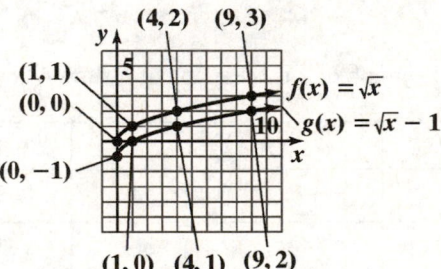

The graph of g is the graph of f shifted down 1 unit.

53.

x	$f(x)=\sqrt{x}$	(x,y)
0	$f(0)=\sqrt{0}=0$	$(0,0)$
1	$f(1)=\sqrt{1}=1$	$(1,1)$
4	$f(4)=\sqrt{4}=2$	$(4,2)$
9	$f(9)=\sqrt{9}=3$	$(9,3)$

x	$g(x)=\sqrt{x-1}$	(x,y)
1	$g(1)=\sqrt{1-1}=0$	$(1,0)$
2	$g(2)=\sqrt{2-1}=1$	$(2,1)$
5	$g(5)=\sqrt{5-1}=2$	$(5,2)$
10	$g(10)=\sqrt{10-1}=3$	$(10,3)$

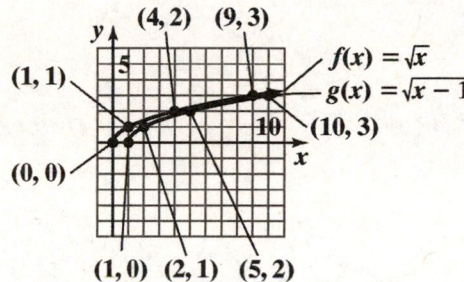

The graph of g is the graph of f shifted right 1 unit.

55. function

57. function

59. not a function

61. function

63. function

65. $f(-2)=-4$

67. $f(4)=4$

69. $f(-3)=0$

71. $g(-4)=2$

73. $g(-10)=2$

75. When $x=-2,\ g(x)=1.$

77.
  a. domain: $(-\infty,\infty)$

  b. range: $[-4,\infty)$

  c. x-intercepts: -3 and 1

  d. y-intercept: -3

  e. $f(-2)=-3$ and $f(2)=5$

79.
  a. domain: $(-\infty,\infty)$

  b. range: $[1,\infty)$

  c. x-intercept: none

  d. y-intercept: 1

  e. $f(-1)=2$ and $f(3)=4$

81.
  a. domain: $[0, 5)$

  b. range: $[-1, 5)$

  c. x-intercept: 2

  d. y-intercept: -1

  e. $f(3) = 1$

83.
  a. domain: $[0,\infty)$

  b. range: $[1,\infty)$

  c. x-intercept: none

  d. y-intercept: 1

  e. $f(4) = 3$

85.
  a. domain: $[-2, 6]$

  b. range: $[-2, 6]$

  c. x-intercept: 4

  d. y-intercept: 4

  e. $f(-1) = 5$

87.
  a. domain: $(-\infty,\infty)$

  b. range: $(-\infty, -2]$

  c. x-intercept: none

  d. y-intercept: -2

  e. $f(-4) = -5$ and $f(4) = -2$

89. **a.** domain: $(-\infty, \infty)$

 b. range: $(0, \infty)$

 c. x-intercept: none

 d. y-intercept: 1.5

 e. $f(4) = 6$

91. **a.** domain: $\{-5, -2, 0, 1, 3\}$

 b. range: $\{2\}$

 c. x-intercept: none

 d. y-intercept: 2

 e. $f(-5) + f(3) = 2 + 2 = 4$

93. $g(1) = 3(1) - 5 = 3 - 5 = -2$

$f(g(1)) = f(-2) = (-2)^2 - (-2) + 4$
$$= 4 + 2 + 4 = 10$$

95. $\sqrt{3 - (-1)} - (-6)^2 + 6 \div (-6) \cdot 4$

$$= \sqrt{3 + 1} - 36 + 6 \div (-6) \cdot 4$$
$$= \sqrt{4} - 36 + -1 \cdot 4$$
$$= 2 - 36 + -4$$
$$= -34 + -4$$
$$= -38$$

97. $f(-x) - f(x)$

$$= (-x)^3 + (-x) - 5 - (x^3 + x - 5)$$
$$= -x^3 - x - 5 - x^3 - x + 5 = -2x^3 - 2x$$

99. **a.** $\{(\text{Iceland}, 9.7), (\text{Finland}, 9.6), (\text{New Zealand}, 9.6), (\text{Denmark}, 9.5)\}$

 b. Yes, the relation is a function. Each element in the domain corresponds to only one element in the range.

 c. $\{(9.7, \text{Iceland}), (9.6, \text{Finland}), (9.6, \text{New Zealand}), (9.5, \text{Denmark})\}$

 d. No, the relation is not a function. 9.6 in the domain corresponds to both Finland and New Zealand in the range.

101. **a.** $f(70) = 83$ which means the chance that a 60-year old will survive to age 70 is 83%.

 b. $g(70) = 76$ which means the chance that a 60-year old will survive to age 70 is 76%.

 c. Function f is the better model.

103. a. $T(x) = -0.125x^2 + 5.25x + 72$

$T(20) = -0.125(20)^2 + 5.25(20) + 72 = 127$

Americans ordered an average of 127 takeout meals per person 20 years after 1984, or 2004. This is represented on the graph by the point (20,127).

b. $R(x) = -0.6x + 94$

$R(0) = -0.6(0) + 94 = 94$

Americans ordered an average of 94 meals in restaurants per person 0 years after 1984, or 1984.
This is represented on the graph by the point (0,94).

c. According to the graphs, the average number of takeout orders approximately equaled the average number of in-restaurant meals 4 years after 1984, or 1988.

$T(x) = -0.125x^2 + 5.25x + 72$

$T(4) = -0.125(4)^2 + 5.25(4) + 72 = 91$

In 1988 Americans ordered an average of 91 takeout meals per person.

$R(x) = -0.6x + 94$

$R(4) = -0.6(4) + 94 = 91.6$

In 1988 Americans ordered an average of 91.6 meals in restaurants per person.

105. $C(x) = 100,000 + 100x$

$C(90) = 100,000 + 100(90) = \$109,000$

It will cost \$109,000 to produce 90 bicycles.

107.

$T(x) = \dfrac{40}{x} + \dfrac{40}{x+30}$

$T(30) = \dfrac{40}{30} + \dfrac{40}{30+30}$

$= \dfrac{80}{60} + \dfrac{40}{60}$

$= \dfrac{120}{60}$

$= 2$

If you travel 30 mph going and 60 mph returning, your total trip will take 2 hours.

109. – 117. Answers may vary.

119. does not make sense; Explanations will vary. Sample explanation: The parentheses used in

function notation, such as $f(x)$, do not imply multiplication.

121. does not make sense; Explanations will vary. Sample explanation: This would not be a function because some elements in the domain would correspond to more than one age in the range.

123. false; Changes to make the statement true will vary. A sample change is: The range is $[-2,2)$.

125. false; Changes to make the statement true will vary. A sample change is: $f(0) = 0.8$

127. Answers may vary.
An example is $\{(1,1),(2,1)\}$

129. $C(t) = 20 + 0.40(t - 60)$

$C(100) = 20 + 0.40(100 - 60)$

$= 20 + 0.40(40)$

$= 20 + 16$

$= 36$

For 100 calling minutes, the monthly cost is \$36.

130. $f(x) = x + 2,\ x \le 1$

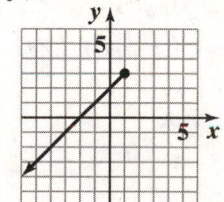

131. $2(x+h)^2 + 3(x+h) + 5 - (2x^2 + 3x + 5)$

$= 2(x^2 + 2xh + h^2) + 3x + 3h + 5 - 2x^2 - 3x - 5$

$= 2x^2 + 4xh + 2h^2 + 3x + 3h + 5 - 2x^2 - 3x - 5$

$= 2x^2 - 2x^2 + 4xh + 2h^2 + 3x - 3x + 3h + 5 - 5$

$= 4xh + 2h^2 + 3h$

Section 1.3

Check Point Exercises

1. The function is increasing on the interval $(-\infty, -1)$, decreasing on the interval $(-1, 1)$, and increasing on the interval $(1, \infty)$.

2. **a.** $f(-x) = (-x)^2 + 6 = x^2 + 6 = f(x)$
   The function is even.

   b. $g(-x) = 7(-x)^3 - (-x) = -7x^3 + x = -f(x)$
   The function is odd.

   c. $h(-x) = (-x)^5 + 1 = -x^5 + 1$
   The function is neither even nor odd.

3. $C(t) = \begin{cases} 20 & \text{if } 0 \le t \le 60 \\ 20 + 0.40(t-60) & \text{if } t > 60 \end{cases}$

   b. Since $0 \le 40 \le 60$, $C(40) = 20$
   With 40 calling minutes, the cost is \$20.
   This is represented by $(40, 20)$.

   c. Since $80 > 60$,
   $C(80) = 20 + 0.40(80 - 60) = 28$
   With 80 calling minutes, the cost is \$28.
   This is represented by $(80, 28)$.

4.

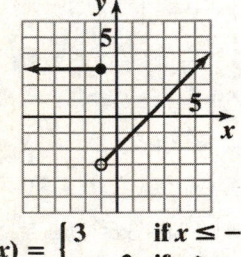

$$f(x) = \begin{cases} 3 & \text{if } x \le -1 \\ x - 2 & \text{if } x > -1 \end{cases}$$

5. **a.** $f(x) = -2x^2 + x + 5$
   $f(x+h) = -2(x+h)^2 + (x+h) + 5$
   $\qquad = -2(x^2 + 2xh + h^2) + x + h + 5$
   $\qquad = -2x^2 - 4xh - 2h^2 + x + h + 5$

b. $\dfrac{f(x+h) - f(x)}{h}$

$= \dfrac{-2x^2 - 4xh - 2h^2 + x + h + 5 - \left(-2x^2 + x + 5\right)}{h}$

$= \dfrac{-2x^2 - 4xh - 2h^2 + x + h + 5 + 2x^2 - x - 5}{h}$

$= \dfrac{-4xh - 2h^2 + h}{h}$

$= \dfrac{h\left(-4x - 2h + 1\right)}{h}$

$= -4x - 2h + 1$

Exercise Set 1.3

1. **a.** increasing: $(-1, \infty)$

   b. decreasing: $(-\infty, -1)$

   c. constant: none

3. **a.** increasing: $(0, \infty)$

   b. decreasing: none

   c. constant: none

5. **a.** increasing: none

   b. decreasing: $(-2, 6)$

   c. constant: none

7. **a.** increasing: $(-\infty, -1)$

   b. decreasing: none

   c. constant: $(-1, \infty)$

9. **a.** increasing: $(-\infty, 0)$ or $(1.5, 3)$

   b. decreasing: $(0, 1.5)$ or $(3, \infty)$

   c. constant: none

11. **a.** increasing: $(-2, 4)$

    b. decreasing: none

    c. constant: $(-\infty, -2)$ or $(4, \infty)$

13. a. $x = 0$, relative maximum $= 4$

    b. $x = -3, 3$, relative minimum $= 0$

15. a. $x = -2$, relative maximum $= 21$

    b. $x = 1$, relative minimum $= -6$

17. $f(x) = x^3 + x$

$f(-x) = (-x)^3 + (-x)$

$f(-x) = -x^3 - x = -(x^3 + x)$

$f(-x) = -f(x), \text{ odd function}$

19. $g(x) = x^2 + x$

$g(-x) = (-x)^2 + (-x)$

$g(-x) = x^2 - x, \text{ neither}$

21. $h(x) = x^2 - x^4$

$h(-x) = (-x)^2 - (-x)^4$

$h(-x) = x^2 - x^4$

$h(-x) = h(x), \text{ even function}$

23. $f(x) = x^2 - x^4 + 1$

$f(-x) = (-x)^2 - (-x)^4 + 1$

$f(-x) = x^2 - x^4 + 1$

$f(-x) = f(x), \text{ even function}$

25. $f(x) = \dfrac{1}{5}x^6 - 3x^2$

$f(-x) = \dfrac{1}{5}(-x)^6 - 3(-x)^2$

$f(-x) = \dfrac{1}{5}x^6 - 3x^2$

$f(-x) = f(x), \text{ even function}$

27. $f(x) = x\sqrt{1 - x^2}$

$f(-x) = -x\sqrt{1 - (-x)^2}$

$f(-x) = -x\sqrt{1 - x^2}$

$\qquad = -\left(x\sqrt{1 - x^2}\right)$

$f(-x) = -f(x), \text{ odd function}$

29. The graph is symmetric with respect to the *y*-axis. The function is even.

31. The graph is symmetric with respect to the origin. The function is odd.

33. a. domain: $(-\infty, \infty)$

    b. range: $[-4, \infty)$

    c. *x*-intercepts: 1, 7

    d. *y*-intercept: 4

    e. $(4, \infty)$

    f. $(0, 4)$

    g. $(-\infty, 0)$

    h. $x = 4$

    i. $y = -4$

    j. $f(-3) = 4$

    k. $f(2) = -2$ and $f(6) = -2$

    l. neither ; $f(-x) \neq x$, $f(-x) \neq -x$

35. a. domain: $(-\infty, 3]$

    b. range: $(-\infty, 4]$

    c. *x*-intercepts: $-3, 3$

    d. $f(0) = 3$

    e. $(-\infty, 1)$

    f. $(1, 3)$

    g. $(-\infty, -3]$

    h. $f(1) = 4$

    i. $x = 1$

    j. positive; $f(-1) = +2$

37. a. $f(-2) = 3(-2) + 5 = -1$

    b. $f(0) = 4(0) + 7 = 7$

    c. $f(3) = 4(3) + 7 = 19$

39. a. $g(0) = 0 + 3 = 3$

b. $g(-6) = -(-6 + 3) = -(-3) = 3$

c. $g(-3) = -3 + 3 = 0$

41. a. $h(5) = \dfrac{5^2 - 9}{5 - 3} = \dfrac{25 - 9}{2} = \dfrac{16}{2} = 8$

b. $h(0) = \dfrac{0^2 - 9}{0 - 3} = \dfrac{-9}{-3} = 3$

c. $h(3) = 6$

43. a.

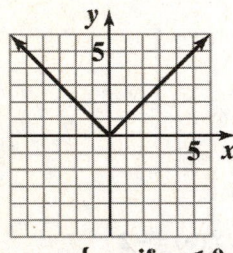

$$f(x) = \begin{cases} -x \text{ if } x < 0 \\ x \text{ if } x \geq 0 \end{cases}$$

b. range: $[0, \infty)$

45. a.

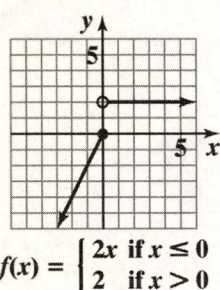

$$f(x) = \begin{cases} 2x \text{ if } x \leq 0 \\ 2 \text{ if } x > 0 \end{cases}$$

b. range: $(-\infty, 0] \cup \{2\}$

47. a.

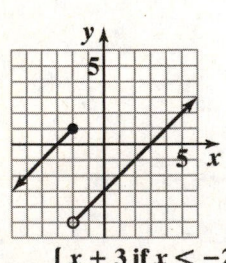

$$f(x) = \begin{cases} x + 3 \text{ if } x < -2 \\ x - 3 \text{ if } x \geq -2 \end{cases}$$

b. range: $(-\infty, \infty)$

49. a.

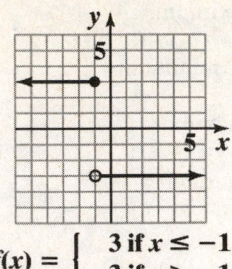

$$f(x) = \begin{cases} 3 \text{ if } x \leq -1 \\ -3 \text{ if } x > -1 \end{cases}$$

b. range: $\{-3, 3\}$

51. a.

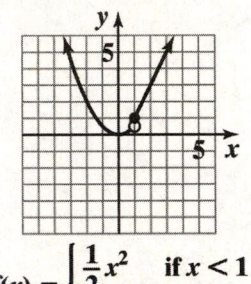

$$f(x) = \begin{cases} \dfrac{1}{2}x^2 & \text{if } x < 1 \\ 2x - 1 \text{ if } x \geq 1 \end{cases}$$

b. range: $[0, \infty)$

53. a.

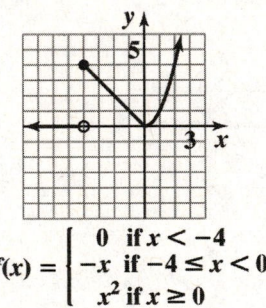

$$f(x) = \begin{cases} 0 \text{ if } x < -4 \\ -x \text{ if } -4 \leq x < 0 \\ x^2 \text{ if } x \geq 0 \end{cases}$$

b. range: $[0, \infty)$

55. $\dfrac{f(x+h) - f(x)}{h}$

$= \dfrac{4(x+h) - 4x}{h}$

$= \dfrac{4x + 4h - 4x}{h}$

$= \dfrac{4h}{h}$

$= 4$

57. $\dfrac{f(x+h)-f(x)}{h}$

$=\dfrac{3(x+h)+7-(3x+7)}{h}$

$=\dfrac{3x+3h+7-3x-7}{h}$

$=\dfrac{3h}{h}$

$=3$

59. $\dfrac{f(x+h)-f(x)}{h}$

$=\dfrac{(x+h)^2-x^2}{h}$

$=\dfrac{x^2+2xh+h^2-x^2}{h}$

$=\dfrac{2xh+h^2}{h}$

$=\dfrac{h(2x+h)}{h}$

$=2x+h$

61. $\dfrac{f(x+h)-f(x)}{h}$

$=\dfrac{(x+h)^2-4(x+h)+3-(x^2-4x+3)}{h}$

$=\dfrac{x^2+2xh+h^2-4x-4h+3-x^2+4x-3}{h}$

$=\dfrac{2xh+h^2-4h}{h}$

$=\dfrac{h(2x+h-4)}{h}$

$=2x+h-4$

63. $\dfrac{f(x+h)-f(x)}{h}$

$=\dfrac{2(x+h)^2+(x+h)-1-(2x^2+x-1)}{h}$

$=\dfrac{2x^2+4xh+2h^2+x+h-1-2x^2-x+1}{h}$

$=\dfrac{4xh+2h^2+h}{h}$

$=\dfrac{h(4x+2h+1)}{h}$

$=4x+2h+1$

65. $\dfrac{f(x+h)-f(x)}{h}$

$=\dfrac{-(x+h)^2+2(x+h)+4-(-x^2+2x+4)}{h}$

$=\dfrac{-x^2-2xh-h^2+2x+2h+4+x^2-2x-4}{h}$

$=\dfrac{-2xh-h^2+2h}{h}$

$=\dfrac{h(-2x-h+2)}{h}$

$=-2x-h+2$

67. $\dfrac{f(x+h)-f(x)}{h}$

$=\dfrac{-2(x+h)^2+5(x+h)+7-(-2x^2+5x+7)}{h}$

$=\dfrac{-2x^2-4xh-2h^2+5x+5h+7+2x^2-5x-7}{h}$

$=\dfrac{-4xh-2h^2+5h}{h}$

$=\dfrac{h(-4x-2h+5)}{h}$

$=-4x-2h+5$

69. $\dfrac{f(x+h)-f(x)}{h}$

$=\dfrac{-2(x+h)^2-(x+h)+3-(-2x^2-x+3)}{h}$

$=\dfrac{-2x^2-4xh-2h^2-x-h+3+2x^2+x-3}{h}$

$=\dfrac{-4xh-2h^2-h}{h}$

$=\dfrac{h(-4x-2h-1)}{h}$

$=-4x-2h-1$

71. $\dfrac{f(x+h)-f(x)}{h}=\dfrac{6-6}{h}=\dfrac{0}{h}=0$

73.

$$\frac{f(x+h)-f(x)}{h}$$

$$=\frac{\dfrac{1}{x+h}-\dfrac{1}{x}}{h}$$

$$=\frac{\dfrac{x}{x(x+h)}+\dfrac{-(x+h)}{x(x+h)}}{h}$$

$$=\frac{\dfrac{x-x-h}{x(x+h)}}{h}$$

$$=\frac{\dfrac{-h}{x(x+h)}}{h}$$

$$=\frac{-h}{x(x+h)}\cdot\frac{1}{h}$$

$$=\frac{-1}{x(x+h)}$$

75.

$$\frac{f(x+h)-f(x)}{h}$$

$$=\frac{\sqrt{x+h}-\sqrt{x}}{h}$$

$$=\frac{\sqrt{x+h}-\sqrt{x}}{h}\cdot\frac{\sqrt{x+h}+\sqrt{x}}{\sqrt{x+h}+\sqrt{x}}$$

$$=\frac{x+h-x}{h\left(\sqrt{x+h}+\sqrt{x}\right)}$$

$$=\frac{h}{h\left(\sqrt{x+h}+\sqrt{x}\right)}$$

$$=\frac{1}{\sqrt{x+h}+\sqrt{x}}$$

77. $\sqrt{f(-1.5)+f(-0.9)}-\left[f(\pi)\right]^2+f(-3)\div f(1)\cdot f(-\pi)$

$=\sqrt{1+0}-\left[-4\right]^2+2\div(-2)\cdot 3$

$=\sqrt{1}-16+(-1)\cdot 3$

$=1-16-3$

$=-18$

79. $30+0.30(t-120)=30+0.3t-36=0.3t-6$

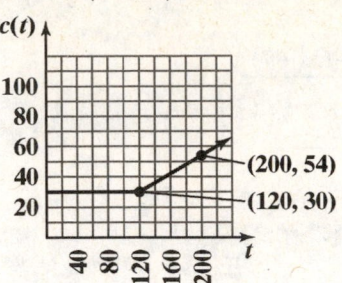

81. $C(t)=\begin{cases}50 & \text{if } 0\le t\le 400 \\ 50+0.30(t-400) & \text{if } t>400\end{cases}$

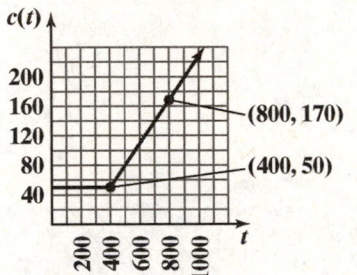

83. increasing: $(25, 55)$; decreasing: $(55, 75)$

85. The percent body fat in women reaches a maximum at age 55. This maximum is 38%.

87. domain: $[25, 75]$; range: $[34, 38]$

89. This model describes percent body fat in men.

91.
$$T(20,000)=782.50+0.15(20,000-7825)$$
$$=2608.75$$
A single taxpayer with taxable income of $20,000 owes $2608.75.

93. $39,148.75+0.33(x-160,850)$

95. $f(3)=0.76$

The cost of mailing a first-class letter weighing 3 ounces is $0.76.

97. The cost to mail a letter weighing 1.5 ounces is $0.59.

99.

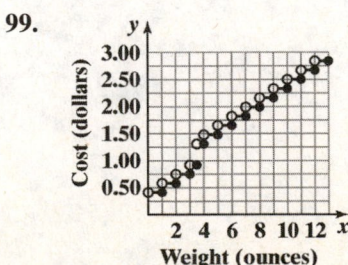

101. – 105. Answers may vary.

107.

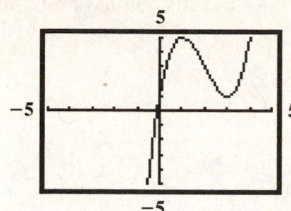

Increasing: $(-\infty, 1)$ or $(3, \infty)$

Decreasing: $(1, 3)$

109.

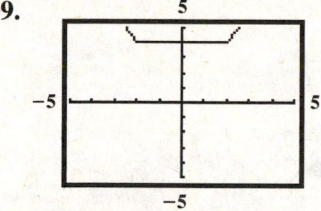

Increasing: $(2, \infty)$

Decreasing: $(-\infty, -2)$

Constant: $(-2, 2)$

111.

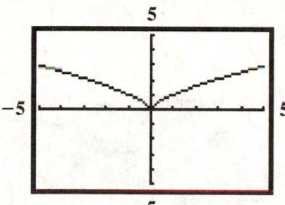

Increasing: $(0, \infty)$

Decreasing: $(-\infty, 0)$

113. a.

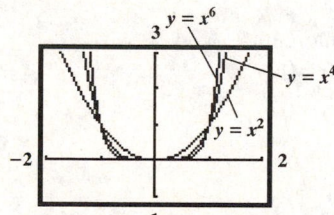

b.

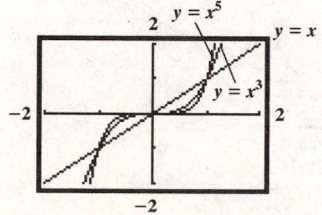

c. Increasing: $(0, \infty)$

Decreasing: $(-\infty, 0)$

d. $f(x) = x^n$ is increasing from $(-\infty, \infty)$ when n is odd.

e.

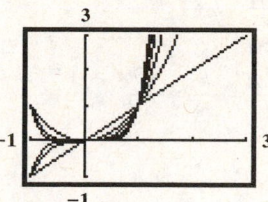

115. makes sense

117. makes sense

119. answers may vary

121. answers may vary

122. $\dfrac{y_2 - y_1}{x_2 - x_1} = \dfrac{4 - 1}{-2 - (-3)} = \dfrac{3}{1} = 3$

123. When $y = 0$:

$4x - 3y - 6 = 0$

$4x - 3(0) - 6 = 0$

$4x - 6 = 0$

$4x = 6$

$x = \dfrac{3}{2}$

The point is $\left(\dfrac{3}{2}, 0\right)$.

When $x = 0$:

$4x - 3y - 6 = 0$

$4(0) - 3y - 6 = 0$

$-3y - 6 = 0$

$-3y = 6$

$x = -2$

The point is $(0, -2)$.

124. $3x + 2y - 4 = 0$

$2y = -3x + 4$

$y = \dfrac{-3x + 4}{2}$

or

$y = -\dfrac{3}{2}x + 2$

Section 1.4

Check Point Exercises

1. a. $m = \dfrac{-2-4}{-4-(-3)} = \dfrac{-6}{-1} = 6$

    b. $m = \dfrac{5-(-2)}{-1-4} = \dfrac{7}{-5} = -\dfrac{7}{5}$

2. $y - y_1 = m(x - x_1)$
    $y - (-5) = 6(x - 2)$
    $y + 5 = 6x - 12$
    $y = 6x - 17$

3. $m = \dfrac{-6-(-1)}{-1-(-2)} = \dfrac{-5}{1} = -5$,

    so the slope is –5. Using the point (–2, –1), we get the point slope equation:
    $y - y_1 = m(x - x_1)$
    $y - (-1) = -5[x - (-2)]$
    $y + 1 = -5(x + 2).$ Solve the equation for y:
    $y + 1 = -5x - 10$
    $y = -5x - 11.$

4. The slope m is $\frac{3}{5}$ and the y-intercept is 1, so one point on the line is (1, 0). We can find a second point on the line by using the slope $m = \frac{3}{5} = \frac{\text{Rise}}{\text{Run}}$: starting at the point (0, 1), move 3 units up and 5 units to the right, to obtain the point (5, 4).

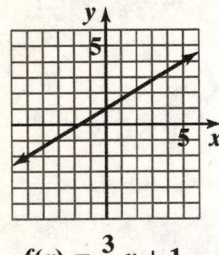

$f(x) = \dfrac{3}{5}x + 1$

5. $y = 3$ is a horizontal line.

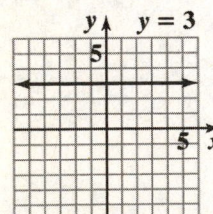

6. All ordered pairs that are solutions of $x = -3$ have a value of x that is always –3. Any value can be used for y.

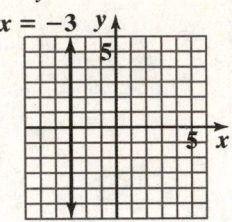

7. $3x + 6y - 12 = 0$
    $6y = -3x + 12$
    $y = \dfrac{-3}{6}x + \dfrac{12}{6}$
    $y = -\dfrac{1}{2}x + 2$

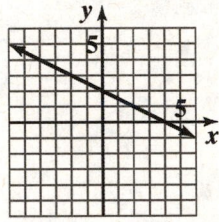

$3x + 6y - 12 = 0$

The slope is $-\dfrac{1}{2}$ and the y-intercept is 2.

8. Find the x-intercept:
    $3x - 2y - 6 = 0$
    $3x - 2(0) - 6 = 0$
    $3x - 6 = 0$
    $3x = 6$
    $x = 2$
    Find the y-intercept:
    $3x - 2y - 6 = 0$
    $3(0) - 2y - 6 = 0$
    $-2y - 6 = 0$
    $-2y = 6$
    $y = -3$

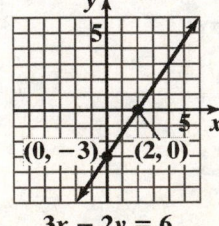

$3x - 2y = 6$

9. First find the slope.

$$m = \frac{\text{Change in } y}{\text{Change in } x} = \frac{57.64 - 57.04}{354 - 317} = \frac{0.6}{37} \approx 0.016$$

Use the point-slope form and then find slope-intercept form.

$$y - y_1 = m(x - x_1)$$
$$y - 57.04 = 0.016(x - 317)$$
$$y - 57.04 = 0.016x - 5.072$$
$$y = 0.016x + 51.968$$
$$f(x) = 0.016x + 52.0$$

Find the temperature at a concentration of 600 parts per million.

$$f(x) = 0.016x + 52.0$$
$$f(600) = 0.016(600) + 52.0$$
$$= 61.6$$

The temperature at a concentration of 600 parts per million would be $61.6°F$.

Exercise Set 1.4

1. $m = \dfrac{10 - 7}{8 - 4} = \dfrac{3}{4}$; rises

3. $m = \dfrac{2 - 1}{2 - (-2)} = \dfrac{1}{4}$; rises

5. $m = \dfrac{2 - (-2)}{3 - 4} = \dfrac{0}{-1} = 0$; horizontal

7. $m = \dfrac{-1 - 4}{-1 - (-2)} = \dfrac{-5}{1} = -5$; falls

9. $m = \dfrac{-2 - 3}{5 - 5} = \dfrac{-5}{0}$ undefined; vertical

11. $m = 2$, $x_1 = 3$, $y_1 = 5$;
point-slope form: $y - 5 = 2(x - 3)$;
slope-intercept form: $y - 5 = 2x - 6$
$$y = 2x - 1$$

13. $m = 6$, $x_1 = -2$, $y_1 = 5$;
point-slope form: $y - 5 = 6(x + 2)$;
slope-intercept form: $y - 5 = 6x + 12$
$$y = 6x + 17$$

15. $m = -3$, $x_1 = -2$, $y_1 = -3$;
point-slope form: $y + 3 = -3(x + 2)$;
slope-intercept form: $y + 3 = -3x - 6$
$$y = -3x - 9$$

17. $m = -4$, $x_1 = -4$, $y_1 = 0$;
point-slope form: $y - 0 = -4(x + 4)$;
slope-intercept form: $y = -4(x + 4)$
$$y = -4x - 16$$

19. $m = -1$, $x_1 = \dfrac{-1}{2}$, $y_1 = -2$;

point-slope form: $y + 2 = -1\left(x + \dfrac{1}{2}\right)$;

slope-intercept form: $y + 2 = -x - \dfrac{1}{2}$

$$y = -x - \dfrac{5}{2}$$

21. $m = \dfrac{1}{2}$, $x_1 = 0$, $y_1 = 0$;

point-slope form: $y - 0 = \dfrac{1}{2}(x - 0)$;

slope-intercept form: $y = \dfrac{1}{2}x$

23. $m = -\dfrac{2}{3}$, $x_1 = 6$, $y_1 = -2$;

point-slope form: $y + 2 = -\dfrac{2}{3}(x - 6)$;

slope-intercept form: $y + 2 = -\dfrac{2}{3}x + 4$

$$y = -\dfrac{2}{3}x + 2$$

25. $m = \dfrac{10 - 2}{5 - 1} = \dfrac{8}{4} = 2$;

point-slope form: $y - 2 = 2(x - 1)$ using
$(x_1, y_1) = (1, 2)$, or $y - 10 = 2(x - 5)$ using
$(x_1, y_1) = (5, 10)$;
slope-intercept form: $y - 2 = 2x - 2$ or
$$y - 10 = 2x - 10,$$
$$y = 2x$$

27. $m = \dfrac{3-0}{0-(-3)} = \dfrac{3}{3} = 1$;

point-slope form: $y - 0 = 1(x+3)$ using $(x_1, y_1) = (-3, 0)$, or $y - 3 = 1(x-0)$ using $(x_1, y_1) = (0, 3)$; slope-intercept form: $y = x + 3$

29. $m = \dfrac{4-(-1)}{2-(-3)} = \dfrac{5}{5} = 1$;

point-slope form: $y + 1 = 1(x+3)$ using $(x_1, y_1) = (-3, -1)$, or $y - 4 = 1(x-2)$ using $(x_1, y_1) = (2, 4)$; slope-intercept form:

$y + 1 = x + 3$ or

$y - 4 = x - 2$

$\quad\ y = x + 2$

31. $m = \dfrac{6-(-2)}{3-(-3)} = \dfrac{8}{6} = \dfrac{4}{3}$;

point-slope form: $y + 2 = \dfrac{4}{3}(x+3)$ using

$(x_1, y_1) = (-3, -2)$, or $y - 6 = \dfrac{4}{3}(x-3)$ using

$(x_1, y_1) = (3, 6)$;

slope-intercept form: $y + 2 = \dfrac{4}{3x} + 4$ or

$\quad\quad y - 6 = \dfrac{4}{3}x - 4$,

$\quad\quad\quad\ y = \dfrac{4}{3}x + 2$

33. $m = \dfrac{-1-(-1)}{4-(-3)} = \dfrac{0}{7} = 0$;

point-slope form: $y + 1 = 0(x+3)$ using $(x_1, y_1) = (-3, -1)$, or $y + 1 = 0(x-4)$ using $(x_1, y_1) = (4, -1)$;

slope-intercept form: $y + 1 = 0$, so

$\quad\quad\quad y = -1$

35. $m = \dfrac{0-4}{-2-2} = \dfrac{-4}{-4} = 1$;

point-slope form: $y - 4 = 1(x-2)$ using $(x_1, y_1) = (2, 4)$, or $y - 0 = 1(x+2)$ using $(x_1, y_1) = (-2, 0)$;

slope-intercept form: $y - 9 = x - 2$, or

$\quad\quad\quad y = x + 2$

37. $m = \dfrac{4-0}{0-\left(-\frac{1}{2}\right)} = \dfrac{4}{\frac{1}{2}} = 8$;

point-slope form: $y - 4 = 8(x-0)$ using $(x_1, y_1) = (0, 4)$, or $y - 0 = 8\left(x+\frac{1}{2}\right)$ using $(x_1, y_1) = \left(-\frac{1}{2}, 0\right)$; or $y - 0 = 8\left(x+\frac{1}{2}\right)$

slope-intercept form: $y = 8x + 4$

39. $m = 2$; $b = 1$

$y = 2x + 1$

41. $m = -2$; $b = 1$

$f(x) = -2x + 1$

43. $m = \dfrac{3}{4}$; $b = -2$

$f(x) = \dfrac{3}{4}x - 2$

45. $m = -\dfrac{3}{5}$; $b = 7$

$y = -\dfrac{3}{5}x + 7$

47. $m = -\dfrac{1}{2}; \ b = 0$

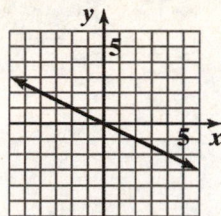

$$g(x) = -\frac{1}{2}x$$

49. $y = -2$

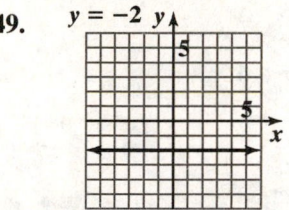

51. $y = -3$

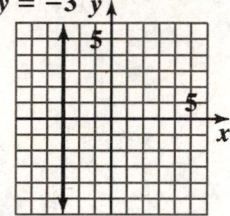

53. $y = 0$

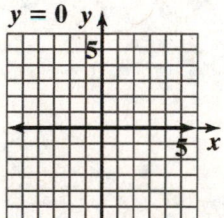

55. $f(x) = 1$

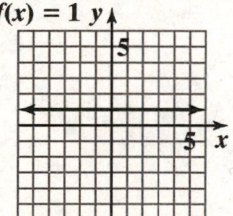

57.
$$3x - 18 = 0$$
$$3x = 18$$
$$x = 6$$

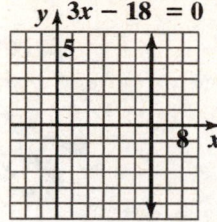

59. a.
$$3x + y - 5 = 0$$
$$y - 5 = -3x$$
$$y = -3x + 5$$

b. $m = -3; \ b = 5$

c.

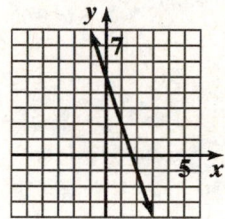

$$3x + y - 5 = 0$$

61. a.
$$2x + 3y - 18 = 0$$
$$2x - 18 = -3y$$
$$-3y = 2x - 18$$
$$y = \frac{2}{-3}x - \frac{18}{-3}$$
$$y = -\frac{2}{3}x + 6$$

b. $m = -\dfrac{2}{3}; \ b = 6$

c.

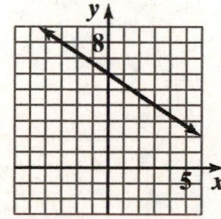

$$2x + 3y - 18 = 0$$

63. a.
$$8x - 4y - 12 = 0$$
$$8x - 12 = 4y$$
$$4y = 8x - 12$$
$$y = \frac{8}{4}x - \frac{12}{4}$$
$$y = 2x - 3$$

b. $m = 2; b = -3$

c.

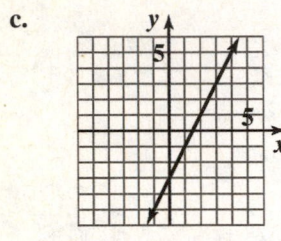

$8x - 4y - 12 = 0$

65. a.
$$3y - 9 = 0$$
$$3y = 9$$
$$y = 3$$

b. $m = 0; b = 3$

c.

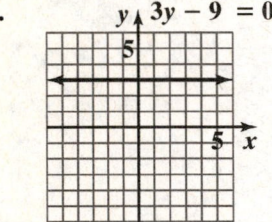

$3y - 9 = 0$

67. Find the x-intercept:
$$6x - 2y - 12 = 0$$
$$6x - 2(0) - 12 = 0$$
$$6x - 12 = 0$$
$$6x = 12$$
$$x = 2$$
Find the y-intercept:
$$6x - 2y - 12 = 0$$
$$6(0) - 2y - 12 = 0$$
$$-2y - 12 = 0$$
$$-2y = 12$$
$$y = -6$$

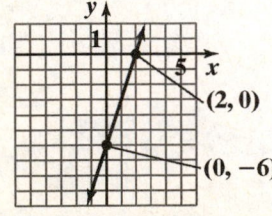

$(2, 0)$

$(0, -6)$

$6x - 2y - 12 = 0$

69. Find the x-intercept:
$$2x + 3y + 6 = 0$$
$$2x + 3(0) + 6 = 0$$
$$2x + 6 = 0$$
$$2x = -6$$
$$x = -3$$
Find the y-intercept:
$$2x + 3y + 6 = 0$$
$$2(0) + 3y + 6 = 0$$
$$3y + 6 = 0$$
$$3y = -6$$
$$y = -2$$

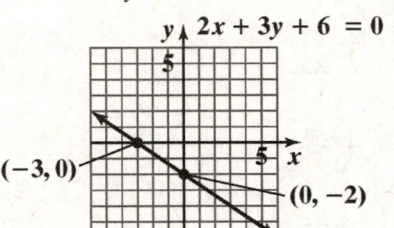

$2x + 3y + 6 = 0$

$(-3, 0)$

$(0, -2)$

71. Find the x-intercept:
$$8x - 2y + 12 = 0$$
$$8x - 2(0) + 12 = 0$$
$$8x + 12 = 0$$
$$8x = -12$$
$$\frac{8x}{8} = \frac{-12}{8}$$
$$x = \frac{-3}{2}$$
Find the y-intercept:
$$8x - 2y + 12 = 0$$
$$8(0) - 2y + 12 = 0$$
$$-2y + 12 = 0$$
$$-2y = -12$$
$$y = -6$$

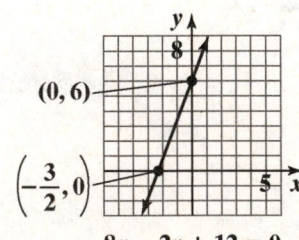

$(0, 6)$

$\left(-\frac{3}{2}, 0\right)$

$8x - 2y + 12 = 0$

73.
$$m = \frac{0-a}{b-0} = \frac{-a}{b} = -\frac{a}{b}$$

Since a and b are both positive, $-\dfrac{a}{b}$ is

negative. Therefore, the line falls.

75.
$$m = \frac{(b+c)-b}{a-a} = \frac{c}{0}$$
The slope is undefined.
The line is vertical.

77. $Ax + By = C$
$$By = -Ax + C$$
$$y = -\frac{A}{B}x + \frac{C}{B}$$

The slope is $-\dfrac{A}{B}$ and the $y-$intercept is $\dfrac{C}{B}$.

79.
$$-3 = \frac{4-y}{1-3}$$
$$-3 = \frac{4-y}{-2}$$
$$6 = 4-y$$
$$2 = -y$$
$$-2 = y$$

81. $3x - 4f(x) = 6$
$$-4f(x) = -3x + 6$$
$$f(x) = \frac{3}{4}x - \frac{3}{2}$$

$3x - 4f(x) - 6 = 0$

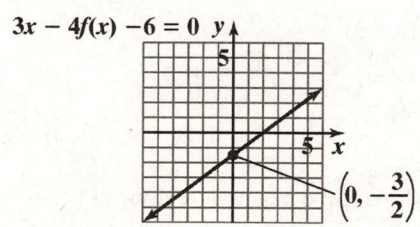

83. Using the slope-intercept form for the equation
of a line:
$$-1 = -2(3) + b$$
$$-1 = -6 + b$$
$$5 = b$$

85. m_1, m_3, m_2, m_4

87. **a.** First, find the slope using $(20, 38.9)$ and
$(10, 31.1)$.
$$m = \frac{38.9 - 31.1}{20 - 10} = \frac{7.8}{10} = 0.78$$
Then use the slope and one of the points to
write the equation in point-slope form.
$$y - y_1 = m(x - x_1)$$
$$y - 31.1 = 0.78(x - 10)$$
or
$$y - 38.9 = 0.78(x - 20)$$

b. $y - 31.1 = 0.78(x - 10)$
$$y - 31.1 = 0.78x - 7.8$$
$$y = 0.78x + 23.3$$
$$f(x) = 0.78x + 23.3$$

c. $f(40) = 0.78(40) + 23.3 = 54.5$

The linear function predicts the percentage of
never married American females, ages 25 – 29,
to be 54.5% in 2020.

89. **a.** **Life Expectancy for United States**
Males, by Year of Birth

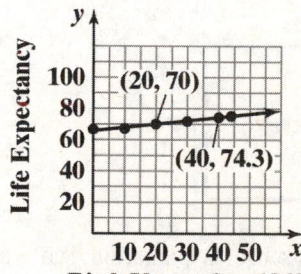

b. $m = \dfrac{\text{Change in } y}{\text{Change in } x} = \dfrac{74.3 - 70.0}{40 - 20} = 0.215$
$$y - y_1 = m(x - x_1)$$
$$y - 70.0 = 0.215(x - 20)$$
$$y - 70.0 = 0.215x - 4.3$$
$$y = 0.215x + 65.7$$
$$E(x) = 0.215x + 65.7$$

c. $E(x) = 0.215x + 65.7$
$$E(60) = 0.215(60) + 65.7$$
$$= 78.6$$
The life expectancy of American men born in
2020 is expected to be 78.6.

91. (10, 230) (60, 110) Points may vary.
$$m = \frac{110-230}{60-10} = -\frac{120}{50} = -2.4$$
$$y - 230 = -2.4(x-10)$$
$$y - 230 = -2.4x + 24$$
$$y = -2.4x + 254$$
Answers may vary for predictions.

93.–99. Answers may vary.

101. Two points are (0, 6) and (10, –24).
$$m = \frac{-24-6}{10-0} = \frac{-30}{10} = -3.$$
Check: $y = mx + b$: $y = -3x + 6$.

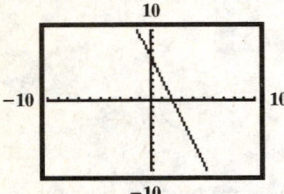

103. Two points are (0, –2) and (10, 5.5).
$$m = \frac{5.5-(-2)}{10-0} = \frac{7.5}{10} = 0.75 \text{ or } \frac{3}{4}.$$
Check: $y = mx + b$: $y = \frac{3}{4}x - 2$.

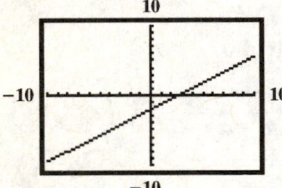

105. does not make sense; Explanations will vary. Sample explanation: Linear functions never change from increasing to decreasing.

107. does not make sense; Explanations will vary. Sample explanation: The slope of line's whose equations are in this form can be determined in several ways. One such way is to rewrite the equation in slope-intercept form.

109. false; Changes to make the statement true will vary. A sample change is: It is possible for *m* to equal *b*.

111. true

113. We are given that the x–intercept is –2 and the y–intercept is 4. We can use the points $(-2,0)$ and $(0,4)$ to find the slope.
$$m = \frac{4-0}{0-(-2)} = \frac{4}{0+2} = \frac{4}{2} = 2$$
Using the slope and one of the intercepts, we can write the line in point-slope form.
$$y - y_1 = m(x - x_1)$$
$$y - 0 = 2(x-(-2))$$
$$y = 2(x+2)$$
$$y = 2x + 4$$
$$-2x + y = 4$$
Find the *x*– and *y*–coefficients for the equation of the line with right-hand-side equal to 12. Multiply both sides of $-2x + y = 4$ by 3 to obtain 12 on the right-hand-side.
$$-2x + y = 4$$
$$3(-2x + y) = 3(4)$$
$$-6x + 3y = 12$$
Therefore, the coefficient of x is –6 and the coefficient of y is 3.

115. Answers may vary.

117. Answers may vary.

118. Since the slope is the same as the slope of $y = 2x + 1$, then $m = 2$.
$$y - y_1 = m(x - x_1)$$
$$y - 1 = 2(x-(-3))$$
$$y - 1 = 2(x+3)$$
$$y - 1 = 2x + 6$$
$$y = 2x + 7$$

119. Since the slope is the negative reciprocal of $-\frac{1}{4}$, then $m = 4$.
$$y - y_1 = m(x - x_1)$$
$$y - (-5) = 4(x-3)$$
$$y + 5 = 4x - 12$$
$$-4x + y + 17 = 0$$
$$4x - y - 17 = 0$$

120. $\dfrac{f(x_2)-f(x_1)}{x_2-x_1}=\dfrac{f(4)-f(1)}{4-1}$

$=\dfrac{4^2-1^2}{4-1}$

$=\dfrac{15}{3}$

$=5$

Section 1.5

Check Point Exercises

1. The slope of the line $y=3x+1$ is 3.

$y-y_1=m(x-x_1)$

$y-5=3\left(x-(-2)\right)$

$y-5=3(x+2)$ point-slope

$y-5=3x+6$

$y=3x+11$ slope-intercept

2. **a.** Write the equation in slope-intercept form:

$x+3y-12=0$

$3y=-x+12$

$y=-\dfrac{1}{3}x+4$

The slope of this line is $-\dfrac{1}{3}$ thus the slope of any line perpendicular to this line is 3.

b. Use $m=3$ and the point $(-2,-6)$ to write the equation.

$y-y_1=m(x-x_1)$

$y-(-6)=3\left(x-(-2)\right)$

$y+6=3(x+2)$

$y+6=3x+6$

$-3x+y=0$

$3x-y=0$ general form

3. $m=\dfrac{\text{Change in } y}{\text{Change in } x}=\dfrac{12.7-9.0}{2005-1990}=\dfrac{3.7}{15}\approx 0.25$

The slope indicates that the number of U.S. men living alone is projected to increase by 0.25 million each year.

4. **a.** $\dfrac{f(x_2)-f(x_1)}{x_2-x_1}=\dfrac{1^3-0^3}{1-0}=1$

b. $\dfrac{f(x_2)-f(x_1)}{x_2-x_1}=\dfrac{2^3-1^3}{2-1}=\dfrac{8-1}{1}=7$

c. $\dfrac{f(x_2)-f(x_1)}{x_2-x_1}=\dfrac{0^3-(-2)^3}{0-(-2)}=\dfrac{8}{2}=4$

5. $\dfrac{f(x_2)-f(x_1)}{x_2-x_1}=\dfrac{f(3)-f(1)}{3-1}=\dfrac{0.05-0.03}{3-1}=0.01$

6. **a.** $s(1)=4(1)^2=4$

$s(2)=4(2)^2=16$

$\dfrac{\Delta s}{\Delta t}=\dfrac{16-4}{2-1}=12$ feet per second

b. $s(1)=4(1)^2=4$

$s(1.5)=4(1.5)^2=9$

$\dfrac{\Delta s}{\Delta t}=\dfrac{9-4}{1.5-1}=10$ feet per second

c. $s(1)=4(1)^2=4$

$s(1.01)=4(1.01)^2=4.0804$

$\dfrac{\Delta s}{\Delta t}=\dfrac{4.0804-4}{1.01-1}=8.04$ feet per second

Exercise Set 1.5

1. Since L is parallel to $y=2x$, we know it will have slope $m=2$. We are given that it passes through $(4,2)$. We use the slope and point to write the equation in point-slope form.

$y-y_1=m(x-x_1)$

$y-2=2(x-4)$

Solve for y to obtain slope-intercept form.

$y-2=2(x-4)$

$y-2=2x-8$

$y=2x-6$

In function notation, the equation of the line is $f(x)=2x-6$.

3. Since L is perpendicular to $y = 2x$, we know it will have slope $m = -\dfrac{1}{2}$. We are given that it passes through $(2, 4)$. We use the slope and point to write the equation in point-slope form.

$$y - y_1 = m(x - x_1)$$

$$y - 4 = -\frac{1}{2}(x - 2)$$

Solve for y to obtain slope-intercept form.

$$y - 4 = -\frac{1}{2}(x - 2)$$

$$y - 4 = -\frac{1}{2}x + 1$$

$$y = -\frac{1}{2}x + 5$$

In function notation, the equation of the line is

$$f(x) = -\frac{1}{2}x + 5.$$

5. $m = -4$ since the line is parallel to
$y = -4x + 3;\ x_1 = -8,\ y_1 = -10;$
point-slope form: $\quad y + 10 = -4(x + 8)$
slope-intercept form: $y + 10 = -4x - 32$
$$y = -4x - 42$$

7. $m = -5$ since the line is perpendicular to
$y = \dfrac{1}{5}x + 6;\ x_1 = 2,\ y_1 = -3;$
point-slope form: $y + 3 = -5(x - 2)$
slope-intercept form: $\ y + 3 = -5x + 10$
$$y = -5x + 7$$

9. $2x - 3y - 7 = 0$
$$-3y = -2x + 7$$
$$y = \frac{2}{3}x - \frac{7}{3}$$

The slope of the given line is $\dfrac{2}{3}$, so $m = \dfrac{2}{3}$ since the lines are parallel.

point-slope form: $y - 2 = \dfrac{2}{3}(x + 2)$

general form: $2x - 3y + 10 = 0$

11. $x - 2y - 3 = 0$
$$-2y = -x + 3$$
$$y = \frac{1}{2}x - \frac{3}{2}$$

The slope of the given line is $\dfrac{1}{2}$, so $m = -2$ since the lines are perpendicular.

point-slope form: $\quad y + 7 = -2(x - 4)$

general form: $2x + y - 1 = 0$

13. $\dfrac{15 - 0}{5 - 0} = \dfrac{15}{5} = 3$

15. $\dfrac{5^2 + 2 \cdot 5 - (3^2 + 2 \cdot 3)}{5 - 3} = \dfrac{25 + 10 - (9 + 6)}{2}$
$$= \frac{20}{2}$$
$$= 10$$

17. $\dfrac{\sqrt{9} - \sqrt{4}}{9 - 4} = \dfrac{3 - 2}{5} = \dfrac{1}{5}$

19.   **a.** $s(3) = 10(3)^2 = 90$
$$s(4) = 10(4)^2 = 160$$
$$\frac{\Delta s}{\Delta t} = \frac{160 - 90}{4 - 3} = 70 \text{ feet per second}$$

   b. $s(3) = 10(3)^2 = 90$
$$s(3.5) = 10(3.5)^2 = 122.5$$
$$\frac{\Delta s}{\Delta t} = \frac{122.5 - 90}{3.5 - 3} = 65 \text{ feet per second}$$

   c. $s(3) = 10(3)^2 = 90$
$$s(3.01) = 10(3.01)^2 = 90.601$$
$$\frac{\Delta s}{\Delta t} = \frac{90.601 - 90}{3.01 - 3} = 60.1 \text{ feet per second}$$

   d. $s(3) = 10(3)^2 = 90$
$$s(3.001) = 10(3.001)^2 = 90.06$$
$$\frac{\Delta s}{\Delta t} = \frac{90.06 - 90}{3.001 - 3} = 60.01 \text{ feet per second}$$

21. Since the line is perpendicular to $x = 6$ which is a vertical line, we know the graph of f is a horizontal line with 0 slope. The graph of f passes through $(-1,5)$, so the equation of f is $f(x) = 5$.

23. First we need to find the equation of the line with x – intercept of 2 and y – intercept of -4. This line will pass through $(2,0)$ and $(0,-4)$. We use these points to find the slope.

$$m = \frac{-4-0}{0-2} = \frac{-4}{-2} = 2$$

Since the graph of f is perpendicular to this line, it will have slope $m = -\frac{1}{2}$.

Use the point $(-6,4)$ and the slope $-\frac{1}{2}$ to find the equation of the line.

$$y - y_1 = m(x - x_1)$$
$$y - 4 = -\frac{1}{2}(x - (-6))$$
$$y - 4 = -\frac{1}{2}(x + 6)$$
$$y - 4 = -\frac{1}{2}x - 3$$
$$y = -\frac{1}{2}x + 1$$
$$f(x) = -\frac{1}{2}x + 1$$

25. First put the equation $3x - 2y - 4 = 0$ in slope-intercept form.
$$3x - 2y - 4 = 0$$
$$-2y = -3x + 4$$
$$y = \frac{3}{2}x - 2$$

The equation of f will have slope $-\frac{2}{3}$ since it is perpendicular to the line above and the same y – intercept -2.

So the equation of f is $f(x) = -\frac{2}{3}x - 2$.

27. $P(x) = -1.2x + 47$

29. $m = \dfrac{1163 - 617}{1998 - 1994} = \dfrac{546}{4} \approx 137$

There was an average increase of approximately 137 discharges per year.

31. **a.** $f(x) = 1.1x^3 - 35x^2 + 264x + 557$

$$f(0) = 1.1(0)^3 - 35(0)^2 + 264(0) + 557 = 557$$
$$f(4) = 1.1(4)^3 - 35(4)^2 + 264(4) + 557 = 1123.4$$
$$m = \frac{1123.4 - 557}{4 - 0} \approx 142$$

b. This overestimates by 5 discharges per year.

33. – 37. Answers may vary.

39.
$$y = \frac{1}{3}x + 1$$
$$y = -3x - 2$$

a. The lines are perpendicular because their slopes are negative reciprocals of each other. This is verified because product of their slopes is -1.

b.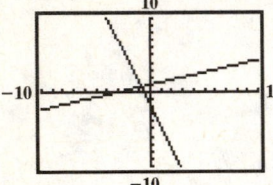

The lines do not appear to be perpendicular.

c.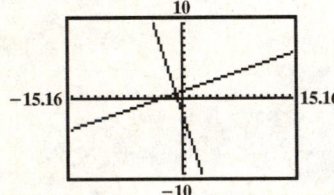

The lines appear to be perpendicular. The calculator screen is rectangular and does not have the same width and height. This causes the scale of the x–axis to differ from the scale on the y–axis despite using the same scale in the window settings. In part (b), this causes the lines not to appear perpendicular when indeed they are. The zoom square feature compensates for this and in part (c), the lines appear to be perpendicular.

41. makes sense

43. makes sense

45. The slope of the line containing $(1, -3)$ and $(-2, 4)$ has slope $m = \dfrac{4 - (-3)}{-2 - 1} = \dfrac{4 + 3}{-3} = \dfrac{7}{-3} = -\dfrac{7}{3}$

Solve $Ax + y - 2 = 0$ for y to obtain slope-intercept form.
$$Ax + y - 2 = 0$$
$$y = -Ax + 2$$
So the slope of this line is $-A$.

This line is perpendicular to the line above so its slope is $\dfrac{3}{7}$. Therefore, $-A = \dfrac{3}{7}$ so $A = -\dfrac{3}{7}$.

46. **a.**

$f(x) = |x|$

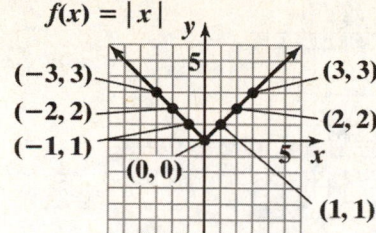

$(-3, 3)$ $(3, 3)$
$(-2, 2)$ $(2, 2)$
$(-1, 1)$
$(0, 0)$ $(1, 1)$

b.

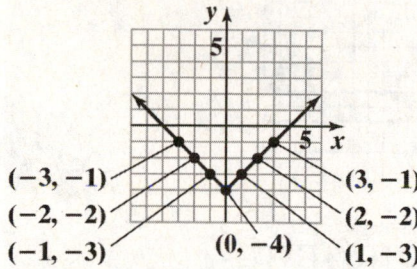

$(-3, -1)$ $(3, -1)$
$(-2, -2)$ $(2, -2)$
$(-1, -3)$ $(0, -4)$ $(1, -3)$

c. The graph in part (b) is the graph in part (a) shifted down 4 units.

47. **a.**

$f(x) = x^2$

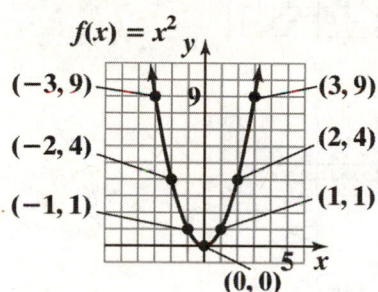

$(-3, 9)$ $(3, 9)$
$(-2, 4)$ $(2, 4)$
$(-1, 1)$ $(1, 1)$
$(0, 0)$

b.

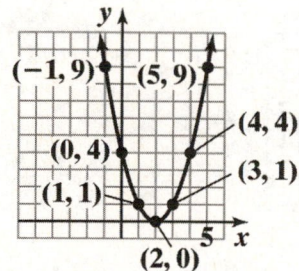

$(-1, 9)$ $(5, 9)$
$(0, 4)$ $(4, 4)$
$(1, 1)$ $(3, 1)$
$(2, 0)$

c. The graph in part (b) is the graph in part (a) shifted to the right 2 units.

48. **a.**

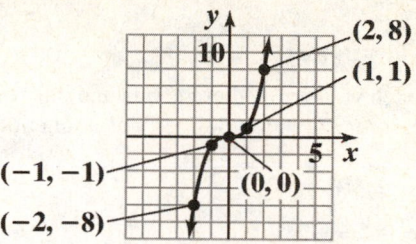

$(2, 8)$
$(1, 1)$
$(-1, -1)$
$(0, 0)$
$(-2, -8)$

b.

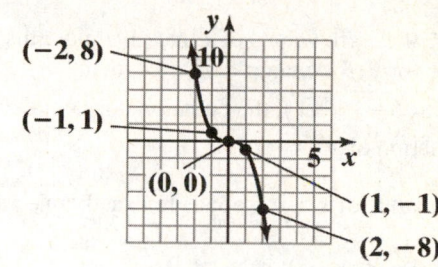

$(-2, 8)$
$(-1, 1)$
$(0, 0)$ $(1, -1)$
$(2, -8)$

c. The graph in part (b) is the graph in part (a) reflected across the *y*-axis.

Mid-Chapter 1 Check Point

1. The relation is not a function.
The domain is $\{1, 2\}$.
The range is $\{-6, 4, 6\}$.

2. The relation is a function.
The domain is $\{0, 2, 3\}$.
The range is $\{1, 4\}$.

3. The relation is a function.
The domain is $\{x \mid -2 \le x < 2\}$.
The range is $\{y \mid 0 \le y \le 3\}$.

4. The relation is not a function.
The domain is $\{x \mid -3 < x \le 4\}$.
The range is $\{y \mid -1 \le y \le 2\}$.

5. The relation is not a function.
The domain is $\{-2, -1, 0, 1, 2\}$.
The range is $\{-2, -1, 1, 3\}$.

6. The relation is a function.
The domain is $\{x \mid x \le 1\}$.
The range is $\{y \mid y \ge -1\}$.

7. $x^2 + y = 5$

$\qquad y = -x^2 + 5$

For each value of x, there is one and only one value for y, so the equation defines y as a function of x.

8. $x + y^2 = 5$

$\qquad y^2 = 5 - x$

$\qquad y = \pm\sqrt{5 - x}$

Since there are values of x that give more than one value for y (for example, if $x = 4$, then $y = \pm\sqrt{5 - 4} = \pm 1$), the equation does not define y as a function of x.

9. Each value of x corresponds to exactly one value of y.

10. Domain: $(-\infty, \infty)$

11. Range: $(-\infty, 4]$

12. x-intercepts: -6 and 2

13. y-intercept: 3

14. increasing: $(-\infty, -2)$

15. decreasing: $(-2, \infty)$

16. $x = -2$

17. $f(-2) = 4$

18. $f(-4) = 3$

19. $f(-7) = -2$ and $f(3) = -2$

20. $f(-6) = 0$ and $f(2) = 0$

21. $(-6, 2)$

22. $f(100)$ is negative.

23. neither; $f(-x) \neq x$ and $f(-x) \neq -x$

24. $\dfrac{f(x_2) - f(x_1)}{x_2 - x_1} = \dfrac{f(4) - f(-4)}{4 - (-4)} = \dfrac{-5 - 3}{4 + 4} = -1$

25. $y = -2x$

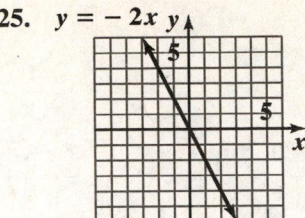

26. $y = -2$

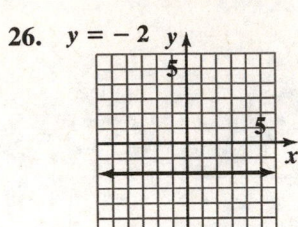

27. $x + y = -2$

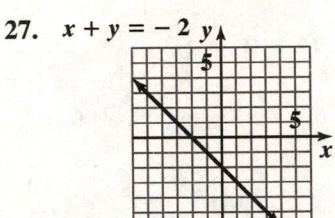

28. $y = \dfrac{1}{3}x - 2$

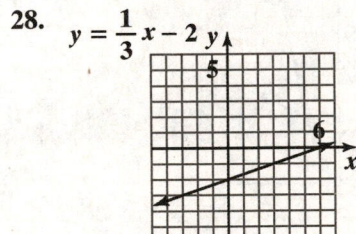

29. $x = 3.5$

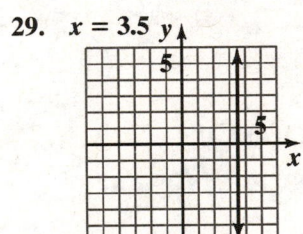

30.

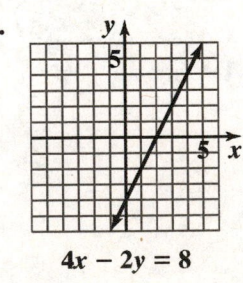

$4x - 2y = 8$

31.

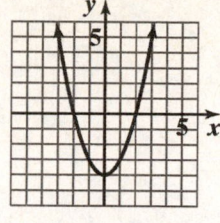

$$f(x) = x^2 - 4$$

32.

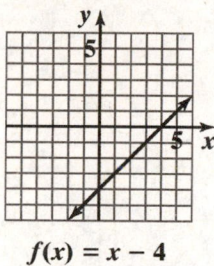

$$f(x) = x - 4$$

33.

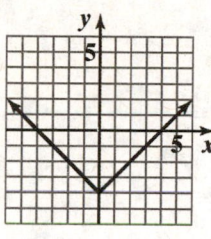

$$f(x) = |x| - 4$$

34. $5y = -3x$

$$y = -\frac{3}{5}x$$

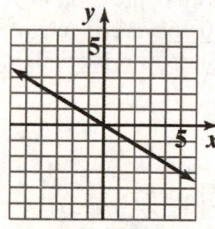

$$5y = -3x$$

35. $5y = 20$

$$y = 4$$

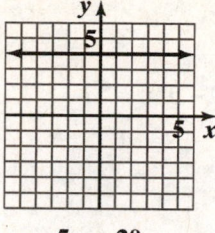

$$5y = 20$$

36.

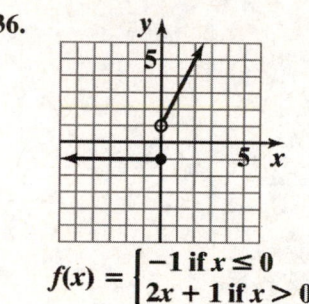

$$f(x) = \begin{cases} -1 \text{ if } x \le 0 \\ 2x + 1 \text{ if } x > 0 \end{cases}$$

37. a. $f(-x) = -2(-x)^2 - x - 5 = -2x^2 - x - 5$

neither; $f(-x) \ne x$ and $f(-x) \ne -x$

b. $\dfrac{f(x+h) - f(x)}{h}$

$$= \frac{-2(x+h)^2 + (x+h) - 5 - (-2x^2 + x - 5)}{h}$$

$$= \frac{-2x^2 - 4xh - 2h^2 + x + h - 5 + 2x^2 - x + 5}{h}$$

$$= \frac{-4xh - 2h^2 + h}{h}$$

$$= \frac{h(-4x - 2h + 1)}{h}$$

$$= -4x - 2h + 1$$

38. $C(x) = \begin{cases} 30 & \text{if } 0 \le t \le 200 \\ 30 + 0.40(t - 200) & \text{if } t > 200 \end{cases}$

  a. $C(150) = 30$

  b. $C(250) = 30 + 0.40(250 - 200) = 50$

39. $y - y_1 = m(x - x_1)$

$y - 3 = -2(x - (-4))$

$y - 3 = -2(x + 4)$

$y - 3 = -2x - 8$

$y = -2x - 5$

$f(x) = -2x - 5$

40. $m = \dfrac{\text{Change in } y}{\text{Change in } x} = \dfrac{1 - (-5)}{2 - (-1)} = \dfrac{6}{3} = 2$

$y - y_1 = m(x - x_1)$

$y - 1 = 2(x - 2)$

$y - 1 = 2x - 4$

$y = 2x - 3$

$f(x) = 2x - 3$

41. $3x - y - 5 = 0$

$-y = -3x + 5$

$y = 3x - 5$

The slope of the given line is 3, and the lines are parallel, so $m = 3$.

$y - y_1 = m(x - x_1)$

$y - (-4) = 3(x - 3)$

$y + 4 = 3x - 9$

$y = 3x - 13$

$f(x) = 3x - 13$

42. $2x - 5y - 10 = 0$

$-5y = -2x + 10$

$\dfrac{-5y}{-5} = \dfrac{-2x}{-5} + \dfrac{10}{-5}$

$y = \dfrac{2}{5}x - 2$

The slope of the given line is $\dfrac{2}{5}$, and the lines are

perpendicular, so $m = -\dfrac{5}{2}$.

$y - y_1 = m(x - x_1)$

$y - (-3) = -\dfrac{5}{2}(x - (-4))$

$y + 3 = -\dfrac{5}{2}x - 10$

$y = -\dfrac{5}{2}x - 13$

$f(x) = -\dfrac{5}{2}x - 13$

43. $m_1 = \dfrac{\text{Change in } y}{\text{Change in } x} = \dfrac{0 - (-4)}{7 - 2} = \dfrac{4}{5}$

$m_2 = \dfrac{\text{Change in } y}{\text{Change in } x} = \dfrac{6 - 2}{1 - (-4)} = \dfrac{4}{5}$

The slope of the lines are equal thus the lines are parallel.

44. **a.** $m = \dfrac{\text{Change in } y}{\text{Change in } x} = \dfrac{42 - 26}{180 - 80} = \dfrac{16}{100} = 0.16$

  b. For each minute of brisk walking, the percentage of patients with depression in remission increased by 0.16%. The rate of change is 0.16% per minute of brisk walking.

45. $\dfrac{f(x_2) - f(x_1)}{x_2 - x_1} = \dfrac{f(2) - f(-1)}{2 - (-1)}$

$= \dfrac{(3(2)^2 - 2) - (3(-1)^2 - (-1))}{2 + 1}$

$= 2$

Section 1.6

Check Point Exercises

1. Shift up vertically 3 units.

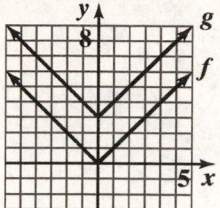

2. Shift to the right 4 units.

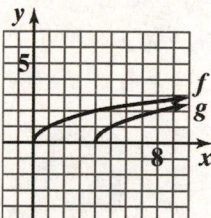

3. Shift to the right 1 unit and down 2 units.

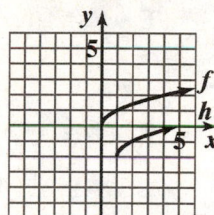

4. Reflect about the *x*-axis.

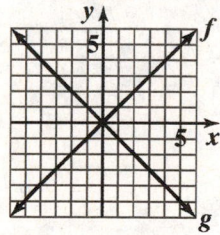

5. Reflect about the *y*-axis.

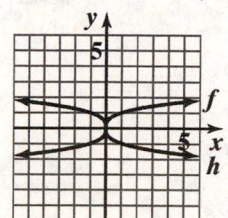

6. Vertically stretch the graph of $f(x) = |x|$.

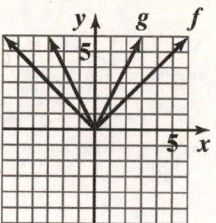

7. a. Horizontally shrink the graph of $y = f(x)$.

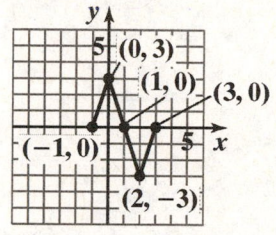

$$g(x) = f(2x)$$

b. Horizontally stretch the graph of $y = f(x)$.

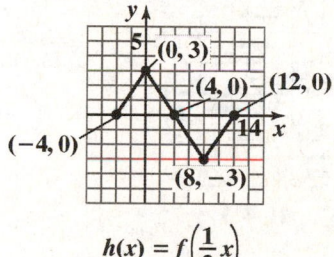

$$h(x) = f\left(\frac{1}{2}x\right)$$

8. The graph of $y = f(x)$ is shifted 1 unit left, shrunk by a factor of $\frac{1}{3}$, reflected about the x-axis, then shifted down 2 units.

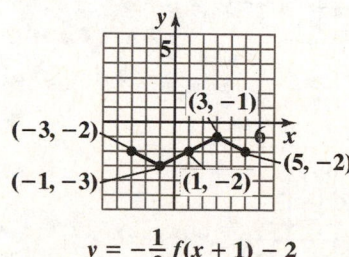

$$y = -\frac{1}{3}f(x+1) - 2$$

9. The graph of $f(x) = x^2$ is shifted 1 unit right, stretched by a factor of 2, then shifted up 3 units.

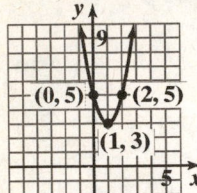

$g(x) = 2(x - 1)^2 + 3$

Exercise Set 1.6

1.

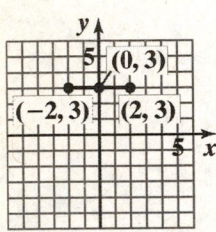

$g(x) = f(x) + 1$

3.

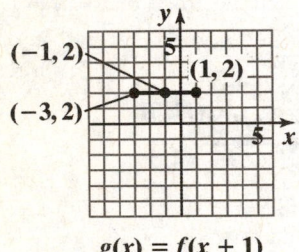

$g(x) = f(x + 1)$

5.

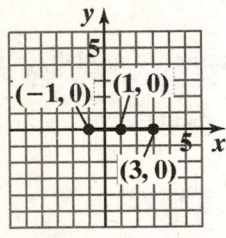

$g(x) = f(x - 1) - 2$

7.

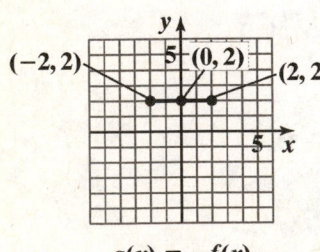

$g(x) = -f(x)$

9.

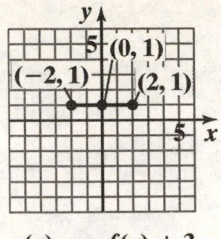

$g(x) = -f(x) + 3$

11.

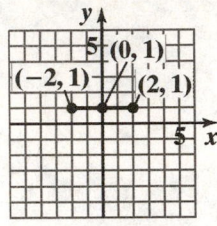

$g(x) = \frac{1}{2} f(x)$

13. $g(x) = f\left(\frac{1}{2} x\right)$

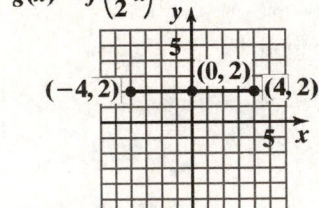

15.

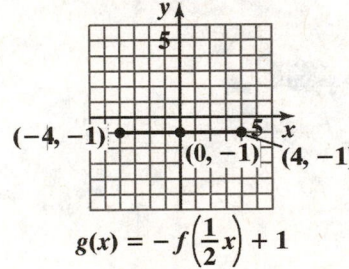

$g(x) = -f\left(\frac{1}{2} x\right) + 1$

17.

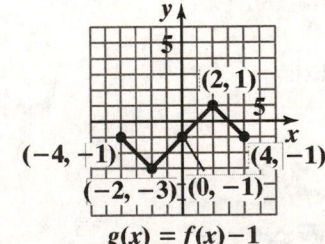

$g(x) = f(x) - 1$

19.

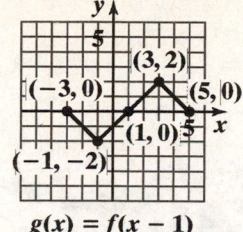

$g(x) = f(x - 1)$

21.

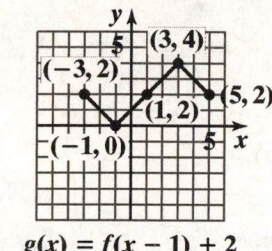

$g(x) = f(x - 1) + 2$

23.

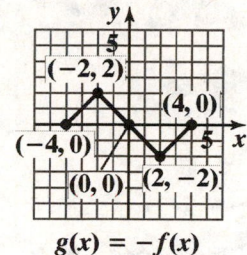

$g(x) = -f(x)$

25.

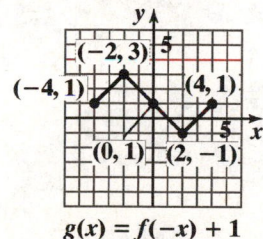

$g(x) = f(-x) + 1$

27.

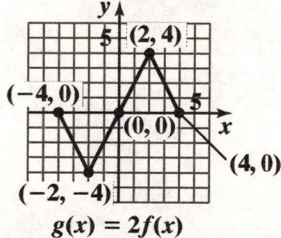

$g(x) = 2f(x)$

29.

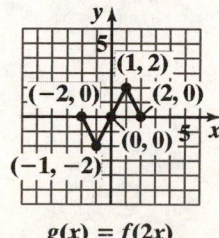

$g(x) = f(2x)$

31.

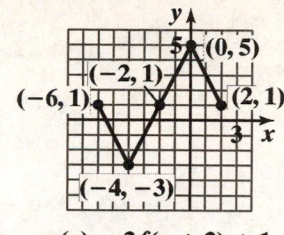

$g(x) = 2f(x + 2) + 1$

33.

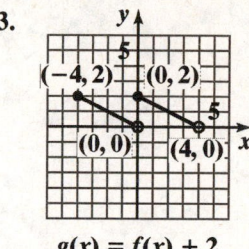

$g(x) = f(x) + 2$

35.

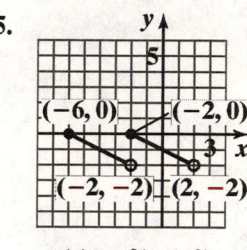

$g(x) = f(x + 2)$

37.

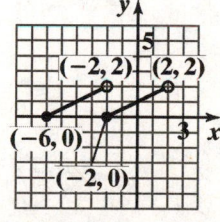

$g(x) = -f(x + 2)$

39.

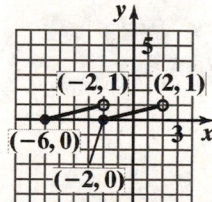

$g(x) = -\dfrac{1}{2} f(x + 2)$

41.

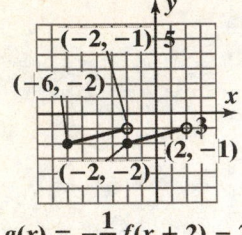

$g(x) = -\frac{1}{2}f(x + 2) - 2$

43.

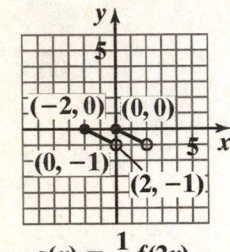

$g(x) = \frac{1}{2}f(2x)$

45.

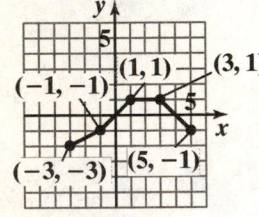

$g(x) = f(x - 1) - 1$

47.

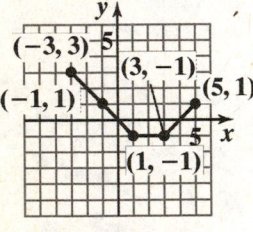

$g(x) = -f(x - 1) + 1$

49.

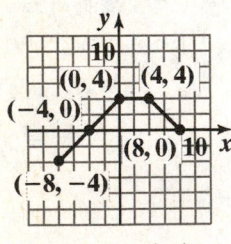

$g(x) = 2f\left(\frac{1}{2}x\right)$

51.

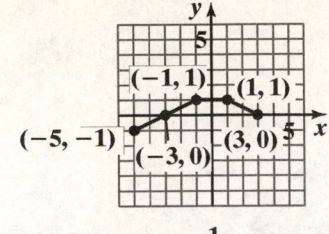

$g(x) = \frac{1}{2}f(x + 1)$

53.

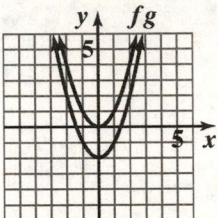

55.

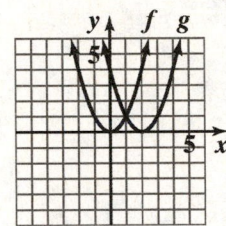

57.

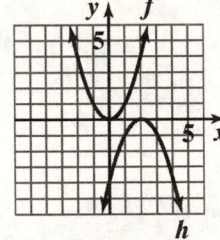

59.

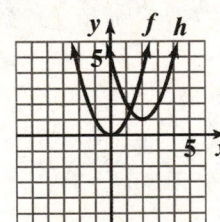

61.

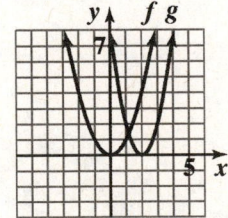

63.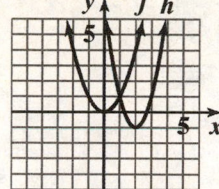

65.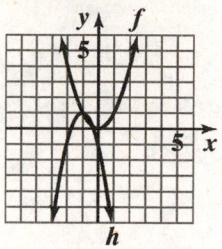

67.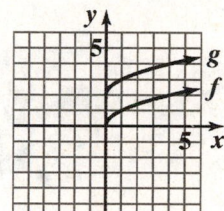

69.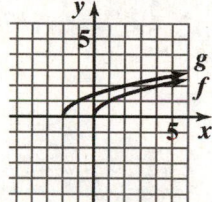

71.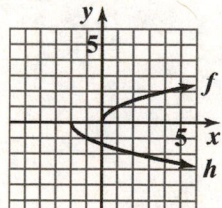

73.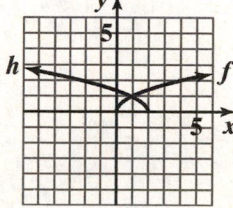

75.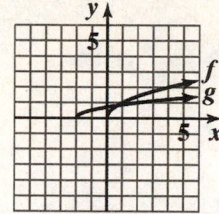

77.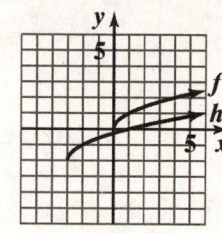

79.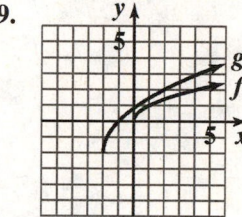

81.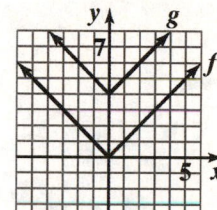

83.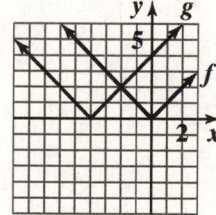

85.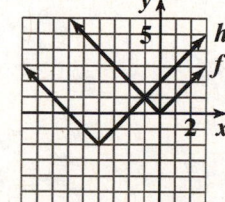

87.

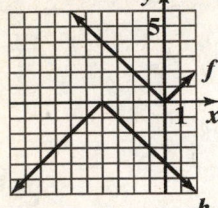

89.

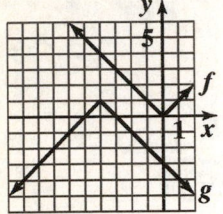

91.

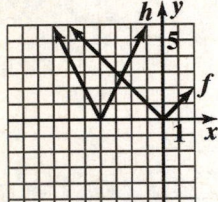

93.

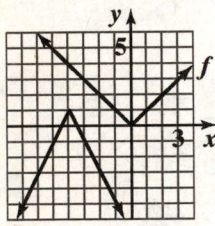

95.

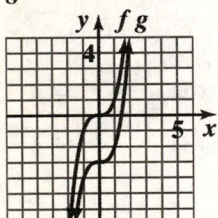

97.

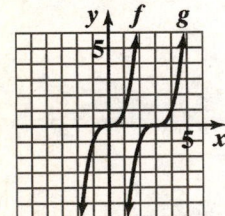

99.

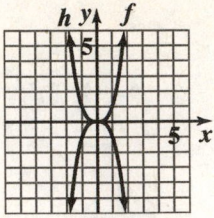

101.

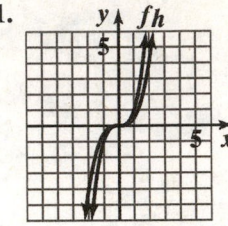

103.

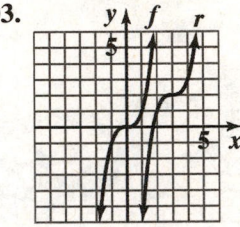

105.

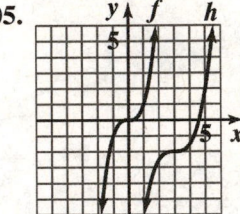

107.

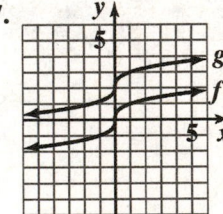

109.

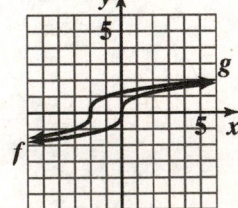

111.

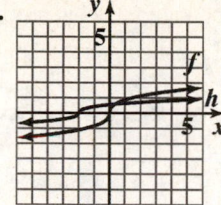

113.

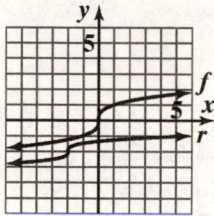

115.

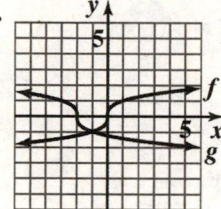

117.

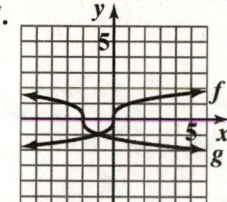

119.

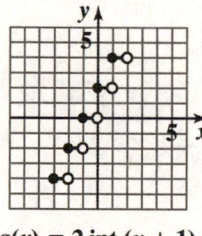

$g(x) = 2 \text{ int } (x + 1)$

121.

$h(x) = \text{int } (-x) + 1$

123. $y = \sqrt{x-2}$

125. $y = (x+1)^2 - 4$

127. a. First, vertically stretch the graph of $f(x) = \sqrt{x}$ by the factor 2.9; then shift the result up 20.1 units.

b. $f(x) = 2.9\sqrt{x} + 20.1$

$f(48) = 2.9\sqrt{48} + 20.1 \approx 40.2$

The model describes the actual data very well.

c. $\dfrac{f(x_2) - f(x_1)}{x_2 - x_1}$

$= \dfrac{f(10) - f(0)}{10 - 0}$

$= \dfrac{\left(2.9\sqrt{10} + 20.1\right) - \left(2.9\sqrt{0} + 20.1\right)}{10 - 0}$

$= \dfrac{29.27 - 20.1}{10}$

≈ 0.9

0.9 inches per month

d. $\dfrac{f(x_2) - f(x_1)}{x_2 - x_1}$

$= \dfrac{f(60) - f(50)}{60 - 50}$

$= \dfrac{\left(2.9\sqrt{60} + 20.1\right) - \left(2.9\sqrt{50} + 20.1\right)}{60 - 50}$

$= \dfrac{42.5633 - 40.6061}{10}$

≈ 0.2

This rate of change is lower than the rate of change in part (c). The relative leveling off of the curve shows this difference.

129. – 133. Answers may vary.

135. a.

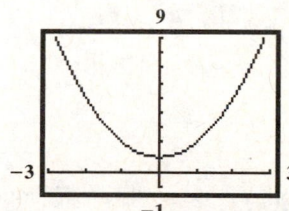

b.

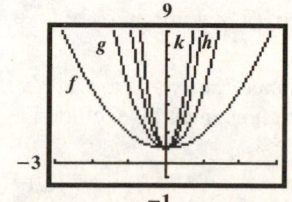

137. makes sense

139. does not make sense; Explanations will vary.
Sample explanation: The reprogram should be
$y = f(t+1)$.

141. false; Changes to make the statement true will vary.
A sample change is: The graph of g is a translation
of f three units to the <u>left</u> and three units upward.

143. false; Changes to make the statement true will vary.
A sample change is: The stretch will be 5 units and
the downward shift will be 10 units.

145. $g(x) = -(x+4)^2$

147. $g(x) = -\sqrt{x-2} + 2$

149. $(-a, b)$

151. $(a+3, b)$

153. $(2x-1)(x^2+x-2) = 2x(x^2+x-2) - 1(x^2+x-2)$
$$= 2x^3 + 2x^2 - 4x - x^2 - x + 2$$
$$= 2x^3 + 2x^2 - x^2 - 4x - x + 2$$
$$= 2x^3 + x^2 - 5x + 2$$

154. $(f(x))^2 - 2f(x) + 6 = (3x-4)^2 - 2(3x-4) + 6$
$$= 9x^2 - 24x + 16 - 6x + 8 + 6$$
$$= 9x^2 - 24x - 6x + 16 + 8 + 6$$
$$= 9x^2 - 30x + 30$$

155. $\dfrac{2}{\dfrac{3}{x}-1} = \dfrac{2x}{\dfrac{3x}{x}-x} = \dfrac{2x}{3-x}$

Section 1.7

Check Point Exercises

1. a. The function $f(x) = x^2 + 3x - 17$ contains
neither division nor an even root. The domain of
f is the set of all real numbers or $(-\infty, \infty)$.

b. The denominator equals zero when $x = 7$ or $x = -7$. These values must be excluded from the
domain.
domain of $g = (-\infty, -7) \cup (-7, 7) \cup (7, \infty)$.

c. Since $h(x) = \sqrt{9x-27}$ contains an even root; the
quantity under the radical must be greater than or
equal to 0.
$9x - 27 \geq 0$
$\quad 9x \geq 27$
$\quad\quad x \geq 3$
Thus, the domain of h is $\{x \mid x \geq 3\}$, or the
interval $[3, \infty)$.

2. a. $(f+g)(x) = f(x) + g(x)$
$$= x - 5 + (x^2 - 1)$$
$$= x - 5 + x^2 - 1$$
$$= -x^2 + x - 6$$

b. $(f-g)(x) = f(x) - g(x)$
$$= x - 5 - (x^2 - 1)$$
$$= x - 5 - x^2 + 1$$
$$= -x^2 + x - 4$$

c. $(fg)(x) = (x-5)(x^2-1)$
$$= x(x^2-1) - 5(x^2-1)$$
$$= x^3 - x - 5x^2 + 5$$
$$= x^3 - 5x^2 - x + 5$$

d. $\left(\dfrac{f}{g}\right)(x) = \dfrac{f(x)}{g(x)}$
$$= \dfrac{x-5}{x^2-1}, \ x \neq \pm 1$$

3. a. $(f+g)(x) = f(x) + g(x)$
$$= \sqrt{x-3} + \sqrt{x+1}$$

b. domain of f: $\quad x - 3 \geq 0$
$\quad\quad\quad\quad\quad\quad x \geq 3$
$\quad\quad\quad\quad\quad [3, \infty)$
domain of g: $\quad x + 1 \geq 0$
$\quad\quad\quad\quad\quad\quad x \geq -1$
$\quad\quad\quad\quad\quad [-1, \infty)$

The domain of $f + g$ is the set of all real
numbers that are common to the domain of f
and the domain of g. Thus, the domain of $f + g$
is $[3, \infty)$.

4. a. $(f \circ g)(x) = f(g(x))$

$= 5(2x^2 - x - 1) + 6$

$= 10x^2 - 5x - 5 + 6$

$= 10x^2 - 5x + 1$

b. $(g \circ f)(x) = g(f(x))$

$= 2(5x + 6)^2 - (5x + 6) - 1$

$= 2(25x^2 + 60x + 36) - 5x - 6 - 1$

$= 50x^2 + 120x + 72 - 5x - 6 - 1$

$= 50x^2 + 115x + 65$

c. $(f \circ g)(x) = 10x^2 - 5x + 1$

$(f \circ g)(-1) = 10(-1)^2 - 5(-1) + 1$

$= 10 + 5 + 1$

$= 16$

5. a. $(f \circ g)(x) = \dfrac{4}{\dfrac{1}{x} + 2} = \dfrac{4x}{1 + 2x}$

b. domain: $\left\{ x \middle| x \neq 0, \ x \neq -\dfrac{1}{2} \right\}$

6. $h(x) = f \circ g$ where $f(x) = \sqrt{x}; \ \ g(x) = x^2 + 5$

Exercise Set 1.7

1. The function contains neither division nor an even root. The domain $= (-\infty, \infty)$

3. The denominator equals zero when $x = 4$. This value must be excluded from the domain.
domain: $(-\infty, 4) \cup (4, \infty)$.

5. The function contains neither division nor an even root. The domain $= (-\infty, \infty)$

7. The values that make the denominator equal zero must be excluded from the domain.
domain: $(-\infty, -3) \cup (-3, 5) \cup (5, \infty)$

9. The values that make the denominators equal zero must be excluded from the domain.
domain: $(-\infty, -7) \cup (-7, 9) \cup (9, \infty)$

11. The first denominator cannot equal zero. The values that make the second denominator equal zero must be excluded from the domain.
domain: $(-\infty, -1) \cup (-1, 1) \cup (1, \infty)$

13. Exclude x for $x = 0$.

Exclude x for $\dfrac{3}{x} - 1 = 0$.

$\dfrac{3}{x} - 1 = 0$

$x\left(\dfrac{3}{x} - 1\right) = x(0)$

$3 - x = 0$

$-x = -3$

$x = 3$

domain: $(-\infty, 0) \cup (0, 3) \cup (3, \infty)$

15. Exclude x for $x - 1 = 0$.

$x - 1 = 0$

$x = 1$

Exclude x for $\dfrac{4}{x - 1} - 2 = 0$.

$\dfrac{4}{x - 1} - 2 = 0$

$(x - 1)\left(\dfrac{4}{x - 1} - 2\right) = (x - 1)(0)$

$4 - 2(x - 1) = 0$

$4 - 2x + 2 = 0$

$-2x + 6 = 0$

$-2x = -6$

$x = 3$

domain: $(-\infty, 1) \cup (1, 3) \cup (3, \infty)$

17. The expression under the radical must not be negative.

$x - 3 \geq 0$

$x \geq 3$

domain: $[3, \infty)$

19. The expression under the radical must be positive.

$x - 3 > 0$

$x > 3$

domain: $(3, \infty)$

21. The expression under the radical must not be negative.
$$5x + 35 \geq 0$$
$$5x \geq -35$$
$$x \geq -7$$
domain: $[-7, \infty)$

23. The expression under the radical must not be negative.
$$24 - 2x \geq 0$$
$$-2x \geq -24$$
$$\frac{-2x}{-2} \leq \frac{-24}{-2}$$
$$x \leq 12$$
domain: $(-\infty, 12]$

25. The expressions under the radicals must not be negative.
$$x - 2 \geq 0 \quad \text{and} \quad x + 3 \geq 0$$
$$x \geq 2 \qquad\qquad x \geq -3$$
To make both inequalities true, $x \geq 2$.
domain: $[2, \infty)$

27. The expression under the radical must not be negative.
$$x - 2 \geq 0$$
$$x \geq 2$$
The denominator equals zero when $x = 5$.
domain: $[2, 5) \cup (5, \infty)$.

29. Find the values that make the denominator equal zero and must be excluded from the domain.
$$x^3 - 5x^2 - 4x + 20$$
$$= x^2(x - 5) - 4(x - 5)$$
$$= (x - 5)(x^2 - 4)$$
$$= (x - 5)(x + 2)(x - 2)$$
-2, 2, and 5 must be excluded.
domain: $(-\infty, -2) \cup (-2, 2) \cup (2, 5) \cup (5, \infty)$

31. $(f + g)(x) = 3x + 2$
domain: $(-\infty, \infty)$
$(f - g)(x) = f(x) - g(x)$
$$= (2x + 3) - (x - 1)$$
$$= x + 4$$
domain: $(-\infty, \infty)$
$(fg)(x) = f(x) \cdot g(x)$
$$= (2x + 3) \cdot (x - 1)$$
$$= 2x^2 + x - 3$$
domain: $(-\infty, \infty)$
$$\left(\frac{f}{g}\right)(x) = \frac{f(x)}{g(x)} = \frac{2x + 3}{x - 1}$$
domain: $(-\infty, 1) \cup (1, \infty)$

33. $(f + g)(x) = 3x^2 + x - 5$
domain: $(-\infty, \infty)$
$(f - g)(x) = -3x^2 + x - 5$
domain: $(-\infty, \infty)$
$(fg)(x) = (x - 5)(3x^2) = 3x^3 - 15x^2$
domain: $(-\infty, \infty)$
$$\left(\frac{f}{g}\right)(x) = \frac{x - 5}{3x^2}$$
domain: $(-\infty, 0) \cup (0, \infty)$

35. $(f + g)(x) = 2x^2 - 2$
domain: $(-\infty, \infty)$
$(f - g)(x) = 2x^2 - 2x - 4$
domain: $(-\infty, \infty)$
$(fg)(x) = (2x^2 - x - 3)(x + 1)$
$$= 2x^3 + x^2 - 4x - 3$$
domain: $(-\infty, \infty)$
$$\left(\frac{f}{g}\right)(x) = \frac{2x^2 - x - 3}{x + 1}$$
$$= \frac{(2x - 3)(x + 1)}{(x + 1)} = 2x - 3$$
domain: $(-\infty, -1) \cup (-1, \infty)$

37. $(f+g)(x) = (3-x^2)+(x^2+2x-15)$
$= 2x-12$
domain: $(-\infty, \infty)$
$(f-g)(x) = (3-x^2)-(x^2+2x-15)$
$= -2x^2-2x+18$
domain: $(-\infty, \infty)$
$(fg)(x) = (3-x^2)(x^2+2x-15)$
$= -x^4-2x^3+18x^2+6x-45$
domain: $(-\infty, \infty)$
$\left(\dfrac{f}{g}\right)(x) = \dfrac{3-x^2}{x^2+2x-15}$
domain: $(-\infty,-5)\cup(-5,3)\cup(3,\infty)$

39. $(f+g)(x) = \sqrt{x}+x-4$
domain: $[0,\infty)$
$(f-g)(x) = \sqrt{x}-x+4$
domain: $[0,\infty)$
$(fg)(x) = \sqrt{x}(x-4)$
domain: $[0,\infty)$
$\left(\dfrac{f}{g}\right)(x) = \dfrac{\sqrt{x}}{x-4}$
domain: $[0,4)\cup(4,\infty)$

41. $(f+g)(x) = 2+\dfrac{1}{x}+\dfrac{1}{x} = 2+\dfrac{2}{x} = \dfrac{2x+2}{x}$
domain: $(-\infty,0)\cup(0,\infty)$
$(f-g)(x) = 2+\dfrac{1}{x}-\dfrac{1}{x} = 2$
domain: $(-\infty,0)\cup(0,\infty)$
$(fg)(x) = \left(2+\dfrac{1}{x}\right)\cdot\dfrac{1}{x} = \dfrac{2}{x}+\dfrac{1}{x^2} = \dfrac{2x+1}{x^2}$
domain: $(-\infty,0)\cup(0,\infty)$
$\left(\dfrac{f}{g}\right)(x) = \dfrac{2+\frac{1}{x}}{\frac{1}{x}} = \left(2+\dfrac{1}{x}\right)\cdot x = 2x+1$
domain: $(-\infty,0)\cup(0,\infty)$

43. $(f+g)(x) = f(x)+g(x)$
$= \dfrac{5x+1}{x^2-9}+\dfrac{4x-2}{x^2-9}$
$= \dfrac{9x-1}{x^2-9}$
domain: $(-\infty,-3)\cup(-3,3)\cup(3,\infty)$
$(f-g)(x) = f(x)-g(x)$
$= \dfrac{5x+1}{x^2-9}-\dfrac{4x-2}{x^2-9}$
$= \dfrac{x+3}{x^2-9}$
$= \dfrac{1}{x-3}$
domain: $(-\infty,-3)\cup(-3,3)\cup(3,\infty)$
$(fg)(x) = f(x)\cdot g(x)$
$= \dfrac{5x+1}{x^2-9}\cdot\dfrac{4x-2}{x^2-9}$
$= \dfrac{(5x+1)(4x-2)}{\left(x^2-9\right)^2}$
domain: $(-\infty,-3)\cup(-3,3)\cup(3,\infty)$
$\left(\dfrac{f}{g}\right)(x) = \dfrac{\frac{5x+1}{x^2-9}}{\frac{4x-2}{x^2-9}}$
$= \dfrac{5x+1}{x^2-9}\cdot\dfrac{x^2-9}{4x-2}$
$= \dfrac{5x+1}{4x-2}$
The domain must exclude -3, 3, and any values that make $4x-2=0$.
$4x-2=0$
$4x=2$
$x=\dfrac{1}{2}$
domain: $(-\infty,-3)\cup\left(-3,\tfrac{1}{2}\right)\cup\left(\tfrac{1}{2},3\right)\cup(3,\infty)$

45. $(f+g)(x) = \sqrt{x+4} + \sqrt{x-1}$
domain: $[1, \infty)$
$(f-g)(x) = \sqrt{x+4} - \sqrt{x-1}$
domain: $[1, \infty)$
$(fg)(x) = \sqrt{x+4} \cdot \sqrt{x-1} = \sqrt{x^2+3x-4}$
domain: $[1, \infty)$
$\left(\dfrac{f}{g}\right)(x) = \dfrac{\sqrt{x+4}}{\sqrt{x-1}}$
domain: $(1, \infty)$

47. $(f+g)(x) = \sqrt{x-2} + \sqrt{2-x}$
domain: $\{2\}$
$(f-g)(x) = \sqrt{x-2} - \sqrt{2-x}$
domain: $\{2\}$
$(fg)(x) = \sqrt{x-2} \cdot \sqrt{2-x} = \sqrt{-x^2+4x-4}$
domain: $\{2\}$
$\left(\dfrac{f}{g}\right)(x) = \dfrac{\sqrt{x-2}}{\sqrt{2-x}}$
domain: $\varnothing$

49. $f(x) = 2x; g(x) = x + 7$

a. $(f \circ g)(x) = 2(x+7) = 2x+14$

b. $(g \circ f)(x) = 2x+7$

c. $(f \circ g)(2) = 2(2)+14 = 18$

51. $f(x) = x + 4; g(x) = 2x + 1$

a. $(f \circ g)(x) = (2x+1)+4 = 2x+5$

b. $(g \circ f)(x) = 2(x+4)+1 = 2x+9$

c. $(f \circ g)(2) = 2(2)+5 = 9$

53. $f(x) = 4x - 3; g(x) = 5x^2 - 2$

a. $(f \circ g)(x) = 4(5x^2-2)-3$
$= 20x^2 - 11$

b. $(g \circ f)(x) = 5(4x-3)^2 - 2$
$= 5(16x^2 - 24x + 9) - 2$
$= 80x^2 - 120x + 43$

c. $(f \circ g)(2) = 20(2)^2 - 11 = 69$

55. $f(x) = x^2 + 2; g(x) = x^2 - 2$

a. $(f \circ g)(x) = (x^2-2)^2 + 2$
$= x^4 - 4x^2 + 4 + 2$
$= x^4 - 4x^2 + 6$

b. $(g \circ f)(x) = (x^2+2)^2 - 2$
$= x^4 + 4x^2 + 4 - 2$
$= x^4 + 4x^2 + 2$

c. $(f \circ g)(2) = 2^4 - 4(2)^2 + 6 = 6$

57. $f(x) = 4 - x; \quad g(x) = 2x^2 + x + 5$

a. $(f \circ g)(x) = 4 - \left(2x^2 + x + 5\right)$
$= 4 - 2x^2 - x - 5$
$= -2x^2 - x - 1$

b. $(g \circ f)(x) = 2(4-x)^2 + (4-x) + 5$
$= 2(16 - 8x + x^2) + 4 - x + 5$
$= 32 - 16x + 2x^2 + 4 - x + 5$
$= 2x^2 - 17x + 41$

c. $(f \circ g)(2) = -2(2)^2 - 2 - 1 = -11$

59. $f(x) = \sqrt{x}; \quad g(x) = x - 1$

a. $(f \circ g)(x) = \sqrt{x-1}$

b. $(g \circ f)(x) = \sqrt{x} - 1$

c. $(f \circ g)(2) = \sqrt{2-1} = \sqrt{1} = 1$

61. $f(x) = 2x - 3; \quad g(x) = \dfrac{x+3}{2}$

a. $(f \circ g)(x) = 2\left(\dfrac{x+3}{2}\right) - 3$
$= x + 3 - 3$
$= x$

b. $(g \circ f)(x) = \dfrac{(2x-3)+3}{2} = \dfrac{2x}{2} = x$

c. $(f \circ g)(2) = 2$

63. $f(x) = \dfrac{1}{x};\quad g(x) = \dfrac{1}{x}$

    a. $(f \circ g)(x) = \dfrac{1}{\frac{1}{x}} = x$

    b. $(g \circ f)(x) = \dfrac{1}{\frac{1}{x}} = x$

    c. $(f \circ g)(2) = 2$

65. **a.** $(f \circ g)(x) = f\left(\dfrac{1}{x}\right) = \dfrac{2}{\frac{1}{x}+3}, x \neq 0$

         $= \dfrac{2(x)}{\left(\frac{1}{x}+3\right)(x)}$

         $= \dfrac{2x}{1+3x}$

    b. We must exclude 0 because it is excluded from g.

        We must exclude $-\dfrac{1}{3}$ because it causes the denominator of $f \circ g$ to be 0.

        domain: $\left(-\infty, -\dfrac{1}{3}\right) \cup \left(-\dfrac{1}{3}, 0\right) \cup (0, \infty)$.

67. **a.** $(f \circ g)(x) = f\left(\dfrac{4}{x}\right) = \dfrac{\frac{4}{x}}{\frac{4}{x}+1}$

         $= \dfrac{\left(\frac{4}{x}\right)(x)}{\left(\frac{4}{x}+1\right)(x)}$

         $= \dfrac{4}{4+x}, x \neq -4$

    b. We must exclude 0 because it is excluded from g.
        We must exclude -4 because it causes the denominator of $f \circ g$ to be 0.
        domain: $(-\infty, -4) \cup (-4, 0) \cup (0, \infty)$.

69. **a.** $f \circ g(x) = f(x-2) = \sqrt{x-2}$

    b. The expression under the radical in $f \circ g$ must not be negative.
        $x - 2 \geq 0$
           $x \geq 2$
        domain: $[2, \infty)$.

71. **a.** $(f \circ g)(x) = f(\sqrt{1-x})$

         $= \left(\sqrt{1-x}\right)^2 + 4$

         $= 1 - x + 4$

         $= 5 - x$

    b. The domain of $f \circ g$ must exclude any values that are excluded from g.
        $1 - x \geq 0$
          $-x \geq -1$
           $x \leq 1$
        domain: $(-\infty, 1]$.

73. $f(x) = x^4;\quad g(x) = 3x - 1$

75. $f(x) = \sqrt[3]{x};\quad g(x) = x^2 - 9$

77. $f(x) = |x|;\quad g(x) = 2x - 5$

79. $f(x) = \dfrac{1}{x};\quad g(x) = 2x - 3$

81. $(f+g)(-3) = f(-3) + g(-3) = 4 + 1 = 5$

83. $(fg)(2) = f(2)g(2) = (-1)(1) = -1$

85. The domain of $f + g$ is $[-4, 3]$.

87. The graph of $f + g$

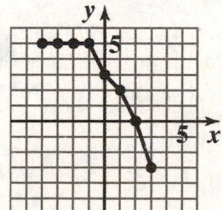

89. $(f \circ g)(-1) = f(g(-1)) = f(-3) = 1$

91. $(g \circ f)(0) = g(f(0)) = g(2) = -6$

93.
$$(f \circ g)(x) = 7$$
$$2(x^2 - 3x + 8) - 5 = 7$$
$$2x^2 - 6x + 16 - 5 = 7$$
$$2x^2 - 6x + 11 = 7$$
$$2x^2 - 6x + 4 = 0$$
$$x^2 - 3x + 2 = 0$$
$$(x - 1)(x - 2) = 0$$
$$x - 1 = 0 \quad \text{or} \quad x - 2 = 0$$
$$x = 1 \qquad\qquad x = 2$$

95. a. $(B - D)(x)$
$$= B(x) - D(x)$$
$$= (7.4x^2 - 15x + 4046) - (-3.5x^2 + 20x + 2405)$$
$$= 7.4x^2 - 15x + 4046 + 3.5x^2 - 20x - 2405$$
$$= 10.9x^2 - 35x + 1641$$

b. $(B - D)(x) = 10.9x^2 - 35x + 1641$
$$(B - D)(3) = 10.9(3)^2 - 35(3) + 1641$$
$$= 1634.1$$
The change in population in the U.S. in 2003 was 1634.1 thousand.

c. $(B - D)(x)$ overestimates the actual change in population in the U.S. in 2003 by 0.1 thousand.

97. $(R - C)(20,000)$
$$= 65(20,000) - (600,000 + 45(20,000))$$
$$= -200,000$$
The company lost $200,000 since costs exceeded revenues.
$(R - C)(30,000)$
$$= 65(30,000) - (600,000 + 45(30,000))$$
$$= 0$$
The company broke even.

99. a. *f* gives the price of the computer after a $400 discount. *g* gives the price of the computer after a 25% discount.

b. $(f \circ g)(x) = 0.75x - 400$
This models the price of a computer after first a 25% discount and then a $400 discount.

c. $(g \circ f)(x) = 0.75(x - 400)$
This models the price of a computer after first a $400 discount and then a 25% discount.

d. The function $f \circ g$ models the greater discount, since the 25% discount is taken on the regular price first.

101. – 105. Answers may vary.

107.

$(f \circ g)(x) = \sqrt{2 - \sqrt{x}}$
The domain of g is $[0, \infty)$.
The expression under the radical in $f \circ g$ must not be negative.
$$2 - \sqrt{x} \geq 0$$
$$-\sqrt{x} \geq -2$$
$$\sqrt{x} \leq 2$$
$$x \leq 4$$
domain: $[0, 4]$

109. makes sense

111. does not make sense; Explanations will vary. Sample explanation: The diagram illustrates $g(f(x)) = x^2 + 4$.

113. false; Changes to make the statement true will vary. A sample change is:
$$f(x) = 2x; g(x) = 3x$$
$$(f \circ g)(x) = f(g(x)) = f(3x) = 2(3x) = 6x$$
$$(g \circ f)(x) = g(f(x)) = g(f(x)) = 3(2x) = 6x$$

115. true

117. Answers may vary.

118. $\{(4, -2), (1, -1), (1, 1), (4, 2)\}$
The element 1 in the domain corresponds to two elements in the range.
Thus, the relation is not a function.

119. $x = \dfrac{5}{y} + 4$

$$y(x) = y\left(\dfrac{5}{y} + 4\right)$$

$$xy = 5 + 4y$$

$$xy - 4y = 5$$

$$y(x - 4) = 5$$

$$y = \dfrac{5}{x - 4}$$

$$x = y^2 - 1$$

$$x + 1 = y^2$$

120. $\sqrt{x+1} = \sqrt{y^2}$

$$\sqrt{x+1} = y$$

$$y = \sqrt{x+1}$$

Section 1.8

Check Point Exercises

1. $f\big(g(x)\big) = 4\left(\dfrac{x+7}{4}\right) - 7 = x$

$$g\big(f(x)\big) = \dfrac{(4x-7)+7}{4} = x$$

$$f\big(g(x)\big) = g\big(f(x)\big) = x$$

2. $f(x) = 2x + 7$

Replace $f(x)$ with y:

$y = 2x + 7$

Interchange x and y:

$x = 2y + 7$

Solve for y:

$$x = 2y + 7$$

$$x - 7 = 2y$$

$$\dfrac{x-7}{2} = y$$

Replace y with $f^{-1}(x)$:

$$f^{-1}(x) = \dfrac{x-7}{2}$$

3. $f(x) = 4x^3 - 1$

Replace $f(x)$ with y:

$y = 4x^3 - 1$

Interchange x and y:

$x = 4y^3 - 1$

Solve for y:

$$x = 4y^3 - 1$$

$$x + 1 = 4y^3$$

$$\dfrac{x+1}{4} = y^3$$

$$\sqrt[3]{\dfrac{x+1}{4}} = y$$

Replace y with $f^{-1}(x)$:

$$f^{-1}(x) = \sqrt[3]{\dfrac{x+1}{4}}$$

Alternative form for answer:

$$f(x)^{-1} = \sqrt[3]{\dfrac{x+1}{4}} = \dfrac{\sqrt[3]{x+1}}{\sqrt[3]{4}}$$

$$= \dfrac{\sqrt[3]{x+1}}{\sqrt[3]{4}} \cdot \dfrac{\sqrt[3]{2}}{\sqrt[3]{2}} = \dfrac{\sqrt[3]{2x+2}}{\sqrt[3]{8}}$$

$$= \dfrac{\sqrt[3]{2x+2}}{2}$$

4. $f(x) = \dfrac{3}{x} - 1$

Replace $f(x)$ with y:

$$y = \dfrac{3}{x} - 1$$

Interchange x and y:

$$x = \dfrac{3}{y} - 1$$

Solve for y:

$$x = \dfrac{3}{y} - 1$$

$$xy = 3 - y$$

$$xy + y = 3$$

$$y(x + 1) = 3$$

$$y = \dfrac{3}{x+1}$$

Replace y with $f^{-1}(x)$:

$$f^{-1}(x) = \dfrac{3}{x+1}$$

5. The graphs of (b) and (c) pass the horizontal line test and thus have an inverse.

6. Find points of f^{-1}.

$f(x)$	$f^{-1}(x)$
$(-2,-2)$	$(-2,-2)$
$(-1,0)$	$(0,-1)$
$(1,2)$	$(2,1)$

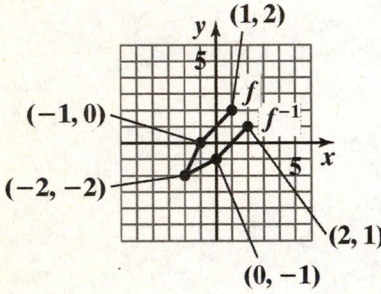

7. $f(x) = x^2 + 1$

Replace $f(x)$ with y:

$y = x^2 + 1$

Interchange x and y:

$x = y^2 + 1$

Solve for y:

$x = y^2 + 1$

$x - 1 = y^2$

$\sqrt{x-1} = y$

Replace y with $f^{-1}(x)$:

$f^{-1}(x) = \sqrt{x-1}$

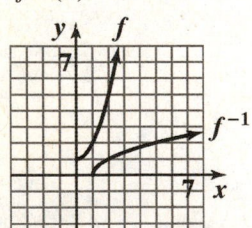

1. $f(x) = 4x;\ g(x) = \dfrac{x}{4}$

$f(g(x)) = 4\left(\dfrac{x}{4}\right) = x$

$g(f(x)) = \dfrac{4x}{4} = x$

f and g are inverses.

3. $f(x) = 3x + 8;\ g(x) = \dfrac{x-8}{3}$

$f(g(x)) = 3\left(\dfrac{x-8}{3}\right) + 8 = x - 8 + 8 = x$

$g(f(x)) = \dfrac{(3x+8)-8}{3} = \dfrac{3x}{3} = x$

f and g are inverses.

5. $f(x) = 5x - 9;\ g(x) = \dfrac{x+5}{9}$

$f(g(x)) = 5\left(\dfrac{x+5}{9}\right) - 9$

$\qquad = \dfrac{5x+25}{9} - 9$

$\qquad = \dfrac{5x-56}{9}$

$g(f(x)) = \dfrac{5x-9+5}{9} = \dfrac{5x-4}{9}$

f and g are not inverses.

7. $f(x) = \dfrac{3}{x-4};\ g(x) = \dfrac{3}{x} + 4$

$f(g(x)) = \dfrac{3}{\frac{3}{x}+4-4} = \dfrac{3}{\frac{3}{x}} = x$

$g(f(x)) = \dfrac{3}{\frac{3}{x-4}} + 4$

$\qquad = 3 \cdot \left(\dfrac{x-4}{3}\right) + 4$

$\qquad = x - 4 + 4$

$\qquad = x$

f and g are inverses.

9. $f(x) = -x;\ g(x) = -x$

$f(g(x)) = -(-x) = x$

$g(f(x)) = -(-x) = x$

f and g are inverses.

11. a.
$$f(x) = x + 3$$
$$y = x + 3$$
$$x = y + 3$$
$$y = x - 3$$
$$f^{-1}(x) = x - 3$$

b.
$$f(f^{-1}(x)) = x - 3 + 3 = x$$
$$f^{-1}(f(x)) = x + 3 - 3 = x$$

13. a.
$$f(x) = 2x$$
$$y = 2x$$
$$x = 2y$$
$$y = \frac{x}{2}$$
$$f^{-1}(x) = \frac{x}{2}$$

b.
$$f(f^{-1}(x)) = 2\left(\frac{x}{2}\right) = x$$
$$f^{-1}(f(x)) = \frac{2x}{2} = x$$

15. a.
$$f(x) = 2x + 3$$
$$y = 2x + 3$$
$$x = 2y + 3$$
$$x - 3 = 2y$$
$$y = \frac{x - 3}{2}$$
$$f^{-1}(x) = \frac{x - 3}{2}$$

b.
$$f(f^{-1}(x)) = 2\left(\frac{x - 3}{2}\right) + 3$$
$$= x - 3 + 3$$
$$= x$$
$$f^{-1}(f(x)) = \frac{2x + 3 - 3}{2} = \frac{2x}{2} = x$$

17. a.
$$f(x) = x^3 + 2$$
$$y = x^3 + 2$$
$$x = y^3 + 2$$
$$x - 2 = y^3$$
$$y = \sqrt[3]{x - 2}$$
$$f^{-1}(x) = \sqrt[3]{x - 2}$$

b.
$$f(f^{-1}(x)) = \left(\sqrt[3]{x - 2}\right)^3 + 2$$
$$= x - 2 + 2$$
$$= x$$
$$f^{-1}(f(x)) = \sqrt[3]{x^3 + 2 - 2} = \sqrt[3]{x^3} = x$$

19. a.
$$f(x) = (x + 2)^3$$
$$y = (x + 2)^3$$
$$x = (y + 2)^3$$
$$\sqrt[3]{x} = y + 2$$
$$y = \sqrt[3]{x} - 2$$
$$f^{-1}(x) = \sqrt[3]{x} - 2$$

b.
$$f(f^{-1}(x)) = \left(\sqrt[3]{x} - 2 + 2\right)^3 = \left(\sqrt[3]{x}\right)^3 = x$$
$$f^{-1}(f(x)) = \sqrt[3]{(x + 2)^3} - 2$$
$$= x + 2 - 2$$
$$= x$$

21. a.
$$f(x) = \frac{1}{x}$$
$$y = \frac{1}{x}$$
$$x = \frac{1}{y}$$
$$xy = 1$$
$$y = \frac{1}{x}$$
$$f^{-1}(x) = \frac{1}{x}$$

b.
$$f(f^{-1}(x)) = \frac{1}{\frac{1}{x}} = x$$
$$f^{-1}(f(x)) = \frac{1}{\frac{1}{x}} = x$$

23. a.
$$f(x) = \sqrt{x}$$
$$y = \sqrt{x}$$
$$x = \sqrt{y}$$
$$y = x^2$$
$$f^{-1}(x) = x^2, x \geq 0$$

b.
$$f(f^{-1}(x)) = \sqrt{x^2} = |x| = x \text{ for } x \geq 0.$$
$$f^{-1}(f(x)) = (\sqrt{x})^2 = x$$

25. a.

$$f(x) = \frac{7}{x} - 3$$

$$y = \frac{7}{x} - 3$$

$$x = \frac{7}{y} - 3$$

$$xy = 7 - 3y$$

$$xy + 3y = 7$$

$$y(x+3) = 7$$

$$y = \frac{7}{x+3}$$

$$f^{-1}(x) = \frac{7}{x+3}$$

b.

$$f\left(f^{-1}(x)\right) = \frac{7}{\frac{7}{x+3}} - 3 = x$$

$$f^{-1}\left(f(x)\right) = \frac{7}{\frac{7}{x} - 3 + 3} = x$$

27. a.

$$f(x) = \frac{2x+1}{x-3}$$

$$y = \frac{2x+1}{x-3}$$

$$x = \frac{2y+1}{y-3}$$

$$x(y-3) = 2y+1$$

$$xy - 3x = 2y + 1$$

$$xy - 2y = 3x + 1$$

$$y(x-2) = 3x + 1$$

$$y = \frac{3x+1}{x-2}$$

$$f^{-1}(x) = \frac{3x+1}{x-2}$$

b.

$$f(f^{-1}(x)) = \frac{2\left(\frac{3x+1}{x-2}\right)+1}{\frac{3x+1}{x-2}-3}$$

$$= \frac{2(3x+1)+x-2}{3x+1-3(x-2)} = \frac{6x+2+x-2}{3x+1-3x+6}$$

$$= \frac{7x}{7} = x$$

$$f^{-1}(f(x)) = \frac{3\left(\frac{2x+1}{x-3}\right)+1}{\frac{2x+1}{x-3}-2}$$

$$= \frac{3(2x+1)+x-3}{2x+1-2(x-3)}$$

$$= \frac{6x+3+x-3}{2x+1-2x+6} = \frac{7x}{7} = x$$

29. The function fails the horizontal line test, so it does not have an inverse function.

31. The function fails the horizontal line test, so it does not have an inverse function.

33. The function passes the horizontal line test, so it does have an inverse function.

35.

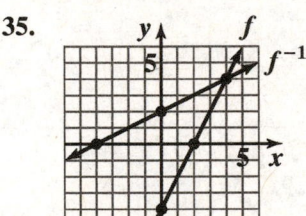

37.

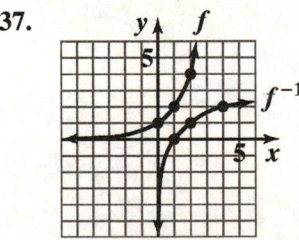

39. a.
$$f(x) = 2x - 1$$
$$y = 2x - 1$$
$$x = 2y - 1$$
$$x + 1 = 2y$$
$$\frac{x+1}{2} = y$$
$$f^{-1}(x) = \frac{x+1}{2}$$

b.

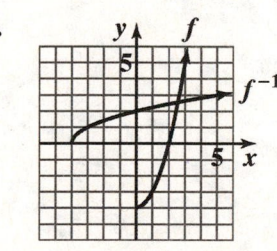

c. domain of f : $(-\infty, \infty)$

range of f : $(-\infty, \infty)$

domain of f^{-1} : $(-\infty, \infty)$

range of f^{-1} : $(-\infty, \infty)$

41. a.
$$f(x) = x^2 - 4$$
$$y = x^2 - 4$$
$$x = y^2 - 4$$
$$x + 4 = y^2$$
$$\sqrt{x+4} = y$$
$$f^{-1}(x) = \sqrt{x+4}$$

b.

c. domain of f : $[0, \infty)$

range of f : $[-4, \infty)$

domain of f^{-1} : $[-4, \infty)$

range of f^{-1} : $[0, \infty)$

43. a.
$$f(x) = (x-1)^2$$
$$y = (x-1)^2$$
$$x = (y-1)^2$$
$$-\sqrt{x} = y - 1$$
$$-\sqrt{x} + 1 = y$$
$$f^{-1}(x) = 1 - \sqrt{x}$$

b.

c. domain of f : $(-\infty, 1]$

range of f : $[0, \infty)$

domain of f^{-1} : $[0, \infty)$

range of f^{-1} : $(-\infty, 1]$

45. a.
$$f(x) = x^3 - 1$$
$$y = x^3 - 1$$
$$x = y^3 - 1$$
$$x + 1 = y^3$$
$$\sqrt[3]{x+1} = y$$
$$f^{-1}(x) = \sqrt[3]{x+1}$$

b.

c. domain of f : $(-\infty, \infty)$

range of f : $(-\infty, \infty)$

domain of f^{-1} : $(-\infty, \infty)$

range of f^{-1} : $(-\infty, \infty)$

47. a.
$$f(x) = (x+2)^3$$
$$y = (x+2)^3$$
$$x = (y+2)^3$$
$$\sqrt[3]{x} = y + 2$$
$$\sqrt[3]{x} - 2 = y$$
$$f^{-1}(x) = \sqrt[3]{x} - 2$$

b.

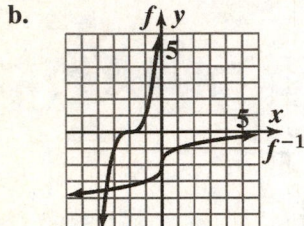

c. domain of f: $(-\infty, \infty)$
range of f: $(-\infty, \infty)$
domain of f^{-1}: $(-\infty, \infty)$
range of f^{-1}: $(-\infty, \infty)$

49. a.
$$f(x) = \sqrt{x-1}$$
$$y = \sqrt{x-1}$$
$$x = \sqrt{y-1}$$
$$x^2 = y - 1$$
$$x^2 + 1 = y$$
$$f^{-1}(x) = x^2 + 1$$

b.

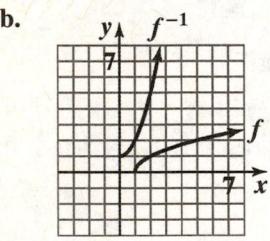

c. domain of f: $[1, \infty)$
range of f: $[0, \infty)$
domain of f^{-1}: $[0, \infty)$
range of f^{-1}: $[1, \infty)$

51. a.
$$f(x) = \sqrt[3]{x} + 1$$
$$y = \sqrt[3]{x} + 1$$
$$x = \sqrt[3]{y} + 1$$
$$x - 1 = \sqrt[3]{y}$$
$$(x-1)^3 = y$$
$$f^{-1}(x) = (x-1)^3$$

b.

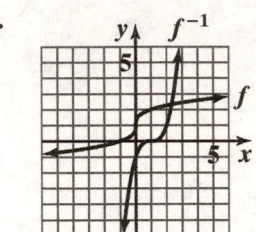

c. domain of f: $(-\infty, \infty)$
range of f: $(-\infty, \infty)$
domain of f^{-1}: $(-\infty, \infty)$
range of f^{-1}: $(-\infty, \infty)$

53. $f(g(1)) = f(1) = 5$

55. $(g \circ f)(-1) = g(f(-1)) = g(1) = 1$

57. $f^{-1}(g(10)) = f^{-1}(-1) = 2$, since $f(2) = -1$.

59.
$$(f \circ g)(0) = f(g(0))$$
$$= f(4 \cdot 0 - 1)$$
$$= f(-1) = 2(-1) - 5 = -7$$

61. Let $f^{-1}(1) = x$. Then
$$f(x) = 1$$
$$2x - 5 = 1$$
$$2x = 6$$
$$x = 3$$
Thus, $f^{-1}(1) = 3$

63.
$$g(f[h(1)]) = g\left(f\left[1^2 + 1 + 2\right]\right)$$
$$= g(f(4))$$
$$= g(2 \cdot 4 - 5)$$
$$= g(3)$$
$$= 4 \cdot 3 - 1 = 11$$

65.   **a.** $\{(17,9.7),(22,8.7),(30,8.4),(40,8.3),(50,8.2),(60,8.3)\}$

     b. $\{(9.7,17),(8.7,22),(8.4,30),(8.3,40),(8.2,50),(8.3,60)\}$

       f is not a one-to-one function because the inverse of f is not a function.

67.   **a.**   It passes the horizontal line test and is one-to-one.

     b. $f^{-1}(0.25) = 15$ If there are 15 people in the room, the probability that 2 of them have the same birthday is 0.25.
       $f^{-1}(0.5) = 21$ If there are 21 people in the room, the probability that 2 of them have the same birthday is 0.5.
       $f^{-1}(0.7) = 30$ If there are 30 people in the room, the probability that 2 of them have the same birthday is 0.7.

69. $f(g(x)) = \dfrac{9}{5}\left[\dfrac{5}{9}(x-32)\right]+32$

             $= x-32+32$

             $= x$

  f and g are inverses.

71. – 75.   Answers may vary.

77.

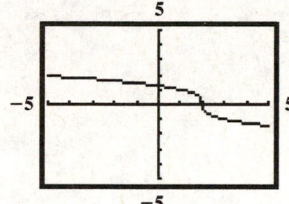

one-to-one

79.

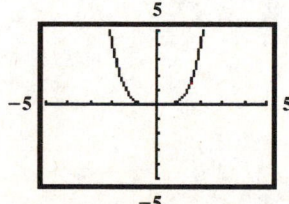

not one-to-one

81.

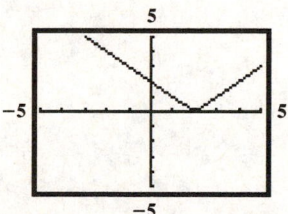

not one-to-one

83.

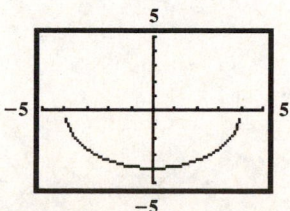

not one-to-one

85.

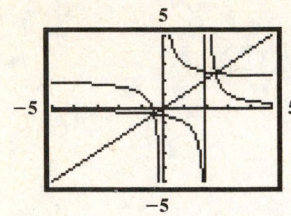

f and g are inverses

87. makes sense

89. makes sense

91. false; Changes to make the statement true will vary. A sample change is: The inverse is $\{(4,1), (7,2)\}$.

93. false; Changes to make the statement true will vary.

A sample change is: $f^{-1}(x) = \dfrac{x}{3}$.

95. $(f \circ g)(x) = 3(x+5) = 3x+15$.

$$y = 3x+15$$
$$x = 3y+15$$
$$y = \dfrac{x-15}{3}$$
$$(f \circ g)^{-1}(x) = \dfrac{x-15}{3}$$

$$g(x) = x+5$$
$$y = x+5$$
$$x = y+5$$
$$y = x-5$$
$$g^{-1}(x) = x-5$$

$$f(x) = 3x$$
$$y = 3x$$
$$x = 3y$$
$$y = \dfrac{x}{3}$$
$$f^{-1}(x) = \dfrac{x}{3}$$

$$\left(g^{-1} \circ f^{-1}\right)(x) = \dfrac{x}{3} - 5 = \dfrac{x-15}{3}$$

97. No, there will be 2 times when the spacecraft is at the same height, when it is going up and when it is coming down.

99. Answers may vary.

100. $\sqrt{(x_2 - x_1)^2 + (y_2 - y_1)^2} = \sqrt{(1-7)^2 + (-1-2)^2}$

$$= \sqrt{(-6)^2 + (-3)^2}$$
$$= \sqrt{36+9}$$
$$= \sqrt{45}$$
$$= 3\sqrt{5}$$

101.

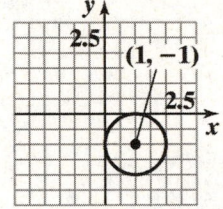

102. $y^2 - 6y - 4 = 0$

$$y^2 - 6y \quad = 4$$
$$y^2 - 6y + 9 = 4 + 9$$
$$(y-3)^2 = 13$$
$$y - 3 = \pm\sqrt{13}$$
$$y = 3 \pm \sqrt{13}$$

Section 1.9

Check Point Exercises

1. $d = \sqrt{(x_2 - x_1)^2 + (y_2 - y_1)^2}$

$d = \sqrt{(1-(-4))^2 + (-3-9)^2}$

$\quad = \sqrt{(5)^2 + (-12)^2}$

$\quad = \sqrt{25 + 144}$

$\quad = \sqrt{169}$

$\quad = 13$

2. $\left(\dfrac{1+7}{2}, \dfrac{2+(-3)}{2}\right) = \left(\dfrac{8}{2}, \dfrac{-1}{2}\right) = \left(4, -\dfrac{1}{2}\right)$

3. $h = 0,\ k = 0,\ r = 4;$

$(x-0)^2 + (y-0)^2 = 4^2$

$\qquad x^2 + y^2 = 16$

4. $h = 0,\ k = -6,\ r = 10;$

$(x-0)^2 + [y-(-6)]^2 = 10^2$

$\quad (x-0)^2 + (y+6)^2 = 100$

$\qquad\quad x^2 + (y+6)^2 = 100$

5. a. $(x+3)^2 + (y-1)^2 = 4$

$[x-(-3)]^2 + (y-1)^2 = 2^2$

So in the standard form of the circle's equation

$(x-h)^2 + (y-k)^2 = r^2,$

we have $h = -3,\ k = 1,\ r = 2.$

center: $(h, k) = (-3, 1)$

radius: $r = 2$

b.

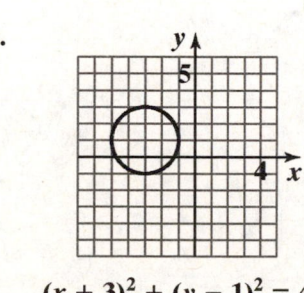

$(x + 3)^2 + (y - 1)^2 = 4$

c. domain: $[-5, -1]$

range: $[-1, 3]$

6. $x^2 + y^2 + 4x - 4y - 1 = 0$

$x^2 + y^2 + 4x - 4y - 1 = 0$

$(x^2 + 4x\quad) + (y^2 - 4y\quad) = 0$

$(x^2 + 4x + 4) + (y^2 + 4y + 4) = 1 + 4 + 4$

$(x+2)^2 + (y-2)^2 = 9$

$[x-(-x)]^2 + (y-2)^2 = 3^2$

So in the standard form of the circle's equation

$(x-h)^2 + (y-k)^2 = r^2$, we have

$h = -2,\ k = 2,\ r = 3$.

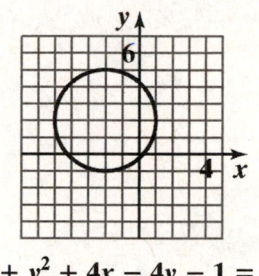

$x^2 + y^2 + 4x - 4y - 1 = 0$

Exercise Set 1.9

1. $d = \sqrt{(14-2)^2 + (8-3)^2}$

$\quad = \sqrt{12^2 + 5^2}$

$\quad = \sqrt{144 + 25}$

$\quad = \sqrt{169}$

$\quad = 13$

3. $d = \sqrt{(-6-4)^2 + (3-(-1))^2}$

$\quad = \sqrt{(-10)^2 + (4)^2}$

$\quad = \sqrt{100 + 16}$

$\quad = \sqrt{116}$

$\quad = 2\sqrt{29}$

$\quad \approx 10.77$

5. $d = \sqrt{(-3-0)^2 + (4-0)^2}$

$\quad = \sqrt{3^2 + 4^2}$

$\quad = \sqrt{9 + 16}$

$\quad = \sqrt{25}$

$\quad = 5$

7. $d = \sqrt{[3-(-2)]^2 + [-4-(-6)]^2}$

$= \sqrt{5^2 + 2^2}$

$= \sqrt{25+4}$

$= \sqrt{29}$

≈ 5.39

9. $d = \sqrt{(4-0)^2 + [1-(-3)]^2}$

$= \sqrt{4^2 + 4^2}$

$= \sqrt{16+16}$

$= \sqrt{32}$

$= 4\sqrt{2}$

≈ 5.66

11. $d = \sqrt{(-.5-3.5)^2 + (6.2-8.2)^2}$

$= \sqrt{(-4)^2 + (-2)^2}$

$= \sqrt{16+4}$

$= \sqrt{20}$

$= 2\sqrt{5}$

≈ 4.47

13. $d = \sqrt{(\sqrt{5}-0)^2 + [0-(-\sqrt{3})]^2}$

$= \sqrt{(\sqrt{5})^2 + (\sqrt{3})^2}$

$= \sqrt{5+3}$

$= \sqrt{8}$

$= 2\sqrt{2}$

≈ 2.83

15. $d = \sqrt{(-\sqrt{3}-3\sqrt{3})^2 + (4\sqrt{5}-\sqrt{5})^2}$

$= \sqrt{(-4\sqrt{3})^2 + (3\sqrt{5})^2}$

$= \sqrt{16(3)+9(5)}$

$= \sqrt{48+45}$

$= \sqrt{93}$

≈ 9.64

17. $d = \sqrt{\left(\dfrac{1}{3}-\dfrac{7}{3}\right)^2 + \left(\dfrac{6}{5}-\dfrac{1}{5}\right)^2}$

$= \sqrt{(-2)^2 + 1^2}$

$= \sqrt{4+1}$

$= \sqrt{5}$

≈ 2.24

19. $\left(\dfrac{6+2}{2}, \dfrac{8+4}{2}\right) = \left(\dfrac{8}{2}, \dfrac{12}{2}\right) = (4,6)$

21. $\left(\dfrac{-2+(-6)}{2}, \dfrac{-8+(-2)}{2}\right)$

$= \left(\dfrac{-8}{2}, \dfrac{-10}{2}\right) = (-4,-5)$

23. $\left(\dfrac{-3+6}{2}, \dfrac{-4+(-8)}{2}\right)$

$= \left(\dfrac{3}{2}, \dfrac{-12}{2}\right) = \left(\dfrac{3}{2}, -6\right)$

25. $\left(\dfrac{\dfrac{-7}{2}+\left(-\dfrac{5}{2}\right)}{2}, \dfrac{\dfrac{3}{2}+\left(-\dfrac{11}{2}\right)}{2}\right)$

$= \left(\dfrac{\dfrac{-12}{2}}{2}, \dfrac{\dfrac{-8}{2}}{2}\right) = \left(-\dfrac{6}{2}, \dfrac{-4}{2}\right) = (-3,-2)$

27. $\left(\dfrac{8+(-6)}{2}, \dfrac{3\sqrt{5}+7\sqrt{5}}{2}\right)$

$= \left(\dfrac{2}{2}, \dfrac{10\sqrt{5}}{2}\right) = \left(1,5\sqrt{5}\right)$

29. $\left(\dfrac{\sqrt{18}+\sqrt{2}}{2}, \dfrac{-4+4}{2}\right)$

$= \left(\dfrac{3\sqrt{2}+\sqrt{2}}{2}, \dfrac{0}{2}\right) = \left(\dfrac{4\sqrt{2}}{2},0\right) = (2\sqrt{2},0)$

31. $(x-0)^2 + (y-0)^2 = 7^2$

$x^2 + y^2 = 49$

33. $(x-3)^2 + (y-2)^2 = 5^2$

$(x-3)^2 + (y-2)^2 = 25$

35. $[x-(-1)]^2+(y-4)^2=2^2$

$(x+1)^2+(y-4)^2=4$

37. $[x-(-3)]^2+[y-(-1)]^2=(\sqrt{3})^2$

$(x+3)^2+(y+1)^2=3$

39. $[x-(-4)]^2+(y-0)^2=10^2$

$(x+4)^2+(y-0)^2=100$

41. $\qquad x^2+y^2=16$

$(x-0)^2+(y-0)^2=y^2$

$h=0,\ k=0,\ r=4;$

center $=(0,0);$ radius $=4$

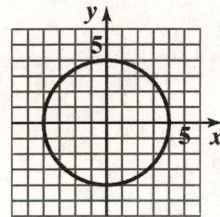

$x^2+y^2=16$

domain: $[-4,4]$

range: $[-4,4]$

43. $(x-3)^2+(y-1)^2=36$

$(x-3)^2+(y-1)^2=6^2$

$h=3,\ k=1,\ r=6;$

center $=(3,1);$ radius $=6$

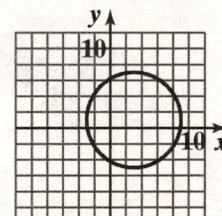

$(x-3)^2+(y-1)^2=36$

domain: $[-3,9]$

range: $[-5,7]$

45. $\qquad (x+3)^2+(y-2)^2=4$

$[x-(-3)]^2+(y-2)^2=2^2$

$h=-3,\ k=2,\ r=2$

center $=(-3,2);$ radius $=2$

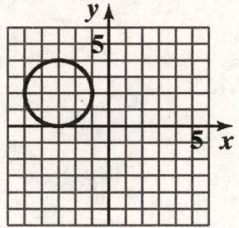

$(x+3)^2+(y-2)^2=4$

domain: $[-5,-1]$

range: $[0,4]$

47. $\qquad (x+2)^2+(y+2)^2=4$

$[x-(-2)]^2+[y-(-2)]^2=2^2$

$h=-2,\ k=-2,\ r=2$

center $=(-2,-2);$ radius $=2$

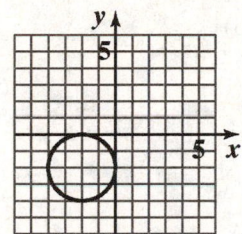

$(x+2)^2+(y+2)^2=4$

domain: $[-4,0]$

range: $[-4,0]$

49. $x^2+(y-1)^2=1$

$h=0,k=1,r=1;$

center $=(0,1);$ radius $=1$

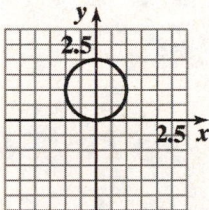

domain: $[-1,1]$

range: $[0,2]$

51. $(x+1)^2 + y^2 = 25$

$h = -1, k = 0, r = 5;$

center $= (-1,0);$ radius $= 5$

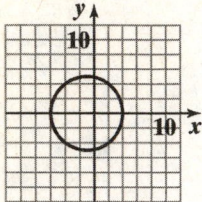

domain: $[-6,4]$

range: $[-5,5]$

53.
$$x^2 + y^2 + 6x + 2y + 6 = 0$$
$$(x^2 + 6x) + (y^2 + 2y) = -6$$
$$(x^2 + 6x + 9) + (y^2 + 2y + 1) = 9 + 1 - 6$$
$$(x+3)^2 + (y+1)^2 = 4$$
$$[x-(-3)]^2 + [9-(-1)]^2 = 2^2$$
center $= (-3, -1);$ radius $= 2$

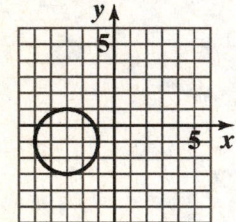

$x^2 + y^2 + 6x + 2y + 6 = 0$

55.
$$x^2 + y^2 - 10x - 6y - 30 = 0$$
$$(x^2 - 10x) + (y^2 - 6y) = 30$$
$$(x^2 - 10x + 25) + (y^2 - 6y + 9) = 25 + 9 + 30$$
$$(x-5)^2 + (y-3)^2 = 64$$
$$(x-5)^2 + (y-3)^2 = 8^2$$
center $= (5, 3);$ radius $= 8$

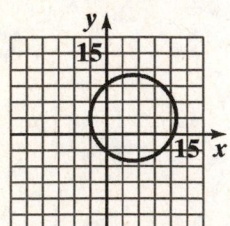

$x^2 + y^2 - 10x - 6y - 30 = 0$

57.
$$x^2 + y^2 + 8x - 2y - 8 = 0$$
$$(x^2 + 8x) + (y^2 - 2y) = 8$$
$$(x^2 + 8x + 16) + (y^2 - 2y + 1) = 16 + 1 + 8$$
$$(x+4)^2 + (y-1)^2 = 25$$
$$[x-(-4)]^2 + (y-1)^2 = 5^2$$
center $= (-4, 1);$ radius $= 5$

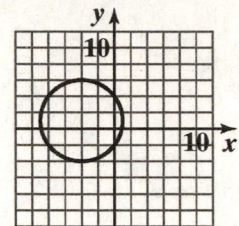

$x^2 + y^2 + 8x - 2y - 8 = 0$

59.
$$x^2 - 2x + y^2 - 15 = 0$$
$$(x^2 - 2x) + y^2 = 15$$
$$(x^2 - 2x + 1) + (y-0)^2 = 1 + 0 + 15$$
$$(x-1)^2 + (y-0)^2 = 16$$
$$(x-1)^2 + (y-0)^2 = 4^2$$
center $= (1, 0);$ radius $= 4$

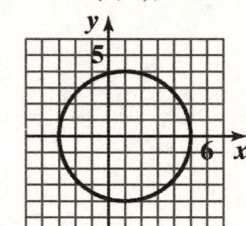

$x^2 - 2x + y^2 - 15 = 0$

142

61.
$$x^2 + y^2 - x + 2y + 1 = 0$$
$$x^2 - x \quad + y^2 + 2y \quad = -1$$
$$x^2 - x + \frac{1}{4} + y^2 + 2y + 1 = -1 + \frac{1}{4} + 1$$
$$\left(x - \frac{1}{2}\right)^2 + (y+1)^2 = \frac{1}{4}$$
center $= \left(\frac{1}{2}, -1\right)$; radius $= \frac{1}{2}$

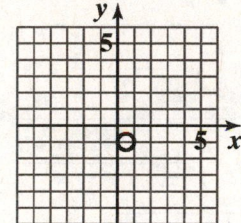

$$x^2 + y^2 - x + 2y + 1 = 0$$

63.
$$x^2 + y^2 + 3x - 2y - 1 = 0$$
$$x^2 + 3x \quad + y^2 - 2y \quad = 1$$
$$x^2 + 3x + \frac{9}{4} + y^2 - 2y + 1 = 1 + \frac{9}{4} + 1$$
$$\left(x + \frac{3}{2}\right)^2 + (y-1)^2 = \frac{17}{4}$$
center $= \left(-\frac{3}{2}, 1\right)$; radius $= \frac{\sqrt{17}}{2}$

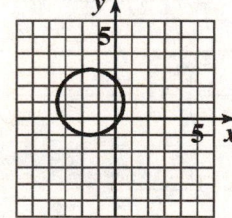

$$x^2 + y^2 + 3x - 2y - 1 = 0$$

65. a. Since the line segment passes through the center, the center is the midpoint of the segment.
$$M = \left(\frac{x_1 + x_2}{2}, \frac{y_1 + y_2}{2}\right)$$
$$= \left(\frac{3+7}{2}, \frac{9+11}{2}\right) = \left(\frac{10}{2}, \frac{20}{2}\right)$$
$$= (5, 10)$$
The center is $(5, 10)$.

b. The radius is the distance from the center to one of the points on the circle. Using the point $(3, 9)$, we get:
$$d = \sqrt{(5-3)^2 + (10-9)^2}$$
$$= \sqrt{2^2 + 1^2} = \sqrt{4+1}$$
$$= \sqrt{5}$$
The radius is $\sqrt{5}$ units.

c.
$$(x-5)^2 + (y-10)^2 = \left(\sqrt{5}\right)^2$$
$$(x-5)^2 + (y-10)^2 = 5$$

67. $x^2 + y^2 = 16$
$x - y = 4$

Intersection points: $(0, -4)$ and $(4, 0)$

Check $(0, -4)$:
$$0^2 + (-4)^2 = 16 \qquad 0 - (-4) = 4$$
$$16 = 16 \text{ true} \qquad 4 = 4 \text{ true}$$

Check $(4, 0)$:
$$4^2 + 0^2 = 16 \qquad 4 - 0 = 4$$
$$16 = 16 \text{ true} \qquad 4 = 4 \text{ true}$$
The solution set is $\{(0, -4), (4, 0)\}$.

69.

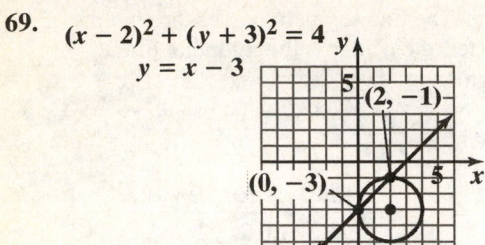

$(x-2)^2 + (y+3)^2 = 4$

$y = x - 3$

Intersection points: $(0,-3)$ and $(2,-1)$

Check $(0,-3)$:

$(0-2)^2 + (-3+3)^2 = 9 \qquad -3 = 0-3$

$(-2)^2 + 0^2 = 4 \qquad\qquad -3 = -3$ true

$4 = 4$

true

Check $(2,-1)$:

$(2-2)^2 + (-1+3)^2 = 4 \qquad -1 = 2-3$

$0^2 + 2^2 = 4 \qquad\qquad\quad -1 = -1$ true

$4 = 4$

true

The solution set is $\{(0,-3),(2,-1)\}$.

71. $d = \sqrt{(8495-4422)^2 + (8720-1241)^2} \cdot \sqrt{0.1}$

$d = \sqrt{72,524,770} \cdot \sqrt{0.1}$

$d \approx 2693$

The distance between Boston and San Francisco is about 2693 miles.

73. If we place L.A. at the origin, then we want the equation of a circle with center at $(-2.4,-2.7)$ and radius 30.

$\left(x-(-2.4)\right)^2 + \left(y-(-2.7)\right)^2 = 30^2$

$(x+2.4)^2 + (y+2.7)^2 = 900$

75. – 81. Answers may vary.

83.

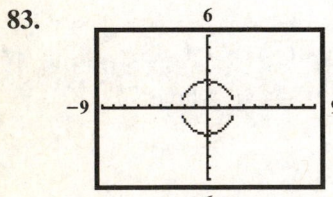

85.

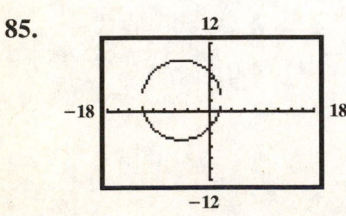

87. makes sense

89. makes sense

91. false; Changes to make the statement true will vary. A sample change is: The center is at $(3, -5)$.

93. false; Changes to make the statement true will vary. A sample change is: Since $r^2 = -36$ this is not the equation of a circle.

95. a. d_1 is distance from (x_1, x_2) to midpoint

$d_1 = \sqrt{\left(\dfrac{x_1+x_2}{2} - x_1\right)^2 + \left(\dfrac{y_1+y_2}{2} - y_1\right)^2}$

$d_1 = \sqrt{\left(\dfrac{x_1+x_2-2x_1}{2}\right)^2 + \left(\dfrac{y_1+y_2-2y_1}{2}\right)^2}$

$d_1 = \sqrt{\left(\dfrac{x_2-x_1}{2}\right)^2 + \left(\dfrac{y_2-y_1}{2}\right)^2}$

$d_1 = \sqrt{\dfrac{x_2 - 2x_1 x_2 + x_1^2}{4} + \dfrac{y_2^2 - 2y_2 y_1 + y_1^2}{4}}$

$d_1 = \sqrt{\dfrac{1}{4}\left(x_2 - 2x_1 x_2 + x_1 + y_2^2 - 2y_2 y_1 + y_1^2\right)}$

$d_1 = \dfrac{1}{2}\sqrt{x_2 - 2x_1 x_2 + x_1 + y_2^2 - 2y_2 y_1 + y_1^2}$

d_2 is distance from midpoint to (x_2, y_2)

$d_2 = \sqrt{\left(\dfrac{x_1+x_2}{2} - x_2\right)^2 + \left(\dfrac{y_1+y_2}{2} - y_2\right)^2}$

$d_2 = \sqrt{\left(\dfrac{x_1+x_2-2x_2}{2}\right)^2 + \left(\dfrac{y_1+y_2-2y_2}{2}\right)^2}$

$d_2 = \sqrt{\left(\dfrac{x_1-x_2}{2}\right)^2 + \left(\dfrac{y_1-y_2}{2}\right)^2}$

$d_2 = \sqrt{\dfrac{x_1^2 - 2x_1 x_2 + x_2^2}{4} + \dfrac{y_1^2 - 2y_2 y_1 + y_2^2}{4}}$

$d_2 = \sqrt{\dfrac{1}{4}\left(x_1^2 - 2x_1 x_2 + x_2^2 + y_1^2 - 2y_2 y_1 + y_2^2\right)}$

$d_2 = \dfrac{1}{2}\sqrt{x_1^2 - 2x_1 x_2 + x_2^2 + y_1^2 - 2y_2 y_1 + y_2^2}$

$d_1 = d_2$

b. d_3 is the distance from (x_1, y_1) to (x_2, y_2)

$d_3 = \sqrt{(x_2-x_1)^2 + (y_2-y_1)^2}$

$d_3 = \sqrt{x_2^2 - 2x_1 x_2 + x_1^2 + y_2^2 - 2y_2 y_1 + y_1^2}$

$d_1 + d_2 = d_3$ because $\dfrac{1}{2}\sqrt{a} + \dfrac{1}{2}\sqrt{a} = \sqrt{a}$

97. The circle is centered at (0,0). The slope of the radius with endpoints (0,0) and (3,–4) is

$m = -\dfrac{-4-0}{3-0} = -\dfrac{4}{3}$. The line perpendicular to the

radius has slope $\dfrac{3}{4}$. The tangent line has slope $\dfrac{3}{4}$ and passes through (3,–4), so its equation is:

$y + 4 = \dfrac{3}{4}(x-3).$

98. $x - 200$

99.   **a.** $p = 2l + 2w = 2(40) + 2(30) = 140$

$A = lw = (40)(30) = 1200$

The perimeter is 140 yd; the area is 1200 sq yd

b. $p = 2l + 2w = 2(50) + 2(20) = 140$

$A = lw = (50)(20) = 1000$

The perimeter is 140 yd; the area is 1000 sq yd

100. $\pi r^2 h = 22$

$h = \dfrac{22}{\pi r^2}$

$2\pi r^2 + 2\pi rh = 2\pi r^2 + 2\pi r\left(\dfrac{22}{\pi r^2}\right) = 2\pi r^2 + \dfrac{44}{r}$

Section 1.10

Check Point Exercises

1.   **a.** $f(x) = 15 + 0.08x$

b. $g(x) = 3 + 0.12x$

c. $15 + 0.08x = 3 + 0.12x$

$12 = 0.04x$

$300 = x$

The plans cost the same for 300 minutes.

2.   **a.** $N(x) = 8000 - 100(x - 100)$

$= 8000 - 100x + 10000$

$= 18,000 - 100x$

b. $R(x) = (18,000 - 100x)x$

$= -100x^2 + 18,000x$

3. $V(x) = (15 - 2x)(8 - 2x)x$

$= (120 - 46x + 4x^2)x$

$= 4x^3 - 46x^2 + 120x$

Since x represents the inches to be cut off, $x > 0$. The smallest side is 8, so must cut less than 4 off each side. The domain of V is $\{x \mid 0 < x < 4\}$ or, in interval notation, $(0, 4)$.

4. $2l + 2w = 200$

$2l = 200 - 2w$

$l = 100 - w$

Let x = width, then length = $100 - x$

$A(x) = x(100 - x)$

$= 100x - x^2$

5. $V = \pi r^2 h$

$1000 = \pi r^2 h$

$\dfrac{1000}{\pi r^2} = h$

$A = 2\pi r^2 + 2\pi rh$

$= 2\pi r^2 + 2\pi r \cdot \dfrac{1000}{\pi r^2}$

$= 2\pi r^2 + \dfrac{2000}{r}$

6. $I(x) = 0.07x + 0.09(25,000 - x)$

7. $d = \sqrt{(x-0)^2 + (y-0)^2}$

$= \sqrt{x^2 + y^2}$

$y = x^3$

$d = \sqrt{x^2 + \left(x^3\right)^2}$

$= \sqrt{x^2 + x^6}$

Exercise Set 1.10

1. **a.** $f(x) = 200 + 0.15x$

b. $320 = 200 + 0.15x$

$120 = 01.5x$

$800 = x$
800 miles

3. **a.** $M(x) = 239.4 - 0.3x$

b. $180 = 239.4 - 0.3x$

$0.3x = 59.4$

$x = 198$
198 years after 1954, in 2152,
someone will run a 3 minute mile.

5. **a.** $f(x) = 1.25x$

b. $g(x) = 21 + 0.5x$

c. $1.25x = 21 + 0.5x$

$0.75x = 21$

$x = 28$

$f(28) = 1.25(28) = 35$

$g(28) = 21 + 0.5(28) = 35$

If a person crosses the bridge 28 times
the cost will be $35 for both options

7. **a.** $f(x) = 100 + 0.8x$

b. $g(x) = 40 + 0.9x$

c. $100 + 0.8x = 40 + 0.9x$

$60 = 0.1x$

$600 = x$
For $600 worth of merchandise,
your cost is $580 for both plans

9. **a.** $N(x) = 30,000 - 500(x - 20)$

$= 30,000 - 500x + 10000$

$= 40,000 - 500x$

b. $R(x) = (40,000 - 500x)x$

$= -500x^2 + 40,000x$

11. **a.** $N(x) = 9000 + 50(150 - x)$

$= 9000 - 50x + 7500$

$= 16500 - 50x$

b. $R(x) = (16500 - 50x)x$

$= -50x^2 + 16500x$

13. **a.** $Y(x) = 320 - 4(x - 50)$

$= 320 - 4x + 200$

$= 520 - 4x$

b. $T(x) = (520 - 4x)x$

$= -4x^2 + 520x$

15. **a.** $V(x) = (24 - 2x)(24 - 2x)x$

$= (576 - 96x + 4x^2)x$

$= 4x^3 - 96x^2 + 576x$

b. $V(2) = 4(2)^3 - 96(2)^2 + 576(2) = 800$ If
2-inch squares are cut off each corner, the
volume will be 800 square inches.

$V(3) = 4(3)^3 - 96(3)^2 + 576(3) = 972$ If 3-
inch squares are cut off each corner, the
volume will be 972 square inches.

$V(4) = 4(4)^3 - 96(4)^2 + 576(4) = 1024$ If
4-inch squares are cut off each corner, the
volume will be 1024 square inches.

$V(5) = 4(5)^3 - 96(5)^2 + 576(5) = 980$ If 5-
inch squares are cut off each corner, the
volume will be 980 square inches.

$V(6) = 4(6)^3 - 96(6)^2 + 576(6) = 864$ If 6-
inch squares are cut off each corner, the
volume will be 864 square inches.

c. If x is the inches to be cut off, $x > 0$.
Since each side is 24, you must cut less
than 12 inches off each end.
$0 < x < 12$

17. $A(x) = x(20 - 2x)$

$= -2x^2 + 20x$

19. $P(x) = x(66 - x)$

$= -x^2 + 66x$

21. $A(x) = x(400 - x)$

$= -x^2 + 400x$

23. $2w + l = 800$

$l = 800 - 2w$

Let $x = w$

$A(x) = x(800 - 2x)$

$= -2x^2 + 800x$

25. $2x + 3y = 1000$

$\qquad 3y = 1000 - 2x$

$\qquad y = \dfrac{1000 - 2x}{3}$

$\qquad A(x) = x\left(\dfrac{1000 - 2x}{3}\right)$

$\qquad\qquad = \dfrac{x(1000 - 2x)}{3}$

27. $\qquad 2x = $ distance around 2 straight sides

$\qquad \pi 2r = $ distance around 2 curved sides

$\quad 2x + 2\pi r = 440$

$\qquad 2x = 440 - 2\pi r$

$\qquad x = 220 - \pi r$

$\quad A(r) = (220 - \pi r)2r + \pi r^2$

$\qquad\quad = 440r - 2\pi r^2 + \pi r^2$

$\qquad\quad = 440r - \pi r^2$

29. $\qquad xy = 4000$

$\qquad y = \dfrac{4000}{x}$

$\quad C(x) = \left[2x + 2\left(\dfrac{4000}{x}\right)\right]175 + 125x$

$\qquad\quad = 350x + \dfrac{1,400,000}{x} + 125x$

$\qquad\quad = 475x + \dfrac{1,400,000}{x}$

31. $\qquad 10 = x^2 y$

$\qquad \dfrac{10}{x^2} = y$

$\quad A(x) = x^2 + 4\left(x \cdot \dfrac{10}{x^2}\right)$

$\qquad\quad = x^2 + \dfrac{40}{x}$

33. $\qquad 300 = y + 4x$

$\quad 300 - 4x = y^2$

$\quad A(x) = x^2(300 - 4x)$

$\qquad\quad = -4x^3 + 300x^2$

35. a. Let $x = $ amount invested at 15%

$\qquad 50000 - x = $ amount invested at 7%

$\qquad I(x) = 0.15x + 0.07(50000 - x)$

b. $\quad 6000 = 0.15x + 0.07x(50000 - x)$

$\qquad 6000 = 0.15x + 3500 - 0.07x$

$\qquad 2500 = 0.08x$

$\qquad 31250 = x$

$\qquad 50000 - 31250 = 18750$

Invest \$31,250 at 15% and \$18,750 at 7%.

37. Let $x = $ amount invested at 12%

$8000 - x = $ amount invested at 5% loss

$I(x) = 0.12x - 0.05(8000 - x)$

39. $d = \sqrt{(x - 0)^2 + (y - 0)^2}$

$\quad = \sqrt{x^2 + y^2}$

$\quad = \sqrt{x^2 + \left(x^2 - 4\right)^2}$

$\quad = \sqrt{x^2 + x^4 - 8x^2 + 16}$

$\quad = \sqrt{x^4 - 7x^2 + 16}$

41. $d = \sqrt{(x - 1)^2 + y^2}$

$\quad = \sqrt{x^2 - 2x + 1 + \left(\sqrt{x}\right)^2}$

$\quad = \sqrt{x^2 - 2x + 1 + x}$

$\quad = \sqrt{x^2 - x + 1}$

43. a. $A(x) = 2xy$

$\qquad\quad = 2x\sqrt{4 - x^2}$

b. $P(x) = 2(2x) + 2y$

$\qquad\quad = 4x + 2\sqrt{4 - x^2}$

45. 6-foot pole

$$c^2 = 6^2 + x^2$$

$$x = \sqrt{36 + x^2}$$

8-foot pole

$$c^2 = 8^2 + (10 - x)^2$$

$$c = \sqrt{64 + 100 - 20x + x^2}$$

$$c = \sqrt{x^2 - 20x + 164}$$

total length

$$f(x) = \sqrt{36 + x^2} + \sqrt{x^2 - 20x + 164}$$

47. $A(x) = \frac{1}{2}x(x - 5) + \frac{1}{2}x(x + 3)$

$$+ (x + 2)\big[(x - 5) + (x + 3)\big]$$

$$A(x) = \frac{1}{2}x^2 - \frac{5}{2}x + \frac{1}{2}x^2 + \frac{3}{2}x + (x + 2)\big[2x - 2\big]$$

$$A(x) = x^2 - x + 2x^2 + 2x - 4$$

$$A(x) = 3x^2 + x - 4$$

49. $V(x) = (x + 5)(2x + 1)(x + 2) - (x + 5)(3)(x)$

$$V(x) = (x + 5)(2x^2 + 5x + 2) - 3x(x + 5)$$

$$V(x) = 2x^3 + 15x^2 + 27x + 10 - 3x^2 - 15x$$

$$V(x) = 2x^3 + 12x^2 + 12x + 10$$

51. – 61. Answers may vary.

63. does not make sense; Explanations will vary. Sample explanation: This model is not reasonable, as it suggests a per minute charge of \$30.

65. does not make sense; Explanations will vary. Sample explanation: The area of a rectangle is not solely determined by its perimeter. For example: A 4 by 6 rectangle and a 3 by 7 rectangle both have perimeters of 20 units, yet their areas are different from each other.

67. Distance and time rowed:

$$d^2 = 2^2 + x^2$$

$$d = \sqrt{4 + x^2}$$

$$rt = d$$

$$2t = \sqrt{4 + x^2}$$

$$t = \frac{\sqrt{4 + x^2}}{2}$$

Distance and time walked:

$$d = 6 - x$$

$$rt = d$$

$$5t = 6 - x$$

$$t = \frac{6 - x}{5}$$

Total time:

$$T(x) = \frac{\sqrt{4 + x^2}}{2} + \frac{6 - x}{5}$$

69.

$$P = 2h + 2r + \frac{1}{2}(\pi 2r)$$

$$12 = 2h + 2r + \pi r$$

$$12 - 2r - \pi r = 2h$$

$$\frac{12 - 2r - \pi r}{2} = h$$

$$A = \frac{12 - 2r - \pi r}{2} \, 2r + \frac{1}{2}\left(\pi r^2\right)$$

$$= 12r - 2r^2 - \pi r^2 + \frac{1}{2}\pi r^2$$

$$= 12r - 2r^2 - \frac{1}{2}\pi r^2$$

71. $(7 - 3x)(-2 - 5x) = -14 - 35x + 6x + 15x^2$

$$= -14 - 29x + 15x^2$$

or

$$= 15x^2 - 29x - 14$$

72. $\sqrt{18} - \sqrt{8} = \sqrt{9 \cdot 2} - \sqrt{4 \cdot 2}$

$$= 3\sqrt{2} - 2\sqrt{2}$$

$$= \sqrt{2}$$

73. $\dfrac{7 + 4\sqrt{2}}{2 - 5\sqrt{2}} \cdot \dfrac{2 + 5\sqrt{2}}{2 + 5\sqrt{2}} = \dfrac{14 + 35\sqrt{2} + 8\sqrt{2} + 40}{4 + 10\sqrt{2} - 10\sqrt{2} - 50}$

$$= \frac{54 + 43\sqrt{2}}{-46}$$

$$= -\frac{54 + 43\sqrt{2}}{46}$$

Chapter 1 Review Exercises

1.

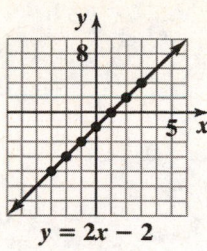

$$y = 2x - 2$$

$x = -3, y = -8$
$x = -2, y = -6$
$x = -1, y = -4$
$x = 0, y = -2$
$x = 1, y = 0$
$x = 2, y = 2$
$x = 3, y = 4$

2.

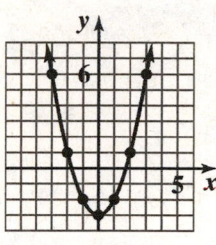

$$y = x^2 - 3$$

$x = -3, y = 6$
$x = -2, y = 1$
$x = -1, y = -2$
$x = 0, y = -3$
$x = 1, y = -2$
$x = 2, y = 1$
$x = 3, y = 6$

3.

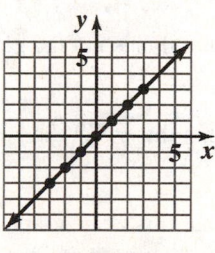

$$y = x$$

$x = -3, y = -3$
$x = -2, y = -2$
$x = -1, y = -1$
$x = 0, y = 0$
$x = 1, y = 1$
$x = 2, y = 2$
$x = 3, y = 3$

4.

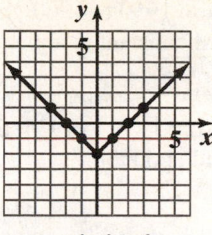

$$y = |x| - 2$$

$x = -3, y = 1$
$x = -2, y = 0$
$x = -1, y = -1$
$x = 0, y = -2$
$x = 1, y = -1$
$x = 2, y = 0$
$x = 3, y = 1$

5. A portion of Cartesian coordinate plane with minimum x-value equal to –20, maximum x-value equal to 40, x-scale equal to 10 and with minimum y-value equal to –5, maximum y-value equal to 5, and y-scale equal to 1.

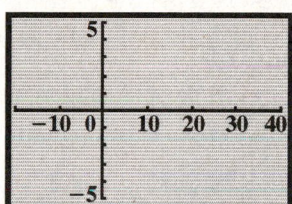

6. x-intercept: –2; The graph intersects the x-axis at (–2, 0).
y-intercept: 2; The graph intersects the y-axis at (0, 2).

7. x-intercepts: 2, –2; The graph intersects the x-axis at (–2, 0) and (2, 0).
y-intercept: –4; The graph intercepts the y-axis at (0, –4).

8. x-intercept: 5; The graph intersects the x-axis at (5, 0).
y-intercept: None; The graph does not intersect the y-axis.

9. The coordinates are (1985, 50%).

10. The top marginal tax rate in 2005 was 35%.

11. The highest marginal tax rate occurred in 1945 and was about 94%.

12. The lowest marginal tax rate occurred in 1990 and was about 28%.

13. During the ten-year period from 1950 to 1960, the top marginal tax rate remained constant at about 91%.

14. During the five-year period from 1930 to 1935, the top marginal tax rate increased about 38%.

15. function
domain: {2, 3, 5}
range: {7}

16. function
domain: {1, 2, 13}
range: {10, 500, π}

17. not a function
domain: {12, 14}
range: {13, 15, 19}

18. $2x + y = 8$

$y = -2x + 8$

Since only one value of y can be obtained for each value of x, y is a function of x.

19. $3x^2 + y = 14$

$y = -3x^2 + 14$

Since only one value of y can be obtained for each value of x, y is a function of x.

20. $2x + y^2 = 6$

$y^2 = -2x + 6$

$y = \pm\sqrt{-2x + 6}$

Since more than one value of y can be obtained from some values of x, y is not a function of x.

21. $f(x) = 5 - 7x$

a. $f(4) = 5 - 7(4) = -23$

b. $f(x + 3) = 5 - 7(x + 3)$

$= 5 - 7x - 21$

$= -7x - 16$

c. $f(-x) = 5 - 7(-x) = 5 + 7x$

22. $g(x) = 3x^2 - 5x + 2$

a. $g(0) = 3(0)^2 - 5(0) + 2 = 2$

b. $g(-2) = 3(-2)^2 - 5(-2) + 2$

$= 12 + 10 + 2$

$= 24$

c. $g(x - 1) = 3(x - 1)^2 - 5(x - 1) + 2$

$= 3(x^2 - 2x + 1) - 5x + 5 + 2$

$= 3x^2 - 11x + 10$

d. $g(-x) = 3(-x)^2 - 5(-x) + 2$

$= 3x^2 + 5x + 2$

23. **a.** $g(13) = \sqrt{13 - 4} = \sqrt{9} = 3$

b. $g(0) = 4 - 0 = 4$

c. $g(-3) = 4 - (-3) = 7$

24. **a.** $f(-2) = \dfrac{(-2)^2 - 1}{-2 - 1} = \dfrac{3}{-3} = -1$

b. $f(1) = 12$

c. $f(2) = \dfrac{2^2 - 1}{2 - 1} = \dfrac{3}{1} = 3$

25. The vertical line test shows that this is not the graph of a function.

26. The vertical line test shows that this is the graph of a function.

27. The vertical line test shows that this is the graph of a function.

28. The vertical line test shows that this is not the graph of a function.

29. The vertical line test shows that this is not the graph of a function.

30. The vertical line test shows that this is the graph of a function.

31. **a.** domain: [-3, 5)

b. range: [-5, 0]

c. x-intercept: -3

d. y-intercept: -2

e.   increasing: $(-2, 0)$ or $(3, 5)$
       decreasing: $(-3, -2)$ or $(0, 3)$

f. $f(-2) = -3$ and $f(3) = -5$

32. **a.**   domain: $(-\infty, \infty)$

       b.   range: $(-\infty, \infty)$

       c. x-intercepts: -2 and 3

       d. y-intercept: 3

       e.   increasing: $(-5, 0)$
              decreasing: $(-\infty, -5)$ or $(0, \infty)$

       f. $f(-2) = 0$ and $f(6) = -3$

33. **a.**   domain: $(-\infty, \infty)$

       b.   range: $[-2, 2]$

       c. x-intercept: 0

       d. y-intercept: 0

       e.   increasing: $(-2, 2)$
              constant: $(-\infty, -2)$ or $(2, \infty)$

       f. $f(-9) = -2$ and $f(14) = 2$

34. **a.**   0, relative maximum -2

       b. $-2, 3$, relative minimum $-3, -5$

35. **a.**   0, relative maximum 3

       b. -5, relative minimum -6

36. $f(x) = x^3 - 5x$
       $f(-x) = (-x)^3 - 5(-x)$
       $\quad\quad = -x^3 + 5x$
       $\quad\quad = -f(x)$
The function is odd. The function is symmetric with respect to the origin.

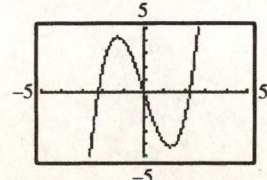

37. $f(x) = x^4 - 2x^2 + 1$
       $f(-x) = (-x)^4 - 2(-x)^2 + 1$
       $\quad\quad = x^4 - 2x^2 + 1$
       $\quad\quad = f(x)$
The function is even. The function is symmetric with respect to the y-axis.

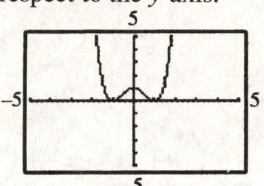

38. $f(x) = 2x\sqrt{1 - x^2}$
       $f(-x) = 2(-x)\sqrt{1 - (-x)^2}$
       $\quad\quad = -2x\sqrt{1 - x^2}$
       $\quad\quad = -f(x)$
The function is odd. The function is symmetric with respect to the origin.

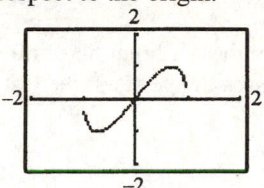

39. **a.**

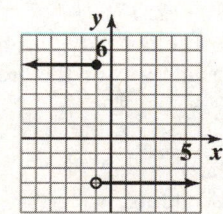

$$f(x) = \begin{cases} 5 \text{ if } x \le -1 \\ -3 \text{ if } x > -1 \end{cases}$$

       b.   range: $\{-3, 5\}$

40. **a.**

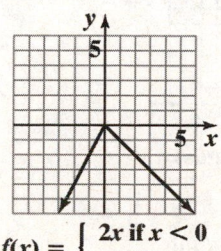

$$f(x) = \begin{cases} 2x \text{ if } x < 0 \\ -x \text{ if } x \ge 0 \end{cases}$$

       b.   range: $\{y \mid y \le 0\}$

41. $\dfrac{8(x+h)-11-(8x-11)}{h}$

$=\dfrac{8x+8h-11-8x+11}{h}$

$=\dfrac{8h}{8}$

$=8$

42. $\dfrac{-2(x+h)^2+(x+h)+10-\left(-2x^2+x+10\right)}{h}$

$=\dfrac{-2\left(x^2+2xh+h^2\right)+x+h+10+2x^2-x-10}{h}$

$=\dfrac{-2x^2-4xh-2h^2+x+h+10+2x^2-x-10}{h}$

$=\dfrac{-4xh-2h^2+h}{h}$

$=\dfrac{h\left(-4x-2h+1\right)}{h}$

$-4x-2h+1$

43. a. Yes, the eagle's height is a function of time since the graph passes the vertical line test.

b. Decreasing: (3, 12)
The eagle descended.

c. Constant: (0, 3) or (12, 17)
The eagle's height held steady during the first 3 seconds and the eagle was on the ground for 5 seconds.

d. Increasing: (17, 30)
The eagle was ascending.

44.

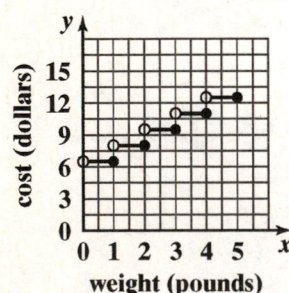

45. $m=\dfrac{1-2}{5-3}=\dfrac{-1}{2}=-\dfrac{1}{2}$; falls

46. $m=\dfrac{-4-(-2)}{-3-(-1)}=\dfrac{-2}{-2}=1$; rises

47. $m=\dfrac{\frac{1}{4}-\frac{1}{4}}{6-(-3)}=\dfrac{0}{9}=0$; horizontal

48. $m=\dfrac{10-5}{-2-(-2)}=\dfrac{5}{0}$ undefined; vertical

49. point-slope form: $y-2=-6(x+3)$
slope-intercept form: $y=-6x-16$

50. $m=\dfrac{2-6}{-1-1}=\dfrac{-4}{-2}=2$
point-slope form: $y-6=2(x-1)$
or $y-2=2(x+1)$
slope-intercept form: $y=2x+4$

51. $3x+y-9=0$
$y=-3x+9$
$m=-3$
point-slope form:
$y+7=-3(x-4)$
slope-intercept form:
$y=-3x+12-7$
$y=-3x+5$

52. perpendicular to $y=\dfrac{1}{3}x+4$

$m=-3$
point-slope form:
$y-6=-3(x+3)$
slope-intercept form:
$y=-3x-9+6$
$y=-3x-3$

53. Write $6x-y-4=0$ in slope intercept form.
$6x-y-4=0$
$-y=-6x+4$
$y=6x-4$
The slope of the perpendicular line is 6, thus the
slope of the desired line is $m=-\dfrac{1}{6}$.
$y-y_1=m(x-x_1)$
$y-(-1)=-\frac{1}{6}\left(x-(-12)\right)$
$y+1=-\frac{1}{6}(x+12)$
$y+1=-\frac{1}{6}x-2$
$6y+6=-x-12$
$x+6y+18=0$

54. slope: $\dfrac{2}{5}$; y-intercept: -1

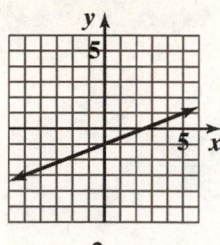

$$y = \frac{2}{5}x - 1$$

55. slope: -4; y-intercept: 5

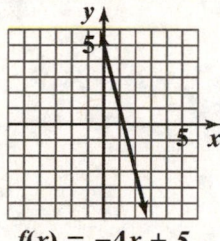

$$f(x) = -4x + 5$$

56. $2x + 3y + 6 = 0$

$$3y = -2x - 6$$

$$y = -\frac{2}{3}x - 2$$

slope: $-\dfrac{2}{3}$; y-intercept: -2

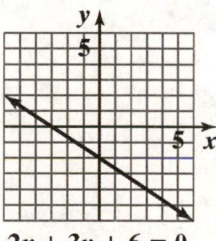

$$2x + 3y + 6 = 0$$

57. $2y - 8 = 0$

$$2y = 8$$

$$y = 4$$

slope: 0; y-intercept: 4

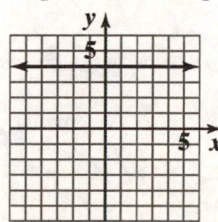

$$2y - 8 = 0$$

58. $2x - 5y - 10 = 0$

Find x-intercept:

$$2x - 5(0) - 10 = 0$$

$$2x - 10 = 0$$

$$2x = 10$$

$$x = 5$$

Find y-intercept:

$$2(0) - 5y - 10 = 0$$

$$-5y - 10 = 0$$

$$-5y = 10$$

$$y = -2$$

$2x - 5y - 10 = 0$

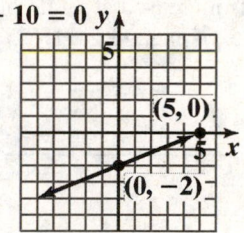

59. $2x - 10 = 0$

$$2x = 10$$

$$x = 5$$

$2x - 10 = 0$

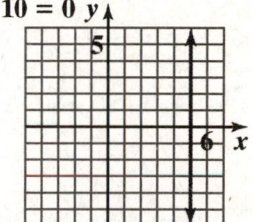

60. **a.** $m = \dfrac{11 - 2.3}{90 - 15} = \dfrac{8.7}{75} = 0.116$

$$y - y_1 = m(x - x_1)$$

$$y - 11 = 0.116(x - 90)$$

or

$$y - 2.3 = 0.116(x - 15)$$

b. $y - 11 = 0.116(x - 90)$

$$y - 11 = 0.116x - 10.44$$

$$y = 0.116x + 0.56$$

$$f(x) = 0.116x + 0.56$$

c. According to the graph, France has about 5 deaths per 100,000 persons.

d. $f(x) = 0.116x + 0.56$

$f(32) = 0.116(32) + 0.56$

$= 4.272$

≈ 4.3

According to the function, France has about 4.3 deaths per 100,000 persons.

This underestimates the value in the graph by 0.7 deaths per 100,000 persons.

The line passes below the point for France.

61. $m = \dfrac{1616 - 886}{2006 - 2002} = \dfrac{730}{4} = 182.5$

Corporate profits increased at a rate of $182.5 billion per year. The rate of change is $182.5 billion per year.

62. $\dfrac{f(x_2) - f(x_1)}{x_2 - x_1} = \dfrac{[9^2 - 4(9)] - [4^2 - 4 \cdot 5]}{9 - 5} = 10$

63.   **a.** $S(0) = -16(0)^2 + 64(0) + 80 = 80$

$S(2) = -16(2)^2 + 64(2) + 80 = 144$

$\dfrac{144 - 80}{2 - 0} = 32$

b. $S(4) = -16(4)^2 + 64(4) + 80 = 80$

$\dfrac{80 - 144}{4 - 2} = -32$

c.   The ball is traveling up until 2 seconds, then it starts to come down.

64. $y = g(x)$

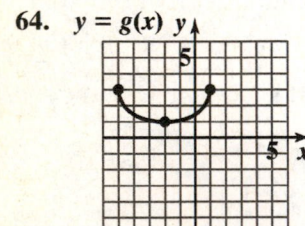

65. $y = g(x)$

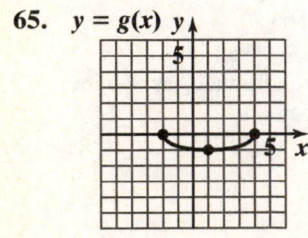

66.

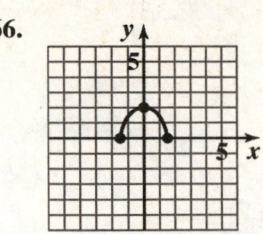

$y = g(x)$

67.

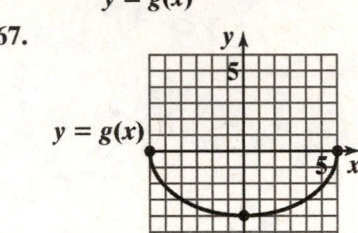

68.

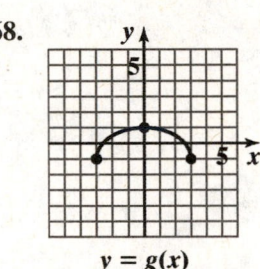

$y = g(x)$

69.

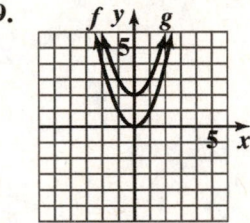

70.

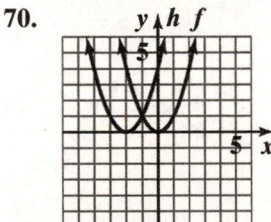

71.

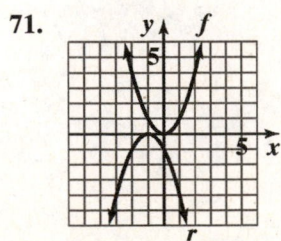

154

72.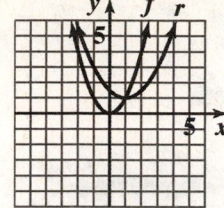

73.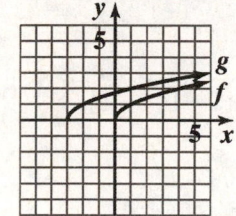

74.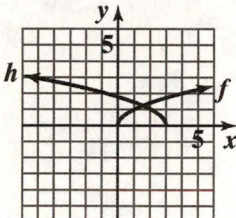

75.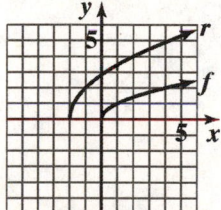

76.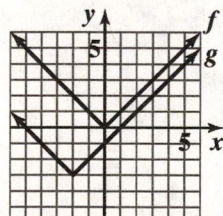

77.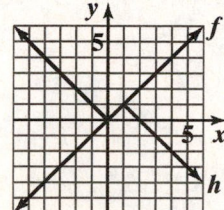

78.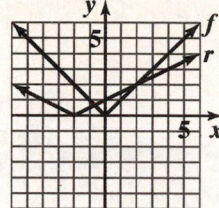

79.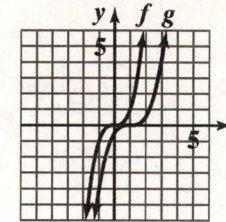

80.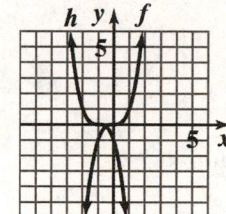

81.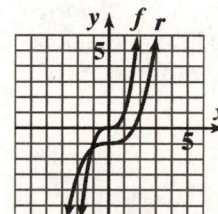

82.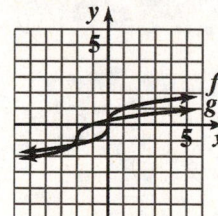

83.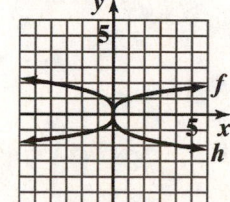

84.

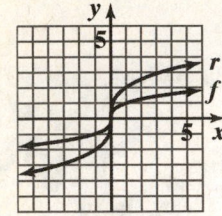

85. domain: $(-\infty, \infty)$

86. The denominator is zero when $x = 7$. The domain is $(-\infty, 7) \cup (7, \infty)$.

87. The expressions under each radical must not be negative.
$8 - 2x \geq 0$
$-2x \geq -8$
$x \leq 4$
domain: $(-\infty, 4]$.

88. The denominator is zero when $x = -7$ or $x = 3$.
domain: $(-\infty, -7) \cup (-7, 3) \cup (3, \infty)$

89. The expressions under each radical must not be negative. The denominator is zero when $x = 5$.
$x - 2 \geq 0$
$x \geq 2$
domain: $[2, 5) \cup (5, \infty)$

90. The expressions under each radical must not be negative.
$x - 1 \geq 0 \quad \text{and} \quad x + 5 \geq 0$
$x \geq 1 \qquad\qquad x \geq -5$
domain: $[1, \infty)$

91. $f(x) = 3x - 1; g(x) = x - 5$
$(f + g)(x) = 4x - 6$
domain: $(-\infty, \infty)$
$(f - g)(x) = (3x - 1) - (x - 5) = 2x + 4$
domain: $(-\infty, \infty)$
$(fg)(x) = (3x - 1)(x - 5) = 3x^2 - 16x + 5$
domain: $(-\infty, \infty)$
$\left(\dfrac{f}{g}\right)(x) = \dfrac{3x - 1}{x - 5}$
domain: $(-\infty, 5) \cup (5, \infty)$

92. $f(x) = x^2 + x + 1; g(x) = x^2 - 1$
$(f + g)(x) = 2x^2 + x$
domain: $(-\infty, \infty)$
$(f - g)(x) = (x^2 + x + 1) - (x^2 - 1) = x + 2$
domain: $(-\infty, \infty)$
$(fg)(x) = (x^2 + x + 1)(x^2 - 1)$
$\qquad\quad = x^4 + x^3 - x - 1$
$\left(\dfrac{f}{g}\right)(x) = \dfrac{x^2 + x + 1}{x^2 - 1}$
domain: $(-\infty, -1) \cup (-1, 1) \cup (1, \infty)$

93. $f(x) = \sqrt{x + 7}; g(x) = \sqrt{x - 2}$
$(f + g)(x) = \sqrt{x + 7} + \sqrt{x - 2}$
domain: $[2, \infty)$
$(f - g)(x) = \sqrt{x + 7} - \sqrt{x - 2}$
domain: $[2, \infty)$
$(fg)(x) = \sqrt{x + 7} \cdot \sqrt{x - 2}$
$\qquad\quad = \sqrt{x^2 + 5x - 14}$
domain: $[2, \infty)$
$\left(\dfrac{f}{g}\right)(x) = \dfrac{\sqrt{x + 7}}{\sqrt{x - 2}}$
domain: $(2, \infty)$

94. $f(x) = x^2 + 3; g(x) = 4x - 1$

a. $(f \circ g)(x) = (4x - 1)^2 + 3$
$\qquad\qquad\quad = 16x^2 - 8x + 4$

b. $(g \circ f)(x) = 4(x^2 + 3) - 1$
$\qquad\qquad\quad = 4x^2 + 11$

c. $(f \circ g)(3) = 16(3)^2 - 8(3) + 4 = 124$

95. $f(x) = \sqrt{x}; \quad g(x) = x + 1$

a. $(f \circ g)(x) = \sqrt{x + 1}$

b. $(g \circ f)(x) = \sqrt{x} + 1$

c. $(f \circ g)(3) = \sqrt{3 + 1} = \sqrt{4} = 2$

96. a.

$$(f \circ g)(x) = f\left(\frac{1}{x}\right)$$

$$= \frac{\frac{1}{x}+1}{\frac{1}{x}-2} = \frac{\left(\frac{1}{x}+1\right)x}{\left(\frac{1}{x}-2\right)x} = \frac{1+x}{1-2x}$$

b. $x \neq 0$ $1-2x \neq 0$

$$x \neq \frac{1}{2}$$

$$(-\infty, 0) \cup \left(0, \frac{1}{2}\right) \cup \left(\frac{1}{2}, \infty\right)$$

97. a. $(f \circ g)(x) = f(x+3) = \sqrt{x+3-1} = \sqrt{x+2}$

b. $x+2 \geq 0$
       $x \geq -2$ $[-2, \infty)$

98. $f(x) = x^4$ $g(x) = x^2 + 2x - 1$

99. $f(x) = \sqrt[3]{x}$ $g(x) = 7x + 4$

100. $f(x) = \frac{3}{5}x + \frac{1}{2}; g(x) = \frac{5}{3}x - 2$

$$f(g(x)) = \frac{3}{5}\left(\frac{5}{3}x - 2\right) + \frac{1}{2}$$

$$= x - \frac{6}{5} + \frac{1}{2}$$

$$= x - \frac{7}{10}$$

$$g(f(x)) = \frac{5}{3}\left(\frac{3}{5}x + \frac{1}{2}\right) - 2$$

$$= x + \frac{5}{6} - 2$$

$$= x - \frac{7}{6}$$

f and *g* are not inverses of each other.

101. $f(x) = 2 - 5x; g(x) = \frac{2-x}{5}$

$$f(g(x)) = 2 - 5\left(\frac{2-x}{5}\right)$$

$$= 2 - (2-x)$$

$$= x$$

$$g(f(x)) = \frac{2-(2-5x)}{5} = \frac{5x}{5} = x$$

f and *g* are inverses of each other.

102. a. $f(x) = 4x - 3$

$$y = 4x - 3$$

$$x = 4y - 3$$

$$y = \frac{x+3}{4}$$

$$f^{-1}(x) = \frac{x+3}{4}$$

b. $f(f^{-1}(x)) = 4\left(\frac{x+3}{4}\right) - 3$

$$= x + 3 - 3$$

$$= x$$

$$f^{-1}(f(x)) = \frac{(4x-3)+3}{4} = \frac{4x}{4} = x$$

103. a. $f(x) = 8x^3 + 1$

$$y = 8x^3 + 1$$

$$x = 8y^3 + 1$$

$$x - 1 = 8y^3$$

$$\frac{x-1}{8} = y^3$$

$$\sqrt[3]{\frac{x-1}{8}} = y$$

$$\frac{\sqrt[3]{x-1}}{2} = y$$

$$f^{-1}(x) = \frac{\sqrt[3]{x-1}}{2}$$

b. $f\left(f^{-1}(x)\right)=8\left(\dfrac{\sqrt[3]{x-1}}{2}\right)^{3}+1$

$\qquad\qquad =8\left(\dfrac{x-1}{8}\right)+1$

$\qquad\qquad =x-1+1$

$\qquad\qquad =x$

$f^{-1}\left(f(x)\right)=\dfrac{\sqrt[3]{\left(8x^{3}+1\right)-1}}{2}$

$\qquad\qquad =\dfrac{\sqrt[3]{8x^{3}}}{2}$

$\qquad\qquad =\dfrac{2x}{2}$

$\qquad\qquad =x$

104. a. $f(x)=\dfrac{2}{x}+5$

$\qquad y=\dfrac{2}{x}+5$

$\qquad x=\dfrac{2}{y}+5$

$\qquad xy=2+5y$

$\qquad xy-5y=2$

$\qquad y(x-5)=2$

$\qquad y=\dfrac{2}{x-5}$

$\qquad f^{-1}(x)=\dfrac{2}{x-5}$

b. $f\left(f^{-1}(x)\right)=\dfrac{2}{\dfrac{2}{x-5}}+5$

$\qquad\qquad =\dfrac{2(x-5)}{2}+5$

$\qquad\qquad =x-5+5$

$\qquad\qquad =x$

$f^{-1}\left(f(x)\right)=\dfrac{2}{\dfrac{2}{x}+5-5}$

$\qquad\qquad =\dfrac{2}{\dfrac{2}{x}}$

$\qquad\qquad =\dfrac{2x}{2}$

$\qquad\qquad =x$

105.   The inverse function exists.

106.   The inverse function does not exist since it does not pass the horizontal line test.

107.   The inverse function exists.

108.   The inverse function does not exist since it does not pass the horizontal line test.

109.

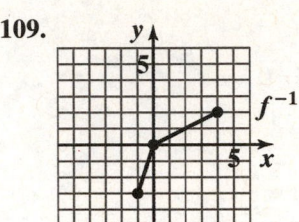

110. $f(x)=1-x^{2}$

$\qquad y=1-x^{2}$

$\qquad x=1-y^{2}$

$\qquad y^{2}=1-x$

$\qquad y=\sqrt{1-x}$

$\qquad f^{-1}(x)=\sqrt{1-x}$

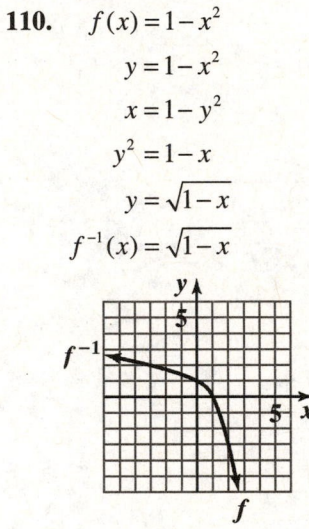

111. $f(x)=\sqrt{x}+1$

$\qquad y=\sqrt{x}+1$

$\qquad x=\sqrt{y}+1$

$\qquad x-1=\sqrt{y}$

$\qquad (x-1)^{2}=y$

$\qquad f^{-1}(x)=(x-1)^{2},\quad x\ge 1$

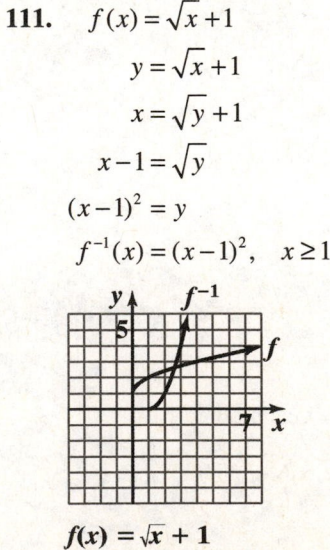

$f(x)=\sqrt{x}+1$
$g(x)=(x-1)^{2}, x\ge 1$

112. $d = \sqrt{[3-(-2)]^2 + [9-(-3)]^2}$

$= \sqrt{5^2 + 12^2}$

$= \sqrt{25 + 144}$

$= \sqrt{169}$

$= 13$

113. $d = \sqrt{[-2-(-4)]^2 + (5-3)^2}$

$= \sqrt{2^2 + 2^2}$

$= \sqrt{4+4}$

$= \sqrt{8}$

$= 2\sqrt{2}$

≈ 2.83

114. $\left(\dfrac{2+(-12)}{2}, \dfrac{6+4}{2}\right) = \left(\dfrac{-10}{2}, \dfrac{10}{2}\right) = (-5,5)$

115. $\left(\dfrac{4+(-15)}{2}, \dfrac{-6+2}{2}\right) = \left(\dfrac{-11}{2}, \dfrac{-4}{2}\right) = \left(\dfrac{-11}{2}, -2\right)$

116. $x^2 + y^2 = 3^2$

$x^2 + y^2 = 9$

117. $(x-(-2))^2 + (y-4)^2 = 6^2$

$(x+2)^2 + (y-4)^2 = 36$

118. center: (0, 0); radius: 1

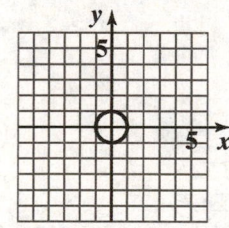

$x^2 + y^2 = 1$

domain: $[-1,1]$

range: $[-1,1]$

119. center: (–2, 3); radius: 3

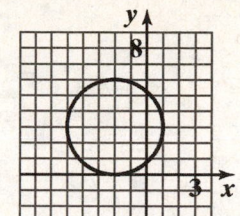

$(x + 2)^2 + (y - 3)^2 = 9$

domain: $[-5,1]$

range: $[0,6]$

120. $x^2 + y^2 - 4x + 2y - 4 = 0$

$x^2 - 4x \quad + y^2 + 2y \quad = 4$

$x^2 - 4x + 4 + y^2 + 2y + 1 = 4 + 4 + 1$

$(x-2)^2 + (y+1)^2 = 9$

center: (2, –1); radius: 3

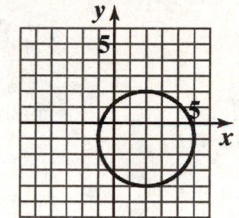

$x^2 + y^2 - 4x + 2y - 4 = 0$

domain: $[-1,5]$

range: $[-4,2]$

121. a. $W(x) = 567 + 15x$

b. $702 = 567 + 15x$

$135 = 15x$

$9 = x$

9 years after 2000, in 2009, the average weekly sales will be $702.

122. a. $f(x) = 15 + 0.05x$

b. $g(x) = 5 + 0.07x$

c. $15 + 0.05x = 5 + 0.07x$

$10 = 0.02x$

$500 = x$

For 500 minutes, the two plans cost the same.

123. **a.** $N(x) = 400 - 2(x - 120)$

$\qquad = 400 - 2x + 240$

$\qquad = 640 - 2x$

b. $R(x) = x(640 - 2x)$

$\qquad = -2x^2 + 640x$

124. **a.** $w = 16 - 2x \qquad l = 24 - 2x$

$V(x) = (16 - 2x)(24 - 2x)x$

b. $0 < x < 8$

125. $2l + 3w = 400$

$\qquad 2l = 400 - 3w$

$\qquad l = \dfrac{400 - 3w}{2}$

Let x = width

$A(x) = x\left(\dfrac{400 - 3w}{2}\right)$

$\qquad = \dfrac{x(400 - 3w)}{2}$

126. $V = lwh$

$\qquad 8 = x \cdot x \cdot h$

$\qquad \dfrac{8}{x^2} = h$

$A(x) = 2x \cdot x + 4hx$

$\qquad = 2x^2 + 4\left(\dfrac{8}{x^2}\right)x$

$\qquad = 2x^2 + \dfrac{32}{x}$

127. $I = 0.08x + 0.12(10,000 - x)$

Chapter 1 Test

1. (b), (c), and (d) are not functions.

2. **a.** $f(4) - f(-3) = 3 - (-2) = 5$

b. domain: $(-5, 6]$

c. range: $[-4, 5]$

d. increasing: $(-1, 2)$

e. decreasing: $(-5, -1)$ or $(2, 6)$

f. $2, f(2) = 5$

g. $(-1, -4)$

h. x-intercepts: $-4, 1,$ and 5.

i. y-intercept: -3

3. **a.** $-2, 2$

b. $-1, 1$

c. 0

d. even; $f(-x) = f(x)$

e. no; f fails the horizontal line test

f. $f(0)$ is a relative minimum.

g.

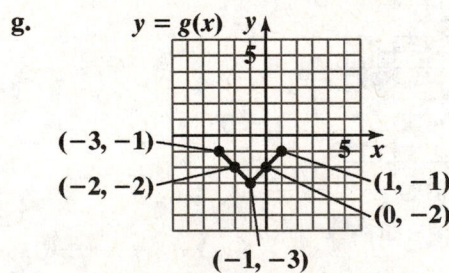

h.

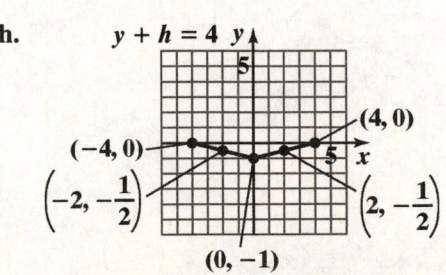

i.

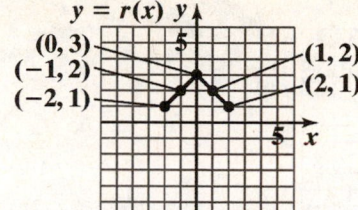

$y = r(x)$
(0, 3)
(−1, 2)
(−2, 1)
(1, 2)
(2, 1)

j. $\dfrac{f(x_2) - f(x_1)}{x_2 - x_1} = \dfrac{-1-0}{1-(-2)} = -\dfrac{1}{3}$

4. $y + y = 4$

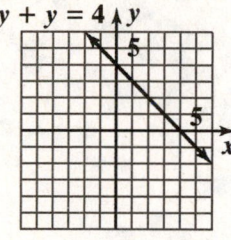

domain: $(-\infty, \infty)$

range: $(-\infty, \infty)$

5. $x^2 + y^2 = 4$

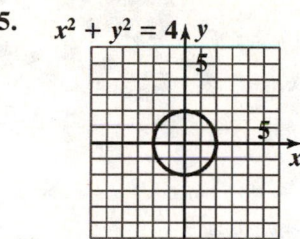

domain: $[-2, 2]$

range: $[-2, 2]$

6. $f(x) = 4$

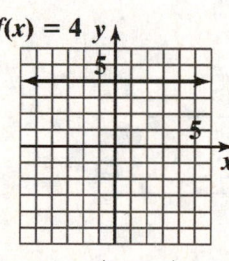

domain: $(-\infty, \infty)$

range: $\{4\}$

7.

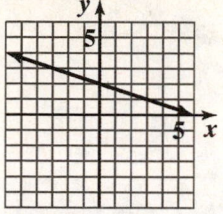

$f(x) = -\dfrac{1}{3}x + 2$

domain: $(-\infty, \infty)$

range: $(-\infty, \infty)$

8.

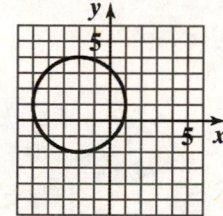

$(x + 2)^2 + (y - 1)^2 = 9$

domain: $[-5, 1]$

range: $[-2, 4]$

9.

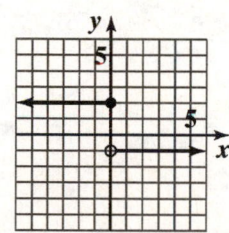

$f(x) = \begin{cases} 2 \text{ if } x \le 0 \\ -1 \text{ if } x > 0 \end{cases}$

domain: $(-\infty, \infty)$

range: $\{-1, 2\}$

10.

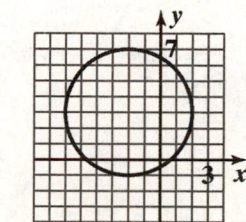

$x^2 + y^2 + 4x + 6y - 3 = 0$

domain: $[-6, 2]$

range: $[-1, 7]$

11.

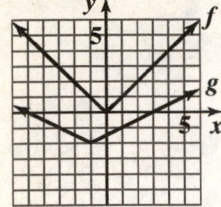

domain of f: $(-\infty, \infty)$

range of f: $[0, \infty)$

domain of g: $(-\infty, \infty)$

range of g: $[-2, \infty)$

12.

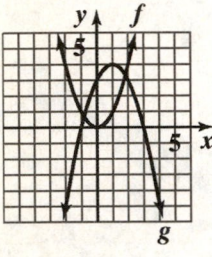

domain of f: $(-\infty, \infty)$

range of f: $[0, \infty)$

domain of g: $(-\infty, \infty)$

range of g: $(-\infty, 4]$

13.

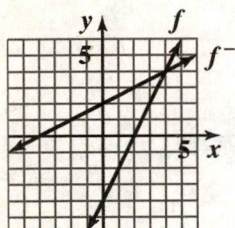

domain of f: $(-\infty, \infty)$

range of f: $(-\infty, \infty)$

domain of f^{-1}: $(-\infty, \infty)$

range of f^{-1}: $(-\infty, \infty)$

14.

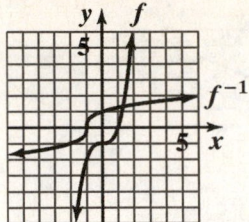

domain of f: $(-\infty, \infty)$

range of f: $(-\infty, \infty)$

domain of f^{-1}: $(-\infty, \infty)$

range of f^{-1}: $(-\infty, \infty)$

15.

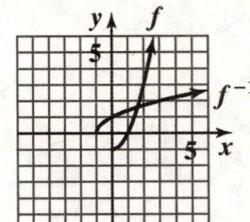

domain of f: $[0, \infty)$

range of f: $[-1, \infty)$

domain of f^{-1}: $[-1, \infty)$

range of f^{-1}: $[0, \infty)$

16. $\quad f(x) = x^2 - x - 4$

$$f(x-1) = (x-1)^2 - (x-1) - 4$$
$$= x^2 - 2x + 1 - x + 1 - 4$$
$$= x^2 - 3x - 2$$

17. $\dfrac{f(x+h) - f(x)}{h}$

$$= \frac{(x+h)^2 - (x+h) - 4 - \left(x^2 - x - 4\right)}{h}$$

$$= \frac{x^2 + 2xh + h^2 - x - h - 4 - x^2 + x + 4}{h}$$

$$= \frac{2xh + h^2 - h}{h}$$

$$= \frac{h(2x + h - 1)}{h}$$

$$= 2x + h - 1$$

18. $\quad (g - f)(x) = 2x - 6 - \left(x^2 - x - 4\right)$

$$= 2x - 6 - x^2 + x + 4$$
$$= -x^2 + 3x - 2$$

19. $\left(\dfrac{f}{g}\right)(x) = \dfrac{x^2 - x - 4}{2x - 6}$

domain: $(-\infty, 3) \cup (3, \infty)$

20. $(f \circ g)(x) = f\left(g(x)\right)$

$= (2x - 6)^2 - (2x - 6) - 4$

$= 4x^2 - 24x + 36 - 2x + 6 - 4$

$= 4x^2 - 26x + 38$

21. $(g \circ f)(x) = g\left(f(x)\right)$

$= 2\left(x^2 - x - 4\right) - 6$

$= 2x^2 - 2x - 8 - 6$

$= 2x^2 - 2x - 14$

22. $g\left(f(-1)\right) = 2\left((-1)^2 - (-1) - 4\right) - 6$

$= 2(1 + 1 - 4) - 6$

$= 2(-2) - 6$

$= -4 - 6$

$= -10$

23. $f(x) = x^2 - x - 4$

$f(-x) = (-x)^2 - (-x) - 4$

$= x^2 + x - 4$

f is neither even nor odd.

24. $m = \dfrac{-8 - 1}{-1 - 2} = \dfrac{-9}{-3} = 3$

point-slope form: $y - 1 = 3(x - 2)$
or $y + 8 = 3(x + 1)$
slope-intercept form: $y = 3x - 5$

25. $y = -\dfrac{1}{4}x + 5$ so $m = 4$

point-slope form: $y - 6 = 4(x + 4)$
slope-intercept form: $y = 4x + 22$

26. Write $4x + 2y - 5 = 0$ in slope intercept form.

$4x + 2y - 5 = 0$

$2y = -4x + 5$

$y = -2x + \dfrac{5}{2}$

The slope of the parallel line is -2, thus the slope of the desired line is $m = -2$.

$y - y_1 = m(x - x_1)$

$y - (-10) = -2\left(x - (-7)\right)$

$y + 10 = -2(x + 7)$

$y + 10 = -2x - 14$

$2x + y + 24 = 0$

27. **a.** First, find the slope using the points
$(2, 476)$ and $(4, 486)$.

$m = \dfrac{486 - 476}{4 - 2} = \dfrac{10}{2} = 5$

Then use the slope and a point to write the equation in point-slope form.

$y - y_1 = m\left(x - x_1\right)$

$y - 486 = 5(x - 4)$

or

$y - 476 = 5(x - 2)$

b. $y - 486 = 5(x - 4)$

$y - 486 = 5x - 20$

$y = 5x + 466$

$f(x) = 5x + 466$

c. $f(10) = 5(10) + 466 = 516$

The function predicts that in 2010 the number of sentenced inmates in the U.S. will be 516 per 100,000 residents.

28. $\dfrac{3(10)^2 - 5 - [3(6)^2 - 5]}{10 - 6}$

$= \dfrac{205 - 103}{4}$

$= \dfrac{192}{4}$

$= 48$

29. $g(-1) = 3 - (-1) = 4$

$g(7) = \sqrt{7 - 3} = \sqrt{4} = 2$

30. The denominator is zero when $x = 1$ or $x = -5$.

domain: $(-\infty, -5) \cup (-5, 1) \cup (1, \infty)$

31. The expressions under each radical must not be negative.

$x + 5 \geq 0$ and $x - 1 \geq 0$

$x \geq -5$ $\qquad x \geq 1$

domain: $[1, \infty)$

32. $(f \circ g)(x) = \dfrac{7}{\dfrac{2}{x} - 4} = \dfrac{7x}{2 - 4x}$

$x \neq 0, \quad 2 - 4x \neq 0$

$\qquad\qquad x \neq \dfrac{1}{2}$

domain: $(-\infty, 0) \cup \left(0, \dfrac{1}{2}\right) \cup \left(\dfrac{1}{2}, \infty\right)$

33. $f(x) = x^7 \qquad g(x) = 2x + 3$

34. $d = \sqrt{(x_2 - x_1)^2 + (y_2 - y_1)^2}$

$d = \sqrt{(x_2 - x_1)^2 + (y_2 - y_1)^2}$

$\quad = \sqrt{(5 - 2)^2 + (2 - (-2))^2}$

$\quad = \sqrt{3^2 + 4^2}$

$\quad = \sqrt{9 + 16}$

$\quad = \sqrt{25}$

$\quad = 5$

$\left(\dfrac{x_1 + x_2}{2}, \dfrac{y_1 + y_2}{2}\right) = \left(\dfrac{2 + 5}{2}, \dfrac{-2 + 2}{2}\right)$

$\qquad\qquad\qquad\quad = \left(\dfrac{7}{2}, 0\right)$

The length is 5 and the midpoint is $\left(\dfrac{7}{2}, 0\right)$.

35. **a.** $T(x) = 41.78 - 0.19x$

b. $35.7 = 41.78 - 0.19x$

$-6.08 = -0.19x$

$32 = x$

32 years after 1980, in 2012, the winning time will be 35.7 seconds.

36. **a.** $Y(x) = 50 - 1.5(x - 30)$

$\qquad\quad = 50 - 1.5x + 45$

$\qquad\quad = 95 - 1.5x$

b. $T(x) = x(95 - 1.5x)$

$\qquad\quad = -1.5x^2 + 95x$

37. $2l + 2w = 600$

$\quad 2l = 600 - 2w$

$\quad\, l = 300 - w$

Let $x = w$

$A(x) = x(300 - x)$

$\qquad = -x^2 + 300x$

38. $V = lwh$

$8000 = x \cdot x \cdot h$

$\dfrac{8000}{x^2} = h$

$A(x) = 2x^2 + 4x\,\dfrac{8000}{x^2}$

$\qquad = 2x^2 + \dfrac{32,000}{x}$

Chapter 2
Polynomial and Rational Functions

Section 2.1

Check Point Exercises

1. **a.** $(5-2i)+(3+3i)$
$= 5-2i+3+3i$
$= (5+3)+(-2+3)i$
$= 8+i$

b. $(2+6i)-(12-i)$
$= 2+6i-12+i$
$= (2-12)+(6+1)i$
$= -10+7i$

2. **a.** $7i(2-9i) = 7i(2)-7i(9i)$
$= 14i-63i^2$
$= 14i-63(-1)$
$= 63+14i$

b. $(5+4i)(6-7i) = 30-35i+24i-28i^2$
$= 30-35i+24i-28(-1)$
$= 30+28-35i+24i$
$= 58-11i$

3. $\dfrac{5+4i}{4-i} = \dfrac{5+4i}{4-i} \cdot \dfrac{4+i}{4+i}$
$= \dfrac{20+5i+16i+4i^2}{16+4i-4i-i^2}$
$= \dfrac{20+21i-4}{16+1}$
$= \dfrac{16+21i}{17}$
$= \dfrac{16}{17}+\dfrac{21}{17}i$

4. **a.** $\sqrt{-27}+\sqrt{-48} = i\sqrt{27}+i\sqrt{48}$
$= i\sqrt{9\cdot3}+i\sqrt{16\cdot3}$
$= 3i\sqrt{3}+4i\sqrt{3}$
$= 7i\sqrt{3}$

b. $(-2+\sqrt{-3})^2 = (-2+i\sqrt{3})^2$
$= (-2)^2+2(-2)(i\sqrt{3})+(i\sqrt{3})^2$
$= 4-4i\sqrt{3}+3i^2$
$= 4-4i\sqrt{3}+3(-1)$
$= 1-4i\sqrt{3}$

c. $\dfrac{-14+\sqrt{-12}}{2} = \dfrac{-14+i\sqrt{12}}{2}$
$= \dfrac{-14+2i\sqrt{3}}{2}$
$= \dfrac{-14}{2}+\dfrac{2i\sqrt{3}}{2}$
$= -7+i\sqrt{3}$

5. $x^2-2x+2 = 0$
$a = 1, b = -2, c = 2$
$x = \dfrac{-b\pm\sqrt{b^2-4ac}}{2a}$
$x = \dfrac{-(-2)\pm\sqrt{(-2)^2-4(1)(2)}}{2(1)}$
$x = \dfrac{2\pm\sqrt{4-8}}{2}$
$x = \dfrac{2\pm\sqrt{-4}}{2}$
$x = \dfrac{2\pm2i}{2}$
$x = 1\pm i$
The solution set is $\{1+i, 1-i\}$.

Polynomial and Rational Functions

Exercise Set 2.1

1. $(7 + 2i) + (1 - 4i) = 7 + 2i + 1 - 4i$
$$= 7 + 1 + 2i - 4i$$
$$= 8 - 2i$$

3. $(3 + 2i) - (5 - 7i) = 3 - 5 + 2i + 7i$
$$= 3 + 2i - 5 + 7i$$
$$= -2 + 9i$$

5. $6 - (-5 + 4i) - (-13 - i) = 6 + 5 - 4i + 13 + i$
$$= 24 - 3i$$

7. $8i - (14 - 9i) = 8i - 14 + 9i$
$$= -14 + 8i + 9i$$
$$= -14 + 17i$$

9. $-3i(7i - 5) = -21i^2 + 15i$
$$= -21(-1) + 15i$$
$$= 21 + 15i$$

11. $(-5 + 4i)(3 + i) = -15 - 5i + 12i + 4i^2$
$$= -15 + 7i - 4$$
$$= -19 + 7i$$

13. $(7 - 5i)(-2 - 3i) = -14 - 21i + 10i + 15i^2$
$$= -14 - 15 - 11i$$
$$= -29 - 11i$$

15. $(3 + 5i)(3 - 5i) = 9 - 15i + 15i - 25i^2$
$$= 9 + 25$$
$$= 34$$

17. $(-5 + i)(-5 - i) = 25 + 5i - 5i - i^2$
$$= 25 + 1$$
$$= 26$$

19. $(2 + 3i)^2 = 4 + 12i + 9i^2$
$$= 4 + 12i - 9$$
$$= -5 + 12i$$

21. $\dfrac{2}{3 - i} = \dfrac{2}{3 - i} \cdot \dfrac{3 + i}{3 + i}$
$$= \dfrac{2(3 + i)}{9 + 1}$$
$$= \dfrac{2(3 + i)}{10}$$
$$= \dfrac{3 + i}{5}$$
$$= \dfrac{3}{5} + \dfrac{1}{5}i$$

23. $\dfrac{2i}{1 + i} = \dfrac{2i}{1 + i} \cdot \dfrac{1 - i}{1 - i} = \dfrac{2i - 2i^2}{1 + 1} = \dfrac{2 + 2i}{2} = 1 + i$

25. $\dfrac{8i}{4 - 3i} = \dfrac{8i}{4 - 3i} \cdot \dfrac{4 + 3i}{4 + 3i}$
$$= \dfrac{32i + 24i^2}{16 + 9}$$
$$= \dfrac{-24 + 32i}{25}$$
$$= -\dfrac{24}{25} + \dfrac{32}{25}i$$

27. $\dfrac{2 + 3i}{2 + i} = \dfrac{2 + 3i}{2 + i} \cdot \dfrac{2 - i}{2 - i}$
$$= \dfrac{4 + 4i - 3i^2}{4 + 1}$$
$$= \dfrac{7 + 4i}{5}$$
$$= \dfrac{7}{5} + \dfrac{4}{5}i$$

29. $\sqrt{-64} - \sqrt{-25} = i\sqrt{64} - i\sqrt{25}$
$$= 8i - 5i = 3i$$

31. $5\sqrt{-16} + 3\sqrt{-81} = 5(4i) + 3(9i)$
$$= 20i + 27i = 47i$$

33. $\left(-2 + \sqrt{-4}\right)^2 = (-2 + 2i)^2$
$$= 4 - 8i + 4i^2$$
$$= 4 - 8i - 4$$
$$= -8i$$

35. $\left(-3-\sqrt{-7}\right)^2 = \left(-3-i\sqrt{7}\right)^2$

$$= 9 + 6i\sqrt{7} + i^2(7)$$
$$= 9 - 7 + 6i\sqrt{7}$$
$$= 2 + 6i\sqrt{7}$$

37. $\dfrac{-8+\sqrt{-32}}{24} = \dfrac{-8+i\sqrt{32}}{24}$

$$= \dfrac{-8+i\sqrt{16\cdot2}}{24}$$
$$= \dfrac{-8+4i\sqrt{2}}{24}$$
$$= -\dfrac{1}{3} + \dfrac{\sqrt{2}}{6}i$$

39. $\dfrac{-6-\sqrt{-12}}{48} = \dfrac{-6-i\sqrt{12}}{48}$

$$= \dfrac{-6-i\sqrt{4\cdot3}}{48}$$
$$= \dfrac{-6-2i\sqrt{3}}{48}$$
$$= -\dfrac{1}{8} - \dfrac{\sqrt{3}}{24}i$$

41. $\sqrt{-8}\left(\sqrt{-3}-\sqrt{5}\right) = i\sqrt{8}(i\sqrt{3}-\sqrt{5})$

$$= 2i\sqrt{2}\left(i\sqrt{3}-\sqrt{5}\right)$$
$$= -2\sqrt{6} - 2i\sqrt{10}$$

43. $\left(3\sqrt{-5}\right)\left(-4\sqrt{-12}\right) = \left(3i\sqrt{5}\right)\left(-8i\sqrt{3}\right)$

$$= -24i^2\sqrt{15}$$
$$= 24\sqrt{15}$$

45. $x^2 - 6x + 10 = 0$

$$x = \dfrac{6\pm\sqrt{(-6)^2-4(1)(10)}}{2(1)}$$
$$x = \dfrac{6\pm\sqrt{36-40}}{2}$$
$$x = \dfrac{6\pm\sqrt{-4}}{2}$$
$$x = \dfrac{6\pm2i}{2}$$
$$x = 3\pm i$$

The solution set is $\left\{3+i, 3-i\right\}$.

47. $4x^2 + 8x + 13 = 0$

$$x = \dfrac{-8\pm\sqrt{8^2-4(4)(13)}}{2(4)}$$
$$= \dfrac{-8\pm\sqrt{64-208}}{8}$$
$$= \dfrac{-8\pm\sqrt{-144}}{8}$$
$$= \dfrac{-8\pm12i}{8}$$
$$= \dfrac{4(-2\pm3i)}{8}$$
$$= \dfrac{-2\pm3i}{2}$$
$$= -1\pm\dfrac{3}{2}i$$

The solution set is $\left\{-1+\dfrac{3}{2}i, -1-\dfrac{3}{2}i\right\}$.

49. $3x^2 - 8x + 7 = 0$

$$x = \frac{-(-8) \pm \sqrt{(-8)^2 - 4(3)(7)}}{2(3)}$$

$$= \frac{8 \pm \sqrt{64 - 84}}{6}$$

$$= \frac{8 \pm \sqrt{-20}}{6}$$

$$= \frac{8 \pm 2i\sqrt{5}}{6}$$

$$= \frac{2(4 \pm i\sqrt{5})}{6}$$

$$= \frac{4 \pm i\sqrt{5}}{3}$$

$$= \frac{4}{3} \pm \frac{\sqrt{5}}{3}i$$

The solution set is $\left\{ \frac{4}{3} + \frac{\sqrt{5}}{3}i, \frac{4}{3} - \frac{\sqrt{5}}{3}i \right\}$.

51. $(2 - 3i)(1 - i) - (3 - i)(3 + i)$

$= (2 - 2i - 3i + 3i^2) - (3^2 - i^2)$

$= 2 - 5i + 3i^2 - 9 + i^2$

$= -7 - 5i + 4i^2$

$= -7 - 5i + 4(-1)$

$= -11 - 5i$

53. $(2 + i)^2 - (3 - i)^2$

$= (4 + 4i + i^2) - (9 - 6i + i^2)$

$= 4 + 4i + i^2 - 9 + 6i - i^2$

$= -5 + 10i$

55. $5\sqrt{-16} + 3\sqrt{-81}$

$= 5\sqrt{16}\sqrt{-1} + 3\sqrt{81}\sqrt{-1}$

$= 5 \cdot 4i + 3 \cdot 9i$

$= 20i + 27i$

$= 47i$ or $0 + 47i$

57. $f(x) = x^2 - 2x + 2$

$f(1 + i) = (1 + i)^2 - 2(1 + i) + 2$

$= 1 + 2i + i^2 - 2 - 2i + 2$

$= 1 + i^2$

$= 1 - 1$

$= 0$

59. $f(x) = \frac{x^2 + 19}{2 - x}$

$f(3i) = \frac{(3i)^2 + 19}{2 - 3i}$

$= \frac{9i^2 + 19}{2 - 3i}$

$= \frac{-9 + 19}{2 - 3i}$

$= \frac{10}{2 - 3i}$

$= \frac{10}{2 - 3i} \cdot \frac{2 + 3i}{2 + 3i}$

$= \frac{20 + 30i}{4 - 9i^2}$

$= \frac{20 + 30i}{4 + 9}$

$= \frac{20 + 30i}{13}$

$= \frac{20}{13} + \frac{30}{13}i$

61. $E = IR = (4 - 5i)(3 + 7i)$

$= 12 + 28i - 15i - 35i^2$

$= 12 + 13i - 35(-1)$

$= 12 + 35 + 13i = 47 + 13i$

The voltage of the circuit is $(47 + 13i)$ volts.

63. Sum:

$(5 + i\sqrt{15}) + (5 - i\sqrt{15})$

$= 5 + i\sqrt{15} + 5 - i\sqrt{15}$

$= 5 + 5$

$= 10$

Product:

$(5 + i\sqrt{15})(5 - i\sqrt{15})$

$= 25 - 5i\sqrt{15} + 5i\sqrt{15} - 15i^2$

$= 25 + 15$

$= 40$

65. – 71. Answers may vary.

73. makes sense

75. does not make sense; Explanations will vary.

Sample explanation: $i = \sqrt{-1}$; It is not a variable in this context.

77. false; Changes to make the statement true will vary. A sample change is: All irrational numbers are complex numbers.

79. false; Changes to make the statement true will vary. A sample change is:

$$\frac{7+3i}{5+3i} = \frac{7+3i}{5+3i} \cdot \frac{5-3i}{5-3i} = \frac{44-6i}{34} = \frac{22}{17} - \frac{3}{17}i$$

81.

$$\frac{4}{(2+i)(3-i)} = \frac{4}{6-2i+3i-i^2}$$

$$= \frac{4}{6+i+1}$$

$$= \frac{4}{7+i}$$

$$= \frac{4}{7+i} \cdot \frac{7-i}{7-i}$$

$$= \frac{28-4i}{49-i^2}$$

$$= \frac{28-4i}{49+1}$$

$$= \frac{28-4i}{50}$$

$$= \frac{28}{50} - \frac{4}{50}i$$

$$= \frac{14}{25} - \frac{2}{25}i$$

83.

$$\frac{8}{1+\frac{2}{i}} = \frac{8}{\frac{i}{i}+\frac{2}{i}}$$

$$= \frac{8}{\frac{2+i}{i}}$$

$$= \frac{8i}{2+i}$$

$$= \frac{8i}{2+i} \cdot \frac{2-i}{2-i}$$

$$= \frac{16i-8i^2}{4-i^2}$$

$$= \frac{16i+8}{4+1}$$

$$= \frac{8+16i}{5}$$

$$= \frac{8}{5} + \frac{16}{5}i$$

84.

$$0 = -2(x-3)^2 + 8$$

$$2(x-3)^2 = 8$$

$$(x-3)^2 = 4$$

$$x-3 = \pm\sqrt{4}$$

$$x = 3 \pm 2$$

$$x = 1, 5$$

85.

$$-x^2 - 2x + 1 = 0$$

$$x^2 + 2x - 1 = 0$$

$$x = \frac{-b \pm \sqrt{b^2 - 4ac}}{2a}$$

$$x = \frac{-(-2) \pm \sqrt{(-2)^2 - 4(1)(-1)}}{2(1)}$$

$$= \frac{2 \pm \sqrt{8}}{2}$$

$$= \frac{2 \pm 2\sqrt{2}}{2}$$

$$= 1 \pm \sqrt{2}$$

The solution set is $\{1 \pm \sqrt{2}\}$.

86. The graph of g is the graph of f shifted 1 unit up and 3 units to the left.

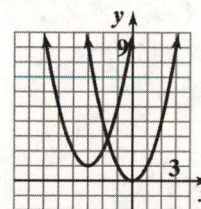

$f(x) = x^2$

$g(x) = (x + 3)^2 + 1$

Section 2.2

Check Point Exercises

1. $f(x) = -(x-1)^2 + 4$

$$f(x) = \overset{a=-1}{-}\left(x - \overset{h=1}{1}\right)^2 + \overset{k=4}{4}$$

Step 1: The parabola opens down because $a < 0$.
Step 2: find the vertex: (1, 4)
Step 3: find the x-intercepts:

$0 = -(x-1)^2 + 4$

$(x-1)^2 = 4$

$x - 1 = \pm 2$

$x = 1 \pm 2$

$x = 3$ or $x = -1$

Step 4: find the y-intercept:

$f(0) = -(0-1)^2 + 4 = 3$

Step 5: The axis of symmetry is $x = 1$.

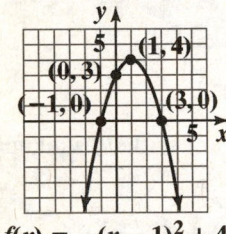

$f(x) = -(x-1)^2 + 4$

2. $f(x) = (x-2)^2 + 1$

Step 1: The parabola opens up because $a > 0$.
Step 2: find the vertex: (2, 1)
Step 3: find the x-intercepts:

$0 = (x-2)^2 + 1$

$(x-2)^2 = -1$

$x - 2 = \sqrt{-1}$

$x = 2 \pm i$

The equation has no real roots, thus the parabola has no x-intercepts.
Step 4: find the y-intercept:

$f(0) = (0-2)^2 + 1 = 5$

Step 5: The axis of symmetry is $x = 2$.

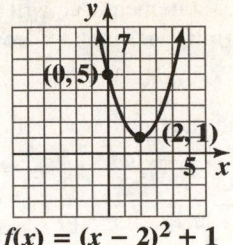

$f(x) = (x-2)^2 + 1$

3. $f(x) = -x^2 + 4x + 1$

Step 1: The parabola opens down because $a < 0$.
Step 2: find the vertex:

$$x = -\frac{b}{2a} = -\frac{4}{2(-1)} = 2$$

$f(2) = -2^2 + 4(2) + 1 = 5$

The vertex is (2, 5).
Step 3: find the x-intercepts:

$0 = -x^2 + 4x + 1$

$$x = \frac{-b \pm \sqrt{b^2 - 4ac}}{2a}$$

$$x = \frac{-4 \pm \sqrt{4^2 - 4(-1)(1)}}{2(-1)}$$

$$x = \frac{-4 \pm \sqrt{20}}{-2}$$

$x = 2 \pm \sqrt{5}$

The x-intercepts are $x \approx -0.2$ and $x \approx -4.2$.
Step 4: find the y-intercept:

$f(0) = -0^2 + 4(0) + 1 = 1$

Step 5: The axis of symmetry is $x = 2$.

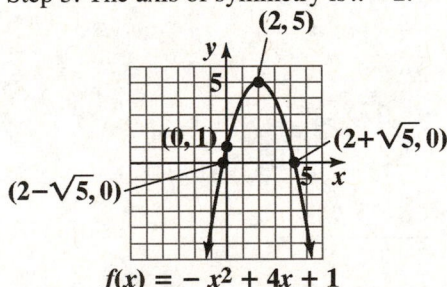

$f(x) = -x^2 + 4x + 1$

4. $f(x) = 4x^2 - 16x + 1000$

a. $a = 4$. The parabola opens upward and has a minimum value.

b. $x = \dfrac{-b}{2a} = \dfrac{16}{8} = 2$

$f(2) = 4(2)^2 - 16(2) + 1000 = 984$

The minimum point is 984 at $x = 2$.

c. domain: $(-\infty, \infty)$ range: $[984, \infty)$

5. $y = -0.005x^2 + 2x + 5$

a. The information needed is found at the vertex.
x-coordinate of vertex

$$x = \frac{-b}{2a} = \frac{-2}{2(-0.005)} = 200$$

y-coordinate of vertex

$$y = -0.005(200)^2 + 2(200) + 5 = 205$$

The vertex is $(200, 205)$.
The maximum height of the arrow is 205 feet.
This occurs 200 feet from its release.

b. The arrow will hit the ground when the height reaches 0.

$$y = -0.005x^2 + 2x + 5$$
$$0 = -0.005x^2 + 2x + 5$$
$$x = \frac{-b \pm \sqrt{b^2 - 4ac}}{2a}$$
$$x = \frac{-2 \pm \sqrt{2^2 - 4(-0.005)(5)}}{2(-0.005)}$$

$x \approx -2$ or $x \approx 402$

The arrow travels 402 feet before hitting the ground.

c. The starting point occurs when $x = 0$. Find the corresponding y-coordinate.

$$y = -0.005(0)^2 + 2(0) + 5 = 5$$

Plot $(0, 5)$, $(402, 0)$, and $(200, 205)$, and connect them with a smooth curve.

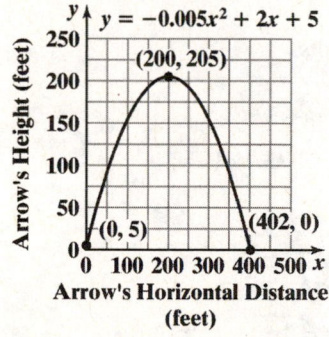

6. Let x = one of the numbers;
$x - 8$ = the other number.

The product is $f(x) = x(x - 8) = x^2 - 8x$
The x-coordinate of the minimum is

$$x = -\frac{b}{2a} = -\frac{-8}{2(1)} = -\frac{-8}{2} = 4.$$

$$f(4) = (4)^2 - 8(4)$$
$$= 16 - 32 = -16$$

The vertex is $(4, -16)$.

The minimum product is -16. This occurs when the two number are 4 and $4 - 8 = -4$.

7. Maximize the area of a rectangle constructed with 120 feet of fencing.
Let x = the length of the rectangle. Let y = the width of the rectangle.
Since we need an equation in one variable, use the perimeter to express y in terms of x.

$$2x + 2y = 120$$
$$2y = 120 - 2x$$
$$y = \frac{120 - 2x}{2} = 60 - x$$

We need to maximize $A = xy = x(60 - x)$. Rewrite A as a function of x.

$$A(x) = x(60 - x) = -x^2 + 60x$$

Since $a = -1$ is negative, we know the function opens downward and has a maximum at

$$x = -\frac{b}{2a} = -\frac{60}{2(-1)} = -\frac{60}{-2} = 30.$$

When the length x is 30, the width y is
$$y = 60 - x = 60 - 30 = 30.$$

The dimensions of the rectangular region with maximum area are 30 feet by 30 feet. This gives an area of $30 \cdot 30 = 900$ square feet.

Exercise Set 2.2

1.   vertex: (1, 1)

    $h(x) = (x-1)^2 + 1$

3.   vertex: (1, –1)

    $j(x) = (x-1)^2 - 1$

5.   The graph is $f(x) = x^2$ translated down one.

    $h(x) = x^2 - 1$

7.   The point (1, 0) is on the graph and

    $g(1) = 0.$ $g(x) = x^2 - 2x + 1$

9. $f(x) = 2(x-3)^2 + 1$

    $h = 3, k = 1$

    The vertex is at (3, 1).

11. $f(x) = -2(x+1)^2 + 5$

    $h = -1, k = 5$

    The vertex is at (–1, 5).

13. $f(x) = 2x^2 - 8x + 3$

    $x = \dfrac{-b}{2a} = \dfrac{8}{4} = 2$

    $f(2) = 2(2)^2 - 8(2) + 3$

    $= 8 - 16 + 3 = -5$

    The vertex is at (2, –5).

15. $f(x) = -x^2 - 2x + 8$

    $x = \dfrac{-b}{2a} = \dfrac{2}{-2} = -1$

    $f(-1) = -(-1)^2 - 2(-1) + 8$

    $= -1 + 2 + 8 = 9$

    The vertex is at (–1, 9).

17. $f(x) = (x-4)^2 - 1$

    vertex: (4, –1)

    x-intercepts:

    $0 = (x-4)^2 - 1$

    $1 = (x-4)^2$

    $\pm 1 = x - 4$

    $x = 3$ or $x = 5$

    y-intercept:

    $f(0) = (0-4)^2 - 1 = 15$

    The axis of symmetry is $x = 4$.

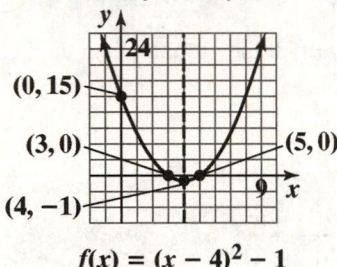

    $f(x) = (x - 4)^2 - 1$

    domain: $(-\infty, \infty)$

    range: $[-1, \infty)$

19. $f(x) = (x-1)^2 + 2$

    vertex: (1, 2)

    x-intercepts:

    $0 = (x-1)^2 + 2$

    $(x-1)^2 = -2$

    $x - 1 = \pm\sqrt{-2}$

    $x = 1 \pm i\sqrt{2}$

    No x-intercepts.

    y-intercept:

    $f(0) = (0-1)^2 + 2 = 3$

    The axis of symmetry is $x = 1$.

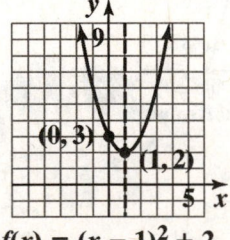

    $f(x) = (x - 1)^2 + 2$

    domain: $(-\infty, \infty)$

    range: $[2, \infty)$

21. $y - 1 = (x - 3)^2$

$y = (x - 3)^2 + 1$

vertex: (3, 1)

x-intercepts:

$0 = (x - 3)^2 + 1$

$(x - 3)^2 = -1$

$x - 3 = \pm i$

$x = 3 \pm i$

No x-intercepts.

y-intercept: 10

$y = (0 - 3)^2 + 1 = 10$

The axis of symmetry is $x = 3$.

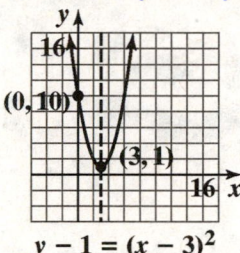

$y - 1 = (x - 3)^2$

domain: $(-\infty, \infty)$

range: $[1, \infty)$

23. $f(x) = 2(x + 2)^2 - 1$

vertex: (−2, −1)

x-intercepts:

$0 = 2(x + 2)^2 - 1$

$2(x + 2)^2 = 1$

$(x + 2)^2 = \dfrac{1}{2}$

$x + 2 = \pm \dfrac{1}{\sqrt{2}}$

$x = -2 \pm \dfrac{1}{\sqrt{2}} = -2 \pm \dfrac{\sqrt{2}}{2}$

y-intercept:

$f(0) = 2(0 + 2)^2 - 1 = 7$

The axis of symmetry is $x = -2$.

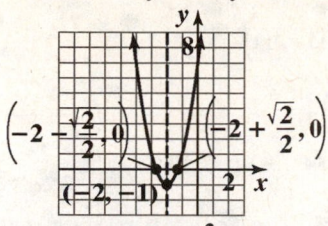

$f(x) = 2(x + 2)^2 - 1$

domain: $(-\infty, \infty)$

range: $[-1, \infty)$

25. $f(x) = 4 - (x - 1)^2$

$f(x) = -(x - 1)^2 + 4$

vertex: (1, 4)

x-intercepts:

$0 = -(x - 1)^2 + 4$

$(x - 1)^2 = 4$

$x - 1 = \pm 2$

$x = -1$ or $x = 3$

y-intercept:

$f(x) = -(0 - 1)^2 + 4 = 3$

The axis of symmetry is $x = 1$.

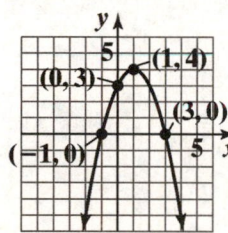

$f(x) = 4 - (x - 1)^2$

domain: $(-\infty, \infty)$

range: $(-\infty, 4]$

27. $f(x) = x^2 - 2x - 3$

$f(x) = (x^2 - 2x + 1) - 3 - 1$

$f(x) = (x-1)^2 - 4$

vertex: $(1, -4)$

x-intercepts:

$0 = (x-1)^2 - 4$

$(x-1)^2 = 4$

$x - 1 = \pm 2$

$x = -1$ or $x = 3$

y-intercept: -3

$f(0) = 0^2 - 2(0) - 3 = -3$

The axis of symmetry is $x = 1$.

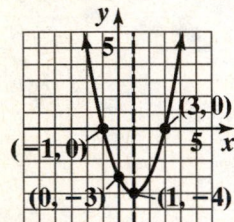

$f(x) = x^2 + 3x - 10$

domain: $(-\infty, \infty)$

range: $[-4, \infty)$

29. $f(x) = x^2 + 3x - 10$

$f(x) = \left(x^2 + 3x + \dfrac{9}{4}\right) - 10 - \dfrac{9}{4}$

$f(x) = \left(x + \dfrac{3}{2}\right)^2 - \dfrac{49}{4}$

vertex: $\left(-\dfrac{3}{2}, -\dfrac{49}{4}\right)$

x-intercepts:

$0 = \left(x + \dfrac{3}{2}\right)^2 - \dfrac{49}{4}$

$\left(x + \dfrac{3}{2}\right)^2 = \dfrac{49}{4}$

$x + \dfrac{3}{2} = \pm \dfrac{7}{2}$

$x = -\dfrac{3}{2} \pm \dfrac{7}{2}$

$x = 2$ or $x = -5$

y-intercept:

$f(x) = 0^2 + 3(0) - 10 = -10$

The axis of symmetry is $x = -\dfrac{3}{2}$.

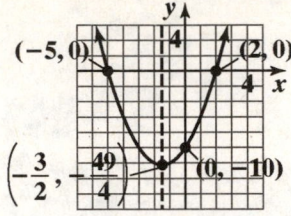

$f(x) = x^2 + 3x - 10$

domain: $(-\infty, \infty)$

range: $\left[-\dfrac{49}{4}, \infty\right)$

31. $f(x) = 2x - x^2 + 3$

$f(x) = -x^2 + 2x + 3$

$f(x) = -(x^2 - 2x + 1) + 3 + 1$

$f(x) = -(x-1)^2 + 4$

vertex: $(1, 4)$

x-intercepts:

$0 = -(x-1)^2 + 4$

$(x-1)^2 = 4$

$x - 1 = \pm 2$

$x = -1$ or $x = 3$

y-intercept:

$f(0) = 2(0) - (0)^2 + 3 = 3$

The axis of symmetry is $x = 1$.

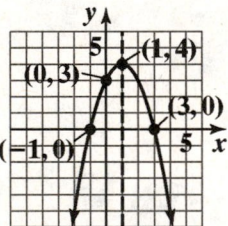

$f(x) = 2x - x^2 + 3$

domain: $(-\infty, \infty)$

range: $(-\infty, 4]$

33. $f(x) = x^2 + 6x + 3$

$f(x) = (x^2 + 6x + 9) + 3 - 9$

$f(x) = (x+3)^2 - 6$

vertex: $(-3, -6)$

x-intercepts:

$0 = (x+3)^2 - 6$

$(x+3)^2 = 6$

$x + 3 = \pm\sqrt{6}$

$x = -3 \pm \sqrt{6}$

y-intercept:

$f(0) = (0)^2 + 6(0) + 3$

$f(0) = 3$

The axis of symmetry is $x = -3$.

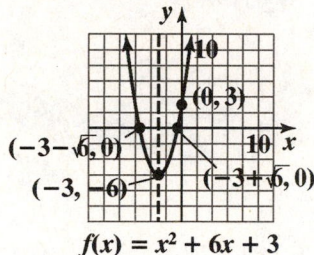

$$f(x) = x^2 + 6x + 3$$

domain: $(-\infty, \infty)$

range: $[-6, \infty)$

35. $f(x) = 2x^2 + 4x - 3$

$f(x) = 2(x^2 + 2x \quad) - 3$

$f(x) = 2(x^2 + 2x + 1) - 3 - 2$

$f(x) = 2(x+1)^2 - 5$

vertex: $(-1, -5)$

x-intercepts:

$0 = 2(x+1)^2 - 5$

$2(x+1)^2 = 5$

$(x+1)^2 = \dfrac{5}{2}$

$x + 1 = \pm\sqrt{\dfrac{5}{2}}$

$x = -1 \pm \dfrac{\sqrt{10}}{2}$

y-intercept:

$f(0) = 2(0)^2 + 4(0) - 3$

$f(0) = -3$

The axis of symmetry is $x = -1$.

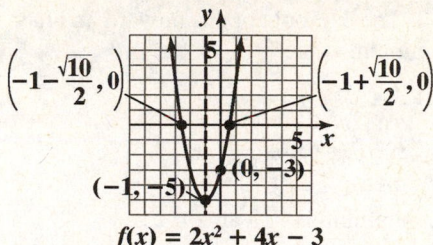

$$f(x) = 2x^2 + 4x - 3$$

domain: $(-\infty, \infty)$

range: $[-5, \infty)$

37. $f(x) = 2x - x^2 - 2$

$f(x) = -x^2 + 2x - 2$

$f(x) = -(x^2 - 2x + 1) - 2 + 1$

$f(x) = -(x-1)^2 - 1$

vertex: $(1, -1)$

x-intercepts:

$0 = -(x-1)^2 - 1$

$(x-1)^2 = -1$

$x - 1 = \pm i$

$x = 1 \pm i$

No x-intercepts.

y-intercept:

$f(0) = 2(0) - (0)^2 - 2 = -2$

The axis of symmetry is $x = 1$.

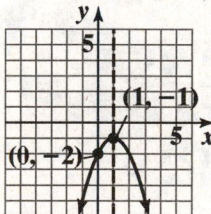

$$f(x) = 2x - x^2 - 2$$

domain: $(-\infty, \infty)$

range: $(-\infty, -1]$

39. $f(x) = 3x^2 - 12x - 1$

   a. $a = 3$. The parabola opens upward and has a minimum value.

   b. $x = \dfrac{-b}{2a} = \dfrac{12}{6} = 2$

       $f(2) = 3(2)^2 - 12(2) - 1$
       $= 12 - 24 - 1 = -13$
       The minimum is -13 at $x = 2$.

   c. domain: $(-\infty, \infty)$ range: $[-13, \infty)$

41. $f(x) = -4x^2 + 8x - 3$

   a. $a = -4$. The parabola opens downward and has a maximum value.

   b. $x = \dfrac{-b}{2a} = \dfrac{-8}{-8} = 1$

       $f(1) = -4(1)^2 + 8(1) - 3$
          $= -4 + 8 - 3 = 1$
       The maximum is 1 at $x = 1$.

   c. domain: $(-\infty, \infty)$ range: $(-\infty, 1]$

43. $f(x) = 5x^2 - 5x$

   a. $a = 5$. The parabola opens upward and has a minimum value.

   b. $x = \dfrac{-b}{2a} = \dfrac{5}{10} = \dfrac{1}{2}$

       $f\left(\dfrac{1}{2}\right) = 5\left(\dfrac{1}{2}\right)^2 - 5\left(\dfrac{1}{2}\right)$

       $= \dfrac{5}{4} - \dfrac{5}{2} = \dfrac{5}{4} - \dfrac{10}{4} = \dfrac{-5}{4}$

       The minimum is $\dfrac{-5}{4}$ at $x = \dfrac{1}{2}$.

   c. domain: $(-\infty, \infty)$ range: $\left[\dfrac{-5}{4}, \infty\right)$

45. Since the parabola opens up, the vertex $(-1, -2)$ is a minimum point.
domain: $(-\infty, \infty)$. range: $[-2, \infty)$

47. Since the parabola has a maximum, it opens down from the vertex $(10, -6)$.
domain: $(-\infty, \infty)$. range: $(-\infty, -6]$

49. $(h, k) = (5, 3)$

   $f(x) = 2(x - h)^2 + k = 2(x - 5)^2 + 3$

51. $(h, k) = (-10, -5)$

   $f(x) = 2(x - h)^2 + k$
       $= 2[x - (-10)]^2 + (-5)$
       $= 2(x + 10)^2 - 5$

53. Since the vertex is a maximum, the parabola opens down and $a = -3$.

   $(h, k) = (-2, 4)$

   $f(x) = -3(x - h)^2 + k$
       $= -3[x - (-2)]^2 + 4$
       $= -3(x + 2)^2 + 4$

55. Since the vertex is a minimum, the parabola opens up and $a = 3$.

   $(h, k) = (11, 0)$

   $f(x) = 3(x - h)^2 + k$
       $= 3(x - 11)^2 + 0$
       $= 3(x - 11)^2$

57. **a.** $y = -0.01x^2 + 0.7x + 6.1$

       $a = -0.01$, $b = 0.7$, $c = 6.1$

       x-coordinate of vertex

       $\dfrac{-b}{2a} = \dfrac{-0.7}{2(-0.01)} = 35$

       y-coordinate of vertex

       $y = -0.01x^2 + 0.7x + 6.1$

       $y = -0.01(35)^2 + 0.7(35) + 6.1 = 18.35$

       The maximum height of the shot is about 18.35 feet. This occurs 35 feet from its point of release.

   b. The ball will reach the maximum horizontal distance when its height returns to 0.

       $y = -0.01x^2 + 0.7x + 6.1$

       $0 = -0.01x^2 + 0.7x + 6.1$

       $a = -0.01$, $b = 0.7$, $c = 6.1$

       $x = \dfrac{-b \pm \sqrt{b^2 - 4ac}}{2a}$

       $x = \dfrac{-0.7 \pm \sqrt{0.7^2 - 4(-0.01)(6.1)}}{2(-0.01)}$

       $x \approx 77.8$ or $x \approx -7.8$

       The maximum horizontal distance is 77.8 feet.

c. The initial height can be found at $x = 0$.

$$y = -0.01x^2 + 0.7x + 6.1$$

$$y = -0.01(0)^2 + 0.7(0) + 6.1 = 6.1$$

The shot was released at a height of 6.1 feet.

59. $f(x) = 0.004x^2 - 0.094x + 2.6$

a. $f(25) = 0.004(25)^2 - 0.094(25) + 2.6$

$$= 2.75$$

According to the function, U.S. adult wine consumption in 2005 was 2.75 gallons per person. This underestimates the graph's value by 0.05 gallon.

b. $year = -\dfrac{b}{2a} = -\dfrac{-0.094}{2(0.004)} \approx 12$

Wine consumption was at a minimum about 12 years after 1980, or 1992.

$$f(12) = 0.004(12)^2 - 0.094(12) + 2.6 \approx 2.048$$

Wine consumption was about 2.048 gallons per U.S. adult in 1992.

This seems reasonable as compared to the values in the graph.

61. Let x = one of the numbers;

$16 - x$ = the other number.

The product is $f(x) = x(16 - x)$

$$= 16x - x^2 = -x^2 + 16x$$

The x-coordinate of the maximum is

$$x = -\dfrac{b}{2a} = -\dfrac{16}{2(-1)} = -\dfrac{16}{-2} = 8.$$

$$f(8) = -8^2 + 16(8) = -64 + 128 = 64$$

The vertex is $(8, 64)$. The maximum product is 64. This occurs when the two number are 8 and $16 - 8 = 8$.

63. Let x = one of the numbers;

$x - 16$ = the other number.

The product is $f(x) = x(x - 16) = x^2 - 16x$

The x-coordinate of the minimum is

$$x = -\dfrac{b}{2a} = -\dfrac{-16}{2(1)} = -\dfrac{-16}{2} = 8.$$

$$f(8) = (8)^2 - 16(8)$$

$$= 64 - 128 = -64$$

The vertex is $(8, -64)$. The minimum product is -64. This occurs when the two number are 8 and $8 - 16 = -8$.

65. Maximize the area of a rectangle constructed along a river with 600 feet of fencing.

Let x = the width of the rectangle;

$600 - 2x$ = the length of the rectangle

We need to maximize.

$$A(x) = x(600 - 2x)$$

$$= 600x - 2x^2 = -2x^2 + 600x$$

Since $a = -2$ is negative, we know the function opens downward and has a maximum at

$$x = -\dfrac{b}{2a} = -\dfrac{600}{2(-2)} = -\dfrac{600}{-4} = 150.$$

When the width is $x = 150$ feet, the length is

$$600 - 2(150) = 600 - 300 = 300 \text{ feet.}$$

The dimensions of the rectangular plot with maximum area are 150 feet by 300 feet. This gives an area of $150 \cdot 300 = 45{,}000$ square feet.

67. Maximize the area of a rectangle constructed with 50 yards of fencing.

Let x = the length of the rectangle. Let y = the width of the rectangle.

Since we need an equation in one variable, use the perimeter to express y in terms of x.

$$2x + 2y = 50$$

$$2y = 50 - 2x$$

$$y = \dfrac{50 - 2x}{2} = 25 - x$$

We need to maximize $A = xy = x(25 - x)$. Rewrite A as a function of x.

$$A(x) = x(25 - x) = -x^2 + 25x$$

Since $a = -1$ is negative, we know the function opens downward and has a maximum at

$$x = -\dfrac{b}{2a} = -\dfrac{25}{2(-1)} = -\dfrac{25}{-2} = 12.5.$$

When the length x is 12.5, the width y is

$$y = 25 - x = 25 - 12.5 = 12.5.$$

The dimensions of the rectangular region with maximum area are 12.5 yards by 12.5 yards. This gives an area of $12.5 \cdot 12.5 = 156.25$ square yards.

69. Maximize the area of the playground with 600 feet of fencing.

Let x = the length of the rectangle. Let y = the width of the rectangle.

Since we need an equation in one variable, use the perimeter to express y in terms of x.

$$2x + 3y = 600$$

$$3y = 600 - 2x$$

$$y = \frac{600 - 2x}{3}$$

$$y = 200 - \frac{2}{3}x$$

We need to maximize $A = xy = x\left(200 - \frac{2}{3}x\right)$.

Rewrite A as a function of x.

$$A(x) = x\left(200 - \frac{2}{3}x\right) = -\frac{2}{3}x^2 + 200x$$

Since $a = -\frac{2}{3}$ is negative, we know the function opens downward and has a maximum at

$$x = -\frac{b}{2a} = -\frac{200}{2\left(-\frac{2}{3}\right)} = -\frac{200}{-\frac{4}{3}} = 150.$$

When the length x is 150, the width y is

$$y = 200 - \frac{2}{3}x = 200 - \frac{2}{3}(150) = 100.$$

The dimensions of the rectangular playground with maximum area are 150 feet by 100 feet. This gives an area of $150 \cdot 100 = 15,000$ square feet.

71. Maximize the cross-sectional area of the gutter:

$$A(x) = x(20 - 2x)$$

$$= 20x - 2x^2 = -2x^2 + 20x.$$

Since $a = -2$ is negative, we know the function opens downward and has a maximum at

$$x = -\frac{b}{2a} = -\frac{20}{2(-2)} = -\frac{20}{-4} = 5.$$

When the height x is 5, the width is

$$20 - 2x = 20 - 2(5) = 20 - 10 = 10.$$

$$A(5) = -2(5)^2 + 20(5)$$

$$= -2(25) + 100 = -50 + 100 = 50$$

The maximum cross-sectional area is 50 square inches. This occurs when the gutter is 5 inches deep and 10 inches wide.

73. x = increase

$$A = (50 + x)(8000 - 100x)$$

$$= 400,000 + 3000x - 100x^2$$

$$x = \frac{-b}{2a} = \frac{-3000}{2(-100)} = 15$$

The maximum price is $50 + 15 = \$65$.
The maximum revenue $= 65(800 - 100 \cdot 15) = \$422,500$.

75. x = increase

$$A = (20 + x)(60 - 2x)$$

$$= 1200 + 20x - 2x^2$$

$$x = \frac{-b}{2a} = \frac{-20}{2(-2)} = 5$$

The maximum number of trees is $20 + 5 = 25$ trees.
The maximum yield is $60 - 2 \cdot 5 = 50$ pounds per tree, $50 \times 25 = 1250$ pounds.

77. – 83. Answers may vary.

85. $y = -0.25x^2 + 40x$

$$x = \frac{-b}{2a} = \frac{-40}{-0.5} = 80$$

$$y = -0.25(80)^2 + 40(80)$$

$$= 1600$$

vertex: (80, 1600)

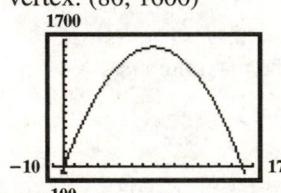

87. $y = 5x^2 + 40x + 600$

$$x = \frac{-b}{2a} = \frac{-40}{10} = -4$$

$$y = 5(-4)^2 + 40(-4) + 600$$

$$= 80 - 160 + 600 = 520$$

vertex: (-4, 520)

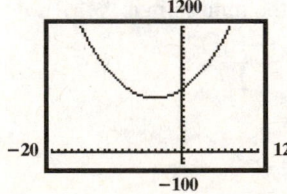

89. a. The values of y increase then decrease.

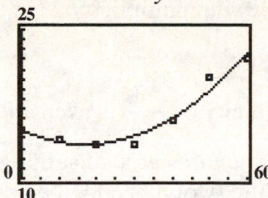

b. $y = 0.005x^2 - 0.170x + 14.817$

c. $x = \dfrac{-(-0.170)}{2(.005)} = 17; \quad 1940 + 17 = 1957$

$y = 0.005(17)^2 - 0.170(17) + 14.817$

≈ 13.372

The worst gas mileage was 13.372 mpg in 1957.

d.

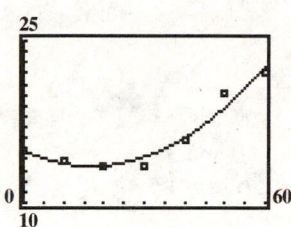

91. makes sense

93. does not make sense; Explanations will vary. Sample explanation: The football's path is better described by a quadratic model.

95. false; Changes to make the statement true will vary. A sample change is: The vertex is $(5, -1)$.

97. false; Changes to make the statement true will vary. A sample change is: The x-coordinate of the maximum is $-\dfrac{b}{2a} = -\dfrac{1}{2(-1)} = -\dfrac{1}{-2} = \dfrac{1}{2}$ and the y-coordinate of the vertex of the parabola is

$f\left(-\dfrac{b}{2a}\right) = f\left(\dfrac{1}{2}\right) = \dfrac{5}{4}.$

The maximum y–value is $\dfrac{5}{4}$.

99. Vertex (3, 2) Axis: $x = 3$
second point (0, 11)

101. We know $(h, k) = (-3, -1)$, so the equation is of the form $f(x) = a(x - h)^2 + k$

$= a\left[x - (-3)\right]^2 + (-1)$

$= a(x + 3)^2 - 1$

We use the point $(-2, -3)$ on the graph to determine the value of a: $f(x) = a(x + 3)^2 - 1$

$-3 = a(-2 + 3)^2 - 1$

$-3 = a(1)^2 - 1$

$-3 = a - 1$

$-2 = a$

Thus, the equation of the parabola is

$f(x) = -2(x + 3)^2 - 1$.

103. $f(x) = (80 + x)(300 - 3x) - 10(300 - 3x)$

$= 24000 + 60x - 3x^2 - 3000 + 30x$

$= -3x^2 + 90x + 21000$

$x = \dfrac{-b}{2a} = \dfrac{-90}{2(-3)} = \dfrac{3}{2} = 15$

The maximum charge is 80 + 15 = \$95.00. the maximum profit is $-3(15)^2 + 9(15) + 21000 =$ \$21,675.

105. Answers may vary.

106. $x^3 + 3x^2 - x - 3 = x^2(x + 3) - 1(x + 3)$

$= (x + 3)(x^2 - 1)$

$= (x + 3)(x + 1)(x - 1)$

107. $f(x) = x^3 - 2x - 5$

$f(2) = (2)^3 - 2(2) - 5 = -1$

$f(3) = (3)^3 - 2(3) - 5 = 16$

The graph passes through (2, –1), which is below the x-axis, and (3, 16), which is above the x-axis. Since the graph of f is continuous, it must cross the x-axis somewhere between 2 and 3 to get from one of these points to the other.

108. $f(x) = x^4 - 2x^2 + 1$

$f(-x) = (-x)^4 - 2(-x)^2 + 1$

$= x^4 - 2x^2 + 1$

Since $f(-x) = f(x)$, the function is even.

Thus, the graph is symmetric with respect to the y-axis.

Polynomial and Rational Functions

Section 2.3

Check Point Exercises

1. Since n is even and $a_n > 0$, the graph rises to the left and to the right.

2. It is not necessary to multiply out the polynomial to determine its degree. We can find the degree of the polynomial by adding the degrees of each of its factors. $f(x) = 2 \overset{\text{degree 3}}{x^3} \overset{\text{degree 1}}{(x-1)} \overset{\text{degree 1}}{(x+5)}$ has degree $3+1+1=5$.

   $f(x) = 2x^3(x-1)(x+5)$ is of odd degree with a positive leading coefficient. Thus its graph falls to the left and rises to the right.

3. Since n is odd and the leading coefficient is negative, the function falls to the right. Since the ratio cannot be negative, the model won't be appropriate.

4. The graph does not show the function's end behavior. Since $a_n > 0$ and n is odd, the graph should fall to the left.

5. $f(x) = x^3 + 2x^2 - 4x - 8$

   $0 = x^2(x+2) - 4(x+2)$

   $0 = (x+2)(x^2 - 4)$

   $0 = (x+2)^2(x-2)$

   $x = 2$ or $x = -2$

   The zeros are 2 and -2.

6. $f(x) = x^4 - 4x^2$

   $x^4 - 4x^2 = 0$

   $x^2(x^2 - 4) = 0$

   $x^2(x+2)(x-2) = 0$

   $x = 0$ or $x = -2$ or $x = 2$

   The zeros are 0, -2, and 2.

7. $f(x) = -4\left(x + \dfrac{1}{2}\right)^2 (x-5)^3$

   $-4\left(x + \dfrac{1}{2}\right)^2 (x-5)^3 = 0$

   $x = -\dfrac{1}{2}$ or $x = 5$

The zeros are $-\dfrac{1}{2}$, with multiplicity 2, and 5, with multiplicity 3.

Because the multiplicity of $-\dfrac{1}{2}$ is even, the graph touches the x-axis and turns around at this zero. Because the multiplicity of 5 is odd, the graph crosses the x-axis at this zero.

8. $f(-3) = 3(-3)^3 - 10(-3) + 9 = -42$

   $f(-2) = 3(-2)^3 - 10(-2) + 9 = 5$

   The sign change shows there is a zero between -3 and -2.

9. $f(x) = x^3 - 3x^2$

   Since $a_n > 0$ and n is odd, the graph falls to the left and rises to the right.

   $x^3 - 3x^2 = 0$

   $x^2(x-3) = 0$

   $x = 0$ or $x = 3$

   The x-intercepts are 0 and 3.

   $f(0) = 0^3 - 3(0)^2 = 0$

   The y-intercept is 0.

   $f(-x) = (-x)^3 - 3(-x)^2 = -x^3 - 3x^2$

   No symmetry.

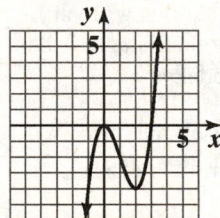

$f(x) = x^3 - 3x^2$

Exercise Set 2.3

1. polynomial function;
   degree: 3

3. polynomial function;
   degree: 5

5. not a polynomial function

7. not a polynomial function

9. not a polynomial function

11. polynomial function

180

13. Not a polynomial function because graph is not continuous.

15. (b)

17. (a)

19. $f(x) = 5x^3 + 7x^2 - x + 9$

Since $a_n > 0$ and n is odd, the graph of $f(x)$ falls to the left and rises to the right.

21. $f(x) = 5x^4 + 7x^2 - x + 9$

Since $a_n > 0$ and n is even, the graph of $f(x)$ rises to the left and to the right.

23. $f(x) = -5x^4 + 7x^2 - x + 9$

Since $a_n < 0$ and n is even, the graph of $f(x)$ falls to the left and to the right.

25. $f(x) = 2(x-5)(x+4)^2$

$x = 5$ has multiplicity 1;
The graph crosses the x-axis.
$x = -4$ has multiplicity 2;
The graph touches the x-axis and turns around.

27. $f(x) = 4(x-3)(x+6)^3$

$x = 3$ has multiplicity 1;
The graph crosses the x-axis.
$x = -6$ has multiplicity 3;
The graph crosses the x-axis.

29. $f(x) = x^3 - 2x^2 + x$

$\qquad = x(x^2 - 2x + 1)$

$\qquad = x(x-1)^2$

$x = 0$ has multiplicity 1;
The graph crosses the x-axis.
$x = 1$ has multiplicity 2;
The graph touches the x-axis and turns around.

31. $f(x) = x^3 + 7x^2 - 4x - 28$

$\qquad = x^2(x+7) - 4(x+7)$

$\qquad = (x^2 - 4)(x+7)$

$\qquad = (x-2)(x+2)(x+7)$

$x = 2$, $x = -2$ and $x = -7$ have multiplicity 1;
The graph crosses the x-axis.

33. $f(x) = x^3 - x - 1$

$f(1) = -1$
$f(2) = 5$
The sign change shows there is a zero between the given values.

35. $f(x) = 2x^4 - 4x^2 + 1$

$f(-1) = -1$
$f(0) = 1$
The sign change shows there is a zero between the given values.

37. $f(x) = x^3 + x^2 - 2x + 1$

$f(-3) = -11$
$f(-2) = 1$
The sign change shows there is a zero between the given values.

39. $f(x) = 3x^3 - 10x + 9$

$f(-3) = -42$
$f(-2) = 5$
The sign change shows there is a zero between the given values.

41. $f(x) = x^3 + 2x^2 - x - 2$

a. Since $a_n > 0$ and n is odd, $f(x)$ rises to the right and falls to the left.

b. $\qquad x^3 + 2x^2 - x - 2 = 0$

$\qquad x^2(x+2) - (x+2) = 0$

$\qquad\qquad (x+2)(x^2 - 1) = 0$

$\qquad (x+2)(x-1)(x+1) = 0$

$\qquad x = -2,\ x = 1,\ x = -1$

The zeros at -2, -1, and 1 have odd multiplicity so $f(x)$ crosses the x-axis at these points.

c. $f(0) = (0)^3 + 2(0)^2 - 0 - 2$

$\qquad = -2$

The y-intercept is -2.

d. $f(-x) = (-x) + 2(-x)^2 - (-x) - 2$

$\qquad = -x^3 + 2x^2 + x - 2$

$-f(x) = -x^3 - 2x^2 + x + 2$

The graph has neither origin symmetry nor y-axis symmetry.

e. The graph has 2 turning points and $2 \le 3 - 1$.

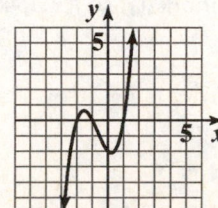

$y = x^3 + 2x^2 - x - 2$

43. $f(x) = x^4 - 9x^2$

a. Since $a_n > 0$ and n is even, $f(x)$ rises to the left and the right.

b.
$$x^4 - 9x^2 = 0$$
$$x^2(x^2 - 9) = 0$$
$$x^2(x - 3)(x + 3) = 0$$
$$x = 0, x = 3, x = -3$$

The zeros at -3 and 3 have odd multiplicity, so $f(x)$ crosses the x-axis at these points. The root at 0 has even multiplicity, so $f(x)$ touches the x-axis at 0.

c. $f(0) = (0)^4 - 9(0)^2 = 0$
The y-intercept is 0.

d. $f(-x) = x^4 - 9x^2$
$f(-x) = f(x)$
The graph has y-axis symmetry.

e. The graph has 3 turning points and $3 \le 4 - 1$.

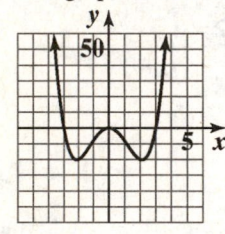

$f(x) = x^4 - 9x^2$

45. $f(x) = -x^4 + 16x^2$

a. Since $a_n < 0$ and n is even, $f(x)$ falls to the left and the right.

b.
$$-x^4 + 16x^2 = 0$$
$$x^2(-x^2 + 16) = 0$$
$$x^2(4 - x)(4 + x) = 0$$
$$x = 0, x = 4, x = -4$$
The zeros at -4 and 4 have odd multiplicity, so $f(x)$ crosses the x-axis at these points. The root at 0 has even multiplicity, so $f(x)$ touches the x-axis at 0.

c. $f(0) = (0)^4 - 9(0)^2 = 0$
The y-intercept is 0.

d. $f(-x) = -x^4 + 16x^2$
$f(-x) = f(x)$
The graph has y-axis symmetry.

e. The graph has 3 turning points and $3 \le 4 - 1$.

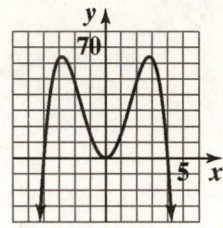

$f(x) = -x^4 + 16x^2$

47. $f(x) = x^4 - 2x^3 + x^2$

a. Since $a_n > 0$ and n is even, $f(x)$ rises to the left and the right.

b.
$$x^4 - 2x^3 + x^2 = 0$$
$$x^2(x^2 - 2x + 1) = 0$$
$$x^2(x - 1)(x - 1) = 0$$
$$x = 0, x = 1$$
The zeros at 1 and 0 have even multiplicity, so $f(x)$ touches the x-axis at 0 and 1.

c. $f(0) = (0)^4 - 2(0)^3 + (0)^2 = 0$
The y-intercept is 0.

d. $f(-x) = x^4 + 2x^3 + x^2$
The graph has neither y-axis nor origin symmetry.

e. The graph has 3 turning points and $3 \le 4 - 1$.

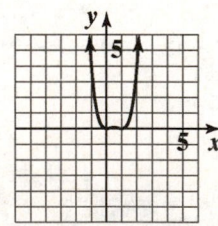

$f(x) = x^4 - 2x^3 + x^2$

49. $f(x) = -2x^4 + 4x^3$

a. Since $a_n < 0$ and n is even, $f(x)$ falls to the left and the right.

b. $-2x^4 + 4x^3 = 0$
$x^3(-2x + 4) = 0$
$x = 0, x = 2$
The zeros at 0 and 2 have odd multiplicity, so $f(x)$ crosses the x-axis at these points.

c. $f(0) = -2(0)^4 + 4(0)^3 = 0$
The y-intercept is 0.

d. $f(-x) = -2x^4 - 4x^3$
The graph has neither y-axis nor origin symmetry.

e. The graph has 1 turning point and $1 \le 4 - 1$.

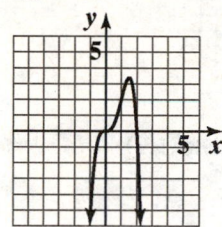

$f(x) = -2x^4 + 4x^3$

51. $f(x) = 6x^3 - 9x - x^5$

a. Since $a_n < 0$ and n is odd, $f(x)$ rises to the left and falls to the right.

b. $-x^5 + 6x^3 - 9x = 0$
$-x(x^4 - 6x^2 + 9) = 0$
$-x(x^2 - 3)(x^2 - 3) = 0$
$x = 0, x = \pm\sqrt{3}$
The root at 0 has odd multiplicity so $f(x)$ crosses the x-axis at $(0, 0)$. The zeros at $-\sqrt{3}$ and $\sqrt{3}$ have even multiplicity so $f(x)$ touches the x-axis at $\sqrt{3}$ and $-\sqrt{3}$.

c. $f(0) = -(0)^5 + 6(0)^3 - 9(0) = 0$
The y-intercept is 0.

d. $f(-x) = x^5 - 6x^3 + 9x$
$f(-x) = -f(x)$
The graph has origin symmetry.

e. The graph has 4 turning point and $4 \le 5 - 1$.

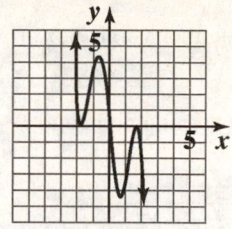

$f(x) = 6x^3 - 9x - x^5$

53. $f(x) = 3x^2 - x^3$

a. Since $a_n < 0$ and n is odd, $f(x)$ rises to the left and falls to the right.

b. $-x^3 + 3x^2 = 0$
$-x^2(x - 3) = 0$
$x = 0, x = 3$
The zero at 3 has odd multiplicity so $f(x)$ crosses the x-axis at that point. The root at 0 has even multiplicity so $f(x)$ touches the axis at $(0, 0)$.

c. $f(0) = -(0)^3 + 3(0)^2 = 0$
The y-intercept is 0.

d. $f(-x) = x^3 + 3x^2$
The graph has neither y-axis nor origin symmetry.

e. The graph has 2 turning point and $2 \le 3 - 1$.

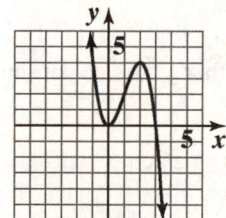

$f(x) = 3x^2 - x^3$

55. $f(x) = -3(x-1)^2(x^2 - 4)$

a. Since $a_n < 0$ and n is even, $f(x)$ falls to the left and the right.

b. $-3(x-1)^2(x^2 - 4) = 0$
$x = 1, x = -2, x = 2$
The zeros at -2 and 2 have odd multiplicity, so $f(x)$ crosses the x-axis at these points. The root at 1 has even multiplicity, so $f(x)$ touches the x-axis at $(1, 0)$.

c. $f(0) = -3(0-1)^2(0^2 - 4)^3$
$\quad = -3(1)(-4) = 12$
The y-intercept is 12.

d. $f(-x) = -3(-x-1)^2 (x^2-4)$
The graph has neither y-axis nor origin symmetry.

e. The graph has 1 turning point and $1 \le 4 - 1$.

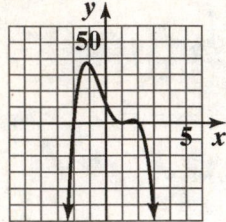

$$f(x) = -3(x-1)^2 (x^2-4)$$

57. $f(x) = x^2 (x-1)^3 (x+2)$

a. Since $a_n > 0$ and n is even, $f(x)$ rises to the left and the right.

b. $x = 0, x = 1, x = -2$
The zeros at 1 and –2 have odd multiplicity so $f(x)$ crosses the x-axis at those points. The root at 0 has even multiplicity so $f(x)$ touches the axis at (0, 0).

c. $f(0) = 0^2 (0-1)^3 (0+2) = 0$
The y-intercept is 0.

d. $f(-x) = x^2 (-x-1)^3 (-x+2)$
The graph has neither y-axis nor origin symmetry.

e. The graph has 2 turning points and $2 \le 6 - 1$.

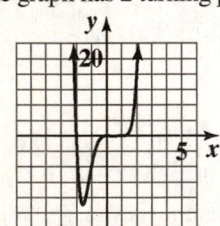

$$f(x) = x^2 (x-1)^3 (x+2)$$

59. $f(x) = -x^2 (x-1)(x+3)$

a. Since $a_n < 0$ and n is even, $f(x)$ falls to the left and the right.

b. $x = 0, x = 1, x = -3$
The zeros at 1 and –3 have odd multiplicity so $f(x)$ crosses the x-axis at those points. The root at 0 has even multiplicity so $f(x)$ touches the axis at (0, 0).

c. $f(0) = -0^2 (0-1)(0+3) = 0$
The y-intercept is 0.

d. $f(-x) = -x^2 (-x-1)(-x+3)$
The graph has neither y-axis nor origin symmetry.

e. The graph has 3 turning points and $3 \le 4 - 1$.

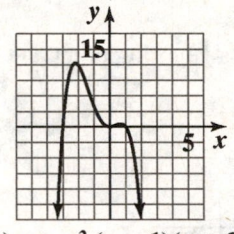

$$f(x) = -x^2 (x-1)(x+3)$$

61. $f(x) = -2x^3 (x-1)^2 (x+5)$

a. Since $a_n < 0$ and n is even, $f(x)$ falls to the left and the right.

b. $x = 0, x = 1, x = -5$
The roots at 0 and –5 have odd multiplicity so $f(x)$ crosses the x-axis at those points. The root at 1 has even multiplicity so $f(x)$ touches the axis at (1, 0).

c. $f(0) = -2(0)^3 (0-1)^2 (0+5) = 0$
The y-intercept is 0.

d. $f(-x) = 2x^3 (-x-1)^2 (-x+5)$
The graph has neither y-axis nor origin symmetry.

e. The graph has 2 turning points and $2 \le 6 - 1$.

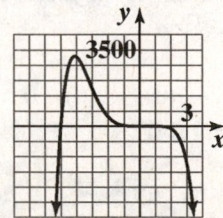

$$f(x) = -2x^3 (x-1)^2 (x+5)$$

63. $f(x) = (x-2)^2(x+4)(x-1)$

  a. Since $a_n > 0$ and n is even, $f(x)$ rises to the left and rises the right.

  b. $x = 2, x = -4, x = 1$
The zeros at -4 and 1 have odd multiplicity so $f(x)$ crosses the x-axis at those points. The root at 2 has even multiplicity so $f(x)$ touches the axis at $(2, 0)$.

  c. $f(0) = (0-2)^2(0+4)(0-1) = -16$
The y-intercept is -16.

  d. $f(-x) = (-x-2)^2(-x+4)(-x-1)$
The graph has neither y-axis nor origin symmetry.

  e. The graph has 3 turning points and $3 \le 4 - 1$.

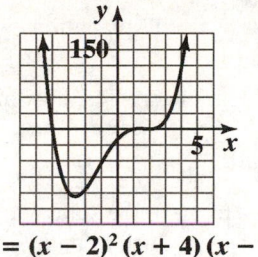

$f(x) = (x-2)^2(x+4)(x-1)$

65. a. The x-intercepts of the graph are -2, 1, and 4, so they are the zeros. Since the graph actually crosses the x-axis at all three places, all three have odd multiplicity.

  b. Since the graph has two turning points, the function must be at least of degree 3. Since -2, 1, and 4 are the zeros, $x+2$, $x-1$, and $x-4$ are factors of the function. The lowest odd multiplicity is 1. From the end behavior, we can tell that the leading coefficient must be positive. Thus, the function is
$f(x) = (x+2)(x-1)(x-4)$.

  c. $f(0) = (0+2)(0-1)(0-4) = 8$

67. a. The x-intercepts of the graph are -1 and 3, so they are the zeros. Since the graph crosses the x-axis at -1, it has odd multiplicity. Since the graph touches the x-axis and turns around at 3, it has even multiplicity.

  b. Since the graph has two turning points, the function must be at least of degree 3. Since -1 and 3 are the zeros, $x+1$ and $x-3$ are factors of the function. The lowest odd multiplicity is 1, and the lowest even multiplicity is 2. From the end behavior, we can tell that the leading coefficient must be positive. Thus, the function is $f(x) = (x+1)(x-3)^2$.

  c. $f(0) = (0+1)(0-3)^2 = 9$

69. a. The x-intercepts of the graph are -3 and 2, so they are the zeros. Since the graph touches the x-axis and turns around at both -3 and 2, both have even multiplicity.

  b. Since the graph has three turning points, the function must be at least of degree 4. Since -3 and 2 are the zeros, $x+3$ and $x-2$ are factors of the function. The lowest even multiplicity is 2. From the end behavior, we can tell that the leading coefficient must be negative. Thus, the function is $f(x) = -(x+3)^2(x-2)^2$.

  c. $f(0) = -(0+3)^2(0-2)^2 = -36$

71. a. The x-intercepts of the graph are -2, -1, and 1, so they are the zeros. Since the graph crosses the x-axis at -1 and 1, they both have odd multiplicity. Since the graph touches the x-axis and turns around at -2, it has even multiplicity.

  b. Since the graph has five turning points, the function must be at least of degree 6. Since -2, -1, and 1 are the zeros, $x+2$, $x+1$, and $x-1$ are factors of the function. The lowest even multiplicity is 2, and the lowest odd multiplicity is 1. However, to reach degree 6, one of the odd multiplicities must be 3. From the end behavior, we can tell that the leading coefficient must be positive. The function is
$f(x) = (x+2)^2(x+1)(x-1)^3$.

  c. $f(0) = (0+2)^2(0+1)(0-1)^3 = -4$

73. **a.** $f(x) = -3402x^2 + 42,203x + 308,453$

$f(3) = -3402(3)^2 + 42,203(3) + 308,453$

$\quad = 404,444$

$g(x) = 2769x^3 - 28,324x^2 + 107,555x + 261,931$

$g(3) = 2769(3)^3 - 28,324(3)^2 + 107,555(3) + 261,931$

$\quad = 404,443$

Function *f* provides a better description of the actual number.

b. Since the degree of *f* is even and the leading coefficient is negative, the graph falls to the right. The function will not be a useful model over an extended period of time because it will eventually give negative values.

75. **a.** The woman's heart rate was increasing from 1 through 4 minutes and from 8 through 10 minutes.

b. The woman's heart rate was decreasing from 4 through 8 minutes and from 10 through 12 minutes.

c. There were 3 turning points during the 12 minutes.

d. Since there were 3 turning points, a polynomial of degree 4 would provide the best fit.

e. The leading coefficient should be negative. The graph falls to the left and to the right.

f. The woman's heart rate reached a maximum of about 116 ± 1 beats per minute. This occurred after 10 minutes.

g. The woman's heart rate reached a minimum of about 64 ± 1 beats per minute. This occurred after 8 minutes.

77. – 93. Answers may vary.

95.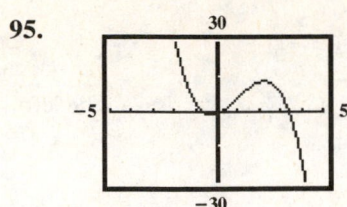

97.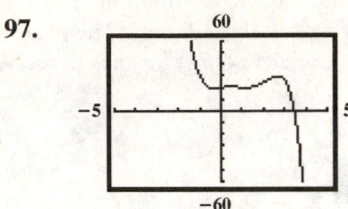

99.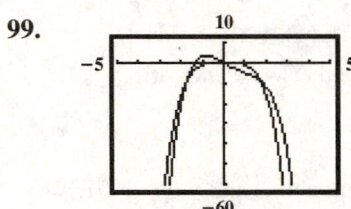

101. does not make sense; Explanations will vary. Sample explanation: Since $(x + 2)$ is raised to an odd power, the graph crosses the *x*-axis at –2.

103. makes sense

105. false; Changes to make the statement true will vary. A sample change is: Such a function falls to the right and will eventually have negative values.

107. false; Changes to make the statement true will vary. A sample change is: A function with origin symmetry either falls to the left and rises to the right, or rises to the left and falls to the right.

109. $f(x) = x^3 - 2x^2$

110. $\dfrac{737}{21} = 35 + \dfrac{2}{21}$

111. $6x^3 - x^2 - 5x + 4$

112. $2x^3 - x^2 - 11x + 6 = (x - 3)(2x^2 + 3x - 2)$

$\qquad\qquad\qquad\qquad = (x - 3)(2x - 1)(x + 2)$

Section 2.4

Check Point Exercises

1.
$$\begin{array}{r} x+5 \\ x+9{\overline{\smash{\big)}\,x^2+14x+45}} \\ \underline{x^2+9x} \\ 5x+45 \\ \underline{5x+45} \\ 0 \end{array}$$
The answer is $x+5$.

2.
$$\begin{array}{r} 2x^2+3x-2 \\ x-3{\overline{\smash{\big)}\,2x^3-3x^2-11x+7}} \\ \underline{2x^3-6x^2} \\ 3x^2-11x \\ \underline{3x^2-9x} \\ -2x+7 \\ \underline{-2x+6} \\ 1 \end{array}$$
The answer is $2x^2+3x-2+\dfrac{1}{x-3}$.

3.
$$\begin{array}{r} 2x^2+7x+14 \\ x^2-2x{\overline{\smash{\big)}\,2x^4+3x^3+0x^2-7x-10}} \\ \underline{2x^4-4x^3} \\ 7x^3+0x^2 \\ \underline{7x^3-14x^2} \\ 14x^2-7x \\ \underline{14x^2-28x} \\ 21x-10 \end{array}$$
The answer is $2x^2+7x+14+\dfrac{21x-10}{x^2-2x}$.

4.
$$\begin{array}{r|rrrr} -2 & 1 & 0 & -7 & -6 \\ & & -2 & 4 & 6 \\ \hline & 1 & -2 & -3 & 0 \end{array}$$
The answer is x^2-2x-3.

5.
$$\begin{array}{r|rrrr} -4 & 3 & 4 & -5 & 3 \\ & & -12 & 32 & -108 \\ \hline & 3 & -8 & 27 & -105 \end{array}$$
$f(-4) = -105$

6.
$$\begin{array}{r|rrrr} -1 & 15 & 14 & -3 & -2 \\ & & -15 & 1 & 2 \\ \hline & 15 & -1 & -2 & 0 \end{array}$$
$15x^2-x-2=0$
$(3x+1)(5x-2)=0$
$x=-\dfrac{1}{3}$ or $x=\dfrac{2}{5}$
The solution set is $\left\{-1,-\dfrac{1}{3},\dfrac{2}{5}\right\}$.

Exercise Set 2.4

1.
$$\begin{array}{r} x+3 \\ x+5{\overline{\smash{\big)}\,x^2+8x+15}} \\ \underline{x^2+5x} \\ 3x+15 \\ \underline{3x+15} \\ 0 \end{array}$$
The answer is $x+3$.

3.
$$\begin{array}{r} x^2+3x+1 \\ x+2{\overline{\smash{\big)}\,x^3+5x^2+7x+2}} \\ \underline{x^3+2x^2} \\ 3x^2+7x \\ \underline{3x^2+6x} \\ x+2 \\ \underline{x+2} \\ 0 \end{array}$$
The answer is x^2+3x+1.

5.
$$\begin{array}{r} 2x^2+3x+5 \\ 3x-1{\overline{\smash{\big)}\,6x^3+7x^2+12x-5}} \\ \underline{6x^3-2x^2} \\ 9x^2+12x \\ \underline{9x^2-3x} \\ 15x-5 \\ \underline{15x-5} \\ 0 \end{array}$$
The answer is $2x^2+3x+5$.

7.

$$\begin{array}{r} 4x+3+\dfrac{2}{3x-2} \\ 3x-2\overline{)12x^2+x-4} \\ \underline{12x^2-8x} \\ 9x-4 \\ \underline{9x-6} \\ 2 \end{array}$$

The answer is $4x+3+\dfrac{2}{3x-2}$.

9.

$$\begin{array}{r} 2x^2+x+6-\dfrac{38}{x+3} \\ x+3\overline{)2x^3+7x^2+9x-20} \\ \underline{2x^3+6x^2} \\ x^2+9x \\ \underline{x^2+3x} \\ 6x-20 \\ \underline{6x+18} \\ -38 \end{array}$$

The answer is $2x^2+x+6-\dfrac{38}{x+3}$.

11.

$$\begin{array}{r} 4x^3+16x^2+60x+246+\dfrac{984}{x-4} \\ x-4\overline{)4x^4-4x^2+6x} \\ \underline{4x^4-16x^3} \\ 16x^3-4x^2 \\ \underline{16x^3-64x^2} \\ 60x^2+6x \\ \underline{60x^2-240x} \\ 246x \\ \underline{246x-984} \\ 984 \end{array}$$

The answer is

$$4x^3+16x^2+60x+246+\dfrac{984}{x-4}.$$

13.

$$\begin{array}{r} 2x+5 \\ 3x^2-x-3\overline{)6x^3+13x^2-11x-15} \\ \underline{6x^3-2x^2-6x} \\ 15x^2-5x-15 \\ \underline{15x^2-5x-15} \\ 0 \end{array}$$

The answer is $2x+5$.

15.

$$\begin{array}{r} 6x^2+3x-1 \\ 3x^2+1\overline{)18x^4+9x^3+3x^2} \\ \underline{18x^4+6x^2} \\ 9x^3-3x^2 \\ \underline{9x^3+3x} \\ -3x^2-3x \\ \underline{-3x^2-1} \\ -3x+1 \end{array}$$

The answer is $6x^2+3x-1-\dfrac{3x-1}{3x^2+1}$.

17. $\left(2x^2+x-10\right)\div(x-2)$

$$\begin{array}{c|rrr} 2 & 2 & 1 & -10 \\ & & 4 & 10 \\ \hline & 2 & 5 & 0 \end{array}$$

The answer is $2x+5$.

19. $\left(3x^2+7x-20\right)\div(x+5)$

$$\begin{array}{c|rrr} -5 & 3 & 7 & -20 \\ & & -15 & 40 \\ \hline & 3 & -8 & 20 \end{array}$$

The answer is $3x-8+\dfrac{20}{x+5}$.

21. $\left(4x^3-3x^2+3x-1\right)\div(x-1)$

$$\begin{array}{c|rrrr} 1 & 4 & -3 & 3 & -1 \\ & & 4 & 1 & 4 \\ \hline & 4 & 1 & 4 & 3 \end{array}$$

The answer is $4x^2+x+4+\dfrac{3}{x-1}$.

23. $\left(6x^5 - 2x^3 + 4x^2 - 3x + 1\right) \div (x - 2)$

$$
\begin{array}{r|rrrrrr}
2 & 6 & 0 & -2 & 4 & -3 & 1 \\
  &   & 12 & 24 & 44 & 96 & 186 \\
\hline
  & 6 & 12 & 22 & 48 & 93 & 187
\end{array}
$$

The answer is

$6x^4 + 12x^3 + 22x^2 + 48x + 93 + \dfrac{187}{x-2}$.

25. $\left(x^2 - 5x - 5x^3 + x^4\right) \div (5 + x) \Rightarrow$

$\left(x^4 - 5x^3 + x^2 - 5x\right) \div (x + 5)$

$$
\begin{array}{r|rrrrr}
-5 & 1 & -5 & 1 & -5 & 0 \\
   &   & -5 & 50 & -255 & 1300 \\
\hline
   & 1 & -10 & 51 & -260 & 1300
\end{array}
$$

The answer is

$x^3 - 10x^2 + 51x - 260 + \dfrac{1300}{x+5}$.

27. $\dfrac{x^5 + x^3 - 2}{x - 1}$

$$
\begin{array}{r|rrrrrr}
1 & 1 & 0 & 1 & 0 & 0 & -2 \\
  &   & 1 & 1 & 2 & 2 & 2 \\
\hline
  & 1 & 1 & 2 & 2 & 2 & 0
\end{array}
$$

The answer is $x^4 + x^3 + 2x^2 + 2x + 2$.

29. $\dfrac{x^4 - 256}{x - 4}$

$$
\begin{array}{r|rrrrr}
4 & 1 & 0 & 0 & 0 & -256 \\
  &   & 4 & 16 & 64 & 256 \\
\hline
  & 1 & 4 & 16 & 64 & 0
\end{array}
$$

The answer is $x^3 + 4x^2 + 16x + 64$.

31. $\dfrac{2x^5 - 3x^4 + x^3 - x^2 + 2x - 1}{x + 2}$

$$
\begin{array}{r|rrrrrr}
-2 & 2 & -3 & 1 & -1 & 2 & -1 \\
   &   & -4 & 14 & -30 & 62 & -128 \\
\hline
   & 2 & -7 & 15 & -31 & 64 & -129
\end{array}
$$

The answer is

$2x^4 - 7x^3 + 15x^2 - 31x + 64 - \dfrac{129}{x+2}$.

33. $f(x) = 2x^3 - 11x^2 + 7x - 5$

$$
\begin{array}{r|rrrr}
4 & 2 & -11 & 7 & -5 \\
  &   & 8 & -12 & -20 \\
\hline
  & 2 & -3 & -5 & -25
\end{array}
$$

$f(4) = -25$

35. $f(x) = 3x^3 - 7x^2 - 2x + 5$

$$
\begin{array}{r|rrrr}
-3 & 3 & -7 & -2 & 5 \\
   &   & -9 & 48 & -138 \\
\hline
   & 3 & -16 & 46 & -133
\end{array}
$$

$f(-3) = -133$

37. $f(x) = x^4 + 5x^3 + 5x^2 - 5x - 6$

$$
\begin{array}{r|rrrr}
3 & 1 & 5 & 5 & -5 & -6 \\
  &   & 3 & 24 & 87 & 246 \\
\hline
  & 1 & 8 & 29 & 82 & 240
\end{array}
$$

$f(3) = 240$

39. $f(x) = 2x^4 - 5x^3 - x^2 + 3x + 2$

$$
\begin{array}{r|rrrrr}
-\frac{1}{2} & 2 & -5 & -1 & 3 & 2 \\
   &   & -1 & 3 & -1 & -1 \\
\hline
   & 2 & -6 & 2 & 2 & 1
\end{array}
$$

$f\left(-\dfrac{1}{2}\right) = 1$

41. Dividend: $x^3 - 4x^2 + x + 6$
Divisor: $x + 1$

$$
\begin{array}{r|rrrr}
-1 & 1 & -4 & 1 & 6 \\
   &   & -1 & 5 & -6 \\
\hline
   & 1 & -5 & 6 & 0
\end{array}
$$

The quotient is $x^2 - 5x + 6$.
$(x+1)(x^2 - 5x + 6) = 0$
$(x+1)(x-2)(x-3) = 0$
$x = -1, x = 2, x = 3$
The solution set is $\{-1, 2, 3\}$.

43. $2x^3 - 5x^2 + x + 2 = 0$

$$
\begin{array}{r|rrrr}
2 & 2 & -5 & 1 & 2 \\
  &   & 4 & -2 & -2 \\
\hline
  & 2 & -1 & -1 & 0
\end{array}
$$

$(x-2)(2x^2 - x - 1) = 0$
$(x-2)(2x+1)(x-1) = 0$

$x = 2, \ x = -\dfrac{1}{2}, \ x = 1$

The solution set is $\left\{-\dfrac{1}{2}, 1, 2\right\}$.

45. $12x^3 + 16x^2 - 5x - 3 = 0$

$$
\begin{array}{r|rrrr}
-\frac{3}{2} & 12 & 16 & -5 & -3 \\
  &   & -18 & 3 & 3 \\
\hline
  & 12 & -2 & -2 & 0
\end{array}
$$

$\left(x + \dfrac{3}{2}\right)(12x^2 - 2x - 2) = 0$

$\left(x + \dfrac{3}{2}\right)2(6x^2 - x - 1) = 0$

$\left(x + \dfrac{3}{2}\right)2(3x + 1)(2x - 1) = 0$

$x = -\dfrac{3}{2}, \ x = -\dfrac{1}{3}, \ x = \dfrac{1}{2}$

The solution set is $\left\{-\dfrac{3}{2}, -\dfrac{1}{3}, \dfrac{1}{2}\right\}$.

47. The graph indicates that 2 is a solution to the equation.

$$
\begin{array}{r|rrrr}
2 & 1 & 2 & -5 & -6 \\
  &   & 2 & 8 & 6 \\
\hline
  & 1 & 4 & 3 & 0
\end{array}
$$

The remainder is 0, so 2 is a solution.
$x^3 + 2x^2 - 5x - 6 = 0$
$(x-2)(x^2 + 4x + 3) = 0$
$(x-2)(x+3)(x+1) = 0$
The solutions are 2, -3, and -1, or $\{-3, -1, 2\}$.

49. The table indicates that 1 is a solution to the equation.

$$
\begin{array}{r|rrrr}
1 & 6 & -11 & 6 & -1 \\
  &   & 6 & -5 & 1 \\
\hline
  & 6 & -5 & 1 & 0
\end{array}
$$

The remainder is 0, so 1 is a solution.
$6x^3 - 11x^2 + 6x - 1 = 0$
$(x-1)(6x^2 - 5x + 1) = 0$
$(x-1)(3x-1)(2x-1) = 0$
The solutions are 1, $\dfrac{1}{3}$, and $\dfrac{1}{2}$, or $\left\{\dfrac{1}{3}, \dfrac{1}{2}, 1\right\}$.

51. a. $14x^3 - 17x^2 - 16x - 177 = 0$

$$
\begin{array}{r|rrrr}
3 & 14 & -17 & -16 & -177 \\
  &   & 42 & 75 & 177 \\
\hline
  & 14 & 25 & 59 & 0
\end{array}
$$

The remainder is 0 so 3 is a solution.
$14x^3 - 17x^2 - 16x - 177$
$= (x-3)(14x^2 + 25x + 59)$

b. $f(x) = 14x^3 - 17x^2 - 16x + 34$
We need to find x when $f(x) = 211$.
$f(x) = 14x^3 - 17x^2 - 16x + 34$
$211 = 14x^3 - 17x^2 - 16x + 34$
$0 = 14x^3 - 17x^2 - 16x - 177$
This is the equation obtained in part **a.** One solution is 3. It can be used to find other solutions (if they exist).
$14x^3 - 17x^2 - 16x - 177 = 0$
$(x-3)(14x^2 + 25x + 59) = 0$
The polynomial $14x^2 + 25x + 59$ cannot be factored, so the only solution is $x = 3$. The female moth's abdominal width is 3 millimeters.

190

53. $A = l \cdot w$ so

$$l = \frac{A}{w} = \frac{0.5x^3 - 0.3x^2 + 0.22x + 0.06}{x + 0.2}$$

$$\begin{array}{r|rrrr} -0.2 & 0.5 & -0.3 & 0.22 & 0.06 \\ & & -0.1 & 0.08 & -0.06 \\ \hline & 0.5 & -0.4 & 0.3 & 0 \end{array}$$

Therefore, the length of the rectangle is $0.5x^2 - 0.4x + 0.3$ units.

55. a.

$$f(30) = \frac{80(30) - 8000}{30 - 110} = 70$$

(30, 70) At a 30% tax rate, the government tax revenue will be \$70 ten billion.

b.
$$\begin{array}{r|rr} 110 & 80 & -8000 \\ & & 8800 \\ \hline & 80 & 800 \end{array}$$

$$f(x) = 80 + \frac{800}{x - 110}$$

$$f(30) = 80 + \frac{800}{80 - 110} = 70$$

(30, 70) same answer as in **a.**

c. $f(x)$ is not a polynomial function. It is a rational function because it is the quotient of two linear polynomials.

57. – 65. Answers may vary.

67. makes sense

69. does not make sense; Explanations will vary. Sample explanation: The zeros of f are the same as the solutions of $f(x) = 0$.

71. true

73. false; Changes to make the statement true will vary. A sample change is: The divisor is a factor of the divided only if the remainder is the whole number 0.

75. $f(x) = d(x) \cdot q(x) + r(x)$

$$2x^2 - 7x + 9 = d(x)(2x - 3) + 3$$

$$2x^2 - 7x + 6 = d(x)(2x - 3)$$

$$\frac{2x^2 - 7x + 6}{2x - 3} = d(x)$$

$$\begin{array}{r} x - 2 \\ 2x-3 \overline{\smash{\big)}\ 2x^2 - 7x + 6} \\ \underline{2x^2 - 3x} \\ -4x + 6 \\ \underline{-4x + 6} \end{array}$$

The polynomial is $x - 2$.

77. $2x - 4 = 2(x - 2)$

Use synthetic division to divide by $x - 2$. Then divide the quotient by 2.

79. $x^2 + 4x - 1 = 0$

$$x = \frac{-b \pm \sqrt{b^2 - 4ac}}{2a}$$

$$x = \frac{-(4) \pm \sqrt{(4)^2 - 4(1)(-1)}}{2(1)}$$

$$x = \frac{-4 \pm \sqrt{20}}{2}$$

$$x = \frac{-4 \pm 2\sqrt{5}}{2}$$

$$x = -2 \pm \sqrt{5}$$

The solution set is $\left\{ -2 \pm \sqrt{5} \right\}$.

80. $x^2 + 4x + 6 = 0$

$$x = \frac{-b \pm \sqrt{b^2 - 4ac}}{2a}$$

$$x = \frac{-(4) \pm \sqrt{(4)^2 - 4(1)(6)}}{2(1)}$$

$$x = \frac{-4 \pm \sqrt{-8}}{2}$$

$$x = \frac{-4 \pm 2i\sqrt{2}}{2}$$

$$x = -2 \pm i\sqrt{2}$$

The solution set is $\left\{ -2 \pm i\sqrt{2} \right\}$.

81. $f(x) = a_n(x^4 - 3x^2 - 4)$

$$f(3) = -150$$

$$a_n\left((3)^4 - 3(3)^2 - 4\right) = -150$$

$$a_n(81 - 27 - 4) = -150$$

$$a_n(50) = -150$$

$$a_n = -3$$

Section 2.5

Check Point Exercises

1. $p: \pm 1, \pm 2, \pm 3, \pm 6$

$q: \pm 1$

$\dfrac{p}{q}: \pm 1, \pm 2, \pm 3, \pm 6$

are the possible rational zeros.

2. $p: \pm 1, \pm 3$

$q: \pm 1, \pm 2, \pm 4$

$\dfrac{p}{q}: \pm 1, \pm 3, \pm \dfrac{1}{2}, \pm \dfrac{1}{4}, \pm \dfrac{3}{2}, \pm \dfrac{3}{4}$

are the possible rational zeros.

3. $\pm 1, \pm 2, \pm 4, \pm 5, \pm 10, \pm 20$ are possible rational zeros

$$
\begin{array}{r|rrrr}
1 & 1 & 8 & 11 & -20 \\
  &   & 1 & 9 & 20 \\
\hline
  & 1 & 9 & 20 & 0
\end{array}
$$

1 is a zero.

$$x^2 + 9x + 20 = 0$$

$$(x+4)(x+5) = 0$$

$$x = -4 \quad \text{or} \quad x = -5$$

The solution set is $\{1, -4, -5\}$.

4. $\pm 1, \pm 2$ are possible rational zeros

$$
\begin{array}{r|rrrr}
2 & 1 & 1 & -5 & -2 \\
  &   & 2 & 6 & 2 \\
\hline
  & 1 & 3 & 1 & 0
\end{array}
$$

2 is a zero.

$$x^2 + 3x + 1 = 0$$

$$x = \dfrac{-b \pm \sqrt{b^2 - 4ac}}{2a}$$

$$x = \dfrac{-3 \pm \sqrt{3^2 - 4(1)(1)}}{2(1)}$$

$$= \dfrac{-3 \pm \sqrt{5}}{2}$$

The solution set is $\left\{2, \dfrac{-3+\sqrt{5}}{2}, \dfrac{-3-\sqrt{5}}{2}\right\}$.

5. $\pm 1, \pm 13$ are possible rational zeros.

$$
\begin{array}{r|rrrrr}
1 & 1 & -6 & 22 & -30 & 13 \\
  &   & 1 & -5 & 17 & -13 \\
\hline
  & 1 & -5 & 17 & -13 & 0
\end{array}
$$

1 is a zero.

$$
\begin{array}{r|rrrr}
1 & 1 & 5 & 17 & -13 \\
  &   & 1 & -4 & 13 \\
\hline
  & 1 & -4 & 13 & 0
\end{array}
$$

1 is a double root.

$$x^2 - 4x + 13 = 0$$

$$x = \dfrac{4 \pm \sqrt{16 - 52}}{2} = \dfrac{4 \pm \sqrt{-36}}{2} = 2 + 3i$$

The solution set is $\{1, 2 + 3i, 2 - 3i\}$.

6. $(x+3)(x-i)(x+i) = (x+3)(x^2+1)$

$$f(x) = a_n(x+3)(x^2+1)$$

$$f(1) = a_n(1+3)(1^2+1) = 8a_n = 8$$

$$a_n = 1$$

$$f(x) = (x+3)(x^2+1) \text{ or } x^3 + 3x^2 + x + 3$$

7. $f(x) = x^4 - 14x^3 + 71x^2 - 154x + 120$

$$f(-x) = x^4 + 14x^3 + 71x^2 + 154x + 120$$

Since $f(x)$ has 4 changes of sign, there are 4, 2, or 0 positive real zeros.

Since $f(-x)$ has no changes of sign, there are no negative real zeros.

Exercise Set 2.5

1. $f(x) = x^3 + x^2 - 4x - 4$

$p: \pm 1, \pm 2, \pm 4$

$q: \pm 1$

$\dfrac{p}{q}: \pm 1, \pm 2, \pm 4$

3. $f(x) = 3x^4 - 11x^3 - x^2 + 19x + 6$

$p: \pm 1, \pm 2, \pm 3, \pm 6$

$q: \pm 1, \pm 3$

$\dfrac{p}{q}: \pm 1, \pm 2, \pm 3, \pm 6, \pm \dfrac{1}{3}, \pm \dfrac{2}{3}$

5. $f(x) = 4x^4 - x^3 + 5x^2 - 2x - 6$

$p: \pm 1, \pm 2, \pm 3, \pm 6$

$q: \pm 1, \pm 2, \pm 4$

$\dfrac{p}{q}: \pm 1, \pm 2, \pm 3, \pm 6, \pm \dfrac{1}{2}, \pm \dfrac{1}{4}, \pm \dfrac{3}{2}, \pm \dfrac{3}{4}$

7. $f(x) = x^5 - x^4 - 7x^3 + 7x^2 - 12x - 12$

$p: \pm 1, \pm 2, \pm 3 \pm 4 \pm 6 \pm 12$

$q: \pm 1$

$\dfrac{p}{q}: \pm 1, \pm 2, \pm 3 \pm 4 \pm 6 \pm 12$

9. $f(x) = x^3 + x^2 - 4x - 4$

a. $p: \pm 1, \pm 2, \pm 4$

$q: \pm 1$

$\dfrac{p}{q}: \pm 1, \pm 2, \pm 4$

b.

$$
\begin{array}{r|rrrr}
2 & 1 & 1 & -4 & -4 \\
  &   & 2 & 6 & 4 \\
\hline
  & 1 & 3 & 2 & 0 \\
\end{array}
$$

2 is a zero.

2, –2, –1 are rational zeros.

c. $x^3 + x^2 - 4x - 4 = 0$

$(x-2)(x^2 + 3x + 2) = 0$

$(x-2)(x+2)(x+1) = 0$

$x - 2 = 0 \quad x + 2 = 0 \quad x + 1 = 0$

$x = 2, \ x = -2, \ x = -1$

The solution set is $\{2, -2, -1\}$.

11. $f(x) = 2x^3 - 3x^2 - 11x + 6$

a. $p: \pm 1, \pm 2, \ \pm 3, \pm 6$

$q: \pm 1, \pm 2$

$\dfrac{p}{q}: \pm 1, \pm 2, \pm 3, \pm 6, \pm \dfrac{1}{2}, \pm \dfrac{3}{2}$

b.

$$
\begin{array}{r|rrrr}
3 & 2 & -3 & -11 & 6 \\
  &   & 6 & 9   & -6 \\
\hline
  & 2 & 3 & -2 & 0 \\
\end{array}
$$

3 is a zero.

$3, \dfrac{1}{2}, -2$ are rational zeros.

c. $2x^3 - 3x^2 - 11x + 6 = 0$

$(x-3)(2x^2 + 3x - 2) = 0$

$(x-3)(2x-1)(x+2) = 0$

$x = 3, \ x = \dfrac{1}{2}, \ x = -2$

The solution set is $\left\{ 3, \dfrac{1}{2}, -2 \right\}$.

13.   **a.** $f(x) = x^3 + 4x^2 - 3x - 6$

$p: \pm 1, \pm 2, \pm 3, \pm 6$

$q: \pm 1$

$\dfrac{p}{q}: \pm 1, \pm 2, \pm 3, \pm 6$

b.

$$
\begin{array}{r|rrrr}
-1 & 1 & 4 & -3 & -6 \\
   &   & -1 & -3 & 6 \\
\hline
   & 1 & 3 & -6 & 0 \\
\end{array}
$$

–1 is a rational zero.

c. $x^2 + 3x - 6 = 0$

$x = \dfrac{-b \pm \sqrt{b^2 - 4ac}}{2a}$

$x = \dfrac{-3 \pm \sqrt{3^2 - 4(1)(-6)}}{2(1)}$

$= \dfrac{-3 \pm \sqrt{33}}{2}$

The solution set is $\left\{ -1, \dfrac{-3 + \sqrt{33}}{2}, \dfrac{-3 - \sqrt{33}}{2} \right\}$.

15. a. $f(x) = 2x^3 + 6x^2 + 5x + 2$

$p: \pm 1, \pm 2$

$q: \pm 1, \pm 2$

$\dfrac{p}{q}: \pm 1, \pm 2, \pm \dfrac{1}{2}$

b.

$$\begin{array}{r|rrrr} -2 & 2 & 6 & 5 & 2 \\ & & -4 & -4 & -2 \\ \hline & 2 & 2 & 1 & 0 \end{array}$$

-2 is a rational zero.

c. $2x^2 + 2x + 1 = 0$

$x = \dfrac{-b \pm \sqrt{b^2 - 4ac}}{2a}$

$x = \dfrac{-2 \pm \sqrt{2^2 - 4(2)(1)}}{2(2)}$

$= \dfrac{-2 \pm \sqrt{-4}}{4}$

$= \dfrac{-2 \pm 2i}{4}$

$= \dfrac{-1 \pm i}{2}$

The solution set is $\left\{ -2, \dfrac{-1+i}{2}, \dfrac{-1-i}{2} \right\}$.

17. $x^3 - 2x^2 - 11x + 12 = 0$

a. $p: \pm 1, \pm 2, \pm 3, \pm 4, \pm 6, \pm 12$

$q: \pm 1$

$\dfrac{p}{q}: \pm 1, \pm 2, \pm 3, \pm 4, \pm 6, \pm 12$

b.

$$\begin{array}{r|rrrr} 4 & 1 & -2 & -11 & 12 \\ & & 4 & 8 & -12 \\ \hline & 1 & 2 & -3 & 0 \end{array}$$

4 is a root.
$-3, 1, 4$ are rational roots.

c. $x^3 - 2x^2 - 11x + 12$

$(x-4)(x^2 + 2x - 3) = 0$

$(x-4)(x+3)(x-1) = 0$

$x - 4 = 0 \quad x + 3 = 0 \quad x - 1 = 0$

$x = 4 \qquad x = -3 \qquad x = 1$

The solution set is $\{-3, 1, 4\}$.

19. $x^3 - 10x - 12 = 0$

a. $p: \pm 1, \pm 2, \pm 3, \pm 4, \pm 6, \pm 12$

$q: \pm 1$

$\dfrac{p}{q}: \pm 1, \pm 2, \pm 3, \pm 4, \pm 6, \pm 12$

b.

$$\begin{array}{r|rrrr} -2 & 1 & 0 & -10 & -12 \\ & & -2 & 4 & 12 \\ \hline & 1 & -2 & -6 & 0 \end{array}$$

-2 is a rational root.

c. $x^3 - 10x - 12 = 0$

$(x+2)(x^2 - 2x - 6) = 0$

$x = \dfrac{2 \pm \sqrt{4 + 24}}{2} = \dfrac{2 \pm \sqrt{28}}{2}$

$= \dfrac{2 \pm 2\sqrt{7}}{2} = 1 \pm \sqrt{7}$

The solution set is $\left\{ -2, 1 + \sqrt{7}, 1 - \sqrt{7} \right\}$.

21. $6x^3 + 25x^2 - 24x + 5 = 0$

a. $p: \pm 1, \pm 5$

$q: \pm 1, \pm 2, \pm 3, \pm 6$

$\dfrac{p}{q}: \pm 1, \pm 5, \pm \dfrac{1}{2}, \pm \dfrac{5}{2}, \pm \dfrac{1}{3}, \pm \dfrac{5}{3}, \pm \dfrac{1}{6}, \pm \dfrac{5}{6}$

b.

$$\begin{array}{r|rrrr} -5 & 6 & 25 & -24 & 5 \\ & & -30 & 25 & -5 \\ \hline & 6 & -5 & 1 & 0 \end{array}$$

-5 is a root.

$-5, \dfrac{1}{2}, \dfrac{1}{3}$ are rational roots.

c. $6x^3 + 25x^2 - 24x + 5 = 0$

$(x+5)(6x^2 - 5x + 1) = 0$

$(x+5)(2x-1)(3x-1) = 0$

$x + 5 = 0 \quad 2x - 1 = 0 \quad 3x - 1 = 0$

$x = -5, \qquad x = \dfrac{1}{2}, \qquad x = \dfrac{1}{3}$

The solution set is $\left\{ -5, \dfrac{1}{2}, \dfrac{1}{3} \right\}$.

23. $x^4 - 2x^3 - 5x^2 + 8x + 4 = 0$

a. $p: \pm 1, \pm 2, \pm 4$

$q: \pm 1$

$\dfrac{p}{q}: \pm 1, \pm 2, \pm 4$

b.

$$\begin{array}{r|rrrrr} 2 & 1 & -2 & -5 & 8 & 4 \\ & & 2 & 0 & -10 & -4 \\ \hline & 1 & 0 & -5 & -2 & 0 \end{array}$$

2 is a root.
$-2, 2$ are rational roots.

c. $x^4 - 2x^3 - 5x^2 + 8x + 4 = 0$

$(x-2)(x^3 - 5x - 2) = 0$

$$\begin{array}{r|rrrr} -2 & 1 & 0 & -5 & -2 \\ & & -2 & 4 & 2 \\ \hline & 1 & -2 & -1 & 0 \end{array}$$

-2 is a zero of $x^3 - 5x - 2 = 0$.

$(x-2)(x+2)(x^2 - 2x - 1) = 0$

$x = \dfrac{2 \pm \sqrt{4+4}}{2} = \dfrac{2 \pm \sqrt{8}}{2} = \dfrac{2 \pm 2\sqrt{2}}{2}$

$= 1 \pm \sqrt{2}$

The solution set is

$\left\{ -2, 2, 1+\sqrt{2}, 1-\sqrt{2} \right\}$.

25.

$(x-1)(x+5i)(x-5i)$

$= (x-1)\left(x^2 + 25 \right)$

$= x^3 + 25x - x^2 - 25$

$= x^3 - x^2 + 25x - 25$

$f(x) = a_n\left(x^3 - x^2 + 25x - 25 \right)$

$f(-1) = a_n(-1 - 1 - 25 - 25)$

$-104 = a_n(-52)$

$a_n = 2$

$f(x) = 2\left(x^3 - x^2 + 25x - 25 \right)$

$f(x) = 2x^3 - 2x^2 + 50x - 50$

27. $(x+5)(x-4-3i)(x-4+3i)$

$= (x+5)\left(x^2 - 4x + 3ix - 4x + 16 - 12i \right.$

$\left. -3ix + 12i - 9i^2 \right)$

$= (x+5)\left(x^2 - 8x + 25 \right)$

$= \left(x^3 - 8x^2 + 25x + 5x^2 - 40x + 125 \right)$

$= x^3 - 3x^2 - 15x + 125$

$f(x) = a_n(x^3 - 3x^2 - 15x + 125)$

$f(2) = a_n\left(2^3 - 3(2)^2 - 15(2) + 125 \right)$

$91 = a_n(91)$

$a_n = 1$

$f(x) = 1\left(x^3 - 3x^2 - 15x + 125 \right)$

$f(x) = x^3 - 3x^2 - 15x + 125$

29. $(x-i)(x+i)(x-3i)(x+3i)$

$= \left(x^2 - i^2 \right)\left(x^2 - 9i^2 \right)$

$= \left(x^2 + 1 \right)\left(x^2 + 9 \right)$

$= x^4 + 10x^2 + 9$

$f(x) = a_n(x^4 + 10x^2 + 9)$

$f(-1) = a_n((-1)^4 + 10(-1)^2 + 9)$

$20 = a_n(20)$

$a_n = 1$

$f(x) = x^4 + 10x^2 + 9$

31. $(x+2)(x-5)(x-3+2i)(x-3-2i)$

$= \left(x^2 - 3x - 10 \right)\left(x^2 - 3x - 2ix - 3x + 9 + 6i + 2ix - 6i - 4i^2 \right)$

$= \left(x^2 - 3x - 10 \right)\left(x^2 - 6x + 13 \right)$

$= x^4 - 6x + 13x^2 - 3x^3 + 18x^2 - 39x - 10x^2 + 60x - 130$

$= x^4 - 9x^3 + 21x^2 + 21x - 130$

$f(x) = a_n\left(x^4 - 9x^3 + 21x^2 + 21x - 130 \right)$

$f(1) = a_n(1 - 9 + 21 + 21 - 130)$

$-96 = a_n(-96)$

$a_n = 1$

$f(x) = x^4 - 9x^3 + 21x^2 + 21x - 130$

33. $f(x) = x^3 + 2x^2 + 5x + 4$

Since $f(x)$ has no sign variations, no positive real roots exist.

$f(-x) = -x^3 + 2x^2 - 5x + 4$

Since $f(-x)$ has 3 sign variations, 3 or 1 negative real roots exist.

35. $f(x) = 5x^3 - 3x^2 + 3x - 1$

Since $f(x)$ has 3 sign variations, 3 or 1 positive real roots exist.

$f(-x) = -5x^3 - 3x^2 - 3x - 1$

Since $f(-x)$ has no sign variations, no negative real roots exist.

37. $f(x) = 2x^4 - 5x^3 - x^2 - 6x + 4$

Since $f(x)$ has 2 sign variations, 2 or 0 positive real roots exist.

$f(-x) = 2x^4 + 5x^3 - x^2 + 6x + 4$

Since $f(-x)$ has 2 sign variations, 2 or 0 negative real roots exist.

39. $f(x) = x^3 - 4x^2 - 7x + 10$

$p : \pm 1, \pm 2, \pm 5, \pm 10$

$q : \pm 1$

$\dfrac{p}{q} : \pm 1, \pm 2, \pm 5, \pm 10$

Since $f(x)$ has 2 sign variations, 0 or 2 positive real zeros exist.

$f(-x) = -x^3 - 4x^2 + 7x + 10$

Since $f(-x)$ has 1 sign variation, exactly one negative real zeros exists.

$$
\begin{array}{r|rrrr}
-2 & 1 & -4 & -7 & 10 \\
   &   & -2 & 12 & -10 \\
\hline
   & 1 & -6 & 5 & 0
\end{array}
$$

-2 is a zero.

$f(x) = (x+2)(x^2 - 6x + 5)$

$\qquad = (x+2)(x-5)(x-1)$

$x = -2,\ x = 5,\ x = 1$

The solution set is $\{-2, 5, 1\}$.

41. $2x^3 - x^2 - 9x - 4 = 0$

$p : \pm 1, \pm 2, \pm 4$

$q : \pm 1, \pm 2$

$\dfrac{p}{q} : \pm 1, \pm 2, \pm 4 \pm \dfrac{1}{2}$

1 positive real root exists.

$f(-x) = -2x^3 - x^2 + 9x - 4$ 2 or no negative real roots exist.

$$
\begin{array}{r|rrrr}
-\dfrac{1}{2} & 2 & -1 & -9 & -4 \\
   &   & -1 & 1 & 4 \\
\hline
   & 2 & -2 & -8 & 0
\end{array}
$$

$-\dfrac{1}{2}$ is a root.

$\left(x + \dfrac{1}{2}\right)(2x^2 - 2x - 8) = 0$

$2\left(x + \dfrac{1}{2}\right)(x^2 - x - 4) = 0$

$x = \dfrac{1 \pm \sqrt{1+16}}{2} = \dfrac{1 \pm \sqrt{17}}{2}$

The solution set is $\left\{ -\dfrac{1}{2}, \dfrac{1+\sqrt{17}}{2}, \dfrac{1-\sqrt{17}}{2} \right\}$.

43. $f(x) = x^4 - 2x^3 + x^2 + 12x + 8$

$p : \pm 1, \pm 2, \pm 4, \pm 8$

$q : \pm 1$

$\dfrac{p}{q} : \pm 1, \pm 2, \pm 4, \pm 8$

Since $f(x)$ has 2 sign changes, 0 or 2 positive roots exist.

$f(-x) = (-x)^4 - 2(-x)^3 + (-x)^2 - 12x + 8$

$\qquad = x^4 + 2x^3 + x^2 - 12x + 8$

Since $f(-x)$ has 2 sign changes, 0 or 2 negative roots exist.

$$
\begin{array}{r|rrrrr}
-1 & 1 & -2 & 1 & 12 & 8 \\
   &   & -1 & 4 & -4 & -8 \\
\hline
   & 1 & -3 & 4 & 8 & 0
\end{array}
$$

$$
\begin{array}{r|rrrr}
-1 & 1 & -3 & 4 & 8 \\
   &   & -1 & 4 & -8 \\
\hline
   & 1 & -4 & 8 & 0
\end{array}
$$

$0 = x^2 - 4x + 8$

$x = \dfrac{-(-4) \pm \sqrt{(-4)^2 - 4(1)(8)}}{2(1)}$

$x = \dfrac{4 \pm \sqrt{16 - 32}}{2}$

$x = \dfrac{4 \pm \sqrt{-16}}{2}$

$x = \dfrac{4 \pm 4i}{2}$

$x = 2 \pm 2i$

The solution set is $\{-1, -1, 2 + 2i, 2 - 2i\}$.

45. $x^4 - 3x^3 - 20x^2 - 24x - 8 = 0$

$p : \pm 1, \pm 2, \pm 4, \pm 8$

$q : \pm 1$

$\dfrac{p}{q} : \pm 1, \pm 2, \pm 4 \pm 8$

1 positive real root exists.

3 or 1 negative real roots exist.

$$
\begin{array}{r|rrrrr}
-1 & 1 & -3 & -20 & -24 & -8 \\
   &   & -1 & 4 & 16 & 8 \\
\hline
   & 1 & -4 & -16 & -8 & 0 \\
\end{array}
$$

$(x + 1)\left(x^3 - 4x^2 - 16x - 8\right) = 0$

$$
\begin{array}{r|rrrr}
-2 & 1 & -4 & -16 & -8 \\
   &   & -2 & 12 & 8 \\
\hline
   & 1 & -6 & -4 & 0 \\
\end{array}
$$

$(x + 1)(x + 2)\left(x^2 - 6x - 4\right) = 0$

$x = \dfrac{6 \pm \sqrt{36 + 16}}{2} = \dfrac{6 \pm \sqrt{52}}{2}$

$= \dfrac{6 \pm 2\sqrt{13}}{2} = \dfrac{3 \pm \sqrt{13}}{2}$

The solution set is

$\left\{-1, -2, 3 \pm \sqrt{13}, 3 - \sqrt{13}\right\}$.

47. $f(x) = 3x^4 - 11x^3 - x^2 + 19x + 6$

$p : \pm 1, \pm 2, \pm 3, \pm 6$

$q : \pm 1, \pm 3$

$\dfrac{p}{q} : \pm 1, \pm 2, \pm 3, \pm 6, \pm \dfrac{1}{3}, \pm \dfrac{2}{3}$

2 or no positive real zeros exists.

$f(-x) = 3x^4 + 11x^3 - x^2 - 19x + 6$

2 or no negative real zeros exist.

$$
\begin{array}{r|rrrrr}
-1 & 3 & -11 & -1 & 19 & 6 \\
   &   & -3 & 14 & -13 & -6 \\
\hline
   & 3 & -14 & 13 & 6 & 0 \\
\end{array}
$$

$f(x) = (x + 1)\left(3x^3 - 14x^2 + 13x + 6\right)$

$$
\begin{array}{r|rrrr}
2 & 3 & -14 & 13 & 6 \\
  &   & 6 & -16 & -6 \\
\hline
  & 3 & -8 & -3 & 0 \\
\end{array}
$$

$f(x) = (x + 1)(x - 2)\left(3x^2 - 8x - 3\right)$

$\quad = (x + 1)(x - 2)(3x + 1)(x - 3)$

$x = -1,\ x = 2\ \ x = -\dfrac{1}{3},\ x = 3$

The solution set is $\left\{-1,\ 2,\ -\dfrac{1}{3},\ 3\right\}$.

49. $4x^4 - x^3 + 5x^2 - 2x - 6 = 0$

$p : \pm 1, \pm 2, \pm 3, \pm 6$

$q : \pm 1, \pm 2, \pm 4$

$\dfrac{p}{q} : \pm 1, \pm 2, \pm 3, \pm 6, \pm \dfrac{1}{2}, \pm \dfrac{3}{2}, \pm \dfrac{1}{4}, \pm \dfrac{3}{4}$

3 or 1 positive real roots exists.

1 negative real root exists.

$$
\begin{array}{r|rrrrr}
1 & 4 & -1 & 5 & -2 & -6 \\
  &   & 4 & 3 & 8 & 6 \\
\hline
  & 4 & 3 & 8 & 6 & 0 \\
\end{array}
$$

$(x - 1)(4x^3 + 3x^2 + 8x + 6) = 0$

$4x^3 + 3x^2 + 8x + 6 = 0$ has no positive real roots.

$$
\begin{array}{r|rrrr}
-\frac{3}{4} & 4 & 3 & 8 & 6 \\
             &   & -3 & 0 & -6 \\
\hline
             & 4 & 0 & 8 & 0 \\
\end{array}
$$

$(x - 1)\left(x + \dfrac{3}{4}\right)\left(4x^2 + 8\right) = 0$

$4(x - 1)\left(x + \dfrac{3}{4}\right)\left(x^2 + 2\right) = 0$

$x^2 + 2 = 0$

$x^2 = -2$

$x = \pm i\sqrt{2}$

The solution set is $\left\{1,\ -\dfrac{3}{4},\ i\sqrt{2},\ -i\sqrt{2}\right\}$.

51. $2x^5 + 7x^4 - 18x^2 - 8x + 8 = 0$

$p: \pm 1, \pm 2, \pm 4, \pm 8$

$q: \pm 1, \pm 2$

$\dfrac{p}{q}: \pm 1, \pm 2, \pm 4, \pm 8, \pm \dfrac{1}{2}$

2 or no positive real roots exists.

3 or 1 negative real root exist.

$$
\begin{array}{r|rrrrrr}
-2 & 2 & 7 & 0 & -18 & -8 & 8 \\
  &   & -4 & -6 & 12 & 12 & -8 \\
\hline
  & 2 & 3 & -6 & -6 & 4 & 0
\end{array}
$$

$(x+2)(2x^4 + 3x^3 - 6x^2 - 6x + 4) = 0$

$4x^3 + 3x^2 + 8x + 6 = 0$ has no positive real roots.

$$
\begin{array}{r|rrrrr}
-2 & 2 & 3 & -6 & -6 & 4 \\
  &   & -4 & 2 & 8 & -4 \\
\hline
  & 2 & -1 & -4 & 2 & 0
\end{array}
$$

$(x+2)^2 (2x^3 - x^2 - 4x + 2)$

$$
\begin{array}{r|rrrr}
\frac{1}{2} & 2 & -1 & -4 & 2 \\
  &   & 1 & 0 & 2 \\
\hline
  & 2 & 0 & -4 & 0
\end{array}
$$

$(x+2)^2 \left(x - \dfrac{1}{2} \right)(2x^2 - 4) = 0$

$2(x+2)^2 \left(x - \dfrac{1}{2} \right)(x^2 - 2) = 0$

$x^2 - 2 = 0$

$x^2 = 2$

$x = \pm\sqrt{2}$

The solution set is $\left\{ -2, \dfrac{1}{2}, \sqrt{2}, -\sqrt{2} \right\}$.

53. $f(x) = -x^3 + x^2 + 16x - 16$

a. From the graph provided, we can see that -4 is an x-intercept and is thus a zero of the function. We verify this below:

$$
\begin{array}{r|rrrr}
-4 & -1 & 1 & 16 & -16 \\
  &   & 4 & -20 & 16 \\
\hline
  & -1 & 5 & -4 & 0
\end{array}
$$

Thus, $-x^3 + x^2 + 16x - 16 = 0$

$(x+4)(-x^2 + 5x - 4) = 0$

$-(x+4)(x^2 - 5x + 4) = 0$

$-(x+4)(x-1)(x-4) = 0$

$x+4 = 0$ or $x-1 = 0$ or $x-4 = 0$

$x = -4 \qquad\quad x = 1 \qquad\quad x = 4$

The zeros are -4, 1, and 4.

b.

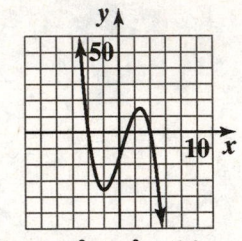

$f(x) = -x^3 + x^2 + 16x - 16$

55. $f(x) = 4x^3 - 8x^2 - 3x + 9$

a. From the graph provided, we can see that -1 is an x-intercept and is thus a zero of the function. We verify this below:

$$
\begin{array}{r|rrrr}
-1 & 4 & -8 & -3 & 9 \\
  &   & -4 & 12 & -9 \\
\hline
  & 4 & -12 & 9 & 0
\end{array}
$$

Thus, $4x^3 - 8x^2 - 3x + 9 = 0$

$(x+1)(4x^2 - 12x + 9) = 0$

$(x+1)(2x-3)^2 = 0$

$x+1 = 0$ or $(2x-3)^2 = 0$

$x = -1 \qquad\quad 2x - 3 = 0$

$\qquad\qquad\qquad 2x = 3$

$\qquad\qquad\qquad x = \dfrac{3}{2}$

The zeros are -1 and $\dfrac{3}{2}$.

b.

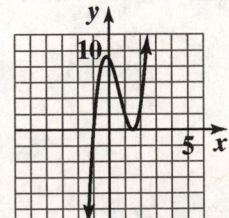

$f(x) = 4x^3 - 8x^2 - 3x + 9$

57. $f(x) = 2x^4 - 3x^3 - 7x^2 - 8x + 6$

    a. From the graph provided, we can see that $\dfrac{1}{2}$ is an

x-intercept and is thus a zero of the function. We verify this below:

$$\begin{array}{r|rrrrr} \frac{1}{2} & 2 & -3 & -7 & -8 & 6 \\ & & 1 & -1 & -4 & -6 \\ \hline & 2 & -2 & -8 & -12 & 0 \end{array}$$

Thus, $\quad 2x^4 - 3x^3 - 7x^2 - 8x + 6 = 0$

$$\left(x - \frac{1}{2}\right)\left(2x^3 - 2x^2 - 8x - 12\right) = 0$$

$$2\left(x - \frac{1}{2}\right)\left(x^3 - x^2 - 4x - 6\right) = 0$$

To factor $x^3 - x^2 - 4x - 6$, we use the Rational Zero Theorem to determine possible rational zeros.

Factors of the constant term -6:
$\pm1, \pm2, \pm3, \pm6$
Factors of the leading coefficient 1: ±1

The possible rational zeros are:

$$\frac{\text{Factors of } -6}{\text{Factors of } 1} = \frac{\pm1, \pm2, \pm3, \pm6}{\pm1}$$
$= \pm1, \pm2, \pm3, \pm6$

We test values from above until we find a zero. One possibility is shown next:

Test 3:

$$\begin{array}{r|rrrr} 3 & 1 & -1 & -4 & -6 \\ & & 3 & 6 & 6 \\ \hline & 1 & 2 & 2 & 0 \end{array}$$

The remainder is 0, so 3 is a zero of f.

$$2x^4 - 3x^3 - 7x^2 - 8x + 6 = 0$$

$$\left(x - \frac{1}{2}\right)\left(2x^3 - 2x^2 - 8x - 12\right) = 0$$

$$2\left(x - \frac{1}{2}\right)\left(x^3 - x^2 - 4x - 6\right) = 0$$

$$2\left(x - \frac{1}{2}\right)(x - 3)\left(x^2 + 2x + 2\right) = 0$$

Note that $x^2 + x + 1$ will not factor, so we use the quadratic formula:
$a = 1 \quad b = 2 \quad c = 2$

$$x = \frac{-2 \pm \sqrt{2^2 - 4(1)(2)}}{2(1)}$$

$$= \frac{-2 \pm \sqrt{-4}}{2} = \frac{-2 \pm 2i}{2} = -1 \pm i$$

The zeros are $\dfrac{1}{2}$, 3, and $-1 \pm i$.

    b.

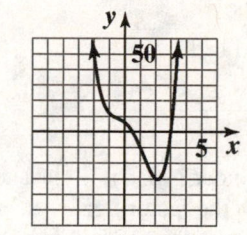

$f(x) = 2x^4 - 3x^3 - 7x^2 - 8x + 6$

59. $f(x) = 3x^5 + 2x^4 - 15x^3 - 10x^2 + 12x + 8$

    a. From the graph provided, we can see that 1 and 2 are x-intercepts and are thus zeros of the function. We verify this below:

$$\begin{array}{r|rrrrrr} 1 & 3 & 2 & -15 & -10 & 12 & 8 \\ & & 3 & 5 & -10 & -20 & -8 \\ \hline & 3 & 5 & -10 & -20 & -8 & 0 \end{array}$$

Thus, $3x^5 + 2x^4 - 15x^3 - 10x^2 + 12x + 8$
$$= (x-1)\left(3x^4 + 5x^3 - 10x^2 - 20x - 8\right)$$

$$\begin{array}{r|rrrrr} 2 & 3 & 5 & -10 & -20 & -8 \\ & & 6 & 22 & 24 & 8 \\ \hline & 3 & 11 & 12 & 4 & 0 \end{array}$$

Thus, $3x^5 + 2x^4 - 15x^3 - 10x^2 + 12x + 8$
$$= (x-1)\left(3x^4 + 5x^3 - 10x^2 - 20x - 8\right)$$
$$= (x-1)(x-2)\left(3x^3 + 11x^2 + 12x + 4\right)$$

To factor $3x^3 + 11x^2 + 12x + 4$, we use the Rational Zero Theorem to determine possible rational zeros.

Factors of the constant term 4: $\pm1, \pm2, \pm4$
Factors of the leading coefficient 3: $\pm1, \pm3$

The possible rational zeros are:

$$\frac{\text{Factors of 4}}{\text{Factors of 3}} = \frac{\pm 1, \ \pm 2, \ \pm 4}{\pm 1, \ \pm 3}$$

$$= \pm 1, \ \pm 2, \ \pm 4, \ \pm \frac{1}{3}, \ \pm \frac{2}{3}, \ \pm \frac{4}{3}$$

We test values from above until we find a zero. One possibility is shown next:

Test -1:

$$\begin{array}{r|rrrr} -1 & 3 & 11 & 12 & 4 \\ & & -3 & -8 & -4 \\ \hline & 3 & 8 & 4 & 0 \end{array}$$

The remainder is 0, so -1 is a zero of f. We can now finish the factoring:

$$3x^5 + 2x^4 - 15x^3 - 10x^2 + 12x + 8 = 0$$

$$(x-1)\left(3x^4 + 5x^3 - 10x^2 - 20x - 8\right) = 0$$

$$(x-1)(x-2)\left(3x^3 + 11x^2 + 12x + 4\right) = 0$$

$$(x-1)(x-2)(x+1)\left(3x^2 + 8x + 4\right) = 0$$

$$(x-1)(x-2)(x+1)(3x+2)(x+2) = 0$$

$$x = 1, \ x = 2, \ x = -1, \ x = -\frac{2}{3}, \ x = -2$$

The zeros are -2, -1, $-\dfrac{2}{3}$, 1 and 2.

b.

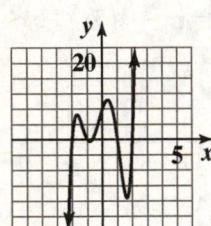

$f(x) = 3x^5 + 2x^4 - 15x^3 - 10x^2 + 12x + 8$

61. $V(x) = x(x+10)(30-2x)$

$$2000 = x(x+10)(30-2x)$$

$$2000 = -2x^3 + 10x^2 + 300x$$

$$2x^3 - 10x^2 - 300x + 2000 = 0$$

$$x^3 - 5x^2 - 150x + 1000 = 0$$

Find the roots.

$$\begin{array}{r|rrrr} 10 & 1 & -5 & -150 & 1000 \\ & & 10 & 50 & -1000 \\ \hline & 1 & 5 & -100 & 0 \end{array}$$

Use the remaining quadratic to find the other 2 roots.

$$x = \frac{-b \pm \sqrt{b^2 - 4ac}}{2a}$$

$$x = \frac{-(5) \pm \sqrt{(5)^2 - 4(1)(-100)}}{2(1)}$$

$$x \approx -12.8, \ 7.8$$

Since the depth must be positive, reject the negative value.

The depth can be 10 inches or 7.8 inches to obtain a volume of 2000 cubic inches.

63. a. The answers correspond to the points $(7.8, 2000)$ and $(10, 2000)$.

b. The range is $(0, 15)$.

65. – 71. Answers may vary.

73. $6x^3 - 19x^2 + 16x - 4 = 0$

p: $\pm 1, \pm 2, \pm 4$

q: $\pm 1, \pm 2, \pm 3, \pm 6$

$\dfrac{p}{q}$: $\pm 1, \pm 2, \pm 4, \pm \dfrac{1}{2}, \pm \dfrac{1}{3}, \pm \dfrac{2}{3}, \pm \dfrac{4}{3}, \pm \dfrac{1}{6}$

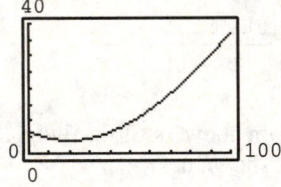

From the graph, we see that the solutions are $\dfrac{1}{2}, \dfrac{2}{3}$ and 2.

75. $4x^4 + 4x^3 + 7x^2 - x - 2 = 0$

p: $\pm 1, \pm 2$

q: $\pm 1, \pm 2, \pm 4$

$\dfrac{p}{q}$: $\pm 1, \pm 2, \pm \dfrac{1}{2}, \pm \dfrac{1}{4}$

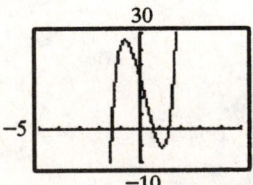

From the graph, we see that the solutions are $-\dfrac{1}{2}$ and $\dfrac{1}{2}$.

77. $f(x) = x^5 - x^4 + x^3 - x^2 + x - 8$

$f(x)$ has 5 sign variations, so either 5, 3, or 1 positive real roots exist.

$f(-x) = -x^5 - x^4 - x^3 - x^2 - x - 8$

$f(-x)$ has no sign variations, so no negative real roots exist.

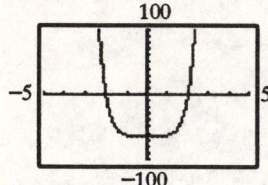

79. $f(x) = x^3 - 6x - 9$

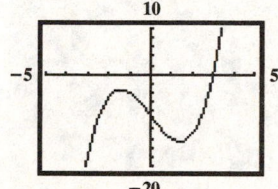

1 real zero
2 nonreal complex zeros

81. $f(x) = 3x^4 + 4x^3 - 7x^2 - 2x - 3$

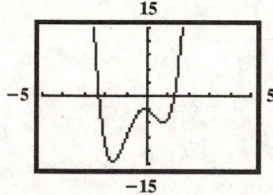

83. makes sense

85. makes sense

87. false; Changes to make the statement true will vary. A sample change is: The equation has 0 sign variations, so no positive roots exist.

89. true

91. $(2x+1)(x+5)(x+2) - 3x(x+5) = 208$

$(2x^2 + 11x + 5)(x+2) - 3x^2 - 15x = 208$

$2x^3 + 4x^2 + 11x^2 + 22x + 5x$

$+ 10 - 3x^2 - 15x = 208$

$2x^3 + 15x^2 + 27x - 3x^2 - 15x - 198 = 0$

$2x^3 + 12x^2 + 12x - 198 = 0$

$2(x^3 + 6x^2 + 6x - 99) = 0$

$$
\begin{array}{r|rrrr}
3 & 1 & 6 & 6 & -99 \\
& & 3 & 27 & 99 \\
\hline
& 1 & 9 & 33 & 0
\end{array}
$$

$x^2 + 9x + 33 = 0$

$b^2 - 4ac = -51$

$x = 3$ in.

93. Because the polynomial has two obvious changes of direction; the smallest degree is 3.

95. Because the polynomial has two obvious changes of direction and two roots have multiplicity 2, the smallest degree is 5.

97. Answers may vary.

98. The function is undefined at $x = 1$ and $x = 2$.

99. The equation of the vertical asymptote is $x = 1$.

100. The equation of the horizontal asymptote is $y = 0$.

Mid-Chapter 2 Check Point

1. $(6 - 2i) - (7 - i) = 6 - 2i - 7 + i = -1 - i$

2. $3i(2 + i) = 6i + 3i^2 = -3 + 6i$

3. $(1+i)(4-3i) = 4 - 3i + 4i - 3i^2$
$= 4 + i + 3 = 7 + i$

4. $\dfrac{1+i}{1-i} = \dfrac{1+i}{1-i} \cdot \dfrac{1+i}{1+i} = \dfrac{1+i+i+i^2}{1-i^2}$

$= \dfrac{1 + 2i - 1}{1 + 1}$

$= \dfrac{2i}{2}$

$= i$

5. $\sqrt{-75} - \sqrt{-12} = 5i\sqrt{3} - 2i\sqrt{3} = 3i\sqrt{3}$

6. $\left(2-\sqrt{-3}\right)^2 = \left(2-i\sqrt{3}\right)^2$

$$= 4-4i\sqrt{3}+3i^2$$
$$= 4-4i\sqrt{3}-3$$
$$= 1-4i\sqrt{3}$$

7. $x(2x-3) = -4$

$2x^2 - 3x = -4$

$2x^2 - 3x + 4 = 0$

$x = \dfrac{-b \pm \sqrt{b^2 - 4ac}}{2a}$

$x = \dfrac{-(-3) \pm \sqrt{(-3)^2 - 4(2)(4)}}{2(2)}$

$x = \dfrac{3 \pm \sqrt{-23}}{4}$

$x = \dfrac{3}{4} \pm \dfrac{\sqrt{23}}{4}i$

8. $f(x) = (x-3)^2 - 4$

The parabola opens up because $a > 0$.
The vertex is (3, −4).
x-intercepts:

$0 = (x-3)^2 - 4$

$(x-3)^2 = 4$

$x - 3 = \pm\sqrt{4}$

$x = 3 \pm 2$

The equation has x-intercepts at $x = 1$ and $x = 5$.
y-intercept:

$f(0) = (0-3)^2 - 4 = 5$

domain: $(-\infty, \infty)$ range: $[-4, \infty)$

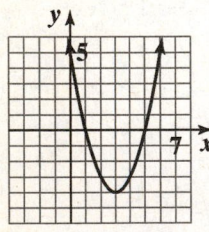

$f(x) = (x - 3)^2 - 4$

9. $f(x) = 5 - (x+2)^2$

The parabola opens down because $a < 0$.
The vertex is (−2, 5).
x-intercepts:

$0 = 5 - (x+2)^2$

$(x+2)^2 = 5$

$x + 2 = \pm\sqrt{5}$

$x = -2 \pm \sqrt{5}$

y-intercept:

$f(0) = 5 - (0+2)^2 = 1$

domain: $(-\infty, \infty)$ range: $(-\infty, 5]$

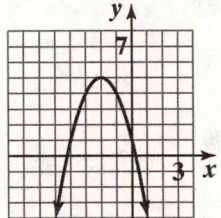

$f(x) = 5 - (x + 2)^2$

10. $f(x) = -x^2 - 4x + 5$

The parabola opens down because $a < 0$.

vertex: $x = -\dfrac{b}{2a} = -\dfrac{-4}{2(-1)} = -2$

$f(-2) = -(-2)^2 - 4(-2) + 5 = 9$

The vertex is (−2, 9).
x-intercepts:

$0 = -x^2 - 4x + 5$

$x = \dfrac{-b \pm \sqrt{b^2 - 4ac}}{2a}$

$x = \dfrac{-(-4) \pm \sqrt{(-4)^2 - 4(-1)(5)}}{2(-1)}$

$x = \dfrac{4 \pm \sqrt{36}}{-2}$

$x = -2 \pm 3$

The x-intercepts are $x = 1$ and $x = -5$.
y-intercept:

$f(0) = -0^2 - 4(0) + 5 = 5$

domain: $(-\infty, \infty)$ range: $(-\infty, 9]$

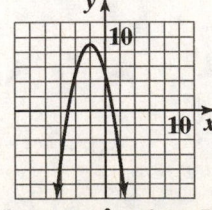

$f(x) = -x^2 - 4x + 5$

11. $f(x) = 3x^2 - 6x + 1$

The parabola opens up because $a > 0$.

vertex: $x = -\dfrac{b}{2a} = -\dfrac{-6}{2(3)} = 1$

$f(1) = 3(1)^2 - 6(1) + 1 = -2$

The vertex is $(1, -2)$.

x-intercepts:

$0 = 3x^2 - 6x + 1$

$x = \dfrac{-b \pm \sqrt{b^2 - 4ac}}{2a}$

$x = \dfrac{-(-6) \pm \sqrt{(-6)^2 - 4(3)(1)}}{2(3)}$

$x = \dfrac{6 \pm \sqrt{24}}{6}$

$x = \dfrac{3 \pm \sqrt{6}}{3}$

y-intercept:

$f(0) = 3(0)^2 - 6(0) + 1 = 1$

domain: $(-\infty, \infty)$ range: $[-2, \infty)$

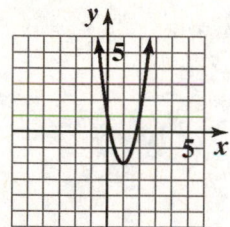

$f(x) = 3x^2 - 6x + 1$

12. $f(x) = (x-2)^2 (x+1)^3$

$(x-2)^2 (x+1)^3 = 0$

Apply the zero-product principle:

$(x-2)^2 = 0$ or $(x+1)^3 = 0$

$x - 2 = 0$ $x + 1 = 0$

$x = 2$ $x = -1$

The zeros are -1 and 2.

The graph of f crosses the x-axis at -1, since the zero has multiplicity 3. The graph touches the x-axis and turns around at 2 since the zero has multiplicity 2.

Since f is an odd-degree polynomial, degree 5, and since the leading coefficient, 1, is positive, the graph falls to the left and rises to the right.

Plot additional points as necessary and construct the graph.

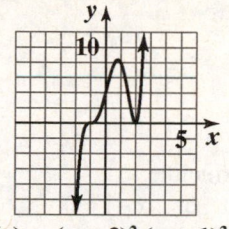

$f(x) = (x - 2)^2 (x + 1)^3$

13. $f(x) = -(x-2)^2 (x+1)^2$

$-(x-2)^2 (x+1)^2 = 0$

Apply the zero-product principle:

$(x-2)^2 = 0$ or $(x+1)^2 = 0$

$x - 2 = 0$ $x + 1 = 0$

$x = 2$ $x = -1$

The zeros are -1 and 2.

The graph touches the x-axis and turns around both at -1 and 2 since both zeros have multiplicity 2.

Since f is an even-degree polynomial, degree 4, and since the leading coefficient, -1, is negative, the graph falls to the left and falls to the right.

Plot additional points as necessary and construct the graph.

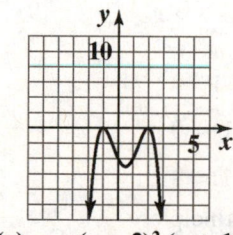

$f(x) = -(x - 2)^2 (x + 1)^2$

14. $f(x) = x^3 - x^2 - 4x + 4$

$$x^3 - x^2 - 4x + 4 = 0$$

$$x^2(x-1) - 4(x-1) = 0$$

$$(x^2 - 4)(x-1) = 0$$

$$(x+2)(x-2)(x-1) = 0$$

Apply the zero-product principle:

$x+2 = 0$ or $x-2 = 0$ or $x-1 = 0$

$\quad x = -2 \qquad\quad x = 2 \qquad\quad x = 1$

The zeros are -2, 1, and 2.

The graph of f crosses the x-axis at all three zeros, -2, 1, and 2, since all have multiplicity 1.

Since f is an odd-degree polynomial, degree 3, and since the leading coefficient, 1, is positive, the graph falls to the left and rises to the right.

Plot additional points as necessary and construct the graph.

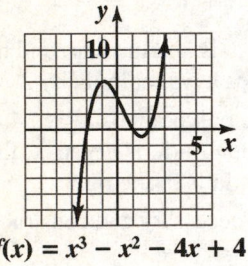

$f(x) = x^3 - x^2 - 4x + 4$

15. $f(x) = x^4 - 5x^2 + 4$

$$x^4 - 5x^2 + 4 = 0$$

$$(x^2 - 4)(x^2 - 1) = 0$$

$$(x+2)(x-2)(x+1)(x-1) = 0$$

Apply the zero-product principle,

$x = -2$, $x = 2$, $x = -1$, $x = 1$

The zeros are -2, -1, 1, and 2.

The graph crosses the x-axis at all four zeros, -2, -1, 1, and 2., since all have multiplicity 1.

Since f is an even-degree polynomial, degree 4, and since the leading coefficient, 1, is positive, the graph rises to the left and rises to the right.

Plot additional points as necessary and construct the graph.

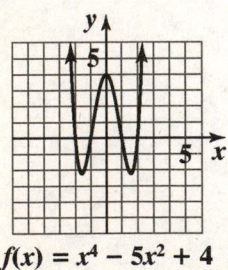

$f(x) = x^4 - 5x^2 + 4$

16. $f(x) = -(x+1)^6$

$$-(x+1)^6 = 0$$

$$(x+1)^6 = 0$$

$$x+1 = 0$$

$$x = -1$$

The zero is are -1.

The graph touches the x-axis and turns around at -1 since the zero has multiplicity 6.

Since f is an even-degree polynomial, degree 6, and since the leading coefficient, -1, is negative, the graph falls to the left and falls to the right.

Plot additional points as necessary and construct the graph.

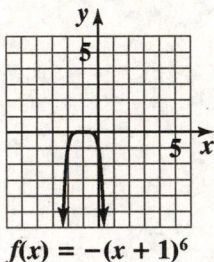

$f(x) = -(x+1)^6$

17. $f(x) = -6x^3 + 7x^2 - 1$

To find the zeros, we use the Rational Zero Theorem:

List all factors of the constant term -1: ± 1

List all factors of the leading coefficient -6:
$\pm 1, \pm 2, \pm 3, \pm 6$

The possible rational zeros are:

$\dfrac{\text{Factors of } -1}{\text{Factors of } -6} = \dfrac{\pm 1}{\pm 1, \pm 2, \pm 3, \pm 6}$

$= \pm 1, \pm \dfrac{1}{2}, \pm \dfrac{1}{3}, \pm \dfrac{1}{6}$

We test values from the above list until we find a zero. One is shown next:

Test 1:

$$\begin{array}{r|rrrr} 1 & -6 & 7 & 0 & -1 \\ & & -6 & 1 & 1 \\ \hline & -6 & 1 & 1 & 0 \end{array}$$

The remainder is 0, so 1 is a zero. Thus,

$-6x^3 + 7x^2 - 1 = 0$

$(x-1)(-6x^2 + x + 1) = 0$

$-(x-1)(6x^2 - x - 1) = 0$

$-(x-1)(3x+1)(2x-1) = 0$

Apply the zero-product property:

$x = 1, \quad x = -\dfrac{1}{3}, \quad x = \dfrac{1}{2}$

The zeros are $-\dfrac{1}{3}$, $\dfrac{1}{2}$, and 1.

The graph of f crosses the x-axis at all three zeros, $-\dfrac{1}{3}$, $\dfrac{1}{2}$, and 1, since all have multiplicity 1.

Since f is an odd-degree polynomial, degree 3, and since the leading coefficient, -6, is negative, the graph rises to the left and falls to the right.

Plot additional points as necessary and construct the graph.

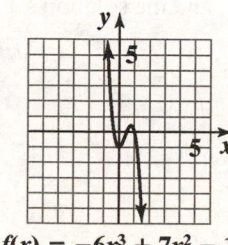

$f(x) = -6x^3 + 7x^2 - 1$

18. $f(x) = 2x^3 - 2x$

$2x^3 - 2x = 0$

$2x(x^2 - 1) = 0$

$2x(x+1)(x-1) = 0$

Apply the zero-product principle:

$x = 0, \qquad x = -1, \qquad x = 1$

The zeros are -1, 0, and 1.

The graph of f crosses the x-axis at all three zeros, -1, 0, and 1, since all have multiplicity 1.

Since f is an odd-degree polynomial, degree 3, and since the leading coefficient, 2, is positive, the graph falls to the left and rises to the right.

Plot additional points as necessary and construct the graph.

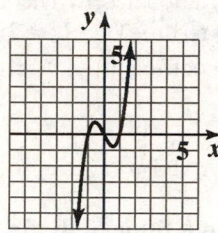

$f(x) = 2x^3 - 2x$

19. $f(x) = x^3 - 2x^2 + 26x$

$x^3 - 2x^2 + 26x = 0$

$x(x^2 - 2x + 26) = 0$

Note that $x^2 - 2x + 26$ does not factor, so we use the quadratic formula:

$x = 0 \quad$ or $\quad x^2 - 2x + 26 = 0$

$a = 1, \quad b = -2, \quad c = 26$

$x = \dfrac{-(-2) \pm \sqrt{(-2)^2 - 4(1)(26)}}{2(1)}$

$= \dfrac{2 \pm \sqrt{-100}}{2} = \dfrac{2 \pm 10i}{2} = 1 \pm 5i$

The zeros are 0 and $1 \pm 5i$.

The graph of f crosses the x-axis at 0 (the only real zero), since it has multiplicity 1.

Since f is an odd-degree polynomial, degree 3, and since the leading coefficient, 1, is positive, the graph falls to the left and rises to the right.

Plot additional points as necessary and construct the graph.

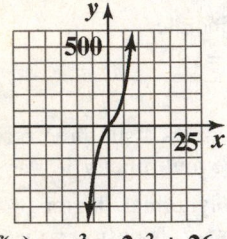

$$f(x) = x^3 - 2x^2 + 26x$$

20. $f(x) = -x^3 + 5x^2 - 5x - 3$

To find the zeros, we use the Rational Zero Theorem:
List all factors of the constant term -3: $\pm 1, \pm 3$
List all factors of the leading coefficient -1: ± 1

The possible rational zeros are:

$$\frac{\text{Factors of } -3}{\text{Factors of } -1} = \frac{\pm 1, \pm 3}{\pm 1} = \pm 1, \pm 3$$

We test values from the previous list until we find a zero. One is shown next:

Test 3:

$$\underline{3|} \quad \begin{array}{cccc} -1 & 5 & -5 & -3 \\ & -3 & 6 & 3 \\ \hline -1 & 2 & 1 & 0 \end{array}$$

The remainder is 0, so 3 is a zero. Thus,

$$-x^3 + 5x^2 - 5x - 3 = 0$$
$$(x - 3)(-x^2 + 2x + 1) = 0$$
$$-(x - 3)(x^2 - 2x - 1) = 0$$

Note that $x^2 - 2x - 1$ does not factor, so we use the quadratic formula:
$x - 3 = 0$ or $x^2 - 2x - 1 = 0$
$\quad x = 3 \qquad a = 1, \ b = -2, \ c = -1$

$$x = \frac{-(-2) \pm \sqrt{(-2)^2 - 4(1)(-1)}}{2(1)}$$

$$= \frac{2 \pm \sqrt{8}}{2} = \frac{2 \pm 2\sqrt{2}}{2} = 1 \pm \sqrt{2}$$

The zeros are 3 and $1 \pm \sqrt{2}$.

The graph of f crosses the x-axis at all three zeros, 3 and $1 \pm \sqrt{2}$, since all have multiplicity 1.

Since f is an odd-degree polynomial, degree 3, and since the leading coefficient, -1, is negative, the graph rises to the left and falls to the right.

Plot additional points as necessary and construct the graph.

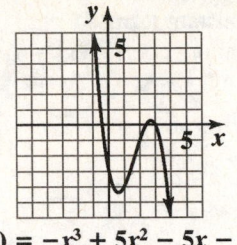

$$f(x) = -x^3 + 5x^2 - 5x - 3$$

21. $x^3 - 3x + 2 = 0$
We begin by using the Rational Zero Theorem to determine possible rational roots.

Factors of the constant term 2: $\pm 1, \pm 2$
Factors of the leading coefficient 1: ± 1

The possible rational zeros are:

$$\frac{\text{Factors of } 2}{\text{Factors of } 1} = \frac{\pm 1, \pm 2}{\pm 1} = \pm 1, \pm 2$$

We test values from above until we find a root. One is shown next:

Test 1:

$$\underline{1|} \quad \begin{array}{cccc} 1 & 0 & -3 & 2 \\ & 1 & 1 & -2 \\ \hline 1 & 1 & -2 & 0 \end{array}$$

The remainder is 0, so 1 is a root of the equation. Thus,

$$x^3 - 3x + 2 = 0$$
$$(x - 1)(x^2 + x - 2) = 0$$
$$(x - 1)(x + 2)(x - 1) = 0$$
$$(x - 1)^2 (x + 2) = 0$$

Apply the zero-product property:
$(x - 1)^2 = 0$ or $x + 2 = 0$
$\quad x - 1 = 0 \qquad\qquad x = -2$
$\quad\quad x = 1$

The solutions are -2 and 1, and the solution set is $\{-2, 1\}$.

22. $6x^3 - 11x^2 + 6x - 1 = 0$

We begin by using the Rational Zero Theorem to determine possible rational roots.

Factors of the constant term -1: ± 1
Factors of the leading coefficient 6:
$\pm 1, \ \pm 2, \ \pm 3, \ \pm 6$

The possible rational zeros are:

$$\frac{\text{Factors of } -1}{\text{Factors of } 6} = \frac{\pm 1}{\pm 1, \ \pm 2, \ \pm 3, \ \pm 6}$$

$$= \pm 1, \ \pm \frac{1}{2}, \ \pm \frac{1}{3}, \ \pm \frac{1}{6}$$

We test values from above until we find a root. One is shown next:

Test 1:

$$\begin{array}{r|rrrr}
1 & 6 & -11 & 6 & -1 \\
  &   & 6 & -5 & 1 \\
\hline
  & 6 & -5 & 1 & 0
\end{array}$$

The remainder is 0, so 1 is a root of the equation. Thus,

$$6x^3 - 11x^2 + 6x - 1 = 0$$

$$(x-1)(6x^2 - 5x + 1) = 0$$

$$(x-1)(3x-1)(2x-1) = 0$$

Apply the zero-product property:

$x - 1 = 0$ or $3x - 1 = 0$ or $2x - 1 = 0$

$x = 1$ $x = \dfrac{1}{3}$ $x = \dfrac{1}{2}$

The solutions are $\dfrac{1}{3}$, $\dfrac{1}{2}$ and 1, and the solution set is

$\left\{ \dfrac{1}{3}, \ \dfrac{1}{2}, \ 1 \right\}$.

23. $(2x+1)(3x-2)^3(2x-7) = 0$

Apply the zero-product property:

$2x + 1 = 0$ or $(3x-2)^3 = 0$ or $2x - 7 = 0$

$x = -\dfrac{1}{2}$ $3x - 2 = 0$ $x = \dfrac{7}{2}$

$\qquad\qquad\qquad x = \dfrac{2}{3}$

The solutions are $-\dfrac{1}{2}$, $\dfrac{2}{3}$ and $\dfrac{7}{2}$, and the solution set

is $\left\{ -\dfrac{1}{2}, \ \dfrac{2}{3}, \ \dfrac{7}{2} \right\}$.

24. $2x^3 + 5x^2 - 200x - 500 = 0$

We begin by using the Rational Zero Theorem to determine possible rational roots.

Factors of the constant term -500:
$\pm 1, \ \pm 2, \ \pm 4, \ \pm 5, \ \pm 10, \ \pm 20, \ \pm 25,$
$\pm 50, \ \pm 100, \ \pm 125, \ \pm 250, \ \pm 500$

Factors of the leading coefficient 2: $\pm 1, \ \pm 2$

The possible rational zeros are:

$$\frac{\text{Factors of } 500}{\text{Factors of } 2} = \pm 1, \ \pm 2, \ \pm 4, \ \pm 5,$$

$$\pm 10, \ \pm 20, \ \pm 25, \ \pm 50, \ \pm 100, \ \pm 125,$$

$$\pm 250, \ \pm 500, \ \pm \frac{1}{2}, \ \pm \frac{5}{2}, \ \pm \frac{25}{2}, \ \pm \frac{125}{2}$$

We test values from above until we find a root. One is shown next:

Test 10:

$$\begin{array}{r|rrrr}
10 & 2 & 5 & -200 & -500 \\
   &   & 20 & 250 & 500 \\
\hline
   & 2 & 25 & 50 & 0
\end{array}$$

The remainder is 0, so 10 is a root of the equation. Thus,

$$2x^3 + 5x^2 - 200x - 500 = 0$$

$$(x-10)(2x^2 + 25x + 50) = 0$$

$$(x-10)(2x+5)(x+10) = 0$$

Apply the zero-product property:

$x - 10 = 0$ or $2x + 5 = 0$ or $x + 10 = 0$

$x = 10$ $x = -\dfrac{5}{2}$ $x = -10$

The solutions are -10, $-\dfrac{5}{2}$, and 10, and the solution

set is $\left\{ -10, \ -\dfrac{5}{2}, \ 10 \right\}$.

25.
$$x^4 - x^3 - 11x^2 = x + 12$$

$$x^4 - x^3 - 11x^2 - x - 12 = 0$$

We begin by using the Rational Zero Theorem to determine possible rational roots.

Factors of the constant term -12:
$\pm 1, \pm 2, \pm 3, \pm 4, \pm 6, \pm 12$

Factors of the leading coefficient 1: ± 1

The possible rational zeros are:

$$\frac{\text{Factors of } -12}{\text{Factors of } 1}$$

$$= \frac{\pm 1, \pm 2, \pm 3, \pm 4, \pm 6, \pm 12}{\pm 1}$$

$$= \pm 1, \pm 2, \pm 3, \pm 4, \pm 6, \pm 12$$

We test values from this list we find a root. One possibility is shown next:

Test -3:

$$
\begin{array}{r|rrrrr}
-3 & 1 & -1 & -11 & -1 & -12 \\
   &   & -3 & 12 & -3 & 12 \\
\hline
   & 1 & -4 & 1 & -4 & 0
\end{array}
$$

The remainder is 0, so -3 is a root of the equation. Using the Factor Theorem, we know that $x - 1$ is a factor. Thus,

$$x^4 - x^3 - 11x^2 - x - 12 = 0$$

$$(x+3)(x^3 - 4x^2 + x - 4) = 0$$

$$(x+3)\left[x^2(x-4) + 1(x-4)\right] = 0$$

$$(x+3)(x-4)(x^2 + 1) = 0$$

As this point we know that -3 and 4 are roots of the equation. Note that $x^2 + 1$ does not factor, so we use the square-root principle: $x^2 + 1 = 0$

$$x^2 = -1$$

$$x = \pm\sqrt{-1} = \pm i$$

The roots are -3, 4, and $\pm i$, and the solution set is $\{-3, 4, \pm i\}$.

26. $2x^4 + x^3 - 17x^2 - 4x + 6 = 0$

We begin by using the Rational Zero Theorem to determine possible rational roots.

Factors of the constant term 6: $\pm 1, \pm 2, \pm 3, \pm 6$

Factors of the leading coefficient 4: $\pm 1, \pm 2$

The possible rational roots are:

$$\frac{\text{Factors of } 6}{\text{Factors of } 2} = \frac{\pm 1, \pm 2, \pm 3, \pm 6}{\pm 1, \pm 2}$$

$$= \pm 1, \pm 2, \pm 3, \pm 6, \pm\frac{1}{2}, \pm\frac{3}{2}$$

We test values from above until we find a root. One possibility is shown next:

Test -3:

$$
\begin{array}{r|rrrrr}
-3 & 2 & 1 & -17 & -4 & 6 \\
   &   & -6 & 15 & 6 & -6 \\
\hline
   & 2 & -5 & -2 & 2 & 0
\end{array}
$$

The remainder is 0, so -3 is a root. Using the Factor Theorem, we know that $x + 3$ is a factor of the polynomial. Thus,

$$2x^4 + x^3 - 17x^2 - 4x + 6 = 0$$

$$(x+3)(2x^3 - 5x^2 - 2x + 2) = 0$$

To solve the equation above, we need to factor $2x^3 - 5x^2 - 2x + 2$. We continue testing potential roots:

Test $\frac{1}{2}$:

$$
\begin{array}{r|rrrr}
\frac{1}{2} & 2 & -5 & -2 & 2 \\
            &   & 1 & -2 & -2 \\
\hline
            & 2 & -4 & -4 & 0
\end{array}
$$

The remainder is 0, so $\frac{1}{2}$ is a zero and $x - \frac{1}{2}$ is a factor.

Summarizing our findings so far, we have

$$2x^4 + x^3 - 17x^2 - 4x + 6 = 0$$

$$(x+3)(2x^3 - 5x^2 - 2x + 2) = 0$$

$$(x+3)\left(x - \frac{1}{2}\right)(2x^2 - 4x - 4) = 0$$

$$2(x+3)\left(x - \frac{1}{2}\right)(x^2 - 2x - 2) = 0$$

At this point, we know that -3 and $\frac{1}{2}$ are roots of the equation. Note that $x^2 - 2x - 2$ does not factor, so we use the quadratic formula:

$$x^2 - 2x - 2 = 0$$

$$a = 1, \ b = -2, \ c = -2$$

$$x = \frac{-(-2) \pm \sqrt{(-2)^2 - 4(1)(-2)}}{2(1)}$$

$$= \frac{2 \pm \sqrt{4+8}}{2} = \frac{2 \pm \sqrt{12}}{2} = \frac{2 \pm 2\sqrt{3}}{2} = 1 \pm \sqrt{3}$$

The solutions are -3, $\frac{1}{2}$, and $1 \pm \sqrt{3}$, and the

solution set is $\left\{ -3, \frac{1}{2}, 1 \pm \sqrt{3} \right\}$.

27. $P(x) = -x^2 + 150x - 4425$

Since $a = -1$ is negative, we know the function opens down and has a maximum at

$$x = -\frac{b}{2a} = -\frac{150}{2(-1)} = -\frac{150}{-2} = 75.$$

$$P(75) = -75^2 + 150(75) - 4425$$
$$= -5625 + 11,250 - 4425 = 1200$$

The company will maximize its profit by manufacturing and selling 75 cabinets per day. The maximum daily profit is $1200.

28. Let $x =$ one of the numbers;
$-18 - x =$ the other number
The product is $f(x) = x(-18 - x) = -x^2 - 18x$
The x-coordinate of the maximum is

$$x = -\frac{b}{2a} = -\frac{-18}{2(-1)} = \frac{-18}{-2} = -9.$$

$$f(-9) = -9 \left[-18 - (-9) \right]$$
$$= -9(-18 + 9) = -9(-9) = 81$$

The vertex is $(-9, 81)$. The maximum product is 81. This occurs when the two number are -9 and $-18 - (-9) = -9$.

29. Let $x =$ height of triangle;
$40 - 2x =$ base of triangle

$$A = \frac{1}{2}bh = \frac{1}{2}x(40 - 2x)$$

$$A(x) = 20x - x^2$$

The height at which the triangle will have

maximum area is $x = -\frac{b}{2a} = -\frac{20}{2(-1)} = 10.$

$$A(10) = 20(10) - (10)^2 = 100$$

The maximum area is 100 squares inches.

30.

$$\begin{array}{r} 2x^2 - x - 3 \\ 3x^2 - 1 \overline{\smash{\big)}\ 6x^4 - 3x^3 - 11x^2 + 2x + 4} \end{array}$$

$$\underline{6x^4 \qquad\quad - 2x^2}$$
$$-3x^3 - 9x^2 + 2x$$
$$\underline{-3x^3 \qquad\quad + x}$$
$$-9x^2 + x + 4$$
$$\underline{-9x^2 \qquad\quad + 3}$$
$$x + 1$$

$$2x^2 - x - 3 + \frac{x+1}{3x^2 - 1}$$

31. $\left(2x^4 - 13x^3 + 17x^2 + 18x - 24 \right) \div (x - 4)$

4	2	−13	17	18	−24
		8	-20	−12	24
	2	−5	−3	6	0

The quotient is $2x^3 - 5x^2 - 3x + 6$.

32. $(x - 1)(x - i)(x + i) = (x - 1)(x^2 + 1)$

$$f(x) = a_n(x - 1)(x^2 + 1)$$

$$f(-1) = a_n(-1 - 1)\left((-1)^2 + 1 \right) = -4a_n = 8$$

$$a_n = -2$$

$$f(x) = -2(x - 1)(x^2 + 1) \text{ or } -2x^3 + 2x^2 - 2x + 2$$

33. $(x - 2)(x - 2)(x - 3i)(x + 3i)$

$$= (x - 2)(x - 2)(x^2 + 9)$$

$$f(x) = a_n(x - 2)(x - 2)(x^2 + 9)$$

$$f(0) = a_n(0 - 2)(0 - 2)(0^2 + 9)$$

$$36 = 36a_n$$

$$a_n = 1$$

$$f(x) = 1(x - 2)(x - 2)(x^2 + 9)$$

$$f(x) = x^4 - 4x^3 + 13x^2 - 36x + 36$$

34. $f(x) = x^3 - x - 5$

$$f(1) = 1^3 - 1 - 5 = -5$$

$$f(2) = 2^3 - 2 - 5 = 1$$

Yes, the function must have a real zero between 1 and 2 because $f(1)$ and $f(2)$ have opposite signs.

Polynomial and Rational Functions

Section 2.6

Check Point Exercises

1. **a.** $x - 5 = 0$

   $x = 5$

   $\{x | x \neq 5\}$

   b. $x^2 - 25 = 0$

   $x^2 = 25$

   $x = \pm 5$

   $\{x | x \neq 5, x \neq -5\}$

   c. The denominator cannot equal zero.
   All real numbers.

2. **a.** $x^2 - 1 = 0$

   $x^2 = 1$

   $x = 1, x = -1$

   b. $g(x) = \dfrac{x-1}{x^2-1} = \dfrac{x-1}{(x-1)(x+1)} = \dfrac{1}{x+1}$

   $x = -1$

   c. The denominator cannot equal zero.
   No vertical asymptotes.

3. **a.** Since $n = m$, $y = \dfrac{9}{3} = 3$

   $y = 3$ is a horizontal asymptote.

   b. Since $n < m$, $y = 0$ is a horizontal asymptote.

   c. Since $n > m$, there is no horizontal asymptote.

4. Begin with the graph of $f(x) = \dfrac{1}{x}$.

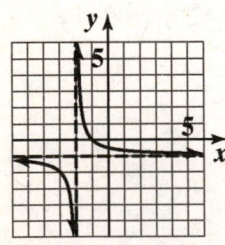

$g(x) = \dfrac{1}{x+2} - 1$

Shift the graph 2 units to the left by subtracting 2 from each x-coordinate. Shift the graph 1 unit down by subtracting 1 from each y-coordinate.

5. $f(x) = \dfrac{3x-3}{x-2}$

   $f(-x) = \dfrac{3(-x)-3}{-x-2} = \dfrac{-3x-3}{-x-2} = \dfrac{3x+3}{x+2}$

   no symmetry

   $f(0) = \dfrac{3(0)-3}{0-2} = \dfrac{3}{2}$

   The y-intercept is $\dfrac{3}{2}$.

   $3x - 3 = 0$

   $3x = 3$

   $x = 1$

   The x-intercept is 1.
   Vertical asymptote:

   $x - 2 = 0$

   $x = 2$

   Horizontal asymptote:

   $y = \dfrac{3}{1} = 3$

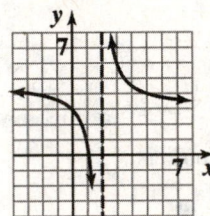

$f(x) = \dfrac{3x-3}{x-2}$

6. $f(x) = \dfrac{2x^2}{x^2-9}$

   $f(-x) = \dfrac{2(-x)^2}{(-x)^2-9} = \dfrac{2x^2}{x^2-9} = f(x)$

   The y-axis symmetry.

   $f(0) = \dfrac{2(0)^2}{0^2-9} = 0$

   The y-intercept is 0.

   $2x^2 = 0$

   $x = 0$

   The x-intercept is 0.

210

vertical asymptotes:

$x^2 - 9 = 0$

$x = 3, x = -3$

horizontal asymptote:

$y = \dfrac{2}{1} = 2$

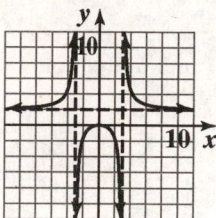

$$f(x) = \dfrac{2x^2}{x^2 - 9}$$

7.　$f(x) = \dfrac{x^4}{x^2 + 2}$

$f(-x) = \dfrac{(-x)^4}{(-x)^2 + 2} = \dfrac{x^4}{x^2 + 2} = f(x)$

y-axis symmetry

$f(0) = \dfrac{0^4}{0^2 + 2} = 0$

The y-intercept is 0.

$x^4 = 0$

$x = 0$

The x-intercept is 0.

vertical asymptotes:

$x^2 + 2 = 0$

$x^2 = -2$

no vertical asymptotes

horizontal asymptote:

Since $n > m$, there is no horizontal asymptote.

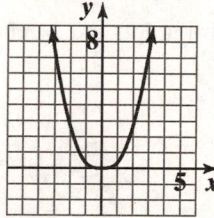

$$f(x) = \dfrac{x^4}{x^2 + 2}$$

8.

$$\begin{array}{r|rrr} 2 & 2 & -5 & 7 \\ & & 4 & -2 \\ \hline & 2 & -1 & 5 \end{array}$$

the equation of the slant asymptote is
$y = 2x - 1$.

9.　a.　$C(x) = 500,000 + 400x$

b.　$\bar{C}(x) = \dfrac{500,000 + 400x}{x}$

c.　$\bar{C}(1000) = \dfrac{500,000 + 400(1000)}{1000}$

$\qquad = 900$

$\bar{C}(10,000) = \dfrac{500,000 + 400(10,000)}{10,000}$

$\qquad = 450$

$\bar{C}(100,000) = \dfrac{500,000 + 400(100,000)}{100,000}$

$\qquad = 405$

The average cost per wheelchair of producing 1000, 10,000, and 100,000 wheelchairs is $900, $450, and $405, respectively.

d.　$y = \dfrac{400}{1} = 400$

The cost per wheelchair approaches $400 as more wheelchairs are produced.

10.　$x - 10 =$ the average velocity on the return trip. The function that expresses the total time required to complete the round trip is

$$T(x) = \dfrac{20}{x} + \dfrac{20}{x - 10}.$$

Exercise Set 2.6

1.　$f(x) = \dfrac{5x}{x - 4}$

$\{x | x \neq 4\}$

3.　$g(x) = \dfrac{3x^2}{(x - 5)(x + 4)}$

$\{x | x \neq 5, x \neq -4\}$

5.　$h(x) = \dfrac{x + 7}{x^2 - 49}$

$x^2 - 49 = (x - 7)(x + 7)$

$\{x | x \neq 7, x \neq -7\}$

7.　$f(x) = \dfrac{x + 7}{x^2 + 49}$

all real numbers

9. $-\infty$

11. $-\infty$

13. 0

15. $+\infty$

17. $-\infty$

19. 1

21. $f(x) = \dfrac{x}{x+4}$

$x+4 = 0$

$x = -4$

vertical asymptote: $x = -4$

23. $g(x) = \dfrac{x+3}{x(x+4)}$

$x(x+4) = 0$

$x = 0, x = -4$

vertical asymptotes: $x = 0$, $x = -4$

25. $h(x) = \dfrac{x}{x(x+4)} = \dfrac{1}{x+4}$

$x+4 = 0$

$x = -4$

vertical asymptote: $x = -4$

27. $r(x) = \dfrac{x}{x^2 + 4}$

$x^2 + 4$ has no real zeros

There are no vertical asymptotes.

29. $f(x) = \dfrac{12x}{3x^2 + 1}$

$n < m$

horizontal asymptote: $y = 0$

31. $g(x) = \dfrac{12x^2}{3x^2 + 1}$

$n = m,$

horizontal asymptote: $y = \dfrac{12}{3} = 4$

33. $h(x) = \dfrac{12x^3}{3x^2 + 1}$

$n > m$

no horizontal asymptote

35. $f(x) = \dfrac{-2x+1}{3x+5}$

$n = m$

horizontal asymptote: $y = -\dfrac{2}{3}$

37. $g(x) = \dfrac{1}{x-1}$

Shift the graph of $f(x) = \dfrac{1}{x}$ 1 unit to the right.

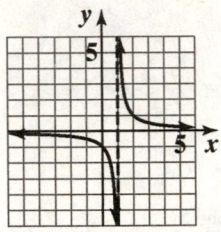

$g(x) = \dfrac{1}{x-1}$

39. $h(x) = \dfrac{1}{x} + 2$

Shift the graph of $f(x) = \dfrac{1}{x}$ 2 units up.

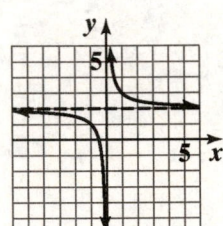

$h(x) = \dfrac{1}{x} + 2$

41. $g(x) = \dfrac{1}{x+1} - 2$

Shift the graph of $f(x) = \dfrac{1}{x}$ 1 unit left and 2 units down.

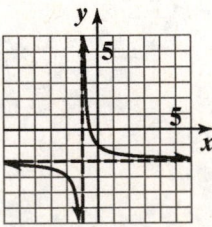

$g(x) = \dfrac{1}{x+1} - 2$

43. $g(x) = \dfrac{1}{(x+2)^2}$

Shift the graph of $f(x) = \dfrac{1}{x^2}$ 2 units left.

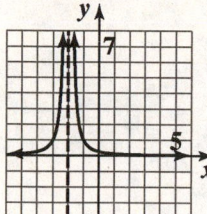

$g(x) = \dfrac{1}{(x+2)^2}$

45. $h(x) = \dfrac{1}{x^2} - 4$

Shift the graph of $f(x) = \dfrac{1}{x^2}$ 4 units down.

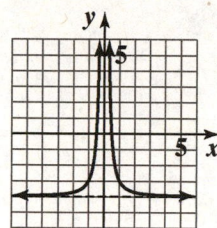

$h(x) = \dfrac{1}{x^2} - 4$

47. $h(x) = \dfrac{1}{(x-3)^2} + 1$

Shift the graph of $f(x) = \dfrac{1}{x^2}$ 3 units right and 1 unit up.

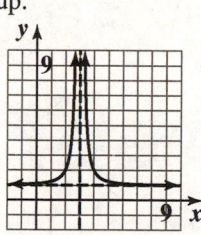

$h(x) = \dfrac{1}{(x-3)^2} + 1$

49. $f(x) = \dfrac{4x}{x-2}$

$f(-x) = \dfrac{4(-x)}{(-x)-2} = \dfrac{4x}{x+2}$

$f(-x) \neq f(x), f(-x) \neq -f(x)$

no symmetry

y-intercept: $y = \dfrac{4(0)}{0-2} = 0$

x-intercept: $4x = 0$

$x = 0$

vertical asymptote:

$x - 2 = 0$

$\quad x = 2$

horizontal asymptote:

$n = m$, so $y = \dfrac{4}{1} = 4$

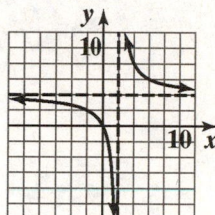

$f(x) = \dfrac{4x}{x-2}$

51. $f(x) = \dfrac{2x}{x^2 - 4}$

$f(-x) = \dfrac{2(-x)}{(-x)^2 - 4} = -\dfrac{2x}{x^2 - 4} = -f(x)$

Origin symmetry

y-intercept: $\dfrac{2(0)}{0^2 - 4} = \dfrac{0}{-4} = 0$

x-intercept:

$2x = 0$

$x = 0$

vertical asymptotes:

$x^2 - 4 = 0$

$x = \pm 2$

horizontal asymptote:

$n < m$ so $y = 0$

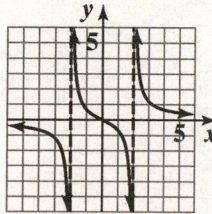

$f(x) = \dfrac{2x}{x^2 - 4}$

53. $f(x) = \dfrac{2x^2}{x^2 - 1}$

$f(-x) = \dfrac{2(-x)^2}{(-x)^2 - 1} = \dfrac{2x^2}{x^2 - 1} = f(x)$

y-axis symmetry

y-intercept: $y = \dfrac{2(0)^2}{0^2 - 1} = \dfrac{0}{1} = 0$

x-intercept:

$2x^2 = 0$

$x = 0$

vertical asymptote:

$x^2 - 1 = 0$

$x^2 = 1$

$x = \pm 1$

horizontal asymptote:

$n = m$, so $y = \dfrac{2}{1} = 2$

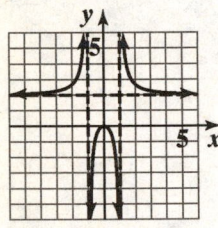

$f(x) = \dfrac{2x^2}{x^2 - 1}$

55. $f(x) = \dfrac{-x}{x + 1}$

$f(-x) = \dfrac{-(-x)}{(-x) + 1} = \dfrac{x}{-x + 1}$

$f(-x) \ne f(x), f(-x) \ne -f(x)$

no symmetry

y-intercept: $y = \dfrac{-(0)}{0 + 1} = \dfrac{0}{1} = 0$

x-intercept:

$-x = 0$

$x = 0$

vertical asymptote:

$x + 1 = 0$

$x = -1$

horizontal asymptote:

$n = m$, so $y = \dfrac{-1}{1} = -1$

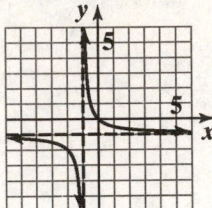

$f(x) = \dfrac{-x}{x + 1}$

57. $f(x) = -\dfrac{1}{x^2 - 4}$

$f(-x) = -\dfrac{1}{(-x)^2 - 4} = -\dfrac{1}{x^2 - 4} = f(x)$

y-axis symmetry

y-intercept: $y = -\dfrac{1}{0^2 - 4} = \dfrac{1}{4}$

x-intercept: $-1 \ne 0$

no *x*-intercept

vertical asymptotes:

$x^2 - 4 = 0$

$x^2 = 4$

$x = \pm 2$

horizontal asymptote:

$n < m$ or $y = 0$

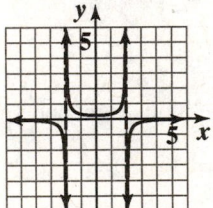

$f(x) = -\dfrac{1}{x^2 - 4}$

59. $f(x) = \dfrac{2}{x^2 + x - 2}$

$f(-x) = -\dfrac{2}{(-x)^2 - x - 2} = \dfrac{2}{x^2 - x - 2}$

$f(-x) \neq f(x), f(-x) \neq -f(x)$

no symmetry

y-intercept: $y = \dfrac{2}{0^2 + 0 - 2} = \dfrac{2}{-2} = -1$

x-intercept: none
vertical asymptotes:

$x^2 + x - 2 = 0$

$(x+2)(x-1) = 0$

$x = -2, x = 1$

horizontal asymptote:
$n < m$ so $y = 0$

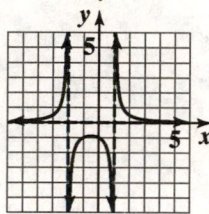

$f(x) = \dfrac{2}{x^2 + x - 2}$

61. $f(x) = \dfrac{2x^2}{x^2 + 4}$

$f(-x) = \dfrac{2(-x)^2}{(-x)^2 + 4} = \dfrac{2x^2}{x^2 + 4} = f(x)$

y axis symmetry

y-intercept: $y = \dfrac{2(0)^2}{0^2 + 4} = 0$

x-intercept: $2x^2 = 0$
$x = 0$
vertical asymptote: none
horizontal asymptote:

$n = m$, so $y = \dfrac{2}{1} = 2$

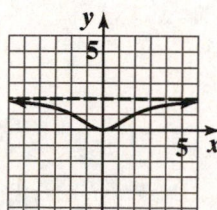

$f(x) = \dfrac{2x^2}{x^2 + 4}$

63. $f(x) = \dfrac{x+2}{x^2 + x - 6}$

$f(-x) = \dfrac{-x+2}{(-x)^2 - (-x) - 6} = \dfrac{-x+2}{x^2 + x - 6}$

$f(-x) \neq f(x), f(-x) \neq -f(x)$

no symmetry

y-intercept: $y = \dfrac{0+2}{0^2 + 0 - 6} = -\dfrac{2}{6} = -\dfrac{1}{3}$

x-intercept:
$x + 2 = 0$
$x = -2$
vertical asymptotes:

$x^2 + x - 6 = 0$

$(x+3)(x-2)$

$x = -3, x = 2$

horizontal asymptote:
$n < m$, so $y = 0$

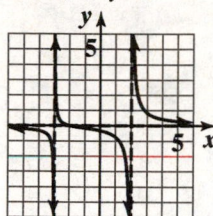

$f(x) = \dfrac{x+2}{x^2 + x - 6}$

65. $f(x) = \dfrac{x^4}{x^2 + 2}$

$f(-x) = \dfrac{(-x)^4}{(-x)^2 + 2} = \dfrac{x^4}{x^2 + 2} = f(x)$

y-axis symmetry

y-intercept: $y = \dfrac{0^4}{0^2 + 2} = 0$

x-intercept: $x^4 = 0$
$x = 0$
vertical asymptote: none
horizontal asymptote:
$n > m$, so none

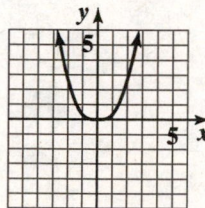

$f(x) = \dfrac{x^4}{x^2 + 2}$

67. $f(x) = \dfrac{x^2 + x - 12}{x^2 - 4}$

$f(-x) = \dfrac{(-x)^2 - x - 12}{(-x)^2 - 4} = \dfrac{x^2 - x - 12}{x^2 - 4}$

$f(-x) \neq f(x), f(-x) \neq -f(x)$

no symmetry

y-intercept: $y = \dfrac{0^2 + 0 - 12}{0^2 - 4} = 3$

x-intercept: $x^2 + x - 12 = 0$

$\quad\quad (x - 3)(x + 4) = 0$

$\quad\quad\quad\quad\quad x = 3, x = -4$

vertical asymptotes:

$\quad x^2 - 4 = 0$

$(x - 2)(x + 2) = 0$

$\quad\quad\quad x = 2, x = -2$

horizontal asymptote:

$n = m$, so $y = \dfrac{1}{1} = 1$

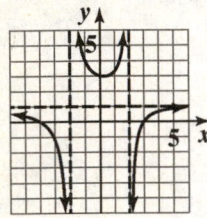

$f(x) = \dfrac{x^2 + x - 12}{x^2 - 4}$

69. $f(x) = \dfrac{3x^2 + x - 4}{2x^2 - 5x}$

$f(-x) = \dfrac{3(-x)^2 - x - 4}{2(-x)^2 + 5x} = \dfrac{3x^2 - x - 4}{2x^2 + 5x}$

$f(-x) \neq f(x), f(-x) \neq -f(x)$

no symmetry

y-intercept: $y = \dfrac{3(0)^2 + 0 - 4}{2(0)^2 - 5(0)} = \dfrac{-4}{0}$

no y-intercept

x-intercepts:

$\quad\quad 3x^2 + x - 4 = 0$

$\quad (3x + 4)(x - 1) = 0$

$\quad\quad 3x + 4 = 0 \quad x - 1 = 0$

$\quad\quad\quad 3x = -4$

$\quad\quad\quad\quad x = -\dfrac{4}{3}, x = 1$

vertical asymptotes:

$2x^2 - 5x = 0$

$x(2x - 5) = 0$

$x = 0, 2x = 5$

$\quad\quad x = \dfrac{5}{2}$

horizontal asymptote:

$n = m$, so $y = \dfrac{3}{2}$

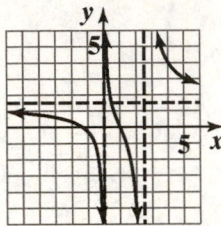

$f(x) = \dfrac{3x^2 + x - 4}{2x^2 - 5x}$

71. a. Slant asymptote:

$f(x) = x - \dfrac{1}{x}$

$y = x$

b. $f(x) = \dfrac{x^2 - 1}{x}$

$f(-x) = \dfrac{(-x)^2 - 1}{(-x)} = \dfrac{x^2 - 1}{-x} = -f(x)$

Origin symmetry

y-intercept: $y = \dfrac{0^2 - 1}{0} = \dfrac{-1}{0}$

no y-intercept

x-intercepts: $x^2 - 1 = 0$

$x = \pm 1$

vertical asymptote: $x = 0$

horizontal asymptote:

$n < m$, so none exist.

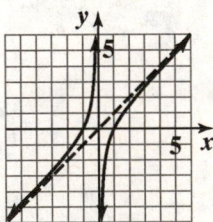

$f(x) = \dfrac{x^2 - 1}{x}$

73. a. Slant asymptote:

$$f(x) = x + \frac{1}{x}$$

$$y = x$$

b. $$f(x) = \frac{x^2 + 1}{x}$$

$$f(-x) = \frac{(-x)^2 + 1}{-x} = \frac{x^2 + 1}{-x} = -f(x)$$

Origin symmetry

y-intercept: $y = \frac{0^2 + 1}{0} = \frac{1}{0}$

no y-intercept

x-intercept:

$x^2 + 1 = 0$

$x^2 = -1$

no x-intercept

vertical asymptote: $x = 0$

horizontal asymptote:

$n > m$, so none exist.

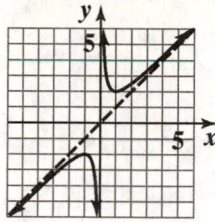

$$f(x) = \frac{x^2 + 1}{x}$$

75. a. Slant asymptote:

$$f(x) = x + 4 + \frac{6}{x - 3}$$

$$y = x + 4$$

b. $$f(x) = \frac{x^2 + x - 6}{x - 3}$$

$$f(-x) = \frac{(-x)^2 + (-x) - 6}{-x - 3} = \frac{x^2 - x - 6}{-x - 3}$$

$f(-x) \neq g(x), g(-x) \neq -g(x)$

No symmetry

y-intercept: $y = \frac{0^2 + 0 - 6}{0 - 3} = \frac{-6}{-3} = 2$

x-intercept:

$x^2 + x - 6 = 0$

$(x + 3)(x - 2) = 0$

$x = -3$ and $x = 2$

vertical asymptote:

$x - 3 = 0$

$x = 3$

horizontal asymptote:

$n > m$, so none exist.

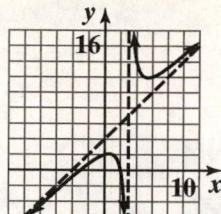

$$f(x) = \frac{x^2 + x - 6}{x - 3}$$

77. $$f(x) = \frac{x^3 + 1}{x^2 + 2x}$$

a. slant asymptote:

$$\begin{array}{r} x - 2 \\ x^2 + 2x\overline{) x^3 \qquad\quad + 1} \\ x^3 + 2x^2 \\ \hline -2x^2 \\ -2x^2 + 4x \\ \hline -4x + 1 \end{array}$$

$$y = x - 2$$

b. $$f(-x) = \frac{(-x)^3 + 1}{(-x)^2 + 2(-x)} = \frac{-x^3 + 1}{x^2 - 2x}$$

$f(-x) \neq f(x), \ f(-x) \neq -f(x)$

no symmetry

y-intercept: $y = \frac{0^3 + 1}{0^2 + 2(0)} = \frac{1}{0}$

no y-intercept

x-intercept: $x^3 + 1 = 0$

$x^3 = -1$

$x = -1$

vertical asymptotes:

$x^2 + 2x = 0$

$x(x + 2) = 0$

$x = 0, x = -2$

horizontal asymptote:

$n > m$, so none

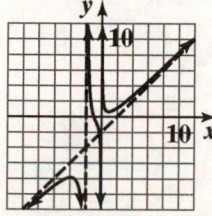

$$f(x) = \frac{x^3 + 1}{x^2 + 2x}$$

79. $\dfrac{5x^2}{x^2-4} \cdot \dfrac{x^2+4x+4}{10x^3}$

$= \dfrac{\cancel{5}\,\cancel{x^2}}{(x+2)(x-2)} \cdot \dfrac{(x+2)^{\cancel{2}}}{\cancel{10}\,x^{\cancel{3}1}}$

$= \dfrac{x+2}{2x(x-2)}$

So, $f(x) = \dfrac{x+2}{2x(x-2)}$

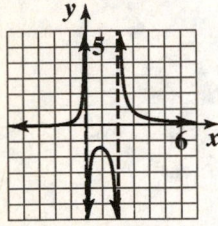

$f(x) = \dfrac{x+2}{2x(x-2)}$

81. $\dfrac{x}{2x+6} - \dfrac{9}{x^2-9}$

$\dfrac{x}{2x+6} - \dfrac{9}{x^2-9}$

$= \dfrac{x}{2(x+3)} - \dfrac{9}{(x+3)(x-3)}$

$= \dfrac{x(x-3)-9(2)}{2(x+3)(x-3)}$

$= \dfrac{x^2-3x-18}{2(x+3)(x-3)}$

$= \dfrac{(x-6)\,\cancel{(x+3)}}{2\,\cancel{(x+3)}(x-3)} = \dfrac{x-6}{2(x-3)}$

So, $f(x) = \dfrac{x-6}{2(x-3)}$

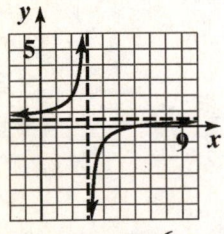

$f(x) = \dfrac{x-6}{2(x-3)}$

83. $\dfrac{1-\dfrac{3}{x+2}}{1+\dfrac{1}{x-2}} = \dfrac{1-\dfrac{3}{x+2}}{1+\dfrac{1}{x-2}} \cdot \dfrac{(x+2)(x-2)}{(x+2)(x-2)}$

$= \dfrac{(x+2)(x-2)-3(x-2)}{(x+2)(x-2)+(x+2)}$

$= \dfrac{x^2-4-3x+6}{x^2-4+x+2}$

$= \dfrac{x^2-3x+2}{x^2+x-2}$

$= \dfrac{(x-2)\,\cancel{(x-1)}}{(x+2)\,\cancel{(x-1)}} = \dfrac{x-2}{x+2}$

So, $f(x) = \dfrac{x-2}{x+2}$

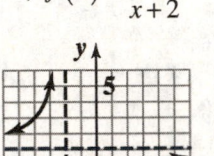

$f(x) = \dfrac{x-2}{x+2}$

85. $g(x) = \dfrac{2x+7}{x+3} = \dfrac{1}{x+3}+2$

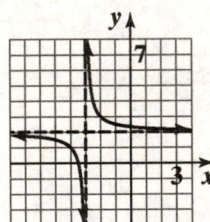

$f(x) = \dfrac{1}{x+3}+2$

87. $g(x) = \dfrac{3x-7}{x-2} = \dfrac{-1}{x-2}+3$

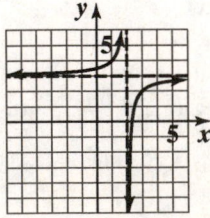

$f(x) = \dfrac{-1}{x-2}+3$

89. a. $C(x) = 100x + 100,000$

b. $\overline{C}(x) = \dfrac{100x + 100,000}{x}$

c. $\overline{C}(500) = \dfrac{100(500) + 100,000}{500} = \$300$

When 500 bicycles are manufactured, it costs $300 to manufacture each.

$\overline{C}(1000) = \dfrac{100(1000) + 100,000}{1000} = \$200$

When 1000 bicycles are manufactured, it costs $200 to manufacture each.

$\overline{C}(2000) = \dfrac{100(2000) + 100,000}{2000} = \$150$

When 2000 bicycles are manufactured, it costs $150 to manufacture each.

$\overline{C}(4000) = \dfrac{100(4000) + 100,000}{4000} = \$125$

When 4000 bicycles are manufactured, it costs $125 to manufacture each.
The average cost decreases as the number of bicycles manufactured increases.

d. $n = m$, so $y = \dfrac{100}{1} = 100$.

As greater numbers of bicycles are manufactured, the average cost approaches $100.

91. a. From the graph the pH level of the human mouth 42 minutes after a person eats food containing sugar will be about 6.0.

b. From the graph, the pH level is lowest after about 6 minutes.

$f(6) = \dfrac{6.5(6)^2 - 20.4(6) + 234}{6^2 + 36}$

$= 4.8$

The pH level after 6 minutes (i.e. the lowest pH level) is 4.8.

c. From the graph, the pH level appears to approach 6.5 as time goes by. Therefore, the normal pH level must be 6.5.

d. $y = 6.5$

Over time, the pH level rises back to the normal level.

e. During the first hour, the pH level drops quickly below normal, and then slowly begins to approach the normal level.

93. $P(10) = \dfrac{100(10 - 1)}{10} = 90 \quad (10, 90)$

For a disease that smokers are 10 times more likely to contact than non-smokers, 90% of the deaths are smoking related.

95. $y = 100$ As incidence of the diseases increases, the percent of death approaches, but never gets to be, 100%.

97. a. $f(x) = \dfrac{11x^2 + 40x + 1040}{12x^2 + 230x + 2190}$

b. According to the graph, $\dfrac{1707.2}{2708.7}$ or about 63% of federal expenditures were spent on human resources in 2006.

c. According to the function,

$f(36) = \dfrac{11(36)^2 + 40(36) + 1040}{12(36)^2 + 230(36) + 2190} = \dfrac{16736}{26022}$ or

about 64% of federal expenditures were spent on human resources in 2006. This overestimates the actual percent found in the graph by 1%.

d. The horizontal asymptote is $y = \dfrac{11}{12}$.

If trends continue, $\dfrac{11}{12}$ or about 92% of federal expenditures will spent on human resources over time.

99. $T(x) = \dfrac{90}{9x} + \dfrac{5}{x} = \dfrac{10}{x} + \dfrac{5}{x}$

The function that expresses the total time for driving and hiking is $T(x) = \dfrac{10}{x} + \dfrac{5}{x}$.

101. $A = lw$

$xy = 50$

$l = y + 2 = \dfrac{50}{x} + 2$

$w = x + 1$

$A = \left(\dfrac{50}{x} + 2\right)(x + 1)$

$= 50 + \dfrac{50}{x} + 2x + 2$

$= 2x + \dfrac{50}{x} + 52$

The total area of the page is

$A(x) = 2x + \dfrac{50}{x} + 52$.

103. – 111. Answers may vary.

113.

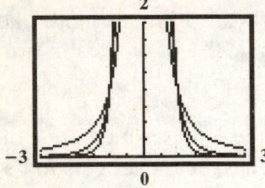

The graph approaches the horizontal asymptote faster and the vertical asymptote slower as *n* increases.

115. a. $f(x) = \dfrac{27725(x-14)}{x^2+9} - 5x$

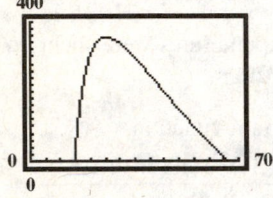

b. The graph increases from late teens until about the age of 25, and then the number of arrests decreases.

c. At age 25 the highest number arrests occurs. There are about 356 arrests for every 100,000 drivers.

117. does not make sense; Explanations will vary. Sample explanation: The function has one vertical asymptote, $x = 2$.

119. does not make sense; Explanations will vary. Sample explanation: As production level increases, the average cost for a company to produce each unit of its product decreases.

121. true

123. true

125. – 127. Answers may vary.

128.
$$2x^2 + x = 15$$
$$2x^2 + x - 15 = 0$$
$$(2x-5)(x+3) = 0$$
$$2x - 5 = 0 \quad \text{or} \quad x + 3 = 0$$
$$x = \frac{5}{2} \qquad\qquad x = -3$$

The solution set is $\left\{-3, \dfrac{5}{2}\right\}$.

129.
$$x^3 + x^2 = 4x + 4$$
$$x^3 + x^2 - 4x - 4 = 0$$
$$x^2(x+1) - 4(x+1) = 0$$
$$(x+1)(x^2 - 4) = 0$$
$$(x+1)(x+2)(x-2) = 0$$

The solution set is $\{-2, -1, 2\}$.

130.
$$\frac{x+1}{x+3} - 2 = \frac{x+1}{x+3} - \frac{2(x+3)}{x+3}$$
$$= \frac{x+1}{x+3} - \frac{2x+6}{x+3}$$
$$= \frac{x+1-2x-6}{x+3}$$
$$= \frac{-x-5}{x+3} \text{ or } -\frac{x+5}{x+3}$$

Section 2.7

Check Point Exercises

1.
$$x^2 - x > 20$$
$$x^2 - x - 20 > 0$$
$$(x+4)(x-5) > 0$$

Solve the related quadratic equation.
$$(x+4)(x-5) = 0$$

Apply the zero product principle.
$$x+4=0 \quad \text{or} \quad x-5=0$$
$$x=-4 \qquad \quad x=5$$

The boundary points are –2 and 4.

Test Interval	Test Number	Test	Conclusion
$(-\infty, -4)$	–5	$(-5)^2 - (-5) > 20$ $30 > 20$, true	$(-\infty, -4)$ belongs to the solution set.
$(-4, 5)$	0	$(0)^2 - (0) > 20$ $0 > 20$, false	$(-4, 5)$ does not belong to the solution set.
$(5, \infty)$	10	$(10)^2 - (10) > 20$ $90 > 20$, true	$(5, \infty)$ belongs to the solution set.

The solution set is $(-\infty, -4) \cup (5, \infty)$ or $\{x | x < -4 \text{ or } x > 5\}$.

2.
$$x^3 + 3x^2 \le x + 3$$
$$x^3 + 3x^2 - x - 3 \le 0$$
$$(x+1)(x-1)(x+3) \le 0$$
$$(x+1)(x-1)(x+3) = 0$$
$$x+1=0 \quad \text{or} \quad x-1=0 \quad \text{or} \quad x+3=0$$
$$x=-1 \qquad \quad x=1 \qquad \quad x=-3$$

Test Interval	Test Number	Test	Conclusion
$(-\infty, -3)$	-4	$(-4)^3 + 3(-4)^2 \le (-4)+3$ $-16 \le -1$ true	$(-\infty, -3)$ belongs to the solution set.
$(-3, -1]$	-2	$(-2)^3 + 3(-2)^2 \le (-2)+3$ $4 \le 1$ false	$(-3, -1]$ does not belong to the solution set.
$[-1, 1]$	0	$(0)^3 + 3(0)^2 \le (0)+3$ $0 \le 3$ true	$[-1, 1]$ belongs to the solution set.
$[1, \infty)$	2	$(6+3)(6-5) > 0$ true	$[1, \infty)$ does not belong to the solution set.

The solution set is $(-\infty, -3] \cup [-1, 1]$ or $\{x | x \le -3 \text{ or } -1 \le x \le 1\}$.

3.

$$\frac{2x}{x+1} \geq 1$$

$$\frac{2x}{x+1} - 1 \geq 0$$

$$\frac{x-1}{x+1} \geq 0$$

$$x - 1 = 0 \quad \text{or} \quad x + 1 = 0$$

$$x = 1 \qquad x = -1$$

Test Interval	Test Number	Test	Conclusion
$(-\infty, -1)$	-2	$\dfrac{2(-2)}{-2+1} \geq 1$ $4 \geq 1$, true	$(-\infty, -1)$ belongs to the solution set.
$(-1, 1]$	0	$\dfrac{2(0)}{0+1} \geq 1$ $0 \geq 1$, false	$(-1, 1]$ does not belong to the solution set.
$[1, \infty)$	2	$\dfrac{2(2)}{2+1} \geq 1$ $\dfrac{4}{3} \geq 1$, true	$[1, \infty)$ belongs to the solution set.

The solution set is $(-\infty, -1) \cup [1, \infty)$ or $\{x \mid x < -1 \text{ or } x \geq 1\}$.

$$\overset{\longleftarrow}{\underset{-5\ -4\ -3\ -2\ -1\ \ 0\ \ 1\ \ 2\ \ 3\ \ 4\ \ 5}{\longrightarrow}}$$

4.

$$-16t^2 + 80t > 64$$

$$-16t^2 + 80t - 64 > 0$$

$$-16(t-1)(t-4) > 0$$

$$t - 1 = 0 \quad \text{or} \quad t - 4 = 0$$

$$t = 1 \qquad t = 4$$

Test Interval	Test Number	Test	Conclusion
$(-\infty, 1)$	0	$-16(0)^2 + 80(0) > 64$ $0 > 64$, false	$(-\infty, 1)$ does not belong to the solution set.
$(1, 4)$	2	$-16(2)^2 + 80(2) > 64$ $96 > 64$, true	$(1, 4)$ belongs to the solution set.
$(4, \infty)$	5	$-16(5)^2 + 80(5) > 64$ $0 > 64$, false	$(4, \infty)$ does not belong to the solution set.

The object will be more than 64 feet above the ground between 1 and 4 seconds.

Exercise Set 2.7

1. $(x-4)(x+2) > 0$
$x = 4$ or $x = -2$

T	F	T
	-2	4

Test -3: $(-3-4)(-3+2) > 0$
$\qquad 7 > 0$ True
Test 0: $(0-4)(0+2) > 0$
$\qquad -8 > 0$ False
Test 5: $(5-4)(5+2) > 0$
$\qquad 7 > 0$ True
$(-\infty, -2)$ or $(4, \infty)$

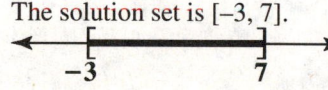

3. $(x-7)(x+3) \le 0$
$x = 7$ or $x = -3$

F	T	F
	-3	7

Test -4: $(-4-7)(-4+3) \le 0$
$\qquad 11 \le 0$ False
Test 0: $(0-7)(0+3) \le 0$
$\qquad -21 \le 0$ True
Test 8: $(8-7)(8+3) \le 0$
$\qquad 11 \le 0$ False
The solution set is $[-3, 7]$.

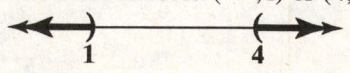

5. $x^2 - 5x + 4 > 0$
$(x-4)(x-1) > 0$
$x = 4$ or $x = 1$

T	F	T
	1	4

Test 0: $0^2 - 5(0) + 4 > 0$
$\qquad 4 > 0$ True
Test 2: $2^2 - 5(2) + 4 > 0$
$\qquad -2 > 0$ False
Test 5: $5^2 - 5(5) + 4 > 0$
$\qquad 4 > 0$ True
The solution set is $(-\infty, 1)$ or $(4, \infty)$.

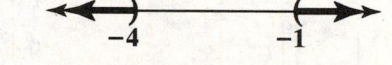

7. $x^2 + 5x + 4 > 0$
$(x+1)(x+4) > 0$
$x = -1$ or $x = -4$

T	F	T
	-4	-1

Test -5: $(-5)^2 + 5(-5) + 4 > 0$
$\qquad 4 > 0$ True
Test -3: $(-3)^2 + 5(-3) + 4 > 0$
$\qquad -2 > 0$ False
Test 0: $0^2 + 5(0) + 4 > 0$
$\qquad 4 > 0$ True
The solution set is $(-\infty, -4)$ or $(-1, \infty)$.

9. $x^2 - 6x + 9 < 0$
$(x-3)(x-3) < 0$
$\qquad\qquad x = 3$

F	F
	3

Test 0: $0^2 - 6(0) + 9 < 0$
$\qquad\qquad 9 < 0$ False
Test 4: $4^2 - 6(4) + 9 < 0$
$\qquad\qquad 1 < 0$ False
The solution set is the empty set, $\varnothing$.

11.
$$3x^2 + 10x - 8 \le 0$$
$$(3x - 2)(x + 4) \le 0$$

$$x = \frac{2}{3} \text{ or } x = -4$$

F		T		F
	-4		$\frac{2}{3}$	

Test -5: $3(-5)^2 + 10(-5) - 8 \le 0$
$$17 \le 0 \text{ False}$$
Test 0: $3(0)^2 + 10(0) - 8 \le 0$
$$8 \le 0 \text{ True}$$
Test 1: $3(1)^2 + 10(1) - 8 \le 0$
$$5 \le 0 \text{ False}$$

The solution set is $\left[-4, \frac{2}{3}\right]$.

13.
$$2x^2 + x < 15$$
$$2x^2 + x - 15 < 0$$
$$(2x - 5)(x + 3) < 0$$
$$2x - 5 = 0 \quad \text{or} \quad x + 3 = 0$$
$$2x = 5$$
$$x = \frac{5}{2} \quad \text{or} \quad x = -3$$

F		T		F
	-3		$\frac{5}{2}$	

Test -4: $2(-4)^2 + (-4) < 15$
$$28 < 15 \text{ False}$$
Test 0: $2(0)^2 + 0 < 15$
$$0 < 15 \text{ True}$$
Test 3: $2(3)^2 + 3 < 15$
$$21 < 15 \text{ False}$$

The solution set is $\left(-3, \frac{5}{2}\right)$.

15.
$$4x^2 + 7x < -3$$
$$4x^2 + 7x + 3 < 0$$
$$(4x + 3)(x + 1) < 0$$
$$4x + 3 = 0 \quad \text{or} \quad x + 1 = 0$$
$$4x - 3 = 0$$
$$x = -\frac{3}{4} \quad \text{or} \quad x = -1$$

F		T		F
	-1		$-\frac{3}{4}$	

Test -2: $4(-2)^2 + 7(-2) < -3$
$$2 < -3 \text{ False}$$
Test $-\frac{7}{8}$: $4\left(-\frac{7}{8}\right)^2 + 7\left(-\frac{7}{8}\right) < -3$
$$\frac{49}{16} - \frac{49}{8} < -3$$
$$-\frac{49}{16} < -3 \text{ True}$$
Test 0: $4(0)^2 + 7(0) < -3$
$$0 < -3 \text{ False}$$

The solution set is $\left(-1, -\frac{3}{4}\right)$.

17.
$$5x \le 2 - 3x^2$$
$$3x^2 + 5x - 2 \le 0$$
$$(3x-1)(x+2) \le 0$$
$$3x - 1 = 0 \text{ or } x + 2 = 0$$
$$3x = 1$$
$$3x - 1 = 0 \quad \text{or} \quad x + 2 = 0$$
$$3x = 1$$
$$x = \frac{1}{3} \quad \text{or} \quad x = -2$$

F		T		F
	-2		$\frac{1}{3}$	

Test -3: $5(-3) \le 2 - 3(-3)^2$

$\quad -15 \le -25$ False

Test 0: $5(0) \le 2 - 3(0)^2$

$\quad 0 \le 2$ True

Test 1: $5(1) \le 2 - 3(1)^2$

$\quad 5 \le -1$ False

The solution set is $\left[-2, \frac{1}{3}\right]$.

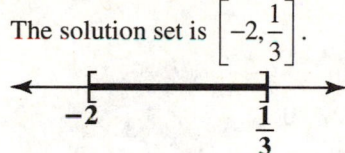

19. $x^2 - 4x \ge 0$
$$x(x-4) \ge 0$$
$$x = 0 \text{ or } x - 4 = 0$$
$$x = 4$$

T		F		T
	0		4	

Test -1: $(-1)^2 - 4(-1) \ge 0$

$\quad 5 \ge 0$ True

Test 1: $(1)^2 - 4(1) \ge 0$

$\quad -3 \ge 0$ False

$\quad 0 \le 2$ True

Test 5: $5^2 - 4(5) \ge 0$

$\quad 5 \ge 0$ True

The solution set is $(-\infty, 0]$ or $[4, \infty)$.

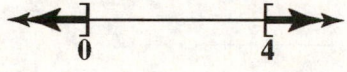

21. $2x^2 + 3x > 0$
$$x(2x + 3) > 0$$
$$x = 0 \text{ or } x = -\frac{3}{2}$$

T		F		T
	$-\frac{3}{2}$		0	

Test -2: $2(-2)^2 + 3(-2) > 0$

$\quad 2 > 0$ True

Test -1: $2(-1)^2 + 3(-1) > 0$

$\quad -1 > 0$ False

Test 1: $2(1)^2 + 3(1) > 0$

$\quad 5 > 0$ True

The solution set is $\left(-\infty, -\frac{3}{2}\right)$ or $(0, \infty)$.

23. $-x^2 + x \ge 0$
$$x^2 - x \le 0$$
$$x(x-1) \le 0$$
$$x = 0 \quad \text{or} \quad x = 1$$

F		T		F
	0		1	

Test -1: $-(-1)^2 + (-1) \ge 0$

$\quad -2 \ge 0$ False

Test $\frac{1}{2}$: $-\left(\frac{1}{2}\right)^2 + \left(\frac{1}{2}\right) \ge 0$

$\quad \frac{1}{4} \ge 0$ True

Test 2: $-(2)^2 + 2 \ge 0$

$\quad -2 \ge 0$ False

The solution set is $[0, 1]$.

25.
$$x^2 \le 4x - 2$$
$$x^2 - 4x + 2 \le 0$$
Solve $x^2 - 4x + 2 = 0$
$$x = \frac{-b \pm \sqrt{b^2 - 4ac}}{2a}$$
$$x = \frac{-(-4) \pm \sqrt{(-4)^2 - 4(1)(2)}}{2(1)}$$
$$= \frac{4 \pm \sqrt{8}}{2}$$
$$= 2 \pm \sqrt{2}$$
$$x \approx 0.59 \text{ or } x \approx 3.41$$

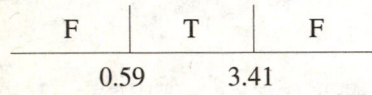

The solution set is $\left[2 - \sqrt{2}, 2 + \sqrt{2}\right]$ or $[0.59, 3.41]$.

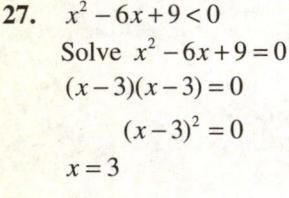

27. $x^2 - 6x + 9 < 0$
Solve $x^2 - 6x + 9 = 0$
$$(x - 3)(x - 3) = 0$$
$$(x - 3)^2 = 0$$
$$x = 3$$

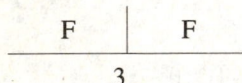

The solution set is the empty set, $\varnothing$.

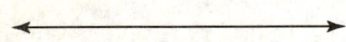

29. $(x - 1)(x - 2)(x - 3) \ge 0$
Boundary points: 1, 2, and 3
Test one value in each interval.

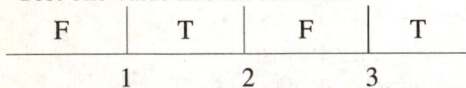

The solution set is $[1, 2] \cup [3, \infty)$.

31. $x(3 - x)(x - 5) \le 0$
Boundary points: 0, 3, and 5
Test one value in each interval.

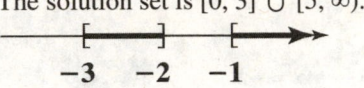

The solution set is $[0, 3] \cup [5, \infty)$.

33. $(2 - x)^2 \left(x - \frac{7}{2}\right) < 0$
Boundary points: 2, and $\frac{7}{2}$
Test one value in each interval.

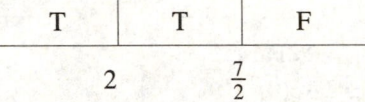

The solution set is $\left(-\infty, 2\right) \cup \left(2, \frac{7}{2}\right)$.

35. $x^3 + 2x^2 - x - 2 \ge 0$
$$x^2(x + 2) - 1(x + 2) \ge 0$$
$$(x + 2)(x^2 - 1) \ge 0$$
$$(x + 2)(x - 1)(x + 1) \ge 0$$
Boundary points: -2, -1, and 2
Test one value in each interval.

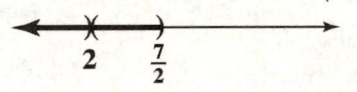

The solution set is $[-2, -1] \cup [1, \infty)$.

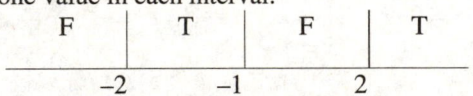

37. $x^3 + 2x^2 - x - 2 \ge 0$
$$x^2(x - 3) - 9(x - 3) \ge 0$$
$$(x - 3)(x^2 - 9) \ge 0$$
$$(x - 3)(x + 3)(x - 3) \ge 0$$
Boundary points: -3 and 3
Test one value in each interval.

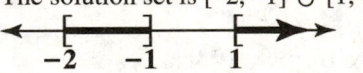

The solution set is $(-\infty, -3]$.

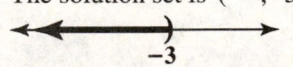

226

39. $x^3 + x^2 + 4x + 4 > 0$

$x^2(x+1) + 4(x+1) \geq 0$

$(x+1)(x^2+4) \geq 0$

Boundary point: -1
Test one value in each interval.

F		T
	-1	

The solution set is $(-1, \infty)$.

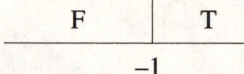

41. $x^3 - 9x^2 \geq 0$

$x^2(x-9) \geq 0$

Boundary points: 0 and 9
Test one value in each interval.

F	F	T
0	9	

The solution set is $[0,0] \cup [9, \infty)$.

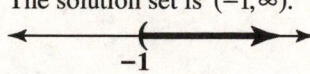

43. $\dfrac{x-4}{x+3} > 0$

$x - 4 = 0 \quad x + 3 = 0$

$x = 4 \qquad x = -3$

T	F	T
-3	4	

The solution set is $(-\infty, -3) \cup (4, \infty)$.

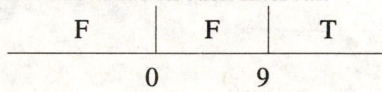

45. $\dfrac{x+3}{x+4} < 0$

$x = -3 \quad$ or $\quad x = -4$

F	T	F
-4	-3	

The solution set is $(-4, -3)$.

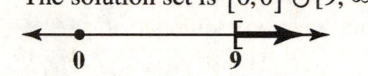

47. $\dfrac{-x+2}{x-4} \geq 0$

$x = 2$ or $x = 4$

F	T	F
2	4	

The solution set is $[2, 4)$.

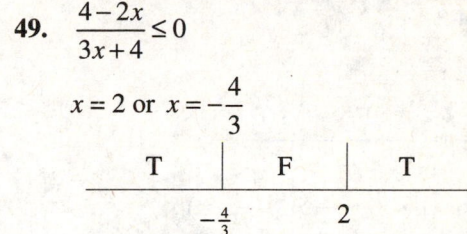

49. $\dfrac{4-2x}{3x+4} \leq 0$

$x = 2$ or $x = -\dfrac{4}{3}$

T	F	T
$-\frac{4}{3}$	2	

The solution set is $\left(-\infty, \dfrac{-4}{3}\right) \cup [2, \infty)$.

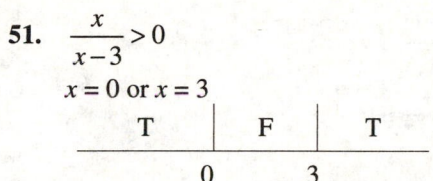

51. $\dfrac{x}{x-3} > 0$

$x = 0$ or $x = 3$

T	F	T
0	3	

The solution set is $(-\infty, 0) \cup (3, \infty)$.

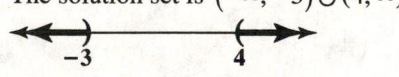

53. $\dfrac{(x+4)(x-1)}{x+2} \leq 0$

$x = -4$ or $x = -2$ or $x = 1$.

T	F	T	F
-4	-2	1	

Values of $x = -4$ or $x = 1$ result in $f(x) = 0$ and, therefore must be included in the solution set.
The solution set is $(-\infty, -4] \cup (-2, 1]$

55.
$$\frac{x+1}{x+3} < 2$$

$$\frac{x+1}{x+3} - 2 < 0$$

$$\frac{x+1-2(x+3)}{x+3} < 0$$

$$\frac{x+1-2x-6}{x+3} < 0$$

$$\frac{-x-5}{x+3} < 0$$

$x =$ or $x = -3$

T		F		T
	-5		-3	

The solution set is $(-\infty, -5) \cup (-3, \infty)$.

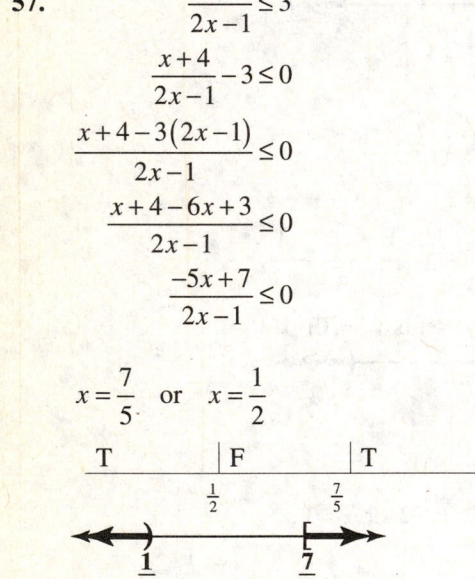

57.
$$\frac{x+4}{2x-1} \le 3$$

$$\frac{x+4}{2x-1} - 3 \le 0$$

$$\frac{x+4-3(2x-1)}{2x-1} \le 0$$

$$\frac{x+4-6x+3}{2x-1} \le 0$$

$$\frac{-5x+7}{2x-1} \le 0$$

$x = \dfrac{7}{5}$ or $x = \dfrac{1}{2}$

T		F		T
	$\frac{1}{2}$		$\frac{7}{5}$	

59.
$$\frac{x-2}{x+2} \le 2$$

$$\frac{x-2}{x+2} - 2 \le 0$$

$$\frac{x-2-2(x+2)}{x+2} \le 0$$

$$\frac{x-2-2x-4}{x+2} \le 0$$

$$\frac{-x-6}{x+2} \le 0$$

$x = -6$ or $x = -2$

T		F		T
	-6		-2	

The solution set is $(-\infty, -6] \cup (-2, \infty)$.

61. $f(x) = \sqrt{2x^2 - 5x + 2}$

The domain of this function requires that
$2x^2 - 5x + 2 \ge 0$
Solve $2x^2 - 5x + 2 = 0$
$(x-2)(2x-1) = 0$

$x = \dfrac{1}{2}$ or $x = 2$

T		F		T
	$\frac{1}{2}$		2	

The domain is $\left(-\infty, \dfrac{1}{2}\right] \cup [2, \infty)$.

63. $f(x) = \sqrt{\dfrac{2x}{x+1} - 1}$

The domain of this function requires that $\dfrac{2x}{x+1} - 1 \geq 0$ or $\dfrac{x-1}{x+1} \geq 0$

$x = -1$ or $x = 1$

T	F	T
-1		1

The value $x = 1$ results in 0 and, thus, it must be included in the domain.

The domain is $(-\infty, -1) \cup [1, \infty)$.

65. $\left| x^2 + 2x - 36 \right| > 12$

Express the inequality without the absolute value symbol:

$x^2 + 2x - 36 < -12$ or $x^2 + 2x - 36 > 12$

$x^2 + 2x - 24 < 0$ $x^2 + 2x - 48 > 0$

Solve the related quadratic equations.

$x^2 + 2x - 24 = 0$ or $x^2 + 2x - 48 = 0$

$(x+6)(x-4) = 0$ $(x+8)(x-6) = 0$

Apply the zero product principle.

$x + 6 = 0$ or $x - 4 = 0$ or $x + 8 = 0$ or $x - 6 = 0$

     $x = -6$ $x = 4$ $x = -8$ $x = 6$

The boundary points are -8, -6, 4 and 6.

Test Interval	Test Number	Test	Conclusion
$(-\infty, -8)$	-9	$\left\lvert (-9)^2 + 2(-9) - 36 \right\rvert > 12$ $27 > 12$, True	$(-\infty, -8)$ belongs to the solution set.
$(-8, -6)$	-7	$\left\lvert (-7)^2 + 2(-7) - 36 \right\rvert > 12$ $1 > 12$, False	$(-8, -6)$ does not belong to the solution set.
$(-6, 4)$	0	$\left\lvert 0^2 + 2(0) - 36 \right\rvert > 12$ $36 > 12$, True	$(-6, 4)$ belongs to the solution set.
$(4, 6)$	5	$\left\lvert 5^2 + 2(5) - 36 \right\rvert > 12$ $1 > 12$, False	$(4, 6)$ does not belong to the solution set.
$(6, \infty)$	7	$\left\lvert 7^2 + 2(7) - 36 \right\rvert > 12$ $27 > 12$, True	$(6, \infty)$ belongs to the solution set.

The solution set is $(-\infty, -8) \cup (-6, 4) \cup (6, \infty)$ or $\{ x \mid x < -8 \text{ or } -6 < x < 4 \text{ or } x > 6 \}$.

67. $\dfrac{3}{x+3} > \dfrac{3}{x-2}$

Express the inequality so that one side is zero.

$$\dfrac{3}{x+3} - \dfrac{3}{x-2} > 0$$

$$\dfrac{3(x-2)}{(x+3)(x-2)} - \dfrac{3(x+3)}{(x+3)(x-2)} > 0$$

$$\dfrac{3x-6-3x-9}{(x+3)(x-2)} < 0$$

$$\dfrac{-15}{(x+3)(x-2)} < 0$$

Find the values of x that make the denominator zero.

$$x+3=0 \qquad x-2=0$$
$$x=-3 \qquad x=2$$

The boundary points are -3 and 2.

Test Interval	Test Number	Test	Conclusion
$(-\infty,-3)$	-4	$\dfrac{3}{-4+3} > \dfrac{3}{-4-2}$ $-3 > \dfrac{1}{2}$, False	$(-\infty,-3)$ does not belong to the solution set.
$(-3,2)$	0	$\dfrac{3}{0+3} > \dfrac{3}{0-2}$ $1 > -\dfrac{3}{2}$, True	$(-3,2)$ belongs to the solution set.
$(2,\infty)$	3	$\dfrac{3}{3+3} > \dfrac{3}{3-2}$ $\dfrac{1}{2} > 3$, False	$(2,\infty)$ does not belong to the solution set.

The solution set is $(-3,2)$ or $\{x\,|\,-3 < x < 2\}$.

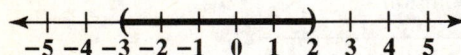

69.

$$\frac{x^2 - x - 2}{x^2 - 4x + 3} > 0$$

Find the values of x that make the numerator and denominator zero.

$$x^2 - x - 2 = 0 \qquad x^2 - 4x + 3 = 0$$

$$(x-2)(x+1) = 0 \qquad (x-3)(x-1) = 0$$

Apply the zero product principle.

$$x - 2 = 0 \quad \text{or} \quad x + 1 = 0 \qquad x - 3 = 0 \quad \text{or} \quad x - 1 = 0$$

$$x = 2 \qquad\qquad x = -1 \qquad\qquad x = 3 \qquad\qquad x = 1$$

The boundary points are -1, 1, 2 and 3.

Test Interval	Test Number	Test	Conclusion
$(-\infty, -1)$	-2	$\dfrac{(-2)^2 - (-2) - 2}{(-2)^2 - 4(-2) + 3} > 0$ $\dfrac{4}{15} > 0$, True	$(-\infty, -1)$ belongs to the solution set.
$(-1, 1)$	0	$\dfrac{0^2 - 0 - 2}{0^2 - 4(0) + 3} > 0$ $-\dfrac{2}{3} > 0$, False	$(-1, 1)$ does not belong to the solution set.
$(1, 2)$	1.5	$\dfrac{1.5^2 - 1.5 - 2}{1.5^2 - 4(1.5) + 3} > 0$ $\dfrac{5}{3} > 0$, True	$(1, 2)$ belongs to the solution set.
$(2, 3)$	2.5	$\dfrac{2.5^2 - 2.5 - 2}{2.5^2 - 4(2.5) + 3} > 0$ $-\dfrac{7}{3} > 0$, False	$(2, 3)$ does not belong to the solution set.
$(3, \infty)$	4	$\dfrac{4^2 - 4 - 2}{4^2 - 4(4) + 3} > 0$ $\dfrac{10}{3} > 0$, True	$(3, \infty)$ belongs to the solution set.

The solution set is $(-\infty, -1) \cup (1, 2) \cup (3, \infty)$ or $\{x \mid x < -1 \text{ or } 1 < x < 2 \text{ or } x > 3\}$.

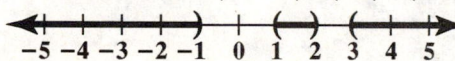

71.

$$2x^3 + 11x^2 \geq 7x + 6$$

$$2x^3 + 11x^2 - 7x - 6 \geq 0$$

The graph of $f(x) = 2x^3 + 11x^2 - 7x - 6$ appears to cross the x-axis at -6, $-\dfrac{1}{2}$, and 1. We verify this numerically by substituting these values into the function:

$$f(-6) = 2(-6)^3 + 11(-6)^2 - 7(-6) - 6 = 2(-216) + 11(36) - (-42) - 6 = -432 + 396 + 42 - 6 = 0$$

$$f\left(-\frac{1}{2}\right) = 2\left(-\frac{1}{2}\right)^3 + 11\left(-\frac{1}{2}\right)^2 - 7\left(-\frac{1}{2}\right) - 6 = 2\left(-\frac{1}{8}\right) + 11\left(\frac{1}{4}\right) - \left(-\frac{7}{2}\right) - 6 = -\frac{1}{4} + \frac{11}{4} + \frac{7}{2} - 6 = 0$$

$$f(1) = 2(1)^3 + 11(1)^2 - 7(1) - 6 = 2(1) + 11(1) - 7 - 6 = 2 + 11 - 7 - 6 = 0$$

Thus, the boundaries are -6, $-\dfrac{1}{2}$, and 1. We need to find the intervals on which $f(x) \geq 0$. These intervals are indicated on the graph where the curve is above the x-axis. Now, the curve is above the x-axis when $-6 < x < -\dfrac{1}{2}$

and when $x > 1$. Thus, the solution set is $\left\{x \,\middle|\, -6 \leq x \leq -\dfrac{1}{2} \text{ or } x \geq 1\right\}$ or $\left[-6, -\dfrac{1}{2}\right] \cup [1, \infty)$.

73.

$$\frac{1}{4(x+2)} \leq -\frac{3}{4(x-2)}$$

$$\frac{1}{4(x+2)} + \frac{3}{4(x-2)} \leq 0$$

Simplify the left side of the inequality:

$$\frac{x-2}{4(x+2)} + \frac{3(x+2)}{4(x-2)} = \frac{x-2+3x+6}{4(x+2)(x-2)} = \frac{4x+4}{4(x+2)(x-2)} = \frac{4(x+1)}{4(x+2)(x-2)} = \frac{x+1}{x^2-4}.$$

The graph of $f(x) = \dfrac{x+1}{x^2-4}$ crosses the x-axis at -1, and has vertical asymptotes at $x = -2$ and $x = 2$. Thus,

the boundaries are -2, -1, and 1. We need to find the intervals on which $f(x) \leq 0$. These intervals are indicated on the graph where the curve is below the x-axis. Now, the curve is below the x-axis when $x < -2$ and when $-1 < x < 2$. Thus, the solution set is $\{x \mid x < -2 \text{ or } -1 \leq x < 2\}$ or $(-\infty, -2) \cup [-1, 2)$.

75. $s(t) = -16t^2 + 8t + 87$

The diver's height will exceed that of the cliff when $s(t) > 87$

$$-16t^2 + 8t + 87 > 87$$

$$-16t^2 + 8t > 0$$

$$-8t(2t-1) > 0$$

The boundaries are 0 and $\dfrac{1}{2}$. Testing each interval shows that the diver will be higher than the cliff for the first half

second after beginning the jump. The interval is $\left(0, \dfrac{1}{2}\right)$.

77. $f(x) = 0.0875x^2 - 0.4x + 66.6$

$g(x) = 0.0875x^2 + 1.9x + 11.6$

a. $f(35) = 0.0875(35)^2 - 0.4(35) + 66.6 \approx 160$ feet

$g(35) = 0.0875(35)^2 + 1.9(35) + 11.6 \approx 185$ feet

b. Dry pavement: graph (b)
Wet pavement: graph (a)

c. The answers to part (a) model the actual stopping distances shown in the figure extremely well. The function values and the data are identical.

d. $0.0875x^2 - 0.4x + 66.6 > 540$

$0.0875x^2 - 0.4x + 473.4 > 0$
Solve the related quadratic equation.
$0.0875x^2 - 0.4x + 473.4 = 0$

$x = \dfrac{-b \pm \sqrt{b^2 - 4ac}}{2a}$

$x = \dfrac{-(-0.4) \pm \sqrt{(-0.4)^2 - 4(0.0875)(473.4)}}{2(0.0875)}$

$x \approx -71$ or 76

Since the function's domain is $x \geq 30$, we must test the following intervals.

Interval	Test Value	Test	Conclusion
$(30, 76)$	50	$0.0875(50)^2 - 0.4(50) + 66.6 > 540$ $265.35 > 540$, False	$(30, 76)$ does not belong to the solution set.
$(76, \infty)$	100	$0.0875(100)^2 - 0.4(100) + 66.6 > 540$ $901.6 > 540$, True	$(76, \infty)$ belongs to the solution set.

On dry pavement, stopping distances will exceed 540 feet for speeds exceeding 76 miles per hour. This is represented on graph (b) to the right of point (76, 540).

79. Let x = the length of the rectangle.

Since Perimeter $= 2(\text{length}) + 2(\text{width})$, we know

$$50 = 2x + 2(\text{width})$$

$$50 - 2x = 2(\text{width})$$

$$\text{width} = \frac{50 - 2x}{2} = 25 - x$$

Now, $A = (\text{length})(\text{width})$, so we have that

$$A(x) \le 114$$

$$x(25 - x) \le 114$$

$$25x - x^2 \le 114$$

Solve the related equation

$$25x - x^2 = 114$$

$$0 = x^2 - 25x + 114$$

$$0 = (x - 19)(x - 6)$$

Apply the zero product principle:

$x - 19 = 0 \quad$ or $\quad x - 6 = 0$

$\qquad x = 19 \qquad\qquad x = 6$

The boundary points are 6 and 19.

Test Interval	Test Number	Test	Conclusion
$(-\infty, 6)$	0	$25(0) - 0^2 \le 114$ $0 \le 114$, True	$(-\infty, 6)$ belongs to the solution set.
$(6, 19)$	10	$25(10) - 10^2 \le 114$ $150 \le 114$, False	$(6, 19)$ does not belong to the solution set.
$(19, \infty)$	20	$25(20) - 20^2 \le 114$ $100 \le 114$, True	$(19, \infty)$ belongs to the solution set.

If the length is 6 feet, then the width is 19 feet. If the length is less than 6 feet, then the width is greater than 19 feet. Thus, if the area of the rectangle is not to exceed 114 square feet, the length of the shorter side must be 6 feet or less.

81. – 85. Answers may vary.

87. Graph $y_1 = 2x^2 + 5x - 3$ in a standard window. The graph is below or equal to the x-axis for $-3 \le x \le \frac{1}{2}$.

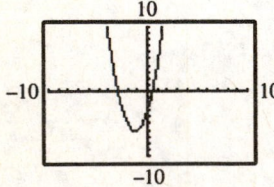

The solution set is $\left\{ x \middle| -3 \le x \le \frac{1}{2} \right\}$ or $\left[-3, \frac{1}{2} \right]$.

89. Graph $y_1 = \dfrac{x-4}{x-1}$ in a standard viewing window. The graph is below the *x*-axis for

$1 < x \le 4$.

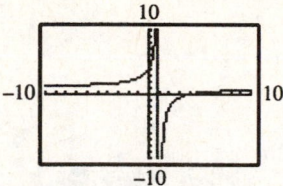

The solution set is (1, 4].

91. Graph $y_1 = \dfrac{1}{x+1}$ and $y_2 = \dfrac{2}{x+4}$

y_1 less than or equal to y_2 for $-4 < x < -1$ or $x \ge 2$.

The solution set is $(-4, -1) \cup [2, \infty)$

93. **a.** $f(x) = 0.1375x^2 + 0.7x + 37.8$

b. $0.1375x^2 + 0.7x + 37.8 > 446$

$0.1375x^2 + 0.7x + 408.2 > 0$
Solve the related quadratic equation.
$0.1375x^2 + 0.7x + 408.2 = 0$

$x = \dfrac{-b \pm \sqrt{b^2 - 4ac}}{2a}$

$x = \dfrac{-(0.7) \pm \sqrt{(0.7)^2 - 4(0.1375)(408.2)}}{2(0.1375)}$

$x \approx -57 \text{ or } 52$

Since the function's domain must be $x \ge 0$, we must test the following intervals.

Interval	Test Value	Test	Conclusion
$(0, 52)$	10	$0.1375(10)^2 + 0.7(10) + 37.8 > 446$ $58.55 > 446$, False	$(0, 52)$ does not belong to the solution set.
$(52, \infty)$	100	$0.1375(100)^2 + 0.7(100) + 37.8 > 446$ $1482.8 > 446$, True	$(52, \infty)$ belongs to the solution set.

On wet pavement, stopping distances will exceed 446 feet for speeds exceeding 52 miles per hour.

95. does not make sense; Explanations will vary. Sample explanation: Polynomials are defined for all values.

97. does not make sense; Explanations will vary. Sample explanation: To solve this inequality you must first subtract 2 from both sides.

99. false; Changes to make the statement true will vary. A sample change is: The inequality cannot be solved by multiplying both sides by $x + 3$. We do not know if $x + 3$ is positive or negative. Thus, we would not know whether or not to reverse the order of the inequality.

101. true

103. One possible solution: $\dfrac{x-3}{x+4} \geq 0$

105. Because any number squared other than zero is positive, the solution includes only 2.

107. Because any number squared other than zero is positive, and the reciprocal of zero is undefined, the solution is all real numbers except 2.

109.
$$\sqrt{27-3x^2} \geq 0$$
$$27-3x^2 \geq 0$$
$$9-x^2 \geq 0$$
$$(3-x)(3+x) \geq 0$$
$$3-x=0 \qquad 3+x=0$$
$$x=3 \text{ or} \qquad x=-3$$

$$\overline{} \underset{-3}{\big|} \quad \overset{T}{} \quad \underset{3}{\big|} \overline{}$$

Test –4: $\sqrt{27-3(-4)^2} \geq 0$
$$\sqrt{27-48} \geq 0$$
$$\sqrt{-21} \geq 0$$
no graph- imaginary

Test 0: $\sqrt{27-3(0)^2} \geq 0$
$$\sqrt{27} \geq 0 \text{ True}$$

Test 4: $\sqrt{27-3(4)^2} \geq 0$
$$\sqrt{27-48} \geq 0$$
$$\sqrt{-21} \geq 0$$
no graph -imaginary
The solution set is [–3, 3].

110. a.
$$y = kx^2$$
$$64 = k \cdot 2^2$$
$$64 = 4k$$
$$16 = k$$

b.
$$y = kx^2$$
$$y = 16x^2$$

c.
$$y = kx^2$$
$$y = 16x^2$$
$$y = 16 \cdot 5^2$$
$$y = 400$$

111. a.
$$y = \frac{k}{x}$$
$$12 = \frac{k}{8}$$
$$96 = k$$

b.
$$y = \frac{k}{x}$$
$$y = \frac{96}{x}$$

c.
$$y = \frac{96}{x}$$
$$y = \frac{96}{3}$$
$$y = 32$$

112.
$$S = \frac{kA}{P}$$
$$12,000 = \frac{k \cdot 60,000}{40}$$
$$\frac{12,000 \cdot 40}{60,000} = k$$
$$8 = k$$

Section 2.8

Check Point Exercises

1. y varies directly as x is expressed as $y = kx$.
The volume of water, W, varies directly as the time, t can be expressed as $W = kt$.
Use the given values to find k.
$$W = kt$$
$$30 = k(5)$$
$$6 = k$$
Substitute the value of k into the equation.
$$W = kt$$
$$W = 6t$$
Use the equation to find W when $t = 11$.
$$W = 6t$$
$$= 6(11)$$
$$= 66$$
A shower lasting 11 minutes will use 66 gallons of water.

2. y varies directly as the cube of x is expressed as $y = kx^3$.

The weight, w, varies directly as the cube of the length, l can be expressed as $w = kl^2$.

Use the given values to find k.

$$w = kl^3$$
$$2025 = k(15)^3$$
$$0.6 = k$$

Substitute the value of k into the equation.

$$w = kl^3$$
$$w = 0.6l^3$$

Use the equation to find w when $l = 25$.

$$w = 0.6l^3$$
$$= 0.6(25)^3$$
$$= 9375$$

The 25-foot long shark was 9375 pounds.

3. y varies inversely as x is expressed as $y = \dfrac{k}{x}$.

The length, L, varies inversely as the frequency, f can be expressed as $L = \dfrac{k}{f}$.

Use the given values to find k.

$$L = \frac{k}{f}$$
$$8 = \frac{k}{640}$$
$$5120 = k$$

Substitute the value of k into the equation.

$$L = \frac{k}{f}$$
$$L = \frac{5120}{f}$$

Use the equation to find f when $L = 10$.

$$L = \frac{5120}{f}$$
$$10 = \frac{5120}{f}$$
$$10f = 5120$$
$$f = 512$$

A 10 inch violin string will have a frequency of 512 cycles per second.

4.   let M represent the number of minutes
let Q represent the number of problems
let P represent the number of people
M varies directly as Q and inversely as P is expressed as $M = \dfrac{kQ}{P}$.

Use the given values to find k.

$$M = \frac{kQ}{P}$$
$$32 = \frac{k(16)}{4}$$
$$8 = k$$

Substitute the value of k into the equation.

$$M = \frac{kQ}{P}$$
$$M = \frac{8Q}{P}$$

Use the equation to find M when $P = 8$ and $Q = 24$.

$$M = \frac{8Q}{P}$$
$$M = \frac{8(24)}{8}$$
$$M = 24$$

It will take 24 minutes for 8 people to solve 24 problems.

5. V varies jointly with h and r^2 and can be modeled as $V = khr^2$.

Use the given values to find k.

$$V = khr^2$$
$$120\pi = k(10)(6)^2$$
$$\frac{\pi}{3} = k$$

Therefore, the volume equation is $V = \dfrac{1}{3}hr^2$.

$$V = \frac{\pi}{3}(2)(12)^2 = 96\pi \text{ cubic feet}$$

Exercise Set 2.8

1. Use the given values to find k.

$y = kx$

$65 = k \cdot 5$

$\dfrac{65}{5} = \dfrac{k \cdot 5}{5}$

$13 = k$

The equation becomes $y = 13x$.

When $x = 12$, $y = 13x = 13 \cdot 12 = 156$.

3. Since y varies inversely with x, we have $y = \dfrac{k}{x}$.

Use the given values to find k.

$y = \dfrac{k}{x}$

$12 = \dfrac{k}{5}$

$5 \cdot 12 = 5 \cdot \dfrac{k}{5}$

$60 = k$

The equation becomes $y = \dfrac{60}{x}$.

When $x = 2$, $y = \dfrac{60}{2} = 30$.

5. Since y varies inversely as x and inversely as the square of z, we have $y = \dfrac{kx}{z^2}$.

Use the given values to find k.

$y = \dfrac{kx}{z^2}$

$20 = \dfrac{k(50)}{5^2}$

$20 = \dfrac{k(50)}{25}$

$20 = 2k$

$10 = k$

The equation becomes $y = \dfrac{10x}{z^2}$.

When $x = 3$ and $z = 6$,

$y = \dfrac{10x}{z^2} = \dfrac{10(3)}{6^2} = \dfrac{10(3)}{36} = \dfrac{30}{36} = \dfrac{5}{6}$.

7. Since y varies jointly as x and z, we have $y = kxz$.

Use the given values to find k.

$y = kxz$

$25 = k(2)(5)$

$25 = k(10)$

$\dfrac{25}{10} = \dfrac{k(10)}{10}$

$\dfrac{5}{2} = k$

The equation becomes $y = \dfrac{5}{2}xz$.

When $x = 8$ and $z = 12$, $y = \dfrac{5}{2}(8)(12) = 240$.

9. Since y varies jointly as a and b and inversely as the square root of c, we have $y = \dfrac{kab}{\sqrt{c}}$.

Use the given values to find k.

$y = \dfrac{kab}{\sqrt{c}}$

$12 = \dfrac{k(3)(2)}{\sqrt{25}}$

$12 = \dfrac{k(6)}{5}$

$12(5) = \dfrac{k(6)}{5}(5)$

$60 = 6k$

$\dfrac{60}{6} = \dfrac{6k}{6}$

$10 = k$

The equation becomes $y = \dfrac{10ab}{\sqrt{c}}$.

When $a = 5$, $b = 3$, $c = 9$,

$y = \dfrac{10ab}{\sqrt{c}} = \dfrac{10(5)(3)}{\sqrt{9}} = \dfrac{150}{3} = 50$.

11. $x = kyz$;

Solving for y:

$x = kyz$

$\dfrac{x}{kz} = \dfrac{kyz}{yz}$.

$y = \dfrac{x}{kz}$

13.
$$x = \frac{kz^3}{y};$$

Solving for y

$$x = \frac{kz^3}{y}$$

$$xy = y \cdot \frac{kz^3}{y}$$

$$xy = kz^3$$

$$\frac{xy}{x} = \frac{kz^3}{x}$$

$$y = \frac{kz^3}{x}$$

15.
$$x = \frac{kyz}{\sqrt{w}};$$

Solving for y:

$$x = \frac{kyz}{\sqrt{w}}$$

$$x\left(\sqrt{w}\right) = \left(\sqrt{w}\right)\frac{kyz}{\sqrt{w}}$$

$$x\sqrt{w} = kyz$$

$$\frac{x\sqrt{w}}{kz} = \frac{kyz}{kz}$$

$$y = \frac{x\sqrt{w}}{kz}$$

17. $x = kz(y+w)$;

Solving for y:

$$x = kz(y+w)$$

$$x = kzy + kzw$$

$$x - kzw = kzy$$

$$\frac{x - kzw}{kz} = \frac{kzy}{kz}$$

$$y = \frac{x - kzw}{kz}$$

19.
$$x = \frac{kz}{y - w};$$

Solving for y:

$$x = \frac{kz}{y - w}$$

$$(y - w)x = (y - w)\frac{kz}{y - w}$$

$$xy - wx = kz$$

$$xy = kz + wx$$

$$\frac{xy}{x} = \frac{kz + wx}{x}$$

$$y = \frac{xw + kz}{x}$$

21. Since T varies directly as B, we have $T = kB$.
Use the given values to find k.

$$T = kB$$

$$3.6 = k(4)$$

$$\frac{3.6}{4} = \frac{k(4)}{4}$$

$$0.9 = k$$

The equation becomes $T = 0.9B$.
When $B = 6$, $T = 0.9(6) = 5.4$.
The tail length is 5.4 feet.

23. Since B varies directly as D, we have $B = kD$.
Use the given values to find k.

$$B = kD$$

$$8.4 = k(12)$$

$$\frac{8.4}{12} = \frac{k(12)}{12}$$

$$k = \frac{8.4}{12} = 0.7$$

The equation becomes $B = 0.7D$.
When $B = 56$,

$$56 = 0.7D$$

$$\frac{56}{0.7} = \frac{0.7D}{0.7}$$

$$D = \frac{56}{0.7} = 80$$

It was dropped from 80 inches.

25. Since a man's weight varies directly as the cube of his height, we have $w = kh^3$. Use the given values to find k.

$$w = kh^3$$
$$170 = k(70)^3$$
$$170 = k(343,000)$$
$$\frac{170}{343,000} = \frac{k(343,000)}{343,000}$$
$$0.000496 = k$$

The equation becomes $w = 0.000496h^3$.
When $h = 107$,
$$w = 0.000496(107)^3$$
$$= 0.000496(1,225,043) \approx 607.$$

Robert Wadlow's weight was approximately 607 pounds.

27. Since the banking angle varies inversely as the turning radius, we have $B = \frac{k}{r}$. Use the given values to find k.

$$B = \frac{k}{r}$$
$$28 = \frac{k}{4}$$
$$28(4) = 28\left(\frac{k}{4}\right)$$
$$112 = k$$

The equation becomes $B = \frac{112}{r}$.

When $r = 3.5$, $B = \frac{112}{r} = \frac{112}{3.5} = 32$.

The banking angle is $32°$ when the turning radius is 3.5 feet.

29. Since intensity varies inversely as the square of the distance, we have pressure, we have
$$I = \frac{k}{d}.$$
Use the given values to find k.

$$I = \frac{k}{d^2}.$$
$$62.5 = \frac{k}{3^2}$$
$$62.5 = \frac{k}{9}$$
$$9(62.5) = 9\left(\frac{k}{9}\right)$$
$$562.5 = k$$

The equation becomes $I = \frac{562.5}{d^2}$.

When $d = 2.5$, $I = \frac{562.5}{2.5^2} = \frac{562.5}{6.25} = 90$

The intensity is 90 milliroentgens per hour.

31. Since index varies directly as weight and inversely as the square of one's height, we have $I = \frac{kw}{h^2}$. Use the given values to find k.

$$I = \frac{kw}{h^2}$$
$$35.15 = \frac{k(180)}{60^2}$$
$$35.15 = \frac{k(180)}{3600}$$
$$(3600)35.15 = \frac{k(180)}{3600}$$
$$126540 = k(180)$$
$$k = \frac{126540}{180} = 703$$

The equation becomes $I = \frac{703w}{h^2}$.

When $w = 170$ and $h = 70$,
$$I = \frac{703(170)}{(70)^2} \approx 24.4.$$

This person has a BMI of 24.4 and is not overweight.

33. Since heat loss varies jointly as the area and temperature difference, we have $L = kAD$. Use the given values to find k.

$$L = kAD$$
$$1200 = k(3 \cdot 6)(20)$$
$$1200 = 360k$$
$$\frac{1200}{360} = \frac{360k}{360}$$
$$k = \frac{10}{3}$$

The equation becomes $L = \frac{10}{3}AD$

When $A = 6 \cdot 9 = 54$, $D = 10$,

$L = \frac{10}{3}(9 \cdot 6)(10) = 1800$.

The heat loss is 1800 Btu.

35. Since intensity varies inversely as the square of the distance from the sound source, we

have $I = \dfrac{k}{d^2}$. If you move to a seat twice as

far, then $d = 2d$. So we have

$I = \dfrac{k}{(2d)^2} = \dfrac{k}{4d^2} = \dfrac{1}{4} \cdot \dfrac{k}{d^2}$. The intensity will

be multiplied by a factor of $\dfrac{1}{4}$. So the sound

intensity is $\dfrac{1}{4}$ of what it was originally.

37. a. Since the average number of phone calls varies jointly as the product of the populations and inversely as the square of the distance, we have

$$C = \frac{kP_1P_2}{d^2}.$$

b. Use the given values to find k.

$$C = \frac{kP_1P_2}{d^2}$$
$$326,000 = \frac{k(777,000)(3,695,000)}{(420)^2}$$
$$326,000 = \frac{k(2.87 \times 10^{12})}{176,400}$$
$$326,000 = 16269841.27k$$
$$0.02 \approx k$$

The equation becomes $C = \dfrac{0.02P_1P_2}{d^2}$.

c. $C = \dfrac{0.02(650,000)(220,000)}{(400)^2}$

$= 17,875$

There are approximately 17,875 daily phone calls.

39. a.

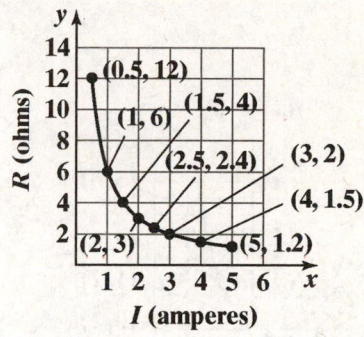

b. Current varies inversely as resistance. Answers will vary.

c. Since the current varies inversely as

resistance we have $R = \dfrac{k}{I}$. Using one of

the given ordered pairs to find k.

$$12 = \frac{k}{0.5}$$
$$12(0.5) = \frac{k}{0.5}(0.5)$$
$$k = 6$$

The equation becomes $R = \dfrac{6}{I}$.

41. – 47. Answers may vary.

49. does not make sense; Explanations will vary. Sample explanation: For an inverse variation, the independent variable can not be zero.

51. makes sense

53. Pressure, P, varies directly as the square of wind velocity, v, can be modeled as $P = kv^2$.

If $v = x$ then $P = k(x)^2 = kx^2$

If $v = 2x$ then $P = k(2x)^2 = 4kx^2$

If the wind speed doubles the pressure is 4 times more destructive.

55. The Heat, H, varies directly as the square of the voltage, v, and inversely as the resistance, r.

$$H = \frac{kv^2}{r}$$

If the voltage remains constant, to triple the heat the resistant must be reduced by a multiple of 3.

57. Answers may vary.

58.

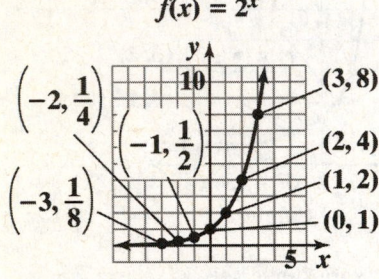

$f(x) = 2^x$

59.

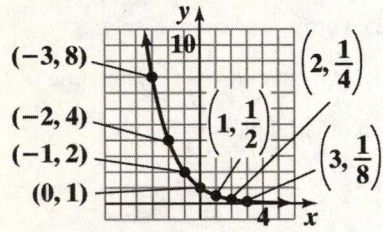

$g(x) = f(-x) = 2^{-x}$

60.

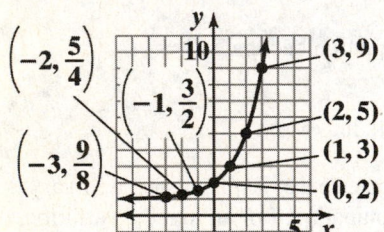

$h(x) = f(x) + 1 = 2^x + 1$

Chapter 2 Review Exercises

1. $(8 - 3i) - (17 - 7i) = 8 - 3i - 17 + 7i$
$$= -9 + 4i$$

2. $4i(3i - 2) = (4i)(3i) + (4i)(-2)$
$$= 12i^2 - 8i$$
$$= -12 - 8i$$

3. $(7 - i)(2 + 3i)$
$$= 7 \cdot 2 + 7(3i) + (-i)(2) + (-i)(3i)$$
$$= 14 + 21i - 2i + 3$$
$$= 17 + 19i$$

4. $(3 - 4i)^2 = 3^2 + 2 \cdot 3(-4i) + (-4i)^2$
$$= 9 - 24i - 16$$
$$= -7 - 24i$$

5. $(7 + 8i)(7 - 8i) = 7^2 + 8^2 = 49 + 64 = 113$

6. $\dfrac{6}{5 + i} = \dfrac{6}{5 + i} \cdot \dfrac{5 - i}{5 - i}$
$$= \frac{30 - 6i}{25 + 1}$$
$$= \frac{30 - 6i}{26}$$
$$= \frac{15 - 3i}{13}$$
$$= \frac{15}{13} - \frac{3}{13}i$$

7. $\dfrac{3 + 4i}{4 - 2i} = \dfrac{3 + 4i}{4 - 2i} \cdot \dfrac{4 + 2i}{4 + 2i}$
$$= \frac{12 + 6i + 16i + 8i^2}{16 - 4i^2}$$
$$= \frac{12 + 22i - 8}{16 + 4}$$
$$= \frac{4 + 22i}{20}$$
$$= \frac{1}{5} + \frac{11}{10}i$$

8. $\sqrt{-32} - \sqrt{-18} = i\sqrt{32} - i\sqrt{18}$
$$= i\sqrt{16 \cdot 2} - i\sqrt{9 \cdot 2}$$
$$= 4i\sqrt{2} - 3i\sqrt{2}$$
$$= (4i - 3i)\sqrt{2}$$
$$= i\sqrt{2}$$

9. $(-2 + \sqrt{-100})^2 = (-2 + i\sqrt{100})^2$
$$= (-2 + 10i)^2$$
$$= 4 - 40i + (10i)^2$$
$$= 4 - 40i - 100$$
$$= -96 - 40i$$

10. $\dfrac{4 + \sqrt{-8}}{2} = \dfrac{4 + i\sqrt{8}}{2} = \dfrac{4 + 2i\sqrt{2}}{2} = 2 + i\sqrt{2}$

11. $x^2 - 2x + 4 = 0$

$$x = \frac{-(-2) \pm \sqrt{(-2)^2 - 4(1)(4)}}{2(1)}$$

$$x = \frac{2 \pm \sqrt{4 - 16}}{2}$$

$$x = \frac{2 \pm \sqrt{-12}}{2}$$

$$x = \frac{2 \pm 2i\sqrt{3}}{2}$$

$$x = 1 \pm i\sqrt{3}$$

The solution set is $\left\{1 - i\sqrt{3},\ 1 + i\sqrt{3}\right\}$

12. $2x^2 - 6x + 5 = 0$

$$x = \frac{-(-6) \pm \sqrt{(-6)^2 - 4(2)(5)}}{2(2)}$$

$$x = \frac{6 \pm \sqrt{36 - 40}}{4}$$

$$x = \frac{6 \pm \sqrt{-4}}{4}$$

$$x = \frac{6 \pm 2i}{4}$$

$$x = \frac{6}{4} \pm \frac{2i}{4}$$

$$= \frac{3}{2} \pm \frac{1}{2}i$$

The solution set is $\left\{\dfrac{3}{2} - \dfrac{1}{2}i,\ \dfrac{3}{2} + \dfrac{1}{2}i\right\}$.

13. $f(x) = -(x+1)^2 + 4$

vertex: (–1, 4)

x-intercepts:

$$0 = -(x+1)^2 + 4$$

$$(x+1)^2 = 4$$

$$x + 1 = \pm 2$$

$$x = -1 \pm 2$$

$$x = -3 \text{ or } x = 1$$

y-intercept:

$$f(0) = -(0+1)^2 + 4 = 3$$

The axis of symmetry is $x = -1$.

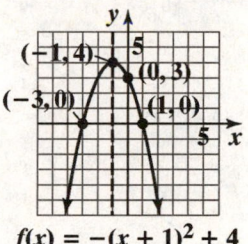

$$f(x) = -(x + 1)^2 + 4$$

domain: $(-\infty, \infty)$ range: $(-\infty, 4]$

14. $f(x) = (x+4)^2 - 2$

vertex: (–4, –2)

x-intercepts:

$$0 = (x+4)^2 - 2$$

$$(x+4)^2 = 2$$

$$x + 4 = \pm\sqrt{2}$$

$$x = -4 \pm \sqrt{2}$$

y-intercept:

$$f(0) = (0+4)^2 - 2 = 14 = -1$$

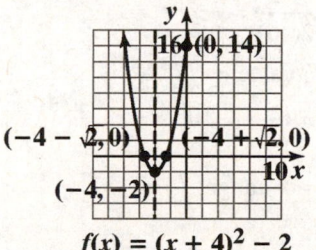

$$f(x) = (x + 4)^2 - 2$$

The axis of symmetry is $x = -4$.

domain: $(-\infty, \infty)$ range: $[-2, \infty)$

15. $f(x) = -x^2 + 2x + 3$

$\qquad = -(x^2 - 2x + 1) + 3 + 1$

$\quad f(x) = -(x-1)^2 + 4$

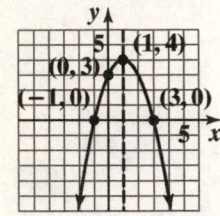

$f(x) = -x^2 + 2x + 3$

domain: $(-\infty, \infty)$ range: $(-\infty, 4]$

16. $f(x) = 2x^2 - 4x - 6$

$\quad f(x) = 2(x^2 - 2x + 1) - 6 - 2$

$\qquad 2(x-1)^2 - 8$

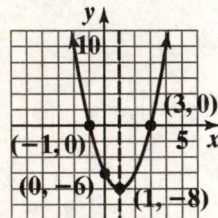

$f(x) = 2x^2 - 4x - 6$

axis of symmetry: $x = 1$

domain: $(-\infty, \infty)$ range: $[-8, \infty)$

17. $f(x) = -x^2 + 14x - 106$

a. Since $a < 0$ the parabola opens down with the maximum value occurring at

$$x = -\frac{b}{2a} = -\frac{14}{2(-1)} = 7.$$

The maximum value is $f(7)$.

$$f(7) = -(7)^2 + 14(7) - 106 = -57$$

b. domain: $(-\infty, \infty)$ range: $(-\infty, -57]$

18. $f(x) = 2x^2 + 12x + 703$

a. Since $a > 0$ the parabola opens up with the minimum value occurring at

$$x = -\frac{b}{2a} = -\frac{12}{2(2)} = -3.$$

The minimum value is $f(-3)$.

$$f(-3) = 2(-3)^2 + 12(-3) + 703 = 685$$

b. domain: $(-\infty, \infty)$ range: $[685, \infty)$

19. a. The maximum height will occur at the vertex.

$$f(x) = -0.025x^2 + x + 6$$

$$x = -\frac{b}{2a} = -\frac{1}{2(-0.025)} = 20$$

$$f(20) = -0.025(20)^2 + (20) + 6 = 16$$

The maximum height of 16 feet occurs when the ball is 20 yards downfield.

b. $f(x) = -0.025x^2 + x + 6$

$$f(0) = -0.025(0)^2 + (0) + 6 = 6$$

The ball was tossed at a height of 6 feet.

c. The ball is at a height of 0 when it hits the ground.

$$f(x) = -0.025x^2 + x + 6$$

$$0 = -0.025x^2 + x + 6$$

$$x = \frac{-b \pm \sqrt{b^2 - 4ac}}{2a}$$

$$x = \frac{-(1) \pm \sqrt{(1)^2 - 4(-0.025)(6)}}{2(-0.025)}$$

$x \approx 45.3, \ -5.3 \,(\text{reject})$

The ball will hit the ground 45.3 yards downfield.

d. The football's path:

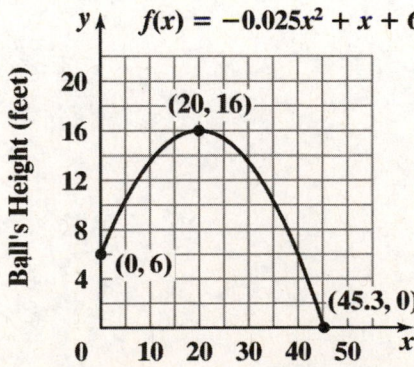

244

20. Maximize the area using $A = lw$.

$A(x) = x(1000 - 2x)$

$A(x) = -2x^2 + 1000x$

Since $a = -2$ is negative, we know the function opens downward and has a maximum at

$x = -\dfrac{b}{2a} = -\dfrac{1000}{2(-2)} = -\dfrac{1000}{-4} = 250$.

The maximum area is achieved when the width is 250 yards. The maximum area is

$A(250) = 250(1000 - 2(250))$

$\qquad\quad = 250(1000 - 500)$

$\qquad\quad = 250(500) = 125,000$.

The area is maximized at 125,000 square yards when the width is 250 yards and the length is $1000 - 2 \cdot 250 = 500$ yards.

21. Let x = one of the numbers
Let $14 + x$ = the other number

We need to minimize the function

$P(x) = x(14 + x)$

$\qquad\quad = 14x + x^2$

$\qquad\quad = x^2 + 14x$.

The minimum is at

$x = -\dfrac{b}{2a} = -\dfrac{14}{2(1)} = -\dfrac{14}{2} = -7$.

The other number is $14 + x = 14 + (-7) = 7$.

The numbers which minimize the product are 7 and -7. The minimum product is $-7 \cdot 7 = -49$.

22. $3x + 4y = 1000$

$4y = 1000 - 3x$

$y = \dfrac{1000 - 3x}{4}$

$A = x \cdot \dfrac{1000 - 3x}{4}$

$\quad = -\dfrac{3}{4}x^2 + 250x$

$x = \dfrac{-b}{2a} = \dfrac{-250}{2\left(-\dfrac{3}{4}\right)} = 125$

$y = \dfrac{1000 - 3(125)}{4} = 166.7$

125 feet by 166.7 feet will maximize the area.

23. $y = (35 + x)(150 - 4x)$

$y = 5250 + 10x - 4x^2$

$x = \dfrac{-b}{2a} = \dfrac{-10}{2(-4)} = \dfrac{5}{4} = 1.25$ or 1 tree

The maximum number of trees should be $35 + 1 = 36$ trees.

The maximum number of trees should be $35 + 1 = 36$ trees.

$y = 36(150 - 4x) = 36(150 - 4 \cdot 1) = 5256$

The maximum yield will be 5256 pounds.

24. $f(x) = -x^3 + 12x^2 - x$

The graph rises to the left and falls to the right and goes through the origin, so graph (c) is the best match.

25. $g(x) = x^6 - 6x^4 + 9x^2$

The graph rises to the left and rises to the right, so graph (b) is the best match.

26. $h(x) = x^5 - 5x^3 + 4x$

The graph falls to the left and rises to the right and crosses the y-axis at zero, so graph (a) is the best match.

27. $f(x) = -x^4 + 1$

$f(x)$ falls to the left and to the right so graph (d) is the best match.

28. The leading coefficient is -0.87 and the degree is 3. This means that the graph will fall to the right. This function is not useful in modeling the number of thefts over an extended period of time. The model predicts that eventually, the number of thefts would be negative. This is impossible.

29. In the polynomial, $f(x) = -x^4 + 21x^2 + 100,$ the leading coefficient is -1 and the degree is 4. Applying the Leading Coefficient Test, we know that even-degree polynomials with negative leading coefficient will fall to the left and to the right. Since the graph falls to the right, we know that the elk population will die out over time.

30. $f(x) = -2(x - 1)(x + 2)^2(x + 5)^3$

$x = 1$, multiplicity 1, the graph crosses the x-axis

$x = -2$, multiplicity 2, the graph touches the x-axis

$x = -5$, multiplicity 5, the graph crosses the x-axis

31. $f(x) = x^3 - 5x^2 - 25x + 125$

$= x^2(x-5) - 25(x-5)$

$= (x^2 - 25)(x-5)$

$= (x+5)(x-5)^2$

$x = -5$, multiplicity 1, the graph crosses the x-axis
$x = 5$, multiplicity 2, the graph touches the x-axis

32. $f(x) = x^3 - 2x - 1$

$f(1) = (1)^3 - 2(1) - 1 = -2$

$f(2) = (2)^3 - 2(2) - 1 = 3$

The sign change shows there is a zero between the given values.

33. $f(x) = x^3 - x^2 - 9x + 9$

a. Since n is odd and $a_n > 0$, the graph falls to the left and rises to the right.

b. $f(-x) = (-x)^3 - (-x)^2 - 9(-x) + 9$

$= -x^3 - x^2 + 9x + 9$

$f(-x) \neq f(x), f(-x) \neq -f(x)$
no symmetry

c. $f(x) = (x-3)(x+3)(x-1)$
zeros: $3, -3, 1$

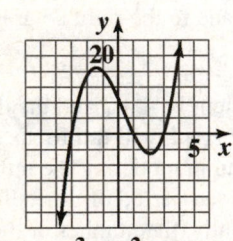

$f(x) = x^3 - x^2 - 9x + 9$

34. $f(x) = 4x - x^3$

a. Since n is odd and $a_n < 0$, the graph rises to the left and falls to the right.

b. $f(-x) = -4x + x^3$

$f(-x) = -f(x)$
origin symmetry

c. $f(x) = x(x^2 - 4) = x(x-2)(x+2)$
zeros: $x = 0, 2, -2$

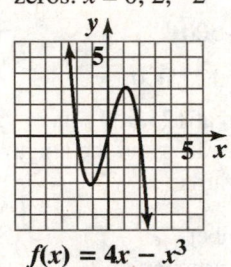

$f(x) = 4x - x^3$

35. $f(x) = 2x^3 + 3x^2 - 8x - 12$

a. Since h is odd and $a_n > 0$, the graph falls to the left and rises to the right.

b. $f(-x) = -2x^3 + 3x^2 + 8x - 12$

$f(-x) \neq f(x), f(-x) = -f(x)$
no symmetry

c. $f(x) = (x-2)(x+2)(2x+3)$

zeros: $x = 2, -2, -\dfrac{3}{2}$

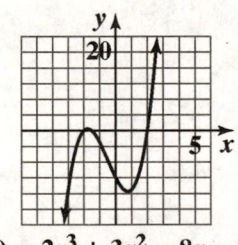

$f(x) = 2x^3 + 3x^2 - 8x - 12$

246

36. $g(x) = -x^4 + 25x^2$

    a. The graph falls to the left and to the right.

    b. $f(-x) = -(-x)^4 + 25(-x)^2$
$$= -x^4 + 25x^2 = f(x)$$
      y-axis symmetry

    c. $\quad -x^4 + 25x^2 = 0$
$$-x^2(x^2 - 25) = 0$$
$$-x^2(x-5)(x+5) = 0$$
      zeros: $x = -5, 0, 5$

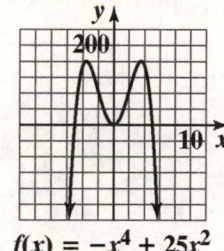

$$f(x) = -x^4 + 25x^2$$

37. $f(x) = -x^4 + 6x^3 - 9x^2$

    a. The graph falls to the left and to the right.

    b. $f(-x) = -(-x)^4 + 6(-x)^3 - 9(-x)$
$$= -x^4 - 6x^3 - 9x^2 \, f(-x) \neq f(x)$$
      $f(-x) \neq -f(x)$
      no symmetry

    c. $\quad = -x^2(x^2 - 6x + 9) = 0$
$$-x^2(x-3)(x-3) = 0$$
      zeros: $x = 0, 3$

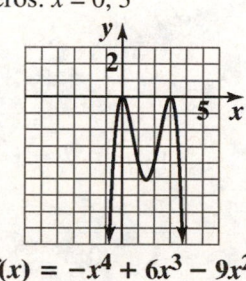

$$f(x) = -x^4 + 6x^3 - 9x^2$$

38. $f(x) = 3x^4 - 15x^3$

    a. The graph rises to the left and to the right.

    b. $f(-x) = 3(-x)^4 - 15(-x)^2 = 3x^4 + 15x^3$
      $f(-x) \neq f(x), \, f(-x) \neq -f(x)$
      no symmetry

    c $\quad 3x^4 - 15x^3 = 0$
$$3x^3(x-5) = 0$$
      zeros: $x = 0, 5$

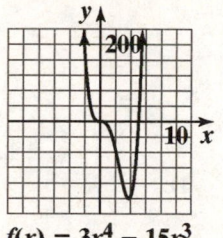

$$f(x) = 3x^4 - 15x^3$$

39. $f(x) = 2x^2(x-1)^3(x+2)$

Since $a_n > 0$ and n is even, $f(x)$ rises to the left and the right.

$x = 0, x = 1, x = -2$

The zeros at 1 and -2 have odd multiplicity so $f(x)$ crosses the x-axis at those points. The root at 0 has even multiplicity so $f(x)$ touches the axis at $(0, 0)$

$$f(0) = 2(0)^2(0-1)^3(0+2) = 0$$

The y-intercept is 0.

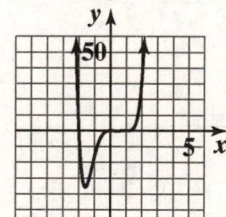

$$f(x) = 2x^2(x-1)^3(x+2)$$

40. $f(x) = -x^3(x+4)^2(x-1)$

Since $a_n < 0$ and n is even, $f(x)$ falls to the left and the right.

$x = 0$, $x = -4$, $x = 1$

The roots at 0 and 1 have odd multiplicity so $f(x)$ crosses the x-axis at those points. The root at -4 has even multiplicity so $f(x)$ touches the axis at $(-4, 0)$

$f(0) = -(0)^3(0+4)^2(0-1) = 0$

The y-intercept is 0.

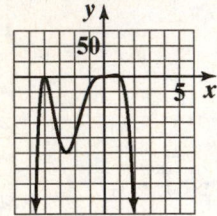

$f(x) = -x^3(x+4)^2(x-1)$

41.

$$\begin{array}{r} 4x^2 - 7x + 5 \\ x+1 \overline{)4x^3 - 3x^2 - 2x + 1} \\ \underline{4x^3 + 4x^2} \\ -7x^2 - 2x \\ \underline{-7x^2 - 7x} \\ 5x + 1 \\ \underline{5x + 5} \\ -4 \end{array}$$

Quotient: $4x^2 - 7x + 5 - \dfrac{4}{x+1}$

42.

$$\begin{array}{r} 2x^2 - 4x + 1 \\ 5x-3 \overline{)10x^3 - 26x^2 + 17x - 13} \\ \underline{10x^3 + 6x^2} \\ -20x^2 + 17x \\ \underline{-20x^2 + 12x} \\ 5x - 13 \\ \underline{5x - 3} \\ -10 \end{array}$$

Quotient: $2x^2 - 4x + 1 - \dfrac{10}{5x-3}$

43.

$$\begin{array}{r} 2x^2 + 3x - 1 \\ 2x^2+1 \overline{)4x^4 + 6x^3 + 3x - 1} \\ \underline{4x^4 + 2x^2} \\ 6x^3 - 2x^2 + 3x \\ \underline{6x^2 + 3x} \\ -2x^2 - 1 \\ \underline{-2x^2 - 1} \\ 0 \end{array}$$

44. $(3x^4 + 11x^3 - 20x^3 + 7x + 35) \div (x+5)$

$$\begin{array}{r|rrrrr} -5 & 3 & 11 & -20 & 7 & 35 \\ & & -15 & 20 & 0 & -35 \\ \hline & 3 & -4 & 0 & 7 & 0 \end{array}$$

Quotient: $3x^3 - 4x^2 + 7$

45. $(3x^4 - 2x^2 - 10x) \div (x-2)$

$$\begin{array}{r|rrrrr} 2 & 3 & 0 & -2 & -10 & 0 \\ & & 6 & 12 & 20 & 20 \\ \hline & 3 & 6 & 10 & 10 & 20 \end{array}$$

Quotient: $3x^3 + 6x^2 + 10x + 10 + \dfrac{20}{x-2}$

46. $f(x) = 2x^3 - 7x^2 + 9x - 3$

$$\begin{array}{r|rrrr} -13 & 2 & -7 & 9 & -3 \\ & & -26 & 429 & -5694 \\ \hline & 2 & -33 & 438 & -5697 \end{array}$$

Quotient: $f(-13) = -5697$

47. $f(x) = 2x^3 + x^2 - 13x + 6$

$$\begin{array}{r|rrrr} 2 & 2 & 1 & -13 & 6 \\ & & 4 & 10 & -6 \\ \hline & 2 & 5 & -3 & 0 \end{array}$$

$f(x) = (x-2)(2x^2 + 5x - 3)$
$\quad\quad = (x-2)(2x-1)(x+3)$

Zeros: $x = 2, \dfrac{1}{2}, -3$

48. $x^3 - 17x + 4 = 0$

$$\begin{array}{r|rrrr} 4 & 1 & 0 & -17 & 4 \\ & & 4 & 16 & -4 \\ \hline & 1 & 4 & -1 & 0 \end{array}$$

$(x-4)(x^2 + 4x - 1) = 0$

$x = \dfrac{-4 \pm \sqrt{16+4}}{2} = \dfrac{-4 \pm 2\sqrt{5}}{2} = -2 \pm \sqrt{5}$

The solution set is $\left\{4, -2+\sqrt{5}, -2-\sqrt{5}\right\}$.

49. $f(x) = x^4 - 6x^3 + 14x^2 - 14x + 5$

$p: \pm 1, \pm 5$

$q: \pm 1$

$\dfrac{p}{q}: \pm 1, \pm 5$

50. $f(x) = 3x^5 - 2x^4 - 15x^3 + 10x^2 + 12x - 8$

$p: \pm 1, \pm 2, \pm 4, \pm 8$

$q: \pm 1, \pm 3$

$\dfrac{p}{q}: \pm 1, \pm 2, \pm 4, \pm 8, \pm\dfrac{8}{3}, \pm\dfrac{4}{3}, \pm\dfrac{2}{3}, \pm\dfrac{1}{3}$

51. $f(x) = 3x^4 - 2x^3 - 8x + 5$

$f(x)$ has 2 sign variations, so $f(x) = 0$ has 2 or 0 positive solutions.

$f(-x) = 3x^4 + 2x^3 + x + 5$

$f(-x)$ has no sign variations, so $f(x) = 0$ has no negative solutions.

52. $f(x) = 2x^5 - 3x^3 - 5x^2 + 3x - 1$

$f(x)$ has 3 sign variations, so $f(x) = 0$ has 3 or 1 positive real roots.

$f(-x) = -2x^5 + 3x^3 - 5x^2 - 3x - 1$

$f(-x)$ has 2 sign variations, so $f(x) = 0$ has 2 or 0 negative solutions.

53. $f(x) = f(-x) = 2x^4 + 6x^2 + 8$

No sign variations exist for either $f(x)$ or $f(-x)$, so no real roots exist.

54. $f(x) = x^3 + 3x^2 - 4$

a. $p: \pm 1, \pm 2, \pm 4$

$q: \pm 1$

$\dfrac{p}{q}: \pm 1, \pm 2, \pm 4$

b.   1 sign variation $\Rightarrow$ 1 positive real zero

$f(-x) = -x^3 + 3x^2 - 4$

2 sign variations $\Rightarrow$ 2 or no negative real zeros

c.
$$\begin{array}{r|rrrr} 1 & 1 & 3 & 0 & -4 \\ & & 1 & 4 & 4 \\ \hline & 1 & 4 & 4 & 0 \end{array}$$

1 is a zero.

1, -2 are rational zeros.

d. $(x-1)(x^2 + 4x + 4) = 0$

$(x-1)(x+2)^2 = 0$

$x = 1$ or $x = -2$

The solution set is $\{1, -2\}$.

55. $f(x) = 6x^3 + x^2 - 4x + 1$

a. $p: \pm 1$

$q: \pm 1, \pm 2, \pm 3, \pm 6$

$\dfrac{p}{q}: \pm 1, \pm\dfrac{1}{2}, \pm\dfrac{1}{3}, \pm\dfrac{1}{6}$

b. $f(x) = 6x^3 + x^2 - 4x + 1$

2 sign variations; 2 or 0 positive real zeros.

$f(-x) = -6x^3 + x^2 + 4x + 1$

1 sign variation; 1 negative real zero.

c.
$$\begin{array}{r|rrrr} -1 & 6 & 1 & -4 & 1 \\ & & -6 & 5 & -1 \\ \hline & 6 & -5 & 1 & 0 \end{array}$$

-1 is a zero.

$-1, \dfrac{1}{3}, \dfrac{1}{2}$ are rational zeros.

d. $6x^3 + x^2 - 4x + 1 = 0$

$(x+1)(6x^2 - 5x + 1) = 0$

$(x+1)(3x-1)(2x-1) = 0$

$x = -1$ or $x = \dfrac{1}{3}$ or $x = \dfrac{1}{2}$

The solution set is $\left\{-1, \dfrac{1}{3}, \dfrac{1}{2}\right\}$.

56. $f(x) = 8x^3 - 36x^2 + 46x - 15$

a. $p: \pm 1, \pm 3, \pm 5, \pm 15$
$q: \pm 1, \pm 2, \pm 4, \pm 8$

$\dfrac{p}{q}: \pm 1, \pm 3, \pm 5, \pm 15, \pm\dfrac{1}{2}, \pm\dfrac{1}{4}, \pm\dfrac{1}{8},$

$\pm\dfrac{3}{2}, \pm\dfrac{3}{4}, \pm\dfrac{3}{8}, \pm\dfrac{5}{2}, \pm\dfrac{5}{4},$

$\pm\dfrac{5}{8}, \pm\dfrac{15}{2}, \pm\dfrac{15}{4}, \pm\dfrac{15}{8}$

b. $f(x) = 8x^3 - 36x^2 + 46x - 15$

3 sign variations; 3 or 1 positive real solutions.
$f(-x) = -8x^3 - 36x^2 - 46x - 15$

0 sign variations; no negative real solutions.

c.

$$\begin{array}{r|rrrr} \frac{1}{2} & 8 & -36 & 46 & -15 \\ & & 4 & -16 & 15 \\ \hline & 8 & -32 & 30 & 0 \end{array}$$

$\dfrac{1}{2}$ is a zero.

$\dfrac{1}{2}, \dfrac{3}{2}, \dfrac{5}{2}$ are rational zeros.

d.

$$8x^3 - 36x^2 + 46x - 15 = 0$$

$$\left(x - \frac{1}{2}\right)(8x^2 - 32x + 30) = 0$$

$$2\left(x - \frac{1}{2}\right)(4x^2 - 16x + 15) = 0$$

$$2\left(x - \frac{1}{2}\right)(2x - 5)(2x - 3) = 0$$

$$x = \frac{1}{2} \text{ or } x = \frac{5}{2} \text{ or } x = \frac{3}{2}$$

The solution set is $\left\{\dfrac{1}{2}, \dfrac{3}{2}, \dfrac{5}{2}\right\}$.

57. $2x^3 + 9x^2 - 7x + 1 = 0$

a. $p: \pm 1$
$q: \pm 1, \pm 2$

$\dfrac{p}{q}: \pm 1, \pm\dfrac{1}{2}$

b. $f(x) = 2x^3 + 9x^2 - 7x + 1$

2 sign variations; 2 or 0 positive real zeros.
$f(-x) = -2x^3 + 9x^2 + 7x + 1$

1 sign variation; 1 negative real zero.

c.

$$\begin{array}{r|rrrr} \frac{1}{2} & 2 & 9 & -7 & 1 \\ & & 1 & 5 & -1 \\ \hline & 2 & 10 & -2 & 0 \end{array}$$

$\dfrac{1}{2}$ is a rational zero.

d.

$$2x^3 + 9x^2 - 7x + 1 = 0$$

$$\left(x - \frac{1}{2}\right)(2x^2 + 10x - 2) = 0$$

$$2\left(x - \frac{1}{2}\right)(x^2 + 5x - 1) = 0$$

Solving $x^2 + 5x - 1 = 0$ using the quadratic

formula gives $x = \dfrac{-5 \pm \sqrt{29}}{2}$

The solution set is $\left\{\dfrac{1}{2}, \dfrac{-5 + \sqrt{29}}{2}, \dfrac{-5 - \sqrt{29}}{2}\right\}$.

58. $x^4 - x^3 - 7x^2 + x + 6 = 0$

a. $p = \dfrac{p}{q}: \pm 1, \pm 2, \pm 3, \pm 6$

b. $f(x) = x^4 - x^3 - 7x^2 + x + 6$

2 sign variations; 2 or 0 positive real zeros.
$f(-x) = x^4 + x^3 - 7x^2 - x + 6$

2 sign variations; 2 or 0 negative real zeros.

c.

$$
\begin{array}{r|rrrrr}
1 & 1 & -1 & -7 & 1 & 6 \\
  &   & 1 & 0 & -7 & -6 \\
\hline
  & 1 & 0 & -7 & -6 & 0
\end{array}
$$

$$
\begin{array}{r|rrrr}
-1 & 1 & 0 & -7 & -6 \\
   &   & -1 & 1 & 6 \\
\hline
   & 1 & -1 & -6 & 0
\end{array}
$$

$-2, -1, 1, 3$ are rational zeros.

d.
$$x^4 - x^3 - 7x^2 + x + 6 = 0$$
$$(x-1)(x+1)(x^2 - x + 6) = 0$$
$$(x-1)(x+1)(x-3)(x+2) = 0$$
The solution set is $\{-2, -1, 1, 3\}$.

59. $4x^4 + 7x^2 - 2 = 0$

a. p: $\pm 1, \pm 2$
q: $\pm 1, \pm 2, \pm 4$
$$\frac{p}{q}: \pm 1, \pm 2, \pm \frac{1}{2}, \pm \frac{1}{4}$$

b. $f(x) = 4x^4 + 7x^2 - 2$
1 sign variation; 1 positive real zero.
$f(-x) = 4x^4 + 7x^2 - 2$
1 sign variation; 1 negative real zero.

c.

$$
\begin{array}{r|rrrrr}
\dfrac{1}{2} & 4 & 0 & 7 & 0 & -2 \\
  &   & 2 & 1 & 4 & 2 \\
\hline
  & 4 & 2 & 8 & 4 & 0
\end{array}
$$

$$
\begin{array}{r|rrrr}
-\dfrac{1}{2} & 4 & 2 & 8 & 4 \\
  &   & -2 & 0 & -4 \\
\hline
  & 4 & 0 & 8 & 0
\end{array}
$$

$-\dfrac{1}{2}, \dfrac{1}{2}$ are rational zeros.

d.
$$4x^4 + 7x^2 - 2 = 0$$
$$\left(x - \frac{1}{2}\right)\left(x + \frac{1}{2}\right)(4x^2 + 8) = 0$$
$$4\left(x - \frac{1}{2}\right)\left(x + \frac{1}{2}\right)(x^2 + 2) = 0$$
Solving $x^2 + 2 = 0$ using the quadratic formula
gives $x = \pm 2i$

The solution set is $\left\{-\dfrac{1}{2}, \dfrac{1}{2}, 2i, -2i\right\}$.

60. $f(x) = 2x^4 + x^3 - 9x^2 - 4x + 4$

a. p: $\pm 1, \pm 2, \pm 4$
q: $\pm 1, \pm 2$
$$\frac{p}{q}: \pm 1, \pm 2, \pm 4, \pm \frac{1}{2}$$

b. $f(x) = 2x^4 + x^3 - 9x^2 - 4x + 4$
2 sign variations; 2 or 0 positive real zeros.
$f(-x) = 2x^4 - x^3 - 9x^2 + 4x + 4$
2 sign variations; 2 or 0 negative real zeros.

c.

$$
\begin{array}{r|rrrrr}
2 & 2 & 1 & -9 & -4 & 4 \\
  &   & 4 & 10 & 2 & -4 \\
\hline
  & 2 & 5 & 1 & -2 & 0
\end{array}
$$

$$
\begin{array}{r|rrrr}
-1 & 2 & 5 & 1 & -2 \\
   &   & -2 & -3 & 2 \\
\hline
   & 2 & 3 & -2 & 0
\end{array}
$$

$-2, -1, \dfrac{1}{2}, 2$ are rational zeros.

d.
$$2x^2 + 3x - 2 = 0$$
$$(2x - 1)(x + 2) = 0$$
$$x = -2 \quad \text{or} \quad x = \frac{1}{2}$$
The solution set is $\left\{-2, -1, \dfrac{1}{2}, 2\right\}$.

61.
$$f(x) = a_n(x-2)(x-2+3i)(x-2-3i)$$
$$f(x) = a_n(x-2)(x^2-4x+13)$$
$$f(1) = a_n(1-2)\left[1^2-4(1)+13\right]$$
$$-10 = -10a_n$$
$$a_n = 1$$
$$f(x) = 1(x-2)(x^2-4x+13)$$
$$f(x) = x^3-4x^2+13x-2x^2+8x-26$$
$$f(x) = x^3-6x^2+21x-26$$

62.
$$f(x) = a_n(x-i)(x+i)(x+3)^2$$
$$f(x) = a_n(x^2+1)(x^2+6x+9)$$
$$f(-1) = a_n\left[(-1)^2+1\right]\left[(-1)^2+6(-1)+9\right]$$
$$16 = 8a_n$$
$$a_n = 2$$
$$f(x) = 2(x^2+1)(x^2+6x+9)$$
$$f(x) = 2(x^4+6x^3+9x^2+x^2+6x+9)$$
$$f(x) = 2x^4+12x^3+20x^2+12x+18$$

63. $f(x) = 2x^4+3x^3+3x-2$

$p: \pm1, \pm2$
$q: \pm1, \pm2$

$\dfrac{p}{q}: \pm1, \pm2, \pm\dfrac{1}{2}$

$$\begin{array}{r|rrrrr}
-2 & 2 & 3 & 0 & 3 & -2 \\
   &   & -4 & 2 & -4 & 2 \\
\hline
   & 2 & -1 & 2 & -1 & 0
\end{array}$$

$$2x^4+3x^3+3x-2 = 0$$
$$(x+2)(2x^3-x^2+2x-1) = 0$$
$$(x+2)[x^2(2x-1)+(2x-1)] = 0$$
$$(x+2)(2x-1)(x^2+1) = 0$$
$$x=-2, \ x=\frac{1}{2} \ \text{or} \ x=\pm i$$

The zeros are -2, $\dfrac{1}{2}$, $\pm i$.

$$f(x) = (x-i)(x+i)(x+2)(2x-1)$$

64. $g(x) = x^4-6x^3+x^2+24x+16$

$p: \pm1, \pm2, \pm4, \pm8, \pm16$
$q: \pm1$

$\dfrac{p}{q}: \pm1, \pm2, \pm4, \pm8, \pm16$

$$\begin{array}{r|rrrrr}
-1 & 1 & -6 & 1 & 24 & 16 \\
   &   & -1 & 7 & -8 & -16 \\
\hline
   & 1 & -7 & 8 & 16 & 0
\end{array}$$

$$x^4-6x^3+x^2+24x+16 = 0$$
$$(x+1)(x^3-7x^2+8x+16) = 0$$

$$\begin{array}{r|rrrr}
-1 & 1 & -7 & 8 & 16 \\
   &   & -1 & 8 & -16 \\
\hline
   & 1 & -8 & 16 & 0
\end{array}$$

$$(x+1)^2(x^2-8x+16) = 0$$
$$(x+1)^2(x-4)^2 = 0$$
$$x=-1 \ \text{or} \ x=4$$
$$g(x) = (x+1)^2(x-4)^2$$

65. 4 real zeros, one with multiplicity two

66. 3 real zeros; 2 nonreal complex zeros

67. 2 real zeros, one with multiplicity two; 2 nonreal complex zeros

68. 1 real zero; 4 nonreal complex zeros

69. $g(x) = \dfrac{1}{(x+2)^2} - 1$

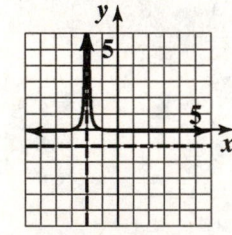

$$g(x) = \frac{1}{(x+2)^2} - 1$$

70. $h(x) = \dfrac{1}{x-1} + 3$

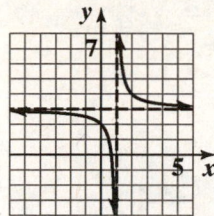

$h(x) = \dfrac{1}{x-1} + 3$

71. $f(x) = \dfrac{2x}{x^2-9}$

Symmetry: $f(-x) = -\dfrac{2x}{x^2-9} = -f(x)$

origin symmetry
x-intercept:

$0 = \dfrac{2x}{x^2-9}$

$2x = 0$

$x = 0$

y-intercept: $y = \dfrac{2(0)}{0^2-9} = 0$

Vertical asymptote:

$x^2 - 9 = 0$

$(x-3)(x+3) = 0$

$x = 3$ and $x = -3$

Horizontal asymptote:
$n < m$, so $y = 0$

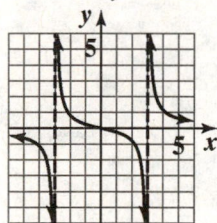

$f(x) = \dfrac{2x}{x^2-9}$

72. $g(x) = \dfrac{2x-4}{x+3}$

Symmetry: $g(-x) = \dfrac{-2x-4}{x+3}$

$g(-x) \neq g(x)$, $g(-x) \neq -g(x)$
No symmetry
x-intercept:
$2x - 4 = 0$
$x = 2$

y-intercept: $y = \dfrac{2(0)-4}{(0)+3} = -\dfrac{4}{3}$

Vertical asymptote:
$x + 3 = 0$
$x = -3$
Horizontal asymptote:

$n = m$, so $y = \dfrac{2}{1} = 2$

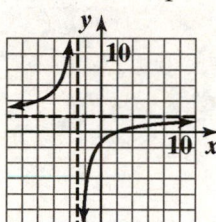

$f(x) = \dfrac{2x-4}{x+3}$

73. $h(x) = \dfrac{x^2 - 3x - 4}{x^2 - x - 6}$

Symmetry: $h(-x) = \dfrac{x^2 + 3x - 4}{x^2 + x - 6}$

$h(-x) \neq h(x), h(-x) \neq -h(x)$
No symmetry
x-intercepts:
$x^2 - 3x - 4 = 0$
$(x - 4)(x + 1)$
$x = 4 \quad x = -1$

y-intercept: $y = \dfrac{0^2 - 3(0) - 4}{0^2 - 0 - 6} = \dfrac{2}{3}$

Vertical asymptotes:
$x^2 - x - 6 = 0$
$(x - 3)(x + 2) = 0$
$x = 3, -2$

Horizontal asymptote:

$n = m$, so $y = \dfrac{1}{1} = 1$

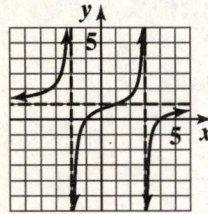

$h(x) = \dfrac{x^2 - 3x - 4}{x^2 - x - 6}$

74. $r(x) = \dfrac{x^2 + 4x + 3}{(x + 2)^2}$

Symmetry: $r(-x) = \dfrac{x^2 - 4x + 3}{(-x + 2)^2}$

$r(-x) \neq r(x), r(-x) \neq -r(x)$
No symmetry
x-intercepts:
$x^2 + 4x + 3 = 0$
$(x + 3)(x + 1) = 0$
$x = -3, -1$

y-intercept: $y = \dfrac{0^2 + 4(0) + 3}{(0 + 2)^2} = \dfrac{3}{4}$

Vertical asymptote:
$x + 2 = 0$
$x = -2$

Horizontal asymptote:

$n = m$, so $y = \dfrac{1}{1} = 1$

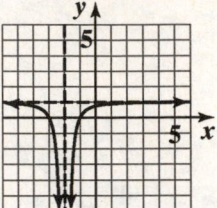

$r(x) = \dfrac{x^2 + 4x + 3}{(x + 4)^2}$

75. $y = \dfrac{x^2}{x + 1}$

Symmetry: $f(-x) = \dfrac{x^2}{-x + 1}$

$f(-x) \neq f(x), f(-x) \neq -f(x)$
No symmetry
x-intercept:
$x^2 = 0$
$x = 0$

y-intercept: $y = \dfrac{0^2}{0 + 1} = 0$

Vertical asymptote:
$x + 1 = 0$
$x = -1$
$n > m$, no horizontal asymptote.
Slant asymptote:

$y = x - 1 + \dfrac{1}{x + 1}$

$y = x - 1$

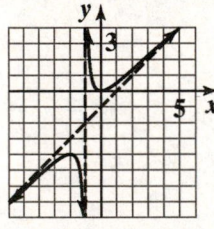

$y = \dfrac{x^2}{x + 1}$

76. $y = \dfrac{x^2 + 2x - 3}{x - 3}$

Symmetry: $f(-x) = \dfrac{x^2 - 2x - 3}{-x - 3}$

$f(-x) \neq f(x), f(-x) \neq -f(x)$
No symmetry
x-intercepts:
$$x^2 + 2x - 3 = 0$$
$$(x + 3)(x - 1) = 0$$
$$x = -3, 1$$

y-intercept: $y = \dfrac{0^2 + 2(0) - 3}{0 - 3} = \dfrac{-3}{-3} = 1$

Vertical asymptote:
$$x - 3 = 0$$
$$x = 3$$
Horizontal asymptote:
$n > m$, so no horizontal asymptote.

Slant asymptote:

$$y = x + 5 + \dfrac{12}{x - 3}$$

$$y = x + 5$$

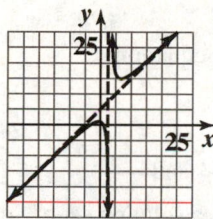

$$f(x) = \dfrac{x^2 + 2x - 3}{x - 3}$$

77. $f(x) = \dfrac{-2x^3}{x^2 + 1}$

Symmetry: $f(-x) = \dfrac{2}{x^2 + 1} = -f(x)$

Origin symmetry
x-intercept:
$$-2x^3 = 0$$
$$x = 0$$

y-intercept: $y = \dfrac{-2(0)^3}{0^2 + 1} = \dfrac{0}{1} = 0$

Vertical asymptote:
$$x^2 + 1 = 0$$
$$x^2 = -1$$
No vertical asymptote.
Horizontal asymptote:
$n > m$, so no horizontal asymptote.
Slant asymptote:

$f(x) = -2x + \dfrac{2x}{x^2 + 1}$

$y = -2x$

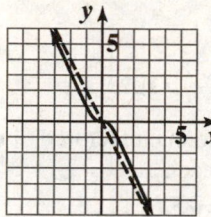

$$f(x) = \dfrac{-2x^3}{x^2 + 1}$$

78. $g(x) = \dfrac{4x^2 - 16x + 16}{2x - 3}$

Symmetry: $g(-x) = \dfrac{4x^2 + 16x + 16}{-2x - 3}$

$g(-x) \neq g(x), g(-x) \neq -g(x)$
No symmetry
x-intercept:
$$4x^2 - 16x + 16 = 0$$
$$4(x - 2)^2 = 0$$
$$x = 2$$
y-intercept:
$$y = \dfrac{4(0)^2 - 16(0) + 16}{2(0) - 3} = -\dfrac{16}{3}$$

Vertical asymptote:
$$2x - 3 = 0$$
$$x = \dfrac{3}{2}$$
Horizontal asymptote:
$n > m$, so no horizontal asymptote.
Slant asymptote:

$g(x) = 2x - 5 + \dfrac{1}{2x - 3}$

$y = 2x - 5$

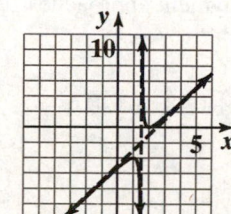

$$g(x) = \dfrac{4x^2 - 16x + 16}{2x - 3}$$

79. **a.** $C(x) = 50,000 + 25x$

b. $\overline{C}(x) = \dfrac{25x + 50,000}{x}$

c. $\overline{C}(50) = \dfrac{25(50) + 50,000}{50} = 1025$

When 50 calculators are manufactured, it costs $1025 to manufacture each.

$\overline{C}(100) = \dfrac{25(100) + 50,000}{100} = 525$

When 100 calculators are manufactured, it costs $525 to manufacture each.

$\overline{C}(1000) = \dfrac{25(1000) + 50,000}{1000} = 75$

When 1,000 calculators are manufactured, it costs $75 to manufacture each.

$\overline{C}(100,000) = \dfrac{25(100,000) + 50,000}{100,000} = 25.5$

When 100,000 calculators are manufactured, it costs $25.50 to manufacture each.

d. $n = m$, so $y = \dfrac{25}{1} = 25$ is the horizontal

asymptote. Minimum costs will approach $25.

80. $f(x) = \dfrac{150x + 120}{0.05x + 1}$

$n = m$, so $y = \dfrac{150}{0.05} = 3000$

The number of fish available in the pond approaches 3000.

81. $P(x) = \dfrac{72,900}{100x^2 + 729}$

$n < m$ so $y = 0$

As the number of years of education increases the percentage rate of unemployment approaches zero.

82. **a.** $P(x) = M(x) + F(x)$

$= 1.58x + 114.4 + 1.48x + 120.6$

$= 3.06x + 235$

b. $R(x) = \dfrac{M(x)}{P(x)} = \dfrac{1.58x + 114.4}{3.06x + 235}$

c. $y = \dfrac{1.58}{3.06} \approx 0.52$

Over time, the percentage of men in the U.S. population will approach 52%.

83. $T(x) = \dfrac{4}{x + 3} + \dfrac{2}{x}$

84. $1000 = lw$

$\dfrac{1000}{w} = l$

$P = 2x + 2\left(\dfrac{1000}{x}\right)$

$P = 2x + \dfrac{2000}{x}$

85. $2x^2 + 5x - 3 < 0$

Solve the related quadratic equation.

$2x^2 + 5x - 3 = 0$

$(2x - 1)(x + 3) = 0$

The boundary points are -3 and $\dfrac{1}{2}$.

Testing each interval gives a solution set of $\left(-3, \dfrac{1}{2}\right)$

86. $2x^2 + 9x + 4 \geq 0$

Solve the related quadratic equation.

$2x^2 + 9x + 4 = 0$

$(2x + 1)(x + 4) = 0$

The boundary points are -4 and $-\frac{1}{2}$.

Testing each interval gives a solution set of

$(-\infty, -4] \cup \left[-\dfrac{1}{2}, \infty\right)$

87. $x^3 + 2x^2 > 3x$

Solve the related equation.

$x^3 + 2x^2 = 3x$

$x^3 + 2x^2 - 3x = 0$

$x(x^2 + 2x - 3) = 0$

$x(x + 3)(x - 1) = 0$

The boundary points are -3, 0, and 1.

Testing each interval gives a solution set of

$(-3, 0) \cup (1, \infty)$

88. $\dfrac{x-6}{x+2} > 0$

Find the values of x that make the numerator and denominator zero.
The boundary points are -2 and 6.
Testing each interval gives a solution set of $(-\infty, -2) \cup (6, \infty)$.

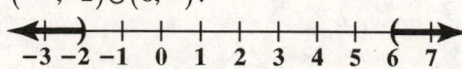

89. $\dfrac{(x+1)(x-2)}{x-1} \geq 0$

Find the values of x that make the numerator and denominator zero.
The boundary points are -1, 1 and 2. We exclude 1 from the solution set, since this would make the denominator zero.
Testing each interval gives a solution set of $[-1, 1) \cup [2, \infty)$.

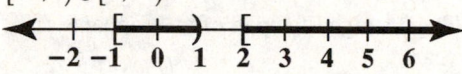

90. $\dfrac{x+3}{x-4} \leq 5$

Express the inequality so that one side is zero.

$$\dfrac{x+3}{x-4} - 5 \leq 0$$

$$\dfrac{x+3}{x-4} - \dfrac{5(x-4)}{x-4} \leq 0$$

$$\dfrac{-4x+23}{x-4} \leq 0$$

Find the values of x that make the numerator and denominator zero.

The boundary points are 4 and $\dfrac{23}{4}$. We exclude 4 from the solution set, since this would make the denominator zero.
Testing each interval gives a solution set of $\left(-\infty, 4\right) \cup \left[\dfrac{23}{4}, \infty\right)$.

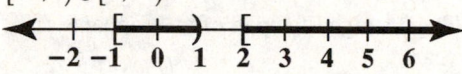

91. **a.** $g(x) = 0.125x^2 + 2.3x + 27$

$g(35) = 0.125(35)^2 + 2.3(35) + 27 \approx 261$

The stopping distance on wet pavement for a motorcycle traveling 35 miles per hour is about 261 feet. This overestimates the distance shown in the graph by 1 foot.

b. $f(x) = 0.125x^2 - 0.8x + 99$

$0.125x^2 - 0.8x + 99 > 267$

$0.125x^2 - 0.8x - 168 > 0$

Solve the related quadratic equation.

$0.125x^2 - 0.8x - 168 = 0$

$$x = \dfrac{-b \pm \sqrt{b^2 - 4ac}}{2a}$$

$$x = \dfrac{-(-0.8) \pm \sqrt{(-0.8)^2 - 4(0.125)(-168)}}{2(0.125)}$$

$x = -33.6,\ 40$

Testing each interval gives a solution set of $(-\infty, -33.6) \cup (40, \infty)$.

Thus, speeds exceeding 40 miles per hour on dry pavement will require over 267 feet of stopping distance.

92. $s = -16t^2 + v_0 t + s_0$

$32 < -16t^2 + 48t + 0$

$0 < -16t^2 + 48t - 32$

$0 < -16\left(t^2 - 3t + 2\right)$

$0 < -16(t-2)(t-1)$

	F		T		F
		1		2	

The projectile's height exceeds 32 feet during the time period from 1 to 2 seconds.

93. $w = ks$

$28 = k \cdot 250$

$0.112 = k$

Thus, $w = 0.112s$.

$w = 0.112(1200) = 134.4$

1200 cubic centimeters of melting snow will produce 134.4 cubic centimeters of water.

94. $d = kt^2$

$144 = k(3)^2$

$k = 16$

$d = 16t^2$

$d = 16(10)^2 = 1{,}600$ ft

95. $p = \dfrac{k}{w}$

$660 = \dfrac{k}{1.6}$

$1056 = k$

Thus, $p = \dfrac{1056}{w}$.

$p = \dfrac{1056}{2.4} = 440$

The pitch is 440 vibrations per second.

96. $l = \dfrac{k}{d^2}$

$28 = \dfrac{k}{8^2}$

$k = 1792$

$l = \dfrac{1792}{d^2}$

$l = \dfrac{1792}{4^2} = 112$ decibels

97. $t = \dfrac{kc}{w}$

$10 = \dfrac{k \cdot 30}{6}$

$10 = 5h$

$h = 2$

$t = \dfrac{2c}{w}$

$t = \dfrac{2(40)}{5} = 16$ hours

98. $V = khB$

$175 = k \cdot 15 \cdot 35$

$k = \dfrac{1}{3}$

$V = \dfrac{1}{3}hB$

$V = \dfrac{1}{3} \cdot 20 \cdot 120 = 800 \text{ ft}^3$

99. **a.** Use $L = \dfrac{k}{R}$ to find k.

$L = \dfrac{k}{R}$

$30 = \dfrac{k}{63}$

$63 \cdot 30 = 63 \cdot \dfrac{k}{63}$

$1890 = k$

Thus, $L = \dfrac{1890}{R}$.

b. This is an approximate model.

c. $L = \dfrac{1890}{R}$

$L = \dfrac{1890}{27} = 70$

The average life span of an elephant is 70 years.

Chapter 2 Test

1. $(6 - 7i)(2 + 5i) = 12 + 30i - 14i - 35i^2$

$\qquad\qquad\qquad\quad = 12 + 16i + 35$

$\qquad\qquad\qquad\quad = 47 + 16i$

2. $\dfrac{5}{2-i} = \dfrac{5}{2-i} \cdot \dfrac{2+i}{2+i}$

$\qquad = \dfrac{5(2+i)}{4+1}$

$\qquad = \dfrac{5(2+i)}{5}$

$\qquad = 2 + i$

3. $2\sqrt{-49} + 3\sqrt{-64} = 2(7i) + 3(8i)$

$\qquad\qquad\qquad\qquad = 14i + 24i$

$\qquad\qquad\qquad\qquad = 38i$

4.
$$x^2 = 4x - 8$$
$$x^2 - 4x + 8 = 0$$
$$x = \frac{-b \pm \sqrt{b^2 - 4ac}}{2a}$$
$$x = \frac{-(-4) \pm \sqrt{(-4)^2 - 4(1)(8)}}{2(1)}$$
$$x = \frac{4 \pm \sqrt{-16}}{2}$$
$$x = \frac{4 \pm 4i}{2}$$
$$x = 2 \pm 2i$$

5. $f(x) = (x+1)^2 + 4$
vertex: $(-1, 4)$
axis of symmetry: $x = -1$
x-intercepts:
$$(x+1)^2 + 4 = 0$$
$$x^2 + 2x + 5 = 0$$
$$x = \frac{-2 \pm \sqrt{4 - 20}}{2} = -1 \pm 2i$$
no x-intercepts
y-intercept:
$$f(0) = (0+1)^2 + 4 = 5$$

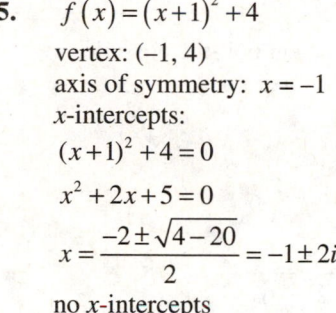
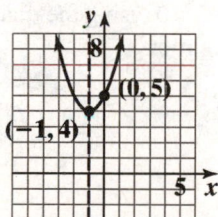

$f(x) = (x + 1)^2 + 4$
domain: $(-\infty, \infty)$; range: $[4, \infty)$

6. $f(x) = x^2 - 2x - 3$
$$x = \frac{-b}{2a} = \frac{2}{2} = 1$$
$$f(1) = 1^2 - 2(1) - 3 = -4$$
vertex: $(1, -4)$
axis of symmetry $x = 1$
x-intercepts:
$$x^2 - 2x - 3 = 0$$
$$(x-3)(x+1) = 0$$
$$x = 3 \text{ or } x = -1$$

y-intercept:
$$f(0) = 0^2 - 2(0) - 3 = -3$$

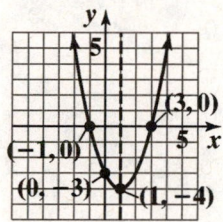

$f(x) = x^2 - 2x - 3$
domain: $(-\infty, \infty)$; range: $[-4, \infty)$

7. $f(x) = -2x^2 + 12x - 16$

Since the coefficient of x^2 is negative, the graph of $f(x)$ opens down and $f(x)$ has a maximum point.
$$x = \frac{-12}{2(-2)} = 3$$

$$f(3) = -2(3)^2 + 12(3) - 16$$
$$= -18 + 36 - 16$$
$$= 2$$
Maximum point: $(3, 2)$
domain: $(-\infty, \infty)$; range: $(-\infty, 2]$

8. $f(x) = -x^2 + 46x - 360$
$$x = -\frac{b}{2a} = \frac{-46}{-2} = 23$$
23 computers will maximize profit.
$$f(23) = -(23)^2 + 46(23) - 360 = 169$$
Maximum daily profit = \$16,900.

9. Let $x =$ one of the numbers;
$14 - x =$ the other number.
The product is $f(x) = x(14 - x)$

$$f(x) = x(14 - x) = -x^2 + 14x$$

The x-coordinate of the maximum is

$$x = -\frac{b}{2a} = -\frac{14}{2(-1)} = -\frac{14}{-2} = 7.$$

$$f(7) = -7^2 + 14(7) = 49$$

The vertex is $(7, 49)$. The maximum product is 49.
This occurs when the two number are 7 and
$14 - 7 = 7$.

10. a. $f(x) = x^3 - 5x^2 - 4x + 20$

$$x^3 - 5x^2 - 4x + 20 = 0$$

$$x^2(x - 5) - 4(x - 5) = 0$$

$$(x - 5)(x - 2)(x + 2) = 0$$

$$x = 5, 2, -2$$

The solution set is $\{5, 2, -2\}$.

b. The degree of the polynomial is odd and the
leading coefficient is positive. Thus the graph
falls to the left and rises to the right.

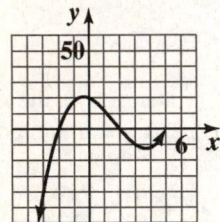

11. $f(x) = x^5 - x$

Since the degree of the polynomial is odd and the
leading coefficient is positive, the graph of f should
fall to the left and rise to the right. The x-intercepts
should be -1 and 1.

12. a. The integral root is 2.

b.

$$
\begin{array}{r|rrrr}
2 & 6 & -19 & 16 & -4 \\
  &   & 12 & -14 & 4 \\
\hline
  & 6 & -7 & 2 & 0
\end{array}
$$

$$6x^2 - 7x + 2 = 0$$
$$(3x - 2)(2x - 1) = 0$$
$$x = \frac{2}{3} \text{ or } x = \frac{1}{2}$$

The other two roots are $\frac{1}{2}$ and $\frac{2}{3}$.

13. $2x^3 + 11x^2 - 7x - 6 = 0$
$p: \pm 1, \pm 2, \pm 3, \pm 6$
$q: \pm 1, \pm 2$

$$\frac{p}{q} : \pm 1, \pm 2, \pm 3, \pm 6, \pm \frac{1}{2}, \pm \frac{3}{2}$$

14. $f(x) = 3x^5 - 2x^4 - 2x^2 + x - 1$

$f(x)$ has 3 sign variations.
$$f(-x) = -3x^5 - 2x^4 - 2x^2 - x - 1$$

$f(-x)$ has no sign variations.
There are 3 or 1 positive real solutions and no
negative real solutions.

15. $x^3 + 9x^2 + 16x - 6 = 0$
Since the leading coefficient is 1, the possible
rational zeros are the factors of 6

$$p = \frac{p}{q} : \pm 1, \pm 2, \pm 3, \pm 6$$

$$
\begin{array}{r|rrrr}
-3 & 1 & 9 & 16 & -6 \\
   &   & -3 & -18 & 6 \\
\hline
   & 1 & 6 & -2 & 0
\end{array}
$$

Thus $x = 3$ is a root.
Solve the quotient $x^2 + 6x - 2 = 0$ using the quadratic
formula to find the remaining roots.

$$x = \frac{-b \pm \sqrt{b^2 - 4ac}}{2a}$$

$$x = \frac{-(6) \pm \sqrt{(6)^2 - 4(1)(-2)}}{2(1)}$$

$$= \frac{-6 \pm \sqrt{44}}{2}$$

$$= -3 \pm \sqrt{11}$$

The zeros are -3, $-3 + \sqrt{11}$, and $-3 - \sqrt{11}$.

16. $f(x) = 2x^4 - x^3 - 13x^2 + 5x + 15$

a. Possible rational zeros are:
$p: \pm 1, \pm 3, \pm 5, \pm 15$
$q: \pm 1, \pm 2$

$$\frac{p}{q} : \pm 1, \pm 3, \pm 5, \pm 15, \pm \frac{1}{2}, \pm \frac{3}{2}, \pm \frac{5}{2}, \pm \frac{15}{2}$$

b. Verify that -1 and $\dfrac{3}{2}$ are zeros as it appears in the graph:

$$\begin{array}{r|rrrrr} -1 & 2 & -1 & -13 & 5 & 15 \\ & & -2 & 3 & 10 & -15 \\ \hline & 2 & -3 & -10 & 15 & 0 \end{array}$$

$$\begin{array}{r|rrrr} \frac{3}{2} & 2 & -3 & -10 & 15 \\ & & 3 & 0 & -15 \\ \hline & 2 & 0 & -10 & 0 \end{array}$$

Thus, -1 and $\dfrac{3}{2}$ are zeros, and the polynomial factors as follows:

$$2x^4 - x^3 - 13x^2 + 5x + 15 = 0$$
$$(x+1)\left(2x^3 - 3x^2 - 10x + 15\right) = 0$$
$$(x+1)\left(x - \dfrac{3}{2}\right)\left(2x^2 - 10\right) = 0$$

Find the remaining zeros by solving:
$$2x^2 - 10 = 0$$
$$2x^2 = 10$$
$$x^2 = 5$$
$$x = \pm\sqrt{5}$$

The zeros are -1, $\dfrac{3}{2}$, and $\pm\sqrt{5}$.

17. $f(x)$ has zeros at -2 and 1. The zero at -2 has multiplicity of 2.
$$x^3 + 3x^2 - 4 = (x-1)(x+2)^2$$

18. $f(x) = a_0(x+1)(x-1)(x+i)(x-i)$
$$= a_0(x^2 - 1)(x^2 + 1)$$
$$= a_0(x^4 - 1)$$
Since $f(3) = 160$, then
$$a_0(3^4 - 1) = 160$$
$$a_0(80) = 160$$
$$a_0 = \dfrac{160}{80}$$
$$a_0 = 2$$
$$f(x) = 2(x^4 - 1) = 2x^4 - 2$$

19. $f(x) = -3x^3 - 4x^2 + x + 2$

The graph shows a root at $x = -1$.
Use synthetic division to verify this root.

$$\begin{array}{r|rrrr} -1 & -3 & -4 & 1 & 2 \\ & & 3 & 1 & -4 \\ \hline & -3 & -1 & 2 & 0 \end{array}$$

Factor the quotient to find the remaining zeros.
$$-3x^2 - x + 2 = 0$$
$$-(3x - 2)(x + 1) = 0$$

The zeros (x-intercepts) are -1 and $\dfrac{2}{3}$.

The y-intercept is $f(0) = 2$

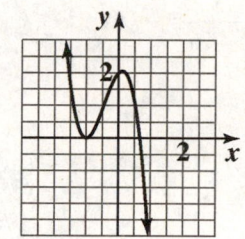

$$f(x) = -3x^3 - 4x^2 + x + 2$$

20. $f(x) = \dfrac{1}{(x+3)^2}$

domain: $\{x \mid x \neq -3\}$ or $(-\infty, -3) \cup (-3, \infty)$

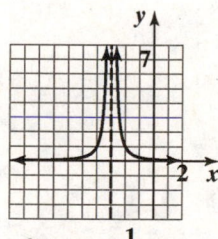

$$f(x) = \dfrac{1}{(x+3)^2}$$

21. $f(x) = \dfrac{1}{x-1} + 2$

domain: $\{x \mid x \neq 1\}$ or $(-\infty, 1) \cup (1, \infty)$

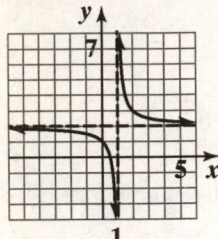

$$f(x) = \frac{1}{x-1} + 2$$

22. $f(x) = \dfrac{x}{x^2 - 16}$

domain: $\{x \mid x \neq 4, x \neq -4\}$

Symmetry: $f(-x) = \dfrac{-x}{x^2 - 16} = -f(x)$

y-axis symmetry

x-intercept: $x = 0$

y-intercept: $y = \dfrac{0}{0^2 - 16} = 0$

Vertical asymptotes:

$x^2 - 16 = 0$

$(x - 4)(x + 4) = 0$

$x = 4, -4$

Horizontal asymptote:

$n < m$, so $y = 0$ is the horizontal asymptote.

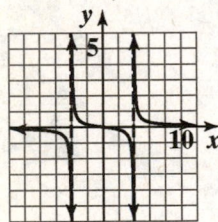

$$f(x) = \frac{x}{x^2 - 16}$$

23. $f(x) = \dfrac{x^2 - 9}{x - 2}$

domain: $\{x \mid x \neq 2\}$

Symmetry: $f(-x) = \dfrac{x^2 - 9}{-x - 2}$

$f(-x) \neq f(x), f(-x) \neq -f(x)$

No symmetry

x-intercepts:

$x^2 - 9 = 0$

$(x - 3)(x + 3) = 0$

$x = 3, -3$

y-intercept: $y = \dfrac{0^2 - 9}{0 - 2} = \dfrac{9}{2}$

Vertical asymptote:

$x - 2 = 0$

$x = 2$

Horizontal asymptote:

$n > m$, so no horizontal asymptote exists.

Slant asymptote: $f(x) = x + 2 - \dfrac{5}{x-2}$

$y = x + 2$

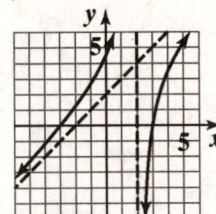

$$f(x) = \frac{x^2 - 9}{x - 2}$$

24. $f(x) = \dfrac{x+1}{x^2 + 2x - 3}$

$x^2 + 2x - 3 = (x+3)(x-1)$

domain: $\{x \mid x \neq -3, x \neq 1\}$

Symmetry: $f(-x) = \dfrac{-x+1}{x^2 - 2x - 3}$

$f(-x) \neq f(x), f(-x) \neq -f(x)$
No symmetry
x-intercept:
$x + 1 = 0$
$x = -1$

y-intercept: $y = \dfrac{0+1}{0^2 + 2(0) - 3} = -\dfrac{1}{3}$

Vertical asymptotes:
$x^2 + 2x - 3 = 0$
$(x + 3)(x - 1) = 0$
$x -3, 1$
Horizontal asymptote:
$n < m$, so $y = 0$ is the horizontal asymptote.

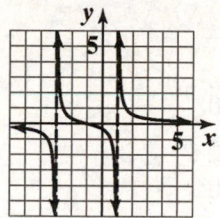

$f(x) = \dfrac{x+1}{x^2 + 2x - 3}$

25. $f(x) = \dfrac{4x^2}{x^2 + 3}$

domain: all real numbers

Symmetry: $f(-x) = \dfrac{4x^2}{x^2 + 3} = f(x)$

y-axis symmetry
x-intercept:
$4x^2 = 0$
$x = 0$

y-intercept: $y = \dfrac{4(0)^2}{0^2 + 3} = 0$

Vertical asymptote:
$x^2 + 3 = 0$
$x^2 = -3$
No vertical asymptote.

Horizontal asymptote:

$n = m$, so $y = \dfrac{4}{1} = 4$ is the horizontal asymptote.

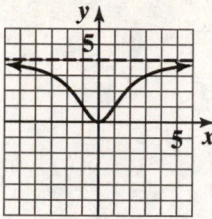

$f(x) = \dfrac{4x^2}{x^2 + 3}$

26. a. $\overline{C}(x) = \dfrac{300,000 + 10x}{x}$

b. Since the degree of the numerator equals the degree of the denominator, the horizontal asymptote is $x = \dfrac{10}{1} = 10$.

This represents the fact that as the number of satellite radio players produced increases, the production cost approaches \$10 per radio.

27. $ x^2 < x + 12$

$x^2 - x - 12 < 0$

$(x + 3)(x - 4) < 0$

Boundary values: -3 and 4
Solution set: $(-3, 4)$

28. $\dfrac{2x+1}{x-3} \leq 3$

$\dfrac{2x+1}{x-3} - 3 \leq 0$

$\dfrac{10 - x}{x - 3} \leq 0$

Boundary values: 3 and 10
Solution set: $(-\infty, 3) \cup [10, \infty)$

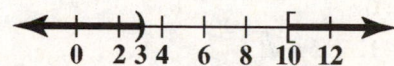

29.

$$i = \frac{k}{d^2}$$

$$20 = \frac{k}{15^2}$$

$$4500 = k$$

$$i = \frac{4500}{d^2} = \frac{4500}{10^2} = 45 \text{ foot-candles}$$

Cumulative Review Exercises (Chapters P–2)

1. domain: $(-2, 2)$ range: $[0, \infty)$

2. The zero at -1 touches the x-axis at turns around so it must have a minimum multiplicity of 2.
The zero at 1 touches the x-axis at turns around so it must have a minimum multiplicity of 2.

3. There is a relative maximum at the point $(0, 3)$.

4. $(f \circ f)(-1) = f\left(f(-1)\right) = f(0) = 3$

5. $f(x) \to \infty$ as $\underline{x \to -2^+}$ or as $\underline{x \to 2^-}$

6.

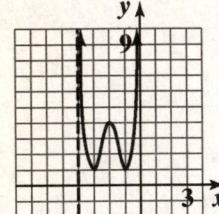

7. $|2x - 1| = 3$

$2x - 1 = 3$

$2x = 4$

$x = 2$

$2x - 1 = -3$

$2x = -2$

$x = -1$

The solution set is $\{2, -1\}$.

8. $3x^2 - 5x + 1 = 0$

$$x = \frac{5 \pm \sqrt{25 - 12}}{6} = \frac{5 \pm \sqrt{13}}{6}$$

The solution set is $\left\{\dfrac{5 + \sqrt{13}}{6}, \dfrac{5 - \sqrt{13}}{6}\right\}$.

9. $9 + \dfrac{3}{x} = \dfrac{2}{x^2}$

$9x^2 + 3x = 2$

$9x^2 + 3x - 2 = 0$

$(3x - 1)(3x + 2) = 0$

$3x - 1 = 0 \qquad 3x + 2 = 0$

$x = \dfrac{1}{3} \quad$ or $\quad x = -\dfrac{2}{3}$

The solution set is $\left\{\dfrac{1}{3}, -\dfrac{2}{3}\right\}$.

10. $x^3 + 2x^2 - 5x - 6 = 0$

$p: \pm 1, \pm 2, \pm 3, \pm 6$

$q: \pm 1$

$\dfrac{p}{q}: \pm 1, \pm 2, \pm 3, \pm 6$

$$
\begin{array}{r|rrrr}
-3 & 1 & 2 & -5 & -6 \\
   &   & -3 & 3 & 6 \\
\hline
   & 1 & -1 & -2 & 0
\end{array}
$$

$x^3 + 2x^2 - 5x - 6 = 0$

$(x + 3)(x^2 - x - 2) = 0$

$(x + 3)(x + 1)(x - 2) = 0$

$x = -3$ or $x = -1$ or $x = 2$

The solution set is $\{-3, -1, 2\}$.

11. $|2x - 5| > 3$

$2x - 5 > 3$

$2x > 8$

$x > 4$

$2x - 5 < -3$

$2x < 2$

$x < 1$

$(-\infty, 1)$ or $(4, \infty)$

12. $\qquad 3x^2 > 2x + 5$

$3x^2 - 2x - 5 > 0$

$3x^2 - 2x - 5 = 0$

$(3x - 5)(x + 1) = 0$

$x = \dfrac{5}{3}$ or $x = -1$

Test intervals are $(-\infty, -1)$, $\left(-1, \dfrac{5}{3}\right)$, $\left(\dfrac{5}{3}, \infty\right)$.

Testing points, the solution is $(-\infty, -1)$ or $\left(\dfrac{5}{3}, \infty\right)$.

13. $f(x) = x^3 - 4x^2 - x + 4$

x-intercepts:

$$x^3 - 4x^2 - x + 4 = 0$$

$$x^2(x-4) - 1(x-4) = 0$$

$$(x-4)(x^2 - 1) = 0$$

$$(x-4)(x+1)(x-1) = 0$$

$$x = -1, 1, 4$$

x-intercepts:

$$f(0) = 0^3 - 4(0)^2 - 0 + 4 = 4$$

The degree of the polynomial is odd and the leading coefficient is positive. Thus the graph falls to the left and rises to the right.

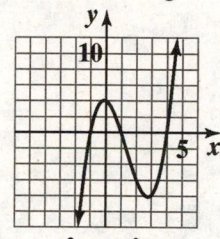

$$f(x) = x^3 - 4x^2 - x + 4$$

14. $f(x) = x^2 + 2x - 8$

$$x = \frac{-b}{2a} = \frac{-2}{2} = -1$$

$$f(-1) = (-1)^2 + 2(-1) - 8$$

$$= 1 - 2 - 8 = -9$$

vertex: $(-1, -9)$

x-intercepts:

$$x^2 + 2x - 8 = 0$$

$$(x+4)(x-2) = 0$$

$$x = -4 \text{ or } x = 2$$

y-intercept: $f(0) = -8$

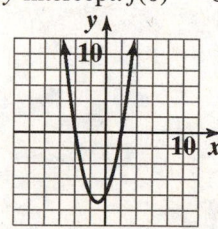

$$f(x) = x^2 + 2x - 8$$

15. $f(x) = x^2(x-3)$

zeros: $x = 0$ (multiplicity 2) and $x = 3$

y-intercept: $y = 0$

$$f(x) = x^3 - 3x^2$$

$n = 3, a_n = 0$ so the graph falls to the left and rises to the right.

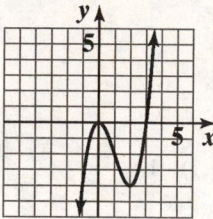

$$f(x) = x^2(x-3)$$

16. $f(x) = \dfrac{x-1}{x-2}$

vertical asymptote: $x = 2$

horizontal asymptote: $y = 1$

x-intercept: $x = 1$

y-intercept: $y = \dfrac{1}{2}$

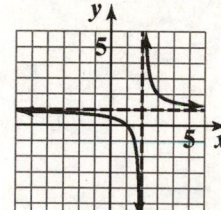

17.

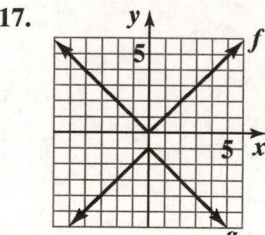

18.

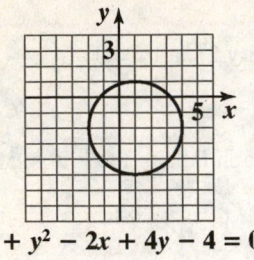

$x^2 + y^2 - 2x + 4y - 4 = 0$

19. $(f \circ g)(x) = f\big(g(x)\big)$

$$= 2(4x-1)^2 - (4x-1) - 1$$
$$= 32x^2 - 20x + 2$$

20. $\dfrac{f(x+h) - f(x)}{h}$

$$= \dfrac{\big[2(x+h)^2 - (x+h) - 1\big] - \big[2x^2 - x - 1\big]}{h}$$

$$= \dfrac{2x^2 + 4hx - x + 2h^2 - h - 1 - 2x^2 + x + 1}{h}$$

$$= \dfrac{4hx + 2h^2 - h}{h}$$

$$= 4x + 2h - 1$$

Chapter 3
Exponential and Logarithmic Functions

Section 3.1

Check Point Exercises

1. $f(x) = 42.2(1.56)^x$

    $f(3) = 42.2(1.56)^3 \approx 160.20876 \approx 160$

    According to the function, the average amount spent after three hours of shopping at the mall is $160. This overestimates the actual amount shown by $11.

2. Begin by setting up a table of coordinates.

x	$f(x) = 3^x$
-3	$f(-3) = 3^{-3} = \frac{1}{27}$
-2	$f(-2) = 3^{-2} = \frac{1}{9}$
-1	$f(-1) = 3^{-1} = \frac{1}{3}$
0	$f(0) = 3^0 = 1$
1	$f(1) = 3^1 = 3$
2	$f(2) = 3^2 = 9$
3	$f(3) = 3^3 = 27$

    Plot these points, connecting them with a continuous curve.

    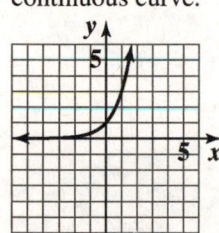

    $f(x) = 3^x$

3. Begin by setting up a table of coordinates.

x	$f(x) = \left(\frac{1}{3}\right)^x$
-2	$\left(\frac{1}{3}\right)^{-2} = 9$
-1	$\left(\frac{1}{3}\right)^{-1} = 3$
0	$\left(\frac{1}{3}\right)^{0} = 1$
1	$\left(\frac{1}{3}\right)^{1} = \frac{1}{3}$
2	$\left(\frac{1}{3}\right)^{2} = \frac{1}{9}$

    Plot these points, connecting them with a continuous curve.

    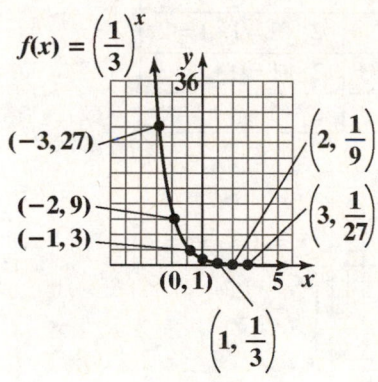

    $f(x) = \left(\frac{1}{3}\right)^x$

4. Note that the function $g(x) = 3^{x-1}$ has the general form $g(x) = b^{x+c}$ where $c = -1$. Because $c < 0$, we graph $g(x) = 3^{x-1}$ by shifting the graph of $f(x) = 3^x$ one unit to the right. Construct a table showing some of the coordinates for f and g.

x	$f(x) = 3^x$	$g(x) = 3^{x-1}$
-2	$3^{-2} = \frac{1}{9}$	$3^{-2-1} = 3^{-3} = \frac{1}{27}$
-1	$3^{-1} = \frac{1}{3}$	$3^{-1-1} = 3^{-2} = \frac{1}{9}$
0	$3^0 = 1$	$3^{0-1} = 3^{-1} = \frac{1}{3}$
1	$3^1 = 3$	$3^{1-1} = 3^0 = 1$
2	$3^2 = 9$	$3^{2-1} = 3^1 = 3$

    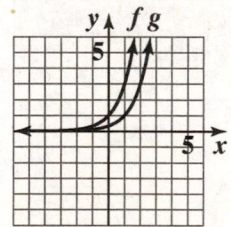

    $f(x) = 3^x$
    $g(x) = 3^{x-1}$

5. Note that the function $g(x) = 2^x + 1$ has the general form $g(x) = b^x + c$ where $c = 1$. Because $c > 0$, we graph $g(x) = 2^x + 1$ by shifting the graph of $f(x) = 2^x$ up one unit. Construct a table showing some of the coordinates for f and g.

x	$f(x) - 2^x$	$g(x) = 2^x + 1$
-2	$2^{-2} = \frac{1}{4}$	$2^{-2} + 1 = \frac{1}{4} + 1 = \frac{5}{4}$
-1	$2^{-1} = \frac{1}{2}$	$2^{-1} + 1 = \frac{1}{2} + 1 = \frac{3}{2}$
0	$2^0 = 1$	$2^0 + 1 = 1 + 1 = 2$
1	$2^1 = 2$	$2^1 + 1 = 2 + 1 = 3$
2	$2^2 = 4$	$2^2 + 1 = 4 + 1 = 5$

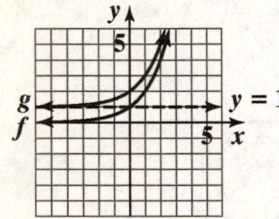

$f(x) = 2^x$
$g(x) = 2^x + 1$

6. 2012 is 34 years after 1978.

$$f(x) = 1066e^{0.042x}$$

$$f(34) = 1066e^{0.042(34)} \approx 4446$$

In 2012 the gray wolf population of the Western Great Lakes is projected to be about 4446.

7. **a.** $A = P\left(1 + \frac{r}{n}\right)^{nt}$

$$A = 10,000\left(1 + \frac{0.08}{4}\right)^{4(5)}$$

$$= \$14,859.47$$

b. $A = Pe^{rt}$

$$A = 10,000e^{0.08(5)}$$

$$= \$14,918.25$$

Exercise Set 3.1

1. $2^{3.4} \approx 10.556$

3. $3^{\sqrt{5}} \approx 11.665$

5. $4^{-1.5} = 0.125$

7. $e^{2.3} \approx 9.974$

9. $e^{-0.95} \approx 0.387$

11.

x	$f(x) = 4^x$
-2	$4^{-2} = \frac{1}{16}$
-1	$4^{-1} = \frac{1}{4}$
0	$4^0 = 1$
1	$4^1 = 4$
2	$4^2 = 16$

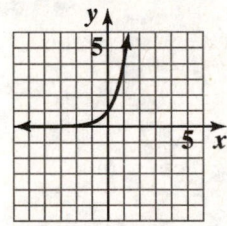

$f(x) = 4^x$

13.

x	$g(x) = \left(\frac{3}{2}\right)^x$
-2	$\left(\frac{3}{2}\right)^{-2} = \frac{4}{9}$
-1	$\left(\frac{3}{2}\right)^{-1} = \frac{2}{3}$
0	$\left(\frac{3}{2}\right)^0 = 1$
1	$\left(\frac{3}{2}\right)^1 = \frac{3}{2}$
2	$\left(\frac{3}{2}\right)^2 = \frac{9}{4}$

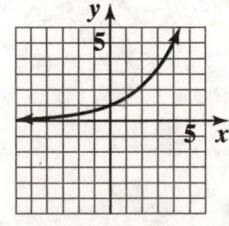

$g(x) = \left(\frac{3}{2}\right)^x$

15.

x	$h(x) = \left(\frac{1}{2}\right)^x$
−2	$\left(\frac{1}{2}\right)^{-2} = 4$
−1	$\left(\frac{1}{2}\right)^{-1} = 2$
0	$\left(\frac{1}{2}\right)^{0} = 1$
1	$\left(\frac{1}{2}\right)^{1} = \frac{1}{2}$
2	$\left(\frac{1}{2}\right)^{2} = \frac{1}{4}$

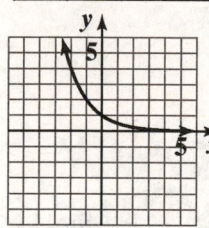

$$h(x) = \left(\frac{1}{2}\right)^x$$

17.

x	$f(x) = (0.6)^x$
−2	$(0.6)^{-2} = 2.\overline{7}$
−1	$(0.6)^{-1} = 1.\overline{6}$
0	$(0.6)^{0} = 1$
1	$(0.6)^{1} = 0.6$
2	$(0.6)^{2} = 0.36$

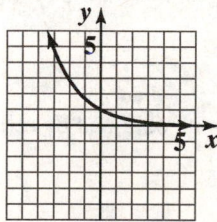

$$f(x) = (0.6)^x$$

19. This is the graph of $f(x) = 3^x$ reflected about the *x*-axis and about the *y*-axis, so the function is $H(x) = -3^{-x}$.

21. This is the graph of $f(x) = 3^x$ reflected about the *x*-axis, so the function is $F(x) = -3^x$.

23. This is the graph of $f(x) = 3^x$ shifted one unit downward, so the function is $h(x) = 3^x - 1$.

25. The graph of $g(x) = 2^{x+1}$ can be obtained by shifting the graph of $f(x) = 2^x$ one unit to the left.

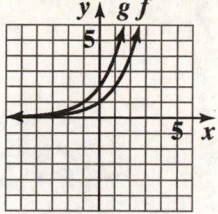

$$f(x) = 2^x$$
$$g(x) = 2^x + 1$$

asymptote: $y = 0$

domain: $(-\infty, \infty)$; range: $(0, \infty)$

27. The graph of $g(x) = 2^x - 1$ can be obtained by shifting the graph of $f(x) = 2^x$ downward one unit.

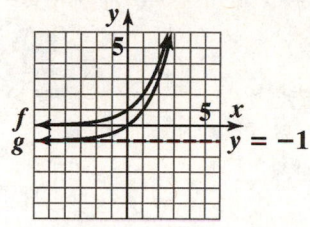

$$f(x) = 2^x$$
$$g(x) = 2^x - 1$$

asymptote: $y = -1$

domain: $(-\infty, \infty)$; range: $(-1, \infty)$

29. The graph of $h(x) = 2^{x+1} - 1$ can be obtained by shifting the graph of $f(x) = 2^x$ one unit to the left and one unit downward.

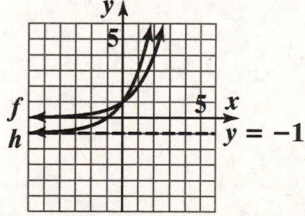

$$f(x) = 2^x$$
$$h(x) = 2^{x+1} - 1$$

asymptote: $y = -1$

domain: $(-\infty, \infty)$; range: $(-1, \infty)$

31. The graph of $g(x) = -2^x$ can be obtained by reflecting the graph of $f(x) = 2^x$ about the x-axis.

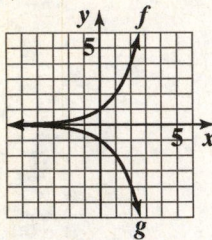

$f(x) = 2^x$
$g(x) = -2^x$

asymptote: $y = 0$

domain: $(-\infty, \infty)$; range: $(-\infty, 0)$

33. The graph of $g(x) = 2 \cdot 2^x$ can be obtained by vertically stretching the graph of $f(x) = 2^x$ by a factor of two.

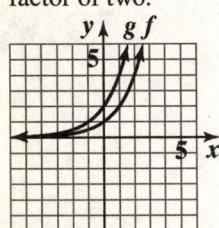

$f(x) = 2^x$
$g(x) = 2 \cdot 2^x$

asymptote: $y = 0$
domain: $(-\infty, \infty)$; range: $(0, \infty)$

35. The graph of $g(x) = e^{x-1}$ can be obtained by moving $f(x) = e^x$ 1 unit right.

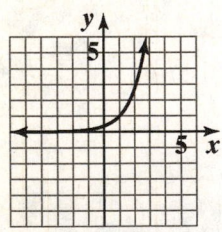

$g(x) = e^{x-1}$

asymptote: $y = 0$
domain: $(-\infty, \infty)$; range: $(0, \infty)$

37. The graph of $g(x) = e^x + 2$ can be obtained by moving $f(x) = e^x$ 2 units up.

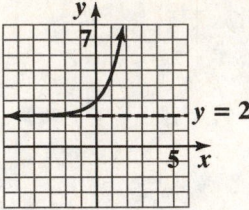

$g(x) = e^x + 2$

asymptote: $y = 2$
domain: $(-\infty, \infty)$; range: $(2, \infty)$

39. The graph of $h(x) = e^{x-1} + 2$ can be obtained by moving $f(x) = e^x$ 1 unit right and 2 units up.

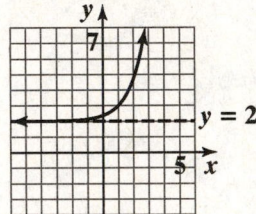

$h(x) = e^{x-1} + 2$

asymptote: $y = 2$
domain: $(-\infty, \infty)$; range: $(2, \infty)$

41. The graph of $h(x) = e^{-x}$ can be obtained by reflecting $f(x) = e^x$ about the y-axis.

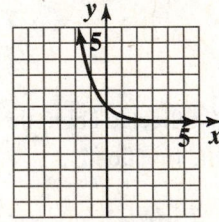

$h(x) = e^{-x}$

asymptote: $y = 0$
domain: $(-\infty, \infty)$; range: $(0, \infty)$

43. The graph of $g(x) = 2e^x$ can be obtained by stretching $f(x) = e^x$ vertically by a factor of 2.

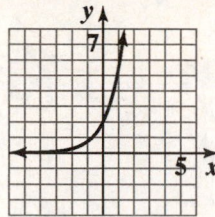

$$g(x) = 2e^x$$

asymptote: $y = 0$

domain: $(-\infty, \infty)$; range: $(0, \infty)$

45. The graph of $h(x) = e^{2x} + 1$ can be obtained by stretching $f(x) = e^x$ horizontally by a factor of 2 and then moving the graph up 1 unit.

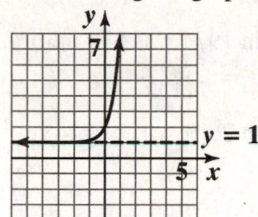

$$h(x) = e^{2x} + 1$$

asymptote: $y = 1$

domain: $(-\infty, \infty)$; range: $(1, \infty)$

47. The graph of $g(x)$ can be obtained by reflecting $f(x)$ about the y-axis.

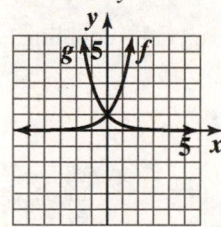

$$f(x) = 3^x$$
$$g(x) = 3^{-x}$$

asymptote of $f(x)$: $y = 0$
asymptote of $g(x)$: $y = 0$

49. The graph of $g(x)$ can be obtained by vertically shrinking $f(x)$ by a factor of $\frac{1}{3}$.

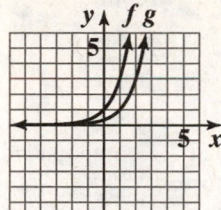

$$f(x) = 3^x$$
$$g(x) = \frac{1}{3} \cdot 3^x$$

asymptote of $f(x)$: $y = 0$
asymptote of $g(x)$: $y = 0$

51. The graph of $g(x)$ can be obtained by moving the graph of $f(x)$ one space to the right and one space up.

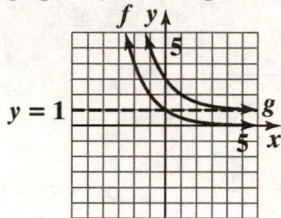

$$f(x) = \left(\frac{1}{2}\right)^x$$
$$g(x) = \left(\frac{1}{2}\right)^{x-1} + 1$$

asymptote of $f(x)$: $y = 0$
asymptote of $g(x)$: $y = 1$

53. a. $A = 10,000\left(1 + \dfrac{0.055}{2}\right)^{2(5)}$

       $\approx \$13,116.51$

b. $A = 10,000\left(1 + \dfrac{0.055}{4}\right)^{4(5)}$

       $\approx \$13,140.67$

c. $A = 10,000\left(1 + \dfrac{0.055}{12}\right)^{12(5)}$

       $\approx \$13,157.04$

d. $A = 10,000e^{0.055(5)}$

       $\approx \$13,165.31$

55. $A = 12,000\left(1 + \dfrac{0.07}{12}\right)^{12(3)}$

$\approx 14,795.11$ (7% yield)

$A = 12,000e^{0.0685(3)}$

$\approx 14,737.67$ (6.85% yield)

Investing \$12,000 for 3 years at 7% compounded monthly yields the greater return.

57.

x	$f(x) = 2^x$	$g(x) = 2^{-x}$
-2	$\dfrac{1}{4}$	4
-1	$\dfrac{1}{2}$	2
0	1	1
1	2	$\dfrac{1}{2}$
2	4	$\dfrac{1}{4}$

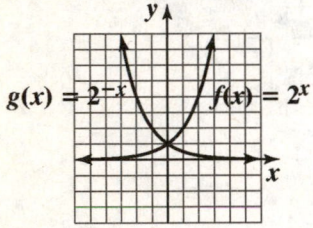

The point of intersection is $(0,1)$.

59.

x	$y = 2^x$
-2	$\dfrac{1}{4}$
-1	$\dfrac{1}{2}$
0	1
1	2
2	4

y	$x = 2^y$
-2	$\dfrac{1}{4}$
-1	$\dfrac{1}{2}$
0	1
1	2
2	4

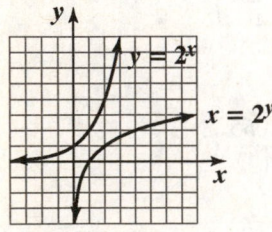

61. The graph is of the form $y = b^x$.

Substitute values from the point $(1, 4)$ to find b.

$y = b^x$

$4 = b^1$

$4 = b$

The equation of the graph is $y = 4^x$

63. The graph is of the form $y = -b^x$.

Substitute values from the point $(1, -e)$ to find b.

$y = -b^x$

$-e = -b^1$

$e = b$

The equation of the graph is $y = -e^x$

65. a. $f(0) = 574(1.026)^0$

$= 574(1) = 574$

India's population in 1974 was 574 million.

b. $f(27) = 574(1.026)^{27} \approx 1148$

India's population in 2001 will be 1148 million.

c. Since $2028 - 1974 = 54$, find

$f(54) = 574(1.026)^{54} \approx 2295$.

India's population in 2028 will be 2295 million.

d. $2055 - 1974 = 81$, find

$f(54) = 574(1.026)^{81} \approx 4590$.

India's population in 2055 will be 4590 million.

e. India's population appears to be doubling every 27 years.

67. $S = 465,000(1 + 0.06)^{10}$

$= 465,000(1.06)^{10} \approx \$832,744$

69. $2^{1.7} \approx 3.249009585$

$2^{1.73} \approx 3.317278183$

$2^{1.732} \approx 3.321880096$

$2^{1.73205} \approx 3.321995226$

$2^{1.7320508} \approx 3.321997068$

$2^{\sqrt{3}} \approx 3.321997085$

The closer the exponent is to $\sqrt{3}$, the closer the value is to $2^{\sqrt{3}}$.

71. **a.** 2005 is 50 years after 1955.
$$f(x) = 0.15x + 1.44$$
$$f(50) = 0.15(50) + 1.44 \approx 8.9$$
According to the linear model, there were about 8.9 million words in the federal tax code in 2005.

b. 2005 is 50 years after 1955.
$$g(x) = 1.87e^{0.0344x}$$
$$g(50) = 1.87e^{0.0344(50)} \approx 10.4$$
According to the exponential model, there were about 10.4 million words in the federal tax code in 2005.

c. The linear model is the better model for the data in 2005.

73. **a.** $f(0) = 80e^{-0.5(0)} + 20$
$$= 80e^0 + 20$$
$$= 80(1) + 20$$
$$= 100$$
100% of the material is remembered at the moment it is first learned.

b. $f(1) = 80e^{-0.5(1)} + 20 \approx 68.5$
68.5% of the material is remembered 1 week after it is first learned.

c. $f(4) = 80e^{-0.5(4)} + 20 \approx 30.8$
30.8% of the material is remembered 4 week after it is first learned.

d. $f(52) = 80e^{-0.5(52)} + 20 \approx 20$
20% of the material is remembered 1 year after it is first learned.

75. $f(x) = 6.19(1.029)^x$
$$f(56) = 6.19(1.029)^{56} \approx 30.7$$
$$g(x) = \frac{37.3}{1 + 6.1e^{-0.052x}}$$
$$g(56) = \frac{37.3}{1 + 6.1e^{-0.052(56)}} \approx 27.9$$
Function $g(x)$ is a better model for the graph's value of 28.0 in 2006.

77. – 79. Answers may vary.

81. **a.**
$$A = 10,000\left(1 + \frac{0.05}{4}\right)^{4t}$$
$$A = 10,000\left(1 + \frac{0.045}{12}\right)^{12t}$$

b. 5% compounded quarterly offers the better return.

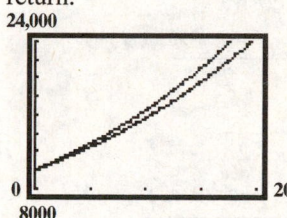

83. does not make sense; Explanations will vary. Sample explanation: The horizontal asymptote is $y = 0$.

85. does not make sense; Explanations will vary. Sample explanation: An exponential model is better than a linear model.

87. false; Changes to make the statement true will vary. A sample change is: The amount of money will not increase without bound.

89. false; Changes to make the statement true will vary. A sample change is: If $f(x) = 2^x$ then $f(a + b) = f(a) \cdot f(b)$.

91. $y = 3^x$ is (d). y increases as x increases, but not as quickly as $y = 5^x$. $y = 5^x$ is (c). $y = \left(\frac{1}{3}\right)^x$ is (a).

$y = \left(\frac{1}{3}\right)^x$ is the same as $y = 3^{-x}$, so it is (d) reflected about the y-axis. $y = \left(\frac{1}{5}\right)^x$ is (b). $y = \left(\frac{1}{5}\right)^x$ is the same as $y = 5^{-x}$, so it is (c) reflected about the y-axis.

93. **a.**
$$\cosh(-x) = \frac{e^{-x} + e^{-(-x)}}{2}$$
$$= \frac{e^{-x} + e^{x}}{2}$$
$$= \frac{e^{x} + e^{-x}}{2}$$
$$= \cosh x$$

b.
$$\sinh(-x) = \frac{e^{-x} - e^{-(-x)}}{2}$$
$$= \frac{e^{-x} - e^{x}}{2}$$
$$= \frac{-\left(-e^{-x} + e^{x}\right)}{2}$$
$$= -\frac{e^{x} - e^{-x}}{2}$$
$$= -\sinh x$$

c.
$$(\cosh x)^2 - (\sinh x)^2 \overset{?}{=} 1$$
$$\left(\frac{e^{x} + e^{-x}}{2}\right)^2 - \left(\frac{e^{x} - e^{-x}}{2}\right)^2 \overset{?}{=} 1$$
$$\frac{e^{2x} + 2 + e^{-2x}}{4} - \frac{e^{2x} - 2 + e^{-2x}}{4} \overset{?}{=} 1$$
$$\frac{e^{2x} + 2 + e^{-2x} - e^{2x} + 2 - e^{-2x}}{4} \overset{?}{=} 1$$
$$\frac{4}{4} \overset{?}{=} 1$$
$$1 = 1$$

94. We do not know how to solve $x = 2^y$ for y.

95. $\frac{1}{2}$; i.e. $25^{1/2} = 5$

96. $(x-3)^2 > 0$

Solving the related equation, $(x-3)^2 > 0$, gives
$x = 3$.
Note that the boundary value $x = 3$ does not satisfy
the inequality.
Testing each interval gives a solution set of
$(-\infty, 3) \cup (3, \infty)$.

Section 3.2

Check Point Exercises

1. **a.** $3 = \log_7 x$ means $7^3 = x$.

   b. $2 = \log_b 25$ means $b^2 = 25$.

   c. $\log_4 26 = y$ means $4^y = 26$.

2. **a.** $2^5 = x$ means $5 = \log_2 x$.

   b. $b^3 = 27$ means $3 = \log_b 27$.

   c. $e^y = 33$ means $y = \log_e 33$.

3. **a.** Question: 10 to what power gives 100?
   $\log_{10} 100 = 2$ because $10^2 = 100$.

   b. Question: 5 to what power gives $\frac{1}{125}$?
   $\log_5 \frac{1}{125} = -3$ because $5^{-3} = \frac{1}{5^3} = \frac{1}{125}$.

   c. Question: 36 to what power gives 6?
   $\log_{36} 6 = \frac{1}{2}$ because $36^{1/2} = \sqrt{36} = 6$

   d. Question: 3 to what power gives $\sqrt[7]{3}$?
   $\log_3 \sqrt[7]{3} = \frac{1}{7}$ because $3^{1/7} = \sqrt[7]{3}$.

4. **a.** Because $\log_b b = 1$, we conclude $\log_9 9 = 1$.

   b. Because $\log_b 1 = 0$, we conclude $\log_8 1 = 0$.

5. **a.** Because $\log_b b^x = x$, we conclude $\log_7 7^8 = 8$.

   b. Because $b^{\log_b x} = x$, we conclude $3^{\log_3 17} = 17$.

6. First, set up a table of coordinates for $f(x) = 3^x$.

x	-2	-1	0	1	2	3
$f(x) = 3^x$	$\frac{1}{9}$	$\frac{1}{3}$	1	3	9	27

Reversing these coordinates gives the coordinates for the inverse function $g(x) = \log_3 x$.

x	$\frac{1}{9}$	$\frac{1}{3}$	1	3	9	27
$g(x) = \log_3 x$	-2	-1	0	1	2	3

The graph of the inverse can also be drawn by reflecting the graph of $f(x) = 3^x$ about the line $y = x$.

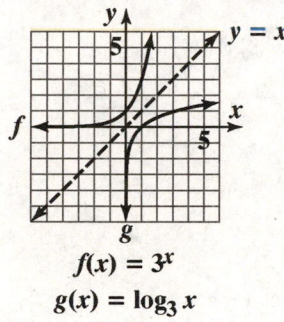

$f(x) = 3^x$
$g(x) = \log_3 x$

7. The domain of h consists of all x for which $x - 5 > 0$. Solving this inequality for x, we obtain $x > 5$. Thus, the domain of h is $(5, \infty)$.

8. Substitute the boy's age, 10, for x and evaluate the function at 10.
$$f(10) = 29 + 48.8 \log(10 + 1)$$
$$= 29 + 48.8 \log(11)$$
$$\approx 80$$
Thus, a 10-year-old boy is approximately 80% of his adult height.

9. Because $I = 10,000 \, I_0$,
$$R = \log \frac{10,000 I_0}{I_0}$$
$$= \log 10,000$$
$$= 4$$
The earthquake registered 4.0 on the Richter scale.

10. a. The domain of f consists of all x for which $4 - x > 0$. Solving this inequality for x, we obtain $x < 4$. Thus, the domain of f is $(-\infty, 4)$

b. The domain of g consists of all x for which $x^2 > 0$. Solving this inequality for x, we obtain $x < 0$ or $x > 0$. Thus the domain of g is $(-\infty, 0) \cup (0, \infty)$.

11. Find the temperature increase after 30 minutes by substituting 30 for x and evaluating the function at 30.
$$f(x) = 13.4 \ln x - 11.6$$
$$f(30) = 13.4 \ln 30 - 11.6$$
$$\approx 34$$
The function models the actual increase shown in the graph quite well.

Exercise Set 3.2

1. $2^4 = 16$

3. $3^2 = x$

5. $b^5 = 32$

7. $6^y = 216$

9. $\log_2 8 = 3$

11. $\log_2 \frac{1}{16} = -4$

13. $\log_8 2 = \frac{1}{3}$

15. $\log_{13} x = 2$

17. $\log_b 1000 = 3$

19. $\log_7 200 = y$

21. $\log_4 16 = 2$ because $4^2 = 16$.

23. $\log_2 64 = 6$ because $2^6 = 64$.

25. $\log_5 \frac{1}{5} = -1$ because $5^{-1} = \frac{1}{5}$.

27. $\log_2 \frac{1}{8} = -3$ because $2^{-3} = \frac{1}{8}$.

29. $\log_7 \sqrt{7} = \frac{1}{2}$ because $7^{\frac{1}{2}} = \sqrt{7}$.

31. $\log_2 \frac{1}{\sqrt{2}} = -\frac{1}{2}$ because $2^{-\frac{1}{2}} = \frac{1}{\sqrt{2}}$.

33. $\log_{64} 8 = \frac{1}{2}$ because $64^{\frac{1}{2}} = \sqrt{64} = 8$.

35. Because $\log_b b = 1$, we conclude $\log_5 5 = 1$.

37. Because $\log_b 1 = 0$, we conclude $\log_4 1 = 0$.

39. Because $\log_b b^x = x$, we conclude $\log_5 5^7 = 7$.

41. Because $b^{\log_b x} = x$, we conclude $8^{\log_8 19} = 19$.

43. First, set up a table of coordinates for $f(x) = 4^x$.

x	-2	-1	0	1	2	3
$f(x) = 4x$	$\frac{1}{16}$	$\frac{1}{4}$	1	4	16	64

Reversing these coordinates gives the coordinates for the inverse function $g(x) = \log_4 x$.

x	$\frac{1}{16}$	$\frac{1}{4}$	1	4	16	64
$g(x) = \log_{4x}$	-2	-1	0	1	2	3

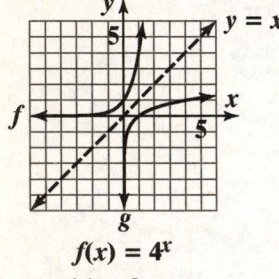

$f(x) = 4^x$

$g(x) = \log_4 x$

45. First, set up a table of coordinates for $f(x) = \left(\frac{1}{2}\right)^x$.

x	-2	-1	0	1	2	3
$f(x) = \left(\frac{1}{2}\right)^x$	4	2	1	$\frac{1}{2}$	$\frac{1}{4}$	$\frac{1}{8}$

Reversing these coordinates gives the coordinates for the inverse function $g(x) = \log_{1/2} x$.

x	4	2	1	$\frac{1}{2}$	$\frac{1}{4}$	$\frac{1}{8}$
$g(x) = \log_{1/2} x$	-2	-1	0	1	2	3

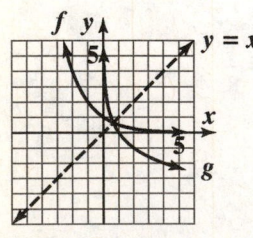

$f(x) = \left(\frac{1}{2}\right)^x$

$g(x) = \log_{1/2} x$

47. This is the graph of $f(x) = \log_3 x$ reflected about the x-axis and shifted up one unit, so the function is
$H(x) = 1 - \log_3 x$.

49. This is the graph of $f(x) = \log_3 x$ shifted down one unit, so the function is $h(x) = \log_3 x - 1$.

51. This is the graph of $f(x) = \log_3 x$ shifted right one unit, so the function is $g(x) = \log_3 (x - 1)$.

53.

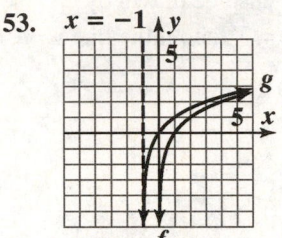

$f(x) = \log_2 x$

$g(x) = \log_2 (x+1)$

vertical asymptote: $x = -1$
domain: $(-1, \infty)$; range: $(-\infty, \infty)$

55.

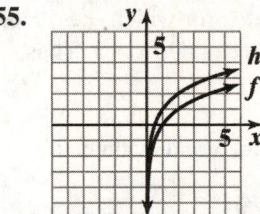

$f(x) = \log_2 x$

$h(x) = 1 + \log_2 x$

vertical asymptote: $x = 0$
domain: $(0, \infty)$; range: $(-\infty, \infty)$

57.

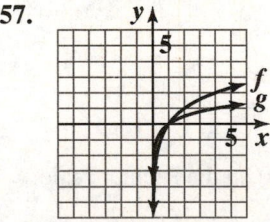

$f(x) = \log_2 x$

$g(x) = \frac{1}{2} \log_2 x$

vertical asymptote: $x = 0$
domain: $(0, \infty)$; range: $(-\infty, \infty)$

59.

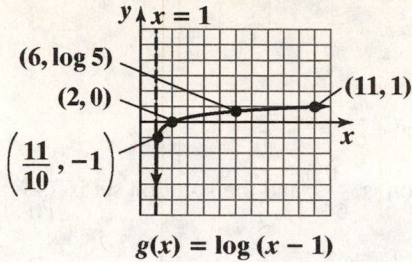

$$g(x) = \log(x - 1)$$

vertical asymptote: $x = 1$
domain: $(1, \infty)$; range: $(-\infty, \infty)$

61.

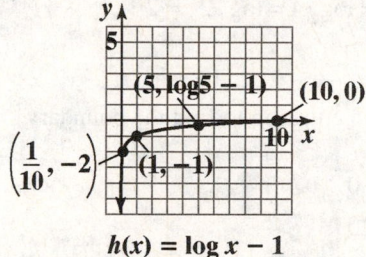

$$h(x) = \log x - 1$$

vertical asymptote: $x = 0$
domain: $(0, \infty)$; range: $(-\infty, \infty)$

63.

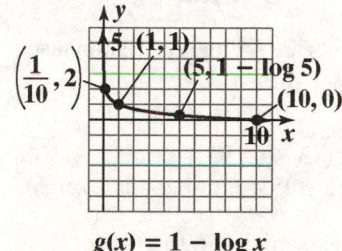

$$g(x) = 1 - \log x$$

vertical asymptote: $x = 0$
domain: $(0, \infty)$; range: $(-\infty, \infty)$

65.

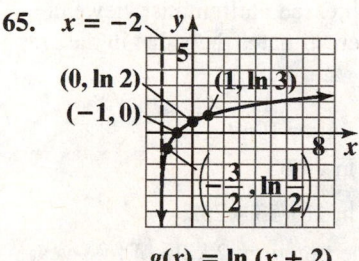

$$g(x) = \ln(x + 2)$$

vertical asymptote: $x = -2$
domain: $(-2, \infty)$; range: $(-\infty, \infty)$

67.

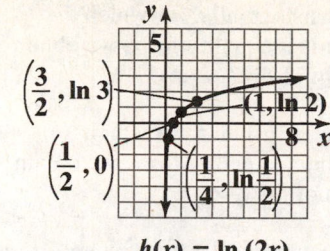

$$h(x) = \ln(2x)$$

vertical asymptote: $x = 0$
domain: $(0, \infty)$; range: $(-\infty, \infty)$

69.

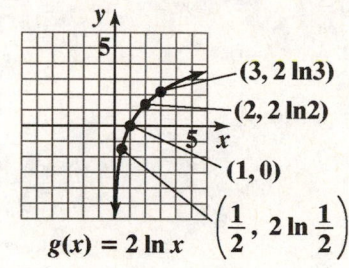

$$g(x) = 2 \ln x$$

vertical asymptote: $x = 0$
domain: $(0, \infty)$; range: $(-\infty, \infty)$

71.

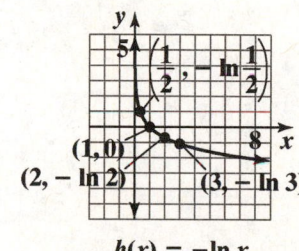

$$h(x) = -\ln x$$

vertical asymptote: $x = 0$
domain: $(0, \infty)$; range: $(-\infty, \infty)$

73.

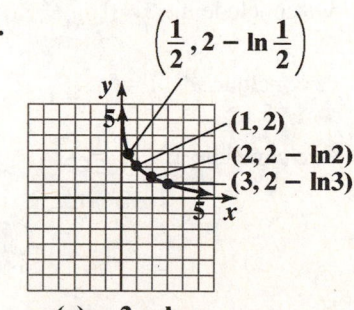

$$g(x) = 2 - \ln x$$

vertical asymptote: $x = 0$
domain: $(0, \infty)$; range: $(-\infty, \infty)$

75. The domain of f consists of all x for which $x + 4 > 0$. Solving this inequality for x, we obtain $x > -4$. Thus, the domain of f is $(-4, \infty)$.

77. The domain of f consists of all x for which $2 - x > 0$. Solving this inequality for x, we obtain $x < 2$. Thus, the domain of f is $(-\infty, 2)$.

79. The domain of f consists of all x for which $(x - 2)^2 > 0$. Solving this inequality for x, we obtain $x < 2$ or $x > 2$. Thus, the domain of f is $(-\infty, 2)$ or $(2, \infty)$.

81. $\log 100 = \log_{10} 100 = 2$

because $10^2 = 100$.

83. Because $\log 10^x = x$, we

conclude $\log 10^7 = 7$.

85. Because $10^{\log x} = x$, we

conclude $10^{\log 33} = 33$.

87. $\ln 1 = 0$ because $e^0 = 1$.

89. Because $\ln e^x = x$, we

conclude $\ln e^6 = 6$.

91. $\ln \dfrac{1}{e^6} = \ln e^{-6}$

Because $\ln e^x = x$ we conclude

$\ln e^{-6} = -6$, so $\ln \dfrac{1}{e^6} = -6$.

93. Because $e^{\ln x} = x$, we conclude $e^{\ln 125} = 125$.

95. Because $\ln e^x = x$, we conclude $\ln e^{9x} = 9x$.

97. Because $e^{\ln x} = x$, we conclude $e^{\ln 5x^2} = 5x^2$.

99. Because $10^{\log x} = x$, we conclude $10^{\log \sqrt{x}} = \sqrt{x}$.

101. $\log_3 (x - 1) = 2$

$$3^2 = x - 1$$
$$9 = x - 1$$
$$10 = x$$

The solution is 10, and the solution set is $\{10\}$.

103. $\log_4 x = -3$

$$4^{-3} = x$$

$$x = \frac{1}{4^3} = \frac{1}{64}$$

The solution is $\dfrac{1}{64}$, and the solution set is $\left\{ \dfrac{1}{64} \right\}$.

105. $\log_3 \left(\log_7 7 \right) = \log_3 1 = 0$

107. $\log_2 \left(\log_3 81 \right) = \log_2 \left(\log_3 3^4 \right)$

$$= \log_2 4 = \log_2 2^2 = 2$$

109. For $f(x) = \ln(x^2 - x - 2)$ to be real, $x^2 - x - 2 > 0$. Solve the related equation to find the boundary points:

$$x^2 - x - 2 = 0$$
$$(x + 1)(x - 2) = 0$$

The boundary points are -1 and 2. Testing each interval gives a domain of $(-\infty, -1) \cup (2, \infty)$.

111. For $f(x) = \ln \left(\dfrac{x + 1}{x - 5} \right)$ to be real, $\dfrac{x + 1}{x - 5} > 0$.

The boundary points are -1 and 5. Testing each interval gives a domain of $(-\infty, -1) \cup (5, \infty)$.

113. $f(13) = 62 + 35 \log(13 - 4) \approx 95.4$
She is approximately 95.4% of her adult height.

115. a. 2004 is 35 years after 1969.

$$f(x) = -7.49 \ln x + 53$$
$$f(35) = -7.49 \ln 35 + 53 \approx 26.4$$

According to the function, 26.4% of first-year college men expressed antifeminist views in 2004. This underestimates the value in the graph by 1%.

b. 2010 is 41 years after 1969.

$$f(x) = -7.49 \ln x + 53$$
$$f(41) = -7.49 \ln 41 + 53 \approx 25.2$$

According to the function, 25.2% of first-year college men will express antifeminist views in 2010.

117. $D = 10 \log \left[10^{12} (6.3 \times 10^6) \right] \approx 188$

Yes, the sound can rupture the human eardrum.

119. a. $f(0) = 88 - 15\ln(0 + 1) = 88$
The average score on the original exam was 88.

b. $f(2) = 88 - 15\ln(2 + 1) = 71.5$
$f(4) = 88 - 15\ln(4 + 1) = 63.9$
$f(6) = 88 - 15\ln(6 + 1) = 58.8$
$f(8) = 88 - 15\ln(8 + 1) = 55$
$f(10) = 88 - 15\ln(10 + 1) = 52$
$f(12) = 88 - 15\ln(12 + 1) = 49.5$

The average score after 2 months was about 71.5, after 4 months was about 63.9, after 6 months was about 58.8, after 8 months was about 55, after 10 months was about 52, and after one year was about 49.5.

121. – 127. Answers may vary.

129.
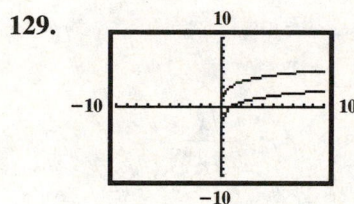

$g(x)$ is $f(x)$ shifted 3 units upward.

131.
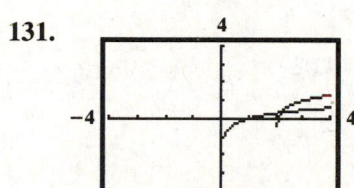

$g(x)$ is $f(x)$ shifted right 2 units and upward 1 unit.

133. a.
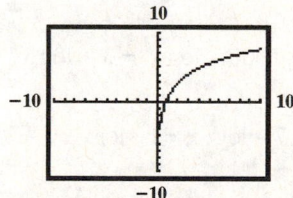

b.
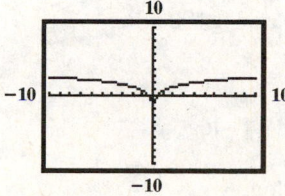

c.
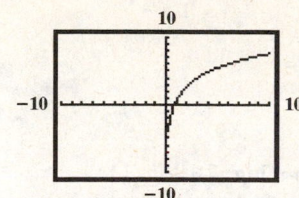

d They are the same.
$\log_b MN = \log_b M + \log_b N$

e. The sum of the logarithms of its factors.

135. makes sense

137. makes sense

139. false; Changes to make the statement true will vary.
A sample change is: $\dfrac{\log_2 8}{\log_2 4} = \dfrac{3}{2}$

141. false; Changes to make the statement true will vary.
A sample change is: The domain of $f(x) = \log_2 x$ is $(0, \infty)$.

143. $\dfrac{\log_3 81 - \log_\pi 1}{\log_{2\sqrt{2}} 8 - \log 0.001} = \dfrac{4 - 0}{2 - (-3)} = \dfrac{4}{5}$

145. $\log_4 60 < \log_4 64 = 3$ so $\log_4 60 < 3$.
$\log_3 40 > \log_3 27 = 3$ so $\log_3 40 > 3$.
$\log_4 60 < 3 < \log_3 40$
$\log_3 40 > \log_4 60$

147. a. $\log_2 32 = \log_2 2^5 = 5$

b. $\log_2 8 + \log_2 4 = \log_2 2^3 + \log_2 2^2 = 3 + 2 = 5$

c. $\log_2(8 \cdot 4) = \log_2 8 + \log_2 4$

148. a. $\log_2 16 = \log_2 2^4 = 4$

b. $\log_2 32 - \log_2 2 = \log_2 2^5 - \log_2 2 = 5 - 1 = 4$

c. $\log_2\left(\dfrac{32}{2}\right) = \log_2 32 - \log_2 2$

149. a. $\log_3 81 = \log_3 3^4 = 4$

b. $2\log_3 9 = 2\log_3 3^2 = 2 \cdot 2 = 4$

c. $\log_3 9^2 = 2\log_3 9$

Section 3.3

Check Point Exercises

1. **a.** $\log_6(7 \cdot 11) = \log_6 7 + \log_6 11$

b. $\log(100x) = \log 100 + \log x$
$= 2 + \log x$

2. **a.** $\log_8\left(\dfrac{23}{x}\right) = \log_8 23 - \log_8 x$

b. $\ln\left(\dfrac{e^5}{11}\right) = \ln e^5 - \ln 11$
$= 5 - \ln 11$

3. **a.** $\log_6 3^9 = 9\log_6 3$

b. $\ln \sqrt[3]{x} = \ln x^{1/3} = \dfrac{1}{3}\ln x$

c. $\log(x+4)^2 = 2\log(x+4)$

4. **a.** $\log_b x^4 \sqrt[3]{y}$
$= \log_b x^4 y^{1/3}$
$= \log_b x^4 + \log_b y^{1/3}$
$= 4\log_b x + \dfrac{1}{3}\log_b y$

b. $\log_5 \dfrac{\sqrt{x}}{25y^3}$
$= \log_5 \dfrac{x^{1/2}}{25y^3}$
$= \log_5 x^{1/2} - \log_5 25y^3$
$= \log_5 x^{1/2} - \left(\log_5 5^2 + \log_5 y^3\right)$
$= \tfrac{1}{2}\log_5 x - \log_5 5^2 - \log_5 y^3$
$= \tfrac{1}{2}\log_5 x - 2\log_5 5 - 3\log_5 y$
$= \tfrac{1}{2}\log_5 x - 2 - 3\log_5 y$

5. **a.** $\log 25 + \log 4 = \log(25 \cdot 4) = \log 100 = 2$

b. $\log(7x+6) - \log x = \log \dfrac{7x+6}{x}$

6. **a.** $\ln x^2 + \dfrac{1}{3}\ln(x+5)$
$= \ln x^2 + \ln(x+5)^{1/3}$
$= \ln x^2(x+5)^{1/3}$
$= \ln x^2 \sqrt[3]{x+5}$

b. $2\log(x-3) - \log x$
$= \log(x-3)^2 - \log x$
$= \log \dfrac{(x-3)^2}{x}$

c. $\dfrac{1}{4}\log_b x - 2\log_b 5 - 10\log_b y$
$= \log_b x^{1/4} - \log_b 5^2 - \log_b y^{10}$
$= \log_b x^{1/4} - \left(\log_b 25 - \log_b y^{10}\right)$
$= \log_b x^{1/4} - \log_b 25y^{10}$
$= \log_b \dfrac{x^{1/4}}{25y^{10}}$ or $\log_b \dfrac{\sqrt[4]{x}}{25y^{10}}$

7. $\log_7 2506 = \dfrac{\log 2506}{\log 7} \approx 4.02$

8. $\log_7 2506 = \dfrac{\ln 2506}{\ln 7} \approx 4.02$

Exercise Set 3.3

1. $\log_5(7 \cdot 3) = \log_5 7 + \log_5 3$

3. $\log_7(7x) = \log_7 7 + \log_7 x = 1 + \log_7 x$

5. $\log(1000x) = \log 1000 + \log x = 3 + \log x$

7. $\log_7\left(\dfrac{7}{x}\right) = \log_7 7 - \log_7 x = 1 - \log_7 x$

9. $\log\left(\dfrac{x}{100}\right) = \log x - \log 100 = \log_x - 2$

11. $\log_4\left(\dfrac{64}{y}\right) = \log_4 64 - \log_4 y$
$= 3 - \log_4 y$

13. $\ln\left(\dfrac{e^2}{5}\right) = \ln e^2 - \ln 5 = 2\ln e - \ln 5 = 2 - \ln 5$

15. $\log_b x^3 = 3\log_b x$

17. $\log N^{-6} = -6\log N$

19. $\ln \sqrt[5]{x} = \ln x^{(1/5)} = \dfrac{1}{5}\ln x$

21. $\log_b x^2 y = \log_b x^2 + \log_b y = 2\log_b x + \log_b y$

23. $\log_4\left(\dfrac{\sqrt{x}}{64}\right) = \log_4 x^{1/2} - \log_4 64 = \dfrac{1}{2}\log_4 x - 3$

25. $\log_6\left(\dfrac{36}{\sqrt{x+1}}\right) = \log_6 36 - \log_6 (x+1)^{1/2}$

$\qquad\qquad = 2 - \dfrac{1}{2}\log_6(x+1)$

27. $\log_b\left(\dfrac{x^2 y}{z^2}\right) = \log_b\left(x^2 y\right) - \log_b z^2$

$\qquad\qquad = \log_b x^2 + \log_b y - \log_b z^2$

$\qquad\qquad = 2\log_b x + \log_b y - 2\log_b z$

29. $\log\sqrt{100x} = \log(100x)^{1/2}$

$\qquad\qquad = \dfrac{1}{2}\log(100x)$

$\qquad\qquad = \dfrac{1}{2}(\log 100 + \log x)$

$\qquad\qquad = \dfrac{1}{2}(2 + \log x)$

$\qquad\qquad = 1 + \dfrac{1}{2}\log x$

31. $\log\sqrt[3]{\dfrac{x}{y}} = \log\left(\dfrac{x}{y}\right)^{1/3}$

$\qquad\qquad = \dfrac{1}{3}\left[\log\left(\dfrac{x}{y}\right)\right]$

$\qquad\qquad = \dfrac{1}{3}(\log x - \log y)$

$\qquad\qquad = \dfrac{1}{3}\log x - \dfrac{1}{3}\log y$

33. $\log_b \dfrac{\sqrt{x}\,y^3}{z^3}$

$\qquad = \log_b x^{1/2} + \log_b y^3 - \log_b z^3$

$\qquad = \dfrac{1}{2}\log_b x + 3\log_b y - 3\log_b z$

35.

$\log_5 \sqrt[3]{\dfrac{x^2 y}{25}}$

$= \log_5 x^{2/3} + \log_5 y^{1/3} - \log_5 25^{1/3}$

$= \dfrac{2}{3}\log_5 x + \dfrac{1}{3}\log_5 y - \log_5 5^{2/3}$

$= \dfrac{2}{3}\log_5 x + \dfrac{1}{3}\log_5 y - \dfrac{2}{3}$

37. $\ln\left[\dfrac{x^3\sqrt{x^2+1}}{(x+1)^4}\right]$

$= \ln x^3 + \ln\sqrt{x^2+1} - \ln(x+1)^4$

$= 3\ln x + \dfrac{1}{2}\ln(x^2+1) - 4\ln(x+1)$

39. $\log\left[\dfrac{10x^2\sqrt[3]{1-x}}{7(x+1)^2}\right]$

$= \log 10 + \log x^2 + \log\sqrt[3]{1-x} - \log 7 - \log(x+1)^2$

$= 1 + 2\log x + \dfrac{1}{3}\log(1-x) - \log 7 - 2\log(x+1)$

41. $\log 5 + \log 2 = \log(5\cdot 2) = \log 10 = 1$

43. $\ln x + \ln 7 = \ln(7x)$

45. $\log_2 96 - \log_2 3 = \log_2\left(\dfrac{96}{3}\right) = \log_2 32 = 5$

47. $\log(2x+5) - \log x = \log\left(\dfrac{2x+5}{x}\right)$

49. $\log x + 3\log y = \log x + \log y^3 = \log(xy^3)$

51. $\dfrac{1}{2}\ln x + \ln y = \ln x^{1/2} + \ln y$

$\qquad\qquad = \ln\left(x^{\frac{1}{2}}y\right) \text{ or } \ln\left(y\sqrt{x}\right)$

53. $2\log_b x + 3\log_b y = \log_b x^2 + \log_b y^3$

$\qquad\qquad = \log_b(x^2 y^3)$

55. $5\ln x - 2\ln y = \ln x^5 - \ln y^2 = \ln\left(\dfrac{x^5}{y^2}\right)$

57. $3\ln x - \dfrac{1}{3}\ln y = \ln x^3 - \ln y^{1/3}$

$$= \ln\left(\dfrac{x^3}{y^{1/3}}\right) \text{ or } \ln\left(\dfrac{x^3}{\sqrt[3]{y}}\right)$$

59. $4\ln(x+6) - 3\ln x = \ln(x+6)^4 - \ln x^3$

$$= \ln\dfrac{(x+6)^4}{x^3}$$

61. $3\ln x + 5\ln y - 6\ln z$

$$= \ln x^3 + \ln y^5 - \ln z^6$$

$$= \ln\dfrac{x^3 y^5}{z^6}$$

63. $\dfrac{1}{2}\left(\log x + \log y\right)$

$$= \dfrac{1}{2}\left(\log xy\right)$$

$$= \log(xy)^{1/2}$$

$$= \log\sqrt{xy}$$

65. $\dfrac{1}{2}\left(\log_5 x + \log_5 y\right) - 2\log_5(x+1)$

$$= \dfrac{1}{2}\log_5 xy - \log_5(x+1)^2$$

$$= \log_5(xy)^{1/2} - \log_5(x+1)^2$$

$$= \log_5\dfrac{(xy)^{1/2}}{(x+1)^2}$$

$$= \log_5\dfrac{\sqrt{xy}}{(x+1)^2}$$

67. $\dfrac{1}{3}[2\ln(x+5) - \ln x - \ln(x^2-4)]$

$$= \dfrac{1}{3}[\ln(x+5)^2 - \ln x - \ln(x^2-4)]$$

$$= \dfrac{1}{3}\left[\ln\dfrac{(x+5)^2}{x(x^2-4)}\right]$$

$$= \ln\left[\dfrac{(x+5)^2}{x(x^2-4)}\right]^{1/3}$$

$$= \ln\sqrt[3]{\dfrac{(x+5)^2}{x(x^2-4)}}$$

69. $\log x + \log(x^2-1) - \log 7 - \log(x+1)$

$$= \log x + \log(x^2-1) - (\log 7 + \log(x+1))$$

$$= \log(x(x^2-1)) - \log(7(x+1))$$

$$= \log\dfrac{x(x^2-1)}{7(x+1)}$$

$$= \log\dfrac{x(x+1)(x-1)}{7(x+1)}$$

$$= \log\dfrac{x(x-1)}{7}$$

71. $\log_5 13 = \dfrac{\log 13}{\log 5} \approx 1.5937$

73. $\log_{14} 87.5 = \dfrac{\ln 87.5}{\ln 14} \approx 1.6944$

75. $\log_{0.1} 17 = \dfrac{\log 17}{\log 0.1} \approx -1.2304$

77. $\log_\pi 63 = \dfrac{\ln 63}{\ln \pi} \approx 3.6193$

79. $y = \log_3 x = \dfrac{\log x}{\log 3}$

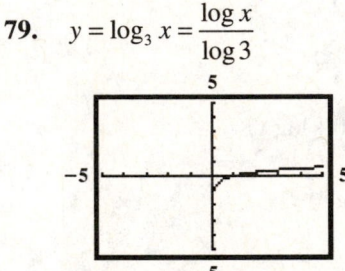

81. $y = \log_2(x+2) = \dfrac{\log(x+2)}{\log 2}$

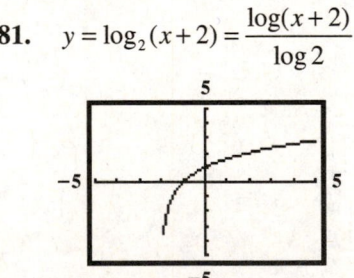

83. $\log_b \dfrac{3}{2} = \log_b 3 - \log_b 2 = C - A$

85. $\log_b 8 = \log_b 2^3 = 3\log_b 2 = 3A$

87. $\log_b \sqrt{\dfrac{2}{27}} = \log_b \left(\dfrac{2}{27}\right)^{\frac{1}{2}}$

$= \dfrac{1}{2}\log_b \left(\dfrac{2}{3^3}\right)$

$= \dfrac{1}{2}\left(\log_b 2 - \log_b 3^3\right)$

$= \dfrac{1}{2}\left(\log_b 2 - 3\log_b 3\right)$

$= \dfrac{1}{2}\log_b 2 - \dfrac{3}{2}\log_b 3$

$= \dfrac{1}{2}A - \dfrac{3}{2}C$

89. false; $\ln e = 1$

91. false; $\log_4 (2x)^3 = 3\log_4 (2x)$

93. true; $x\log 10^x = x \cdot x = x^2$

95. true; $\ln(5x) + \ln 1 = \ln 5x + 0 = \ln 5x$

97. false; $\log(x+3) - \log(2x) = \log\dfrac{x+3}{2x}$

99. true; quotient rule

101. true; $\log_3 7 = \dfrac{\log 7}{\log 3} = \dfrac{1}{\frac{\log 3}{\log 7}} = \dfrac{1}{\log_7 3}$

103. a. $D = 10\log\left(\dfrac{I}{I_0}\right)$

b. $D_1 = 10\log\left(\dfrac{100I}{I_0}\right)$

$= 10\log\left(100I - I_0\right)$

$= 10\log 100 + 10\log I - 10\log I_0$

$= 10(2) + 10\log I - 10\log I_0$

$= 20 + 10\log\left(\dfrac{I}{I_0}\right)$

This is 20 more than the loudness level of the softer sound. This means that the 100 times louder sound will be 20 decibels louder.

105. – 111. Answers may vary.

113. a. $y = \log_3 x = \dfrac{\ln x}{\ln 3}$

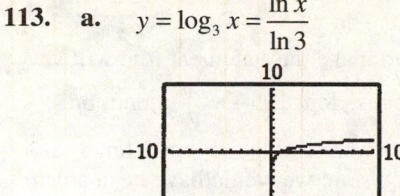

b.

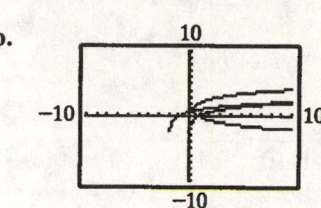

To obtain the graph of $y = 2 + \log 3x$, shift the graph of $y = \log 3x$ two units upward. To obtain the graph of $y = \log 3(x + 2)$, shift the graph of $y = \log 3x$ two units left. To obtain the graph of $y = -\log 3x$, reflect the graph of $y = \log 3x$ about the x-axis.

115. $\log_3 x = \dfrac{\log x}{\log 3}$;

$\log_{25} x = \dfrac{\log x}{\log 25}$;

$\log_{100} x = \dfrac{\log x}{\log 100}$

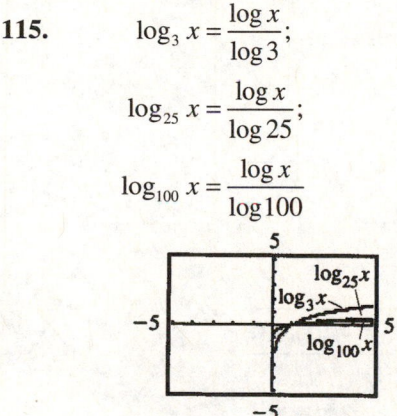

a. top graph: $y = \log_{100} x$
 bottom graph: $y = \log_3 x$

b. top graph: $y = \log_3 x$
 bottom graph: $y = \log_{100} x$

c. Comparing graphs of $\log_b x$ for $b > 1$, the graph of the equation with the largest b will be on the top in the interval $(0, 1)$ and on the bottom in the interval $(1, \infty)$.

117. – 119. Answers may vary.

121. makes sense

123. makes sense

125. true

127. false; Changes to make the statement true will vary.

A sample change is: $\log_b\left(x^3 + y^3\right)$ cannot be simplified. If we were taking the logarithm of a product and not a sum, we would have been able to simplify as follows.

$$\log_b\left(x^3 y^3\right) = \log_b x^3 + \log_b y^3$$
$$= 3\log_b x + 3\log_b y$$

129. $\log e = \log_{10} e = \dfrac{\ln e}{\ln 10} = \dfrac{1}{\ln 10}$

131. $e^{\ln 8x^5 - \ln 2x^2} = e^{\ln\left(\frac{8x^5}{2x^2}\right)} = e^{\ln\left(4x^3\right)} = 4x^3$

133. $a(x-2) = b(2x+3)$
$ax - 2a = 2bx + 3b$
$ax - 2bx = 2a + 3b$
$x(a - 2b) = 2a + 3b$
$x = \dfrac{2a + 3b}{a - 2b}$

134. $x(x-7) = 3$
$x^2 - 7x = 3$
$x^2 - 7x - 3 = 0$
$x = \dfrac{-b \pm \sqrt{b^2 - 4ac}}{2a}$
$x = \dfrac{-(-7) \pm \sqrt{(-7)^2 - 4(1)(-3)}}{2(1)}$
$x = \dfrac{7 \pm \sqrt{61}}{2}$

The solution set is $\left\{\dfrac{7 \pm \sqrt{61}}{2}\right\}$.

135.
$$\dfrac{x+2}{4x+3} = \dfrac{1}{x}$$
$$x(4x+3)\left(\dfrac{x+2}{4x+3}\right) = x(4x+3)\left(\dfrac{1}{x}\right)$$
$$x(x+2) = 4x+3$$
$$x^2 + 2x = 4x+3$$
$$x^2 - 2x - 3 = 0$$
$$(x+1)(x-3) = 0$$
$$x + 1 = 0 \quad \text{or} \quad x - 3 = 0$$
$$x = -1 \qquad\qquad x = 3$$

The solution set is $\{-1, 3\}$.

Mid-Chapter 3 Check Point

1.

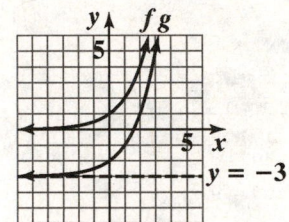

$f(x) = 2^x$
$g(x) = 2^x - 3$

asymptote of f: $y = 0$
asymptote of g: $y = -3$
domain of f = domain of $g = (-\infty, \infty)$
range of $f = (0, \infty)$
range of $g = (-3, \infty)$

2.

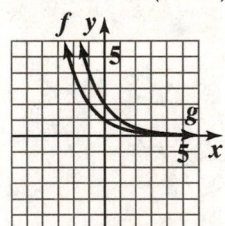

$f(x) = \left(\dfrac{1}{2}\right)^x$

$g(x) = \left(\dfrac{1}{2}\right)^{x-1}$

asymptote of f: $y = 0$
asymptote of g: $y = 0$
domain of f = domain of $g = (-\infty, \infty)$
range of f = range of $g = (0, \infty)$

284

3.

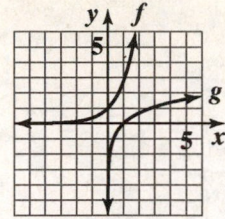

$$f(x) = e^x$$
$$g(x) = \ln x$$

asymptote of f: $y = 0$

asymptote of g: $x = 0$

domain of f = range of g = $(-\infty, \infty)$

range of f = domain of g = $(0, \infty)$

4.

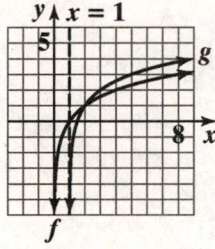

$$f(x) = \log_2 x$$
$$g(x) = \log_2 (x - 1) + 1$$

asymptote of f: $x = 0$

asymptote of g: $x = 1$

domain of f = $(0, \infty)$

domain of g = $(1, \infty)$

range of f = range of g = $(-\infty, \infty)$

5.

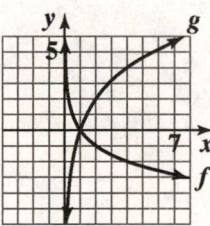

$$f(x) = \log_{1/2} x$$
$$g(x) = -2 \log_{1/2} x$$

asymptote of f: $x = 0$

asymptote of g: $x = 0$

domain of f = domain of g = $(0, \infty)$

range of f = range of g = $(-\infty, \infty)$

6. $f(x) = \log_3 (x + 6)$

The argument of the logarithm must be positive:
$x + 6 > 0$

$x > -6$

domain: $\{x \mid x > -6\}$ or $(-6, \infty)$.

7. $f(x) = \log_3 x + 6$

The argument of the logarithm must be positive:
$x > 0$

domain: $\{x \mid x > 0\}$ or $(0, \infty)$.

8. $\log_3 (x + 6)^2$

The argument of the logarithm must be positive.

Now $(x + 6)^2$ is always positive, except when

$x = -6$

domain: $\{x \mid x \neq 0\}$ or $(-\infty, -6) \cup (-6, \infty)$.

9. $f(x) = 3^{x+6}$

domain: $\{x \mid x \text{ is a real number}\}$ or $(-\infty, \infty)$.

10. $\log_2 8 + \log_5 25 = \log_2 2^3 + \log_5 5^2$
$$= 3 + 2 = 5$$

11. $\log_3 \dfrac{1}{9} = \log_3 \dfrac{1}{3^2} = \log_3 3^{-2} = -2$

12. Let $\log_{100} 10 = y$
$$100^y = 10$$
$$(10^2)^y = 10^1$$
$$10^{2y} = 10^1$$
$$2y = 1$$
$$y = \dfrac{1}{2}$$

13. $\log \sqrt[3]{10} = \log 10^{\frac{1}{3}} = \dfrac{1}{3}$

14. $\log_2 (\log_3 81) = \log_2 (\log_3 3^4)$
$$= \log_2 4 = \log_2 2^2 = 2$$

15. $\log_3\left(\log_2\dfrac{1}{8}\right) = \log_3\left(\log_2\dfrac{1}{2^3}\right)$

$\qquad\qquad = \log_3\left(\log_2 2^{-3}\right)$

$\qquad\qquad = \log_3(-3)$

$\qquad\qquad = \text{not possible}$

This expression is impossible to evaluate.

16. $6^{\log_6 5} = 5$

17. $\ln e^{\sqrt{7}} = \sqrt{7}$

18. $10^{\log 13} = 13$

19. $\log_{100} 0.1 = y$

$\qquad 100^y = 0.1$

$\qquad \left(10^2\right)^y = \dfrac{1}{10}$

$\qquad 10^{2y} = 10^{-1}$

$\qquad 2y = -1$

$\qquad y = -\dfrac{1}{2}$

20. $\log_\pi \pi^{\sqrt{\pi}} = \sqrt{\pi}$

21. $\log\left(\dfrac{\sqrt{xy}}{1000}\right) = \log\left(\sqrt{xy}\right) - \log 1000$

$\qquad\qquad = \log(xy)^{\frac{1}{2}} - \log 10^3$

$\qquad\qquad = \dfrac{1}{2}\log(xy) - 3$

$\qquad\qquad = \dfrac{1}{2}(\log x + \log y) - 3$

$\qquad\qquad = \dfrac{1}{2}\log x + \dfrac{1}{2}\log y - 3$

22. $\ln\left(e^{19}x^{20}\right) = \ln e^{19} + \ln x^{20}$

$\qquad\qquad = 19 + 20\ln x$

23. $8\log_7 x - \dfrac{1}{3}\log_7 y = \log_7 x^8 - \log_7 y^{\frac{1}{3}}$

$\qquad\qquad = \log_7\left(\dfrac{x^8}{y^{\frac{1}{3}}}\right)$

$\qquad\qquad = \log_7\left(\dfrac{x^8}{\sqrt[3]{y}}\right)$

24. $7\log_5 x + 2\log_5 x = \log_5 x^7 + \log_5 x^2$

$\qquad\qquad = \log_5\left(x^7 \cdot x^2\right)$

$\qquad\qquad = \log_5 x^9$

25. $\dfrac{1}{2}\ln x - 3\ln y - \ln(z-2)$

$\qquad = \ln x^{\frac{1}{2}} - \ln y^3 - \ln(z-2)$

$\qquad = \ln\sqrt{x} - \left[\ln y^3 + \ln(z-2)\right]$

$\qquad = \ln\sqrt{x} - \ln\left[y^3(z-2)\right]$

$\qquad = \ln\left[\dfrac{\sqrt{x}}{y^3(z-2)}\right]$

26. Continuously: $A = 8000e^{0.08(3)}$

$\qquad\qquad\qquad \approx 10{,}170$

Monthly: $A = 8000\left(1+\dfrac{0.08}{12}\right)^{12\cdot 3}$

$\qquad\qquad\quad \approx 10{,}162$

$10{,}170 - 10{,}162 = 8$

Interest returned will be \$8 more if compounded continuously.

Section 3.4

Check Point Exercises

1. **a.** $5^{3x-6} = 125$

$5^{3x-6} = 5^3$

$3x - 6 = 3$

$3x = 9$

$x = 3$

b. $8^{x+2} = 4^{x-3}$

$\left(2^3\right)^{x+2} = \left(2^2\right)^{x-3}$

$2^{3x+6} = 2^{2x-6}$

$3x + 6 = 2x - 6$

$x = -12$

2. **a.** $5^x = 134$

$\ln 5^x = \ln 134$

$x \ln 5 = \ln 134$

$x = \dfrac{\ln 134}{\ln 5} \approx 3.04$

The solution set is $\left\{ \dfrac{\ln 134}{\ln 5} \right\}$,

approximately 3.04.

b. $10^x = 8000$

$\log 10^x = \log 8000$

$x \log 10 = \log 8000$

$x = \log 8000 \approx 3.90$

The solution set is $\{\log 8000\}$, approximately 3.90.

3. $7e^{2x} = 63$

$e^{2x} = 9$

$\ln e^{2x} = \ln 9$

$2x = \ln 9$

$x = \dfrac{\ln 9}{2} \approx 1.10$

The solution set is $\left\{ \dfrac{\ln 9}{2} \right\}$,

approximately 1.10.

4. $3^{2x-1} = 7^{x+1}$

$\ln 3^{2x-1} = \ln 7^{x+1}$

$(2x - 1)\ln 3 = (x + 1)\ln 7$

$2x \ln 3 - \ln 3 = x \ln 7 + \ln 7$

$2x \ln 3 - x \ln 7 = \ln 3 + \ln 7$

$x(2\ln 3 - \ln 7) = \ln 3 + \ln 7$

$x = \dfrac{\ln 3 + \ln 7}{2\ln 3 - \ln 7}$

$x \approx 12.11$

5. $e^{2x} - 8e^x + 7 = 0$

$\left(e^x - 7\right)\left(e^x - 1\right) = 0$

$e^x - 7 = 0 \qquad \text{or} \quad e^x - 1 = 0$

$e^x = 7 \qquad\qquad e^x = 1$

$\ln e^x = \ln 7 \qquad \ln e^x = \ln 1$

$x = \ln 7 \qquad\qquad x = 0$

The solution set is $\{0, \ln 7\}$. The solutions are 0 and (approximately) 1.95.

6. **a.** $\log_2 (x - 4) = 3$

$2^3 = x - 4$

$8 = x - 4$

$12 = x$

Check:

$\log_2 (x - 4) = 3$

$\log_2 (12 - 4) = 3$

$\log_2 8 = 3$

$3 = 3$

The solution set is $\{12\}$.

b.
$$4\ln 3x = 8$$
$$\ln 3x = 2$$
$$e^{\ln 3x} = e^2$$
$$3x = e^2$$
$$x = \frac{e^2}{3} \approx 2.46$$

Check
$$4\ln 3x = 8$$
$$4\ln 3\left(\frac{e^2}{3}\right) = 8$$
$$4\ln e^2 = 8$$
$$4(2) = 8$$
$$8 = 8$$

The solution set is $\left\{\dfrac{e^2}{3}\right\}$,

approximately 2.46.

7.
$$\log x + \log(x-3) = 1$$
$$\log x(x-3) = 1$$
$$10^1 = x(x-3)$$
$$10 = x^2 - 3x$$
$$0 = x^2 - 3x - 10$$
$$0 = (x-5)(x+2)$$
$$x - 5 = 0 \quad \text{or} \quad x + 2 = 0$$
$$x = 5 \quad \text{or} \quad x = -2$$

Check
Checking 5:
$$\log 5 + \log(5-3) = 1$$
$$\log 5 + \log 2 = 1$$
$$\log(5 \cdot 2) = 1$$
$$\log 10 = 1$$
$$1 = 1$$

Checking –2:
$$\log x + \log(x-3) = 1$$
$$\log(-2) + \log(-2-3) \ 0 \ 1$$

Negative numbers do not have logarithms so
–2 does not check.
The solution set is {5}.

8.
$$\ln(x-3) = \ln(7x - 23) - \ln(x+1)$$
$$\ln(x-3) = \ln\frac{7x-23}{x+1}$$
$$x - 3 = \frac{7x-23}{x+1}$$
$$(x-3)(x+1) = 7x - 23$$
$$x^2 - 2x - 3 = 7x - 23$$
$$x^2 - 9x + 20 = 0$$
$$(x-4)(x-5) = 0$$
$$x = 4 \quad \text{or} \quad x = 5$$

Both values produce true statements.
The solution set is {4, 5}

9. For a risk of 7%, let $R = 7$ in
$$R = 6e^{12.77x}$$
$$6e^{12.77x} = 7$$
$$e^{12.77x} = \frac{7}{6}$$
$$\ln e^{12.77x} = \ln\left(\frac{7}{6}\right)$$
$$12.77x = \ln\left(\frac{7}{6}\right)$$
$$x = \frac{\ln\left(\frac{7}{6}\right)}{12.77} \approx 0.01$$

For a blood alcohol concentration of 0.01, the risk
of a car accident is 7%.

10.
$$A = P\left(1 + \frac{r}{n}\right)^{nt}$$
$$3600 = 1000\left(1 + \frac{0.08}{4}\right)^{4t}$$
$$1000\left(1 - \frac{0.08}{4}\right)^{4t} = 3600$$
$$1000(1 + 0.02)^{4t} = 3600$$
$$1000(1.02)^{4t} = 3600$$
$$(1.02)^{4t} = \ln 3.6$$
$$4t\ln(1.02) = \ln 3.6$$
$$t = \frac{\ln 3.6}{4\ln 1.02}$$
$$\approx 16.2$$

After approximately 16.2 years, the $1000 will
grow to an accumulated value of $3600.

11. $f(x) = 54.8 - 12.3 \ln x$

Solve equation when $f(x) = 25$.

$$54.8 - 12.3 \ln x = 25$$
$$-12.3 \ln x = -29.8$$
$$\ln x = \frac{-29.8}{-12.3}$$
$$\log_e x = \frac{29.8}{12.3}$$
$$x = e^{\frac{29.8}{12.3}}$$
$$x \approx 11.277$$

An annual income of approximately \$11,000 corresponds to 25% of Americans reporting fair or poor health.

Exercise Set 3.4

1. $2^x = 64$

$$2^x = 2^6$$
$$x = 6$$

The solution is 6, and the solution set is $\{6\}$.

3. $5^x = 125$

$$5^x = 5^3$$
$$x = 3$$

The solution is 3, and the solution set is $\{3\}$.

5. $2^{2x-1} = 32$

$$2^{2x-1} = 2^5$$
$$2x - 1 = 5$$
$$2x = 6$$
$$x = 3$$

The solution is 3, and the solution set is $\{3\}$.

7. $4^{2x-1} = 64$

$$4^{2x-1} = 4^3$$
$$2x - 1 = 3$$
$$2x = 4$$
$$x = 2$$

The solution is 2, and the solution set is $\{2\}$.

9. $32^x = 8$

$$\left(2^5\right)^x = 2^3$$
$$2^{5x} = 2^3$$
$$5x = 3$$
$$x = \frac{3}{5}$$

The solution is $\frac{3}{5}$, and the solution set is $\left\{\frac{3}{5}\right\}$.

11. $9^x = 27$

$$\left(3^2\right)^x = 3^3$$
$$3^{2x} = 3^3$$
$$2x = 3$$
$$x = \frac{3}{2}$$

The solution is $\frac{3}{2}$, and the solution set is $\left\{\frac{3}{2}\right\}$.

13. $3^{1-x} = \frac{1}{27}$

$$3^{1-x} = \frac{1}{3^3}$$
$$3^{1-x} = 3^{-3}$$
$$1 - x = -3$$
$$-x = -4$$
$$x = 4$$

The solution set is $\{4\}$.

15. $6^{\frac{x-3}{4}} = \sqrt{6}$

$$6^{\frac{x-3}{4}} = 6^{\frac{1}{2}}$$
$$\frac{x-3}{4} = \frac{1}{2}$$
$$2(x-3) = 4(1)$$
$$2x - 6 = 4$$
$$2x = 10$$
$$x = 5$$

The solution is 5, and the solution set is $\{5\}$.

17. $4^x = \dfrac{1}{\sqrt{2}}$

$\left(2^2\right)^x = \dfrac{1}{2^{\frac{1}{2}}}$

$2^{2x} = 2^{-\frac{1}{2}}$

$2x = -\dfrac{1}{2}$

$x = \dfrac{1}{2}\left(-\dfrac{1}{2}\right) = -\dfrac{1}{4}$

The solution is $-\dfrac{1}{4}$, and the solution set is $\left\{-\dfrac{1}{4}\right\}$.

19. $8^{x+3} = 16^{x-1}$

$\left(2^3\right)^{x+3} = \left(2^4\right)^{x-1}$

$2^{3x+9} = 2^{4x-4}$

$3x + 9 = 4x - 4$

$13 = x$

The solution set is $\{13\}$.

21. $e^{x+1} = \dfrac{1}{e}$

$e^{x+1} = e^{-1}$

$x + 1 = -1$

$x = -2$

The solution set is $\{-2\}$.

23. $10^x = 3.91$

$\ln 10^x = \ln 3.91$

$x \ln 10 = \ln 3.91$

$x = \dfrac{\ln 3.91}{\ln 10} \approx 0.59$

25. $e^x = 5.7$

$\ln e^x = 5.7$

$x = \ln 5.7 \approx 1.74$

27. $5^x = 17$

$\ln 5^x = \ln 17$

$x \ln 5 = \ln 17$

$x = \dfrac{\ln 17}{\ln 5} \approx 1.76$

29. $5e^x = 23$

$e^x = \dfrac{23}{5}$

$\ln e^x = \ln \dfrac{23}{5}$

$x = \ln \dfrac{23}{5} \approx 1.53$

31. $3e^{5x} = 1977$

$e^{5x} = 659$

$\ln e^{5x} = \ln 659$

$x = \dfrac{\ln 659}{5} \approx 1.30$

33. $e^{1-5x} = 793$

$\ln e^{1-5x} = \ln 793$

$(1-5x)(\ln e) = \ln 793$

$1 - 5x = \ln 793$

$5x = 1 - \ln 793$

$x = \dfrac{1 - \ln 793}{5} \approx -1.14$

35. $e^{5x-3} - 2 = 10,476$

$e^{5x-3} = 10,478$

$\ln e^{5x-3} = \ln 10,478$

$(5x-3)\ln e = \ln 10,478$

$5x - 3 = \ln 10,478$

$5x = \ln 10,478 + 3$

$x = \dfrac{\ln 10,478 + 3}{5} \approx 2.45$

37. $7^{x+2} = 410$

$\ln 7^{x+2} = \ln 410$

$(x+2)\ln 7 = \ln 410$

$x + 2 = \dfrac{\ln 410}{\ln 7}$

$x = \dfrac{\ln 410}{\ln 7} - 2 \approx 1.09$

39. $7^{0.3x} = 813$

$\ln 7^{0.3x} = \ln 813$

$0.3x \ln 7 = \ln 813$

$x = \dfrac{\ln 813}{0.3 \ln 7} \approx 11.48$

41. $5^{2x+3} = 3^{x-1}$

$\ln 5^{2x+3} = \ln 3^{x-1}$

$(2x+3)\ln 5 = (x-1)\ln 3$

$2x\ln 5 + 3\ln 5 = x\ln 3 - \ln 3$

$3\ln 5 + \ln 3 = x\ln 3 - 2x\ln 5$

$3\ln 5 + \ln 3 = x(\ln 3 - 2\ln 5)$

$\dfrac{3\ln 5 + \ln 3}{\ln 3 - 2\ln 5} = x$

$-2.80 \approx x$

43. $e^{2x} - 3e^x + 2 = 0$

$\left(e^x - 2\right)\left(e^x - 1\right) = 0$

$e^x - 2 = 0 \quad$ or $\quad e^x - 1 = 0$

$e^x = 2 \qquad\qquad e^x = 1$

$\ln e^x = \ln 2 \qquad \ln e^x = \ln 1$

$x = \ln 2 \qquad\qquad x = 0$

The solution set is {0, ln 2}. The solutions are 0 and approximately 0.69.

45. $e^{4x} + 5e^{2x} - 24 = 0$

$\left(e^{2x} + 8\right)\left(e^{2x} - 3\right) = 0$

$e^{2x} + 8 = 0 \qquad$ or $\quad e^{2x} - 3 = 0$

$e^{2x} = -8 \qquad\qquad e^{2x} = 3$

$\ln e^{2x} = \ln(-8) \qquad \ln e^{2x} = \ln 3$

$2x = \ln(-8) \qquad\qquad 2x = \ln 3$

$\ln(-8)$ does not exist $\qquad x = \dfrac{\ln 3}{2}$

$x = \dfrac{\ln 3}{2} \approx 0.55$

47. $3^{2x} + 3^x - 2 = 0$

$(3^x + 2)(3^x - 1) = 0$

$3^x + 2 = 0 \qquad\qquad 3^x - 1 = 0$

$3^x = -2 \qquad\qquad\quad 3^x = 1$

$\log 3^x = \log(-2) \qquad \log 3^x = \log 1$

does not exist $\qquad\qquad \log 3 = 0$

$\qquad\qquad\qquad\qquad x = \dfrac{0}{\log 3}$

$\qquad\qquad\qquad\qquad x = 0$

The solution set is {0}.

49. $\log_3 x = 4$

$3^4 = x$

$81 = x$

51. $\ln x = 2$

$e^2 = x$

$7.39 \approx x$

53. $\log_4 (x+5) = 3$

$4^3 = x+5$

$59 = x$

55. $\log_3 (x-4) = -3$

$3^{-3} = x-4$

$\dfrac{1}{27} = x-4$

$4\dfrac{1}{27} = x$

$4.04 \approx x$

57. $\log_4 (3x+2) = 3$

$4^3 = 3x+2$

$64 = 3x+2$

$62 = 3x$

$\dfrac{62}{3} = x$

$20.67 \approx x$

59. $5\ln 2x = 20$

$\ln 2x = 4$

$e^{\ln 2x} = e^4$

$2x = e^4$

$x = \dfrac{e^4}{2} \approx 27.30$

61. $6 + 2\ln x = 5$

$2\ln x = -1$

$\ln x = -\dfrac{1}{2}$

$e^{\ln x} = e^{-1/2}$

$x = e^{-1/2} \approx 0.61$

63. $\ln \sqrt{x+3} = 1$

$e^{\ln \sqrt{x+3}} = e^1$

$\sqrt{x+3} = e$

$x+3 = e^2$

$x = e^2 - 3 \approx 4.39$

65. $\log_5 x + \log_5 (4x - 1) = 1$

$\log_5 (4x^2 - x) = 1$

$4x^2 - x = 5$

$4x^2 - x - 5 = 0$

$(4x - 5)(x + 1) = 0$

$x = \dfrac{5}{4}$ or $x = -1$

$x = -1$ does not check because $\log_5 (-1)$ does not exist.

The solution set is $\left\{ \dfrac{5}{4} \right\}$.

67. $\log_3 (x - 5) + \log_3 (x + 3) = 2$

$\log_3 \left[(x - 5)(x + 3) \right] = 2$

$(x - 5)(x + 3) = 3^2$

$x^2 - 2x - 15 = 9$

$x^2 - 2x - 24 = 0$

$(x - 6)(x + 4) = 0$

$x = 6$ or $x = -4$

$x = -4$ does not check because $\log_3 (-4 - 5)$ does not exist. The solution set is $\{6\}$.

69. $\log_2 (x + 2) - \log_2 (x - 5) = 3$

$\log_2 \left(\dfrac{x + 2}{x - 5} \right) = 3$

$\dfrac{x + 2}{x - 5} = 2^3$

$\dfrac{x + 2}{x - 5} = 8$

$x + 2 = 8(x - 5)$

$x + 2 = 8x - 40$

$7x = 42$

$x = 6$

71. $2 \log_3 (x + 4) = \log_3 9 + 2$

$2 \log_3 (x + 4) = 2 + 2$

$2 \log_3 (x + 4) = 4$

$\log_3 (x + 4) = 2$

$3^2 = x + 4$

$9 = x + 4$

$5 = x$

73. $\log_2 (x - 6) + \log_2 (x - 4) - \log_2 x = 2$

$\log_2 \dfrac{(x - 6)(x - 4)}{x} = 2$

$\dfrac{(x - 6)(x - 4)}{x} = 2^2$

$x^2 - 10x + 24 = 4x$

$x^2 - 14x + 24 = 0$

$(x - 12)(x - 2) = 0$

$x - 12 = 0 \quad$ or $\quad x - 2 = 0$

$x = 12 \qquad\qquad x = 2$

The solution set is $\{12\}$ since $\log_2 (2 - 6) = \log_2 (-4)$ is not possible.

75. $\log(x + 4) = \log x + \log 4$

$\log(x + 4) = \log 4x$

$x + 4 = 4x$

$4 = 3x$

$x = \dfrac{4}{3}$

This value is rejected. The solution set is $\left\{ \dfrac{4}{3} \right\}$.

77. $\log(3x - 3) = \log(x + 1) + \log 4$

$\log(3x - 3) = \log(4x + 4)$

$3x - 3 = 4x + 4$

$-7 = x$

This value is rejected. The solution set is $\{\ \}$.

79. $2 \log x = \log 25$

$\log x^2 = \log 25$

$x^2 = 25$

$x = \pm 5$

-5 is rejected. The solution set is $\{5\}$.

81. $\log(x + 4) - \log 2 = \log(5x + 1)$

$\log \dfrac{x + 4}{2} = \log(5x + 1)$

$\dfrac{x + 4}{2} = 5x + 1$

$x + 4 = 10x + 2$

$-9x = -2$

$x = \dfrac{2}{9}$

$x \approx 0.22$

83. $2\log x - \log 7 = \log 112$

$\log x^2 - \log 7 = \log 112$

$\log \dfrac{x^2}{7} = \log 112$

$\dfrac{x^2}{7} = 112$

$x^2 = 784$

$x = \pm 28$

-28 is rejected. The solution set is $\{28\}$.

85. $\log x + \log(x+3) = \log 10$

$\log(x^2 + 3x) = \log 10$

$x^2 + 3x = 10$

$x^2 + 3x - 10 = 0$

$(x+5)(x-2) = 0$

$x = -5$ or $x = 2$

-5 is rejected. The solution set is $\{2\}$.

87. $\ln(x-4) + \ln(x+1) = \ln(x-8)$

$\ln(x^2 - 3x - 4) = \ln(x-8)$

$x^2 - 3x - 4 = x - 8$

$x^2 - 4x + 4 = 0$

$(x-2)(x-2) = 0$

$x = 2$

2 is rejected. The solution set is { }.

89. $\ln(x-2) - \ln(x+3) = \ln(x-1) - \ln(x+7)$

$\ln \dfrac{x-2}{x+3} = \ln \dfrac{x-1}{x+7}$

$\dfrac{x-2}{x+3} = \dfrac{x-1}{x+7}$

$(x-2)(x+7) = (x+3)(x-1)$

$x^2 + 5x - 14 = x^2 + 2x - 3$

$3x = 11$

$x = \dfrac{11}{3}$

$x \approx 3.67$

91. $5^{2x} \cdot 5^{4x} = 125$

$5^{2x+4x} = 5^3$

$5^{6x} = 5^3$

$6x = 3$

$x = \dfrac{1}{2}$

93. $2|\ln x| - 6 = 0$

$2|\ln x| = 6$

$|\ln x| = 3$

$\ln x = 3$ or $\ln x = -3$

$x = e^3$ $x = e^{-3}$

$x \approx 20.09$ $x \approx 0.05$

95. $3^{x^2} = 45$

$\ln 3^{x^2} = \ln 45$

$x^2 \ln 3 = \ln 45$

$x^2 = \dfrac{\ln 45}{\ln 3}$

$x = \pm\sqrt{\dfrac{\ln 45}{\ln 3}} \approx \pm 1.86$

97. $\ln(2x+1) + \ln(x-3) - 2\ln x = 0$

$\ln(2x+1) + \ln(x-3) - \ln x^2 = 0$

$\ln \dfrac{(2x+1)(x-3)}{x^2} = 0$

$\dfrac{(2x+1)(x-3)}{x^2} = e^0$

$\dfrac{2x^2 - 5x - 3}{x^2} = 1$

$2x^2 - 5x - 3 = x^2$

$x^2 - 5x - 3 = 0$

$x = \dfrac{-b \pm \sqrt{b^2 - 4ac}}{2a}$

$x = \dfrac{-(-5) \pm \sqrt{(-5)^2 - 4(1)(-3)}}{2(1)}$

$x = \dfrac{5 \pm \sqrt{37}}{2}$

$x = \dfrac{5 + \sqrt{37}}{2} \approx 5.54$

$x = \dfrac{5 - \sqrt{37}}{2} \approx -0.54$ (rejected)

The solution set is $\left\{ \dfrac{5 + \sqrt{37}}{2} \right\}$.

99.
$$5^{x^2-12} = 25^{2x}$$
$$5^{x^2-12} = \left(5^2\right)^{2x}$$
$$5^{x^2-12} = 5^{4x}$$
$$x^2 - 12 = 4x$$
$$x^2 - 4x - 12 = 0$$
$$(x-6)(x+2) = 0$$

Apply the zero product property:
$$x - 6 = 0 \quad \text{or} \quad x + 2 = 0$$
$$x = 6 \qquad\qquad x = -2$$

The solution set is $\{-2,\ 6\}$.

101. a. 2005 is 0 years after 2005.
$$A = 36.1e^{0.0126t}$$
$$A = 36.1e^{0.0126(0)} = 36.1$$
The population of California was 36.1 million in 2005.

b.
$$A = 36.1e^{0.0126t}$$
$$40 = 36.1e^{0.0126t}$$
$$\frac{40}{36.1} = e^{0.0126t}$$
$$\ln\frac{40}{36.1} = \ln e^{0.0126t}$$
$$0.0126t = \ln\frac{40}{36.1}$$
$$t = \frac{\ln\dfrac{40}{36.1}}{0.0126} \approx 8$$

The population of California will reach 40 million about 8 years after 2005, or 2013

103. $f(x) = 20(0.975)^x$
$$1 = 20(0.975)^x$$
$$\frac{1}{20} = 0.975^x$$
$$\ln\frac{1}{20} = \ln 0.975^x$$
$$\ln\frac{1}{20} = x\ln 0.975$$
$$x = \frac{\ln\dfrac{1}{20}}{\ln 0.975}$$
$$x \approx 118$$

There is 1% of surface sunlight at 118 feet. This is represented by the point $(118,1)$.

105. $20,000 = 12,500\left(1 + \dfrac{0.0575}{4}\right)^{4t}$
$$12,500(1.014375)^{4t} = 20,000$$
$$(1.014375)^{4t} = 1.6$$
$$\ln(1.014375)^{4t} = \ln 1.6$$
$$4t\ln(1.014375) = \ln 1.6$$
$$t = \frac{\ln 1.6}{4\ln 1.014375} \approx 8.2$$

8.2 years

107.
$$1400 = 1000\left(1 + \frac{r}{360}\right)^{360\cdot 2}$$
$$\left(1 + \frac{r}{360}\right)^{720} = 1.4$$
$$\ln\left(1 + \frac{r}{360}\right)^{720} = \ln 1.4$$
$$720\ln\left(1 + \frac{r}{360}\right) = \ln 1.4$$
$$\ln\left(1 + \frac{r}{360}\right) = \frac{\ln 1.4}{720}$$
$$e^{\ln(1+r/360)} = e^{(\ln 1.4)/720}$$
$$1 + \frac{r}{360} = e^{(\ln 1.4)/720} - 1$$
$$r = 360(e^{(\ln 1.4)/720} - 1)$$
$$\approx 0.168$$

16.8%

109. accumulated amount $= 2(8000) = 16,000$
$$16,000 = 8000e^{0.08t}$$
$$e^{0.08t} = 2$$
$$\ln e^{0.08t} = \ln 2$$
$$0.08t = \ln 2$$
$$t = \frac{\ln 2}{0.08}$$
$$t \approx 8.7$$

The amount would double in 8.7 years.

111. accumulated amount $= 3(2350) = 7050$
$$7050 = 2350e^{r\cdot 7}$$
$$e^{7r} = 3$$
$$\ln e^{7r} = \ln 3$$
$$7r = \ln 3$$
$$r = \frac{\ln 3}{7} \approx 0.157$$

15.7%

113. a. 2007 is 5 years after 2002.

$f(x) = 8 + 38\ln x$

$f(5) = 8 + 38\ln 5 \approx 69$

According to the function, 69% of new cellphones will have cameras in 2007. This overestimates the value shown in the graph by 1%.

b. $f(x) = 8 + 38\ln x$

$87 = 8 + 38\ln x$

$79 = 38\ln x$

$\dfrac{79}{38} = \ln x$

$x = e^{\frac{79}{38}}$

$x \approx 8$

If the trend continues, 87% of new cellphones will have cameras 8 years after 2002, or 2010.

115. $P(x) = 95 - 30\log_2 x$

$40 = 95 - 30\log_2 x$

$30\log_2 x = 45$

$\log_2 x = 1.5$

$x = 2^{1.5} \approx 2.8$

Only half the students recall the important features of the lecture after 2.8 days.
(2.8, 50)

117. a. $\text{pH} = -\log x$

$5.6 = -\log x$

$-5.6 = \log x$

$x = 10^{-5.6}$

The hydrogen ion concentration is $10^{-5.6}$ mole per liter.

b. $\text{pH} = -\log x$

$2.4 = -\log x$

$-2.4 = \log x$

$x = 10^{-2.4}$

The hydrogen ion concentration is $10^{-2.4}$ mole per liter.

c. $\dfrac{10^{-2.4}}{10^{-5.6}} = 10^{-2.4-(-5.6)} = 10^{3.2}$

The concentration of the acidic rainfall in part (b) is $10^{3.2}$ times greater than the normal rainfall in part (a).

119. – 121.   Answers may vary.

123.

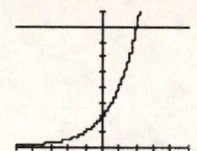

The intersection point is (2, 8).
Verify: $x = 2$

$2^{x+1} = 8$

$2^{2+1} = 2$

$2^3 = 8$

$8 = 8$

The solution set is $\{2\}$.

125.

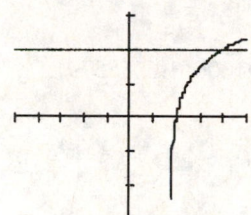

The intersection point is (4, 2).
Verify: $x = 4$

$\log_3(4 \cdot 4 - 7) = 2$

$\log_3 9 = 2$

$2 = 2$

The solution set is $\{4\}$.

127.

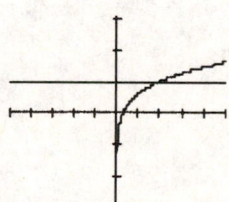

The intersection point is (2, 1).
Verify: $x = 2$

$\log(2 + 3) + \log 2 = 1$

$\log 5 + \log 2 = 1$

$\log(5 \cdot 2) = 1$

$\log 10 = 1$

$1 = 1$

The solution set is $\{2\}$.

129.

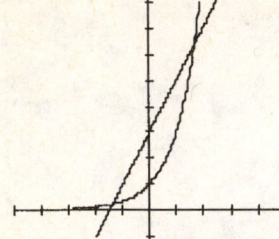

There are 2 points of intersection, approximately
$(-1.391606, 0.21678798)$ and
$(1.6855579, 6.3711158)$.
Verify $x \approx -1.391606$
$3^x = 2x + 3$
$3^{-1.391606} \approx 2(-1.391606) + 3$
$0.2167879803 \approx 0.216788$
Verify $x \approx 1.6855579$
$3^x = 2x + 3$
$3^{1.6855579} \approx 2(1.6855579) + 3$
$6.37111582 \approx 6.371158$
The solution set is $\{-1.391606, 1.6855579\}$.

131.

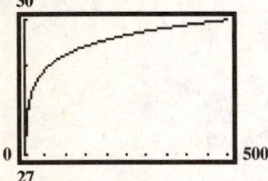

As the distance from the eye increases, barometric air pressure increases, leveling off at about 30 inches of mercury.

133.

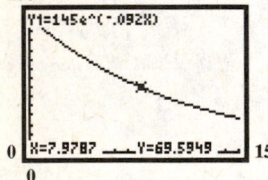

When $P = 70$, $t \approx 7.9$, so it will take about 7.9 minutes.
Verify:
$70 = 45e^{-0.092(7.9)}$

$70 \approx 70.10076749$
The runner's pulse will be 70 beats per minute after about 7.9 minutes.

135. does not make sense; Explanations will vary. Sample explanation: $2^x = 15$ requires logarithms. $2^x = 16$ can be solved by rewriting 16 as 2^4.

$2^x = 15$
$\ln 2^x = \ln 15$
$x \ln 2 = \ln 15$
$x = \dfrac{\ln 15}{\ln 2}$

$2^x = 16$
$2^x = 2^4$
$x = 4$

137. makes sense

139. false; Changes to make the statement true will vary. A sample change is: If $\log(x+3) = 2$, then $10^2 = x + 3$.

141. true

143. Account paying 3% interest:
$A = 4000\left(1 + \dfrac{0.03}{1}\right)^{1 \cdot t}$
Account paying 5% interest:
$A = 2000\left(1 + \dfrac{0.05}{1}\right)^{1 \cdot t}$
The two accounts will have the same balance when
$4000(1.03)^t = 2000(1.05)^t$
$(1.03)^t = 0.5(1.05)^t$
$\left(\dfrac{1.03}{1.05}\right)^t = 0.5$
$\ln\left(\dfrac{1.03}{1.05}\right)^t = \ln 0.5$
$t \ln\left(\dfrac{1.03}{1.05}\right) = \ln 0.5$
$t = \dfrac{\ln 0.5}{\ln\left(\dfrac{1.03}{1.05}\right)} \approx 36$

The accounts will have the same balance in about 36 years.

145. $(\log x)(2\log x + 1) = 6$

$2(\log x)^2 + \log x - 6 = 0$

$(2\log x - 3)(\log x + 2) = 0$

$2\log x - 3 = 0$ or $\log x + 2 = 0$

$2\log x = 3 \qquad \log x = -2$

$\log x = \dfrac{3}{2} \qquad x = 10^{-2}$

$x = 10^{3/2} \qquad x = \dfrac{1}{100}$

$x = 10\sqrt{10}$

The solution set is $\left\{ \dfrac{1}{100}, 10\sqrt{10} \right\}$.

Check by direct substitution:

$\text{Check:} x = 10\sqrt{10} = 10^{3/2}$

$(\log x)(2\log x + 1) = 6$

$\left(\log 10^{3/2}\right)\left(2\log 10^{3/2} + 1\right) = 6$

$\left(\dfrac{3}{2}\right)\left(2 \cdot \dfrac{3}{2} + 1\right) = 6$

$\left(\dfrac{3}{2}\right)(3 + 1) = 6$

$\left(\dfrac{3}{2}\right)(4) = 6$

$6 = 6$

147. $A = 10e^{-0.003t}$

a. 2006: $A = 10e^{-0.003(0)} = 10$ million

2007: $A = 10e^{-0.003(1)} \approx 9.97$ million

2008: $A = 10e^{-0.003(2)} \approx 9.94$ million

2009: $A = 10e^{-0.003(3)} \approx 9.91$ million

b. The population is decreasing.

148. An exponential function is the best choice.

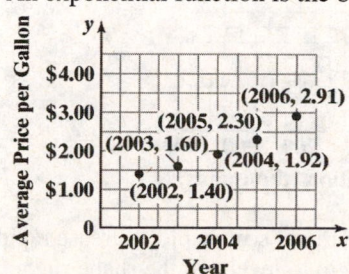

149. a. $e^{\ln 3} = 3$

b. $e^{\ln 3} = 3$

$\left(e^{\ln 3}\right)^x = 3^x$

$e^{(\ln 3)x} = 3^x$

Section 3.5

Check Point Exercises

1. a. $A_0 = 643$. Since 2006 is 16 years after 1990, when $t = 16$, $A = 906$.

$A = A_0 e^{kt}$

$906 = 643e^{k(16)}$

$\dfrac{906}{643} = e^{16k}$

$\ln\left(\dfrac{906}{643}\right) = \ln e^{16k}$

$\ln\left(\dfrac{906}{643}\right) = 16k$

$k = \dfrac{\ln\left(\dfrac{906}{643}\right)}{16} \approx 0.021$

Thus, the growth function is $A = 643e^{0.021t}$.

b. $A = 643e^{0.021t}$

$2000 = 643e^{0.021t}$

$\dfrac{2000}{643} = e^{0.021t}$

$\ln\left(\dfrac{2000}{643}\right) = \ln e^{0.021t}$

$\ln\left(\dfrac{2000}{643}\right) = 0.021t$

$t = \dfrac{\ln\left(\dfrac{2000}{643}\right)}{0.021} \approx 54$

Africa's population will reach 2000 million approximately 54 years after 1990, or 2044.

2. a. In the exponential decay model $A = A_0 e^{kt}$, substitute $\dfrac{A_0}{2}$ for A since the amount present after 28 years is half the original amount.

$$\frac{A_0}{2} = A_0 e^{k \cdot 28}$$

$$e^{28k} = \frac{1}{2}$$

$$\ln e^{28k} = \ln \frac{1}{2}$$

$$28k = \ln \frac{1}{2}$$

$$k = \frac{\ln^{1/2}}{28} \approx -0.0248$$

So the exponential decay model is
$$A = A_0 e^{-0.0248t}$$

b. Substitute 60 for A_0 and 10 for A in the model from part (a) and solve for t.

$$10 = 60 e^{-0.0248t}$$

$$e^{-0.0248t} = \frac{1}{6}$$

$$\ln e^{-0.0248t} = \ln \frac{1}{6}$$

$$-0.0248t = \ln \frac{1}{6}$$

$$t = \frac{\ln \frac{1}{6}}{-0.0248} \approx 72$$

The strontium-90 will decay to a level of 10 grams about 72 years after the accident.

3. a. The time prior to learning trials corresponds to $t = 0$.

$$f(0) = \frac{0.8}{1 + e^{-0.2(0)}} = 0.4$$

The proportion of correct responses prior to learning trials was 0.4.

b. Substitute 10 for t in the model:

$$f(10) = \frac{0.8}{1 + e^{-0.2(10)}} \approx 0.7$$

The proportion of correct responses after 10 learning trials was 0.7.

c. In the logistic growth model, $f(t) = \dfrac{c}{1 + ae^{-bt}}$, the constant c represents the limiting size that $f(t)$ can attain. The limiting size of the proportion of correct responses as continued learning trials take place is 0.8.

4. a.
$$T = C + (T_o - C)e^{kt}$$
$$80 = 30 + (100 - 30)e^{k5}$$
$$80 = 30 + 70 e^{5k}$$
$$50 = 70 e^{5k}$$
$$\frac{5}{7} = e^{5k}$$
$$\ln \frac{5}{7} = \ln e^{5k}$$
$$\ln \frac{5}{7} = 5k$$
$$\frac{\ln \frac{5}{7}}{5} = k$$
$$-0.0673 \approx k$$
$$T = 30 + 70 e^{-0.0673t}$$

b. $T = 30 + 70 e^{-0.0673(20)} \approx 48°$
After 20 minutes, the temperature will be 48°.

c.
$$35 = 30 + 70 e^{-0.0673t}$$
$$5 = 70 e^{-0.0673t}$$
$$\frac{1}{14} = e^{-0.0673t}$$
$$\ln \frac{1}{14} = \ln e^{-0.0673t}$$
$$\ln \frac{1}{14} = -0.0673t$$
$$\frac{\ln \frac{1}{14}}{-0.0673} = t$$
$$39 \approx t$$
The temperature will reach 35° after 39 min.

5. Scatter plot:

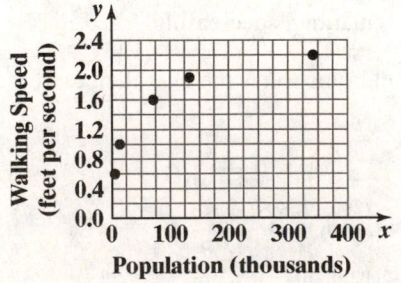

Because the data in the scatter plot increase rapidly at first and then begin to level off, the shape suggests that a logarithmic function is a good choice for modeling the data.

6.   Scatter plot:

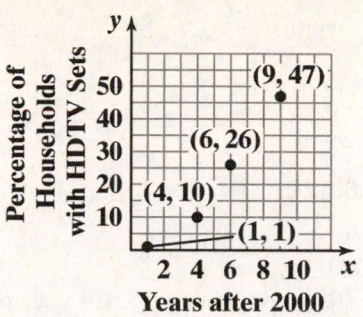

Because the data in the scatter plot appear to increase more and more rapidly, the shape suggests that an exponential function is a good choice for modeling the data.

7. $y = ab^x$ is equivalent to $y = ae^{(\ln b)x}$.

For $y = 4(7.8)^x$, $a = 4$, $b = 7.8$.

Thus, $y = 4(7.8)^x$ is equivalent to $y = 4e^{(\ln 7.8)x}$ in terms of a natural logarithm. Rounded to three decimal places, the model is approximately equivalent to $y = 4e^{2.054x}$.

Exercise Set 3.5

1.   Since 2006 is 0 years after 2006, find A when $t = 0$:

$$A = 127.5e^{0.001t}$$

$$A = 127.5e^{0.001(0)}$$

$$A = 127.5e^0$$

$$A = 127.5(1)$$

$$A = 127.5$$

In 2006, the population of Japan was 127.5 million.

3.   Iraq has the greatest growth rate at 2.7% per year.

5.   Substitute $A = 1238$ into the model for India and solve for t:

$$1238 = 1095.4e^{0.014t}$$

$$\frac{1238}{1095.4} = e^{0.014t}$$

$$\ln \frac{1238}{1095.4} = \ln e^{0.014t}$$

$$\ln \frac{1238}{1095.4} = 0.014t$$

$$t = \frac{\ln \dfrac{1238}{1095.4}}{0.014} \approx 9$$

The population of India will be 1238 million approximately 9 years after 2006, or 2015.

7.   **a.** $A_0 = 6.04$. Since 2050 is 50 years after 2000, when $t = 50$, $A = 10$.

$$A = A_0 e^{kt}$$

$$10 = 6.04e^{k(50)}$$

$$\frac{10}{6.04} = e^{50k}$$

$$\ln\left(\frac{10}{6.04}\right) = \ln e^{50k}$$

$$\ln\left(\frac{10}{6.04}\right) = 50k$$

$$k = \frac{\ln\left(\dfrac{10}{6.04}\right)}{50} \approx 0.01$$

Thus, the growth function is $A = 6.04e^{0.01t}$.

b. $9 = 6.04e^{0.01t}$

$$\frac{9}{6.04} = e^{0.01t}$$

$$\ln\left(\frac{9}{6.04}\right) = \ln e^{0.01t}$$

$$\ln\left(\frac{9}{6.04}\right) = 0.01t$$

$$t = \frac{\ln\left(\dfrac{9}{6.04}\right)}{0.01} \approx 40$$

Now, $2000 + 40 = 2040$, so the population will be 9 million is approximately the year 2040

9. $P(x) = 91.1e^{0.0147t}$

$$P(18) = 91.1e^{0.0147(18)}$$

$$P(18) = 91.1e^{0.0147(18)} \approx 118.7$$

The population is projected to be 118.7 million in 2025.

11.
$$P(x) = 44.4e^{kt}$$
$$55.2 = 44.4e^{18k}$$
$$\frac{55.2}{44.4} = e^{18k}$$
$$\ln\left(\frac{55.2}{44.4}\right) = \ln e^{18k}$$
$$\ln\left(\frac{55.2}{44.4}\right) = 18k$$
$$\frac{\ln\left(\frac{55.2}{44.4}\right)}{18} = k$$
$$k \approx 0.0121$$
The growth rate is 0.0121.

13.
$$P(x) = 44.0e^{kt}$$
$$40.0 = 44.0e^{18k}$$
$$\frac{40.0}{44.0} = e^{18k}$$
$$\ln\left(\frac{40.0}{44.0}\right) = \ln e^{18k}$$
$$\ln\left(\frac{40.0}{44.0}\right) = 18k$$
$$\frac{\ln\left(\frac{40.0}{44.0}\right)}{18} = k$$
$$k \approx -0.0053$$
The growth rate is –0.0053.

15.
$$A = 16e^{-0.000121t}$$
$$A = 16e^{-0.000121(5715)}$$
$$A = 16e^{-0.691515}$$
$$A \approx 8.01$$
Approximately 8 grams of carbon-14 will be present in 5715 years.

17. After 10 seconds, there will be $16 \cdot \frac{1}{2} = 8$ grams present. After 20 seconds, there will be $8 \cdot \frac{1}{2} = 4$ grams present. After 30 seconds, there will be $4 \cdot \frac{1}{2} = 2$ grams present. After 40 seconds, there will be $2 \cdot \frac{1}{2} = 1$ grams present. After 50 seconds, there will be $1 \cdot \frac{1}{2} = \frac{1}{2}$ gram present.

19.
$$A = A_0 e^{-0.000121t}$$
$$15 = 100e^{-0.000121t}$$
$$\frac{15}{100} = e^{-0.000121t}$$
$$\ln 0.15 = \ln e^{-0.000121t}$$
$$\ln 0.15 = -0.000121t$$
$$t = \frac{\ln 0.15}{-0.000121} \approx 15,679$$
The paintings are approximately 15,679 years old.

21.
$$0.5 = e^{kt}$$
$$0.5 = e^{-0.055t}$$
$$\ln 0.5 = \ln e^{-0.055t}$$
$$\ln 0.5 = -0.055t$$
$$\frac{\ln 0.5}{-0.055} = t$$
$$t \approx 12.6$$
The half-life is 12.6 years.

23.
$$0.5 = e^{kt}$$
$$0.5 = e^{1620k}$$
$$\ln 0.5 = \ln e^{1620k}$$
$$\ln 0.5 = 1620k$$
$$\frac{\ln 0.5}{1620} = k$$
$$k \approx -0.000428$$
The decay rate is 0.0428% per year.

25.
$$0.5 = e^{kt}$$
$$0.5 = e^{17.5k}$$
$$\ln 0.5 = \ln e^{17.5k}$$
$$\ln 0.5 = 17.5k$$
$$\frac{\ln 0.5}{17.5} = k$$
$$k \approx -0.039608$$
The decay rate is 3.9608% per day.

27. a.

$$\frac{1}{2} = 1e^{k1.31}$$

$$\ln\frac{1}{2} = \ln e^{1.31k}$$

$$\ln\frac{1}{2} = 1.31k$$

$$k = \frac{\ln\frac{1}{2}}{1.31} \approx -0.52912$$

The exponential model is given by $A = A_0 e^{-0.52912t}$.

b.

$$A = A_0 e^{-0.52912t}$$

$$0.945A_0 = A_0 e^{-0.52912t}$$

$$0.945 = e^{-0.52912t}$$

$$\ln 0.945 = \ln e^{-0.52912t}$$

$$\ln 0.945 = -0.52912t$$

$$t = \frac{\ln 0.945}{-0.52912} \approx 0.1069$$

The age of the dinosaur ones is approximately 0.1069 billion or 106,900,000 years old.

29. First find the decay equation.

$$0.5 = e^{kt}$$

$$0.5 = e^{22k}$$

$$\ln 0.5 = \ln e^{22k}$$

$$\ln 0.5 = 22k$$

$$\frac{\ln 0.5}{22} = k$$

$$k \approx -0.031507$$

$$A = e^{-0.031507t}$$

Next use the decay equation answer question.

$$A = e^{-0.031507t}$$

$$0.8 = e^{-0.031507t}$$

$$\ln 0.8 = \ln e^{-0.031507t}$$

$$\ln 0.8 = -0.031507t$$

$$\frac{\ln 0.8}{-0.031507} = t$$

$$t \approx 7.1$$

It will take 7.1 years.

31. First find the decay equation.

$$0.5 = e^{kt}$$

$$0.5 = e^{36k}$$

$$\ln 0.5 = \ln e^{36k}$$

$$\ln 0.5 = 36k$$

$$\frac{\ln 0.5}{36} = k$$

$$k \approx -0.019254$$

$$A = e^{-0.019254t}$$

Next use the decay equation answer question.

$$A = e^{-0.019254t}$$

$$0.9 = e^{-0.019254t}$$

$$\ln 0.9 = \ln e^{-0.019254t}$$

$$\ln 0.9 = -0.019254t$$

$$\frac{\ln 0.9}{-0.019254} = t$$

$$t \approx 5.5$$

It will take 5.5 hours.

33.

$$2A_0 = A_0 e^{kt}$$

$$2 = e^{kt}$$

$$\ln 2 = \ln e^{kt}$$

$$\ln 2 = kt$$

$$t = \frac{\ln 2}{k}$$

The population will double in $t = \dfrac{\ln 2}{k}$ years.

35. $A = 4.1e^{0.01t}$

a. $k = 0.01$, so New Zealand's growth rate is 1%.

b.

$$A = 4.1e^{0.01t}$$

$$2 \cdot 4.1 = 4.1e^{0.01t}$$

$$2 = e^{0.01t}$$

$$\ln 2 = \ln e^{0.01t}$$

$$\ln 2 = 0.01t$$

$$t = \frac{\ln 2}{0.01} \approx 69$$

New Zealand's population will double in approximately 69 years.

37. a. When the epidemic began, $t = 0$.

$$f(0) = \frac{100,000}{1 + 5000e^0} \approx 20$$

Twenty people became ill when the epidemic began.

b. $$f(4) = \frac{100,000}{1 + 5,000e^{-4}} \approx 1080$$

About 1080 people were ill at the end of the fourth week.

c. In the logistic growth model,

$$f(t) = \frac{c}{1 + ae^{-bt}},$$

the constant c represents the limiting size that $f(t)$ can attain. The limiting size of the population that becomes ill is 100,000 people.

39. $$f(x) = \frac{11.82}{1 + 3.81e^{-0.027(x)}}$$

$$f(54) = \frac{11.82}{1 + 3.81e^{-0.027(57)}} \approx 6.5$$

The function models the data very well.

41. $$f(x) = \frac{11.82}{1 + 3.81e^{-0.027(x)}}$$

$$8 = \frac{11.82}{1 + 3.81e^{-0.027(x)}}$$

$$8\left(1 + 3.81e^{-0.027(x)}\right) = 11.82$$

$$8 + 30.48e^{-0.027(x)} = 11.82$$

$$30.48e^{-0.027(x)} = 3.82$$

$$e^{-0.027(x)} = \frac{3.82}{30.48}$$

$$\ln e^{-0.027(x)} = \ln \frac{3.82}{30.48}$$

$$-0.027x = \ln \frac{3.82}{30.48}$$

$$x = \frac{\ln \frac{3.82}{30.48}}{-0.027}$$

$$x \approx 77$$

The world population will reach 8 billion 77 years after 1949, or 2026.

43. $$P(20) = \frac{90}{1 + 271e^{-0.122(20)}} \approx 3.7$$

The probability that a 20-year-old has some coronary heart disease is about 3.7%.

45. $$0.5 = \frac{0.9}{1 + 271e^{-0.122t}}$$

$$0.5\left(1 + 271e^{-0.122t}\right) = 0.9$$

$$1 + 271e^{-0.122t} = 1.8$$

$$271e^{-0.122t} = 0.8$$

$$e^{-0.122t} = \frac{0.8}{271}$$

$$\ln e^{-0.122t} = \ln \frac{0.8}{271}$$

$$-0.122t = \ln \frac{0.8}{271}$$

$$t = \frac{\ln \frac{0.8}{271}}{-0.122} \approx 48$$

The probability of some coronary heart disease is 50% at about age 48.

47. a. $$55 = 45 + (70 - 45)e^{k10}$$

$$10 = 25e^{10k}$$

$$\frac{2}{5} = e^{10k}$$

$$\ln \frac{2}{5} = \ln e^{10k}$$

$$\ln \frac{2}{5} = 10k$$

$$\frac{\ln \frac{2}{5}}{10} = k$$

$$-0.0916 \approx k$$

$$T = 45 + 25e^{-0.0916t}$$

b. $T = 45 + 25e^{-0.0916(15)} \approx 51°$

After 15 minutes, the temperature will be 51°.

c. $$50 = 45 + 25e^{-0.0916t}$$

$$5 = 25e^{-0.0916t}$$

$$\frac{1}{5} = e^{-0.0916t}$$

$$\ln \frac{1}{5} = \ln e^{-0.0916t}$$

$$\ln \frac{1}{5} = -0.0916t$$

$$\frac{\ln \frac{1}{5}}{-0.0916} = t$$

$$18 \approx t$$

The temperature will reach 50° after 18 min.

49.

$$T = C + (T_o - C)e^{kt}$$

$$38 = 75 + (28 - 75)e^{k10}$$

$$-37 = -47e^{10k}$$

$$\frac{-37}{-47} = e^{10k}$$

$$\ln\frac{37}{47} = \ln e^{10k}$$

$$\ln\frac{37}{47} = 10k$$

$$\frac{\ln\frac{37}{47}}{10} = k$$

$$-0.0239 \approx k$$

$$T = 75 - 47e^{-0.0239t}$$

$$50 = 75 - 47e^{-0.0239t}$$

$$-25 = -47e^{-0.0239t}$$

$$\frac{-25}{-47} = e^{-0.0239t}$$

$$\ln\frac{25}{47} = \ln e^{-0.0239t}$$

$$\ln\frac{25}{47} = -0.0239t$$

$$\frac{\ln\frac{17}{47}}{-0.0239} = t$$

$$26 = t$$

The temperature will reach $50°$ after 26 min.

51. a.

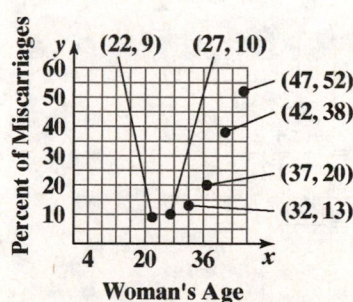

b. An exponential function appears to be the best choice for modeling the data.

53. a.

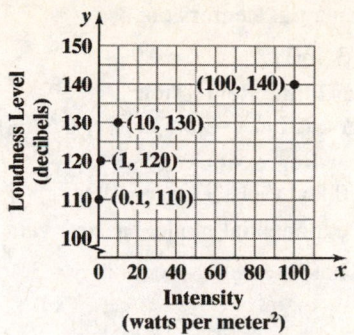

b. A logarithmic function appears to be the best choice for modeling the data.

55. $y = 100(4.6)^x$ is equivalent to
$y = 100e^{(\ln 4.6)x}$;
Using $\ln 4.6 \approx 1.526$,
$y = 100e^{1.526x}$.

57. $y = 2.5(0.7)^x$ is equivalent to
$y = 2.5e^{(\ln 0.7)x}$;
Using $\ln 0.7 \approx -0.357$,
$y = 2.5e^{-0.357x}$.

59. – 67. Answers may vary.

69. The logarithmic model is $y = 193.16 + 23.574\ln x$.
Since $r = 0.878$ is fairly close to 1, the model fits the data okay, but not great.

71. The power regression model is $y = 195.871x^{0.097}$.
Since $r = 0.901$, the model fits the data fairly well.

73. a. Exponential Regression:

$y = 3.46(1.02)^x$; $r \approx 0.994$

Logarithmic Regression:
$y = 14.752 \ln x - 26.512$; $r \approx 0.673$

Linear Regression:
$y = 0.557x - 10.972$; $r \approx 0.947$

The exponential model has an r value closer to 1. Thus, the better model is $y = 3.46(1.02)^x$.

b. $y = 3.46(1.02)^x$

$y = 3.46e^{(\ln 1.02)x}$

$y = 3.46e^{0.02x}$

The 65-and-over population is increasing by approximately 2% each year.

75. does not make sense; Explanations will vary. Sample explanation: Since the car's value is decreasing (depreciating), the growth rate is negative.

77. makes sense

79. true

81. true

83. Use data to find k.

$$827 = 70 + (85.6 - 70)e^{k30}$$

$$12.7 = 15.6e^{30k}$$

$$\frac{12.7}{15.6} = e^{30k}$$

$$\ln \frac{12.7}{15.6} = \ln e^{30k}$$

$$\ln \frac{12.7}{15.6} = 30k$$

$$\frac{\ln \frac{12.7}{15.6}}{30} = k$$

$$-0.0069 \approx k$$

Use k to write equation.

$$85.6 = 70 + (98.6 - 70)e^{-0.0069t}$$

$$15.6 = 28.6e^{-0.0069t}$$

$$\frac{15.6}{28.6} = e^{-0.0069t}$$

$$\ln \frac{15.6}{28.6} = \ln e^{-0.0069t}$$

$$\ln \frac{15.6}{28.6} = -0.0069$$

$$\frac{\ln \frac{15.6}{28.6}}{-0.0069} = t$$

$$88 \approx t$$

The death occurred at 88 minutes before 9:30, or 8:22 am.

85.

$$\frac{5\pi}{4} = 2\pi x$$

$$\frac{5\pi}{4 \cdot 2\pi} = \frac{2\pi x}{2\pi}$$

$$\frac{5}{8} = x$$

The solution set is $\left\{ \frac{5}{8} \right\}$.

86.

$$\frac{17\pi}{6} - 2\pi = \frac{17\pi}{6} - \frac{12\pi}{6}$$

$$= \frac{17\pi - 12\pi}{6}$$

$$= \frac{5\pi}{6}$$

87.

$$-\frac{\pi}{12} + 2\pi = -\frac{\pi}{12} + \frac{24\pi}{12}$$

$$= \frac{-\pi + 24\pi}{12}$$

$$= \frac{23\pi}{12}$$

Chapter 3 Review Exercises

1. This is the graph of $f(x) = 4^x$ reflected about the y-axis, so the function is $g(x) = 4^{-x}$.

2. This is the graph of $f(x) = 4^x$ reflected about the x-axis and about the y-axis, so the function is $h(x) = -4^{-x}$.

3. This is the graph of $f(x) = 4^x$ reflected about the x-axis and about the y-axis then shifted upward 3 units, so the function is $r(x) = -4^{-x} + 3$.

4. This is the graph of $f(x) = 4^x$.

5. The graph of $g(x)$ shifts the graph of $f(x)$ one unit to the right.

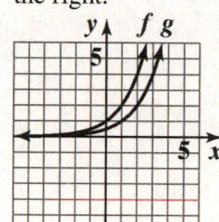

$f(x) = 2^x$
$g(x) = 2^x - 1$

asymptote of f: $y = 0$
asymptote of g: $y = 0$
domain of f = domain of g = $(-\infty, \infty)$
range of f = range of g = $(0, \infty)$

6. The graph of $g(x)$ shifts the graph of $f(x)$ one unit down.

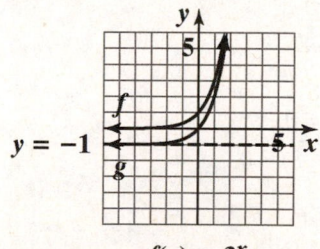

$f(x) = 3^x$
$g(x) = 3^x - 1$

asymptote of f: $y = 0$
asymptote of g: $y = -1$
domain of f = domain of g = $(-\infty, \infty)$
range of f = $(0, \infty)$
range of g = $(-1, \infty)$

7. The graph of $g(x)$ reflects the graph of $f(x)$ about the y – axis.

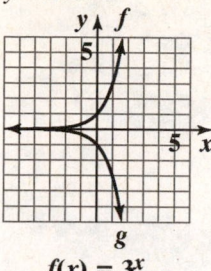

$f(x) = 3^x$
$g(x) = -3^x$

asymptote of f: $y = 0$
asymptote of g: $y = 0$
domain of f = domain of g = $(-\infty, \infty)$
range of f = $(0, \infty)$
range of g = $(-\infty, 0)$

8. The graph of $g(x)$ reflects the graph of $f(x)$ about the x – axis.

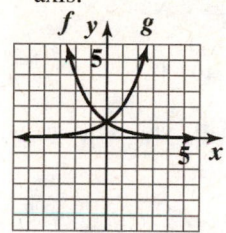

$f(x) = \left(\dfrac{1}{2}\right)^x$

$g(x) = \left(\dfrac{1}{2}\right)^{-x}$

asymptote of f: $y = 0$
asymptote of g: $y = 0$
domain of f = domain of g = $(-\infty, \infty)$
range of f = range of g = $(0, \infty)$

9. The graph of $g(x)$ vertically stretches the graph of $f(x)$ by a factor of 2.

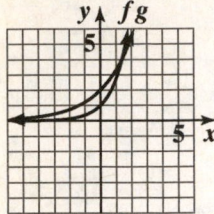

$$f(x) = e^x$$
$$g(x) = 2e^{x/2}$$

asymptote of f: $y = 0$

asymptote of g: $y = 0$

domain of f = domain of $g = (-\infty, \infty)$

range of f = range of $g = (0, \infty)$

10. 5.5% compounded semiannually:

$$A = 5000\left(1 + \frac{0.055}{2}\right)^{2 \cdot 5} \approx 6558.26$$

5.25% compounded monthly:

$$A = 5000\left(1 + \frac{0.0525}{12}\right)^{12 \cdot 5} \approx 6497.16$$

5.5% compounded semiannually yields the greater return.

11. 7% compounded monthly:

$$A = 14,000\left(1 + \frac{0.07}{12}\right)^{12 \cdot 10} \approx 28,135.26$$

6.85% compounded continuously:

$$A = 14,000e^{0.0685(10)} \approx 27,772.81$$

7% compounded monthly yields the greater return.

12. a. When first taken out of the microwave, the temperature of the coffee was 200°.

   b. After 20 minutes, the temperature of the coffee was about 120°.
   $$T = 70 + 130e^{-0.04855(20)} \approx 119.23$$
   Using a calculator, the temperature is about 119°.

   c. The coffee will cool to about 70°;
   The temperature of the room is 70°.

13. $49^{1/2} = 7$

14. $4^3 = x$

15. $3^y = 81$

16. $\log_6 216 = 3$

17. $\log_b 625 = 4$

18. $\log_{13} 874 = y$

19. $\log_4 64 = 3$ because $4^3 = 64$.

20. $\log_5 \frac{1}{25} = -2$ because $5^{-2} = \frac{1}{25}$.

21. $\log_3(-9)$ cannot be evaluated since $\log_b x$ is defined only for $x > 0$.

22. $\log_{16} 4 = \frac{1}{2}$ because $16^{1/2} = \sqrt{16} = 4$.

23. Because $\log_b b = 1$,
we conclude $\log_{17} 17 = 1$.

24. Because $\log_b b^x = x$,
we conclude $\log_3 3^8 = 8$.

25. Because $\ln e^x = x$,
we conclude $\ln e^5 = 5$.

26. $\log_3 \frac{1}{\sqrt{3}} = \log_3 \frac{1}{3^{\frac{1}{2}}} = \log_3 3^{-\frac{1}{2}} = -\frac{1}{2}$

27. $\ln \frac{1}{e^2} = \ln e^{-2} = -2$

28. $\log \frac{1}{1000} = \log \frac{1}{10^3} = \log 10^{-3} = -3$

29. Because $\log_b = 1$,
we conclude $\log_8 8 = 1$.
So, $\log_3(\log_8 8) = \log_3 1$.
Because $\log_b 1 = 0$
we conclude $\log_3 1 = 0$.
Therefore, $\log_3(\log_8 8) = 0$.

30.

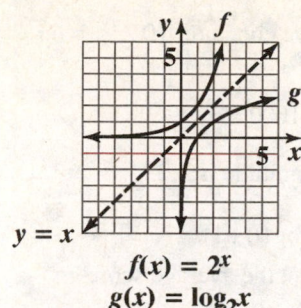

$$f(x) = 2^x$$
$$g(x) = \log_2 x$$

domain of f = range of $g = (-\infty, \infty)$

range of f = domain of $g = (0, \infty)$

31.

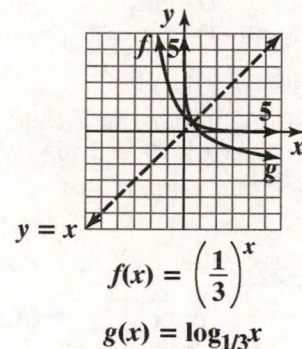

$$f(x) = \left(\frac{1}{3}\right)^x$$
$$g(x) = \log_{1/3} x$$

domain of f = range of $g = (-\infty, \infty)$

range of f = domain of $g = (0, \infty)$

32. This is the graph of $f(x) = \log x$ reflected about the
y-axis, so the function is
$g(x) = \log(-x)$.

33. This is the graph of $f(x) = \log x$
shifted left 2 units, reflected about the
y-axis, then shifted upward one unit, so the function
is $r(x) = 1 + \log(2 - x)$.

34. This is the graph of $f(x) = \log x$
shifted left 2 units then reflected about the
y-axis, so the function is $h(x) = \log(2 - x)$.

35. This is the graph of $f(x) = \log x$.

36.

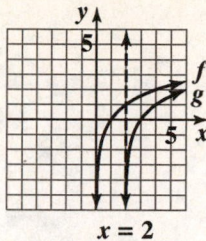

$$x = 2$$

$$f(x) = \log_2 x$$
$$g(x) = \log_2(x - 2)$$

x-intercept: (3, 0)
vertical asymptote: $x = 2$
domain: $(2, \infty)$

range: $(-\infty, \infty)$

37.

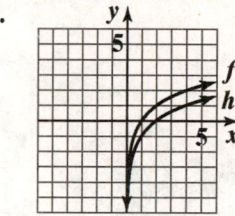

$$f(x) = \log_2 x$$
$$h(x) = -1 + \log_2 x$$

x-intercept: (2, 0)
vertical asymptote: $x = 0$
domain: $(0, \infty)$

range: $(-\infty, \infty)$

38.

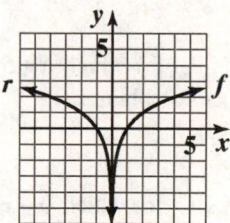

$$f(x) = \log_2 x$$
$$r(x) = \log_2(-x)$$

x-intercept: (-1, 0)
vertical asymptote: $x = 0$
domain: $(-\infty, 0)$

range: $(-\infty, \infty)$

39. $x = -3$

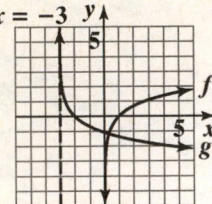

$f(x) = \log x$

$g(x) = -\log(x + 3)$

asymptote of f: $x = 0$

asymptote of g: $x = -3$

domain of $f = (0, \infty)$

domain of $g = (-3, \infty)$

range of f = range of $g = (-\infty, \infty)$

40.

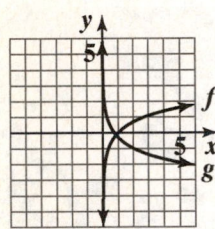

$f(x) = \ln x$

$g(x) = -\ln(2x)$

asymptote of f: $x = 0$

asymptote of g: $x = 0$

domain of f = domain of $g = (0, \infty)$

range of f = range of $g = (-\infty, \infty)$

41. The domain of f consists of all x for which $x + 5 > 0$. Solving this inequality for x, we obtain $x > -5$. Thus the domain of f is $(-5, \infty)$

42. The domain of f consists of all x for which $3 - x > 0$. Solving this inequality for x, we obtain $x < 3$. Thus, the domain of f is $(-\infty, 3)$.

43. The domain of f consists of all x for which $(x-1)^2 > 0$.

Solving this inequality for x, we obtain $x < 1$ or $x > 1$.

Thus, the domain of f is $(-\infty, 1) \cup (1, \infty)$.

44. Because $\ln e^x = x$, we conclude $\ln e^{6x} = 6x$.

45. Because $e^{\ln x} = x$, we conclude $e^{\ln \sqrt{x}} = \sqrt{x}$.

46. Because $10^{\log x} = x$, we conclude $10^{\log 4x^2} = 4x^2$.

47. $R = \log \dfrac{1000 I_0}{I_0} = \log 1000 = 3$

The Richter scale magnitude is 3.0.

48. a. $f(0) = 76 - 18\log(0+1) = 76$

When first given, the average score was 76.

b. $f(2) = 76 - 18\log(2+1) \approx 67$

$f(4) = 76 - 18\log(4+1) \approx 63$

$f(6) = 76 - 18\log(6+1) \approx 61$

$f(8) = 76 - 18\log(8+1) \approx 59$

$f(12) = 76 - 18\log(12+1) \approx 56$

After 2, 4, 6, 8, and 12 months, the average scores are about 67, 63, 61, 59, and 56, respectively.

c.

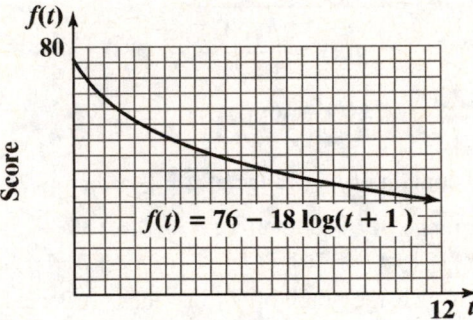

Time (months)

Retention decreases as time passes.

49. $t = \dfrac{1}{0.06}\ln\left(\dfrac{12}{12-5}\right) \approx 8.98$

It will take about 9 weeks.

50. $\log_6\left(36x^3\right)$

$= \log_6 36 + \log_6 x^3$

$= \log_6 36 + 3\log_6 x$

$= 2 + 3\log_6 x$

51. $\log_4 \dfrac{\sqrt{x}}{64} = \log_4 x^{1/2} - \log_4 64$

$= \dfrac{1}{2}\log_4 x - 3$

52. $\log_2 \dfrac{xy^2}{64} = \log_2 xy^2 - \log_2 64$

$= \log_2 x + \log_2 y^2 - \log_2 64$

$= \log_2 x + 2\log_2 y - 6$

53. $\ln \sqrt[3]{\dfrac{x}{e}}$

$= \ln\left(\dfrac{x}{e}\right)^{\!\frac{1}{3}}$

$= \dfrac{1}{3}\big[\ln x - \ln e\big]$

$= \dfrac{1}{3}\ln x - \dfrac{1}{3}\ln e$

$= \dfrac{1}{3}\ln x - \dfrac{1}{3}$

54. $\log_b 7 + \log_b 3$

$= \log_b (7 \cdot 3)$

$= \log_b 21$

55. $\log 3 - 3\log x$

$= \log 3 - \log x^3$

$= \log \dfrac{3}{x^3}$

56. $3\ln x + 4\ln y$

$= \ln x^3 + \ln y^4$

$= \ln\left(x^3 y^4\right)$

57. $\dfrac{1}{2}\ln x - \ln y$

$= \ln x^{\frac{1}{2}} - \ln y$

$= \ln \dfrac{\sqrt{x}}{y}$

58. $\log_6 72{,}348 = \dfrac{\log 72{,}348}{\log 6} \approx 6.2448$

59. $\log_4 0.863 = \dfrac{\ln 0.863}{\ln 4} \approx -0.1063$

60. true; $(\ln x)(\ln 1) = (\ln x)(0) = 0$

61. false; $\log(x+9) - \log(x+1) = \log\dfrac{(x+9)}{(x+1)}$

62. false; $\log_2 x^4 = 4\log_2 x$

63. true; $\ln e^x = x\ln e$

64. $\quad 2^{4x-2} = 64$

$\quad 2^{4x-2} = 2^6$

$\quad 4x - 2 = 6$

$\quad 4x = 8$

$\quad x = 2$

65. $\quad 125^x = 25$

$\quad \left(5^3\right)^x = 5^2$

$\quad 5^{3x} = 5^2$

$\quad 3x = 2$

$\quad x = \dfrac{2}{3}$

66. $\quad 10^x = 7000$

$\quad \log 10^x = \log 7000$

$\quad x\log 10 = \log 7000$

$\quad x = \log 7000$

$\quad x \approx 3.85$

67. $\quad 9^{x+2} = 27^{-x}$

$\quad \left(3^2\right)^{x+2} = \left(3^3\right)^{-x}$

$\quad 3^{2x+4} = 3^{-3x}$

$\quad 2x + 4 = -3x$

$\quad 5x = -4$

$\quad x = -\dfrac{4}{5}$

68. $\quad 8^x = 12{,}143$

$\quad \ln 8^x = \ln 12{,}143$

$\quad x\ln 8 = \ln 12{,}143$

$\quad x = \dfrac{\ln 12{,}143}{\ln 8} \approx 4.52$

69. $\quad 9e^{5x} = 1269$

$\quad e^{5x} = 141$

$\quad \ln e^{5x} = \ln 141$

$\quad 5x = \ln 141$

$\quad x = \dfrac{\ln 141}{5} \approx 0.99$

70. $e^{12-5x} - 7 = 123$

$e^{12-5x} = 130$

$\ln e^{12-5x} = \ln 130$

$12 - 5x = \ln 130$

$5x = 12 - \ln 130$

$x = \dfrac{12 - \ln 130}{5} \approx 1.43$

71. $5^{4x+2} = 37,500$

$\ln 5^{4x+2} = \ln 37,500$

$(4x+2)\ln 5 = \ln 37,500$

$4x \ln 5 + 2 \ln 5 = \ln 37,500$

$4x \ln 5 = \ln 37,500 - 2 \ln 5$

$x = \dfrac{\ln 37,500 - 2 \ln 5}{4 \ln 5} \approx 1.14$

72. $3^{x+4} = 7^{2x-1}$

$\ln 3^{x+4} = \ln 7^{2x-1}$

$(x+4)\ln 3 = (2x-1)\ln 7$

$x \ln 3 + 4 \ln 3 = 2x \ln 7 - \ln 7$

$x \ln 3 - 2x \ln 7 = -4 \ln 3 - \ln 7$

$x(\ln 3 - 2 \ln 7) = -4 \ln 3 - \ln 7$

$x = \dfrac{-4 \ln 3 - \ln 7}{\ln 3 - 2 \ln 7}$

$x = \dfrac{4 \ln 3 + \ln 7}{2 \ln 7 - \ln 3}$

$x \approx 2.27$

73. $e^{2x} - e^x - 6 = 0$

$(e^x - 3)(e^x + 2) = 0$

$e^x - 3 = 0 \ \text{ or } \ e^x + 2 = 0$

$e^x = 3 \qquad\qquad e^x = -2$

$\ln e^x = \ln 3 \quad \ln e^x = \ln(-2)$

$x = \ln 3 \qquad\quad x = \ln(-2)$

$x = \ln 3 \approx 1.099 \ \ \ln(-2)$ does not exist.

The solution set is $\{\ln 3\}$,

approximately 1.10.

74. $\log_4 (3x - 5) = 3$

$3x - 5 = 4^3$

$3x - 5 = 64$

$3x = 69$

$x = 23$

The solutions set is $\{23\}$.

75. $3 + 4 \ln(2x) = 15$

$4 \ln(2x) = 12$

$\ln(2x) = 3$

$2x = e^3$

$x = \dfrac{e^3}{2}$

$x \approx 10.04$

The solutions set is $\left\{ \dfrac{e^3}{2} \right\}$.

76. $\log_2 (x+3) + \log_2 (x-3) = 4$

$\log_2 (x+3)(x-3) = 4$

$\log_2 (x^2 - 9) = 4$

$x^2 - 9 = 2^4$

$x^2 - 9 = 16$

$x^2 = 25$

$x = \pm 5$

$x = -5$ does not check because $\log_2(-5+3)$ does not exist.

The solution set is $\{5\}$.

77. $\log_3 (x-1) - \log_3 (x+2) = 2$

$\log_3 \dfrac{x-1}{x+2} = 2$

$\dfrac{x-1}{x-2} = 3^2$

$\dfrac{x-1}{x+2} = 9$

$x - 1 = 9(x+2)$

$x - 1 = 9x + 18$

$8x = -19$

$x = -\dfrac{19}{8}$

$x = -\dfrac{19}{8}$ does not check because $\log_3 \left(-\dfrac{19}{8} - 1 \right)$

does not exist.

The solution set is $\varnothing$.

78. $\ln(x+4)-\ln(x+1)=\ln x$

$$\ln\frac{x+4}{x+1}=\ln x$$

$$\frac{x+4}{x+1}=x$$

$$x(x+1)=x+4$$

$$x^2+x=x+4$$

$$x^2=4$$

$$x=\pm2$$

$x=-2$ does not check and must be rejected.
The solution set is $\{2\}$.

79. $\log_4(2x+1)=\log_4(x-3)+\log_4(x+5)$

$\log_4(2x+1)=\log_4(x-3)+\log_4(x+5)$

$\log_4(2x+1)=\log_4(x^2+2x-15)$

$$2x+1=x^2+2x-15$$

$$16=x^2$$

$$x^2=16$$

$$x=\pm4$$

$x=-4$ does not check and must be rejected.
The solution set is $\{4\}$.

80. $P(x)=14.7e^{-0.21x}$

$$4.6=14.7e^{-0.21x}$$

$$\frac{4.6}{14.7}=e^{-0.21x}$$

$$\ln\frac{4.6}{14.7}=\ln e^{-0.21x}$$

$$\ln\frac{4.6}{14.7}=-0.21x$$

$$t=\frac{\ln\frac{4.6}{14.7}}{-0.21}\approx5.5$$

The peak of Mt. Everest is about 5.5 miles above sea level.

81. $f(t)=364(1.005)^t$

$$560=364(1.005)^t$$

$$\frac{560}{364}=(1.005)^t$$

$$\ln\frac{560}{364}=\ln(1.005)^t$$

$$\ln\frac{560}{364}=t\ln1.005$$

$$t=\frac{\ln\frac{560}{364}}{\ln1.005}\approx86.4$$

The carbon dioxide concentration will be double the pre-industrial level approximately 86 years after the year 2000 in the year 2086.

82. $W(x)=0.37\ln x+0.05$

$$3.38=0.37\ln x+0.05$$

$$3.33=0.37\ln x$$

$$\frac{3.33}{0.37}=\ln x$$

$$9=\ln x$$

$$e^9=e^{\ln x}$$

$$x=e^9\approx8103$$

The population of New York City is approximately 8103 thousand, or 8,103,000.

83. $20,000=12,500\left(1+\frac{0.065}{4}\right)^{4t}$

$$12,500(1.01625)^{4t}=20,000$$

$$(1.01625)^{4t}=1.6$$

$$\ln(1.01625)^{4t}=\ln1.6$$

$$4t\ln1.01625=\ln1.6$$

$$t=\frac{\ln1.6}{4\ln1.01625}\approx7.3$$

It will take about 7.3 years.

84. $3\cdot50,000=50,000e^{0.075t}$

$$50,000e^{0.075t}=150,000$$

$$e^{0.075}=3$$

$$\ln e^{0.075t}=\ln3$$

$$0.075t=\ln3$$

$$t=\frac{\ln3}{0.075}\approx14.6$$

It will take about 14.6 years.

Exponential and Logarithmic Functions

85. When an investment value triples, $A = 3P$.

$3P = Pe^{5r}$

$e^{5r} = 3$

$\ln e^{5r} = \ln 3$

$5r = \ln 3$

$r = \dfrac{\ln 3}{5} \approx 0.2197$

The interest rate would need to be about 22%

86. **a.** $35.3 = 22.4e^{k10}$

$\dfrac{35.3}{22.4} = e^{10k}$

$\ln \dfrac{35.3}{22.4} = \ln e^{10k}$

$\ln \dfrac{35.3}{22.4} = 10k$

$\dfrac{\ln \dfrac{35.3}{22.4}}{10} = k$

$0.045 \approx k$

$A = 22.4e^{0.045t}$

b. $A = 22.4e^{0.045(20)} \approx 55.1$

In 2010, the population will be about 55.1 million.

c. $60 = 22.4e^{0.045t}$

$\dfrac{60}{22.4} = e^{0.045t}$

$\ln \dfrac{60}{22.4} = \ln e^{0.045t}$

$\ln \dfrac{60}{22.4} = 0.045t$

$\dfrac{\ln \dfrac{60}{22.4}}{0.045} = t$

$22 \approx t$

The population will reach 60 million about 22 years after 1990, in 2012.

87. Use the half-life of 140 days to find k.

$A = A_0 e^{kt}$

$\tfrac{1}{2} = e^{k \cdot 140}$

$\tfrac{1}{2} = e^{140k}$

$\ln \tfrac{1}{2} = \ln e^{140k}$

$\ln \tfrac{1}{2} = 140k$

$\dfrac{\ln \tfrac{1}{2}}{140} = k$

$k \approx -0.004951$

Use $A = A_0 e^{kt}$ to find t.

$A = A_0 e^{-0.004951t}$

$0.2 = e^{-0.004951t}$

$\ln 0.2 = \ln e^{-0.004951t}$

$\ln 0.2 = -0.004951t$

$t = \dfrac{\ln 0.2}{-0.004951}$

$t \approx 325$

It will take about 325 days for the substance to decay to 20% of its original amount.

88. **a.** $f(0) = \dfrac{500,000}{1 + 2499e^{-0.92(0)}} = 200$

200 people became ill when the epidemic began.

b. $f(6) = \dfrac{500,000}{1 + 2499e^{-0.92(6)}} = 45,411$

45,410 were ill after 6 weeks.

c. 500,000 people

89. a.

$$T = C + (T_o - C)e^{kt}$$

$$150 = 65 + (185 - 65)e^{k2}$$

$$90 = 120e^{2k}$$

$$\frac{90}{120} = e^{2k}$$

$$\ln\frac{3}{4} = \ln e^{2k}$$

$$\ln\frac{3}{4} = 2k$$

$$\frac{\ln\frac{3}{4}}{2} = k$$

$$-0.1438 \approx k$$

$$T = 65 + 120e^{-0.1438t}$$

b.

$$105 = 65 + 120e^{-0.1438t}$$

$$40 = 120e^{-0.1438t}$$

$$\frac{1}{3} = e^{-0.1438t}$$

$$\ln\frac{1}{3} = \ln e^{-0.1438t}$$

$$\ln\frac{1}{3} = -0.1438t$$

$$\frac{\ln\frac{1}{3}}{-0.1438} = t$$

$$7.6 \approx t$$

The temperature will reach 105° after 8 min.

90. a.

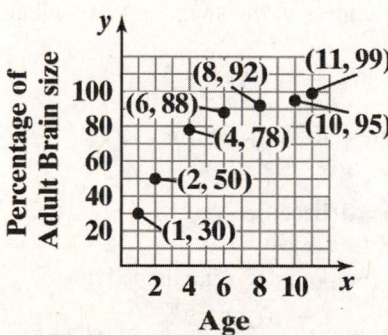

b. A logarithmic function appears to be the better choice for modeling the data.

Chapter 3 Test

1.

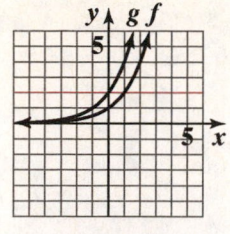

$$f(x) = 2^x$$
$$g(x) = 2^x + 1$$

2.

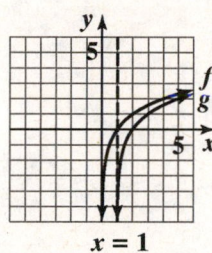

$$x = 1$$
$$f(x) = \log_2 x$$
$$g(x) = \log_2(x - 1)$$

3. $5^3 = 125$

4. $\log_{36} 6 = \dfrac{1}{2}$

5. The domain of f consists of all x for which $3 - x > 0$. Solving this inequality for x, we obtain $x < 3$.
Thus, the domain of f is $(-\infty, 3)$.

6. $\log_4\left(64x^5\right) = \log_4 64 + \log_4 x^5$
$$= 3 + 5\log_4 x$$

7. $\log_3 \dfrac{\sqrt[3]{x}}{81} = \log_3 x^{\frac{1}{3}} - \log_3 81$
$$= \frac{1}{3}\log_3 x - 4$$

8. $6\log x + 2\log y = \log x^6 + \log y^2$
$$= \log\left(x^6 y^2\right)$$

9. $\ln 7 - 3\ln x = \ln 7 - \ln x^3$
$$= \ln\frac{7}{x^3}$$

10. $\log_{15} 71 = \dfrac{\log 71}{\log 15} \approx 1.5741$

11. $3^{x-2} = 9^{x+4}$

$3^{x-2} = \left(3^2\right)^{x+4}$

$3^{x-2} = 3^{2x+8}$

$x - 2 = 2x + 8$

$-x = 10$

$x = -10$

12. $5^x = 1.4$

$\ln 5^x = \ln 1.4$

$x \ln 5 = \ln 1.4$

$x = \dfrac{\ln 1.4}{\ln 5} \approx 0.2091$

13. $400e^{0.005x} = 1600$

$e^{0.005x} = 4$

$\ln e^{0.005x} = \ln 4$

$0.005x = \ln 4$

$x = \dfrac{\ln 4}{0.005} \approx 277.2589$

14. $e^{2x} - 6e^x + 5 = 0$

$\left(e^x - 5\right)\left(e^x - 1\right) = 0$

$e^x - 5 = 0 \qquad$ or $\qquad e^x - 1 = 0$

$\qquad e^x = 5 \qquad\qquad\qquad e^x = 1$

$\ln e^x = \ln 5 \qquad\qquad \ln e^x = \ln 1$

$\qquad x = \ln 5 \qquad\qquad\qquad x = \ln 1$

$\qquad x \approx 1.6094 \qquad\qquad\quad x = 0$

The solution set is $\{0, \ln 5\}$; $\ln \approx 1.6094$.

15. $\log_6 (4x - 1) = 3$

$4x - 1 = 6^3$

$4x - 1 = 216$

$4x = 217$

$x = \dfrac{217}{4} = 54.25$

16. $2 \ln 3x = 8$

$\ln 3x = 4$

$3x = e^4$

$x = \dfrac{e^4}{3} \approx 18.1994$

17. $\log x + \log (x + 15) = 2$

$\log \left(x^2 + 15x\right) = 2$

$x^2 + 15x = 10^2$

$x^2 + 15x - 100 = 0$

$(x + 20)(x - 5) = 0$

$x + 20 = 0$ or $x - 5 = 0$

$x = -20 \qquad x = 5$

$x = -20$ does not check because $\log(-20)$ does not exist.

The solution set is $\{5\}$.

18. $\ln (x - 4) - \ln (x + 1) = \ln 6$

$\ln \dfrac{x - 4}{x + 1} = \ln 6$

$\dfrac{x - 4}{x + 1} = 6$

$6(x + 1) = x - 4$

$6x + 6 = x - 4$

$5x = -10$

$x = -2$

$x = -2$ does not check and must be rejected.

The solution set is $\{\ \}$.

19. $D = 10 \log \dfrac{10^{12} I_0}{I_0}$

$\quad = 10 \log 10^{12}$

$\quad = 10 \cdot 12$

$\quad = 120$

The loudness of the sound is 120 decibels.

20. Since $\ln e^x = x$, $\ln e^{5x} = 5x$.

21. $\log_b b = 1$ because $b^1 = b$.

22. $\log_6 1 = 0$ because $6^0 = 1$.

23. 6.5% compounded semiannually:

$A = 3,000 \left(1 + \dfrac{0.065}{2}\right)^{2(10)} \approx \$5,687.51$

6% compounded continuously:

$A = 3,000 e^{0.06(10)} \approx \$5,466.36$

6.5% compounded semiannually yields about \$221 more than 6% compounded continuously.

24.

$$8000 = 4000\left(1 + \frac{0.05}{4}\right)^{4t}$$

$$\frac{8000}{4000} = (1 + 0.0125)^{4t}$$

$$2 = (1.0125)^{4t}$$

$$\ln 2 = \ln(1.0125)^{4t}$$

$$\ln 2 = 4t\ln(1.0125)$$

$$\frac{\ln 2}{4\ln(1.0125)} = \frac{4t\ln(1.0125)}{4\ln(1.0125)}$$

$$t = \frac{\ln 2}{4\ln(1.0125)} \approx 13.9$$

It will take approximately 13.9 years for the money to grow to $8000.

25.

$$2 = 1e^{r10}$$

$$2 = e^{10r}$$

$$\ln 2 = \ln e^{10r}$$

$$\ln 2 = 10r$$

$$r = \frac{\ln 2}{10} \approx 0.069$$

The money will double in 10 years with an interest rate of approximately 6.9%.

26. a.

$$A = 82.4e^{-0.002(x)}$$

$$A = 82.4e^{-0.002(0)} \approx 82.4$$

In 2006, the population of Germany was 82.4 million.

b. The population of Germany is decreasing. We can tell because the model has a negative $k = -0.002$.

c.

$$81.5 = 82.4e^{-0.002t}$$

$$\frac{81.5}{82.4} = e^{-0.002t}$$

$$\ln \frac{81.5}{82.4} = \ln e^{-0.002t}$$

$$\ln \frac{81.5}{82.4} = -0.002t$$

$$t = \frac{\ln \frac{81.5}{82.4}}{-0.002} \approx 5$$

The population of Germany will be 81.5 million approximately 5 years after 2006 in the year 2011.

27. In 1990, $t = 0$ and $A_0 = 509$
In 2000, $t = 2000 - 1990 = 10$ and $A = 729$.

$$729 = 509e^{k10}$$

$$\frac{729}{509} = e^{10k}$$

$$\ln \frac{729}{509} = \ln e^{10k}$$

$$\ln \frac{729}{509} = 10k$$

$$\frac{\ln \frac{729}{509}}{10} = k$$

$$0.036 \approx k$$

The exponential growth function is
$$A = 509e^{0.036t}.$$

28. First find the decay equation.

$$0.5 = e^{kt}$$

$$0.5 = e^{7.2k}$$

$$\ln 0.5 = \ln e^{7.2k}$$

$$\ln 0.5 = 7.2k$$

$$\frac{\ln 0.5}{7.2} = k$$

$$k \approx -0.096270$$

$$A = e^{-0.096270t}$$

Next use the decay equation answer question.

$$A = e^{-0.096270t}$$

$$0.3 = e^{-0.096270t}$$

$$\ln 0.3 = \ln e^{-0.096270t}$$

$$\ln 0.3 = -0.096270t$$

$$\frac{\ln 0.3}{-0.096270} = t$$

$$t \approx 12.5$$

It will take 12.5 days.

29. a. $f(0) = \dfrac{140}{1+9e^{-0.165(0)}} = 14$

Fourteen elk were initially introduced to the habitat.

b. $f(10) = \dfrac{140}{1+9e^{-0.165(10)}} \approx 51$

After 10 years, about 51 elk are expected.

c. In the logistic growth model,

$$f(t) = \dfrac{c}{1+ae^{-bt}},$$

the constant c represents the limiting size that $f(t)$ can attain. The limiting size of the elk population is 140 elk.

30. Plot the ordered pairs.

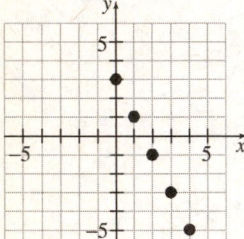

The values appear to belong to a linear function.

31. Plot the ordered pairs.

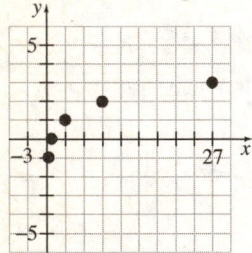

The values appear to belong to a logarithmic function.

32. Plot the ordered pairs.

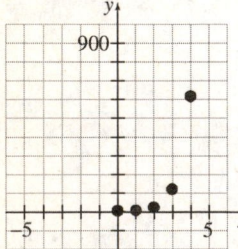

The values appear to belong to an exponential function.

33. Plot the ordered pairs.

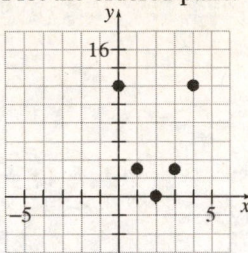

The values appear to belong to a quadratic function.

34.

$$y = 96(0.38)^x$$

$$y = 96e^{(\ln 0.38)x}$$

$$y = 96e^{-0.968x}$$

Cumulative Review Exercises (Chapters P–3)

1. $|3x - 4| = 2$

$3x - 4 = 2$ or $\quad 3x - 4 = -2$

$3x = 6 \qquad\qquad 3x = 2$

$x = 2 \qquad\qquad x = \dfrac{2}{3}$

The solution set is $\left\{\dfrac{2}{3}, 2\right\}$.

2. $x^2 + 2x + 5 = 0$

$x = \dfrac{-b \pm \sqrt{b^2 - 4ac}}{2a}$

$x = \dfrac{-(2) \pm \sqrt{(2)^2 - 4(1)(5)}}{2(1)}$

$x = \dfrac{-2 \pm \sqrt{-16}}{2}$

$x = \dfrac{-2 \pm 4i}{2}$

$x = -1 \pm 2i$

The solution set is $\{-1 \pm 2i\}$.

3. $x^4 + x^3 - 3x^2 - x + 2 = 0$

$p: \pm 1, \pm 2$

$q: \pm 1$

$\dfrac{p}{q}: \pm 1, \pm 2$

$$\begin{array}{r|rrrrr} -2 & 1 & 1 & -3 & -1 & 2 \\ & & -2 & 2 & 2 & -2 \\ \hline & 1 & -1 & -1 & 1 & 0 \end{array}$$

$(x+2)(x^3 - x^2 - x + 1) = 0$

$(x+2)[x^2(x-1) - (x-1)] = 0$

$(x+2)(x^2 - 1)(x-1) = 0$

$(x+2)(x+1)(x-1)(x-1) = 0$

$(x+2)(x+1)(x-1)^2 = 0$

$x+2 = 0 \quad$ or $\quad x+1 = 0 \quad$ or $\quad x-1 = 0$

$x = -2 \qquad\qquad x = -1 \qquad\qquad x = 1$

The solution set is $\{-2, -1, 1\}$.

4. $e^{5x} - 32 = 96$

$e^{5x} = 128$

$\ln e^{5x} = \ln 128$

$5x = \ln 128$

$x = \dfrac{\ln 128}{5} \approx 0.9704$

The solution set is $\left\{ \dfrac{\ln 128}{5} \right\}$, approximately 0.9704.

5. $\log_2(x+5) + \log_2(x-1) = 4$

$\log_2[(x+5)(x-1)] = 4$

$(x+5)(x-1) = 2^4$

$x^2 + 4x - 5 = 16$

$x^2 + 4x - 21 = 0$

$(x+7)(x-3) = 0$

$x+7 = 0 \quad$ or $\quad x-3 = 0$

$x = -7 \qquad\qquad x = 3$

$x = -7$ does not check because $\log_2(-7+5)$ does not exist.

The solution set is $\{3\}$.

6. $\ln(x+4) + \ln(x+1) = 2\ln(x+3)$

$\ln((x+4)(x+1)) = \ln(x+3)^2$

$(x+4)(x+1) = (x+3)^2$

$x^2 + 5x + 4 = x^2 + 6x + 9$

$5x + 4 = 6x + 9$

$-x = 5$

$x = -5$

$x = -5$ does not check and must be rejected.

The solution set is { }.

7. $14 - 5x \geq -6$

$-5x \geq -20$

$x \leq 4$

The solution set is $(-\infty, 4]$.

8. $|2x - 4| \leq 2$

$2x - 4 \leq 2$ and $2x - 4 \geq -2$

$2x \leq 6 \qquad\qquad 2x \geq 2$

$x \leq 3 \qquad$ and $\quad x \geq 1$

The solution set is $[1, 3]$.

9. Circle with center: $(3, -2)$ and radius of 2

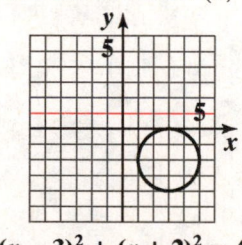

$(x - 3)^2 + (y + 2)^2 = 4$

10. Parabola with vertex: $(2, -1)$

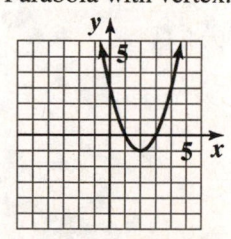

$f(x) = (x - 2)^2 - 1$

11. x-intercepts:

$$x^2 - 1 = 0$$
$$x^2 = 1$$
$$x = \pm 1$$

The x-intercepts are $(1, 0)$ and $(-1, 0)$.

vertical asymptotes:

$$x^2 - 4 = 0$$
$$x^2 = 4$$
$$x = \pm 2$$

The vertical asymptotes are $x = 2$ and $x = -2$.

Horizontal asymptote: y 5 1

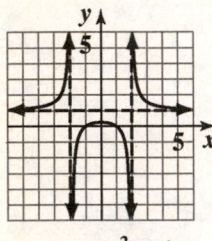

$$f(x) = \frac{x^2 - 1}{x^2 - 4}$$

12. x-intercepts:

$$x - 2 = 0 \quad \text{or} \quad x + 1 = 0$$
$$x = 2 \quad \text{or} \quad x = -1$$

The x-intercepts are $(2, 0)$ and $(-1, 0)$.

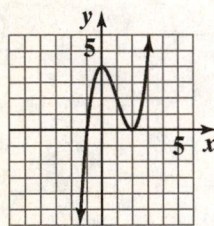

$$f(x) = (x - 2)^2(x + 1)$$

13.

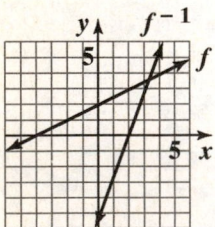

$$f(x) = 2x - 4$$
$$f^{-1}(x) = \frac{x + 4}{2}$$

14.

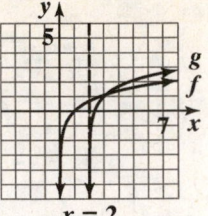

$$f(x) = \ln x$$
$$g(x) = \ln (x - 2) + 1$$

15. $m = \dfrac{3 - (-3)}{1 - 3} = \dfrac{6}{-2} = -3$

Using $(1, 3)$ point-slope form:

$$y - 3 = -3(x - 1)$$

slope-intercept form:

$$y - 3 = -3(x - 1)$$
$$y - 3 = -3x + 3$$
$$y = -3x + 6$$

16. $(f \circ g)(x) = f(x + 2)$
$$= (x + 2)^2$$
$$= x^2 + 4x + 4$$

$(g \circ f)(x) = g(x^2)$
$$= x^2 + 2$$

17. y varies inversely as the square of x is expressed as

$y = \dfrac{k}{x^2}$.

The hours, H, vary inversely as the square of the number of cups of coffee, C can be expressed

as $H = \dfrac{k}{C^2}$.

Use the given values to find k.

$H = \dfrac{k}{C^2}$

$8 = \dfrac{k}{2^2}$

$32 = k$

Substitute the value of k into the equation.

$H = \dfrac{k}{C^2}$

$H = \dfrac{32}{C^2}$

Use the equation to find H when $C = 4$.

$H = \dfrac{32}{C^2}$

$H = \dfrac{32}{4^2}$

$H = 2$

If 4 cups of coffee are consumed you should expect to sleep 2 hours.

18. $s(t) = -16t^2 + 64t + 5$

The ball reaches its maximum height at

$t = \dfrac{-b}{2a} = \dfrac{-(64)}{2(-16)} = 2$ seconds.

The maximum height is $s(2)$.

$s(2) = -16(2)^2 + 64(2) + 5 = 69$ feet.

19. $s(t) = -16t^2 + 64t + 5$

Let $s(t) = 0$:

$0 = -16t^2 + 64t + 5$

Use the quadratic formula to solve.

$t = \dfrac{-b \pm \sqrt{b^2 - 4ac}}{2a}$

$t = \dfrac{-(64) \pm \sqrt{(64)^2 - 4(-16)(5)}}{2(-16)}$

$t \approx 4.1, \quad t \approx -0.1$

The negative value is rejected.

The ball hits the ground after about 4.1 seconds.

20. $40x + 10(1.5x) = 660$

$40x + 15x = 660$

$55x = 660$

$x = 12$

Your normal hourly salary is \$12 per hour.

Chapter 4
Trigonometric Functions

Section 4.1

Check Point Exercises

1. The radian measure of a central angle is the length of the intercepted arc, s, divided by the circle's radius, r. The length of the intercepted arc is 42 feet: $s = 42$ feet. The circle's radius is 12 feet: $r = 12$ feet. Now use the formula for radian measure to find the radian measure of θ.

$$\theta = \frac{s}{r} = \frac{42 \text{ feet}}{12 \text{ feet}} = 3.5$$

Thus, the radian measure of θ is 3.5

2. a. $60° = 60° \cdot \dfrac{\pi \text{ radians}}{180°} = \dfrac{60\pi}{180}$ radians

$\qquad = \dfrac{\pi}{3}$ radians

b. $270° = 270° \cdot \dfrac{\pi \text{ radians}}{180°} = \dfrac{270\pi}{180}$ radians

$\qquad = \dfrac{3\pi}{2}$ radians

c. $-300° = -300° \cdot \dfrac{\pi \text{ radians}}{180°} = \dfrac{-300\pi}{180}$ radians

$\qquad = -\dfrac{5\pi}{3}$ radians

3. a. $\dfrac{\pi}{4}$ radians $= \dfrac{\pi \text{ radians}}{4} \cdot \dfrac{180°}{\pi \text{ radians}}$

$\qquad = \dfrac{180°}{4} = 45°$

b. $-\dfrac{4\pi}{3}$ radians $= -\dfrac{4\pi \text{ radians}}{3} \cdot \dfrac{180°}{\pi}$

$\qquad = -\dfrac{4 \cdot 180°}{3} = -240°$

c. 6 radians $= 6$ radians $\cdot \dfrac{180°}{\pi \text{ radians}}$

$\qquad = \dfrac{6 \cdot 180°}{\pi} \approx 343.8°$

4. a.

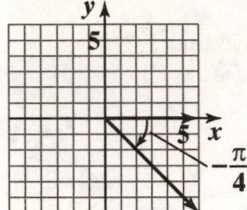

b.

c.

d.

5. a. For a 400° angle, subtract 360° to find a positive coterminal angle.
$400° - 360° = 40°$

b. For a −135° angle, add 360° to find a positive coterminal angle.
$-135° + 360° = 225°$

6. a. $\dfrac{13\pi}{5} - 2\pi = \dfrac{13\pi}{5} - \dfrac{10\pi}{5} = \dfrac{3\pi}{5}$

b. $-\dfrac{\pi}{15} + 2\pi = -\dfrac{\pi}{15} + \dfrac{30\pi}{15} = \dfrac{29\pi}{15}$

7. a. $855° - 360° \cdot 2 = 855° - 720° = 135°$

    b. $\dfrac{17\pi}{3} - 2\pi \cdot 2 = \dfrac{17\pi}{3} - 4\pi$

          $= \dfrac{17\pi}{3} - \dfrac{12\pi}{3} = \dfrac{5\pi}{3}$

    c. $-\dfrac{25\pi}{6} + 2\pi \cdot 3 = -\dfrac{25\pi}{6} + 6\pi$

          $= -\dfrac{25\pi}{6} + \dfrac{36\pi}{6} = \dfrac{11\pi}{6}$

8. The formula $s = r\theta$ can only be used when θ is expressed in radians. Thus, we begin by converting $45°$ to radians. Multiply by $\dfrac{\pi \text{ radians}}{180°}$.

    $45° = 45° \cdot \dfrac{\pi \text{ radians}}{180°} = \dfrac{45}{180}\pi \text{ radians}$

    $= \dfrac{\pi}{4} \text{ radians}$

    Now we can use the formula $s = r\theta$ to find the length of the arc. The circle's radius is 6 inches : $r = 6$ inches. The measure of the central angle in radians is $\dfrac{\pi}{4} : \theta = \dfrac{\pi}{4}$. The length of the arc intercepted by this central angle is

    $s = r\theta = (6 \text{ inches})\left(\dfrac{\pi}{4}\right) = \dfrac{6\pi}{4} \text{ inches} \approx 4.71 \text{ inches}.$

9. We are given ω, the angular speed.
    $\omega = 45$ revolutions per minute
    We use the formula $v = r\omega$ to find v, the linear speed. Before applying the formula, we must express ω in radians per minute.

    $\omega = \dfrac{45 \text{ revolutions}}{1 \text{ minute}} \cdot \dfrac{2\pi \text{ radians}}{1 \text{ revolution}}$

    $= \dfrac{90\pi \text{ radians}}{1 \text{ minute}}$

    The angular speed of the propeller is 90π radians per minute. The linear speed is

    $v = r\omega = 1.5 \text{ inches} \cdot \dfrac{90\pi}{1 \text{ minute}} = \dfrac{135\pi \text{ inches}}{\text{minute}}$

    The linear speed is 135π inches per minute, which is approximately 424 inches per minute.

Exercise Set 4.1

1. obtuse

3. acute

5. straight

7. $\theta = \dfrac{s}{r} = \dfrac{40 \text{ inches}}{10 \text{ inches}} = 4 \text{ radians}$

9. $\theta = \dfrac{s}{r} = \dfrac{8 \text{ yards}}{6 \text{ yards}} = \dfrac{4}{3} \text{ radians}$

11. $\theta = \dfrac{s}{r} = \dfrac{400 \text{ centimeters}}{100 \text{ centimeters}} = 4 \text{ radians}$

13. $45° = 45° \cdot \dfrac{\pi \text{ radians}}{180°}$

    $= \dfrac{45\pi}{180} \text{ radians}$

    $= \dfrac{\pi}{4} \text{ radians}$

15. $135° = 135° \cdot \dfrac{\pi \text{ radians}}{180°}$

    $= \dfrac{135\pi}{180} \text{ radians}$

    $= \dfrac{3\pi}{4} \text{ radians}$

17. $300° = 300° \cdot \dfrac{\pi \text{ radians}}{180°}$

    $= \dfrac{300\pi}{180} \text{ radians}$

    $= \dfrac{5\pi}{3} \text{ radians}$

19. $-225° = -225° \cdot \dfrac{\pi \text{ radians}}{180°}$

    $= -\dfrac{225\pi}{180} \text{ radians}$

    $= -\dfrac{5\pi}{4} \text{ radians}$

21. $\dfrac{\pi}{2} \text{ radians} = \dfrac{\pi \text{ radians}}{2} \cdot \dfrac{180°}{\pi \text{ radians}}$

    $= \dfrac{180°}{2}$

    $= 90°$

23. $\dfrac{2\pi}{3}$ radians $= \dfrac{2\pi \text{ radians}}{3} \cdot \dfrac{180^{\circ}}{\pi \text{ radians}}$

$\qquad = \dfrac{2 \cdot 180^{\circ}}{3}$

$\qquad = 120^{\circ}$

25. $\dfrac{7\pi}{6}$ radians $= \dfrac{7\pi \text{ radians}}{6} \cdot \dfrac{180^{\circ}}{\pi \text{ radians}}$

$\qquad = \dfrac{7 \cdot 180^{\circ}}{6}$

$\qquad = 210^{\circ}$

27. -3π radians $= -3\pi \text{ radians} \cdot \dfrac{180^{\circ}}{\pi \text{ radians}}$

$\qquad = -3 \cdot 180^{\circ}$

$\qquad = -540^{\circ}$

29. $18^{\circ} = 18^{\circ} \cdot \dfrac{\pi \text{ radians}}{180^{\circ}}$

$\qquad = \dfrac{18\pi}{180}$ radians

$\qquad \approx 0.31$ radians

31. $-40^{\circ} = -40^{\circ} \cdot \dfrac{\pi \text{ radians}}{180^{\circ}}$

$\qquad = -\dfrac{40\pi}{180}$ radians

$\qquad \approx -0.70$ radians

33. $200^{\circ} = 200^{\circ} \cdot \dfrac{\pi \text{ radians}}{180^{\circ}}$

$\qquad = \dfrac{200\pi}{180}$ radians

$\qquad \approx 3.49$ radians

35. 2 radians $= 2 \text{ radians} \cdot \dfrac{180^{\circ}}{\pi \text{ radians}}$

$\qquad = \dfrac{2 \cdot 180^{\circ}}{\pi}$

$\qquad \approx 114.59^{\circ}$

37. $\dfrac{\pi}{13}$ radians $= \dfrac{\pi \text{ radians}}{13} \cdot \dfrac{180^{\circ}}{\pi \text{ radians}}$

$\qquad = \dfrac{180^{\circ}}{13}$

$\qquad \approx 13.85^{\circ}$

39. -4.8 radians $= -4.8 \text{ radians} \cdot \dfrac{180^{\circ}}{\pi \text{ radians}}$

$\qquad = \dfrac{-4.8 \cdot 180^{\circ}}{\pi}$

$\qquad \approx -275.02^{\circ}$

41.

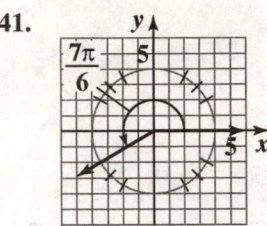

43.

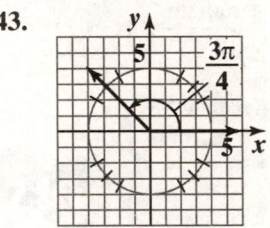

45.

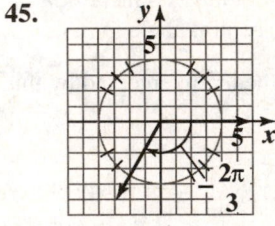

47.

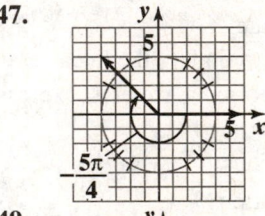

49.

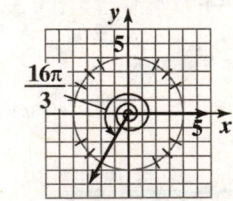

51.

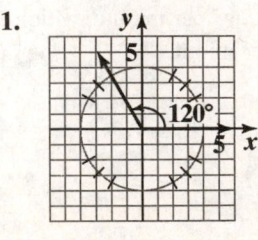

53.

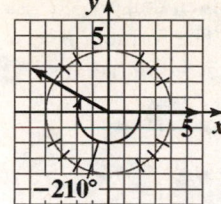

55.

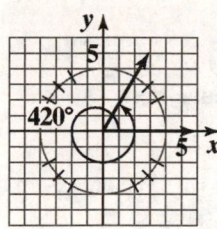

57. $395° - 360° = 35°$

59. $-150° + 360° = 210°$

61. $-765° + 360° \cdot 3 = -765° + 1080° = 315°$

63. $\dfrac{19\pi}{6} - 2\pi = \dfrac{19\pi}{6} - \dfrac{12\pi}{6} = \dfrac{7\pi}{6}$

65. $\dfrac{23\pi}{5} - 2\pi \cdot 2 = \dfrac{23\pi}{5} - 4\pi = \dfrac{23\pi}{5} - \dfrac{20\pi}{5} = \dfrac{3\pi}{5}$

67. $-\dfrac{\pi}{50} + 2\pi = -\dfrac{\pi}{50} + \dfrac{100\pi}{50} = \dfrac{99\pi}{50}$

69. $-\dfrac{31\pi}{7} + 2\pi \cdot 3 = -\dfrac{31\pi}{7} + 6\pi$

$= -\dfrac{31\pi}{7} + \dfrac{42\pi}{7} = \dfrac{11\pi}{7}$

71. $r = 12$ inches, $\theta = 45°$

Begin by converting 45° to radians, in order to use the formula $s = r\theta$.

$45° = 45° \cdot \dfrac{\pi \text{ radians}}{180°} = \dfrac{\pi}{4}$ radians

Now use the formula $s = r\theta$.

$s = r\theta = 12 \cdot \dfrac{\pi}{4} = 3\pi$ inches ≈ 9.42 inches

73. $r = 8$ feet, $\theta = 225°$

Begin by converting 225° to radians, in order to use the formula $s = r\theta$.

$225° = 225° \cdot \dfrac{\pi \text{ radians}}{180°} = \dfrac{5\pi}{4}$ radians

Now use the formula $s = r\theta$.

$s = r\theta = 8 \cdot \dfrac{5\pi}{4} = 10\pi$ feet ≈ 31.42 feet

75. 6 revolutions per second

$= \dfrac{6 \text{ revolutions}}{1 \text{ second}} \cdot \dfrac{2\pi \text{ radians}}{1 \text{ revolutions}} = \dfrac{12\pi \text{ radians}}{1 \text{ seconds}}$

$= 12\pi$ radians per second

77. $-\dfrac{4\pi}{3}$ and $\dfrac{2\pi}{3}$

79. $-\dfrac{3\pi}{4}$ and $\dfrac{5\pi}{4}$

81. $-\dfrac{\pi}{2}$ and $\dfrac{3\pi}{2}$

83. $\dfrac{55}{60} \cdot 2\pi = \dfrac{11\pi}{6}$

85. 3 minutes and 40 seconds equals 220 seconds.

$\dfrac{220}{60} \cdot 2\pi = \dfrac{22\pi}{3}$

87. First, convert to degrees.

$\dfrac{1}{6}$ revolution $= \dfrac{1}{6}$ revolution $\cdot \dfrac{360°}{1 \text{ revolution}}$

$= \dfrac{1}{6} \cdot 360° = 60°$

Now, convert 60° to radians.

$60° = 60° \cdot \dfrac{\pi \text{ radians}}{180°} = \dfrac{60\pi}{180}$ radians

$= \dfrac{\pi}{3}$ radians

Therefore, $\dfrac{1}{6}$ revolution is equivalent to 60° or $\dfrac{\pi}{3}$ radians.

89. The distance that the tip of the minute hand moves is given by its arc length, s. Since $s = r\theta$, we begin by finding r and θ. We are given that $r = 8$ inches. The minute hand moves from 12 to 2 o'clock, or $\dfrac{1}{6}$ of a complete revolution. The formula $s = r\theta$ can only be used when θ is expressed in radians. We must convert $\dfrac{1}{6}$ revolution to radians.

$$\frac{1}{6}\text{ revolution} = \frac{1}{6}\text{ revolution} \cdot \frac{2\pi\text{ radians}}{1\text{ revolution}}$$

$$= \frac{\pi}{3}\text{ radians}$$

The distance the tip of the minute hand moves is

$$s = r\theta = (8\text{ inches})\left(\frac{\pi}{3}\right) = \frac{8\pi}{3}\text{ inches}$$

$$\approx 8.38\text{ inches.}$$

91. The length of each arc is given by $s = r\theta$. We are given that $r = 24$ inches and $\theta = 90°$. The formula $s = r\theta$ can only be used when θ is expressed in radians.

$$90° = 90° \cdot \frac{\pi\text{ radians}}{180°} = \frac{90\pi}{180}\text{ radians}$$

$$= \frac{\pi}{2}\text{ radians}$$

The length of each arc is

$$s = r\theta = (24\text{ inches})\left(\frac{\pi}{2}\right) = 12\pi\text{ inches}$$

$$\approx 37.70\text{ inches.}$$

93. Recall that $\theta = \dfrac{s}{r}$. We are given that $s = 8000$ miles and $r = 4000$ miles.

$$\theta = \frac{s}{r} = \frac{8000\text{ miles}}{4000\text{ miles}} = 2\text{ radians}$$

Now, convert 2 radians to degrees.

$$2\text{ radians} = 2\text{ radians} \cdot \frac{180°}{\pi\text{ radians}} \approx 114.59°$$

95. Recall that $s = r\theta$. We are given that $r = 4000$ miles and $\theta = 30°$. The formula $s = r\theta$ can only be used when θ is expressed in radians.

$$30° = 30° \cdot \frac{\pi\text{ radians}}{180°} = \frac{30\pi}{180}\text{ radians}$$

$$= \frac{\pi}{6}\text{ radians}$$

$$s = r\theta = (4000\text{ miles})\left(\frac{\pi}{6}\right) \approx 2094\text{ miles}$$

To the nearest mile, the distance from A to B is 2094 miles.

97. Linear speed is given by $v = r\omega$. We are given that

$\omega = \dfrac{\pi}{12}$ radians per hour and

$r = 4000$ miles. Therefore,

$$v = r\omega = (4000\text{ miles})\left(\frac{\pi}{12}\right)$$

$$= \frac{4000\pi}{12}\text{ miles per hour}$$

$$\approx 1047\text{ miles per hour}$$

The linear speed is about 1047 miles per hour.

99. Linear speed is given by $v = r\omega$. We are given that $r = 12$ feet and the wheel rotates at 20 revolutions per minute.

20 revolutions per minute

$$= 20\text{ revolutions per minute} \cdot \frac{2\pi\text{ radians}}{1\text{ revolution}}$$

$$= 40\pi\text{ radians per minute}$$

$$v = r\omega = (12\text{ feet})(40\pi)$$

$$\approx 1508\text{ feet per minute}$$

The linear speed of the wheel is about 1508 feet per minute.

101. – 111. Answers may vary.

113.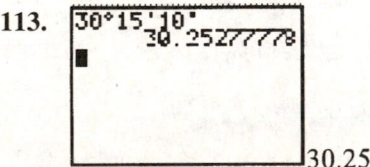

30.25°

115.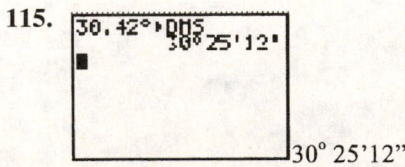

30° 25'12"

117. does not make sense; Explanations will vary. Sample explanation: Angles greater than π will exceed a straight angle.

119. makes sense

121. A right angle measures 90° and

$90° = \dfrac{\pi}{2}$ radians ≈ 1.57 radians.

If $\theta = \dfrac{3}{2}$ radians $= 1.5$ radians, θ is smaller than a right angle.

123. $s = r\theta$

Begin by changing $\theta = 26°$ to radians.

$26° = 26° \cdot \dfrac{\pi}{180°} = \dfrac{13\pi}{90}$ radians

$s = 4000 \cdot \dfrac{13\pi}{90}$

≈ 1815 miles

To the nearest mile, Miami, Florida is 1815 miles north of the equator.

124.

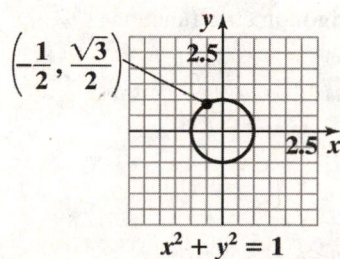

$x^2 + y^2 = 1$

125. domain: $\left\{x \mid -1 \le x \le 1\right\}$ or $[-1,1]$

range: $\left\{y \mid -1 \le y \le 1\right\}$ or $[-1,1]$

126. $x = -\dfrac{1}{2}; \quad y = \dfrac{\sqrt{3}}{2}$

$\dfrac{x}{y} = \dfrac{-\dfrac{1}{2}}{\dfrac{\sqrt{3}}{2}} = -\dfrac{1}{\sqrt{3}} = -\dfrac{1}{\sqrt{3}}\dfrac{\sqrt{3}}{\sqrt{3}} = -\dfrac{\sqrt{3}}{3}$

Section 4.2

Check Point Exercises

1. $P\left(\dfrac{\sqrt{3}}{2}, \dfrac{1}{2}\right)$

$\sin t = y = \dfrac{1}{2}$

$\cos t = x = \dfrac{\sqrt{3}}{2}$

$\tan t = \dfrac{y}{x} = \dfrac{\frac{1}{2}}{\frac{\sqrt{3}}{2}} = \dfrac{\sqrt{3}}{3}$

$\csc t = \dfrac{1}{y} = 2$

$\sec t = \dfrac{1}{x} = \dfrac{2\sqrt{3}}{3}$

$\cot t = \dfrac{x}{y} = \sqrt{3}$

2. The point P on the unit circle that corresponds to $t = \pi$ has coordinates $(-1, 0)$. Use $x = -1$ and $y = 0$ to find the values of the trigonometric functions.

$\sin \pi = y = 0$

$\cos \pi = x = -1$

$\tan \pi = \dfrac{y}{x} = \dfrac{0}{-1} = 0$

$\sec \pi = \dfrac{1}{x} = \dfrac{1}{-1} = -1$

$\cot \pi = \dfrac{x}{y} = \dfrac{-1}{0} = $ undefined

$\csc \pi = \dfrac{1}{y} = \dfrac{1}{0} = $ undefined

3. $t = \dfrac{\pi}{4}, P\left(\dfrac{1}{\sqrt{2}}, \dfrac{1}{\sqrt{2}}\right)$

$\csc \dfrac{\pi}{4} = \dfrac{1}{y} = \sqrt{2}$

$\sec \dfrac{\pi}{4} = \dfrac{1}{x} = \sqrt{2}$

$\cot \dfrac{\pi}{4} = \dfrac{x}{y} = \dfrac{\frac{1}{y}}{\frac{1}{\sqrt{2}}} = 1$

Trigonometric Functions

4. a. $\sec\left(-\frac{\pi}{4}\right) = \sec\left(\frac{\pi}{4}\right) = \sqrt{2}$

b. $\sin\left(-\frac{\pi}{4}\right) = -\sin\left(\frac{\pi}{4}\right) = -\frac{\sqrt{2}}{2}$

5. $\tan\theta = \frac{\sin\theta}{\cos\theta} = \frac{\frac{2}{3}}{\frac{\sqrt{5}}{3}}$

$= \frac{2}{3}\cdot\frac{3}{\sqrt{5}} = \frac{2}{\sqrt{5}}$

$= \frac{2}{\sqrt{5}}\cdot\frac{\sqrt{5}}{\sqrt{5}} = \frac{2\sqrt{5}}{5}$

$\csc\theta = \frac{1}{\sin\theta} = \frac{1}{\frac{2}{3}} = \frac{3}{2}$

$\sec\theta = \frac{1}{\cos\theta} = \frac{1}{\frac{\sqrt{5}}{3}} = \frac{3}{\sqrt{5}}$

$= \frac{3}{\sqrt{5}}\cdot\frac{\sqrt{5}}{\sqrt{5}} = \frac{3\sqrt{5}}{5}$

$\cot\theta = \frac{1}{\tan\theta} = \frac{1}{\frac{2}{\sqrt{5}}} = \frac{\sqrt{5}}{2}$

6. $\sin t = \frac{1}{2}, 0 \le t < \frac{\pi}{2}$

$\sin^2 t + \cos^2 t = 1$

$\left(\frac{1}{2}\right)^2 + \cos^2 t = 1$

$\cos^2 t = 1 - \frac{1}{4}$

$\cos t = \sqrt{\frac{3}{4}} = \frac{\sqrt{3}}{2}$

Because $0 \le t < \frac{\pi}{2}$, $\cos t$ is positive.

7. a. $\cot\frac{5\pi}{4} = \cot\left(\frac{\pi}{4}+\pi\right) = \cot\frac{\pi}{4} = 1$

b. $\cos\left(-\frac{9\pi}{4}\right) = \cos\left(-\frac{9\pi}{4}+4\pi\right)$

$= \cos\frac{7\pi}{4}$

$= \frac{\sqrt{2}}{2}$

8. a. $\sin\frac{\pi}{4} \approx 0.7071$

b. $\csc 1.5 \approx 1.0025$

Exercise Set 4.2

1. The point P on the unit circle has coordinates $\left(-\frac{15}{17}, \frac{8}{17}\right)$. Use $x = -\frac{15}{17}$ and $y = \frac{8}{17}$ to find the values of the trigonometric functions.

$\sin t = y = \frac{8}{17}$

$\cos t = x = -\frac{15}{17}$

$\tan t = \frac{y}{x} = \frac{\frac{8}{17}}{-\frac{15}{17}} = -\frac{8}{15}$

$\csc t = \frac{1}{y} = \frac{17}{8}$

$\sec t = \frac{1}{x} = -\frac{17}{15}$

$\cot t = \frac{x}{y} = -\frac{15}{8}$

3. The point P on the unit circle that corresponds to $t = -\frac{\pi}{4}$ has coordinates $\left(\frac{\sqrt{2}}{2}, -\frac{\sqrt{2}}{2}\right)$. Use $x = \frac{\sqrt{2}}{2}$ and $y = -\frac{\sqrt{2}}{2}$ to find the values of the trigonometric functions.

$\sin t = y = -\frac{\sqrt{2}}{2}$

$\cos t = x = \frac{\sqrt{2}}{2}$

$\tan t = \frac{y}{x} = \frac{-\frac{\sqrt{2}}{2}}{\frac{\sqrt{2}}{2}} = -1$

$\csc t = \frac{1}{y} = -\sqrt{2}$

$\sec t = \frac{1}{x} = \sqrt{2}$

$\cot t = \frac{x}{y} = -1$

5. $\sin\frac{\pi}{6} = \frac{1}{2}$

7. $\cos\frac{5\pi}{6} = -\frac{\sqrt{3}}{2}$

9. $\tan\pi = \frac{0}{-1} = 0$

11. $\csc\frac{7\pi}{6} = \frac{1}{-\frac{1}{2}} = -2$

13. $\sec\frac{11\pi}{6} = \frac{1}{\frac{\sqrt{3}}{2}} = \frac{2\sqrt{3}}{3}$

15. $\sin\frac{3\pi}{2} = -1$

17. $\sec\frac{3\pi}{2} = $ undefined

19. a. $\cos\frac{\pi}{6} = \frac{\sqrt{3}}{2}$

    b. $\cos\left(-\frac{\pi}{6}\right) = \cos\frac{\pi}{6} = \frac{\sqrt{3}}{2}$

21. a. $\sin\frac{5\pi}{6} = \frac{1}{2}$

    b. $\sin\left(-\frac{5\pi}{6}\right) = -\sin\frac{5\pi}{6} = -\frac{1}{2}$

23. a. $\tan\frac{5\pi}{3} = \frac{-\frac{\sqrt{3}}{2}}{\frac{1}{2}} = -\sqrt{3}$

    b. $\tan\left(-\frac{5\pi}{3}\right) = -\tan\frac{5\pi}{3} = \sqrt{3}$

25. $\sin t = \frac{8}{17}, \cos t = \frac{15}{17}$

    $\tan t = \frac{\frac{8}{17}}{\frac{15}{17}} = \frac{8}{15}$

    $\csc t = \frac{17}{8}$

    $\sec t = \frac{17}{15}$

    $\cot t = \frac{15}{8}$

27. $\sin t = \frac{1}{3}, \cos t = \frac{2\sqrt{2}}{3}$

    $\tan t = \frac{\frac{1}{3}}{\frac{2\sqrt{2}}{3}} = \frac{\sqrt{2}}{4}$

    $\csc t = 3$

    $\sec t = \frac{3\sqrt{2}}{4}$

    $\cot t = 2\sqrt{2}$

29. $\sin t = \frac{6}{7}, 0 \le t < \frac{\pi}{2}$

    $\sin^2 t + \cos^2 t = 1$

    $\left(\frac{6}{7}\right)^2 + \cos^2 t = 1$

    $\cos^2 t = 1 - \frac{36}{49}$

    $\cos t = \sqrt{\frac{13}{49}} = \frac{\sqrt{13}}{7}$

    Because $0 \le t < \frac{\pi}{2}, \cos t$ is positive.

31. $\sin t = \dfrac{\sqrt{39}}{8}, 0 \le t < \dfrac{\pi}{2}$

$\sin^2 t + \cos^2 t = 1$

$\left(\dfrac{\sqrt{39}}{8}\right)^2 + \cos^2 t = 1$

$\cos^2 t = 1 - \dfrac{39}{64}$

$\cos t = \sqrt{\dfrac{25}{64}} = \dfrac{5}{8}$

Because $0 \le t < \dfrac{\pi}{2}, \cos t$ is positive.

33. $\sin 1.7 \csc 1.7 = \sin 1.7 \left(\dfrac{1}{\sin 1.7}\right) = 1$

35. $\sin^2 \dfrac{\pi}{6} + \cos^2 \dfrac{\pi}{2} = 1$ by the Pythagorean identity.

37. $\sec^2 \dfrac{\pi}{3} - \tan^2 \dfrac{\pi}{3} = 1$ because $1 + \tan^2 t = \sec^2 t$.

39. $\cos \dfrac{9\pi}{4} = \cos\left(\dfrac{\pi}{4} + 2\pi\right) = \cos \dfrac{\pi}{4} = \dfrac{\sqrt{2}}{2}$

41. $\sin\left(-\dfrac{9\pi}{4}\right) = \sin\left(-\dfrac{9\pi}{4} + 4\pi\right) = \sin \dfrac{7\pi}{4} = -\dfrac{\sqrt{2}}{2}$

43. $\tan \dfrac{5\pi}{4} = \tan\left(\dfrac{\pi}{4} + \pi\right) = \tan \dfrac{\pi}{4} = 1$

45. $\cot\left(-\dfrac{5\pi}{4}\right) = \cot\left(\dfrac{3\pi}{4} - 2\pi\right) = \cot \dfrac{3\pi}{4} = -1$

47. $-\tan\left(\dfrac{\pi}{4} + 15\pi\right) = -\tan \dfrac{\pi}{4} = -1$

49. $\sin\left(-\dfrac{\pi}{4} - 1000\pi\right) = \sin\left(-\dfrac{\pi}{4} + 2\pi\right)$

$= \sin \dfrac{7\pi}{4}$

$= -\dfrac{\sqrt{2}}{2}$

51. $\cos\left(-\dfrac{\pi}{4} - 1000\pi\right) = \cos\left(-\dfrac{\pi}{4} + 2\pi\right)$

$= \cos \dfrac{7\pi}{4}$

$= \dfrac{\sqrt{2}}{2}$

53. a. $\sin \dfrac{3\pi}{4} = \dfrac{\sqrt{2}}{2}$

b. $\sin \dfrac{11\pi}{4} = \sin\left(\dfrac{3\pi}{4} + 2\pi\right) = \sin \dfrac{3\pi}{4} = \dfrac{\sqrt{2}}{2}$

55. a. $\cos \dfrac{\pi}{2} = 0$

b. $\cos \dfrac{9\pi}{2} = \cos\left(\dfrac{\pi}{2} + 4\pi\right)$

$= \cos\left[\dfrac{\pi}{2} + 2(2\pi)\right]$

$= \cos \dfrac{\pi}{2}$

$= 0$

57. a. $\tan \pi = \dfrac{0}{-1} = 0$

b. $\tan 17\pi = \tan(\pi + 16\pi)$

$= \tan[\pi + 8(2\pi)]$

$= \tan \pi$

$= 0$

59. a. $\sin \dfrac{7\pi}{4} = -\dfrac{\sqrt{2}}{2}$

b. $\sin \dfrac{47\pi}{4} = \sin\left(\dfrac{7\pi}{4} + 10\pi\right)$

$= \sin\left[\dfrac{7\pi}{4} + 5(2\pi)\right]$

$= \sin \dfrac{7\pi}{4}$

$= -\dfrac{\sqrt{2}}{2}$

61. $\sin 0.8 \approx 0.7174$

63. $\tan 3.4 \approx 0.2643$

65. $\csc 1 \approx 1.1884$

67. $\cos \dfrac{\pi}{10} \approx 0.9511$

69. $\cot \dfrac{\pi}{12} \approx 3.7321$

71. $\sin(-t) - \sin t = -\sin t - \sin t = -2\sin t = -2a$

73. $4\cos(-t) - \cos t = 4\cos t - \cos t = 3\cos t = 3b$

75. $\sin(t + 2\pi) - \cos(t + 4\pi) + \tan(t + \pi)$
$= \sin(t) - \cos(t) + \tan(t)$
$= a - b + c$

77. $\sin(-t - 2\pi) - \cos(-t - 4\pi) - \tan(-t - \pi)$
$= -\sin(t + 2\pi) - \cos(t + 4\pi) + \tan(t + \pi)$
$= -\sin(t) - \cos(t) + \tan(t)$
$= -a - b + c$

79. $\cos t + \cos(t + 1000\pi) - \tan t - \tan(t + 999\pi)$
$\qquad\qquad\qquad - \sin t + 4\sin(t - 1000\pi)$
$= \cos t + \cos t - \tan t - \tan t - \sin t + 4\sin t$
$= 2\cos t - 2\tan t + 3\sin t$
$= 3a + 2b - 2c$

81. a. $H = 12 + 8.3\sin\left[\dfrac{2\pi}{365}(80 - 80)\right]$
$\qquad = 12 + 8.3\sin 0 = 12 + 8.3(0)$
$\qquad = 12$
There are 12 hours of daylight in Fairbanks on March 21.

   b. $H = 12 + 8.3\sin\left[\dfrac{2\pi}{365}(172 - 80)\right]$
$\qquad \approx 12 + 8.3\sin 1.5837$
$\qquad \approx 20.3$
There are about 20.3 hours of daylight in Fairbanks on June 21.

   c. $H = 12 + 8.3\sin\left[\dfrac{2\pi}{365}(355 - 80)\right]$
$\qquad \approx 12 + 8.3\sin 4.7339$
$\qquad \approx 3.7$
There are about 3.7 hours of daylight in Fairbanks on December 21.

83. a. For $t = 7$,
$$E = \sin\dfrac{\pi}{14} \cdot 7 = \sin\dfrac{\pi}{2} = 1$$
For $t = 14$,
$$E = \sin\dfrac{\pi}{14} \cdot 14 = \sin\pi = 0$$
For $t = 21$,
$$E = \sin\dfrac{\pi}{14} \cdot 21 = \sin\dfrac{3\pi}{2} = -1$$
For $t = 28$,
$$E = \sin\dfrac{\pi}{14} \cdot 28 = \sin 2\pi = \sin 0 = 0$$
For $t = 35$,
$$E = \sin\dfrac{\pi}{14} \cdot 35 = \sin\dfrac{5\pi}{2} = \sin\dfrac{\pi}{2} = 1$$
Observations may vary.

   b. Because $E(35) = E(7) = 1$, the period is $35 - 7 = 28$ or 28 days.

85. – 95. Answers may vary.

97. makes sense

99. does not make sense; Explanations will vary. Sample explanation: Cosine is not an odd function.

101. t is in the third quadrant therefore $\sin t < 0$, $\tan t > 0$, and $\cot t > 0$.
Thus, only choice (c) is true.

103. $f(x) = \sin x$ and $f(a) = \dfrac{1}{4}$
$f(a) + 2f(-a) = f(a) - 2f(a)$
$\qquad\qquad = \dfrac{1}{4} - 2\left(\dfrac{1}{4}\right)$
$\qquad\qquad = -\dfrac{1}{4}$
$f(-a) = -f(a)$ because $\sin(-x) = -\sin x$.
Sine is an odd function.

105. First find the hypotenuse.

$$c^2 = a^2 + b^2$$

$$c^2 = 5^2 + 12^2$$

$$c^2 = 25 + 144$$

$$c^2 = 169$$

$$c = 13$$

Next write the ratio.

$$\frac{a}{c} = \frac{5}{13}$$

106. First find the hypotenuse.

$$c^2 = a^2 + b^2$$

$$c^2 = 1^2 + 1^2$$

$$c^2 = 1 + 1$$

$$c^2 = 2$$

$$c = \sqrt{2}$$

Next write the ratio and simplify.

$$\frac{a}{c} = \frac{1}{\sqrt{2}}$$

$$= \frac{1}{\sqrt{2}} \cdot \frac{\sqrt{2}}{\sqrt{2}}$$

$$= \frac{\sqrt{2}}{2}$$

107. $\left(\dfrac{a}{c}\right)^2 + \left(\dfrac{b}{c}\right)^2 = \dfrac{a^2}{c^2} + \dfrac{b^2}{c^2}$

$$= \frac{a^2 + b^2}{c^2}$$

Since $c^2 = a^2 + b^2$, continue simplifying by substituting c^2 for $a^2 + b^2$.

$$\left(\frac{a}{c}\right)^2 + \left(\frac{b}{c}\right)^2 = \frac{a^2}{c^2} + \frac{b^2}{c^2}$$

$$= \frac{a^2 + b^2}{c^2}$$

$$= \frac{\overbrace{a^2 + b^2}^{c^2}}{c^2}$$

$$= \frac{c^2}{c^2}$$

$$= 1$$

Section 4.3

Check Point Exercises

1. Use the Pythagorean Theorem, $c^2 = a^2 + b^2$, to find c.

$$a = 3, b = 4$$

$$c^2 = a^2 + b^2 = 3^2 + 4^2 = 9 + 16 = 25$$

$$c = \sqrt{25} = 5$$

Referring to these lengths as opposite, adjacent, and hypotenuse, we have

$$\sin\theta = \frac{\text{opposite}}{\text{hypotenuse}} = \frac{3}{5}$$

$$\cos\theta = \frac{\text{adjacent}}{\text{hypotenuse}} = \frac{4}{5}$$

$$\tan\theta = \frac{\text{opposite}}{\text{adjacent}} = \frac{3}{4}$$

$$\csc\theta = \frac{\text{hypotenuse}}{\text{opposite}} = \frac{5}{3}$$

$$\sec\theta = \frac{\text{hypotenuse}}{\text{adjacent}} = \frac{5}{4}$$

$$\cot\theta = \frac{\text{adjacent}}{\text{opposite}} = \frac{4}{3}$$

2. Use the Pythagorean Theorem, $c^2 = a^2 + b^2$, to find b.

$$a^2 + b^2 = c^2$$

$$1^2 + b^2 = 5^2$$

$$1 + b^2 = 25$$

$$b^2 = 24$$

$$b = \sqrt{24} = 2\sqrt{6}$$

Note that side a is opposite θ and side b is adjacent to θ.

$$\sin\theta = \frac{\text{opposite}}{\text{hypotenuse}} = \frac{1}{5}$$

$$\cos\theta = \frac{\text{adjacent}}{\text{hypotenuse}} = \frac{2\sqrt{6}}{5}$$

$$\tan\theta = \frac{\text{opposite}}{\text{adjacent}} = \frac{1}{2\sqrt{6}} = \frac{\sqrt{6}}{12}$$

$$\csc\theta = \frac{\text{hypotenuse}}{\text{opposite}} = \frac{5}{1} = 5$$

$$\sec\theta = \frac{\text{hypotenuse}}{\text{adjacent}} = \frac{5}{2\sqrt{6}} = \frac{5\sqrt{6}}{12}$$

$$\cot\theta = \frac{\text{adjacent}}{\text{opposite}} = \frac{2\sqrt{6}}{1} = 2\sqrt{6}$$

3.   Apply the definitions of these three trigonometric functions.

$$\csc 45° = \frac{\text{length of hypotenuse}}{\text{length of side opposite } 45°}$$

$$= \frac{\sqrt{2}}{1} = \sqrt{2}$$

$$\sec 45° = \frac{\text{length of hypotenuse}}{\text{length of side adjacent to } 45°}$$

$$= \frac{\sqrt{2}}{1} = \sqrt{2}$$

$$\cot 45° = \frac{\text{length of side adjacent to } 45°}{\text{length of side opposite } 45°}$$

$$= \frac{1}{1} = 1$$

4.

$$\tan 60° = \frac{\text{length of side opposite } 60°}{\text{length of side adjacent to } 60°}$$

$$= \frac{\sqrt{3}}{1} = \sqrt{3}$$

$$\tan 30° = \frac{\text{length of side opposite } 30°}{\text{length of side adjacent to } 30°}$$

$$= \frac{1}{\sqrt{3}} = \frac{1}{\sqrt{3}} \cdot \frac{\sqrt{3}}{3} = \frac{\sqrt{3}}{3}$$

5.   **a.** $\sin 46° = \cos(90° - 46°) = \cos 44°$

   b. $\cot \dfrac{\pi}{12} = \tan\left(\dfrac{\pi}{2} - \dfrac{\pi}{12}\right)$

$$= \tan\left(\frac{6\pi}{12} - \frac{\pi}{12}\right)$$

$$= \tan \frac{5\pi}{12}$$

6.   Because we have a known angle, an unknown opposite side, and a known adjacent side, we select the tangent function.

$$\tan 24° = \frac{a}{750}$$

$$a = 750 \tan 24°$$

$$a \approx 750(0.4452) \approx 334$$

The distance across the lake is approximately 334 yards.

7. $\tan\theta = \dfrac{\text{side opposite}}{\text{side adjacent}} = \dfrac{14}{10}$

Use a calculator in degree mode to find θ.

Many Scientific Calculators	**Many Graphing Calculators**
$\boxed{\text{TAN}^{-1}}\ \boxed{(}\ \boxed{14}\ \boxed{\div}\ \boxed{10}\ \boxed{)}\ \boxed{\text{ENTER}}$	$\boxed{\text{TAN}}\ \boxed{(}\ \boxed{14}\ \boxed{\div}\ \boxed{10}\ \boxed{)}\ \boxed{\text{ENTER}}$

The display should show approximately 54. Thus, the angle of elevation of the sun is approximately 54°.

Exercise Set 4.3

1. $c^2 = 9^2 + 12^2 = 225$

$c = \sqrt{225} = 15$

$\sin\theta = \dfrac{\text{opposite}}{\text{hypotenuse}} = \dfrac{9}{15} = \dfrac{3}{5}$

$\cos\theta = \dfrac{\text{adjacent}}{\text{hypotenuse}} = \dfrac{12}{15} = \dfrac{4}{5}$

$\tan\theta = \dfrac{\text{opposite}}{\text{adjacent}} = \dfrac{9}{12} = \dfrac{3}{4}$

$\csc\theta = \dfrac{\text{hypotenuse}}{\text{opposite}} = \dfrac{15}{9} = \dfrac{5}{3}$

$\sec\theta = \dfrac{\text{hypotenuse}}{\text{adjacent}} = \dfrac{15}{12} = \dfrac{5}{4}$

$\cot\theta = \dfrac{\text{adjacent}}{\text{opposite}} = \dfrac{12}{9} = \dfrac{4}{3}$

3. $a^2 + 21^2 = 29^2$

$a^2 = 841 - 441 = 400$

$a = \sqrt{400} = 20$

$\sin\theta = \dfrac{\text{opposite}}{\text{hypotenuse}} = \dfrac{20}{29}$

$\cos\theta = \dfrac{\text{adjacent}}{\text{hypotenuse}} = \dfrac{21}{29}$

$\tan\theta = \dfrac{\text{opposite}}{\text{adjacent}} = \dfrac{20}{21}$

$\csc\theta = \dfrac{\text{hypotenuse}}{\text{opposite}} = \dfrac{29}{20}$

$\sec\theta = \dfrac{\text{hypotenuse}}{\text{adjacent}} = \dfrac{29}{21}$

$\cot\theta = \dfrac{\text{adjacent}}{\text{opposite}} = \dfrac{21}{20}$

5. $10^2 + b^2 = 26^2$

$\qquad b^2 = 676 - 100 = 576$

$\qquad b = \sqrt{576} = 24$

$\sin\theta = \dfrac{\text{opposite}}{\text{hypotenuse}} = \dfrac{10}{26} = \dfrac{5}{13}$

$\cos\theta = \dfrac{\text{adjacent}}{\text{hypotenuse}} = \dfrac{24}{26} = \dfrac{12}{13}$

$\tan\theta = \dfrac{\text{opposite}}{\text{adjacent}} = \dfrac{10}{24} = \dfrac{5}{12}$

$\csc\theta = \dfrac{\text{hypotenuse}}{\text{opposite}} = \dfrac{26}{10} = \dfrac{13}{5}$

$\sec\theta = \dfrac{\text{hypotenuse}}{\text{adjacent}} = \dfrac{26}{24} = \dfrac{13}{12}$

$\cot\theta = \dfrac{\text{adjacent}}{\text{opposite}} = \dfrac{24}{10} = \dfrac{12}{5}$

7. $a^2 + 21^2 = 35^2$

$\qquad a^2 = 1225 - 441 = 784$

$\qquad a = \sqrt{784} = 28$

$\sin\theta = \dfrac{\text{opposite}}{\text{hypotenuse}} = \dfrac{28}{35} = \dfrac{4}{5}$

$\cos\theta = \dfrac{\text{adjacent}}{\text{hypotenuse}} = \dfrac{21}{35} = \dfrac{3}{5}$

$\tan\theta = \dfrac{\text{opposite}}{\text{adjacent}} = \dfrac{28}{21} = \dfrac{4}{3}$

$\csc\theta = \dfrac{\text{hypotenuse}}{\text{opposite}} = \dfrac{35}{28} = \dfrac{5}{4}$

$\sec\theta = \dfrac{\text{hypotenuse}}{\text{adjacent}} = \dfrac{35}{21} = \dfrac{5}{3}$

$\cot\theta = \dfrac{\text{adjacent}}{\text{opposite}} = \dfrac{21}{28} = \dfrac{3}{4}$

9. $\cos 30° = \dfrac{\text{length of side adjacent to } 30°}{\text{length of hypotenuse}}$

$\qquad = \dfrac{\sqrt{3}}{2}$

11. $\sec 45° = \dfrac{\text{length of hypotenuse}}{\text{length of side adjacent to } 45°}$

$\qquad = \dfrac{\sqrt{2}}{1} = \sqrt{2}$

13. $\tan\dfrac{\pi}{3} = \tan 60°$

$\qquad = \dfrac{\text{length of side opposite } 60°}{\text{length of side adjacent to } 60°}$

$\qquad = \dfrac{\sqrt{3}}{1} = \sqrt{3}$

15. $\sin\dfrac{\pi}{4} - \cos\dfrac{\pi}{4} = \sin 45° - \cos 45°$

$\qquad = \dfrac{1}{\sqrt{2}} - \dfrac{1}{\sqrt{2}} = 0$

17. $\sin\dfrac{\pi}{3}\cos\dfrac{\pi}{4} - \tan\dfrac{\pi}{4} = \left(\dfrac{\sqrt{3}}{2}\right)\left(\dfrac{\sqrt{2}}{2}\right) - 1$

$\qquad = \dfrac{\sqrt{6}}{4} - 1$

$\qquad = \dfrac{\sqrt{6} - 4}{4}$

19. $2\tan\dfrac{\pi}{3} + \cos\dfrac{\pi}{4}\tan\dfrac{\pi}{6} = 2\left(\sqrt{3}\right) + \left(\dfrac{\sqrt{2}}{2}\right)\left(\dfrac{\sqrt{3}}{3}\right)$

$\qquad = 2\sqrt{3} + \dfrac{\sqrt{6}}{6}$

$\qquad = \dfrac{12\sqrt{3} + \sqrt{6}}{6}$

21. $\sin 7° = \cos(90° - 7°) = \cos 83°$

23. $\csc 25° = \sec(90° - 25°) = \sec 65°$

25. $\tan\dfrac{\pi}{9} = \cot\left(\dfrac{\pi}{2} - \dfrac{\pi}{9}\right)$

$\qquad = \cot\left(\dfrac{9\pi}{18} - \dfrac{2\pi}{18}\right)$

$\qquad = \cot\dfrac{7\pi}{18}$

27. $\cos\dfrac{2\pi}{5} = \sin\left(\dfrac{\pi}{2} - \dfrac{2\pi}{5}\right)$

$\qquad = \sin\left(\dfrac{5\pi}{10} - \dfrac{4\pi}{10}\right)$

$\qquad = \sin\dfrac{\pi}{10}$

29. $\tan 37° = \dfrac{a}{250}$

$a = 250 \tan 37°$

$a \approx 250(0.7536) \approx 188$ cm

31. $\cos 34° = \dfrac{b}{220}$

$b = 220 \cos 34°$

$b \approx 220(0.8290) \approx 182$ in.

33. $\sin 23° = \dfrac{16}{c}$

$c = \dfrac{16}{\sin 23°} \approx \dfrac{16}{0.3907} \approx 41$ m

35.

Scientific Calculator	Graphing Calculator	Display (rounded to the nearest degree)
.2974 $\boxed{\text{SIN}^{-1}}$	$\boxed{\text{SIN}^{-1}}$ .2974 $\boxed{\text{ENTER}}$	17

If $\sin \theta = 0.2974$, then $\theta \approx 17°$.

37.

Scientific Calculator	Graphing Calculator	Display (rounded to the nearest degree)
4.6252 $\boxed{\text{TAN}^{-1}}$	$\boxed{\text{TAN}^{-1}}$ 4.6252 $\boxed{\text{ENTER}}$	78

If $\tan \theta = 4.6252$, then $\theta \approx 78°$.

39.

Scientific Calculator	Graphing Calculator	Display (rounded to three places)
.4112 $\boxed{\text{COS}^{-1}}$	$\boxed{\text{COS}^{-1}}$ .4112 $\boxed{\text{ENTER}}$	1.147

If $\cos \theta = 0.4112$, then $\theta \approx 1.147$ radians.

41.

Scientific Calculator	Graphing Calculator	Display (rounded to three places)
.4169 $\boxed{\text{TAN}^{-1}}$	$\boxed{\text{TAN}^{-1}}$ .4169 $\boxed{\text{ENTER}}$	.395

If $\tan \theta = 0.4169$, then $\theta \approx 0.395$ radians.

43.

$$\dfrac{\tan \dfrac{\pi}{3}}{2} - \dfrac{1}{\sec \dfrac{\pi}{6}} = \dfrac{\sqrt{3}}{2} - \dfrac{1}{\dfrac{1}{\cos \frac{\pi}{6}}}$$

$$= \dfrac{\sqrt{3}}{2} - \dfrac{1}{\dfrac{1}{\frac{\sqrt{3}}{2}}}$$

$$= \dfrac{\sqrt{3}}{2} - \dfrac{\sqrt{3}}{2}$$

$$= 0$$

45. $1 + \sin^2 40° + \sin^2 50°$

   $= 1 + \sin^2(90° - 50°) + \sin^2 50°$

   $= 1 + \cos^2 50° + \sin^2 50°$

   $= 1 + 1$

   $= 2$

47. $\csc 37° \sec 53° - \tan 53° \cot 37°$

   $= \sec 53° \sec 53° - \tan 53° \tan 53°$

   $= \sec^2 53° - \tan^2 53°$

   $= 1$

49. $f(\theta) = 2\cos\theta - \cos 2\theta$

   $f\left(\dfrac{\pi}{6}\right) = 2\cos\dfrac{\pi}{6} - \cos\left(2 \cdot \dfrac{\pi}{6}\right)$

   $\qquad = 2\left(\dfrac{\sqrt{3}}{2}\right) - \cos\left(\dfrac{\pi}{3}\right)$

   $\qquad = \dfrac{2\sqrt{3}}{2} - \dfrac{1}{2}$

   $\qquad = \dfrac{2\sqrt{3} - 1}{2}$

51. $\tan\left(\dfrac{\pi}{2} - \theta\right) = \cot\theta = \dfrac{1}{4}$

53. $\tan 40° = \dfrac{a}{630}$

   $a = 630 \tan 40°$

   $a \approx 630(0.8391) \approx 529$

   The distance across the lake is approximately 529 yards.

55. $\tan\theta = \dfrac{125}{172}$

   Use a calculator in degree mode to find θ.

Many Scientific Calculators	Many Graphing Calculators
125 ÷ 172 = TAN⁻¹	TAN⁻¹ (125 ÷ 172) ENTER

   The display should show approximately 36. Thus, the angle of elevation of the sun is approximately 36°.

57. $\sin 10° = \dfrac{500}{c}$

   $c = \dfrac{500}{\sin 10°} \approx \dfrac{500}{0.1736} \approx 2880$

   The plane has flown approximately 2880 feet.

59. $\cos\theta = \dfrac{60}{75}$

   Use a calculator in degree mode to find θ.

Many Scientific Calculators	Many Graphing Calculators
60 ÷ 75 = COS⁻¹	COS⁻¹ (60 ÷ 75) ENTER

The display should show approximately 37. Thus, the angle between the wire and the pole is approximately 37°.

61. – 67. Answers may vary.

Many Scientific Calculators	Many Graphing Calculators
55 ÷ 80 = COS⁻¹	COS⁻¹ (55 ÷ 80) ENTER

The display should show approximately 47. Thus, the angle between the wire and the pole is approximately 47°.

69.

θ	0.4	0.3	0.2	0.1	0.01	0.001	0.0001	0.00001
$\cos\theta$	0.92106	0.95534	0.98007	0.99500	0.99995	0.9999995	0.999999995	1
$\dfrac{\cos\theta-1}{\theta}$	–0.19735	–0.148878	–0.099667	–0.04996	–0.005	–0.0005	–0.00005	0

$\dfrac{\cos\theta-1}{\theta}$ approaches 0 as θ approaches 0.

71. does not make sense; Explanations will vary. Sample explanation: This value is irrational. Irrational numbers are rounded on calculators.

73. makes sense

75. true

77. true

79. Use a calculator in degree mode to generate the following table. Then use the table to describe what happens to the tangent of an acute angle as the angle gets close to 90°.

θ	60	70	80	89	89.9	89.99	89.999	89.9999
$\tan\theta$	1.7321	2.7475	5.6713	57	573	5730	57,296	572,958

As θ approaches 90°, $\tan\theta$ increases without bound. At 90°, $\tan\theta$ is undefined.

81.   **a.** $\dfrac{y}{r}$

     b.   First find r: $r = \sqrt{x^2 + y^2}$

$$r = \sqrt{(-3)^2 + 4^2}$$

$$r = 5$$

$\dfrac{y}{r} = \dfrac{4}{5}$, which is positive.

82.   **a.** $\dfrac{x}{r}$

     b.   First find r: $r = \sqrt{x^2 + y^2}$

$$r = \sqrt{(-3)^2 + 5^2}$$

$$r = \sqrt{34}$$

$\dfrac{x}{r} = \dfrac{-3}{\sqrt{34}} = \dfrac{-3}{\sqrt{34}} \cdot \dfrac{\sqrt{34}}{\sqrt{34}} = \dfrac{-3\sqrt{34}}{34}$, which is

negative.

83.   **a.** $\theta' = 360° - 345° = 15°$

     b. $\theta' = \pi - \dfrac{5\pi}{6} = \dfrac{6\pi}{6} - \dfrac{5\pi}{6} = \dfrac{\pi}{6}$

Section 4.4

Check Point Exercises

1. $r = \sqrt{x^2 + y^2}$

$$r = \sqrt{1^2 + (-3)^2} = \sqrt{1 + 9} = \sqrt{10}$$

Now that we know *x*, *y*, and *r*, we can find the six trigonometric functions of θ.

$$\sin\theta = \frac{y}{r} = \frac{-3}{\sqrt{10}} = -\frac{3\sqrt{10}}{10}$$

$$\cos\theta = \frac{x}{r} = \frac{1}{\sqrt{10}} = \frac{\sqrt{10}}{10}$$

$$\tan\theta = \frac{y}{x} = \frac{-3}{1} = -3$$

$$\csc\theta = \frac{r}{y} = \frac{\sqrt{10}}{-3} = -\frac{\sqrt{10}}{3}$$

$$\sec\theta = \frac{r}{x} = \frac{\sqrt{10}}{1} = \sqrt{10}$$

$$\cot\theta = \frac{x}{y} = \frac{1}{-3} = -\frac{1}{3}$$

2.   **a.** $\theta = 0° = 0$ radians

The terminal side of the angle is on the positive *x*-axis. Select the point
$P = (1,0)$: $x = 1$, $y = 0$, $r = 1$

Apply the definitions of the cosine and cosecant functions.

$$\cos 0° = \cos 0 = \frac{x}{r} = \frac{1}{1} = 1$$

$$\csc 0° = \csc 0 = \frac{r}{y} = \frac{1}{0}, \text{ undefined}$$

     b. $\theta = 90° = \dfrac{\pi}{2}$ radians

The terminal side of the angle is on the positive *y*-axis. Select the point
$P = (0,1)$: $x = 0$, $y = 1$, $r = 1$

Apply the definitions of the cosine and cosecant functions.

$$\cos 90° = \cos \frac{\pi}{2} = \frac{x}{r} = \frac{0}{1} = 0$$

$$\csc 90° = \csc \frac{\pi}{2} = \frac{r}{y} = \frac{1}{1} = 1$$

     c. $\theta = 180° = \pi$ radians

The terminal side of the angle is on the negative *x*-axis. Select the point
$P = (-1,0)$: $x = -1$, $y = 0$, $r = 1$

Apply the definitions of the cosine and cosecant functions.

$$\cos 180° = \cos \pi = \frac{x}{r} = \frac{-1}{1} = -1$$

$$\csc 180° = \csc \pi = \frac{r}{y} = \frac{1}{0}, \text{ undefined}$$

     d. $\theta = 270° = \dfrac{3\pi}{2}$ radians

The terminal side of the angle is on the negative *y*-axis. Select the point
$P = (0,-1)$: $x = 0$, $y = -1$, $r = 1$

Apply the definitions of the cosine and cosecant functions.

$$\cos 270° = \cos \frac{3\pi}{2} = \frac{x}{r} = \frac{0}{1} = 0$$

$$\csc 270° = \csc \frac{3\pi}{2} = \frac{r}{y} = \frac{1}{-1} = -1$$

3. Because $\sin\theta < 0$, θ cannot lie in quadrant I; all the functions are positive in quadrant I. Furthermore, θ cannot lie in quadrant II; $\sin\theta$ is positive in quadrant II. Thus, with $\sin\theta < 0$, θ lies in quadrant III or quadrant IV. We are also given that $\cos\theta < 0$. Because quadrant III is the only quadrant in which cosine is negative and the sine is negative, we conclude that θ lies in quadrant III.

4. Because the tangent is negative and the cosine is negative, θ lies in quadrant II. In quadrant II, x is negative and y is positive. Thus,

$$\tan\theta = -\frac{1}{3} = \frac{y}{x} = \frac{1}{-3}$$

$x = -3,\ y = 1$

Furthermore,

$$r = \sqrt{x^2 + y^2} = \sqrt{(-3)^2 + 1^2} = \sqrt{9 + 1} = \sqrt{10}$$

Now that we know x, y, and r, we can find $\sin\theta$ and $\sec\theta$.

$$\sin\theta = \frac{y}{r} = \frac{1}{\sqrt{10}} = \frac{1}{\sqrt{10}} \cdot \frac{\sqrt{10}}{\sqrt{10}} = \frac{\sqrt{10}}{10}$$

$$\sec\theta = \frac{r}{x} = \frac{\sqrt{10}}{-3} = -\frac{\sqrt{10}}{3}$$

5. a. Because $210°$ lies between $180°$ and $270°$, it is in quadrant III. The reference angle is $\theta' = 210° - 180° = 30°$.

b. Because $\dfrac{7\pi}{4}$ lies between $\dfrac{3\pi}{2} = \dfrac{6\pi}{4}$ and

$2\pi = \dfrac{8\pi}{4}$, it is in quadrant IV. The reference

angle is $\theta' = 2\pi - \dfrac{7\pi}{4} = \dfrac{8\pi}{4} - \dfrac{7\pi}{4} = \dfrac{\pi}{4}$.

c. Because $-240°$ lies between $-180°$ and $-270°$, it is in quadrant II. The reference angle is $\theta = 240 - 180 = 60°$.

d. Because 3.6 lies between $\pi \approx 3.14$ and

$\dfrac{3\pi}{2} \approx 4.71$, it is in quadrant III. The reference

angle is $\theta' = 3.6 - \pi \approx 0.46$.

6. a. $665° - 360° = 305°$
This angle is in quadrant IV, thus the reference angle is $\theta' = 360° - 305° = 55°$.

b. $\dfrac{15\pi}{4} - 2\pi = \dfrac{15\pi}{4} - \dfrac{8\pi}{4} = \dfrac{7\pi}{4}$
This angle is in quadrant IV, thus the reference
angle is $\theta' = 2\pi - \dfrac{7\pi}{4} = \dfrac{8\pi}{4} - \dfrac{7\pi}{4} = \dfrac{\pi}{4}$.

c. $-\dfrac{11\pi}{3} + 2 \cdot 2\pi = -\dfrac{11\pi}{3} + \dfrac{12\pi}{3} = \dfrac{\pi}{3}$
This angle is in quadrant I, thus the reference
angle is $\theta' = \dfrac{\pi}{3}$.

7. a. $300°$ lies in quadrant IV. The reference angle is $\theta' = 360° - 300° = 60°$.

$$\sin 60° = \frac{\sqrt{3}}{2}$$

Because the sine is negative in quadrant IV,

$$\sin 300° = -\sin 60° = -\frac{\sqrt{3}}{2}.$$

b. $\dfrac{5\pi}{4}$ lies in quadrant III. The reference angle is

$$\theta' = \frac{5\pi}{4} - \pi = \frac{5\pi}{4} - \frac{4\pi}{4} = \frac{\pi}{4}.$$

$$\tan\frac{\pi}{4} = 1$$

Because the tangent is positive in quadrant III,

$$\tan\frac{5\pi}{4} = +\tan\frac{\pi}{4} = 1.$$

c. $-\dfrac{\pi}{6}$ lies in quadrant IV. The reference angle is

$$\theta' = \frac{\pi}{6}.$$

$$\sec\frac{\pi}{6} = \frac{2\sqrt{3}}{3}$$

Because the secant is positive in quadrant IV,

$$\sec\left(-\frac{\pi}{6}\right) = +\sec\frac{\pi}{6} = \frac{2\sqrt{3}}{3}.$$

8. **a.** $\dfrac{17\pi}{6} - 2\pi = \dfrac{17\pi}{6} - \dfrac{12\pi}{6} = \dfrac{5\pi}{6}$ lies in quadrant II. The reference angle is $\theta' = \pi - \dfrac{5\pi}{6} = \dfrac{\pi}{6}$.

The function value for the reference angle is

$\cos \dfrac{\pi}{6} = \dfrac{\sqrt{3}}{2}$.

Because the cosine is negative in quadrant II,

$\cos \dfrac{17\pi}{6} = \cos \dfrac{5\pi}{6} = -\cos \dfrac{\pi}{6} = -\dfrac{\sqrt{3}}{2}$.

b. $\dfrac{-22\pi}{3} + 8\pi = \dfrac{-22\pi}{3} + \dfrac{24\pi}{3} = \dfrac{2\pi}{3}$ lies in quadrant II. The reference angle is

$\theta' = \pi - \dfrac{2\pi}{3} = \dfrac{\pi}{3}$.

The function value for the reference angle is

$\sin \dfrac{\pi}{3} = \dfrac{\sqrt{3}}{2}$.

Because the sine is positive in quadrant II,

$\sin \dfrac{-22\pi}{3} = \sin \dfrac{2\pi}{3} = \sin \dfrac{\pi}{3} = \dfrac{\sqrt{3}}{2}$.

Exercise Set 4.4

1. We need values for *x*, *y*, and *r*. Because $P = (-4, 3)$ is a point on the terminal side of θ, $x = -4$ and $y = 3$. Furthermore,

$r = \sqrt{x^2 + y^2} = \sqrt{(-4)^2 + 3^2} = \sqrt{16 + 9} = \sqrt{25} = 5$ Now that we know *x*, *y*, and *r*, we can find the six trigonometric functions of θ.

$\sin \theta = \dfrac{y}{r} = \dfrac{3}{5}$

$\cos \theta = \dfrac{x}{r} = \dfrac{-4}{5} = -\dfrac{4}{5}$

$\tan \theta = \dfrac{y}{x} = \dfrac{3}{-4} = -\dfrac{3}{4}$

$\csc \theta = \dfrac{r}{y} = \dfrac{5}{3}$

$\sec \theta = \dfrac{r}{x} = \dfrac{5}{-4} = -\dfrac{5}{4}$

$\cot \theta = \dfrac{x}{y} = \dfrac{-4}{3} = -\dfrac{4}{3}$

3. We need values for *x*, *y*, and *r*. Because $P = (2, 3)$ is a point on the terminal side of θ, $x = 2$ and $y = 3$. Furthermore,

$r = \sqrt{x^2 + y^2} = \sqrt{2^2 + 3^2} = \sqrt{4 + 9} = \sqrt{13}$

Now that we know *x*, *y*, and *r*, we can find the six trigonometric functions of θ.

$\sin \theta = \dfrac{y}{r} = \dfrac{3}{\sqrt{13}} = \dfrac{3}{\sqrt{13}} \cdot \dfrac{\sqrt{13}}{\sqrt{13}} = \dfrac{3\sqrt{13}}{13}$

$\cos \theta = \dfrac{x}{r} = \dfrac{2}{\sqrt{13}} = \dfrac{2}{\sqrt{13}} \cdot \dfrac{\sqrt{13}}{\sqrt{13}} = \dfrac{2\sqrt{13}}{13}$

$\tan \theta = \dfrac{y}{x} = \dfrac{3}{2}$

$\csc \theta = \dfrac{r}{y} = \dfrac{\sqrt{13}}{3}$

$\sec \theta = \dfrac{r}{x} = \dfrac{\sqrt{13}}{2}$

$\cot \theta = \dfrac{x}{y} = \dfrac{2}{3}$

5. We need values for *x*, *y*, and *r*. Because $P = (3, -3)$ is a point on the terminal side of θ, $x = 3$ and $y = -3$.

Furthermore, $r = \sqrt{x^2 + y^2} = \sqrt{3^2 + (-3)^2} = \sqrt{9 + 9}$

$= \sqrt{18} = 3\sqrt{2}$

Now that we know *x*, *y*, and *r*, we can find the six trigonometric functions of θ.

$\sin \theta = \dfrac{y}{r} = \dfrac{-3}{3\sqrt{2}} = \dfrac{-1}{\sqrt{2}} \cdot \dfrac{\sqrt{2}}{\sqrt{2}} = -\dfrac{\sqrt{2}}{2}$

$\cos \theta = \dfrac{x}{r} = \dfrac{3}{3\sqrt{2}} = \dfrac{1}{\sqrt{2}} \cdot \dfrac{\sqrt{2}}{\sqrt{2}} = \dfrac{\sqrt{2}}{2}$

$\tan \theta = \dfrac{y}{x} = \dfrac{-3}{3} = -1$

$\csc \theta = \dfrac{r}{y} = \dfrac{3\sqrt{2}}{-3} = -\sqrt{2}$

$\sec \theta = \dfrac{r}{x} = \dfrac{3\sqrt{2}}{3} = \sqrt{2}$

$\cot \theta = \dfrac{x}{y} = \dfrac{3}{-3} = -1$

7. We need values for *x, y,* and *r.* Because $P = (-2, -5)$ is a point on the terminal side of θ, $x = -2$ and $y = -5$. Furthermore,

$r = \sqrt{x^2 + y^2} = \sqrt{(-2)^2 + (-5)^2} = \sqrt{4 + 25} = \sqrt{29}$ Now that we know *x, y,* and *r,* we can find the six trigonometric functions of θ.

$\sin\theta = \dfrac{y}{r} = \dfrac{-5}{\sqrt{29}} = \dfrac{-5}{\sqrt{29}} \cdot \dfrac{\sqrt{29}}{\sqrt{29}} = -\dfrac{5\sqrt{29}}{29}$

$\cos\theta = \dfrac{x}{r} = \dfrac{-2}{\sqrt{29}} = \dfrac{-2}{\sqrt{29}} \cdot \dfrac{\sqrt{29}}{\sqrt{29}} = -\dfrac{2\sqrt{29}}{29}$

$\tan\theta = \dfrac{y}{x} = \dfrac{-5}{-2} = \dfrac{5}{2}$

$\csc\theta = \dfrac{r}{y} = \dfrac{\sqrt{29}}{-5} = -\dfrac{\sqrt{29}}{5}$

$\sec\theta = \dfrac{r}{x} = \dfrac{\sqrt{29}}{-2} = -\dfrac{\sqrt{29}}{2}$

$\cot\theta = \dfrac{x}{y} = \dfrac{-2}{-5} = \dfrac{2}{5}$

9. $\theta = \pi$ radians
The terminal side of the angle is on the negative *x*-axis. Select the point $P = (-1, 0)$:
$x = -1$, $y = 0$, $r = 1$ Apply the definition of the cosine function.

$\cos\pi = \dfrac{x}{r} = \dfrac{-1}{1} = -1$

11. $\theta = \pi$ radians
The terminal side of the angle is on the negative *x*-axis. Select the point $P = (-1, 0)$:
$x = -1$, $y = 0$, $r = 1$ Apply the definition of the secant function.

$\sec\pi = \dfrac{r}{x} = \dfrac{1}{-1} = -1$

13. $\theta = \dfrac{3\pi}{2}$ radians
The terminal side of the angle is on the negative *y*-axis. Select the point $P = (0, -1)$:
$x = 0$, $y = -1$, $r = 1$ Apply the definition of the tangent function. $\tan\dfrac{3\pi}{2} = \dfrac{y}{x} = \dfrac{-1}{0}$, undefined

15. $\theta = \dfrac{\pi}{2}$ radians
The terminal side of the angle is on the positive *y*-axis. Select the point $P = (0, 1)$:
$x = 0$, $y = 1$, $r = 1$ Apply the definition of the cotangent function. $\cot\dfrac{\pi}{2} = \dfrac{x}{y} = \dfrac{0}{1} = 0$

17. Because $\sin\theta > 0$, θ cannot lie in quadrant III or quadrant IV; the sine function is negative in those quadrants. Thus, with $\sin\theta > 0$, θ lies in quadrant I or quadrant II. We are also given that $\cos\theta > 0$. Because quadrant I is the only quadrant in which the cosine is positive and sine is positive, we conclude that θ lies in quadrant I.

19. Because $\sin\theta < 0$, θ cannot lie in quadrant I or quadrant II; the sine function is positive in those two quadrants. Thus, with $\sin\theta < 0$, θ lies in quadrant III or quadrant IV. We are also given that $\cos\theta < 0$. Because quadrant III is the only quadrant in which the cosine is positive and the sine is negative, we conclude that θ lies in quadrant III.

21. Because $\tan\theta < 0$, θ cannot lie in quadrant I or quadrant III; the tangent function is positive in those quadrants. Thus, with $\tan\theta < 0$, θ lies in quadrant II or quadrant IV. We are also given that $\cos\theta < 0$. Because quadrant II is the only quadrant in which the cosine is negative and the tangent is negative, we conclude that θ lies in quadrant II.

23. In quadrant III x is negative and y is negative. Thus,

$\cos\theta = -\dfrac{3}{5} = \dfrac{x}{r} = \dfrac{-3}{5}$, $x = -3$, $r = 5$. Furthermore,

$r^2 = x^2 + y^2$

$5^2 = (-3)^2 + y^2$

$y^2 = 25 - 9 = 16$

$y = -\sqrt{16} = -4$

Now that we know x, y, and r, we can find the remaining trigonometric functions of θ.

$\sin\theta = \dfrac{y}{r} = \dfrac{-4}{5} = -\dfrac{4}{5}$

$\tan\theta = \dfrac{y}{x} = \dfrac{-4}{-3} = \dfrac{4}{3}$

$\csc\theta = \dfrac{r}{y} = \dfrac{5}{-4} = -\dfrac{5}{4}$

$\sec\theta = \dfrac{r}{x} = \dfrac{5}{-3} = -\dfrac{5}{3}$

$\cot\theta = \dfrac{x}{y} = \dfrac{-3}{-4} = \dfrac{3}{4}$

25. In quadrant II x is negative and y is positive. Thus,

$\sin\theta = \dfrac{5}{13} = \dfrac{y}{r}$, $y = 5$, $r = 13$. Furthermore,

$x^2 + y^2 = r^2$

$x^2 + 5^2 = 13^2$

$x^2 = 169 - 25 = 144$

$x = -\sqrt{144} = -12$

Now that we know x, y, and r, we can find the remaining trigonometric functions of θ.

$\cos\theta = \dfrac{x}{r} = \dfrac{-12}{13} = -\dfrac{12}{13}$

$\tan\theta = \dfrac{y}{x} = \dfrac{5}{-12} = -\dfrac{5}{12}$

$\csc\theta = \dfrac{r}{y} = \dfrac{13}{5}$

$\sec\theta = \dfrac{r}{x} = \dfrac{13}{-12} = -\dfrac{13}{12}$

$\cot\theta = \dfrac{x}{y} = \dfrac{-12}{5} = -\dfrac{12}{5}$

27. Because $270° < \theta < 360°$, θ is in quadrant IV. In quadrant IV x is positive and y is negative. Thus,

$\cos\theta = \dfrac{8}{17} = \dfrac{x}{r}$, $x = 8$,

$r = 17$. Furthermore

$x^2 + y^2 = r^2$

$8^2 + y^2 = 17^2$

$y^2 = 289 - 64 = 225$

$y = -\sqrt{225} = -15$

Now that we know x, y, and r, we can find the remaining trigonometric functions of θ.

$\sin\theta = \dfrac{y}{r} = \dfrac{-15}{17} = -\dfrac{15}{17}$

$\tan\theta = \dfrac{y}{x} = \dfrac{-15}{8} = -\dfrac{15}{8}$

$\csc\theta = \dfrac{r}{y} = \dfrac{17}{-15} = -\dfrac{17}{15}$

$\sec\theta = \dfrac{r}{x} = \dfrac{17}{8}$

$\cot\theta = \dfrac{x}{y} = \dfrac{8}{-15} = -\dfrac{8}{15}$

29. Because the tangent is negative and the sine is positive, θ lies in quadrant II. In quadrant II, x is negative and y is positive. Thus,

$\tan\theta = -\dfrac{2}{3} = \dfrac{y}{x} = \dfrac{2}{-3}$, $x = -3$, $y = 2$. Furthermore,

$r = \sqrt{x^2 + y^2} = \sqrt{(-3)^2 + 2^2} = \sqrt{9+4} = \sqrt{13}$

Now that we know x, y, and r, we can find the remaining trigonometric functions of θ.

$\sin\theta = \dfrac{y}{r} = \dfrac{2}{\sqrt{13}} = \dfrac{2}{\sqrt{13}} \cdot \dfrac{\sqrt{13}}{\sqrt{13}} = \dfrac{2\sqrt{13}}{13}$

$\cos\theta = \dfrac{x}{r} = \dfrac{-3}{\sqrt{13}} = \dfrac{-3}{\sqrt{13}} \cdot \dfrac{\sqrt{13}}{\sqrt{13}} = -\dfrac{3\sqrt{13}}{13}$

$\csc\theta = \dfrac{r}{y} = \dfrac{\sqrt{13}}{2}$

$\sec\theta = \dfrac{r}{x} = \dfrac{\sqrt{13}}{-3} = -\dfrac{\sqrt{13}}{3}$

$\cot\theta = \dfrac{x}{y} = \dfrac{-3}{2} = -\dfrac{3}{2}$

31. Because the tangent is positive and the cosine is negative, θ lies in quadrant III. In quadrant III, x is negative and y is negative. Thus, $\tan\theta = \dfrac{4}{3} = \dfrac{y}{x} = \dfrac{-4}{-3}$,

$x = -3$, $y = -4$. Furthermore,

$r = \sqrt{x^2 + y^2} = \sqrt{(-3)^2 + (-4)^2} = \sqrt{9+16}$

$\quad = \sqrt{25} = 5$

Now that we know x, y, and r, we can find the remaining trigonometric functions of θ.

$\sin\theta = \dfrac{y}{r} = \dfrac{-4}{5} = -\dfrac{4}{5}$

$\cos\theta = \dfrac{x}{r} = \dfrac{-3}{5} = -\dfrac{3}{5}$

$\csc\theta = \dfrac{r}{y} = \dfrac{5}{-4} = -\dfrac{5}{4}$

$\sec\theta = \dfrac{r}{x} = \dfrac{5}{-3} = -\dfrac{5}{3}$

$\cot\theta = \dfrac{x}{y} = \dfrac{-3}{-4} = \dfrac{3}{4}$

33. Because the secant is negative and the tangent is positive, θ lies in quadrant III. In quadrant III, x is negative and y is negative. Thus,

$\sec\theta = -3 = \dfrac{r}{x} = \dfrac{3}{-1}$, $x = -1$, $r = 3$. Furthermore,

$x^2 + y^2 = r^2$

$(-1)^2 + y^2 = 3^2$

$\quad y^2 = 9 - 1 = 8$

$\quad y = -\sqrt{8} = -2\sqrt{2}$

Now that we know x, y, and r, we can find the remaining trigonometric functions of θ.

$\sin\theta = \dfrac{y}{r} = \dfrac{-2\sqrt{2}}{3} = -\dfrac{2\sqrt{2}}{3}$

$\cos\theta = \dfrac{x}{r} = \dfrac{-1}{3} = -\dfrac{1}{3}$

$\tan\theta = \dfrac{y}{x} = \dfrac{-2\sqrt{2}}{-1} = 2\sqrt{2}$

$\csc\theta = \dfrac{r}{y} = \dfrac{3}{-2\sqrt{2}} = \dfrac{3}{-2\sqrt{2}} \cdot \dfrac{\sqrt{2}}{\sqrt{2}} = -\dfrac{3\sqrt{2}}{4}$

$\cot\theta = \dfrac{x}{y} = \dfrac{-1}{-2\sqrt{2}} = \dfrac{1}{2\sqrt{2}} \cdot \dfrac{\sqrt{2}}{\sqrt{2}} = \dfrac{\sqrt{2}}{4}$

35. Because $160°$ lies between $90°$ and $180°$, it is in quadrant II. The reference angle is $\theta' = 180° - 160° = 20°$.

37. Because $205°$ lies between $180°$ and $270°$, it is in quadrant III. The reference angle is $\theta' = 205° - 180° = 25°$.

39. Because $355°$ lies between $270°$ and $360°$, it is in quadrant IV. The reference angle is $\theta' = 360° - 355° = 5°$.

41. Because $\dfrac{7\pi}{4}$ lies between $\dfrac{3\pi}{2} = \dfrac{6\pi}{4}$ and $2\pi = \dfrac{8\pi}{4}$, it is in quadrant IV. The reference angle is

$\theta' = 2\pi - \dfrac{7\pi}{4} = \dfrac{8\pi}{4} - \dfrac{7\pi}{4} = \dfrac{\pi}{4}$.

43. Because $\dfrac{5\pi}{6}$ lies between $\dfrac{\pi}{2} = \dfrac{3\pi}{6}$ and $\pi = \dfrac{6\pi}{6}$, it is in quadrant II. The reference angle is

$\theta' = \pi - \dfrac{5\pi}{6} = \dfrac{6\pi}{6} - \dfrac{5\pi}{6} = \dfrac{\pi}{6}$.

45. $-150° + 360° = 210°$

Because the angle is in quadrant III, the reference angle is $\theta' = 210° - 180° = 30°$.

47. $-335° + 360° = 25°$

Because the angle is in quadrant I, the reference angle is $\theta' = 25°$.

49. Because 4.7 lies between $\pi \approx 3.14$ and $\frac{3\pi}{2} \approx 4.71$, it is in quadrant III. The reference angle is $\theta' = 4.7 - \pi \approx 1.56$.

51. $565° - 360° = 205°$

Because the angle is in quadrant III, the reference angle is $\theta' = 205° - 180° = 25°$.

53. $\frac{17\pi}{6} - 2\pi = \frac{17\pi}{6} - \frac{12\pi}{6} = \frac{5\pi}{6}$

Because the angle is in quadrant II, the reference angle is $\theta' = \pi - \frac{5\pi}{6} = \frac{\pi}{6}$.

55. $\frac{23\pi}{4} - 4\pi = \frac{23\pi}{4} - \frac{16\pi}{4} = \frac{7\pi}{4}$

Because the angle is in quadrant IV, the reference angle is $\theta' = 2\pi - \frac{7\pi}{4} = \frac{\pi}{4}$.

57. $-\frac{11\pi}{4} + 4\pi = -\frac{11\pi}{4} + \frac{16\pi}{4} = \frac{5\pi}{4}$

Because the angle is in quadrant III, the reference angle is $\theta' = \frac{5\pi}{4} - \pi = \frac{\pi}{4}$.

59. $-\frac{25\pi}{6} + 6\pi = -\frac{25\pi}{6} + \frac{36\pi}{6} = \frac{11\pi}{6}$

Because the angle is in quadrant IV, the reference angle is $\theta' = 2\pi - \frac{11\pi}{6} = \frac{\pi}{6}$.

61. $225°$ lies in quadrant III. The reference angle is $\theta' = 225° - 180° = 45°$.

$\cos 45° = \frac{\sqrt{2}}{2}$

Because the cosine is negative in quadrant III,

$\cos 225° = -\cos 45° = -\frac{\sqrt{2}}{2}$.

63. $210°$ lies in quadrant III. The reference angle is $\theta' = 210° - 180° = 30°$.

$\tan 30° = \frac{\sqrt{3}}{3}$

Because the tangent is positive in quadrant III,

$\tan 210° = \tan 30° = \frac{\sqrt{3}}{3}$.

65. $420°$ lies in quadrant I. The reference angle is $\theta' = 420° - 360° = 60°$.

$\tan 60° = \sqrt{3}$

Because the tangent is positive in quadrant I,

$\tan 420° = \tan 60° = \sqrt{3}$.

67. $\frac{2\pi}{3}$ lies in quadrant II. The reference angle is

$\theta' = \pi - \frac{2\pi}{3} = \frac{3\pi}{3} - \frac{2\pi}{3} = \frac{\pi}{3}$.

$\sin \frac{\pi}{3} = \frac{\sqrt{3}}{2}$

Because the sine is positive in quadrant II,

$\sin \frac{2\pi}{3} = \sin \frac{\pi}{3} = \frac{\sqrt{3}}{2}$.

69. $\frac{7\pi}{6}$ lies in quadrant III. The reference angle is

$\theta' = \frac{7\pi}{6} - \pi = \frac{7\pi}{6} - \frac{6\pi}{6} = \frac{\pi}{6}$.

$\csc \frac{\pi}{6} = 2$

Because the cosecant is negative in quadrant III,

$\csc \frac{7\pi}{6} = -\csc \frac{\pi}{6} = -2$.

71. $\frac{9\pi}{4}$ lies in quadrant I. The reference angle is

$\theta' = \frac{9\pi}{4} - 2\pi = \frac{9\pi}{4} - \frac{8\pi}{4} = \frac{\pi}{4}$.

$\tan \frac{\pi}{4} = 1$

Because the tangent is positive in quadrant I,

$\tan \frac{9\pi}{4} = \tan \frac{\pi}{4} = 1$

73. $-240°$ lies in quadrant II. The reference angle is $\theta' = 240° - 180° = 60°$.

$$\sin 60° = \frac{\sqrt{3}}{2}$$

Because the sine is positive in quadrant II,

$$\sin(-240°) = \sin 60° = \frac{\sqrt{3}}{2}.$$

75. $-\dfrac{\pi}{4}$ lies in quadrant IV. The reference angle is

$$\theta' = \frac{\pi}{4}.$$

$$\tan \frac{\pi}{4} = 1$$

Because the tangent is negative in quadrant IV,

$$\tan\left(-\frac{\pi}{4}\right) = -\tan \frac{\pi}{4} = -1$$

77. $\sec 495° = \sec 135° = -\sqrt{2}$

79. $\cot \dfrac{19\pi}{6} = \cot \dfrac{7\pi}{6} = \sqrt{3}$

81. $\cos \dfrac{23\pi}{4} = \cos \dfrac{7\pi}{4} = \dfrac{\sqrt{2}}{2}$

83. $\tan\left(-\dfrac{17\pi}{6}\right) = \tan \dfrac{7\pi}{6} = \dfrac{\sqrt{3}}{3}$

85. $\sin\left(-\dfrac{17\pi}{3}\right) = \sin \dfrac{\pi}{3} = \dfrac{\sqrt{3}}{2}$

87. $\sin \dfrac{\pi}{3} \cos \pi - \cos \dfrac{\pi}{3} \sin \dfrac{3\pi}{2}$

$$= \left(\frac{\sqrt{3}}{2}\right)(-1) - \left(\frac{1}{2}\right)(-1)$$

$$= -\frac{\sqrt{3}}{2} + \frac{1}{2}$$

$$= \frac{1 - \sqrt{3}}{2}$$

89. $\sin \dfrac{11\pi}{4} \cos \dfrac{5\pi}{6} + \cos \dfrac{11\pi}{4} \sin \dfrac{5\pi}{6}$

$$= \left(\frac{\sqrt{2}}{2}\right)\left(-\frac{\sqrt{3}}{2}\right) + \left(-\frac{\sqrt{2}}{2}\right)\left(\frac{1}{2}\right)$$

$$= -\frac{\sqrt{6}}{4} - \frac{\sqrt{2}}{4}$$

$$= -\frac{\sqrt{6} + \sqrt{2}}{4}$$

91. $\sin \dfrac{3\pi}{2} \tan\left(-\dfrac{15\pi}{4}\right) - \cos\left(-\dfrac{5\pi}{3}\right)$

$$= (-1)(1) - \left(\frac{1}{2}\right)$$

$$= -1 - \frac{1}{2}$$

$$= -\frac{2}{2} - \frac{1}{2}$$

$$= -\frac{3}{2}$$

93. $f\left(\dfrac{4\pi}{3} + \dfrac{\pi}{6}\right) + f\left(\dfrac{4\pi}{3}\right) + f\left(\dfrac{\pi}{6}\right)$

$$= \sin\left(\frac{4\pi}{3} + \frac{\pi}{6}\right) + \sin \frac{4\pi}{3} + \sin \frac{\pi}{6}$$

$$= \sin \frac{3\pi}{2} + \sin \frac{4\pi}{3} + \sin \frac{\pi}{6}$$

$$= (-1) + \left(-\frac{\sqrt{3}}{2}\right) + \left(\frac{1}{2}\right)$$

$$= -\frac{\sqrt{3} + 1}{2}$$

95. $(h \circ g)\left(\dfrac{17\pi}{3}\right) = h\left(g\left(\dfrac{17\pi}{3}\right)\right)$

$$= 2\left(\cos\left(\frac{17\pi}{3}\right)\right)$$

$$= 2\left(\frac{1}{2}\right)$$

$$= 1$$

97. The average rate of change is the slope of the line through the points $(x_1, f(x_1))$ and $(x_2, f(x_2))$

$$m = \frac{f(x_2) - f(x_1)}{x_2 - x_1}$$

$$= \frac{\sin\left(\dfrac{3\pi}{2}\right) - \sin\left(\dfrac{5\pi}{4}\right)}{\dfrac{3\pi}{2} - \dfrac{5\pi}{4}}$$

$$= \frac{-1 - \left(-\dfrac{\sqrt{2}}{2}\right)}{\dfrac{\pi}{4}}$$

$$= \frac{-1 + \dfrac{\sqrt{2}}{2}}{\dfrac{\pi}{4}}$$

$$= \frac{4\left(-1 + \dfrac{\sqrt{2}}{2}\right)}{4\left(\dfrac{\pi}{4}\right)}$$

$$= \frac{2\sqrt{2} - 4}{\pi}$$

99. $\sin\theta = \dfrac{\sqrt{2}}{2}$ when the reference angle is $\dfrac{\pi}{4}$ and θ is in quadrants I or II.

QI	QII
$\theta = \dfrac{\pi}{4}$	$\theta = \pi - \dfrac{\pi}{4}$
	$= \dfrac{3\pi}{4}$

$$\theta = \frac{\pi}{4}, \frac{3\pi}{4}$$

101. $\sin\theta = -\dfrac{\sqrt{2}}{2}$ when the reference angle is $\dfrac{\pi}{4}$ and θ is in quadrants III or IV.

QIII	QIV
$\theta = \pi + \dfrac{\pi}{4}$	$\theta = 2\pi - \dfrac{\pi}{4}$
$= \dfrac{5\pi}{4}$	$= \dfrac{7\pi}{4}$

$$\theta = \frac{5\pi}{4}, \frac{7\pi}{4}$$

103. $\tan\theta = -\sqrt{3}$ when the reference angle is $\dfrac{\pi}{3}$ and θ is in quadrants II or IV.

QII	QIV
$\theta = \pi - \dfrac{\pi}{3}$	$\theta = 2\pi - \dfrac{\pi}{3}$
$= \dfrac{2\pi}{3}$	$= \dfrac{5\pi}{3}$

$$\theta = \frac{2\pi}{3}, \frac{5\pi}{3}$$

105. – 109. Answers may vary.

111. does not make sense; Explanations will vary. Sample explanation: Sine and cosecant have the same sign within any quadrant because they are reciprocals of each other.

113. makes sense

114. $y = \dfrac{1}{2}\cos(4x + \pi)$

x	$-\dfrac{\pi}{4}$	$-\dfrac{\pi}{8}$	0	$\dfrac{\pi}{8}$	$\dfrac{\pi}{4}$
y	$\dfrac{1}{2}$	0	$-\dfrac{1}{2}$	0	$\dfrac{1}{2}$

115. $y = 4\sin\left(2x - \dfrac{2\pi}{3}\right)$

x	$\dfrac{\pi}{3}$	$\dfrac{7\pi}{12}$	$\dfrac{5\pi}{6}$	$\dfrac{13\pi}{12}$	$\dfrac{4\pi}{3}$
y	0	4	0	-4	0

116. $y = 3\sin\dfrac{\pi}{2}x$

x	0	$\dfrac{1}{3}$	1	$\dfrac{5}{3}$	2	$\dfrac{7}{3}$	3	$\dfrac{11}{3}$	4
y	0	$\dfrac{3}{2}$	3	$\dfrac{3}{2}$	0	$-\dfrac{3}{2}$	-3	$-\dfrac{3}{2}$	0

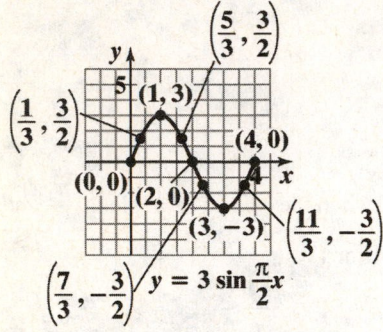

Mid-Chapter 4 Check Point

1. $10° = 10° \cdot \dfrac{\pi \text{ radians}}{180°} = \dfrac{10\pi}{180}$ radians

$\qquad = \dfrac{\pi}{18}$ radians

2. $-105° = -105° \cdot \dfrac{\pi \text{ radians}}{180°} = -\dfrac{105\pi}{180}$ radians

$\qquad = -\dfrac{7\pi}{12}$ radians

3. $\dfrac{5\pi}{12}$ radians $= \dfrac{5\pi \text{ radians}}{12} \cdot \dfrac{180^{O}}{\pi \text{ radians}} = 75^{O}$

4. $-\dfrac{13\pi}{20}$ radians $= -\dfrac{13\pi \text{ radians}}{20} \cdot \dfrac{180^{O}}{\pi \text{ radians}}$

$\qquad\qquad = -117^{O}$

5. **a.** $\dfrac{11\pi}{3} - 2\pi = \dfrac{11\pi}{3} - \dfrac{6\pi}{3} = \dfrac{5\pi}{3}$

b.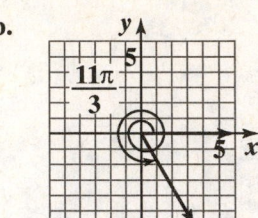

c. Since $\dfrac{5\pi}{3}$ is in quadrant IV, the reference angle

is $2\pi - \dfrac{5\pi}{3} = \dfrac{6\pi}{3} - \dfrac{5\pi}{3} = \dfrac{\pi}{3}$

6. **a.** $-\dfrac{19\pi}{4} + 6\pi = -\dfrac{19\pi}{4} + \dfrac{24\pi}{4} = \dfrac{5\pi}{4}$

b.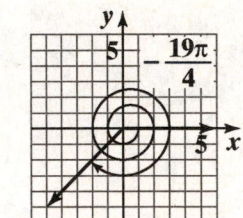

c. Since $\dfrac{5\pi}{4}$ is in quadrant III, the reference angle

is $\dfrac{5\pi}{4} - \pi = \dfrac{5\pi}{4} - \dfrac{4\pi}{4} = \dfrac{\pi}{4}$

7. **a.** $510° - 360° = 150°$

b.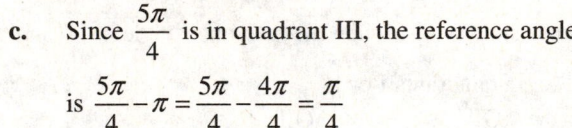

c. Since $150°$ is in quadrant II, the reference angle is $180° - 150° = 30°$

8. $r = \sqrt{x^2 + y^2}$

$$r = \sqrt{\left(-\frac{3}{5}\right)^2 + \left(-\frac{4}{5}\right)^2} = \sqrt{\frac{9}{25} + \frac{16}{25}} = \sqrt{\frac{25}{25}} = 1$$

Now that we know x, y, and r, we can find the six trigonometric functions of θ.

$$\sin\theta = \frac{y}{r} = \frac{-\frac{4}{5}}{1} = -\frac{4}{5}$$

$$\cos\theta = \frac{x}{r} = \frac{-\frac{3}{5}}{1} = -\frac{3}{5}$$

$$\tan\theta = \frac{y}{x} = \frac{-\frac{4}{5}}{-\frac{3}{5}} = \frac{4}{3}$$

$$\csc\theta = \frac{r}{y} = \frac{1}{-\frac{4}{5}} = -\frac{5}{4}$$

$$\sec\theta = \frac{r}{x} = \frac{1}{-\frac{3}{5}} = -\frac{5}{3}$$

$$\cot\theta = \frac{x}{y} = \frac{-\frac{3}{5}}{-\frac{4}{5}} = \frac{3}{4}$$

9.   Use the Pythagorean theorem to find b.

$$a^2 + b^2 = c^2$$
$$5^2 + b^2 = 6^2$$
$$25 + b^2 = 36$$
$$b^2 = 11$$
$$b = \sqrt{11}$$

$$\sin\theta = \frac{\text{opposite}}{\text{hypotenuse}} = \frac{5}{6}$$

$$\cos\theta = \frac{\text{adjacent}}{\text{hypotenuse}} = \frac{\sqrt{11}}{6}$$

$$\tan\theta = \frac{\text{opposite}}{\text{adjacent}} = \frac{5\sqrt{11}}{11}$$

$$\csc\theta = \frac{\text{hypotenuse}}{\text{opposite}} = \frac{6}{5}$$

$$\sec\theta = \frac{\text{hypotenuse}}{\text{adjacent}} = \frac{6}{\sqrt{11}} = \frac{6\sqrt{11}}{11}$$

$$\cot\theta = \frac{\text{adjacent}}{\text{opposite}} = \frac{\sqrt{11}}{5}$$

10. $r = \sqrt{x^2 + y^2}$

$$r = \sqrt{3^2 + (-2)^2} = \sqrt{9 + 4} = \sqrt{13}$$

Now that we know x, y, and r, we can find the six trigonometric functions of θ.

$$\sin\theta = \frac{y}{r} = \frac{-2}{\sqrt{13}} = -\frac{2\sqrt{13}}{13}$$

$$\cos\theta = \frac{x}{r} = \frac{3}{\sqrt{13}} = \frac{3\sqrt{13}}{13}$$

$$\tan\theta = \frac{y}{x} = \frac{-2}{3} = -\frac{2}{3}$$

$$\csc\theta = \frac{r}{y} = \frac{\sqrt{13}}{-2} = -\frac{\sqrt{13}}{2}$$

$$\sec\theta = \frac{r}{x} = \frac{\sqrt{13}}{3}$$

$$\cot\theta = \frac{x}{y} = \frac{3}{-2} = -\frac{3}{2}$$

11.   Because the tangent is negative and the cosine is negative, θ is in quadrant II. In quadrant II, x is negative and y is positive. Thus,

$$\tan\theta = -\frac{3}{4} = \frac{x}{y}, \quad x = -4, \ y = 3. \text{ Furthermore,}$$

$$r^2 = x^2 + y^2$$
$$r^2 = (-3)^2 + 4^2$$
$$r^2 = 9 + 16 = 25$$
$$r = 5$$

Now that we know x, y, and r, we can find the remaining trigonometric functions of θ.

$$\sin\theta = \frac{y}{r} = \frac{3}{5}$$

$$\cos\theta = \frac{x}{r} = \frac{-4}{5} = -\frac{4}{5}$$

$$\csc\theta = \frac{r}{y} = \frac{5}{3}$$

$$\sec\theta = \frac{r}{x} = \frac{5}{-3} = -\frac{5}{4}$$

$$\cot\theta = \frac{x}{y} = \frac{-3}{4} = -\frac{4}{3}$$

12. Since $\cos\theta = \dfrac{3}{7} = \dfrac{x}{r}$, $x = 3$, $r = 7$. Furthermore,

$$x^2 + y^2 = r^2$$
$$3^2 + y^2 = 7^2$$
$$9 + y^2 = 49$$
$$y^2 = 40$$
$$y = \pm\sqrt{40} = \pm 2\sqrt{10}$$

Because the cosine is positive and the sine is negative, θ is in quadrant IV. In quadrant IV, x is positive and y is negative.

Therefore $y = -2\sqrt{10}$

Use x, y, and r to find the remaining trigonometric functions of θ.

$$\sin\theta = \frac{y}{r} = \frac{-2\sqrt{10}}{7} = -\frac{2\sqrt{10}}{7}$$

$$\tan\theta = \frac{y}{x} = \frac{-2\sqrt{10}}{3} = -\frac{2\sqrt{10}}{3}$$

$$\csc\theta = \frac{r}{y} = \frac{7}{-2\sqrt{10}} = -\frac{7\sqrt{10}}{20}$$

$$\sec\theta = \frac{r}{x} = \frac{7}{3}$$

$$\cot\theta = \frac{x}{y} = \frac{3}{-2\sqrt{10}} = -\frac{3\sqrt{10}}{20}$$

13. $\tan\theta = \dfrac{\text{side opposite }\theta}{\text{side adjacent }\theta}$

$$\tan 41° = \frac{a}{60}$$
$$a = 60\tan 41°$$
$$a \approx 52 \text{ cm}$$

14. $\cos\theta = \dfrac{\text{side adjacent }\theta}{\text{hypotenuse}}$

$$\cos 72° = \frac{250}{c}$$
$$c = \frac{250}{\cos 72°}$$
$$c \approx 809 \text{ m}$$

15. Since $\cos\theta = \dfrac{1}{6} = \dfrac{x}{r}$, $x = 1$, $r = 6$. Furthermore,

$$x^2 + y^2 = r^2$$
$$1^2 + y^2 = 6^2$$
$$1 + y^2 = 36$$
$$y^2 = 35$$
$$y = \pm\sqrt{35}$$

Since θ is acute, $y = +\sqrt{35} = \sqrt{35}$

$$\cot\left(\frac{\pi}{2} - \theta\right) = \tan\theta = \frac{y}{x} = \frac{\sqrt{35}}{1} = \sqrt{35}$$

16. $\tan 30° = \dfrac{\sqrt{3}}{3}$

17. $\cot 120° = \dfrac{1}{\tan 120°} = \dfrac{1}{-\tan 60°} = \dfrac{1}{-\sqrt{3}} = -\dfrac{\sqrt{3}}{3}$

18. $\cos 240° = -\cos 60° = -\dfrac{1}{2}$

19. $\sec\dfrac{11\pi}{6} = \dfrac{1}{\cos\dfrac{11\pi}{6}} = \dfrac{1}{\cos\dfrac{\pi}{6}} = \dfrac{1}{\dfrac{\sqrt{3}}{2}} = \dfrac{2}{\sqrt{3}} = \dfrac{2\sqrt{3}}{3}$

20. $\sin^2\dfrac{\pi}{7} + \cos^2\dfrac{\pi}{7} = 1$

21. $\sin\left(-\dfrac{2\pi}{3}\right) = \sin\left(-\dfrac{2\pi}{3} + 2\pi\right)$

$$= \sin\frac{4\pi}{3} = -\sin\frac{\pi}{3}$$

$$= -\frac{\sqrt{3}}{2}$$

22. $\csc\left(\dfrac{22\pi}{3}\right) = \csc\left(\dfrac{22\pi}{3} - 6\pi\right) = \csc\dfrac{4\pi}{3}$

$$= \frac{1}{\sin\dfrac{4\pi}{3}} = \frac{1}{-\sin\dfrac{\pi}{3}} = \frac{1}{-\dfrac{\sqrt{3}}{2}}$$

$$= -\frac{2}{\sqrt{3}} = -\frac{2\sqrt{3}}{3}$$

23. $\cos 495° = \cos\left(495° - 360°\right) = \cos 135°$

$$= -\cos 45° = -\frac{\sqrt{2}}{2}$$

24. $\tan\left(-\dfrac{17\pi}{6}\right) = \tan\left(-\dfrac{17\pi}{6}+4\pi\right) = \tan\dfrac{7\pi}{6}$

$\qquad\qquad = \tan\dfrac{\pi}{6} = \dfrac{\sqrt{3}}{3}$

25. $\sin^2\dfrac{\pi}{2}-\cos\pi = (1)^2 -(-1) = 1+1 = 2$

26. $\cos\left(\dfrac{5\pi}{6}+2\pi n\right)+\tan\left(\dfrac{5\pi}{6}+n\pi\right)$

$\qquad = \cos\dfrac{5\pi}{6} + \tan\dfrac{5\pi}{6} = -\cos\dfrac{\pi}{6} - \tan\dfrac{\pi}{6}$

$\qquad = -\dfrac{\sqrt{3}}{2} - \dfrac{\sqrt{3}}{3} = -\dfrac{3\sqrt{3}}{6} - \dfrac{2\sqrt{3}}{6}$

$\qquad = -\dfrac{5\sqrt{3}}{6}$

27. Begin by converting from degrees to radians.

$36° = 36° \cdot \dfrac{\pi \text{ radians}}{180°} = \dfrac{\pi}{5} \text{ radians}$

$s = r\theta = 40\cdot\dfrac{\pi}{5} = 8\pi \approx 25.13 \text{ cm}$

28. Linear speed is given by $v = r\omega$. It is given that $r = 10$ feet and the merry-go-round rotates at 8 revolutions per minute. Convert 8 revolutions per minute to radians per minute.
8 revolutions per minute

$= 8 \text{ revolutions per minute} \cdot \dfrac{2\pi \text{ radians}}{1 \text{ revolution}}$

$= 16\pi \text{ radians per minute}$

$v = r\omega = (10)(16\pi) = 160\pi \approx 502.7 \text{ feet per minute}$

The linear speed of the horse is about 502.7 feet per minute.

29. $\sin\theta = \dfrac{\text{side opposite }\theta}{\text{hypotenuse}}$

$\sin 6° = \dfrac{h}{5280}$

$h = 5280\sin 6°$

$h \approx 551.9 \text{ feet}$

30. $\tan\theta = \dfrac{\text{side opposite }\theta}{\text{side adjacent }\theta}$

$\tan\theta = \dfrac{50}{60}$

$\theta = \tan^{-1}\left(\dfrac{50}{60}\right)$

$\theta \approx 40°$

Section 4.5

Check Point Exercises

1. The equation $y = 3\sin x$ is of the form $y = A\sin x$ with $A = 3$. Thus, the amplitude is $|A| = |3| = 3$ The period for both $y = 3\sin x$ and $y = \sin x$ is 2π. We find the three x–intercepts, the maximum point, and the minimum point on the interval $[0, 2\pi]$ by dividing the period, 2π, by 4, $\dfrac{\text{period}}{4} = \dfrac{2\pi}{4} = \dfrac{\pi}{2}$, then by adding quarter-periods to generate x-values for each of the key points. The five x-values are

$x = 0$

$x = 0 + \dfrac{\pi}{2} = \dfrac{\pi}{2}$

$x = \dfrac{\pi}{2} + \dfrac{\pi}{2} = \pi$

$x = \pi + \dfrac{\pi}{2} = \dfrac{3\pi}{2}$

$x = \dfrac{3\pi}{2} + \dfrac{\pi}{2} = 2\pi$

Evaluate the function at each value of x.

x	$y = 3\sin x$	coordinates
0	$y = 3\sin 0 = 3\cdot 0 = 0$	$(0, 0)$
$\dfrac{\pi}{2}$	$y = 3\sin\dfrac{\pi}{2} = 3\cdot 1 = 3$	$\dfrac{\pi}{2}, 3$
π	$y = 3\sin x = 3\cdot 0 = 0$	$(\pi, 0)$
$\dfrac{3\pi}{2}$	$y = 3\sin\dfrac{3\pi}{2}$ $= 3(-1) = -3$	$\dfrac{3\pi}{2}, -3$
2π	$y = 3\sin 2\pi = 3\cdot 0 = 0$	$(2\pi, 0)$

Connect the five points with a smooth curve and graph one complete cycle of the given function with the graph of $y = \sin x$.

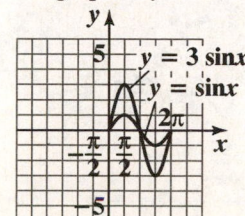

2. The equation $y = -\dfrac{1}{2}\sin x$ is of the form $y = A\sin x$

with $A = -\dfrac{1}{2}$. Thus, the amplitude is

$|A| = \left|-\dfrac{1}{2}\right| = \dfrac{1}{2}$. The period for both $y = -\dfrac{1}{2}\sin x$

and $y = \sin x$ is 2π.

Find the x–values for the five key points by dividing

the period, 2π, by 4, $\dfrac{\text{period}}{4} = \dfrac{2\pi}{4} = \dfrac{\pi}{2}$, then by

adding quarter- periods. The five x-values are

$x = 0$

$x = 0 + \dfrac{\pi}{2} = \dfrac{\pi}{2}$

$x = \dfrac{\pi}{2} + \dfrac{\pi}{2} = \pi$

$x = \pi + \dfrac{\pi}{2} = \dfrac{3\pi}{2}$

$x = \dfrac{3\pi}{2} + \dfrac{\pi}{2} = 2\pi$

Evaluate the function at each value of x.

x	$y = -\dfrac{1}{2}\sin x$	coordinates
0	$y = -\dfrac{1}{2}\sin 0$ $= -\dfrac{1}{2}\cdot 0 = 0$	$(0, 0)$
$\dfrac{\pi}{2}$	$y = -\dfrac{1}{2}\sin\dfrac{\pi}{2}$ $= -\dfrac{1}{2}\cdot 1 = -\dfrac{1}{2}$	$\dfrac{\pi}{2}, -\dfrac{1}{2}$
π	$y = -\dfrac{1}{2}\sin \pi$ $= -\dfrac{1}{2}\cdot 0 = 0$	$(\pi, 0)$
$\dfrac{3\pi}{2}$	$y = -\dfrac{1}{2}\sin\dfrac{3\pi}{2}$ $= -\dfrac{1}{2}(-1) = \dfrac{1}{2}$	$\dfrac{3\pi}{2}, \dfrac{1}{2}$
2π	$y = -\dfrac{1}{2}\sin 2\pi$ $= -\dfrac{1}{2}\cdot 0 = 0$	$(2\pi, 0)$

Connect the five key points with a smooth curve and graph one complete cycle of the given function with the graph of $y = \sin x$. Extend the pattern of each graph to the left and right as desired.

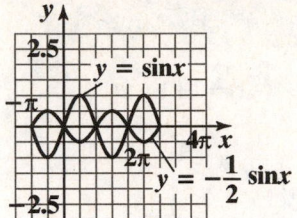

3. The equation $y = 2\sin\dfrac{1}{2}x$ is of the form

$y = A\sin Bx$ with $A = 2$ and $B = \dfrac{1}{2}$.

The amplitude is $|A| = |2| = 2$.

The period is $\dfrac{2\pi}{B} = \dfrac{2\pi}{\frac{1}{2}} = 4\pi$.

Find the x–values for the five key points by dividing

the period, 4π, by 4, $\dfrac{\text{period}}{4} = \dfrac{4\pi}{4} = \pi$, then by

adding quarter-periods.
The five x-values are

$x = 0$

$x = 0 + \pi = \pi$

$x = \pi + \pi = 2\pi$

$x = 2\pi + \pi = 3\pi$

$x = 3\pi + \pi = 4\pi$

Evaluate the function at each value of x.

x	$y = 2\sin\dfrac{1}{2}x$	coordinates
0	$y = 2\sin\dfrac{1}{2}\cdot 0$ $= 2\sin 0$ $= 2\cdot 0 = 0$	$(0, 0)$
π	$y = 2\sin\dfrac{1}{2}\cdot\pi$ $= 2\sin\dfrac{\pi}{2} = 2\cdot 1 = 2$	$(\pi, 2)$
2π	$y = 2\sin\dfrac{1}{2}\cdot 2\pi$ $= 2\sin\pi = 2\cdot 0 = 0$	$(2\pi, 0)$

3π	$\begin{aligned} y &= 2\sin\frac{1}{2}\cdot 3\pi \\ &= 2\sin\frac{3\pi}{2} \\ &= 2\cdot(-1) = -2 \end{aligned}$	$(3\pi, -2)$
4π	$\begin{aligned} y &= 2\sin\frac{1}{2}\cdot 4\pi \\ &= 2\sin 2\pi = 2\cdot 0 = 0 \end{aligned}$	$(4\pi, 0)$

Connect the five key points with a smooth curve and graph one complete cycle of the given function. Extend the pattern of the graph another full period to the right.

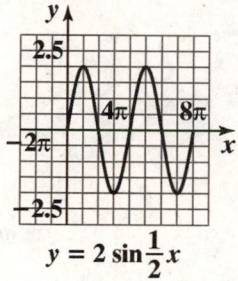

$y = 2\sin\dfrac{1}{2}x$

4. The equation $y = 3\sin\left(2x - \dfrac{\pi}{3}\right)$ is of the form $y = A\sin(Bx - C)$ with $A = 3$, $B = 2$, and $C = \dfrac{\pi}{3}$. The amplitude is $|A| = |3| = 3$.

The period is $\dfrac{2\pi}{B} = \dfrac{2\pi}{2} = \pi$.

The phase shift is $\dfrac{C}{B} = \dfrac{\frac{\pi}{3}}{2} = \dfrac{\pi}{3}\cdot\dfrac{1}{2} = \dfrac{\pi}{6}$.

Find the x-values for the five key points by dividing the period, π, by 4, $\dfrac{\text{period}}{4} = \dfrac{\pi}{4}$, then by adding quarter-periods to the value of x where the cycle begins, $x = \dfrac{\pi}{6}$.

The five x-values are

$$x = \frac{\pi}{6}$$

$$x = \frac{\pi}{6} + \frac{\pi}{4} = \frac{2\pi}{12} + \frac{3\pi}{12} = \frac{5\pi}{12}$$

$$x = \frac{5\pi}{12} + \frac{\pi}{4} = \frac{5\pi}{12} + \frac{3\pi}{12} = \frac{8\pi}{12} = \frac{2\pi}{3}$$

$$x = \frac{2\pi}{3} + \frac{\pi}{4} = \frac{8\pi}{12} + \frac{3\pi}{12} = \frac{11\pi}{12}$$

$$x = \frac{11\pi}{12} + \frac{\pi}{4} = \frac{11\pi}{12} + \frac{3\pi}{12} = \frac{14\pi}{12} = \frac{7\pi}{6}$$

Evaluate the function at each value of x.

x	$y = 3\sin\left(2x - \dfrac{\pi}{3}\right)$	coordinates
$\dfrac{\pi}{6}$	$\begin{aligned} y &= 3\sin\left(2\cdot\frac{\pi}{6} - \frac{\pi}{3}\right) \\ &= 3\sin 0 = 3\cdot 0 = 0 \end{aligned}$	$\dfrac{\pi}{6}, 0$
$\dfrac{5\pi}{12}$	$\begin{aligned} y &= 3\sin\left(2\cdot\frac{5\pi}{12} - \frac{\pi}{3}\right) \\ &= 3\sin\frac{3\pi}{6} = 3\sin\frac{\pi}{2} \\ &= 3\cdot 1 = 3 \end{aligned}$	$\dfrac{5\pi}{12}, 3$
$\dfrac{2\pi}{3}$	$\begin{aligned} y &= 3\sin\left(2\cdot\frac{2\pi}{3} - \frac{\pi}{3}\right) \\ &= 3\sin\frac{3\pi}{3} = 3\sin\pi \\ &= 3\cdot 0 = 0 \end{aligned}$	$\dfrac{2\pi}{3}, 0$
$\dfrac{11\pi}{12}$	$\begin{aligned} y &= 3\sin\left(2\cdot\frac{11\pi}{12} - \frac{\pi}{3}\right) \\ &= 3\sin\frac{9\pi}{6} = 3\sin\frac{3\pi}{2} \\ &= 3(-1) = -3 \end{aligned}$	$\dfrac{11\pi}{12}, -3$
$\dfrac{7\pi}{6}$	$\begin{aligned} y &= 3\sin\left(2\cdot\frac{7\pi}{6} - \frac{\pi}{3}\right) \\ &= 3\sin\frac{6\pi}{3} = 3\sin 2\pi \\ &= 3\cdot 0 = 0 \end{aligned}$	$\dfrac{7\pi}{6}, 0$

Connect the five key points with a smooth curve and graph one complete cycle of the given graph.

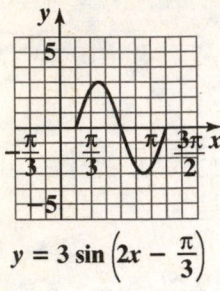

$$y = 3 \sin \left(2x - \frac{\pi}{3} \right)$$

5. The equation $y = -4 \cos \pi x$ is of the form $y = A \cos Bx$ with $A = -4$, and $B = \pi$.

Thus, the amplitude is $|A| = |-4| = 4$.

The period is $\frac{2\pi}{B} = \frac{2\pi}{\pi} = 2$.

Find the x-values for the five key points by dividing the period, 2, by 4, $\frac{\text{period}}{4} = \frac{2}{4} = \frac{1}{2}$, then by adding quarter periods to the value of x where the cycle begins. The five x-values are

$x = 0$

$x = 0 + \frac{1}{2} = \frac{1}{2}$

$x = \frac{1}{2} + \frac{1}{2} = 1$

$x = 1 + \frac{1}{2} = \frac{3}{2}$

$x = \frac{3}{2} + \frac{1}{2} = 2$

Evaluate the function at each value of x.

x	$y = -4 \cos \pi x$	coordinates
0	$y = -4 \cos (\pi \cdot 0)$ $= -4 \cos 0 = -4$	$(0, -4)$
$\frac{1}{2}$	$y = -4 \cos \left(\pi \cdot \frac{1}{2} \right)$ $= -4 \cos \frac{\pi}{2} = 0$	$\frac{1}{2}, 0$
1	$y = -4 \cos (\pi \cdot 1)$ $= -4 \cos \pi = 4$	$(1, 4)$
$\frac{3}{2}$	$y = -4 \cos \left(\pi \cdot \frac{3}{2} \right)$ $= -4 \cos \frac{3\pi}{2} = 0$	$\frac{3}{2}, 0$
2	$y = -4 \cos (\pi \cdot 2)$ $= -4 \cos 2\pi = -4$	$(2, -4)$

Connect the five key points with a smooth curve and graph one complete cycle of the given function. Extend the pattern of the graph another full period to the left.

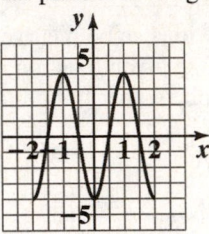

$$y = -4 \cos \pi x$$

6. $y = \frac{3}{2} \cos(2x + \pi) = \frac{3}{2} \cos(2x - (-\pi))$

The equation is of the form $y = A \cos(Bx - C)$ with

$A = \frac{3}{2}$, $B = 2$, and $C = -\pi$.

Thus, the amplitude is $|A| = \left| \frac{3}{2} \right| = \frac{3}{2}$.

The period is $\frac{2\pi}{B} = \frac{2\pi}{2} = \pi$.

The phase shift is $\frac{C}{B} = \frac{-\pi}{2} = -\frac{\pi}{2}$.

Find the x-values for the five key points by dividing the period, π, by 4, $\frac{\text{period}}{4} = \frac{\pi}{4}$, then by adding quarter-periods to the value of x where the cycle begins, $x = -\frac{\pi}{2}$.

The five x-values are

$x = -\frac{\pi}{2}$

$x = -\frac{\pi}{2} + \frac{\pi}{4} = -\frac{\pi}{4}$

$x = -\frac{\pi}{4} + \frac{\pi}{4} = 0$

$x = 0 + \frac{\pi}{4} = \frac{\pi}{4}$

$x = \frac{\pi}{4} + \frac{\pi}{4} = \frac{\pi}{2}$

Evaluate the function at each value of x.

x	$y = \frac{3}{2}\cos(2x + \pi)$	coordinates
$-\dfrac{\pi}{2}$	$y = \dfrac{3}{2}\cos(-\pi + \pi)$ $= \dfrac{3}{2}\cdot 1 = \dfrac{3}{2}$	$-\dfrac{\pi}{2}, \dfrac{3}{2}$
$-\dfrac{\pi}{4}$	$y = \dfrac{3}{2}\cos\left(-\dfrac{\pi}{2} + \pi\right)$ $= \dfrac{3}{2}\cdot 0 = 0$	$-\dfrac{\pi}{4}, 0$
0	$y = \dfrac{3}{2}\cos(0 + \pi)$ $= \dfrac{3}{2}\cdot -1 = -\dfrac{3}{2}$	$0, -\dfrac{3}{2}$
$\dfrac{\pi}{4}$	$y = \dfrac{3}{2}\cos\left(\dfrac{\pi}{2} + \pi\right)$ $= \dfrac{3}{2}\cdot 0 = 0$	$\dfrac{\pi}{4}, 0$
$\dfrac{\pi}{2}$	$y = \dfrac{3}{2}\cos(\pi + \pi)$ $= \dfrac{3}{2}\cdot 1 = \dfrac{3}{2}$	$\dfrac{\pi}{2}, \dfrac{3}{2}$

Connect the five key points with a smooth curve and graph one complete cycle of the given graph.

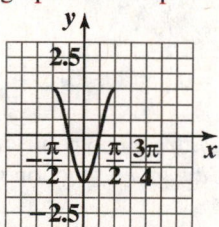

$y = \dfrac{3}{2}\cos(2x + \pi)$

7. The graph of $y = 2\cos x + 1$ is the graph of $y = 2\cos x$ shifted one unit upwards. The period for both functions is 2π. The quarter-period is $\dfrac{2\pi}{4}$ or $\dfrac{\pi}{2}$. The cycle begins at $x = 0$. Add quarter-periods to generate x-values for the key points.

$x = 0$

$x = 0 + \dfrac{\pi}{2} = \dfrac{\pi}{2}$

$x = \dfrac{\pi}{2} + \dfrac{\pi}{2} = \pi$

$x = \pi + \dfrac{\pi}{2} = \dfrac{3\pi}{2}$

$x = \dfrac{3\pi}{2} + \dfrac{\pi}{2} = 2\pi$

Evaluate the function at each value of x.

x	$y = 2\cos x + 1$	coordinates
0	$y = 2\cos 0 + 1$ $= 2\cdot 1 + 1 = 3$	$(0, 3)$
$\dfrac{\pi}{2}$	$y = 2\cos\dfrac{\pi}{2} + 1$ $= 2\cdot 0 + 1 = 1$	$\dfrac{\pi}{2}, 1$
π	$y = 2\cos\pi + 1$ $= 2\cdot(-1) + 1 = -1$	$(\pi, -1)$
$\dfrac{3\pi}{2}$	$y = 2\cos\dfrac{3\pi}{2} + 1$ $= 2\cdot 0 + 1 = 1$	$\dfrac{3\pi}{2}, 1$
2π	$y = 2\cos 2\pi + 1$ $= 2\cdot 1 + 1 = 3$	$(2\pi, 3)$

By connecting the points with a smooth curve, we obtain one period of the graph.

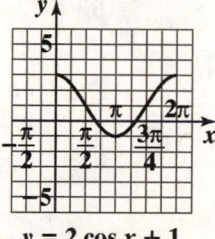

$y = 2\cos x + 1$

8. *A*, the amplitude, is the maximum value of *y*. The graph shows that this maximum value is 4, Thus, $A = 4$. The period is $\dfrac{\pi}{2}$, and period $= \dfrac{2\pi}{B}$. Thus,

$$\frac{\pi}{2} = \frac{2\pi}{B}$$
$$\pi B = 4\pi$$
$$B = 4$$

Substitute these values into $y = A \sin Bx$.
The graph is modeled by $y = 4 \sin 4x$.

9. Because the hours of daylight ranges from a minimum of 10 hours to a maximum of 14 hours, the curve oscillates about the middle value, 12 hours. Thus, $D = 12$. The maximum number of hours is 2 hours above 12 hours. Thus, $A = 2$. The graph shows that one complete cycle occurs in 12–0, or 12 months. The period is 12. Thus, $12 = \dfrac{2\pi}{B}$

$$12B = 2\pi$$
$$B = \frac{2\pi}{12} = \frac{\pi}{6}$$

The graph shows that the starting point of the cycle is shifted from 0 to 3. The phase shift, $\dfrac{C}{B}$, is 3.

$$3 = \frac{C}{B}$$
$$3 = \frac{C}{\frac{\pi}{6}}$$
$$\frac{\pi}{2} = C$$

Substitute these values into $y = A \sin(Bx - C) + D$.
The number of hours of daylight is modeled by

$$y = 2\sin\left(\frac{\pi}{6}x - \frac{\pi}{2}\right) + 12.$$

Exercise Set 4.5

1. The equation $y = 4 \sin x$ is of the form $y = A \sin x$ with $A = 4$. Thus, the amplitude is $|A| = |4| = 4$.

The period is 2π. The quarter-period is $\dfrac{2\pi}{4}$ or $\dfrac{\pi}{2}$.

The cycle begins at $x = 0$. Add quarter-periods to generate *x*-values for the key points.
$x = 0$

$$x = 0 + \frac{\pi}{2} = \frac{\pi}{2}$$

$$x = \frac{\pi}{2} + \frac{\pi}{2} = \pi$$

$$x = \pi + \frac{\pi}{2} = \frac{3\pi}{2}$$

$$x = \frac{3\pi}{2} + \frac{\pi}{2} = 2\pi$$

Evaluate the function at each value of *x*.

x	$y = 4\sin x$	coordinates
0	$y = 4\sin 0 = 4 \cdot 0 = 0$	$(0, 0)$
$\dfrac{\pi}{2}$	$y = 4\sin\dfrac{\pi}{2} = 4 \cdot 1 = 4$	$\left(\dfrac{\pi}{2}, 4\right)$
π	$y = 4\sin \pi = 4 \cdot 0 = 0$	$(\pi, 0)$
$\dfrac{3\pi}{2}$	$y = 4\sin\dfrac{3\pi}{2}$ $= 4(-1) = -4$	$\left(\dfrac{3\pi}{2}, -4\right)$
2π	$y = 4\sin 2\pi = 4 \cdot 0 = 0$	$(2\pi, 0)$

Connect the five key points with a smooth curve and graph one complete cycle of the given function with the graph of $y = \sin x$.

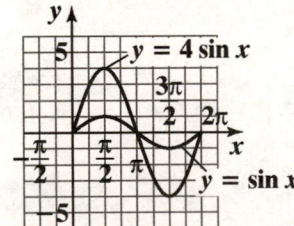

3. The equation $y = \frac{1}{3}\sin x$ is of the form $y = A\sin x$ with $A = \frac{1}{3}$. Thus, the amplitude is $|A| = \left|\frac{1}{3}\right| = \frac{1}{3}$.

The period is 2π. The quarter-period is $\frac{2\pi}{4}$ or $\frac{\pi}{2}$.

The cycle begins at $x = 0$. Add quarter-periods to generate x-values for the key points.

$x = 0$

$x = 0 + \frac{\pi}{2} = \frac{\pi}{2}$

$x = \frac{\pi}{2} + \frac{\pi}{2} = \pi$

$x = \pi + \frac{\pi}{2} = \frac{3\pi}{2}$

$x = \frac{3\pi}{2} + \frac{\pi}{2} = 2\pi$

Evaluate the function at each value of x.

x	$y = \frac{1}{3}\sin x$	coordinates
0	$y = \frac{1}{3}\sin 0 = \frac{1}{3}\cdot 0 = 0$	$(0, 0)$
$\frac{\pi}{2}$	$y = \frac{1}{3}\sin\frac{\pi}{2} = \frac{1}{3}\cdot 1 = \frac{1}{3}$	$\frac{\pi}{2}, \frac{1}{3}$
π	$y = \frac{1}{3}\sin \pi = \frac{1}{3}\cdot 0 = 0$	$(\pi, 0)$
$\frac{3\pi}{2}$	$y = \frac{1}{3}\sin\frac{3\pi}{2}$ $= \frac{1}{3}(-1) = -\frac{1}{3}$	$\frac{3\pi}{2}, -\frac{1}{3}$
2π	$y = \frac{1}{3}\sin 2\pi = \frac{1}{3}\cdot 0 = 0$	$(2\pi, 0)$

Connect the five key points with a smooth curve and graph one complete cycle of the given function with the graph of $y = \sin x$.

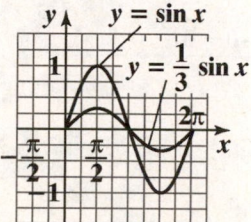

5. The equation $y = -3\sin x$ is of the form $y = A\sin x$ with $A = -3$. Thus, the amplitude is $|A| = |-3| = 3$.

The period is 2π. The quarter-period is $\frac{2\pi}{4}$ or $\frac{\pi}{2}$.

The cycle begins at $x = 0$. Add quarter-periods to generate x-values for the key points.

$x = 0$

$x = 0 + \frac{\pi}{2} = \frac{\pi}{2}$

$x = \frac{\pi}{2} + \frac{\pi}{2} = \pi$

$x = \pi + \frac{\pi}{2} = \frac{3\pi}{2}$

$x = \frac{3\pi}{2} + \frac{\pi}{2} = 2\pi$

Evaluate the function at each value of x.

x	$y = -3\sin x$	coordinates
0	$y = -3\sin x$ $= -3\cdot 0 = 0$	$(0, 0)$
$\frac{\pi}{2}$	$y = -3\sin\frac{\pi}{2}$ $= -3\cdot 1 = -3$	$\frac{\pi}{2}, -3$
π	$y = -3\sin \pi$ $= -3\cdot 0 = 0$	$(\pi, 0)$
$\frac{3\pi}{2}$	$y = -3\sin\frac{3\pi}{2}$ $= -3(-1) = 3$	$\frac{3\pi}{2}, 3$
2π	$y = -3\sin 2\pi$ $= -3\cdot 0 = 0$	$(2\pi, 0)$

Connect the five key points with a smooth curve and graph one complete cycle of the given function with the graph of $y = \sin x$.

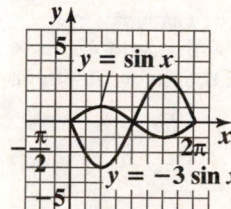

7. The equation $y = \sin 2x$ is of the form $y = A \sin Bx$ with $A = 1$ and $B = 2$. The amplitude is $|A| = |1| = 1$. The period is $\dfrac{2\pi}{B} = \dfrac{2\pi}{2} = \pi$. The quarter-period is $\dfrac{\pi}{4}$. The cycle begins at $x = 0$. Add quarter-periods to generate x-values for the key points.

$x = 0$

$x = 0 + \dfrac{\pi}{4}$

$x = \dfrac{\pi}{4} + \dfrac{\pi}{4} = \dfrac{\pi}{2}$

$x = \dfrac{\pi}{2} + \dfrac{\pi}{4} = \dfrac{3\pi}{4}$

$x = \dfrac{3\pi}{4} + \dfrac{\pi}{4} = \pi$

Evaluate the function at each value of x.

x	$y = \sin 2x$	coordinates
0	$y = \sin 2 \cdot 0 = \sin 0 = 0$	$(0, 0)$
$\dfrac{\pi}{4}$	$y = \sin 2 \cdot \dfrac{\pi}{4}$ $= \sin \dfrac{\pi}{2} = 1$	$\dfrac{\pi}{4}, 1$
$\dfrac{\pi}{2}$	$y = \sin 2 \cdot \dfrac{\pi}{2}$ $= \sin \pi = 0$	$\dfrac{\pi}{2}, 0$
$\dfrac{3\pi}{4}$	$y = \sin 2 \cdot \dfrac{3\pi}{4}$ $= \sin \dfrac{3\pi}{2} = -1$	$\dfrac{3\pi}{4}, -1$
π	$y = \sin(2 \cdot \pi)$ $= \sin 2\pi = 0$	$(\pi, 0)$

Connect the five key points with a smooth curve and graph one complete cycle of the given function.

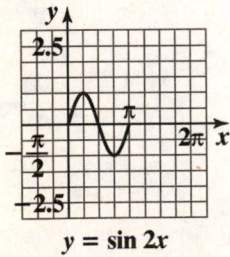

$y = \sin 2x$

9. The equation $y = 3\sin \dfrac{1}{2}x$ is of the form $y = A \sin Bx$ with $A = 3$ and $B = \dfrac{1}{2}$. The amplitude is $|A| = |3| = 3$. The period is $\dfrac{2\pi}{B} = \dfrac{2\pi}{\frac{1}{2}} = 2\pi \cdot 2 = 4\pi$. The quarter-period is $\dfrac{4\pi}{4} = \pi$. The cycle begins at $x = 0$. Add quarter-periods to generate x-values for the key points.

$x = 0$

$x = 0 + \pi = \pi$

$x = \pi + \pi = 2\pi$

$x = 2\pi + \pi = 3\pi$

$x = 3\pi + \pi = 4\pi$

Evaluate the function at each value of x.

x	$y = 3\sin \dfrac{1}{2}x$	coordinates
0	$y = 3\sin \dfrac{1}{2} \cdot 0$ $= 3\sin 0 = 3 \cdot 0 = 0$	$(0, 0)$
π	$y = 3\sin \dfrac{1}{2} \cdot \pi$ $= 3\sin \dfrac{\pi}{2} = 3 \cdot 1 = 3$	$(\pi, 3)$
2π	$y = 3\sin \dfrac{1}{2} \cdot 2\pi$ $= 3\sin \pi = 3 \cdot 0 = 0$	$(2\pi, 0)$
3π	$y = 3\sin \dfrac{1}{2} \cdot 3\pi$ $= 3\sin \dfrac{3\pi}{2}$ $= 3(-1) = -3$	$(3\pi, -3)$
4π	$y = 3\sin \dfrac{1}{2} \cdot 4\pi$ $= 3\sin 2\pi = 3 \cdot 0 = 0$	$(4\pi, 0)$

Connect the five points with a smooth curve and graph one complete cycle of the given function.

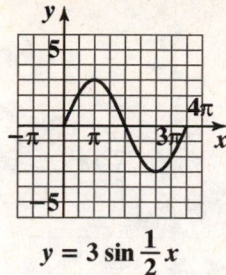

$$y = 3\sin\frac{1}{2}x$$

Connect the five points with a smooth curve and graph one complete cycle of the given function.

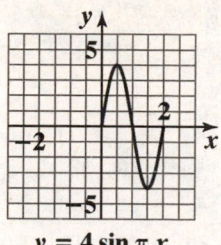

$$y = 4\sin\pi x$$

11. The equation $y = 4\sin\pi x$ is of the form $y = A\sin Bx$ with $A = 4$ and $B = \pi$. The amplitude is $|A| = |4| = 4$. The period is $\dfrac{2\pi}{B} = \dfrac{2\pi}{\pi} = 2$. The quarter-period is $\dfrac{2}{4} = \dfrac{1}{2}$. The cycle begins at $x = 0$. Add quarter-periods to generate x-values for the key points.

$x = 0$

$x = 0 + \dfrac{1}{2} = \dfrac{1}{2}$

$x = \dfrac{1}{2} + \dfrac{1}{2} = 1$

$x = 1 + \dfrac{1}{2} = \dfrac{3}{2}$

$x = \dfrac{3}{2} + \dfrac{1}{2} = 2$

Evaluate the function at each value of x.

x	$y = 4\sin\pi x$	coordinates
0	$y = 4\sin(\pi \cdot 0)$ $= 4\sin 0 = 4 \cdot 0 = 0$	$(0, 0)$
$\dfrac{1}{2}$	$y = 4\sin\left(\pi \cdot \dfrac{1}{2}\right)$ $= 4\sin\dfrac{\pi}{2} = 4(1) = 4$	$\left(\dfrac{1}{2}, 4\right)$
1	$y = 4\sin(\pi \cdot 1)$ $= 4\sin\pi = 4 \cdot 0 = 0$	$(1, 0)$
$\dfrac{3}{2}$	$y = 4\sin\left(\pi \cdot \dfrac{3}{2}\right)$ $= 4\sin\dfrac{3\pi}{2}$ $= 4(-1) = -4$	$\left(\dfrac{3}{2}, -4\right)$
2	$y = 4\sin(\pi \cdot 2)$ $= 4\sin 2\pi = 4 \cdot 0 = 0$	$(2, 0)$

13. The equation $y = -3\sin 2\pi x$ is of the form $y = A\sin Bx$ with $A = -3$ and $B = 2\pi$. The amplitude is $|A| = |-3| = 3$. The period is $\dfrac{2\pi}{B} = \dfrac{2\pi}{2\pi} = 1$. The quarter-period is $\dfrac{1}{4}$. The cycle begins at $x = 0$. Add quarter-periods to generate x-values for the key points.

$x = 0$

$x = 0 + \dfrac{1}{4} = \dfrac{1}{4}$

$x = \dfrac{1}{4} + \dfrac{1}{4} = \dfrac{1}{2}$

$x = \dfrac{1}{2} + \dfrac{1}{4} = \dfrac{3}{4}$

$x = \dfrac{3}{4} + \dfrac{1}{4} = 1$

Evaluate the function at each value of x.

x	$y = -3\sin 2\pi x$	coordinates
0	$y = -3\sin(2\pi \cdot 0)$ $= -3\sin 0$ $= -3 \cdot 0 = 0$	$(0, 0)$
$\dfrac{1}{4}$	$y = -3\sin\left(2\pi \cdot \dfrac{1}{4}\right)$ $= -3\sin\dfrac{\pi}{2}$ $= -3 \cdot 1 = -3$	$\dfrac{1}{4}, -3$
$\dfrac{1}{2}$	$y = -3\sin\left(2\pi \cdot \dfrac{1}{2}\right)$ $= -3\sin\pi$ $= -3 \cdot 0 = 0$	$\dfrac{1}{2}, 0$

$\dfrac{3}{4}$	$y = -3\sin 2\pi \cdot \dfrac{3}{4}$ $= -3\sin \dfrac{3\pi}{2}$ $= -3(-1) = 3$	$\dfrac{3}{4}, 3$
1	$y = -3\sin(2\pi \cdot 1)$ $= -3\sin 2\pi$ $= -3 \cdot 0 = 0$	$(1, 0)$

Connect the five points with a smooth curve and graph one complete cycle of the given function.

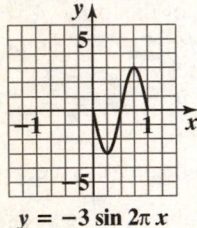

$y = -3 \sin 2\pi x$

15. The equation $y = -\sin \dfrac{2}{3}x$ is of the form $y = A\sin Bx$

with $A = -1$ and $B = \dfrac{2}{3}$.

The amplitude is $|A| = |-1| = 1$.

The period is $\dfrac{2\pi}{B} = \dfrac{2\pi}{\frac{2}{3}} = 2\pi \cdot \dfrac{3}{2} = 3\pi$.

The quarter-period is $\dfrac{3\pi}{4}$. The cycle begins at $x = 0$. Add quarter-periods to generate x-values for the key points.

$x = 0$

$x = 0 + \dfrac{3\pi}{4} = \dfrac{3\pi}{4}$

$x = \dfrac{3\pi}{4} + \dfrac{3\pi}{4} = \dfrac{3\pi}{2}$

$x = \dfrac{3\pi}{2} + \dfrac{3\pi}{4} = \dfrac{9\pi}{4}$

$x = \dfrac{9\pi}{4} + \dfrac{3\pi}{4} = 3\pi$

Evaluate the function at each value of x.

x	$y = -\sin \dfrac{2}{3}x$	coordinates
0	$y = -\sin \dfrac{2}{3} \cdot 0$ $= -\sin 0 = 0$	$(0, 0)$

$\dfrac{3\pi}{4}$	$y = -\sin \dfrac{2}{3} \cdot \dfrac{3\pi}{4}$ $= -\sin \dfrac{\pi}{2} = -1$	$\dfrac{3\pi}{4}, -1$
$\dfrac{3\pi}{2}$	$y = -\sin \dfrac{2}{3} \cdot \dfrac{3\pi}{2}$ $= -\sin \pi = 0$	$\dfrac{3\pi}{2}, 0$
$\dfrac{9\pi}{4}$	$y = -\sin \dfrac{2}{3} \cdot \dfrac{9\pi}{4}$ $= -\sin \dfrac{3\pi}{2}$ $= -(-1) = 1$	$\dfrac{9\pi}{4}, 1$
3π	$y = -\sin \dfrac{2}{3} \cdot 3\pi$ $= -\sin 2\pi = 0$	$(3\pi, 0)$

Connect the five points with a smooth curve and graph one complete cycle of the given function.

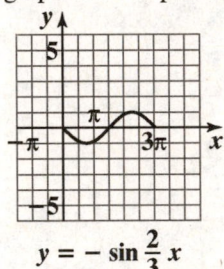

$y = -\sin \dfrac{2}{3}x$

17. The equation $y = \sin(x - \pi)$ is of the form $y = A\sin(Bx - C)$ with $A = 1$, $B = 1$, and $C = \pi$. The amplitude is $|A| = |1| = 1$. The period is $\dfrac{2\pi}{B} = \dfrac{2\pi}{1} = 2\pi$. The phase shift is $\dfrac{C}{B} = \dfrac{\pi}{1} = \pi$. The quarter-period is $\dfrac{2\pi}{4} = \dfrac{\pi}{2}$. The cycle begins at $x = \pi$.

Add quarter-periods to generate x-values for the key points.

$x = \pi$

$x = \pi + \dfrac{\pi}{2} = \dfrac{3\pi}{2}$

$x = \dfrac{3\pi}{2} + \dfrac{\pi}{2} = 2\pi$

$x = 2\pi + \dfrac{\pi}{2} = \dfrac{5\pi}{2}$

$x = \dfrac{5\pi}{2} + \dfrac{\pi}{2} = 3\pi$

Evaluate the function at each value of x.

x	$y = \sin(x - \pi)$	coordinates
π	$y = \sin(\pi - \pi)$ $= \sin 0 = 0$	$(\pi, 0)$
$\dfrac{3\pi}{2}$	$y = \sin\left(\dfrac{3\pi}{2} - \pi\right)$ $= \sin\dfrac{\pi}{2} = 1$	$\left(\dfrac{3\pi}{2}, 1\right)$
2π	$y = \sin(2\pi - \pi)$ $= \sin \pi = 0$	$(2\pi, 0)$
$\dfrac{5\pi}{2}$	$y = \sin\left(\dfrac{5\pi}{2} - \pi\right)$ $= \sin\dfrac{3\pi}{2} = -1$	$\left(\dfrac{5\pi}{2}, -1\right)$
3π	$y = \sin(3\pi - \pi)$ $= \sin 2\pi = 0$	$(3\pi, 0)$

Connect the five points with a smooth curve and graph one complete cycle of the given function.

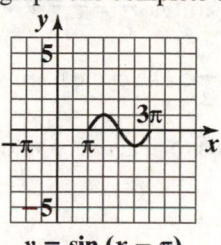

$y = \sin (x - \pi)$

19. The equation $y = \sin(2x - \pi)$ is of the form
$y = A\sin(Bx - C)$ with $A = 1$, $B = 2$, and $C = \pi$. The
amplitude is $|A| = |1| = 1$. The period is
$\dfrac{2\pi}{B} = \dfrac{2\pi}{2} = \pi$. The phase shift is $\dfrac{C}{B} = \dfrac{\pi}{2}$. The quarter-
period is $\dfrac{\pi}{4}$. The cycle begins at $x = \dfrac{\pi}{2}$. Add quarter-
periods to generate x-values for the key points.

$x = \dfrac{\pi}{2}$

$x = \dfrac{\pi}{2} + \dfrac{\pi}{4} = \dfrac{3\pi}{4}$

$x = \dfrac{3\pi}{4} + \dfrac{\pi}{4} = \pi$

$x = \pi + \dfrac{\pi}{4} = \dfrac{5\pi}{4}$

$x = \dfrac{5\pi}{4} + \dfrac{\pi}{4} = \dfrac{3\pi}{2}$

Evaluate the function at each value of x.

x	$y = \sin(2x - \pi)$	coordinates
$\dfrac{\pi}{2}$	$y = \sin\left(2 \cdot \dfrac{\pi}{2} - \pi\right)$ $= \sin(\pi - \pi)$ $= \sin 0 = 0$	$\left(\dfrac{\pi}{2}, 0\right)$
$\dfrac{3\pi}{4}$	$y = \sin\left(2 \cdot \dfrac{3\pi}{4} - \pi\right)$ $= \sin\left(\dfrac{3\pi}{2} - \pi\right)$ $= \sin\dfrac{\pi}{2} = 1$	$\left(\dfrac{3\pi}{4}, 1\right)$
π	$y = \sin(2 \cdot \pi - \pi)$ $= \sin(2\pi - \pi)$ $= \sin \pi = 0$	$(\pi, 0)$
$\dfrac{5\pi}{4}$	$y = \sin\left(2 \cdot \dfrac{5\pi}{4} - \pi\right)$ $= \sin\left(\dfrac{5\pi}{2} - \pi\right)$ $= \sin\dfrac{3\pi}{2} = -1$	$\left(\dfrac{5\pi}{4}, -1\right)$
$\dfrac{3\pi}{2}$	$y = \sin\left(2 \cdot \dfrac{3\pi}{2} - \pi\right)$ $= \sin(3\pi - \pi)$ $= \sin 2\pi = 0$	$\left(\dfrac{3\pi}{2}, 0\right)$

Connect the five points with a smooth curve and graph one complete cycle of the given function.

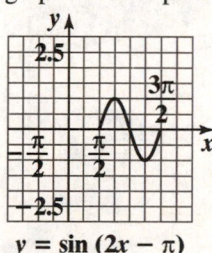

$y = \sin (2x - \pi)$

21. The equation $y = 3\sin(2x - \pi)$ is of the form $y = A\sin(Bx - C)$ with $A = 3$, $B = 2$, and $C = \pi$. The amplitude is $|A| = |3| = 3$. The period is $\frac{2\pi}{B} = \frac{2\pi}{2} = \pi$. The phase shift is $\frac{C}{B} = \frac{\pi}{2}$. The quarter-period is $\frac{\pi}{4}$. The cycle begins at $x = \frac{\pi}{2}$. Add quarter-periods to generate x-values for the key points.

$$x = \frac{\pi}{2}$$

$$x = \frac{\pi}{2} + \frac{\pi}{4} = \frac{3\pi}{4}$$

$$x = \frac{3\pi}{4} + \frac{\pi}{4} = \pi$$

$$x = \pi + \frac{\pi}{4} = \frac{5\pi}{4}$$

$$x = \frac{5\pi}{4} + \frac{\pi}{4} = \frac{3\pi}{2}$$

Evaluate the function at each value of x.

x	$y = 3\sin(2x - \pi)$	coordinates
$\frac{\pi}{2}$	$y = 3\sin\left(2 \cdot \frac{\pi}{2} - \pi\right)$ $= 3\sin(\pi - \pi)$ $= 3\sin 0 = 3 \cdot 0 = 0$	$\frac{\pi}{2}, 0$
$\frac{3\pi}{4}$	$y = 3\sin\left(2 \cdot \frac{3\pi}{4} - \pi\right)$ $= 3\sin\left(\frac{3\pi}{2} - \pi\right)$ $= 3\sin\frac{\pi}{2} = 3 \cdot 1 = 3$	$\frac{3\pi}{4}, 3$
π	$y = 3\sin(2 \cdot \pi - \pi)$ $= 3\sin(2\pi - \pi)$ $= 3\sin\pi = 3 \cdot 0 = 0$	$(\pi, 0)$
$\frac{5\pi}{4}$	$y = 3\sin\left(2 \cdot \frac{5\pi}{4} - \pi\right)$ $= 3\sin\left(\frac{5\pi}{2} - \pi\right)$ $= 3\sin\frac{3\pi}{2}$ $= 3(-1) = -3$	$\frac{5\pi}{4}, -3$
$\frac{3\pi}{2}$	$y = 3\sin\left(2 \cdot \frac{3\pi}{2} - \pi\right)$ $= 3\sin(3\pi - \pi)$ $= 3\sin 2\pi = 3 \cdot 0 = 0$	$\frac{3\pi}{2}, 0$

Connect the five points with a smooth curve and graph one complete cycle of the given function.

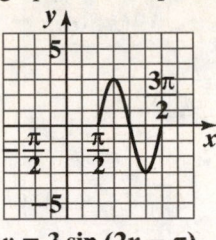

$y = 3\sin(2x - \pi)$

23. $y = \frac{1}{2}\sin\left(x + \frac{\pi}{2}\right) = \frac{1}{2}\sin\left(x - \left(-\frac{\pi}{2}\right)\right)$

The equation $y = \frac{1}{2}\sin\left(x - \left(-\frac{\pi}{2}\right)\right)$ is of the form $y = A\sin(Bx - C)$ with $A = \frac{1}{2}$, $B = 1$, and $C = -\frac{\pi}{2}$.

The amplitude is $|A| = \left|\frac{1}{2}\right| = \frac{1}{2}$. The period is $\frac{2\pi}{B} = \frac{2\pi}{1} = 2\pi$. The phase shift is $\frac{C}{B} = \frac{-\frac{\pi}{2}}{1} = -\frac{\pi}{2}$.

The quarter-period is $\frac{2\pi}{4} = \frac{\pi}{2}$. The cycle begins at $x = -\frac{\pi}{2}$. Add quarter-periods to generate x-values for the key points.

$$x = -\frac{\pi}{2}$$

$$x = -\frac{\pi}{2} + \frac{\pi}{2} = 0$$

$$x = 0 + \frac{\pi}{2} = \frac{\pi}{2}$$

$$x = \frac{\pi}{2} + \frac{\pi}{2} = \pi$$

$$x = \pi + \frac{\pi}{2} = \frac{3\pi}{2}$$

Evaluate the function at each value of x.

x	$y = \frac{1}{2}\sin\left(x + \frac{\pi}{2}\right)$	coordinates
$-\dfrac{\pi}{2}$	$y = \dfrac{1}{2}\sin\left(-\dfrac{\pi}{2} + \dfrac{\pi}{2}\right)$ $= \dfrac{1}{2}\sin 0 = \dfrac{1}{2}\cdot 0 = 0$	$\left(-\dfrac{\pi}{2}, 0\right)$
0	$y = \dfrac{1}{2}\sin\left(0 + \dfrac{\pi}{2}\right)$ $= \dfrac{1}{2}\sin\dfrac{\pi}{2} = \dfrac{1}{2}\cdot 1 = \dfrac{1}{2}$	$\left(0, \dfrac{1}{2}\right)$
$\dfrac{\pi}{2}$	$y = \dfrac{1}{2}\sin\left(\dfrac{\pi}{2} + \dfrac{\pi}{2}\right)$ $= \dfrac{1}{2}\sin \pi = \dfrac{1}{2}\cdot 0 = 0$	$\left(\dfrac{\pi}{2}, 0\right)$
π	$y = \dfrac{1}{2}\sin\left(\pi + \dfrac{\pi}{2}\right)$ $= \dfrac{1}{2}\sin\dfrac{3\pi}{2}$ $= \dfrac{1}{2}\cdot(-1) = -\dfrac{1}{2}$	$\left(\pi, -\dfrac{1}{2}\right)$
$\dfrac{3\pi}{2}$	$y = \dfrac{1}{2}\sin\left(\dfrac{3\pi}{2} + \dfrac{\pi}{2}\right)$ $= \dfrac{1}{2}\sin 2\pi$ $= \dfrac{1}{2}\cdot 0 = 0$	$\left(\dfrac{3\pi}{2}, 0\right)$

Connect the five points with a smooth curve and graph one complete cycle of the given function.

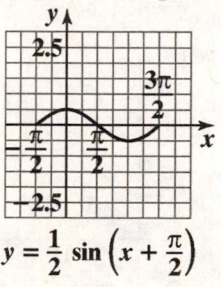

$$y = \frac{1}{2}\sin\left(x + \frac{\pi}{2}\right)$$

25. $y = -2\sin\left(2x + \dfrac{\pi}{2}\right) = -2\sin\left(2x - \left(-\dfrac{\pi}{2}\right)\right)$

The equation $y = -2\sin\left(2x - \left(-\dfrac{\pi}{2}\right)\right)$ is of the form $y = A\sin(Bx - C)$ with $A = -2$, $B = 2$, and $C = -\dfrac{\pi}{2}$. The amplitude is $|A| = |-2| = 2$. The period is $\dfrac{2\pi}{B} = \dfrac{2\pi}{2} = \pi$. The phase shift is $\dfrac{C}{B} = \dfrac{-\frac{\pi}{2}}{2} = -\dfrac{\pi}{2}\cdot\dfrac{1}{2} = -\dfrac{\pi}{4}$. The quarter-period is $\dfrac{\pi}{4}$. The cycle begins at $x = -\dfrac{\pi}{4}$. Add quarter-periods to generate x-values for the key points.

$x = -\dfrac{\pi}{4}$

$x = -\dfrac{\pi}{4} + \dfrac{\pi}{4} = 0$

$x = 0 + \dfrac{\pi}{4} = \dfrac{\pi}{4}$

$x = \dfrac{\pi}{4} + \dfrac{\pi}{4} = \dfrac{\pi}{2}$

$x = \dfrac{\pi}{2} + \dfrac{\pi}{4} = \dfrac{3\pi}{4}$

Evaluate the function at each value of x.

x	$y = -2\sin\left(2x + \dfrac{\pi}{2}\right)$	coordinates
$-\dfrac{\pi}{4}$	$y = -2\sin\left(2\cdot\left(-\dfrac{\pi}{4}\right) + \dfrac{\pi}{2}\right)$ $= -2\sin\left(-\dfrac{\pi}{2} + \dfrac{\pi}{2}\right)$ $= -2\sin 0 = -2\cdot 0 = 0$	$\left(-\dfrac{\pi}{4}, 0\right)$
0	$y = -2\sin\left(2\cdot 0 + \dfrac{\pi}{2}\right)$ $= -2\sin\left(0 + \dfrac{\pi}{2}\right)$ $= -2\sin\dfrac{\pi}{2}$ $= -2\cdot 1 = -2$	$(0, -2)$

$\dfrac{\pi}{4}$	$y = -2\sin\left(2\cdot\dfrac{\pi}{4}+\dfrac{\pi}{2}\right)$ $= -2\sin\left(\dfrac{\pi}{2}+\dfrac{\pi}{2}\right)$ $= -2\sin\pi$ $= -2\cdot 0 = 0$	$\dfrac{\pi}{4},\ 0$
$\dfrac{\pi}{2}$	$y = -2\sin\left(2\cdot\dfrac{\pi}{2}+\dfrac{\pi}{2}\right)$ $= -2\sin\left(\pi+\dfrac{\pi}{2}\right)$ $= -2\sin\dfrac{3\pi}{2}$ $= -2(-1) = 2$	$\dfrac{\pi}{2},\ 2$
$\dfrac{3\pi}{4}$	$y = -2\sin\left(2\cdot\dfrac{3\pi}{4}+\dfrac{\pi}{2}\right)$ $= -2\sin\left(\dfrac{3\pi}{2}+\dfrac{\pi}{2}\right)$ $= -2\sin 2\pi$ $= -2\cdot 0 = 0$	$\dfrac{3\pi}{4},\ 0$

Connect the five points with a smooth curve and graph one complete cycle of the given function.

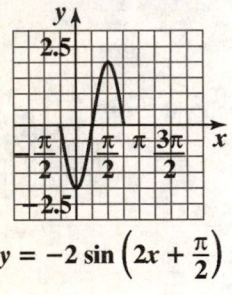

$$y = -2\sin\left(2x+\dfrac{\pi}{2}\right)$$

27. $y = 3\sin(\pi x + 2)$

The equation $y = 3\sin(\pi x - (-2))$ is of the form $y = A\sin(Bx - C)$ with $A = 3$, $B = \pi$, and $C = -2$. The amplitude is $|A| = |3| = 3$. The period is $\dfrac{2\pi}{B} = \dfrac{2\pi}{\pi} = 2$. The phase shift is $\dfrac{C}{B} = \dfrac{-2}{\pi} = -\dfrac{2}{\pi}$. The quarter-period is $\dfrac{2}{4} = \dfrac{1}{2}$. The cycle begins at $x = -\dfrac{2}{\pi}$. Add quarter-periods to generate x-values for the key points.

$$x = -\dfrac{2}{\pi}$$
$$x = -\dfrac{2}{\pi}+\dfrac{1}{2} = \dfrac{\pi-4}{2\pi}$$
$$x = \dfrac{\pi-4}{2\pi}+\dfrac{1}{2} = \dfrac{\pi-2}{\pi}$$
$$x = \dfrac{\pi-2}{\pi}+\dfrac{1}{2} = \dfrac{3\pi-4}{2\pi}$$
$$x = \dfrac{3\pi-4}{2\pi}+\dfrac{1}{2} = \dfrac{2\pi-2}{\pi}$$

Evaluate the function at each value of x.

x	$y = 3\sin(\pi x + 2)$	coordinates
$-\dfrac{2}{\pi}$	$y = 3\sin\left(\pi\left(-\dfrac{2}{\pi}\right)+2\right)$ $= 3\sin(-2+2)$ $= 3\sin 0 = 3\cdot 0 = 0$	$-\dfrac{2}{\pi},\ 0$
$\dfrac{\pi-4}{2\pi}$	$y = 3\sin\left(\pi\left(\dfrac{\pi-4}{2\pi}\right)+2\right)$ $= 3\sin\left(\dfrac{\pi-4}{2}+2\right)$ $= 3\sin\left(\dfrac{\pi}{2}-2+2\right)$ $= 3\sin\dfrac{\pi}{2}$ $= 3\cdot 1 = 3$	$\dfrac{\pi-4}{2\pi},\ 3$
$\dfrac{\pi-2}{\pi}$	$y = 3\sin\left(\pi\left(\dfrac{\pi-2}{\pi}\right)+2\right)$ $= 3\sin(\pi-2+2)$ $= 3\sin\pi = 3\cdot 0 = 0$	$\dfrac{\pi-2}{\pi},\ 0$

$\dfrac{3\pi-4}{2\pi}$	$y=3\sin\ \pi\ \dfrac{3\pi-4}{2\pi}+2$ $=3\sin\ \dfrac{3\pi-4}{2}+2$ $=3\sin\ \dfrac{3\pi}{2}-2+2$ $=3\sin\dfrac{3\pi}{2}$ $=3(-1)=-3$	$\dfrac{5\pi}{4},-3$
$\dfrac{2\pi-2}{\pi}$	$y=3\sin\ \pi\ \dfrac{2\pi-2}{\pi}+2$ $=3\sin(2\pi-2+2)$ $=3\sin 2\pi=3\cdot 0=0$	$\dfrac{2\pi-2}{\pi},0$

Connect the five points with a smooth curve and graph one complete cycle of the given function.

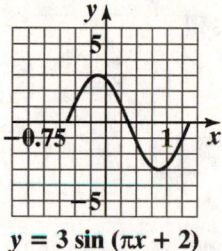

$y = 3 \sin (\pi x + 2)$

29. $y=-2\sin(2\pi x+4\pi)=-2\sin(2\pi x-(-4\pi))$

The equation $y=-2\sin(2\pi x-(-4\pi))$ is of the form $y=A\sin(Bx-C)$ with $A=-2$, $B=2\pi$, and $C=-4\pi$. The amplitude is $|A|=|-2|=2$. The period is $\dfrac{2\pi}{B}=\dfrac{2\pi}{2\pi}=1$. The phase shift is $\dfrac{C}{B}=\dfrac{-4\pi}{2\pi}=-2$. The quarter-period is $\dfrac{1}{4}$. The cycle begins at $x=-2$. Add quarter-periods to generate x-values for the key points.

$x=-2$

$x=-2+\dfrac{1}{4}=-\dfrac{7}{4}$

$x=-\dfrac{7}{4}+\dfrac{1}{4}=-\dfrac{3}{2}$

$x=-\dfrac{3}{2}+\dfrac{1}{4}=-\dfrac{5}{4}$

$x=-\dfrac{5}{4}+\dfrac{1}{4}=-1$

Evaluate the function at each value of x.

x	$y=-2\sin(2\pi x+4\pi)$	coordinates
-2	$y=-2\sin(2\pi(-2)+4\pi)$ $=-2\sin(-4\pi+4\pi)$ $=-2\sin 0$ $=-2\cdot 0=0$	$(-2,0)$
$-\dfrac{7}{4}$	$y=-2\sin\ 2\pi\ -\dfrac{7}{4}\ +4\pi$ $=-2\sin\ -\dfrac{7\pi}{2}+4\pi$ $=-2\sin\dfrac{\pi}{2}=-2\cdot 1=-2$	$-\dfrac{7}{4},-2$
$-\dfrac{3}{2}$	$y=-2\sin\ 2\pi\ -\dfrac{3}{2}\ +4\pi$ $=-2\sin(-3\pi+4\pi)$ $=-2\sin\pi=-2\cdot 0=0$	$-\dfrac{3}{2},0$
$-\dfrac{5}{4}$	$y=-2\sin\ 2\pi\ -\dfrac{5}{4}\ +4\pi$ $=-2\sin\ -\dfrac{5\pi}{2}+4\pi$ $=-2\sin\dfrac{3\pi}{2}$ $=-2(-1)=2$	$-\dfrac{5}{4},2$
-1	$y=-2\sin(2\pi(-1)+4\pi)$ $=-2\sin(-2\pi+4\pi)$ $=-2\sin 2\pi$ $=-2\cdot 0=0$	$(-1,0)$

Connect the five points with a smooth curve and graph one complete cycle of the given function.

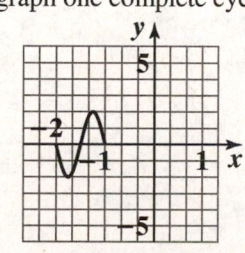

$y = -2 \sin (2\pi x + 4\pi)$

31. The equation $y = 2\cos x$ is of the form $y = A\cos x$ with $A = 2$. Thus, the amplitude is $|A| = |2| = 2$. The period is 2π. The quarter-period is $\dfrac{2\pi}{4}$ or $\dfrac{\pi}{2}$.

The cycle begins at $x = 0$. Add quarter-periods to generate x-values for the key points.

$x = 0$

$x = 0 + \dfrac{\pi}{2} = \dfrac{\pi}{2}$

$x = \dfrac{\pi}{2} + \dfrac{\pi}{2} = \pi$

$x = \pi + \dfrac{\pi}{2} = \dfrac{3\pi}{2}$

$x = \dfrac{3\pi}{2} + \dfrac{\pi}{2} = 2\pi$

Evaluate the function at each value of x.

x	$y = 2\cos x$	coordinates
0	$y = 2\cos 0$ $= 2 \cdot 1 = 2$	$(0, 2)$
$\dfrac{\pi}{2}$	$y = 2\cos \dfrac{\pi}{2}$ $= 2 \cdot 0 = 0$	$\dfrac{\pi}{2}, 0$
π	$y = 2\cos \pi$ $= 2 \cdot (-1) = -2$	$(\pi, -2)$
$\dfrac{3\pi}{2}$	$y = 2\cos \dfrac{3\pi}{2}$ $= 2 \cdot 0 = 0$	$\dfrac{3\pi}{2}, 0$
2π	$y = 2\cos 2\pi$ $= 2 \cdot 1 = 2$	$(2\pi, 2)$

Connect the five points with a smooth curve and graph one complete cycle of the given function with the graph of $y = 2\cos x$.

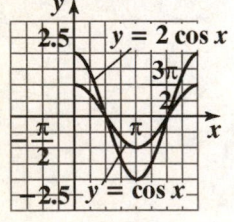

33. The equation $y = -2\cos x$ is of the form $y = A\cos x$ with $A = -2$. Thus, the amplitude is $|A| = |-2| = 2$. The period is 2π. The quarter-period is $\dfrac{2\pi}{4}$ or $\dfrac{\pi}{2}$. The cycle begins at $x = 0$. Add quarter-periods to generate x-values for the key points.

$x = 0$

$x = 0 + \dfrac{\pi}{2} = \dfrac{\pi}{2}$

$x = \dfrac{\pi}{2} + \dfrac{\pi}{2} = \pi$

$x = \pi + \dfrac{\pi}{2} = \dfrac{3\pi}{2}$

$x = \dfrac{3\pi}{2} + \dfrac{\pi}{2} = 2\pi$

Evaluate the function at each value of x.

x	$y = -2\cos x$	coordinates
0	$y = -2\cos 0$ $= -2 \cdot 1 = -2$	$(0, -2)$
$\dfrac{\pi}{2}$	$y = -2\cos \dfrac{\pi}{2}$ $= -2 \cdot 0 = 0$	$\dfrac{\pi}{2}, 0$
π	$y = -2\cos \pi$ $= -2 \cdot (-1) = 2$	$(\pi, 2)$
$\dfrac{3\pi}{2}$	$y = -2\cos \dfrac{3\pi}{2}$ $= -2 \cdot 0 = 0$	$\dfrac{3\pi}{2}, 0$
2π	$y = -2\cos 2\pi$ $= -2 \cdot 1 = -2$	$(2\pi, -2)$

Connect the five points with a smooth curve and graph one complete cycle of the given function with the graph of $y = \cos x$.

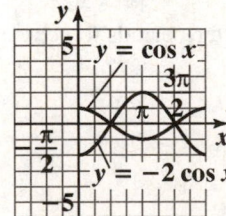

35. The equation $y = \cos 2x$ is of the form $y = A\cos Bx$ with $A = 1$ and $B = 2$. Thus, the amplitude is $|A| = |1| = 1$. The period is $\frac{2\pi}{B} = \frac{2\pi}{2} = \pi$. The quarter-period is $\frac{\pi}{4}$. The cycle begins at $x = 0$. Add quarter-periods to generate x-values for the key points.

$x = 0$

$x = 0 + \dfrac{\pi}{4} = \dfrac{\pi}{4}$

$x = \dfrac{\pi}{4} + \dfrac{\pi}{4} = \dfrac{\pi}{2}$

$x = \dfrac{\pi}{2} + \dfrac{\pi}{4} = \dfrac{3\pi}{4}$

$x = \dfrac{3\pi}{4} + \dfrac{\pi}{4} = \pi$

Evaluate the function at each value of x.

x	$y = \cos 2x$	coordinates
0	$y = \cos(2 \cdot 0)$ $= \cos 0 = 1$	$(0, 1)$
$\dfrac{\pi}{4}$	$y = \cos\left(2 \cdot \dfrac{\pi}{4}\right)$ $= \cos\dfrac{\pi}{2} = 0$	$\left(\dfrac{\pi}{4}, 0\right)$
$\dfrac{\pi}{2}$	$y = \cos\left(2 \cdot \dfrac{\pi}{2}\right)$ $= \cos\pi = -1$	$\left(\dfrac{\pi}{2}, -1\right)$
$\dfrac{3\pi}{4}$	$y = \cos\left(2 \cdot \dfrac{3\pi}{4}\right)$ $= \cos\dfrac{3\pi}{2} = 0$	$\left(\dfrac{3\pi}{4}, 0\right)$
π	$y = \cos(2 \cdot \pi)$ $= \cos 2\pi = 1$	$(\pi, 1)$

Connect the five points with a smooth curve and graph one complete cycle of the given function.

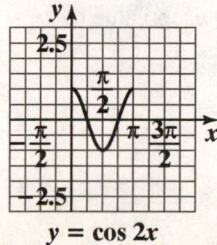

$y = \cos 2x$

37. The equation $y = 4\cos 2\pi x$ is of the form $y = A\cos Bx$ with $A = 4$ and $B = 2\pi$. Thus, the amplitude is $|A| = |4| = 4$. The period is $\frac{2\pi}{B} = \frac{2\pi}{2\pi} = 1$. The quarter-period is $\frac{1}{4}$. The cycle begins at $x = 0$. Add quarter-periods to generate x-values for the key points.

$x = 0$

$x = 0 + \dfrac{1}{4} = \dfrac{1}{4}$

$x = \dfrac{1}{4} + \dfrac{1}{4} = \dfrac{1}{2}$

$x = \dfrac{1}{2} + \dfrac{1}{4} = \dfrac{3}{4}$

$x = \dfrac{3}{4} + \dfrac{1}{4} = 1$

Evaluate the function at each value of x.

x	$y = 4\cos 2\pi x$	coordinates
0	$y = 4\cos(2\pi \cdot 0)$ $= 4\cos 0$ $= 4 \cdot 1 = 4$	$(0, 4)$
$\dfrac{1}{4}$	$y = 4\cos\left(2\pi \cdot \dfrac{1}{4}\right)$ $= 4\cos\dfrac{\pi}{2}$ $= 4 \cdot 0 = 0$	$\left(\dfrac{1}{4}, 0\right)$
$\dfrac{1}{2}$	$y = 4\cos\left(2\pi \cdot \dfrac{1}{2}\right)$ $= 4\cos\pi$ $= 4 \cdot (-1) = -4$	$\left(\dfrac{1}{2}, -4\right)$
$\dfrac{3}{4}$	$y = 4\cos\left(2\pi \cdot \dfrac{3}{4}\right)$ $= 4\cos\dfrac{3\pi}{2}$ $= 4 \cdot 0 = 0$	$\left(\dfrac{3}{4}, 0\right)$
1	$y = 4\cos(2\pi \cdot 1)$ $= 4\cos 2\pi$ $= 4 \cdot 1 = 4$	$(1, 4)$

Connect the five points with a smooth curve and graph one complete cycle of the given function.

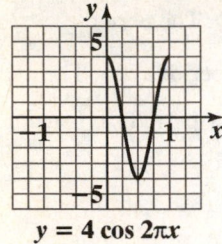

$y = 4 \cos 2\pi x$

39. The equation $y = -4 \cos \frac{1}{2} x$ is of the form

$y = A \cos Bx$ with $A = -4$ and $B = \frac{1}{2}$. Thus, the

amplitude is $|A| = |-4| = 4$. The period is

$\frac{2\pi}{B} = \frac{2\pi}{\frac{1}{2}} = 2\pi \cdot 2 = 4\pi$. The quarter-period is

$\frac{4\pi}{4} = \pi$. The cycle begins at $x = 0$. Add quarter-periods to generate x-values for the key points.

$x = 0$

$x = 0 + \pi = \pi$

$x = \pi + \pi = 2\pi$

$x = 2\pi + \pi = 3\pi$

$x = 3\pi + \pi = 4\pi$

Evaluate the function at each value of x.

x	$y = -4 \cos \frac{1}{2} x$	coordinates
0	$y = -4 \cos \frac{1}{2} \cdot 0$ $= -4 \cos 0$ $= -4 \cdot 1 = -4$	$(0, -4)$
π	$y = -4 \cos \frac{1}{2} \cdot \pi$ $= -4 \cos \frac{\pi}{2}$ $= -4 \cdot 0 = 0$	$(\pi, 0)$
2π	$y = -4 \cos \frac{1}{2} \cdot 2\pi$ $= -4 \cos \pi$ $= -4 \cdot (-1) = 4$	$(2\pi, 4)$

3π	$y = -4 \cos \frac{1}{2} \cdot 3\pi$ $= -4 \cos \frac{3\pi}{2}$ $= -4 \cdot 0 = 0$	$(3\pi, 0)$
4π	$y = -4 \cos \frac{1}{2} \cdot 4\pi$ $= -4 \cos 2\pi$ $= -4 \cdot 1 = -4$	$(4\pi, -4)$

Connect the five points with a smooth curve and graph one complete cycle of the given function.

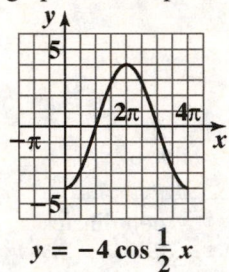

$y = -4 \cos \frac{1}{2} x$

41. The equation $y = -\frac{1}{2} \cos \frac{\pi}{3} x$ is of the form

$y = A \cos Bx$ with $A = -\frac{1}{2}$ and $B = \frac{\pi}{3}$. Thus, the

amplitude is $|A| = \left| -\frac{1}{2} \right| = \frac{1}{2}$. The period is

$\frac{2\pi}{B} = \frac{2\pi}{\frac{\pi}{3}} = 2\pi \cdot \frac{3}{\pi} = 6$. The quarter-period is $\frac{6}{4} = \frac{3}{2}$.

The cycle begins at $x = 0$. Add quarter-periods to generate x-values for the key points.

$x = 0$

$x = 0 + \frac{3}{2} = \frac{3}{2}$

$x = \frac{3}{2} + \frac{3}{2} = 3$

$x = 3 + \frac{3}{2} = \frac{9}{2}$

$x = \frac{9}{2} + \frac{3}{2} = 6$

Evaluate the function at each value of x.

x	$y = -\dfrac{1}{2}\cos\dfrac{\pi}{3}x$	coordinates
0	$y = -\dfrac{1}{2}\cos\dfrac{\pi}{3}\cdot 0$ $= -\dfrac{1}{2}\cos 0$ $= -\dfrac{1}{2}\cdot 1 = -\dfrac{1}{2}$	$0, -\dfrac{1}{2}$
$\dfrac{3}{2}$	$y = -\dfrac{1}{2}\cos\dfrac{\pi}{3}\cdot\dfrac{3}{2}$ $= -\dfrac{1}{2}\cos\dfrac{\pi}{2}$ $= -\dfrac{1}{2}\cdot 0 = 0$	$\dfrac{3}{2}, 0$
3	$y = -\dfrac{1}{2}\cos\dfrac{\pi}{3}\cdot 3$ $= -\dfrac{1}{2}\cos\pi$ $= -\dfrac{1}{2}\cdot(-1) = \dfrac{1}{2}$	$3, \dfrac{1}{2}$
$\dfrac{9}{2}$	$y = -\dfrac{1}{2}\cos\dfrac{\pi}{3}\cdot\dfrac{9}{2}$ $= -\dfrac{1}{2}\cos\dfrac{3\pi}{2}$ $= -\dfrac{1}{2}\cdot 0 = 0$	$\dfrac{9}{2}, 0$
6	$y = -\dfrac{1}{2}\cos\dfrac{\pi}{3}\cdot 6$ $= -\dfrac{1}{2}\cos 2\pi$ $= -\dfrac{1}{2}\cdot 1 = -\dfrac{1}{2}$	$6, -\dfrac{1}{2}$

Connect the five points with a smooth curve and graph one complete cycle of the given function.

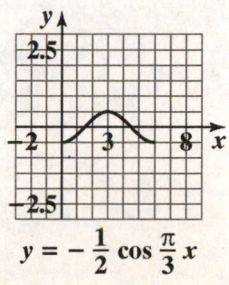

$$y = -\frac{1}{2}\cos\frac{\pi}{3}x$$

43. The equation $y = \cos\left(x - \dfrac{\pi}{2}\right)$ is of the form

$y = A\cos(Bx - C)$ with $A = 1$, and $B = 1$, and $C = \dfrac{\pi}{2}$.

Thus, the amplitude is $|A| = |1| = 1$. The period is

$\dfrac{2\pi}{B} = \dfrac{2\pi}{1} = 2\pi$. The phase shift is $\dfrac{C}{B} = \dfrac{\frac{\pi}{2}}{1} = \dfrac{\pi}{2}$. The

quarter-period is $\dfrac{2\pi}{4} = \dfrac{\pi}{2}$. The cycle begins at $x = \dfrac{\pi}{2}$.

Add quarter-periods to generate x-values for the key points.

$x = \dfrac{\pi}{2}$

$x = \dfrac{\pi}{2} + \dfrac{\pi}{2} = \pi$

$x = \pi + \dfrac{\pi}{2} = \dfrac{3\pi}{2}$

$x = \dfrac{3\pi}{2} + \dfrac{\pi}{2} = 2\pi$

$x = 2\pi + \dfrac{\pi}{2} = \dfrac{5\pi}{2}$

Evaluate the function at each value of x.

x	coordinates
$\dfrac{\pi}{2}$	$\left(\dfrac{\pi}{2}, 1\right)$
π	$(\pi, 0)$
$\dfrac{3\pi}{2}$	$\left(\dfrac{3\pi}{2}, -1\right)$
2π	$(2\pi, 0)$
$\dfrac{5\pi}{2}$	$\left(\dfrac{5\pi}{2}, 1\right)$

Connect the five points with a smooth curve and graph one complete cycle of the given function

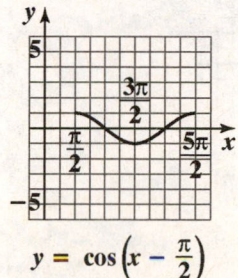

$$y = \cos\left(x - \frac{\pi}{2}\right)$$

45. The equation $y = 3\cos(2x - \pi)$ is of the form $y = A\cos(Bx - C)$ with $A = 3$, and $B = 2$, and $C = \pi$. Thus, the amplitude is $|A| = |3| = 3$. The period is $\dfrac{2\pi}{B} = \dfrac{2\pi}{2} = \pi$. The phase shift is $\dfrac{C}{B} = \dfrac{\pi}{2}$.

The quarter-period is $\dfrac{\pi}{4}$. The cycle begins at $x = \dfrac{\pi}{2}$. Add quarter-periods to generate x-values for the key points.

$x = \dfrac{\pi}{2}$

$x = \dfrac{\pi}{2} + \dfrac{\pi}{4} = \dfrac{3\pi}{4}$

$x = \dfrac{3\pi}{4} + \dfrac{\pi}{4} = \pi$

$x = \pi + \dfrac{\pi}{4} = \dfrac{5\pi}{4}$

$x = \dfrac{5\pi}{4} + \dfrac{\pi}{4} = \dfrac{3\pi}{2}$

Evaluate the function at each value of x.

x	coordinates
$\dfrac{\pi}{2}$	$\dfrac{\pi}{2}, 3$
$\dfrac{3\pi}{4}$	$\dfrac{3\pi}{4}, 0$
π	$(\pi, -3)$
$\dfrac{5\pi}{4}$	$\dfrac{5\pi}{4}, 0$
$\dfrac{3\pi}{2}$	$\dfrac{3\pi}{2}, 3$

Connect the five points with a smooth curve and graph one complete cycle of the given function

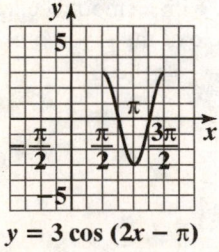

$y = 3\cos(2x - \pi)$

47. $y = \dfrac{1}{2}\cos\ 3x + \dfrac{\pi}{2} = \dfrac{1}{2}\cos\ 3x - -\dfrac{\pi}{2}$

The equation $y = \dfrac{1}{2}\cos\ 3x - -\dfrac{\pi}{2}$ is of the form

$y = A\cos(Bx - C)$ with $A = \dfrac{1}{2}$, and $B = 3$, and

$C = -\dfrac{\pi}{2}$. Thus, the amplitude is $|A| = \left|\dfrac{1}{2}\right| = \dfrac{1}{2}$.

The period is $\dfrac{2\pi}{B} = \dfrac{2\pi}{3}$. The phase shift is

$\dfrac{C}{B} = \dfrac{-\dfrac{\pi}{2}}{3} = -\dfrac{\pi}{2} \cdot \dfrac{1}{3} = -\dfrac{\pi}{6}$. The quarter-period is

$\dfrac{\frac{2\pi}{3}}{4} = \dfrac{2\pi}{3} \cdot \dfrac{1}{4} = \dfrac{\pi}{6}$. The cycle begins at $x = -\dfrac{\pi}{6}$. Add

quarter-periods to generate x-values for the key points.

$x = -\dfrac{\pi}{6}$

$x = -\dfrac{\pi}{6} + \dfrac{\pi}{6} = 0$

$x = 0 + \dfrac{\pi}{6} = \dfrac{\pi}{6}$

$x = \dfrac{\pi}{6} + \dfrac{\pi}{6} = \dfrac{\pi}{3}$

$x = \dfrac{\pi}{3} + \dfrac{\pi}{6} = \dfrac{\pi}{2}$

Evaluate the function at each value of x.

x	coordinates
$-\dfrac{\pi}{6}$	$-\dfrac{\pi}{6}, \dfrac{1}{2}$
0	$(0, 0)$
$\dfrac{\pi}{6}$	$\dfrac{\pi}{6}, -\dfrac{1}{2}$
$\dfrac{\pi}{3}$	$\dfrac{\pi}{3}, 0$
$\dfrac{\pi}{2}$	$\dfrac{\pi}{2}, \dfrac{1}{2}$

Connect the five points with a smooth curve and graph one complete cycle of the given function

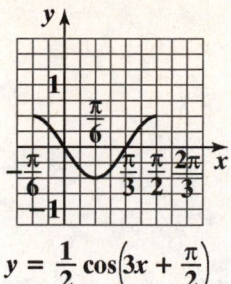

$$y = \frac{1}{2}\cos\left(3x + \frac{\pi}{2}\right)$$

49. The equation $y = -3\cos\left(2x - \frac{\pi}{2}\right)$ is of the form

$y = A\cos(Bx - C)$ with $A = -3$, and

$B = 2$, and $C = \frac{\pi}{2}$. Thus, the amplitude is

$|A| = |-3| = 3$. The period is $\frac{2\pi}{B} = \frac{2\pi}{2} = \pi$. The

phase shift is $\frac{C}{B} = \frac{\frac{\pi}{2}}{2} = \frac{\pi}{2} \cdot \frac{1}{2} = \frac{\pi}{4}$. The quarter-period

is $\frac{\pi}{4}$. The cycle begins at $x = \frac{\pi}{4}$. Add quarter-

periods to generate x-values for the key points.

$x = \frac{\pi}{4}$

$x = \frac{\pi}{4} + \frac{\pi}{4} = \frac{\pi}{2}$

$x = \frac{\pi}{2} + \frac{\pi}{4} = \frac{3\pi}{4}$

$x = \frac{3\pi}{4} + \frac{\pi}{4} = \pi$

$x = \pi + \frac{\pi}{4} = \frac{5\pi}{4}$

Evaluate the function at each value of x.

x	coordinates
$\frac{\pi}{4}$	$\frac{\pi}{4}, -3$
$\frac{\pi}{2}$	$\frac{\pi}{2}, 0$
$\frac{3\pi}{4}$	$\frac{3\pi}{4}, 3$
π	$(\pi, 0)$

$\frac{5\pi}{4}$	$\frac{5\pi}{4}, -3$

Connect the five points with a smooth curve and graph one complete cycle of the given function

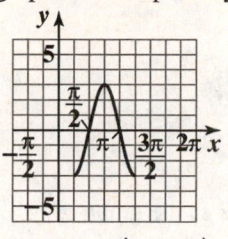

$$y = -3\cos\left(2x - \frac{\pi}{2}\right)$$

51. $y = 2\cos(2\pi x + 8\pi) = 2\cos(2\pi x - (-8\pi))$

The equation $y = 2\cos(2\pi x - (-8\pi))$ is of the form
$y = A\cos(Bx - C)$ with $A = 2$, $B = 2\pi$, and $C = -8\pi$.
Thus, the amplitude is $|A| = |2| = 2$. The period is

$\frac{2\pi}{B} = \frac{2\pi}{2\pi} = 1$. The phase shift is $\frac{C}{B} = \frac{-8\pi}{2\pi} = -4$. The

quarter-period is $\frac{1}{4}$. The cycle begins at $x = -4$. Add

quarter-periods to generate x-values for the key points.

$x = -4$

$x = -4 + \frac{1}{4} = -\frac{15}{4}$

$x = -\frac{15}{4} + \frac{1}{4} = -\frac{7}{2}$

$x = -\frac{7}{2} + \frac{1}{4} = -\frac{13}{4}$

$x = -\frac{13}{4} + \frac{1}{4} = -3$

Evaluate the function at each value of x.

x	coordinates
-4	$(-4, 2)$
$-\frac{15}{4}$	$-\frac{15}{4}, 0$
$-\frac{7}{2}$	$-\frac{7}{2}, -2$
$-\frac{13}{4}$	$-\frac{13}{4}, 0$
-3	$(-3, 2)$

Connect the five points with a smooth curve and graph one complete cycle of the given function

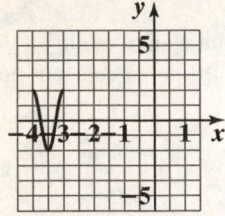

$y = 2 \cos (2\pi x + 8\pi)$

By connecting the points with a smooth curve we obtain one period of the graph.

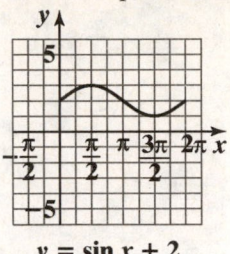

$y = \sin x + 2$

53. The graph of $y = \sin x + 2$ is the graph of $y = \sin x$ shifted up 2 units upward. The period for both functions is 2π. The quarter-period is $\frac{2\pi}{4}$ or $\frac{\pi}{2}$. The cycle begins at $x = 0$. Add quarter-periods to generate x-values for the key points.

$x = 0$

$x = 0 + \dfrac{\pi}{2} = \dfrac{\pi}{2}$

$x = \dfrac{\pi}{2} + \dfrac{\pi}{2} = \pi$

$x = \pi + \dfrac{\pi}{2} = \dfrac{3\pi}{2}$

$x = \dfrac{3\pi}{2} + \dfrac{\pi}{2} = 2\pi$

Evaluate the function at each value of x.

x	$y = \sin x + 2$	coordinates
0	$y = \sin 0 + 2$ $= 0 + 2 = 2$	$(0, 2)$
$\dfrac{\pi}{2}$	$y = \sin \dfrac{\pi}{2} + 2$ $= 1 + 2 = 3$	$\dfrac{\pi}{2}, 3$
π	$y = \sin \pi + 2$ $= 0 + 2 = 2$	$(\pi, 2)$
$\dfrac{3\pi}{2}$	$y = \sin \dfrac{3\pi}{2} + 2$ $= -1 + 2 = 1$	$\dfrac{3\pi}{2}, 1$
2π	$y = \sin 2\pi + 2$ $= 0 + 2 = 2$	$(2\pi, 2)$

55. The graph of $y = \cos x - 3$ is the graph of $y = \cos x$ shifted 3 units downward. The period for both functions is 2π. The quarter-period is $\frac{2\pi}{4}$ or $\frac{\pi}{2}$. The cycle begins at $x = 0$. Add quarter-periods to generate x-values for the key points.

$x = 0$

$x = 0 + \dfrac{\pi}{2} = \dfrac{\pi}{2}$

$x = \dfrac{\pi}{2} + \dfrac{\pi}{2} = \pi$

$x = \pi + \dfrac{\pi}{2} = \dfrac{3\pi}{2}$

$x = \dfrac{3\pi}{2} + \dfrac{\pi}{2} = 2\pi$

Evaluate the function at each value of x.

x	$y = \cos x - 3$	coordinates
0	$y = \cos 0 - 3$ $= 1 - 3 = -2$	$(0, -2)$
$\dfrac{\pi}{2}$	$y = \cos \dfrac{\pi}{2} - 3$ $= 0 - 3 = -3$	$\dfrac{\pi}{2}, -3$
π	$y = \cos \pi - 3$ $= -1 - 3 = -4$	$(\pi, -4)$
$\dfrac{3\pi}{2}$	$y = \cos \dfrac{3\pi}{2} - 3$ $= 0 - 3 = -3$	$\dfrac{3\pi}{2}, -3$
2π	$y = \cos 2\pi - 3$ $= 1 - 3 = -2$	$(2\pi, -2)$

By connecting the points with a smooth curve we obtain one period of the graph.

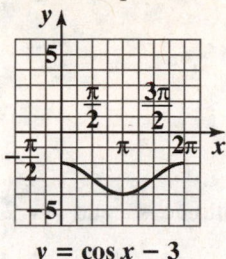

$y = \cos x - 3$

57. The graph of $y = 2\sin\frac{1}{2}x + 1$ is the graph of $y = 2\sin\frac{1}{2}x$ shifted one unit upward. The amplitude for both functions is $\mid 2 \mid = 2$. The period for both functions is $\dfrac{2\pi}{\frac{1}{2}} = 2\pi \cdot 2 = 4\pi$. The quarter-period is $\dfrac{4\pi}{4} = \pi$. The cycle begins at $x = 0$. Add quarter-periods to generate x-values for the key points.

$x = 0$

$x = 0 + \pi = \pi$

$x = \pi + \pi = 2\pi$

$x = 2\pi + \pi = 3\pi$

$x = 3\pi + \pi = 4\pi$

Evaluate the function at each value of x.

x	$y = 2\sin\frac{1}{2}x + 1$	coordinates
0	$y = 2\sin\left(\frac{1}{2}\cdot 0\right) + 1$ $= 2\sin 0 + 1$ $= 2\cdot 0 + 1 = 0 + 1 = 1$	$(0, 1)$
π	$y = 2\sin\left(\frac{1}{2}\cdot \pi\right) + 1$ $= 2\sin\frac{\pi}{2} + 1$ $= 2\cdot 1 + 1 = 2 + 1 = 3$	$(\pi, 3)$
2π	$y = 2\sin\left(\frac{1}{2}\cdot 2\pi\right) + 1$ $= 2\sin \pi + 1$ $= 2\cdot 0 + 1 = 0 + 1 = 1$	$(2\pi, 1)$

3π	$y = 2\sin\left(\frac{1}{2}\cdot 3\pi\right) + 1$ $= 2\sin\frac{3\pi}{2} + 1$ $= 2\cdot(-1) + 1$ $= -2 + 1 = -1$	$(3\pi, -1)$
4π	$y = 2\sin\left(\frac{1}{2}\cdot 4\pi\right) + 1$ $= 2\sin 2\pi + 1$ $= 2\cdot 0 + 1 = 0 + 1 = 1$	$(4\pi, 1)$

By connecting the points with a smooth curve we obtain one period of the graph.

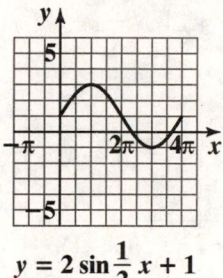

$y = 2\sin\frac{1}{2}x + 1$

59. The graph of $y = -3\cos 2\pi x + 2$ is the graph of $y = -3\cos 2\pi x$ shifted 2 units upward. The amplitude for both functions is $\mid -3 \mid = 3$. The period for both functions is $\dfrac{2\pi}{2\pi} = 1$. The quarter-period is $\dfrac{1}{4}$. The cycle begins at $x = 0$. Add quarter-periods to generate x-values for the key points.

$x = 0$

$x = 0 + \dfrac{1}{4} = \dfrac{1}{4}$

$x = \dfrac{1}{4} + \dfrac{1}{4} = \dfrac{1}{2}$

$x = \dfrac{1}{2} + \dfrac{1}{4} = \dfrac{3}{4}$

$x = \dfrac{3}{4} + \dfrac{1}{4} = 1$

Evaluate the function at each value of x.

x	$y = -3\cos 2\pi x + 2$	coordinates
0	$y = -3\cos(2\pi \cdot 0) + 2$ $= -3\cos 0 + 2$ $= -3\cdot 1 + 2$ $= -3 + 2 = -1$	$(0, -1)$

$\dfrac{1}{4}$	$y = -3 \cos\left[2\pi \cdot \dfrac{1}{4}\right] + 2$ $= -3 \cos\dfrac{\pi}{2} + 2$ $= -3 \cdot 0 + 2$ $= 0 + 2 = 2$	$\dfrac{1}{4}, 2$
$\dfrac{1}{2}$	$y = -3 \cos\left[2\pi \cdot \dfrac{1}{2}\right] + 2$ $= -3 \cos\pi + 2$ $= -3 \cdot(-1) + 2$ $= 3 + 2 = 5$	$\dfrac{1}{2}, 5$
$\dfrac{3}{4}$	$y = -3 \cos\left[2\pi \cdot \dfrac{3}{4}\right] + 2$ $= -3 \cos\dfrac{3\pi}{2} + 2$ $= -3 \cdot 0 + 2$ $= 0 + 2 = 2$	$\dfrac{3}{4}, 2$
1	$y = -3 \cos(2\pi \cdot 1) + 2$ $= -3 \cos 2\pi + 2$ $= -3 \cdot 1 + 2$ $= -3 + 2 = -1$	$(1, -1)$

By connecting the points with a smooth curve we obtain one period of the graph.

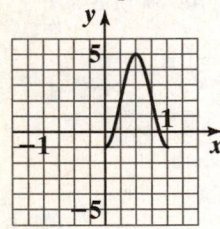

$y = -3 \cos 2\pi x + 2$

61. Using $y = A\cos Bx$ the amplitude is 3 and $A = 3$, The period is 4π and thus

$$B = \frac{2\pi}{\text{period}} = \frac{2\pi}{4\pi} = \frac{1}{2}$$

$y = A\cos Bx$

$$y = 3\cos\frac{1}{2}x$$

63. Using $y = A\sin Bx$ the amplitude is 2 and $A = -2$, The period is π and thus

$$B = \frac{2\pi}{\text{period}} = \frac{2\pi}{\pi} = 2$$

$y = A\sin Bx$

$y = -2\sin 2x$

65. Using $y = A\sin Bx$ the amplitude is 2 and $A = 2$, The period is 4 and thus

$$B = \frac{2\pi}{\text{period}} = \frac{2\pi}{4} = \frac{\pi}{2}$$

$y = A\sin Bx$

$$y = 2\sin\left(\frac{\pi}{2}x\right)$$

67.

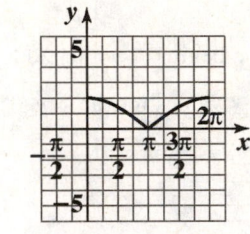

$$y = \left|2\cos\frac{x}{2}\right|$$

69.

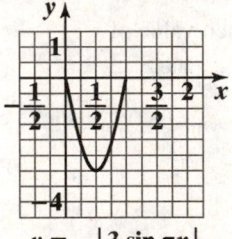

$$y = -\left|3\sin\pi x\right|$$

71.

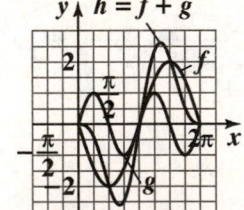

73.

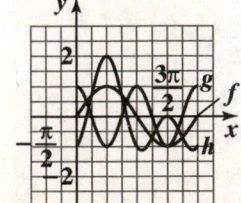

75. The period of the physical cycle is 33 days.

77. The period of the intellectual cycle is 23 days.

79. In the month of March, March 21 would be the best day to meet an on-line friend for the first time, because the emotional cycle is at a maximum.

81. Answers may vary.

83. The information gives the five key point of the graph.
(0, 14) corresponds to June,
(3, 12) corresponds to September,
(6, 10) corresponds to December,
(9, 12) corresponds to March,
(12, 14) corresponds to June
By connecting the five key points with a smooth curve we graph the information from June of one year to June of the following year.

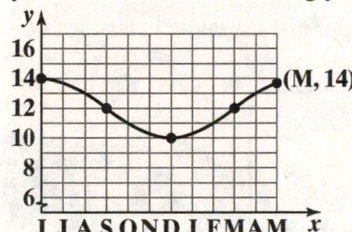

85. The function $y = 3\sin\dfrac{2\pi}{365}(x - 79) + 12$ is of the form

$$y = A\sin B\left(x - \frac{C}{B}\right) + D \text{ with}$$

$$A = 3 \text{ and } B = \frac{2\pi}{365}.$$

a. The amplitude is $\mid A \mid = \mid 3 \mid = 3$.

b. The period is $\dfrac{2\pi}{B} = \dfrac{2\pi}{\frac{2\pi}{365}} = 2\pi \cdot \dfrac{365}{2\pi} = 365$.

c. The longest day of the year will have the most hours of daylight. This occurs when the sine function equals 1.

$$y = 3\sin\frac{2\pi}{365}(x - 79) + 12$$

$$y = 3(1) + 12$$

$$y = 15$$

There will be 15 hours of daylight.

d. The shortest day of the year will have the least hours of daylight. This occurs when the sine function equals –1.

$$y = 3\sin\frac{2\pi}{365}(x - 79) + 12$$

$$y = 3(-1) + 12$$

$$y = 9$$

There will be 9 hours of daylight.

e. The amplitude is 3. The period is 365. The phase shift is $\dfrac{C}{B} = 79$. The quarter-period is

$\dfrac{365}{4} = 91.25$. The cycle begins at $x = 79$. Add quarter-periods to find the x-values of the key points.

$x = 79$

$x = 79 + 91.25 = 170.25$

$x = 170.25 + 91.25 = 261.5$

$x = 261.5 + 91.25 = 352.75$

$x = 352.75 + 91.25 = 444$

Because we are graphing for $0 \le x \le 365$, we will evaluate the function for the first four x-values along with $x = 0$ and $x = 365$. Using a calculator we have the following points.
(0, 9.07) (79, 12) (170.25, 15)
(261.5, 12) (352.75, 9) (365, 9.07)

By connecting the points with a smooth curve we obtain one period of the graph, starting on January 1.

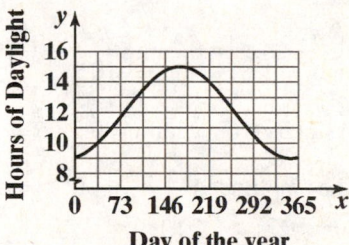

87. Because the depth of the water ranges from a minimum of 6 feet to a maximum of 12 feet, the curve oscillates about the middle value, 9 feet. Thus, $D = 9$. The maximum depth of the water is 3 feet above 9 feet. Thus, $A = 3$. The graph shows that one complete cycle occurs in 12-0, or 12 hours. The period is 12. Thus,

$$12 = \frac{2\pi}{B}$$

$$12B = 2\pi$$

$$B = \frac{2\pi}{12} = \frac{\pi}{6}$$

Substitute these values into $y = A\cos Bx + D$. The

depth of the water is modeled by $y = 3\cos\dfrac{\pi x}{6} + 9$.

89. – 99. Answers may vary.

101. The function $y = 3\sin(2x + \pi) = 3\sin(2x - (-\pi))$ is of the form $y = A\sin(Bx - C)$ with $A = 3$, $B = 2$, and $C = -\pi$. The amplitude is $|A| = |3| = 3$. The

period is $\dfrac{2\pi}{B} = \dfrac{2\pi}{2} = \pi$. The cycle begins at

$x = \dfrac{C}{B} = \dfrac{-\pi}{2} = -\dfrac{\pi}{2}$. We choose $-\dfrac{\pi}{2} \le x \le \dfrac{3\pi}{2}$, and

$-4 \le y \le 4$ for our graph.

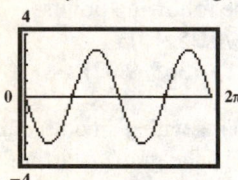

103. The function

$y = 0.2\sin\left(\dfrac{\pi}{10}x + \pi\right) = 0.2\sin\left(\dfrac{\pi}{10}x - (-\pi)\right)$ is of the

form $y = A\sin(Bx - C)$ with $A = 0.2$, $B = \dfrac{\pi}{10}$, and

$C = -\pi$. The amplitude is $|A| = |0.2| = 0.2$. The

period is $\dfrac{2\pi}{B} = \dfrac{2\pi}{\frac{\pi}{10}} = 2\pi \cdot \dfrac{10}{\pi} = 20$. The cycle begins

at $x = \dfrac{C}{B} = \dfrac{-\pi}{\frac{\pi}{10}} = -\pi \cdot \dfrac{10}{\pi} = -10$. We choose

$-10 \le x \le 30$, and $-1 \le y \le 1$ for our graph.

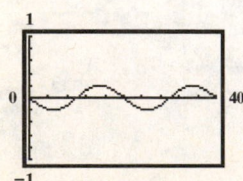

105.

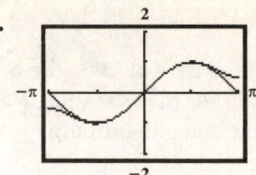

The graphs appear to be the same from $-\dfrac{\pi}{2}$ to $\dfrac{\pi}{2}$.

107.

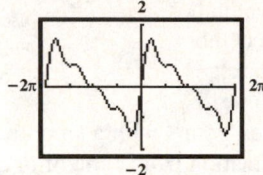

The graph is similar to $y = \sin x$, except the amplitude is greater and the curve is less smooth.

109. a.

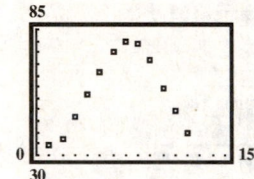

b. $y = 22.61\sin(0.50x - 2.04) + 57.17$

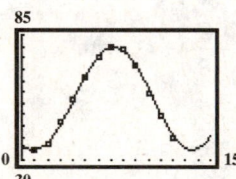

111. makes sense

113. makes sense

115. a. Since $A = 3$ and $D = -2$, the maximum will occur at $3 - 2 = 1$ and the minimum will occur at $-3 - 2 = -5$. Thus the range is $[-5, 1]$

Viewing rectangle: $\left[-\dfrac{\pi}{6}, \dfrac{23\pi}{6}, \dfrac{\pi}{6}\right]$ by $[-5, 1, 1]$

b. Since $A = 1$ and $D = -2$, the maximum will occur at $1 - 2 = -1$ and the minimum will occur at $-1 - 2 = -3$. Thus the range is $[-3, -1]$

Viewing rectangle: $\left[-\dfrac{\pi}{6}, \dfrac{7\pi}{6}, \dfrac{\pi}{6}\right]$ by $[-3, -1, 1]$

117. $y = \sin^2 x = \dfrac{1}{2} - \dfrac{1}{2}\cos 2x$

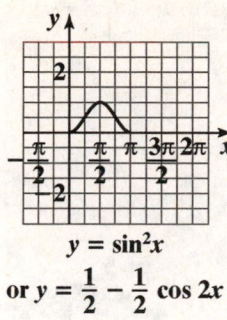

$y = \sin^2 x$

or $y = \dfrac{1}{2} - \dfrac{1}{2}\cos 2x$

119. Answers may vary.

120.
$$-\frac{\pi}{2} < x + \frac{\pi}{4} < \frac{\pi}{2}$$

$$-\frac{\pi}{2} - \frac{\pi}{4} < x + \frac{\pi}{4} - \frac{\pi}{4} < \frac{\pi}{2} - \frac{\pi}{4}$$

$$-\frac{2\pi}{4} - \frac{\pi}{4} < x < \frac{2\pi}{4} - \frac{\pi}{4}$$

$$-\frac{3\pi}{4} < x < \frac{\pi}{4}$$

$$\left\{ x \middle| -\frac{3\pi}{4} < x < \frac{\pi}{4} \right\} \text{ or } \left(-\frac{3\pi}{4}, \frac{\pi}{4} \right)$$

121. $\dfrac{-\dfrac{3\pi}{4} + \dfrac{\pi}{4}}{2} = \dfrac{-\dfrac{2\pi}{4}}{2} = \dfrac{-\dfrac{\pi}{2}}{2} = -\dfrac{\pi}{4}$

122. a.

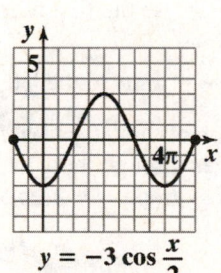

$y = -3\cos\dfrac{x}{2}$

b. The reciprocal function is undefined.

Section 4.6

Check Point Exercises

1. Solve the equations $2x = -\dfrac{\pi}{2}$ and $2x = \dfrac{\pi}{2}$

$$x = -\frac{\pi}{4} \qquad\qquad x = \frac{\pi}{4}$$

Thus, two consecutive asymptotes occur at $x = -\dfrac{\pi}{4}$

and $x = \dfrac{\pi}{4}$. Midway between these asymptotes is $x =$

0. An x-intercept is 0 and the graph passes through
(0, 0). Because the coefficient of the tangent is 3, the
points on the graph midway between an x-intercept
and the asymptotes have y-coordinates of
-3 and 3. Use the two asymptotes, the x-intercept,
and the points midway between to graph one period

of $y = 3\tan 2x$ from $-\dfrac{\pi}{4}$ to $\dfrac{\pi}{4}$. In order to graph for

$-\dfrac{\pi}{4} < x < \dfrac{3\pi}{4}$, Continue the pattern and extend the

graph another full period to the right.

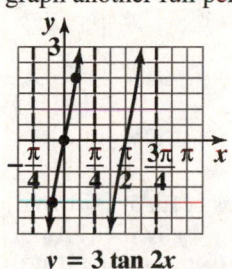

$y = 3\tan 2x$

2. Solve the equations

$$x - \frac{\pi}{2} = -\frac{\pi}{2} \qquad \text{and} \qquad x - \frac{\pi}{2} = \frac{\pi}{2}$$

$$x = \frac{\pi}{2} - \frac{\pi}{2} \qquad\qquad x = \frac{\pi}{2} + \frac{\pi}{2}$$

$$x = 0 \qquad\qquad\qquad x = \pi$$

Thus, two consecutive asymptotes occur at
$x = 0$ and $x = \pi$.

x-intercept $= \dfrac{0 + \pi}{2} = \dfrac{\pi}{2}$

An x-intercept is $\dfrac{\pi}{2}$ and the graph passes through

$\left(\dfrac{\pi}{2}, 0\right)$. Because the coefficient of the tangent is 1,

the points on the graph midway between an x-intercept and the asymptotes have y-coordinates of -1 and 1. Use the two consecutive asymptotes, $x = 0$ and $x = \pi$, to graph one full period of

$y = \tan\left(x - \dfrac{\pi}{2}\right)$ from 0 to π. Continue the pattern

and extend the graph another full period to the right.

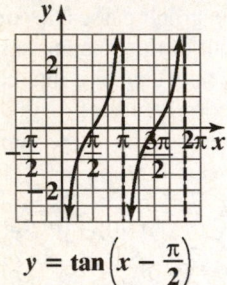

$$y = \tan\left(x - \frac{\pi}{2}\right)$$

3. Solve the equations

$$\dfrac{\pi}{2}x = 0 \quad \text{and} \quad \dfrac{\pi}{2}x = \pi$$

$$x = 0 \qquad\qquad x = \dfrac{\pi}{\frac{\pi}{2}}$$

$$x = 2$$

Two consecutive asymptotes occur at $x = 0$ and $x = 2$. Midway between $x = 0$ and $x = 2$ is $x = 1$. An x-intercept is 1 and the graph passes through $(1, 0)$.

Because the coefficient of the cotangent is $\dfrac{1}{2}$, the

points on the graph midway between an x-intercept

and the asymptotes have y-coordinates of $-\dfrac{1}{2}$ and

$\dfrac{1}{2}$. Use the two consecutive asymptotes, $x = 0$ and x

$= 2$, to graph one full period of $y = \dfrac{1}{2}\cot\dfrac{\pi}{2}x$. The

curve is repeated along the x-axis one full period as shown.

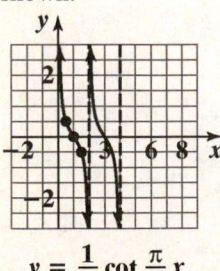

$$y = \frac{1}{2}\cot\frac{\pi}{2}x$$

4. The x-intercepts of $y = \sin\left(x + \dfrac{\pi}{4}\right)$ correspond to

vertical asymptotes of $y = \csc\left(x + \dfrac{\pi}{4}\right)$.

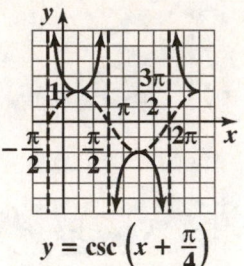

$$y = \csc\left(x + \frac{\pi}{4}\right)$$

5. Graph the reciprocal cosine function, $y = 2\cos 2x$.

The equation is of the form $y = A\cos Bx$ with $A = 2$ and $B = 2$.

amplitude: $|A| = |2| = 2$

period: $\dfrac{2\pi}{B} = \dfrac{2\pi}{2} = \pi$

Use quarter-periods, $\dfrac{\pi}{4}$, to find x-values for the five

key points. Starting with $x = 0$, the x-values are

$0, \dfrac{\pi}{4}, \dfrac{\pi}{2}, \dfrac{3\pi}{4}$, and π. Evaluating the function at each

value of x, the key points are

$(0, 2), \left(\dfrac{\pi}{4}, 0\right), \left(\dfrac{\pi}{2}, -2\right), \left(\dfrac{3\pi}{4}, 0\right), (\pi, 2)$. In order to

graph for $-\dfrac{3\pi}{4} \le x \le \dfrac{3\pi}{4}$, Use the first four points

and extend the graph $-\dfrac{3\pi}{4}$ units to the left. Use the

graph to obtain the graph of the reciprocal function. Draw vertical asymptotes through the x-intercepts, and use them as guides to graph $y = 2\sec 2x$.

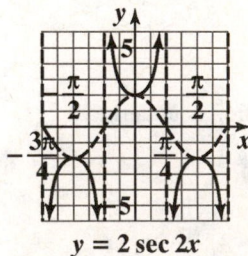

$$y = 2\sec 2x$$

Exercise Set 4.6

1. The graph has an asymptote at $x = -\dfrac{\pi}{2}$.

The phase shift, $\dfrac{C}{B}$, from $\dfrac{\pi}{2}$ to $-\dfrac{\pi}{2}$ is $-\pi$ units.

Thus, $\dfrac{C}{B} = \dfrac{C}{1} = -\pi$

$C = -\pi$

The function with $C = -\pi$ is $y = \tan(x + \pi)$.

3. The graph has an asymptote at $x = \pi$.

$\pi = \dfrac{\pi}{2} + C$

$C = \dfrac{\pi}{2}$

The function is $y = -\tan\left(x - \dfrac{\pi}{2}\right)$.

$$\dfrac{x}{4} = -\dfrac{\pi}{2} \quad \text{and} \quad \dfrac{x}{4} = \dfrac{\pi}{2}$$

5. Solve the equations $x = \left(-\dfrac{\pi}{2}\right)4 \qquad x = \left(\dfrac{\pi}{2}\right)4$

$\qquad\qquad x = -2\pi \qquad\qquad x = 2\pi$

Thus, two consecutive asymptotes occur at $x = -2\pi$ and $x = 2\pi$.

x-intercept $= \dfrac{-2\pi + 2\pi}{2} = \dfrac{0}{2} = 0$

An x-intercept is 0 and the graph passes through (0, 0). Because the coefficient of the tangent is 3, the points on the graph midway between an x-intercept and the asymptotes have y-coordinates of -3 and 3. Use the two consecutive asymptotes, $x = -2\pi$ and $x = 2\pi$, to graph one full period of $y = 3\tan\dfrac{x}{4}$ from -2π to 2π.

Continue the pattern and extend the graph another full period to the right.

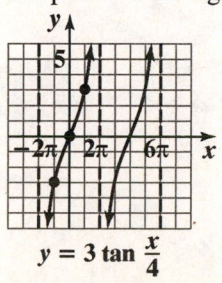

$y = 3\tan\dfrac{x}{4}$

7. Solve the equations $2x = -\dfrac{\pi}{2}$ and $2x = \dfrac{\pi}{2}$

$\qquad x = \dfrac{-\frac{\pi}{2}}{2} \qquad\qquad x = \dfrac{\frac{\pi}{2}}{2}$

$\qquad x = -\dfrac{\pi}{4} \qquad\qquad x = \dfrac{\pi}{4}$

Thus, two consecutive asymptotes occur at $x = -\dfrac{\pi}{4}$ and $x = \dfrac{\pi}{4}$.

x-intercept $= \dfrac{-\frac{\pi}{4} + \frac{\pi}{4}}{2} = \dfrac{0}{2} = 0$

An x-intercept is 0 and the graph passes through (0, 0). Because the coefficient of the tangent is $\dfrac{1}{2}$, the points on the graph midway between an x-intercept and the asymptotes have y-coordinates of $-\dfrac{1}{2}$ and $\dfrac{1}{2}$. Use the two consecutive asymptotes, $x = -\dfrac{\pi}{4}$ and $x = \dfrac{\pi}{4}$, to graph one full period of $y = \dfrac{1}{2}\tan 2x$ from $-\dfrac{\pi}{4}$ to $\dfrac{\pi}{4}$. Continue the pattern and extend the graph another full period to the right.

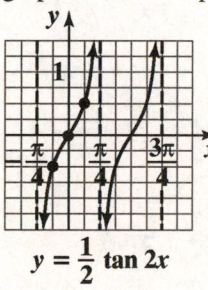

$y = \dfrac{1}{2}\tan 2x$

9. Solve the equations

$$\frac{1}{2}x = -\frac{\pi}{2} \qquad \text{and} \qquad \frac{1}{2}x = \frac{\pi}{2}$$

$$x = \left(-\frac{\pi}{2}\right)2 \qquad\qquad x = \left(\frac{\pi}{2}\right)2$$

$$x = -\pi \qquad\qquad\qquad x = \pi$$

Thus, two consecutive asymptotes occur at $x = -\pi$ and $x = \pi$.

$$x\text{-intercept} = \frac{-\pi + \pi}{2} = \frac{0}{2} = 0$$

An x-intercept is 0 and the graph passes through (0, 0). Because the coefficient of the tangent is –2, the points on the graph midway between an x-intercept and the asymptotes have y-coordinates of 2 and –2. Use the two consecutive asymptotes, $x = -\pi$ and

$x = \pi$, to graph one full period of $y = -2\tan\dfrac{1}{2}x$

from $-\pi$ to π. Continue the pattern and extend the graph another full period to the right.

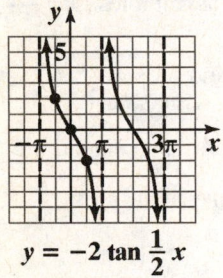

$$y = -2\tan\frac{1}{2}x$$

11. Solve the equations

$$x - \pi = -\frac{\pi}{2} \qquad \text{and} \qquad x - \pi = \frac{\pi}{2}$$

$$x = -\frac{\pi}{2} + \pi \qquad\qquad x = \frac{\pi}{2} + \pi$$

$$x = \frac{\pi}{2} \qquad\qquad\qquad x = \frac{3\pi}{2}$$

Thus, two consecutive asymptotes occur at $x = \dfrac{\pi}{2}$

and $x = \dfrac{3\pi}{2}$.

$$x\text{-intercept} = \frac{\frac{\pi}{2} + \frac{3\pi}{2}}{2} = \frac{\frac{4\pi}{2}}{2} = \frac{4\pi}{4} = \pi$$

An x-intercept is π and the graph passes through $(\pi, 0)$. Because the coefficient of the tangent is 1, the points on the graph midway between an x-intercept and the asymptotes have y-coordinates of –1 and 1.

Use the two consecutive asymptotes, $x = \dfrac{\pi}{2}$ and

$x = \dfrac{3\pi}{2}$, to graph one full period of $y = \tan(x - \pi)$

from $\dfrac{\pi}{2}$ to $\dfrac{3\pi}{2}$. Continue the pattern and extend the graph another full period to the right.

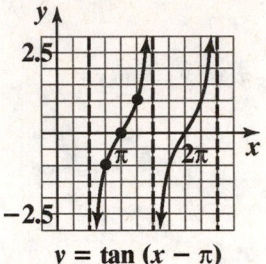

$$y = \tan(x - \pi)$$

13. There is no phase shift. Thus,

$$\frac{C}{B} = \frac{C}{1} = 0$$

$$C = 0$$

Because the points on the graph midway between an x-intercept and the asymptotes have y-coordinates of –1 and 1, $A = -1$. The function with $C = 0$ and $A = -1$ is $y = -\cot x$.

15. The graph has an asymptote at $-\dfrac{\pi}{2}$. The phase shift,

$\dfrac{C}{B}$, from 0 to $-\dfrac{\pi}{2}$ is $-\dfrac{\pi}{2}$ units. Thus, $\dfrac{C}{B} = \dfrac{C}{1} = -\dfrac{\pi}{2}$

$$C = -\frac{\pi}{2}$$

The function with $C = -\dfrac{\pi}{2}$ is $y = \cot\left(x + \dfrac{\pi}{2}\right)$.

17. Solve the equations $x = 0$ and $x = \pi$. Two consecutive asymptotes occur at $x = 0$ and $x = \pi$.

$$x\text{-intercept} = \frac{0 + \pi}{2} = \frac{\pi}{2}$$

An x-intercept is $\frac{\pi}{2}$ and the graph passes through

$\left(\frac{\pi}{2}, 0\right)$. Because the coefficient of the cotangent is 2, the points on the graph midway between an x-intercept and the asymptotes have y-coordinates of 2 and –2. Use the two consecutive asymptotes, $x = 0$ and $x = \pi$, to graph one full period of $y = 2\cot x$. The curve is repeated along the x-axis one full period as shown.

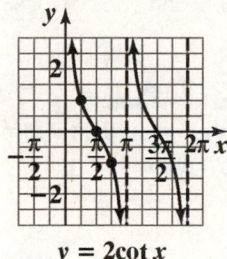

$y = 2\cot x$

19. Solve the equations $2x = 0$ and $2x = \pi$

$$x = 0 \qquad x = \frac{\pi}{2}$$

Two consecutive asymptotes occur at $x = 0$ and $x = \frac{\pi}{2}$.

$$x\text{-intercept} = \frac{0 + \frac{\pi}{2}}{2} = \frac{\frac{\pi}{2}}{2} = \frac{\pi}{4}$$

An x-intercept is $\frac{\pi}{4}$ and the graph passes through

$\left(\frac{\pi}{4}, 0\right)$. Because the coefficient of the cotangent is $\frac{1}{2}$, the points on the graph midway between an x-intercept and the asymptotes have y-coordinates of $\frac{1}{2}$ and $-\frac{1}{2}$.

Use the two consecutive asymptotes, $x = 0$ and $x = \frac{\pi}{2}$,

to graph one full period of $y = \frac{1}{2}\cot 2x$. The curve is repeated along the x-axis one full period as shown.

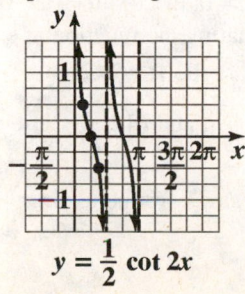

$y = \frac{1}{2}\cot 2x$

21. Solve the equations $\frac{\pi}{2}x = 0$ and $\frac{\pi}{2}x = \pi$

$$x = 0 \qquad x = \frac{\pi}{\frac{\pi}{2}}$$
$$x = 2$$

Two consecutive asymptotes occur at $x = 0$ and $x = 2$.

$$x\text{-intercept} = \frac{0 + 2}{2} = \frac{2}{2} = 1$$

An x-intercept is 1 and the graph passes through $(1, 0)$. Because the coefficient of the cotangent is –3, the points on the graph midway between an x-intercept and the asymptotes have y-coordinates of –3 and 3. Use the two consecutive asymptotes, $x = 0$ and $x = 2$, to graph one full period of

$y = -3\cot\frac{\pi}{2}x$. The curve is repeated along the x-axis one full period as shown.

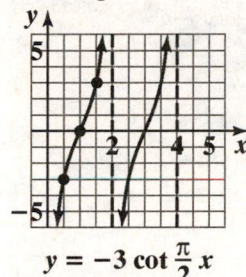

$y = -3\cot\frac{\pi}{2}x$

23. Solve the equations

$$x + \frac{\pi}{2} = 0 \qquad \text{and} \qquad x + \frac{\pi}{2} = \pi$$
$$x = 0 - \frac{\pi}{2} \qquad x = \pi - \frac{\pi}{2}$$
$$x = -\frac{\pi}{2} \qquad x = \frac{\pi}{2}$$

Two consecutive asymptotes occur at $x = -\frac{\pi}{2}$ and

$x = \frac{\pi}{2}$.

$$x\text{-intercept} = \frac{-\frac{\pi}{2} + \frac{\pi}{2}}{2} = \frac{0}{2} = 0$$

An x-intercept is 0 and the graph passes through (0, 0). Because the coefficient of the cotangent is 3, the points on the graph midway between an x-intercept and the asymptotes have y-coordinates of 3 and –3.

Use the two consecutive asymptotes, $x = -\frac{\pi}{2}$ and

$x = \frac{\pi}{2}$, to graph one full period of $y = 3\cot\left(x + \frac{\pi}{2}\right)$.

The curve is repeated along the *x*-axis one full period as shown.

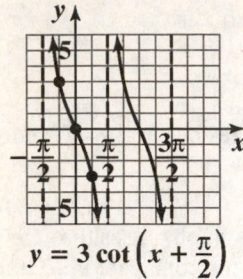

$$y = 3 \cot\left(x + \frac{\pi}{2}\right)$$

25. The *x*-intercepts of $y = -\frac{1}{2}\sin\frac{x}{2}$ corresponds to vertical asymptotes of $y = -\frac{1}{2}\csc\frac{x}{2}$. Draw the vertical asymptotes, and use them as a guide to sketch the graph of $y = -\frac{1}{2}\csc\frac{x}{2}$.

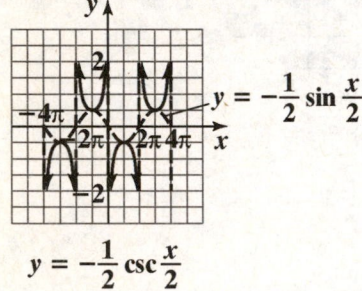

$$y = -\frac{1}{2}\csc\frac{x}{2}$$

27. The *x*-intercepts of $y = \frac{1}{2}\cos 2\pi x$ corresponds to vertical asymptotes of $y = \frac{1}{2}\sec 2\pi x$. Draw the vertical asymptotes, and use them as a guide to sketch the graph of $y = \frac{1}{2}\sec 2\pi x$.

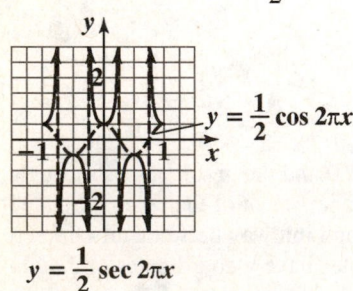

$$y = \frac{1}{2}\sec 2\pi x$$

29. Graph the reciprocal sine function, $y = 3\sin x$. The equation is of the form $y = A\sin Bx$ with $A = 3$ and $B = 1$.

amplitude: $|A| = |3| = 3$

period: $\frac{2\pi}{B} = \frac{2\pi}{1} = 2\pi$

Use quarter-periods, $\frac{\pi}{2}$, to find *x*-values for the five key points. Starting with $x = 0$, the *x*-values are $0, \frac{\pi}{2}, \pi, \frac{3\pi}{2}$, and 2π. Evaluating the function at each value of *x*, the key points are $(0, 0)$, $\left(\frac{\pi}{2}, 3\right)$, $(\pi, 0)$, $\left(\frac{3\pi}{2}, -3\right)$, and $(2\pi, 0)$. Use these key points to graph $y = 3\sin x$ from 0 to 2π. Extend the graph one cycle to the right.

Use the graph to obtain the graph of the reciprocal function. Draw vertical asymptotes through the *x*-intercepts, and use them as guides to graph $y = 3\csc x$.

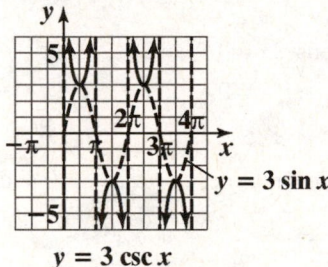

$$y = 3\csc x$$

31. Graph the reciprocal sine function, $y = \frac{1}{2}\sin\frac{x}{2}$. The equation is of the form $y = A\sin Bx$ with $A = \frac{1}{2}$ and $B = \frac{1}{2}$.

amplitude: $|A| = \left|\frac{1}{2}\right| = \frac{1}{2}$

period: $\frac{2\pi}{B} = \frac{2\pi}{\frac{1}{2}} = 2\pi \cdot 2 = 4\pi$

Use quarter-periods, π, to find *x*-values for the five key points. Starting with $x = 0$, the *x*-values are 0, $\pi, 2\pi, 3\pi$, and 4π. Evaluating the function at each value of *x*, the key points are $(0, 0)$, $\left(\pi, \frac{1}{2}\right)$, $(2\pi, 0)$, $\left(3\pi, -\frac{1}{2}\right)$, and $(4\pi, 0)$. Use these key points to graph $y = \frac{1}{2}\sin\frac{x}{2}$ from 0 to 4π.

Extend the graph one cycle to the right. Use the graph to obtain the graph of the reciprocal function. Draw vertical asymptotes through the x-intercepts, and use them as guides to graph $y = \dfrac{1}{2}\csc\dfrac{x}{2}$.

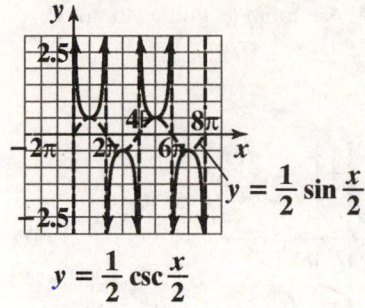

$$y = \dfrac{1}{2}\csc\dfrac{x}{2}$$

33. Graph the reciprocal cosine function, $y = 2\cos x$.
The equation is of the form $y = A\cos Bx$ with $A = 2$ and $B = 1$.
amplitude: $\mid A\mid = \mid 2\mid = 2$
period: $\dfrac{2\pi}{B} = \dfrac{2\pi}{1} = 2\pi$

Use quarter-periods, $\dfrac{\pi}{2}$, to find x-values for the five key points. Starting with $x = 0$, the x-values are 0, $\dfrac{\pi}{2}$, π, $\dfrac{3\pi}{2}$, 2π. Evaluating the function at each value of x, the key points are $(0, 2)$, $\left(\dfrac{\pi}{2}, 0\right)$, $(\pi, -2)$, $\left(\dfrac{3\pi}{2}, 0\right)$, and $(2\pi, 2)$. Use these key points to graph $y = 2\cos x$ from 0 to 2π. Extend the graph one cycle to the right. Use the graph to obtain the graph of the reciprocal function. Draw vertical asymptotes through the x-intercepts, and use them as guides to graph $y = 2\sec x$.

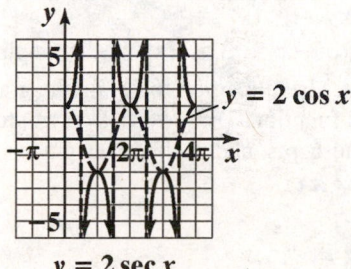

$$y = 2\sec x$$

35. Graph the reciprocal cosine function, $y = \cos\dfrac{x}{3}$. The equation is of the form $y = A\cos Bx$ with $A = 1$ and $B = \dfrac{1}{3}$.
amplitude: $\mid A\mid = \mid 1\mid = 1$
period: $\dfrac{2\pi}{B} = \dfrac{2\pi}{\frac{1}{3}} = 2\pi\cdot 3 = 6\pi$

Use quarter-periods, $\dfrac{6\pi}{4} = \dfrac{3\pi}{2}$, to find x-values for the five key points. Starting with $x = 0$, the x-values are 0, $\dfrac{3\pi}{2}$, 3π, $\dfrac{9\pi}{2}$, and 6π. Evaluating the function at each value of x, the key points are $(0, 1)$, $\left(\dfrac{3\pi}{2}, 0\right)$, $(3\pi, -1)$, $\left(\dfrac{9\pi}{2}, 0\right)$, and $(6\pi, 1)$. Use these key points to graph $y = \cos\dfrac{x}{3}$ from 0 to 6π. Extend the graph one cycle to the right. Use the graph to obtain the graph of the reciprocal function. Draw vertical asymptotes through the x-intercepts, and use them as guides to graph $y = \sec\dfrac{x}{3}$.

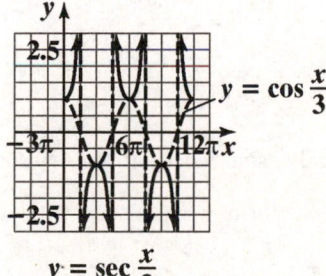

$$y = \sec\dfrac{x}{3}$$

37. Graph the reciprocal sine function, $y = -2\sin \pi x$.
The equation is of the form $y = A\sin Bx$ with $A = -2$
and $B = \pi$.
amplitude: $|A| = |-2| = 2$

period: $\dfrac{2\pi}{B} = \dfrac{2\pi}{\pi} = 2$

Use quarter-periods, $\dfrac{2}{4} = \dfrac{1}{2}$, to find

x-values for the five key points. Starting with $x = 0$,

the x-values are 0, $\dfrac{1}{2}$, 1, $\dfrac{3}{2}$, and 2. Evaluating the

function at each value of x, the key points are $(0, 0)$,

$\left(\dfrac{1}{2}, -2\right)$, $(1, 0)$, $\left(\dfrac{3}{2}, 2\right)$, and $(2, 0)$. Use these key

points to graph $y = -2\sin \pi x$ from 0 to 2. Extend the
graph one cycle to the right. Use the graph to obtain
the graph of the reciprocal function.

Draw vertical asymptotes through the x-intercepts,
and use them as guides to graph $y = -2\csc \pi x$.

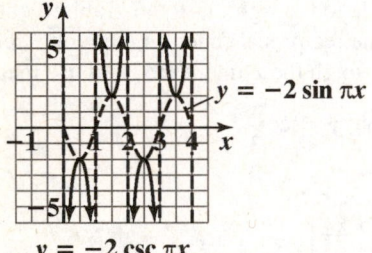

$y = -2 \csc \pi x$

39. Graph the reciprocal cosine function, $y = -\dfrac{1}{2}\cos \pi x$.

The equation is of the form $y = A\cos Bx$ with

$A = -\dfrac{1}{2}$ and $B = \pi$.

amplitude: $|A| = \left| -\dfrac{1}{2} \right| = \dfrac{1}{2}$

period: $\dfrac{2\pi}{B} = \dfrac{2\pi}{\pi} = 2$

Use quarter-periods, $\dfrac{2}{4} = \dfrac{1}{2}$, to find x-values for the

five key points. Starting with $x = 0$, the x-values are

0, $\dfrac{1}{2}$, 1, $\dfrac{3}{2}$, and 2. Evaluating the function at each

value of x, the key points are $\left(0, -\dfrac{1}{2} \right)$,

$\left(\dfrac{1}{2}, 0\right)$, $\left(1, \dfrac{1}{2}\right)$, $\left(\dfrac{3}{2}, 0\right)$, $\left(2, -\dfrac{1}{2}\right)$. Use these key

points to graph $y = -\dfrac{1}{2}\cos \pi x$ from 0 to 2. Extend

the graph one cycle to the right. Use the graph to
obtain the graph of the reciprocal function. Draw
vertical asymptotes through the
x-intercepts, and use them as guides to graph

$y = -\dfrac{1}{2}\sec \pi x$.

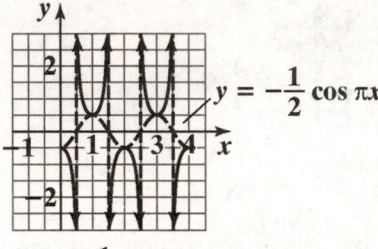

$y = -\dfrac{1}{2} \sec \pi x$

41. Graph the reciprocal sine function, $y = \sin(x - \pi)$.
The equation is of the form $y = A\sin(Bx - C)$ with A
$= 1$, and $B = 1$, and $C = \pi$.
amplitude: $|A| = |1| = 1$

period: $\dfrac{2\pi}{B} = \dfrac{2\pi}{1} = 2\pi$

phase shift: $\dfrac{C}{B} = \dfrac{\pi}{1} = \pi$

Use quarter-periods, $\dfrac{2\pi}{4} = \dfrac{\pi}{2}$, to find

x-values for the five key points. Starting with $x = \pi$,

the x-values are π, $\dfrac{3\pi}{2}$, 2π, $\dfrac{5\pi}{2}$, and 3π.

Evaluating the function at each value of x, the key

points are $(\pi, 0)$, $\left(\dfrac{3\pi}{2}, 1\right)$, $(2\pi, 0)$,

$\left(\dfrac{5\pi}{2}, -1\right)$, $(3\pi, 0)$. Use these key points to graph

$y = \sin(x - \pi)$ from π to 3π. Extend the graph one
cycle to the right. Use the graph to obtain the graph
of the reciprocal function. Draw vertical asymptotes
through the x-intercepts, and use them as guides to
graph $y = \csc(x - \pi)$.

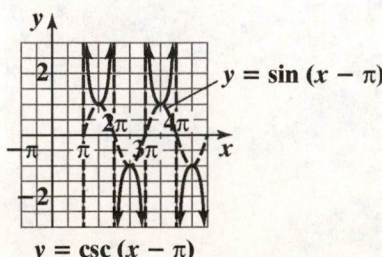

$y = \csc (x - \pi)$

43. Graph the reciprocal cosine function,

$y = 2\cos(x + \pi)$. The equation is of the form

$y = A\cos(Bx + C)$ with $A = 2$, $B = 1$, and $C = -\pi$.

amplitude: $|A| = |2| = 2$

period: $\dfrac{2\pi}{B} = \dfrac{2\pi}{1} = 2\pi$

phase shift: $\dfrac{C}{B} = \dfrac{-\pi}{1} = -\pi$

Use quarter-periods, $\dfrac{2\pi}{4} = \dfrac{\pi}{2}$, to find x-values for the

five key points. Starting with $x = -\pi$, the x-values

are $-\pi$, $-\dfrac{\pi}{2}$, 0, $\dfrac{\pi}{2}$, and π. Evaluating the function

at each value of x, the key points are $(-\pi, 2)$,

$\left(-\dfrac{\pi}{2}, 0\right)$, $(0, -2)$, $\left(\dfrac{\pi}{2}, 0\right)$, and $(\pi, 2)$. Use these

key points to graph $y = 2\cos(x + \pi)$ from $-\pi$ to π.

Extend the graph one cycle to the right. Use the graph
to obtain the graph of the reciprocal function. Draw
vertical asymptotes through the x-intercepts, and use
them as guides to graph $y = 2\sec(x + \pi)$.

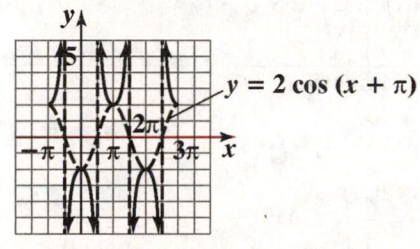

$y = 2\cos(x + \pi)$

$y = 2\sec(x + \pi)$

45.

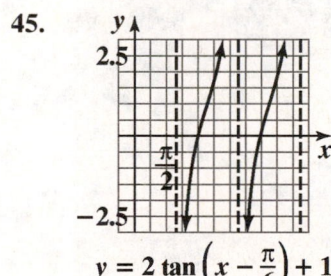

$y = 2\tan\left(x - \dfrac{\pi}{6}\right) + 1$

47.

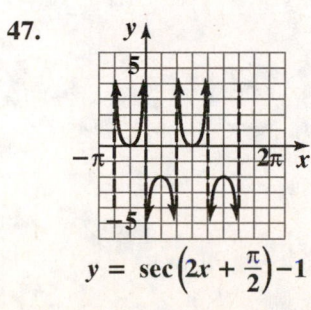

$y = \sec\left(2x + \dfrac{\pi}{2}\right) - 1$

49.

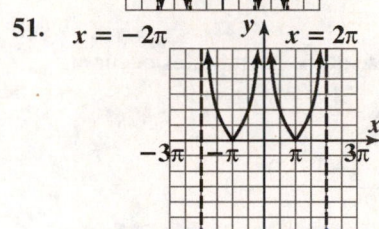

51.

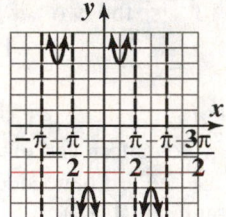

53. $y = (f \circ h)(x) = f(h(x)) = 2\sec\left(2x - \dfrac{\pi}{2}\right)$

55. Use a graphing utility with $y_1 = \tan x$ and $y_2 = -1$.

For the window use $\text{Xmin} = -2\pi$, $\text{Xmax} = 2\pi$,

$\text{Ymin} = -2$, and $\text{Ymax} = 2$.

$x = -\dfrac{5\pi}{4}$, $-\dfrac{\pi}{4}$, $\dfrac{3\pi}{4}$, $\dfrac{7\pi}{4}$

$x \approx -3.93$, -0.79, 2.36, 5.50

57. Use a graphing utility with $y_1 = 1/\sin x$ and $y_2 = 1$.

For the window use $\text{Xmin} = -2\pi$, $\text{Xmax} = 2\pi$,

$\text{Ymin} = -2$, and $\text{Ymax} = 2$.

$x = -\dfrac{3\pi}{2}$, $\dfrac{\pi}{2}$

$x \approx -4.71$, 1.57

59. $d = 12 \tan 2\pi t$

   a. Solve the equations

$$2\pi t = -\frac{\pi}{2} \quad \text{and} \quad 2\pi t = \frac{\pi}{2}$$

$$t = \frac{-\frac{\pi}{2}}{2\pi} \qquad\qquad t = \frac{\frac{\pi}{2}}{2\pi}$$

$$t = -\frac{1}{4} \qquad\qquad t = \frac{1}{4}$$

Thus, two consecutive asymptotes occur at

$$x = -\frac{1}{4} \text{ and } x = \frac{1}{4}.$$

$$x\text{-intercept} = \frac{-\frac{1}{4} + \frac{1}{4}}{2} = \frac{0}{2} = 0$$

An x-intercept is 0 and the graph passes through $(0, 0)$. Because the coefficient of the tangent is 12, the points on the graph midway between an x-intercept and the asymptotes have y-coordinates of -12 and 12. Use the two consecutive asymptotes, $x = -\frac{1}{4}$ and $x = \frac{1}{4}$, to graph one full period of $d = 12 \tan 2\pi t$. To graph on $[0, 2]$, continue the pattern and extend the graph to 2. (Do not use the left hand side of the first period of the graph on $[0, 2]$.)

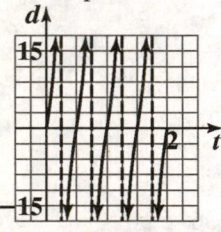

$d = 12 \tan 2\pi t$

   b. The function is undefined for $t = 0.25, 0.75, 1.25,$ and 1.75.
The beam is shining parallel to the wall at these times.

61. Use the function that relates the acute angle with the hypotenuse and the adjacent leg, the secant function.

$$\sec x = \frac{d}{10}$$

$$d = 10 \sec x$$

Graph the reciprocal cosine function, $y = 10 \cos x$.
The equation is of the form $y = A \cos Bx$ with $A = 10$ and $B = 1$.
amplitude: $|A| = |10| = 10$
period: $\frac{2\pi}{B} = \frac{2\pi}{1} = 2\pi$

For $-\frac{\pi}{2} < x < \frac{\pi}{2}$, use the x-values $-\frac{\pi}{2}$, 0, and $\frac{\pi}{2}$ to find the key points $\left(-\frac{\pi}{2}, 0\right)$, $(0, 10)$, and $\left(\frac{\pi}{2}, 0\right)$.

Connect these points with a smooth curve, then draw vertical asymptotes through the x-intercepts, and use them as guides to graph $d = 10 \sec x$ on $\left[-\frac{\pi}{2}, \frac{\pi}{2}\right]$.

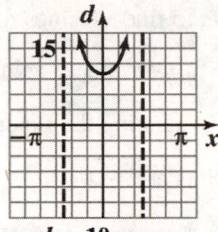

$d = 10 \sec x$

63.

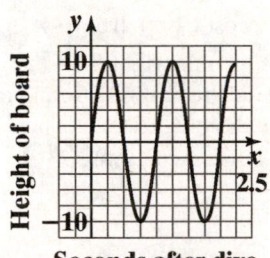

Seconds after dive

65. – 75. Answers may vary.

77. period: $\frac{\pi}{B} = \frac{\pi}{\frac{1}{4}} = \pi \cdot 4 = 4\pi$

Graph $y = \tan \frac{x}{4}$ for $0 \le x \le 8\pi$.

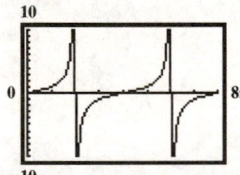

79. period: $\frac{\pi}{B} = \frac{\pi}{2}$

Graph $y = \cot 2x$ for $0 \le x \le \pi$.

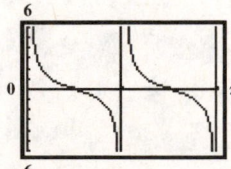

81. period: $\dfrac{\pi}{B} = \dfrac{\pi}{\pi} = 1$

Graph $y = \dfrac{1}{2}\tan \pi x$ for $0 \le x \le 2$.

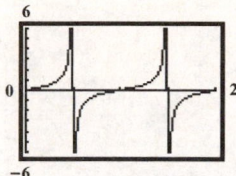

83. period: $\dfrac{2\pi}{B} = \dfrac{2\pi}{\frac{1}{2}} = 2\pi \cdot 2 = 4\pi$

Graph the functions for $0 \le x \le 8\pi$.

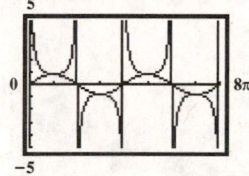

85. period: $\dfrac{2\pi}{B} = \dfrac{2\pi}{2} = \pi$

phase shift: $\dfrac{C}{B} = \dfrac{\frac{\pi}{6}}{2} = \dfrac{\pi}{12}$

Thus, we include $\dfrac{\pi}{12} \le x \le \dfrac{25\pi}{12}$ in our graph, and

graph for $0 \le x \le \dfrac{5\pi}{2}$.

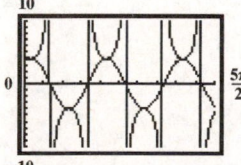

87.

The graph shows that carbon dioxide concentration rises and falls each year, but over all the concentration increased from 1990 to 2008.

89. makes sense

91. does not make sense; Explanations will vary.
Sample explanation: To obtain a cosecant graph, you can use a sine graph.

93. The graph has the shape of a cotangent function with consecutive asymptotes at

$x = 0$ and $x = \dfrac{2\pi}{3}$. The period is $\dfrac{2\pi}{3} - 0 = \dfrac{2\pi}{3}$. Thus,

$\dfrac{\pi}{B} = \dfrac{2\pi}{3}$

$2\pi B = 3\pi$

$B = \dfrac{3\pi}{2\pi} = \dfrac{3}{2}$

The points on the graph midway between an x-intercept and the asymptotes have y-coordinates of 1 and -1. Thus, $A = 1$. There is no phase shift. Thus,

$C = 0$. An equation for this graph is $y = \cot \dfrac{3}{2}x$.

95. The range shows that $A = 2$.

Since the period is 3π, the coefficient of x is given

by B where $\dfrac{2\pi}{B} = 3\pi$

$\dfrac{2\pi}{B} = 3\pi$

$3B\pi = 2\pi$

$B = \dfrac{2}{3}$

Thus, $y = 2\csc \dfrac{2x}{3}$

97. a. Since $A = 1$, the range is $(-\infty, -1] \cup [1, \infty)$

Viewing rectangle: $\left[-\dfrac{\pi}{6}, \pi, \dfrac{7\pi}{6}\right]$ by $[-3, 3, 1]$

b. Since $A = 3$, the range is $(-\infty, -3] \cup [3, \infty)$

Viewing rectangle: $\left[-\dfrac{1}{2}, \dfrac{7}{2}, 1\right]$ by $[-6, 6, 1]$

99. a.

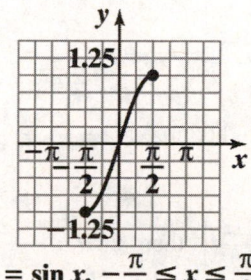

$y = \sin x,\ -\dfrac{\pi}{2} \le x \le \dfrac{\pi}{2}$

b. yes; Explanations will vary.

c. The angle is $-\dfrac{\pi}{6}$.

This is represented by the point $\left(-\dfrac{\pi}{6}, -\dfrac{1}{2}\right)$.

100. a.

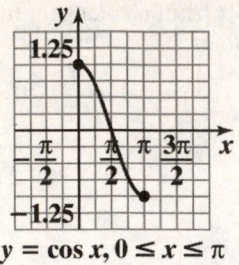

$y = \cos x, \, 0 \le x \le \pi$

b. yes; Explanations will vary.

c. The angle is $\dfrac{5\pi}{6}$.

This is represented by the point $\left(\dfrac{5\pi}{6}, -\dfrac{\sqrt{3}}{2}\right)$.

101. a.

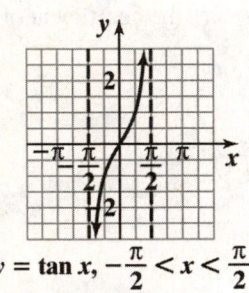

$y = \tan x, \, -\dfrac{\pi}{2} < x < \dfrac{\pi}{2}$

b. yes; Explanations will vary.

c. The angle is $-\dfrac{\pi}{3}$.

This is represented by the point $\left(-\dfrac{\pi}{3}, -\sqrt{3}\right)$.

Section 4.7

Check Point Exercises

1. Let $\theta = \sin^{-1}\dfrac{\sqrt{3}}{2}$, then $\sin\theta = \dfrac{\sqrt{3}}{2}$.

The only angle in the interval $\left[-\dfrac{\pi}{2}, \dfrac{\pi}{2}\right]$ that satisfies $\sin\theta = \dfrac{\sqrt{3}}{2}$ is $\dfrac{\pi}{3}$. Thus, $\theta = \dfrac{\pi}{3}$, or $\sin^{-1}\dfrac{\sqrt{3}}{2} = \dfrac{\pi}{3}$.

2. Let $\theta = \sin^{-1}\left(-\dfrac{\sqrt{2}}{2}\right)$, then $\sin\theta = -\dfrac{\sqrt{2}}{2}$.

The only angle in the interval $\left[-\dfrac{\pi}{2}, \dfrac{\pi}{2}\right]$ that satisfies $\cos\theta = -\dfrac{\sqrt{2}}{2}$ is $-\dfrac{\pi}{4}$. Thus $\theta = -\dfrac{\pi}{4}$, or $\sin^{-1}\left(-\dfrac{\sqrt{2}}{2}\right) = -\dfrac{\pi}{4}$.

3. Let $\theta = \cos^{-1}\left(-\dfrac{1}{2}\right)$, then $\cos\theta = -\dfrac{1}{2}$. The only angle in the interval $[0, \pi]$ that satisfies $\cos\theta = -\dfrac{1}{2}$ is $\dfrac{2\pi}{3}$. Thus,

$\theta = \dfrac{2\pi}{3}$, or $\cos^{-1}\left(-\dfrac{1}{2}\right) = \dfrac{2\pi}{3}$.

4. Let $\theta = \tan^{-1}(-1)$, then $\tan\theta = -1$. The only angle in the interval $\left(-\dfrac{\pi}{2}, \dfrac{\pi}{2}\right)$ that satisfies $\tan\theta = -1$ is $-\dfrac{\pi}{4}$. Thus

$\theta = -\dfrac{\pi}{4}$ or $\tan^{-1}\theta = -\dfrac{\pi}{4}$.

5.

			Scientific Calculator Solution	
	Function	**Mode**	**Keystrokes**	**Display** (rounded to four places)
a.	$\cos^{-1}\left(\dfrac{1}{3}\right)$	Radian	$1\ \boxed{\div}\ 3\ \boxed{=}\ \boxed{\text{COS}^{-1}}$	1.2310
b.	$\tan^{-1}(-35.85)$	Radian	$35.85\ \boxed{+\!/\!-}\ \boxed{\text{TAN}^{-1}}$	−1.5429

			Graphing Calculator Solution	
	Function	**Mode**	**Keystrokes**	**Display** (rounded to four places)
a.	$\cos^{-1}\left(\dfrac{1}{3}\right)$	Radian	$\boxed{\text{COS}^{-1}}\ \boxed{(}\ 1\ \boxed{\div}\ 3\ \boxed{)}\ \boxed{\text{ENTER}}$	1.2310
b.	$\tan^{-1}(-35.85)$	Radian	$\boxed{\text{TAN}^{-1}}\ \boxed{-}\ 35.85\ \boxed{\text{ENTER}}$	−1.5429

6. **a.** $\cos\left(\cos^{-1} 0.7\right)$

$x = 0.7$, x is in $[-1,1]$ so $\cos(\cos^{-1} 0.7) = 0.7$

b. $\sin^{-1}(\sin\pi)$

$x = \pi$, x is not in $\left[-\dfrac{\pi}{2}, \dfrac{\pi}{2}\right]$. x is in the domain of $\sin x$, so $\sin^{-1}(\sin\pi) = \sin^{-1}(0) = 0$

c. $\cos\left(\cos^{-1}\pi\right)$

$x = \pi$, x is not in $[-1,1]$ so $\cos\left(\cos^{-1}\pi\right)$ is not defined.

7. Let $\theta = \tan^{-1}\left(\dfrac{3}{4}\right)$, then $\tan\theta = \dfrac{3}{4}$. Because $\tan\theta$ is positive, θ is in the first quadrant.

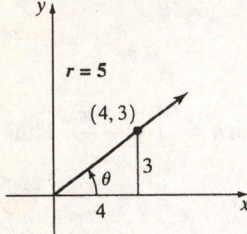

Use the Pythagorean Theorem to find r.

$r^2 = 3^2 + 4^2 = 9 + 16 = 25$

$r = \sqrt{25} = 5$

Use the right triangle to find the exact value.

$\sin\left(\tan^{-1}\dfrac{3}{4}\right) = \sin\theta = \dfrac{\text{side opposite } \theta}{\text{hypotenuse}} = \dfrac{3}{5}$

8. Let $\theta = \sin^{-1}\left(-\dfrac{1}{2}\right)$, then $\sin\theta = -\dfrac{1}{2}$. Because $\sin\theta$ is negative, θ is in quadrant IV.

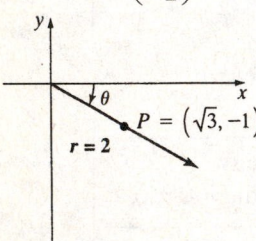

Use the Pythagorean Theorem to find x.

$x^2 + (-1)^2 = 2^2$

$x^2 + 1 = 4$

$x^2 = 3$

$x = \sqrt{3}$

Use values for x and r to find the exact value.

$\cos\left[\sin^{-1}\left(-\dfrac{1}{2}\right)\right] = \cos\theta = \dfrac{x}{r} = \dfrac{\sqrt{3}}{2}$

9. Let $\theta = \tan^{-1} x$, then $\tan\theta = x = \dfrac{x}{1}$.

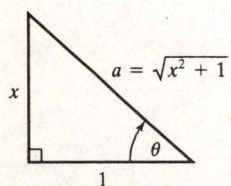

Use the Pythagorean Theorem to find the third side, a.

$a^2 = x^2 + 1^2$

$a = \sqrt{x^2 + 1}$

Use the right triangle to write the algebraic expression.

$$\sec\left(\tan^{-1} x\right) = \sec\theta = \frac{\sqrt{x^2 + 1}}{1} = \sqrt{x^2 + 1}$$

Exercise Set 4.7

1. Let $\theta = \sin^{-1}\dfrac{1}{2}$, then $\sin\theta = \dfrac{1}{2}$. The only angle in the interval $\left[-\dfrac{\pi}{2}, \dfrac{\pi}{2}\right]$ that satisfies $\sin\theta = \dfrac{1}{2}$ is $\dfrac{\pi}{6}$. Thus, $\theta = \dfrac{\pi}{6}$, or

    $\sin^{-1}\dfrac{1}{2} = \dfrac{\pi}{6}$.

3. Let $\theta = \sin^{-1}\dfrac{\sqrt{2}}{2}$, then $\sin\theta = \dfrac{\sqrt{2}}{2}$. The only angle in the interval $\left[-\dfrac{\pi}{2}, \dfrac{\pi}{2}\right]$ that satisfies $\sin\theta = \dfrac{\sqrt{2}}{2}$ is $\dfrac{\pi}{4}$. Thus

    $\theta = \dfrac{\pi}{4}$, or $\sin^{-1}\dfrac{\sqrt{2}}{2} = \dfrac{\pi}{4}$.

5. Let $\theta = \sin^{-1}\left(-\dfrac{1}{2}\right)$, then $\sin\theta = -\dfrac{1}{2}$. The only angle in the interval $\left[-\dfrac{\pi}{2}, \dfrac{\pi}{2}\right]$ that satisfies $\sin\theta = -\dfrac{1}{2}$ is $-\dfrac{\pi}{6}$. Thus

    $\theta = -\dfrac{\pi}{6}$, or $\sin^{-1}\left(-\dfrac{1}{2}\right) = -\dfrac{\pi}{6}$.

7. Let $\theta = \cos^{-1}\dfrac{\sqrt{3}}{2}$, then $\cos\theta = \dfrac{\sqrt{3}}{2}$. The only angle in the interval $[0, \pi]$ that satisfies $\cos\theta = \dfrac{\sqrt{3}}{2}$ is $\dfrac{\pi}{6}$. Thus $\theta = \dfrac{\pi}{6}$,

    or $\cos^{-1}\dfrac{\sqrt{3}}{2} = \dfrac{\pi}{6}$.

9. Let $\theta = \cos^{-1}\left(-\dfrac{\sqrt{2}}{2}\right)$, then $\cos\theta = -\dfrac{\sqrt{2}}{2}$. The only angle in the interval $[0, \pi]$ that satisfies $\cos\theta = -\dfrac{\sqrt{2}}{2}$ is $\dfrac{3\pi}{4}$. Thus

    $\theta = \dfrac{3\pi}{4}$, or $\cos^{-1}\left(-\dfrac{\sqrt{2}}{2}\right) = \dfrac{3\pi}{4}$.

11. Let $\theta = \cos^{-1} 0$, then $\cos\theta = 0$. The only angle in the interval $[0, \pi]$ that satisfies $\cos\theta = 0$ is $\dfrac{\pi}{2}$.

    Thus $\theta = \dfrac{\pi}{2}$, or $\cos^{-1} 0 = \dfrac{\pi}{2}$.

13. Let $\theta = \tan^{-1}\dfrac{\sqrt{3}}{3}$, then $\tan\theta = \dfrac{\sqrt{3}}{3}$. The only angle in the interval $\left(-\dfrac{\pi}{2}, \dfrac{\pi}{2}\right)$ that satisfies $\tan\theta = \dfrac{\sqrt{3}}{3}$ is $\dfrac{\pi}{6}$. Thus

    $\theta = \dfrac{\pi}{6}$, or $\tan^{-1}\dfrac{\sqrt{3}}{3} = \dfrac{\pi}{6}$.

15. Let $\theta = \tan^{-1} 0$, then $\tan\theta = 0$. The only angle in the interval $\left(-\dfrac{\pi}{2}, \dfrac{\pi}{2}\right)$ that satisfies $\tan\theta = 0$ is 0. Thus $\theta = 0$, or

    $\tan^{-1} 0 = 0$.

17. Let $\theta = \tan^{-1}\left(-\sqrt{3}\right)$, then $\tan\theta = -\sqrt{3}$. The only angle in the interval $\left(-\dfrac{\pi}{2}, \dfrac{\pi}{2}\right)$ that satisfies $\tan\theta = -\sqrt{3}$ is $-\dfrac{\pi}{3}$.

Thus $\theta = -\dfrac{\pi}{3}$, or $\tan^{-1}\left(-\sqrt{3}\right) = -\dfrac{\pi}{3}$.

19.

Scientific Calculator Solution			
Function	**Mode**	**Keystrokes**	**Display** (rounded to two places)
$\sin^{-1} 0.3$	Radian	0.3 $\boxed{\text{SIN}^{-1}}$	0.30

Graphing Calculator Solution			
Function	**Mode**	**Keystrokes**	**Display** (rounded to two places)
$\sin^{-1} 0.3$	Radian	$\boxed{\text{SIN}^{-1}}$ 0.3 $\boxed{\text{ENTER}}$	0.30

21.

Scientific Calculator Solution			
Function	**Mode**	**Keystrokes**	**Display** (rounded to two places)
$\sin^{-1}(-0.32)$	Radian	0.32 $\boxed{+/-}$ $\boxed{\text{SIN}^{-1}}$	−0.33

Graphing Calculator Solution			
Function	**Mode**	**Keystrokes**	**Display** (rounded to two places)
$\sin^{-1}(-0.32)$	Radian	$\boxed{\text{SIN}^{-1}}$ $\boxed{-}$ 0.32 $\boxed{\text{ENTER}}$	−0.33

23.

Scientific Calculator Solution			
Function	**Mode**	**Keystrokes**	**Display** (rounded to two places)
$\cos^{-1}\left(\dfrac{3}{8}\right)$	Radian	3 $\boxed{\div}$ 8 $\boxed{=}$ $\boxed{\text{COS}^{-1}}$	1.19

Graphing Calculator Solution			
Function	**Mode**	**Keystrokes**	**Display** (rounded to two places)
$\cos^{-1}\left(\dfrac{3}{8}\right)$	Radian	$\boxed{\text{COS}^{-1}}$ $\boxed{(}$ 3 $\boxed{\div}$ 8 $\boxed{)}$ $\boxed{\text{ENTER}}$	1.19

25.

Scientific Calculator Solution			
Function	**Mode**	**Keystrokes**	**Display** (rounded to two places)
$\cos^{-1}\dfrac{\sqrt{5}}{7}$	Radian	5 $\boxed{\sqrt{}}$ $\boxed{\div}$ 7 $\boxed{=}$ $\boxed{\text{COS}^{-1}}$	1.25

Graphing Calculator Solution			
Function	**Mode**	**Keystrokes**	**Display** (rounded to two places)
$\cos^{-1}\dfrac{\sqrt{5}}{7}$	Radian	$\boxed{\text{COS}^{-1}}$ $\boxed{(}$ $\boxed{\sqrt{}}$ 5 $\boxed{\div}$ 7 $\boxed{)}$ $\boxed{\text{ENTER}}$	1.25

27.

Scientific Calculator Solution			
Function	**Mode**	**Keystrokes**	**Display** (rounded to two places)
$\tan^{-1}(-20)$	Radian	20 $\boxed{+/-}$ $\boxed{\text{TAN}^{-1}}$	-1.52

Graphing Calculator Solution			
Function	**Mode**	**Keystrokes**	**Display** (rounded to two places)
$\tan^{-1}(-20)$	Radian	$\boxed{\text{TAN}^{-1}}$ $\boxed{-}$ 20 $\boxed{\text{ENTER}}$	-1.52

29.

Scientific Calculator Solution			
Function	**Mode**	**Keystrokes**	**Display** (rounded to two places)
$\tan^{-1}\left(-\sqrt{473}\right)$	Radian	473 $\boxed{\sqrt{\ }}$ $\boxed{+/-}$ $\boxed{\text{TAN}^{-1}}$	-1.52

Graphing Calculator Solution			
Function	**Mode**	**Keystrokes**	**Display** (rounded to two places)
$\tan^{-1}\left(-\sqrt{473}\right)$	Radian	$\boxed{\text{TAN}^{-1}}$ $\boxed{(}$ $\boxed{-}$ $\boxed{\sqrt{\ }}$ $\boxed{473}$ $\boxed{)}$ $\boxed{\text{ENTER}}$	-1.52

31. $\sin\left(\sin^{-1} 0.9\right)$

$x = 0.9$, x is in $[-1, 1]$, so $\sin(\sin^{-1} 0.9) = 0.9$

33. $\sin^{-1}\left(\sin\dfrac{\pi}{3}\right)$

$x = \dfrac{\pi}{3}$, x is in $\left[-\dfrac{\pi}{2}, \dfrac{\pi}{2}\right]$, so $\sin^{-1}\left(\sin\dfrac{\pi}{3}\right) = \dfrac{\pi}{3}$

35. $\sin^{-1}\left(\sin\dfrac{5\pi}{6}\right)$

$x = \dfrac{5\pi}{6}$, x is not in $\left[-\dfrac{\pi}{2}, \dfrac{\pi}{2}\right]$, x is in the domain of $\sin x$, so $\sin^{-1}\left(\sin\dfrac{5\pi}{6}\right) = \sin^{-1}\left(\dfrac{1}{2}\right) = \dfrac{\pi}{6}$

37. $\tan\left(\tan^{-1} 125\right)$

$x = 125$, x is a real number, so $\tan\left(\tan^{-1} 125\right) = 125$

39. $\tan^{-1}\left[\tan\left(-\dfrac{\pi}{6}\right)\right]$

$x = -\dfrac{\pi}{6}$, x is in $\left(-\dfrac{\pi}{2}, \dfrac{\pi}{2}\right)$, so $\tan^{-1}\left[\tan\left(-\dfrac{\pi}{6}\right)\right] = -\dfrac{\pi}{6}$

41. $\tan^{-1}\left(\tan\dfrac{2\pi}{3}\right)$

$x=\dfrac{2\pi}{3}$, x is not in $\left(-\dfrac{\pi}{2},\dfrac{\pi}{2}\right)$, x is in the domain of

$\tan x$, so $\tan^{-1}\left(\tan\dfrac{2\pi}{3}\right)=\tan^{-1}\left(-\sqrt{3}\right)=-\dfrac{\pi}{3}$

43. $\sin^{-1}(\sin\pi)$

$x=\pi$, x is not in $\left[-\dfrac{\pi}{2},\dfrac{\pi}{2}\right]$,

x is in the domain of $\sin x$, so

$\sin^{-1}(\sin\pi)=\sin^{-1}0=0$

45. $\sin\left(\sin^{-1}\pi\right)$

$x=\pi$, x is not in $[-1,1]$, so $\sin\left(\sin^{-1}\pi\right)$ is not

defined.

47. Let $\theta=\sin^{-1}\dfrac{4}{5}$, then $\sin\theta=\dfrac{4}{5}$. Because $\sin\theta$ is

positive, θ is in the first quadrant.

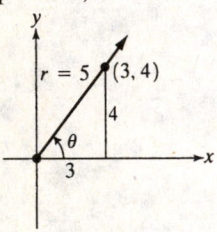

$x^2+y^2=r^2$

$x^2+4^2=5^2$

$\quad x^2=25-16=9$

$\quad\; x=3$

$\cos\left(\sin^{-1}\dfrac{4}{5}\right)=\cos\theta=\dfrac{x}{r}=\dfrac{3}{5}$

49. Let $\theta = \cos^{-1}\dfrac{5}{13}$, then $\cos\theta = \dfrac{5}{13}$. Because $\cos\theta$ is positive, θ is in the first quadrant.

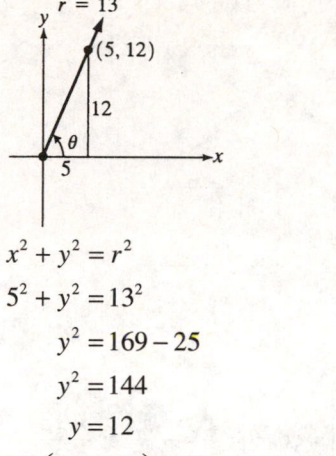

$$x^2 + y^2 = r^2$$
$$5^2 + y^2 = 13^2$$
$$y^2 = 169 - 25$$
$$y^2 = 144$$
$$y = 12$$
$$\tan\left(\cos^{-1}\dfrac{5}{13}\right) = \tan\theta = \dfrac{y}{x} = \dfrac{12}{5}$$

51. Let $\theta = \sin^{-1}\left(-\dfrac{3}{5}\right)$, then $\sin\theta = -\dfrac{3}{5}$. Because $\sin\theta$ is negative, θ is in quadrant IV.

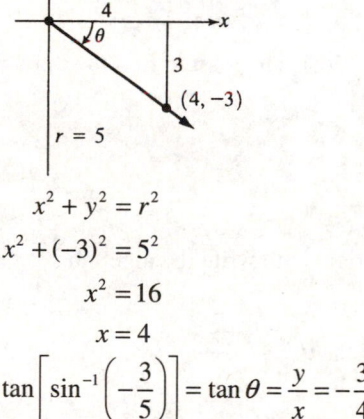

$$x^2 + y^2 = r^2$$
$$x^2 + (-3)^2 = 5^2$$
$$x^2 = 16$$
$$x = 4$$
$$\tan\left[\sin^{-1}\left(-\dfrac{3}{5}\right)\right] = \tan\theta = \dfrac{y}{x} = -\dfrac{3}{4}$$

53. Let, $\theta = \cos^{-1}\dfrac{\sqrt{2}}{2}$, then $\cos\theta = \dfrac{\sqrt{2}}{2}$. Because $\cos\theta$ is positive, θ is in the first quadrant.

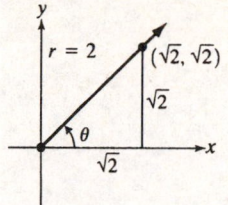

$$x^2 + y^2 = r^2$$
$$\left(\sqrt{2}\right)^2 + y^2 = 2^2$$
$$y^2 = 2$$
$$y = \sqrt{2}$$
$$\sin\left(\cos^{-1}\dfrac{\sqrt{2}}{2}\right) = \sin\theta = \dfrac{y}{r} = \dfrac{\sqrt{2}}{2}$$

55. Let $\theta = \sin^{-1}\left(-\dfrac{1}{4}\right)$, then $\sin\theta = -\dfrac{1}{4}$. Because $\sin\theta$ is negative, θ is in quadrant IV.

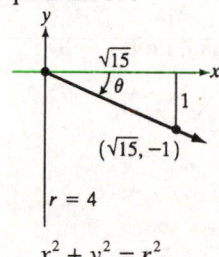

$$x^2 + y^2 = r^2$$
$$x^2 + (-1)^2 = 4^2$$
$$x^2 = 15$$
$$x = \sqrt{15}$$
$$\sec\left[\sin^{-1}\left(-\dfrac{1}{4}\right)\right] = \sec\theta = \dfrac{r}{x} = \dfrac{4}{\sqrt{15}} = \dfrac{4\sqrt{15}}{15}$$

57. Let $\theta = \cos^{-1}\left(-\dfrac{1}{3}\right)$, then $\cos\theta = -\dfrac{1}{3}$. Because

$\cos\theta$ is negative, θ is in quadrant II.

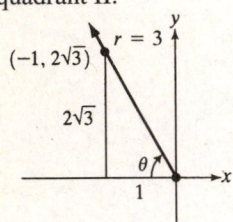

$$x^2 + y^2 = r^2$$
$$(-1)^2 + y^2 = 3^2$$
$$y^2 = 8$$
$$y = \sqrt{8}$$
$$y = 2\sqrt{2}$$

Use the right triangle to find the exact value.

$$\tan\left[\cos^{-1}\left(-\frac{1}{3}\right)\right] = \tan\theta = \frac{y}{x} = \frac{2\sqrt{2}}{-1} = -2\sqrt{2}$$

59. Let $\theta = \cos^{-1}\left(-\dfrac{\sqrt{3}}{2}\right)$, then $\cos\theta = -\dfrac{\sqrt{3}}{2}$. Because

$\cos\theta$ is negative, θ is in quadrant II.

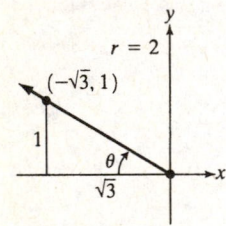

$$x^2 + y^2 = r^2$$
$$\left(-\sqrt{3}\right)^2 + y^2 = 2^2$$
$$y^2 = 1$$
$$y = 1$$

$$\csc\left[\cos^{-1}\left(-\frac{\sqrt{3}}{2}\right)\right] = \csc\theta = \frac{r}{y} = \frac{2}{1} = 2$$

61. Let $\theta = \tan^{-1}\left(-\dfrac{2}{3}\right)$, then $\tan\theta = -\dfrac{2}{3}$.

Because $\tan\theta$ is negative, θ is in quadrant IV.

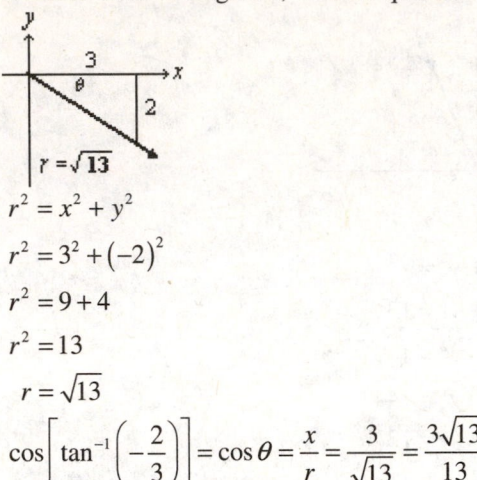

$$r^2 = x^2 + y^2$$
$$r^2 = 3^2 + (-2)^2$$
$$r^2 = 9 + 4$$
$$r^2 = 13$$
$$r = \sqrt{13}$$

$$\cos\left[\tan^{-1}\left(-\frac{2}{3}\right)\right] = \cos\theta = \frac{x}{r} = \frac{3}{\sqrt{13}} = \frac{3\sqrt{13}}{13}$$

63. Let $\theta = \cos^{-1} x$, then $\cos\theta = x = \dfrac{x}{1}$.

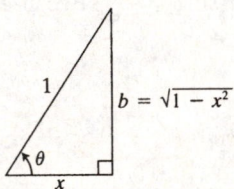

Use the Pythagorean Theorem to find the third side, b.

$$x^2 + b^2 = 1^2$$
$$b^2 = 1 - x^2$$
$$b = \sqrt{1 - x^2}$$

Use the right triangle to write the algebraic expression.

$$\tan\left(\cos^{-1} x\right) = \tan\theta = \frac{\sqrt{1 - x^2}}{x}$$

65. Let $\theta = \sin^{-1} 2x$, then $\sin\theta = 2x$
$y = 2x, r = 1$
Use the Pythagorean Theorem to find x.
$$x^2 + (2x)^2 = 1^2$$
$$x^2 = 1 - 4x^2$$
$$x = \sqrt{1 - 4x^2}$$
$$\cos(\sin^{-1} 2x) = \sqrt{1 - 4x^2}$$

67. Let $\theta = \sin^{-1}\dfrac{1}{x}$, then $\sin\theta = \dfrac{1}{x}$.

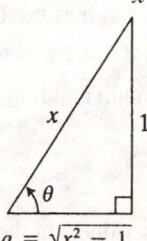

$a = \sqrt{x^2 - 1}$

Use the Pythagorean Theorem to find the third side, a.
$$a^2 + 1^2 = x^2$$
$$a^2 = x^2 - 1$$
$$a = \sqrt{x^2 - 1}$$
Use the right triangle to write the algebraic expression.
$$\cos\left(\sin^{-1}\frac{1}{x}\right) = \cos\theta = \frac{\sqrt{x^2 - 1}}{x}$$

69. $\cot\left(\tan^{-1}\dfrac{x}{\sqrt{3}}\right) = \dfrac{\sqrt{3}}{x}$

71. Let $\theta = \sin^{-1}\dfrac{x}{\sqrt{x^2 + 4}}$, then $\sin\theta = \dfrac{x}{\sqrt{x^2 + 4}}$.

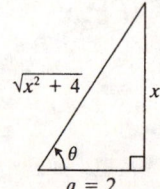

$a = 2$

Use the Pythagorean Theorem to find the third side, a.
$$a^2 + x^2 = \left(\sqrt{x^2 + 4}\right)^2$$
$$a^2 = x^2 + 4 - x^2 = 4$$
$$a = 2$$
Use the right triangle to write the algebraic expression.
$$\sec\left(\sin^{-1}\frac{x}{\sqrt{x^2 + 4}}\right) = \sec\theta = \frac{\sqrt{x^2 + 4}}{2}$$

73. a. $y = \sec x$ is the reciprocal of $y = \cos x$. The x-values for the key points in the interval $[0, \pi]$ are $0, \dfrac{\pi}{4}, \dfrac{\pi}{2}, \dfrac{3\pi}{4}$, and π. The key points are

$(0, 1), \left(\dfrac{\pi}{4}, \dfrac{\sqrt{2}}{2}\right), \left(\dfrac{\pi}{2}, 0\right), \left(\dfrac{3\pi}{4}, -\dfrac{\sqrt{2}}{2}\right)$, and

$(\pi, -1)$, Draw a vertical asymptote at $x = \dfrac{\pi}{2}$.

Now draw our graph from $(0, 1)$ through

$\left(\dfrac{\pi}{4}, \sqrt{2}\right)$ to ∞ on the left side of the

asymptote. From $-\infty$ on the right side of the

asymptote through $\left(\dfrac{3\pi}{4}, -\sqrt{2}\right)$ to $(\pi, -1)$.

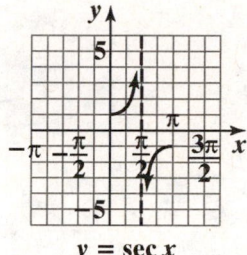

$y = \sec x$

b. With this restricted domain, no horizontal line intersects the graph of $y = \sec x$ more than once, so the function is one-to-one and has an inverse function.

c. Reflecting the graph of the restricted secant function about the line $y = x$, we get the graph of $y = \sec^{-1} x$.

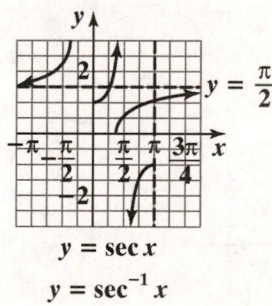

$y = \sec x$
$y = \sec^{-1} x$

75.

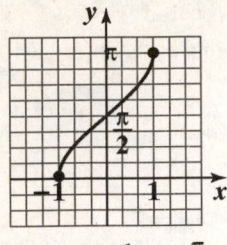

$$f(x) = \sin^{-1} x + \frac{\pi}{2}$$

domain: $[-1, 1]$;
range: $[0, \pi]$

77.

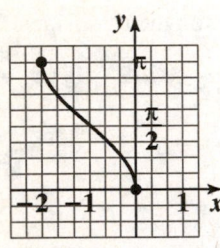

$$g(x) = \cos^{-1}(x + 1)$$

domain: $[-2, 0]$;
range: $[0, \pi]$

79.

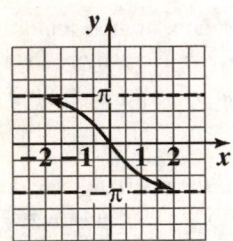

$$f(x) = -2\tan^{-1} x$$

domain: $(-\infty, \infty)$;
range: $(-\pi, \pi)$

81.

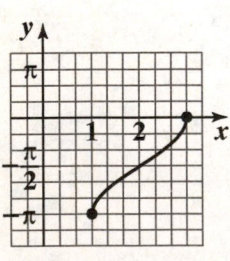

$$f(x) = \sin^{-1}(x - 2) - \frac{\pi}{2}$$

domain: $(1, 3)$;
range: $[-\pi, 0]$

83.

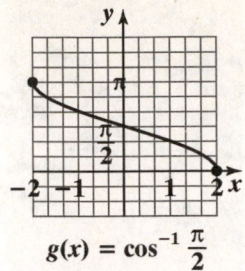

$$g(x) = \cos^{-1} \frac{\pi}{2}$$

domain: $[-2, 2]$;
range: $[0, \pi]$

85. The inner function, $\sin^{-1} x$, accepts values on the interval $[-1,1]$. Since the inner and outer functions are inverses of each other, the domain and range are as follows.
domain: $[-1,1]$; range: $[-1,1]$

87. The inner function, $\cos x$, accepts values on the interval $(-\infty,\infty)$. The outer function returns values on the interval $[0,\pi]$
domain: $(-\infty,\infty)$; range: $[0,\pi]$

89. The inner function, $\cos x$, accepts values on the interval $(-\infty,\infty)$. The outer function returns values on the interval $\left[-\frac{\pi}{2},\frac{\pi}{2}\right]$
domain: $(-\infty,\infty)$; range: $\left[-\frac{\pi}{2},\frac{\pi}{2}\right]$

91. The functions $\sin^{-1} x$ and $\cos^{-1} x$ accept values on the interval $[-1,1]$. The sum of these values is always $\frac{\pi}{2}$.
domain: $[-1,1]$; range: $\left\{\frac{\pi}{2}\right\}$

93. $\theta = \tan^{-1}\dfrac{33}{x} - \tan^{-1}\dfrac{8}{x}$

x	θ
5	$\tan^{-1}\dfrac{33}{5} - \tan^{-1}\dfrac{8}{5} \approx 0.408$ radians
10	$\tan^{-1}\dfrac{33}{10} - \tan^{-1}\dfrac{8}{10} \approx 0.602$ radians
15	$\tan^{-1}\dfrac{33}{15} - \tan^{-1}\dfrac{8}{15} \approx 0.654$ radians
20	$\tan^{-1}\dfrac{33}{20} - \tan^{-1}\dfrac{8}{20} \approx 0.645$ radians
25	$\tan^{-1}\dfrac{33}{25} - \tan^{-1}\dfrac{8}{25} \approx 0.613$ radians

95. $\theta = 2\tan^{-1}\dfrac{21.634}{28} \approx 1.3157$ radians;

$1.3157\left(\dfrac{180}{\pi}\right) \approx 75.4°$

97. $\tan^{-1} b - \tan^{-1} a = \tan^{-1} 2 - \tan^{-1} 0$
≈ 1.1071 square units

99. – 109. Answers may vary.

111. The domain of $y = \cos^{-1} x$ is the interval
$[-1, 1]$, and the range is the interval $[0, \pi]$. Because
the second equation is the first equation with 1
subtracted from the variable, we will move our x
max to π, and graph in a $\left[-\dfrac{\pi}{2}, \pi, \dfrac{\pi}{4}\right]$ by $[0, 4, 1]$
viewing rectangle.

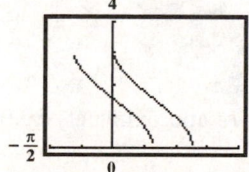

The graph of the second equation is the graph of the
first equation shifted right 1 unit.

113. The domain of $y = \sin^{-1} x$ is the interval
$[-1, 1]$, and the range is $\left[-\dfrac{\pi}{2}, \dfrac{\pi}{2}\right]$. Because the
second equation is the first equation plus 1, and with
2 added to the variable, we will move our y max to 3,
and move our x min to $-\pi$, and graph in a
$\left[-\pi, \dfrac{\pi}{2}, \dfrac{\pi}{2}\right]$ by
$[-2, 3, 1]$ viewing rectangle.

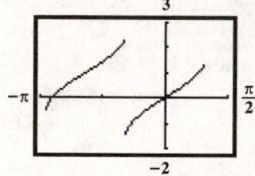

The graph of the second equation is the
graph of the first equation shifted left
2 units and up 1 unit.

115.

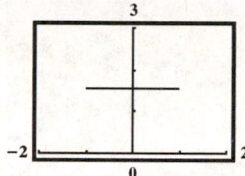

It seems $\sin^{-1} x + \cos^{-1} x = \dfrac{\pi}{2}$ for $-1 \le x \le 1$.

117. does not make sense; Explanations will vary.
Sample explanation: Though this restriction works
for tangent, it is not selected simply because it is
easier to remember. Rather the restrictions are based
on which intervals will have inverses.

119. does not make sense; Explanations will vary.
Sample explanation:

$\sin^{-1}\left(\sin\dfrac{5\pi}{4}\right) = \sin^{-1}\left(-\dfrac{\sqrt{2}}{2}\right) = -\dfrac{\pi}{4}$

121. $2\sin^{-1} x = \dfrac{\pi}{4}$

$\sin^{-1} x = \dfrac{\pi}{8}$

$x = \sin\dfrac{\pi}{8}$

123. Let α equal the acute angle in the smaller right triangle.

$$\tan\alpha = \frac{8}{x}$$

so $\tan^{-1}\dfrac{8}{x} = \alpha$

$$\tan(\alpha + \theta) = \frac{33}{x}$$

so $\tan^{-1}\dfrac{33}{x} = \alpha + \theta$

$$\theta = \alpha + \theta - \alpha = \tan^{-1}\frac{33}{x} - \tan^{-1}\frac{8}{x}$$

124. $\tan A = \dfrac{a}{b}$

$$\tan 22.3° = \frac{a}{12.1}$$

$$a = 12.1\tan 22.3°$$

$$a \approx 4.96$$

$$\cos A = \frac{b}{c}$$

$$\cos 22.3° = \frac{12.1}{c}$$

$$c = \frac{12.1}{\cos 22.3°}$$

$$c \approx 13.08$$

125. $\tan\theta = \dfrac{\text{opposite}}{\text{adjacent}}$

$$\tan\theta = \frac{18}{25}$$

$$\theta = \tan^{-1}\left(\frac{18}{25}\right)$$

$$\theta \approx 35.8°$$

126. $10\cos\left(\dfrac{\pi}{6}x\right)$

amplitude: $|10| = 10$

period: $\dfrac{2\pi}{\frac{\pi}{6}} = 2\pi \cdot \dfrac{6}{\pi} = 12$

Section 4.8

Check Point Exercises

1. We begin by finding the measure of angle B. Because $C = 90°$ and the sum of a triangle's angles is $180°$, we see that $A + B = 90°$. Thus, $B = 90° - A = 90° - 62.7° = 27.3°$.

Now we find b. Because we have a known angle, a known opposite side, and an unknown adjacent side, use the tangent function.

$$\tan 62.7° = \frac{8.4}{b}$$

$$b = \frac{8.4}{\tan 62.7°} \approx 4.34$$

Finally, we need to find c. Because we have a known angle, a known opposite side and an unknown hypotenuse, use the sine function.

$$\sin 62.7° = \frac{8.4}{c}$$

$$c = \frac{8.4}{\sin 62.7} \approx 9.45$$

In summary, $B = 27.3°$, $b \approx 4.34$, and $c \approx 9.45$.

2. Using a right triangle, we have a known angle, an unknown opposite side, a, and a known adjacent side. Therefore, use the tangent function.

$$\tan 85.4° = \frac{a}{80}$$

$$a = 80\tan 85.4° \approx 994$$

The Eiffel tower is approximately 994 feet high.

3. Using a right triangle, we have an unknown angle, A, a known opposite side, and a known hypotenuse. Therefore, use the sine function.

$$\sin A = \frac{6.7}{13.8}$$

$$A = \sin^{-1}\frac{6.7}{13.8} \approx 29.0°$$

The wire makes an angle of approximately 29.0° with the ground.

4. Using two right triangles, a smaller right triangle corresponding to the smaller angle of elevation drawn inside a larger right triangle corresponding to the larger angle of elevation, we have a known angle, an unknown opposite side, a in the smaller triangle, b in the larger triangle, and a known adjacent side in each triangle. Therefore, use the tangent function.

$$\tan 32° = \frac{a}{800}$$

$$a = 800 \tan 32° \approx 499.9$$

$$\tan 35° = \frac{b}{800}$$

$$b = 800 \tan 35° \approx 560.2$$

The height of the sculpture of Lincoln's face is 560.2 – 499.9, or approximately 60.3 feet.

5. **a.** We need the acute angle between ray *OD* and the north-south line through *O*.
The measurement of this angle is given to be 25°. The angle is measured from the south side of the north-south line and lies east of the north-south line. Thus, the bearing from *O* to *D* is S 25°E.

   b. We need the acute angle between ray *OC* and the north-south line through *O*.
This angle measures $90° - 75° = 15°$.
This angle is measured from the south side of the north-south line and lies west of the north-south line. Thus the bearing from *O* to *C* is S 15° W.

6. **a.** Your distance from the entrance to the trail system is represented by the hypotenuse, *c*, of a right triangle. Because we know the length of the two sides of the right triangle, we find *c* using the Pythagorean Theorem.
We have
$$c^2 = a^2 + b^2 = (2.3)^2 + (3.5)^2 = 17.54$$
$$c = \sqrt{17.54} \approx 4.2$$
You are approximately 4.2 miles from the entrance to the trail system.

   b. To find your bearing from the entrance to the trail system, consider a north-south line passing through the entrance. The acute angle from this line to the ray on which you lie is $31° + \theta$. Because we are measuring the angle from the south side of the line and you are west of the entrance, your bearing from the entrance is $S(31° + \theta)$ W. To find θ, Use a right triangle and the tangent function.
$$\tan \theta = \frac{3.5}{2.3}$$
$$\theta = \tan^{-1} \frac{3.5}{2.3} \approx 56.7°$$
Thus, $31° + \theta = 31° + 56.7° = 87.7°$. Your bearing from the entrance to the trail system is S 87.7° W.

7. When the object is released ($t = 0$), the ball's distance, *d*, from its rest position is 6 inches down. Because it is down, *d* is negative: when $t = 0$, $d = -6$. Notice the greatest distance from rest position occurs at $t = 0$. Thus, we will use the equation with the cosine function, $y = a \cos \omega t$, to model the ball's motion. Recall that $|a|$ is the maximum distance. Because the ball initially moves down, $a = -6$. The value of ω can be found using the formula for the period.

$$period = \frac{2\pi}{\omega} = 4$$
$$2\pi = 4\omega$$
$$\omega = \frac{2\pi}{4} = \frac{\pi}{2}$$

Substitute these values into $d = a \cos wt$. The equation for the ball's simple harmonic motion is

$$d = -6 \cos \frac{\pi}{2} t.$$

8. We begin by identifying values for *a* and ω.
$$d = 12 \cos \frac{\pi}{4} t, \ a = 12 \text{ and } \omega = \frac{\pi}{4}.$$

   a. The maximum displacement from the rest position is the amplitude. Because $a = 12$, the maximum displacement is 12 centimeters.

   b. The frequency, *f*, is
$$f = \frac{\omega}{2\pi} = \frac{\frac{\pi}{4}}{2\pi} = \frac{\pi}{4} \cdot \frac{1}{2\pi} = \frac{1}{8}$$
The frequency is $\frac{1}{8}$ cycle per second.

   c. The time required for one cycle is the period.
$$period = \frac{2\pi}{\omega} = \frac{2\pi}{\frac{\pi}{4}} = 2\pi \cdot \frac{4}{\pi} = 8$$
The time required for one cycle is 8 seconds.

Exercise Set 4.8

1. Find the measure of angle B. Because
$C = 90°$, $A + B = 90°$. Thus,
$B = 90° - A = 90° - 23.5° = 66.5°$.
Because we have a known angle, a known adjacent side, and an unknown opposite side, use the tangent function.

$$\tan 23.5° = \frac{a}{10}$$

$$a = 10\tan 23.5° \approx 4.35$$

Because we have a known angle, a known adjacent side, and an unknown hypotenuse, use the cosine function.

$$\cos 23.5° = \frac{10}{c}$$

$$c = \frac{10}{\cos 23.5°} \approx 10.90$$

In summary, $B = 66.5°$, $a \approx 4.35$, and $c \approx 10.90$.

3. Find the measure of angle B. Because
$C = 90°$, $A + B = 90°$.
Thus, $B = 90° - A = 90° - 52.6° = 37.4°$.
Because we have a known angle, a known hypotenuse, and an unknown opposite side, use the sine function.

$$\sin 52.6 = \frac{a}{54}$$

$$a = 54\sin 52.6° \approx 42.90$$

Because we have a known angle, a known hypotenuse, and an unknown adjacent side, use the cosine function.

$$\cos 52.6° = \frac{b}{54}$$

$$b = 54\cos 52.6° \approx 32.80$$

In summary, $B = 37.4°$, $a \approx 42.90$, and $b \approx 32.80$.

5. Find the measure of angle A. Because
$C = 90°$, $A + B = 90°$.
Thus, $A = 90° - B = 90° - 16.8° = 73.2°$.
Because we have a known angle, a known opposite side and an unknown adjacent side, use the tangent function.

$$\tan 16.8° = \frac{30.5}{a}$$

$$a = \frac{30.5}{\tan 16.8°} \approx 101.02$$

Because we have a known angle, a known opposite side, and an unknown hypotenuse, use the sine function.

$$\sin 16.8° = \frac{30.5}{c}$$

$$c = \frac{30.5}{\sin 16.8°} \approx 105.52$$

In summary, $A = 73.2°$, $a \approx 101.02$, and $c \approx 105.52$.

7. Find the measure of angle A. Because we have a known hypotenuse, a known opposite side, and an unknown angle, use the sine function.

$$\sin A = \frac{30.4}{50.2}$$

$$A = \sin^{-1}\left(\frac{30.4}{50.2}\right) \approx 37.3°$$

Find the measure of angle B. Because
$C = 90°$, $A + B = 90°$. Thus,
$B = 90° - A \approx 90° - 37.3° = 52.7°$.
Use the Pythagorean Theorem.

$$a^2 + b^2 = c^2$$

$$(30.4)^2 + b^2 = (50.2)^2$$

$$b^2 = (50.2)^2 - (30.4)^2 = 1595.88$$

$$b = \sqrt{1595.88} \approx 39.95$$

In summary, $A \approx 37.3°$, $B \approx 52.7°$, and $b \approx 39.95$.

9. Find the measure of angle A. Because we have a known opposite side, a known adjacent side, and an unknown angle, use the tangent function.

$$\tan A = \frac{10.8}{24.7}$$

$$A = \tan^{-1}\left(\frac{10.8}{24.7}\right) \approx 23.6°$$

Find the measure of angle B. Because $C = 90°$, $A + B = 90°$.
Thus, $B = 90° - A \approx 90° - 23.6° = 66.4°$.
Use the Pythagorean Theorem.

$$c^2 = a^2 + b^2 = (10.8)^2 + (24.7)^2 = 726.73$$

$$c = \sqrt{726.73} \approx 26.96$$

In summary, $A \approx 23.6°$, $B \approx 66.4°$, and $c \approx 26.96$.

11. Find the measure of angle A. Because we have a known hypotenuse, a known adjacent side, and unknown angle, use the cosine function.

$$\cos A = \frac{2}{7}$$

$$A = \cos^{-1}\left(\frac{2}{7}\right) \approx 73.4°$$

Find the measure of angle B. Because $C = 90°$, $A + B = 90°$.
Thus, $B = 90° - A \approx 90° - 73.4° = 16.6°$.
Use the Pythagorean Theorem.

$$a^2 + b^2 = c^2$$

$$a^2 + (2)^2 = (7)^2$$

$$a^2 = (7)^2 - (2)^2 = 45$$

$$a = \sqrt{45} \approx 6.71$$

In summary, $A \approx 73.4°$, $B \approx 16.6°$, and $a \approx 6.71$.

13. We need the acute angle between ray OA and the north-south line through O. This angle measure $90° - 75° = 15°$. This angle is measured from the north side of the north-south line and lies east of the north-south line. Thus, the bearing from O and A is N 15° E.

15. The measurement of this angle is given to be 80°. The angle is measured from the south side of the north-south line and lies west of the north-south line. Thus, the bearing from O to C is S 80° W.

17. When the object is released $(t = 0)$, the object's distance, d, from its rest position is 6 centimeters down. Because it is down, d is negative: When $t = 0$, $d = -6$. Notice the greatest distance from rest position occurs at $t = 0$. Thus, we will use the equation with the cosine function, $y = a\cos\omega t$ to model the object's motion. Recall that $|a|$ is the maximum distance. Because the object initially moves down, $a = -6$. The value of ω can be found using the formula for the period.

$$\text{period} = \frac{2\pi}{\omega} = 4$$

$$2\pi = 4\omega$$

$$\omega = \frac{2\pi}{4} = \frac{\pi}{2}$$

Substitute these values into $d = a\cos\omega t$. The equation for the object's simple harmonic motion is

$$d = -6\cos\frac{\pi}{2}t.$$

19. When $t = 0$, $d = 0$. Therefore, we will use the equation with the sine function, $y = a\sin\omega t$, to model the object's motion. Recall that $|a|$ is the maximum distance. Because the object initially moves down, and has an amplitude of 3 inches, $a = -3$. The value of ω can be found using the formula for the period.

$$\text{period} = \frac{2\pi}{\omega} = 1.5$$

$$2\pi = 1.5\omega$$

$$\omega = \frac{2\pi}{1.5} = \frac{4\pi}{3}$$

Substitute these values into $d = a\sin\omega t$. The equation for the object's simple harmonic motion is

$$d = -3\sin\frac{4\pi}{3}t.$$

21. We begin by identifying values for a and ω.

$d = 5\cos\dfrac{\pi}{2}t$, $a = 5$ and $\omega = \dfrac{\pi}{2}$

a. The maximum displacement from the rest position is the amplitude. Because $a = 5$, the maximum displacement is 5 inches.

b. The frequency, f, is

$$f = \frac{\omega}{2\pi} = \frac{\frac{\pi}{2}}{2\pi} = \frac{\pi}{2} \cdot \frac{1}{2\pi} = \frac{1}{4}.$$

The frequency is $\dfrac{1}{4}$ inch per second.

c. The time required for one cycle is the period.

$$\text{period} = \frac{2\pi}{\omega} = \frac{2\pi}{\frac{\pi}{2}} = 2\pi \cdot \frac{2}{\pi} = 4$$

The time required for one cycle is 4 seconds.

23. We begin by identifying values for a and ω.
$d = -6\cos 2\pi t$, $a = -6$ and $\omega = 2\pi$

a. The maximum displacement from the rest position is the amplitude.
Because $a = -6$, the maximum displacement is 6 inches.

b. The frequency, f, is

$$f = \frac{\omega}{2\pi} = \frac{2\pi}{2\pi} = 1.$$

The frequency is 1 inch per second.

c. The time required for one cycle is the period.

$$\text{period} = \frac{2\pi}{\omega} = \frac{2\pi}{2\pi} = 1$$

The time required for one cycle is
1 second.

25. We begin by identifying values for a and ω.

$d = \dfrac{1}{2}\sin 2t$, $a = \dfrac{1}{2}$ and $\omega = 2$

a. The maximum displacement from the rest position is the amplitude.

Because $a = \dfrac{1}{2}$, the maximum displacement is

$\dfrac{1}{2}$ inch.

b. The frequency, f, is

$$f = \frac{\omega}{2\pi} = \frac{2}{2\pi} = \frac{1}{\pi} \approx 0.32.$$

The frequency is approximately 0.32 cycle per second.

c. The time required for one cycle is the period.

$$\text{period} = \frac{2\pi}{\omega} = \frac{2\pi}{2} = \pi \approx 3.14$$

The time required for one cycle is approximately 3.14 seconds.

27. We begin by identifying values for a and ω.

$d = -5\sin\dfrac{2\pi}{3}t$, $a = -5$ and $\omega = \dfrac{2\pi}{3}$

a. The maximum displacement from the rest position is the amplitude.
Because $a = -5$, the maximum displacement is 5 inches.

b. The frequency, f, is

$$f = \frac{\omega}{2\pi} = \frac{\frac{2\pi}{3}}{2\pi} = \frac{2\pi}{3} \cdot \frac{1}{2\pi} = \frac{1}{3}.$$

The frequency is $\dfrac{1}{3}$ cycle per second.

c. The time required for one cycle is the period.

$$\text{period} = \frac{2\pi}{\omega} = \frac{2\pi}{\frac{2\pi}{3}} = 2\pi \cdot \frac{3}{2\pi} = 3$$

The time required for one cycle is 3 seconds.

29. $x = 500\tan 40° + 500\tan 25°$

$x \approx 653$

31. $x = 600\tan 28° - 600\tan 25°$

$x \approx 39$

33. $x = \dfrac{300}{\tan 34°} - \dfrac{300}{\tan 64°}$

$x \approx 298$

35. $x = \dfrac{400 \tan 40° \tan 20°}{\tan 40° - \tan 20°}$

$x \approx 257$

37. $d = 4\cos\left(\pi t - \dfrac{\pi}{2}\right)$

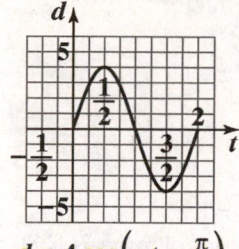

$d = 4\cos\left(\pi t - \dfrac{\pi}{2}\right)$

a. 4 in.

b. $\dfrac{1}{2}$ in. per sec

c. 2 sec

d. $\dfrac{1}{2}$

39. $d = -2\sin\left(\dfrac{\pi t}{4} + \dfrac{\pi}{2}\right)$

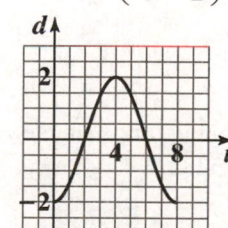

$d = -2\sin\left(\dfrac{\pi}{4}t + \dfrac{\pi}{2}\right)$

a. 2 in.

b. $\dfrac{1}{8}$ in. per sec

c. 8 sec

d. -2

41. Using a right triangle, we have a known angle, an unknown opposite side, a, and a known adjacent side. Therefore, use tangent function.

$\tan 21.3° = \dfrac{a}{5280}$

$a = 5280 \tan 21.3° \approx 2059$

The height of the tower is approximately 2059 feet.

43. Using a right triangle, we have a known angle, a known opposite side, and an unknown adjacent side, a. Therefore, use the tangent function.

$\tan 23.7° = \dfrac{305}{a}$

$a = \dfrac{305}{\tan 23.7°} \approx 695$

The ship is approximately 695 feet from the statue's base.

45. The angle of depression from the helicopter to point P is equal to the angle of elevation from point P to the helicopter. Using a right triangle, we have a known angle, a known opposite side, and an unknown adjacent side, d. Therefore, use the tangent function.

$\tan 36° = \dfrac{1000}{d}$

$d = \dfrac{1000}{\tan 36°} \approx 1376$

The island is approximately 1376 feet off the coast.

47. Using a right triangle, we have an unknown angle, A, a known opposite side, and a known hypotenuse. Therefore, use the sine function.

$\sin A = \dfrac{6}{23}$

$A = \sin^{-1}\left(\dfrac{6}{23}\right) \approx 15.1°$

The ramp makes an angle of approximately 15.1° with the ground.

49. Using the two right triangles, we have a known angle, an unknown opposite side, a in the smaller triangle, b in the larger triangle, and a known adjacent side in each triangle. Therefore, use the tangent function.

$\tan 19.2° = \dfrac{a}{125}$

$a = 125 \tan 19.2° \approx 43.5$

$\tan 31.7° = \dfrac{b}{125}$

$b = 125 \tan 31.7° \approx 77.2$

The balloon rises approximately $77.2 - 43.5$ or 33.7 feet.

51. Using a right triangle, we have a known angle, a known hypotenuse, and unknown sides. To find the opposite side, a, use the sine function.

$$\sin 53° = \frac{a}{150}$$
$$a = 150 \sin 53° \approx 120$$

To find the adjacent side, b, use the cosine function.

$$\cos 53° = \frac{b}{150}$$
$$b = 150 \cos 53° \approx 90$$

The boat has traveled approximately 90 miles north and 120 miles east.

53. The bearing from the fire to the second ranger is N 28° E. Using a right triangle, we have a known angle, a known opposite side, and an unknown adjacent side, b. Therefore, use the tangent function.

$$\tan 28° = \frac{7}{b}$$
$$b = \frac{7}{\tan 28°} \approx 13.2$$

The first ranger is 13.2 miles from the fire, to the nearest tenth of a mile.

55. Using a right triangle, we have a known adjacent side, a known opposite side, and an unknown angle, A. Therefore, use the tangent function.

$$\tan A = \frac{1.5}{2}$$
$$A = \tan\left(\frac{1.5}{2}\right) \approx 37°$$

We need the acute angle between the ray that runs from your house through your location, and the north-south line through your house. This angle measures approximately $90° - 37° = 53°$. This angle is measured from the north side of the north-south line and lies west of the north-south line. Thus, the bearing from your house to you is N 53° W.

57. To find the jet's bearing from the control tower, consider a north-south line passing through the tower. The acute angle from this line to the ray on which the jet lies is $35° + \theta$. Because we are measuring the angle from the north side of the line and the jet is east of the tower, the jet's bearing from the tower is N $(35° + \theta)$ E. To find θ, use a right triangle and the tangent function.

$$\tan \theta = \frac{7}{5}$$
$$\theta = \tan^{-1}\left(\frac{7}{5}\right) \approx 54.5°$$

Thus, $35° + \theta = 35° + 54.5° = 89.5°$.
The jet's bearing from the control tower is N 89.5° E.

59. The frequency, f, is $f = \frac{\omega}{2\pi}$, so

$$\frac{1}{2} = \frac{\omega}{2\pi}$$
$$\omega = \frac{1}{2} \cdot 2\pi = \pi$$

Because the amplitude is 6 feet, $a = 6$. Thus, the equation for the object's simple harmonic motion is $d = 6 \sin \pi t$.

61. The frequency, f, is $f = \frac{\omega}{2\pi}$, so

$$264 = \frac{\omega}{2\pi}$$
$$\omega = 264 \cdot 2\pi = 528\pi$$

Thus, the equation for the tuning fork's simple harmonic motion is $d = \sin 528\pi t$.

63. – 69. Answers may vary.

71. $y = -6e^{-0.09x} \cos 2\pi x$

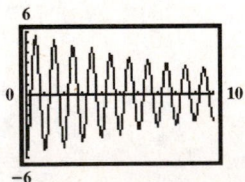

10 complete oscillations occur.

73. does not make sense; Explanations will vary. Sample explanation: When using bearings, the angle must be less than 90°.

75. does not make sense; Explanations will vary. Sample explanation: Frequency and Period are inverses of each other. If the period is 10 seconds then the frequency is $\dfrac{1}{10} = 0.1$ oscillations per second.

77. Let d be the adjacent side to the 40° angle. Using the right triangles, we have a known angle and unknown sides in both triangles. Use the tangent function.

$$\tan 20° = \frac{h}{75 + d}$$
$$h = (75 + d)\tan 20°$$

Also, $\tan 40° = \dfrac{h}{d}$

$$h = d\tan 40°$$

Using the transitive property we have

$$(75 + d)\tan 20° = d\tan 40°$$
$$75\tan 20° + d\tan 20° = d\tan 40°$$
$$d\tan 40° - d\tan 20° = 75\tan 20°$$
$$d(\tan 40° - \tan 20°) = 75\tan 20°$$
$$d = \frac{75\tan 20°}{\tan 40° - \tan 20°}$$

Thus, $h = d\tan 40°$

$$= \frac{75\tan 20°}{\tan 40° - \tan 20°}\tan 40° \approx 48$$

The height of the building is approximately 48 feet.

79. $\sec x \cot x = \dfrac{1}{\cos x} \cdot \dfrac{\cos x}{\sin x} = \dfrac{1}{\sin x}$ or $\csc x$

80. $\tan x \csc x \cos x = \dfrac{\sin x}{\cos x} \cdot \dfrac{1}{\sin x} \cdot \dfrac{\cos x}{1} = 1$

81. $\sec x + \tan x = \dfrac{1}{\cos x} + \dfrac{\sin x}{\cos x} = \dfrac{1 + \sin x}{\cos x}$

Chapter 4 Review Exercises

1. The radian measure of a central angle is the length of the intercepted arc divided by the circle's radius.

$$\theta = \frac{27}{6} = 4.5 \text{ radians}$$

2. $15° = 15° \cdot \dfrac{\pi \text{ radians}}{180°} = \dfrac{15\pi}{180} \text{ radian}$

$$= \frac{\pi}{12} \text{ radian}$$

3. $120° = 120° \cdot \dfrac{\pi \text{ radians}}{180°} = \dfrac{120\pi}{180} \text{ radians}$

$$= \frac{2\pi}{3} \text{ radians}$$

4. $315° = 315° \cdot \dfrac{\pi \text{ radians}}{180°} = \dfrac{315\pi}{180} \text{ radians}$

$$= \frac{7\pi}{4} \text{ radians}$$

5. $\dfrac{5\pi}{3} \text{ radians} = \dfrac{5\pi}{3} \text{ radians} \cdot \dfrac{180°}{\pi \text{ radians}}$

$$= \frac{5 \cdot 180°}{3} = 300°$$

6. $\dfrac{7\pi}{5} \text{ radians} = \dfrac{7\pi}{5} \text{ radians} \cdot \dfrac{180°}{\pi \text{ radians}}$

$$= \frac{7 \cdot 180°}{5} = 252°$$

7. $-\dfrac{5\pi}{6} \text{ radians} = -\dfrac{5\pi}{6} \text{ radians} \cdot \dfrac{180°}{\pi \text{ radians}}$

$$= -\frac{5 \cdot 180°}{6} = -150°$$

8.

9.

10.

11.

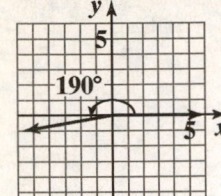

190°

12.

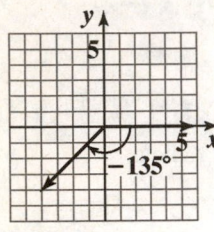

−135°

13. $400° − 360° = 40°$

14. $−445° + (2)360° = 275°$

15. $\dfrac{13\pi}{4} − 2\pi = \dfrac{13\pi}{4} − \dfrac{8\pi}{4} = \dfrac{5\pi}{4}$

16. $\dfrac{31\pi}{6} − (2)2\pi = \dfrac{31\pi}{6} − \dfrac{24\pi}{6} = \dfrac{7\pi}{6}$

17. $−\dfrac{8\pi}{3} + (2)2\pi = −\dfrac{8\pi}{3} + \dfrac{12\pi}{3} = \dfrac{4\pi}{3}$

18. $135° = 135° \cdot \dfrac{\pi \text{ radians}}{180°} = \dfrac{135 \cdot \pi}{180}$ radians

$= \dfrac{3\pi}{4}$ radians

$s = r\theta$

$s = (10 \text{ ft})\left(\dfrac{3\pi}{4}\right) = \dfrac{15\pi}{2}$ ft ≈ 23.56 ft

19. $\dfrac{10.3 \text{ revolutions}}{1 \text{ minute}} \cdot \dfrac{2\pi \text{ radians}}{1 \text{ revolution}}$

$= \dfrac{20.6\pi \text{ radians}}{1 \text{ minute}} = 20.6\pi$ radians per minute

20. Use $v = r\omega$ where v is the linear speed and ω is the angular speed in radians per minute.

$\omega = \dfrac{2250 \text{ revolutions}}{1 \text{ minute}} \cdot \dfrac{2\pi \text{ radians}}{1 \text{ revolution}}$

$= 4500\pi$ radians per minute

$v = 3 \text{ feet} \cdot \dfrac{4500\pi}{\text{minute}} = \dfrac{13,500\pi \text{ feet}}{\text{min}}$

$\approx 42,412$ ft per min

21. $P\left(−\dfrac{4}{5}, −\dfrac{3}{5}\right)$

$\sin t = y = −\dfrac{3}{5}$

$\cos t = x = −\dfrac{4}{5}$

$\tan t = \dfrac{y}{x} = \dfrac{−\frac{3}{5}}{−\frac{4}{5}} = \dfrac{3}{4}$

$\csc t = \dfrac{1}{y} = −\dfrac{5}{3}$

$\sec t = \dfrac{1}{x} = −\dfrac{5}{4}$

$\cot t = \dfrac{x}{y} = \dfrac{4}{3}$

22. $P\left(\dfrac{8}{17}, −\dfrac{15}{17}\right)$

$\sin t = y = −\dfrac{15}{17}$

$\cos t = x = \dfrac{8}{17}$

$\tan t = \dfrac{y}{x} = \dfrac{−\frac{15}{17}}{\frac{8}{17}} = −\dfrac{15}{8}$

$\csc t = \dfrac{1}{y} = −\dfrac{17}{15}$

$\sec t = \dfrac{1}{x} = \dfrac{17}{8}$

$\cot t = \dfrac{x}{y} = −\dfrac{8}{15}$

23. $\sec \dfrac{5\pi}{6} = \dfrac{1}{−\frac{\sqrt{3}}{2}} = −\dfrac{2\sqrt{3}}{3}$

24. $\tan \dfrac{4\pi}{3} = \dfrac{−\frac{\sqrt{3}}{2}}{−\frac{1}{2}} = \sqrt{3}$

25. $\sec \dfrac{\pi}{2}$ is undefined.

26. $\cot \pi$ is undefined.

27. $\sin t = \dfrac{2}{\sqrt{7}}, 0 \le t < \dfrac{\pi}{2}$

$\sin^2 t + \cos^2 t = 1$

$\left(\dfrac{2}{\sqrt{7}}\right)^2 + \cos^2 t = 1$

$\cos^2 t = 1 - \dfrac{4}{7}$

$\cos t = \sqrt{\dfrac{3}{7}} = \dfrac{\sqrt{21}}{7}$

Because $0 \le t < \dfrac{\pi}{2}, \cos t$ is positive.

$\tan t = \dfrac{\frac{2}{\sqrt{7}}}{\sqrt{\frac{3}{7}}} = \dfrac{2\sqrt{3}}{3}$

$\csc t = \dfrac{\sqrt{7}}{2}$

$\sec t = \dfrac{\sqrt{21}}{3}$

$\cot t = \dfrac{\sqrt{3}}{2}$

28. $\tan 4.7 \cot 4.7 = \tan 4.7 \left(\dfrac{1}{\tan 4.7}\right) = 1$

29. $\sin^2 \dfrac{\pi}{17} + \cos^2 \dfrac{\pi}{17} = 1$ because

$\sin^2 t + \cos^2 t = 1.$

30. $\tan^2 1.4 - \sec^2 1.4 = -\left(\sec^2 1.4 - \tan^2 1.4\right)$

$= -1$

31. Use the Pythagorean Theorem to find the hypotenuse, c.

$c^2 = a^2 + b^2$

$c = \sqrt{8^2 + 5^2} = \sqrt{64 + 25} = \sqrt{89}$

$\sin \theta = \dfrac{5}{\sqrt{89}} = \dfrac{5\sqrt{89}}{\sqrt{89}}$

$\cos \theta = \dfrac{8}{\sqrt{89}} = \dfrac{8\sqrt{89}}{\sqrt{89}}$

$\tan \theta = \dfrac{5}{8}$

$\csc \theta = \dfrac{\sqrt{89}}{5}$

$\sec \theta = \dfrac{\sqrt{89}}{8}$

$\cot \theta = \dfrac{3}{5}$

32. $\sin \dfrac{\pi}{6} + \tan^2 \dfrac{\pi}{3} = \dfrac{1}{2} + \left(\sqrt{3}\right)^2$

$= \dfrac{1}{2} + 3$

$= \dfrac{7}{2}$

33. $\cos^2 \dfrac{\pi}{4} + \tan^2 \dfrac{\pi}{4} = \left(\dfrac{\sqrt{2}}{2}\right)^2 - (1)^2$

$= \dfrac{1}{2} - 1$

$= -\dfrac{1}{2}$

34. $\sec^2 \dfrac{\pi}{5} - \tan^2 \dfrac{\pi}{5} = 1$

35. $\cos \dfrac{2\pi}{9} \sec \dfrac{2\pi}{9} = 1$

36. $\sin 70° = \cos\left(90° - 70°\right) = \cos 20°$

37. $\cos \dfrac{\pi}{2} = \sin\left(\dfrac{\pi}{2} - \dfrac{\pi}{2}\right) = \sin 0$

38. $\tan 23° = \dfrac{a}{100}$

$a = 100 \tan 23°$

$a \approx 100(0.4245) \approx 42 \, \text{mm}$

39. $\sin 61° = \dfrac{20}{c}$

$c = \dfrac{20}{\sin 61°}$

$c \approx \dfrac{20}{0.8746} \approx 23\,\text{cm}$

40. $\sin 48° = \dfrac{a}{50}$

$a = 50\sin 48°$

$a \approx 50(0.7431) \approx 37\,\text{in.}$

41. $\sin\theta = \dfrac{y}{r} = \dfrac{1}{4}$

$x^2 + y^2 = r^2$

$x^2 + 1^2 = 4^2$

$x^2 = 15$

$x = \sqrt{15}$

$\tan\left(\dfrac{\pi}{2} - \theta\right) = \cot\theta = \dfrac{x}{y} = \dfrac{\sqrt{15}}{1} = \sqrt{15}$

42. $\dfrac{1}{2}\,\text{mi.} = \dfrac{1}{2} \cdot 5280\,\text{ft} = 2640\,\text{ft}$

$\sin 17° = \dfrac{a}{2640}$

$a = 2640 \cdot \sin 17°$

$a \approx 2640(0.2924) \approx 772$

The hiker gains 772 feet of altitude.

43. $\tan 32° = \dfrac{d}{50}$

$d = 50\tan 32°$

$d \approx 50(0.6249) \approx 31$

The distance across the lake is about 31 meters.

44. $\tan\theta = \dfrac{6}{4}$

Use a calculator in degree mode to find θ.

Scientific Calculator
6 ÷ 4 = TAN⁻¹

Graphing Calculator
TAN⁻¹ (6 ÷ 4) ENTER

The display should show approximately 56. Thus, the angle of elevation of the sun is approximately 56°.

45. We need values for x, y, and r. Because $P = (-1, -5)$ is a point on the terminal side of θ, $x = -1$ and $y = -5$. Furthermore,

$r = \sqrt{(-1)^2 + (-5)^2}$

$= \sqrt{1 + 25} = \sqrt{26}$

Now that we know x, y, and r, we can find the six trigonometric functions of θ.

$\sin\theta = \dfrac{y}{r} = \dfrac{-5}{\sqrt{26}} = \dfrac{-5\sqrt{26}}{\sqrt{26} \cdot \sqrt{26}} = -\dfrac{5\sqrt{26}}{26}$

$\cos\theta = \dfrac{x}{r} = \dfrac{-1}{\sqrt{26}} = \dfrac{-1\sqrt{26}}{\sqrt{26} \cdot \sqrt{26}} = -\dfrac{\sqrt{26}}{26}$

$\tan\theta = \dfrac{y}{x} = \dfrac{-5}{-1} = 5$

$\csc\theta = \dfrac{r}{y} = \dfrac{\sqrt{26}}{-5} = -\dfrac{\sqrt{26}}{5}$

$\sec\theta = \dfrac{r}{x} = \dfrac{\sqrt{26}}{-1} = -\sqrt{26}$

$\cot\theta = \dfrac{x}{y} = \dfrac{-1}{-5} = \dfrac{1}{5}$

46. We need values for x, y, and r. Because $P = (0, -1)$ is a point on the terminal side of θ, $x = 0$ and $y = -1$. Furthermore,

$r = \sqrt{x^2 + y^2} = \sqrt{0^2 + (-1)^2}$

$= \sqrt{0 + 1} = \sqrt{1} = 1$

Now that we know x, y, and r, we can find the six trigonometric functions of θ.

$\sin\theta = \dfrac{y}{r} = \dfrac{-1}{1} = -1$

$\cos\theta = \dfrac{x}{r} = \dfrac{0}{1} = 0$

$\tan\theta = \dfrac{y}{x} = \dfrac{-1}{0}$, undefined

$\csc\theta = \dfrac{r}{y} = \dfrac{1}{-1} = -1$

$\sec\theta = \dfrac{r}{x} = \dfrac{1}{0}$, undefined

$\cot\theta = \dfrac{x}{y} = \dfrac{0}{-1} = 0$

47. Because $\tan\theta > 0$, θ cannot lie in quadrant II and quadrant IV; the tangent function is negative in those two quadrants. Thus, with $\tan\theta > 0$, θ lies in quadrant I or quadrant III. We are also given that $\sec\theta > 0$. Because quadrant I is the only quadrant in which the tangent is positive and the secant is positive, we conclude that θ lies in quadrant I.

48. Because $\tan\theta > 0$, θ cannot lie in quadrant II and quadrant IV; the tangent function is negative in those two quadrants. Thus, with $\tan\theta > 0$, θ lies in quadrant I or quadrant III. We are also given that $\cos\theta < 0$. Because quadrant III is the only quadrant in which the tangent is positive and the cosine is negative, we conclude that θ lies in quadrant III.

49. Because the cosine is positive and the sine is negative, θ lies in quadrant IV. In quadrant IV, x is positive and y is negative. Thus, $\cos\theta = \dfrac{2}{5} = \dfrac{x}{r}$, $x = 2$, $r = 5$. Furthermore,
$$x^2 + y^2 = r^2$$
$$2^2 + y^2 = 5^2$$
$$y^2 = 25 - 4 = 21$$
$$y = -\sqrt{21}$$

Now that we know x, y, and r, we can find the six trigonometric functions of θ.

$$\sin\theta = \frac{y}{r} = \frac{-\sqrt{21}}{5} = -\frac{\sqrt{21}}{5}$$

$$\tan\theta = \frac{y}{x} = \frac{-\sqrt{21}}{2} = -\frac{\sqrt{21}}{2}$$

$$\csc\theta = \frac{r}{y} = \frac{5}{-\sqrt{21}} = -\frac{5\cdot\sqrt{21}}{\sqrt{21}\cdot\sqrt{21}} = -\frac{5\sqrt{21}}{21}$$

$$\sec\theta = \frac{r}{x} = \frac{5}{2}$$

$$\cot\theta = \frac{x}{y} = \frac{2}{-\sqrt{21}} = -\frac{2\sqrt{21}}{\sqrt{21}\cdot\sqrt{21}} = -\frac{2\sqrt{21}}{21}$$

50. Because the tangent is negative and the sine is positive, θ lies in quadrant II. In quadrant II x is negative and y is positive. Thus,
$$\tan\theta = -\frac{1}{3} = \frac{y}{x} = \frac{1}{-3}, \quad x = -3, \; y = 1.$$
Furthermore,
$$r = \sqrt{x^2 + y^2} = \sqrt{(-3)^2 + 1^2} = \sqrt{9+1} = \sqrt{10}$$
Now that we know x, y, and r, we can find the six trigonometric functions of θ.

$$\sin\theta = \frac{y}{r} = \frac{1}{\sqrt{10}} = \frac{1\cdot\sqrt{10}}{\sqrt{10}\cdot\sqrt{10}} = \frac{\sqrt{10}}{10}$$

$$\cos\theta = \frac{x}{r} = \frac{-3}{\sqrt{10}} = -\frac{3\sqrt{10}}{\sqrt{10}\cdot\sqrt{10}} = -\frac{3\sqrt{10}}{10}$$

$$\csc\theta = \frac{r}{y} = \frac{\sqrt{10}}{1} = \sqrt{10}$$

$$\sec\theta = \frac{r}{x} = \frac{\sqrt{10}}{-3} = -\frac{\sqrt{10}}{3}$$

$$\cot\theta = \frac{x}{y} = \frac{-3}{1} = -3$$

51. Because the cotangent is positive and the cosine is negative, θ lies in quadrant III. In quadrant III x and y are both negative. Thus,
$$\cot\theta = \frac{3}{1} = \frac{x}{y} = \frac{-3}{-1}, \quad x = -3, \; y = -1.$$
Furthermore,
$$r = \sqrt{x^2 + y^2} = \sqrt{(-3)^2 + (-1)^2} = \sqrt{9+1} = \sqrt{10}$$
Now that we know x, y, and r, we can find the six trigonometric functions of θ.

$$\sin\theta = \frac{y}{r} = \frac{-1}{\sqrt{10}} = -\frac{\sqrt{10}}{10}$$

$$\cos\theta = \frac{x}{r} = \frac{-3}{\sqrt{10}} = -\frac{3\sqrt{10}}{10}$$

$$\tan\theta = \frac{y}{x} = \frac{-1}{-3} = \frac{1}{3}$$

$$\csc\theta = \frac{r}{y} = \frac{\sqrt{10}}{-1} = -\sqrt{10}$$

$$\sec\theta = \frac{r}{x} = \frac{\sqrt{10}}{-3} = -\frac{\sqrt{10}}{3}$$

52. Because $265°$ lies between $180°$ and $270°$, it is in quadrant III.
The reference angle is $\theta' = 265° - 180° = 85°$.

53. Because $\dfrac{5\pi}{8}$ lies between $\dfrac{\pi}{2} = \dfrac{4\pi}{8}$ and $\pi = \dfrac{8\pi}{8}$, it is in quadrant II.

The reference angle is $\theta' = \pi - \dfrac{5\pi}{8} = \dfrac{8\pi}{8} - \dfrac{5\pi}{8} = \dfrac{3\pi}{8}$.

54. Find the coterminal angle: $-410° + (2)360° = 310°$

Find the reference angle: $360° - 310° = 50°$

55. Find the coterminal angle: $\dfrac{17\pi}{6} - 2\pi = \dfrac{5\pi}{6}$

Find the reference angle: $2\pi - \dfrac{5\pi}{6} = \dfrac{\pi}{6}$

56. Find the coterminal angle: $-\dfrac{11\pi}{3} + 4\pi = \dfrac{\pi}{3}$

Find the reference angle: $\dfrac{\pi}{3}$

57. $240°$ lies in quadrant III.
The reference angle is
$\theta' = 240° - 180° = 60°$.

$\sin 60° = \dfrac{\sqrt{3}}{2}$

In quadrant III, $\sin \theta < 0$, so

$\sin 240° = -\sin 60° = -\dfrac{\sqrt{3}}{2}$.

58. $120°$ lies in quadrant II.
The reference angle is
$\theta' = 180° - 120° = 60°$.

$\tan 60° = \sqrt{3}$

In quadrant II, $\tan \theta < 0$, so
$\tan 120° = -\tan 60° = -\sqrt{3}$.

59. $\dfrac{7\pi}{4}$ lies in quadrant IV.
The reference angle is

$\theta' = 2\pi - \dfrac{7\pi}{4} = \dfrac{8\pi}{4} - \dfrac{7\pi}{4} = \dfrac{\pi}{4}$.

$\sec \dfrac{\pi}{4} = \sqrt{2}$

In quadrant IV, $\sec \theta > 0$, so

$\sec \dfrac{7\pi}{4} = \sec \dfrac{\pi}{4} = \sqrt{2}$.

60. $\dfrac{11\pi}{6}$ lies in quadrant IV.
The reference angle is

$\theta' = 2\pi - \dfrac{11\pi}{6} = \dfrac{12\pi}{6} - \dfrac{11\pi}{6} = \dfrac{\pi}{6}$.

$\cos \dfrac{\pi}{6} = \dfrac{\sqrt{3}}{2}$

In quadrant IV, $\cos \theta > 0$, so $\cos \dfrac{11\pi}{6} = \cos \dfrac{\pi}{6} = \dfrac{\sqrt{3}}{2}$.

61. $-210°$ lies in quadrant II.
The reference angle is
$\theta' = 210° - 180° = 30°$.

$\cot 30° = \sqrt{3}$
In quadrant II, $\cot \theta < 0$, so
$\cot(-210°) = -\cot 30° = -\sqrt{3}$.

62. $-\dfrac{2\pi}{3}$ lies in quadrant III.

The reference angle is

$\theta' = \pi + \dfrac{-2\pi}{3} = \dfrac{3\pi}{3} - \dfrac{2\pi}{3} = \dfrac{\pi}{3}$.

$\csc\left(\dfrac{\pi}{3}\right) = \dfrac{2\sqrt{3}}{3}$

In quadrant III, $\csc \theta < 0$, so

$\csc\left(-\dfrac{2\pi}{3}\right) = -\csc\left(\dfrac{\pi}{3}\right) = -\dfrac{2\sqrt{3}}{3}$.

63. $-\dfrac{\pi}{3}$ lies in quadrant IV.

The reference angle is

$\theta' = \dfrac{\pi}{3}$.

$\sin\left(\dfrac{\pi}{3}\right) = \dfrac{\sqrt{3}}{2}$

In quadrant IV, $\sin \theta < 0$, so

$\sin\left(-\dfrac{\pi}{3}\right) = -\sin\left(\dfrac{\pi}{3}\right) = -\dfrac{\sqrt{3}}{2}$.

64. $495°$ lies in quadrant II.
$495° - 360° = 135°$
The reference angle is
$\theta' = 180° - 135° = 45°$.
$\sin 45° = \dfrac{\sqrt{2}}{2}$
In quadrant II, $\sin \theta > 0$, so
$\sin 495° = \sin 45° = \dfrac{\sqrt{2}}{2}$.

65. $\dfrac{13\pi}{4}$ lies in quadrant III.
$\dfrac{13\pi}{4} - 2\pi = \dfrac{13\pi}{4} - \dfrac{8\pi}{4} = \dfrac{5\pi}{4}$
The reference angle is
$\theta' = \dfrac{5\pi}{4} - \pi = \dfrac{5\pi}{4} - \dfrac{4\pi}{4} = \dfrac{\pi}{4}$.
$\tan \dfrac{\pi}{4} = 1$

In quadrant III, $\tan \theta > 0$, so $\tan \dfrac{13\pi}{4} = \tan \dfrac{\pi}{4} = 1$.

66. $\sin \dfrac{22\pi}{3} = \sin \left(\dfrac{22\pi}{3} - 6\pi \right)$

$= \sin \dfrac{4\pi}{3}$

$= -\sin \dfrac{\pi}{3}$

$= -\dfrac{\sqrt{3}}{2}$

67. $\cos \left(-\dfrac{35\pi}{6} \right) = \cos \left(-\dfrac{35\pi}{6} + 6\pi \right)$

$= \cos \dfrac{\pi}{6}$

$= \dfrac{\sqrt{3}}{2}$

68. The equation $y = 3\sin 4x$ is of the form $y = A\sin Bx$ with $A = 3$ and $B = 4$. The amplitude is $|A| = |3| = 3$.

The period is $\dfrac{2\pi}{B} = \dfrac{2\pi}{4} = \dfrac{\pi}{2}$. The quarter-period is

$\dfrac{\frac{\pi}{2}}{4} = \dfrac{\pi}{2} \cdot \dfrac{1}{4} = \dfrac{\pi}{8}$. The cycle begins at $x = 0$. Add quarter-periods to generate x-values for the key points.
$x = 0$

$x = 0 + \dfrac{\pi}{8} = \dfrac{\pi}{8}$

$x = \dfrac{\pi}{8} + \dfrac{\pi}{8} = \dfrac{\pi}{4}$

$x = \dfrac{\pi}{4} + \dfrac{\pi}{8} = \dfrac{3\pi}{8}$

$x = \dfrac{3\pi}{8} + \dfrac{\pi}{8} = \dfrac{\pi}{2}$

Evaluate the function at each value of x.

x	coordinates
0	$(0, 0)$
$\dfrac{\pi}{8}$	$\left(\dfrac{\pi}{8}, 3 \right)$
$\dfrac{\pi}{4}$	$\left(\dfrac{\pi}{4}, 0 \right)$
$\dfrac{3\pi}{8}$	$\left(\dfrac{3\pi}{8}, -3 \right)$
$\dfrac{\pi}{2}$	$(2\pi, 0)$

Connect the five key points with a smooth curve and graph one complete cycle of the given function.

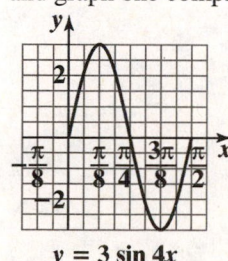

$y = 3 \sin 4x$

69. The equation $y = -2\cos 2x$ is of the form $y = A\cos Bx$ with $A = -2$ and $B = 2$. The amplitude is $|A| = |-2| = 2$. The period is $\dfrac{2\pi}{B} = \dfrac{2\pi}{2} = \pi$. The quarter-period is $\dfrac{\pi}{4}$. The cycle begins at $x = 0$. Add quarter-periods to generate x-values for the key points.

$x = 0$

$x = 0 + \dfrac{\pi}{4} = \dfrac{\pi}{4}$

$x = \dfrac{\pi}{4} + \dfrac{\pi}{4} = \dfrac{\pi}{2}$

$x = \dfrac{\pi}{2} + \dfrac{\pi}{4} = \dfrac{3\pi}{4}$

$x = \dfrac{3\pi}{4} + \dfrac{\pi}{4} = \pi$

Evaluate the function at each value of x.

x	coordinates
0	$(0, -2)$
$\dfrac{\pi}{4}$	$\left(\dfrac{\pi}{4}, 0\right)$
$\dfrac{\pi}{2}$	$\left(\dfrac{\pi}{2}, 2\right)$
$\dfrac{3\pi}{4}$	$\left(\dfrac{3\pi}{4}, 0\right)$
π	$(\pi, -2)$

Connect the five key points with a smooth curve and graph one complete cycle of the given function.

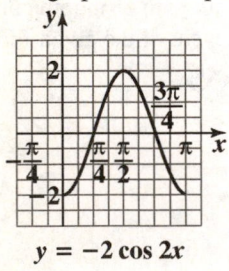

$y = -2\cos 2x$

70. The equation $y = 2\cos\dfrac{1}{2}x$ is of the form $y = A\cos Bx$ with $A = 2$ and $B = \dfrac{1}{2}$. The amplitude is $|A| = |2| = 2$. The period is $\dfrac{2\pi}{B} = \dfrac{2\pi}{\frac{1}{2}} = 2\pi \cdot 2 = 4\pi$.

The quarter-period is $\dfrac{4\pi}{4} = \pi$. The cycle begins at $x = 0$. Add quarter-periods to generate x-values for the key points.

$x = 0$

$x = 0 + \pi = \pi$

$x = \pi + \pi = 2\pi$

$x = 2\pi + \pi = 3\pi$

$x = 3\pi + \pi = 4\pi$

Evaluate the function at each value of x.

x	coordinates
0	$(0, 2)$
π	$(\pi, 0)$
2π	$(2\pi, -2)$
3π	$(3\pi, 0)$
4π	$(4\pi, 2)$

Connect the five key points with a smooth curve and graph one complete cycle of the given function.

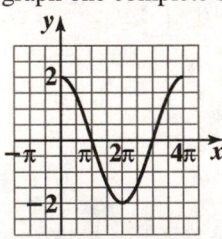

$y = 2\cos\dfrac{1}{2}x$

71. The equation $y = \dfrac{1}{2}\sin\dfrac{\pi}{3}x$ is of the form

$y = A\sin Bx$ with $A = \dfrac{1}{2}$ and $B = \dfrac{\pi}{3}$. The amplitude

is $|A| = \left|\dfrac{1}{2}\right| = \dfrac{1}{2}$. The period is $\dfrac{2\pi}{B} = \dfrac{2\pi}{\frac{\pi}{3}} = 2\pi \cdot \dfrac{3}{\pi} = 6.$

The quarter-period is $\dfrac{6}{4} = \dfrac{3}{2}$. The cycle begins at $x =$

0. Add quarter-periods to generate x-values for the key points.

$x = 0$

$x = 0 + \dfrac{3}{2} = \dfrac{3}{2}$

$x = \dfrac{3}{2} + \dfrac{3}{2} = 3$

$x = 3 + \dfrac{3}{2} = \dfrac{9}{2}$

$x = \dfrac{9}{2} + \dfrac{3}{2} = 6$

Evaluate the function at each value of x.

x	coordinates
0	$(0, 0)$
$\dfrac{3}{2}$	$\left(\dfrac{3}{2}, \dfrac{1}{2}\right)$
3	$(3, 0)$
$\dfrac{9}{2}$	$\left(\dfrac{9}{2}, -\dfrac{1}{2}\right)$
6	$(6, 0)$

Connect the five key points with a smooth curve and graph one complete cycle of the given function.

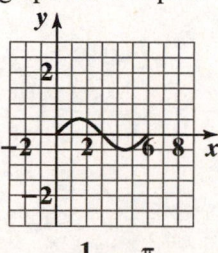

$y = \dfrac{1}{2}\sin\dfrac{\pi}{3}x$

72. The equation $y = -\sin\pi x$ is of the form

$y = A\sin Bx$ with $A = -1$ and $B = \pi$. The amplitude

is $|A| = |-1| = 1$. The period is $\dfrac{2\pi}{B} = \dfrac{2\pi}{\pi} = 2.$ The

quarter-period is $\dfrac{2}{4} = \dfrac{1}{2}$. The cycle begins at $x = 0$.

Add quarter-periods to generate x-values for the key points.

$x = 0$

$x = 0 + \dfrac{1}{2} = \dfrac{1}{2}$

$x = \dfrac{1}{2} + \dfrac{1}{2} = 1$

$x = 1 + \dfrac{1}{2} = \dfrac{3}{2}$

$x = \dfrac{3}{2} + \dfrac{1}{2} = 2$

Evaluate the function at each value of x.

x	coordinates
0	$(0, 0)$
$\dfrac{1}{2}$	$\left(\dfrac{1}{2}, -1\right)$
1	$(1, 0)$
$\dfrac{3}{2}$	$\left(\dfrac{3}{2}, 1\right)$
2	$(2, 0)$

Connect the five key points with a smooth curve and graph one complete cycle of the given function.

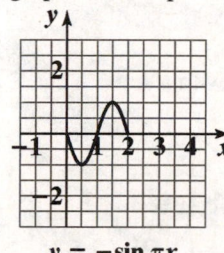

$y = -\sin\pi x$

73. The equation $y = 3\cos\dfrac{x}{3}$ is of the form $y = A\cos Bx$ with $A = 3$ and $B = \dfrac{1}{3}$. The amplitude is $|A| = |3| = 3$.

The period is $\dfrac{2\pi}{B} = \dfrac{2\pi}{\frac{1}{3}} = 2\pi \cdot 3 = 6\pi$. The quarter-period is $\dfrac{6\pi}{4} = \dfrac{3\pi}{2}$. The cycle begins at $x = 0$. Add quarter-periods to generate x-values for the key points.

$x = 0$

$x = 0 + \dfrac{3\pi}{2} = \dfrac{3\pi}{2}$

$x = \dfrac{3\pi}{2} + \dfrac{3\pi}{2} = 3\pi$

$x = 3\pi + \dfrac{3\pi}{2} = \dfrac{9\pi}{2}$

$x = \dfrac{9\pi}{2} + \dfrac{3\pi}{2} = 6\pi$

Evaluate the function at each value of x.

x	coordinates
0	$(0, 3)$
$\dfrac{3\pi}{2}$	$\left(\dfrac{3\pi}{2}, 0\right)$
3π	$(3\pi, -3)$
$\dfrac{9\pi}{2}$	$\left(\dfrac{9\pi}{2}, 0\right)$
6π	$(6\pi, 3)$

Connect the five key points with a smooth curve and graph one complete cycle of the given function.

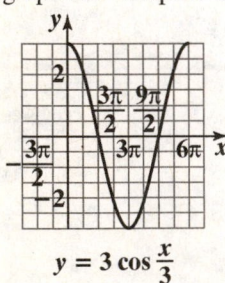

$y = 3\cos\dfrac{x}{3}$

74. The equation $y = 2\sin(x - \pi)$ is of the form $y = A\sin(Bx - C)$ with $A = 2$, $B = 1$, and $C = \pi$. The amplitude is $|A| = |2| = 2$. The period is $\dfrac{2\pi}{B} = \dfrac{2\pi}{1} = 2\pi$. The phase shift is $\dfrac{C}{B} = \dfrac{\pi}{1} = \pi$. The quarter-period is $\dfrac{2\pi}{4} = \dfrac{\pi}{2}$.

The cycle begins at $x = \pi$. Add quarter-periods to generate x-values for the key points.

$x = \pi$

$x = \pi + \dfrac{\pi}{2} = \dfrac{3\pi}{2}$

$x = \dfrac{3\pi}{2} + \dfrac{\pi}{2} = 2\pi$

$x = 2\pi + \dfrac{\pi}{2} = \dfrac{5\pi}{2}$

$x = \dfrac{5\pi}{2} + \dfrac{\pi}{2} = 3\pi$

Evaluate the function at each value of x.

x	coordinates
π	$(\pi, 0)$
$\dfrac{3\pi}{2}$	$\left(\dfrac{3\pi}{2}, 2\right)$
2π	$(2\pi, 0)$
$\dfrac{5\pi}{2}$	$\left(\dfrac{5\pi}{2}, -2\right)$
3π	$(3\pi, 0)$

Connect the five key points with a smooth curve and graph one complete cycle of the given function.

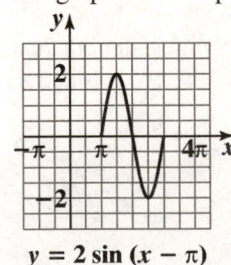

$y = 2\sin(x - \pi)$

75. $y = -3\cos(x + \pi) = -3\cos(x - (-\pi))$

The equation $y = -3\cos(x - (-\pi))$ is of the form

$y = A\cos(Bx - C)$ with $A = -3$, $B = 1$, and $C = -\pi$.

The amplitude is $|A| = |-3| = 3$.

The period is $\dfrac{2\pi}{B} = \dfrac{2\pi}{1} = 2\pi$. The phase shift is

$\dfrac{C}{B} = \dfrac{-\pi}{1} = -\pi$. The quarter-period is $\dfrac{2\pi}{4} = \dfrac{\pi}{2}$. The

cycle begins at $x = -\pi$. Add quarter-periods to generate x-values for the key points.

$x = -\pi$

$x = -\pi + \dfrac{\pi}{2} = -\dfrac{\pi}{2}$

$x = -\dfrac{\pi}{2} + \dfrac{\pi}{2} = 0$

$x = 0 + \dfrac{\pi}{2} = \dfrac{\pi}{2}$

$x = \dfrac{\pi}{2} + \dfrac{\pi}{2} = \pi$

Evaluate the function at each value of x.

x	coordinates
$-\pi$	$(-\pi, -3)$
$-\dfrac{\pi}{2}$	$\left(-\dfrac{\pi}{2}, 0\right)$
0	$(0, 3)$
$\dfrac{\pi}{2}$	$\left(\dfrac{\pi}{2}, 0\right)$
π	$(\pi, -3)$

Connect the five key points with a smooth curve and graph one complete cycle of the given function.

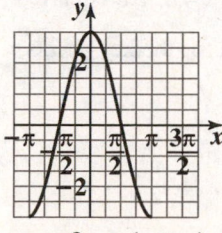

$y = -3\cos(x + \pi)$

76. $y = \dfrac{3}{2}\cos\left(2x + \dfrac{\pi}{4}\right) = \dfrac{3}{2}\cos\left(2x - \left(-\dfrac{\pi}{4}\right)\right)$

The equation $y = \dfrac{3}{2}\cos\left(2x - \left(-\dfrac{\pi}{4}\right)\right)$ is of

the form $y = A\cos(Bx - C)$ with $A = \dfrac{3}{2}$,

$B = 2$, and $C = -\dfrac{\pi}{4}$. The amplitude is

$|A| = \left|\dfrac{3}{2}\right| = \dfrac{3}{2}$.

The period is $\dfrac{2\pi}{B} = \dfrac{2\pi}{2} = \pi$. The phase shift is

$\dfrac{C}{B} = \dfrac{-\frac{\pi}{4}}{2} = -\dfrac{\pi}{4} \cdot \dfrac{1}{2} = -\dfrac{\pi}{8}$. The quarter-period is $\dfrac{\pi}{4}$.

The cycle begins at $x = -\dfrac{\pi}{8}$. Add quarter-periods to

generate x-values for the key points.

$x = -\dfrac{\pi}{8}$

$x = -\dfrac{\pi}{8} + \dfrac{\pi}{4} = \dfrac{\pi}{8}$

$x = \dfrac{\pi}{8} + \dfrac{\pi}{4} = \dfrac{3\pi}{8}$

$x = \dfrac{3\pi}{8} + \dfrac{\pi}{4} = \dfrac{5\pi}{8}$

$x = \dfrac{5\pi}{8} + \dfrac{\pi}{4} = \dfrac{7\pi}{8}$

Evaluate the function at each value of x.

x	coordinates
$-\dfrac{\pi}{8}$	$\left(-\dfrac{\pi}{8}, \dfrac{3}{2}\right)$
$\dfrac{\pi}{8}$	$\left(\dfrac{\pi}{8}, 0\right)$
$\dfrac{3\pi}{8}$	$\left(\dfrac{3\pi}{8}, -\dfrac{3}{2}\right)$
$\dfrac{5\pi}{8}$	$\left(\dfrac{5\pi}{8}, 0\right)$
$\dfrac{7\pi}{8}$	$\left(\dfrac{7\pi}{8}, \dfrac{3}{2}\right)$

Connect the five key points with a smooth curve and graph one complete cycle of the given function.

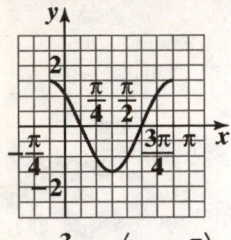

$$y = \frac{3}{2}\cos\left(2x + \frac{\pi}{4}\right)$$

77. $y = \frac{5}{2}\sin\left(2x + \frac{\pi}{2}\right) = \frac{5}{2}\sin\left(2x - \left(-\frac{\pi}{2}\right)\right)$

The equation $y = \frac{5}{2}\sin\left(2x - \left(-\frac{\pi}{2}\right)\right)$ is of

the form $y = A\sin(Bx - C)$ with $A = \frac{5}{2}$,

$B = 2$, and $C = -\frac{\pi}{2}$. The amplitude is

$|A| = \left|\frac{5}{2}\right| = \frac{5}{2}$.

The period is $\frac{2\pi}{B} = \frac{2\pi}{2} = \pi$. The phase shift is

$\frac{C}{B} = \frac{-\frac{\pi}{2}}{2} = -\frac{\pi}{2}\cdot\frac{1}{2} = -\frac{\pi}{4}$. The quarter-period is $\frac{\pi}{4}$.

The cycle begins at $x = -\frac{\pi}{4}$. Add quarter-periods to generate x-values for the key points.

$x = -\frac{\pi}{4}$

$x = -\frac{\pi}{4} + \frac{\pi}{4} = 0$

$x = 0 + \frac{\pi}{4} = \frac{\pi}{4}$

$x = \frac{\pi}{4} + \frac{\pi}{4} = \frac{\pi}{2}$

$x = \frac{\pi}{2} + \frac{\pi}{4} = \frac{3\pi}{4}$

Evaluate the function at each value of x.

x	coordinates
$-\dfrac{\pi}{4}$	$\left(-\dfrac{\pi}{4}, 0\right)$
0	$\left(0, \dfrac{5}{2}\right)$
$\dfrac{\pi}{4}$	$\left(\dfrac{\pi}{4}, 0\right)$
$\dfrac{\pi}{2}$	$\left(\dfrac{\pi}{2}, -\dfrac{5}{2}\right)$
$\dfrac{3\pi}{4}$	$\left(\dfrac{3\pi}{4}, 0\right)$

Connect the five key points with a smooth curve and graph one complete cycle of the given function.

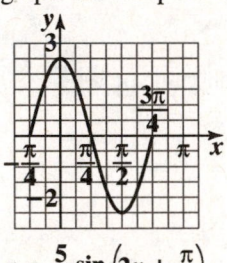

$$y = \frac{5}{2}\sin\left(2x + \frac{\pi}{2}\right)$$

78. The equation $y = -3\sin\left(\frac{\pi}{3}x - 3\pi\right)$ is of

the form $y = A\sin(Bx - C)$ with $A = -3$,

$B = \frac{\pi}{3}$, and $C = 3\pi$. The amplitude is $|A| = |-3| = 3$.

The period is $\frac{2\pi}{B} = \frac{2\pi}{\frac{\pi}{3}} = 2\pi\cdot\frac{3}{\pi} = 6$. The phase shift

is $\frac{C}{B} = \frac{3\pi}{\frac{\pi}{3}} = 3\pi\cdot\frac{3}{\pi} = 9$. The quarter-period is

$\frac{6}{4} = \frac{3}{2}$. The cycle begins at $x = 9$. Add quarter-

periods to generate x-values for the key points.

$x = 9$

$x = 9 + \dfrac{3}{2} = \dfrac{21}{2}$

$x = \dfrac{21}{2} + \dfrac{3}{2} = 12$

$x = 12 + \dfrac{3}{2} = \dfrac{27}{2}$

$x = \dfrac{27}{2} + \dfrac{3}{2} = 15$

Evaluate the function at each value of x.

x	coordinates
9	$(9, 0)$
$\dfrac{21}{2}$	$\left(\dfrac{21}{2}, -3\right)$
12	$(12, 0)$
$\dfrac{27}{2}$	$\left(\dfrac{27}{2}, 3\right)$
15	$(15, 0)$

Connect the five key points with a smooth curve and graph one complete cycle of the given function.

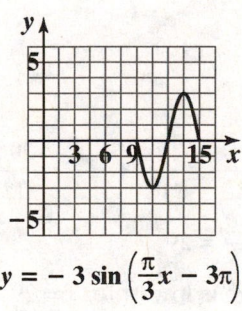

$y = -3\sin\left(\dfrac{\pi}{3}x - 3\pi\right)$

79. The graph of $y = \sin 2x + 1$ is the graph of $y = \sin 2x$ shifted one unit upward. The period for both functions is $\dfrac{2\pi}{2} = \pi$. The quarter-period is $\dfrac{\pi}{4}$. The cycle begins at $x = 0$. Add quarter-periods to generate x-values for the key points.

$x = 0$

$x = 0 + \dfrac{\pi}{4} = \dfrac{\pi}{4}$

$x = \dfrac{\pi}{4} + \dfrac{\pi}{4} = \dfrac{\pi}{2}$

$x = \dfrac{\pi}{2} + \dfrac{\pi}{4} = \dfrac{3\pi}{4}$

$x = \dfrac{3\pi}{4} + \dfrac{\pi}{4} = \pi$

Evaluate the function at each value of x.

x	coordinates
0	$(0, 1)$
$\dfrac{\pi}{4}$	$\left(\dfrac{\pi}{4}, 2\right)$
$\dfrac{\pi}{2}$	$\left(\dfrac{\pi}{2}, 1\right)$
$\dfrac{3\pi}{4}$	$\left(\dfrac{3\pi}{4}, 0\right)$
π	$(\pi, 1)$

By connecting the points with a smooth curve we obtain one period of the graph.

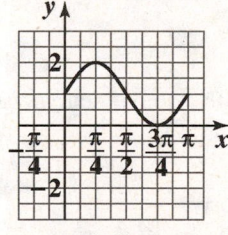

$y = \sin 2x + 1$

80. The graph of $y = 2\cos\dfrac{1}{3}x - 2$ is the graph of

$y = 2\cos\dfrac{1}{3}x$ shifted two units downward. The

period for both functions is $\dfrac{2\pi}{\frac{1}{3}} = 2\pi \cdot 3 = 6\pi$. The

quarter-period is $\dfrac{6\pi}{4} = \dfrac{3\pi}{2}$. The cycle begins at $x =$

0. Add quarter-periods to generate x-values for the
key points.

$x = 0$

$x = 0 + \dfrac{3\pi}{2} = \dfrac{3\pi}{2}$

$x = \dfrac{3\pi}{2} + \dfrac{3\pi}{2} = 3\pi$

$x = 3\pi + \dfrac{3\pi}{2} = \dfrac{9\pi}{2}$

$x = \dfrac{9\pi}{2} + \dfrac{3\pi}{2} = 6\pi$

Evaluate the function at each value of x.

x	coordinates
0	$(0, 0)$
$\dfrac{3\pi}{2}$	$\left(\dfrac{3\pi}{2}, -2\right)$
3π	$(3\pi, -4)$
$\dfrac{9\pi}{2}$	$\left(\dfrac{9\pi}{2}, -2\right)$
6π	$(6\pi, 0)$

By connecting the points with a smooth curve we
obtain one period of the graph.

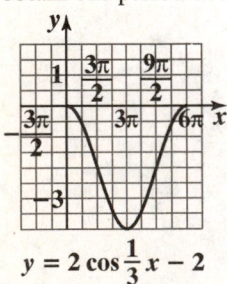

$y = 2\cos\dfrac{1}{3}x - 2$

81. a. At midnight $x = 0$. Thus,

$$y = 98.6 + 0.3\sin\left(\dfrac{\pi}{12} \cdot 0 - \dfrac{11\pi}{12}\right)$$

$$= 98.6 + 0.3\sin\left(-\dfrac{11\pi}{12}\right)$$

$$\approx 98.6 + 0.3(-0.2588) \approx 98.52$$

The body temperature is about 98.52°F.

b. period: $\dfrac{2\pi}{B} = \dfrac{2\pi}{\frac{\pi}{12}} = 2\pi \cdot \dfrac{12}{\pi} = 24$ hours

c. Solve the equation

$$\dfrac{\pi}{12}x - \dfrac{11\pi}{12} = \dfrac{\pi}{2}$$

$$\dfrac{\pi}{12}x = \dfrac{\pi}{2} + \dfrac{11\pi}{12} = \dfrac{6\pi}{12} + \dfrac{11\pi}{12} = \dfrac{17\pi}{12}$$

$$x = \dfrac{17\pi}{12} \cdot \dfrac{12}{\pi} = 17$$

The body temperature is highest for
$x = 17$.

$$y = 98.6 + 0.3\sin\left(\dfrac{\pi}{12} \cdot 17 - \dfrac{11\pi}{12}\right)$$

$$= 98.6 + 0.3\sin\dfrac{\pi}{2} = 98.6 + 0.3 = 98.9$$

17 hours after midnight, which is
5 P.M., the body temperature is 98.9°F.

d. Solve the equation

$$\dfrac{\pi}{12}x - \dfrac{11\pi}{12} = \dfrac{3\pi}{2}$$

$$\dfrac{\pi}{12}x = \dfrac{3\pi}{2} + \dfrac{11\pi}{12} = \dfrac{18\pi}{12} + \dfrac{11\pi}{12} = \dfrac{29\pi}{12}$$

$$x = \dfrac{29\pi}{12} \cdot \dfrac{12}{\pi} = 29$$

The body temperature is lowest for $x = 29$.

$$y = 98.6 + 0.3\sin\left(\dfrac{\pi}{12} \cdot 29 - \dfrac{11\pi}{12}\right)$$

$$= 98.6 + 0.3\sin\left(\dfrac{3\pi}{2}\right)$$

$$= 98.6 + 0.3(-1) = 98.3°$$

29 hours after midnight or 5 hours after
midnight, at 5 A.M., the body temperature is
98.3°F.

e. The graph of $y = 98.6 + 0.3 \sin\left(\dfrac{\pi}{12}x - \dfrac{11\pi}{12}\right)$ is

of the form $y = D + A\sin(Bx - C)$ with $A = 0.3$,

$B = \dfrac{\pi}{12}$, $C = \dfrac{11\pi}{12}$, and $D = 98.6$. The amplitude

is $|A| = |0.3| = 0.3$. The period from part (b)

is 24. The quarter-period is $\dfrac{24}{4} = 6$. The phase

shift is $\dfrac{C}{B} = \dfrac{\frac{11\pi}{12}}{\frac{\pi}{12}} = \dfrac{11\pi}{12} \cdot \dfrac{12}{\pi} = 11$. The cycle

begins at $x = 11$. Add quarter-periods to
generate x-values for the key points.

$x = 11$

$x = 11 + 6 = 17$

$x = 17 + 6 = 23$

$x = 23 + 6 = 29$

$x = 29 + 6 = 35$

Evaluate the function at each value of x. The
key points are (11, 98.6), (17, 98.9), (23, 98.6),
(29, 98.3), (35, 98.6). Extend the pattern to the
left, and graph the function for $0 \le x \le 24$.

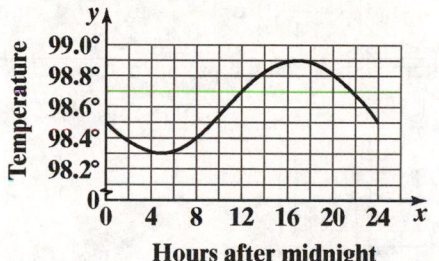

82. Blue:
This is a sine wave with a period of 480.
Since the amplitude is 1, $A = 1$.

$B = \dfrac{2\pi}{\text{period}} = \dfrac{2\pi}{480} = \dfrac{\pi}{240}$

The equation is $y = \sin\dfrac{\pi}{240}x$.

Red:
This is a sine wave with a period of 640.
Since the amplitude is 1, $A = 1$.

$B = \dfrac{2\pi}{\text{period}} = \dfrac{2\pi}{640} = \dfrac{\pi}{320}$

The equation is $y = \sin\dfrac{\pi}{320}x$.

83. Solve the equations

$2x = -\dfrac{\pi}{2}$ and $2x = \dfrac{\pi}{2}$

$x = \dfrac{-\frac{\pi}{2}}{2}$ $\qquad x = \dfrac{\frac{\pi}{2}}{2}$

$x = -\dfrac{\pi}{4}$ $\qquad x = \dfrac{\pi}{4}$

Thus, two consecutive asymptotes occur at

$x = -\dfrac{\pi}{4}$ and $x = \dfrac{\pi}{4}$.

x-intercept $= \dfrac{-\frac{\pi}{4} + \frac{\pi}{4}}{2} = \dfrac{0}{2} = 0$

An x-intercept is 0 and the graph passes through (0,
0). Because the coefficient of the tangent is 4, the
points on the graph midway between an x-intercept
and the asymptotes have y-coordinates of –4 and 4.

Use the two consecutive asymptotes. $x = -\dfrac{\pi}{4}$ and

$x = \dfrac{\pi}{4}$, to graph one full period of $y = 4\tan 2x$ from

$-\dfrac{\pi}{4}$ to $\dfrac{\pi}{4}$.

Continue the pattern and extend the graph another
full period to the right.

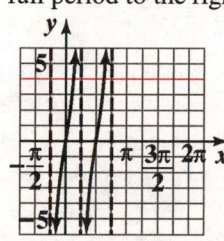

$y = 4 \tan 2x$

84. Solve the equations

$\dfrac{\pi}{4}x = -\dfrac{\pi}{2}$ and $\dfrac{\pi}{4}x = \dfrac{\pi}{2}$

$x = -\dfrac{\pi}{2} \cdot \dfrac{4}{\pi}$ $\qquad x = \dfrac{\pi}{2} \cdot \dfrac{4}{\pi}$

$x = -2$ $\qquad x = 2$

Thus, two consecutive asymptotes occur at
$x = -2$ and $x = 2$.

x-intercept $= \dfrac{-2 + 2}{2} = \dfrac{0}{2} = 0$

An x-intercept is 0 and the graph passes through
(0, 0). Because the coefficient of the tangent is –2,
the points on the graph midway between an x-
intercept and the asymptotes have y-coordinates of 2
and –2. Use the two consecutive asymptotes, $x = -2$

and $x = 2$, to graph one full period of $y = -2 \tan \dfrac{\pi}{4} x$

from -2 to 2. Continue the pattern and extend the graph another full period to the right.

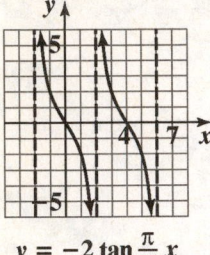

$$y = -2 \tan \frac{\pi}{4} x$$

85. Solve the equations

$$x + \pi = -\frac{\pi}{2} \quad \text{and} \quad x + \pi = \frac{\pi}{2}$$

$$x = -\frac{\pi}{2} - \pi \qquad x = \frac{\pi}{2} - \pi$$

$$x = -\frac{3\pi}{2} \qquad x = -\frac{\pi}{2}$$

Thus, two consecutive asymptotes occur at

$$x = -\frac{3\pi}{2} \text{ and } x = -\frac{\pi}{2}.$$

$$x\text{-intercept} = \frac{-\frac{3\pi}{2} - \frac{\pi}{2}}{2} = \frac{-2\pi}{2} = -\pi$$

An x-intercept is $-\pi$ and the graph passes through $(-\pi, 0)$. Because the coefficient of the tangent is 1, the points on the graph midway between an x-intercept and the asymptotes have y-coordinates of -1 and 1. Use the two consecutive asymptotes,

$$x = -\frac{3\pi}{2} \text{ and } x = -\frac{\pi}{2}, \text{ to graph one full period of}$$

$$y = \tan(x + \pi) \text{ from } -\frac{3\pi}{2} \text{ to } -\frac{\pi}{2}.$$

Continue the pattern and extend the graph another full period to the right.

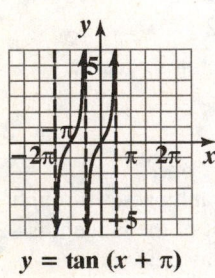

$$y = \tan(x + \pi)$$

86. Solve the equations

$$x - \frac{\pi}{4} = -\frac{\pi}{2} \quad \text{and} \quad x - \frac{\pi}{4} = \frac{\pi}{2}$$

$$x = -\frac{\pi}{2} + \frac{\pi}{4} \qquad x = \frac{\pi}{2} + \frac{\pi}{4}$$

$$x = -\frac{\pi}{4} \qquad x = \frac{3\pi}{4}$$

Thus, two consecutive asymptotes occur at

$$x = -\frac{\pi}{4} \text{ and } x = -\frac{3\pi}{4}.$$

$$x\text{-intercept} = \frac{-\frac{\pi}{4} - \frac{3\pi}{4}}{2} = \frac{\frac{\pi}{2}}{2} = \frac{\pi}{4}$$

An x-intercept is $\dfrac{\pi}{4}$ and the graph passes through

$\left(\dfrac{\pi}{4}, 0 \right)$. Because the coefficient of the tangent is -1,

the points on the graph midway between an x-intercept and the asymptotes have y-coordinates of 1 and -1. Use the two consecutive asymptotes,

$$x = -\frac{\pi}{4} \text{ and } x = \frac{3\pi}{4}, \text{ to graph one full period of}$$

$$y = -\tan\left(x - \frac{\pi}{4}\right) \text{ from } -\frac{\pi}{4} \text{ to } \frac{3\pi}{4}. \text{ Continue the}$$

pattern and extend the graph another full period to the right.

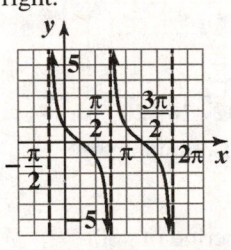

$$y = -\tan\left(x - \frac{\pi}{4}\right)$$

87. Solve the equations

$$3x = 0 \quad \text{and} \quad 3x = \pi$$

$$x = 0 \qquad x = \frac{\pi}{3}$$

Thus, two consecutive asymptotes occur at

$$x = 0 \text{ and } x = \frac{\pi}{3}.$$

$$x\text{-intercept} = \frac{0 + \frac{\pi}{3}}{2} = \frac{\frac{\pi}{3}}{2} = \frac{\pi}{6}$$

An x-intercept is $\dfrac{\pi}{6}$ and the graph passes through

$\left(\dfrac{\pi}{6}, 0 \right)$.

Because the coefficient of the tangent is 2, the points on the graph midway between an x-intercept and the asymptotes have y-coordinates of 2 and -2. Use the two consecutive asymptotes, $x = 0$ and $x = \dfrac{\pi}{3}$, to graph one full period of $y = 2\cot 3x$ from 0 to $\dfrac{\pi}{3}$. Continue the pattern and extend the graph another full period to the right.

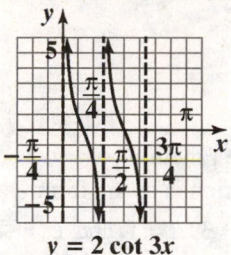

$$y = 2\cot 3x$$

88. Solve the equations

$$\frac{\pi}{2}x = 0 \quad \text{and} \quad \frac{\pi}{2}x = \pi$$

$$x = 0 \qquad\qquad x = \pi \cdot \frac{2}{\pi}$$

$$x = 2$$

Thus, two consecutive asymptotes occur at $x = 0$ and $x = 2$.

$$x\text{-intercept} = \frac{0+2}{2} = \frac{2}{2} = 1$$

An x-intercept is 1 and the graph passes through $(1, 0)$. Because the coefficient of the cotangent is $-\dfrac{1}{2}$, the points on the graph midway between an x-intercept and the asymptotes have y-coordinates of $-\dfrac{1}{2}$ and $\dfrac{1}{2}$. Use the two consecutive asymptotes, $x = 0$ and $x = 2$, to graph one full period of $y = -\dfrac{1}{2}\cot\dfrac{\pi}{2}x$ from 0 to 2. Continue the pattern and extend the graph another full period to the right.

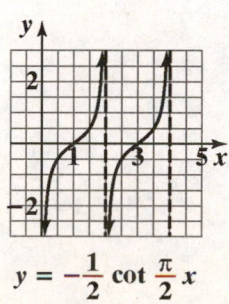

$$y = -\frac{1}{2}\cot\frac{\pi}{2}x$$

89. Solve the equations

$$x + \frac{\pi}{2} = 0 \qquad \text{and} \qquad x + \frac{\pi}{2} = \pi$$

$$x = 0 - \frac{\pi}{2} \qquad\qquad x = \pi - \frac{\pi}{2}$$

$$x = -\frac{\pi}{2} \qquad\qquad x = \frac{\pi}{2}$$

Thus, two consecutive asymptotes occur at

$$x = -\frac{\pi}{2} \quad \text{and} \quad x = \frac{\pi}{2}.$$

$$x\text{-intercept} = \frac{-\frac{\pi}{2} + \frac{\pi}{2}}{2} = \frac{0}{2} = 0$$

An x-intercept is 0 and the graph passes through $(0, 0)$. Because the coefficient of the cotangent is 2, the points on the graph midway between an x-intercept and the asymptotes have y-coordinates of 2 and -2.

Use the two consecutive asymptotes, $x = -\dfrac{\pi}{2}$ and $x = \dfrac{\pi}{2}$, to graph one full period of $y = 2\cot\left(x + \dfrac{\pi}{2}\right)$ from $-\dfrac{\pi}{2}$ to $\dfrac{\pi}{2}$. Continue the pattern and extend the graph another full period to the right.

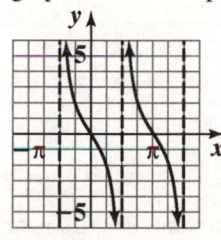

$$y = 2\cot\left(x + \frac{\pi}{2}\right)$$

90. Graph the reciprocal cosine function, $y = 3\cos 2\pi x$.
The equation is of the form $y = A\cos Bx$ with $A = 3$
and $B = 2\pi$.
amplitude: $|A| = |3| = 3$

period: $\dfrac{2\pi}{B} = \dfrac{2\pi}{2\pi} = 1$

Use quarter-periods, $\dfrac{1}{4}$, to find x-values for the five

key points. Starting with $x = 0$, the x-values are 0, $\dfrac{1}{4}$,

$\dfrac{1}{2}$, $\dfrac{3}{4}$, 1. Evaluating the function at each value of x,

the key points are $(0, 3)$,

$\left(\dfrac{1}{4}, 0\right), \left(\dfrac{1}{2}, -3\right), \left(\dfrac{3}{4}, 0\right)$, $(1, 3)$ Use these key

points to graph $y = 3\cos 2\pi x$ from 0 to 1. Extend the
graph one cycle to the right. Use the graph to obtain
the graph of the reciprocal function. Draw vertical
asymptotes through the x-intercepts, and use them as
guides to graph $y = 3\sec 2\pi x$.

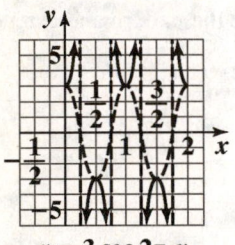

$y = 3 \sec 2\pi\, x$

91. Graph the reciprocal sine function, $y = -2\sin \pi x$.
The equation is of the form $y = A\sin Bx$ with
$A = -2$ and $B = \pi$.
amplitude: $|A| = |-2| = 2$

period: $\dfrac{2\pi}{B} = \dfrac{2\pi}{\pi} = 2$

Use quarter-periods, $\dfrac{2}{4} = \dfrac{1}{2}$, to find

x-values for the five key points. Starting with

$x = 0$, the x-values are 0, $\dfrac{1}{2}$, 1, $\dfrac{3}{2}$, 2. Evaluating the

function at each value of x, the key points are $(0, 0)$,

$\left(\dfrac{1}{2}, -2\right), (1, 0), \left(\dfrac{3}{2}, 2\right), (2, 0)$. Use these key

points to graph $y = -2\sin \pi x$ from 0 to 2.

Extend the graph one cycle to the right. Use the graph
to obtain the graph of the reciprocal function. Draw
vertical asymptotes through the x-intercepts, and use
them as guides to graph $y = -2\csc \pi x$.

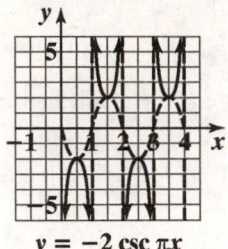

$y = -2\,\mathbf{csc}\,\pi x$

92. Graph the reciprocal cosine function,
$y = 3\cos(x + \pi)$. The equation is of the form
$y = A\cos(Bx - C)$ with $A = 3$, $B = 1$, and $C = -\pi$.
amplitude: $|A| = |3| = 3$

period: $\dfrac{2\pi}{B} = \dfrac{2\pi}{1} = 2\pi$

phase shift: $\dfrac{C}{B} = \dfrac{-\pi}{1} = -\pi$

Use quarter-periods, $\dfrac{2\pi}{4} = \dfrac{\pi}{2}$, to find

x-values for the five key points. Starting with

$x = -\pi$, the x-values are $-\pi$, $-\dfrac{\pi}{2}$, 0, $\dfrac{\pi}{2}$, π.

Evaluating the function at each value of x, the key

points are $(-\pi, 3), \left(-\dfrac{\pi}{2}, 0\right), (0, -3)$,

$\left(\dfrac{\pi}{2}, 0\right), (\pi, 3)$. Use these key points to graph

$y = 3\cos(x + \pi)$ from $-\pi$ to π. Extend the graph
one cycle to the right. Use the graph to obtain the
graph of the reciprocal function. Draw vertical
asymptotes through the x-intercepts, and use them as
guides to graph $y = 3\sec(x + \pi)$.

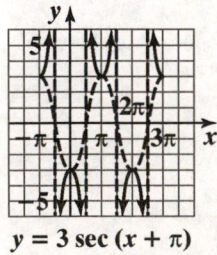

$y = 3 \sec (x + \pi)$

93. Graph the reciprocal sine function, $y = \dfrac{5}{2}\sin(x - \pi)$.

The equation is of the form $y = A\sin(Bx - C)$ with

$A = \dfrac{5}{2}$, $B = 1$, and $C = \pi$.

amplitude: $|A| = \left|\dfrac{5}{2}\right| = \dfrac{5}{2}$

period: $\dfrac{2\pi}{B} = \dfrac{2\pi}{1} = 2\pi$

phase shift: $\dfrac{C}{B} = \dfrac{\pi}{1} = \pi$

Use quarter-periods, $\dfrac{2\pi}{4} = \dfrac{\pi}{2}$, to find

x-values for the five key points. Starting with $x = \pi$,

the x-values are π, $\dfrac{3\pi}{2}$, 2π, $\dfrac{5\pi}{2}$, 3π. Evaluating the

function at each value of x, the key points

are $(\pi, 0)$, $\left(\dfrac{3\pi}{2}, \dfrac{5}{2}\right)$, $(2\pi, 0)$, $\left(\dfrac{5\pi}{2}, -\dfrac{5}{2}\right)$, $(3\pi, 0)$.

Use these key points to graph $y = \dfrac{5}{2}\sin(x - \pi)$ from

π to 3π. Extend the graph one cycle to the right.
Use the graph to obtain the graph of the reciprocal
function. Draw vertical asymptotes through the x-
intercepts, and use them as guides to graph

$y = \dfrac{5}{2}\csc(x - \pi)$.

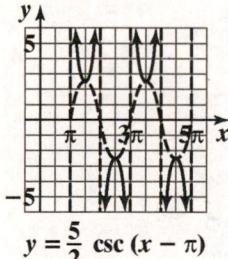

$y = \dfrac{5}{2}\csc(x - \pi)$

94. Let $\theta = \sin^{-1}1$, then $\sin\theta = 1$.

The only angle in the interval $\left[-\dfrac{\pi}{2}, \dfrac{\pi}{2}\right]$ that satisfies

$\sin\theta = 1$ is $\dfrac{\pi}{2}$. Thus $\theta = \dfrac{\pi}{2}$, or $\sin^{-1}1 = \dfrac{\pi}{2}$.

95. Let $\theta = \cos^{-1}1$, then $\cos\theta = 1$.
The only angle in the interval $[0, \pi]$ that satisfies
$\cos\theta = 1$ is 0. Thus $\theta = 0$, or $\cos^{-1}1 = 0$.

96. Let $\theta = \tan^{-1}1$, then $\tan\theta = 1$.

The only angle in the interval $\left(-\dfrac{\pi}{2}, \dfrac{\pi}{2}\right)$ that

satisfies $\tan\theta = 1$ is $\dfrac{\pi}{4}$. Thus $\theta = \dfrac{\pi}{4}$, or

$\tan^{-1}1 = \dfrac{\pi}{4}$.

97. Let $\theta = \sin^{-1}\left(-\dfrac{\sqrt{3}}{2}\right)$, then $\sin\theta = -\dfrac{\sqrt{3}}{2}$.

The only angle in the interval $\left(-\dfrac{\pi}{2}, \dfrac{\pi}{2}\right)$ that

satisfies $\sin\theta = -\dfrac{\sqrt{3}}{2}$ is $-\dfrac{\pi}{3}$. Thus $\theta = -\dfrac{\pi}{3}$, or

$\sin^{-1}\left(-\dfrac{\sqrt{3}}{2}\right) = -\dfrac{\pi}{3}$.

98. Let $\theta = \cos^{-1}\left(-\dfrac{1}{2}\right)$, then $\cos\theta = -\dfrac{1}{2}$.

The only angle in the interval $[0, \pi]$ that satisfies

$\cos\theta = -\dfrac{1}{2}$ is $\dfrac{2\pi}{3}$. Thus $\theta = \dfrac{2\pi}{3}$, or

$\cos^{-1}\left(-\dfrac{1}{2}\right) = \dfrac{2\pi}{3}$.

99. Let $\theta = \tan^{-1}\left(-\dfrac{\sqrt{3}}{3}\right)$, then $\tan\theta = -\dfrac{\sqrt{3}}{3}$.

The only angle in the interval $\left(-\dfrac{\pi}{2}, \dfrac{\pi}{2}\right)$ that

satisfies $\tan\theta = -\dfrac{\sqrt{3}}{3}$ is $-\dfrac{\pi}{6}$.

Thus $\theta = -\dfrac{\pi}{6}$, or $\tan^{-1}\left(-\dfrac{\sqrt{3}}{3}\right) = -\dfrac{\pi}{6}$.

100. Let $\theta = \sin^{-1}\dfrac{\sqrt{2}}{2}$, then $\sin\theta = \dfrac{\sqrt{2}}{2}$. The only angle

in the interval $\left[-\dfrac{\pi}{2}, \dfrac{\pi}{2}\right]$ that satisfies $\sin\theta = \dfrac{\sqrt{2}}{2}$ is

$\dfrac{\pi}{4}$.

Thus, $\cos\left(\sin^{-1}\dfrac{\sqrt{2}}{2}\right) = \cos\dfrac{\pi}{4} = \dfrac{\sqrt{2}}{2}$.

101. Let $\theta = \cos^{-1} 0$, then $\cos \theta = 0$. The only angle in the interval $[0, \pi]$ that satisfies $\cos \theta = 0$ is $\dfrac{\pi}{2}$.

Thus, $\sin\left(\cos^{-1} 0\right) = \sin \dfrac{\pi}{2} = 1$.

102. Let $\theta = \sin^{-1}\left(-\dfrac{1}{2}\right)$, then $\sin \theta = -\dfrac{1}{2}$. The only

angle in the interval $\left[-\dfrac{\pi}{2}, \dfrac{\pi}{2}\right]$ that satisfies

$\sin \theta = -\dfrac{1}{2}$ is $-\dfrac{\pi}{6}$.

Thus, $\tan\left[\sin^{-1}\left(-\dfrac{1}{2}\right)\right] = \tan\left(-\dfrac{\pi}{6}\right) = -\dfrac{\sqrt{3}}{3}$.

103. Let $\theta = \cos^{-1}\left(-\dfrac{\sqrt{3}}{2}\right)$, then $\cos \theta = -\dfrac{\sqrt{3}}{2}$. The only

angle in the interval $[0, \pi]$ that satisfies

$\cos \theta = -\dfrac{\sqrt{3}}{2}$ is $\dfrac{5\pi}{6}$.

Thus, $\tan\left[\cos^{-1}\left(-\dfrac{\sqrt{3}}{2}\right)\right] = \tan \dfrac{5\pi}{6} = -\dfrac{\sqrt{3}}{3}$.

104. Let $\theta = \tan^{-1} \dfrac{\sqrt{3}}{3}$, then $\tan \theta = \dfrac{\sqrt{3}}{3}$.

The only angle in the interval $\left(-\dfrac{\pi}{2}, \dfrac{\pi}{2}\right)$ that

satisfies $\tan \theta = \dfrac{\sqrt{3}}{3}$ is $\dfrac{\pi}{6}$.

Thus $\csc\left(\tan^{-1} \dfrac{\sqrt{3}}{3}\right) = \csc \dfrac{\pi}{6} = 2$.

105. Let $\theta = \tan^{-1} \dfrac{3}{4}$, then $\tan \theta = \dfrac{3}{4}$.

Because $\tan \theta$ is positive, θ is in the first quadrant.

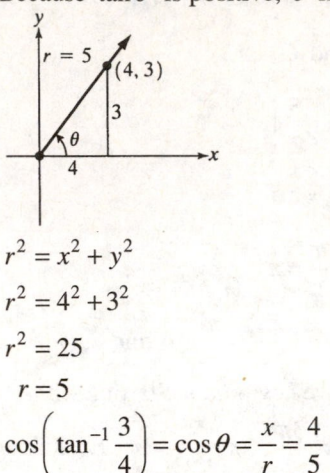

$r^2 = x^2 + y^2$

$r^2 = 4^2 + 3^2$

$r^2 = 25$

$r = 5$

$\cos\left(\tan^{-1} \dfrac{3}{4}\right) = \cos \theta = \dfrac{x}{r} = \dfrac{4}{5}$

106. Let $\theta = \cos^{-1} \dfrac{3}{5}$, then $\cos \theta = \dfrac{3}{5}$.

Because $\cos \theta$ is positive, θ is in the first quadrant.

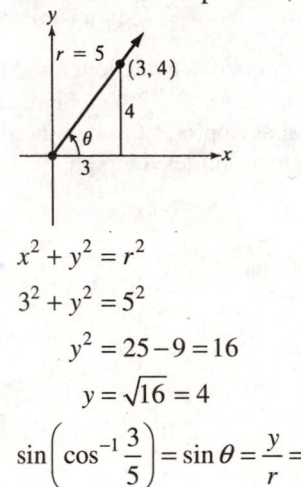

$x^2 + y^2 = r^2$

$3^2 + y^2 = 5^2$

$y^2 = 25 - 9 = 16$

$y = \sqrt{16} = 4$

$\sin\left(\cos^{-1} \dfrac{3}{5}\right) = \sin \theta = \dfrac{y}{r} = \dfrac{4}{5}$

107. Let $\theta = \sin^{-1}\left(-\dfrac{3}{5}\right)$, then $\sin\theta = -\dfrac{3}{5}$.

Because $\sin\theta$ is negative, θ is in quadrant IV.

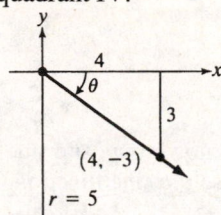

$$x^2 + (-3)^2 = 5^2$$
$$x^2 + y^2 = r^2$$
$$x^2 = 25 - 9 = 16$$
$$x = \sqrt{16} = 4$$
$$\tan\left[\sin^{-1}\left(-\frac{3}{5}\right)\right] = \tan\theta = \frac{y}{x} = -\frac{3}{4}$$

108. Let $\theta = \cos^{-1}\left(-\dfrac{4}{5}\right)$, then $\cos\theta = -\dfrac{4}{5}$.

Because $\cos\theta$ is negative, θ is in quadrant II.

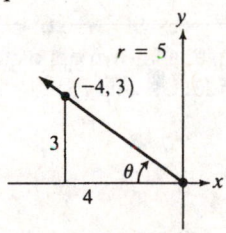

$$x^2 + y^2 = r^2$$
$$(-4)^2 + y^2 = 5^2$$
$$y^2 = 25 - 16 = 9$$
$$y = \sqrt{9} = 3$$

Use the right triangle to find the exact value.

$$\tan\left[\cos^{-1}\left(-\frac{4}{5}\right)\right] = \tan\theta = -\frac{3}{4}$$

109. Let $\theta = \tan^{-1}\left(-\dfrac{1}{3}\right)$,

Because $\tan\theta$ is negative, θ is in quadrant IV and $x = 3$ and $y = -1$.

$$r^2 = x^2 + y^2$$
$$r^2 = 3^2 + (-1)^2$$
$$r^2 = 10$$
$$r = \sqrt{10}$$
$$\sin\left[\tan^{-1}\left(-\frac{4}{5}\right)\right] = \sin\theta = \frac{y}{r} = \frac{-1}{\sqrt{10}} = -\frac{\sqrt{10}}{10}$$

110. $x = \dfrac{\pi}{3}$, x is in $\left[-\dfrac{\pi}{2}, \dfrac{\pi}{2}\right]$, so $\sin^{-1}\left(\sin\dfrac{\pi}{3}\right) = \dfrac{\pi}{3}$

111. $x = \dfrac{2\pi}{3}$, x is not in $\left[-\dfrac{\pi}{2}, \dfrac{\pi}{2}\right]$. x is in the domain of $\sin x$, so

$$\sin^{-1}\left(\sin\frac{2\pi}{3}\right) = \sin^{-1}\frac{\sqrt{3}}{2} = \frac{\pi}{3}$$

112. $\sin^{-1}\left(\cos\dfrac{2\pi}{3}\right) = \sin^{-1}\left(-\dfrac{1}{2}\right)$

Let $\theta = \sin^{-1}\left(-\dfrac{1}{2}\right)$, then $\sin\theta = -\dfrac{1}{2}$. The only

angle in the interval $\left[-\dfrac{\pi}{2}, \dfrac{\pi}{2}\right]$ that satisfies

$\sin\theta = -\dfrac{1}{2}$ is $-\dfrac{\pi}{6}$. Thus, $\theta = -\dfrac{\pi}{6}$, or

$$\sin^{-1}\left(\cos\frac{2\pi}{3}\right) = \sin^{-1}\left(-\frac{1}{2}\right) = -\frac{\pi}{6}.$$

113. Let $\theta = \tan^{-1}\dfrac{x}{2}$, then $\tan\theta = \dfrac{x}{2}$.

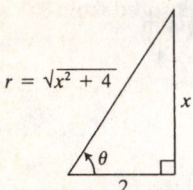

$r^2 = x^2 + 2^2$

$r^2 = x^2 + y^2$

$r = \sqrt{x^2 + 4}$

Use the right triangle to write the algebraic expression.

$$\cos\left(\tan^{-1}\frac{x}{2}\right) = \cos\theta = \frac{2}{\sqrt{x^2+4}} = \frac{2\sqrt{x^2+4}}{x^2+4}$$

114. Let $\theta = \sin^{-1}\dfrac{1}{x}$, then $\sin\theta = \dfrac{1}{x}$.

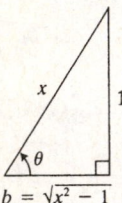

Use the Pythagorean theorem to find the third side, b.

$1^2 + b^2 = x^2$

$b^2 = x^2 - 1$

$b = \sqrt{x^2 - 1}$

Use the right triangle to write the algebraic expression.

$$\sec\left(\sin^{-1}\frac{1}{x}\right) = \sec\theta = \frac{x}{\sqrt{x^2-1}} = \frac{x\sqrt{x^2-1}}{x^2-1}$$

115. Find the measure of angle B. Because $C = 90°$, $A + B = 90°$. Thus, $B = 90° - A = 90° - 22.3° = 67.7°$
We have a known angle, a known hypotenuse, and an unknown opposite side. Use the sine function.

$\sin 22.3° = \dfrac{a}{10}$

$a = 10\sin 22.3° \approx 3.79$

We have a known angle, a known hypotenuse, and an unknown adjacent side. Use the cosine function.

$\cos 22.3° = \dfrac{b}{10}$

$b = 10\cos 22.3° \approx 9.25$

In summary, $B = 67.7°$, $a \approx 3.79$, and $b \approx 9.25$.

116. Find the measure of angle A. Because $C = 90°$, $A + B = 90°$. Thus, $A = 90° - B = 90° - 37.4° = 52.6°$
We have a known angle, a known opposite side, and an unknown adjacent side. Use the tangent function.

$\tan 37.4° = \dfrac{6}{a}$

$a = \dfrac{6}{\tan 37.4°} \approx 7.85$

We have a known angle, a known opposite side, and an unknown hypotenuse. Use the sine function.

$\sin 37.4° = \dfrac{6}{c}$

$c = \dfrac{6}{\sin 37.4°} \approx 9.88$

In summary, $A = 52.6°$, $a \approx 7.85$, and $c \approx 9.88$.

117. Find the measure of angle A. We have a known hypotenuse, a known opposite side, and an unknown angle. Use the sine function.

$\sin A = \dfrac{2}{7}$

$A = \sin^{-1}\left(\dfrac{2}{7}\right) \approx 16.6°$

Find the measure of angle B. Because $C = 90°$, $A + B = 90°$. Thus, $B = 90° - A \approx 90° - 16.6° = 73.4°$
We have a known hypotenuse, a known opposite side, and an unknown adjacent side. Use the Pythagorean theorem.

$a^2 + b^2 = c^2$

$2^2 + b^2 = 7^2$

$b^2 = 7^2 - 2^2 = 45$

$b = \sqrt{45} \approx 6.71$

In summary, $A \approx 16.6°$, $B \approx 73.4°$, and $b \approx 6.71$.

118. Find the measure of angle A. We have a known opposite side, a known adjacent side, and an unknown angle. Use the tangent function.

$\tan A = \dfrac{1.4}{3.6}$

$A = \tan^{-1}\left(\dfrac{1.4}{3.6}\right) \approx 21.3°$

Find the measure of angle B. Because $C = 90°$, $A + B = 90°$. Thus, $B = 90° - A \approx 90° - 21.3° = 68.7°$
We have a known opposite side, a known adjacent side, and an unknown hypotenuse.
Use the Pythagorean theorem.

$c^2 = a^2 + b^2 = (1.4)^2 + (3.6)^2 = 14.92$

$c = \sqrt{14.92} \approx 3.86$

In summary, $A \approx 21.3°$, $B \approx 68.7°$, and $c \approx 3.86$.

119. Using a right triangle, we have a known angle, an unknown opposite side, h, and a known adjacent side. Therefore, use the tangent function.

$$\tan 25.6° = \frac{h}{80}$$

$$h = 80 \tan 25.6°$$

$$\approx 38.3$$

The building is about 38 feet high.

120. Using a right triangle, we have a known angle, an unknown opposite side, h, and a known adjacent side. Therefore, use the tangent function.

$$\tan 40° = \frac{h}{60}$$

$$h = 60 \tan 40° \approx 50 \text{ yd}$$

The second building is 50 yds taller than the first. Total height $= 40 + 50 = 90$ yd .

121. Using two right triangles, a smaller right triangle corresponding to the smaller angle of elevation drawn inside a larger right triangle corresponding to the larger angle of elevation, we have a known angle, a known opposite side, and an unknown adjacent side, d, in the smaller triangle. Therefore, use the tangent function.

$$\tan 68° = \frac{125}{d}$$

$$d = \frac{125}{\tan 68°} \approx 50.5$$

We now have a known angle, a known adjacent side, and an unknown opposite side, h, in the larger triangle. Again, use the tangent function.

$$\tan 71° = \frac{h}{50.5}$$

$$h = 50.5 \tan 71° \approx 146.7$$

The height of the antenna is $146.7 - 125$, or 21.7 ft, to the nearest tenth of a foot.

122. We need the acute angle between ray OA and the north-south line through O. This angle measures $90° - 55° = 35°$. This angle measured from the north side of the north-south line and lies east of the north-south line. Thus the bearing from O to A is N35°E.

123. We need the acute angle between ray OA and the north-south line through O. This angle measures $90° - 55° = 35°$. This angle measured from the south side of the north-south line and lies west of the north-south line. Thus the bearing from O to A is S35°W.

124. Using a right triangle, we have a known angle, a known adjacent side, and an unknown opposite side, d. Therefore, use the tangent function.

$$\tan 64° = \frac{d}{12}$$

$$d = 12 \tan 64° \approx 24.6$$

The ship is about 24.6 miles from the lighthouse.

125.

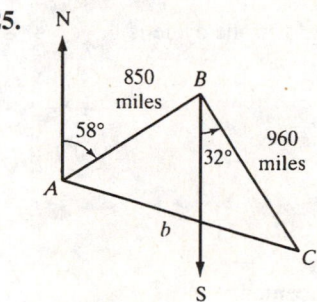

a. Using the figure,
$$B = 58° + 32° = 90°$$
Thus, use the Pythagorean Theorem to find the distance from city A to city C.

$$850^2 + 960^2 = b^2$$

$$b^2 = 722,500 + 921,600$$

$$b^2 = 1,644,100$$

$$b = \sqrt{1,644,100} \approx 1282.2$$

The distance from city A to city B is about 1282.2 miles.

b. Using the figure,

$$\tan A = \frac{\text{opposite}}{\text{adjacent}} = \frac{960}{850} \approx 1.1294$$

$$A \approx \tan^{-1}(1.1294) \approx 48°$$

$$180° - 58° - 48° = 74°$$

The bearing from city A to city C is S74°E.

126. $d = 20\cos\dfrac{\pi}{4}t$

$a = 20$ and $\omega = \dfrac{\pi}{4}$

a.　maximum displacement:
$|a| = |20| = 20\,\text{cm}$

b.　$f = \dfrac{\omega}{2\pi} = \dfrac{\frac{\pi}{4}}{2\pi} = \dfrac{\pi}{4} \cdot \dfrac{1}{2\pi} = \dfrac{1}{8}$

frequency: $\dfrac{1}{8}$ cycle per second

c.　period: $\dfrac{2\pi}{\omega} = \dfrac{2\pi}{\frac{\pi}{4}} = 2\pi \cdot \dfrac{4}{\pi} = 8$

The time required for one cycle is 8 seconds.

127. $d = \dfrac{1}{2}\sin 4t$

$a = \dfrac{1}{2}$ and $\omega = 4$

a.　maximum displacement:
$|a| = \left|\dfrac{1}{2}\right| = \dfrac{1}{2}\,\text{cm}$

b.　$f = \dfrac{\omega}{2\pi} = \dfrac{4}{2\pi} = \dfrac{2}{\pi} \approx 0.64$

frequency: 0.64 cycle per second

c.　period: $\dfrac{2\pi}{\omega} = \dfrac{2\pi}{4} = \dfrac{\pi}{2} \approx 1.57$

The time required for one cycle is about 1.57 seconds.

128. Because the distance of the object from the rest position at $t = 0$ is a maximum, use the form

$d = a\cos\omega t$. The period is $\dfrac{2\pi}{\omega}$ so,

$2 = \dfrac{2\pi}{\omega}$

$\omega = \dfrac{2\pi}{2} = \pi$

Because the amplitude is 30 inches, $|a| = 30$.

because the object starts below its rest position $a = -30$. the equation for the object's simple harmonic motion is $d = -30\cos\pi t$.

129. Because the distance of the object from the rest position at $t = 0$ is 0, use the form $d = a\sin\omega t$. The period is $\dfrac{2\pi}{\omega}$ so

$5 = \dfrac{2\pi}{\omega}$

$\omega = \dfrac{2\pi}{5}$

Because the amplitude is $\dfrac{1}{4}$ inch, $|a| = \dfrac{1}{4}$. a is negative since the object begins pulled down. The equation for the object's simple harmonic motion is

$d = -\dfrac{1}{4}\sin\dfrac{2\pi}{5}t$.

Chapter 4 Test

1.　$135° = 135° \cdot \dfrac{\pi \text{ radians}}{180°}$

$= \dfrac{135\pi}{180}$ radians

$= \dfrac{3\pi}{4}$ radians

2.　$75° = 75° \cdot \dfrac{\pi \text{ radians}}{180°} = \dfrac{75\pi}{180}$ radians

$= \dfrac{5\pi}{12}$ radians

$s = r\theta$

$s = 20\left(\dfrac{5\pi}{12}\right) = \dfrac{25\pi}{3}$ ft ≈ 26.18 ft

3.　**a.**　$\dfrac{16\pi}{3} - 4\pi = \dfrac{16\pi}{3} - \dfrac{12\pi}{3} = \dfrac{4\pi}{3}$

b.　$\dfrac{16\pi}{3}$ is coterminal with $\dfrac{4\pi}{3}$.

$\dfrac{4\pi}{3} - \pi = \dfrac{4\pi}{3} - \dfrac{3\pi}{3} = \dfrac{\pi}{3}$

4. $P = (-2, 5)$ is a point on the terminal side of θ, $x = -2$ and $y = 5$. Furthermore,

$$r = \sqrt{x^2 + y^2} = \sqrt{(-2)^2 + (5)^2}$$
$$= \sqrt{4 + 25} = \sqrt{29}$$

Use *x, y,* and *r,* to find the six trigonometric functions of θ.

$$\sin\theta = \frac{y}{r} = \frac{5}{\sqrt{29}} = \frac{5\sqrt{29}}{\sqrt{29}\sqrt{29}} = \frac{5\sqrt{29}}{29}$$

$$\cos\theta = \frac{x}{r} = \frac{-2}{\sqrt{29}} = -\frac{2\sqrt{29}}{\sqrt{29}\sqrt{29}} = -\frac{2\sqrt{29}}{29}$$

$$\tan\theta = \frac{y}{x} = \frac{5}{-2} = -\frac{5}{2}$$

$$\csc\theta = \frac{r}{y} = \frac{\sqrt{29}}{5}$$

$$\sec\theta = \frac{r}{x} = \frac{\sqrt{29}}{-2} = -\frac{\sqrt{29}}{2}$$

$$\cot\theta = \frac{x}{y} = \frac{-2}{5} = -\frac{2}{5}$$

5. Because $\cos\theta < 0$, θ cannot lie in quadrant I and quadrant IV; the cosine function is positive in those two quadrants. Thus, with $\cos\theta < 0$, θ lies in quadrant II or quadrant III. We are also given that $\cot\theta > 0$. Because quadrant III is the only quadrant in which the cosine is negative and the cotangent is positive, θ lies in quadrant III.

6. Because the cosine is positive and the tangent is negative, θ lies in quadrant IV. In quadrant IV *x* is positive and *y* is negative. Thus,

$$\cos\theta = \frac{1}{3} = \frac{x}{r}, \; x = 1, \; r = 3. \text{ Furthermore,}$$

$$x^2 + y^2 = r^3$$

$$1^2 + y^2 = 3^2$$

$$y^2 = 9 - 1 = 8$$

$$y = -\sqrt{8} = -2\sqrt{2}$$

Use *x, y,* and *r,* to find the six trigonometric functions of θ.

$$\sin\theta = \frac{y}{r} = \frac{-2\sqrt{2}}{3} = -\frac{2\sqrt{2}}{3}$$

$$\tan\theta = \frac{y}{x} = \frac{-2\sqrt{2}}{1} = -2\sqrt{2}$$

$$\csc\theta = \frac{r}{y} = \frac{3}{-2\sqrt{2}} = -\frac{3\sqrt{2}}{2\sqrt{2}\cdot\sqrt{2}} = -\frac{3\sqrt{2}}{4}$$

$$\sec\theta = \frac{r}{x} = \frac{3}{1} = 3$$

$$\cot\theta = \frac{x}{y} = \frac{1}{-2\sqrt{2}} = -\frac{1\cdot\sqrt{2}}{2\sqrt{2}\sqrt{2}} = -\frac{\sqrt{2}}{4}$$

7. $\tan\dfrac{\pi}{6}\cos\dfrac{\pi}{3} - \cos\dfrac{\pi}{2} = \dfrac{\sqrt{3}}{3}\cdot\dfrac{1}{2} - 0 = \dfrac{\sqrt{3}}{6}$

8. $300°$ lies in quadrant IV.
The reference angle is
$$\theta' = 360° - 300° = 60°$$
$$\tan 60° = \sqrt{3}$$
In quadrant IV, $\tan\theta < 0$, so
$$\tan 300° = -\tan 60 = -\sqrt{3}.$$

9. $\dfrac{7\pi}{4}$ lies in quadrant IV.
The reference angle is
$$\theta' = 2\pi - \frac{7\pi}{4} = \frac{8\pi}{4} - \frac{7\pi}{4} = \frac{\pi}{4}$$
$$\sin\frac{\pi}{4} = \frac{\sqrt{2}}{2}$$
In quadrant IV, $\sin\theta < 0$, so
$$\sin\frac{7\pi}{4} = -\sin\frac{\pi}{4} = -\frac{\sqrt{2}}{2}.$$

10. $\sec\dfrac{22\pi}{3} = \sec\dfrac{4\pi}{3} = -\sec\dfrac{\pi}{3}$

$$= \frac{1}{-\cos\dfrac{\pi}{3}} = \frac{1}{-\dfrac{1}{2}} = -2$$

11. $\cot\left(-\dfrac{8\pi}{3}\right) = \cot\left(\dfrac{4\pi}{3}\right) = \cot\dfrac{\pi}{3}$

$$= \frac{1}{\tan\dfrac{\pi}{3}} = \frac{1}{\sqrt{3}} = \frac{\sqrt{3}}{3}$$

12. $\tan\left(\dfrac{7\pi}{3} + n\pi\right) = \tan\dfrac{7\pi}{3} = \tan\dfrac{\pi}{3} = \sqrt{3}$

13. a. $\sin(-\theta) + \cos(-\theta) = -\sin(\theta) + \cos(\theta)$
$$= -a + b$$

   b. $\tan\theta - \sec\theta = \dfrac{\sin\theta}{\cos\theta} - \dfrac{1}{\cos\theta}$
$$= \dfrac{a}{b} - \dfrac{1}{b}$$
$$= \dfrac{a-1}{b}$$

14.   The equation $y = 3\sin 2x$ is of the form $y = A\sin Bx$ with $A = 3$ and $B = 2$. The amplitude is $|A| = |3| = 3$.
The period is $\dfrac{2\pi}{B} = \dfrac{2\pi}{2} = \pi$. The quarter-period is $\dfrac{\pi}{4}$.
The cycle begins at $x = 0$. Add quarter-periods to generate x-values for the key points.
$x = 0$
$x = 0 + \dfrac{\pi}{4} = \dfrac{\pi}{4}$
$x = \dfrac{\pi}{4} + \dfrac{\pi}{4} = \dfrac{\pi}{2}$
$x = \dfrac{\pi}{2} + \dfrac{\pi}{4} = \dfrac{3\pi}{4}$
$x = \dfrac{3\pi}{4} + \dfrac{\pi}{4} = \pi$
Evaluate the function at each value of x.

x	coordinates
0	$(0, 0)$
$\dfrac{\pi}{4}$	$\left(\dfrac{\pi}{4}, 3\right)$
$\dfrac{\pi}{2}$	$\left(\dfrac{\pi}{2}, 0\right)$
$\dfrac{3\pi}{4}$	$\left(\dfrac{3\pi}{4}, -3\right)$
π	$(\pi, 0)$

Connect the five key points with a smooth curve and graph one complete cycle of the given function.

$y = 3\sin 2x$

15.   The equation $y = -2\cos\left(x - \dfrac{\pi}{2}\right)$ is of the form
$y = A\cos(Bx - C)$ with $A = -2$, $B = 1$, and
$C = \dfrac{\pi}{2}$. The amplitude is $|A| = |-2| = 2$.
The period is $\dfrac{2\pi}{B} = \dfrac{2\pi}{1} = 2\pi$. The phase shift is
$\dfrac{C}{B} = \dfrac{\frac{\pi}{2}}{1} = \dfrac{\pi}{2}$. The quarter-period is $\dfrac{2\pi}{4} = \dfrac{\pi}{2}$.
The cycle begins at $x = \dfrac{\pi}{2}$. Add quarter-periods to generate x-values for the key points.
$x = \dfrac{\pi}{2}$
$x = \dfrac{\pi}{2} + \dfrac{\pi}{2} = \pi$
$x = \pi + \dfrac{\pi}{2} = \dfrac{3\pi}{2}$
$x = \dfrac{3\pi}{2} + \dfrac{\pi}{2} = 2\pi$
$x = 2\pi + \dfrac{\pi}{2} = \dfrac{5\pi}{2}$
Evaluate the function at each value of x.

x	coordinates
$\dfrac{\pi}{2}$	$\left(\dfrac{\pi}{2}, -2\right)$
π	$(\pi, 0)$
$\dfrac{3\pi}{2}$	$\left(\dfrac{3\pi}{2}, 2\right)$
2π	$(2\pi, 0)$
$\dfrac{5\pi}{2}$	$\left(\dfrac{5\pi}{2}, -2\right)$

Connect the five key points with a smooth curve and graph one complete cycle of the given function.

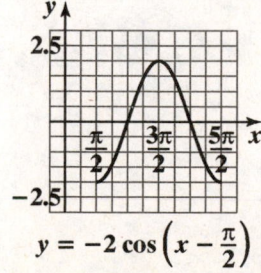

$y = -2\cos\left(x - \dfrac{\pi}{2}\right)$

16. Solve the equations

$$\frac{x}{2} = -\frac{\pi}{2} \quad \text{and} \quad \frac{x}{2} = \frac{\pi}{2}$$

$$x = -\frac{\pi}{2} \cdot 2 \qquad x = \frac{\pi}{2} \cdot 2$$

$$x = -\pi \qquad\qquad x = \pi$$

Thus, two consecutive asymptotes occur at $x = -\pi$ and $x = \pi$.

$$x\text{-intercept} = \frac{-\pi + \pi}{2} = \frac{0}{2} = 0$$

An x-intercept is 0 and the graph passes through (0, 0). Because the coefficient of the tangent is 2, the points on the graph midway between an x-intercept and the asymptotes have y-coordinates of –2 and 2. Use the two consecutive asymptotes, $x = -\pi$ and $x = \pi$, to graph one

full period of $y = 2\tan\dfrac{x}{2}$ from $-\pi$ to π.

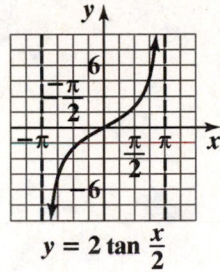

$$y = 2\tan\frac{x}{2}$$

17. Graph the reciprocal sine function, $y = -\dfrac{1}{2}\sin\pi x$.

The equation is of the form $y = A\sin Bx$ with $A = -\dfrac{1}{2}$ and $B = \pi$.

amplitude: $|A| = \left| -\dfrac{1}{2} \right| = \dfrac{1}{2}$

period: $\dfrac{2\pi}{B} = \dfrac{2\pi}{\pi} = 2$

Use quarter-periods, $\dfrac{2}{4} = \dfrac{1}{2}$, to find x-values for the five key points. Starting with $x = 0$, the x-values are $0, \dfrac{1}{2}, 1, \dfrac{3}{2}, 2$. Evaluating the function at each value of x, the key points are

$(0, 3), \left(\dfrac{1}{2}, -\dfrac{1}{2}\right), (1, 0), \left(\dfrac{3}{2}, \dfrac{1}{2}\right), (2, 0)$.

Use these key points to graph $y = -\dfrac{1}{2}\sin\pi x$ from 0 to 2. Use the graph to obtain the graph of the

reciprocal function. Draw vertical asymptotes through the x-intercepts, and use them as guides to graph $y = -\dfrac{1}{2}\csc\pi x$.

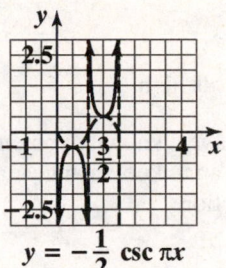

$$y = -\frac{1}{2}\csc\pi x$$

18. Let $\theta = \cos^{-1}\left(-\dfrac{1}{2}\right)$, then $\cos\theta = -\dfrac{1}{2}$.

Because $\cos\theta$ is negative, θ is in quadrant II.

$$x^2 + y^2 = r^2$$

$$(-1)^2 + y^2 = 2^2$$

$$y^2 = 4 - 1 = 3$$

$$y = \sqrt{3}$$

$$\tan\left[\cos^{-1}\left(-\frac{1}{2}\right)\right] = \tan\theta = \frac{y}{x} = \frac{\sqrt{3}}{-1} = -\sqrt{3}$$

19. Let $\theta = \cos^{-1}\left(\dfrac{x}{3}\right)$, then $\cos\theta = \dfrac{x}{3}$.

Because $\cos\theta$ is positive, θ is in quadrant I.

$$x^2 + y^2 = r^2$$

$$x^2 + y^2 = 3^2$$

$$y^2 = 9 - x^2$$

$$y = \sqrt{9 - x^2}$$

$$\sin\left[\cos^{-1}\left(\frac{x}{3}\right)\right] = \sin\theta = \frac{y}{r} = \frac{\sqrt{9-x^2}}{3}$$

20. Find the measure of angle B. Because $C = 90°$, $A + B = 90°$.
Thus, $B = 90° - A = 90° - 21° = 69°$.
We have a known angle, a known hypotenuse, and an unknown opposite side. Use the sine function.

$$\sin 21° = \frac{a}{13}$$

$$a = 13\sin 21° \approx 4.7$$

We have a known angle, a known hypotenuse, and an unknown adjacent side. Use the cosine function.

$$\cos 21° = \frac{b}{13}$$

$$b = 13\cos 21° \approx 12.1$$

In summary, $B = 69°$, $a \approx 4.7$, and $b \approx 12.1$.

21. Using a right triangle, we have a known angle, an unknown opposite side, h, and a known adjacent side. Therefore, use the tangent function.

$$\tan 37° = \frac{h}{30}$$

$$h = 30 \tan 37° \approx 23$$

The building is about 23 yards high.

22. Using a right triangle, we have a known hypotenuse, a known opposite side, and an unknown angle. Therefore, use the sine function.

$$\sin \theta = \frac{43}{73}$$

$$\theta = \sin^{-1}\left(\frac{43}{73}\right) \approx 36.1°$$

The rope makes an angle of about 36.1° with the pole.

23. We need the acute angle between ray OP and the north-south line through O. This angle measures 90° – 10°. This angle is measured from the north side of the north-south line and lies west of the north-south line. Thus the bearing from O to P is N80°W.

24. $d = -6\cos\pi t$

$a = -6$ and $\omega = \pi$

a. maximum displacement: $|a| = |-6| = 6$ in.

b. $f = \dfrac{\omega}{2\pi} = \dfrac{\pi}{2\pi} = \dfrac{1}{2}$

frequency: $\dfrac{1}{2}$ cycle per second

c. period $= \dfrac{2\pi}{\omega} = \dfrac{2\pi}{\pi} = 2$

The time required for one cycle is 2 seconds.

25. Trigonometric functions are periodic.

Cumulative Review Exercises (Chapters P-4)

1. $x^2 = 18 + 3x$

$x^2 - 3x - 18 = 0$

$(x-6)(x+3) = 0$

$x - 6 = 0$ or $x + 3 = 0$

$x = 6 \qquad\qquad x = -3$

The solution set is $\{-3, 6\}$.

2. $x^3 + 5x^2 - 4x - 20 = 0$

$x^2(x+5) - 4(x+5) = 0$

$(x^2 - 4)(x+5) = 0$

$(x-2)(x+2)(x+5) = 0$

$x - 2 = 0$ or $x + 2 = 0$ or $x + 5 = 0$

$x = 2 \qquad\quad x = -2 \qquad\quad x = -5$

The solution set is $\{-5, -2, 2\}$.

3. $\log_2 x + \log_2(x-2) = 3$

$\log_2 x(x-2) = 3$

$x(x-2) = 2^3$

$x^2 - 2x = 2^3$

$x^2 - 2x - 8 = 0$

$(x-4)(x+2) = 0$

$x - 4 = 0$ or $x + 2 = 0$

$x = 4 \qquad\qquad x = -2$

$x = -2$ is extraneous

The solution set is $\{4\}$

4. $\sqrt{x-3} + 5 = x$

$\sqrt{x-3} = x - 5$

$\left(\sqrt{x-3}\right)^2 = (x-5)^2$

$x - 3 = x^2 - 10x + 25$

$x^2 - 11x + 28 = 0$

$(x-4)(x-7) = 0$

$x - 4 = 0$ or $x - 7 = 0$

$x = 4 \qquad\qquad x = 7$

$\sqrt{4-3} + 5 = 4$

$\sqrt{1} + 5 = 4$

$1 + 5 = 4$ false

$x = 4$ is not a solution

$\sqrt{7-3} + 5 = 7$

$\sqrt{4} + 5 = 7$

$2 + 5 = 7$ true

The solution set is $\{7\}$.

5. $x^3 - 4x^2 + x + 6 = 0$

$p: \pm 1, \pm 2, \pm 3, \pm 6$

$q: \pm 1$

$\dfrac{p}{q}: \pm 1, \pm 2, \pm 3, \pm 6$

$$\begin{array}{r|rrrr}
2 & 1 & -4 & 1 & 6 \\
  &   & 2 & -4 & -6 \\
\hline
  & 1 & -2 & -3 & 0
\end{array}$$

$x^3 - 4x^2 + x + 6 = (x - 2)(x^2 - 2x - 3)$

Thus,

$x^3 - 4x^2 + x + 6 = 0$

$(x - 2)(x^2 - 2x - 3) = 0$

$(x - 2)(x - 3)(x + 1) = 0$

$x - 2 = 0$ or $x - 3 = 0$ or $x + 1 = 0$

$\quad x = 2 \qquad\qquad x = 3 \qquad\qquad x = -1$

The solution set is $\{-1, 2, 3\}$

6. $|2x - 5| \le 11$

$-11 \le 2x - 5 \le 11$

$-6 \le 2x \le 16$

$-3 \le x \le 8$

The solution set is $\{x \mid -3 \le x \le 8\}$

7. $f(x) = \sqrt{x - 6}$

$\quad x = \sqrt{y - 6}$

$\quad x^2 = y - 6$

$\quad y = x^2 + 6$

$f^{-1}(x) = x^2 + 6$

8.

$$
\require{enclose}
\begin{array}{r}
4x^2 - \dfrac{14}{5}x - \dfrac{17}{25} \\[2pt]
5x + 2\overline{\smash{\big)}\,20x^3 - 6x^2 - 9x + 10} \\[2pt]
\underline{20x^3 + 8x^2} \\[2pt]
-14x^2 - 9x \\[2pt]
\underline{-14x^2 - \dfrac{28}{5}x} \\[2pt]
-\dfrac{17}{5}x + 10 \\[2pt]
\underline{-\dfrac{17}{5}x - \dfrac{34}{25}} \\[2pt]
\dfrac{284}{25}
\end{array}
$$

The quotient is $4x^2 - \dfrac{14}{5}x - \dfrac{17}{25} + \dfrac{284}{125x + 50}$.

9. $\log 25 + \log 40 = \log(25 \cdot 40)$

$\qquad\qquad\qquad\quad = \log 1000$

$\qquad\qquad\qquad\quad = \log 10^3$

$\qquad\qquad\qquad\quad = 3$

10. $\dfrac{14\pi}{9}$ radians $= \dfrac{14\pi}{9}$ radians $\cdot \dfrac{180°}{\pi \text{ radians}}$

$\qquad\qquad\qquad = \dfrac{14 \cdot 180°}{9} = 280°$

11. $3x^4 - 2x^3 + 5x^2 + x - 9 = 0$

The sign changes 3 times so the equation has at most 3 positive real roots;

$f(-x) = 3x^4 + 2x^3 + 5x^2 - x - 9$

The sign changes 1 time, so the equation has at most 1 negative real root.

12. $f(x) = \dfrac{x}{x^2 - 1}$

vertical asymptotes: $x^2 - 1 = 0$, $x = 1$ and $x = -1$

horizontal asymptote: $m = 1$ and $n = 2$ so $m < n$ and the x-axis is a horizontal asymptote.

x-intercept: $(0, 0)$

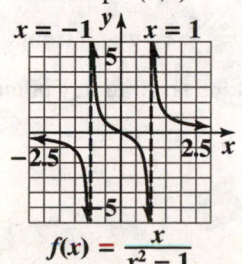

$f(x) = \dfrac{x}{x^2 - 1}$

13. $(x-2)^2 + y^2 = 1$

The graph is a circle with center $(2,0)$ and $r = 1$.

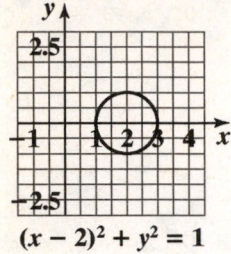

$(x - 2)^2 + y^2 = 1$

14. $y = (x-1)(x+2)^2$

x-intercepts: $(1,0)$ and $(-2,0)$

y-intercept: $y = (-1)(2)^2 = -4$

$(0,-4)$

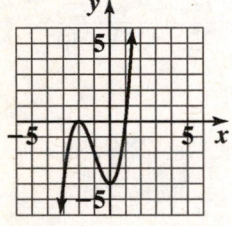

$y = (x - 1)(x + 2)^2$

15. $y = \sin\left(2x + \dfrac{\pi}{2}\right) = \sin\left(2x - \left(-\dfrac{\pi}{2}\right)\right)$

The equation $y = \sin\left(2x - \left(-\dfrac{\pi}{2}\right)\right)$ is of the form

$y = A\sin(Bx - C)$ with $A = 1$, $B = 2$, and $C = -\dfrac{\pi}{2}$.

The amplitude is $|A| = |1| = 1$

The period is $\dfrac{2\pi}{B} = \dfrac{2\pi}{2} = \pi$. The phase shift is

$\dfrac{C}{B} = \dfrac{-\frac{\pi}{2}}{2} = -\dfrac{\pi}{2} \cdot \dfrac{1}{2} = -\dfrac{\pi}{4}$. The quarter-period is $\dfrac{\pi}{4}$.

The cycle begins at $x = -\dfrac{\pi}{4}$. Add quarter-periods to

generate x-values for the key points.

$x = -\dfrac{\pi}{4}$, $x = -\dfrac{\pi}{4} + \dfrac{\pi}{4} = 0$, $x = 0 + \dfrac{\pi}{4} = \dfrac{\pi}{4}$,

$x = \dfrac{\pi}{4} + \dfrac{\pi}{4} = \dfrac{\pi}{2}$, $x = \dfrac{\pi}{2} + \dfrac{\pi}{4} = \dfrac{3\pi}{4}$ To graph from 0 to

π, evaluate the function at the last four key points

and at $x = \pi$.

x	coordinates
0	$(0, 1)$
$\dfrac{\pi}{4}$	$\left(\dfrac{\pi}{4}, 0\right)$
$\dfrac{\pi}{2}$	$\left(\dfrac{\pi}{2}, -1\right)$
$\dfrac{3\pi}{4}$	$\left(\dfrac{3\pi}{4}, 0\right)$
π	$(\pi, 1)$

Connect the points with a smooth curve and extend the graph one cycle to the right to graph from 0 to 2π.

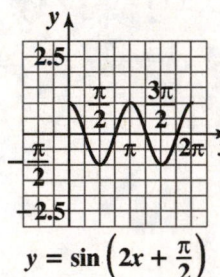

$y = \sin\left(2x + \dfrac{\pi}{2}\right)$

16. Solve the equations

$$3x = -\frac{\pi}{2} \quad \text{and} \quad 3x = \frac{\pi}{2}$$

$$x = \frac{-\frac{\pi}{2}}{3} \qquad\qquad x = \frac{\frac{\pi}{2}}{3}$$

$$x = -\frac{\pi}{6} \qquad\qquad x = \frac{\pi}{6}$$

Thus, two consecutive asymptotes occur at $x = -\frac{\pi}{6}$

and $x = \frac{\pi}{6}$.

$$x\text{-intercept} = \frac{-\frac{\pi}{6} + \frac{\pi}{6}}{2} = \frac{0}{2} = 0$$

An x-intercept is 0 and the graph passes through $(0,0)$. Because the coefficient of the tangent is 2, the points on the graph midway between an x-intercept and the asymptotes have y-coordinates of -2 and 2. Use the two consecutive asymptotes, $x = -\frac{\pi}{6}$ and $x = \frac{\pi}{6}$, to graph one full period of $y = 2\tan 3x$ from $-\frac{\pi}{6}$ to $\frac{\pi}{6}$. Extend the pattern to the right to graph two complete cycles.

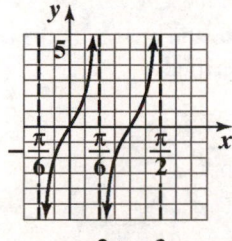

$$y = 2\tan 3x$$

17. $C(p) = 30,000 + 2500p$

$R(p) = 3125p$

$30,000 + 2500p = 3125p$

$\qquad 30,000 = 625p$

$\qquad\qquad p = 48$

48 performances must be played for you to break even.

18. a. Let t be the number of years after 2000.

$$A = A_0 e^{kt}$$

$$A = 110e^{kt}$$

$$233 = 110e^{k(6)}$$

$$\frac{233}{110} = e^{k(6)}$$

$$\ln\frac{233}{110} = \ln e^{k(6)}$$

$$\ln\frac{233}{110} = 6k$$

$$\frac{\ln\frac{233}{110}}{6} = k$$

$$k \approx 0.1251$$

Thus, $A = 110e^{0.1251t}$

b. $\qquad A = 110e^{0.2151t}$

$$300 = 110e^{0.1251t}$$

$$\frac{300}{110} = e^{0.1251t}$$

$$\ln\frac{300}{110} = \ln e^{0.1251t}$$

$$\ln\frac{300}{110} = 0.1251t$$

$$\frac{\ln\frac{300}{110}}{0.1251} = t$$

$$t \approx 8$$

There will be 300 million cell phone subscribers in the United States 8 years after 2000, or 2008.

19. $2200 = \dfrac{k}{3.5}$

$k = 7700$

$h = \dfrac{7700}{5} = 1540$

The rate of heat loss is 1540 Btu per hour.

20. Using a right triangle, we have a known opposite side, a known adjacent side, and an unknown angle. Therefore, use the tangent function.

$$\tan\theta = \frac{200}{50} = 4$$

$$\theta = \tan^{-1}(4) \approx 76°$$

The angle of elevation is about $76°$.

Chapter 5
Analytic Trigonometry

Check Point Exercises

1. $\csc x \tan x = \dfrac{1}{\sin x} \cdot \dfrac{\sin x}{\cos x}$

    $= \dfrac{1}{\cos x}$

    $= \sec x$

    We worked with the left side and arrived at the right side. Thus, the identity is verified.

2. $\cos x \cot x + \sin x = \cos x \cdot \dfrac{\cos x}{\sin x} + \sin x$

    $= \dfrac{\cos^2 x}{\sin x} + \sin x$

    $= \dfrac{\cos^2 x}{\sin x} + \sin x \cdot \dfrac{\sin x}{\sin x}$

    $= \dfrac{\cos^2 x}{\sin x} + \dfrac{\sin^2 x}{\sin x}$

    $= \dfrac{\cos^2 x + \sin^2 x}{\sin x}$

    $= \dfrac{1}{\sin x}$

    $= \csc x$

    We worked with the left side and arrived at the right side. Thus, the identity is verified.

3. $\sin x - \sin x \cos^2 x = \sin x \left(1 - \cos^2 x\right)$

    $= \sin x \cdot \sin^2 x$

    $= \sin^3 x$

    We worked with the left side and arrived at the right side. Thus, the identity is verified.

4. $\dfrac{1 + \cos \theta}{\sin \theta} = \dfrac{1}{\sin \theta} + \dfrac{\cos \theta}{\sin \theta}$

    $= \csc \theta + \cot \theta$

    We worked with the left side and arrived at the right side. Thus, the identity is verified.

5. $\dfrac{\sin x}{1 + \cos x} + \dfrac{1 + \cos x}{\sin x}$

    $= \dfrac{\sin x(\sin x)}{(1 + \cos x) \sin x} + \dfrac{(1 + \cos x)(1 + \cos x)}{\sin x(1 + \cos x)}$

    $= \dfrac{\sin^2 x}{(1 + \cos x) \sin x} + \dfrac{1 + 2\cos x + \cos^2 x}{(1 + \cos x) \sin x}$

    $= \dfrac{\sin^2 x + \cos^2 x + 2\cos x + 1}{(1 + \cos x) \sin x}$

    $= \dfrac{1 + 1 + 2\cos x}{(1 + \cos x) \sin x}$

    $= \dfrac{2 + 2\cos x}{(1 + \cos x) \sin x}$

    $= \dfrac{2(1 + \cos x)}{(1 + \cos x) \sin x}$

    $= \dfrac{2}{\sin x}$

    $= 2 \csc x$

    We worked with the left side and arrived at the right side. Thus, the identity is verified.

6. $\dfrac{\cos x}{1 + \sin x} = \dfrac{\cos x}{(1 + \sin x)} \cdot \dfrac{1 - \sin x}{1 - \sin x}$

    $= \dfrac{\cos x(1 - \sin x)}{1 - \sin^2 x}$

    $= \dfrac{\cos x(1 - \sin x)}{\cos^2 x}$

    $= \dfrac{1 - \sin x}{\cos x}$

    We worked with the left side and arrived at the right side. Thus, the identity is verified.

7. $\dfrac{\sin x}{1+\cos x}+\dfrac{1+\cos}{\sin x}$

$=\dfrac{\sin x(\sin x)}{(1+\cos x)\sin x}+\dfrac{(1+\cos x)(1+\cos x)}{\sin x(1+\cos x)}$

$=\dfrac{\sin^2 x}{(1+\cos x)\sin x}+\dfrac{1+2\cos x+\cos^2 x}{(1+\cos x)\sin x}$

$=\dfrac{\sin^2 x+\cos^2 x+2\cos x+1}{(1+\cos x)\sin x}$

$=\dfrac{1+1+2\cos x}{(1+\cos x)\sin x}$

$=\dfrac{2+2+\cos x}{(1+\cos x)\sin x}$

$=\dfrac{2(1+\cos x)}{(1+\cos x)\sin x}$

$=\dfrac{2}{\sin x}$

$=2\csc x$

We worked with the left side and arrived at the right side. Thus, the identity is verified.

8. Left side:

$\dfrac{1}{1+\sin\theta}+\dfrac{1}{1-\sin\theta}$

$=\dfrac{1(1-\sin\theta)}{(1+\sin\theta)(1-\sin\theta)}+\dfrac{1(1+\sin\theta)}{(1-\sin\theta)(1+\sin\theta)}$

$=\dfrac{1-\sin\theta+1+\sin\theta}{(1+\sin\theta)(1-\sin\theta)}$

$=\dfrac{2}{(1+\sin\theta)(1-\sin\theta)}$

$=\dfrac{2}{1-\sin^2\theta}$

Right side:

$2+2\tan^2\theta=2+2\left(\dfrac{\sin^2\theta}{\cos^2\theta}\right)$

$=\dfrac{2\cos^2\theta}{\cos^2\theta}+\dfrac{2\sin^2\theta}{\cos^2\theta}$

$=\dfrac{2\cos^2\theta+2\sin^2\theta}{\cos^2\theta}$

$=\dfrac{2}{\cos^2\theta}=\dfrac{2}{1-\sin^2\theta}$

The identity is verified because both sides are equal to $\dfrac{2}{1-\sin^2\theta}$.

Exercise Set 5.1

1. $\sin x\sec x=\sin x\cdot\dfrac{1}{\cos x}$

$=\dfrac{\sin x}{\cos x}$

$=\tan x$

3. $\tan(-x)\cdot\cos x=-\tan x\cdot\cos x$

$=-\dfrac{\sin x}{\cos x}\cdot\cos x$

$=-\sin x$

5. $\tan x\csc x\cos x=\dfrac{\sin x}{\cos x}\cdot\dfrac{1}{\sin x}\cos x$

$=1$

7. $\sec x-\sec x\sin^2 x=\sec x(1-\sin^2 x)$

$=\dfrac{1}{\cos x}\cdot\cos^2 x$

$=\cos x$

9. $\cos^2 x-\sin^2 x=\left(1-\sin^2 x\right)-\sin^2 x$

$=1-\sin^2 x-\sin^2 x$

$=1-2\sin^2 x$

11. $\csc\theta-\sin\theta=\dfrac{1}{\sin\theta}-\sin\theta$

$=\dfrac{1}{\sin\theta}-\dfrac{\sin^2\theta}{\sin\theta}$

$=\dfrac{1-\sin^2\theta}{\sin\theta}$

$=\dfrac{\cos^2\theta}{\sin\theta}$

$=\dfrac{\cos\theta}{\sin\theta}\cdot\cos\theta$

$=\cot\theta\cos\theta$

13. $\dfrac{\tan\theta\cot\theta}{\csc\theta} = \dfrac{\dfrac{\sin\theta}{\cos\theta}\cdot\dfrac{\cos\theta}{\sin\theta}}{\dfrac{1}{\sin\theta}}$

$\qquad = \dfrac{1}{\dfrac{1}{\sin\theta}}$

$\qquad = 1 \div \dfrac{1}{\sin\theta}$

$\qquad = 1 \cdot \dfrac{\sin\theta}{1}$

$\qquad = \sin\theta$

15. $\sin^2\theta(1+\cot^2\theta) = \sin^2\theta(\csc^2\theta)$

$\qquad\qquad\qquad = \sin^2\theta\cdot\dfrac{1}{\sin^2\theta}$

$\qquad\qquad\qquad = 1$

17. $\dfrac{1-\cos^2 t}{\cos t} = \dfrac{\sin^2 t}{\cos t}$

$\qquad = \sin t\cdot\dfrac{\sin t}{\cos t}$

$\qquad = \sin t\tan t$

19. $\dfrac{\csc^2 t}{\cot t} = \dfrac{\dfrac{1}{\sin^2 t}}{\dfrac{\cos t}{\sin t}}$

$\qquad = \dfrac{1}{\sin^2 t}\div\dfrac{\cos t}{\sin t}$

$\qquad = \dfrac{1}{\sin^2 t}\cdot\dfrac{\sin t}{\cos t}$

$\qquad = \dfrac{1}{\sin t}\cdot\dfrac{1}{\cos t}$

$\qquad = \csc t\sec t$

21. $\dfrac{\tan^2 t}{\sec t} = \dfrac{\sec^2 t - 1}{\sec t}$

$\qquad = \dfrac{\sec^2 t}{\sec t} - \dfrac{1}{\sec t}$

$\qquad = \sec t - \cos t$

23. $\dfrac{1-\cos\theta}{\sin\theta} = \dfrac{1}{\sin\theta} - \dfrac{\cos\theta}{\sin\theta}$

$\qquad\qquad = \csc\theta - \cot\theta$

25. $\dfrac{\sin t}{\csc t} + \dfrac{\cos t}{\sec t} = \dfrac{\sin t}{\dfrac{1}{\sin t}} + \dfrac{\cos t}{\dfrac{1}{\cos t}}$

$\qquad = \sin t \div \dfrac{1}{\sin t} + \cos t \div \dfrac{1}{\cos t}$

$\qquad = \sin t\cdot\dfrac{\sin t}{1} + \cos t\cdot\dfrac{\cos t}{1}$

$\qquad = \sin^2 t + \cos^2 t$

$\qquad = 1$

27. $\tan t + \dfrac{\cos t}{1+\sin t}$

$\qquad = \dfrac{\sin t}{\cos t} + \dfrac{\cos t}{1+\sin t}$

$\qquad = \dfrac{\sin t}{\cos t}\cdot\dfrac{1+\sin t}{1+\sin t} + \dfrac{\cos t}{1+\sin t}\cdot\dfrac{\cos t}{\cos t}$

$\qquad = \dfrac{\sin t + \sin^2 t}{\cos t(1+\sin t)} + \dfrac{\cos^2 t}{\cos t(1+\sin t)}$

$\qquad \dfrac{\sin t + \sin^2 t + \cos^2 t}{\cos t(1+\sin t)}$

$\qquad = \dfrac{1+\sin t}{\cos t(1+\sin t)}$

$\qquad = \dfrac{1}{\cos t}$

$\qquad = \sec t$

29. $1 - \dfrac{\sin^2 x}{1+\cos x} = 1 - \dfrac{\sin^2 x}{1+\cos x}\cdot\dfrac{1-\cos x}{1-\cos x}$

$\qquad = 1 - \dfrac{\sin^2 x(1-\cos x)}{1-\cos^2 x}$

$\qquad = 1 - \dfrac{\sin^2 x(1-\cos x)}{\sin^2 x}$

$\qquad = 1 - 1 + \cos x$

$\qquad = \cos x$

31. $\dfrac{\cos x}{1-\sin x}+\dfrac{1-\sin x}{\cos x}$

$=\dfrac{\cos x}{1-\sin x}\cdot\dfrac{1+\sin x}{1+\sin x}+\dfrac{1-\sin x}{\cos x}$

$=\dfrac{\cos x(1+\sin x)}{1-\sin^2 x}+\dfrac{1-\sin x}{\cos x}$

$=\dfrac{\cos x(1+\sin x)}{\cos^2 x}+\dfrac{1-\sin x}{\cos x}$

$=\dfrac{1+\sin x}{\cos x}+\dfrac{1-\sin x}{\cos x}$

$=\dfrac{2}{\cos x}$

$=2\cdot\dfrac{1}{\cos x}$

$=2\sec x$

33. $\sec^2 x\csc^2 x=(1+\tan^2 x)\csc^2 x$

$=\csc^2 x+\tan^2 x\csc^2 x$

$=\csc^2 x+\dfrac{\sin^2 x}{\cos^2 x}\cdot\dfrac{1}{\sin^2 x}$

$=\csc^2 x+\dfrac{1}{\cos^2 x}$

$=\csc^2 x+\sec^2 x$

$=\sec^2 x+\csc^2 x$

35. $\dfrac{\sec x-\csc x}{\sec x+\csc x}=\dfrac{\dfrac{1}{\cos x}-\dfrac{1}{\sin x}}{\dfrac{1}{\cos x}+\dfrac{1}{\sin x}}$

$=\dfrac{\dfrac{1}{\cos x}-\dfrac{1}{\sin x}}{\dfrac{1}{\cos x}+\dfrac{1}{\sin x}}\cdot\dfrac{\sin x}{\sin x}$

$=\dfrac{\dfrac{\sin x}{\cos x}-1}{\dfrac{\sin x}{\cos x}+1}$

$=\dfrac{\tan x-1}{\tan x+1}$

37. $\dfrac{\sin^2 x-\cos^2 x}{\sin x+\cos x}=\dfrac{(\sin x+\cos x)(\sin x-\cos x)}{\sin x+\cos x}$

$=\sin x-\cos x$

39. $\tan^2 2x+\sin^2 2x+\cos^2 2x=\tan^2 2x+1$

$=\sec^2 2x$

41. $\dfrac{\tan 2\theta+\cot 2\theta}{\csc 2\theta}=\dfrac{\dfrac{\sin 2\theta}{\cos 2\theta}+\dfrac{\cos 2\theta}{\sin 2\theta}}{\dfrac{1}{\sin 2\theta}}$

$=\dfrac{\dfrac{\sin 2\theta}{\cos 2\theta}\cdot\dfrac{\sin 2\theta}{\sin 2\theta}+\dfrac{\cos 2\theta}{\sin 2\theta}\cdot\dfrac{\cos 2\theta}{\cos 2\theta}}{\dfrac{1}{\sin 2\theta}}$

$=\dfrac{\dfrac{\sin^2 2\theta+\cos 2\theta}{\cos 2\theta\sin 2\theta}}{\dfrac{1}{\sin 2\theta}}$

$=\dfrac{1}{\cos 2\theta\sin 2\theta}\div\dfrac{1}{\sin 2\theta}$

$=\dfrac{1}{\cos 2\theta\sin 2\theta}\cdot\dfrac{\sin 2\theta}{1}$

$=\dfrac{1}{\cos 2\theta}=\sec 2\theta$

43. $\dfrac{\tan x+\tan y}{1-\tan x\tan y}=\dfrac{\dfrac{\sin x}{\cos x}+\dfrac{\sin y}{\cos y}}{1-\dfrac{\sin x}{\cos y}\cdot\dfrac{\sin y}{\cos y}}\cdot\dfrac{\cos x\cos y}{\cos x\cos y}$

$=\dfrac{\sin x\cos y+\cos x\sin y}{\cos x\cos y-\sin x\sin y}$

45. Left side:

$$(\sec x - \tan x)^2 = \left(\frac{1}{\cos x} - \frac{\sin x}{\cos x}\right)^2$$

$$= \left(\frac{1-\sin x}{\cos x}\right)^2$$

$$= \frac{(1-\sin x)^2}{\cos^2 x}$$

Right side:

$$\frac{1-\sin x}{1+\sin x} = \frac{1-\sin x}{1+\sin x} \cdot \frac{1-\sin x}{1-\sin x}$$

$$= \frac{(1-\sin x)^2}{1-\sin^2 x}$$

$$= \frac{(1-\sin x)^2}{\cos^2 x}$$

The identity is verified because both sides are equal
to $\dfrac{(1-\sin x)^2}{\cos^2 x}$.

47. $\dfrac{\tan t}{\sec t - 1} = \dfrac{\tan t}{\sec t - 1} \cdot \dfrac{\sec t + 1}{\sec t + 1}$

$$= \frac{\tan t(\sec t + 1)}{\sec^2 t - 1}$$

$$= \frac{\tan t(\sec t + 1)}{\tan^2 t}$$

$$= \frac{\sec t + 1}{\tan t}$$

49. Left side:

$$\frac{1+\cos t}{1-\cos t} = \frac{1+\cos t}{1-\cos t} \cdot \frac{1+\cos t}{1+\cos t}$$

$$= \frac{(1+\cos t)^2}{1-\cos^2 t}$$

$$= \frac{(1+\cos t)^2}{\sin^2 t}$$

Right side:

$$(\csc t + \cot t)^2 = \left(\frac{1}{\sin t} + \frac{\cos t}{\sin t}\right)^2$$

$$= \left(\frac{1+\cos t}{\sin t}\right)^2$$

$$= \frac{(1+\cos t)^2}{\sin^2 t}$$

The identity is verified because both sides are equal
to $\dfrac{(1+\cos t)^2}{\sin^2 t}$.

51. $\cos^4 t - \sin^4 t = \left(\cos^2 t - \sin^2 t\right)\left(\cos^2 t + \sin^2 t\right)$

$$= \left(\cos^2 t - \sin^2 t\right) \cdot 1$$

$$= 1 - \sin^2 t - \sin^2 t$$

$$= 1 - 2\sin^2 t$$

53. $\dfrac{\sin\theta - \cos\theta}{\sin\theta} + \dfrac{\cos\theta - \sin\theta}{\cos\theta}$

$$= \frac{(\sin\theta - \cos\theta)\cos\theta}{\cos\theta\sin\theta} + \frac{(\cos\theta - \sin\theta)\sin\theta}{\cos\theta\sin\theta}$$

$$= \frac{\sin\theta\cos\theta - \cos^2\theta + \sin\theta\cos\theta - \sin^2\theta}{\sin\theta\cos\theta}$$

$$= \frac{2\sin\theta\cos\theta - \left(\cos^2\theta + \sin^2\theta\right)}{\sin\theta\cos\theta}$$

$$= \frac{2\sin\theta\cos\theta - 1}{\sin\theta\cos\theta}$$

$$= \frac{2\sin\theta\cos\theta}{\sin\theta\cos\theta} - \frac{1}{\sin\theta\cos\theta}$$

$$= 2 - \frac{1}{\sin\theta} \cdot \frac{1}{\cos\theta}$$

$$= 2 - \csc\theta\sec\theta$$

$$= 2 - \sec\theta\csc\theta$$

55. $\left(\tan^2\theta + 1\right)\left(\cos^2\theta + 1\right)$

$$= \tan^2\theta\cos^2\theta + \tan^2\theta + \cos^2\theta + 1$$

$$= \frac{\sin^2\theta}{\cos^2\theta} \cdot \cos^2\theta + \tan^2\theta + \cos^2\theta + 1$$

$$= \sin^2\theta + \tan^2\theta + \cos^2\theta + 1$$

$$= \sin^2\theta + \cos^2\theta + \tan^2\theta + 1$$

$$= 1 + \tan^2\theta + 1$$

$$= \tan^2\theta + 2$$

57. $(\cos\theta - \sin\theta)^2 + (\cos\theta + \sin\theta)^2$

$= \cos^2\theta - 2\cos\theta\sin\theta + \sin^2\theta + \cos^2\theta + 2\cos\theta\sin\theta + \sin^2\theta$

$= \cos^2\theta + \sin^2\theta + \cos^2\theta + \sin^2\theta$

$= 1 + 1 = 2$

59. $\dfrac{\cos^2 x - \sin^2 x}{1 - \tan^2 x}$

$= \dfrac{\cos^2 x - \sin^2 x}{1 - \dfrac{\sin^2 x}{\cos^2 x}} = \dfrac{\cos^2 x - \sin^2 x}{\dfrac{\cos^2 x - \sin^2 x}{\cos^2 x}}$

$= \dfrac{\cos^2 x - \sin^2 x}{1} \div \dfrac{\cos^2 x - \sin^2 x}{\cos^2 x}$

$= \dfrac{\cos^2 x - \sin^2 x}{1} \cdot \dfrac{\cos^2 x}{\cos^2 x - \sin^2 x} = \cos^2 x$

61. Conjecture: left side is equal to $\cos x$

$\dfrac{(\sec x + \tan x)(\sec x - \tan x)}{\sec x} = \dfrac{\sec^2 x - \tan^2 x}{\sec x}$

$= \dfrac{1}{\sec x}$

$= \cos x$

63. Conjecture: left side is equal to $2\sin x$

$\dfrac{\cos x + \cot x \sin x}{\cot x} = \dfrac{\cos x}{\cot x} + \dfrac{\cot x \sin x}{\cot x}$

$= \dfrac{\cos x}{\dfrac{\cos x}{\sin x}} + \dfrac{\cot x \sin x}{\cot x}$

$= \dfrac{\cos x \sin x}{\cos x} + \sin x$

$= \sin x + \sin x$

$= 2\sin x$

65. Conjecture: left side is equal to $2\sec x$

$\dfrac{1}{\sec x + \tan x} + \dfrac{1}{\sec x - \tan x} = \dfrac{\sec^2 x - \tan^2 x}{\sec x + \tan x} + \dfrac{\sec^2 x - \tan^2 x}{\sec x - \tan x}$

$= \dfrac{(\sec x + \tan x)(\sec x - \tan x)}{\sec x + \tan x} + \dfrac{(\sec x + \tan x)(\sec x - \tan x)}{\sec x - \tan x}$

$= \sec x - \tan x + \sec x + \tan x$

$= 2\sec x$

67. $\dfrac{\tan x + \cot x}{\csc x} = \dfrac{\dfrac{\sin x}{\cos x} + \dfrac{\cos x}{\sin x}}{\dfrac{1}{\sin x}}$

$= \left(\dfrac{\sin x}{\cos x} + \dfrac{\cos x}{\sin x} \right) \dfrac{\sin x}{1}$

$= \dfrac{\sin^2 x}{\cos x} + \dfrac{\sin x \cos x}{\sin x}$

$= \dfrac{1 - \cos^2 x}{\cos x} + \dfrac{\cos x}{1}$

$= \dfrac{1 - \cos^2 x}{\cos x} + \dfrac{\cos^2 x}{\cos x}$

$= \dfrac{1}{\cos x}$

69. $\dfrac{\cos x}{1 + \sin x} + \tan x = \dfrac{\cos x}{1 + \sin x} \cdot \dfrac{\cos x}{\cos x} + \dfrac{\sin x}{\cos x} \cdot \dfrac{1 + \sin x}{1 + \sin x}$

$= \dfrac{\cos^2 x}{(1 + \sin x)(\cos x)} + \dfrac{\sin x + \sin^2 x}{(1 + \sin x)(\cos x)}$

$= \dfrac{\cos^2 x + \sin x + \sin^2 x}{(1 + \sin x)(\cos x)}$

$= \dfrac{\sin x + \cos^2 x + \sin^2 x}{(1 + \sin x)(\cos x)}$

$= \dfrac{\sin x + 1}{(1 + \sin x)(\cos x)}$

$= \dfrac{1}{\cos x}$

71. $\dfrac{1}{1 - \cos x} - \dfrac{\cos x}{1 + \cos x} = \dfrac{1}{1 - \cos x} \cdot \dfrac{1 + \cos x}{1 + \cos x} - \dfrac{\cos x}{1 + \cos x} \cdot \dfrac{1 - \cos x}{1 - \cos x}$

$= \dfrac{1 + \cos x}{1 - \cos^2 x} - \dfrac{\cos x - \cos^2 x}{1 - \cos^2 x}$

$= \dfrac{1 + \cos x - \cos x + \cos^2 x}{1 - \cos^2 x}$

$= \dfrac{1 + \cos^2 x}{\sin^2 x}$

$= \dfrac{1}{\sin^2 x} + \dfrac{\cos^2 x}{\sin^2 x}$

$= \csc^2 x + \cot^2 x$

$= \csc^2 x + \csc^2 x - 1$

$= 2 \csc^2 x - 1$

73.

$$\frac{1}{\csc x - \sin x} = \frac{1}{\dfrac{1}{\sin x} - \sin x}$$

$$= \frac{1}{\dfrac{1}{\sin x} - \sin x}$$

$$= \frac{1}{\dfrac{1}{\sin x} - \dfrac{\sin^2 x}{\sin x}}$$

$$= \frac{1}{\dfrac{1 - \sin^2 x}{\sin x}}$$

$$= \frac{1}{\dfrac{\cos^2 x}{\sin x}}$$

$$= \frac{\sin x}{\cos^2 x}$$

$$= \frac{1}{\cos x} \cdot \frac{\sin x}{\cos x}$$

$$= \sec x \tan x$$

75. – 77. Answers may vary.

79.

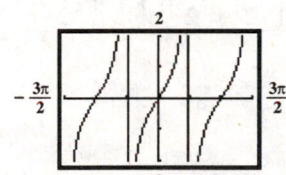

$$\sec x(\sin x - \cos x) + 1 = \frac{1}{\cos x}(\sin x - \cos x) + 1$$

$$= \frac{\sin x}{\cos x} - \frac{\cos x}{\cos x} + 1$$

$$= \tan x - 1 + 1$$

$$= \tan x$$

81.

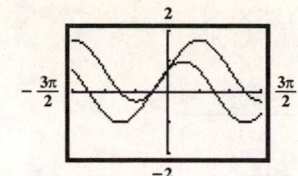

The graphs do not coincide.
Values for x may vary.

83.

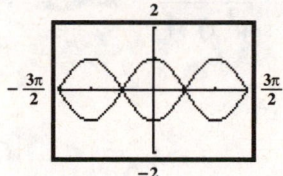

The graphs do not coincide.
Values for x may vary.

85.

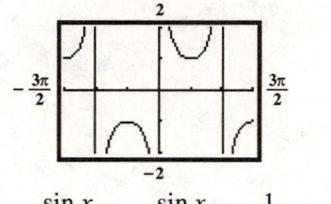

$$\frac{\sin x}{1 - \cos^2 x} = \frac{\sin x}{\sin^2 x} = \frac{1}{\sin x} = \csc x$$

87.

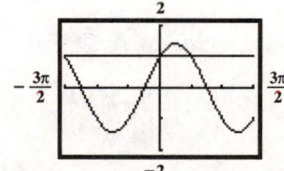

The graphs do not coincide.
Values for x may vary.

89. makes sense

91. does not make sense; Explanations will vary.
Sample explanation: The most efficient way to
simplify the identity is to multiply out the numerator
and then use a Pythagorean identity.

93. $\dfrac{\sin x - \cos x + 1}{\sin x + \cos x - 1}$

$= \dfrac{\sin x - \cos x + 1}{\sin x + \cos x - 1} \cdot \dfrac{\sin x - \cos x - 1}{\sin x - \cos x - 1}$

$= \dfrac{\sin^2 x - 2\cos x \sin x + \cos^2 x - 1}{\sin^2 x - 2\sin x - \cos^2 x + 1}$

$= \dfrac{\sin^2 x + \cos^2 x - 2\cos x \sin x - 1}{\sin x^2 - 2\sin x - (1 - \sin^2 x) + 1}$

$= \dfrac{1 - 2\cos x \sin x - 1}{\sin^2 x - 2\sin x + \sin x^2}$

$= \dfrac{-2\cos x \sin x}{2\sin^2 x - 2\sin x}$

$= \dfrac{-2\sin x \cos x}{2\sin x(\sin x - 1)}$

$= \dfrac{-\cos x}{\sin x - \cos x}$

$= \dfrac{-\cos x}{\sin x - 1} \cdot \dfrac{\sin x + 1}{\sin x + 1}$

$= \dfrac{-\cos x(\sin x + 1)}{\sin^2 x - 1}$

$= \dfrac{-\cos x(\sin x + 1)}{\cos^2 x - 1 - \cos^2 x}$

$= \dfrac{-\cos x(\sin x + 1)}{\cos^2 x}$

$= \dfrac{\sin x + 1}{\cos x}$

95. $\ln e^{\tan^2 x - \sec^2 x}$

$= \tan^2 x - \sec^2 x$

$= -(-\tan^2 x + \sec^2 x)$

$= -(\sec^2 x - \tan^2 x)$

$= -1$

97. Answers may vary.

98. $\cos 30° = \dfrac{\sqrt{3}}{2}$

$\sin 30° = \dfrac{1}{2}$

$\cos 60° = \dfrac{1}{2}$

$\sin 60° = \dfrac{\sqrt{3}}{2}$

$\cos 90° = 0$

$\sin 90° = 1$

99. **a.** No, they are not equal.

$\cos(30° + 60°) \neq \cos 30° + \cos 60°$

$\cos 90° \neq \dfrac{\sqrt{3}}{2} + \dfrac{1}{2}$

$0 \neq \dfrac{1 + \sqrt{3}}{2}$

b. Yes, they are equal.

$\cos(30° + 60°) = \cos 30° \cos 60° - \sin 30° \sin 60°$

$\cos 90° = \left(\dfrac{\sqrt{3}}{2}\right)\left(\dfrac{1}{2}\right) - \left(\dfrac{1}{2}\right)\left(\dfrac{\sqrt{3}}{2}\right)$

$0 = \dfrac{\sqrt{3}}{4} - \dfrac{\sqrt{3}}{4}$

$0 = 0$

100. **a.** No, they are not equal.

$\sin(30° + 60°) \neq \sin 30° + \sin 60°$

$\sin 90° \neq \dfrac{1}{2} + \dfrac{\sqrt{3}}{2}$

$1 \neq \dfrac{1 + \sqrt{3}}{2}$

b. Yes, they are equal.

$\sin(30° + 60°) = \sin 30° \cos 60° + \cos 30° \sin 60°$

$\sin 90° = \left(\dfrac{1}{2}\right)\left(\dfrac{1}{2}\right) + \left(\dfrac{\sqrt{3}}{2}\right)\left(\dfrac{\sqrt{3}}{2}\right)$

$1 = \dfrac{1}{4} + \dfrac{3}{4}$

$1 = 1$

Section 5.2

Check Point Exercises

1.　$\cos 30° = \cos(90° - 60°)$

$\qquad = \cos 90° \cos 60° + \sin 90° \sin 60°$

$\qquad = 0 \cdot \dfrac{1}{2} + 1 \cdot \dfrac{\sqrt{3}}{2}$

$\qquad = 0 + \dfrac{\sqrt{3}}{2}$

$\qquad = \dfrac{\sqrt{3}}{2}$

2.　$\cos 70° \cos 40° + \sin 70° \sin 40°$

$\qquad = \cos(70 - 40°)$

$\qquad = \cos 30°$

$\qquad = \dfrac{\sqrt{3}}{2}$

3.　$\dfrac{\cos(\alpha - \beta)}{\cos \alpha \cos \beta} = \dfrac{\cos \alpha \cos \beta + \sin \alpha \sin \beta}{\cos \alpha \cos \beta}$

$\qquad\qquad = \dfrac{\cos \alpha}{\cos \alpha} \cdot \dfrac{\cos \beta}{\cos \beta} + \dfrac{\sin \alpha}{\cos \alpha} \cdot \dfrac{\sin \beta}{\cos \beta}$

$\qquad\qquad = 1 \cdot 1 + \tan \alpha \cdot \tan \beta$

$\qquad\qquad = 1 + \tan \alpha \tan \beta$

We worked with the left side and arrived at the right side. Thus, the identity is verified.

4.　$\sin \dfrac{5\pi}{12} = \sin\left(\dfrac{\pi}{6} + \dfrac{\pi}{4}\right)$

$\qquad = \sin \dfrac{\pi}{6} \cos \dfrac{\pi}{4} + \cos \dfrac{\pi}{6} \sin \dfrac{\pi}{4}$

$\qquad = \dfrac{1}{2} \cdot \dfrac{\sqrt{2}}{2} + \dfrac{\sqrt{3}}{2} \cdot \dfrac{\sqrt{2}}{2}$

$\qquad = \dfrac{\sqrt{2}}{4} + \dfrac{\sqrt{6}}{4}$

$\qquad = \dfrac{\sqrt{2} + \sqrt{6}}{4}$

5.　**a.**　$\sin \alpha = \dfrac{4}{5} = \dfrac{y}{r}$

Find x:

$x^2 + y^2 = r^2$

$x^2 + 4^2 = 5^2$

$x^2 + 16 = 25$

$x^2 = 9$

Because α is in Quadrant II, x is negative.

$x = -\sqrt{9} = -3$

$\cos \alpha = \dfrac{x}{r} = \dfrac{-3}{5} = -\dfrac{3}{5}$

b.　$\sin \beta = \dfrac{1}{2} = \dfrac{y}{r}$

Find x:

$x^2 + y^2 = r^2$

$x^2 + 1^2 = 2^2$

$x^2 + 1 = 4$

$x^2 = 3$

Because β is in Quadrant I, x is positive.

$x = \sqrt{3}$

$\cos \beta = \dfrac{x}{r} = \dfrac{\sqrt{3}}{2}$

c.　$\cos(\alpha + \beta) = \cos \alpha \cos \beta - \sin \alpha \sin \beta$

$\qquad = -\dfrac{3}{5} \cdot \dfrac{\sqrt{3}}{2} - \dfrac{4}{5} \cdot \dfrac{1}{2}$

$\qquad = \dfrac{-3\sqrt{3}}{10} - \dfrac{4}{10}$

$\qquad = \dfrac{-3\sqrt{3} - 4}{10}$

d.　$\sin(\alpha + \beta) = \sin \alpha \cos \beta + \cos \alpha \sin \beta$

$\qquad = \dfrac{4}{5} \cdot \dfrac{\sqrt{3}}{2} + \dfrac{-3}{5} \cdot \dfrac{1}{2}$

$\qquad = \dfrac{4\sqrt{3}}{10} + \dfrac{-3}{10}$

$\qquad = \dfrac{4\sqrt{3} - 3}{10}$

6. **a.** The graph appears to be the sine curve,
$y = \sin x$.

It cycles through intercept, maximum, intercept, minimum and back to intercept. Thus, $y = \sin x$ also describes the graph.

b. $\cos\left(x + \dfrac{3\pi}{2}\right) = \cos x \cos\dfrac{3\pi}{2} - \sin x \sin\dfrac{3\pi}{2}$

$= \cos x \cdot 0 - \sin x \cdot (-1)$

$= \sin x$

This verifies our observation that

$y = \cos\left(x + \dfrac{3\pi}{2}\right)$ and $y = \sin x$ describe the

same graph.

7. $\tan(x + \pi) = \dfrac{\tan x + \tan \pi}{1 - \tan x \tan \pi}$

$= \dfrac{\tan x + 0}{1 - \tan x \cdot 0}$

$= \dfrac{\tan x}{1}$

$= \tan x$

Exercise Set 5.2

1. $\cos(45° - 30°) = \cos 45° \cos 30° + \sin 45° \sin 30°$

$= \dfrac{\sqrt{2}}{2} \cdot \dfrac{\sqrt{3}}{2} + \dfrac{\sqrt{2}}{2} \cdot \dfrac{1}{2}$

$= \dfrac{\sqrt{6}}{4} + \dfrac{\sqrt{2}}{4}$

$= \dfrac{\sqrt{6} + \sqrt{2}}{4}$

3. $\cos\left(\dfrac{3\pi}{4} - \dfrac{\pi}{6}\right) = \cos\dfrac{3\pi}{4}\cos\dfrac{\pi}{6} + \sin\dfrac{3\pi}{4}\sin\dfrac{\pi}{6}$

$= -\dfrac{\sqrt{2}}{2} \cdot \dfrac{\sqrt{3}}{2} + \dfrac{\sqrt{2}}{2} \cdot \dfrac{1}{2}$

$= -\dfrac{\sqrt{6}}{4} + \dfrac{\sqrt{2}}{4}$

$= \dfrac{\sqrt{2} - \sqrt{6}}{4}$

5. **a.** $\cos 50° \cos 20° + \sin 50° \sin 20°$

$= \cos\alpha\cos\beta + \sin\alpha\sin\beta$

Thus, $\alpha = 50°$ and $\beta = 20°$.

b. $\cos 50° \cos 20° + \sin 50° \sin 20°$

$= \cos(50° - 20°)$

$= \cos 30°$

c. $\cos 30° = \dfrac{\sqrt{3}}{2}$

7. **a.** $\cos\dfrac{5\pi}{12}\cos\dfrac{\pi}{12} + \sin\dfrac{5\pi}{12}\sin\dfrac{\pi}{12}$

$= \cos\alpha\cos\beta + \sin\alpha\sin\beta$

Thus, $\alpha = \dfrac{5\pi}{12}$ and $\beta = \dfrac{\pi}{12}$.

b. $\cos\dfrac{5\pi}{12}\cos\dfrac{\pi}{12} + \sin\dfrac{5\pi}{12}\sin\dfrac{\pi}{12}$

$= \cos\left(\dfrac{5\pi}{12} - \dfrac{\pi}{12}\right)$

$= \cos\dfrac{4\pi}{12}$

$= \cos\dfrac{\pi}{3}$

c. $\cos\dfrac{\pi}{3} = \dfrac{1}{2}$

9. $\dfrac{\cos(\alpha - \beta)}{\cos\alpha\sin\beta} = \dfrac{\cos\alpha\cos\beta - \sin\alpha\sin\beta}{\cos\alpha\sin\beta}$

$= \dfrac{\cos\alpha}{\cos\alpha} \cdot \dfrac{\cos\beta}{\sin\beta} - \dfrac{\sin\alpha}{\cos\alpha} \cdot \dfrac{\sin\beta}{\sin\beta}$

$= 1 \cdot \cot\beta + \tan\alpha \cdot 1$

$= \tan\alpha + \cot\beta$

11. $\cos\left(x - \dfrac{\pi}{4}\right) = \cos x \cos\dfrac{\pi}{4} + \sin x \sin\dfrac{\pi}{4}$

$= \cos x \cdot \dfrac{\sqrt{2}}{2} + \sin x \cdot \dfrac{\sqrt{2}}{2}$

$= \dfrac{\sqrt{2}}{2}(\cos x + \sin x)$

446

13. $\sin(45° - 30°) = \sin 45° \cos 30° - \cos 45° \sin 30°$

$$= \frac{\sqrt{2}}{2} \cdot \frac{\sqrt{3}}{2} - \frac{\sqrt{2}}{2} \cdot \frac{1}{2}$$

$$= \frac{\sqrt{6}}{4} - \frac{\sqrt{2}}{4}$$

$$= \frac{\sqrt{6} - \sqrt{2}}{4}$$

15. $\sin(105°) = \sin(60° + 45°)$

$$= \sin 60° \cos 45° + \cos 60° \sin 45°$$

$$= \frac{\sqrt{3}}{2} \cdot \frac{\sqrt{2}}{2} + \frac{1}{2} \cdot \frac{\sqrt{2}}{2}$$

$$= \frac{\sqrt{6}}{4} + \frac{\sqrt{2}}{4}$$

$$= \frac{\sqrt{6} + \sqrt{2}}{4}$$

17. $\cos(135° + 30°) = \cos 135° \cos 30° - \sin 135° \sin 30°$

$$= \cos(90° + 45°)\cos 30° - \sin(90° + 45°)\sin 30°$$

$$= (\cos 90° \cos 45° - \sin 90° \sin 45°)\cos 30° - (\sin 90° \cos 45° + \cos 90° \sin 45°)\sin 30°$$

$$= \left(0 \cdot \frac{\sqrt{2}}{2} - 1 \cdot \frac{\sqrt{2}}{2}\right)\frac{\sqrt{3}}{2} - \left(1 \cdot \frac{\sqrt{2}}{2} + 0 \cdot \frac{\sqrt{2}}{2}\right)\frac{1}{2}$$

$$= \left(-\frac{\sqrt{2}}{2}\right)\frac{\sqrt{3}}{2} - \left(\frac{\sqrt{2}}{2}\right)\frac{1}{2}$$

$$= -\frac{\sqrt{6}}{4} - \frac{\sqrt{2}}{4}$$

$$= -\frac{\sqrt{6} + \sqrt{2}}{4}$$

19. $\cos 75° = \cos(45° + 30°)$

$$= \cos 45° \cos 30° - \sin 45° \sin 30°$$

$$= \frac{\sqrt{2}}{2} \cdot \frac{\sqrt{3}}{2} - \frac{\sqrt{2}}{2} \cdot \frac{1}{2}$$

$$= \frac{\sqrt{6}}{4} - \frac{\sqrt{2}}{4}$$

$$= \frac{\sqrt{6} - \sqrt{2}}{4}$$

21. $\tan\left(\dfrac{\pi}{6}+\dfrac{\pi}{4}\right)=\dfrac{\tan\dfrac{\pi}{6}+\tan\dfrac{\pi}{4}}{1-\tan\dfrac{\pi}{6}\tan\dfrac{\pi}{4}}$

$=\dfrac{\dfrac{\sqrt{3}}{3}+1}{1-\dfrac{\sqrt{3}}{3}\cdot 1}$

$=\dfrac{\dfrac{\sqrt{3}}{3}+\dfrac{3}{3}}{\dfrac{3}{3}-\dfrac{\sqrt{3}}{3}}$

$=\dfrac{\sqrt{3}+3}{3-\sqrt{3}}$

$=\dfrac{3+\sqrt{3}}{3-\sqrt{3}}\cdot\dfrac{3+\sqrt{3}}{3+\sqrt{3}}$

$=\dfrac{9+6\sqrt{3}+3}{9-3}$

$=\dfrac{12+6\sqrt{3}}{6}$

$=2+\sqrt{3}$

23. $\tan\left(\dfrac{4\pi}{3}-\dfrac{\pi}{4}\right)=\dfrac{\tan\dfrac{4\pi}{3}-\tan\dfrac{\pi}{4}}{1+\tan\dfrac{4\pi}{3}\tan\dfrac{\pi}{4}}$

$=\dfrac{\sqrt{3}-1}{1+\sqrt{3}\cdot 1}$

$=\dfrac{-1+\sqrt{3}}{1+\sqrt{3}}$

$=\dfrac{-1+\sqrt{3}}{1+\sqrt{3}}\cdot\dfrac{1-\sqrt{3}}{1-\sqrt{3}}$

$=\dfrac{-1+2\sqrt{3}-3}{1-3}$

$=\dfrac{-4+2\sqrt{3}}{-2}$

$=2-\sqrt{3}$

25. $\sin 25°\cos 5°+\cos 25°\sin 5°=\sin(25°+5°)$

$=\sin 30°$

$=\dfrac{1}{2}$

27. $\dfrac{\tan 10°+\tan 35°}{1-\tan 10°\tan 35°}=\tan(10°+35°)$

$=\tan 45°$

$=1$

29. $\sin\dfrac{5\pi}{12}\cos\dfrac{\pi}{4}-\cos\dfrac{5\pi}{12}\sin\dfrac{\pi}{4}=\sin\left(\dfrac{5\pi}{12}-\dfrac{\pi}{4}\right)$

$=\sin\left(\dfrac{2\pi}{12}\right)$

$=\sin\left(\dfrac{\pi}{6}\right)$

$=\dfrac{1}{2}$

31. $\dfrac{\tan\frac{\pi}{5}-\tan\frac{\pi}{30}}{1+\tan\frac{\pi}{5}\tan\frac{\pi}{30}}=\tan\left(\dfrac{\pi}{5}-\dfrac{\pi}{30}\right)$

$=\tan\left(\dfrac{5\pi}{30}\right)=\tan\left(\dfrac{\pi}{6}\right)$

$=\dfrac{\sqrt{3}}{3}$

33. $\sin\left(x+\dfrac{\pi}{2}\right)=\sin x\cos\dfrac{\pi}{2}+\cos x\sin\dfrac{\pi}{2}$

$=\sin x\cdot 0+\cos x\cdot 1$

$=\cos x$

35. $\cos\left(x-\dfrac{\pi}{2}\right)=\cos x\cos\dfrac{\pi}{2}+\sin x\sin\dfrac{\pi}{2}$

$=\cos x\cdot 0+\sin x\cdot 1$

$=\sin x$

37. $\tan(2\pi-x)=\dfrac{\tan 2\pi-\tan x}{1+\tan 2\pi\tan x}$

$=\dfrac{0-\tan x}{1+0\cdot\tan x}$

$=-\tan x$

39. $\sin(\alpha+\beta)+\sin(\alpha-\beta)$

$=\sin\alpha\cos\beta+\cos\alpha\sin\beta$

$+\sin\alpha\cos\beta-\cos\alpha\sin\beta$

$=2\sin\alpha\cos\beta$

41. $\dfrac{\sin(\alpha-\beta)}{\cos\alpha\cos\beta}=\dfrac{\sin\alpha\cos\beta-\cos\alpha\sin\beta}{\cos\alpha\cos\beta}$

$=\dfrac{\sin\alpha\cos\beta}{\cos\alpha\cos\beta}-\dfrac{\cos\alpha\sin\beta}{\cos\alpha\cos\beta}$

$=\tan\alpha\cdot 1-1\cdot\tan\beta$

$=\tan\alpha-\tan\beta$

43. $\tan\left(\theta+\dfrac{\pi}{4}\right) = \dfrac{\tan\theta+\tan\frac{\pi}{4}}{1-\tan\theta\tan\frac{\pi}{4}}$

$= \dfrac{\tan\theta+1}{1-\tan\theta}$

$= \dfrac{\frac{\sin\theta}{\cos\theta}+\frac{\cos\theta}{\cos\theta}}{\frac{\cos\theta}{\cos\theta}-\frac{\sin\theta}{\cos\theta}}$

$= \dfrac{\frac{\sin\theta+\cos\theta}{\cos\theta}}{\frac{\cos\theta-\sin\theta}{\cos\theta}}$

$= \dfrac{\sin\theta+\cos\theta}{\cos\theta}\cdot\dfrac{\cos\theta}{\cos\theta-\sin\theta}$

$= \dfrac{\sin\theta+\cos\theta}{\cos\theta-\sin\theta}$

$= \dfrac{\cos\theta+\sin\theta}{\cos\theta-\sin\theta}$

45. $\cos(\alpha+\beta)\cos(\alpha-\beta)$

$= (\cos\alpha\cos\beta-\sin\alpha\sin\beta)$

$\cdot(\cos\alpha\cos\beta+\sin\alpha\sin\beta)$

$= \cos^2\alpha\cos^2\beta-\sin^2\alpha\sin^2\beta$

$= \left(1-\sin^2\alpha\right)\cos^2\beta-\sin^2\alpha\left(1-\cos^2\beta\right)$

$= \cos^2\beta-\sin^2\alpha\cos^2\beta$

$-\sin^2\alpha+\sin^2\alpha\cos^2\beta$

$= \cos^2\beta-\sin^2\alpha$

47. $\dfrac{\sin(\alpha+\beta)}{\sin(\alpha-\beta)}$

$= \dfrac{\sin\alpha\cos\beta+\cos\alpha\sin\beta}{\sin\alpha\cos\beta-\cos\alpha\sin\beta}$

$= \dfrac{\sin\alpha\cos\beta+\cos\alpha\sin\beta}{\sin\alpha\cos\beta-\cos\alpha\sin\beta}\cdot\dfrac{\frac{1}{\cos\alpha\cos\beta}}{\frac{1}{\cos\alpha\cos\beta}}$

$= \dfrac{\frac{\sin\alpha\cos\beta+\cos\alpha\sin\beta}{\cos\alpha\cos\beta}}{\frac{\sin\alpha\cos\beta-\cos\alpha\sin\beta}{\cos\alpha\cos\beta}}$

$= \dfrac{\frac{\sin\alpha\cos\beta}{\cos\alpha\cos\beta}+\frac{\cos\alpha\sin\beta}{\cos\alpha\cos\beta}}{\frac{\sin\alpha\cos\beta}{\cos\alpha\cos\beta}-\frac{\cos\alpha\sin\beta}{\cos\alpha\cos\beta}}$

$= \dfrac{\tan\alpha\cdot1+1\cdot\tan\beta}{\tan\alpha\cdot1-1\cdot\tan\beta}$

$= \dfrac{\tan\alpha+\tan\beta}{\tan\alpha-\tan\beta}$

49. $\dfrac{\cos(x+h)-\cos x}{h}$

$= \dfrac{\cos x\cos h-\sin x\sin h-\cos x}{h}$

$= \dfrac{\cos x\cos h-\cos x-\sin x\sin h}{h}$

$= \dfrac{\cos x(\cos h-1)-\sin x\sin h}{h}$

$= \cos x\cdot\dfrac{\cos h-1}{h}-\sin x\cdot\dfrac{\sin h}{h}$

51. $\sin2\alpha = \sin(\alpha+\alpha)$

$= \sin\alpha\cos\alpha+\cos\alpha\sin\alpha$

$= 2\sin\alpha\cos\alpha$

53. $\tan2\alpha = \tan(\alpha+\alpha)$

$= \dfrac{\tan\alpha+\tan\alpha}{1-\tan\alpha\tan\alpha}$

$= \dfrac{2\tan\alpha}{1-\tan^2\alpha}$

55. $\tan(\alpha+\beta) = \dfrac{\sin(\alpha+\beta)}{\cos(\alpha+\beta)}$

$= \dfrac{\sin\alpha\cos\beta+\cos\alpha\sin\beta}{\cos\alpha\cos\beta-\sin\alpha\sin\beta}$

$= \dfrac{\sin\alpha\cos\beta+\cos\alpha\sin\beta}{\cos\alpha\cos\beta-\sin\alpha\sin\beta}\cdot\dfrac{\frac{1}{\cos\alpha\cos\beta}}{\frac{1}{\cos\alpha\cos\beta}}$

$= \dfrac{\frac{\sin\alpha\cos\beta+\cos\alpha\sin\beta}{\cos\alpha\cos\beta}}{\frac{\cos\alpha\cos\beta-\sin\alpha\sin\beta}{\cos\alpha\cos\beta}}$

$= \dfrac{\frac{\sin\alpha\cos\beta}{\cos\alpha\cos\beta}+\frac{\cos\alpha\sin\beta}{\cos\alpha\cos\beta}}{\frac{\cos\alpha\cos\beta}{\cos\alpha\cos\beta}-\frac{\sin\alpha\sin\beta}{\cos\alpha\cos\beta}}$

$= \dfrac{\tan\alpha+\tan\beta}{1-\tan\alpha\tan\beta}$

57. $\sin\alpha = \dfrac{3}{5} = \dfrac{y}{r}$

$x^2 + y^2 = r^2$

$x^2 + 3^2 = 5^2$

$x^2 + 9 = 25$

$x^2 = 16$

Because α lies in quadrant I, x is positive.

$x = 4$

Thus, $\cos\alpha = \dfrac{x}{r} = \dfrac{4}{5}$, and

$\tan\alpha = \dfrac{\sin\alpha}{\cos\alpha} = \dfrac{\frac{3}{5}}{\frac{4}{5}} = \dfrac{3}{4}.$

$\sin\beta = \dfrac{5}{13} = \dfrac{y}{r}$

$x^2 + y^2 = r^2$

$x^2 + 5^2 = 13^2$

$x^2 + 25 = 169$

$x^2 = 144$

Because β lies in quadrant II, x is negative.

$x = -12$

Thus, $\cos\beta = \dfrac{x}{r} = \dfrac{-12}{13} = -\dfrac{12}{13}$, and

$\tan\beta = \dfrac{\sin\beta}{\cos\beta} = \dfrac{\frac{5}{13}}{-\frac{12}{13}} = -\dfrac{5}{12}.$

a. $\cos(\alpha + \beta) = \cos\alpha\cos\beta - \sin\alpha\sin\beta$

$= \dfrac{4}{5}\cdot\left(-\dfrac{12}{13}\right) - \dfrac{3}{5}\cdot\dfrac{5}{13} = -\dfrac{63}{65}$

b. $\sin(\alpha + \beta) = \sin\alpha\cos\beta + \cos\alpha\sin\beta$

$= \dfrac{3}{5}\cdot\left(-\dfrac{12}{13}\right) + \dfrac{4}{5}\cdot\dfrac{5}{13} = -\dfrac{16}{65}$

c. $\tan(\alpha + \beta) = \dfrac{\tan\alpha + \tan\beta}{1 - \tan\alpha\tan\beta}$

$= \dfrac{\frac{3}{4} + \left(-\frac{5}{12}\right)}{1 - \frac{3}{4}\cdot\left(-\frac{5}{12}\right)} = \dfrac{\frac{4}{12}}{\frac{63}{48}} = \dfrac{16}{63}$

59. $\tan\alpha = -\dfrac{3}{4} = \dfrac{3}{-4} = \dfrac{y}{x}$

$x^2 + y^2 = r^2$

$(-4)^2 + 3^2 = r^2$

$16 + 9 = r^2$

$25 = r^2$

Because r is a distance, it is positive.

$r = 5$

Thus, $\cos\alpha = \dfrac{x}{r} = \dfrac{-4}{5} = -\dfrac{4}{5}$, and

$\sin\alpha = \dfrac{y}{r} = \dfrac{3}{5}.$

$\cos\beta = \dfrac{1}{3} = \dfrac{x}{r}$

$x^2 + y^2 = r^2$

$1^2 + y^2 = 3^2$

$1 + y^2 = 9$

$y^2 = 8$

Because β lies in quadrant I, y is positive.

$y = \sqrt{8} = 2\sqrt{2}$

Thus, $\sin\beta = \dfrac{y}{r} = \dfrac{2\sqrt{2}}{3}$, and

$\tan\beta = \dfrac{\sin\beta}{\cos\beta} = \dfrac{\frac{2\sqrt{2}}{3}}{\frac{1}{3}} = 2\sqrt{2}.$

a. $\cos(\alpha + \beta) = \cos\alpha\cos\beta - \sin\alpha\sin\beta$

$= \left(-\dfrac{4}{5}\right)\cdot\dfrac{1}{3} - \dfrac{3}{5}\cdot\dfrac{2\sqrt{2}}{3}$

$= -\dfrac{4}{15} - \dfrac{6\sqrt{2}}{15}$

$= \dfrac{-4 - 6\sqrt{2}}{15}$

$= -\dfrac{4 + 6\sqrt{2}}{15}$

b. $\sin(\alpha + \beta) = \sin\alpha\cos\beta + \cos\alpha\sin\beta$

$= \dfrac{3}{5}\cdot\dfrac{1}{3} + \left(-\dfrac{4}{5}\right)\cdot\dfrac{2\sqrt{2}}{3}$

$= \dfrac{3}{15} - \dfrac{8\sqrt{2}}{15}$

$= \dfrac{3 - 8\sqrt{2}}{15}$

c. $\tan(\alpha+\beta) = \dfrac{\tan\alpha+\tan\beta}{1-\tan\alpha\tan\beta}$

$= \dfrac{-\frac{3}{4}+2\sqrt{2}}{1-\left(-\frac{3}{4}\right)\left(2\sqrt{2}\right)}$

$= \dfrac{\frac{-3+8\sqrt{2}}{4}}{\frac{4+6\sqrt{2}}{4}}$

$= \dfrac{-3+8\sqrt{2}}{4+6\sqrt{2}}\cdot\dfrac{\left(4-6\sqrt{2}\right)}{\left(4-6\sqrt{2}\right)}$

$= \dfrac{-108+50\sqrt{2}}{-56}$

$= \dfrac{54-25\sqrt{2}}{28}$

61. $\cos\alpha = \dfrac{8}{17} = \dfrac{x}{r}$

$x^2+y^2 = r^2$

$8^2+y^2 = 17^2$

$64+y^2 = 289$

$y^2 = 225$

Because α lies in quadrant IV, y is negative.

$y = -15$

Thus, $\sin\alpha = \dfrac{y}{r} = \dfrac{-15}{17} = -\dfrac{15}{17}$, and

$\tan\alpha = \dfrac{\sin\alpha}{\cos\alpha} = \dfrac{-\frac{15}{17}}{\frac{8}{17}} = -\dfrac{15}{8}$.

$\sin\beta = -\dfrac{1}{2} = \dfrac{-1}{2} = \dfrac{y}{r}$

$x^2+y^2 = r^2$

$x^2+(-1)^2 = 2^2$

$x^2+1 = 4$

$x^2 = 3$

Because β lies in quadrant III, x is negative.

$x = -\sqrt{3}$

Thus, $\cos\beta = \dfrac{x}{r} = \dfrac{-\sqrt{3}}{2} = -\dfrac{\sqrt{3}}{2}$, and

$\tan\beta = \dfrac{\sin\beta}{\cos\beta} = \dfrac{-\frac{1}{2}}{-\frac{\sqrt{3}}{2}} = \dfrac{1}{\sqrt{3}} = \dfrac{\sqrt{3}}{3}$.

a. $\cos(\alpha+\beta) = \cos\alpha\cos\beta - \sin\alpha\sin\beta$

$= \dfrac{8}{17}\cdot\left(-\dfrac{\sqrt{3}}{2}\right) - \left(-\dfrac{15}{17}\right)\cdot\left(-\dfrac{1}{2}\right)$

$= \dfrac{-8\sqrt{3}-15}{34}$

$= -\dfrac{8\sqrt{3}+15}{34}$

b. $\sin(\alpha+\beta) = \sin\alpha\cos\beta + \cos\alpha\sin\beta$

$= \left(-\dfrac{15}{17}\right)\cdot\left(-\dfrac{\sqrt{3}}{2}\right) + \dfrac{8}{17}\cdot\left(-\dfrac{1}{2}\right)$

$= \dfrac{15\sqrt{3}-8}{34}$

c. $\tan(\alpha+\beta) = \dfrac{\tan\alpha+\tan\beta}{1-\tan\alpha\tan\beta}$

$= \dfrac{-\dfrac{15}{8}+\dfrac{\sqrt{3}}{3}}{1-\left(-\dfrac{15}{8}\right)\left(\dfrac{\sqrt{3}}{3}\right)}$

$= \dfrac{\frac{-45+8\sqrt{3}}{24}}{\frac{24+15\sqrt{3}}{24}}$

$= \dfrac{-45+8\sqrt{3}}{24+15\sqrt{3}}\cdot\dfrac{24-15\sqrt{3}}{24-15\sqrt{3}}$

$= \dfrac{-1440+867\sqrt{3}}{-99}$

$= \dfrac{489-289\sqrt{3}}{33}$

63. $\tan\alpha = \dfrac{3}{4} = \dfrac{y}{x}$

Because α lies in quadrant III, x and y are negative.

$r^2 = x^2 + y^2$

$r^2 = (-4)^2 + (-3)^2$

$r^2 = 25$

$r = 5$

$\sin\alpha = \dfrac{y}{r} = \dfrac{-3}{5} = -\dfrac{3}{5}$

$\cos\alpha = \dfrac{x}{r} = \dfrac{-4}{5} = -\dfrac{4}{5}$

$\cos\beta = \dfrac{1}{4} = \dfrac{x}{r}$

Because β lies in quadrant IV, y is negative.

$x^2 + y^2 = r^2$

$1^2 + y^2 = 4^2$

$y^2 = 15$

$y = -\sqrt{15}$

$\sin\beta = \dfrac{y}{r} = \dfrac{-\sqrt{15}}{4} = -\dfrac{\sqrt{15}}{4}$

$\tan\beta = \dfrac{y}{x} = \dfrac{-\sqrt{15}}{1} = -\sqrt{15}$

a. $\cos(\alpha+\beta) = \cos\alpha\cos\beta - \sin\alpha\sin\beta$

$= -\dfrac{4}{5}\cdot\left(\dfrac{1}{4}\right) - \left(-\dfrac{3}{5}\right)\left(-\dfrac{\sqrt{15}}{4}\right)$

$= -\dfrac{4}{20} - \dfrac{3\sqrt{15}}{20}$

$= -\dfrac{4+3\sqrt{15}}{20}$

b. $\sin(\alpha+\beta) = \sin\alpha\cos\beta + \cos\alpha\sin\beta$

$= \left(-\dfrac{3}{5}\right)\left(\dfrac{1}{4}\right) + \left(-\dfrac{4}{5}\right)\left(-\dfrac{\sqrt{15}}{4}\right)$

$= -\dfrac{3}{20} + \dfrac{4\sqrt{15}}{20}$

$= \dfrac{-3+4\sqrt{15}}{20}$

c. $\tan(\alpha+\beta) = \dfrac{\tan\alpha + \tan\beta}{1 - \tan\alpha\tan\beta}$

$= \dfrac{\dfrac{3}{4} + \left(-\sqrt{15}\right)}{1 - \dfrac{3}{4}\left(-\sqrt{15}\right)}$

$= \dfrac{\dfrac{3}{4} - \dfrac{4\sqrt{15}}{4}}{\dfrac{4}{4} + \dfrac{3\sqrt{15}}{4}}$

$= \dfrac{3-4\sqrt{15}}{4+3\sqrt{15}}$

$= \dfrac{3-4\sqrt{15}}{4+3\sqrt{15}} \cdot \dfrac{4-3\sqrt{15}}{4-3\sqrt{15}}$

$= \dfrac{12-9\sqrt{15}-16\sqrt{15}+180}{16-135}$

$= \dfrac{192-25\sqrt{15}}{-119}$

$= \dfrac{-192+25\sqrt{15}}{119}$

65. a. The graph appears to be the sine curve, $y = \sin x$. It cycles through intercept, maximum, minimum and back to intercept. Thus, $y = \sin x$ also describes the graph.

b. $\sin(\pi - x) = \sin\pi\cos x - \cos\pi\sin x$

$= 0\cdot\cos x - (-1)\cdot\sin x$

$= \sin x$

This verifies our observation that $y = \sin(\pi - x)$ and $y = \sin x$ describe the same graph.

67. a. The graph appears to be 2 times the cosine curve, $y = 2\cos x$. It cycles through maximum, intercept, minimum, intercept and back to maximum. Thus $y = 2\cos x$ also describes the graph.

b. $\sin\left(x + \dfrac{\pi}{2}\right) + \sin\left(\dfrac{\pi}{2} - x\right)$

$= \sin x \cos\dfrac{\pi}{2} + \cos x \sin\dfrac{\pi}{2} + \sin\dfrac{\pi}{2}\cos x$

$\quad - \cos\dfrac{\pi}{2}\sin x$

$= \sin x \cdot 0 + \cos x \cdot 1 + 1 \cdot \cos x - 0 \cdot \sin x$

$= \cos x + \cos x$

$= 2\cos x$

This verifies our observation that

$y = \sin\left(x + \dfrac{\pi}{2}\right) + \sin\left(\dfrac{\pi}{2} - x\right)$ and $y = 2\cos x$

describe the same graph.

69. $\cos(\alpha + \beta)\cos\beta + \sin(\alpha + \beta)\sin\beta$

$= \cos\left[(\alpha + \beta) - \beta\right]$

$= \cos\alpha$

71. $\dfrac{\sin(\alpha + \beta) - \sin(\alpha - \beta)}{\cos(\alpha + \beta) + \cos(\alpha - \beta)}$

$= \dfrac{(\sin\alpha\cos\beta + \cos\alpha\sin\beta) - (\sin\alpha\cos\beta - \cos\alpha\sin\beta)}{(\cos\alpha\cos\beta - \sin\alpha\sin\beta) + (\cos\alpha\cos\beta + \sin\alpha\sin\beta)}$

$= \dfrac{\sin\alpha\cos\beta + \cos\alpha\sin\beta - \sin\alpha\cos\beta + \cos\alpha\sin\beta}{\cos\alpha\cos\beta - \sin\alpha\sin\beta + \cos\alpha\cos\beta + \sin\alpha\sin\beta}$

$= \dfrac{\cos\alpha\sin\beta + \cos\alpha\sin\beta}{\cos\alpha\cos\beta + \cos\alpha\cos\beta}$

$= \dfrac{2\cos\alpha\sin\beta}{2\cos\alpha\cos\beta}$

$= \dfrac{\sin\beta}{\cos\beta}$

$= \tan\beta$

73. $\cos\left(\dfrac{\pi}{6} + \alpha\right)\cos\left(\dfrac{\pi}{6} - \alpha\right) - \sin\left(\dfrac{\pi}{6} + \alpha\right)\sin\left(\dfrac{\pi}{6} - \alpha\right)$

$= \cos\left[\left(\dfrac{\pi}{6} + \alpha\right) + \left(\dfrac{\pi}{6} - \alpha\right)\right]$

$= \cos\left[\dfrac{\pi}{6} + \alpha + \dfrac{\pi}{6} - \alpha\right]$

$= \cos\dfrac{\pi}{3}$

$= \dfrac{1}{2}$

75. Conjecture: the left side is equal to $\cos 3x$.

$\cos 2x\cos 5x + \sin 2x\sin 5x$

$= \cos(2x - 5x)$

$= \cos(-3x)$

$= \cos 3x$

77. Conjecture: the left side is equal to $\sin\dfrac{x}{2}$.

$\sin\dfrac{5x}{2}\cos 2x - \cos\dfrac{5x}{2}\sin 2x$

$= \sin\left(\dfrac{5x}{2} - 2x\right)$

$= \sin\left(\dfrac{5x}{2} - \dfrac{4x}{2}\right)$

$= \sin\dfrac{x}{2}$

79. $\tan\theta = \dfrac{3}{2} = \dfrac{y}{x}$

$x^2 + y^2 = r^2$

$2^2 + 3^3 = r^2$

$4 + 9 = r^2$

$13 = r^2$

Because r is a distance, it is positive.

$r = \sqrt{13}$

Thus, $\sin\theta = \dfrac{y}{r} = \dfrac{3}{\sqrt{13}}$

and $\cos\theta = \dfrac{x}{r} = \dfrac{2}{\sqrt{13}}$

$\sqrt{13}\cos(t-\theta) = \sqrt{13}(\cos t\cos\theta + \sin t\sin\theta)$

$\qquad = \sqrt{13}\left(\cos t\cdot\dfrac{2}{\sqrt{13}} + \sin t\cdot\dfrac{3}{\sqrt{13}}\right)$

$\qquad = \cos t\cdot 2 + \sin t\cdot 3$

$\qquad = 2\cos t + 3\sin t$

For the equation $y = \sqrt{13}\cos(t-\theta)$, the amplitude is

$\left|\sqrt{13}\right| = \sqrt{13}$, and the period is $\dfrac{2\pi}{1} = 2\pi$.

81. – 87. Answers may vary.

89.

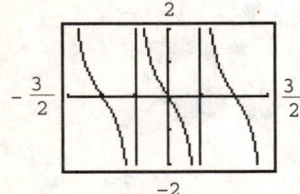

$\tan(\pi - x) = \dfrac{\tan\pi - \tan x}{1 + \tan\pi\tan x}$

$\qquad = \dfrac{0 - \tan x}{1 + 0\cdot\tan x}$

$\qquad = \dfrac{-\tan x}{1}$

$\qquad = -\tan x$

91.

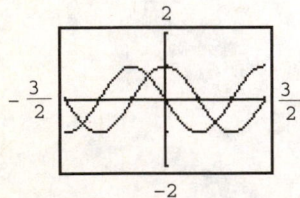

The graphs do not coincide.
Values for x may vary.

93.

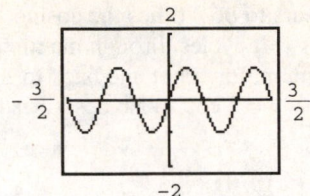

$\sin 1.2x\cos 0.8x + \cos 1.2x\sin 0.8x$

$\sin(1.2x + 0.8x)$

$\sin 2x$

95. makes sense

97. makes sense

99. $\cos^{-1}\dfrac{1}{2}$

$x = 1$

$y = \sqrt{3}$

$r = 2$

$\sin^{-1}\dfrac{3}{5}$

$x = 4$

$y = 3$

$r = 5$

$\sin\left(\cos^{-1}\dfrac{1}{2} + \sin^{-1}\dfrac{3}{5}\right)$

$= \sin\cos^{-1}\dfrac{1}{2}\cos\sin^{-1}\dfrac{3}{5}$

$\quad + \cos\cos^{-1}\dfrac{1}{2}\sin\sin^{-1}\dfrac{3}{5}$

$= \dfrac{\sqrt{3}}{2}\cdot\dfrac{4}{5} + \dfrac{1}{2}\cdot\dfrac{3}{5}$

$= \dfrac{4\sqrt{3} + 3}{10}$

101. $\tan^{-1}\dfrac{4}{3}$

$x = 3$

$y = 4$

$r = 5$

$\cos^{-1}\dfrac{5}{13}$

$x = 5$

$y = 12$

$r = 13$

$\cos\left(\tan^{-1}\dfrac{4}{3} + \cos^{-1}\dfrac{5}{13}\right)$

$= \cos\tan^{-1}\dfrac{4}{3}\cos\cos^{-1}\dfrac{5}{13}$

$\quad - \sin\tan^{-1}\dfrac{4}{3}\sin\cos^{-1}\dfrac{5}{13}$

$= \dfrac{3}{5}\cdot\dfrac{5}{13} - \dfrac{4}{5}\cdot\dfrac{12}{13}$

$= -\dfrac{33}{65}$

103. Let $\alpha = \sin^{-1}x$, where $-\dfrac{\pi}{2} \le \alpha \le \dfrac{\pi}{2}$. $\sin\alpha = x$

Because x is positive, $\sin\alpha$ is positive. Thus α is in quadrant I. Using a right triangle in quadrant I with $\sin\alpha = x$, the third side can be found using the Pythagorean Theorem.

$a^2 + x^2 = 1^2$

$\qquad a^2 = 1 - x^2$

$\qquad a = \sqrt{1 - x^2}$

Thus $\cos\alpha = \dfrac{\sqrt{1 - x^2}}{1} = \sqrt{1 - x^2}$

Because y is positive, $\cos\beta$ is positive. Thus β is in quadrant I. Using a right triangle in quadrant I with $\cos\beta = y$, the third side can be found using the Pythagorean Theorem.

$b^2 + y^2 = 1^2$

$\qquad b^2 = 1 - x^2$

$\qquad a = \sqrt{1 - y^2}$

Thus $\cos\alpha = \dfrac{\sqrt{1 - y^2}}{1} = \sqrt{1 - y^2}$

$\cos(\sin^{-1}x - \cos^{-1}y) = \cos(\alpha - \beta)$

$\qquad = \cos\alpha\cos\beta + \sin\alpha\sin\beta$

$\qquad = \sqrt{1 - x^2}\,y + x\sqrt{1 - y^2}$

$\qquad = y\sqrt{1 - x^2} + x\sqrt{1 - y^2}$

105. $\sin^{-1}$

$x = \sqrt{1 - x^2}$

$y = x$

$r = 1$

$\cos^{-1}y$

$x = y$

$y = \sqrt{1 - y^2}$

$r = 1$

$\tan(\sin^{-1}x + \cos^{-1}y)$

$= \dfrac{\tan\sin^{-1}x + \tan\cos^{-1}y}{1 - \tan\sin^{-1}x \cdot \tan\cos^{-1}y}$

$= \dfrac{\dfrac{x}{\sqrt{1 - x^2}} + \dfrac{\sqrt{1 - y^2}}{y}}{1 - \dfrac{x}{\sqrt{1 - x^2}}\cdot\dfrac{\sqrt{1 - y^2}}{y}}$

$= \dfrac{xy + \sqrt{1 - y^2}\sqrt{1 - x^2}}{y\sqrt{1 - x^2} - x\sqrt{1 - y^2}}$

107. $\sin 30° = \dfrac{1}{2}$

$\cos 30° = \dfrac{\sqrt{3}}{2}$

$\sin 60° = \dfrac{\sqrt{3}}{2}$

$\cos 60° = \dfrac{1}{2}$

108. a. No, they are not equal.

$\sin(2 \cdot 30°) \ne 2 \sin 30°$

$\sin 60° \ne 2 \cdot \dfrac{1}{2}$

$\dfrac{\sqrt{3}}{2} \ne 1$

b. Yes, they are equal.

$\sin(2 \cdot 30°) = 2 \sin 30° \cos 30°$

$\sin 60° = 2 \cdot \dfrac{1}{2} \cdot \dfrac{\sqrt{3}}{2}$

$\dfrac{\sqrt{3}}{2} = \dfrac{\sqrt{3}}{2}$

109. a. No, they are not equal.

$\cos(2 \cdot 30°) \ne 2 \cos 30°$

$\cos 60° \ne 2 \cdot \dfrac{\sqrt{3}}{2}$

$\dfrac{1}{2} \ne \sqrt{3}$

b. Yes, they are equal.

$\cos(2 \cdot 30°) = \cos^2 30° - \sin^2 30°$

$\cos 60° = \left(\dfrac{\sqrt{3}}{2}\right)^2 - \left(\dfrac{1}{2}\right)^2$

$\dfrac{1}{2} = \dfrac{3}{4} - \dfrac{1}{4}$

$\dfrac{1}{2} = \dfrac{1}{2}$

Section 5.3

Check Point Exercises

1. $\sin\theta = \dfrac{4}{5} = \dfrac{y}{r}$

Because θ lies in quadrant II, x is negative.

$x^2 + y^2 = r^2$

$x^2 + 4^2 = 5^2$

$x^2 = 5^2 - 4^2 = 9$

$x = -\sqrt{9} = -3$

Now we use values for x, y, and r to find the required values.

a. $\sin 2\theta = 2 \sin\theta \cos\theta$

$= 2\left(\dfrac{4}{5}\right)\left(-\dfrac{3}{5}\right) = -\dfrac{24}{25}$

b. $\cos 2\theta = \cos^2\theta - \sin^2\theta$

$= \left(-\dfrac{3}{5}\right)^2 - \left(\dfrac{4}{5}\right)^2 = \dfrac{9}{25} - \dfrac{16}{25}$

$= -\dfrac{7}{25}$

c. $\tan 2\theta = \dfrac{2\tan\theta}{1 - \tan^2\theta}$

$= \dfrac{2\left(-\frac{4}{3}\right)}{1 - \left(-\frac{4}{3}\right)^2} = \dfrac{-\frac{8}{3}}{1 - \frac{16}{9}} = \dfrac{-\frac{8}{3}}{-\frac{7}{9}}$

$= \left(-\dfrac{8}{3}\right)\left(-\dfrac{9}{7}\right) = \dfrac{24}{7}$

2. The given expression is the right side of the formula for $\cos 2\theta$ with $\theta = 15°$.

$\cos^2 15° - \sin^2 15° = \cos(2 \cdot 15°)$

$= \cos 30° = \dfrac{\sqrt{3}}{2}$

3. $\sin 3\theta = \sin(2\theta + \theta)$

$= \sin 2\theta \cos \theta + \cos 2\theta \sin \theta$

$= 2\sin \theta \cos \theta \cos \theta + (2\cos^2 \theta - 1)\sin \theta$

$= 2\sin \theta \cos^2 \theta + 2\sin \theta \cos^2 \theta - \sin \theta$

$= 4\sin \theta \cos^2 \theta - \sin \theta$

$= 4\sin \theta(1 - \sin^2 \theta) - \sin \theta$

$= 4\sin \theta - 4\sin^3 \theta - \sin \theta$

$= 3\sin \theta - 4\sin^3 \theta$

By working with the left side and expressing it in a form identical to the right side, we have verified the identity.

4. $\sin^4 x = \left(\sin^2 x\right)^2$

$= \left(\dfrac{1 - \cos 2x}{2}\right)^2$

$= \dfrac{1 - 2\cos 2x + \cos^2 2x}{4}$

$= \dfrac{1}{4} - \dfrac{1}{2}\cos 2x + \dfrac{1}{4}\cos^2 2x$

$= \dfrac{1}{4} - \dfrac{1}{2}\cos 2x + \dfrac{1}{4}\left(\dfrac{1 + \cos 2(2x)}{2}\right)$

$= \dfrac{1}{4} - \dfrac{1}{2}\cos 2x + \dfrac{1}{8} + \dfrac{1}{8}\cos 4x$

$= \dfrac{3}{8} - \dfrac{1}{2}\cos 2x + \dfrac{1}{8}\cos 4x$

5. Because $105°$ lies in quadrant II, $\cos 105° < 0$.

$\cos 105° = \cos\left(\dfrac{210°}{2}\right)$

$= -\sqrt{\dfrac{1 + \cos 210°}{2}}$

$= -\sqrt{\dfrac{1 - \frac{\sqrt{3}}{2}}{2}}$

$= -\sqrt{\dfrac{2 - \sqrt{3}}{4}}$

$= -\dfrac{\sqrt{2 - \sqrt{3}}}{2}$

6. $\dfrac{\sin 2\theta}{1 + \cos 2\theta} = \dfrac{2\sin \theta \cos \theta}{1 + \left(1 - 2\sin^2 \theta\right)}$

$= \dfrac{2\sin \theta \cos \theta}{2 - 2\sin^2 \theta}$

$= \dfrac{2\sin \theta \cos \theta}{2\left(1 - \sin^2 \theta\right)}$

$= \dfrac{2\sin \theta \cos \theta}{2\cos^2 \theta}$

$= \dfrac{\sin \theta}{\cos \theta} = \tan \theta$

The right side simplifies to $\tan \theta$, the expression on the left side. Thus, the identity is verified.

7. $\dfrac{\sec \alpha}{\sec \alpha \csc \alpha + \csc \alpha} = \dfrac{\frac{1}{\cos \alpha}}{\frac{1}{\cos \alpha} \cdot \frac{1}{\sin \alpha} + \frac{1}{\sin \alpha}}$

$= \dfrac{\frac{1}{\cos \alpha}}{\frac{1}{\cos \alpha \sin \alpha} + \frac{\cos \alpha}{\cos \alpha \sin \alpha}}$

$= \dfrac{\frac{1}{\cos \alpha}}{\frac{1 + \cos \alpha}{\cos \alpha \sin \alpha}}$

$= \dfrac{1}{\cos \alpha} \cdot \dfrac{\cos \alpha \sin \alpha}{1 + \cos \alpha}$

$= \dfrac{\sin \alpha}{1 + \cos \alpha}$

$= \tan \dfrac{\alpha}{2}$

We worked with the right side and arrived at the left side. Thus, the identity is verified.

Analytic Trigonometry

Exercise Set 5.3

1. $\sin 2\theta = 2\sin\theta\cos\theta = 2\left(\dfrac{3}{5}\right)\left(\dfrac{4}{5}\right) = \dfrac{24}{25}$

3. $\tan 2\theta = \dfrac{2\tan\theta}{1-\tan^2\theta}$

$= \dfrac{2\left(\frac{3}{4}\right)}{1-\left(\frac{3}{4}\right)^2} = \dfrac{\frac{3}{2}}{1-\frac{9}{16}}$

$= \dfrac{\frac{3}{2}}{\frac{7}{16}} = \left(\dfrac{3}{2}\right)\left(\dfrac{16}{7}\right) = \dfrac{24}{7}$

Use this information to solve problems 4, 5, and 6.

$\tan\alpha = \dfrac{7}{24} = \dfrac{y}{x}$

Because r is a distance it is positive.

$x^2 + y^2 = r^2$

$24^2 + 7^2 = r^2$

$576 + 49 = r^2$

$625 = r^2$

$r = 25$

$\sin\alpha = \dfrac{y}{r} = \dfrac{7}{25}$

$\cos\alpha = \dfrac{x}{r} = \dfrac{24}{25}$

5. $\cos 2\alpha = \cos^2\alpha - \sin^2\alpha$

$= \left(\dfrac{24}{25}\right)^2 - \left(\dfrac{7}{25}\right)^2 = \dfrac{576}{625} - \dfrac{49}{625}$

$= \dfrac{527}{625}$

7. $\sin\theta = \dfrac{15}{17} = \dfrac{y}{r}$

Because θ lies in quadrant II, x is negative.

$x^2 + y^2 = r^2$

$x^2 + 15^2 = 17^2$

$x^2 = 17^2 - 15^2 = 64$

$x = -\sqrt{64} = -8$

Now we use values for x, y, and r to find the required values.

a. $\sin 2\theta = 2\sin\theta\cos\theta$

$= 2\left(\dfrac{15}{17}\right)\left(-\dfrac{8}{17}\right) = -\dfrac{240}{289}$

b. $\cos 2\theta = \cos^2\theta - \sin^2\theta$

$= \left(-\dfrac{8}{17}\right)^2 - \left(\dfrac{15}{17}\right)^2 = \dfrac{64}{289} - \dfrac{225}{289}$

$= -\dfrac{161}{289}$

c. $\tan 2\theta = \dfrac{2\tan\theta}{1-\tan^2\theta}$

$= \dfrac{2\left(-\frac{15}{8}\right)}{1-\left(-\frac{15}{8}\right)^2} = \dfrac{-\frac{15}{4}}{1-\frac{225}{64}} = \dfrac{-\frac{15}{4}}{-\frac{161}{64}}$

$= \left(-\dfrac{15}{4}\right)\left(-\dfrac{64}{161}\right) = \dfrac{240}{161}$

9. $\cos\theta = \dfrac{24}{25} = \dfrac{x}{r}$

Because θ lies in quadrant IV, y is negative.

$x^2 + y^2 = r^2$

$24^2 + y^2 = 25^2$

$y^2 = 25^2 - 24^2 = 49$

$y = -\sqrt{49} = -7$

Now we use values for x, y, and r to find the required values.

a. $\sin 2\theta = 2\sin\theta\cos\theta$

$= 2\left(-\dfrac{7}{25}\right)\left(\dfrac{24}{25}\right) = -\dfrac{336}{625}$

b. $\cos 2\theta = \cos^2\theta - \sin^2\theta$

$= \left(\dfrac{24}{25}\right)^2 - \left(-\dfrac{7}{25}\right)^2$

$= \dfrac{576}{625} - \dfrac{49}{625} = \dfrac{527}{625}$

c. $\tan 2\theta = \dfrac{2\tan\theta}{1-\tan^2\theta}$

$= \dfrac{2\left(-\frac{7}{24}\right)}{1-\left(-\frac{7}{24}\right)^2} = \dfrac{-\frac{7}{12}}{1-\frac{49}{576}} = \dfrac{-\frac{7}{12}}{\frac{527}{576}}$

$= \left(-\dfrac{7}{12}\right)\left(\dfrac{576}{527}\right) = -\dfrac{336}{527}$

11. $\cot\theta = 2 = \dfrac{-2}{-1} = \dfrac{x}{y}$

Because r is a distance, it is positive.

$r^2 = x^2 + y^2$

$r^2 = (-2)^2 + (-1)^2$

$r^2 = 5$

$r = \sqrt{5}$

Now we use values for x, y, and r to find the required values.

a. $\sin 2\theta = 2\sin\theta\cos\theta$

$= 2\left(-\dfrac{1}{\sqrt{5}}\right)\left(-\dfrac{2}{\sqrt{5}}\right) = \dfrac{4}{5}$

b. $\cos 2\theta = \cos^2\theta - \sin^2\theta$

$= \left(-\dfrac{2}{\sqrt{5}}\right)^2 - \left(-\dfrac{1}{\sqrt{5}}\right)^2$

$= \dfrac{4}{5} - \dfrac{1}{5} = \dfrac{3}{5}$

c. $\tan 2\theta = \dfrac{2\tan\theta}{1-\tan^2\theta}$

$= \dfrac{2\left(\frac{1}{2}\right)}{1-\left(\frac{1}{2}\right)^2} = \dfrac{1}{1-\frac{1}{4}} = \dfrac{1}{\frac{3}{4}}$

$= (1)\left(\dfrac{4}{3}\right) = \dfrac{4}{3}$

13. $\sin\theta = -\dfrac{9}{41} = \dfrac{-9}{41} = \dfrac{y}{r}$

Because θ lies in quadrant III, x is negative.

$x^2 + y^2 = r^2$

$x^2 + (-9)^2 = 41^2$

$x^2 = 1600$

$x = -\sqrt{1600}$

$x = -40$

Now we use values for x, y, and r to find the required values.

a. $\sin 2\theta = 2\sin\theta\cos\theta$

$= 2\left(-\dfrac{9}{41}\right)\left(-\dfrac{40}{41}\right) = \dfrac{720}{1681}$

b. $\cos 2\theta = \cos^2\theta - \sin^2\theta$

$= \left(-\dfrac{40}{41}\right)^2 - \left(-\dfrac{9}{41}\right)^2$

$= \dfrac{1600}{1681} - \dfrac{81}{1681}$

$= \dfrac{1519}{1681}$

c. $\tan 2\theta = \dfrac{2\tan\theta}{1-\tan^2\theta}$

$= \dfrac{2\left(\frac{9}{40}\right)}{1-\left(\frac{9}{40}\right)^2} = \dfrac{\frac{9}{20}}{1-\frac{81}{1600}} = \dfrac{\frac{9}{20}}{\frac{1519}{1600}}$

$= \left(\dfrac{9}{20}\right)\left(\dfrac{1600}{1519}\right) = \dfrac{720}{1519}$

15. The given expression is the right side of the formula for $\sin 2\theta$ with $\theta = 15°$.

$2\sin 15°\cos 15° = \sin(2\cdot 15°)$

$= \sin 30° = \dfrac{1}{2}$

17. The given expression is the right side of the formula for $\cos 2\theta$ with $\theta = 75°$.

$\cos^2 75° - \sin^2 75° = \cos(2\cdot 75°)$

$= \cos 150° = -\dfrac{\sqrt{3}}{2}$

19. The given expression is the right side of the formula for $\cos 2\theta$ with $\theta = \dfrac{\pi}{8}$.

$2\cos^2\dfrac{\pi}{8} - 1 = \cos\left(2\cdot\dfrac{\pi}{8}\right)$

$= \cos\dfrac{\pi}{4} = \dfrac{\sqrt{2}}{2}$

21. The given expression is the right side of the formula for $\tan 2\theta$ with $\theta = \dfrac{\pi}{12}$.

$\dfrac{2\tan\frac{\pi}{12}}{1-\tan^2\frac{\pi}{12}} = \tan\left(2\cdot\dfrac{\pi}{12}\right) = \tan\dfrac{\pi}{6} = \dfrac{\sqrt{3}}{3}$

23. $\dfrac{2\tan\theta}{1+\tan^2\theta} = \dfrac{2\cdot\frac{\sin\theta}{\cos\theta}}{\frac{\cos^2\theta}{\cos^2\theta}+\frac{\sin^2\theta}{\cos^2\theta}}$

$= \dfrac{\frac{2\sin\theta}{\cos\theta}}{\frac{\cos^2\theta+\sin^2\theta}{\cos^2\theta}}$

$= \dfrac{\frac{2\sin\theta}{\cos\theta}}{\frac{1}{\cos^2\theta}}$

$= \dfrac{2\sin\theta}{\cos\theta}\cdot\dfrac{\cos^2\theta}{1}$

$= 2\sin\theta\cos\theta$

$= \sin 2\theta$

25. $(\sin\theta+\cos\theta)^2 = \sin^2\theta+2\sin\theta\cos\theta+\cos^2\theta$

$= \sin^2\theta+\cos^2\theta+2\sin\theta\cos\theta$

$= 1+2\sin\theta\cos\theta$

$= 1+\sin 2\theta$

27. $\sin^2 x+\cos 2x = \sin^2 x+\cos^2 x-\sin^2 x$

$= \cos^2 x$

29. $\dfrac{\sin 2x}{1-\cos 2x} = \dfrac{2\sin x\cos x}{1-\left(\cos^2 x-\sin^2 x\right)}$

$= \dfrac{2\sin x\cos x}{1-\cos^2 x+\sin^2 x}$

$= \dfrac{2\sin x\cos x}{\sin^2+\sin^2 x}$

$= \dfrac{2\sin x\cos x}{2\sin^2 x}$

$= \dfrac{\cos x}{\sin x}$

$= \cot x$

31. $\tan t\cos 2t = \dfrac{\sin t}{\cos t}\cdot\left(2\cos^2 t-1\right)$

$= \dfrac{2\sin t\cos^2 t}{\cos t}-\dfrac{\sin t}{\cos t}$

$= 2\sin t\cos t-\tan t$

$= \sin 2t-\tan t$

33. $\sin 4t = \sin(2t+2t)$

$= \sin 2t\cos 2t+\cos 2t\sin 2t$

$= \cos 2t(\sin 2t+\sin 2t)$

$= \cos 2t\cdot 2\sin 2t$

$= \left(\cos^2 t-\sin^2 t\right)\cdot 2\cdot 2\sin t\cos t$

$= 4\sin t\cos^3 t-4\sin^3 t\cos t$

35. $6\sin^4 x$

$= 6\left(\dfrac{1-\cos 2x}{2}\right)^2$

$= 6\left(\dfrac{1-2\cos 2x+\cos^2 2x}{4}\right)$

$= \dfrac{6-12\cos 2x+6\cos^2 2x}{4}$

$= \dfrac{3}{4}-3\cos 2x+\dfrac{3}{2}\cos^2 2x$

$= \dfrac{3}{4}-3\cos 2x+\dfrac{3}{2}\left(\dfrac{1+\cos 4x}{2}\right)$

$= \dfrac{3}{4}-3\cos 2x+\dfrac{3}{2}\left(\dfrac{1}{2}+\dfrac{\cos 4x}{2}\right)$

$= \dfrac{3}{4}-3\cos 2x+\dfrac{3}{4}+\dfrac{3}{4}\cos 4x$

$= \dfrac{9}{4}-3\cos 2x+\dfrac{3}{4}\cos 4x$

37. $\sin^2 x\cos^2 x = \left(\dfrac{1-\cos 2x}{2}\right)\left(\dfrac{1+\cos 2x}{2}\right)$

$= \dfrac{1-\cos^2 2x}{4}$

$= \dfrac{1}{4}-\dfrac{1}{4}\cos^2 2x$

$= \dfrac{1}{4}-\dfrac{1}{4}\left(\dfrac{1+\cos(2\cdot 2x)}{2}\right)$

$= \dfrac{1}{4}-\dfrac{1}{8}(1+\cos 4x)$

$= \dfrac{1}{4}-\dfrac{1}{8}-\dfrac{1}{8}\cos 4x$

$= \dfrac{1}{8}-\dfrac{1}{8}\cos 4x$

39. Because $15°$ lies in quadrant I, $\sin 15° > 0$.

$$\sin 15° = \sin \frac{30°}{2}$$

$$= \sqrt{\frac{1 - \cos 30°}{2}} = \sqrt{\frac{1 - \frac{\sqrt{3}}{2}}{2}}$$

$$= \sqrt{\frac{2 - \sqrt{3}}{4}} = \frac{\sqrt{2 - \sqrt{3}}}{2}$$

41. Because $157.5°$ lies in quadrant II, $\cos 157.5° < 0$.

$$\cos 157.5° = \cos \frac{315°}{2} = -\sqrt{\frac{1 + \cos 315°}{2}}$$

$$= -\sqrt{\frac{1 + \frac{\sqrt{2}}{2}}{2}} = -\sqrt{\frac{2 + \sqrt{2}}{4}}$$

$$= -\frac{\sqrt{2 + \sqrt{2}}}{2}$$

43. Because $75°$ lies in quadrant I, $\tan 75° > 0$.

$$\tan 75° = \tan \frac{150°}{2} = \frac{1 - \cos 150°}{\sin 150°}$$

$$= \frac{1 - \left(-\frac{\sqrt{3}}{2}\right)}{\frac{1}{2}} = 2 + \sqrt{3}$$

45. Because $\frac{7\pi}{8}$ lies in quadrant II, $\tan \frac{7\pi}{8} < 0$.

$$\tan \frac{7\pi}{8} = \tan\left(\frac{\frac{7\pi}{4}}{2}\right) = \frac{1 - \cos \frac{7\pi}{4}}{\sin \frac{7\pi}{4}}$$

$$= \frac{1 - \frac{\sqrt{2}}{2}}{-\frac{\sqrt{2}}{2}} = -\frac{2}{\sqrt{2}} + 1$$

$$= -\sqrt{2} + 1$$

47. $\sin \dfrac{\theta}{2} = \sqrt{\dfrac{1 - \cos\theta}{2}}$

$$= \sqrt{\frac{1 - \frac{4}{5}}{2}} = \sqrt{\frac{1}{10}}$$

$$= \frac{1}{\sqrt{10}} = \frac{\sqrt{10}}{10}$$

49. $\tan \dfrac{\theta}{2} = \dfrac{1 - \cos\theta}{\sin\theta}$

$$= \frac{1 - \frac{4}{5}}{\frac{3}{5}}$$

$$= \frac{1}{3}$$

Use this information to solve problems 50, 51, 52 and 54.

$$\tan \alpha = \frac{7}{24} = \frac{y}{x}$$

Because r is a distance, it is positive.

$$r^2 = x^2 + y^2$$
$$r^2 = 24^2 + 7^2$$
$$r^2 = 625$$
$$r = 25$$

$$\sin \alpha = \frac{y}{r} = \frac{7}{25}$$

$$\cos \alpha = \frac{x}{r} = \frac{24}{25}$$

51. $\cos \dfrac{\alpha}{2} = \sqrt{\dfrac{1 + \cos\alpha}{2}} = \sqrt{\dfrac{1 + \frac{24}{25}}{2}} = \sqrt{\dfrac{49}{50}}$

$$= \frac{7}{5\sqrt{2}} = \frac{7\sqrt{2}}{10}$$

53. $2\sin \dfrac{\theta}{2}\cos \dfrac{\theta}{2} = 2 \cdot \sqrt{\dfrac{1 - \cos\theta}{2}} \cdot \sqrt{\dfrac{1 + \cos\theta}{2}}$

$$= 2\sqrt{\frac{1 - \frac{4}{5}}{2}} \cdot \sqrt{\frac{1 + \frac{4}{5}}{2}}$$

$$= 2 \cdot \sqrt{\frac{1}{10}} \cdot \sqrt{\frac{9}{10}}$$

$$= 2 \cdot \frac{1}{\sqrt{10}} \cdot \frac{3}{\sqrt{10}}$$

$$= \frac{6}{10} = \frac{3}{5}$$

55. $\tan\alpha = \dfrac{4}{3} = \dfrac{-4}{-3} = \dfrac{y}{x}$

Because r is a distance, it is positive.

$r^2 = x^2 + y^2$

$r^2 = (-4)^2 + (-3)^2$

$r^2 = 25$

$r = 5$

Since $180° < \alpha < 270°$, then $90° < \dfrac{\alpha}{2} < 135°$.

Therefore $\dfrac{\alpha}{2}$ lies in quadrant II.

Thus, $\sin\dfrac{\alpha}{2} > 0$, $\cos\dfrac{\alpha}{2} < 0$, and $\tan\dfrac{\alpha}{2} < 0$.

a. $\sin\dfrac{\alpha}{2} = \sqrt{\dfrac{1-\cos\alpha}{2}} = \sqrt{\dfrac{1-\left(-\dfrac{3}{5}\right)}{2}}$

$= \sqrt{\dfrac{\frac{8}{5}}{2}} = \sqrt{\dfrac{4}{5}} = \dfrac{2}{\sqrt{5}} = \dfrac{2\sqrt{5}}{5}$

b. $\cos\dfrac{\alpha}{2} = -\sqrt{\dfrac{1+\cos\alpha}{2}} = -\sqrt{\dfrac{1+\left(-\dfrac{3}{5}\right)}{2}}$

$= -\sqrt{\dfrac{\frac{2}{5}}{2}} = -\sqrt{\dfrac{1}{5}} = -\dfrac{1}{\sqrt{5}} = -\dfrac{\sqrt{5}}{5}$

c. $\tan\dfrac{\alpha}{2} = \dfrac{1-\cos\alpha}{\sin\alpha} = \dfrac{1-\left(-\dfrac{3}{5}\right)}{-\dfrac{4}{5}}$

$= \dfrac{\frac{8}{5}}{-\frac{4}{5}} = \dfrac{8}{-4} = -2$

57. $\sec\alpha = -\dfrac{13}{5} = \dfrac{13}{-5} = \dfrac{r}{x}$

Because α lies in quadrant II, y is positive.

$x^2 + y^2 = r^2$

$(-5)^2 + y^2 = (13)^2$

$y^2 = 144$

$y = 12$

Since $\dfrac{\pi}{2} < \alpha < \pi$, then $\dfrac{\pi}{4} < \dfrac{\alpha}{2} < \dfrac{\pi}{2}$. Therefore $\dfrac{\alpha}{2}$ lies in quadrant I.

Thus, $\sin\dfrac{\alpha}{2} > 0$, $\cos\dfrac{\alpha}{2} > 0$, and $\tan\dfrac{\alpha}{2} > 0$.

a. $\sin\dfrac{\alpha}{2} = \sqrt{\dfrac{1-\cos\alpha}{2}} = \sqrt{\dfrac{1-\left(-\dfrac{5}{13}\right)}{2}}$

$= \sqrt{\dfrac{18}{26}} = \sqrt{\dfrac{9}{13}} = \dfrac{3}{\sqrt{13}}$

$= \dfrac{3\sqrt{13}}{13}$

b. $\cos\dfrac{\alpha}{2} = \sqrt{\dfrac{1+\cos\alpha}{2}} = \sqrt{\dfrac{1+\left(-\dfrac{5}{13}\right)}{2}}$

$= \sqrt{\dfrac{8}{26}} = \sqrt{\dfrac{4}{13}} = \dfrac{2}{\sqrt{13}}$

$= \dfrac{2\sqrt{13}}{13}$

c. $\tan\dfrac{\alpha}{2} = \dfrac{1-\cos\alpha}{\sin\alpha} = \dfrac{1-\left(-\dfrac{5}{13}\right)}{\frac{12}{13}}$

$= \dfrac{13+5}{12} = \dfrac{18}{12} = \dfrac{3}{2}$

59. $\sin^2\dfrac{\theta}{2} = \dfrac{1-\cos 2\left(\frac{\theta}{2}\right)}{2}$

$= \dfrac{1-\cos\theta}{2} \cdot \dfrac{\frac{1}{\cos\theta}}{\frac{1}{\cos\theta}}$

$= \dfrac{\frac{1-\cos\theta}{\cos\theta}}{\frac{2}{\cos\theta}}$

$= \dfrac{\frac{1}{\cos\theta} - \frac{\cos\theta}{\cos\theta}}{2 \cdot \frac{1}{\cos\theta}}$

$= \dfrac{\sec\theta - 1}{2\sec\theta}$

61. $\cos^2\dfrac{\theta}{2} = \dfrac{1+\cos 2\left(\frac{\theta}{2}\right)}{2}$

$\qquad = \dfrac{1+\cos\theta}{2}$

$\qquad = \dfrac{1+\cos\theta}{2} \cdot \dfrac{\frac{\sin\theta}{\cos\theta}}{\frac{\sin\theta}{\cos\theta}}$

$\qquad = \dfrac{\frac{\sin\theta}{\cos\theta}+\sin\theta}{2\cdot\frac{\sin\theta}{\cos\theta}}$

$\qquad = \dfrac{\tan\theta+\sin\theta}{2\tan\theta}$

$\qquad = \dfrac{\sin\theta+\tan\theta}{2\tan\theta}$

63. $\tan\dfrac{\alpha}{2} = \dfrac{\sin\alpha}{1+\cos\alpha}$

$\qquad = \dfrac{\sin\alpha}{1+\cos\alpha} \cdot \dfrac{\frac{1}{\cos\alpha}}{\frac{1}{\cos\alpha}}$

$\qquad = \dfrac{\frac{\sin\alpha}{\cos\alpha}}{\frac{1+\cos\alpha}{\cos\alpha}}$

$\qquad = \dfrac{\tan\alpha}{\frac{1}{\cos\alpha}+\frac{\cos\alpha}{\cos\alpha}}$

$\qquad = \dfrac{\tan\alpha}{\sec\alpha+1}$

65. $\dfrac{\sin x}{1-\cos x} = \dfrac{\sin x}{1-\cos x}\cdot\dfrac{\frac{1}{\sin x}}{\frac{1}{\sin x}}$

$\qquad = \dfrac{\frac{\sin x}{\sin x}}{\frac{1-\cos x}{\sin x}}$

$\qquad = \dfrac{1}{\tan\frac{x}{2}}$

$\qquad = \cot\dfrac{x}{2}$

67. $\tan\dfrac{x}{2}+\cot\dfrac{x}{2} = \dfrac{1-\cos x}{\sin x}+\dfrac{1}{\tan\frac{x}{2}}$

$\qquad = \dfrac{1-\cos x}{\sin x}+\dfrac{1}{\frac{\sin x}{1+\cos x}}$

$\qquad = \dfrac{1-\cos x}{\sin x}+\dfrac{1+\cos x}{\sin x}$

$\qquad = \dfrac{1-\cos x+1+\cos x}{\sin x}$

$\qquad = \dfrac{2}{\sin x} = 2\csc x$

69. Conjecture: The left side is equal to $\cos 2x$.

$\dfrac{\cot x-\tan x}{\cot x+\tan x} = \dfrac{\dfrac{\cos x}{\sin x}-\dfrac{\sin x}{\cos x}}{\dfrac{\cos x}{\sin x}+\dfrac{\sin x}{\cos x}}$

$\qquad = \dfrac{\dfrac{\cos x}{\sin x}\cdot\dfrac{\cos x}{\cos x}-\dfrac{\sin x}{\cos x}\cdot\dfrac{\sin x}{\sin x}}{\dfrac{\cos x}{\sin x}\cdot\dfrac{\cos x}{\cos x}+\dfrac{\sin x}{\cos x}\cdot\dfrac{\sin x}{\sin x}}$

$\qquad = \dfrac{\dfrac{\cos^2 x}{\sin x\cos x}-\dfrac{\sin^2 x}{\sin x\cos x}}{\dfrac{\cos^2 x}{\sin x\cos x}+\dfrac{\sin^2 x}{\sin x\cos x}}$

$\qquad = \dfrac{\cos^2 x-\sin^2 x}{\cos^2 x+\sin^2 x}$

$\qquad = \dfrac{\cos 2x}{1}$

$\qquad = \cos 2x$

71. Conjecture: The left side is equal to $\sin x+1$.

$\left(\sin\dfrac{x}{2}+\cos\dfrac{x}{2}\right)^2 = \sin^2\dfrac{x}{2}+2\sin\dfrac{x}{2}\cos\dfrac{x}{2}+\cos^2\dfrac{x}{2}$

$\qquad = 2\sin\dfrac{x}{2}\cos\dfrac{x}{2}+\sin^2\dfrac{x}{2}+\cos^2\dfrac{x}{2}$

$\qquad = \left[2\sin\dfrac{x}{2}\cos\dfrac{x}{2}\right]+\left[\sin^2\dfrac{x}{2}+\cos^2\dfrac{x}{2}\right]$

$\qquad = \sin\left(2\cdot\dfrac{x}{2}\right)+1$

$\qquad = \sin x+1$

73. Conjecture: The left side is equal to $\sec x$.

$$\frac{\sin 2x}{\sin x} - \frac{\cos 2x}{\cos x} = \frac{2\sin x \cos x}{\sin x} - \frac{2\cos^2 x - 1}{\cos x}$$

$$= 2\cos x - \frac{2\cos^2 x}{\cos x} + \frac{1}{\cos x}$$

$$= 2\cos x - 2\cos x + \sec x$$

$$= \sec x$$

75. Conjecture: The left side is equal to $2\csc 2x$.

$$\frac{\csc^2 x}{\cot x} = \frac{\dfrac{1}{\sin^2 x}}{\dfrac{\cos x}{\sin x}}$$

$$= \frac{1}{\sin^2 x} \frac{\sin x}{\cos x}$$

$$= \frac{1}{\sin x \cos x}$$

$$= \frac{2}{2\sin x \cos x}$$

$$= \frac{2}{\sin 2x}$$

$$= 2\csc 2x$$

77. Conjecture: The left side is equal to $\sin 3x$.

$$\sin x \left(4\cos^2 x - 1\right) = \sin x \left(2\cos^2 x + 2\cos^2 x - 1\right)$$

$$= \sin x \left(2\cos^2 x + \cos 2x\right)$$

$$= 2\sin x \cos^2 x + \sin x \cos 2x$$

$$= 2\sin x \cos x \cos x + \sin x \cos 2x$$

$$= \sin 2x \cos x + \sin x \cos 2x$$

$$= \sin(2x + x)$$

$$= \sin 3x$$

79. **a.** $\dfrac{v_o^2}{16}\sin\theta\cos\theta = \dfrac{v_o^2}{32} \cdot 2\sin\theta\cos\theta$

$$= \frac{v_o^2}{32}\cdot\sin 2\theta$$

b. $\sin\alpha$ is at a maximum in the interval $[0, 2\pi]$

when $\alpha = \dfrac{\pi}{2}$, so $\sin 2\theta$ is at a maximum when

$2\theta = \dfrac{\pi}{2}$ or $\theta = \dfrac{\pi}{4}$.

81. $\theta = \dfrac{\pi}{4}$

$$\sin\frac{\theta}{2} = \sqrt{\frac{1-\cos\theta}{2}}$$

$$= \sqrt{\frac{1-\cos\frac{\pi}{4}}{2}}$$

$$= \sqrt{\frac{1-\frac{\sqrt{2}}{2}}{2}}$$

$$= \sqrt{\frac{2-\sqrt{2}}{4}}$$

$$= \frac{\sqrt{2-\sqrt{2}}}{2}$$

$$\sin\frac{\theta}{2} = \frac{1}{M}$$

$$\frac{\sqrt{2-\sqrt{2}}}{2} = \frac{1}{M}$$

$$M = \frac{2}{\sqrt{2-\sqrt{2}}}$$

$$= \frac{2\sqrt{2-\sqrt{2}}}{2-\sqrt{2}}$$

$$= \frac{2\sqrt{2-\sqrt{2}}}{2-\sqrt{2}} \cdot \frac{2+\sqrt{2}}{2+\sqrt{2}}$$

$$= \frac{4\sqrt{2-\sqrt{2}} + 2\sqrt{2}\sqrt{2-\sqrt{2}}}{4-2}$$

$$= \frac{2\left(2\sqrt{2-\sqrt{2}} + \sqrt{2}\sqrt{2-\sqrt{2}}\right)}{2}$$

$$= 2\sqrt{2-\sqrt{2}} + \sqrt{2}\sqrt{2-\sqrt{2}}$$

$$= \sqrt{2-\sqrt{2}} \cdot \left(2+\sqrt{2}\right) \approx 2.6$$

83. – 93. Answers may vary.

95.

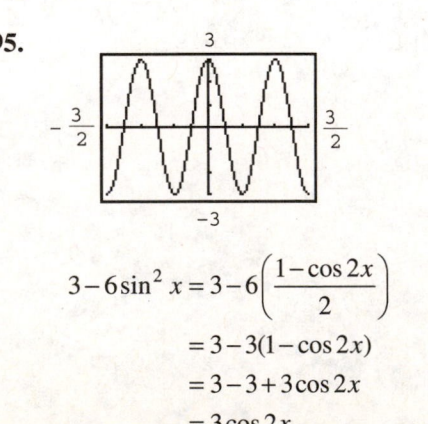

$$3 - 6\sin^2 x = 3 - 6\left(\frac{1-\cos 2x}{2}\right)$$

$$= 3 - 3(1-\cos 2x)$$

$$= 3 - 3 + 3\cos 2x$$

$$= 3\cos 2x$$

97.

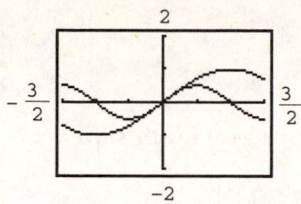

The graphs do not coincide.
Values for *x* may vary.

99.

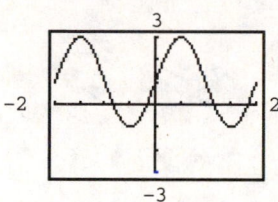

a. The graph appears to be the sum of 1 and 2 times the sine curve, $y = 1 + 2\sin x$. If you subtract 1 from the graph, it cycles through intercept, maximum, intercept, minimum, and back to intercept. Thus, $y = 1 + 2\sin x$ also describes the graph.

b. $\dfrac{1 - 2\cos 2x}{2\sin x - 1}$

$= \dfrac{1 - 2(1 - 2\sin^2 x)}{2\sin x - 1} = \dfrac{1 - 2 + 4\sin^2 x}{2\sin x - 1}$

$= \dfrac{4\sin^2 x - 1}{2\sin x - 1} = \dfrac{(2\sin x - 1)(2\sin x + 1)}{2\sin x - 1}$

$= 2\sin x + 1 = 1 + 2\sin x$

This verifies our observation that

$y = \dfrac{1 - 2\cos 2x}{2\sin x - 1}$ and $y = 1 + 2\sin x$ describe the same graph.

101.

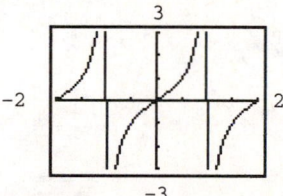

a. The graph appears to be the tangent of half the angle. It cycles from negative infinity through intercept to positive infinity. Thus, $y = \tan\dfrac{x}{2}$ also describes the graph.

b. $\tan\dfrac{x}{2} = \dfrac{1 - \cos x}{\sin x} = \dfrac{1}{\sin x} - \dfrac{\cos x}{\sin x} = \csc x - \cot x$

This verifies our observation that

$y = \csc x - \cot x$ and $y = \tan\dfrac{x}{2}$ describe the same graph.

103. does not make sense; Explanations will vary. Sample explanation: That procedure is not algebraically sound.

105. does not make sense; Explanations will vary. Sample explanation: That method will not work well because $200°$ is not an angle with known trigonometric values.

107. $\sin\left(2\sin^{-1}\dfrac{\sqrt{3}}{2}\right) = \sin\left(2 \cdot \dfrac{\pi}{3}\right)$

$= \sin\dfrac{2\pi}{3}$

$= \dfrac{\sqrt{3}}{2}$

109. $\cos^2\left[\dfrac{1}{2}\sin^{-1}\dfrac{3}{5}\right]$

Let $\theta = \sin^{-1}\dfrac{3}{5}$, then $\sin\theta = \dfrac{y}{r} = \dfrac{3}{5}$

Since θ is in quadrant I, *x* is positive.

$x^2 + y^2 = r^2$

$x^2 + 3^2 = 5^2$

$x^2 = 16$

$x = 4$

$\cos^2\left[\dfrac{1}{2}\sin^{-1}\dfrac{3}{5}\right] = \cos^2\left[\dfrac{1}{2}\theta\right]$

$= \dfrac{1 + \cos\left(2 \cdot \dfrac{1}{2}\theta\right)}{2}$

$= \dfrac{1 + \cos\theta}{2}$

$= \dfrac{1 + \dfrac{4}{5}}{2}$

$= \dfrac{1 + \dfrac{x}{r}}{2}$

$= \dfrac{9}{10}$

111 Let $\alpha = \sin^{-1} x$ where $-\dfrac{\pi}{2} \le \alpha \le \dfrac{\pi}{2}$.

$\sin \alpha = x$

Because x is positive, $\sin \alpha$ is positive. Thus, α is in quadrant I. Using a right triangle in quadrant I with $\sin \alpha = x = \dfrac{x}{1}$

the third side a can be found using the Pythagorean Theorem.

$\alpha^2 + x^2 = 1^2$

$a^2 = 1 - x^2$

$a = \sqrt{1 - x^2}$

$\cos \alpha = \dfrac{\sqrt{1 - x^2}}{1} = \sqrt{1 - x^2}$

$\sin(2 \sin^{-1} x) = \sin 2\alpha = 2x\sqrt{1 - x^2}$

113. $\sin 60° \sin 30° = \dfrac{1}{2}\Big[\cos(60° - 30°) - \cos(60° + 30°)\Big]$

$\qquad \dfrac{\sqrt{3}}{2} \cdot \dfrac{1}{2} = \dfrac{1}{2}\Big[\cos 30° - \cos 90°\Big]$

$\qquad \dfrac{\sqrt{3}}{4} = \dfrac{1}{2}\Big[\dfrac{\sqrt{3}}{2} - 0\Big]$

$\qquad \dfrac{\sqrt{3}}{4} = \dfrac{\sqrt{3}}{4}$

114. $\cos \dfrac{\pi}{2} \cos \dfrac{\pi}{3} = \dfrac{1}{2}\Big[\cos\Big(\dfrac{\pi}{2} - \dfrac{\pi}{3}\Big) + \cos\Big(\dfrac{\pi}{2} + \dfrac{\pi}{3}\Big)\Big]$

$\qquad 0 \cdot \dfrac{1}{2} = \dfrac{1}{2}\Big[\cos\Big(\dfrac{\pi}{6}\Big) + \cos\Big(\dfrac{5\pi}{6}\Big)\Big]$

$\qquad 0 = \dfrac{1}{2}\Big[\dfrac{\sqrt{3}}{2} - \dfrac{\sqrt{3}}{2}\Big]$

$\qquad 0 = \dfrac{1}{2}[0]$

$\qquad 0 = 0$

115. $\sin \pi \cos \dfrac{\pi}{2} = \dfrac{1}{2}\Big[\sin\Big(\pi + \dfrac{\pi}{2}\Big) + \sin\Big(\pi - \dfrac{\pi}{2}\Big)\Big]$

$\qquad 0 \cdot 0 = \dfrac{1}{2}\Big[\sin\Big(\dfrac{3\pi}{2}\Big) + \sin\Big(\dfrac{\pi}{2}\Big)\Big]$

$\qquad 0 \cdot 0 = \dfrac{1}{2}[-1 + 1]$

$\qquad 0 = \dfrac{1}{2}[0]$

$\qquad 0 = 0$

Mid-Chapter 5 Check Point

1. $\cos x(\tan x + \cot x)$

$$= \cos x\left(\frac{\sin x}{\cos x} + \frac{\cos x}{\sin x}\right)$$

$$= \cos x\left(\frac{\sin x}{\cos x} \cdot \frac{\sin x}{\sin x} + \frac{\cos x}{\sin x} \cdot \frac{\cos x}{\cos x}\right)$$

$$= \cos x\left(\frac{\sin^2 x}{\sin x \cos x} + \frac{\cos^2 x}{\sin x \cos x}\right)$$

$$= \cos x\left(\frac{\sin^2 x + \cos^2 x}{\sin x \cos x}\right)$$

$$= \cos x\left(\frac{1}{\sin x \cos x}\right)$$

$$= \frac{\cos x}{\sin x \cos x}$$

$$= \frac{1}{\sin x}$$

$$= \csc x$$

2. $\dfrac{\sin(x+\pi)}{\cos\left(x+\dfrac{3\pi}{2}\right)} = \dfrac{-\sin x}{\cos\left((x+\pi)+\dfrac{\pi}{2}\right)}$

$$= \frac{-\sin x}{-\sin(x+\pi)}$$

$$= \frac{-\sin x}{\sin x}$$

$$= -1$$

$$= -\left(\sec^2 x - \tan^2 x\right)$$

$$= \tan^2 x - \sec^2 x$$

3. $(\sin\theta + \cos\theta)^2 + (\sin\theta - \cos\theta)^2$

$$= \sin^2\theta + 2\sin\theta\cos\theta + \cos^2\theta$$

$$\quad + \sin^2\theta - 2\sin\theta\cos\theta + \cos^2\theta$$

$$= \sin^2\theta + \cos^2\theta + \sin^2\theta + \cos^2\theta$$

$$= 1+1$$

$$= 2$$

4. $\dfrac{\sin t - 1}{\cos t} = \dfrac{\sin t - 1}{\cos t} \cdot \dfrac{\cot t}{\cot t}$

$$= \frac{\sin t \cot t - \cot t}{\cos t \cot t}$$

$$= \frac{\sin t \cdot \dfrac{\cos t}{\sin t} - \cot t}{\cos t \cot t}$$

$$= \frac{\cos t - \cot t}{\cos t \cot t}$$

5. $\dfrac{1 - \cos 2x}{\sin 2x} = \dfrac{1 - 2\cos^2 x - 1}{2\sin x \cos x}$

$$= \frac{2\cos^2 x}{2\sin x \cos x}$$

$$= \frac{\cos x}{\sin x}$$

$$= \tan x$$

6. $\sin\theta\cos\theta + \cos^2\theta$

$$= \cos\theta(\sin\theta + \cos\theta)$$

$$= \cos\theta(\sin\theta + \cos\theta) \cdot \frac{\csc\theta}{\csc\theta}$$

$$= \frac{\cos\theta(\sin\theta\csc\theta + \cos\theta\csc\theta)}{\csc\theta}$$

$$= \frac{\cos\theta\left(\sin\theta \cdot \dfrac{1}{\sin\theta} + \cos\theta \cdot \dfrac{1}{\sin\theta}\right)}{\csc\theta}$$

$$= \frac{\cos\theta(1 + \tan\theta)}{\csc\theta}$$

7. $\dfrac{\sin x}{\tan x} + \dfrac{\cos x}{\cot x} = \dfrac{\sin x}{\frac{\sin x}{\cos x}} + \dfrac{\cos x}{\frac{\cos x}{\sin x}}$

$\qquad\qquad = \sin x \cdot \dfrac{\cos x}{\sin x} + \cos x \cdot \dfrac{\sin x}{\cos x}$

$\qquad\qquad = \cos x + \sin x$

$\qquad\qquad = \sin x + \cos x$

8. $\sin^2 \dfrac{t}{2} = \left(\sin \dfrac{t}{2} \right)^2$

$\qquad\quad = \left(\pm \sqrt{\dfrac{1 - \cos t}{2}} \right)^2$

$\qquad\quad = \dfrac{1 - \cos t}{2}$

$\qquad\quad = \dfrac{1 - \cos t}{2} \cdot \dfrac{\tan t}{\tan t}$

$\qquad\quad = \dfrac{\tan t - \cos t \tan t}{2 \tan t}$

$\qquad\quad = \dfrac{\tan t - \cos t \cdot \frac{\sin t}{\cos t}}{2 \tan t}$

$\qquad\quad = \dfrac{\tan t - \sin t}{2 \tan t}$

9. $\dfrac{1}{2} \big[\sin(\alpha + \beta) + \sin(\alpha - \beta) \big]$

$\quad = \dfrac{1}{2} \big[\sin \alpha \cos \beta + \cos \alpha \sin \beta +$

$\qquad\qquad \sin \alpha \cos \beta - \cos \alpha \sin \beta \big]$

$\quad = \dfrac{1}{2} \big[2 \sin \alpha \cos \beta \big]$

$\quad = \sin \alpha \cos \beta$

10. $\dfrac{1 + \csc x}{\sec x} - \cot x = \dfrac{1 + \frac{1}{\sin x}}{\frac{1}{\cos x}} - \dfrac{\cos x}{\sin x}$

$\qquad\qquad\qquad = \cos x \left(1 + \dfrac{1}{\sin x} \right) - \dfrac{\cos x}{\sin x}$

$\qquad\qquad\qquad = \cos x + \dfrac{\cos x}{\sin x} - \dfrac{\cos x}{\sin x}$

$\qquad\qquad\qquad = \cos x$

11. $\dfrac{\cot x - 1}{\cot x + 1} = \dfrac{\frac{\cot x}{\cot x} - \frac{1}{\cot x}}{\frac{\cot x}{\cot x} + \frac{1}{\cot x}}$

$\qquad\qquad = \dfrac{1 - \tan x}{1 + \tan x}$

12. $2 \sin^3 \theta \cos \theta + 2 \sin \theta \cos^3 \theta$

$\quad = 2 \sin \theta \cos \theta \left(\sin^2 \theta + \cos^2 \theta \right)$

$\quad = 2 \sin \theta \cos \theta$

$\quad = \sin 2\theta$

13. $\dfrac{\sin t + \cos t}{\sec t + \csc t} = \dfrac{\sin t + \cos t}{\frac{1}{\cos t} + \frac{1}{\sin t}}$

$\qquad\qquad = \dfrac{\sin t + \cos t}{\frac{1}{\cos t} \cdot \frac{\sin t}{\sin t} + \frac{1}{\sin t} \cdot \frac{\cos t}{\cos t}}$

$\qquad\qquad = \dfrac{\sin t + \cos t}{\frac{\sin t + \cos t}{\sin t \cos t}}$

$\qquad\qquad = (\sin t + \cos t) \dfrac{\sin t \cos t}{\sin t + \cos t}$

$\qquad\qquad = \sin t \cos t$

$\qquad\qquad = \sin t \dfrac{1}{\sec t}$

$\qquad\qquad = \dfrac{\sin t}{\sec t}$

14. $\dfrac{\sec^2 x}{2 - \sec^2 x} = \dfrac{\frac{\sec^2 x}{\sec^2 x}}{\frac{2}{\sec^2 x} - \frac{\sec^2 x}{\sec^2 x}}$

$\qquad\qquad = \dfrac{1}{2 \cos^2 x - 1}$

$\qquad\qquad = \dfrac{1}{\cos 2x}$

$\qquad\qquad = \sec 2x$

15. $\tan(\alpha+\beta)\tan(\alpha-\beta)$

$= \tan(\alpha+\beta)\tan(\alpha-\beta)$

$= \dfrac{\tan\alpha+\tan\beta}{1-\tan\alpha\tan\beta}\cdot\dfrac{\tan\alpha-\tan\beta}{1+\tan\alpha\tan\beta}$

$= \dfrac{\tan^2\alpha-\tan^2\beta}{1-\tan^2\alpha\tan^2\beta}$

16. $\dfrac{\sin\theta}{1-\cos\theta} = \dfrac{\sin\theta}{1-\cos\theta}\cdot\dfrac{1+\cos\theta}{1+\cos\theta}$

$= \dfrac{\sin\theta+\sin\theta\cos\theta}{1-\cos^2\theta}$

$= \dfrac{\sin\theta+\sin\theta\cos\theta}{\sin^2\theta}$

$= \dfrac{\sin\theta}{\sin^2\theta}+\dfrac{\sin\theta\cos\theta}{\sin^2\theta}$

$= \dfrac{1}{\sin\theta}+\dfrac{\cos\theta}{\sin\theta}$

$= \csc\theta+\cot\theta$

17. $\dfrac{1}{\csc 2x} = \sin 2x$

$= 2\sin x\cos x$

$= 2\sin x\cos x\cdot\dfrac{\cos x}{\cos x}$

$= \dfrac{2\sin x\cos^2 x}{\cos x}$

$= \dfrac{2\sin x}{\cos x}\cdot\cos^2 x$

$= 2\tan x\cdot\dfrac{1}{\sec^2 x}$

$= \dfrac{2\tan x}{\sec^2 x}$

$= \dfrac{2\tan x}{1+\tan^2 x}$

18. $\dfrac{\sec t-1}{t\sec t} = \dfrac{\sec t-1}{t\sec t}\cdot\dfrac{\cos t}{\cos t}$

$= \dfrac{\sec t\cos t-\cos t}{t\sec t\cos t}$

$= \dfrac{\dfrac{1}{\cos t}\cos t-\cos t}{t\dfrac{1}{\cos t}\cos t}$

$= \dfrac{1-\cos t}{t}$

19. Use $\sin\alpha = \dfrac{3}{5} = \dfrac{y}{r}$ to find $\cos\alpha$ and $\tan\alpha$.

Because α is in Quadrant II, x is negative.

$x^2+y^2 = r^2$

$x^2+3^2 = 5^2$

$x^2 = 16$

$x = -\sqrt{16}$

$x = -4$

Thus, $\cos\alpha = \dfrac{-4}{5} = -\dfrac{4}{5}$ and $\tan\alpha = \dfrac{-3}{4} = -\dfrac{3}{4}$.

Use $\cos\beta = \dfrac{-12}{13} = \dfrac{x}{r}$ to find $\sin\beta$ and $\tan\beta$.

Because β is in Quadrant III, x and y are negative.

$x^2+y^2 = r^2$

$(-12)^2+y^2 = 13^2$

$y^2 = 25$

$y = -\sqrt{25}$

$y = -5$

Thus, $\sin\beta = \dfrac{-5}{13} = -\dfrac{5}{13}$ and $\tan\beta = \dfrac{-5}{-12} = \dfrac{5}{12}$.

$\cos(\alpha-\beta) = \cos\alpha\cos\beta+\sin\alpha\sin\beta$

$= \left(-\dfrac{4}{5}\right)\left(-\dfrac{12}{13}\right)+\left(\dfrac{3}{5}\right)\left(-\dfrac{5}{13}\right)$

$= \dfrac{33}{65}$

20. In exercise 19 it was shown that

$\tan\alpha = -\dfrac{3}{4}$ and $\tan\beta = \dfrac{5}{12}$. Thus,

$\tan(\alpha+\beta) = \dfrac{\tan\alpha+\tan\beta}{1-\tan\alpha\tan\beta}$

$= \dfrac{-\dfrac{3}{4}+\dfrac{5}{12}}{1-\left(-\dfrac{3}{4}\right)\dfrac{5}{12}}$

$= -\dfrac{16}{63}$

21. In exercise 19 it was shown that

$\sin\alpha = \dfrac{3}{5}$ and $\cos\alpha = -\dfrac{4}{5}$.

Thus, $\sin 2\alpha = 2\sin\alpha\cos\alpha$

$$= 2\cdot\frac{3}{5}\left(-\frac{4}{5}\right)$$

$$= -\frac{24}{25}$$

22. $\cos\beta = -\dfrac{12}{13}$

Since β is in quadrant III, $\dfrac{\beta}{2}$ is in quadrant II.

The cosine is negative in quadrant II.

$$\cos\frac{\beta}{2} = -\sqrt{\frac{1+\cos\beta}{2}}$$

$$= -\sqrt{\frac{1+\left(-\dfrac{12}{13}\right)}{2}}$$

$$= -\sqrt{\frac{1}{26}}$$

$$= -\frac{1}{\sqrt{26}}$$

$$= -\frac{\sqrt{26}}{26}$$

23. $\sin\left(\dfrac{3\pi}{4}+\dfrac{5\pi}{6}\right)$

$$= \sin\frac{3\pi}{4}\cos\frac{5\pi}{6}+\cos\frac{3\pi}{4}\sin\frac{5\pi}{6}$$

$$= \left(\frac{\sqrt{2}}{2}\right)\left(-\frac{\sqrt{3}}{2}\right)+\left(-\frac{\sqrt{2}}{2}\right)\left(\frac{1}{2}\right)$$

$$= -\frac{\sqrt{6}}{4}-\frac{\sqrt{2}}{4}$$

$$= -\frac{\sqrt{6}+\sqrt{2}}{4}$$

24. $\cos^2 15° - \sin^2 15° = \cos(2\cdot 15°)$

$$= \cos 30°$$

$$= \frac{\sqrt{3}}{2}$$

25. $\cos\dfrac{5\pi}{12}\cos\dfrac{\pi}{12}+\sin\dfrac{5\pi}{12}\sin\dfrac{\pi}{12} = \cos\left(\dfrac{5\pi}{12}-\dfrac{\pi}{12}\right)$

$$= \cos\frac{4\pi}{12}$$

$$= \cos\frac{\pi}{3}$$

$$= \frac{1}{2}$$

26. $\tan 22.5° = \tan\dfrac{45°}{2}$

$$= \frac{\sin 45°}{1+\cos 45°}$$

$$= \frac{\dfrac{\sqrt{2}}{2}}{1+\dfrac{\sqrt{2}}{2}}$$

$$= \frac{\sqrt{2}}{2+\sqrt{2}}$$

$$= \frac{\sqrt{2}}{2+\sqrt{2}}\cdot\frac{2-\sqrt{2}}{2-\sqrt{2}}$$

$$= \frac{2\sqrt{2}-2}{4-2}$$

$$= \frac{2\sqrt{2}-2}{2}$$

$$= \sqrt{2}-1$$

Section 5.4

Check Point Exercises

1. a. $\sin 5x\sin 2x$

$$= \frac{1}{2}[\cos(5x-2x)-\cos(5x+2x)]$$

$$= \frac{1}{2}[\cos 3x-\cos 7x]$$

b. $\cos 7x\cos x$

$$= \frac{1}{2}[\cos(7x-x)+\cos(7x+x)]$$

$$= \frac{1}{2}[\cos 6x+\cos 8x]$$

2. **a.** $\sin 7x + \sin 3x$

$$= 2\sin\left(\frac{7x+3x}{2}\right)\cos\left(\frac{7x-3x}{2}\right)$$

$$= 2\sin\left(\frac{10x}{2}\right)\cos\left(\frac{4x}{2}\right)$$

$$= 2\sin 5x \cos 2x$$

b. $\cos 3x + \cos 2x$

$$= 2\cos\left(\frac{3x+2x}{2}\right)\cos\left(\frac{3x-2x}{2}\right)$$

$$= 2\cos\left(\frac{5x}{2}\right)\cos\left(\frac{x}{2}\right)$$

3. $\dfrac{\cos 3x - \cos x}{\sin 3x + \sin x} = \dfrac{-2\sin\left(\dfrac{3x+x}{2}\right)\sin\left(\dfrac{3x-2x}{2}\right)}{2\sin\dfrac{3x+x}{2}\cos\left(\dfrac{3x-x}{2}\right)}$

$$= \frac{-2\sin 2x \sin x}{2\sin 2x \cos x}$$

$$= \frac{-\sin x}{\cos x}$$

$$= -\tan x$$

We worked with the left side and arrived at the right side. Thus, the identity is verified.

Exercise Set 5.4

1. The given formula can be used to change a <u>product</u> of two sines into the <u>difference</u> of two cosine expressions.

3. The given formula can be used to change a <u>product</u> of a sine and a cosine into the <u>sum</u> of two sine expressions.

5. $\sin 6x \sin 2x = \dfrac{1}{2}\left[\cos(6x-2x) - \cos(6x+2x)\right]$

$$= \frac{1}{2}\left[\cos 4x - \cos 8x\right]$$

7. $\cos 7x \cos 3x = \dfrac{1}{2}\left[\cos(7x-3x) + \cos(7x+3x)\right]$

$$= \frac{1}{2}\left[\cos 4x + \cos 10x\right]$$

9. $\sin x \cos 2x = \dfrac{1}{2}\left[\sin(x+2x) + \sin(x-2x)\right]$

$$= \frac{1}{2}\left[\sin 3x + \sin(-x)\right]$$

$$= \frac{1}{2}\left[\sin 3x - \sin x\right]$$

11. $\cos\dfrac{3x}{2}\sin\dfrac{x}{2} = \dfrac{1}{2}\left[\sin\left(\dfrac{3x}{2}+\dfrac{x}{2}\right) - \sin\left(\dfrac{3x}{2}-\dfrac{x}{2}\right)\right]$

$$= \frac{1}{2}\left[\sin\left(\frac{4x}{2}\right) - \sin\left(\frac{2x}{2}\right)\right]$$

$$= \frac{1}{2}\left[\sin 2x - \sin x\right]$$

13. The given formula can be used to change a <u>sum</u> of two sines into the <u>product</u> of a sine and a cosine expression.

15. The given formula can be used to change a <u>sum</u> of two cosines into the <u>product</u> of two cosine the expressions.

17. $\sin 6x + \sin 2x = 2\sin\left(\dfrac{6x+2x}{2}\right)\cos\left(\dfrac{6x-2x}{2}\right)$

$$= 2\sin\left(\frac{8x}{2}\right)\cos\left(\frac{4x}{2}\right)$$

$$= 2\sin 4x \cos 2x$$

19. $\sin 7x - \sin 3x = 2\sin\left(\dfrac{7x-3x}{2}\right)\cos\left(\dfrac{7x+3x}{2}\right)$

$$= 2\sin\left(\frac{4x}{2}\right)\cos\left(\frac{10x}{2}\right)$$

$$= 2\sin 2x \cos 5x$$

21. $\cos 4x + \cos 2x = 2\cos\left(\dfrac{4x+2x}{2}\right)\cos\left(\dfrac{4x-2x}{2}\right)$

$$= 2\cos\left(\frac{6x}{2}\right)\cos\left(\frac{2x}{2}\right)$$

$$= 2\cos 3x \cos x$$

23. $\sin x + \sin 2x = 2\sin\left(\dfrac{x+2x}{2}\right)\cos\left(\dfrac{x-2x}{2}\right)$

$$= 2\sin\left(\frac{3x}{2}\right)\cos\left(\frac{-x}{2}\right)$$

$$= 2\sin\frac{3x}{2}\cos\frac{x}{2}$$

25. $\cos\dfrac{3x}{2} + \cos\dfrac{x}{2} = 2\cos\left(\dfrac{\frac{3x}{2}+\frac{x}{2}}{2}\right)\cos\left(\dfrac{\frac{3x}{2}-\frac{x}{2}}{2}\right)$

$$= 2\cos\left(\dfrac{4x}{4}\right)\cos\left(\dfrac{2x}{4}\right)$$

$$= 2\cos x \cos\dfrac{x}{2}$$

27. $\sin 75° + \sin 15°$

$$= 2\sin\left(\dfrac{75°+15°}{2}\right)\cos\left(\dfrac{75°-15°}{2}\right)$$

$$= 2\sin(45°)\cos(30°)$$

$$= 2\left(\dfrac{\sqrt{2}}{2}\right)\left(\dfrac{\sqrt{3}}{2}\right)$$

$$= \dfrac{\sqrt{6}}{2}$$

29. $\sin\dfrac{\pi}{12} - \sin\dfrac{5\pi}{12}$

$$= 2\sin\left(\dfrac{\frac{\pi}{12}-\frac{5\pi}{12}}{2}\right)\cos\left(\dfrac{\frac{\pi}{12}+\frac{5\pi}{12}}{2}\right)$$

$$= 2\sin\left(-\dfrac{4\pi}{24}\right)\cos\left(\dfrac{6\pi}{24}\right)$$

$$= -2\sin\dfrac{\pi}{6}\cos\dfrac{\pi}{4}$$

$$= -2\left(\dfrac{1}{2}\right)\left(\dfrac{\sqrt{2}}{2}\right)$$

$$= -\dfrac{\sqrt{2}}{2}$$

31. $\dfrac{\sin 3x - \sin x}{\cos 3x - \cos x}$

$$= \dfrac{2\sin\left(\dfrac{3x-x}{2}\right)\cos\left(\dfrac{3x+x}{2}\right)}{-2\sin\left(\dfrac{3x+x}{2}\right)\sin\left(\dfrac{3x-x}{2}\right)}$$

$$= \dfrac{2\sin\left(\dfrac{2x}{2}\right)\cos\left(\dfrac{4x}{2}\right)}{-2\sin\left(\dfrac{4x}{2}\right)\sin\left(\dfrac{2x}{2}\right)}$$

$$= \dfrac{2\sin x\cos 2x}{-2\sin 2x\sin x}$$

$$= -\dfrac{\cos 2x}{\sin 2x} = -\cot 2x$$

33. $\dfrac{\sin 2x + \sin 4x}{\cos 2x + \cos 4x}$

$$= \dfrac{2\sin\left(\dfrac{2x+4x}{2}\right)\cos\left(\dfrac{2x-4x}{2}\right)}{2\cos\left(\dfrac{2x+4x}{2}\right)\cos\left(\dfrac{2x-4x}{2}\right)}$$

$$= \dfrac{2\sin\left(\dfrac{6x}{2}\right)\cos\left(\dfrac{-2x}{2}\right)}{2\cos\left(\dfrac{6x}{2}\right)\cos\left(\dfrac{-2x}{2}\right)}$$

$$= \dfrac{2\sin 3x\cos(-x)}{2\cos 3x\cos(-x)}$$

$$= \dfrac{\sin 3x}{\cos 3x}$$

$$= \tan 3x$$

35. $\dfrac{\sin x - \sin y}{\sin x + \sin y} = \dfrac{2\sin\left(\dfrac{x-y}{2}\right)\cos\left(\dfrac{x+y}{2}\right)}{2\sin\left(\dfrac{x+y}{2}\right)\cos\left(\dfrac{x-y}{2}\right)}$

$$= \dfrac{\sin\left(\dfrac{x-y}{2}\right)}{\cos\left(\dfrac{x-y}{2}\right)}\cdot\dfrac{\cos\left(\dfrac{x+y}{2}\right)}{\sin\left(\dfrac{x+y}{2}\right)}$$

$$= \tan\dfrac{x-y}{2}\cot\dfrac{x+y}{2}$$

37. $\dfrac{\sin x + \sin y}{\cos x + \cos y} = \dfrac{2\sin\left(\dfrac{x+y}{2}\right)\cos\left(\dfrac{x-y}{2}\right)}{2\cos\left(\dfrac{x+y}{2}\right)\cos\left(\dfrac{x-y}{2}\right)}$

$\qquad\qquad = \dfrac{\sin\left(\dfrac{x+y}{2}\right)}{\cos\left(\dfrac{x+y}{2}\right)}$

$\qquad\qquad = \tan\dfrac{x+y}{2}$

39. a. $y = \cos x$ also describes the graph.

b. $\dfrac{\sin x + \sin 3x}{2\sin 2x} = \dfrac{2\sin\left(\dfrac{x+3x}{2}\right)\cos\left(\dfrac{x-3x}{2}\right)}{2\sin 2x} = \dfrac{2\sin\left(\dfrac{4x}{2}\right)\cos\left(\dfrac{-2x}{2}\right)}{2\sin 2x}$

$\qquad\qquad = \dfrac{2\sin 2x\cos(-x)}{2\sin 2x} = \cos(-x) = \cos x$

41. a. $y = \tan 2x$ also describes the graph.

b. $\dfrac{\cos x - \cos 5x}{\sin x + \sin 5x} = \dfrac{-2\sin\left(\dfrac{x+5x}{2}\right)\sin\left(\dfrac{x-5x}{2}\right)}{2\sin\left(\dfrac{x+5x}{2}\right)\cos\left(\dfrac{x-5x}{2}\right)} = \dfrac{-2\sin\left(\dfrac{6x}{2}\right)\sin\left(\dfrac{-4x}{2}\right)}{2\sin\left(\dfrac{6x}{2}\right)\cos\left(\dfrac{-4x}{2}\right)}$

$\qquad\qquad = \dfrac{-2\sin 3x\sin(-2x)}{2\sin 3x\cos(-2x)} = \dfrac{-\sin(-2x)}{\cos 2x} = \dfrac{\sin 2x}{\cos 2x} = \tan 2x$

43. a. $y = -\cot 2x$ also describes the graph.

b. $\dfrac{\sin x - \sin 3x}{\cos x - \cos 3x} = \dfrac{2\sin\left(\dfrac{x-3x}{2}\right)\cos\left(\dfrac{x+3x}{2}\right)}{-2\sin\left(\dfrac{x+3x}{2}\right)\sin\left(\dfrac{x-3x}{2}\right)}$

$\qquad\qquad = \dfrac{2\sin(-x)\cos 2x}{-2\sin 2x\sin(-x)}$

$\qquad\qquad = \dfrac{\cos 2x}{-\sin 2x}$

$\qquad\qquad = -\cot 2x$

45. a.   The low frequency is $l = 852$ cycles per second and the high frequency is $h = 1209$ cycles per second. The sound produced by touching 7 is described by $y = \sin 2\pi(852)t + \sin 2\pi(1209)t$, or $y = \sin 1704\pi t + \sin 2418\pi t$.

b. $y = \sin 1704\pi t + \sin 2418\pi t$

$\qquad = 2\sin\left(\dfrac{1704\pi t + 2418\pi t}{2}\right)\cdot\cos\left(\dfrac{1704\pi t - 2418\pi t}{2}\right)$

$\qquad = 2\sin 2061\pi t \cdot \cos(-357\pi t)$

$\qquad = 2\sin 2061\pi t \cdot \cos 357\pi t$

47. – 51. Answers may vary.

53.

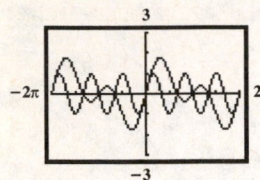

The graphs do not coincide.
Values for x may vary.

55.

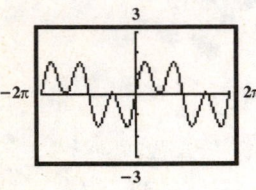

$$\sin x + \sin 3x = 2\sin\left(\frac{x+3x}{2}\right)\cos\left(\frac{x-3x}{2}\right)$$

$$= 2\sin 2x\cos(-x)$$

$$= 2\sin 2x\cos x$$

We worked with the left side and arrived at
the right side. Thus, the identity is verified.

57.

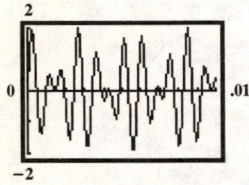

59. a.

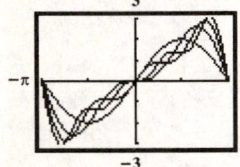

Answers may vary.

b.

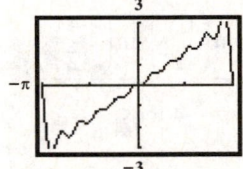

Answers may vary.

c. When $x = \dfrac{\pi}{2}$,

$$\frac{\pi}{2} = 2\left(\frac{\sin\frac{\pi}{2}}{1} - \frac{\sin\left(2\cdot\frac{\pi}{2}\right)}{2} + \frac{\sin\left(3\cdot\frac{\pi}{2}\right)}{3} - \frac{\sin\left(4\cdot\frac{\pi}{2}\right)}{4} + \frac{\sin\left(5\cdot\frac{\pi}{2}\right)}{5} - \frac{\sin\left(6\cdot\frac{\pi}{2}\right)}{6} + \frac{\sin\left(7\cdot\frac{\pi}{2}\right)}{7} - \frac{\sin\left(8\cdot\frac{\pi}{2}\right)}{8} + \cdots\right)$$

$$= 2\left(1 - 0 + \left(-\frac{1}{3}\right) - 0 + \frac{1}{5} - 0 + \left(-\frac{1}{7}\right) + \cdots\right)$$

$$= 2 - \frac{2}{3} + \frac{2}{5} - \frac{2}{7} + \cdots$$

Multiplying both sides by 2 gives: $\pi = 4 - \dfrac{4}{3} + \dfrac{4}{5} - \dfrac{4}{7} + \cdots$

61. makes sense

63. makes sense

65.

$$\sin(\alpha+\beta) = \sin\alpha\cos\beta + \cos\alpha\sin\beta$$

$$\underline{-[\sin(\alpha-\beta) = \sin\alpha\cos\beta - \cos\alpha\sin\beta]}$$

$$\sin(\alpha+\beta) - \sin(\alpha-\beta) = 2\cos\alpha\sin\beta$$

Solve for $\cos\alpha\sin\beta$ by multiplying both sides by $\dfrac{1}{2}$: $\dfrac{1}{2}[\sin(\alpha+\beta) - \sin(\alpha-\beta)] = \cos\alpha\sin\beta$

67. $2\cos\dfrac{\alpha+\beta}{2}\cos\dfrac{\alpha-\beta}{2} = 2\cdot\dfrac{1}{2}\left[\cos\left(\dfrac{\alpha+\beta}{2}-\dfrac{\alpha-\beta}{2}\right)+\cos\left(\dfrac{\alpha+\beta}{2}+\dfrac{\alpha-\beta}{2}\right)\right]$

$\qquad\qquad\qquad\qquad = \cos\left(\dfrac{2\beta}{2}\right)+\cos\left(\dfrac{2\alpha}{2}\right)$

$\qquad\qquad\qquad\qquad = \cos\beta+\cos\alpha$

$\qquad\qquad\qquad\qquad = \cos\alpha+\cos\beta$

69. $\sin 2x+\sin 4x+\sin 6x = \sin 4x+(\sin 2x+\sin 6x)$

$\qquad\qquad\qquad\qquad\quad = \sin 4x+2\sin\left(\dfrac{2x+6x}{2}\right)\cos\left(\dfrac{2x-6x}{2}\right)$

$\qquad\qquad\qquad\qquad\quad = \sin 4x+2\sin\left(\dfrac{8x}{2}\right)\cos\left(\dfrac{-4x}{2}\right)$

$\qquad\qquad\qquad\qquad\quad = \sin 4x+2\sin 4x\cos(-2x)$

$\qquad\qquad\qquad\qquad\quad = \sin 4x+2\sin 4x\cos 2x$

$\qquad\qquad\qquad\qquad\quad = \sin(2\cdot 2x)+2\sin 4x\cos 2x$

$\qquad\qquad\qquad\qquad\quad = 2\sin 2x\cos 2x+2\sin 4x\cos 2x$

$\qquad\qquad\qquad\qquad\quad = 2\cos 2x(\sin 2x+\sin 4x)$

$\qquad\qquad\qquad\qquad\quad = 2\cos 2x\left(2\sin\left(\dfrac{2x+4x}{2}\right)\cos\left(\dfrac{2x-4x}{2}\right)\right)$

$\qquad\qquad\qquad\qquad\quad = 2\cos 2x\cdot 2\sin\left(\dfrac{6x}{2}\right)\cos\left(\dfrac{-2x}{2}\right)$

$\qquad\qquad\qquad\qquad\quad = 2\cos 2x\cdot 2\sin 3x\cos(-x)$

$\qquad\qquad\qquad\qquad\quad = 4\cos 2x\sin 3x\cos x$

$\qquad\qquad\qquad\qquad\quad = 4\cos x\cos 2x\sin 3x$

71. $2(1-u^2)+3u = 0$

$\quad 2-2u^2+3u = 0$

$\quad 2u^2-3u-2 = 0$

$\quad (2u+1)(u-2) = 0$

$\quad 2u+1 = 0 \qquad\text{and}\qquad u-2 = 0$

$\qquad 2u = -1 \qquad\qquad\qquad u = 2$

$\qquad u = -\dfrac{1}{2}$

The solution set is $\left\{-\dfrac{1}{2}, 2\right\}$.

72. $u^3-3u = 0$

$\quad u(u^2-3) = 0$

$\quad u = 0 \quad\text{or}\quad u^2-3 = 0$

$\qquad\qquad\qquad\qquad u^2 = 3$

$\qquad\qquad\qquad\qquad u = \pm\sqrt{3}$

The solution set is $\left\{-\sqrt{3}, 0, \sqrt{3}\right\}$.

73. $u^2-u-1 = 0$

$\quad a = 1, \quad b = -1, \quad c = -1$

$\quad x = \dfrac{-b\pm\sqrt{b^2-4ac}}{2a}$

$\quad x = \dfrac{-(-1)\pm\sqrt{(-1)^2-4(1)(-1)}}{2(1)}$

$\quad x = \dfrac{1\pm\sqrt{5}}{2}$

The solution set is $\left\{\dfrac{1-\sqrt{5}}{2}, \dfrac{1+\sqrt{5}}{2}\right\}$.

Section 5.5

Check Point Exercises

1.
$$5\sin x = 3\sin x + \sqrt{3}$$
$$5\sin x - 3\sin x = 3\sin x - 3\sin x + \sqrt{3}$$
$$2\sin x = \sqrt{3}$$
$$\sin x = \frac{\sqrt{3}}{2}$$

Because $\sin\dfrac{\pi}{3} = \dfrac{\sqrt{3}}{2}$, the solutions for $\sin x = \dfrac{\sqrt{3}}{2}$

in $[0, 2\pi)$ are

$$x = \frac{\pi}{3}$$

$$x = \pi - \frac{\pi}{3} = \frac{3\pi}{3} - \frac{\pi}{3} = \frac{2\pi}{3}.$$

Because the period of the sine function is $2\pi,$ the solutions are given by

$$x = \frac{\pi}{3} + 2n\pi \quad \text{or}$$

$$x = \frac{2\pi}{3} + 2n\pi$$

where n is any integer.

2. The period of the tangent function is π. In the interval $[0, \pi)$, the only value for which the tangent

function is $\sqrt{3}$ is $\dfrac{\pi}{3}$. All the solutions to

$\tan 2x = \sqrt{3}$ are given by

$$2x = \frac{\pi}{3} + n\pi$$

$$x = \frac{\pi}{6} + \frac{n\pi}{2}$$

where n is any integer. In the interval $[0, 2\pi)$, we obtain solutions as follows:

Let $n = 0.$ $x = \dfrac{\pi}{6} + \dfrac{0\pi}{2}$

$$= \frac{\pi}{6}$$

Let $n = 1.$ $x = \dfrac{\pi}{6} + \dfrac{1\pi}{2}$

$$= \frac{\pi}{6} + \frac{3\pi}{6} = \frac{2\pi}{3}$$

Let $n = 2.$ $x = \dfrac{\pi}{6} + \dfrac{2\pi}{2}$

$$= \frac{\pi}{6} + \frac{6\pi}{6} = \frac{7\pi}{6}$$

Let $n = 3.$ $x = \dfrac{\pi}{6} + \dfrac{3\pi}{2}$

$$= \frac{\pi}{6} + \frac{9\pi}{6} = \frac{5\pi}{3}$$

In the interval $[0, 2\pi)$, the solutions are

$$\frac{\pi}{6}, \frac{2\pi}{3}, \frac{7\pi}{6}, \text{ and } \frac{5\pi}{3}.$$

3. The period of the sine function is $2\pi.$
In the interval $[0, 2\pi)$, there are two values at which

the sine function is $\dfrac{1}{2}$. One is $\dfrac{\pi}{6}$. The sine is

positive in quadrant II. Thus, the other value is

$\pi - \dfrac{\pi}{6} = \dfrac{5\pi}{6}$. All the solutions to $\sin\dfrac{x}{3} = \dfrac{1}{2}$ are

given by

$$\frac{x}{3} = \frac{\pi}{6} + 2n\pi$$

$$x = \frac{\pi}{2} + 6n\pi$$

or

$$\frac{x}{3} = \frac{5\pi}{6} + 2n\pi$$

$$x = \frac{5\pi}{2} + 6n\pi$$

where n is any integer. In the interval $[0, 2\pi)$, we obtain solutions as follows:

Let $n = 0.$ $x = \dfrac{\pi}{2}$ or $x = \dfrac{5\pi}{2}$

If we let $n = 1$, we are adding 6π to each of these expressions. These values of x exceed $2\pi.$ Thus in

the interval $[0, 2\pi)$, the solution set is $\left\{\dfrac{\pi}{2}, \dfrac{5\pi}{2}\right\}$

4. The given equation is in quadratic form
$2t^2 - 3t + 1 = 0$ with $t = \sin x$.

$2\sin^2 x - 3\sin x + 1 = 0$

$(2\sin x - 1)(\sin x - 1) = 0$

$2\sin x - 1 = 0$ or $\sin x - 1 = 0$

$2\sin x = 1$ $\sin x = 1$

$\sin x = \dfrac{1}{2}$

$x = \dfrac{\pi}{6}$ $x = \dfrac{\pi}{2}$

$x = \pi - \dfrac{\pi}{6} = \dfrac{5\pi}{6}$

The solutions in the interval $[0, 2\pi)$ are $\dfrac{\pi}{6}, \dfrac{\pi}{2}$,

and $\dfrac{5\pi}{6}$.

5. $4\cos^2 x - 3 = 0$

$\cos^2 x = \dfrac{3}{4}$

$\cos x = \pm\sqrt{\dfrac{3}{4}}$

$\cos x = \pm\dfrac{\sqrt{3}}{2}$

$\cos x = \dfrac{\sqrt{3}}{2}$ or $\cos x = -\dfrac{\sqrt{3}}{2}$

$x = \dfrac{\pi}{6}, \dfrac{11\pi}{6}$ $x = \dfrac{5\pi}{6}, \dfrac{7\pi}{6}$

The solutions in the interval $[0, 2\pi)$ are

$\dfrac{\pi}{6}, \dfrac{5\pi}{6}, \dfrac{7\pi}{6}$, and $\dfrac{11\pi}{6}$.

6. $\sin x \tan x = \sin x$

$\sin x \tan x - \sin x = 0$

$\sin x(\tan x - 1) = 0$

$\sin x = 0$ or $\tan x - 1 = 0$

$x = 0$ $x = \pi$ $\tan x = 1$

$x = \dfrac{\pi}{4}$

$x = \pi + \dfrac{\pi}{4} = \dfrac{5\pi}{4}$

The solutions in the interval $[0, 2\pi)$ are

$0, \dfrac{\pi}{4}, \pi$, and $\dfrac{5\pi}{4}$.

7. $2\sin^2 x - 3\cos x = 0$

$2(1 - \cos^2 x) - 3\cos x = 0$

$2 - 2\cos^2 x - 3\cos x = 0$

$-2\cos^2 x - 3\cos x + 2 = 0$

$2\cos^2 x + 3\cos x - 2 = 0$

$(2\cos x - 1)(\cos x + 2) = 0$

$2\cos x - 1 = 0$ or $\cos x + 2 = 0$

$\cos x = \dfrac{1}{2}$ $\cos x = -2$

                   ~~$\cos x = -2$~~

$x = \dfrac{\pi}{3}, \dfrac{5\pi}{3}$ This equation has no solution.

The solutions in the interval $[0, 2\pi)$ are $\dfrac{\pi}{3}$ and $\dfrac{5\pi}{3}$.

8. $\cos 2x + \sin x = 0$

$1 - 2\sin^2 x + \sin x = 0$

$-2\sin^2 x + \sin x + 1 = 0$

$2\sin^2 x - \sin x - 1 = 0$

$(2\sin x + 1)(\sin x - 1) = 0$

$2\sin x + 1 = 0$ or $\sin x - 1 = 0$

$2\sin x = -1$ $\sin x = 1$

$\sin x = -\dfrac{1}{2}$ $x = \dfrac{\pi}{2}$

$x = \pi + \dfrac{\pi}{6} = \dfrac{7\pi}{6}$ or

$x = 2\pi - \dfrac{\pi}{6} = \dfrac{11\pi}{6}$

The solutions in the interval $[0, 2\pi)$ are

$\dfrac{\pi}{2}, \dfrac{7\pi}{6}$, and $\dfrac{11\pi}{6}$.

9.
$$\sin x \cos x = -\frac{1}{2}$$
$$2 \sin x \cos x = -1$$
$$\sin 2x = -1$$

The period of the sine function is 2π. In the interval $[0, 2\pi)$, the sine function is -1 at $\frac{3\pi}{2}$. All the solutions to $\sin 2x$ are given by

$$2x = \frac{3\pi}{2} + 2n\pi$$
$$x = \frac{3\pi}{4} + n\pi,$$

where n is any integer. The solutions in the interval $[0, 2\pi)$ are obtained by letting $n = 0$ and $n = 1$. The solutions are $\frac{3\pi}{4}$ and $\frac{7\pi}{4}$.

10.
$$\cos x - \sin x = -1$$
$$(\cos x - \sin x)^2 = (-1)^2$$
$$\cos^2 x - 2\cos x \sin x + \sin^2 x = 1$$
$$\cos^2 x + \sin^2 x - 2\cos x \sin x = 1$$
$$1 - 2\cos x \sin x = 1$$
$$-2\cos x \sin x = 0$$
$$\cos x \sin x = 0$$

$\cos x = 0 \qquad$ or $\qquad \sin x = 0$

$x = \dfrac{\pi}{2} \qquad\qquad\qquad x = 0$

$x = \dfrac{3\pi}{2} \qquad\qquad\qquad x = \pi$

We check these proposed solutions to see if any are extraneous.

Check 0: $\cos 0 - \sin 0 = -1$
$$1 - 0 = -1$$
$$1 = -1, \text{ false}$$

Check $\dfrac{\pi}{2}$: $\cos\dfrac{\pi}{2} - \sin\dfrac{\pi}{2} = -1$
$$0 - 1 = -1$$
$$-1 = -1, \text{ true}$$

Check π: $\cos\dfrac{\pi}{2} - \sin\dfrac{\pi}{2} = -1$
$$-1 - 0 = -1$$
$$-1 = -1, \text{ true}$$

Check $\dfrac{3\pi}{2}$: $\cos\dfrac{3\pi}{2} - \sin\dfrac{3\pi}{2} = -1$
$$0 - (-1) = -1$$
$$1 = -1, \text{ false}$$

The actual solutions in the interval $[0, 2\pi)$ are $\dfrac{\pi}{2}$ and π.

11. **a.** $\tan x = 3.1044$

Be sure calculator is in radian mode and find the inverse tangent of 3.1044. This gives the first quadrant reference angle.
$$\theta = \tan^{-1} 3.1044 \approx 1.2592$$
The tangent is positive in quadrants I and III thus,
$$x \approx 1.2592 \quad \text{or} \quad x \approx \pi + 1.2592$$
$$x \approx 4.4008$$

b. $\sin x = -0.2315$

Be sure calculator is in radian mode and find the inverse sine of $+0.2315$. This gives the first quadrant reference angle.
$$\theta = \sin^{-1}(0.2315) \approx 0.2336$$
The sine is negative in quadrants III and IV thus,
$$x \approx \pi + 0.2336 \quad \text{or} \quad x \approx 2\pi - 1.2592$$
$$x \approx 3.3752 \qquad\qquad x \approx 6.0496$$

12. $\cos^2 x + 5\cos x + 3 = 0$

Use the quadratic formula to solve for $\cos x$.
$$\cos x = \frac{-b \pm \sqrt{b^2 - 4ac}}{2a}$$
$$\cos x = \frac{-(5) \pm \sqrt{(5)^2 - 4(1)(3)}}{2(1)}$$
$$\cos x = \frac{-5 \pm \sqrt{13}}{2}$$
$$\cos x \approx -0.6972 \quad \text{or} \quad \cos x \approx -4.3028$$
$$\cancel{\cos x \approx -4.3028}$$
This equation has no solution.

Be sure calculator is in radian mode and find the inverse cosine of $+0.6972$. This gives the first quadrant reference angle.
$$\theta = \cos^{-1} 0.6972 \approx 0.7993$$
The cosine is negative in quadrants II and III thus,
$$x \approx \pi - 0.7993 \quad \text{or} \quad x \approx \pi + 0.7993$$
$$x \approx 2.3423 \qquad\qquad x \approx 3.9409$$

Exercise Set 5.5

1. $\cos\dfrac{\pi}{4} = \dfrac{\sqrt{2}}{2}$

$\dfrac{\sqrt{2}}{2} = \dfrac{\sqrt{2}}{2}$ is true.

Thus, $\dfrac{\pi}{4}$ is a solution.

3. $\sin\dfrac{\pi}{6} = \dfrac{\sqrt{3}}{2}$

$\dfrac{1}{2} = \dfrac{\sqrt{3}}{2}$ is false.

Thus, $\dfrac{\pi}{6}$ is not a solution.

5. $\cos\dfrac{2\pi}{3} = -\dfrac{1}{2}$

$-\dfrac{1}{2} = -\dfrac{1}{2}$ is true.

Thus, $\dfrac{2\pi}{3}$ is a solution.

7. $\tan\left(2 \cdot \dfrac{5\pi}{12}\right) = -\dfrac{\sqrt{3}}{3}$

$\tan\dfrac{5\pi}{6} = -\dfrac{\sqrt{3}}{3}$

$-\dfrac{\sqrt{3}}{3} = -\dfrac{\sqrt{3}}{3}$ is true.

Thus, $\dfrac{5\pi}{12}$ is a solution.

9. $\cos\dfrac{\pi}{3} = \dfrac{\sqrt{3}}{2}$

$\dfrac{1}{2} = \dfrac{\sqrt{3}}{2}$ is false.

Thus, $\dfrac{\pi}{3}$ is not a solution.

11. $\sin x = \dfrac{\sqrt{3}}{2}$

Because $\sin\dfrac{\pi}{3} = \dfrac{\sqrt{3}}{2}$, the solutions

for $\sin x = \dfrac{\sqrt{3}}{2}$ in $[0, 2\pi)$ are

$x = \dfrac{\pi}{3}$

$x = \pi - \dfrac{\pi}{3} = \dfrac{3\pi}{3} - \dfrac{\pi}{3} = \dfrac{2\pi}{3}.$

Because the period of the sine function is 2π, the solutions are given by

$x = \dfrac{\pi}{3} + 2n\pi \quad$ or $\quad x = \dfrac{2\pi}{3} + 2n\pi$

where n is any integer.

13. $\tan x = 1$

Because $\tan\dfrac{\pi}{4} = 1$, the solution

for $\tan x = 1$ in $[0, \pi)$ is

$x = \dfrac{\pi}{4}.$

Because the period of the tangent function is π, the solutions are given by

$x = \dfrac{\pi}{4} + n\pi$

where n is any integer.

15. $\cos x = -\dfrac{1}{2}$

Because $\cos\dfrac{\pi}{3} = \dfrac{1}{2}$, the solutions

for $\cos x = -\dfrac{1}{2}$ in $[0, 2\pi)$ are

$x = \pi - \dfrac{\pi}{3} = \dfrac{3\pi}{3} - \dfrac{\pi}{3} = \dfrac{2\pi}{3}$

$x = \pi + \dfrac{\pi}{3} = \dfrac{3\pi}{3} + \dfrac{\pi}{3} = \dfrac{4\pi}{3}.$

Because the period of the cosine function is 2π, the solutions are given by

$x = \dfrac{2\pi}{3} + 2n\pi \quad$ or $\quad x = \dfrac{4\pi}{3} + 2n\pi$

where n is any integer.

17. $\tan x = 0$

Because $\tan 0 = 0$, the solution

for $\tan x = 0$ in $[0, \pi)$ is

$x = 0$.

Because the period of the tangent function is π, the solutions are given by

$x = 0 + n\pi = n\pi$

where n is any integer.

19. $2\cos x + \sqrt{3} = 0$

$2\cos x = -\sqrt{3}$

$\cos x = -\dfrac{\sqrt{3}}{2}$

Because $\cos\dfrac{\pi}{6} = \dfrac{\sqrt{3}}{2}$, the solutions

for $\cos x = -\dfrac{\sqrt{3}}{2}$ in $[0, 2\pi)$ are

$x = \pi - \dfrac{\pi}{6} = \dfrac{6\pi}{6} - \dfrac{\pi}{6} = \dfrac{5\pi}{6}$

$x = \pi + \dfrac{\pi}{6} = \dfrac{6\pi}{6} + \dfrac{\pi}{6} = \dfrac{7\pi}{6}$.

Because the period of the cosine function is 2π, the solutions are given by

$x = \dfrac{5\pi}{6} + 2n\pi$ or $x = \dfrac{7\pi}{6} + 2n\pi$

where n is any integer.

21. $4\sin\theta - 1 = 2\sin\theta$

$4\sin\theta - 2\sin\theta = 1$

$2\sin\theta = 1$

$\sin\theta = \dfrac{1}{2}$

Because $\sin\dfrac{\pi}{6} = \dfrac{1}{2}$, the solutions

for $\sin\theta = \dfrac{1}{2}$ in $[0, 2\pi)$ are

$\theta = \dfrac{\pi}{6}$

$\theta = \pi - \dfrac{\pi}{6} = \dfrac{6\pi}{6} - \dfrac{\pi}{6} = \dfrac{5\pi}{6}$.

Because the period of the sine function is 2π, the solutions are given by

$\theta = \dfrac{\pi}{6} + 2n\pi$ or $\theta = \dfrac{5\pi}{6} + 2n\pi$

where n is any integer.

23. $3\sin\theta + 5 = -2\sin\theta$

$3\sin\theta + 2\sin\theta = -5$

$5\sin\theta = -5$

$\sin\theta = -1$

Because $\sin\dfrac{\pi}{2} = 1$, the solutions

for $\sin\theta = -1$ in $[0, 2\pi)$ are

$\theta = \pi + \dfrac{\pi}{2} = \dfrac{2\pi}{2} + \dfrac{\pi}{2} = \dfrac{3\pi}{2}$

$\theta = 2\pi - \dfrac{\pi}{2} = \dfrac{4\pi}{2} - \dfrac{\pi}{2} = \dfrac{3\pi}{2}$.

Because the period of the sine function is 2π, the solutions are given by

$\theta = \dfrac{3\pi}{2} + 2n\pi$

where n is any integer.

25. The period of the sine function is 2π. In the interval $[0, 2\pi)$, there are two values at which the sine

function is $\dfrac{\sqrt{3}}{2}$. One is $\dfrac{\pi}{3}$. The sine is positive in

quadrant II; thus, the other value is $\pi - \dfrac{\pi}{3} = \dfrac{2\pi}{3}$. All

the solutions to $\sin 2x = \dfrac{\sqrt{3}}{2}$ are given by

$2x = \dfrac{\pi}{3} + 2n\pi$ or $2x = \dfrac{2\pi}{3} + 2n\pi$

$x = \dfrac{\pi}{6} + n\pi$ $\qquad x = \dfrac{\pi}{3} + n\pi$

Where n is any integer.

The solutions in the interval $[0, 2\pi)$ are obtained by letting $n = 0$ and $n = 1$.

The solutions are $\dfrac{\pi}{6}, \dfrac{\pi}{3}, \dfrac{7\pi}{6}$, and $\dfrac{4\pi}{3}$.

27. The period of the cosine function is 2π. In the interval $[0, 2\pi)$, there are two values at which the cosine function is $-\dfrac{\sqrt{3}}{2}$. One is $\dfrac{5\pi}{6}$. The cosine is negative in quadrant III; thus, the other value is $2\pi - \dfrac{5\pi}{6} = \dfrac{7\pi}{6}$. All the solutions to

$$\cos 4x = -\dfrac{\sqrt{3}}{2}$$ are given by

$$4x = \dfrac{5\pi}{6} + 2n\pi \quad \text{or} \quad 4x = \dfrac{7\pi}{6} + 2n\pi$$

$$x = \dfrac{5\pi}{24} + \dfrac{n\pi}{2} \qquad\qquad x = \dfrac{7\pi}{24} + \dfrac{n\pi}{2}$$

where n is any integer.
The solutions in the interval $[0, 2\pi)$ are obtained by letting $n = 0$, $n = 1$, $n = 2$, and $n = 3$.

The solutions are $\dfrac{5\pi}{24}, \dfrac{7\pi}{24}, \dfrac{17\pi}{24}, \dfrac{19\pi}{24},$

$\dfrac{29\pi}{24}, \dfrac{31\pi}{24}, \dfrac{41\pi}{24}$ and $\dfrac{43\pi}{24}$.

29. The period of the tangent function is π. In the interval $[0, \pi)$, the only value for which the tangent function is $\dfrac{\sqrt{3}}{3}$ is $\dfrac{\pi}{6}$.

All the solutions to $\tan 3x = \dfrac{\sqrt{3}}{3}$ are given by

$$3x = \dfrac{\pi}{6} + n\pi$$

$$x = \dfrac{\pi}{18} + \dfrac{n\pi}{3}$$

where n is any integer.
The solutions in the interval $[0, 2\pi)$ are obtained by letting $n = 0$, $n = 1$, $n = 2$, $n = 3$, $n = 4$, and $n = 5$.
The solutions are

$\dfrac{\pi}{18}, \dfrac{7\pi}{18}, \dfrac{13\pi}{18}, \dfrac{19\pi}{18}, \dfrac{25\pi}{18},$ and $\dfrac{31\pi}{18}$.

31. The period of the tangent function is π. In the interval $[0, \pi)$, the only value for which the tangent function is $\sqrt{3}$ is $\dfrac{\pi}{3}$.

All the solutions to $\tan \dfrac{x}{2} = \sqrt{3}$ are given by

$$\dfrac{x}{2} = \dfrac{\pi}{3} + n\pi$$

$$x = \dfrac{2\pi}{3} + 2n\pi \text{ where } n \text{ is any integer. The solution}$$

in the interval $[0, 2\pi)$ is obtained by letting $n = 0$.

The only solution is $\dfrac{2\pi}{3}$.

33. The period of the sine function is 2π. In the interval $[0, 2\pi)$, the only value for which the sine function is -1 is $\dfrac{3\pi}{2}$.

All the solutions to $\sin \dfrac{2\theta}{3} = -1$ are given by

$$\dfrac{2\theta}{3} = \dfrac{3\pi}{2} + 2n\pi$$

$$\theta = \dfrac{9\pi}{4} + 3n\pi \text{ where } n \text{ is any integer. All values of}$$

θ exceed 2π or are less than zero.
Thus, in the interval $[0, 2\pi)$ there is no solution.

35. The period of the secant function is 2π. In the interval $[0, 2\pi)$, there are two values at which the secant function is -2. One is $\dfrac{2\pi}{3}$. The secant is negative in quadrant III; thus, the other value is $2\pi - \dfrac{2\pi}{3} = \dfrac{4\pi}{3}$. All the solutions to $\sec \dfrac{3\theta}{2} = -2$ are given by

$$\dfrac{3\theta}{2} = \dfrac{2\pi}{3} + 2n\pi \quad \text{or} \quad \dfrac{3\theta}{2} = \dfrac{4\pi}{3} + 2n\pi$$

$$\theta = \dfrac{4\pi}{9} + \dfrac{4n\pi}{3} \qquad\qquad \theta = \dfrac{8\pi}{9} + \dfrac{4n\pi}{3}$$

where n is any integer. The solutions in the interval $[0, 2\pi)$ are obtained by letting $n = 0$ and $n = 1$.

Since $\dfrac{20\pi}{9}$ is not in $[0, 2\pi)$, the solutions are

$\dfrac{4\pi}{9}, \dfrac{8\pi}{9},$ and $\dfrac{16\pi}{9}$.

37. The period of the sine function is 2π. In the interval $[0, 2\pi)$, there are two values at which the sine function is $\frac{1}{2}$. One is $\frac{\pi}{6}$. The sine is positive in quadrant II; Thus, the other value is $\pi - \frac{\pi}{6} = \frac{5\pi}{6}$.

All the solutions to $\sin\left(2x + \frac{\pi}{6}\right) = \frac{1}{2}$ are given by

$$2x + \frac{\pi}{6} = \frac{\pi}{6} + 2n\pi$$
$$2x = 2n\pi$$
$$x = n\pi \qquad \text{or}$$
$$2x + \frac{\pi}{6} = \frac{5\pi}{6} + 2n\pi$$
$$2x = \frac{4\pi}{6} + 2n\pi$$
$$x = \frac{2\pi}{6} + n\pi$$
$$x = \frac{\pi}{3} + n\pi$$

where n is any integer. The solutions in the interval $[0, 2\pi)$ are obtained by letting $n = 0$ and $n = 1$.

The solutions are 0, $\frac{\pi}{3}$, π, and $\frac{4\pi}{3}$.

39.
$$2\sin^2 x - \sin x - 1 = 0$$
$$(2\sin x + 1)(\sin x - 1) = 0$$

$2\sin x + 1 = 0$ or $\sin x - 1 = 0$
$2\sin x = -1$ $\sin x = 1$
$\sin x = -\frac{1}{2}$

$x = \frac{7\pi}{6}$ $x = \frac{11\pi}{6}$ $x = \frac{\pi}{2}$

The solutions in the interval $[0, 2\pi)$ are $\frac{\pi}{2}$, $\frac{7\pi}{6}$, and $\frac{11\pi}{6}$.

41.
$$2\cos^2 x + 3\cos x + 1 = 0$$
$$(2\cos x + 1)(\cos x + 1) = 0$$

$2\cos x + 1 = 0$ or $\cos x + 1 = 0$
$2\cos x = -1$ $\cos x = -1$
$\cos x = -\frac{1}{2}$

$x = \frac{2\pi}{3}$ $x = \frac{4\pi}{3}$ $x = \pi$

The solutions in the interval $[0, 2\pi)$ are $\frac{2\pi}{3}$, π, and $\frac{4\pi}{3}$.

43.
$$2\sin^2 x = \sin x + 3$$
$$2\sin^2 x - \sin x - 3 = 0$$
$$(2\sin x - 3)(\sin x + 1) = 0$$

$2\sin x - 3 = 0$ or $\sin x + 1 = 0$
$2\sin x = 3$ $\sin x = -1$
$\sin x = \frac{3}{2}$ $x = \frac{3\pi}{2}$

$\sin x$ cannot be greater than 1.

The solution in the interval $[0, 2\pi)$ is $\frac{3\pi}{2}$.

45.
$$\sin^2 \theta - 1 = 0$$
$$(\sin\theta - 1)(\sin\theta + 1) = 0$$

$\sin\theta - 1 = 0$ or $\sin\theta + 1 = 0$
$\sin\theta = 1$ $\sin\theta = -1$
$\theta = \frac{\pi}{2}$ $\theta = \frac{3\pi}{2}$

The solutions in the interval $[0, 2\pi)$ are $\frac{\pi}{2}$ and $\frac{3\pi}{2}$.

47.
$$4\cos^2 x - 1 = 0$$
$$(2\cos x + 1)(2\cos x - 1) = 1$$

$2\cos x + 1 = 0$ or $2\cos x - 1 = 0$
$\cos x = -\frac{1}{2}$ $\cos x = \frac{1}{2}$

$x = \frac{2\pi}{3}, \frac{4\pi}{3}$ $x = \frac{\pi}{3}, \frac{5\pi}{3}$

The solutions in the interval $[0, 2\pi)$ are

$$\frac{\pi}{3}, \frac{2\pi}{3}, \frac{4\pi}{3}, \text{ and } \frac{5\pi}{3}.$$

49. $9\tan^2 x - 3 = 0$

$$\tan^2 x = \frac{3}{9}$$

$$\tan x = \pm\sqrt{\frac{3}{9}}$$

$$\tan x = \pm\frac{\sqrt{3}}{3}$$

$\tan x = \frac{\sqrt{3}}{3}$ or $\tan x = -\frac{\sqrt{3}}{3}$

$x = \frac{\pi}{6}, \frac{7\pi}{6}$ $\qquad$ $x = \frac{5\pi}{6}, \frac{11\pi}{6}$

The solutions in the interval $[0, 2\pi)$ are

$\frac{\pi}{6}, \frac{5\pi}{6}, \frac{7\pi}{6},$ and $\frac{11\pi}{6}$.

51. $\sec^2 x - 2 = 0$

$$\sec^2 x = 2$$

$$\cos^2 x = \frac{1}{2}$$

$$\cos x = \pm\sqrt{\frac{1}{2}}$$

$$\cos x = \pm\frac{\sqrt{2}}{2}$$

$\cos x = \frac{\sqrt{2}}{2}$ or $\cos x = -\frac{\sqrt{2}}{2}$

$x = \frac{\pi}{4}, \frac{7\pi}{4}$ $\qquad$ $x = \frac{3\pi}{4}, \frac{5\pi}{4}$

The solutions in the interval $[0, 2\pi)$ are

$\frac{\pi}{4}, \frac{3\pi}{4}, \frac{5\pi}{4},$ and $\frac{7\pi}{4}$.

53. $(\tan x - 1)(\cos x + 1) = 0$

$\tan x - 1 = 0$ or $\cos x + 1 = 0$

$\tan x = 1$ $\qquad$ $\cos x = -1$

$x = \frac{\pi}{4}$ $x = \frac{5\pi}{4}$ $\qquad$ $x = \pi$

The solutions in the interval $[0, 2\pi)$ are

$\frac{\pi}{4}, \pi,$ and $\frac{5\pi}{4}$.

55. $\left(2\cos x + \sqrt{3}\right)\left(2\sin x + 1\right) = 0$

$2\cos x + \sqrt{3} = 0$ or $2\sin x + 1 = 0$

$2\cos x = -\sqrt{3}$ $\qquad$ $2\sin x = -1$

$\cos x = -\frac{\sqrt{3}}{2}$ $\qquad$ $\sin x = -\frac{1}{2}$

$x = \frac{5\pi}{6}$ $x = \frac{7\pi}{6}$ $\qquad$ $x = \frac{7\pi}{6}$ $x = \frac{11\pi}{6}$

The solutions in the interval $[0, 2\pi)$ are

$\frac{5\pi}{6}, \frac{7\pi}{6},$ and $\frac{11\pi}{6}$.

57. $\cot x(\tan x - 1) = 0$

$\cot x = 0$ or $\tan x - 1 = 0$

$\qquad\qquad\qquad$ $\tan x = 1$

$x = \frac{\pi}{2}$ $x = \frac{3\pi}{2}$ $x = \frac{\pi}{4}$ $x = \frac{5\pi}{4}$

The solutions in the interval $[0, 2\pi)$ are $\frac{\pi}{4}$ and $\frac{5\pi}{4}$

since tan is undefined for $\frac{\pi}{2}$ and $\frac{3\pi}{2}$.

59. $\sin x + 2\sin x\cos x = 0$

$\sin x(1 + 2\cos x) = 0$

$\sin x = 0$ or $1 + 2\cos x = 0$

$\qquad\qquad\qquad$ $2\cos x = -1$

$\qquad\qquad\qquad$ $\cos x = -\frac{1}{2}$

$x = 0$ $x = \pi$ $x = \frac{2\pi}{3}$ $x = \frac{4\pi}{3}$

The solutions in the interval $[0, 2\pi)$ are

$0, \frac{2\pi}{3}, \pi,$ and $\frac{4\pi}{3}$.

61. $\tan^2 x\cos x = \tan^2 x$

$\tan^2 x\cos x - \tan^2 x = 0$

$\tan^2 x(\cos x - 1) = 0$

$\tan^2 x = 0$ $\qquad$ or $\qquad$ $\cos x - 1 = 0$

$\tan x = 0$ $\qquad\qquad\qquad$ $\cos x = 1$

$x = 0$ $x = \pi$ $\qquad\qquad$ $x = 0$

The solutions in the interval $[0, 2\pi)$ are 0 and π.

63.

$$2\cos^2 x + \sin x - 1 = 0$$
$$2\left(1 - \sin^2 x\right) + \sin x - 1 = 0$$
$$2 - 2\sin^2 x + \sin x - 1 = 0$$
$$-2\sin^2 x + \sin x + 1 = 0$$
$$2\sin^2 x - \sin x - 1 = 0$$
$$(2\sin x + 1)(\sin x - 1) = 0$$

$$2\sin x + 1 = 0 \qquad \text{or} \qquad \sin x - 1 = 0$$
$$2\sin x = -1 \qquad\qquad\qquad \sin x = 1$$
$$\sin x = -\frac{1}{2}$$

$$x = \frac{7\pi}{6} \quad x = \frac{11\pi}{6} \qquad\qquad x = \frac{\pi}{2}$$

The solutions in the interval $[0, 2\pi)$ are

$$\frac{\pi}{2}, \frac{7\pi}{6}, \text{ and } \frac{11\pi}{6}.$$

65.

$$\sin^2 x - 2\cos x - 2 = 0$$
$$1 - \cos^2 x - 2\cos x - 2 = 0$$
$$-\cos^2 x - 2\cos x - 1 = 0$$
$$\cos^2 x + 2\cos x + 1 = 0$$
$$(\cos x + 1)(\cos x + 1) = 0$$
$$\cos x + 1 = 0$$
$$\cos x = -1$$
$$x = \pi$$

The solution in the interval $[0, 2\pi)$ is π.

67.

$$4\cos^2 x = 5 - 4\sin x$$
$$4\cos^2 x + 4\sin x - 5 = 0$$
$$4\left(1 - \sin^2 x\right) + 4\sin x - 5 = 0$$
$$4 - 4\sin^2 x + 4\sin x - 5 = 0$$
$$-4\sin^2 x + 4\sin x - 1 = 0$$
$$4\sin^2 x - 4\sin x + 1 = 0$$
$$(2\sin x - 1)(2\sin x - 1) = 0$$
$$2\sin x - 1 = 0$$
$$2\sin x = 1$$
$$\sin x = \frac{1}{2}$$
$$x = \frac{\pi}{6} \quad x = \frac{5\pi}{6}$$

The solutions in the interval $[0, 2\pi)$ are $\frac{\pi}{6}$ and $\frac{5\pi}{6}$.

69.

$$\sin 2x = \cos x$$
$$2\sin x \cos x = \cos x$$
$$2\sin x \cos x - \cos x = 0$$
$$\cos x(2\sin x - 1) = 0$$

$$\cos x = 0 \qquad \text{or} \qquad 2\sin x - 1 = 0$$
$$2\sin x = 1$$
$$\sin x = \frac{1}{2}$$

$$x = \frac{\pi}{2} \quad x = \frac{3\pi}{2} \qquad\qquad x = \frac{\pi}{6} \quad x = \frac{5\pi}{6}$$

The solutions in the interval $[0, 2\pi)$ are

$$\frac{\pi}{6}, \frac{\pi}{2}, \frac{5\pi}{6}, \text{and } \frac{3\pi}{2}.$$

71.

$$\cos 2x = \cos x$$
$$2\cos^2 x - 1 = \cos x$$
$$2\cos^2 x - 1 - \cos x = 0$$
$$2\cos^2 x - \cos x - 1 = 0$$
$$(2\cos x + 1)(\cos x - 1) = 0$$

$$2\cos x + 1 = 0 \qquad\qquad \text{or} \qquad \cos x - 1 = 0$$
$$2\cos x = -1 \qquad\qquad\qquad\qquad \cos x = 1$$
$$\cos x = -\frac{1}{2} \qquad\qquad\qquad\qquad \cos x = 1$$

$$x = \frac{2\pi}{3} \quad x = \frac{4\pi}{3} \qquad\qquad\qquad x = 0$$

The solutions in the interval $[0, 2\pi)$ are

$$0, \frac{2\pi}{3}, \text{and } \frac{4\pi}{3}.$$

73.

$$\cos 2x + 5\cos x + 3 = 0$$
$$2\cos^2 x - 1 + 5\cos x + 3 = 0$$
$$2\cos^2 x + 5\cos x + 2 = 0$$
$$(2\cos x + 1)(\cos x + 2) = 0$$

$$2\cos x + 1 = 0 \qquad \text{or} \quad \cos x + 2 = 0$$
$$2\cos x = -1 \qquad\qquad \cos x = -2$$
$$\cos x = -\frac{1}{2}$$

$$x = \frac{2\pi}{3} \quad x = \frac{4\pi}{3} \qquad \begin{array}{l} \cos x \text{ cannot} \\ \text{be less than } -1 \end{array}$$

The solutions in the interval $[0, 2\pi)$ are $\frac{2\pi}{3}$ and $\frac{4\pi}{3}$.

75.

$$\sin x \cos x = \frac{\sqrt{2}}{4}$$

$$2\sin x \cos x = \frac{\sqrt{2}}{2}$$

$$\sin 2x = \frac{\sqrt{2}}{2}$$

The period of the sine function is 2π. In the interval $[0, 2\pi)$, there are two values at which the sine

function is $\frac{\sqrt{2}}{2}$. One is $\frac{\pi}{4}$. The sine is positive in

quadrant II; thus, the other value is $\pi - \frac{\pi}{4} = \frac{3\pi}{4}$.

All the solutions to $\sin 2x = \frac{\sqrt{2}}{2}$ are given by

$$2x = \frac{\pi}{4} + 2n\pi \quad \text{or} \quad 2x = \frac{3\pi}{4} + 2n\pi$$

$$x = \frac{\pi}{8} + n\pi \qquad\qquad x = \frac{3\pi}{8} + n\pi$$

where n is any integer.
The solutions in the interval $[0, 2\pi)$ are obtained by
letting $n = 0$ and $n = 1$.

The solutions are $\frac{\pi}{8}, \frac{3\pi}{8}, \frac{9\pi}{8},$ and $\frac{11\pi}{8}$.

77.

$$\sin x + \cos x = 1$$

$$(\sin x + \cos x)^2 = 1^2$$

$$\sin^2 x + 2\sin x \cos x + \cos^2 x = 1$$

$$\sin^2 x + \cos^2 x + 2\sin x \cos x = 1$$

$$1 + 2\sin x \cos x = 1$$

$$2\sin x \cos x = 0$$

$$\sin x \cos x = 0$$

$$\sin x = 0 \quad \text{or} \quad \cos x = 0$$

$$x = 0 \qquad\qquad x = \frac{\pi}{2}$$

$$x = \pi \qquad\qquad x = \frac{3\pi}{2}$$

After checking these proposed solutions,
the actual solutions in the interval $[0, 2\pi)$ are 0 and

$\frac{\pi}{2}$.

79.

$$\sin\left(x + \frac{\pi}{4}\right) + \sin\left(x - \frac{\pi}{4}\right) = 1$$

$$\frac{1}{2}\left[\sin\left(x + \frac{\pi}{4}\right) + \sin\left(x - \frac{\pi}{4}\right)\right] = 1 \cdot \frac{1}{2}$$

$$\sin x \cos \frac{\pi}{4} = \frac{1}{2}$$

$$\sin x \cdot \frac{\sqrt{2}}{2} = \frac{1}{2}$$

$$\sin x = \frac{1}{\sqrt{2}}$$

$$\sin x = \frac{\sqrt{2}}{2}$$

$$x = \frac{\pi}{4} \quad \text{or} \quad x = \frac{3\pi}{4}$$

The solutions in the interval $[0, 2\pi)$ are $\frac{\pi}{4}$ and $\frac{3\pi}{4}$.

81.

$$\sin 2x \cos x + \cos 2x \sin x = \frac{\sqrt{2}}{2}$$

$$\sin(2x + x) = \frac{\sqrt{2}}{2}$$

$$\sin 3x = \frac{\sqrt{2}}{2}$$

The period of the sine function is 2π. In the interval $[0, 2\pi)$, there are two values at which the sine function

is $\frac{\sqrt{2}}{2}$. One is $\frac{\pi}{4}$. The sine function is positive in

quadrant II; thus, the other value is $\pi - \frac{\pi}{4} = \frac{3\pi}{4}$.

All the solutions to $\sin 3x = \frac{\sqrt{2}}{2}$ are given by

$$3x = \frac{\pi}{4} + 2n\pi \quad \text{or} \quad 3x = \frac{3\pi}{4} + 2n\pi$$

$$x = \frac{\pi}{12} + \frac{2n\pi}{3} \qquad\qquad x = \frac{\pi}{4} + \frac{2n\pi}{3}$$

where n is any integer. The solutions in the interval $[0, 2\pi)$ are obtained by letting $n = 0$, $n = 1$, and $n = 2$.

The solutions are $\frac{\pi}{12}, \frac{\pi}{4}, \frac{3\pi}{4}, \frac{11\pi}{12}, \frac{17\pi}{12},$ and $\frac{19\pi}{12}$.

83.
$$\tan x + \sec x = 1$$
$$\tan x - 1 = -\sec x$$
$$(\tan x - 1)^2 = (-\sec x)^2$$
$$\tan^2 x - 2\tan x + 1 = \sec^2 x$$
$$\tan^2 x - 2\tan x + 1 = 1 + \tan^2 x$$
$$-2\tan x = 0$$
$$\tan x = 0$$
$$x = 0 \quad x = \pi$$

We check these proposed solutions to see if any are extraneous.

Check 0: $\tan 0 \; + \; \sec 0 \quad 0 \quad 1$

$\quad\quad\quad\quad 0 \; + \; 1 \quad\quad 0 \quad 1$ True

Check π: $\tan \pi \; + \; \sec \pi \quad 0 \quad 1$

$\quad\quad\quad\quad 0 \quad\;\; + \; (-1) \quad 0 \quad 1$ False

The actual solution in the interval $[0, 2\pi)$ is 0.

85. $\sin x = 0.8246$

Be sure calculator is in radian mode and find the inverse sine of 0.8246. This gives the first quadrant reference angle.

$$\theta = \sin^{-1} 0.8246 \approx 0.9695$$

The sine is positive in quadrants I and II thus,

$$x \approx 0.9695 \quad \text{or} \quad x \approx \pi - 0.9695$$
$$x \approx 2.1721$$

87. $\cos x = -\dfrac{2}{5}$

Be sure calculator is in radian mode and find the inverse cosine of $+\dfrac{2}{5}$. This gives the first quadrant reference angle.

$$\theta = \cos^{-1}\dfrac{2}{5} \approx 1.1593$$

The cosine is negative in quadrants II and III thus,

$$x \approx \pi - 1.1593 \quad \text{or} \quad x \approx \pi + 1.1593$$
$$x \approx 1.9823 \quad\quad\quad\quad x \approx 4.3009$$

89. $\tan x = -3$

Be sure calculator is in radian mode and find the inverse tangent of $+3$. This gives the first quadrant reference angle.

$$\theta = \tan^{-1} 3 \approx 1.2490$$

The tangent is negative in quadrants II and IV thus,

$$x \approx \pi - 1.2490 \quad \text{or} \quad x \approx 2\pi - 1.2490$$
$$x \approx 1.8925 \quad\quad\quad\quad x \approx 5.0341$$

91. $\cos^2 x - \cos x - 1 = 0$

Use the quadratic formula to solve for $\cos x$.

$$\cos x = \frac{-b \pm \sqrt{b^2 - 4ac}}{2a}$$
$$\cos x = \frac{-(-1) \pm \sqrt{(-1)^2 - 4(1)(-1)}}{2(1)}$$
$$\cos x = \frac{1 \pm \sqrt{5}}{2}$$
$$\cos x \approx -0.6180 \quad \text{or} \quad \cos x \approx 1.6180$$

$\quad\quad\quad\quad\quad\quad\quad\quad \cancel{\cos x \approx 1.6180}$

This equation has no solution.

Be sure calculator is in radian mode and find the inverse cosine of $+0.6180$. This gives the first quadrant reference angle.

$$\theta = \cos^{-1} 0.6180 \approx 0.9046$$

The cosine is negative in quadrants II and III thus,

$$x \approx \pi - 0.9046 \quad \text{or} \quad x \approx \pi + 0.9046$$
$$x \approx 2.2370 \quad\quad\quad\quad x \approx 4.0462$$

93.
$$4\tan^2 x - 8\tan x + 3 = 0$$
$$(2\tan x - 1)(2\tan x - 3) = 0$$
$$2\tan x - 1 = 0 \quad \text{or} \quad 2\tan x - 3 = 0$$
$$\tan x = \frac{1}{2} \quad\quad\quad\quad \tan x = \frac{3}{2}$$
$$x \approx 0.4636, \; 3.6052 \quad x \approx 0.9828, \; 4.1244$$

95.
$$7\sin^2 x - 1 = 0$$
$$\sin^2 x = \frac{1}{7}$$
$$\sin x = \pm\sqrt{\frac{1}{7}}$$
$$\sin x = \pm\frac{\sqrt{7}}{7}$$
$$\sin x \approx 0.3780 \quad \text{or} \quad \sin x \approx -0.3780$$
$$x \approx 0.3876, \; 2.7540 \quad x \approx 3.5292, \; 5.8956$$

97. $2\cos 2x + 1 = 0$

$$\cos 2x = -\frac{1}{2}$$

The period of the cosine function is 2π. On the interval $[0, 2\pi)$ the cosine function equals $-\frac{1}{2}$ at $\frac{2\pi}{3}$ and $\frac{4\pi}{3}$. This means that $2x = \frac{2\pi}{3}$ or $2x = \frac{4\pi}{3}$. Because the period is 2π, all the solutions of the equation are given by

$$2x = \frac{2\pi}{3} + 2n\pi \quad \text{or} \quad 2x = \frac{4\pi}{3} + 2n\pi$$

$$x = \frac{\pi}{3} + n\pi \qquad\qquad x = \frac{2\pi}{3} + n\pi$$

Use all values of n that result in x values on the interval $[0, 2\pi)$. Thus,

$$x = \frac{\pi}{3}, \frac{5\pi}{3} \quad \text{or} \quad x = \frac{2\pi}{3}, \frac{4\pi}{3}$$

99. $\sin 2x + \sin x = 0$

$2\sin x\cos x + \sin x = 0$

$\sin x(2\cos x + 1) = 0$

$\sin x = 0 \quad \text{or} \quad 2\cos x + 1 = 0$

$x = 0, \pi \qquad\qquad \cos x = -\frac{1}{2}$

$$x = \frac{2\pi}{3}, \frac{4\pi}{3}$$

Thus, $x = 0, \dfrac{2\pi}{3}, \pi, \text{ and } \dfrac{4\pi}{3}$.

101. $3\cos x - 6\sqrt{3} = \cos x - 5\sqrt{3}$

$$2\cos x = \sqrt{3}$$

$$\cos x = \frac{\sqrt{3}}{2}$$

$$x = \frac{\pi}{6}, \frac{11\pi}{6}$$

103. $\tan x = -4.7143$

Be sure calculator is in radian mode and find the inverse tangent of +4.7143. This gives the first quadrant reference angle. $\theta = \tan^{-1} 4.7143 \approx 1.3618$
The tangent is negative in quadrants II and IV thus,

$x \approx \pi - 1.3618 \quad \text{or} \quad x \approx 2\pi - 1.3618$

$x \approx 1.7798 \qquad\qquad x \approx 4.9214$

105. $2\sin^2 x = 3 - \sin x$

$2\sin^2 x + \sin x - 3 = 0$

$(\sin x - 1)(2\sin x + 3) = 0$

$\sin x - 1 = 0 \quad \text{or} \quad 2\sin x + 3 = 0$

$\sin x = 1 \qquad\qquad \sin x = -\frac{3}{2}$

$x = \dfrac{\pi}{2} \qquad\qquad \sin x = -\frac{3}{2}$

107. $\cos x \csc x = 2\cos x$

$\cos x \csc x - 2\cos x = 0$

$\cos x(\csc x - 2) = 0$

$\cos x = 0 \qquad \text{or} \quad \csc x - 2 = 0$

$\sin x = \dfrac{1}{2} \qquad\qquad \csc x = 2$

$x = \dfrac{\pi}{2}, \dfrac{3\pi}{2} \qquad\qquad \sin x = \dfrac{1}{2}$

$$x = \frac{\pi}{6}, \frac{5\pi}{6}$$

Thus, $x = \dfrac{\pi}{6}, \dfrac{\pi}{2}, \dfrac{5\pi}{6}, \text{ and } \dfrac{3\pi}{2}$.

109. $5\cot^2 x - 15 = 0$

$$\cot^2 x = 3$$

$$\tan^2 x = \frac{1}{3}$$

$$\tan x = \pm\sqrt{\frac{1}{3}}$$

$$\tan x = \pm\frac{\sqrt{3}}{3}$$

$\tan x = \dfrac{\sqrt{3}}{3} \qquad \text{or} \quad \tan x = -\dfrac{\sqrt{3}}{3}$

$x = \dfrac{\pi}{6}, \dfrac{7\pi}{6} \qquad\qquad x = \dfrac{5\pi}{6}, \dfrac{11\pi}{6}$

Thus, $x = \dfrac{\pi}{6}, \dfrac{5\pi}{6}, \dfrac{7\pi}{6}, \text{ and } \dfrac{11\pi}{6}$.

111. $\cos^2 x + 2\cos x - 2 = 0$

Use the quadratic formula to solve for $\cos x$.

$$\cos x = \frac{-b \pm \sqrt{b^2 - 4ac}}{2a}$$

$$\cos x = \frac{-(2) \pm \sqrt{(2)^2 - 4(1)(-2)}}{2(1)}$$

$$\cos x = \frac{-2 \pm \sqrt{12}}{2}$$

$$\cos x = \frac{-2 \pm 2\sqrt{3}}{2}$$

$$\cos x = -1 \pm \sqrt{3}$$

$\cos x \approx 0.7321$ or $\cos x \approx -2.7321$

~~$\cos x \approx -2.7321$~~

This equation
has no solution.

Be sure calculator is in radian mode and find the inverse cosine of 0.7321. This gives the first quadrant reference angle.

$\theta = \cos^{-1} 0.7321 \approx 0.7495$

The cosine is positive in quadrants I and IV thus,

$x \approx 0.7495$ or $x \approx 2\pi - 0.7495$

$$x \approx 5.5337$$

113.
$$5\sin x = 2\cos^2 x - 4$$
$$5\sin x = 2\left(1 - \sin^2 x\right) - 4$$
$$5\sin x = 2 - 2\sin^2 x - 4$$
$$2\sin^2 x + 5\sin x + 2 = 0$$
$$\left(2\sin x + 1\right)\left(\sin x + 2\right) = 0$$

$2\sin x + 1 = 0$ or $\sin x + 2 = 0$

$\sin x = -\dfrac{1}{2}$ $\qquad$ $\sin x = -2$

$\qquad\qquad\qquad$ ~~$\sin x = -2$~~

$x = \dfrac{7\pi}{6}, \dfrac{11\pi}{6}$

115. $2\tan^2 x + 5\tan x + 3 = 0$

$\left(\tan x + 1\right)\left(2\tan x + 3\right) = 0$

$\tan x + 1 = 0$ or $2\tan x + 3 = 0$

$\tan x = -1$ $\qquad\qquad$ $\tan x = -\dfrac{3}{2}$

$x = \dfrac{3\pi}{4}, \dfrac{7\pi}{4}$ $\qquad$ $x \approx 2.1588, \ 5.3004$

117.

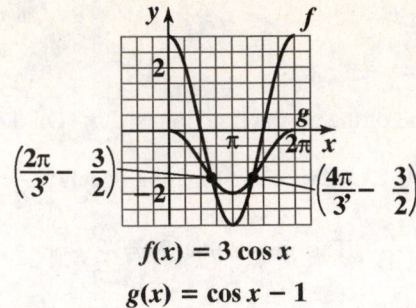

$f(x) = 3\cos x$

$g(x) = \cos x - 1$

$3\cos x = \cos x - 1$

$2\cos x = -1$

$\cos x = -\dfrac{1}{2}$

$x = \dfrac{2\pi}{3}, \dfrac{4\pi}{3}$

$\left(\dfrac{2\pi}{3}, -\dfrac{3}{2}\right), \left(\dfrac{4\pi}{3}, -\dfrac{3}{2}\right)$

119.

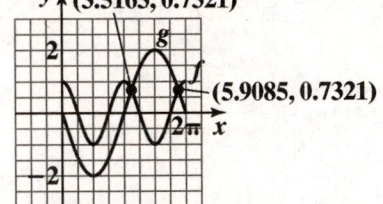

$f(x) = \cos 2x$

$g(x) = -2\sin x$

$\cos 2x = -2\sin x$

$1 - 2\sin^2 x = -2\sin x$

$-2\sin^2 x + 2\sin x + 1 = 0$

$2\sin^2 x - 2\sin x - 1 = 0$

$$\sin x = \frac{-b \pm \sqrt{b^2 - 4ac}}{2a}$$

$$\sin x = \frac{-(-2) \pm \sqrt{(-2)^2 - 4(2)(-1)}}{2(2)}$$

$$\sin x = \frac{2 \pm \sqrt{12}}{4}$$

$$\sin x = \frac{2 \pm 2\sqrt{3}}{4}$$

$$\sin x = \frac{1 \pm \sqrt{3}}{2}$$

$$\sin x \approx -0.3660 \qquad \text{or} \qquad \sin x \approx 1.3660$$

$$\sin x \approx 1.3660$$

$$x \approx 3.5163, \ 5.9085 \qquad \text{This equation}$$
has no solution.

$$(3.5163, 0.7321), (5.9085, 0.7321)$$

121. $\left|\cos x\right| = \dfrac{\sqrt{3}}{2}$

$$\cos x = \frac{\sqrt{3}}{2} \qquad \text{or} \qquad \cos x = -\frac{\sqrt{3}}{2}$$

$$x = \frac{\pi}{6}, \ \frac{11\pi}{6} \qquad\qquad x = \frac{5\pi}{6}, \ \frac{7\pi}{6}$$

123. $\qquad 10\cos^2 x + 3\sin x - 9 = 0$

$$10\left(1 - \sin^2 x\right) + 3\sin x - 9 = 0$$

$$10 - 10\sin^2 x + 3\sin x - 9 = 0$$

$$-10\sin^2 x + 3\sin x + 1 = 0$$

$$10\sin^2 x - 3\sin x - 1 = 0$$

$$\left(2\sin x - 1\right)\left(5\sin x + 1\right) = 0$$

$$2\sin x - 1 = 0 \qquad \text{or} \qquad 5\sin x + 1 = 0$$

$$\sin x = \frac{1}{2} \qquad\qquad\qquad \sin x = -\frac{1}{5}$$

$$x = \frac{\pi}{6}, \ \frac{5\pi}{6} \qquad\qquad x = 3.3430, \ 6.0818$$

125. $2\cos^3 x + \cos^2 x - 2\cos x - 1 = 0$

$$\cos^2 x\left(2\cos x + 1\right) - 1\left(2\cos x + 1\right) = 0$$

$$\left(2\cos x + 1\right)\left(\cos^2 x - 1\right) = 0$$

$$\left(2\cos x + 1\right)\left(\cos x + 1\right)\left(\cos x - 1\right) = 0$$

$$\cos x = -\frac{1}{2} \quad \text{or} \quad \cos x = -1 \quad \text{or} \quad \cos x = 1$$

$$x = 0, \ \frac{2\pi}{3}, \ \pi, \ \frac{4\pi}{3}$$

127. 0.3649, 1.2059, 3.5064, 4.3475
This matches graph a.

129. Substitute $y = 0.3$ into the equation and solve
for x:

$$0.3 = 0.6\sin\frac{2\pi}{5}x$$

$$\frac{0.3}{0.6} = \frac{0.6\sin\frac{2\pi}{5}x}{0.6}$$

$$\frac{1}{2} = \sin\frac{2\pi}{5}x$$

$$\sin\frac{2\pi}{5}x = \frac{1}{2}$$

The period of the sine function is 2π. In the interval
$[0, 2\pi)$, there are two values at which the sine

function is $\dfrac{1}{2}$. One is $\dfrac{\pi}{6}$. The sine is positive in

quadrant II; thus, the

other value is $\pi - \dfrac{\pi}{6} = \dfrac{5\pi}{6}$. All of the solutions to

$\sin\dfrac{2\pi}{5}x = \dfrac{1}{2}$ are given by

$$\frac{2\pi}{5}x = \frac{\pi}{6} + 2n\pi$$

$$x = \frac{5}{12} + 5n$$

or

$$\frac{2\pi}{5}x = \frac{5\pi}{6} + 2n\pi$$

$$x = \frac{25}{12} + 5n$$

where n is any integer. In the interval $[0, 5]$ we
obtain solutions when $n = 0$. The solutions are

$\dfrac{5}{12}$ and $\dfrac{25}{12}$.

Therefore, we are inhaling at 0.3 liter per second at

$x = \dfrac{5}{12} \approx 0.4$ second and at $x = \dfrac{25}{12} \approx 2.1$ seconds.

131. Substitute $y = 10.5$ into the equation and solve for x:

$$10.5 = 3\sin\left[\frac{2\pi}{365}(x-79)\right] + 12$$

$$-1.5 = 3\sin\left[\frac{2\pi}{365}(x-79)\right]$$

$$\frac{-1.5}{3} = \frac{3\sin\left[\frac{2\pi}{365}(x-79)\right]}{3}$$

$$-\frac{1}{2} = \sin\left[\frac{2\pi}{365}(x-79)\right]$$

$$\sin\left[\frac{2\pi}{365}(x-79)\right] = -\frac{1}{2}$$

The period of the sine function is 2π. In the interval $[0, 2\pi)$, there are two values at which the sine function is $-\frac{1}{2}$. One is $\pi + \frac{\pi}{6} = \frac{7\pi}{6}$. The other is $2\pi - \frac{\pi}{6} = \frac{11\pi}{6}$. All the solutions to $\sin\left[\frac{2\pi}{365}(x-79)\right] = -\frac{1}{2}$ are given by

$$\frac{2\pi}{365}(x-79) = \frac{7\pi}{6} + 2n\pi$$

$$x - 79 = \frac{2555}{12} + 365n$$

$$x = \frac{3503}{12} + 365n$$

or

$$\frac{2\pi}{365}(x-79) = \frac{11\pi}{6} + 2n\pi$$

$$x - 79 = \frac{4015}{12} + 365n$$

$$x = \frac{4963}{12} + 365n$$

where n is any integer.
Substitute various integers for n in the two equations. In the interval $[0, 365]$ we obtain values of 49 and 292 days. Thus, Boston has 10.5 hours of daylight 49 and 292 days after January 1.

133. Substitute $d = 2$ into the equation and solve for t:

$$2 = -4\cos\frac{\pi}{3}t$$

$$\frac{2}{-4} = \frac{-4\cos\frac{\pi}{3}t}{-4}$$

$$-\frac{1}{2} = \cos\frac{\pi}{3}t$$

$$\cos\frac{\pi}{3}t = -\frac{1}{2}$$

The period of the cosine function is 2π. In the interval $[0, 2\pi)$, there are two values at which the cosine function is $-\frac{1}{2}$. One is $\frac{2\pi}{3}$. The cosine function is negative in quadrant III; thus, the other value is $2\pi - \frac{2\pi}{3} = \frac{4\pi}{3}$. All solutions to $\cos\frac{\pi}{3}t = -\frac{1}{2}$ are given by

$$\frac{\pi}{3}t = \frac{2\pi}{3} + 2n\pi$$

$$t = 2 + 6n$$

or

$$\frac{\pi}{3}t = \frac{4\pi}{3} + 2n\pi$$

$$t = 4 + 6n$$

where n is any nonnegative integer.

490

135. Substitute $v_0 = 90$ and $d = 170$, and solve for θ:

$$170 = \frac{90^2}{16}\sin\theta\cos\theta$$

$$\frac{136}{405} = \sin\theta\cos\theta$$

$$2 \cdot \frac{136}{405} = 2\sin\theta\cos\theta$$

$$\frac{272}{405} = \sin 2\theta$$

$$\sin 2\theta = \frac{272}{405}$$

The period of the sine function is $360°$. In the interval $[0, 360°]$, there are two values at which the sine function is $\frac{272}{405}$.

One is $\sin^{-1}\left(\frac{272}{405}\right) \approx 42.19°$. The sine function is positive in quadrant II; Thus, the other value is $180° - 42.19° = 137.81°$. All solutions to $\sin 2\theta = \frac{272}{405}$ are given by

$2\theta = 42.19° + 360°n$

$\theta = 21.095° + 180°n$

or

$2\theta = 137.81° + 360°n$

$\theta = 68.905° + 180°n$

where n is any integer.
In the interval $[0, 90°)$ we obtain the solutions by letting $n = 0$. The solutions are approximately $21°$ and $69°$. Therefore, the angle of elevation should be $21°$ or $69°$.

137. – 145. Answers may vary.

147.

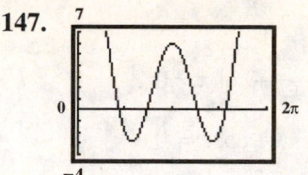

$x = 1.37,$ $x = 2.30$

$x = 3.98,$ or $x = 4.91$

149.

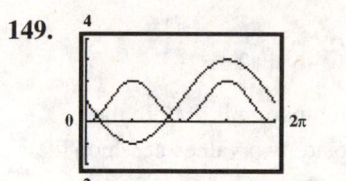

$x = 0.37, 2.77$

151.

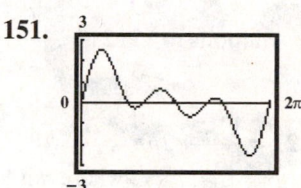

$x = 0,$ $x = 1.57,$ $x = 2.09,$

$x = 3.14,$ $x = 4.19,$ or $x = 4.71$

153. makes sense

155. does not make sense; Explanations will vary. Sample explanation: You do not need to solve a trigonometric equation. You nee to find a trigonometric value of an angle ang simplify using arithmetic.

157. false; Changes to make the statement true will vary. A sample change is: The equation has an infinite number of solutions

159. false; Changes to make the statement true will vary. A sample change is: Over this interval, the first equation has two solutions and the second equation has 4 solutions.

161. $\sin 3x + \sin x + \cos x = 0$

$$2\sin\left(\frac{3x+x}{2}\right)\cos\left(\frac{3x-x}{2}\right) + \cos x = 0$$

$$2\sin 2x \cos x + \cos x = 0$$

$$\cos x(2\sin 2x + 1) = 0$$

$$\cos x = 0 \quad\text{or}\quad 2\sin 2x + 1 = 0$$

$$x = \frac{\pi}{2} \quad x = \frac{3\pi}{2} \qquad 2\sin 2x = -1$$

$$\sin 2x = -\frac{1}{2}$$

The period of the sine function is 2π. In the interval $[0, 2\pi)$, there are two values at which the sine function is $-\frac{1}{2}$. One is $\pi + \frac{\pi}{6} = \frac{7\pi}{6}$. The other is $2\pi - \frac{\pi}{6} = \frac{11\pi}{6}$. All the solutions to $\sin 2x = -\frac{1}{2}$ are given

$$2x = \frac{7\pi}{6} + 2n\pi \quad\text{or}\quad 2x = \frac{11\pi}{6} + 2n\pi$$

by $\qquad\qquad\qquad\qquad\qquad$ where n is

$$x = \frac{7\pi}{12} + n\pi \qquad x = \frac{11\pi}{12} + n\pi$$

any integer. The solutions in the interval $[0, 2\pi)$ are obtained by letting $n = 0$ and $n = 1$. The solutions are $\frac{\pi}{2}, \frac{3\pi}{2}, \frac{7\pi}{12}, \frac{11\pi}{12}, \frac{19\pi}{12}$, and $\frac{23\pi}{12}$.

163. $\dfrac{a}{\sin 46°} = \dfrac{56}{\sin 63°}$

$$a \sin 63° = 56\sin 46°$$

$$a = \frac{56\sin 46°}{\sin 63°}$$

$$a \approx 45.2°$$

164. $\dfrac{81}{\sin 43°} = \dfrac{62}{\sin B}$

$$81\sin B = 62\sin 43°$$

$$\sin B = \frac{62\sin 43°}{81}$$

$$\sin B \approx 0.522023436$$

$$B \approx \sin^{-1}(0.522023436)$$

$$B \approx 31.5°$$

165. $\dfrac{51}{\sin 75°} = \dfrac{71}{\sin B}$

$$51\sin B = 71\sin 75°$$

$$\sin B = \frac{71\sin 75°}{51}$$

$$\sin B \approx 1.344720268$$

No solution.

Chapter 5 Review Exercises

1. $\sec x - \cos x = \dfrac{1}{\cos x} - \cos x$

$$= \frac{1}{\cos x} - \frac{\cos x}{1} \cdot \frac{\cos x}{\cos x}$$

$$= \frac{1}{\cos x} - \frac{\cos^2 x}{\cos x}$$

$$= \frac{1 - \cos^2 x}{\cos x}$$

$$= \frac{\sin^2 x}{\cos x}$$

$$= \frac{\sin x}{\cos x} \cdot \sin x$$

$$= \tan x \sin x$$

2. $\cos x + \sin x \tan x$

$$= \frac{\cos x}{\cos x} \cdot \cos x + \sin x \cdot \frac{\sin x}{\cos x}$$

$$= \frac{\cos^2 x}{\cos x} + \frac{\sin^2 x}{\cos x}$$

$$= \frac{\cos^2 x + \sin^2 x}{\cos x}$$

$$= \frac{1}{\cos x} = \sec x$$

3. $\sin^2 \theta(1 + \cot^2 \theta) = \sin^2 \theta + \sin^2 \theta \cot^2 \theta$

$$= \sin^2 \theta + \sin^2 \theta \cdot \frac{\cos^2 \theta}{\sin^2 \theta}$$

$$= \sin^2 \theta + \cos^2 \theta$$

$$= 1$$

4. $(\sec\theta - 1)(\sec\theta + 1) = \sec^2 \theta - 1$

$$= 1 + \tan^2 \theta - 1$$

$$= \tan^2 \theta$$

5. $\dfrac{1-\tan x}{\sin x} = \dfrac{1}{\sin x} - \dfrac{\tan x}{\sin x}$

$\qquad = \csc x - \dfrac{\sin x}{\cos x} \cdot \dfrac{1}{\sin x}$

$\qquad = \csc x - \dfrac{1}{\cos x}$

$\qquad = \csc x - \sec x$

6. $\dfrac{1}{\sin t - 1} + \dfrac{1}{\sin t + 1}$

$= \dfrac{1}{\sin t - 1} \cdot \dfrac{\sin t + 1}{\sin t + 1} + \dfrac{1}{\sin t + 1} \cdot \dfrac{\sin t - 1}{\sin t - 1}$

$= \dfrac{\sin t + 1}{\sin^2 t - 1} + \dfrac{\sin t - 1}{\sin^2 t - 1}$

$= \dfrac{\sin t + 1 + \sin t - 1}{\sin^2 t - 1}$

$= \dfrac{2\sin t}{\sin^2 t - 1}$

$= \dfrac{2\sin t}{-\cos^2 t}$

$= -2 \cdot \dfrac{\sin t}{\cos t} \cdot \dfrac{1}{\cos t}$

$= -2\tan t \sec t$

7. $\dfrac{1+\sin t}{\cos^2 t} = \dfrac{1}{\cos^2 t} + \dfrac{\sin t}{\cos^2 t}$

$\qquad = \sec^2 t + \dfrac{\sin t}{\cos t} \cdot \dfrac{1}{\cos t}$

$\qquad = \tan^2 t + 1 + \tan t \sec t$

8. $\dfrac{\cos x}{1-\sin x} = \dfrac{\cos x}{1-\sin x} \cdot \dfrac{1+\sin x}{1+\sin x}$

$\qquad = \dfrac{\cos x(1+\sin x)}{1-\sin^2 x}$

$\qquad = \dfrac{\cos x(1+\sin x)}{\cos^2 x}$

$\qquad = \dfrac{1+\sin x}{\cos x}$

9. $1 - \dfrac{\sin^2 x}{1+\cos x} = 1 - \dfrac{1-\cos^2 x}{1+\cos x}$

$\qquad = 1 - \dfrac{(1+\cos x)(1-\cos x)}{1+\cos x}$

$\qquad = 1 - (1-\cos x)$

$\qquad = 1 - 1 + \cos x$

$\qquad = \cos x$

10. $(\tan\theta + \cot\theta)^2$

$= \tan^2\theta + 2\tan\theta\cot\theta + \cot^2\theta$

$= \sec^2\theta - 1 + 2\dfrac{\sin\theta}{\cos\theta} \cdot \dfrac{\cos\theta}{\sin\theta} + \csc^2\theta - 1$

$= \sec^2\theta - 1 + 2 + \csc^2\theta - 1$

$= \sec^2\theta + \csc^2\theta$

11. $\dfrac{1}{\sin\theta + \cos\theta} + \dfrac{1}{\sin\theta - \cos\theta}$

$= \dfrac{\sin\theta - \cos\theta}{\sin\theta - \cos\theta} \cdot \dfrac{1}{\sin\theta + \cos\theta}$

$\quad + \dfrac{\sin\theta + \cos\theta}{\sin\theta + \cos\theta} \cdot \dfrac{1}{\sin\theta - \cos\theta}$

$= \dfrac{\sin\theta - \cos\theta}{\sin^2\theta - \cos^2\theta} + \dfrac{\sin\theta + \cos\theta}{\sin^2\theta - \cos^2\theta}$

$= \dfrac{2\sin\theta}{\sin^2\theta - \cos^2\theta}$

$= \dfrac{2\sin\theta}{\sin^2\theta - \cos^2\theta} \cdot \dfrac{\sin^2\theta + \cos^2\theta}{\sin^2\theta + \cos^2\theta}$

$= \dfrac{2\sin\theta \cdot 1}{\sin^4\theta - \cos^4\theta}$

$= \dfrac{2\sin\theta}{\sin^4\theta - \cos^4\theta}$

12. $\dfrac{\cos t}{\cot t - 5\cos t} = \dfrac{\cos t}{\cot t - 5\cos t} \cdot \dfrac{\frac{1}{\cos t}}{\frac{1}{\cos t}}$

$= \dfrac{\frac{\cos t}{\cos t}}{\frac{\cot t - 5\cos t}{\cos t}}$

$= \dfrac{1}{\frac{\cot t}{\cos t} - 5}$

$= \dfrac{1}{\frac{\frac{\cos t}{\sin t}}{\cos t} - 5}$

$= \dfrac{1}{\frac{\cos t}{\sin t} \cdot \frac{1}{\cos t} - 5}$

$= \dfrac{1}{\frac{1}{\sin t} - 5}$

$= \dfrac{1}{\csc t - 5}$

13. $\dfrac{1-\cos t}{1+\cos t} = \dfrac{1-\cos t}{1+\cos t} \cdot \dfrac{1-\cos t}{1-\cos t}$

$\qquad = \dfrac{(1-\cos t)^2}{1-\cos^2 t}$

$\qquad = \dfrac{(1-\cos t)^2}{\sin^2 t}$

$\qquad = \left(\dfrac{1-\cos t}{\sin t}\right)^2$

$\qquad = \left(\dfrac{1}{\sin t} - \dfrac{\cos t}{\sin t}\right)^2$

$\qquad = (\csc t - \cot t)^2$

14. $\cos(45° + 30°)$

$\qquad = \cos 45° \cos 30° - \sin 45° \sin 30°$

$\qquad = \dfrac{\sqrt{2}}{2} \cdot \dfrac{\sqrt{3}}{2} - \dfrac{\sqrt{2}}{2} \cdot \dfrac{1}{2}$

$\qquad = \dfrac{\sqrt{6}}{4} - \dfrac{\sqrt{2}}{4}$

$\qquad = \dfrac{\sqrt{6} - \sqrt{2}}{4}$

15. $\sin 195° = \sin(135° + 60°)$

$\qquad = \sin 135° \cos 60° + \cos 135° \sin 60°$

$\qquad = \dfrac{\sqrt{2}}{2} \cdot \dfrac{1}{2} + \left(-\dfrac{\sqrt{2}}{2}\right) \cdot \dfrac{\sqrt{3}}{2}$

$\qquad = \dfrac{\sqrt{2}}{4} - \dfrac{\sqrt{6}}{4}$

$\qquad = \dfrac{\sqrt{2} - \sqrt{6}}{4}$

16. $\tan\left(\dfrac{4\pi}{3} - \dfrac{\pi}{4}\right) = \dfrac{\tan\frac{4\pi}{3} - \tan\frac{\pi}{4}}{1 + \tan\frac{4\pi}{3} \cdot \tan\frac{\pi}{4}} = \dfrac{\sqrt{3} - 1}{1 + \sqrt{3} \cdot (1)}$

$\qquad = \dfrac{(\sqrt{3}-1)(1-\sqrt{3})}{(1+\sqrt{3})(1-\sqrt{3})} = \dfrac{-(1-\sqrt{3})^2}{1-3}$

$\qquad = \dfrac{-(1-2\sqrt{3}+3)}{-2} = \dfrac{1-2\sqrt{3}+3}{2}$

$\qquad = \dfrac{4-2\sqrt{3}}{2} = \dfrac{2(2-\sqrt{3})}{2} = 2 - \sqrt{3}$

17. $\tan\dfrac{5\pi}{12} = \tan\left(\dfrac{2\pi}{12} + \dfrac{3\pi}{12}\right)$

$\qquad = \tan\left(\dfrac{\pi}{6} + \dfrac{\pi}{4}\right)$

$\qquad = \dfrac{\tan\frac{\pi}{6} + \tan\frac{\pi}{4}}{1 - \tan\frac{\pi}{6}\tan\frac{\pi}{4}}$

$\qquad = \dfrac{\frac{\sqrt{3}}{3} + 1}{1 - \frac{\sqrt{3}}{3}\cdot 1} = \dfrac{\frac{\sqrt{3}}{3}+1}{1-\frac{\sqrt{3}}{3}} \cdot \dfrac{\left(1+\frac{\sqrt{3}}{3}\right)}{\left(1+\frac{\sqrt{3}}{3}\right)}$

$\qquad = \dfrac{\frac{2\sqrt{3}}{3}+1+\frac{1}{3}}{1-\frac{1}{3}} = \dfrac{\frac{2\sqrt{3}}{3}+\frac{4}{3}}{\frac{2}{3}}$

$\qquad = \left(\dfrac{2\sqrt{3}}{3} + \dfrac{4}{3}\right)\cdot\dfrac{3}{2} = \sqrt{3} + 2$

18. $\cos 65° \cos 5° + \sin 65° \sin 5°$

$\qquad = \cos(65° - 5°)$

$\qquad = \cos 60°$

$\qquad = \dfrac{1}{2}$

19. $\sin 80° \cos 50° - \cos 80° \sin 50°$

$\qquad = \sin(80° - 50°)$

$\qquad = \sin 30°$

$\qquad = \dfrac{1}{2}$

20. $\sin\left(x + \dfrac{\pi}{6}\right) - \cos\left(x + \dfrac{\pi}{3}\right)$

$\qquad = \sin x \cos\dfrac{\pi}{6} + \cos x \sin\dfrac{\pi}{6}$

$\qquad\quad - \left(\cos x \cos\dfrac{\pi}{3} - \sin x \sin\dfrac{\pi}{3}\right)$

$\qquad = \sin x \cdot \dfrac{\sqrt{3}}{2} + \cos x \cdot \dfrac{1}{2}$

$\qquad\quad - \left(\cos x \cdot \dfrac{1}{2} - \sin x \cdot \dfrac{\sqrt{3}}{2}\right)$

$\qquad = 2 \cdot \dfrac{\sqrt{3}}{2} \cdot \sin x$

$\qquad = \sqrt{3} \sin x$

21. $\tan\left(x+\dfrac{3\pi}{4}\right)=\dfrac{\tan x+\tan\frac{3\pi}{4}}{1-\tan x\tan\frac{3\pi}{4}}$

$\qquad\qquad\quad =\dfrac{\tan x+(-1)}{1-\tan x(-1)}$

$\qquad\qquad\quad =\dfrac{\tan x-1}{1+\tan x}$

22. $\sec(\alpha+\beta)=\dfrac{1}{\cos(\alpha+\beta)}$

$\qquad\qquad\quad =\dfrac{1}{\cos\alpha\cos\beta-\sin\alpha\sin\beta}$

$\qquad\qquad\quad =\dfrac{1}{\cos\alpha\cos\beta-\sin\alpha\sin\beta}\cdot\dfrac{\frac{1}{\cos\alpha\cos\beta}}{\frac{1}{\cos\alpha\cos\beta}}$

$\qquad\qquad\quad =\dfrac{\frac{1}{\cos\alpha\cos\beta}}{\frac{\cos\alpha\cos\beta-\sin\alpha\sin\beta}{\cos\alpha\cos\beta}}$

$\qquad\qquad\quad =\dfrac{\frac{1}{\cos\alpha}\cdot\frac{1}{\cos\beta}}{\frac{\cos\alpha\cos\beta}{\cos\alpha\cos\beta}-\frac{\sin\alpha\sin\beta}{\cos\alpha\cos\beta}}$

$\qquad\qquad\quad =\dfrac{\sec\alpha\sec\beta}{1-\frac{\sin\alpha\sin\beta}{\cos\alpha\cos\beta}}$

$\qquad\qquad\quad =\dfrac{\sec\alpha\sec\beta}{1-\tan\alpha\tan\beta}$

23. $\dfrac{\cos(\alpha-\beta)}{\cos\alpha\cos\beta}=\dfrac{\cos\alpha\cos\beta+\sin\alpha\sin\beta}{\cos\alpha\cos\beta}$

$\qquad\qquad\quad =\dfrac{\cos\alpha\cos\beta}{\cos\alpha\cos\beta}+\dfrac{\sin\alpha\sin\beta}{\cos\alpha\cos\beta}$

$\qquad\qquad\quad =1+\tan\alpha\tan\beta$

24. $\cos^4 t-\sin^4 t=\left(\cos^2 t-\sin^2 t\right)\left(\cos^2 t+\sin^2 t\right)$

$\qquad\qquad\quad =(\cos 2t)\cdot(1)$

$\qquad\qquad\quad =\cos 2t$

25. $\sin t-\cos 2t=\sin t-\left(1-2\sin^2 t\right)$

$\qquad\qquad\quad =\sin t-1+2\sin^2 t$

$\qquad\qquad\quad =2\sin^2 t+\sin t-1$

$\qquad\qquad\quad =(2\sin t-1)(\sin t+1)$

26. $\dfrac{\sin 2\theta-\sin\theta}{\cos 2\theta+\cos\theta}=\dfrac{2\sin\theta\cos\theta-\sin\theta}{2\cos^2\theta-1+\cos\theta}$

$\qquad\qquad\quad =\dfrac{\sin\theta(2\cos\theta-1)}{2\cos^2\theta+\cos\theta-1}$

$\qquad\qquad\quad =\dfrac{\sin\theta(2\cos\theta-1)}{(2\cos\theta-1)(\cos\theta+1)}$

$\qquad\qquad\quad =\dfrac{\sin\theta}{\cos\theta+1}$

$\qquad\qquad\quad =\dfrac{\sin\theta}{\cos\theta+1}\cdot\dfrac{\cos\theta-1}{\cos\theta-1}$

$\qquad\qquad\quad =\dfrac{\sin\theta(\cos\theta-1)}{\cos^2\theta-1}$

$\qquad\qquad\quad =\dfrac{\sin\theta(\cos\theta-1)}{-\sin^2\theta}$

$\qquad\qquad\quad =\dfrac{-(\cos\theta-1)}{\sin\theta}$

$\qquad\qquad\quad =\dfrac{1-\cos\theta}{\sin\theta}$

27. $\dfrac{\sin 2\theta}{1-\sin^2\theta}=\dfrac{2\sin\theta\cos\theta}{\cos^2\theta}$

$\qquad\qquad\quad =\dfrac{2\sin\theta}{\cos\theta}\cdot\dfrac{\cos\theta}{\cos\theta}$

$\qquad\qquad\quad =2\tan\theta$

28. $2\sin t\cos t\sec 2t=\sin 2t\cdot\sec 2t$

$\qquad\qquad\quad =\sin 2t\cdot\dfrac{1}{\cos 2t}$

$\qquad\qquad\quad =\dfrac{\sin 2t}{\cos 2t}$

$\qquad\qquad\quad =\tan 2t$

29. $\cos 4t=\cos(2\cdot 2t)$

$\qquad\qquad\quad =1-2\sin^2 2t$

$\qquad\qquad\quad =1-2(\sin 2t)^2$

$\qquad\qquad\quad =1-2\cdot(2\sin t\cos t)^2$

$\qquad\qquad\quad =1-2\cdot 4\sin^2 t\cos^2 t$

$\qquad\qquad\quad =1-8\sin^2 t\cos^2 t$

30. $\tan\dfrac{x}{2}(1+\cos x)=\dfrac{\sin x}{1+\cos x}\cdot(1+\cos x)=\sin x$

31. $\tan\dfrac{x}{2} = \dfrac{1-\cos x}{\sin x}$

$= \dfrac{1-\cos x}{\sin x} \cdot \dfrac{\frac{1}{\cos x}}{\frac{1}{\cos x}}$

$= \dfrac{\frac{1-\cos x}{\cos x}}{\frac{\sin x}{\cos x}}$

$= \dfrac{\frac{1}{\cos x} - \frac{\cos x}{\cos x}}{\tan x}$

$= \dfrac{\sec x - 1}{\tan x}$

32. **a.** The graph appears to be the cosine curve, $y = \cos x$. It cycles through maximum, intercept, minimum, intercept and back to maximum. Thus, $y = \cos x$ also describes the graph.

b. $\sin\left(x - \dfrac{3\pi}{2}\right) = \sin x \cos\dfrac{3\pi}{2} - \cos x \sin\dfrac{3\pi}{2}$

$= \sin x \cdot 0 - \cos x \cdot (-1)$

$= \cos x$

33. **a.** The graph appears to be the negative of the sine curve, $y = -\sin x$. It cycles through intercept, minimum, intercept, maximum and back to intercept. Thus, $y = -\sin x$ also describes the graph.

b. $\cos\left(x + \dfrac{\pi}{2}\right) = \cos x \cos\dfrac{\pi}{2} - \sin x \sin\dfrac{\pi}{2}$

$= \cos x \cdot 0 - \sin x \cdot 1$

$= -\sin x$

34. **a.** The graph appears to be the tangent curve, $y = \tan x$. It cycles through intercept to positive infinity, then from negative infinity through the intercept. Thus, $y = \tan x$ also describes the graph.

b. $y = \dfrac{\tan x - 1}{1 - \cot x}$

$= \dfrac{\frac{\cos x - 1}{\cos x}}{1 - \frac{\cos x}{\sin x}}$

$= \dfrac{\frac{\sin x - \cos x}{\cos x}}{\frac{\sin x - \cos x}{\sin x}}$

$= \dfrac{\sin x - \cos x}{\cos x} \cdot \dfrac{\sin x}{\sin x - \cos x}$

$= \dfrac{\sin x}{\cos x}$

$= \tan x$

35. $\sin\alpha = \dfrac{3}{5} = \dfrac{y}{r}$

Because α lies in quadrant I, x is positive.

$x^2 + 3^2 = 5^2$

$x^2 = 5^2 - 3^2 = 16$

$x = \sqrt{16} = 4$

Thus, $\cos\alpha = \dfrac{x}{r} = \dfrac{4}{5}$, and $\tan\alpha = \dfrac{y}{x} = \dfrac{3}{4}$.

$\sin\beta = \dfrac{12}{13} = \dfrac{y}{r}$

Because β lies in quadrant II, x is negative.

$x^2 + 12^2 = 13^2$

$x^2 = 13^2 - 12^2 = 25$

$x = -\sqrt{25} = -5$

Thus, $\cos\beta = \dfrac{x}{r} = \dfrac{-5}{13} = -\dfrac{5}{13}$, and

$\tan\beta = \dfrac{y}{x} = \dfrac{12}{-5} = -\dfrac{12}{5}$.

a. $\sin(\alpha + \beta) = \sin\alpha\cos\beta + \cos\alpha\sin\beta$

$= \dfrac{3}{5}\cdot\left(-\dfrac{5}{13}\right) + \dfrac{4}{5}\cdot\dfrac{12}{13} = \dfrac{33}{65}$

b. $\cos(\alpha - \beta) = \cos\alpha\cos\beta + \sin\alpha\sin\beta$

$= \dfrac{4}{5}\cdot\left(-\dfrac{5}{13}\right) + \dfrac{3}{5}\cdot\dfrac{12}{13} = \dfrac{16}{65}$

c. $\tan(\alpha + \beta) = \dfrac{\tan \alpha + \tan \beta}{1 - \tan \alpha \tan \beta}$

$= \dfrac{\frac{3}{4} + \left(-\frac{12}{5}\right)}{1 - \frac{3}{4}\left(-\frac{12}{5}\right)}$

$= \dfrac{-\frac{33}{20}}{1 + \frac{36}{20}} = \dfrac{-\frac{33}{20}}{\frac{56}{20}}$

$= -\dfrac{33}{56}$

d. $\sin 2\alpha = 2 \sin \alpha \cos \alpha = 2 \cdot \dfrac{3}{5} \cdot \dfrac{4}{5} = \dfrac{24}{25}$

e. $\cos \dfrac{\beta}{2} = \sqrt{\dfrac{1 + \cos \beta}{2}} = \sqrt{\dfrac{1 - \frac{5}{13}}{2}}$

$= \sqrt{\dfrac{8}{26}} = \sqrt{\dfrac{4}{13}} = \dfrac{2}{\sqrt{13}} = \dfrac{2\sqrt{13}}{13}$

36. $\tan \alpha = \dfrac{4}{3} = \dfrac{-4}{-3} = \dfrac{y}{x}$

Because r is a distance, it is positive.

$r^2 = (-4)^2 + (-3)^2 = 25$

$r = \sqrt{25} = 5$

Thus, $\sin \alpha = \dfrac{y}{r} = \dfrac{-4}{5} = -\dfrac{4}{5}$, and

$\cos \alpha = \dfrac{x}{r} = \dfrac{-3}{5} = -\dfrac{3}{5}$.

$\tan \beta = \dfrac{5}{12} = \dfrac{y}{x}$

Because r is a distance, it is positive.

$r^2 = 5^2 + 12^2 = 169$

$r = \sqrt{169} = 13$

Thus, $\sin \beta = \dfrac{y}{r} = \dfrac{5}{13}$, and $\cos \beta = \dfrac{x}{r} = \dfrac{12}{13}$.

a. $\sin(\alpha + \beta) = \sin \alpha \cos \beta + \cos \alpha \sin \beta$

$= -\dfrac{4}{5} \cdot \dfrac{12}{13} + \left(-\dfrac{3}{5}\right) \cdot \dfrac{5}{13}$

$= -\dfrac{63}{65}$

b. $\cos(\alpha - \beta) = \cos \alpha \cos \beta + \sin \alpha \sin \beta$

$= -\dfrac{3}{5} \cdot \dfrac{12}{13} + \left(-\dfrac{4}{5}\right) \cdot \dfrac{5}{13}$

$= -\dfrac{56}{65}$

c. $\tan(\alpha + \beta) = \dfrac{\tan \alpha + \tan \beta}{1 - \tan \alpha \tan \beta}$

$= \dfrac{\frac{4}{3} + \frac{5}{12}}{1 - \frac{4}{3} \cdot \frac{5}{12}} = \dfrac{\frac{21}{12}}{1 - \frac{20}{36}} = \dfrac{\frac{21}{12}}{\frac{16}{36}}$

$= \dfrac{21}{12} \cdot \dfrac{36}{16} = \dfrac{63}{16}$

d. $\sin 2\alpha = 2 \sin \alpha \sin \alpha$

$= 2\left(-\dfrac{4}{5}\right)\left(-\dfrac{4}{5}\right) = \dfrac{24}{25}$

e. $\cos \dfrac{\beta}{2} = \sqrt{\dfrac{1 + \cos \beta}{2}} = \sqrt{\dfrac{1 + \frac{12}{13}}{2}}$

$= \sqrt{\dfrac{25}{26}} = \dfrac{5}{\sqrt{26}} = \dfrac{5\sqrt{26}}{26}$

37.. $\tan \alpha = -3 = \dfrac{3}{-1} = \dfrac{y}{x}$

Because r is a distance, it is positive.

$r^2 = 3^2 + (-1)^2$

$r^2 = 10$

$r = \sqrt{10}$

$\sin \alpha = \dfrac{3}{\sqrt{10}} = \dfrac{3\sqrt{10}}{10}$

$\cos \alpha = \dfrac{-1}{\sqrt{10}} = -\dfrac{\sqrt{10}}{10}$

$\cot \beta = -3 = \dfrac{3}{-1} = \dfrac{x}{y}$

Because r is a distance, it is positive.

$r^2 = 3^2 + (-1)^2$

$r^2 = 10$

$r = \sqrt{10}$

$\sin \beta = \dfrac{-1}{\sqrt{10}} = -\dfrac{\sqrt{10}}{10}$

$\cos \beta = \dfrac{3}{\sqrt{10}} = \dfrac{3\sqrt{10}}{10}$

a. $\sin(\alpha + \beta) = \sin\alpha\cos\beta + \cos\alpha\sin\beta$

$$= \frac{3\sqrt{10}}{10} \cdot \frac{3\sqrt{10}}{10} + \left(-\frac{\sqrt{10}}{10}\right)\left(-\frac{\sqrt{10}}{10}\right)$$

$$= \frac{90}{100} + \frac{10}{100}$$

$$= \frac{100}{100}$$

$$= 1$$

b. $\cos(\alpha - \beta)$

$$= \cos\alpha\cos\beta + \sin\alpha\sin\beta$$

$$= \left(-\frac{\sqrt{10}}{10}\right)\left(\frac{3\sqrt{10}}{10}\right) + \frac{3\sqrt{10}}{10}\left(\frac{-\sqrt{10}}{10}\right)$$

$$= -\frac{60}{100}$$

$$= -\frac{3}{5}$$

c. $\tan(\alpha + \beta) = \dfrac{\tan\alpha + \tan\beta}{1 - \tan\alpha\tan\beta}$

$$= \frac{-3 + \left(\dfrac{-1}{3}\right)}{1 - (-3)\left(-\dfrac{1}{3}\right)}$$

$$= \frac{\dfrac{-10}{3}}{0}$$

Since this value is undefined, the tangent function is undefined at $\alpha + \beta$.

d. $\sin 2\alpha = 2\sin\alpha\cos\alpha$

$$= 2\left(\frac{3\sqrt{10}}{10}\right)\left(\frac{-\sqrt{10}}{10}\right)$$

$$= -\frac{3}{5}$$

e. $\cos\dfrac{\beta}{2} = \sqrt{\dfrac{1 + \cos\beta}{2}}$

$$= \sqrt{\frac{1 + \dfrac{3\sqrt{10}}{10}}{2}}$$

$$= \sqrt{\frac{10 + 3\sqrt{10}}{20}}$$

$$= \frac{\sqrt{10 + 3\sqrt{10}}}{2\sqrt{5}}$$

38. $\sin\alpha = -\dfrac{1}{3} = \dfrac{-1}{3} = \dfrac{y}{r}$

Because α is in quadrant II, x is negative.

$$x^2 + (-1)^2 = 3^2$$

$$x^2 + 1 = 9$$

$$x^2 = 8$$

$$x = -\sqrt{8} = -2\sqrt{2}$$

$$\cos\alpha = \frac{-2\sqrt{2}}{3}$$

$$\tan\alpha = \frac{-1}{-2\sqrt{2}} = \frac{\sqrt{2}}{4}$$

$$\cos\beta = -\frac{1}{3} = \frac{-1}{3} = \frac{x}{r}$$

Because β is in quadrant III, y is negative.

$$(-1)^2 + y^2 = 3^2$$

$$y^2 = 8$$

$$y = -\sqrt{8} = -2\sqrt{2}$$

$$\sin\beta = \frac{-2\sqrt{2}}{3}$$

$$\tan\beta = \frac{-2\sqrt{2}}{-1} = 2\sqrt{2}$$

a. $\sin(\alpha + \beta)$

$$= \sin\alpha\cos\beta + \cos\alpha\sin\beta$$

$$= -\frac{1}{3} \cdot -\frac{1}{3} + \left(-\frac{2\sqrt{2}}{3}\right)\cdot\left(-\frac{2\sqrt{2}}{3}\right)$$

$$= \frac{9}{9} = 1$$

b. $\cos(\alpha - \beta)$

$$= \cos\alpha\cos\beta + \sin\alpha\sin\beta$$

$$= -\frac{2\sqrt{2}}{3}\cdot\left(-\frac{1}{3}\right) + \left(-\frac{1}{3}\right)\cdot\left(-\frac{2\sqrt{2}}{3}\right)$$

$$= \frac{4\sqrt{2}}{9}$$

c. $\tan(\alpha + \beta) = \dfrac{\tan\alpha + \tan\beta}{1 - \tan\alpha\tan B}$

$= \dfrac{\frac{\sqrt{2}}{4} + 2\sqrt{2}}{1 - \left(\frac{\sqrt{2}}{4}\right)\left(2\sqrt{2}\right)}$

$= \dfrac{\frac{9\sqrt{2}}{4}}{0}$

Since this value is undefined, the tangent function is undefined at $\alpha + \beta$.

d. $\sin 2\alpha = 2\sin\alpha\cos\alpha$

$= 2\left(-\dfrac{1}{3}\right)\left(-\dfrac{2\sqrt{2}}{3}\right) = \dfrac{4\sqrt{2}}{9}$

e. $\cos\dfrac{\beta}{2} = -\sqrt{\dfrac{1 + \cos\beta}{2}}$

$= -\sqrt{\dfrac{1 + \left(-\frac{1}{3}\right)}{2}}$

$= -\sqrt{\dfrac{\frac{2}{3}}{2}} = -\sqrt{\dfrac{1}{3}}$

$= -\dfrac{1}{\sqrt{3}}$

$= -\dfrac{\sqrt{3}}{3}$

39. The given expression is the right side of the formula for $\cos 2\theta$ with $\theta = 15°$.

$\cos^2 15° - \sin^2 15° = \cos(2 \cdot 15°)$

$= \cos 30°$

$= \dfrac{\sqrt{3}}{2}$

40. The given expression is the right side of the formula for $\tan 2\theta$ with $\theta = \dfrac{5\pi}{12}$.

$\dfrac{2\tan\frac{5\pi}{12}}{1 - \tan^2\frac{5\pi}{12}} = \tan\left(2 \cdot \dfrac{5\pi}{12}\right)$

$= \tan\dfrac{5\pi}{6}$

$= -\dfrac{\sqrt{3}}{3}$

41. Because $22.5°$ lies in quadrant I, $\sin 22.5° > 0$.

$\sin 22.5° = \sin\dfrac{45°}{2}$

$= \sqrt{\dfrac{1 - \cos 45°}{2}} = \sqrt{\dfrac{1 - \frac{\sqrt{2}}{2}}{2}}$

$= \sqrt{\dfrac{2 - \sqrt{2}}{4}} = \dfrac{\sqrt{2 - \sqrt{2}}}{2}$

42. Because $\dfrac{\pi}{12}$ lies in quadrant I, $\tan\dfrac{\pi}{12} > 0$.

$\tan\dfrac{\pi}{12} = \tan\dfrac{\frac{\pi}{6}}{2}$

$= \dfrac{1 - \cos\frac{\pi}{6}}{\sin\frac{\pi}{6}} = \dfrac{1 - \frac{\sqrt{3}}{2}}{\frac{1}{2}}$

$= 2 - \sqrt{3}$

43. $\sin 6x\sin 4x$

$= \dfrac{1}{2}\left[\cos(6x - 4x) - \cos(6x + 4x)\right]$

$= \dfrac{1}{2}\left[\cos 2x - \cos 10x\right]$

44. $\sin 7x\cos 3x$

$= \dfrac{1}{2}\left[\sin(7x + 3x) + \sin(7x - 3x)\right]$

$= \dfrac{1}{2}\left[\sin 10x + \sin 4x\right]$

45. $\sin 2x - \sin 4x$

$= 2\sin\left(\dfrac{2x - 4x}{2}\right)\cos\left(\dfrac{2x + 4x}{2}\right)$

$= 2\sin(-x)\cos 3x$

$= -2\sin x\cos 3x$

46. $\cos 75° + \cos 15°$

$= 2\cos\left(\dfrac{75° + 15°}{2}\right)\cos\left(\dfrac{75° - 15°}{2}\right)$

$= 2\cos 45°\cos 30°$

$= 2\left(\dfrac{\sqrt{2}}{2}\right)\left(\dfrac{\sqrt{3}}{2}\right) = \dfrac{\sqrt{6}}{2}$

47. $\dfrac{\cos 3x + \cos 5x}{\cos 3x - \cos 5x} = \dfrac{2\cos\left(\dfrac{3x+5x}{2}\right)\cos\left(\dfrac{3x-5x}{2}\right)}{-2\sin\left(\dfrac{3x+5x}{2}\right)\sin\left(\dfrac{3x-5x}{2}\right)}$

$= \dfrac{2\cos\left(\dfrac{8x}{2}\right)\cos\left(\dfrac{-2x}{2}\right)}{-2\sin\left(\dfrac{8x}{2}\right)\sin\left(\dfrac{-2x}{2}\right)}$

$= \dfrac{2\cos 4x \cos(-x)}{-2\sin 4x \sin(-x)}$

$= \dfrac{2\cos 4x \cos x}{2\sin 4x \sin x}$

$= \dfrac{\cos 4x}{\sin 4x} \cdot \dfrac{\cos x}{\sin x}$

$= \cot 4x \cot x$

$= \cot x \cot 4x$

48. $\dfrac{\sin 2x + \sin 6x}{\sin 2x - \sin 6x} = \dfrac{2\sin\left(\dfrac{2x+6x}{2}\right)\cos\left(\dfrac{2x-6x}{2}\right)}{2\sin\left(\dfrac{2x-6x}{2}\right)\cos\left(\dfrac{2x+6x}{2}\right)}$

$= \dfrac{2\sin\left(\dfrac{8x}{2}\right)\cos\left(\dfrac{-4x}{2}\right)}{2\sin\left(\dfrac{-4x}{2}\right)\cos\left(\dfrac{8x}{2}\right)}$

$= \dfrac{\sin 4x \cos(-2x)}{\sin(-2x)\cos 4x}$

$= -\dfrac{\sin 4x \cos 2x}{\sin 2x \cos 4x}$

$= -\dfrac{\sin 4x}{\cos 4x} \cdot \dfrac{\cos 2x}{\sin 2x}$

$= -\tan 4x \cot 2x$

49. **a.** The graph appears to be the cotangent curve, $y = \cot x$. It cycles from positive infinity through the intercept to negative infinity. Thus, $y = \cot x$ also describes the graph.

b. $\dfrac{\cos 3x + \cos x}{\sin 3x - \sin x} = \dfrac{2\cos\left(\dfrac{3x+x}{2}\right)\cos\left(\dfrac{3x-x}{2}\right)}{2\sin\left(\dfrac{3x-x}{2}\right)\cos\left(\dfrac{3x+x}{2}\right)}$

$= \dfrac{2\cos\left(\dfrac{4x}{2}\right)\cos\left(\dfrac{2x}{2}\right)}{2\sin\left(\dfrac{2x}{2}\right)\cos\left(\dfrac{4x}{2}\right)}$

$= \dfrac{2\cos 2x \cos x}{2\sin x \cos 2x}$

$= \dfrac{\cos x}{\sin x}$

$= \cot x$

This verifies our observation that

$y = \dfrac{\cos 3x + \cos x}{\sin 3x - \sin x}$ and $y = \cot x$ describe the same graph.

50. $\cos x = -\dfrac{1}{2}$

Because $\cos\dfrac{\pi}{3} = \dfrac{1}{2}$, the solutions for $\cos x = -\dfrac{1}{2}$ in $[0, 2\pi)$ are $x = \pi - \dfrac{\pi}{3} = \dfrac{2\pi}{3}$ $x = \pi + \dfrac{\pi}{3} = \dfrac{4\pi}{3}$.

Because the period of the cosine function is 2π, the solutions are given by

$x = \dfrac{2\pi}{3} + 2n\pi$ or $x = \dfrac{4\pi}{3} + 2n\pi$ where n is any integer.

51. $\sin x = \dfrac{\sqrt{2}}{2}$

Because $\sin\dfrac{\pi}{4} = \dfrac{\sqrt{2}}{2}$, the solutions for $\sin x = \dfrac{\sqrt{2}}{2}$ in $[0, 2\pi)$ are $x = \dfrac{\pi}{4}$ and $x = \pi - \dfrac{\pi}{4} = \dfrac{3\pi}{4}$. Because the period of the cosine function is 2π, the solutions are given by $x = \dfrac{\pi}{4} + 2n\pi$ or $x = \dfrac{3\pi}{4} + 2n\pi$ where n is any integer.

500

52. $2\sin x + 1 = 0$

$2\sin x = -1$

$\sin x = -\dfrac{1}{2}$

Because $\sin\dfrac{\pi}{6} = \dfrac{1}{2}$, the solutions for

$\sin x = -\dfrac{1}{2}$ in $[0, 2\pi)$ are

$x = \pi + \dfrac{\pi}{6} = \dfrac{7\pi}{6}$

$x = 2\pi - \dfrac{\pi}{6} = \dfrac{11\pi}{6}$.

Because the period of the sine function is 2π, the solutions are given by

$x = \dfrac{7\pi}{6} + 2n\pi$ or $x = \dfrac{11\pi}{6} + 2n\pi$

where n is any integer

53. $\sqrt{3}\tan x - 1 = 0$

$\sqrt{3}\tan x = 1$

$\tan x = \dfrac{1}{\sqrt{3}}$

Because $\tan\dfrac{\pi}{6} = \dfrac{1}{\sqrt{3}}$, the solution for

$\tan x = \dfrac{1}{\sqrt{3}}$ in $[0, \pi)$ is $x = \dfrac{\pi}{6}$.

Because the period of the tangent function is π, the solutions are given by

$x = \dfrac{\pi}{6} + n\pi$ where n is any integer.

54. The period of the cosine function is 2π. In the interval $[0, 2\pi)$, the only value at which the cosine function is -1 is π. All the solutions to $\cos 2x = -1$ are given by

$2x = \pi + 2n\pi$

$x = \dfrac{\pi}{2} + n\pi$ where n is any integer.

The solutions in the interval $[0, 2\pi)$ are obtained by letting $n = 0$ and $n = 1$.

The solutions are $\dfrac{\pi}{2}$ and $\dfrac{3\pi}{2}$.

55. The period of the sine function is 2π. In the interval $[0, 2\pi)$, the only value at which the sine function is

1 is $\dfrac{\pi}{2}$. All the solutions to $\sin 3x = 1$ are given by

$3x = \dfrac{\pi}{2} + 2n\pi$

$x = \dfrac{\pi}{6} + \dfrac{2n\pi}{3}$

where n is any integer. The solutions in the interval $[0, 2\pi)$ are obtained by letting $n = 0$, $n = 1$, and $n = 2$.

The solutions are $\dfrac{\pi}{6}, \dfrac{5\pi}{6}$, and $\dfrac{9\pi}{6}$.

56. The period of the tangent function is π. In the interval $[0, \pi)$, the only value for which the tangent

function is -1 is $\dfrac{3\pi}{4}$. All the solutions to

$\tan\dfrac{x}{2} = -1$ are given by

$\dfrac{x}{2} = \dfrac{3\pi}{4} + n\pi$

$x = \dfrac{3\pi}{2} + 2n\pi$

where n is any integer. The solution in the interval $[0, 2\pi)$ is obtained by letting $n = 0$.

The solution is $\dfrac{3\pi}{2}$.

57. $\tan x = 2\cos x \tan x$

$\tan x - 2\cos x \tan x = 0$

$\tan x(1 - 2\cos x) = 0$

$\tan x = 0$ or $1 - 2\cos x = 0$

$x = 0 \quad x = \pi$ $-2\cos x = -1$

$\cos x = \dfrac{1}{2}$

$x = \dfrac{\pi}{3} \quad x = \dfrac{5\pi}{3}$

The solutions in the interval $[0, 2\pi)$ are 0,

$\dfrac{\pi}{3}$, π, and $\dfrac{5\pi}{3}$.

58. The given equation is in quadratic form
$t^2 - 2t = 3$ with $t = \cos x$.

$$\cos^2 x - 2\cos x = 3$$

$$\cos^2 x - 2\cos x - 3 = 0$$

$$(\cos x + 1)(\cos x - 3) = 0$$

$\cos x + 1 = 0$ or $\cos x - 3 = 0$

$\cos x = -1 \qquad\qquad \cos x = 3$

$\quad x = \pi \qquad\qquad \cos x$ cannot be

greater than 1.

The solution in the interval $[0, 2\pi)$ is π.

59.
$$2\cos^2 x - \sin x = 1$$

$$2\left(1 - \sin^2 x\right) - \sin x = 1$$

$$2 - 2\sin^2 x - \sin x - 1 = 0$$

$$-2\sin^2 x - \sin x + 1 = 0$$

$$2\sin^2 x + \sin x - 1 = 0$$

$$(2\sin x - 1)(\sin x + 1) = 0$$

$2\sin x - 1 = 0$ or $\sin x + 1 = 0$

$\quad 2\sin x = 1 \qquad\qquad \sin x = -1$

$$\sin x = \frac{1}{2} \qquad\qquad x = \frac{3\pi}{2}$$

$$x = \frac{\pi}{6}, \frac{5\pi}{6}$$

The solutions in the interval $[0, 2\pi)$ are $\dfrac{\pi}{6}, \dfrac{5\pi}{6}$, and

$\dfrac{3\pi}{2}$.

60. The given equation is in quadratic form
$4t^2 = 1$ with $t = \sin x$.

$$4\sin^2 x = 1$$

$$4\sin^2 x - 1 = 0$$

$$(2\sin x - 1)(2\sin x + 1) = 0$$

$2\sin x - 1 = 0$ or $2\sin x + 1 = 0$

$\quad 2\sin x = 1 \qquad\qquad 2\sin x = -1$

$$\sin x = \frac{1}{2} \qquad\qquad \sin x = -\frac{1}{2}$$

$$x = \frac{\pi}{6} \quad x = \frac{5\pi}{6} \qquad x = \frac{7\pi}{6} \quad x = \frac{11\pi}{6}$$

The solutions in the interval $[0, 2\pi)$ are

$\dfrac{\pi}{6}, \dfrac{5\pi}{6}, \dfrac{7\pi}{6}$, and $\dfrac{11\pi}{6}$.

61.
$$\cos 2x - \sin x = 1$$

$$2\cos^2 x - 1 - \sin x = 1$$

$$2\left(1 - \sin^2 x\right) - \sin x - 2 = 0$$

$$2 - 2\sin^2 x - \sin x - 2 = 0$$

$$-2\sin^2 x - \sin x = 0$$

$$2\sin^2 x + \sin x = 0$$

$$\sin x(2\sin x + 1) = 0$$

$\sin x = 0 \qquad\qquad 2\sin x + 1 = 0$

$\quad x = 0, \pi \qquad\qquad \sin x = -\dfrac{1}{2}$

$$x = \frac{7\pi}{6}, \frac{11\pi}{6}$$

The solutions in the interval $[0, 2\pi)$ are

$0, \pi, \dfrac{7\pi}{6}$, and $\dfrac{11\pi}{6}$..

62.
$$\sin 2x = \sqrt{3}\sin x$$

$$2\sin x\cos x = \sqrt{3}\sin x$$

$$2\sin x\cos x - \sqrt{3}\sin x = 0$$

$$\sin x\left(2\cos x - \sqrt{3}\right) = 0$$

$\sin x = 0$ or $2\cos x - \sqrt{3} = 0$

$x = 0 \; x = \pi \qquad\qquad 2\cos x = \sqrt{3}$

$$\cos x = \frac{\sqrt{3}}{2}$$

$$x = \frac{\pi}{6} \qquad x = \frac{11\pi}{6}$$

The solutions in the interval $[0, 2\pi)$ are 0,

$\dfrac{\pi}{6}, \pi$, and $\dfrac{11\pi}{6}$.

63.
$$\sin x = \tan x$$

$$\sin x = \frac{\sin x}{\cos x}$$

$$\sin x \cdot \cos x = \sin x$$

$$\sin x\cos x - \sin x = 0$$

$$\sin x(\cos x - 1) = 0$$

$\sin x = 0$ or $\cos x - 1 = 0$

$x = 0 \; x = \pi \qquad\qquad \cos x = 1$

$$x = 0$$

The solutions in the interval $[0, 2\pi)$ are 0 and π.

64. $\sin x = -0.6031$

Be sure calculator is in radian mode and find the inverse sine of $+0.6031$. This gives the first quadrant reference angle.

$\theta = \sin^{-1}(0.6031) \approx 0.6474$

The sine is negative in quadrants III and IV thus,

$x \approx \pi + 0.6474$ or $x \approx 2\pi - 0.6474$

$x \approx 3.7890$ $x \approx 5.6358$

65. $5\cos^2 x - 3 = 0$

$\cos^2 x = \dfrac{3}{5}$

$\cos x = \pm\sqrt{\dfrac{3}{5}}$

$\cos x = \pm\dfrac{\sqrt{15}}{5}$

$\cos x \approx 0.7746$ or $\cos x \approx -0.7746$

$x \approx 0.6847, \ 5.5985$ $x \approx 2.4569, \ 3.8263$

66.

$1 + \tan^2 x = 4\tan x - 2$

$\tan^2 x - 4\tan x + 3 = 0$

$(\tan x - 1)(\tan x - 3) = 0$

$\tan x = 1$ or $\tan x = 3$

$x = \dfrac{\pi}{4}, \dfrac{5\pi}{4}$ $x \approx 1.2490, \ 4.3906$

67. $2\sin^2 x + \sin x - 2 = 0$

$\sin x = \dfrac{-b \pm \sqrt{b^2 - 4ac}}{2a}$

$\sin x = \dfrac{-(1) \pm \sqrt{(1)^2 - 4(2)(-2)}}{2(2)}$

$\sin x = \dfrac{-1 \pm \sqrt{17}}{4}$

$\sin x = 0.7808$ or $\sin x = -1.2808$

$x = 0.8959, \ 2.2457$ $\cancel{\sin x = -1.2808}$

68. Substitute $d = -3$ into the equation and solve for t:

$-3 = -6\cos\dfrac{\pi}{2}t$

$\dfrac{-3}{-6} = \dfrac{-6\cos\frac{\pi}{2}t}{-6}$

$\dfrac{1}{2} = \cos\dfrac{\pi}{2}t$

$\cos\dfrac{\pi}{2}t = \dfrac{1}{2}$

The period of the cosine function is 2π. In the interval $[0, 2\pi)$, there are two values at which the cosine function is $\dfrac{1}{2}$. One is $\dfrac{\pi}{3}$. The cosine function is positive in quadrant IV. Thus, the other value is

$2\pi - \dfrac{\pi}{3} = \dfrac{5\pi}{3}$.

All solutions to $\cos\dfrac{\pi}{2}t = \dfrac{1}{2}$ are given by

$\dfrac{\pi}{2}t = \dfrac{\pi}{3} + 2n\pi$

$\dfrac{\pi}{2}t = \dfrac{5\pi}{3} + 2n\pi$

$t = \dfrac{2}{3} + 4n$ or

$t = \dfrac{10}{3} + 4n$

where n is any integer.

69. Substitute $v_0 = 90$ and $d = 100$, and solve for θ:

$100 = \dfrac{90^2}{16}\sin\theta\cos\theta$

$\dfrac{16}{81} = \sin\theta\cos\theta$

$2 \cdot \dfrac{16}{81} = 2\sin\theta\cos\theta$

$\dfrac{32}{81} = \sin 2\theta$

$\sin 2\theta = \dfrac{32}{81}$

The period of the sine function is $360°$. In the interval $[0, 360°)$, there are two values at which the sine function is $\dfrac{32}{81}$. One is $\sin^{-1}\left(\dfrac{32}{81}\right) \approx 23.27°$. The sine function is positive in quadrant II. Thus, the other value is $180° - 23.27° = 156.73°$. All solutions to $\sin 2\theta = \dfrac{32}{81}$ are given by

$2\theta = 23.27° + 360°n$

$\theta = 11.635° + 180°n$

or

$2\theta = 156.73° + 360°n$

$\theta = 78.365° + 180°n$

where n is any integer.

In the interval $[0, 90°)$ we obtain the solutions by letting $n = 0$. The solutions are approximately $12°$ and $78°$. Therefore, the angle of elevation should be $12°$ or $78°$.

Chapter 5 Test

For problems 1–4: $\sin\alpha = \dfrac{4}{5} = \dfrac{y}{r}$

Because α lies in quadrant II, x is negative.

$x^2 + 4^2 = 5^2$

$x^2 = 5^2 - 4^2 = 9$

$x = -\sqrt{9} = -3$

Thus, $\cos\alpha = \dfrac{x}{r} = \dfrac{-3}{5} = -\dfrac{3}{5}$, and

$\tan\alpha = \dfrac{y}{x} = \dfrac{4}{-3} = -\dfrac{4}{3}$.

$\cos\beta = \dfrac{5}{13} = \dfrac{x}{r}$

Because β lies in quadrant I, y is positive.

$5^2 + y^2 = 13^2$

$y^2 = 13^2 - 5^2 = 144$

$y = \sqrt{144} = 12$

Thus, $\sin\beta = \dfrac{y}{r} = \dfrac{12}{13}$, and $\tan\beta = \dfrac{y}{x} = \dfrac{12}{5}$.

1. $\cos(\alpha+\beta) = \cos\alpha\cos\beta - \sin\alpha\sin\beta$

$= -\dfrac{3}{5}\cdot\dfrac{5}{13} - \dfrac{4}{5}\cdot\dfrac{12}{13} = -\dfrac{63}{65}$

2. $\tan(\alpha-\beta) = \dfrac{\tan\alpha - \tan\beta}{1 + \tan\alpha\tan\beta}$

$= \dfrac{-\frac{4}{3} - \frac{12}{5}}{1 + \left(-\frac{4}{3}\right)\cdot\frac{12}{5}} = \dfrac{-\frac{56}{15}}{-\frac{33}{15}} = \dfrac{56}{33}$

3. $\sin 2\alpha = 2\sin\alpha\cos\alpha = 2\left(\dfrac{4}{5}\right)\left(-\dfrac{3}{5}\right) = -\dfrac{24}{25}$

4. $\cos\dfrac{\beta}{2} = \sqrt{\dfrac{1+\cos\beta}{2}} = \sqrt{\dfrac{1+\frac{5}{13}}{2}} = \sqrt{\dfrac{18}{26}}$

$= \dfrac{3\sqrt{2}}{\sqrt{26}} = \dfrac{3\sqrt{52}}{26} = \dfrac{3\cdot2\sqrt{13}}{26}$

$= \dfrac{3\sqrt{13}}{13}$

5. $\sin 105° = \sin(135° - 30°)$

$= \sin 135°\cos 30° - \cos 135°\sin 30°$

$= \dfrac{\sqrt{2}}{2}\cdot\dfrac{\sqrt{3}}{2} - \left(-\dfrac{\sqrt{2}}{2}\right)\dfrac{1}{2}$

$= \dfrac{\sqrt{6}}{4} + \dfrac{\sqrt{2}}{4} = \dfrac{\sqrt{6}+\sqrt{2}}{4}$

6. $\cos x \csc x = \cos x\cdot\dfrac{1}{\sin x} = \dfrac{\cos x}{\sin x} = \cot x$

7. $\dfrac{\sec x}{\cot x + \tan x} = \dfrac{\frac{1}{\cos x}}{\frac{\cos x}{\sin x} + \frac{\sin x}{\cos x}}$

$= \dfrac{\frac{1}{\cos x}}{\frac{\cos x}{\sin x}\cdot\frac{\cos x}{\cos x} + \frac{\sin x}{\cos x}\cdot\frac{\sin x}{\sin x}}$

$= \dfrac{\frac{1}{\cos x}}{\frac{\cos^2 x + \sin^2 x}{\sin x\cos x}} = \dfrac{\frac{1}{\cos x}}{\frac{1}{\sin x\cos x}}$

$= \dfrac{1}{\cos x}\cdot\dfrac{\sin x\cos x}{1}$

$= \sin x$

8. $1 - \dfrac{\cos^2 x}{1+\sin x} = 1 - \dfrac{\left(1-\sin^2 x\right)}{1+\sin x}$

$= 1 - \dfrac{(1+\sin x)(1-\sin x)}{1+\sin x}$

$= 1 - (1-\sin x)$

$= \sin x$

9. $\cos\left(\theta + \dfrac{\pi}{2}\right) = \cos\theta\cos\dfrac{\pi}{2} - \sin\theta\sin\dfrac{\pi}{2}$

$= \cos\theta\cdot 0 - \sin\theta\cdot 1$

$= -\sin\theta$

10. $\dfrac{\sin(\alpha-\beta)}{\sin\alpha\cos\beta} = \dfrac{\sin\alpha\cos\beta - \cos\alpha\sin\beta}{\sin\alpha\cos\beta}$

$= \dfrac{\sin\alpha\cos\beta}{\sin\alpha\cos\beta} - \dfrac{\cos\alpha\sin\beta}{\sin\alpha\cos\beta}$

$= 1 - \cot\alpha\tan\beta$

11. $\sin t\cos t(\tan t + \cot t) = \sin t\cos t\left(\dfrac{\sin t}{\cos t} + \dfrac{\cos t}{\sin t}\right)$

$= \dfrac{\sin^2 t\cos t}{\cos t} + \dfrac{\sin t\cos^2 t}{\sin t}$

$= \sin^2 t + \cos^2 t$

$= 1$

12. The period of the sine function is 2π. In the interval $[0, 2\pi)$, there are two values at which the sine function is $-\dfrac{1}{2}$. One is $\pi + \dfrac{\pi}{6} = \dfrac{7\pi}{6}$. The other is $2\pi - \dfrac{\pi}{6} = \dfrac{11\pi}{6}$. All the solutions to $\sin 3x = -\dfrac{1}{2}$ are given by $3x = \dfrac{7\pi}{6} + 2n\pi$

$$x = \dfrac{7\pi}{18} + \dfrac{2n\pi}{3}$$

or

$$3x = \dfrac{11\pi}{6} + 2n\pi$$

$$x = \dfrac{11\pi}{18} + \dfrac{2n\pi}{3}$$

where n is any integer. The solutions in the interval $[0, 2\pi)$ are obtained by letting $n = 0$, $n = 1$, and $n = 2$.

The solutions are $\dfrac{7\pi}{18}, \dfrac{11\pi}{18}, \dfrac{19\pi}{18}, \dfrac{23\pi}{18},$

$\dfrac{31\pi}{18},$ and $\dfrac{35\pi}{18}.$

13.
$$\sin 2x + \cos x = 0$$
$$2\sin x \cos x + \cos x = 0$$
$$\cos x(2\sin x + 1) = 0$$

$\cos x = 0$ or $2\sin x + 1 = 0$

$x = \dfrac{\pi}{2} \; x = \dfrac{3\pi}{2}$ $2\sin x = -1$

$$\sin x = -\dfrac{1}{2}$$

$$x = \dfrac{7\pi}{6} \; x = \dfrac{11\pi}{6}$$

The solutions in the interval $[0, 2\pi)$ are $\dfrac{\pi}{2},$

$\dfrac{7\pi}{6}, \dfrac{3\pi}{2},$ and $\dfrac{11\pi}{6}.$

14.
$$2\cos^2 x - 3\cos x + 1 = 0$$
$$(2\cos x - 1)(\cos x - 1) = 0$$

$2\cos x - 1 = 0$ or $\cos x - 1 = 0$

$2\cos x = 1$ $\cos x = 1$

$\cos x = \dfrac{1}{2}$ $x = 0$

$x = \dfrac{\pi}{3} \quad x = \dfrac{5\pi}{3}$

The solutions in the interval $[0, 2\pi)$ are

$0, \dfrac{\pi}{3},$ and $\dfrac{5\pi}{3}.$

15.
$$2\sin^2 x + \cos x = 1$$
$$2(1 - \cos^2 x) + \cos x - 1 = 0$$
$$2 - 2\cos^2 x + \cos x - 1 = 0$$
$$-2\cos^2 x + \cos x + 1 = 0$$
$$2\cos^2 x - \cos x - 1 = 0$$
$$(2\cos x + 1)(\cos x - 1) = 0$$

$2\cos x + 1 = 0$ or $\cos x - 1 = 0$

$2\cos x = -1$ $\cos x = 1$

$\cos x = -\dfrac{1}{2}$ $x = 0$

$x = \dfrac{2\pi}{3} \quad x = \dfrac{4\pi}{3}$

The solutions in the interval $[0, 2\pi)$ are $0, \dfrac{2\pi}{3},$ and $\dfrac{4\pi}{3}.$

16. $\cos x = -0.8092$

Be sure calculator is in radian mode and find the inverse cosine of $+0.8092$. This gives the first quadrant reference angle.

$\theta = \cos^{-1}(0.8092) \approx 0.6280$

The cosine is negative in quadrants II and III thus,

$x \approx \pi - 0.6280$ or $x \approx \pi + 0.6280$

$x \approx 2.5136$ $x \approx 3.7696$

17.
$$\tan x \sec x = 3\tan x$$
$$\tan x \sec x - 3\tan x = 0$$
$$\tan x(\sec x - 3) = 0$$

$\tan x = 0$ or $\sec x - 3 = 0$

$x = 0, \; \pi$ $\sec x = 3$

$$\cos x = \dfrac{1}{3}$$

$$x \approx 1.2310, \; 5.0522$$

18. $\tan^2 x - 3\tan x - 2 = 0$

$$\tan x = \frac{-b \pm \sqrt{b^2 - 4ac}}{2a}$$

$$\tan x = \frac{-(-3) \pm \sqrt{(-3)^2 - 4(1)(-2)}}{2(1)}$$

$$\tan x = \frac{3 \pm \sqrt{17}}{2}$$

$\tan x \approx 3.5616 \qquad$ or $\quad \tan x \approx -0.5616$

$x \approx 1.2971, \ 4.4387 \qquad x \approx 2.6299, \ 5.7715$

Cumulative Review Exercises (Chapters 1–5)

1. $x^3 + x^2 - x + 15 = 0$

The possible rational zeros are: $\pm 1, \pm 3, \pm 5, \pm 15$.

Synthetic division shows that -3 is a zero:

$$
\begin{array}{r|rrrr}
-3 & 1 & 1 & -1 & 15 \\
   &   & -3 & 6 & -15 \\
\hline
   & 1 & -2 & 5 & 0
\end{array}
$$

The quotient is $x^2 - 2x + 5$. The remaining zeros are found using the quadratic formula:

$$x = \frac{-(-2) \pm \sqrt{(-2)^2 - 4(1)(5)}}{2(1)}$$

$$= \frac{2 \pm \sqrt{4 - 20}}{2}$$

$$= \frac{2 \pm \sqrt{-16}}{2}$$

$$= \frac{2 \pm 4i}{2}$$

$$= 1 \pm 2i$$

All solutions are: -3, $1 + 2i$ and $1 - 2i$.

2.

$$11^{x-1} = 125$$

$$\log 11^{x-1} = \log 125$$

$$(x-1)\log 11 = \log 125$$

$$x - 1 = \frac{\log 125}{\log 11}$$

$$x = \frac{\log 125}{\log 11} + 1$$

or $x \approx 3.01$

3. $x^2 + 2x - 8 > 0$

$(x-2)(x+4) > 0$

zero points are $x = 2$ and $x = -4$.

Test Interval	Representative Number	Substitute into $x^2 + 2x - 8 > 0$	Conclusion
$(-\infty, -4)$	-5	$(-5)^2 + 2(-5) - 8 = 25 - 10 - 8 = 7 > 0$	$(-\infty, -4)$ belongs to the solution set.
$(-4, 2)$	0	$0^2 + 2(0) - 8 = -8 > 0$	$(-4, 2)$ does not belong to the solution set.
$(2, \infty)$	3	$3^2 + 2(3) - 8 = 9 + 6 - 8 = 7 > 0$	$(2, \infty)$ belongs to the solution set.

The solution intervals are $(-\infty, -4) \cup (2, \infty)$.

4. $\cos 2x + 3 = 5\cos x$

$2\cos^2 x - 1 + 3 = 5\cos x$

$2\cos^2 x - 5\cos x + 2 = 0$

$(2\cos x - 1)(\cos x - 2) = 0$

$2\cos x - 1 = 0$ or $\cos x - 2 = 0$

$2\cos x = 1$ $\cos x = 2$

$\cos x = \dfrac{1}{2}$ $\cos x$ cannot
                                    be greater
                                    than 1.

$x = \dfrac{\pi}{3}$ $x = \dfrac{5\pi}{3}$

The solutions in the interval $[0, 2\pi)$ are $\dfrac{\pi}{3}$ and $\dfrac{5\pi}{3}$.

5. $\tan x + \sec^2 x = 3$

$\tan x + 1 + \tan^2 x = 3$

$\tan^2 x + \tan x - 2 = 0$

$(\tan x - 1)(\tan x + 2) = 0$

$\tan x - 1 = 0$ or $\tan x + 2 = 0$

$\tan x = 1$ $\tan x = -2$

$x = \dfrac{\pi}{4}, \dfrac{5\pi}{4}$ $x \approx 2.0344, \ 5.1760$

6.

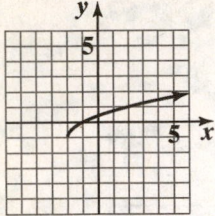

$$y = \sqrt{x + 2} - 1$$

Shift the graph of $y = \sqrt{x}$ left 2 units and down 1 unit.

7.

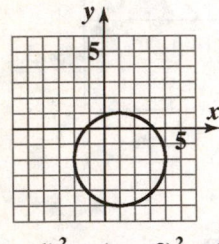

$$(x - 1)^2 + (y + 2)^2 = 9$$

8.

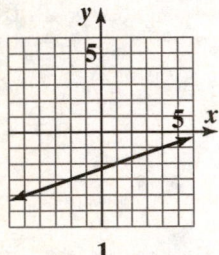

$$y + 2 = \frac{1}{3}(x - 1)$$

9.

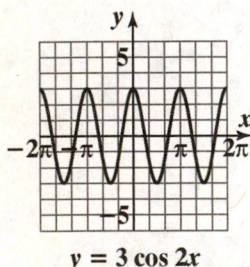

$$y = 3 \cos 2x$$

10.

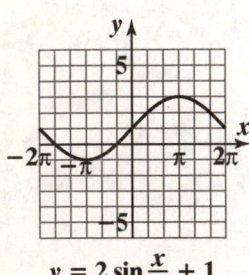

$$y = 2 \sin \frac{x}{2} + 1$$

11.

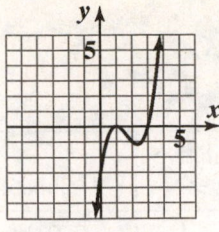

$$f(x) = (x - 1)^2 (x - 3)$$

12. $f(x) = x^2 + 3x - 1$

$$\frac{f(a+h) - f(a)}{h}$$

$$= \frac{(a+h)^2 + 3(a+h) - 1 - \left(a^2 + 3a - 1\right)}{h}$$

$$= \frac{a^2 + 2ah + h^2 + 3a + 3h - 1 - a^2 - 3a + 1}{h}$$

$$= \frac{2ah + h^2 + 3h}{h}$$

$$= 2a + h + 3$$

13. $\sin 225° = \sin(180° + 45°)$

$$= \sin 180° \cos 45° + \cos 180° \sin 45°$$

$$= 0 \cdot \frac{\sqrt{2}}{2} + (-1) \cdot \frac{\sqrt{2}}{2}$$

$$= -\frac{\sqrt{2}}{2}$$

14. $\sec^4 x - \sec^2 x$

$$= \sec^2 x \cdot \sec^2 x - \sec^2 x$$

$$= \left(1 + \tan^2 x\right)\left(1 + \tan^2 x\right) - \left(1 + \tan^2 x\right)$$

$$= 1 + 2\tan^2 x + \tan^4 x - 1 - \tan^2 x$$

$$= \tan^4 x + \tan^2 x$$

We worked with the left side and arrived at the right side. Thus, the identity is verified.

15. $320° \times \dfrac{\pi}{180°} = \dfrac{16}{9}\pi$ or 5.59 radians

508

16.
$$A = Pe^{rt}$$
$$3P = Pe^{0.0575t}$$
$$3 = e^{0.0575t}$$
$$\ln 3 = \ln e^{0.0575t}$$
$$\ln 3 = 0.0575t$$
$$\frac{\ln 3}{0.0575} = t$$
$$t \approx 19.1 \text{ years}$$

17.
$$f(x) = \frac{2x+1}{x-3}$$
$$y = \frac{2x+1}{x-3}$$
$$x = \frac{2y+1}{y-3}$$
$$x(y-3) = 2y+1$$
$$xy - 3x = 2y+1$$
$$xy - 2y = 3x+1$$
$$y(x-2) = 3x+1$$
$$y = \frac{3x+1}{x-2}$$
$$f^{-1}(x) = \frac{3x+1}{x-2}$$

18. The third angle is:
$$B = 180° - 90° - 23° = 67°.$$

Since $\sin\theta = \dfrac{\text{opposite}}{\text{hypotenuse}}$,

$$\sin A = \sin 23° = \frac{12}{c}$$
$$c = \frac{12}{\sin 23°} \approx 30.71 \text{ and}$$
$$\sin B = \sin 67° = \frac{b}{30.71}$$
$$b = 30.71 \cdot \sin 67° \approx 28.27$$

The angles are 90°, 23°, and 67°.
The sides are 12, 30.71, and 28.27.

19. Solve $8.5 = \dfrac{12}{150} \cdot a$

where a is the adult dose.
$$a = \frac{(8.5) \cdot 150}{12}$$
$$= 106.25 \text{ mg}$$
$$a \approx 106 \text{ mg}$$

20.. Let h be the height of the flagpole.
Then $\tan 53° = \dfrac{h}{12}$
$$h = 12 \cdot \tan 53°$$
$$h \approx 15.9 \text{ feet}$$

Chapter 6
Additional Topics in Trigonometry

Section 6.1

Check Point Exercises

1. Begin by finding B, the third angle of the triangle.

$$A + B + C = 180°$$
$$64° + B + 82° = 180°$$
$$146° + B = 180°$$
$$B = 34°$$

In this problem, we are given c and C:
$c = 14$ and $C = 82°$. Thus, use the ratio

$\dfrac{c}{\sin C}$, or $\dfrac{14}{\sin 82°}$, to find the other two sides. Use

the Law of Sines to find a.

$$\frac{a}{\sin A} = \frac{c}{\sin C}$$
$$\frac{a}{\sin 64°} = \frac{14}{\sin 82°}$$
$$a = \frac{14 \sin 64°}{\sin 82°}$$
$$a \approx 13 \text{ centimeters}$$

Use the Law of Sines again, this time to find b.

$$\frac{b}{\sin B} = \frac{c}{\sin C}$$
$$\frac{b}{\sin 34°} = \frac{14}{\sin 82°}$$
$$b = \frac{14 \sin 34°}{\sin 82°}$$
$$b \approx 8 \text{ centimeters}$$

The solution is $B = 34°$, $a \approx 13$ centimeters, and $b \approx 8$ centimeters.

2. Begin by finding B.

$$A + B + C = 180°$$
$$40° + B + 22.5° = 180°$$
$$62.5° + B = 180°$$
$$B = 117.5°$$

In this problem, we are given that $b = 12$ and we find that $B = 117.5°$. Thus, use the ratio

$\dfrac{b}{\sin B}$, or $\dfrac{12}{\sin 117.5°}$, to find the other two sides. Use

the Law of Sines to find a.

$$\frac{a}{\sin A} = \frac{b}{\sin B}$$
$$\frac{a}{\sin 40°} = \frac{12}{\sin 117.5°}$$
$$a = \frac{12 \sin 40°}{\sin 117.5°} \approx 9$$

Use the Law of Sines again, this time to find c.

$$\frac{c}{\sin C} = \frac{b}{\sin B}$$
$$\frac{c}{\sin 22.5°} = \frac{12}{\sin 117.5°}$$
$$c = \frac{12 \sin 22.5°}{\sin 117.5°} \approx 5$$

The solution is $B = 117.5°$, $a \approx 9$, and $c \approx 5$.

3. The known ratio is $\dfrac{a}{\sin A}$, or $\dfrac{33}{\sin 57°}$. Because side b is given, Use the Law of Sines to find angle B.

$$\frac{a}{\sin A} = \frac{b}{\sin B}$$
$$\frac{33}{\sin 57°} = \frac{26}{\sin B}$$
$$33 \sin B = 26 \sin 57°$$
$$\sin B = \frac{26 \sin 57°}{33} \approx 0.6608$$
$$\sin B \approx 0.6608$$
$$B \approx 41°$$

$180° - 41° = 139°$ also has this sine value, but, the sum of $57°$ and $139°$ exceeds $180°$, so B cannot have this value.

$C = 180° - B - A = 180° - 41° - 57° = 82°$.

Use the law of sines to find C.

$$\frac{a}{\sin A} = \frac{c}{\sin C}$$
$$\frac{33}{\sin 57°} = \frac{c}{\sin 82°}$$
$$c = \frac{33 \sin 82°}{\sin 57°}$$
$$c \approx 39$$

Thus, $B \approx 41°$, $C \approx 82°$, $c \approx 39$.

4. The known ratio is $\dfrac{a}{\sin A}$, or $\dfrac{10}{\sin 50^\circ}$. Because side b is given, Use the Law of Sines to find angle B.

$$\frac{a}{\sin A} = \frac{b}{\sin B}$$

$$\frac{10}{\sin 50^\circ} = \frac{20}{\sin B}$$

$$10\sin B = 20\sin 50^\circ$$

$$\sin B = \frac{20\sin 50^\circ}{10} \approx 1.53$$

Because the sine can never exceed 1, there is no angle B for which $\sin B \approx 1.53$. There is no triangle with the given measurements.

5. The known ratio is $\dfrac{a}{\sin A}$, or $\dfrac{12}{\sin 35^\circ}$. Because side b is given, Use the Law of Sines to find angle B.

$$\frac{a}{\sin A} = \frac{b}{\sin B}$$

$$\frac{12}{\sin 35^\circ} = \frac{16}{\sin B}$$

$$12\sin B = 16\sin 35^\circ$$

$$\sin B = \frac{16\sin 35^\circ}{12} \approx 0.7648$$

There are two angles possible:
$B_1 \approx 50^\circ$, $B_2 \approx 180^\circ - 50^\circ = 130^\circ$
There are two triangles:
$C_1 = 180^\circ - A - B_1 \approx 180^\circ - 35^\circ - 50^\circ = 95^\circ$
$C_2 = 180^\circ - A - B_2 \approx 180^\circ - 35^\circ - 130^\circ = 15^\circ$
Use the Law of Sines to find c_1 and c_2.

$$\frac{c_1}{\sin C_1} = \frac{a}{\sin A}$$

$$\frac{c_1}{\sin 95^\circ} = \frac{12}{\sin 35^\circ}$$

$$c_1 = \frac{12\sin 95^\circ}{\sin 35^\circ} \approx 21$$

$$\frac{c_2}{\sin C_2} = \frac{a}{\sin A}$$

$$\frac{c_2}{\sin 15^\circ} = \frac{12}{\sin 35^\circ}$$

$$c_2 = \frac{12\sin 15^\circ}{\sin 35^\circ} \approx 5$$

In one triangle, the solution is $B_1 \approx 50^\circ$, $C_1 \approx 95^\circ$, and $c_1 \approx 21$. In the other triangle, $B_2 \approx 130^\circ$, $C_2 \approx 15^\circ$, and $c_2 \approx 5$.

6. The area of the triangle is half the product of the lengths of the two sides times the sine of the included angle.

$$\text{Area} = \frac{1}{2}(8)(12)(\sin 135^\circ) \approx 34$$

The area of the triangle is approximately 34 square meters.

7.

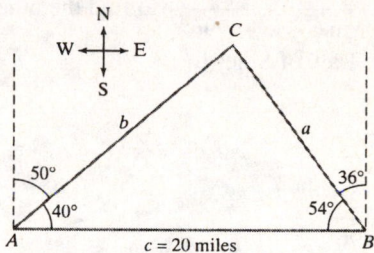

Using a north-south line, the interior angles are found as follows:
$A = 90^\circ - 35^\circ = 55^\circ$

$B = 90^\circ - 49^\circ = 41^\circ$
Find angle C using a 180° angle sum in the triangle.
$C = 180^\circ - A - B = 180^\circ - 55^\circ - 41^\circ = 84^\circ$

The ratio $\dfrac{c}{\sin C}$, or $\dfrac{13}{\sin 84^\circ}$ is now known. Use this ratio and the Law of Sines to find a.

$$\frac{a}{\sin A} = \frac{c}{\sin C}$$

$$\frac{a}{\sin 55^\circ} = \frac{13}{\sin 84^\circ}$$

$$a = \frac{13\sin 55^\circ}{\sin 84^\circ} \approx 11$$

The fire is approximately 11 miles from station B.

Additional Topics in Trigonometry

Exercise Set 6.1

1. Begin by finding B.
$$A + B + C = 180°$$
$$42° + B + 96° = 180°$$
$$138° + B = 180°$$
$$B = 42°$$

Use the ratio $\dfrac{c}{\sin C}$, or $\dfrac{12}{\sin 96°}$, to find the other two sides. Use the Law of Sines to find a.
$$\frac{a}{\sin A} = \frac{c}{\sin C}$$
$$\frac{a}{\sin 42°} = \frac{12}{\sin 96°}$$
$$a = \frac{12\sin 42°}{\sin 96°}$$
$$a \approx 8.1$$

Use the Law of Sines again, this time to find b.
$$\frac{b}{\sin B} = \frac{c}{\sin C}$$
$$\frac{b}{\sin 42°} = \frac{12}{\sin 96°}$$
$$b = \frac{12\sin 42°}{\sin 96°}$$
$$b \approx 8.1$$

The solution is $B = 42°$, $a \approx 8.1$, and $b \approx 8.1$.

3. Begin by finding A.
$$A + B + C = 180°$$
$$A + 54° + 82° = 180°$$
$$A + 136° = 180°$$
$$A = 44°$$

Use the ratio $\dfrac{a}{\sin A}$, or $\dfrac{16}{\sin 44°}$, to find the other two sides. Use the Law of Sines to find b.
$$\frac{b}{\sin B} = \frac{a}{\sin A}$$
$$\frac{b}{\sin 54°} = \frac{16}{\sin 44°}$$
$$b = \frac{16\sin 54°}{\sin 44°}$$
$$b \approx 18.6$$

Use the Law of Sines again, this time to find c.

$$\frac{c}{\sin C} = \frac{a}{\sin A}$$
$$\frac{c}{\sin 82°} = \frac{16}{\sin 44°}$$
$$c = \frac{16\sin 82°}{\sin 44°}$$
$$c \approx 22.8$$

The solution is $A = 44°$, $b \approx 18.6$, and $c \approx 22.8$.

5. Begin by finding C.
$$A + B + C = 180°$$
$$48° + 37° + C = 180°$$
$$85° + C = 180°$$
$$C = 95°$$

Use the ratio $\dfrac{a}{\sin A}$, or $\dfrac{100}{\sin 48°}$, to find the other two sides. Use the Law of Sines to find b.
$$\frac{b}{\sin B} = \frac{a}{\sin A}$$
$$\frac{b}{\sin 37°} = \frac{100}{\sin 48°}$$
$$b = \frac{100\sin 37°}{\sin 48°}$$
$$b \approx 81.0$$

Use the Law of Sines again, this time to find c.
$$\frac{c}{\sin C} = \frac{a}{\sin A}$$
$$\frac{c}{\sin 95°} = \frac{100}{\sin 48°}$$
$$c = \frac{100\sin 95°}{\sin 48°}$$
$$c \approx 134.1$$

The solution is $C = 95°$, $b \approx 81.0$, and $c \approx 134.1$.

7. Begin by finding B.
$$A + B + C = 180°$$
$$38° + B + 102° = 180°$$
$$B + 140° = 180°$$
$$B = 40°$$

Use the ratio $\dfrac{a}{\sin A}$, or $\dfrac{20}{\sin 38°}$, to find the other two sides. Use the Law of Sines to find b.

$$\frac{b}{\sin B} = \frac{a}{\sin A}$$

$$\frac{b}{\sin 40°} = \frac{20}{\sin 38°}$$

$$b = \frac{20 \sin 40°}{\sin 38°}$$

$$b \approx 20.9$$

Use the Law of Sines again, this time to find c.

$$\frac{c}{\sin C} = \frac{a}{\sin A}$$

$$\frac{c}{\sin 102°} = \frac{20}{\sin 38°}$$

$$c = \frac{20 \sin 102°}{\sin 38°}$$

$$c \approx 31.8$$

The solution is $B = 40°$, $b \approx 20.9$, and $c \approx 31.8$.

9. Begin by finding C.

$$A + B + C = 180°$$

$$44° + 25° + C = 180°$$

$$69° + C = 180°$$

$$C = 111°$$

Use the ratio $\dfrac{a}{\sin A}$, or $\dfrac{12}{\sin 44°}$, to find the other two sides. Use the Law of Sines to find b.

$$\frac{b}{\sin B} = \frac{a}{\sin A}$$

$$\frac{b}{\sin 25°} = \frac{12}{\sin 44°}$$

$$b = \frac{12 \sin 25°}{\sin 44°}$$

$$b \approx 7.3$$

Use the Law of Sines again, this time to find c.

$$\frac{c}{\sin C} = \frac{a}{\sin A}$$

$$\frac{c}{\sin 111°} = \frac{12}{\sin 44°}$$

$$c = \frac{12 \sin 111°}{\sin 44°}$$

$$c \approx 16.1$$

The solution is $C = 111°$, $b \approx 7.3$, and $c \approx 16.1$.

11. Begin by finding A.

$$A + B + C = 180°$$

$$A + 85° + 15° = 180°$$

$$A + 100° = 180°$$

$$A = 80°$$

Use the ratio $\dfrac{b}{\sin B}$, or $\dfrac{40}{\sin 85°}$, to find the other two sides. Use the Law of Sines to find a.

$$\frac{a}{\sin A} = \frac{b}{\sin B}$$

$$\frac{a}{\sin 80°} = \frac{40}{\sin 85°}$$

$$a = \frac{40 \sin 80°}{\sin 85°}$$

$$a \approx 39.5$$

Use the Law of Sines again, this time to find c.

$$\frac{c}{\sin C} = \frac{b}{\sin B}$$

$$\frac{c}{\sin 15°} = \frac{40}{\sin 85°}$$

$$c = \frac{40 \sin 15°}{\sin 85°}$$

$$c \approx 10.4$$

The solution is $A = 80°$, $a \approx 39.5$, and $c \approx 10.4$.

13. Begin by finding B.

$$A + B + C = 180°$$

$$115° + B + 35° = 180°$$

$$B + 150° = 180°$$

$$B = 30°$$

Use the ratio $\dfrac{c}{\sin C}$, or $\dfrac{200}{\sin 35°}$, to find the other two sides. Use the Law of Sines to find a.

$$\frac{a}{\sin A} = \frac{c}{\sin C}$$

$$\frac{a}{\sin 115°} = \frac{200}{\sin 35°}$$

$$a = \frac{200 \sin 115°}{\sin 35°}$$

$$a \approx 316.0$$

Use the Law of Sines again, this time to find b.

$$\frac{b}{\sin B} = \frac{c}{\sin C}$$

$$\frac{b}{\sin 30°} = \frac{200}{\sin 35°}$$

$$b = \frac{200 \sin 30°}{\sin 35°}$$

$$b \approx 174.3$$

The solution is $B = 30°$, $a \approx 316.0$, and $b \approx 174.3$.

15. Begin by finding C.

$$A + B + C = 180°$$
$$65° + 65° + C = 180°$$
$$130° + C = 180°$$
$$C = 50°$$

Use the ratio $\dfrac{c}{\sin C}$, or $\dfrac{6}{\sin 50°}$, to find the other two sides. Use the Law of Sines to find a.

$$\frac{a}{\sin A} = \frac{c}{\sin C}$$

$$\frac{a}{\sin 65°} = \frac{6}{\sin 50°}$$

$$a = \frac{6 \sin 65°}{\sin 50°}$$

$$a \approx 7.1$$

Use the Law of Sines to find angle B.

$$\frac{b}{\sin B} = \frac{c}{\sin C}$$

$$\frac{b}{\sin 65°} = \frac{6}{\sin 50°}$$

$$b = \frac{6 \sin 65°}{\sin 50°}$$

$$b \approx 7.1$$

The solution is $C = 50°$, $a \approx 7.1$, and $b \approx 7.1$.

17. The known ratio is $\dfrac{a}{\sin A}$, or $\dfrac{20}{\sin 40°}$.

Use the Law of Sines to find angle B.

$$\frac{a}{\sin A} = \frac{b}{\sin B}$$

$$\frac{20}{\sin 40°} = \frac{15}{\sin B}$$

$$20 \sin B = 15 \sin 40°$$

$$\sin B = \frac{15 \sin 40°}{20}$$

$$\sin B \approx 0.4821$$

There are two angles possible:

$B_1 \approx 29°$, $B_2 \approx 180° - 29° = 151°$

B_2 is impossible, since $40° + 151° = 191°$.

We find C using B_1 and the given information $A = 40°$.

$$C = 180° - B_1 - A \approx 180° - 29° - 40° = 111°$$

Use the Law of Sines to find side c.

$$\frac{c}{\sin C} = \frac{a}{\sin A}$$

$$\frac{c}{\sin 111°} = \frac{20}{\sin 40°}$$

$$c = \frac{20 \sin 111°}{\sin 40°} \approx 29.0$$

There is one triangle and the solution is

B_1 (or B) $\approx 29°$, $C \approx 111°$, and $c \approx 29.0$.

19. The known ratio is $\dfrac{a}{\sin A}$, or $\dfrac{10}{\sin 63°}$.

Use the Law of Sines to find angle C.

$$\frac{a}{\sin A} = \frac{c}{\sin C}$$

$$\frac{10}{\sin 63°} = \frac{8.9}{\sin C}$$

$$10 \sin C = 8.9 \sin 63°$$

$$\sin C = \frac{8.9 \sin 63°}{10}$$

$$\sin C \approx 0.7930$$

There are two angles possible:

$C_1 \approx 52°$, $C_2 \approx 180° - 52° = 128°$

C_2 is impossible, since $63° + 128° = 191°$.

We find B using C_1 and the given information $A = 63°$.

$$B = 180° - C_1 - A \approx 180° - 52° - 63° = 65°$$

Use the Law of Sines to find side b.

$$\frac{b}{\sin B} = \frac{a}{\sin A}$$

$$\frac{b}{\sin 65°} = \frac{10}{\sin 63°}$$

$$b = \frac{10 \sin 65°}{\sin 63°} \approx 10.2$$

There is one triangle and the solution is

C_1 (or C) $\approx 52°$, $B \approx 65°$, and $b \approx 10.2$.

21. The known ratio is $\dfrac{a}{\sin A}$, or $\dfrac{42.1}{\sin 112°}$.

Use the Law of Sines to find angle C.

$$\frac{a}{\sin A} = \frac{c}{\sin C}$$

$$\frac{42.1}{\sin 112°} = \frac{37}{\sin C}$$

$$42.1 \sin C = 37 \sin 112°$$

$$\sin C = \frac{37 \sin 112°}{42.1}$$

$$\sin C \approx 0.8149$$

There are two angles possible:

$C_1 \approx 55°$, $C_2 \approx 180° - 55° = 125°$

C_2 is impossible, since $112° + 125° = 237°$.

We find B using C_1 and the given information $A = 112°$.

$B = 180° - C_1 - A \approx 180° - 55° - 112° = 13°$

Use the Law of Sines to find b.

$$\frac{b}{\sin B} = \frac{a}{\sin A}$$

$$\frac{b}{\sin 13°} = \frac{42.1}{\sin 112°}$$

$$b = \frac{42.1 \sin 13°}{\sin 112°} \approx 10.2$$

There is one triangle and the solution is

C_1 (or C) $\approx 55°$, $B \approx 13°$, and $b \approx 10.2$.

23. The known ratio is $\dfrac{a}{\sin A}$, or $\dfrac{10}{\sin 30°}$.

Use the Law of Sines to find angle B.

$$\frac{a}{\sin A} = \frac{b}{\sin B}$$

$$\frac{10}{\sin 30°} = \frac{40}{\sin B}$$

$$10 \sin B = 40 \sin 30°$$

$$\sin B = \frac{40 \sin 30°}{10} = 2$$

Because the sine can never exceed 1, there is no angle B for which $\sin B = 2$. There is no triangle with the given measurements.

25. The known ratio is $\dfrac{a}{\sin A}$, or $\dfrac{16}{\sin 60°}$.

Use the Law of Sines to find angle B.

$$\frac{a}{\sin A} = \frac{b}{\sin B}$$

$$\frac{16}{\sin 60°} = \frac{18}{\sin B}$$

$$16 \sin B = 18 \sin 60°$$

$$\sin B = \frac{18 \sin 60°}{16}$$

$$\sin B \approx 0.9743$$

There are two angles possible:

$B_1 \approx 77°$, $B_2 \approx 180° - 77° = 103°$

There are two triangles:

$C_1 = 180° - B_1 - A \approx 180° - 77° - 60° = 43°$

$C_2 = 180° - B_2 - A \approx 180° - 103° - 60° = 17°$ Use the Law of Sines to find c_1 and c_2.

$$\frac{c_1}{\sin C_1} = \frac{a}{\sin A}$$

$$\frac{c_1}{\sin 43°} = \frac{16}{\sin 60°}$$

$$c_1 = \frac{16 \sin 43°}{\sin 60°} \approx 12.6$$

$$\frac{c_2}{\sin C_2} = \frac{a}{\sin A}$$

$$\frac{c_2}{\sin 17°} = \frac{16}{\sin 60°}$$

$$c_2 = \frac{16 \sin 17°}{\sin 60°} \approx 5.4$$

In one triangle, the solution is

$B_1 \approx 77°$, $C_1 \approx 43°$, and $c_1 \approx 12.6$.

In the other triangle,

$B_2 \approx 103°$, $C_2 \approx 17°$, and $c_2 \approx 5.4$.

27. The known ratio is $\dfrac{a}{\sin A}$, or $\dfrac{12}{\sin 37°}$.

Use the Law of Sines to find angle B.

$$\frac{a}{\sin A} = \frac{b}{\sin B}$$

$$\frac{12}{\sin 37°} = \frac{16.1}{\sin B}$$

$$12 \sin B = 16.1 \sin 37°$$

$$\sin B = \frac{16.1 \sin 37°}{12}$$

$$\sin B \approx 0.8074$$

There are two angles possible:

$B_1 \approx 54°$, $B_2 \approx 180° - 54° = 126°$

Additional Topics in Trigonometry

There are two triangles:
$C_1 = 180° - B_1 - A \approx 180° - 54° - 37° = 89°$
$C_2 = 180° - B_2 - A \approx 180° - 126° - 37° = 17°$ Use the
Law of Sines to find c_1 and c_2.

$$\frac{c_1}{\sin C_1} = \frac{a}{\sin A}$$

$$\frac{c_1}{\sin 89°} = \frac{12}{\sin 37°}$$

$$c_1 = \frac{12 \sin 89°}{\sin 37°} \approx 19.9$$

$$\frac{c_2}{\sin C_2} = \frac{a}{\sin A}$$

$$\frac{c_2}{\sin 17°} = \frac{12}{\sin 37°}$$

$$c_2 = \frac{12 \sin 17°}{\sin 37°} \approx 5.8$$

In one triangle, the solution is
$B_1 \approx 54°$, $C_1 \approx 89°$, and $c_1 \approx 19.9$.
In the other triangle,
$B_2 \approx 126°$, $C_2 \approx 17°$, and $c_2 \approx 5.8$.

29. The known ratio is $\dfrac{a}{\sin A}$, or $\dfrac{22}{\sin 58°}$.
Use the Law of Sines to find angle C.

$$\frac{a}{\sin A} = \frac{c}{\sin C}$$

$$\frac{22}{\sin 58°} = \frac{24.1}{\sin C}$$

$$22 \sin C = 24.1 \sin 58°$$

$$\sin C = \frac{24.1 \sin 58°}{22}$$

$$\sin C \approx 0.9290$$

There are two angles possible:
$C_1 \approx 68°$, $C_2 \approx 180° - 68° = 112°$

There are two triangles:
$B_1 = 180° - C_1 - A \approx 180° - 68° - 58° = 54°$
$B_2 = 180° - C_2 - A \approx 180° - 112° - 58° = 10°$ Use the
Law of Sines to find b_1 and b_2.

$$\frac{b_1}{\sin B_1} = \frac{a}{\sin A}$$

$$\frac{b_1}{\sin 54°} = \frac{22}{\sin 58°}$$

$$b_1 = \frac{22 \sin 54°}{\sin 58°} \approx 21.0$$

$$\frac{b_2}{\sin B_2} = \frac{a}{\sin A}$$

$$\frac{b_2}{\sin 10°} = \frac{22}{\sin 58°}$$

$$b_2 = \frac{22 \sin 10°}{\sin 58°} \approx 4.5$$

In one triangle, the solution is
$C_1 \approx 68°$, $B_1 \approx 54°$, and $b_1 \approx 21.0$.
In the other triangle,
$C_2 \approx 112°$, $B_2 \approx 10°$, and $b_2 \approx 4.5$.

31. The known ratio is $\dfrac{a}{\sin A}$, or $\dfrac{9.3}{\sin 18°}$.
Use the Law of Sines to find angle B.

$$\frac{a}{\sin A} = \frac{b}{\sin B}$$

$$\frac{9.3}{\sin 18°} = \frac{41}{\sin B}$$

$$9.3 \sin B = 41 \sin 18°$$

$$\sin B = \frac{41 \sin 18°}{9.3} \approx 1.36$$

Because the sine can never exceed 1, there is no
angle B for which $\sin B = 1.36$. There is no triangle
with the given measurements.

33. Area $= \dfrac{1}{2} bc \sin A = \dfrac{1}{2}(20)(40)(\sin 48°) \approx 297$
The area of the triangle is approximately
297 square feet.

35. Area $= \dfrac{1}{2} ac \sin B = \dfrac{1}{2}(3)(6)(\sin 36°) \approx 5$
The area of the triangle is approximately
5 square yards.

37. Area $= \dfrac{1}{2} ab \sin C = \dfrac{1}{2}(4)(6)(\sin 124°) \approx 10$
The area of the triangle is approximately
10 square meters.

39. $\angle ABC = 180° - 67° = 113°$

$\angle ACB = 180° - 43° - 113° = 24°$

Use the law of sines to find $\overline{BC}$.

$$\frac{\overline{BC}}{\sin 43°} = \frac{312}{\sin 24°}$$

$$\overline{BC} = \frac{312\sin 43°}{\sin 24°}$$

$$\overline{BC} \approx 523.1$$

Use the law of sines to find h.

$$\frac{h}{\sin 67°} = \frac{523.1}{\sin 90°}$$

$$h = \frac{523.1\sin 67°}{\sin 90°}$$

$$h \approx 481.5$$

41. Begin by finding the six angles inside the two triangles. Then use the law of sines.

$$\frac{a}{\sin 4°} = \frac{\dfrac{450\sin 145°}{\sin 34°}}{\sin 30°}$$

$$a \approx 64.4$$

43.

$$\frac{a}{\sin A} = \frac{b}{\sin B}$$

$$\frac{300}{\sin 2\theta} = \frac{200}{\sin \theta}$$

$$200\sin 2\theta = 300\sin \theta$$

$$400\sin \theta \cos \theta = 300\sin \theta$$

$$\cos \theta = \frac{300\sin \theta}{400\sin \theta}$$

$$\cos \theta = \frac{3}{4}$$

$$\theta \approx 41°$$

$$2\theta \approx 82°$$

$$A \approx 82°, B \approx 41°, C \approx 57°, c \approx 255.7$$

45.

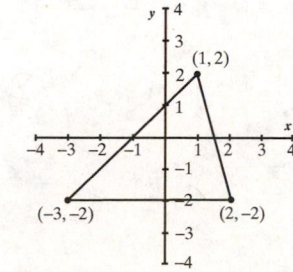

$$A = \frac{1}{2}bh$$

$$= \frac{1}{2}(5)(4)$$

$$= 10$$

47.

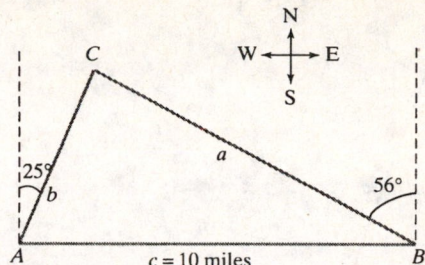

Using a north-south line, the interior angles are found as follows:

$A = 90° - 25° = 65°$

$B = 90° - 56° = 34°$

Find angle C using a $180°$ angle sum in the triangle.

$C = 180° - A - B = 180° - 65° - 34° = 81°$

The ratio $\dfrac{c}{\sin C}$, or $\dfrac{10}{\sin 81°}$, is now known. Use this ratio and the Law of Sines to find b and a.

$$\frac{b}{\sin B} = \frac{c}{\sin C}$$

$$\frac{b}{\sin 34°} = \frac{10}{\sin 81°}$$

$$b = \frac{10\sin 34°}{\sin 81°} \approx 6$$

Station A is about 6 miles from the fire.

$$\frac{a}{\sin A} = \frac{c}{\sin C}$$

$$\frac{a}{\sin 65°} = \frac{10}{\sin 81°}$$

$$a = \frac{10\sin 65°}{\sin 81°} \approx 9$$

Station B is about 9 miles from the fire.

49.

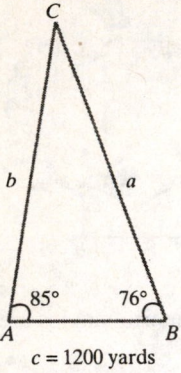

Using the figure,
$C = 180° - A - B = 180° - 85° - 76° = 19°$

The ratio $\dfrac{c}{\sin C}$, or $\dfrac{1200}{\sin 19°}$, is now known. Use this

ratio and the Law of Sines to find a and b.

$$\frac{a}{\sin A} = \frac{c}{\sin C}$$

$$\frac{a}{\sin 85°} = \frac{1200}{\sin 19°}$$

$$a = \frac{1200 \sin 85°}{\sin 19°} \approx 3672$$

$$\frac{b}{\sin B} = \frac{c}{\sin C}$$

$$\frac{b}{\sin 76°} = \frac{1200}{\sin 19°}$$

$$b = \frac{1200 \sin 76°}{\sin 19°} \approx 3576$$

The platform is about 3672 yards from one end of the beach and 3576 yards from the other.

51. According to the figure,
$C = 180° - A - B = 180° - 84.7° - 50° = 45.3°$ The

ratio $\dfrac{c}{\sin C}$, or $\dfrac{171}{\sin 45.3°}$, is now known. Use this

ratio and the Law of Sines to find b.

$$\frac{b}{\sin B} = \frac{c}{\sin C}$$

$$\frac{b}{\sin 50°} = \frac{171}{\sin 45.3°}$$

$$b = \frac{171 \sin 50°}{\sin 45.3°} \approx 184$$

The distance is about 184 feet.

53. The ratio $\dfrac{b}{\sin B}$, or $\dfrac{562}{\sin 85.3°}$, is known.

Use this ratio, the figure, and the Law of Sines to find c.

$$\frac{c}{\sin C} = \frac{b}{\sin B}$$

$$\frac{c}{\sin 5.7°} = \frac{562}{\sin 85.3°}$$

$$c = \frac{562 \sin 5.7°}{\sin 85.3°} \approx 56$$

The toss was about 56 feet.

55.

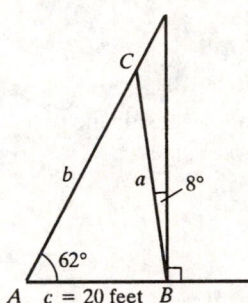

Using the figure,
$B = 90° - 8° = 82°$

$C = 180° - A - B = 180° - 62° - 82° = 36°$

The ratio $\dfrac{c}{\sin C}$, or $\dfrac{20}{\sin 36°}$, is now known. Use this

ratio and the Law of Sines to find a.

$$\frac{a}{\sin A} = \frac{c}{\sin C}$$

$$\frac{a}{\sin 62°} = \frac{20}{\sin 36°}$$

$$a = \frac{20 \sin 62°}{\sin 36°} \approx 30$$

The length of the pole is about 30 feet.

57. a. Using the figure and the measurements shown,
$B = 180° - 44° = 136°$

$C = 180° - B - A = 180° - 136° - 37° = 7°$

The ratio $\dfrac{c}{\sin C}$, or $\dfrac{100}{\sin 7°}$, is now known. Use this ratio and the Law of Sines to find a.

$$\frac{a}{\sin A} = \frac{c}{\sin C}$$

$$\frac{a}{\sin 37°} = \frac{100}{\sin 7°}$$

$$a = \frac{100 \sin 37°}{\sin 7°} \approx 494$$

To the nearest foot, $a = 494$ feet.

Let $a = 494$ be the hypotenuse of the right triangle. Then if h represents the height of the tree,

$$\frac{h}{\sin 44°} = \frac{494}{\sin 90°}$$

$$h = \frac{494 \sin 44°}{\sin 90°} \approx 343$$

A typical redwood tree is about 343 feet.

59.

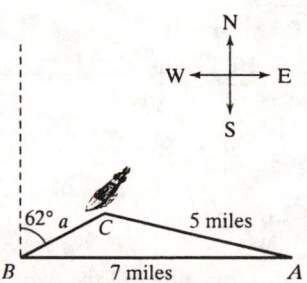

Using the figure,
$B = 90° - 62° = 28°$

The known ratio is $\dfrac{b}{\sin B}$, or $\dfrac{5}{\sin 28°}$.

Use the Law of Sines to find angle C.

$$\frac{b}{\sin B} = \frac{c}{\sin C}$$

$$\frac{5}{\sin 28°} = \frac{7}{\sin C}$$

$$5 \sin C = 7 \sin 28°$$

$$\sin C = \frac{7 \sin 28°}{5} \approx 0.6573$$

There are two angles possible:
$C_1 \approx 41°$, $C_2 \approx 180° - 41° = 139°$

There are two triangles:

$A_1 = 180° - C_1 - B \approx 180° - 41° - 28° = 111°$

$A_2 = 180° - C_2 - B \approx 180° - 139° - 28° = 13°$ Use the

Law of Sines to find a_1 and a_2.

$$\frac{a_1}{\sin A_1} = \frac{b}{\sin B}$$

$$\frac{a_1}{\sin 111°} = \frac{5}{\sin 28°}$$

$$a_1 = \frac{5 \sin 111°}{\sin 28°} \approx 9.9$$

$$\frac{a_2}{\sin A_2} = \frac{b}{\sin B}$$

$$\frac{a_2}{\sin 13°} = \frac{5}{\sin 28°}$$

$$a_2 = \frac{5 \sin 13°}{\sin 28°} \approx 2.4$$

The boat is either 9.9 miles or 2.4 miles from lighthouse B, to the nearest tenth of a mile.

61. – 69. Answers may vary.

71. does not make sense; Explanations will vary. Sample explanation: The law of cosines would be appropriate for this situation.

73. does not make sense; Explanations will vary. Sample explanation: The calculator will give you the acute angle. The obtuse angle is the supplement of the acute angle.

75. No. Explanations may vary.

77.

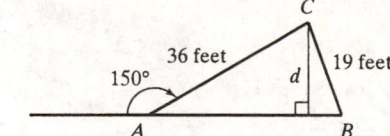

Using the figure,
$A = 180° - 150° = 30°$

Using the Law of Sines we have,

$$\frac{d}{\sin A} = \frac{36}{\sin 90°}$$

$$\frac{d}{\sin 30°} = \frac{36}{\sin 90°}$$

$$d = \frac{36 \sin 30°}{\sin 90°} = 18$$

$CC' = 18 + 5 + 18 = 41$

The wingspan CC' is 41 feet.

78. $\cos B = \dfrac{6^2 + 4^2 - 9^2}{2 \cdot 6 \cdot 4}$

$\cos B = \dfrac{-29}{48}$

$\cos B = \dfrac{-29}{48}$

$B = \cos^{-1}\left(\dfrac{-29}{48}\right)$

$B \approx 127°$

79. $\sqrt{26(26-12)(26-16)(26-24)}$

$= \sqrt{26(14)(10)(2)}$

$= \sqrt{7280}$

$= 4\sqrt{455}$

≈ 85

80.

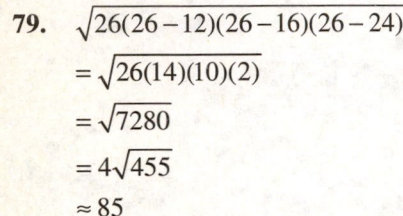

Section 6.2

Check Point Exercises

1. Apply the three-step procedure for solving a SAS triangle. Use the Law of Cosines to find the side opposite the given angle. Thus, we will find a.

$a^2 = b^2 + c^2 - 2bc \cos A$

$a^2 = 7^2 + 8^2 - 2(7)(8)\cos 120°$

$\quad = 49 + 64 - 112(-0.5)$

$\quad = 169$

$a = \sqrt{169} = 13$

Use the Law of Sines to find the angle opposite the shorter of the two sides. Thus, we will find acute angle B.

$\dfrac{b}{\sin B} = \dfrac{a}{\sin A}$

$\dfrac{7}{\sin B} = \dfrac{13}{\sin 120°}$

$13 \sin B = 7 \sin 120°$

$\sin B = \dfrac{7 \sin 120°}{13} \approx 0.4663$

$B \approx 28°$

Find the third angle.

$C = 180° - A - B \approx 180° - 120° - 28° = 32°$

The solution is $a = 13, B \approx 28°,$ and $C \approx 32°$.

2. Apply the three-step procedure for solving a SSS triangle. Use the Law of Cosines to find the angle opposite the longest side.

Thus, we will find angle B.

$b^2 = a^2 + c^2 - 2ac \cos B$

$2ac \cos B = a^2 + c^2 - b^2$

$\cos B = \dfrac{a^2 + c^2 - b^2}{2ac}$

$\cos B = \dfrac{8^2 + 5^2 - 10^2}{2 \cdot 8 \cdot 5} = -\dfrac{11}{80}$

$\cos^{-1}\left(\dfrac{11}{80}\right) \approx 82.1°$

B is obtuse, since $\cos B$ is negative.

$B \approx 180° - 82.1° = 97.9°$

Use the Law of Sines to find either of the two remaining acute angles. We will find angle A.

$\dfrac{a}{\sin A} = \dfrac{b}{\sin B}$

$\dfrac{8}{\sin A} = \dfrac{10}{\sin 97.9°}$

$10 \sin A = 8 \sin 97.9°$

$\sin A = \dfrac{8 \sin 97.9°}{10} \approx 0.7924$

$A \approx 52.4°$

Find the third angle.

$C = 180° - A - B \approx 180° - 52.4° - 97.9°$

$\quad = 29.7°$

The solution is $B \approx 97.9°, A \approx 52.4°,$ and $C \approx 29.7°$

3. The plane flying 400 miles per hour travels $400 \cdot 2 = 800$ miles in 2 hours. Similarly, the other plane travels 700 miles.

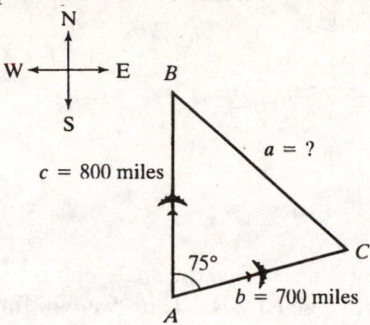

Use the figure and the Law of Cosines to find a in this SAS situation.

$a^2 = b^2 + c^2 - 2bc \cos A$

$a^2 = 700^2 + 800^2 - 2(700)(800)\cos 75°$

$\approx 840,123$

$a \approx \sqrt{840,123} \approx 917$

After 2 hours, the planes are approximately 917 miles apart.

4. Begin by calculating one-half the perimeter:

$s = \dfrac{1}{2}(a+b+c) = \dfrac{1}{2}(6+16+18) = 20$

Use Heron's formula to find the area.

$\text{Area} = \sqrt{s(s-a)(s-b)(s-c)}$

$= \sqrt{20(20-6)(20-16)(20-18)}$

$= \sqrt{2240} \approx 47$

The area of the triangle is approximately 47 square meters.

Exercise Set 6.2

1. Apply the three-step procedure for solving a SAS triangle. Use the Law of Cosines to find the side opposite the given angle.
Thus, we will find a.

$a^2 = b^2 + c^2 - 2bc \cos A$

$a^2 = 4^2 + 8^2 - 2(4)(8)\cos 46°$

$a^2 = 16 + 64 - 64(\cos 46°)$

$a^2 \approx 35.54$

$a \approx \sqrt{35.54} \approx 6.0$

Use the Law of Sines to find the angle opposite the shorter of the two given sides. Thus, we will find acute angle B.

$\dfrac{b}{\sin B} = \dfrac{a}{\sin A}$

$\dfrac{4}{\sin B} = \dfrac{\sqrt{35.54}}{\sin 46°}$

$\sqrt{35.54} \sin B = 4 \sin 46°$

$\sin B = \dfrac{4\sin 46°}{\sqrt{35.54}} \approx 0.4827$

$B \approx 29°$

Find the third angle.

$C = 180° - A - B \approx 180° - 46° - 29° = 105°$

The solution is $a \approx 6.0, B \approx 29°,$ and $C \approx 105°$.

3. Apply the three-step procedure for solving a SAS triangle. Use the Law of Cosines to find the side opposite the given angle.
Thus, we will find c.

$c^2 = a^2 + b^2 - 2ab \cos C$

$c^2 = 6^2 + 4^2 - 2(6)(4)\cos 96°$

$c^2 = 36 + 16 - 48(\cos 96°)$

$c^2 \approx 57.02$

$c \approx \sqrt{57.02} \approx 7.6$

Use the Law of Sines to find the angle opposite the shorter of the two given sides. Thus, we will find acute angle B.

$\dfrac{b}{\sin B} = \dfrac{c}{\sin C}$

$\dfrac{4}{\sin B} = \dfrac{\sqrt{57.02}}{\sin 96°}$

$\sqrt{57.02} \sin B = 4 \sin 96°$

$\sin B = \dfrac{4\sin 96°}{\sqrt{57.02}} \approx 0.5268$

$B \approx 32°$

Find the third angle.

$A = 180° - B - C \approx 180° - 32° - 96° = 52°$

The solution is $c \approx 7.6, A \approx 52°,$ and $B \approx 32°$.

5. Apply the three-step procedure for solving a SSS triangle. Use the Law of Cosines to find the angle opposite the longest side. Since two sides have length 8, we can begin by finding angle B or C.

$b^2 = a^2 + c^2 - 2ac \cos B$

$\cos B = \dfrac{a^2 + c^2 - b^2}{2ac}$

$\cos B = \dfrac{6^2 + 8^2 - 8^2}{2 \cdot 6 \cdot 8} = \dfrac{36}{96} = \dfrac{3}{8}$

$B \approx 68°$

Use the Law of Sines to find either of the two remaining acute angles. We will find angle A.

$$\frac{a}{\sin A} = \frac{b}{\sin B}$$

$$\frac{6}{\sin A} = \frac{8}{\sin 68°}$$

$$8\sin A = 6\sin 68°$$

$$\sin A = \frac{6\sin 68°}{8} \approx 0.6954$$

$$A \approx 44°$$

Find the third angle.

$$C = 180° - B - A \approx 180° - 68° - 44° = 68°$$

The solution is $A \approx 44°, B \approx 68°,$ and $C \approx 68°$.

7. Apply the three-step procedure for solving a SSS triangle. Use the Law of Cosines to find the angle opposite the longest side. Thus, we will find angle A

$$a^2 = b^2 + c^2 - 2bc\cos A$$

$$\cos A = \frac{b^2 + c^2 - a^2}{2bc}$$

$$\cos A = \frac{4^2 + 3^2 - 6^2}{2 \cdot 4 \cdot 3} = -\frac{11}{24}$$

A is obtuse, since $\cos A$ is negative.

$$\cos^{-1}\left(\frac{11}{24}\right) \approx 63°$$

$$A \approx 180° - 63° = 117°$$

Use the Law of Sines to find either of the two remaining acute angles. We will find angle B.

$$\frac{b}{\sin B} = \frac{a}{\sin A}$$

$$\frac{4}{\sin B} = \frac{6}{\sin 117°}$$

$$6\sin B = 4\sin 117°$$

$$\sin B = \frac{4\sin 117°}{6} \approx 0.5940$$

$$B \approx 36°$$

Find the third angle.

$$C = 180° - B - A \approx 180° - 36° - 117° = 27°$$

The solution is $A \approx 117°, B \approx 36°,$ and $C \approx 27°$.

9. Apply the three-step procedure for solving a SAS triangle. Use the Law of Cosines to find the side opposite the given angle.
Thus, we will find c.

$$c^2 = a^2 + b^2 - 2ab\cos C$$

$$c^2 = 5^2 + 7^2 - 2(5)(7)\cos 42°$$

$$c^2 = 25 + 49 - 70(\cos 42°)$$

$$c^2 \approx 21.98$$

$$c \approx \sqrt{21.98} \approx 4.7$$

Use the Law of Sines to find the angle opposite the shorter of the two given sides. Thus, we will find acute angle A.

$$\frac{a}{\sin A} = \frac{c}{\sin C}$$

$$\frac{5}{\sin A} = \frac{\sqrt{21.98}}{\sin 42°}$$

$$\sqrt{21.98}\sin A = 5\sin 42°$$

$$\sin A = \frac{5\sin 42°}{\sqrt{21.98}} \approx 0.7136$$

$$A \approx 46°$$

Find the third angle.

$$B = 180° - C - A \approx 180° - 42° - 46° = 92°$$

The solution is $c \approx 4.7, A \approx 46°,$ and $B \approx 92°$.

11. Apply the three-step procedure for solving a SAS triangle. Use the Law of Cosines to find the side opposite the given angle.
Thus, we will find a.

$$a^2 = b^2 + c^2 - 2bc\cos A$$

$$a^2 = 5^2 + 3^2 - 2(5)(3)\cos 102°$$

$$a^2 = 25 + 9 - 30(\cos 102°)$$

$$a^2 \approx 40.24$$

$$a \approx \sqrt{40.24} \approx 6.3$$

Use the Law of Sines to find the angle opposite the shorter of the two given sides. Thus, we will find acute angle C.

$$\frac{c}{\sin C} = \frac{a}{\sin A}$$

$$\frac{3}{\sin C} = \frac{\sqrt{40.24}}{\sin 102°}$$

$$\sqrt{40.24}\sin C = 3\sin 102°$$

$$\sin C = \frac{3\sin 102°}{\sqrt{40.24}} \approx 0.4626$$

$$C \approx 28°$$

Find the third angle.

$$B = 180° - C - A \approx 180° - 28° - 102° = 50°$$

The solution is $a \approx 6.3, C \approx 28°,$ and $B \approx 50°$.

13. Apply the three-step procedure for solving a SAS triangle. Use the Law of Cosines to find the side opposite the given angle.
Thus, we will find b.

$b^2 = a^2 + c^2 - 2ac \cos B$

$b^2 = 6^2 + 5^2 - 2(6)(5) \cos 50°$

$b^2 = 36 + 25 - 60(\cos 50°)$

$b^2 \approx 22.43$

$b \approx \sqrt{22.43} \approx 4.7$

Use the Law of Sines to find the angle opposite the shorter of the two given sides. Thus, we will find acute angle C.

$\dfrac{c}{\sin C} = \dfrac{b}{\sin B}$

$\dfrac{5}{\sin C} = \dfrac{\sqrt{22.43}}{\sin 50°}$

$\sqrt{22.43} \sin C = 5 \sin 50°$

$\sin C = \dfrac{5 \sin 50°}{\sqrt{22.43}} \approx 0.8087$

$C \approx 54°$

Find the third angle.
$A = 180° - C - B \approx 180° - 54° - 50° = 76°$
The solution is $b \approx 4.7, C \approx 54°,$ and $A \approx 76°$.

15. Apply the three-step procedure for solving a SAS triangle. Use the Law of Cosines to find the side opposite the given angle.
Thus, we will find b.

$b^2 = a^2 + c^2 - 2ac \cos 90°$

$b^2 = 5^2 + 2^2 - 2(5)(2) \cos 90°$

$b^2 = 25 + 4 - 20 \cos 90°$

$b^2 = 29$

$b = \sqrt{29} \approx 5.4$

(use exact value of b from previous step) Use the Law of Sines to find the angle opposite the shorter of the two given sides. Thus, we will find acute angle C.

$\dfrac{c}{\sin C} = \dfrac{b}{\sin B}$

$\dfrac{2}{\sin C} = \dfrac{\sqrt{29}}{\sin 90°}$

$\sqrt{29} \sin C = 2 \sin 90°$

$\sin C = \dfrac{2 \sin 90°}{\sqrt{29}} \approx 0.3714$

$C \approx 22°$

Find the third angle.
$A = 180° - C - B \approx 180° - 22° - 90° = 68°$
The solution is $b \approx 5.4, C \approx 22°,$ and $A \approx 68°$.

17. Apply the three-step procedure for solving a SSS triangle. Use the Law of Cosines to find the angle opposite the longest side. Thus, we will find C.

$c^2 = a^2 + b^2 - 2ab \cos C$

$\cos C = \dfrac{a^2 + b^2 - c^2}{2ab}$

$\cos C = \dfrac{5^2 + 7^2 - 10^2}{2 \cdot 5 \cdot 7} = -\dfrac{13}{35}$

C is obtuse, since cos C is negative.

$\cos^{-1}\left(\dfrac{13}{35}\right) \approx 68°$

$C \approx 180° - 68° = 112°$

Use the Law of Sines to find either of the two remaining angles. We will find angle A.

$\dfrac{a}{\sin A} = \dfrac{c}{\sin C}$

$\dfrac{5}{\sin A} = \dfrac{10}{\sin 112°}$

$10 \sin A = 5 \sin 112°$

$\sin A = \dfrac{5 \sin 112°}{10} \approx 0.4636$

$A \approx 28°$

Find the third angle.
$B = 180° - C - A \approx 180° - 112° - 28° = 40°$
The solution is $C \approx 112°, A \approx 28°,$ and $B \approx 40°$.

19. Apply the three-step procedure for solving a SSS triangle. Use the Law of Cosines to find the angle opposite the longest side. Thus, we will find B.

$b^2 = a^2 + c^2 - 2ac \cos B$

$\cos B = \dfrac{a^2 + c^2 - b^2}{2ac}$

$\cos B = \dfrac{3^2 + 8^2 - 9^2}{2 \cdot 3 \cdot 8} = -\dfrac{1}{6}$

B is obtuse, since cos B is negative.

$\cos^{-1}\left(\dfrac{1}{6}\right) \approx 80°$

$B \approx 180° - 80° = 100°$

Use the Law of Sines to find either of the two remaining angles. We will find angle A.

$$\frac{a}{\sin A} = \frac{b}{\sin B}$$

$$\frac{3}{\sin A} = \frac{9}{\sin 100°}$$

$$9 \sin A = 3 \sin 100°$$

$$\sin A = \frac{3 \sin 100°}{9} \approx 0.3283$$

$$A \approx 19°$$

Find the third angle.

$$C = 180° - B - A \approx 180° - 100° - 19° = 61°$$

The solution is $B \approx 100°, A \approx 19°$, and $C \approx 61°$.

21. Apply the three-step procedure for solving a SSS triangle. Use the Law of Cosines to find any of the three angles, since each side has the same measure.

$$a^2 = b^2 + c^2 - 2bc \cos A$$

$$\cos A = \frac{b^2 + c^2 - a^2}{2bc}$$

$$\cos A = \frac{3^2 + 3^2 - 3^2}{2 \cdot 3 \cdot 3} = \frac{1}{2}$$

$$A = 60°$$

Use the Law of Sines to find either of the two remaining angles. We will find angle B.

$$\frac{b}{\sin B} = \frac{a}{\sin A}$$

$$\frac{3}{\sin B} = \frac{3}{\sin 60°}$$

$$3 \sin B = 3 \sin 60°$$

$$\sin B = \sin 60°$$

$$B = 60°$$

Find the third angle.

$$C = 180° - A - B = 180° - 60° - 60° = 60°$$

The solution is $A = 60°, B = 60°$, and $C = 60°$.

23. Apply the three-step procedure for solving a SSS triangle. Use the Law of Cosines to find the angle opposite the longest side. Thus, we will find A.

$$a^2 = b^2 + c^2 - 2bc \cos A$$

$$\cos A = \frac{b^2 + c^2 - a^2}{2bc}$$

$$\cos A = \frac{22^2 + 50^2 - 63^2}{2 \cdot 22 \cdot 50} = -\frac{985}{2200}$$

$$A \approx 117°$$

Use the Law of Sines to find either of the two remaining angles. We will find angle B.

$$\frac{b}{\sin B} = \frac{a}{\sin A}$$

$$\frac{22}{\sin B} = \frac{63}{\sin 117°}$$

$$63 \sin B = 22 \sin 117°$$

$$\sin B = \frac{22 \sin 117°}{63}$$

$$B = 18°$$

Find the third angle.

$$C = 180° - A - B = 180° - 117° - 18° = 45°$$

The solution is $A = 117°, B = 18°$, and $C = 45°$.

25. $s = \dfrac{1}{2}(a + b + c) = \dfrac{1}{2}(4 + 4 + 2) = 5$

Area $= \sqrt{s(s-a)(s-b)(s-c)}$

$\qquad = \sqrt{5(5-4)(5-4)(5-2)}$

$\qquad = \sqrt{15} \approx 4$

The area of the triangle is approximately 4 square feet.

27. $s = \dfrac{1}{2}(a + b + c) = \dfrac{1}{2}(14 + 12 + 4) = 15$

Area $= \sqrt{s(s-a)(s-b)(s-c)}$

$\qquad = \sqrt{15(15-14)(15-12)(15-11)}$

$\qquad = \sqrt{495} \approx 22$

The area of the triangle is approximately 22 square meters.

29. $s = \dfrac{1}{2}(a + b + c) = \dfrac{1}{2}(11 + 9 + 7) = 13.5$

Area $= \sqrt{s(s-a)(s-b)(s-c)}$

$\qquad = \sqrt{13.5(13.5-11)(13.5-9)(13.5-7)}$

$\qquad = \sqrt{987.1875} \approx 31$

The area of the triangle is approximately 31 square yards.

31. $C = 180° - 15° - 35° = 130°$

$c^2 = b^2 + c^2 - 2bc \cos C$

$c^2 = 8^2 + 13^2 - 2(8)(13) \cos 130°$

$c^2 \approx 366.6998$

$c \approx 19.1$

Use the law of sines to find the solution is
$A \approx 31°, B \approx 19°, C \approx 130°,$ and $c \approx 19.1..$

33. Use the given radii to determine that
$a = 7.5, b = 8.5,$ and $c = 9.0$.

$c^2 = a^2 + b^2 - 2ab \cos C$

$9^2 = 7.5^2 + 8.5^2 - 2(7.5)(8.5) \cos C$

$\cos C \approx 0.3725$

$C \approx 68°$

Use the law of sines to find the solution is
$A \approx 51°, B \approx 61°,$ and $C \approx 68°$.

35. Use the distance formula to determine that
$a = \sqrt{61} \approx 7.8, b = \sqrt{10} \approx 3.2,$ and $c = 5$.

$a^2 = b^2 + c^2 - 2bc \cos A$

$\sqrt{61}^2 = \sqrt{10}^2 + 5^2 - 2\left(\sqrt{10}\right)(5) \cos A$

$\cos A \approx -0.8222$

$A \approx 145°$

Use the law of sines to find the solution is
$A \approx 145°, B \approx 13°,$ and $C \approx 22°$.

37. Use the law of cosines.

$c^2 = a^2 + b^2 - 2ab \cos C$

$5.78^2 = 2.9^2 + 3.0^2 - 2(2.9)(3.0) \cos \theta$

$\cos \theta \approx -0.9194$

$\theta \approx 157°$

This dinosaur was an efficient walker.

39. Let b = the distance between the ships after three hours.
After three hours, the ship traveling 14 miles per hour has gone $3 \cdot 14$ or 42 miles. Similarly, the ship traveling 10 miles per hour has gone 30 miles.

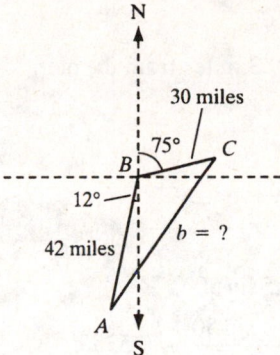

Using the figure,

$B = 180° - 75° + 12° = 117°$

$b^2 = a^2 + c^2 - 2ac \cos B$

$b^2 = 30^2 + 42^2 - 2(30)(42) \cos 117° \approx 3808$

$b \approx 61.7$

After three hours, the ships will be about 61.7 miles apart.

41. Let b = the distance across the lake.

$b^2 = a^2 + c^2 - 2ac \cos B$

$b^2 = 160^2 + 140^2 - 2(160)(140) \cos 80°$

$\approx 37,421$

$b \approx \sqrt{37,421} \approx 193$

The distance across the lake is about 193 yards.

43. Assume that Island B is due east of Island A. Let A = angle at Island A.

$a^2 = b^2 + c^2 - 2bc \cos A$

$\cos A = \dfrac{b^2 + c^2 - a^2}{2bc}$

$\cos A = \dfrac{5^2 + 6^2 - 7^2}{2 \cdot 5 \cdot 6} = \dfrac{1}{5}$

$A \approx 78°$

Since $90° - 78° = 12°$, you should navigate on a bearing of N12°E.

45. a. Using the figure,
$$B = 90° - 40° = 50°$$
$$b^2 = a^2 + c^2 - 2ac\cos B$$
$$b^2 = 13.5^2 + 25^2 - 2(13.5)(25)\cos 50°$$
$$\approx 373$$
$$b \approx \sqrt{373} \approx 19.3$$
You are about 19.3 miles from the pier.

b.
$$\frac{a}{\sin A} = \frac{b}{\sin B}$$
$$\frac{13.5}{\sin A} = \frac{\sqrt{373}}{\sin 50°}$$
$$\sqrt{373}\sin A = 13.5\sin 50°$$
$$\sin A = \frac{13.5\sin 50°}{\sqrt{373}} \approx 0.5355$$
$$A \approx 32°$$
Since $90° - 32° = 58°$, the original bearing could have been S58°E.

47.

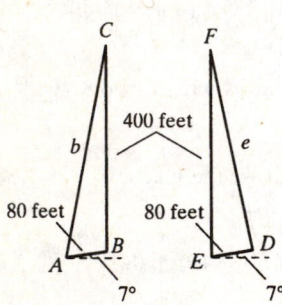

In the figure, b = the guy wire anchored downhill, e = the guy wire anchored uphill.
$$B = 90° + 7° = 97°$$
$$E = 90° - 7° = 83°$$
$$b^2 = a^2 + c^2 - 2ac\cos B$$
$$b^2 = 400^2 + 80^2 - 2(400)(80)\cos 97°$$
$$\approx 174,200$$
$$b \approx \sqrt{174,200} \approx 417.4$$
$$e^2 = d^2 + f^2 - 2df\cos E$$
$$e^2 = 400^2 + 80^2 - 2(400)(80)\cos 83°$$
$$\approx 158,600$$
$$e \approx \sqrt{158.600} \approx 398.2$$
The guy wire anchored downhill is about 417.4 feet long. The one anchored uphill is about 398.2 feet long.

49.

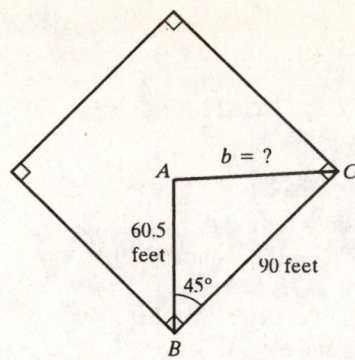

Using the figure,
$$B = 90° \div 2 = 45° \text{ (using symmetry)}$$
$$b^2 = a^2 + c^2 - 2ac\cos B$$
$$b^2 = 90^2 + 60.5^2 - 2(90)(60.5)\cos 45°$$
$$\approx 4060$$
$$b \approx \sqrt{4060} \approx 63.7$$
It is about 63.7 feet from the pitcher's mound to first base.

51. First, find the area using Heron's formula.
$$s = \frac{1}{2}(a+b+c) = \frac{1}{2}(240+300+420) = 480$$
$$\text{Area} = \sqrt{s(s-a)(s-b)(s-c)}$$
$$= \sqrt{480(480-240)(480-300)(480-420)}$$
$$= \sqrt{1,244,160,000} \approx 35,272.65$$
Now multiply by the price per square foot.
$$(35,272.65)(3.50) \approx 123,454$$
The cost is $123,454, to the nearest dollar.

53. – 59. Answers may vary.

61. makes sense

63. makes sense

65. If we call the lower left point D, and the lower right point E, then the Law of Cosines will give all three angles in triangle ADE and triangle ABE. That allows us find $A \approx 29°$, $B \approx 87°$, and $C \approx 64°$. The Law of Sines will then allow us to find $a \approx 11.6$ and $b \approx 23.9$.

Let d = the distance between the tips of the hands.
$$d^2 = m^2 + h^2 - 2mh\cos 60°$$
$$= m^2 + h^2 - 2mh\left(\frac{1}{2}\right)$$
$$= m^2 + h^2 - mh$$
$$d = \sqrt{m^2 + h^2 - mh}$$

67. Answers may vary.

68. $y = 3$ is a horizontal line through (0, 3).

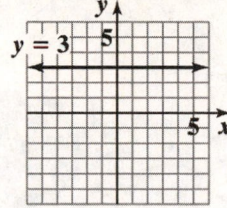

69. $x^2 + (y-1)^2 = 1$ is a circle centered at (0, 1) with a radius of 1.

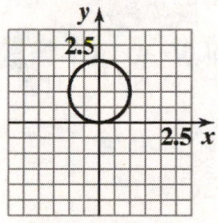

$$x^2 + (y-1)^2 = 1$$

70.

$$x^2 + 6x + y^2 = 0$$

$$x^2 + 6x \quad + y^2 = 0$$

$$x^2 + 6x + 9 + y^2 = 0 + 9$$

$$(x+3)^2 + y^2 = 9$$

$(x+3)^2 + y^2 = 9$ is a circle centered at $(-3, 0)$ with a radius of 3.

Section 6.3

Check Point Exercises

1. **a.** $(r, \theta) = (3, 315°)$

Because 315° is a positive angle, draw $\theta = 315°$ counterclockwise from the polar axis. Because $r > 0$, plot the point by going out 3 units on the terminal side of θ.

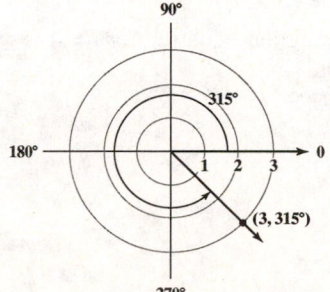

b. $(r, \theta) = (-2, \pi)$

Because π is a positive angle, draw $\theta = \pi$ counterclockwise from the polar axis. Because $r < 0$, plot the point by going out 2 units along the ray opposite the terminal side of θ.

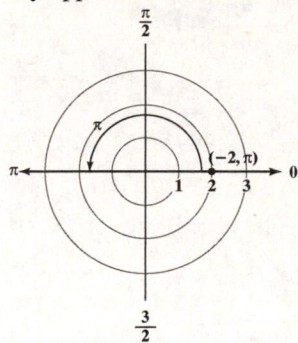

c. $(r, \theta) = \left(-1, -\dfrac{\pi}{2}\right)$

Because $-\dfrac{\pi}{2}$ is a negative angle, draw $\theta = -\dfrac{\pi}{2}$ clockwise from the polar axis. Because $r < 0$, plot the point by going out one unit along the ray opposite the terminal side of θ.

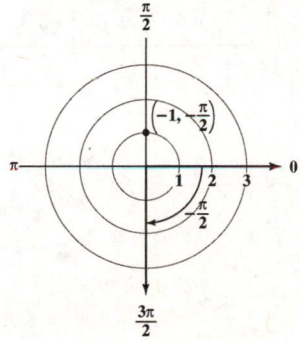

2. **a.** Add 2π to the angle and do not change r.

$$\left(5, \frac{\pi}{4}\right) = \left(5, \frac{\pi}{4} + 2\pi\right) = \left(5, \frac{\pi}{4} + \frac{8\pi}{4}\right)$$

$$= \left(5, \frac{9\pi}{4}\right)$$

b. Add π to the angle and replace r by $-r$.

$$\left(5, \frac{\pi}{4}\right) = \left(-5, \frac{\pi}{4} + \pi\right) = \left(-5, \frac{\pi}{4} + \frac{4\pi}{4}\right)$$

$$= \left(-5, \frac{5\pi}{4}\right)$$

c. Subtract 2π from the angle and do not change r.

$$\left(5, \frac{\pi}{4}\right) = \left(5, \frac{\pi}{4} - 2\pi\right) = \left(5, \frac{\pi}{4} - \frac{8\pi}{4}\right)$$

$$= \left(5, -\frac{7\pi}{4}\right)$$

3. a. $(r, \theta) = (3, \pi)$

$x = r\cos\theta = 3\cos\pi = 3(-1) = -3$

$y = r\sin\theta = 3\sin\pi = 3(0) = 0$

The rectangular coordinates of $(3, \pi)$ are $(-3, 0)$.

b. $(r, \theta) = \left(-10, \frac{\pi}{6}\right)$

$$x = r\cos\theta = -10\cos\frac{\pi}{6} = -10\left(\frac{\sqrt{3}}{2}\right)$$

$$= -5\sqrt{3}$$

$$y = r\sin\theta = -10\sin\frac{\pi}{6} = -10\left(\frac{1}{2}\right) = -5$$

The rectangular coordinates of $\left(-10, \frac{\pi}{6}\right)$ are

$\left(-5\sqrt{3}, -5\right)$.

4.

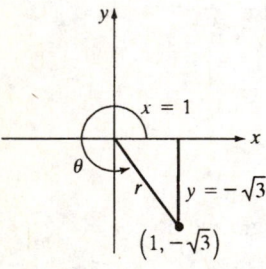

$$r = \sqrt{x^2 + y^2} = \sqrt{1^2 + \left(-\sqrt{3}\right)^2}$$

$$= \sqrt{1+3} = \sqrt{4} = 2$$

$$\tan\theta = \frac{y}{x} = \frac{-\sqrt{3}}{1} = -\sqrt{3}$$

Because $\tan\frac{\pi}{3} = \sqrt{3}$ and θ lies in quadrant IV,

$$\theta = 2\pi - \frac{\pi}{3} = \frac{6\pi}{3} - \frac{\pi}{3} = \frac{5\pi}{3}$$

The polar coordinates of $\left(1, -\sqrt{3}\right)$ are

$$(r, \theta) = \left(2, \frac{5\pi}{3}\right)$$

5.

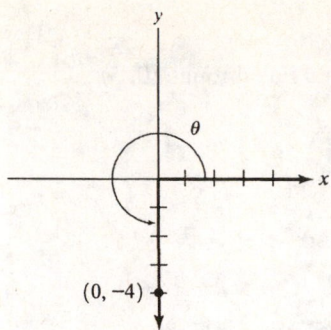

$$r = \sqrt{x^2 + y^2} = \sqrt{(0)^2 + (-4)^2} = \sqrt{16} = 4$$

The point $(0, -4)$ is on the negative y-axis. Thus,

$\theta = \frac{3\pi}{2}$. Polar coordinates of $(0, -4)$ are $\left(4, \frac{3\pi}{2}\right)$.

6. a.
$$3x - y = 6$$
$$3r\cos\theta - r\sin\theta = 6$$
$$r(3\cos\theta - \sin\theta) = 6$$
$$r = \frac{6}{3\cos\theta - \sin\theta}$$

b.
$$x^2 + (y+1)^2 = 1$$
$$(r\cos\theta)^2 + (r\sin\theta + 1)^2 = 1$$
$$r^2\cos^2\theta + r^2\sin^2\theta + 2r\sin\theta + 1 = 1$$
$$r^2 + 2r\sin\theta = 0$$
$$r(r + 2\sin\theta) = 0$$
$$r = 0 \quad \text{or} \quad r + 2\sin\theta = 0$$
$$r = -2\sin\theta$$

7. a. Use $r^2 = x^2 + y^2$ to convert to a rectangular equation.
$$r = 4$$
$$r^2 = 16$$
$$x^2 + y^2 = 16$$
The rectangular equation for $r = 4$ is
$$x^2 + y^2 = 16.$$

b.   Use $\tan\theta = \dfrac{y}{x}$ to convert to a rectangular equation in x and y.

$$\theta = \frac{3\pi}{4}$$

$$\tan\theta = \tan\frac{3\pi}{4}$$

$$\tan\theta = -1$$

$$\frac{y}{x} = -1$$

$$y = -x$$

The rectangular equation for $\theta = \dfrac{3\pi}{4}$ is $y = -x$.

c. $r = -2\sec\theta$

$$r = \frac{-2}{\cos\theta}$$

$$r\cos\theta = -2$$

$$x = -2$$

d. $r = 10\sin\theta$

$$r^2 = 10r\sin\theta$$

$$x^2 + y^2 = 10y$$

$$x^2 + y^2 - 10y = 0$$

$$x^2 + y^2 - 10y + 25 = 25$$

$$x^2 + (y-5)^2 = 25$$

Exercise Set 6.3

1.   225° is in the third quadrant.
C

3. $\dfrac{5\pi}{4} = 225°$ is in the third quadrant. Since r is

negative, the point lies along the ray opposite the terminal side of θ, in the first quadrant.
A

5. $\pi = 180°$ lies on the negative x-axis.
B

7.   −135° is measured clockwise 135° from the positive x-axis. The point lies in the third quadrant.
C

9. $-\dfrac{3\pi}{4} = -135°$ is measured clockwise 135° from the positive x-axis. Since r is negative, the point lies along the ray opposite the terminal side of θ, in the first quadrant.
A

11.   Draw $\theta = 45°$ counterclockwise, since θ is positive, from the polar axis. Go out 2 units on the terminal side of θ, since $r > 0$.

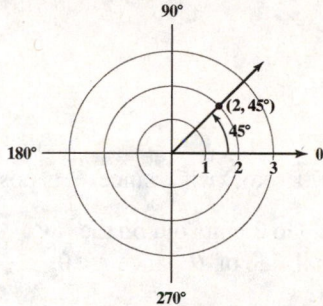

13.   Draw $\theta = 90°$ counterclockwise, since θ is positive, from the polar axis. Go out 3 units on the terminal side of θ, since $r > 0$.

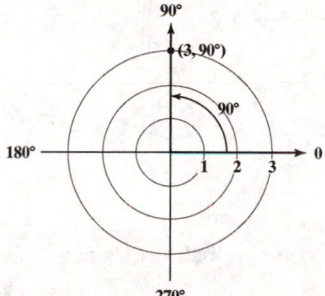

15.   Draw $\theta = \dfrac{4\pi}{3} = 240°$ counterclockwise, since θ is positive, from polar axis. Go out 3 units on the terminal side of θ, since $r > 0$.

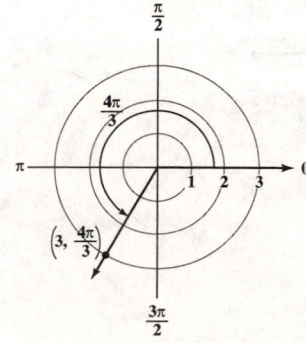

17. Draw $\theta = \pi = 180°$ counterclockwise, since θ is positive, from the polar axis. Go one unit out on the ray opposite the terminal side of θ, since $r < 0$.

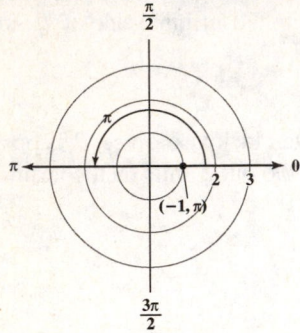

a. Add 2π to the angle and do not change r.

$$\left(5, \frac{\pi}{6}\right) = \left(5, \frac{\pi}{6} + 2\pi\right) = \left(5, \frac{13\pi}{6}\right)$$

b. Add π to the angle and replace r by $-r$.

$$\left(5, \frac{\pi}{6}\right) = \left(-5, \frac{\pi}{6} + \pi\right) = \left(-5, \frac{7\pi}{6}\right)$$

c. Subtract 2π from the angle and do not change r.

$$\left(5, \frac{\pi}{6}\right) = \left(5, \frac{\pi}{6} - 2\pi\right) = \left(5, -\frac{11\pi}{6}\right)$$

19. Draw $\theta = -\frac{\pi}{2} = -90°$ clockwise, since θ is positive, from the polar axis. Go 2 units out on the ray opposite the terminal side of θ, since $r < 0$.

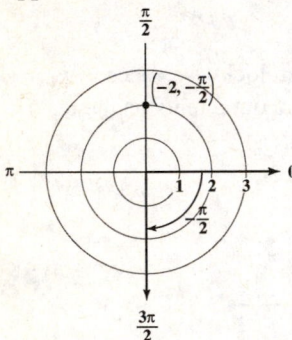

23. Draw $\theta = \frac{3\pi}{4} = 135°$ counterclockwise, since θ is positive, from the polar axis. Go out 10 units on the terminal side of θ, since $r > 0$.

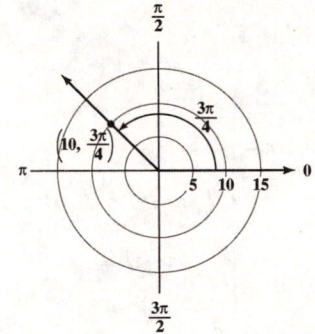

a. Add 2π to the angle and do not change r.

$$\left(10, \frac{3\pi}{4}\right) = \left(10, \frac{3\pi}{4} + 2\pi\right) = \left(10, \frac{11\pi}{4}\right)$$

b. Add π to the angle and replace r by $-r$.

$$\left(10, \frac{3\pi}{4}\right) = \left(-10, \frac{3\pi}{4} + \pi\right) = \left(-10, \frac{7\pi}{4}\right)$$

c. Subtract 2π from the angle and do not change r.

$$\left(10, \frac{3\pi}{4}\right) = \left(10, \frac{3\pi}{4} - 2\pi\right) = \left(10, \frac{-5\pi}{4}\right)$$

21. Draw $\theta = \frac{\pi}{6} = 30°$ counterclockwise, since θ is positive, from the polar axis. Go 5 units out on the terminal side of θ, since $r > 0$.

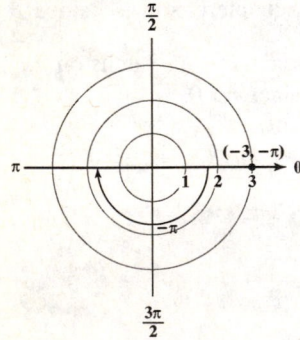

25. Draw $\theta = \dfrac{\pi}{2} = 90°$ counterclockwise, since θ is positive, from the polar axis. Go 4 units out on the terminal side of θ, since $r > 0$.

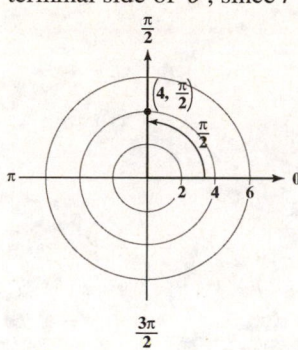

a. Add 2π to the angle and do not change r.

$$\left(4, \frac{\pi}{2}\right) = \left(4, \frac{\pi}{2} + 2\pi\right) = \left(4, \frac{5\pi}{2}\right)$$

b. Add π to the angle and replace r by $-r$.

$$\left(4, \frac{\pi}{2}\right) = \left(-4, \frac{\pi}{2} + \pi\right) = \left(-4, \frac{3\pi}{2}\right)$$

c. Subtract 2π from the angle and do not change r.

$$\left(4, \frac{\pi}{2}\right) = \left(4, \frac{\pi}{2} - 2\pi\right) = \left(4, -\frac{3\pi}{2}\right)$$

27. a, b, d

29. b, d

31. a, b

33. The rectangular coordinates of $(4, 90°)$ are $(0, 4)$.

35. $x = r\cos\theta = 2\cos\dfrac{\pi}{3} = 2\left(\dfrac{1}{2}\right) = 1$

$y = r\sin\theta = 2\sin\dfrac{\pi}{3} = 2\left(\dfrac{\sqrt{3}}{2}\right) = \sqrt{3}$

The rectangular coordinates of $\left(2, \dfrac{\pi}{3}\right)$ are $\left(1, \sqrt{3}\right)$.

37. $x = r\cos\theta = -4\cos\dfrac{\pi}{2} = -4 \cdot 0 = 0$

$y = r\sin\theta = -4\sin\dfrac{\pi}{2} = -4(1) = -4$

The rectangular coordinates of $\left(-4, \dfrac{\pi}{2}\right)$ are $(0, -4)$.

39. $x = r\cos\theta = 7.4\cos 2.5 \approx 7.4(-0.80) \approx -5.9$

$y = r\sin\theta = 7.4\sin 2.5 \approx 7.4(0.60) \approx 4.4$

The rectangular coordinates of $(7.4, 2.5)$ are approximately $(-5.9, 4.4)$.

41. $r = \sqrt{x^2 + y^2} = \sqrt{(-2)^2 + 2^2}$
$= \sqrt{4 + 4} = \sqrt{8} = 2\sqrt{2}$

$\tan\theta = \dfrac{y}{x} = \dfrac{2}{-2} = -1$

Because $\tan\theta = -1$ and θ lies in quadrant II, $\theta = \dfrac{3\pi}{4}$.

The polar coordinates of $(-2, 2)$ are $(r, \theta) = \left(\sqrt{8}, \dfrac{3\pi}{4}\right)$.

43. $r = \sqrt{x^2 + y^2} = \sqrt{(2)^2 + \left(-2\sqrt{3}\right)^2}$
$= \sqrt{4 + 12} = \sqrt{16} = 4$

$\tan\theta = \dfrac{y}{x} = \dfrac{-2\sqrt{3}}{2} = -\sqrt{3}$

Because $\tan\dfrac{\pi}{3} = \sqrt{3}$ and θ lies in quadrant IV, $\theta = 2\pi - \dfrac{\pi}{3} = \dfrac{5\pi}{3}$.

The polar coordinates of $\left(2, -2\sqrt{3}\right)$ are $(r, \theta) = \left(4, \dfrac{5\pi}{3}\right)$.

45.
$$r = \sqrt{x^2 + y^2} = \sqrt{\left(-\sqrt{3}\right)^2 + (-1)^2}$$
$$= \sqrt{3+1} = \sqrt{4} = 2$$
$$\tan\theta = \frac{y}{x} = \frac{-1}{-\sqrt{3}} = \frac{1}{\sqrt{3}}$$

Because $\tan\dfrac{\pi}{6} = \dfrac{1}{\sqrt{3}}$ and θ lies in quadrant III,

$$\theta = \pi + \frac{\pi}{6} = \frac{7\pi}{6}.$$

The polar coordinates of $\left(-\sqrt{3}, -1\right)$ are

$$(r, \theta) = \left(2, \frac{7\pi}{6}\right).$$

47.
$$r = \sqrt{x^2 + y^2} = \sqrt{(5)^2 + (0)^2} = \sqrt{25} = 5$$
$$\tan\theta = \frac{y}{x} = \frac{0}{5} = 0$$

Because $\tan 0 = 0$ and θ lies on the polar axis,
$\theta = 0$.
The polar coordinates of $(5, 0)$ are
$(r, \theta) = (5, 0)$.

49.
$$3x + y = 7$$
$$3r\cos\theta + r\sin\theta = 7$$
$$r(3\cos\theta + \sin\theta) = 7$$
$$r = \frac{7}{3\cos\theta + \sin\theta}$$

51.
$$x = 7$$
$$r\cos\theta = 7$$
$$r = \frac{7}{\cos\theta}$$

53.
$$x^2 + y^2 = 9$$
$$r^2 = 9$$
$$r = 3$$

55.
$$(x-2)^2 + y^2 = 4$$
$$(r\cos\theta - 2)^2 + (r\sin\theta)^2 = 4$$
$$r^2\cos^2\theta - 4r\cos\theta + 4 + r^2\sin^2\theta^2 = 4$$
$$r^2 - 4r\cos\theta = 0$$
$$r^2 = 4r\cos\theta$$
$$r = 4\cos\theta$$

57.
$$y^2 = 6x$$
$$(r\sin\theta)^2 = 6r\cos\theta$$
$$r^2\sin^2\theta = 6r\cos\theta$$
$$r\sin^2\theta = 6\cos\theta$$
$$r = \frac{6\cos\theta}{\sin^2\theta}$$

59.
$$r = 8$$
$$r^2 = 64$$
$$x^2 + y^2 = 64$$

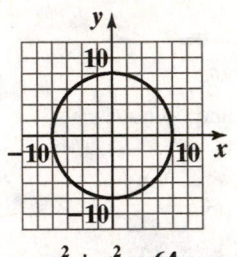

$$x^2 + y^2 = 64$$

61.
$$\theta = \frac{\pi}{2}$$
$$\tan\theta = \tan\frac{\pi}{2}$$
$\tan\theta$ is undefined
$\dfrac{y}{x}$ is undefined
$x = 0$

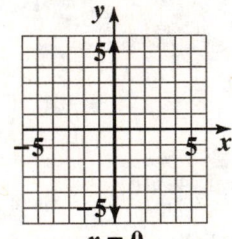

$$x = 0$$

63.
$$r\sin\theta = 3$$
$$y = 3$$

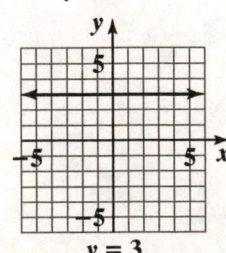

$$y = 3$$

65.
$$r = 4\csc\theta$$
$$r = \frac{4}{\sin\theta}$$
$$r\sin\theta = 4$$
$$y = 4$$

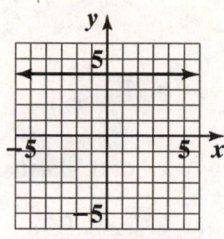

$y = 4$

67.
$$r = \sin\theta$$
$$r \cdot r = r \cdot \sin\theta$$
$$r^2 = r\sin\theta$$
$$x^2 + y^2 = y$$

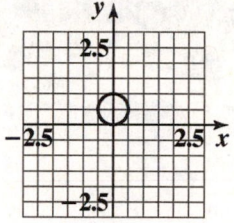

$x^2 + y^2 = y$

69.
$$r = 12\cos\theta$$
$$r^2 = 12r\cos\theta$$
$$x^2 + y^2 = 12x$$
$$x^2 - 12x + y^2 = 0$$
$$x^2 - 12x + 36 + y^2 = 36$$
$$(x - 6)^2 + y^2 = 36$$

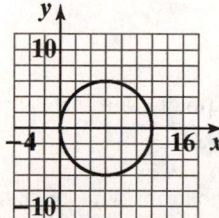

$(x - 6)^2 + y^2 = 36$

71.
$$r = 6\cos\theta + 4\sin\theta$$
$$r \cdot r = r(6\cos\theta + 4\sin\theta)$$
$$r^2 = 6r\cos\theta + 4r\sin\theta$$
$$x^2 + y^2 = 6x + 4y$$

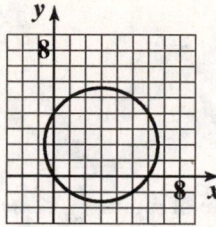

$x^2 + y^2 = 6x + 4y$

73.
$$r^2 \sin 2\theta = 2$$
$$r^2(2\sin\theta\cos\theta) = 2$$
$$2r\sin\theta\, r\cos\theta = 2$$
$$2yx = 2$$
$$xy = 1$$
$$y = \frac{1}{x}$$

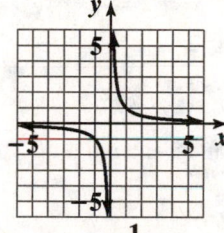

$y = \dfrac{1}{x}$

75.
$$r = a\sec\theta$$
$$r = \frac{a}{\cos\theta}$$
$$r\cos\theta = a$$
$$x = a$$
This is the equation of a vertical line.

77.
$$r = a\sin\theta$$
$$r^2 = ar\sin\theta$$
$$x^2 + y^2 = ay$$
$$x^2 + y^2 - ay = 0$$
$$x^2 + y^2 - ay + \frac{a^2}{4} = \frac{a^2}{4}$$
$$x^2 + \left(y - \frac{a}{2}\right)^2 = \left(\frac{a}{2}\right)^2$$

This is the equation of a circle of radius $\frac{a}{2}$ centered

at $\left(0, \frac{a}{2}\right)$.

79.
$$r\sin\left(\theta - \frac{\pi}{4}\right) = 2$$
$$r\left(\sin\theta\cos\frac{\pi}{4} - \cos\theta\sin\frac{\pi}{4}\right) = 2$$
$$r\sin\theta \cdot \frac{\sqrt{2}}{2} - r\cos\theta \cdot \frac{\sqrt{2}}{2} = 2$$
$$y \cdot \frac{\sqrt{2}}{2} - x \cdot \frac{\sqrt{2}}{2} = 2$$
$$y = x + 2\sqrt{2}$$

$y = x + 2\sqrt{2}$ has slope of 1 and y-intercept of $2\sqrt{2}$.

81. $x_1 = r\cos\theta = 2\cos\frac{2\pi}{3} = -1$

$y_1 = r\sin\theta = 2\sin\frac{2\pi}{3} = \sqrt{3}$

$\left(-1, \sqrt{3}\right)$

$x_2 = r\cos\theta = 4\cos\frac{\pi}{6} = 2\sqrt{3}$

$y_2 = r\sin\theta = 4\sin\frac{\pi}{6} = 2$

$\left(2\sqrt{3}, 2\right)$

$d = \sqrt{\left(x_2 - x_1\right)^2 + \left(y_2 - y_1\right)^2}$

$d = \sqrt{\left(2\sqrt{3} + 1\right)^2 + \left(2 - \sqrt{3}\right)^2}$

$d = 2\sqrt{5}$

83. The angle is measured counterclockwise from the polar axis.

$\theta = \frac{2}{3}(360°) = 240°$ or $\frac{4\pi}{3}$.

The distance from the inner circle's center to the outer circle is
$r = 6 + 3(3) = 6 + 9 = 15$

The polar coordinates are $(r, \theta) = \left(15, \frac{4\pi}{3}\right)$.

85. (6.3, 50°) represents a sailing speed of 6.3 knots at an angle of 50° to the wind.

87. Out of the four points in this 10-knot-wind situation, you would recommend a sailing angle of 105°. A sailing speed of 7.5 knots is achieved at this angle.

89. – 95. Answers may vary.

97.

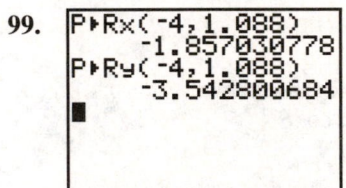

To three decimal places, the rectangular coordinates are (–2, 3.464).

99.
```
P▸Rx(-4,1.088)
      -1.857030778
P▸Ry(-4,1.088)
      -3.542800684
■
```

To three decimal places, the rectangular coordinates are (–1.857, –3.543).

101.
```
R▸Pr(√(5),2)
              3
R▸Pθ(√(5),2)
      .7297276562
```

To three decimal places, the polar coordinates are $(r, \theta) = (3, 0.730)$.

103. does not make sense; Explanations will vary. Sample explanation: There are multiple polar representations for a given point.

105. makes sense

107. Use the distance formula for rectangular coordinates, $d = \sqrt{(x_2 - x_1)^2 + (y_2 - y_1)^2}$.

Let $x_1 = r_1 \cos\theta_1, y_1 = r_1 \sin\theta_1,$

$x_2 = r_2 \cos\theta_2, y_2 = r_2 \sin\theta_2$

$$d = \sqrt{(r_2 \cos\theta_2 - r_1 \cos\theta_1)^2 + (r_2 \sin\theta_2 - r_1 \sin\theta_1)^2}$$

$$= \sqrt{r_2^2 \cos^2\theta_2 - 2r_1 r_2 \cos\theta_1 \cos\theta_2 + r_1^2 \cos^2\theta_1 + r_2^2 \sin^2\theta_2 - 2r_1 r_2 \sin\theta_1 \sin\theta_2 + r_1^2 \sin^2\theta_1}$$

$$= \sqrt{r_2^2 \left(\cos^2\theta_2 + \sin^2\theta_2\right) + r_1^2 \left(\cos^2\theta_1 + \sin^2\theta_1\right) - 2r_1 r_2 \left(\cos\theta_1 \cos\theta_2 + \sin\theta_1 \sin\theta_2\right)}$$

$$= \sqrt{r_2^2 (1) + r_1^2 (1) - 2r_1 r_2 \left(\cos(\theta_2 - \theta_1)\right)}$$

$$= \sqrt{r_1^2 + r_2^2 - 2r_1 r_2 \cos(\theta_2 - \theta_1)}$$

109.

θ	0	$\dfrac{\pi}{6}$	$\dfrac{\pi}{3}$	$\dfrac{\pi}{2}$	$\dfrac{2\pi}{3}$	$\dfrac{5\pi}{6}$	π
$r = 1 - \cos\theta$	0	$\dfrac{2 - \sqrt{3}}{2}$ ≈ 0.13	$\dfrac{1}{2}$ $= 0.5$	1	$\dfrac{3}{2}$ $= 1.5$	$\dfrac{2 + \sqrt{3}}{2}$ ≈ 1.87	0

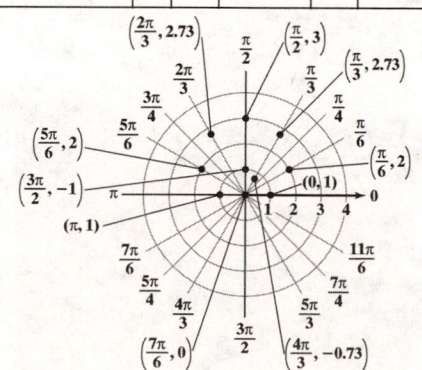

110.

θ	0	$\dfrac{\pi}{6}$	$\dfrac{\pi}{3}$	$\dfrac{\pi}{2}$	$\dfrac{2\pi}{3}$	$\dfrac{5\pi}{6}$	π	$\dfrac{7\pi}{6}$	$\dfrac{4\pi}{3}$	$\dfrac{3\pi}{2}$
$r = 1 + 2\sin\theta$	1	2	$1 + \sqrt{3}$ ≈ 2.73	3	$1 + \sqrt{3}$ ≈ 2.73	2	1	0	$1 - \sqrt{3}$ ≈ -0.73	-1

111.

θ	0	$\dfrac{\pi}{6}$	$\dfrac{\pi}{4}$	$\dfrac{\pi}{3}$	$\dfrac{\pi}{2}$	$\dfrac{2\pi}{3}$	$\dfrac{3\pi}{4}$	$\dfrac{5\pi}{6}$	π
$r = 4\sin 2\theta$	0	$2\sqrt{3}$ ≈ 3.46	4	$2\sqrt{3}$ ≈ 3.46	0	$-2\sqrt{3}$ ≈ -3.46	-4	$-2\sqrt{3}$ ≈ -3.46	0

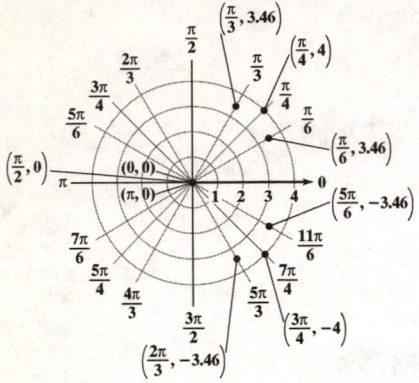

Section 6.4

Check Point Exercises

1. Construct a partial table of coordinates using multiples of $\dfrac{\pi}{6}$. Then plot the points and join them with a smooth curve.

$r = 4\sin\theta$

θ	r	(r, θ)
0	$4\sin 0 = 4\cdot 0 = 0$	$(0, 0)$
$\dfrac{\pi}{6}$	$4\sin\dfrac{\pi}{6} = 4\cdot\dfrac{1}{2} = 2$	$\left(2, \dfrac{\pi}{6}\right)$
$\dfrac{\pi}{3}$	$4\sin\dfrac{\pi}{3} = 4\cdot\dfrac{\sqrt{3}}{2} = 2\sqrt{3}$	$\left(2\sqrt{3}, \dfrac{\pi}{3}\right)$
$\dfrac{\pi}{2}$	$4\sin\dfrac{\pi}{2} = 4\cdot 1 = 4$	$\left(4, \dfrac{\pi}{2}\right)$
$\dfrac{2\pi}{3}$	$4\sin\dfrac{2\pi}{3} = 4\cdot\dfrac{\sqrt{3}}{2} = 2\sqrt{3}$	$\left(2\sqrt{3}, \dfrac{2\pi}{3}\right)$
$\dfrac{5\pi}{6}$	$4\sin\dfrac{5\pi}{6} = 4\cdot\dfrac{1}{2} = 2$	$\left(2, \dfrac{5\pi}{6}\right)$
π	$4\sin\pi = 4\cdot 0 = 0$	$(0, \pi)$

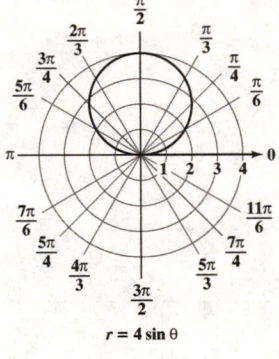

$r = 4\sin\theta$

2.   Polar Axis: Replace θ by $-\theta$ in $r = 1 + \cos\theta$.

$$r = 1 + \cos(-\theta)$$
$$r = 1 + \cos\theta$$

Because the polar equation does not change when θ is replaced by $-\theta$, the graph is symmetric with respect to the polar axis.

The Line $\theta = \dfrac{\pi}{2}$: Replace (r, θ) by $(-r, -\theta)$ in $r = 1 + \cos\theta$.

$$-r = 1 + \cos(-\theta)$$
$$-r = 1 + \cos\theta$$
$$r = -1 - \cos\theta$$

Because the polar equation changes when (r, θ) is replaced by $(-r, -\theta)$, the equation fails the symmetry test. The graph may or may not be symmetric with respect to the line $\theta = \dfrac{\pi}{2}$.

The Pole: Replace r by $-r$ in $r = 1 + \cos\theta$.

$$-r = 1 + \cos\theta$$
$$r = -1 - \cos\theta$$

Because the polar equation changes when r is replaced by $-r$, the equation fails the symmetry test. The graph may or may not be symmetric with respect to the pole.

Because the period of the cosine function is 2π, and the graph is symmetric with respect to the polar axis, begin by finding vales of r for values of θ from 0 to π. then graph $r = 1 + \cos\theta$ for these values and reflect the graph about the polar axis.

θ	0	$\dfrac{\pi}{6}$	$\dfrac{\pi}{3}$	$\dfrac{\pi}{2}$	$\dfrac{2\pi}{3}$	$\dfrac{5\pi}{6}$	π
r	2	1.87	1.5	1	0.5	0.13	0

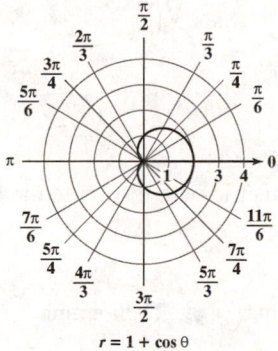

$r = 1 + \cos\theta$

3. $r = 1 - 2\sin\theta$

Check for symmetry:

Polar Axis	The Line $\theta = \dfrac{\pi}{2}$	The Pole
$r = 1 - 2\sin(-\theta)$	$-r = 1 - 2\sin(-\theta)$	$-r = 1 - 2\sin\theta$
$r = 1 - 2(-\sin\theta)$	$-r = 1 + 2\sin\theta$	
$r = 1 + 2\sin\theta$	$r = -1 - 2\sin\theta$	$r = -1 + 2\sin\theta$

There may be no symmetry, since each equation is not equivalent to $r = 1 - 2\sin\theta$. Because the period of the sine function is 2π, we need not consider values of θ beyond 2π.

θ	0	$\dfrac{\pi}{6}$	$\dfrac{\pi}{3}$	$\dfrac{\pi}{2}$	$\dfrac{2\pi}{3}$	$\dfrac{5\pi}{6}$	π	$\dfrac{7\pi}{6}$	$\dfrac{4\pi}{3}$	$\dfrac{3\pi}{2}$	$\dfrac{5\pi}{6}$	$\dfrac{11\pi}{6}$	2π
r	1	0	-0.73	-1	-0.73	0	1	2	2.73	3	2.73	2	1

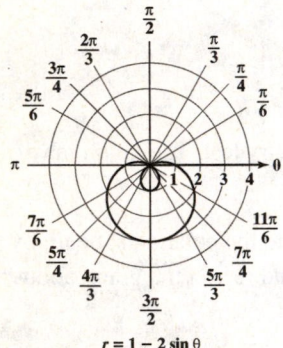

$r = 1 - 2\sin\theta$

4. $r = 3\cos 2\theta$

Check for symmetry:

Polar Axis	The Line $\theta = \dfrac{\pi}{2}$	The Pole
$r = 3\cos 2(-\theta)$	$-r = 3\cos 2(-\theta)$	$-r = 3\cos 2\theta$
	$-r = 3\cos 2\theta$	
$r = 3\cos 2\theta$	$r = -3\cos 2\theta$	$r = -3\cos 2\theta$

The graph has symmetry with respect to the polar axis. The graph may or may not be symmetric with respect to the line $\theta = \dfrac{\pi}{2}$ or the pole.

Since the graph is symmetric with respect to the polar axis, calculate values of r for θ from 0 to π. Then, graph $r = 3\cos 2\theta$ for these values and reflect the graph about the polar axis.

θ	0	$\dfrac{\pi}{6}$	$\dfrac{\pi}{3}$	$\dfrac{\pi}{2}$	$\dfrac{2\pi}{3}$	$\dfrac{5\pi}{6}$	π
r	3	1.5	-1.5	-3	-1.5	1.5	3

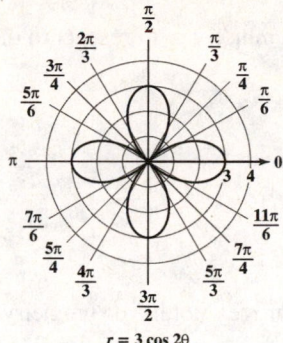

$r = 3\cos 2\theta$

5. $r^2 = 4\cos 2\theta$

Check for symmetry:

Polar Axis	The Line $\theta = \dfrac{\pi}{2}$	The Pole
$r^2 = 4\cos 2(-\theta)$	$(-r)^2 = 4\cos 2(-\theta)$	$(-r)^2 = 4\cos 2\theta$
$r^2 = 4\cos(-2\theta)$	$r^2 = 4\cos(-2\theta)$	
$r^2 = 4\cos 2\theta$	$r^2 = 4\cos 2\theta$	$r^2 = 4\cos 2\theta$

The graph has symmetry with respect to the polar axis, the line $\theta = \dfrac{\pi}{2}$, and the pole.

Calculate values of r for θ from 0 to $\dfrac{\pi}{2}$, and use symmetry to obtain the graph.

θ	0	$\dfrac{\pi}{6}$	$\dfrac{\pi}{4}$	$\dfrac{\pi}{3}$	$\dfrac{\pi}{2}$
r	± 2	± 1.41	0	undef.	undef.

Use symmetry to obtain the graph.

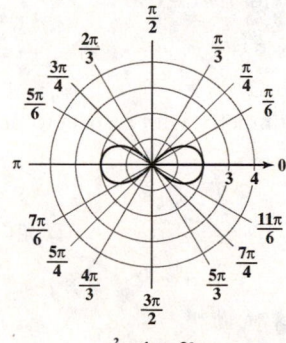

$r^2 = 4\cos 2\theta$

Exercise Set 6.4

1. heart-shaped limaçon or cardioid

    $\dfrac{a}{b} = 1$

    $r = 0$ when $\theta = \dfrac{\pi}{2}$

    The polar equation is $r = 1 - \sin\theta$.

3. circle

    $r = 2$ when $\theta = 0$

    The polar equation is $r = 2\cos\theta$.

5. rose curve

    3 petals $\Rightarrow n = 3$

    The polar equation is $r = 3\sin 3\theta$.

7. **a.** $r = \sin\theta$

    Replace θ with $-\theta$.

    $r = \sin(-\theta)$

    $r = -\sin\theta$

    The graph may or may not have symmetry with respect to the polar axis.

    b. $r = \sin\theta$

    Replace (r, θ) with $(-r, -\theta)$.

    $-r = \sin(-\theta)$

    $-r = -\sin\theta$

    $r = \sin\theta$

    The graph has symmetry with respect to the line $\theta = \dfrac{\pi}{2}$.

    c. $r = \sin\theta$

    Replace r with $-r$.

    $-r = \sin\theta$

    $r = -\sin\theta$

    The graph may or may not have symmetry about the pole.

9. **a.** $r = 4 + 3\cos\theta$

    Replace θ with $-\theta$.

    $r = 4 + 3\cos(-\theta)$

    $r = 4 + 3\cos\theta$

    The graph has symmetry with respect to the polar axis.

    b. $r = 4 + 3\cos\theta$

    Replace (r, θ) with $(-r, -\theta)$.

    $-r = 4 + 3\cos(-\theta)$

    $-r = 4 + 3\cos\theta$

    $r = -4 - 3\cos\theta$

    The graph may or may not have symmetry with respect to the line $\theta = \dfrac{\pi}{2}$.

    c. $r = 4 + 3\cos\theta$

    Replace r with $-r$.

    $-r = 4 + 3\cos\theta$

    $r = -4 - 3\cos\theta$

    The graph may or may not have symmetry about the pole.

11. **a.** $r^2 = 16\cos 2\theta$

    Replace θ with $-\theta$.

    $r^2 = 16\cos 2(-\theta)$

    $r^2 = 16\cos(-2\theta)$

    $r^2 = 16\cos 2\theta$

    The graph has symmetry with respect to the polar axis.

    b. $r^2 = 16\cos 2\theta$

    Replace (r, θ) with $(-r, -\theta)$.

    $(-r)^2 = 16\cos 2(-\theta)$

    $r^2 = 16\cos 2\theta$

    The graph has symmetry with respect to the line $\theta = \dfrac{\pi}{2}$.

    c. $r^2 = 16\cos 2\theta$

    Replace r with $-r$.

    $(-r)^2 = 16\cos 2\theta$

    $r^2 = 16\cos 2\theta$

    The graph has symmetry about the pole.

13. $r = 2\cos\theta$
Check for symmetry:

Polar Axis	The Line $\theta = \dfrac{\pi}{2}$	The Pole
$r = 2\cos(-\theta)$	$-r = 2\cos(-\theta)$	$-r = 2\cos\theta$
	$-r = 2\cos\theta$	
$r = 2\cos\theta$	$r = -2\cos\theta$	$r = -2\cos\theta$

The graph is symmetric with respect to the polar axis. The graph may or may not be symmetric with respect to the line $\theta = \dfrac{\pi}{2}$ or the pole.

Calculate values of r for θ from 0 to π and use symmetry to obtain the graph.

θ	0	$\dfrac{\pi}{6}$	$\dfrac{\pi}{3}$	$\dfrac{\pi}{2}$
r	2	1.73	1	0

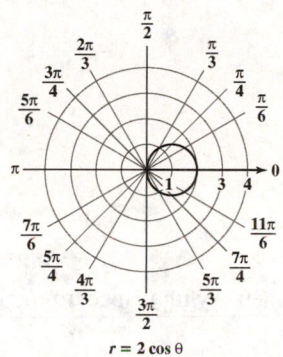

$r = 2\cos\theta$

Notice that there are no points in quadrants II or III. Because the cosine is negative in quadrants II and III, r is negative here. This places the points in quadrants IV and I respectively.

15. $r = 1 - \sin\theta$
Check for symmetry:

Polar Axis	The Line $\theta = \dfrac{\pi}{2}$	The Pole
$r = 1 - \sin(-\theta)$	$-r = 1 - \sin(-\theta)$	$-r = 1 - \sin\theta$
	$-r = 1 + \sin\theta$	
$r = 1 + \sin\theta$	$r = -1 - \sin\theta$	$r = -1 + \sin\theta$

There may be no symmetry since each equation is not equivalent to $r = 1 - \sin\theta$. Because the period of the sine function is 2π, we need not consider values of θ beyond 2π.

θ	0	$\dfrac{\pi}{6}$	$\dfrac{\pi}{3}$	$\dfrac{\pi}{2}$	$\dfrac{2\pi}{3}$	$\dfrac{5\pi}{6}$	π	$\dfrac{7\pi}{6}$	$\dfrac{4\pi}{3}$	$\dfrac{3\pi}{2}$	$\dfrac{5\pi}{6}$	$\dfrac{11\pi}{6}$	2π
r	1	0.5	0.13	0	0.13	0.5	1	1.5	1.87	2	1.87	1.5	1

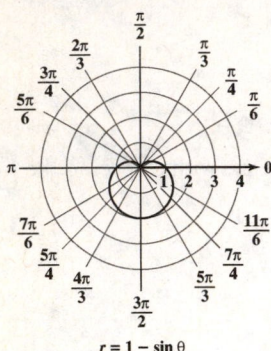

$r = 1 - \sin \theta$

17. $r = 2 + 2\cos\theta$

Check for symmetry:

Polar Axis	**The Line** $\theta = \dfrac{\pi}{2}$	**The Pole**
$r = 2 + 2\cos(-\theta)$	$-r = 2 + 2\cos(-\theta)$	$-r = 2 + 2\cos\theta$
	$-r = 2 + 2\cos\theta$	
$r = 2 + 2\cos\theta$	$r = -2 - 2\cos\theta$	$r = -2 - 2\cos\theta$

The graph is symmetric with respect to the polar axis. The graph may or may not be symmetric with respect to the line $\theta = \dfrac{\pi}{2}$ or the pole.

Calculate values of r for θ from 0 to π and use symmetry to obtain the graph.

θ	0	$\dfrac{\pi}{6}$	$\dfrac{\pi}{3}$	$\dfrac{\pi}{2}$	$\dfrac{2\pi}{3}$	$\dfrac{5\pi}{6}$	π
r	4	3.73	3	2	1	0.27	0

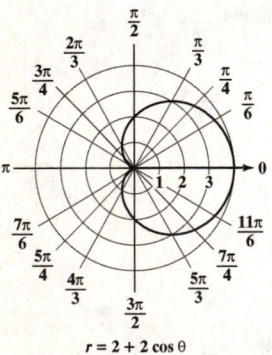

$r = 2 + 2\cos\theta$

19. $r = 2 + \cos\theta$
Check for symmetry:

Polar Axis	The Line $\theta = \dfrac{\pi}{2}$	The Pole
$r = 2 + \cos(-\theta)$	$-r = 2 + \cos(-\theta)$	$-r = 2 + \cos\theta$
	$-r = 2 + \cos\theta$	
$r = 2 + \cos\theta$	$r = -2 - \cos\theta$	$r = -2 - \cos\theta$

The graph is symmetric with respect to the polar axis. The graph may or may not be symmetric with respect to the line
$\theta = \dfrac{\pi}{2}$ or the pole.

Calculate values of r for θ from 0 to π and use symmetry to obtain the graph.

θ	0	$\dfrac{\pi}{6}$	$\dfrac{\pi}{3}$	$\dfrac{\pi}{2}$	$\dfrac{2\pi}{3}$	$\dfrac{5\pi}{6}$	π
r	3	2.87	2.5	2	1.5	1.13	1

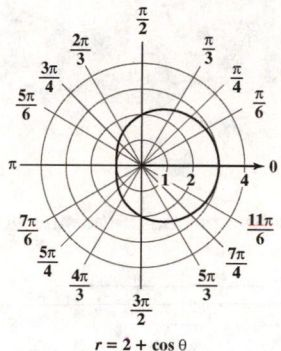

$r = 2 + \cos\theta$

21. $r = 1 + 2\cos\theta$
Check for symmetry:

Polar Axis	The Line $\theta = \dfrac{\pi}{2}$	The Pole
$r = 1 + 2\cos(-\theta)$	$-r = 1 + 2\cos(-\theta)$	$-r = 1 + 2\cos\theta$
	$-r = 1 + 2\cos\theta$	
$r = 1 + 2\cos\theta$	$r = -1 - 2\cos\theta$	$r = -1 - 2\cos\theta$

The graph is symmetric with respect to the polar axis. The graph may or may not be symmetric with respect to the line
$\theta = \dfrac{\pi}{2}$ or the pole.

Calculate values of r for θ from 0 to π and use symmetry to obtain the graph.

θ	0	$\dfrac{\pi}{6}$	$\dfrac{\pi}{3}$	$\dfrac{\pi}{2}$	$\dfrac{2\pi}{3}$	$\dfrac{5\pi}{6}$	π
r	3	2.73	2	1	0	-0.73	-1

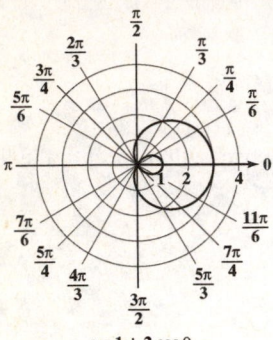

$r = 1 + 2\cos\theta$

23. $r = 2 - 3\sin\theta$

Check for symmetry:

Polar Axis	The Line $\theta = \dfrac{\pi}{2}$	The Pole
$r = 2 - 3\sin(-\theta)$	$-r = 2 - 3\sin(-\theta)$	$-r = 2 - 3\sin\theta$
	$-r = 2 + 3\sin\theta$	
$r = 2 + 3\sin\theta$	$r = -2 - 3\sin\theta$	$r = -2 + 3\sin\theta$

There may be no symmetry since each equation is not equivalent to $r = 2 - 3\sin\theta$. Because the period of the sine function is 2π, we need not consider values of θ beyond 2π.

θ	0	$\dfrac{\pi}{6}$	$\dfrac{\pi}{3}$	$\dfrac{\pi}{2}$	$\dfrac{2\pi}{3}$	$\dfrac{5\pi}{6}$	π	$\dfrac{7\pi}{6}$	$\dfrac{4\pi}{3}$	$\dfrac{3\pi}{2}$	$\dfrac{5\pi}{6}$	$\dfrac{11\pi}{6}$	2π
r	2	0.5	-0.60	-1	-0.60	0.5	2	3.5	4.6	5	4.6	3.5	2

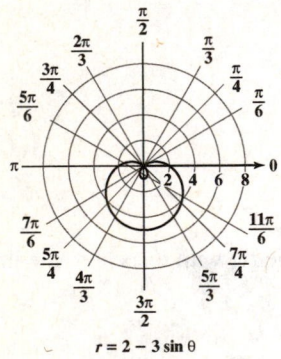

$r = 2 - 3\sin\theta$

25. $r = 2\cos 2\theta$

Check for symmetry:

Polar Axis	The Line $\theta = \dfrac{\pi}{2}$	The Pole
$r = 2\cos 2(-\theta)$	$-r = 2\cos 2(-\theta)$	$-r = 2\cos 2\theta$
$r = 2\cos(-2\theta)$	$-r = 2\cos(-2\theta)$	
	$-r = 2\cos 2\theta$	
$r = 2\cos 2\theta$	$r = -2\cos 2\theta$	$r = -2\cos 2\theta$

The graph is symmetric with respect to the polar axis. The graph may or may not be symmetric with respect to the line $\theta = \dfrac{\pi}{2}$ or the pole.

Calculate values of r for θ from 0 to π and use symmetry to obtain the graph.

θ	0	$\dfrac{\pi}{6}$	$\dfrac{\pi}{3}$	$\dfrac{\pi}{2}$	$\dfrac{2\pi}{3}$	$\dfrac{5\pi}{6}$	π
r	2	1	-1	-2	-1	1	2

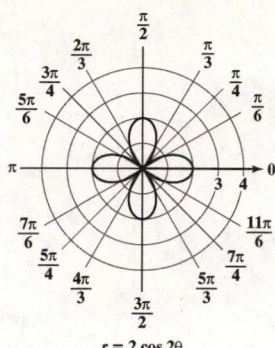

$r = 2\cos 2\theta$

27. $r = 4\sin 3\theta$
Check for symmetry:

Polar Axis	The Line $\theta = \dfrac{\pi}{2}$	The Pole
$r = 4\sin 3(-\theta)$	$-r = 4\sin 3(-\theta)$	$-r = 4\sin 3\theta$
$r = 4\sin(-3\theta)$	$-r = 4\sin(-3\theta)$	
	$-r = -4\sin 3\theta$	
$r = -4\sin 3\theta$	$r = 4\sin 3\theta$	$r = -4\sin 3\theta$

The graph is symmetric with respect to the line $\theta = \dfrac{\pi}{2}$. The graph may or may not be symmetric with respect to the polar axis or the poles.

Calculate values of r for θ from 0 to $\dfrac{\pi}{2}$ and for θ from π to $\dfrac{3\pi}{2}$. Then, use symmetry to obtain the graph.

θ	0	$\dfrac{\pi}{6}$	$\dfrac{\pi}{4}$	$\dfrac{\pi}{3}$	$\dfrac{\pi}{2}$	π	$\dfrac{7\pi}{6}$	$\dfrac{5\pi}{4}$	$\dfrac{4\pi}{3}$	$\dfrac{3\pi}{2}$
r	0	4	2.83	0	-4	0	-4	-2.83	0	4

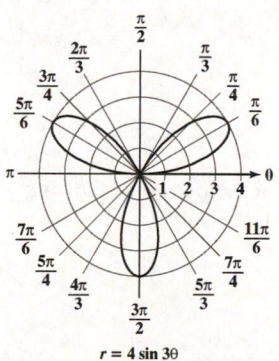

$r = 4\sin 3\theta$

29. $r^2 = 9\cos 2\theta$
Check for symmetry:

Polar Axis	The Line $\theta = \dfrac{\pi}{2}$	The Pole
$r^2 = 9\cos 2(-\theta)$	$(-r)^2 = 9\cos 2(-\theta)$	$(-r)^2 = 9\cos 2\theta$
$r^2 = 9\cos(-2\theta)$	$r^2 = 9\cos(-2\theta)$	
$r^2 = 9\cos 2\theta$	$r^2 = 9\cos 2\theta$	$r^2 = 9\cos 2\theta$

The graph is symmetric with respect to the polar axis, the line $\theta = \dfrac{\pi}{2}$, and the pole.

Note that since $\cos 2\theta$ is negative for $\dfrac{\pi}{4} < \theta < \dfrac{3\pi}{4}$, there is no graph there.

Calculate values of r for θ from 0 to $\dfrac{\pi}{4}$ and use symmetry to obtain the graph.

θ	0	$\dfrac{\pi}{6}$	$\dfrac{\pi}{4}$
r	± 3	± 2.12	0

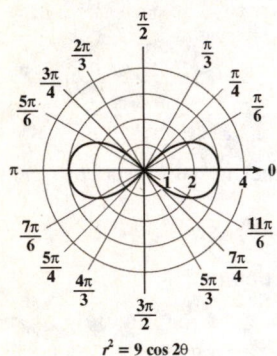

$r^2 = 9\cos 2\theta$

31. $r = 1 - 3\sin\theta$
Check for symmetry:

Polar Axis	The Line $\theta = \dfrac{\pi}{2}$	The Pole
$r = 1 - 3\sin(-\theta)$	$-r = 1 - 3\sin(-\theta)$	$-r = 1 - 3\sin\theta$
	$-r = 1 + 3\sin\theta$	
$r = 1 + 3\sin\theta$	$r = -1 - 3\sin\theta$	$r = -1 + 3\sin\theta$

There may be no symmetry. Since each equation is not equivalent to $r = 1 - 3\sin\theta$. Because the period of the sine function is 2π, we need not consider values of θ beyond 2π.

θ	0	$\dfrac{\pi}{6}$	$\dfrac{\pi}{3}$	$\dfrac{\pi}{2}$	$\dfrac{2\pi}{3}$	$\dfrac{5\pi}{6}$	π	$\dfrac{7\pi}{6}$	$\dfrac{4\pi}{3}$	$\dfrac{3\pi}{2}$	$\dfrac{5\pi}{6}$	$\dfrac{11\pi}{6}$	2π
r	1	−0.5	−1.6	−2	−1.6	−0.5	1	2.5	3.6	4	3.6	2.5	1

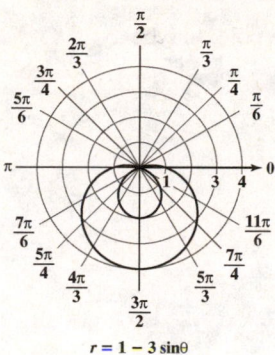

$r = 1 - 3\sin\theta$

33. $r\cos\theta = -3$

$$r = \frac{-3}{\cos\theta} = -3\sec\theta$$

Check for symmetry:

Polar Axis	**The Line** $\theta = \dfrac{\pi}{2}$	**The Pole**
$r = -3\sec(-\theta)$	$-r = -3\sec(-\theta)$	$-r = -3\sec\theta$
	$-r = -3\sec\theta$	
$r = -3\sec\theta$	$r = 3\sec\theta$	$r = 3\sec\theta$

The graph is symmetric with respect to the polar axis. The graph may or may not be symmetric with respect to the line

$\theta = \dfrac{\pi}{2}$ or the pole.

Calculate values of r for θ from 0 to π. Then, use symmetry to obtain the graph.

θ	0	$\dfrac{\pi}{6}$	$\dfrac{\pi}{3}$	$\dfrac{\pi}{2}$	$\dfrac{2\pi}{3}$	$\dfrac{5\pi}{6}$	π
r	−3	−3.46	−6	Undef.	6	3.46	3

Note that since $\sec\theta$ is undefined when $\theta = \dfrac{\pi}{2}$ and $\theta = \dfrac{3\pi}{2}$, r increases without bound as θ approaches these angles.

$r\cos\theta = -3$ is equivalent to $x = -3$.

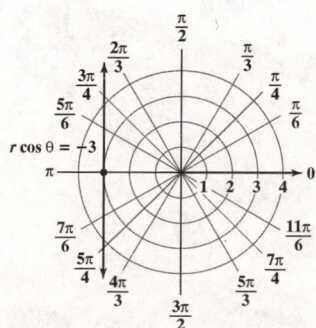

$r\cos\theta = -3$

35. $r = \cos\dfrac{\theta}{2}$

Symmetrical about the polar axis: yes

Symmetrical about the line $\theta = \dfrac{\pi}{2}$: yes

Symmetrical about the pole: yes

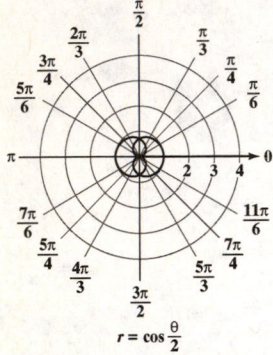

$r = \cos\dfrac{\theta}{2}$

37. $r = \sin\theta + \cos\theta$

Symmetrical about the polar axis: maybe

Symmetrical about the line $\theta = \dfrac{\pi}{2}$: maybe

Symmetrical about the pole: maybe

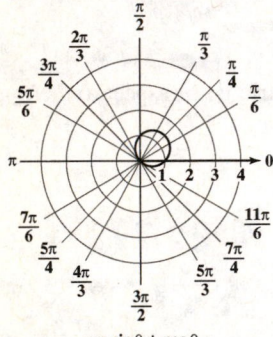

$r = \sin\theta + \cos\theta$

39. $r = \dfrac{1}{1 - \cos\theta}$

Symmetrical about the polar axis: yes

Symmetrical about the line $\theta = \dfrac{\pi}{2}$: maybe

Symmetrical about the pole: maybe

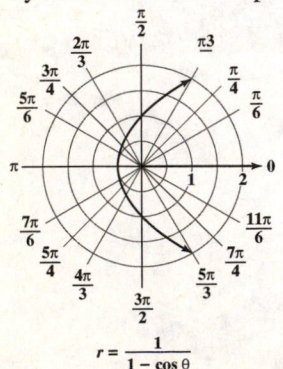

$r = \dfrac{1}{1 - \cos\theta}$

41. $r = \sin\theta\cos^2\theta$

Symmetrical about the polar axis: maybe

Symmetrical about the line $\theta = \dfrac{\pi}{2}$: yes

Symmetrical about the pole: maybe

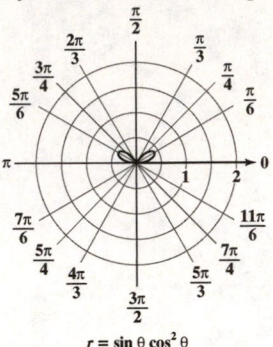

$r = \sin\theta\cos^2\theta$

43. $r = 2 + 3\sin 2\theta$

Symmetrical about the polar axis: maybe

Symmetrical about the line $\theta = \dfrac{\pi}{2}$: maybe

Symmetrical about the pole: yes

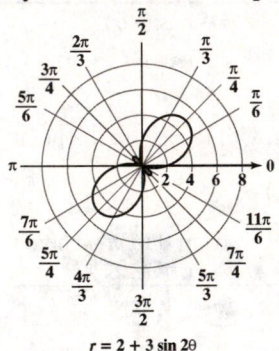

$r = 2 + 3\sin 2\theta$

45. Using the graph, sailing at a 60° angle to the wind gives a speed of about 6 knots (to the nearest knot).

47. Using the graph, sailing at a 90° angle to the wind gives a speed of about 8 knots (to the nearest knot).

49. It appears that an angle of 90° gives a maximum speed of about $7\frac{1}{2}$ knots.

51. – 57. Answers may vary.

59.

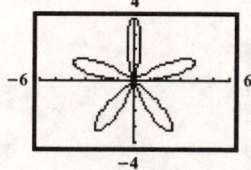

61.

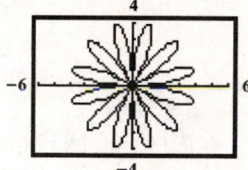

63.

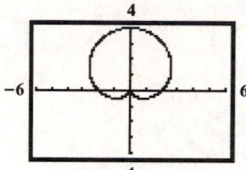

65.

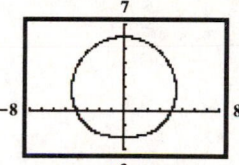

67.

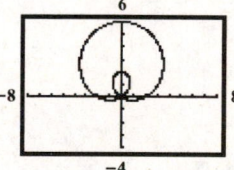

69.

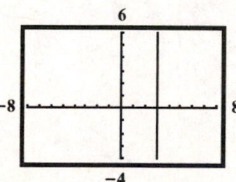

71.

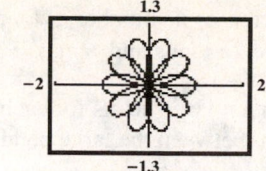

73.

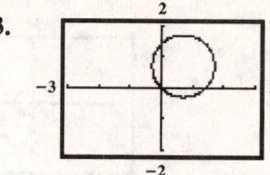

75.

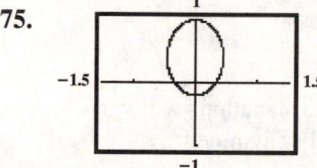

77. If $\theta \max = 2\pi$, the graph is drawn once.

79. θ step $= 0.1$

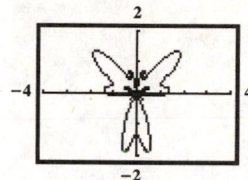

81. θ step $= 0.1$

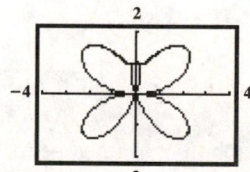

83. As n increases, $\sin n\theta$ increases its number of loops. If n is odd, there are n loops and θ max $= \pi$ traces the graph once, while if n is even, there are $2n$ loops and θ max $= 2\pi$ traces the graph once.

85. There are n small petals and n large petals for each value of n. For odd values of n, the small petals are inside the large petals. For even n, they are between the large petals.

87. θ min $= 0$, θ max $= 2\pi$ θ min $= 0$, θ max $= 4\pi$ θ min $= 0$, θ max $= 8\pi$

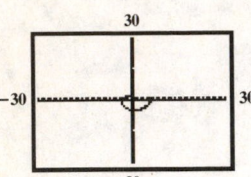 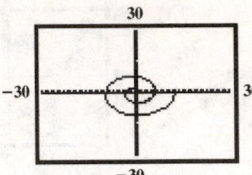 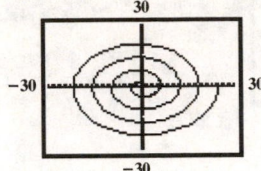

89. does not make sense; Explanations will vary. Sample explanation: If a polar equation fails a symmetry test, its graph may still have that kind of symmetry.

91. makes sense

93.

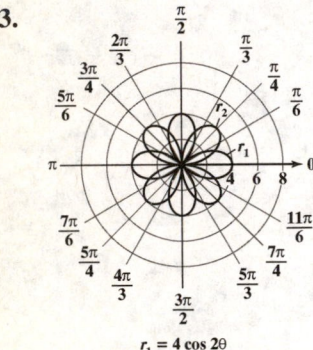

$r_1 = 4 \cos 2\theta$

$r_2 = 4 \cos 2\left(\theta - \dfrac{\pi}{4}\right)$

The graph of r_2 is the graph of r_1 rotated $\dfrac{\pi}{4}$ or $45°$.

95. Answers may vary.

96. $(1+i)(2+2i) = 2 + 2i + 2i + 2i^2$
$$= 2 + 4i - 2$$
$$= 4i$$

97. $(-1+i\sqrt{3})(-1+i\sqrt{3})(-1+i\sqrt{3}) = (-1+i\sqrt{3})(1-i\sqrt{3}-i\sqrt{3}+3i^2)$
$$= (-1+i\sqrt{3})(1-2i\sqrt{3}-3)$$
$$= (-1+i\sqrt{3})(-2-2i\sqrt{3})$$
$$= 2 + 2i\sqrt{3} - 2i\sqrt{3} - 6i^2$$
$$= 2 + 6$$
$$= 8$$

98. $\dfrac{2+2i}{1+i} = \dfrac{2+2i}{1+i} \cdot \dfrac{1-i}{1-i}$
$$= \dfrac{2-2i+2i-2i^2}{1-i+i-i^2}$$
$$= \dfrac{2+2}{1+1}$$
$$= 2$$

Mid-Chapter 6 Check Point

1. $C = 180° - 32° - 41° = 107°$
Use the Law of Sines to find b.

$$\dfrac{a}{\sin A} = \dfrac{b}{\sin B}$$

$$\dfrac{20}{\sin 32°} = \dfrac{b}{\sin 41°}$$

$$b = \dfrac{20 \sin 41°}{\sin 32°}$$

$$b \approx 24.8$$

Use the Law of Sines to find c.

$$\dfrac{a}{\sin A} = \dfrac{c}{\sin C}$$

$$\dfrac{20}{\sin 32°} = \dfrac{c}{\sin 107°}$$

$$c = \dfrac{20 \sin 107°}{\sin 32°}$$

$$c \approx 36.1$$

The solution is $C = 107°$, $b \approx 24.8$, and $c \approx 36.1$.

2. Use the Law of Sines to find B.

$$\frac{a}{\sin A} = \frac{b}{\sin B}$$

$$\frac{63}{\sin 42°} = \frac{57}{\sin B}$$

$$\sin B = \frac{57 \sin 42°}{63}$$

$$\sin B \approx 0.6054$$

There are two angles possible:

$B_1 \approx 37°$, $B_2 \approx 180° - 37° = 143°$

B_2 is impossible, since $42° + 143° = 185°$.

$C = 180° - B_1 - A \approx 180° - 37° - 42° = 101°$

Use the Law of Sines to find c.

$$\frac{c}{\sin C} = \frac{a}{\sin A}$$

$$\frac{c}{\sin 101°} = \frac{63}{\sin 42°}$$

$$c = \frac{63 \sin 101°}{\sin 42°}$$

$$c \approx 92.4$$

There is one triangle and the solution is
B_1 (or B) $\approx 37°$, $C \approx 101°$, and $c \approx 92.4$.

3. Use the Law of Sines to find angle B.

$$\frac{a}{\sin A} = \frac{b}{\sin B}$$

$$\frac{6}{\sin 65°} = \frac{7}{\sin B}$$

$$\sin B = \frac{7 \sin 65°}{6}$$

$$\sin B \approx 1.0574$$

The sine can never exceed 1. There is no triangle
with the given measurements.

4. Use the Law of Cosines to find b.

$$b^2 = a^2 + c^2 - 2ac \cos B$$

$$b^2 = 10^2 + 16^2 - 2(10)(16) \cos 110°$$

$$b^2 \approx 465.4464$$

$$b \approx 21.6$$

Use the Law of Sines to find A.

$$\frac{b}{\sin B} = \frac{a}{\sin A}$$

$$\frac{21.6}{\sin 110°} = \frac{10}{\sin A}$$

$$\sin A = \frac{10 \sin 110°}{21.6}$$

$$\sin A \approx 0.4350$$

$$A \approx 26°$$

Find the third angle.

$C = 180° - A - B \approx 180° - 26° - 110° = 44°$

The solution is $A \approx 26°$, $C \approx 44°$, and $b \approx 21.6$.

5. Use the Law of Sines to find angle A.

$$\frac{a}{\sin A} = \frac{c}{\sin C}$$

$$\frac{16}{\sin A} = \frac{13}{\sin 42°}$$

$$\sin A = \frac{16 \sin 42°}{13}$$

$$\sin A \approx 0.8235$$

There are two angles possible:

$A_1 \approx 55°$, $A_2 \approx 180° - 55° = 125°$

There are two triangles:

$B_1 = 180° - C - A_1 \approx 180° - 42° - 55° = 83°$

$B_2 = 180° - C - A_2 \approx 180° - 42° - 125° = 13°$

Use the Law of Sines to find b_1 and b_2.

$$\frac{b_1}{\sin B_1} = \frac{c}{\sin C}$$

$$\frac{b_1}{\sin 83°} = \frac{13}{\sin 42°}$$

$$b_1 = \frac{13 \sin 83°}{\sin 42°} \approx 19.3$$

$$\frac{b_2}{\sin B_2} = \frac{c}{\sin C}$$

$$\frac{b_2}{\sin 13°} = \frac{13}{\sin 42°}$$

$$b_2 = \frac{13 \sin 13°}{\sin 42°} \approx 4.4$$

In one triangle, the solution is
$A_1 \approx 55°$, $B_1 \approx 83°$, $b_1 \approx 19.3$.

In the other triangle, $A_2 \approx 125°$, $B_2 \approx 13°$, $b_2 \approx 4.4$.

6. Use the Law of Cosines to find the angle opposite the longest side.

Thus, find angle C.

$$c^2 = a^2 + b^2 - 2ab\cos C$$

$$\cos C = \frac{a^2 + b^2 - c^2}{2ab}$$

$$\cos C = \frac{5^2 + 7.2^2 - 10.1^2}{2 \cdot 5 \cdot 7.2}$$

$$\cos C \approx -0.3496$$

$$C \approx 110°$$

Use the Law of Sines to find angle A.

$$\frac{a}{\sin A} = \frac{c}{\sin C}$$

$$\frac{5}{\sin A} = \frac{10.1}{\sin 110°}$$

$$\sin A = \frac{5\sin 110°}{10.1}$$

$$\sin A \approx 0.4652$$

$$A \approx 28°$$

Find the third angle.

$$B = 180° - A - C = 180° - 28° - 110° = 42°$$

The solution is $A \approx 28°, B \approx 42°$, and $C \approx 110°$

7. The area of the triangle is half the product of the lengths of the two sides times the sine of the included angle.

$$\text{Area} = \frac{1}{2}(5)(7)(\sin 36°) \approx 10$$

The area of the triangle is approximately 10 square feet.

8. Begin by calculating one-half the perimeter:

$$s = \frac{1}{2}(a+b+c) = \frac{1}{2}(7+9+12) = 14$$

Use Heron's formula to find the area.

$$\begin{aligned}\text{Area} &= \sqrt{s(s-a)(s-b)(s-c)} \\ &= \sqrt{14(14-7)(14-9)(14-12)} \\ &= \sqrt{980} \approx 31\end{aligned}$$

The area of the triangle is approximately 31 square meters.

9. The first train traveled 100 miles, the second train traveled 80 miles.

Use the Law of Cosines to find the distance.

$$c^2 = a^2 + b^2 - 2ab\cos C$$

$$c^2 = 100^2 + 80^2 - 2(100)(80)\cos 110°$$

$$c^2 \approx 21872.32229$$

$$c \approx 148$$

The two trains are 148 miles apart.

10. Let the fire be at point C.

$$A = 90° - 56° = 34°$$

$$B = 90° - 23° = 67°$$

$$C = 180° - 34° - 67° = 79°$$

Use the law of sines to find b.

$$\frac{b}{\sin B} = \frac{c}{\sin C}$$

$$\frac{b}{\sin 67°} = \frac{16}{\sin 79°}$$

$$b = \frac{16\sin 67°}{\sin 79°}$$

$$b \approx 15.0$$

The fire is 15.0 miles from station A

11. Let point A be where the angle of elevation is $66°$.

Let point B be where the angle of elevation is $50°$.

Let point C be at the top of the tree.

$$C = 180° - A - B = 180° - 66° - 50° = 64°$$

Use the law of sines to find a.

$$\frac{a}{\sin A} = \frac{c}{\sin C}$$

$$\frac{a}{\sin 66°} = \frac{420}{\sin 64°}$$

$$a = \frac{420\sin 66°}{\sin 64°}$$

$$a \approx 427$$

The height of the tree, h, is given by

$$h = a\sin B$$

$$h = 427\sin 50°$$

$$h \approx 327$$

The tree is 327 feet tall.

12. $$x = r\cos\theta = -3\cos\frac{5\pi}{4} = \frac{3\sqrt{2}}{2}$$

$$y = r\sin\theta = -3\sin\frac{5\pi}{4} = -\frac{3\sqrt{2}}{2}$$

Ordered pair: $\left(\dfrac{3\sqrt{2}}{2}, -\dfrac{3\sqrt{2}}{2}\right)$

13. $$x = r\cos\theta = 6\cos\left(-\frac{\pi}{2}\right) = 0$$

$$y = r\sin\theta = 6\sin\left(-\frac{\pi}{2}\right) = -6$$

Ordered pair: $(0, -6)$

14. $r = \sqrt{x^2 + y^2} = \sqrt{2^2 + \left(-2\sqrt{3}\right)^2} = 4$

$\tan \theta = \dfrac{y}{x} = \dfrac{-2\sqrt{3}}{2} = -\sqrt{3}$

Because θ lies in quadrant IV, $\theta = 2\pi - \dfrac{\pi}{3} = \dfrac{5\pi}{3}$

Polar coordinates: $\left(4, \dfrac{5\pi}{3}\right)$

15. $r = \sqrt{x^2 + y^2} = \sqrt{(-6)^2 + 0^2} = 6$

$\tan \theta = \dfrac{y}{x} = \dfrac{0}{-6} = 0$

$\theta = \pi$

Polar coordinates: $(6, \pi)$

16.

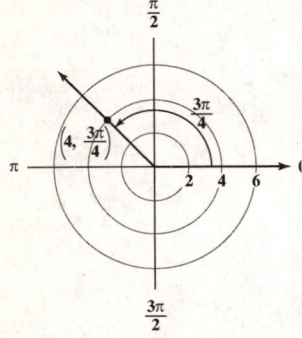

a. $\left(4, \dfrac{11\pi}{4}\right)$

b. $\left(-4, \dfrac{7\pi}{4}\right)$

c. $\left(4, \dfrac{5\pi}{4}\right)$

17.

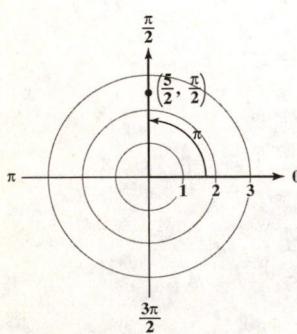

a. $\left(\dfrac{5}{2}, \dfrac{5\pi}{2}\right)$

b. $\left(-\dfrac{5}{2}, \dfrac{3\pi}{2}\right)$

c. $\left(\dfrac{5}{2}, -\dfrac{3\pi}{2}\right)$

18.
$$5x - y = 7$$
$$5r\cos\theta - r\sin\theta = 7$$
$$r(5\cos\theta - \sin\theta) = 7$$
$$r = \dfrac{7}{5\cos\theta - \sin\theta}$$

19.
$$y = -7$$
$$r\sin\theta = -7$$
$$r = \dfrac{-7}{\sin\theta}$$
$$r = -7\csc\theta$$

20.
$$(x+1)^2 + y^2 = 1$$
$$(r\cos\theta + 1)^2 + (r\sin\theta)^2 = 1$$
$$r^2\cos^2\theta + 2r\cos\theta + 1 + r^2\sin^2\theta = 1$$
$$r^2 + 2r\cos\theta = 0$$
$$r^2 = -2r\cos\theta$$
$$r = -2\cos\theta$$

21.
$$r = 6$$
$$r^2 = 36$$
$$x^2 + y^2 = 36$$

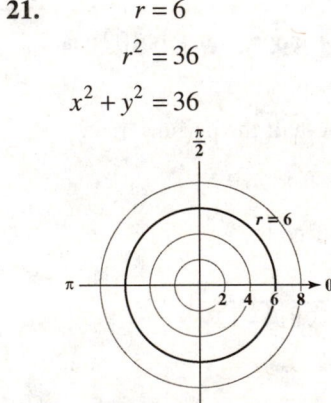

22.
$$\theta = \frac{\pi}{3}$$
$$\tan \theta = \tan \frac{\pi}{3}$$
$$\tan \theta = \sqrt{3}$$
$$\frac{y}{x} = \sqrt{3}$$
$$y = \sqrt{3}x$$

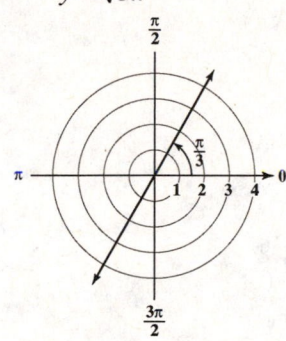

23.
$$r = -3\csc \theta$$
$$r = \frac{-3}{\sin \theta}$$
$$r \sin \theta = -3$$
$$y = -3$$

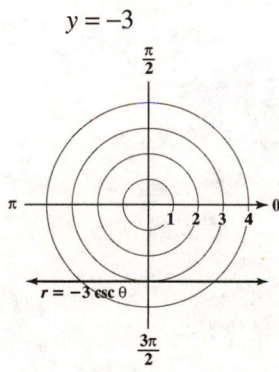

24.
$$r = -10\cos \theta$$
$$r^2 = -10r\cos \theta$$
$$x^2 + y^2 = -10x$$
$$x^2 + 10x + y^2 = 0$$
$$x^2 + 10x + 25 + y^2 = 25$$
$$(x+5)^2 + y^2 = 25$$

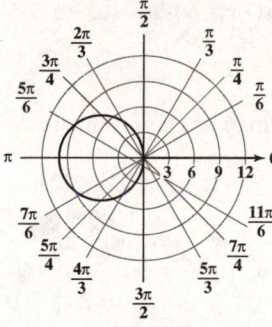

$$r = -10\cos \theta$$

25.
$$r = 4\sin \theta \sec^2 \theta$$
$$r = \frac{4\sin \theta}{\cos^2 \theta}$$
$$r\cos^2 \theta = 4\sin \theta$$
$$r^2 \cos^2 \theta = 4r\sin \theta$$
$$x^2 = 4y$$
$$y = \frac{1}{4}x^2$$

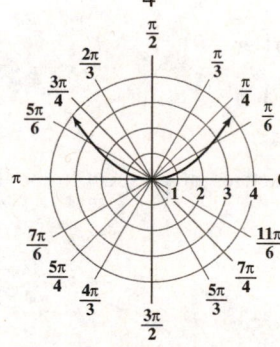

$$r = 4\sin \theta \sec^2 \theta$$

26. $r = 1 - 4\cos\theta$

    a. Replace θ with $-\theta$.

        $r = 1 - 4\cos(-\theta)$

        $r = 1 - 4\cos\theta$

        The graph is symmetric with respect to the polar axis.

    b. Replace (r, θ) with $(-r, -\theta)$.

        $-r = 1 - 4\cos(-\theta)$

        $r = -1 + 4\cos\theta$

        The graph may or may not be symmetric with respect to the line $\theta = \dfrac{\pi}{2}$.

    c. Replace r with $-r$.

        $-r = 1 - 4\cos\theta$

        $r = -1 + 4\cos\theta$

        The graph may or may not be symmetric with respect to the polar axis.

27. $r^2 = 4\cos 2\theta$

    a. Replace θ with $-\theta$.

        $r^2 = 4\cos(-2\theta)$

        $r^2 = 4\cos 2\theta$

        The graph is symmetric with respect to the polar axis.

    b. Replace (r, θ) with $(-r, -\theta)$.

        $(-r)^2 = 4\cos(-2\theta)$

        $r^2 = 4\cos 2\theta$

        The graph is symmetric with respect to the line $\theta = \dfrac{\pi}{2}$.

    c. Replace r with $-r$.

        $(-r)^2 = 4\cos 2\theta$

        $r^2 = 4\cos 2\theta$

        The graph is symmetric with respect to the polar axis.

28.

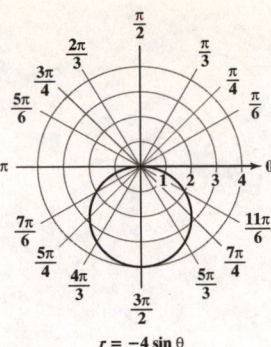

$r = -4\sin\theta$

29.

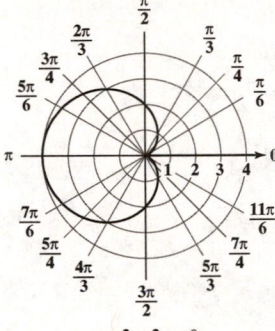

$r = 2 - 2\cos\theta$

30.

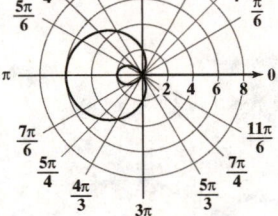

$r = 2 - 4\cos\theta$

31.

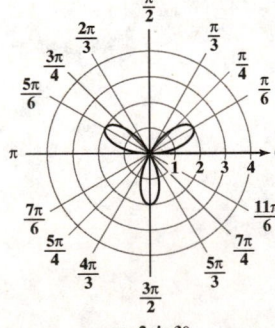

$r = 2\sin 3\theta$

32.

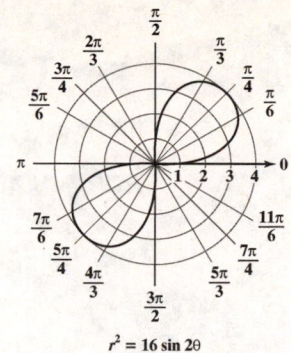

$r^2 = 16 \sin 2\theta$

Section 6.5

Check Point Exercises

1. a. $z = 2 + 3i$ corresponds to the point
(2, 3). Plot the complex number by moving two units to the right on the
real axis and 3 units up parallel to the imaginary axis.

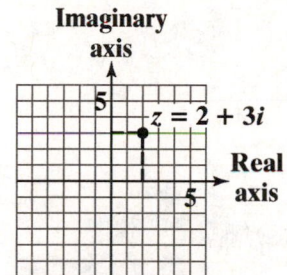

b. $z = -3 - 5i$ corresponds to the point
(–3, –5). Plot the complex number by moving three units to the left on the
real axis and five units down parallel to the imaginary axis.

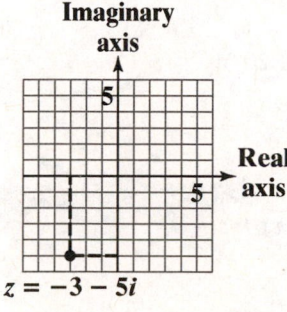

c. Because $z = -4 = -4 + 0i$, this complex number corresponds to the point (–4, 0). Plot the complex number by moving four units to the left on the real axis.

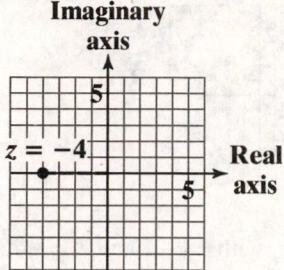

d. Because $z = -i = 0 - i$, this complex number corresponds to the point (0, –1). Plot the complex number by moving one unit down on the imaginary axis.

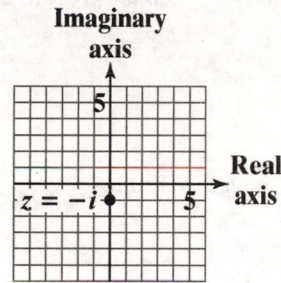

2 a. $z = 5 + 12i$

$a = 5, b = 12$

$|z| = \sqrt{5^2 + 12^2} = \sqrt{25 + 144} = \sqrt{169} = 13$

b. $z = 2 - 3i$

$a = 2, b = -3$

$|z| = \sqrt{2^2 + (-3)^2} = \sqrt{4 + 9} = \sqrt{13}$

3. $z = -1 - i\sqrt{3}$ corresponds to the point $\left(-1, -\sqrt{3}\right)$.

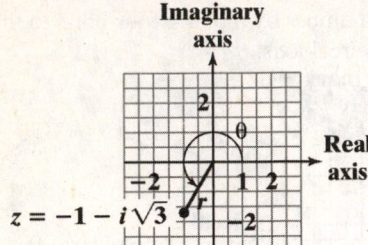

Use $r = \sqrt{a^2 + b^2}$ with $a = -1$ and $b = -\sqrt{3}$ to find r.

$$r = \sqrt{a^2 + b^2} = \sqrt{(-1)^2 + \left(-\sqrt{3}\right)^2}$$

$$= \sqrt{1 + 3} = \sqrt{4} = 2$$

Use $\tan\theta = \dfrac{b}{a}$ with $a = -1$ and $b = -\sqrt{3}$ to find θ.

$$\tan\theta = \frac{b}{a} = \frac{-\sqrt{3}}{-1} = \sqrt{3}$$

Because $\tan\dfrac{\pi}{3} = \sqrt{3}$ and θ lies in

quadrant III, $\theta = \pi + \dfrac{\pi}{3} = \dfrac{3\pi}{3} + \dfrac{\pi}{3} = \dfrac{4\pi}{3}$.

The polar form of $z = -1 - i\sqrt{3}$ is

$$z = r(\cos\theta + i\sin\theta) = 2\left(\cos\frac{4\pi}{3} + i\sin\frac{4\pi}{3}\right).$$

4. The complex number $z = 4(\cos 30° + i\sin 30°)$ is in polar form, with $r = 4$ and $\theta = 30°$. We use exact values for $\cos 30°$ and $\sin 30°$ to write the number in rectangular form.

$$4(\cos 30° + i\sin 30°) = 4\left(\frac{\sqrt{3}}{2} + i\frac{1}{2}\right) = 2\sqrt{3} + 2i$$

The rectangular form of $z = 4(\cos 30° + i\sin 30°)$ is $z = 2\sqrt{3} + 2i$.

5. $z_1 z_2 = [6(\cos 40° + i\sin 40°)][5(\cos 20° + i\sin 20°)] = (6 \cdot 5)[(\cos(40° + 20°) + i\sin(40° + 20°)]$
$\qquad = 30(\cos 60° + i\sin 60°)$

6. $\dfrac{z_1}{z_2} = \dfrac{50\left(\cos\dfrac{4\pi}{3} + i\sin\dfrac{4\pi}{3}\right)}{5\left(\cos\dfrac{\pi}{3} + i\sin\dfrac{\pi}{3}\right)} = \dfrac{50}{5}\left[\cos\left(\dfrac{4\pi}{3} - \dfrac{\pi}{3}\right) + i\sin\left(\dfrac{4\pi}{3} - \dfrac{\pi}{3}\right)\right] = 10(\cos\pi + i\sin\pi)$

7. $\left[2(\cos 30° + i\sin 30°)\right]^5 = 2^5\left[\cos(5 \cdot 30°) + i\sin(5 \cdot 30°)\right] = 32(\cos 150° + i\sin 150°) = 32\left(-\dfrac{\sqrt{3}}{2} + i\dfrac{1}{2}\right)$

$$= -16\sqrt{3} + 16i$$

8. Write $1+i$ in $r(\cos\theta + i\sin\theta)$ form.

$$r = \sqrt{a^2 + b^2} = \sqrt{1^2 + 1^2} = \sqrt{2}$$

$$\tan\theta = \frac{b}{a} = \frac{1}{1} = 1 \text{ and } \theta = \frac{\pi}{4}$$

$$1+i = r(\cos\theta + i\sin\theta) = \sqrt{2}\left(\cos\frac{\pi}{4} + i\sin\frac{\pi}{4}\right)$$

Use DeMoivre's Theorem to raise $1 + i$ to the fourth power.

$$(1+i)^4 = \left[\sqrt{2}\left(\cos\frac{\pi}{4} + i\sin\frac{\pi}{4}\right)\right]^4 = \left(\sqrt{2}\right)^4\left[\cos\left(4\cdot\frac{\pi}{4}\right) + i\sin\left(4\cdot\frac{\pi}{4}\right)\right] = 4(\cos\pi + i\sin\pi) = 4(-1 + 0i) = -4$$

9. From DeMoivre's Theorem for finding complex roots, the fourth roots of $16(\cos 60° + i\sin 60°)$ are

$$z_k = \sqrt[4]{16}\left[\cos\left(\frac{60° + 360°k}{4}\right) + i\sin\left(\frac{60° + 360°k}{4}\right)\right], k = 0,1,2,3.$$

Substitute 0, 1, 2, and 3 for k in the above expression for z_k.

$$z_0 = \sqrt[4]{16}\left[\cos\left(\frac{60° + 360°\cdot 0}{4}\right) + i\sin\left(\frac{60° + 360°\cdot 0}{4}\right)\right] = \sqrt[4]{16}\left[\cos\frac{60°}{4} + i\sin\frac{60°}{4}\right] = 2(\cos 15° + i\sin 15°)$$

$$z_1 = \sqrt[4]{16}\left[\cos\left(\frac{60° + 360°\cdot 1}{4}\right) + i\sin\left(\frac{60° + 360°\cdot 1}{4}\right)\right] = \sqrt[4]{16}\left[\cos\frac{420°}{4} + i\sin\frac{420°}{4}\right] = 2(\cos 105° + i\sin 105°)$$

$$z_2 = \sqrt[4]{16}\left[\cos\left(\frac{60° + 360°\cdot 2}{4}\right) + i\sin\left(\frac{60° + 360°\cdot 2}{4}\right)\right] = \sqrt[4]{16}\left[\cos\frac{780°}{4} + i\sin\frac{780°}{4}\right] = 2(\cos 195° + i\sin 195°)$$

$$z_3 = \sqrt[4]{16}\left[\cos\left(\frac{60° + 360°\cdot 3}{4}\right) + i\sin\left(\frac{60° + 360°\cdot 3}{4}\right)\right] = \sqrt[4]{16}\left[\cos\frac{1140°}{4} + i\sin\frac{1140°}{4}\right] = 2(\cos 285° + i\sin 285°)$$

10. First, write 27 in polar form. $27 = r(\cos\theta + i\sin\theta) = 27(\cos 0 + \sin 0)$. From DeMoivre's theorem for finding complex roots, the cube roots of 27 are

$$z_k = \sqrt[3]{27}\left[\cos\left(\frac{0 + 2\pi k}{3}\right) + i\sin\left(\frac{0 + 2\pi k}{3}\right)\right], k = 0,1,2.$$

$$z_0 = \sqrt[3]{27}\left[\cos\left(\frac{0 + 2\pi\cdot 0}{3}\right) + i\sin\left(\frac{0 + 2\pi\cdot 0}{3}\right)\right] = 3(\cos 0 + i\sin 0) = 3(1 + i\cdot 0) = 3$$

$$z_1 = \sqrt[3]{27}\left[\cos\left(\frac{0 + 2\pi\cdot 1}{3}\right) + i\sin\left(\frac{0 + 2\pi\cdot 1}{3}\right)\right] = 3\left(\cos\frac{2\pi}{3} + i\sin\frac{2\pi}{3}\right) = 3\left(-\frac{1}{2} + i\cdot\frac{\sqrt{3}}{2}\right) = -\frac{3}{2} + \frac{3\sqrt{3}}{2}i$$

$$z_2 = \sqrt[3]{27}\left[\cos\left(\frac{0 + 2\pi\cdot 2}{3}\right) + i\sin\left(\frac{0 + 2\pi\cdot 2}{3}\right)\right] = 3\left(\cos\frac{4\pi}{3} + i\sin\frac{4\pi}{3}\right) = 3\left(-\frac{1}{2} + i\cdot\left(-\frac{\sqrt{3}}{2}\right)\right) = -\frac{3}{2} - \frac{3\sqrt{3}}{2}i$$

Exercise Set 6.5

1. Because $z = 4i = 0 + 4i$, this complex number corresponds to the point $(0, 4)$.

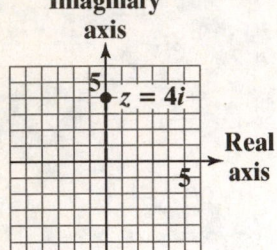

With $a = 0$ and $b = 4$, $|z| = \sqrt{0^2 + 4^2} = \sqrt{16} = 4$.

3. Because $z = 3 = 3 + 0i$, this complex number corresponds to the point $(3, 0)$.

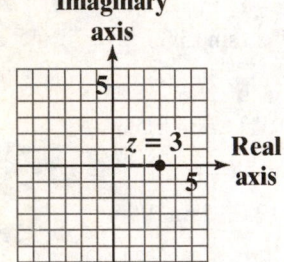

With $a = 3$ and $b = 0$, $|z| = \sqrt{3^2 + 0^2} = \sqrt{9} = 3$.

5. $z = 3 + 2i$ corresponds to the point $(3, 2)$.

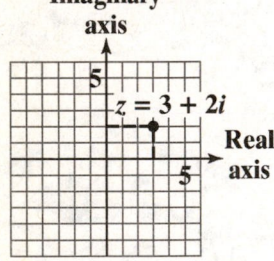

With $a = 3$ and $b = 2$,

$|z| = \sqrt{3^2 + 2^2} = \sqrt{9 + 4} = \sqrt{13}$.

7. $z = 3 - i$ corresponds to the point $(3, -1)$.

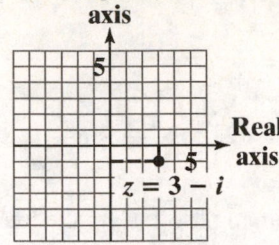

With $a = 3$ and $b = -1$,

$|z| = \sqrt{3^2 + (-1)^2} = \sqrt{9 + 1} = \sqrt{10}$.

9. $z = -3 + 4i$ corresponds to the point $(-3, 4)$.

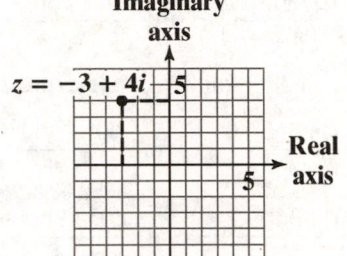

With $a = -3$ and $b = 4$,

$|z| = \sqrt{(-3)^2 + 4^2} = \sqrt{9 + 16} = \sqrt{25} = 5$.

11. $z = 2 + 2i$ corresponds to the point $(2, 2)$.

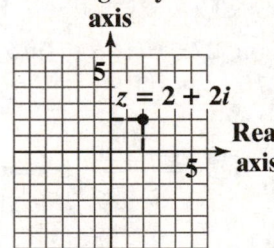

Use $r = \sqrt{a^2 + b^2}$ and $\tan\theta = \dfrac{b}{a}$, with $a = 2$ and $b = 2$,

to find r and θ.

$r = \sqrt{2^2 + 2^2} = \sqrt{4 + 4} = \sqrt{8} = 2\sqrt{2}$

$\tan\theta = \dfrac{2}{2} = 1$

Because $\tan\dfrac{\pi}{4} = 1$ and θ lies in quadrant I, $\theta = \dfrac{\pi}{4}$.

$z = 2 + 2i = r(\cos\theta + i\sin\theta)$

$\quad = 2\sqrt{2}\left(\cos\dfrac{\pi}{4} + i\sin\dfrac{\pi}{4}\right)$

or $2\sqrt{2}(\cos 45° + i\sin 45°)$

13. $z = -1 - i$ corresponds to the point $(-1, -1)$.

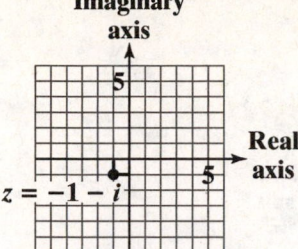

$z = -1 - i$

Use $r = \sqrt{a^2 + b^2}$ and $\tan\theta = \dfrac{b}{a}$, with

$a = -1$ and $b = -1$, to find r and θ.

$r = \sqrt{(-1)^2 + (-1)^2} = \sqrt{1+1} = \sqrt{2}$

$\tan\theta = \dfrac{-1}{-1} = 1$

Because $\tan\dfrac{\pi}{4} = 1$ and θ lies in

quadrant III, $\theta = \pi + \dfrac{\pi}{4} = \dfrac{5\pi}{4}$.

$z = -1 - i = r(\cos\theta + i\sin\theta)$

$= \sqrt{2}\left(\cos\dfrac{5\pi}{4} + i\sin\dfrac{5\pi}{4}\right)$

or $\sqrt{2}(\cos 225° + i\sin 225°)$

15. $z = -4i$ corresponds to the point $(0, -4)$.

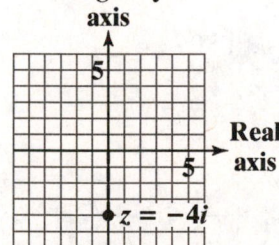

$z = -4i$

Use $r = \sqrt{a^2 + b^2}$ and $\tan\theta = \dfrac{b}{a}$, with

$a = 0$ and $b = -4$, to find r and θ.

$r = \sqrt{0^2 + (-4)^2} = \sqrt{16} = 4$

$\tan\theta = \dfrac{-4}{0}$ is undefined.

Because $\tan\dfrac{\pi}{2}$ is undefined and θ lies on the

negative y-axis, $\theta = \dfrac{\pi}{2} + \pi = \dfrac{3\pi}{2}$.

$z = -4i = r(\cos\theta + i\sin\theta)$

$= 4\left(\cos\dfrac{3\pi}{2} + i\sin\dfrac{3\pi}{2}\right)$

or $4(\cos 270° + i\sin 270°)$

17. $z = 2\sqrt{3} - 2i$ corresponds to the point $\left(2\sqrt{3}, -2\right)$.

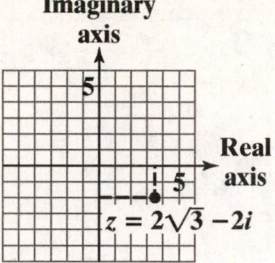

$z = 2\sqrt{3} - 2i$

Use $r = \sqrt{a^2 + b^2}$ and $\tan\theta = \dfrac{b}{a}$, with

$a = 2\sqrt{3}$ and $b = -2$, to find r and θ.

$r = \sqrt{\left(2\sqrt{3}\right)^2 + (-2)^2} = \sqrt{12+4} = \sqrt{16} = 4$

$\tan\theta = \dfrac{-2}{2\sqrt{3}} = -\dfrac{1}{\sqrt{3}}$

Because $\tan\dfrac{\pi}{6} = \dfrac{1}{\sqrt{3}}$ and θ lies in

quadrant IV, $\theta = 2\pi - \dfrac{\pi}{6} = \dfrac{11\pi}{6}$.

$z = 2\sqrt{3} - 2i = r(\cos\theta + i\sin\theta)$

$= 4\left(\cos\dfrac{11\pi}{6} + i\sin\dfrac{11\pi}{6}\right)$

or $4(\cos 330° + i\sin 330°)$

19. $z = -3$ corresponds to the point $(-3, 0)$.

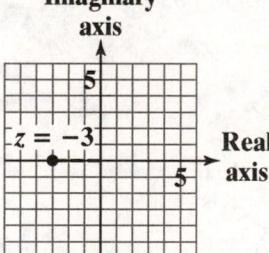

$z = -3$

Use $r = \sqrt{a^2 + b^2}$ and $\tan\theta = \dfrac{b}{a}$, with

$a = -3$ and $b = 0$, to find r and θ.

$r = \sqrt{(-3)^2 + 0^2} = \sqrt{9} = 3$

$\tan\theta = \dfrac{0}{-3} = 0$

Because $\tan 0 = 0$ and θ lies on the negative x-axis,

$\theta = 0 + \pi = \pi$.

$z = -3 = r(\cos\theta + i\sin\theta)$

$= 3\left(\cos\pi + i\sin\pi\right)$

or $3(\cos 180° + i\sin 180°)$

21. $z = -3\sqrt{2} - 3i\sqrt{3}$ corresponds to the point $\left(-3\sqrt{2}, -3\sqrt{3}\right)$.

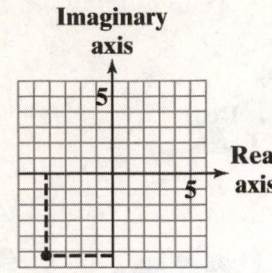

$z = -3\sqrt{2} - 3i\sqrt{3}$

Use $r = \sqrt{a^2 + b^2}$ and $\tan\theta = \dfrac{b}{a}$, with

$a = -3\sqrt{2}$ and $b = -3\sqrt{3}$, to find r and θ.

$r = \sqrt{\left(-3\sqrt{2}\right)^2 + \left(-3\sqrt{3}\right)^2} = \sqrt{18 + 27}$

$= \sqrt{45} = 3\sqrt{5}$

$\tan\theta = \dfrac{-3\sqrt{3}}{-3\sqrt{2}} = \dfrac{\sqrt{3}}{\sqrt{2}} = \dfrac{\sqrt{6}}{2}$

Because θ lies in quadrant III,

$\theta = 180° + \tan^{-1}\left(\dfrac{\sqrt{6}}{2}\right) \approx 180° + 50.8°$

$= 230.8°$

$z = -3\sqrt{2} - 3i\sqrt{3} = r(\cos\theta + i\sin\theta)$

$\approx 3\sqrt{5}(\cos 230.8° + i\sin 230.8°)$

23. $z = -3 + 4i$ corresponds to the point $(-3, 4)$.

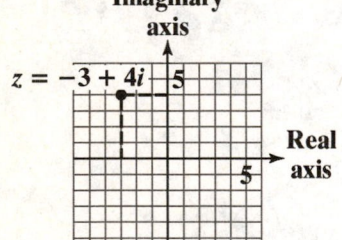

Use $r = \sqrt{a^2 + b^2}$ and $\tan\theta = \dfrac{b}{a}$, with

$a = -3$ and $b = 4$, to find r and θ.

$r = \sqrt{(-3)^2 + (4)^2} = \sqrt{9 + 16} = \sqrt{25} = 5$

$\tan\theta = \dfrac{4}{-3} = -\dfrac{4}{3}$

Because θ lies in quadrant II,

$\theta = 180° - \tan^{-1}\left(\dfrac{4}{3}\right) \approx 180° - 53.1° = 126.9°$.

$z = -3 + 4i = r(\cos\theta + i\sin\theta)$

$\approx 5(\cos 126.9° + i\sin 126.9°)$

25. $z = 2 - i\sqrt{3}$ corresponds to the point $\left(2, -\sqrt{3}\right)$.

Use $r = \sqrt{a^2 + b^2}$ and $\tan\theta = \dfrac{b}{a}$, with

$a = 2$ and $b = -\sqrt{3}$, to find r and θ.

$r = \sqrt{2^2 + \left(-\sqrt{3}\right)^2} = \sqrt{4 + 3} = \sqrt{7}$

$\tan\theta = \dfrac{-\sqrt{3}}{2} = -\dfrac{\sqrt{3}}{2}$

Because θ lies in quadrant IV,

$\theta = 360° - \tan^{-1}\left(\dfrac{\sqrt{3}}{2}\right) \approx 360° - 40.9° = 319.1°$

$z = 2 - i\sqrt{3} = r(\cos\theta + i\sin\theta)$

$\approx \sqrt{7}(\cos 319.1° + i\sin 319.1°)$

27. $6(\cos 30° + i\sin 30°) = 6\left(\dfrac{\sqrt{3}}{2} + i\dfrac{1}{2}\right)$

$= 3\sqrt{3} + 3i$

The rectangular form of

$z = 6(\cos 30° + i\sin 30°)$ is $z = 3\sqrt{3} + 3i$.

29. $4(\cos 240° + i\sin 240°) = 4\left(-\dfrac{1}{2} + i\left(-\dfrac{\sqrt{3}}{2}\right)\right)$

$= -2 - 2i\sqrt{3}$

The rectangular form of

$z = 4(\cos 240° + i\sin 240°)$ is $z = -2 - 2i\sqrt{3}$.

31. $8\left(\cos\dfrac{7\pi}{4} + i\sin\dfrac{7\pi}{4}\right) = 8\left(\dfrac{\sqrt{2}}{2} + i\left(-\dfrac{\sqrt{2}}{2}\right)\right)$

$= 4\sqrt{2} - 4i\sqrt{2}$

The rectangular form of

$8\left(\cos\dfrac{7\pi}{4} + i\sin\dfrac{7\pi}{4}\right)$ is $z = 4\sqrt{2} - 4i\sqrt{2}$.

33. $5\left(\cos\dfrac{\pi}{2}+i\sin\dfrac{\pi}{2}\right)=5\left(0+i(1)\right)$

$$=5i$$

The rectangular form of

$z=5\left(\cos\dfrac{\pi}{2}+i\sin\dfrac{\pi}{2}\right)$ is $z=5i$.

35. $20\left(\cos 205°+i\sin 205°\right)$

$\approx 20\left(-0.91+i(-0.42)\right)=-18.2-8.4i$

The rectangular form of

$z=20\left(\cos 205°+i\sin 205°\right)$ is $z\approx -18.2-8.5i$.

37. $z_1 z_2$

$=\left[6(\cos 20°+i\sin 20°)\right]\left[5(\cos 50°+i\sin 50°)\right]$

$=(6\cdot 5)\left[\cos(20°+50°)+i\sin(20°+50°)\right]$

$=30(\cos 70°+i\sin 70°)$

39. $z_1 z_2 =\left[3\left(\cos\dfrac{\pi}{5}+i\sin\dfrac{\pi}{5}\right)\right]\left[4\left(\cos\dfrac{\pi}{10}+i\sin\dfrac{\pi}{10}\right)\right]=(3\cdot 4)\left[\cos\left(\dfrac{\pi}{5}+\dfrac{\pi}{10}\right)+i\sin\left(\dfrac{\pi}{5}+\dfrac{\pi}{10}\right)\right]$

$=12\left(\cos\dfrac{3\pi}{10}+i\sin\dfrac{3\pi}{10}\right)$

41. $z_1 z_2 =\left[\cos\dfrac{\pi}{4}+i\sin\dfrac{\pi}{4}\right]\left[\cos\dfrac{\pi}{3}+i\sin\dfrac{\pi}{3}\right]=\cos\left(\dfrac{\pi}{4}+\dfrac{\pi}{3}\right)+i\sin\left(\dfrac{\pi}{4}+\dfrac{\pi}{3}\right)=\cos\left(\dfrac{3\pi}{12}+\dfrac{4\pi}{12}\right)+i\sin\left(\dfrac{3\pi}{12}+\dfrac{4\pi}{12}\right)$

$=\cos\dfrac{7\pi}{12}+i\sin\dfrac{7\pi}{12}$

43. Begin by converting $z_1 =1+i$ and $z_2 =-1+i$ to polar form.

For z_1: $a=1$ and $b=1$

$r=\sqrt{a^2+b^2}=\sqrt{1^2+1^2}=\sqrt{2}$

$\tan\theta=\dfrac{b}{a}=\dfrac{1}{1}=1$ and $\theta=\dfrac{\pi}{4}$.

$z_1 =r(\cos\theta+i\sin\theta)=\sqrt{2}\left(\cos\dfrac{\pi}{4}+i\sin\dfrac{\pi}{4}\right)$

For z_2: $a=-1$ and $b=1$

$r=\sqrt{a^2+b^2}=\sqrt{(-1)^2+1^2}=\sqrt{2}$

$\tan\theta=\dfrac{b}{a}=\dfrac{1}{-1}=-1$

Because $\tan\dfrac{\pi}{4}=1$ and θ lies in quadrant II, $\theta=\pi-\dfrac{\pi}{4}=\dfrac{3\pi}{4}$.

$z_2 =r(\cos\theta+i\sin\theta)=\sqrt{2}\left(\cos\dfrac{3\pi}{4}+i\sin\dfrac{3\pi}{4}\right)$

Now, find the product.

$z_1 z_2 = (1+i)(-1+i)$

$$= \left[\sqrt{2}\left(\cos\frac{\pi}{4} + i\sin\frac{\pi}{4}\right)\right]\left[\sqrt{2}\left(\cos\frac{3\pi}{4} + i\sin\frac{3\pi}{4}\right)\right] = \left(\sqrt{2}\cdot\sqrt{2}\right)\left[\cos\left(\frac{\pi}{4} + \frac{3\pi}{4}\right) + i\sin\left(\frac{\pi}{4} + \frac{3\pi}{4}\right)\right]$$

$$= 2\left(\cos\pi + i\sin\pi\right)$$

45. $\dfrac{z_1}{z_2} = \dfrac{20(\cos 75° + i\sin 75°)}{4(\cos 25° + i\sin 25°)} = \dfrac{20}{4}\left[\cos(75° - 25°) + i\sin(75° - 25°)\right]$

$= 5(\cos 50° + i\sin 50°)$

47. $\dfrac{z_1}{z_2} = \dfrac{3\left(\cos\dfrac{\pi}{5} + i\sin\dfrac{\pi}{5}\right)}{4\left(\cos\dfrac{\pi}{10} + i\sin\dfrac{\pi}{10}\right)} = \dfrac{3}{4}\left[\cos\left(\dfrac{\pi}{5} - \dfrac{\pi}{10}\right) + i\sin\left(\dfrac{\pi}{5} - \dfrac{\pi}{10}\right)\right] = \dfrac{3}{4}\left(\cos\dfrac{\pi}{10} + i\sin\dfrac{\pi}{10}\right)$

49. $\dfrac{z_1}{z_2} = \dfrac{\cos 80° + i\sin 80°}{\cos 200° + i\sin 200°} = \cos(80° - 200°) + i\sin(80° - 200°) = \cos(-120°) + i\sin(-120°)$

$= \cos 240° + i\sin 240°$

51. Begin by converting $z_1 = 2 + 2i$ and $z_2 = 1 + i$ to polar form.

For z_1: $a = 2$ and $b = 2$

$r = \sqrt{a^2 + b^2} = \sqrt{2^2 + 2^2} = \sqrt{8} = 2\sqrt{2}$

$\tan\theta = \dfrac{b}{a} = \dfrac{2}{2} = 1$ and $\theta = \dfrac{\pi}{4}$

$z_1 = r(\cos\theta + i\sin\theta) = 2\sqrt{2}\left(\cos\dfrac{\pi}{4} + i\sin\dfrac{\pi}{4}\right)$

For z_2: $a = 1$ and $b = 1$

$r = \sqrt{a^2 + b^2} = \sqrt{1^2 + 1^2} = \sqrt{2}$

$\tan\theta = \dfrac{b}{a} = \dfrac{1}{1} = 1$ and $\theta = \dfrac{\pi}{4}$

$z_2 = r(\cos\theta + i\sin\theta) = \sqrt{2}\left(\cos\dfrac{\pi}{4} + i\sin\dfrac{\pi}{4}\right).$

Now, find the quotient.

$$\dfrac{z_1}{z_2} = \dfrac{2+2i}{1+i} = \dfrac{2\sqrt{2}\left(\cos\dfrac{\pi}{4} + i\sin\dfrac{\pi}{4}\right)}{\sqrt{2}\left(\cos\dfrac{\pi}{4} + i\sin\dfrac{\pi}{4}\right)} = 2\left[\cos\left(\dfrac{\pi}{4} - \dfrac{\pi}{4}\right) + i\sin\left(\dfrac{\pi}{4} - \dfrac{\pi}{4}\right)\right] = 2(\cos 0 + i\sin 0)$$

53. $\left[4(\cos 15° + i\sin 15°)\right]^3 = (4)^3\left[\cos(3\cdot 15°) + i\sin(3\cdot 15°)\right] = 64(\cos 45° + i\sin 45°) = 64\left(\dfrac{\sqrt{2}}{2} + i\dfrac{\sqrt{2}}{2}\right)$

$= 32\sqrt{2} + 32i\sqrt{2}$

55. $\left[2(\cos 80° + i \sin 80°)\right]^3 = (2)^3 \left[\cos(3 \cdot 80°) + i \sin(3 \cdot 80°)\right] = 8(\cos 240° + i \sin 240°)$

$$= 8\left(-\frac{1}{2} + i\left(-\frac{\sqrt{3}}{2}\right)\right) = -4 - 4i\sqrt{3}$$

57. $\left[\frac{1}{2}\left(\cos\frac{\pi}{12} + i\sin\frac{\pi}{12}\right)\right]^6 = \left(\frac{1}{2}\right)^6 \left[\cos\left(6 \cdot \frac{\pi}{12}\right) + i\sin\left(6 \cdot \frac{\pi}{12}\right)\right] = \frac{1}{64}\left(\cos\frac{\pi}{2} + i\sin\frac{\pi}{2}\right) = \frac{1}{64}(0 + i) = \frac{1}{64}i$

59. $\left[\sqrt{2}\left(\cos\frac{5\pi}{6} + i\sin\frac{5\pi}{6}\right)\right]^4 = \left(\sqrt{2}\right)^4 \left[\cos\left(4 \cdot \frac{5\pi}{6}\right) + i\sin\left(4 \cdot \frac{5\pi}{6}\right)\right]$

$$= 4\left(\cos\frac{20\pi}{6} + i\sin\frac{20\pi}{6}\right) = 4\left(\cos\frac{4\pi}{3} + i\sin\frac{4\pi}{3}\right)$$

$$= 4\left(-\frac{1}{2} + i\left(-\frac{\sqrt{3}}{2}\right)\right) = -2 - 2i\sqrt{3}$$

61. Write $1 + i$ in $r(\cos\theta + i\sin\theta)$ form.

$r = \sqrt{a^2 + b^2} = \sqrt{1^2 + 1^2} = \sqrt{2}$

$\tan\theta = \frac{b}{a} = \frac{1}{1} = 1$ and $\theta = \frac{\pi}{4}$

$1 + i = r(\cos\theta + i\sin\theta) = \sqrt{2}\left(\cos\frac{\pi}{4} + i\sin\frac{\pi}{4}\right)$

Use DeMoivre's Theorem to raise $1 + i$ to the fifth power.

$(1 + i)^5 = \left[\sqrt{2}\left(\cos\frac{\pi}{4} + i\sin\frac{\pi}{4}\right)\right]^5 = \left(\sqrt{2}\right)^5 \left[\cos\left(5 \cdot \frac{\pi}{4}\right) + i\sin\left(5 \cdot \frac{\pi}{4}\right)\right]$

$$= 4\sqrt{2}\left(\cos\frac{5\pi}{4} + i\sin\frac{5\pi}{4}\right) = 4\sqrt{2}\left(-\frac{\sqrt{2}}{2} + i\left(-\frac{\sqrt{2}}{2}\right)\right) = -4 - 4i$$

63. Write $\sqrt{3} - i$ in $r(\cos\theta + i\sin\theta)$ form.

$r = \sqrt{a^2 + b^2} = \sqrt{\left(\sqrt{3}\right)^2 + (-1)^2} = \sqrt{4} = 2$

$\tan\theta = \frac{b}{a} = \frac{-1}{\sqrt{3}} = -\frac{1}{\sqrt{3}}$

Because $\tan 30° = \frac{1}{\sqrt{3}}$ and θ lies in quadrant IV, $\theta = 360° - 30° = 330°$.

$\sqrt{3} - i = r(\cos\theta + i\sin\theta) = 2(\cos 330° + i\sin 330°)$

Use DeMoivre's Theorem to raise $\sqrt{3} - i$ to the sixth power.

$(\sqrt{3} - i)^6 = \left[2(\cos 330° + i\sin 330°)\right]^6 = (2)^6 \left[\cos(6 \cdot 330°) + i\sin(6 \cdot 330°)\right]$

$$= 64(\cos 1980° + i\sin 1980°) = 64(\cos 180° + i\sin 180°)$$

$$= 64(-1 + 0i) = -64$$

65. $9(\cos 30° + i\sin 30°)$

$$z_k = \sqrt[2]{9}\left[\cos\left(\frac{30°+360°k}{2}\right) + i\sin\left(\frac{30°+360°k}{2}\right)\right], \ k=0,1$$

$$z_0 = \sqrt{9}\left[\cos\left(\frac{30°+360°\cdot 0}{2}\right) + i\sin\left(\frac{30°+360°\cdot 0}{2}\right)\right] = \sqrt{9}\left[\cos\left(\frac{30°}{2}\right) + i\sin\left(\frac{30°}{2}\right)\right] = 3(\cos 15° + i\sin 15°)$$

$$z_1 = \sqrt{9}\left[\cos\left(\frac{30°+360°\cdot 1}{2}\right) + i\sin\left(\frac{30°+360°\cdot 1}{2}\right)\right] = \sqrt{9}\left[\cos\left(\frac{390°}{2}\right) + i\sin\left(\frac{390°}{2}\right)\right] = 3(\cos 195° + i\sin 195°)$$

67. $8(\cos 210° + i\sin 210°)$

$$z_k = \sqrt[3]{8}\left[\cos\left(\frac{210°+360°k}{3}\right) + i\sin\left(\frac{210°+360°k}{3}\right)\right], \ k=0,1,2$$

$$z_0 = \sqrt[3]{8}\left[\cos\left(\frac{210°+360°\cdot 0}{3}\right) + i\sin\left(\frac{210°+360°\cdot 0}{3}\right)\right] = \sqrt[3]{8}\left[\cos\left(\frac{210°}{3}\right) + i\sin\left(\frac{210°}{3}\right)\right]$$

$$= 2(\cos 70° + i\sin 70°)$$

$$z_1 = \sqrt[3]{8}\left[\cos\left(\frac{210°+360°\cdot 1}{3}\right) + i\sin\left(\frac{210°+360°\cdot 1}{3}\right)\right] = \sqrt[3]{8}\left[\cos\left(\frac{570°}{3}\right) + i\sin\left(\frac{570°}{3}\right)\right]$$

$$= 2(\cos 190° + i\sin 190°)$$

$$z_2 = \sqrt[3]{8}\left[\cos\left(\frac{210°+360°\cdot 2}{3}\right) + i\sin\left(\frac{210°+360°\cdot 2}{3}\right)\right] = \sqrt[3]{8}\left[\cos\left(\frac{930°}{3}\right) + i\sin\left(\frac{930°}{3}\right)\right]$$

$$= 2(\cos 310° + i\sin 310°)$$

69. $81\left(\cos\frac{4\pi}{3} + i\sin\frac{4\pi}{3}\right)$

$$z_k = \sqrt[4]{81}\left[\cos\left(\frac{\frac{4\pi}{3}+2\pi k}{4}\right) + i\sin\left(\frac{\frac{4\pi}{3}+2\pi k}{4}\right)\right], \ k=0,1,2,3$$

$$z_0 = \sqrt[4]{81}\left[\cos\left(\frac{\frac{4\pi}{3}+2\pi\cdot 0}{4}\right) + i\sin\left(\frac{\frac{4\pi}{3}+2\pi\cdot 0}{4}\right)\right] = \sqrt[4]{81}\left(\cos\frac{\pi}{3} + i\sin\frac{\pi}{3}\right) = 3\left(\frac{1}{2}+i\frac{\sqrt{3}}{2}\right) = \frac{3}{2}+\frac{3\sqrt{3}}{2}i$$

$$z_1 = \sqrt[4]{81}\left[\cos\left(\frac{\frac{4\pi}{3}+2\pi\cdot 1}{4}\right) + i\sin\left(\frac{\frac{4\pi}{3}+2\pi\cdot 1}{4}\right)\right] = \sqrt[4]{81}\left(\cos\frac{5\pi}{6} + i\sin\frac{5\pi}{6}\right) = 3\left(-\frac{\sqrt{3}}{2}+i\frac{1}{2}\right) = -\frac{3\sqrt{3}}{2}+\frac{3}{2}i$$

$$z_2 = \sqrt[4]{81}\left[\cos\left(\frac{\frac{4\pi}{3}+2\pi\cdot 2}{4}\right) + i\sin\left(\frac{\frac{4\pi}{3}+2\pi\cdot 2}{4}\right)\right] = \sqrt[4]{81}\left(\cos\frac{4\pi}{3} + i\sin\frac{4\pi}{3}\right)$$

$$= 3\left(-\frac{1}{2}+i\left(-\frac{\sqrt{3}}{2}\right)\right) = -\frac{3}{2}-\frac{3\sqrt{3}}{2}i$$

$$z_3 = \sqrt[4]{81}\left[\cos\left(\frac{\frac{4\pi}{3}+2\pi\cdot 3}{4}\right) + i\sin\left(\frac{\frac{4\pi}{3}+2\pi\cdot 3}{4}\right)\right] = \sqrt[4]{81}\left(\cos\frac{11\pi}{6} + i\sin\frac{11\pi}{6}\right)$$

$$= 3\left(\frac{\sqrt{3}}{2}+i\left(-\frac{1}{2}\right)\right) = \frac{3\sqrt{3}}{2}-\frac{3}{2}i$$

71. $32 = 32(\cos 0° + i\sin 0°)$

$z_k = \sqrt[5]{32}\left[\cos\left(\dfrac{0° + 360°k}{5}\right) + i\sin\left(\dfrac{0° + 360°k}{5}\right)\right], \ k = 0, 1, 2, 3, 4$

$z_0 = \sqrt[5]{32}\left[\cos\left(\dfrac{0° + 360°\cdot 0}{5}\right) + i\sin\left(\dfrac{0° + 360°\cdot 0}{5}\right)\right] = \sqrt[5]{32}(\cos 0° + i\sin 0°) = 2(1 + 0i) = 2$

$z_1 = \sqrt[5]{32}\left[\cos\left(\dfrac{0° + 360°\cdot 1}{5}\right) + i\sin\left(\dfrac{0° + 360°\cdot 1}{5}\right)\right] = \sqrt[5]{32}(\cos 72° + i\sin 72°) \approx 2(0.31 + i(0.95))$

$\approx 0.6 + 1.9i$

$z_2 = \sqrt[5]{32}\left[\cos\left(\dfrac{0° + 360°\cdot 2}{5}\right) + i\sin\left(\dfrac{0° + 360°\cdot 2}{5}\right)\right] = \sqrt[5]{32}(\cos 144° + i\sin 144°) \approx 2(-0.81 + i(0.59))$

$\approx -1.6 + 1.2i$

$z_3 = \sqrt[5]{32}\left[\cos\left(\dfrac{0° + 360°\cdot 3}{5}\right) + i\sin\left(\dfrac{0° + 360°\cdot 3}{5}\right)\right] = \sqrt[5]{32}(\cos 216° + i\sin 216°) \approx 2(-0.81 + i(-0.59))$

$\approx -1.6 - 1.2i$

$z_4 = \sqrt[5]{32}\left[\cos\left(\dfrac{0° + 360°\cdot 4}{5}\right) + i\sin\left(\dfrac{0° + 360°\cdot 4}{5}\right)\right] = \sqrt[5]{32}(\cos 288° + i\sin 288°) \approx 2(0.31 + i(-0.95))$

$\approx 0.6 - 1.9i$

73. $1 = 1(\cos 0° + i\sin 0°)$

$z_k = \sqrt[3]{1}\left[\cos\left(\dfrac{0° + 360°k}{3}\right) + i\sin\left(\dfrac{0° + 360°k}{3}\right)\right], \ k = 0, 1, 2$

$z_0 = \sqrt[3]{1}\left[\cos\left(\dfrac{0° + 360°\cdot 0}{3}\right) + i\sin\left(\dfrac{0° + 360°\cdot 0}{3}\right)\right] = \sqrt[3]{1}\left(\cos 0° + i\sin 0°\right) = 1(1 + 0i) = 1$

$z_1 = \sqrt[3]{1}\left[\cos\left(\dfrac{0° + 360°\cdot 1}{3}\right) + i\sin\left(\dfrac{0° + 360°\cdot 1}{3}\right)\right] = \sqrt[3]{1}\left(\cos 120° + i\sin 120°\right) = 1\left(-\dfrac{1}{2} + i\dfrac{\sqrt{3}}{2}\right) = -\dfrac{1}{2} + \dfrac{\sqrt{3}}{2}i$

$z_2 = \sqrt[3]{1}\left[\cos\left(\dfrac{0° + 360°\cdot 2}{3}\right) + i\sin\left(\dfrac{0° + 360°\cdot 2}{3}\right)\right] = \sqrt[3]{1}\left(\cos 240° + i\sin 240°\right) = 1\left(-\dfrac{1}{2} + i\left(-\dfrac{\sqrt{3}}{2}\right)\right)$

$= -\dfrac{1}{2} - \dfrac{\sqrt{3}}{2}i$

75. $1 + i = \sqrt{2}\left(\cos 45° + i\sin 45°\right)$

$z_k = \sqrt[4]{\sqrt{2}}\left[\cos\left(\dfrac{45° + 360°k}{4}\right) + i\sin\left(\dfrac{45° + 360°k}{4}\right)\right], \ k = 0, 1, 2, 3$

$z_0 = \sqrt[4]{\sqrt{2}}\left[\cos\left(\dfrac{45° + 360°\cdot 0}{4}\right) + i\sin\left(\dfrac{45° + 360°\cdot 0}{4}\right)\right] = \sqrt[4]{\sqrt{2}}\left(\cos 11.25° + i\sin 11.25°\right) \approx 1.1 + 0.2i$

$z_1 = \sqrt[4]{\sqrt{2}}\left[\cos\left(\dfrac{45° + 360°\cdot 1}{4}\right) + i\sin\left(\dfrac{45° + 360°\cdot 1}{4}\right)\right] = \sqrt[4]{\sqrt{2}}\left(\cos 101.25° + i\sin 101.25°\right) \approx -0.2 + 1.1i$

$z_2 = \sqrt[4]{\sqrt{2}}\left[\cos\left(\dfrac{45° + 360°\cdot 2}{4}\right) + i\sin\left(\dfrac{45° + 360°\cdot 2}{4}\right)\right] = \sqrt[4]{\sqrt{2}}\left(\cos 191.25° + i\sin 191.25°\right) \approx -1.1 - 0.2i$

$z_3 = \sqrt[4]{\sqrt{2}}\left[\cos\left(\dfrac{45° + 360°\cdot 3}{4}\right) + i\sin\left(\dfrac{45° + 360°\cdot 3}{4}\right)\right] = \sqrt[4]{\sqrt{2}}\left(\cos 281.25° + i\sin 281.25°\right) \approx 0.2 - 1.1i$

77. $i(2+2i)(-\sqrt{3}+i)$

$$= \left[1(\cos 90° + i\sin 90°)\right]\left[2\sqrt{2}(\cos 45° + i\sin 45°)\right]\left[2(\cos 150° + i\sin 150°)\right]$$

$$= 4\sqrt{2}(\cos 285° + i\sin 285°)$$

$$\approx 1.4641 - 5.4641i$$

79. $\dfrac{\left(1+i\sqrt{3}\right)(1-i)}{2\sqrt{3}-2i}$

$$= \frac{\left[2(\cos 60° + i\sin 60°)\right]\left[\sqrt{2}(\cos 315° + i\sin 315°)\right]}{\left[4(\cos 330° + i\sin 330°)\right]}$$

$$= \frac{\sqrt{2}}{2}(\cos 45° + i\sin 45°)$$

$$= \frac{1}{2} + \frac{1}{2}i$$

81. $x^6 - 1 = 0$

$$x^6 = 1$$

$$x = \sqrt[6]{1}$$

$$x = \sqrt[6]{1 + 0i}$$

$$x = \sqrt[6]{\cos 0° + i\sin 0°}$$

$$z_k = \sqrt[6]{1}\left[\cos\left(\frac{0° + 360° \cdot k}{6}\right) + i\sin\left(\frac{0° + 360° \cdot k}{6}\right)\right], \; k = 0,1,2,3,4,5$$

$$z_0 = \sqrt[6]{1}\left[\cos\left(\frac{0° + 360° \cdot 0}{6}\right) + i\sin\left(\frac{0° + 360° \cdot 0}{6}\right)\right] = \cos 0° + i\sin 0° = 1 + 0i = 1$$

$$z_1 = \sqrt[6]{1}\left[\cos\left(\frac{0° + 360° \cdot 1}{6}\right) + i\sin\left(\frac{0° + 360° \cdot 1}{6}\right)\right] = \cos 60° + i\sin 60° = \frac{1}{2} + \frac{\sqrt{3}}{2}i$$

$$z_2 = \sqrt[6]{1}\left[\cos\left(\frac{0° + 360° \cdot 2}{6}\right) + i\sin\left(\frac{0° + 360° \cdot 2}{6}\right)\right] = \cos 120° + i\sin 120° = -\frac{1}{2} + \frac{\sqrt{3}}{2}i$$

$$z_3 = \sqrt[6]{1}\left[\cos\left(\frac{0° + 360° \cdot 3}{6}\right) + i\sin\left(\frac{0° + 360° \cdot 3}{6}\right)\right] = \cos 180° + i\sin 180° = -1 + 0i = -1$$

$$z_4 = \sqrt[6]{1}\left[\cos\left(\frac{0° + 360° \cdot 4}{6}\right) + i\sin\left(\frac{0° + 360° \cdot 4}{6}\right)\right] = \cos 240° + i\sin 240° = -\frac{1}{2} + i\left(-\frac{\sqrt{3}}{2}\right) = -\frac{1}{2} - \frac{\sqrt{3}}{2}i$$

$$z_5 = \sqrt[6]{1}\left[\cos\left(\frac{0° + 360° \cdot 5}{6}\right) + i\sin\left(\frac{0° + 360° \cdot 5}{6}\right)\right] = \cos 300° + i\sin 300° = \frac{1}{2} + i\left(-\frac{\sqrt{3}}{2}\right) = \frac{1}{2} - \frac{\sqrt{3}}{2}i$$

83. $x^4 + 16i = 0$

$$x^4 = -16i$$

$$x = \sqrt[4]{-16i}$$

$$x = \sqrt[4]{0 - 16i}$$

$$x = \sqrt[4]{16\left(\cos 270° + i\sin 270°\right)}$$

$$z_k = \sqrt[4]{16}\left[\cos\left(\frac{270° + 360°k}{4}\right) + i\sin\left(\frac{270° + 360°k}{4}\right)\right], \; k = 0,1,2,3$$

$$z_0 = \sqrt[4]{16}\left[\cos\left(\frac{270° + 360° \cdot 0}{4}\right) + i\sin\left(\frac{270° + 360° \cdot 0}{4}\right)\right] = 2\left(\cos 67.5° + i\sin 67.5°\right) \approx 0.7654 + 1.8478i$$

$$z_1 = \sqrt[4]{16}\left[\cos\left(\frac{270° + 360° \cdot 1}{4}\right) + i\sin\left(\frac{270° + 360° \cdot 1}{4}\right)\right] = 2\left(\cos 157.5° + i\sin 157.5°\right) \approx -1.8478 + 0.7654i$$

$$z_2 = \sqrt[4]{16}\left[\cos\left(\frac{270° + 360° \cdot 2}{4}\right) + i\sin\left(\frac{270° + 360° \cdot 2}{4}\right)\right] = 2\left(\cos 247.5° + i\sin 247.5°\right) \approx -0.7654 - 1.8478i$$

$$z_3 = \sqrt[4]{16}\left[\cos\left(\frac{270° + 360° \cdot 3}{4}\right) + i\sin\left(\frac{270° + 360° \cdot 3}{4}\right)\right] = 2\left(\cos 337.5° + i\sin 337.5°\right) \approx 1.8478 - 0.7654i$$

85. $x^3 - \left(1 + i\sqrt{3}\right) = 0$

$$x^3 = 1 + i\sqrt{3}$$

$$x = \sqrt[3]{1 + i\sqrt{3}}$$

$$x = \sqrt[3]{2(\cos 60° + i\sin 60°)}$$

$$z_k = \sqrt[3]{2}\left[\cos\left(\frac{60° + 360°k}{3}\right) + i\sin\left(\frac{60° + 360°k}{3}\right)\right], \; k = 0,1,2$$

$$z_0 = \sqrt[3]{2}\left[\cos\left(\frac{60° + 360° \cdot 0}{3}\right) + i\sin\left(\frac{60° + 360° \cdot 0}{3}\right)\right] = \sqrt[3]{2}\left(\cos 20° + i\sin 20°\right) \approx 1.1839 + 0.4309i$$

$$z_1 = \sqrt[3]{2}\left[\cos\left(\frac{60° + 360° \cdot 1}{3}\right) + i\sin\left(\frac{60° + 360° \cdot 1}{3}\right)\right] = \sqrt[3]{2}\left(\cos 140° + i\sin 140°\right) \approx -0.9652 + 0.8099i$$

$$z_2 = \sqrt[3]{2}\left[\cos\left(\frac{60° + 360° \cdot 2}{3}\right) + i\sin\left(\frac{60° + 360° \cdot 2}{3}\right)\right] = \sqrt[3]{2}\left(\cos 260° + i\sin 260°\right) \approx -0.2188 - 1.2408i$$

87. $e^{\frac{\pi i}{4}} = \cos\frac{\pi}{4} + i\sin\frac{\pi}{4}$

$= \frac{\sqrt{2}}{2} + \frac{\sqrt{2}}{2}i$

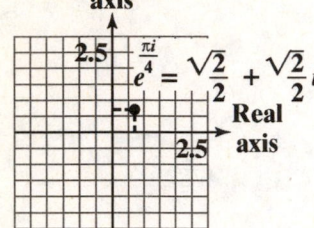

89. $-e^{-\pi i} = -1\left(\cos(-\pi) + i\sin(-\pi)\right)$

$= -(-1) - i(0)$

$= 1 + 0i$

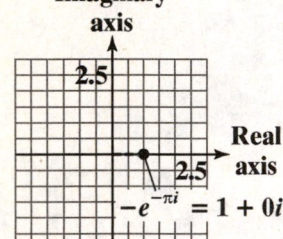

91. $z = i$

a. $z_1 = z = i$

$z_2 = z^2 + z = (i)^2 + i = -1 + i$

$z_3 = \left(z^2 + z\right)^2 + z = z_2^2 + z = (-1+i)^2 + i = -i$

$z_4 = \left[\left(z^2 + z\right)^2 + z\right]^2 + z = z_3^2 + z = (-i)^2 + i = -1 + i$

$z_5 = z_4^2 + z = (-1+i)^2 + i = -i$

$z_6 = z_5^2 + z = (-i)^2 + i = -1 + i$

b. $\left|-1+i\right| = \sqrt{(-1)^2 + 1^2} = \sqrt{2}$

$\left|i\right| = \sqrt{0^2 + 1^2} = 1$

The absolute values of the terms in the sequence are 1 and $\sqrt{2}$.

Choose a complex number with absolute value less than 1, and another with absolute value greater than $\sqrt{2}$. Complex numbers may vary.

93. – 105. Answers may vary.

107. does not make sense; Explanations will vary. Sample explanation: This process involves four multiplications.

109. does not make sense; Explanations will vary. Sample explanation: $-1-i\sqrt{3}$ and $-1+i\sqrt{3}$ are the other 2 cube roots of 8.

111. $1 = 1\left(\cos 0° + i\sin 0°\right)$

$$z_k = \sqrt[4]{1}\left[\cos\left(\frac{0°+360°k}{4}\right) + i\sin\left(\frac{0°+360°k}{4}\right)\right], \ k = 0,1,2,3$$

$$z_0 = \sqrt[4]{1}\left[\cos\left(\frac{0°+360°\cdot 0}{4}\right) + i\sin\left(\frac{0°+360°\cdot 0}{4}\right)\right] = \sqrt[4]{1}\left(\cos 0° + i\sin 0\right) = 1(1+0i) = 1$$

$$z_1 = \sqrt[4]{1}\left[\cos\left(\frac{0°+360°\cdot 1}{4}\right) + i\sin\left(\frac{0°+360°\cdot 1}{4}\right)\right] = \sqrt[4]{1}\left(\cos 90° + i\sin 90°\right) = 1(0+i(1)) = i$$

$$z_2 = \sqrt[4]{1}\left[\cos\left(\frac{0°+360°\cdot 2}{4}\right) + i\sin\left(\frac{0°+360°\cdot 2}{4}\right)\right] = \sqrt[4]{1}\left(\cos 180° + i\sin 180°\right) = 1(-1+0i) = -1$$

$$z_3 = \sqrt[4]{1}\left[\cos\left(\frac{0°+360°\cdot 3}{4}\right) + i\sin\left(\frac{0°+360°\cdot 3}{4}\right)\right] = \sqrt[4]{1}\left(\cos 270° + i\sin 270°\right) = 1(0+i(-1)) = -i$$

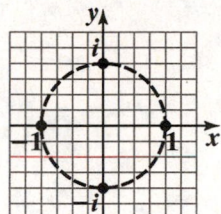

112. Answers may vary.

113. Find the distance from $(-3,-3)$ and $(0,3)$.

$$d = \sqrt{(x_1 - x_2)^2 + (y_1 - y_2)^2}$$
$$= \sqrt{(3+3)^2 + (0+3)^2}$$
$$= \sqrt{45}$$
$$= 3\sqrt{5}$$

Find the distance from $(0,0)$ and $(3,6)$.

$$d = \sqrt{(x_1 - x_2)^2 + (y_1 - y_2)^2}$$
$$= \sqrt{(6-0)^2 + (3-0)^2}$$
$$= \sqrt{45}$$
$$= 3\sqrt{5}$$

The line segments have the same length.

Section 6.6

Check Point Exercises

1. First, we show that **u** and **v** have the same magnitude.

$$\|\mathbf{u}\| = \sqrt{(x_2 - x_1)^2 + (y_2 - y_1)^2}$$
$$= \sqrt{(-2 - (-5))^2 + (6 - 2)^2}$$
$$= \sqrt{3^2 + 4^2}$$
$$= \sqrt{9 + 16}$$
$$= \sqrt{25}$$
$$= 5$$

$$\|\mathbf{v}\| = \sqrt{(x_2 - x_1)^2 + (y_2 - y_1)^2}$$
$$= \sqrt{(5 - 2)^2 + (6 - 2)^2}$$
$$= \sqrt{3^2 + 4^2}$$
$$= \sqrt{9 + 16}$$
$$= \sqrt{25}$$
$$= 5$$

Thus, **u** and **v** have the same magnitude: $\|\mathbf{u}\| = \|\mathbf{v}\|$.

Next, we show that **u** and **v** have the same direction. the line on which **u** lies has slope

$$m = \frac{y_2 - y_1}{x_2 - x_1} = \frac{6 - 2}{-2 - (-5)} = \frac{4}{3}.$$

The line on which **v** lies has slope

$$m = \frac{y_2 - y_1}{x_2 - x_1} = \frac{6 - 2}{5 - 2} = \frac{4}{3}.$$

Because **u** and **v** are both directed toward

the upper right on lines having the same slope, $\frac{4}{3}$,

they have the same direction. Thus, **u** and **v** have the same magnitude and direction, and **u** = **v**.

2. For the given vector $\mathbf{v} = 3\mathbf{i} - 3\mathbf{j}$, $a = 3$ and $b = -3$. The vector's initial point is the origin, $(0, 0)$. The vector's terminal point is $(a, b) = (3, -3)$. We sketch the vector by drawing an arrow from $(0, 0)$ to $(3, -3)$.

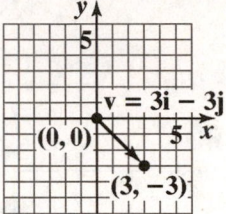

We determine the magnitude of the vector by using the distance formula. Thus, the magnitude is

$$\|\mathbf{v}\| = \sqrt{a^2 + b^2}$$
$$= \sqrt{3^2 + (-3)^2}$$
$$= \sqrt{9 + 9}$$
$$= \sqrt{18}$$
$$= 3\sqrt{2}.$$

3. We identify the values for the variables in the formula.

$$P_1 = (-1, 3) \quad P_2 = (2, 7)$$

$$\begin{array}{cc} \uparrow\uparrow & \uparrow\uparrow \\ x_1\ y_1 & x_2\ y_2 \end{array}$$

Using these values, we write **v** in terms of **i** and **j** as follows:

$$\mathbf{v} = (x_2 - x_1)\mathbf{i} + (y_2 - y_1)\mathbf{j}$$
$$= (2 - (-1))\mathbf{i} + (7 - 3)\mathbf{j}$$
$$= 3\mathbf{i} + 4\mathbf{j}$$

4. a. $\mathbf{v} + \mathbf{w} = (7\mathbf{i} + 3\mathbf{j}) + (4\mathbf{i} - 5\mathbf{j})$
$$= (7 + 4)\mathbf{i} + (3 - 5)\mathbf{j}$$
$$= 11\mathbf{i} - 2\mathbf{j}$$

   b. $\mathbf{v} - \mathbf{w} = (7\mathbf{i} + 3\mathbf{j}) - (4\mathbf{i} - 5\mathbf{j})$
$$= (7 - 4)\mathbf{i} + (3 - (-5))\mathbf{j}$$
$$= 3\mathbf{i} + 8\mathbf{j}$$

5. **a.** $8\mathbf{v} = 8(7\mathbf{i} + 10\mathbf{j})$
$= (8 \cdot 7)\mathbf{i} + (8 \cdot 10)\mathbf{j}$
$= 56\mathbf{i} + 80\mathbf{j}$

b. $-5\mathbf{v} = -5(7\mathbf{i} + 10\mathbf{j})$
$= (-5 \cdot 7)\mathbf{i} + (-5 \cdot 10)\mathbf{j}$
$= -35\mathbf{i} - 50\mathbf{j}$

6. $6\mathbf{v} - 3\mathbf{w} = 6(7\mathbf{i} + 3\mathbf{j}) - 3(4\mathbf{i} - 5\mathbf{j})$
$= 42\mathbf{i} + 18\mathbf{j} - 12\mathbf{i} + 15\mathbf{j}$
$= (42 - 12)\mathbf{i} + (18 + 15)\mathbf{j}$
$= 30\mathbf{i} + 33\mathbf{j}$

7. First, find the magnitude of **v**.
$\|\mathbf{v}\| = \sqrt{a^2 + b^2}$
$= \sqrt{4^2 + (-3)^2}$
$= \sqrt{16 + 9}$
$= \sqrt{25}$
$= 5$

A unit vector in the same direction as **v** is
$\dfrac{\mathbf{v}}{\|\mathbf{v}\|} = \dfrac{4\mathbf{i} - 3\mathbf{j}}{5} = \dfrac{4}{5}\mathbf{i} - \dfrac{3}{5}\mathbf{j}$

Now, we must verify that the magnitude of the vector is 1. The magnitude of $\dfrac{4}{5}\mathbf{i} - \dfrac{3}{5}\mathbf{j}$ is

$\sqrt{\left(\dfrac{4}{5}\right)^2 + \left(-\dfrac{3}{5}\right)^2} = \sqrt{\dfrac{16}{25} + \dfrac{9}{25}} = \sqrt{\dfrac{25}{25}} = 1.$

8. $60 \cos 45° \, \mathbf{i} + 60 \sin 45° \, \mathbf{j}$
$= 60 \cdot \dfrac{\sqrt{2}}{2}\mathbf{i} + 60 \cdot \dfrac{\sqrt{2}}{2}\mathbf{j}$
$= 30\sqrt{2}\mathbf{i} + 30\sqrt{2}\mathbf{j}$

9. We need to find $\|\mathbf{F}\|$ and θ.

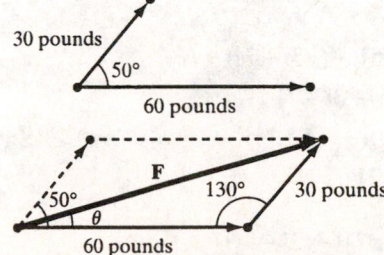

Use the Law of Cosines to find the magnitude of **F**.

$\|\mathbf{F}\|^2 = 60^2 + 30^2 - 2(60)(30)\cos 130° \approx 6814$ The
$\|\mathbf{F}\| \approx \sqrt{6814} \approx 82.5$
magnitude of the resultant force is about 82.5 pounds.

To find θ, the direction of the resultant force, we use the Law of Sines.

$\dfrac{82.5}{\sin 130°} = \dfrac{30}{\sin \theta}$
$82.5 \sin \theta = 30 \sin 130°$
$\sin \theta = \dfrac{30 \sin 130°}{82.5}$
$\theta = \sin^{-1}\left(\dfrac{30 \sin 130°}{82.5}\right) \approx 16.2°$

The two given forces are equivalent to a single force of about 82.5 pounds in the direction of approximately 16.2° relative to the 60-pound force.

Exercise Set 6.6

1. **a.** $\|\mathbf{u}\| = \sqrt{(x_2 - x_1)^2 + (y_2 - y_1)^2}$
$= \sqrt{(4 - (-1))^2 + (6 - 2)^2}$
$= \sqrt{5^2 + 4^2}$
$= \sqrt{25 + 16}$
$= \sqrt{41}$

b. $\|\mathbf{v}\| = \sqrt{(x_2 - x_1)^2 + (y_2 - y_1)^2}$
$= \sqrt{(5 - 0)^2 + (4 - 0)^2}$
$= \sqrt{5^2 + 4^2}$
$= \sqrt{25 + 16}$
$= \sqrt{41}$

c. Since $\|\mathbf{u}\| = \|\mathbf{v}\|$, and **u** and **v** have the same direction, we can conclude that $\mathbf{u} = \mathbf{v}$.

3. **a.** $\|\mathbf{u}\| = \sqrt{(x_2 - x_1)^2 + (y_2 - y_1)^2}$

$= \sqrt{(5 - (-1))^2 + (1 - 1)^2}$

$= \sqrt{6^2 + 0^2}$

$= \sqrt{36 + 0}$

$= \sqrt{36}$

$= 6$

b. $\|\mathbf{v}\| = \sqrt{(x_2 - x_1)^2 + (y_2 - y_1)^2}$

$= \sqrt{(4 - (-2))^2 + (-1 - (-1))^2}$

$= \sqrt{6^2 + 0^2}$

$= \sqrt{36 + 0}$

$= \sqrt{36}$

$= 6$

c. Since $\|\mathbf{u}\| = \|\mathbf{v}\|$, and $\mathbf{u}$ and $\mathbf{v}$ have the same direction, we can conclude that

$\mathbf{u} = \mathbf{v}$.

5.

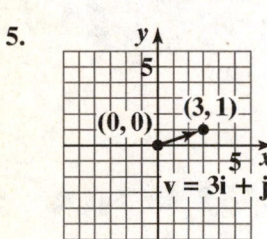

$\|\mathbf{v}\| = \sqrt{3^2 + 1^2} = \sqrt{9 + 1} = \sqrt{10}$

7.

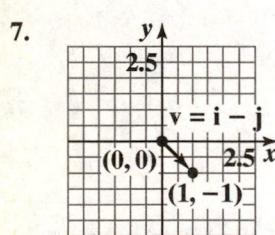

$\|\mathbf{v}\| = \sqrt{1^2 + (-1)^2} = \sqrt{1 + 1} = \sqrt{2}$

9.

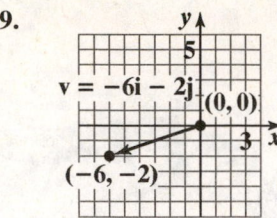

$\|\mathbf{v}\| = \sqrt{(-6)^2 + (-2)^2}$

$= \sqrt{36 + 4}$

$= \sqrt{40}$

$= 2\sqrt{10}$

11.

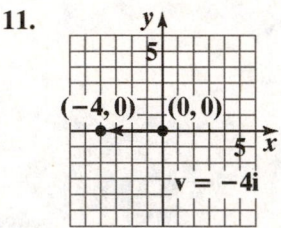

$\|\mathbf{v}\| = \sqrt{(-4)^2 + 0^2} = \sqrt{16 + 0} = \sqrt{16} = 4$

13. $\mathbf{v} = (x_2 - x_1)\mathbf{i} + (y_2 - y_1)\mathbf{j}$

$\mathbf{v} = (6 - (-4))\mathbf{i} + (2 - (-4))\mathbf{j} = 10\mathbf{i} + 6\mathbf{j}$

15. $\mathbf{v} = (x_2 - x_1)\mathbf{i} + (y_2 - y_1)\mathbf{j}$

$\mathbf{v} = (-2 - (-8))\mathbf{i} + (3 - 6)\mathbf{j} = 6\mathbf{i} - 3\mathbf{j}$

17. $\mathbf{v} = (x_2 - x_1)\mathbf{i} + (y_2 - y_1)\mathbf{j}$

$\mathbf{v} = (-7 - (-1))\mathbf{i} + (-7 - 7)\mathbf{j} = -6\mathbf{i} - 14\mathbf{j}$

19. $\mathbf{v} = (x_2 - x_1)\mathbf{i} + (y_2 - y_1)\mathbf{j}$

$\mathbf{v} = (6 - (-3))\mathbf{i} + (4 - 4)\mathbf{j} = 9\mathbf{i} + 0\mathbf{j} = 9\mathbf{i}$

21. $\mathbf{u} + \mathbf{v} = (2\mathbf{i} - 5\mathbf{j}) + (-3\mathbf{i} + 7\mathbf{j})$

$= (2 - 3)\mathbf{i} + (-5 + 7)\mathbf{j}$

$= -\mathbf{i} + 2\mathbf{j}$

23. $\mathbf{u} - \mathbf{v} = (2\mathbf{i} - 5\mathbf{j}) - (-3\mathbf{i} + 7\mathbf{j})$

$= 2\mathbf{i} - 5\mathbf{j} + 3\mathbf{i} - 7\mathbf{j}$

$= (2 + 3)\mathbf{i} + (-5 - 7)\mathbf{j}$

$= 5\mathbf{i} - 12\mathbf{j}$

25. $\mathbf{v} - \mathbf{u} = (-3\mathbf{i} + 7\mathbf{j}) - (2\mathbf{i} - 5\mathbf{j})$

$= -3\mathbf{i} + 7\mathbf{j} - 2\mathbf{i} + 5\mathbf{j}$

$= (-3 - 2)\mathbf{i} + (7 + 5)\mathbf{j}$

$= -5\mathbf{i} + 12\mathbf{j}$

27. $5\mathbf{v} = 5(-3\mathbf{i} + 7\mathbf{j}) = -15\mathbf{i} + 35\mathbf{j}$

29. $-4\mathbf{w} = -4(-\mathbf{i} - 6\mathbf{j}) = 4\mathbf{i} + 24\mathbf{j}$

31. $3\mathbf{w} + 2\mathbf{v} = 3(-\mathbf{i} - 6\mathbf{j}) + 2(-3\mathbf{i} + 7\mathbf{j})$
$= -3\mathbf{i} - 18\mathbf{j} - 6\mathbf{i} + 14\mathbf{j}$
$= (-3 - 6)\mathbf{i} + (-18 + 14)\mathbf{j}$
$= -9\mathbf{i} - 4\mathbf{j}$

33. $3\mathbf{v} - 4\mathbf{w} = 3(-3\mathbf{i} + 7\mathbf{j}) - 4(-\mathbf{i} - 6\mathbf{j})$
$= -9\mathbf{i} + 21\mathbf{j} + 4\mathbf{i} + 24\mathbf{j}$
$= (-9 + 4)\mathbf{i} + (21 + 24)\mathbf{j}$
$= -5\mathbf{i} + 45\mathbf{j}$

35. $\|2\mathbf{u}\| = \|2(2\mathbf{i} - 5\mathbf{j})\|$
$= \|4\mathbf{i} - 10\mathbf{j}\|$
$= \sqrt{4^2 + (-10)^2}$
$= \sqrt{16 + 100}$
$= \sqrt{116}$
$= 2\sqrt{29}$

37. $\|\mathbf{w} - \mathbf{u}\| = \|(-\mathbf{i} - 6\mathbf{j}) - (2\mathbf{i} - 5\mathbf{j})\|$
$= \|-\mathbf{i} - 6\mathbf{j} - 2\mathbf{i} + 5\mathbf{j}\|$
$= \|(-1 - 2)\mathbf{i} + (-6 + 5)\mathbf{j}\|$
$= \|-3\mathbf{i} - \mathbf{j}\|$
$= \sqrt{(-3)^2 + (-1)^2}$
$= \sqrt{9 + 1}$
$= \sqrt{10}$

39. $\dfrac{\mathbf{v}}{\|\mathbf{v}\|} = \dfrac{6\mathbf{i}}{\sqrt{6^2 + 0^2}} = \dfrac{6\mathbf{i}}{\sqrt{36}} = \dfrac{6\mathbf{i}}{6} = \mathbf{i}$

41. $\dfrac{\mathbf{v}}{\|\mathbf{v}\|} = \dfrac{3\mathbf{i} - 4\mathbf{j}}{\sqrt{3^2 + (-4)^2}}$
$= \dfrac{3\mathbf{i} - 4\mathbf{j}}{\sqrt{9 + 16}}$
$= \dfrac{3\mathbf{i} - 4\mathbf{j}}{\sqrt{25}}$
$= \dfrac{3\mathbf{i} - 4\mathbf{j}}{5}$
$= \dfrac{3}{5}\mathbf{i} - \dfrac{4}{5}\mathbf{j}$

43. $\dfrac{\mathbf{v}}{\|\mathbf{v}\|} = \dfrac{3\mathbf{i} - 2\mathbf{j}}{\sqrt{3^2 + (-2)^2}}$
$= \dfrac{3\mathbf{i} - 2\mathbf{j}}{\sqrt{9 + 4}}$
$= \dfrac{3\mathbf{i} - 2\mathbf{j}}{\sqrt{13}}$
$= \dfrac{3}{\sqrt{13}}\mathbf{i} - \dfrac{2}{\sqrt{13}}\mathbf{j}$

45. $\dfrac{\mathbf{v}}{\|\mathbf{v}\|} = \dfrac{\mathbf{i} + \mathbf{j}}{\sqrt{1^2 + 1^2}}$
$= \dfrac{\mathbf{i} + \mathbf{j}}{\sqrt{2}}$
$= \dfrac{\mathbf{i}}{\sqrt{2}} + \dfrac{\mathbf{j}}{\sqrt{2}}$
$= \dfrac{\sqrt{2}}{2}\mathbf{i} + \dfrac{\sqrt{2}}{2}\mathbf{j}$

47. $\mathbf{v} = \|\mathbf{v}\|\cos\theta\,\mathbf{i} + \|\mathbf{v}\|\sin\theta\,\mathbf{j}$
$= 6\cos 30°\,\mathbf{i} + 6\sin 30°\,\mathbf{j}$
$= 6\left(\dfrac{\sqrt{3}}{2}\right)\mathbf{i} + 6\left(\dfrac{1}{2}\right)\mathbf{j}$
$= 3\sqrt{3}\mathbf{i} + 3\mathbf{j}$

49. $\mathbf{v} = \|\mathbf{v}\|\cos\theta\,\mathbf{i} + \|\mathbf{v}\|\sin\theta\,\mathbf{j}$
$= 12\cos 225°\,\mathbf{i} + 12\sin 225°\,\mathbf{j}$
$= 12\left(-\dfrac{\sqrt{2}}{2}\right)\mathbf{i} + 12\left(-\dfrac{\sqrt{2}}{2}\right)\mathbf{j}$
$= -6\sqrt{2}\mathbf{i} - 6\sqrt{2}\mathbf{j}$

51. $\mathbf{v} = \|\mathbf{v}\|\cos\theta\,\mathbf{i} + \|\mathbf{v}\|\sin\theta\,\mathbf{j}$
$= \dfrac{1}{2}\cos 113°\,\mathbf{i} + \dfrac{1}{2}\sin 113°\,\mathbf{j}$
$\approx \dfrac{1}{2}(-0.39)\mathbf{i} + \dfrac{1}{2}(0.92)\mathbf{j}$
$\approx -0.20\mathbf{i} + 0.46\mathbf{j}$

53. $4\mathbf{u} - (2\mathbf{v} - \mathbf{w}) = 4(-2\mathbf{i} + 3\mathbf{j}) - \left[2(6\mathbf{i} - \mathbf{j}) - (-3\mathbf{i})\right]$

$\quad\quad\quad\quad\quad = -8\mathbf{i} + 12\mathbf{j} - \left[12\mathbf{i} - 2\mathbf{j} + 3\mathbf{i})\right]$

$\quad\quad\quad\quad\quad = -8\mathbf{i} + 12\mathbf{j} - 12\mathbf{i} + 2\mathbf{j} - 3\mathbf{i})$

$\quad\quad\quad\quad\quad = -23\mathbf{i} + 14\mathbf{j}$

55. $\|\mathbf{u} + \mathbf{v}\|^2 - \|\mathbf{u} - \mathbf{v}\|^2$

$= \|-2\mathbf{i} + 3\mathbf{j} + 6\mathbf{i} - \mathbf{j}\|^2 - \|-2\mathbf{i} + 3\mathbf{j} - (6\mathbf{i} - \mathbf{j})\|^2$

$= \|4\mathbf{i} + 2\mathbf{j}\|^2 - \|-8\mathbf{i} + 4\mathbf{j}\|^2$

$= \left(\sqrt{4^2 + 2^2}\right)^2 - \left(\sqrt{(-8)^2 + 4^2}\right)^2$

$= 16 + 4 - (64 + 16)$

$= 20 - 80$

$= -60$

57. $\quad\quad\quad\quad\quad\quad \mathbf{u} + \mathbf{v} = \mathbf{v} + \mathbf{u}$

$\left(a_1\mathbf{i} + b_1\mathbf{j}\right) + \left(a_2\mathbf{i} + b_2\mathbf{j}\right) = \left(a_2\mathbf{i} + b_2\mathbf{j}\right) + \left(a_1\mathbf{i} + b_1\mathbf{j}\right)$

$\left(a_1 + a_2\right)\mathbf{i} + \left(b_1 + b_2\right)\mathbf{j} = \left(a_1 + a_2\right)\mathbf{i} + \left(b_1 + b_2\right)\mathbf{j}$

This demonstrates the commutative property of vectors.

59. $\quad\quad\quad\quad\quad\quad c(\mathbf{u} + \mathbf{v}) = c\mathbf{u} + c\mathbf{v}$

$c\left(\left(a_1\mathbf{i} + b_1\mathbf{j}\right) + \left(a_2\mathbf{i} + b_2\mathbf{j}\right)\right) = c\left(a_1\mathbf{i} + b_1\mathbf{j}\right) + c\left(a_2\mathbf{i} + b_2\mathbf{j}\right)$

$\left(ca_1 + ca_2\right)\mathbf{i} + \left(cb_1 + cb_2\right)\mathbf{j} = \left(ca_1 + ca_2\right)\mathbf{i} + \left(cb_1 + cb_2\right)\mathbf{j}$

This demonstrates a distributive property of vectors.

61. $\|\mathbf{v}\| = \sqrt{(-10)^2 + 15^2} = \sqrt{325} \approx 18.03$

$\theta = \tan^{-1}\left(\dfrac{15}{-10}\right) \approx 123.7°$

63. $\mathbf{v} = \left(4\mathbf{i} - 2\mathbf{j}\right) - \left(4\mathbf{i} - 8\mathbf{j}\right) = 6\mathbf{j}$

$\|\mathbf{v}\| = 6$

$\theta = 90°$

65. $\mathbf{v} = \|\mathbf{v}\|\cos\theta\mathbf{i} + \|\mathbf{v}\|\sin\theta\mathbf{j}$

$\quad = 44\cos 30°\mathbf{i} + 44\sin 30°\mathbf{j}$

$\quad = 44\left(\dfrac{\sqrt{3}}{2}\right)\mathbf{i} + 44\left(\dfrac{1}{2}\right)\mathbf{j}$

$\quad = 22\sqrt{3}\mathbf{i} + 22\mathbf{j}$

67. $\mathbf{v} = \|\mathbf{v}\|\cos\theta\mathbf{i} + \|\mathbf{v}\|\sin\theta\mathbf{j}$

$\quad = 150\cos 8°\mathbf{i} + 150\sin 8°\mathbf{j}$

$\quad \approx 148.5\mathbf{i} + 20.9\mathbf{j}$

69. $\mathbf{v} = \|\mathbf{v}\|\cos\theta\mathbf{i} + \|\mathbf{v}\|\sin\theta\mathbf{j}$

$= 1.5\cos 25°\mathbf{i} + 1.5\sin 25°\mathbf{j}$

$\approx 1.4\mathbf{i} + 0.6\mathbf{j}$

The length of the shadow is $|1.4| = 1.4$ inches.

71. $\mathbf{F_1} = \|\mathbf{F_1}\|\cos\theta\mathbf{i} + \|\mathbf{F_1}\|\sin\theta\mathbf{j}$

$= 70\cos 326°\mathbf{i} + 70\sin 326°\mathbf{j}$

$= 58\mathbf{i} - 39.1\mathbf{j}$

$\mathbf{F_2} = \|\mathbf{F_2}\|\cos\theta\mathbf{i} + \|\mathbf{F_2}\|\sin\theta\mathbf{j}$

$= 50\cos 18°\mathbf{i} + 50\sin 18°\mathbf{j}$

$= 47.6\mathbf{i} + 15.5\mathbf{j}$

$\mathbf{F} = \mathbf{F_1} + \mathbf{F_2} = (58\mathbf{i} - 39.1\mathbf{j}) + (47.6\mathbf{i} + 15.5\mathbf{j})$

$= 105.6\mathbf{i} - 23.6\mathbf{j}$

$\|F\| = \sqrt{105.6^2 + (-23.6)^2} = 108.2$ pounds

$\cos\theta = \dfrac{a}{\|F\|}$

$\theta = \cos^{-1}\dfrac{105.6}{108.2} = 12.6°, 90 - 12.6 = S77.4°E$

73. $\mathbf{F_1} = 1610\cos 125°\mathbf{i} + 1610\sin 125°\mathbf{j}$

$= -923.46\mathbf{i} + 1318.83\mathbf{j}$

$\mathbf{F_2} = 1250\cos 215°\mathbf{i} + 1250\sin 215°\mathbf{j}$

$= -1023.94\mathbf{i} - 716.97\mathbf{j}$

$\mathbf{F} = (-923.46 - 1023.94)\mathbf{i} + (1318.83 - 716.97)\mathbf{j}$

$= -1947.40\mathbf{i} + 601.86\mathbf{j}$

$\|\mathbf{F}\| = \sqrt{(-1947.40)^2 + 601.86^2} \approx 2038.28$

2038 kilograms

$\cos\theta = \dfrac{-1947.40}{2038.28}$

$\theta = 162.8°$

75. $\mathbf{F_1} = 70\cos 326°\mathbf{i} + 70\sin 326°\mathbf{j}$

$= -100\mathbf{j}$

To find the length of the BC: $\cos 18° = \dfrac{a}{100}$

$a \approx 95$

$\mathbf{F_2} = 95\cos 2888° + 95\sin 288°$

$= 29.4\mathbf{i} - 90.4\mathbf{j}$

$\mathbf{F} = \mathbf{F_1} - \mathbf{F_2} = (-100\mathbf{j}) - (29.4\mathbf{i} - 90.4\mathbf{j})$

$= -29.4\mathbf{i} - 9.6\mathbf{j}$

$\sqrt{(-29.4)^2 + (-9.6)^2} \approx 30.9$

The force required to pull the weight is 30.9 pounds.

77.   **a.**   335 lb

b.   3484 lb

79.   **a.** $\mathbf{F_1} + \mathbf{F_2} = (3 + 6)\mathbf{i} + (-5 + 2)\mathbf{j} = 9\mathbf{i} - 3\mathbf{j}$

b. $-9\mathbf{i} + 3\mathbf{j}$

81.   **a.** $\mathbf{F_1} = -3\mathbf{i} \quad (-3, 0)$

$\mathbf{F_2} = -\mathbf{i} + 4\mathbf{j} \quad (-1, 4)$

$\mathbf{F_3} = 4\mathbf{i} - 2\mathbf{i} \quad (4, -2)$

$\mathbf{F_4} = -4\mathbf{j} \quad (0, -4)$

$\mathbf{F_1} + \mathbf{F_2} + \mathbf{F_2} + \mathbf{F_2} = (-3 - 1 + 4)\mathbf{i}$

$+ (4 - 2 - 4)\mathbf{j} = -2\mathbf{j}$

b. $2\mathbf{j}$

83.   **a.** $\mathbf{v} = 180\cos 40°\,\mathbf{i} + 180\sin 40°\,\mathbf{j}$

$= 137.88\mathbf{i} + 115.7\mathbf{j}$

$\mathbf{w} = 40\cos 0°\,\mathbf{i} + 40\sin 0°\,\mathbf{j}$

$= 40\mathbf{i}$

b. $\mathbf{v} + \mathbf{w} = (137.88 + 40)\mathbf{i} + 115.7\mathbf{j}$

$= 177.88\mathbf{i} + 115.7\mathbf{j}$

c. $\sqrt{177.88^2 + 115.7^2} \approx 212$ mph

d. $\cos\theta = \dfrac{177.88}{212}$

$\theta = 33°$

$90° - 33° = N57°E$

85. $\mathbf{v} = 320\cos 20°\,\mathbf{i} + 320\sin 20°\,\mathbf{j}$

$= 300.7\mathbf{i} + 109.5\mathbf{j}$

$\mathbf{w} = 370\cos 30°\,\mathbf{i} + 370\sin 30°\,\mathbf{j}$

$= 320.4\mathbf{i} + 185\mathbf{j}$

$\mathbf{w} - \mathbf{v} = (320.4 - 300.7)\mathbf{i}$

$+ (115.7 - 109.5)\mathbf{j}$

$= 19.7\mathbf{i} + 75.6\mathbf{j}$

$\sqrt{19.7^2 + 75.6^2} \approx 78$ mph

$\cos\theta = \dfrac{19.7}{78}$

$\theta = 75.4°$

87. – 103.   Answers may vary.

105.   does not make sense; Explanations will vary. Sample explanation: A vector represents a distance and a direction. A rate of change does not represent a distance and a direction.

107. does not make sense; Explanations will vary. Sample explanation: The resultant force will have a magnitude less than two pounds unless both forces are in the same direction.

109. true

111. true

113.

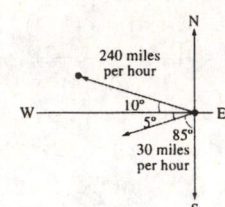

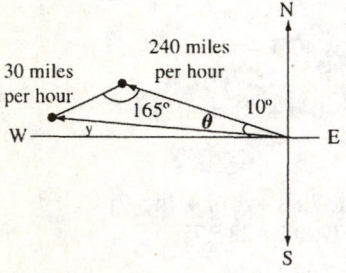

To find the magnitude of **v**, we use the Law of Cosines.

$$\|\mathbf{v}\|^2 = 30^2 + 240^2 - 2(30)(240)\cos 165°$$

$$\approx 72,409.3$$

$$\|\mathbf{v}\| \approx \sqrt{72,409.3} \approx 269.1$$

The plane's true speed relative to the ground is about 269.1 miles per hour. To find the compass heading, relative to the ground, use the Law of Sines.

$$\frac{269.1}{\sin 165°} = \frac{30}{\sin \theta}$$

$$269.1 \sin \theta = 30 \sin 165°$$

$$\sin \theta = \frac{30 \sin 165°}{269.1}$$

$$\theta = \sin^{-1}\left(\frac{30 \sin 165°}{269.1}\right)$$

$$\theta \approx 1.7$$

The compass heading relative to the ground, is approximately $270° + (10° - 1.7°) = 278.3°$.

115. **a.**
$$\mathbf{a} = 310 \cos \theta° \, \mathbf{i} + 310 \sin \theta° \, \mathbf{j}$$
$$\mathbf{w} = 75 \cos 0° \, \mathbf{i} + 75 \sin 0° \, \mathbf{j}$$
$$= 75\mathbf{i}$$
$$310 \cos \theta + 75 = 0$$
$$\cos \theta = \frac{-75}{310}$$
$$\theta = 104°$$
$$180° - 104° = 76°$$

b. increase

116.
$$\cos \theta = \frac{3(-1) + (-2)(4)}{\|\mathbf{v}\| \|\mathbf{w}\|}$$

$$\cos \theta = \frac{-3 - 8}{\sqrt{3^2 + (-2)^2}\sqrt{(-1)^2 + 4^2}}$$

$$\cos \theta = \frac{-11}{\sqrt{13}\sqrt{17}}$$

$$\cos \theta = \frac{-11}{\sqrt{221}}$$

$$\theta = \cos^{-1}\left(\frac{-11}{\sqrt{221}}\right)$$

$$\theta \approx 137.7°$$

117.
$$\frac{2(-2) + 4(-6)}{\|\mathbf{w}\|^2}\mathbf{w} = \frac{-4 - 24}{\sqrt{(-2)^2 + 6^2}^2}(-2\mathbf{i} + 6\mathbf{j})$$

$$= \frac{-28}{40}(-2\mathbf{i} + 6\mathbf{j})$$

$$= \frac{7}{5}\mathbf{i} - \frac{21}{5}\mathbf{j}$$

118. **a.** $\|\mathbf{u}\|^2 = \|\mathbf{v}\|^2 + \|\mathbf{w}\|^2 - 2\|\mathbf{v}\|\|\mathbf{w}\|\cos \theta$

b. $\|\mathbf{u}\| = \sqrt{(a_1 - a_2)^2 + (b_1 - b_2)^2}$

$\|\mathbf{u}\|^2 = (a_1 - a_2)^2 + (b_1 - b_2)^2$

$\|\mathbf{v}\| = \sqrt{(a_1 - 0)^2 + (b_1 - 0)^2} = \sqrt{a_1^2 + b_1^2}$

$\|\mathbf{v}\|^2 = a_1^2 + b_1^2$

$\|\mathbf{w}\| = \sqrt{(0 - a_2)^2 + (0 - b_2)^2} = \sqrt{a_2^2 + b_2^2}$

$\|\mathbf{w}\|^2 = a_2^2 + b_2^2$

Section 6.7

Check Point Exercises

1.　**a.**　$\mathbf{v \cdot w} = 7(2) + (-4)(-1) = 14 + 4 = 18$

　　b.　$\mathbf{w \cdot v} = 2(7) + (-1)(-4) = 14 + 4 = 18$

　　c.　$\mathbf{w \cdot w} = 2(2) + (-1)(-1) = 4 + 1 = 5$

2.　$\cos\theta = \dfrac{\mathbf{v \cdot w}}{\|\mathbf{v}\| \cdot \|\mathbf{w}\|}$

　　$= \dfrac{(4\mathbf{i} - 3\mathbf{j}) \cdot (\mathbf{i} + 2\mathbf{j})}{\sqrt{4^2 + (-3)^2}\sqrt{1^2 + 2^2}}$

　　$= \dfrac{4(1) + (-3)(2)}{\sqrt{25}\sqrt{5}}$

　　$= \dfrac{2}{\sqrt{125}}$

　　The angle θ between the vectors is

　　$\theta = \cos^{-1}\left(-\dfrac{2}{\sqrt{125}}\right) \approx 100.3°.$

3.　$\mathbf{v \cdot w} = (6\mathbf{i} - 3\mathbf{j}) \cdot (\mathbf{i} + 2\mathbf{j})$

　　$= 6(1) + (-3)(2) = 6 - 6 = 0$

　　The dot product is zero.

　　Thus, the given vectors are orthogonal.

4.　$\text{proj}_{\mathbf{w}}\mathbf{v} = \dfrac{\mathbf{v \cdot w}}{\|\mathbf{w}\|^2}\mathbf{w}$

　　$= \dfrac{(2\mathbf{i} - 5\mathbf{j}) \cdot (\mathbf{i} - \mathbf{j})}{\left(\sqrt{1^2 + (-1)^2}\right)^2}\mathbf{w}$

　　$= \dfrac{2(1) + (-5)(-1)}{\left(\sqrt{2}\right)^2}\mathbf{w}$

　　$= \dfrac{7}{2}\mathbf{w}$

　　$= \dfrac{7}{2}(\mathbf{i} - \mathbf{j})$

　　$= \dfrac{7}{2}\mathbf{i} - \dfrac{7}{2}\mathbf{j}$

5.　$\mathbf{v}_1 = \text{proj}_{\mathbf{w}}\mathbf{v} = \dfrac{7}{2}\mathbf{i} - \dfrac{7}{2}\mathbf{j}$

　　$\mathbf{v}_2 = \mathbf{v} - \mathbf{v}_1$

　　$= (2\mathbf{i} - 5\mathbf{j}) - \left(\dfrac{7}{2}\mathbf{i} - \dfrac{7}{2}\mathbf{j}\right)$

　　$= -\dfrac{3}{2}\mathbf{i} - \dfrac{3}{2}\mathbf{j}$

6.　$W = \|\mathbf{F}\|\,\|\overrightarrow{AB}\|\cos\theta = (20)(150)\cos 30°$

　　≈ 2598

　　The work done is approximately 2598 foot-pounds.

Exercise Set 6.7

1.　$\mathbf{v \cdot w} = 3(1) + 1(3) = 3 + 3 = 6$
　　$\mathbf{v \cdot v} = 3(3) + 1(1) = 9 + 1 = 10$

3.　$\mathbf{v \cdot w} = 5(-2) + (-4)(-1) = -10 + 4 = -6$
　　$\mathbf{v \cdot v} = 5(5) + (-4)(-4) = 25 + 16 = 41$

5.　$\mathbf{v \cdot w} = -6(-10) + (-5)(-8) = 60 + 40 = 100$
　　$\mathbf{v \cdot v} = -6(-6) + (-5)(-5) = 36 + 25 = 61$

7.　$\mathbf{v \cdot w} = 5(0) + 0(1) = 0 + 0 = 0$
　　$\mathbf{v \cdot v} = 5(5) + 0(0) = 25 + 0 = 25$

9.　$\mathbf{v} \cdot (\mathbf{v} + \mathbf{w}) = (2\mathbf{i} - \mathbf{j})[(3\mathbf{i} + \mathbf{j}) + (\mathbf{i} + 4\mathbf{j})]$

　　$= (2\mathbf{i} - \mathbf{j})[(3+1)\mathbf{i} + (1+4)\mathbf{j})]$

　　$= (2\mathbf{i} - \mathbf{j})(4\mathbf{i} + 5\mathbf{j})$

　　$= 2(4) + (-1)(5)$

　　$= 8 - 5$

　　$= 3$

11.　$\mathbf{u \cdot v} + \mathbf{u \cdot w}$

　　$= (2\mathbf{i} - \mathbf{j}) \cdot (3\mathbf{i} + \mathbf{j}) + (2\mathbf{i} - \mathbf{j})(\mathbf{i} + 4\mathbf{j})$

　　$= (2)(3) + (-1)(1) + 2(1) + (-1)(4)$

　　$= 6 - 1 + 2 - 4$

　　$= 3$

13.　$(4\mathbf{u}) \cdot \mathbf{v}$

　　$= [(4(2\mathbf{i} - \mathbf{j})] \cdot (3\mathbf{i} + \mathbf{j})$

　　$= (8\mathbf{i} - 4\mathbf{j}) \cdot (3\mathbf{i} + \mathbf{j})$

　　$= (8)(3) + (-4)(1)$

　　$= 24 - 4$

　　$= 20$

15. $4(\mathbf{u \cdot v})$
$= 4[(2\mathbf{i} - \mathbf{j}) \cdot (3\mathbf{i} + \mathbf{j})]$
$= 4[2(3) + (-1)1]$
$= 4[6-1]$
$= 4[5]$
$= 20$

17. $\cos\theta \dfrac{\mathbf{v \cdot w}}{\|\mathbf{v}\| \ \|\mathbf{w}\|}$

$= \dfrac{(2\mathbf{i} - \mathbf{j}) \cdot (3\mathbf{i} + 4\mathbf{j})}{\sqrt{2^2 + (-1)^2} \sqrt{3^2 + 4^2}}$

$= \dfrac{2(3) + (-1)(4)}{\sqrt{5}\sqrt{25}}$

$= \dfrac{6-4}{\sqrt{125}}$

$= \dfrac{2}{\sqrt{125}}$

The angle θ between the vectors is

$\theta = \cos^{-1}\left(\dfrac{2}{\sqrt{125}}\right) \approx 79.7°.$

19. $\cos\theta \dfrac{\mathbf{v \cdot w}}{\|\mathbf{v}\| \ \|\mathbf{w}\|}$

$= \dfrac{(-3\mathbf{i} + 2\mathbf{j}) \cdot (4\mathbf{i} - \mathbf{j})}{\sqrt{(-3)^2 + 2^2} \sqrt{4^2 + (-1)^2}}$

$= \dfrac{-3(4) + 2(-1)}{\sqrt{13}\sqrt{17}}$

$= \dfrac{-14}{\sqrt{221}}$

The angle θ between the vectors is

$\theta = \cos^{-1}\left(-\dfrac{14}{\sqrt{221}}\right) \approx 160.3°.$

21. $\cos\theta \dfrac{\mathbf{v \cdot w}}{\|\mathbf{v}\| \ \|\mathbf{w}\|}$

$= \dfrac{(6\mathbf{i} + 0\mathbf{j}) \cdot (5\mathbf{i} + 4\mathbf{j})}{\sqrt{6^2 + 0^2} \sqrt{5^2 + 4^2}}$

$= \dfrac{6(5) + 0(4)}{\sqrt{36}\sqrt{41}}$

$= \dfrac{30}{\sqrt{1476}}$

The angle θ between the vectors is

$\theta = \cos^{-1}\left(\dfrac{30}{\sqrt{1476}}\right) \approx 38.7°.$

23. $\mathbf{v \cdot w} = (\mathbf{i} + \mathbf{j}) \cdot (\mathbf{i} - \mathbf{j}) = (1)(1) + 1(-1) = 1 - 1 = 0$
The dot product is zero. Thus, the given vectors are orthogonal.

25. $\mathbf{v \cdot w} = (2\mathbf{i} + 8\mathbf{j}) \cdot (4\mathbf{i} - \mathbf{j})$
$= 2(4) + (8)(-1)$
$= 8 - 8$
$= 0$
The dot product is zero. Thus, the given vectors are orthogonal.

27. $\mathbf{v \cdot w} = (2\mathbf{i} - 2\mathbf{j}) \cdot (-\mathbf{i} + \mathbf{j})$
$= 2(-1) + (-2)(1)$
$= -2 - 2$
$= -4$
The dot product is not zero. Thus, the given vectors are not orthogonal.

29. $\mathbf{v \cdot w} = (3\mathbf{i} + 0\mathbf{j}) \cdot (-4\mathbf{i} + 0\mathbf{j})$
$= 3(-4) + 0(0)$
$= -12 + 0$
$= -12$
The dot product is not zero. Thus, the given vectors are not orthogonal.

31. $\mathbf{v \cdot w} = (3\mathbf{i} + 0\mathbf{j}) \cdot (0\mathbf{i} - 4\mathbf{j})$
$= 3(0) + (0)(-4)$
$= 0 + 0$
$= 0$
The dot product is zero. Thus, the given vectors are orthogonal.

33. $\text{proj}_{\mathbf{W}}\mathbf{v} = \dfrac{\mathbf{v}\cdot\mathbf{w}}{\|\mathbf{w}\|^2}\mathbf{w}$

$\qquad = \dfrac{(3\mathbf{i}-2\mathbf{j})\cdot(\mathbf{i}-\mathbf{j})}{\left(\sqrt{1^2+(-1)^2}\right)^2}\mathbf{w}$

$\qquad = \dfrac{3(1)+(-2)(-1)}{\left(\sqrt{2}\right)^2}$

$\qquad = \dfrac{5}{2}\mathbf{w}$

$\qquad = \dfrac{5}{2}(\mathbf{i}-\mathbf{j})$

$\qquad = \dfrac{5}{2}\mathbf{i}-\dfrac{5}{2}\mathbf{j}$

$\mathbf{v}_1 = \text{proj}_{\mathbf{w}}\mathbf{v} = \dfrac{5}{2}\mathbf{i}-\dfrac{5}{2}\mathbf{j}$

$\mathbf{v}_2 = \mathbf{v}-\mathbf{v}_1 = (3\mathbf{i}-2\mathbf{j})-\left(\dfrac{5}{2}\mathbf{i}-\dfrac{5}{2}\mathbf{j}\right)$

$\qquad = \dfrac{1}{2}\mathbf{i}+\dfrac{1}{2}\mathbf{j}$

35. $\text{proj}_{\mathbf{W}}\mathbf{v} = \dfrac{\mathbf{v}\cdot\mathbf{w}}{\|\mathbf{w}\|}\mathbf{w}$

$\qquad = \dfrac{(\mathbf{i}+3\mathbf{j})\cdot(-2\mathbf{i}+5\mathbf{j})}{\sqrt{1^2+(-1)^2}}\mathbf{w}$

$\qquad = \dfrac{1(-2)+3(5)}{\left(\sqrt{(-2)^2+5^2}\right)^2}\mathbf{w}$

$\qquad = \dfrac{13}{\left(\sqrt{29}\right)^2}\mathbf{w}$

$\qquad = \dfrac{13}{29}\mathbf{w}$

$\qquad = \dfrac{13}{29}(-2\mathbf{i}+5\mathbf{j})$

$\qquad = \dfrac{-26}{29}\mathbf{i}+\dfrac{65}{29}\mathbf{j}$

$\mathbf{v}_1 = \text{proj}_{\mathbf{w}}\mathbf{v} = -\dfrac{26}{29}\mathbf{i}+\dfrac{65}{29}\mathbf{j}$

$\mathbf{v}_2 = \mathbf{v}-\mathbf{v}_1$

$\qquad = (\mathbf{i}+3\mathbf{j})-\left(-\dfrac{26}{29}\mathbf{i}+\dfrac{65}{29}\mathbf{j}\right)$

$\qquad = \dfrac{55}{29}\mathbf{i}+\dfrac{22}{29}\mathbf{j}$

$\mathbf{v}_1 = \text{proj}_w\mathbf{v} = -\dfrac{6}{5}\mathbf{i}+\dfrac{12}{5}\mathbf{j}$

$\mathbf{v}_2 = \mathbf{v}-\mathbf{v}_1 = (2\mathbf{i}+4\mathbf{j})-\left(-\dfrac{6}{5}\mathbf{i}+\dfrac{12}{5}\mathbf{j}\right)$

$\qquad = \dfrac{16}{5}\mathbf{i}+\dfrac{8}{5}\mathbf{j}$

37. $\text{proj}_{\mathbf{W}}\mathbf{v} = \dfrac{\mathbf{v}\cdot\mathbf{w}}{\|\mathbf{w}\|^2}\mathbf{w}$

$\qquad = \dfrac{(\mathbf{i}+2\mathbf{j})\cdot(3\mathbf{i}+6\mathbf{j})}{\left(\sqrt{3^2+6^2}\right)^2}\mathbf{w}$

$\qquad = \dfrac{1(3)+2(6)}{\sqrt{45}}$

$\qquad = \dfrac{15}{45}\mathbf{w}$

$\qquad = \dfrac{1}{3}\mathbf{w}$

$\qquad = \dfrac{1}{3}(3\mathbf{i}+6\mathbf{j})$

$\qquad = \mathbf{i}+2\mathbf{j}$

$\mathbf{v}_1 = \text{proj}_w\mathbf{v} = \mathbf{i}+2\mathbf{j}$

$\mathbf{v}_2 = \mathbf{v}-\mathbf{v}_1$

$\qquad = (\mathbf{i}+2\mathbf{j})-\left(\mathbf{i}+2\mathbf{j}\right)$

$\qquad = 0\mathbf{i}+0\mathbf{j}$

$\qquad = \mathbf{0}$

39. $5\mathbf{u} \cdot (3\mathbf{v} - 4\mathbf{w}) = 15\mathbf{u} \cdot \mathbf{v} - 20\mathbf{u} \cdot \mathbf{w}$

$$= 15\big[(-1)(3) + (1)(-2)\big] - 20\big[(-1)(0) + (1)(-5)\big]$$

$$= 15\big[-5\big] - 20\big[-5\big]$$

$$= 25$$

41. $\text{proj}_{\mathbf{u}}(\mathbf{v} + \mathbf{w}) = \dfrac{(\mathbf{v} + \mathbf{w}) \cdot \mathbf{u}}{\|u\|^2}\mathbf{u}$

$$= \dfrac{(3\mathbf{i} - 2\mathbf{j} - 5\mathbf{j}) \cdot (-\mathbf{i} + \mathbf{j})}{\|-\mathbf{i} + \mathbf{j}\|^2}(-\mathbf{i} + \mathbf{j})$$

$$= \dfrac{(3\mathbf{i} - 7\mathbf{j}) \cdot (-\mathbf{i} + \mathbf{j})}{\left(\sqrt{(-1)^2 + 1^2}\right)^2}(-\mathbf{i} + \mathbf{j})$$

$$= \dfrac{-3 - 7}{2}(-\mathbf{i} + \mathbf{j})$$

$$= -5(-\mathbf{i} + \mathbf{j})$$

$$= 5\mathbf{i} - 5\mathbf{j}$$

43. $\cos\theta = \dfrac{\mathbf{v} \cdot \mathbf{w}}{\|\mathbf{v}\|\|\mathbf{w}\|}$

$$\cos\theta = \dfrac{\left(2\cos\dfrac{4\pi}{3}\right)\left(3\cos\dfrac{3\pi}{2}\right) + \left(2\sin\dfrac{4\pi}{3}\right)\left(3\sin\dfrac{3\pi}{2}\right)}{\sqrt{\left(2\cos\dfrac{4\pi}{3}\right)^2 + \left(2\sin\dfrac{4\pi}{3}\right)^2}\sqrt{\left(3\cos\dfrac{3\pi}{2}\right)^2 + \left(3\sin\dfrac{3\pi}{2}\right)^2}}$$

$$\cos\theta = \dfrac{3\sqrt{3}}{6}$$

$$\cos\theta = \dfrac{\sqrt{3}}{2}$$

$$\theta = 30°$$

45. $\cos\theta = \dfrac{\mathbf{v} \cdot \mathbf{w}}{\|\mathbf{v}\|\|\mathbf{w}\|}$

$$\cos\theta = \dfrac{(3)(6) + (-5)(-10)}{\sqrt{(3)^2 + (-5)^2}\sqrt{(6)^2 + (-10)^2}}$$

$$\cos\theta = \dfrac{68}{68}$$

$$\cos\theta = 1$$

$$\theta = 0°$$

The vectors are parallel.

$v \cdot w = 68$

$v \cdot w \neq 0$

The vectors are not orthogonal.

47. $\cos\theta = \dfrac{\mathbf{v}\cdot\mathbf{w}}{\|\mathbf{v}\|\|\mathbf{w}\|}$

$\cos\theta = \dfrac{(3)(6)+(-5)(10)}{\sqrt{(3)^2+(-5)^2}\sqrt{(6)^2+(10)^2}}$

$\cos\theta = \dfrac{-32}{68}$

$\theta \approx 118°$

The vectors are not parallel.

$\mathbf{v}\cdot\mathbf{w} = -32$

$\mathbf{v}\cdot\mathbf{w} \neq 0$

The vectors are not orthogonal.

49. $\mathbf{v}\cdot\mathbf{w} = (3)(6)+(-5)\left(\dfrac{18}{5}\right) = 0$

The vectors are orthogonal.

51. $\mathbf{v}\cdot\mathbf{w} = (240\mathbf{i}+300\mathbf{j})\cdot(2.90\mathbf{i}+3.07\mathbf{j})$

$= 240(2.90)+300(3.07)$

$= 696+921$

$= 1617$

$\mathbf{v}\cdot\mathbf{w} = 1617$ means \$1617 in revenue
was generated on Monday by the sale of
240 gallons of regular gas at \$2.90 per gallon and 300 gallons of premium gas at \$3.07 per gallon.

53. Since the car is pushed along a level road, the angle between the force and the direction of motion is $\theta = 0$. The work done

$W = \|\mathbf{F}\|\left\|\overrightarrow{AB}\right\|\cos\theta$

$= (95)(80)\cos 0°$

$= 7600.$

The work done is 7600 foot-pounds.

55. $W = \|\mathbf{F}\|\left\|\overrightarrow{AB}\right\|\cos\theta$

$= (40)(100)\cos 32°$

≈ 3392

The work done is approximately 3392
foot-pounds.

57. $\mathbf{w} = \mathbf{F}\cdot\overrightarrow{AB}$

$= 60(20)\cos(38°-12°)$

$= 1200\cos 26°$

≈ 1079 foot-pounds

59. $\mathbf{w} = \mathbf{F} \cdot \overrightarrow{\mathbf{AB}}$

$= (3, 2) \cdot [(10, 20) - (4, 9)]$

$= (3, 2) \cdot (6, 11)$

$= 18 + 22$

$= 40$ foot-pounds

61. $\mathbf{w} = \mathbf{F} \cdot \overrightarrow{\mathbf{AB}}$

$= (4 \cos 50^\circ, 4 \sin 50^\circ) \cdot [(8, 10) - (3, 7)]$

$= (4 \cos 50^\circ, 4 \sin 50^\circ) \cdot (5, 3)$

$= 20 \cos 50^\circ + 12 \sin 50^\circ$

≈ 22.05 foot-pounds

63. a. $\cos 30^\circ \mathbf{i} + \sin 30^\circ \mathbf{j} = \dfrac{\sqrt{3}}{2} \mathbf{i} + \dfrac{1}{2} \mathbf{j}$

b. $\text{proj}_{\mathbf{u}} \mathbf{F} = \dfrac{(0, -700) \cdot \left(\dfrac{\sqrt{3}}{2}, \dfrac{1}{2} \right)}{\| \mathbf{u} \|^2} \left(\dfrac{\sqrt{3}}{2}, \dfrac{1}{2} \right)$

$= -350 \left(\dfrac{\sqrt{3}}{2}, \dfrac{1}{2} \right) = -175\sqrt{3}\mathbf{i} - 175\mathbf{j}$

c. $\sqrt{ \left(-175\sqrt{3} \right)^2 + \left(-175 \right)^2 }$

$= \sqrt{122,500} = 350$

A force of 350 pounds is required to keep the boat from rolling down the ramp.

65. – 73. Answers may vary.

75. makes sense

77. makes sense

79. $\mathbf{u} \cdot \mathbf{v} = (a_1 \mathbf{i} + b_1 \mathbf{j}) \cdot (a_2 \mathbf{i} + b_2 \mathbf{j})$

$= a_1 a_2 + b_1 b_2$

$= a_2 a_1 + b_2 b_1$

$= (a_2 \mathbf{i} + b_2 \mathbf{j}) \cdot (a_1 \mathbf{i} + b_1 \mathbf{j})$

$= \mathbf{v} \cdot \mathbf{u}$

Thus $\mathbf{u} \cdot \mathbf{v} = \mathbf{v} \cdot \mathbf{u}$.

81.
$$\mathbf{u} \cdot (\mathbf{v} + \mathbf{w}) = (a_1\mathbf{i} + b_1\mathbf{j}) \cdot [(a_2\mathbf{i} + b_2\mathbf{j}) + (a_3\mathbf{i} + a_3\mathbf{j})]$$
$$= (a_1\mathbf{i} + b_1\mathbf{j}) \cdot [(a_2 + a_3)\mathbf{i} + (b_2 + b_3)\mathbf{j}]$$
$$= a_1(a_2 + a_3) + b_1(b_2 + b_3)$$
$$= a_1a_2 + a_1a_3 + b_1b_2 + b_1b_3$$
$$= a_1a_2 + b_1b_2 + a_1a_3 + b_1b_3$$
$$= (a_1\mathbf{i} + b_1\mathbf{j}) \cdot (a_2\mathbf{i} + b_2\mathbf{j}) + (a_1\mathbf{i} + b_1\mathbf{j}) \cdot (a_3\mathbf{i} + b_3\mathbf{j})$$
$$= \mathbf{u} \cdot \mathbf{v} + \mathbf{u} \cdot \mathbf{w}$$

83. Let $\mathbf{v} = 15\mathbf{i} - 3\mathbf{j}$ and $\mathbf{w} = -4\mathbf{i} + b\mathbf{j}$. The vectors $\mathbf{v}$ and $\mathbf{w}$ are orthogonal if $\mathbf{u} \cdot \mathbf{w} = 0$.

$\mathbf{v} \cdot \mathbf{w} = (15\mathbf{i} - 3\mathbf{j}) \cdot (-4\mathbf{i} + b\mathbf{j}) = 15(-4) + (-3)b = -60 - 3b$

$\mathbf{v} \cdot \mathbf{w} = 0$ if $-60 - 3b = 0$. Solving the equation for b, we find $b = -20$.

85. We know that $\operatorname{proj}_{\mathbf{w}}\mathbf{v} = \dfrac{\mathbf{v} \cdot \mathbf{w}}{\|\mathbf{w}\|^2}\mathbf{w}$ If the projection of $\mathbf{v}$ onto $\mathbf{w}$ is $\mathbf{v}$, then $\mathbf{v} = \dfrac{\mathbf{v} \cdot \mathbf{w}}{\|\mathbf{w}\|^2}\mathbf{w}$.

Since $\dfrac{\mathbf{v} \cdot \mathbf{w}}{\|\mathbf{w}\|^2}$ is a scalar for all $\mathbf{v}$ and $\mathbf{w}$, let $k = \dfrac{\mathbf{v} \cdot \mathbf{w}}{\|\mathbf{w}\|^2}$. Substituting, we have $\mathbf{v} = k\mathbf{w}$.

When one vector can be expressed as a scalar multiple of another, the vectors have the same direction. Thus, the projection of $\mathbf{v}$ onto $\mathbf{w}$ is $\mathbf{v}$ only if $\mathbf{v}$ and $\mathbf{w}$ have the same direction. Thus, any two vectors, $\mathbf{v}$ and $\mathbf{w}$, having the same direction will satisfy the condition that the projection of $\mathbf{v}$ onto $\mathbf{w}$ is $\mathbf{v}$.

87. **a.** $x + 2y = 2$

$\qquad 4 + 2(-1) = 2$

$\qquad\qquad 2 = 2, \text{ true}$

Yes, $(4, -1)$ satisfies the equation.

b. $x - 2y = 6$

$\qquad 4 - 2(-1) = 6$

$\qquad\qquad 6 = 6, \text{ true}$

Yes, $(4, -1)$ satisfies the equation.

88. The graphs intersect at $(4, -1)$.

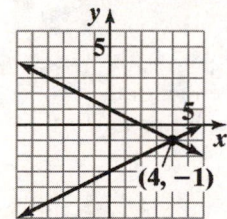

89. $5(2x - 3) - 4x = 9$

$\qquad 10x - 15 - 4x = 9$

$\qquad\qquad 6x - 15 = 9$

$\qquad\qquad\qquad 6x = 24$

$\qquad\qquad\qquad\quad x = 4$

The solution set is $\{4\}$.

Additional Topics in Trigonometry

Chapter 6 Review Exercises

1. Begin by finding C.
$$A + B + C = 180°$$
$$70° + 55° + C = 180°$$
$$125° + C = 180°$$
$$C = 55°$$

Use the ratio $\dfrac{a}{\sin A}$, or $\dfrac{12}{\sin 70°}$, to find the other two sides. Use the Law of Sines to find b.
$$\frac{b}{\sin B} = \frac{a}{\sin A}$$
$$\frac{b}{\sin 55°} = \frac{12}{\sin 70°}$$
$$b = \frac{12\sin 55°}{\sin 70°} \approx 10.5$$

Use the Law of Sines again, this time to find c.
$$\frac{c}{\sin C} = \frac{a}{\sin A}$$
$$\frac{c}{\sin 55°} = \frac{12}{\sin 70°}$$
$$c = \frac{12\sin 55°}{\sin 70°} \approx 10.5$$
The solution is $C = 55°$, $b \approx 10.5$, $c \approx 10.5$.

2. Begin by finding A.
$$A + B + C = 180°$$
$$A + 107° + 30° = 180°$$
$$A + 137° = 180°$$
$$A = 43°$$

Use the ratio $\dfrac{c}{\sin C}$, or $\dfrac{126}{\sin 30°}$, to find the other two sides. Use the Law of Sines to find a.
$$\frac{a}{\sin A} = \frac{c}{\sin C}$$
$$\frac{a}{\sin 43°} = \frac{126}{\sin 30°}$$
$$a = \frac{126\sin 43°}{\sin 30°} \approx 171.9$$
Use the Law of Sines again, this time to find b.
$$\frac{b}{\sin B} = \frac{c}{\sin C}$$
$$\frac{b}{\sin 107°} = \frac{126}{\sin 30°}$$
$$b = \frac{126\sin 107°}{\sin 30°} \approx 241.0$$
The solution is $A = 43°$, $a \approx 171.9$, and $b \approx 241.0$.

3. Apply the three-step procedure for solving a SAS triangle. Use the Law of Cosines to find the side opposite the given angle. Thus, we will find b.
$$b^2 = a^2 + c^2 - 2ac\cos B$$
$$b^2 = 17^2 + 12^2 - 2(17)(12)\cos 66°$$
$$b^2 = 289 + 144 - 408(\cos 66°)$$
$$b^2 \approx 267.05$$
$$b \approx \sqrt{267.05} \approx 16.3$$
Use the Law of Sines to find the angle opposite the shorter of the two given sides. Thus, we will find acute angle C.
$$\frac{c}{\sin C} = \frac{b}{\sin B}$$
$$\frac{12}{\sin C} = \frac{\sqrt{267.05}}{\sin 66°}$$
$$\sqrt{267.05}\sin C = 12\sin 66°$$
$$\sin C = \frac{12\sin 66°}{\sqrt{267.05}} \approx 0.6708$$
$$C \approx 42°$$
$$A = 180° - B - C = 180° - 66° - 42° = 72°$$
The solution is $b \approx 16.3$, $A \approx 72°$, and $C \approx 42°$.

4. Apply the three-step procedure for solving a SSS triangle. Use the Law of Cosines to find the angle opposite the longest side. Thus, we will find angle C.
$$c^2 = a^2 + b^2 - 2ab\cos C$$
$$\cos C = \frac{a^2 + b^2 - c^2}{2ab}$$
$$= \frac{117^2 + 66^2 - 142^2}{2 \cdot 117 \cdot 66} \approx -0.1372$$
C is obtuse because $\cos C$ is negative.
$$\cos^{-1}(0.1372) \approx 82°$$
$$C \approx 180° - 82° = 98°$$
Use the Law of Sines to find either of the two remaining acute angles. We will find angle A.
$$\frac{a}{\sin A} = \frac{c}{\sin C}$$
$$\frac{117}{\sin A} = \frac{142}{\sin 98°}$$
$$142\sin A = 117\sin 98°$$
$$\sin A = \frac{117\sin 98°}{142} \approx 0.8159$$
$$A \approx 55°$$
$$B = 180° - A - C \approx 180° - 55° - 98° = 27°$$
The solution is $C \approx 98°$, $A \approx 55°$, and $B \approx 27°$.

5. Begin by finding C.

$$A + B + C = 180°$$
$$35° + 25° + C = 180°$$
$$60° + C = 180°$$
$$C = 120°$$

Use the ratio $\dfrac{c}{\sin C}$, or $\dfrac{68}{\sin 120°}$, to find the other two sides. Use the Law of Sines to find a.

$$\frac{a}{\sin A} = \frac{c}{\sin C}$$
$$\frac{a}{\sin 35°} = \frac{68}{\sin 120°}$$
$$a = \frac{68 \sin 35°}{\sin 120°} \approx 45.0$$

Use the Law of Sines again, this time to find b.

$$\frac{b}{\sin B} = \frac{c}{\sin C}$$
$$\frac{b}{\sin 25°} = \frac{68°}{\sin 120°}$$
$$b = \frac{68 \sin 25°}{\sin 120°} \approx 33.2$$

The solution is $C = 120°$, $a \approx 45.0$, and $b \approx 33.2$.

6. The known ratio is $\dfrac{a}{\sin A}$, or $\dfrac{20}{\sin 39°}$. Because side b is given, we used the Law of Sines to find angle B.

$$\frac{b}{\sin B} = \frac{a}{\sin A}$$
$$\frac{26}{\sin B} = \frac{20}{\sin 39°}$$
$$\sin B = \frac{26 \sin 39°}{20} \approx 0.8181$$
$$B_1 \approx 55°, \ B_2 \approx 180° - 55° = 125°$$
$$C_1 = 180° - A - B_1 \approx 180° - 39° - 55° = 86°$$
$$C_2 = 180° - A - B_2 \approx 180° - 39° - 125° = 16°$$

Use the Law of Sines to find c_1 and c_2.

$$\frac{c_1}{\sin C_1} = \frac{a}{\sin A}$$
$$\frac{c_1}{\sin 86°} = \frac{20}{\sin 39°}$$
$$c_1 = \frac{20 \sin 86°}{\sin 39°} \approx 31.7$$
$$\frac{c_2}{\sin C_2} = \frac{a}{\sin A}$$
$$\frac{c_2}{\sin 16°} = \frac{20}{\sin 39°}$$
$$c_2 = \frac{20 \sin 16°}{\sin 39°} \approx 8.8$$

There are two triangles. In one triangle, the solution is $B_1 \approx 55°$, $C_1 \approx 86°$, and $c_1 \approx 31.7$. In the other triangle, $B_2 \approx 125°$, $C_2 \approx 16°$, and $c_2 \approx 8.8$.

7. The known ration is $\dfrac{c}{\sin C}$, or $\dfrac{1}{\sin 50°}$. Because side a is given, we used the Law of Sines to find angle A.

$$\frac{a}{\sin A} = \frac{c}{\sin C}$$
$$\frac{3}{\sin A} = \frac{1}{\sin 50°}$$
$$\sin A = \frac{3 \sin 50°}{1} \approx 2.30$$

Because the sine can never exceed 1, there is no triangle with the given measurements.

8. Apply the three-step procedure for solving a SAS triangle. Use the Law of Cosines to find the side opposite the given angle. Thus, we will find a.

$a^2 = b^2 + c^2 - 2bc \cos A$ Use

$a^2 = (11.2)^2 + (48.2)^2 - 2(11.2)(48.2) \cos 162°$

$\quad \approx 3475.5$

$a \approx \sqrt{3475.5} \approx 59.0$

the Law of Sines to find the angle opposite the shorter of the two given sides. Thus, we will find acute angle B.

$\dfrac{b}{\sin B} = \dfrac{a}{\sin A}$

$\dfrac{11.2}{\sin B} = \dfrac{\sqrt{3475.5}}{\sin 162°}$

$\sin B = \dfrac{11.2 \sin 162°}{\sqrt{3475.5}} \approx 0.0587$

$B \approx 3°$

$C = 180° - A - B \approx 180° - 162° - 3° = 15°$

The solution is $a \approx 59.0$, $B \approx 3°$, and $C \approx 15°$.

9. Apply the three-step procedure for solving a SSS triangle. Use the Law of Cosines to find the angle opposite the longest side. Thus, we will find angle B.

$\cos B = \dfrac{a^2 + c^2 - b^2}{2ac}$

$\cos B = \dfrac{(26.1)^2 + (36.5)^2 - (40.2)^2}{2 \cdot 26.1 \cdot 36.5}$

$\quad \approx 0.2086$

$B \approx 78°$

Use the Law of Sines to find either of the two remaining acute angles. We will find angle A.

$\dfrac{a}{\sin A} = \dfrac{b}{\sin B}$

$\dfrac{26.1}{\sin A} = \dfrac{40.2}{\sin 78°}$

$\sin A = \dfrac{26.1 \sin 78°}{40.2} \approx 0.6351$

$A \approx 39°$

$C = 180° - A - B \approx 180° - 39° - 78° = 63°$

The solution is $B \approx 78°$, $A \approx 39°$, and $C \approx 63°$.

10. The known ratio is $\dfrac{a}{\sin A}$, or $\dfrac{6}{\sin 40°}$. Because side b is given, we used the Law of Sines to find angle B.

$\dfrac{b}{\sin B} = \dfrac{a}{\sin A}$

$\dfrac{4}{\sin B} = \dfrac{6}{\sin 40°}$

$\sin B = \dfrac{4 \sin 40°}{6} \approx 0.4285$

$B_1 \approx 25°$, $B_2 \approx 180° - 25° = 155°$

B_2 is impossible, since $40° + 155° = 195°$.

$C = 180° - A - B_1 \approx 180° - 40° - 25° = 115°$

Use the Law of Sines to find c.

$\dfrac{c}{\sin C} = \dfrac{a}{\sin A}$

$\dfrac{c}{\sin 115°} = \dfrac{6}{\sin 40°}$

$c = \dfrac{6 \sin 115°}{\sin 40°} \approx 8.5$

The solution is B_1 (or B) $\approx 25°$, $C \approx 115°$, and $c \approx 8.5$.

11. The known ratio is $\dfrac{b}{\sin B}$, or $\dfrac{8.7}{\sin 37°}$. Because side a is given, we use the Law of Sines to find angle A.

$\dfrac{a}{\sin A} = \dfrac{b}{\sin B}$

$\dfrac{12.4}{\sin A} = \dfrac{8.7}{\sin 37°}$

$\sin A = \dfrac{12.4 \sin 37°}{8.7} \approx 0.8578$

$A_1 \approx 59°$, $A_2 \approx 180° - 59° = 121°$

$C_1 = 180° - A_1 - B$

$\quad \approx 180° - 59° - 37° = 84°$

$C_2 = 180° - A_2 - B$

$\quad \approx 180° - 121° - 37° = 22°$

Use the Law of Sines to find c_1 and c_2.

$$\frac{c_1}{\sin C_1} = \frac{b}{\sin B}$$

$$\frac{c_1}{\sin 84°} = \frac{8.7}{\sin 37°}$$

$$c_1 = \frac{8.7 \sin 84°}{\sin 37°} \approx 14.4$$

$$\frac{c_2}{\sin C_2} = \frac{b}{\sin B}$$

$$\frac{c_2}{\sin 22°} = \frac{8.7}{\sin 37°}$$

$$c_2 = \frac{8.7 \sin 22°}{\sin 37°} \approx 5.4$$

There are two triangles. In one triangle, the solution is $A_1 \approx 59°$, $C_1 \approx 84°$, and $c_1 \approx 14.4$. In the other triangle, $A_2 \approx 121°$, $C_2 \approx 22°$, and $c_2 \approx 5.4$.

12. The known ratio is $\dfrac{a}{\sin A}$, or $\dfrac{54.3}{\sin 23°}$. Because side b is given, we used the Law of Sines to find angle B.

$$\frac{b}{\sin B} = \frac{a}{\sin A}$$

$$\frac{22.1}{\sin B} = \frac{54.3}{\sin 23°}$$

$$\sin B = \frac{22.1 \sin 23°}{54.3} \approx 0.1590$$

$B_1 \approx 9°$, $B_2 \approx 180° - 9° = 171°$

B_2 is impossible, since $23° + 171° = 194°$.

$C = 180° - A - B_1 \approx 180° - 23° - 9° = 148°$

Use the Law of Sines to find c.

$$\frac{c}{\sin C} = \frac{a}{\sin A}$$

$$\frac{c}{\sin 148°} = \frac{54.3}{\sin 23°}$$

$$c = \frac{54.3 \sin 148°}{\sin 23°} \approx 73.6$$

The solution is B_1 (or B) $\approx 9°$, $C \approx 148°$, and $c \approx 73.6$.

13. Area $= \dfrac{1}{2} ab \sin C$

$$= \frac{1}{2}(4)(6) \sin 42°$$

$$\approx 8$$

The area of the triangle is approximately 8 square feet.

14. Area $= \dfrac{1}{2} bc \sin A$

$$= \frac{1}{2}(4)(5) \sin 22°$$

$$\approx 4$$

The area of the triangle is approximately 4 square feet.

15.

$$s = \frac{1}{2}(a + b + c) = \frac{1}{2}(2 + 4 + 5) = \frac{11}{2}$$

$$\text{Area} = \sqrt{s(s-a)(s-b)(s-c)}$$

$$= \sqrt{\frac{11}{2}\left(\frac{11}{2} - 2\right)\left(\frac{11}{2} - 4\right)\left(\frac{11}{2} - 5\right)}$$

$$= \sqrt{\frac{231}{16}} \approx 4$$

The area of the triangle is approximately 4 square meters.

16.

$$s = \frac{1}{2}(a + b + c) = \frac{1}{2}(2 + 2 + 2) = 3$$

$$\text{Area} = \sqrt{s(s-a)(s-b)(s-c)}$$

$$= \sqrt{3(3-2)(3-2)(3-2)}$$

$$= \sqrt{3} \approx 2$$

The area of the triangle is approximately 2 square meters.

17.

Using the figure, $C = 180° - 60° - 60° = 60°$

Use the Law of Sines to find a.

$$\frac{a}{\sin 60°} = \frac{35}{\sin 60°}$$

$$a = 35$$

The length of the roof is 35 feet.

18. One car travels 60 miles per hour for 30 minutes (half an hour), or $60\left(\frac{1}{2}\right) = 30$ miles. Similarly, the other car travels 25 miles.

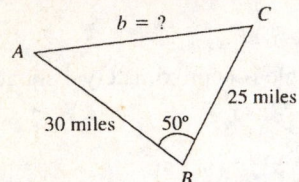

Using the figure,
$$b^2 = a^2 + c^2 - 2ac \cos B$$
$$= 25^2 + 30^2 - 2(25)(30)\cos 80° \approx 1264.53$$
$$b \approx \sqrt{1264.53} \approx 35.6$$
The cars will be about 35.6 miles apart.

19. The first plane travels 325 miles per hour for 2 hours, or $325 \cdot 2 = 650$ miles. Similarly, the other plane travels $300 \cdot 2 = 600$ miles.

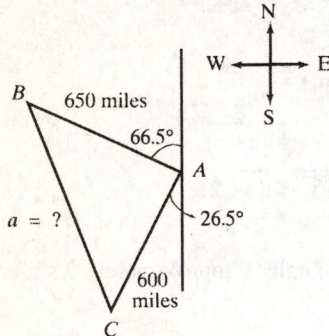

Using the figure,
$A = 180° - 66.5° - 26.5° = 87°$
Use the Law of Cosines to find a.
$$a^2 = b^2 + c^2 - 2bc \cos A$$
$$= 600^2 + 650^2 - 2(600)(650)\cos 87°$$
$$\approx 741,678$$
$$a \approx \sqrt{741,678} \approx 861$$
The planes are about 861 miles apart.

20.

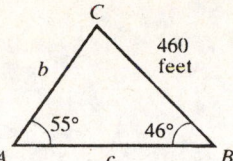

Using the figure,
$C = 180° - A - B = 180° - 55° - 46° = 79°$
Use the Law of Sines to find b.
$$\frac{b}{\sin B} = \frac{a}{\sin A}$$
$$\frac{b}{\sin 46°} = \frac{460}{\sin 55°}$$
$$b = \frac{460 \sin 46°}{\sin 55°} \approx 404$$
Use the Law of Sines again, this time to find c.
$$\frac{c}{\sin C} = \frac{a}{\sin A}$$
$$\frac{c}{\sin 79°} = \frac{460}{\sin 55°}$$
$$c = \frac{460 \sin 79°}{\sin 55°} \approx 551$$
The lengths are about 404 feet and 551 feet.

21. $s = \frac{1}{2}(a + b + c) = \frac{1}{2}(260 + 320 + 450) = 515$
$$\text{Area} = \sqrt{s(s-a)(s-b)(s-c)}$$
$$= \sqrt{515(515-260)(515-320)(515-450)}$$
$$= \sqrt{1,664,544,375} \approx 40,798.83$$
$$\text{cost} \approx (5.25)(40,798.83) \approx 214,194$$
The cost is approximately \$214,194.

22. Draw $\theta = 60°$ counterclockwise, since, θ is positive, from the polar axis. Go 4 units out on the terminal side of θ, since $r > 0$.

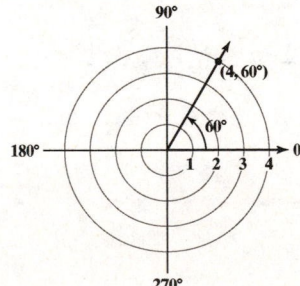

$$x = r \cos \theta = 4 \cos 60° = 4\left(\frac{1}{2}\right) = 2$$

$$y = r \sin \theta = 4 \sin 60° = 4\left(\frac{\sqrt{3}}{2}\right) = 2\sqrt{3}$$

The rectangular coordinates of $(4, 60°)$ are $\left(2, 2\sqrt{3}\right)$.

23. Draw $\theta = 150°$ counterclockwise, since θ is positive, from the polar axis. Go 3 units out on the terminal side of θ, since $r > 0$.

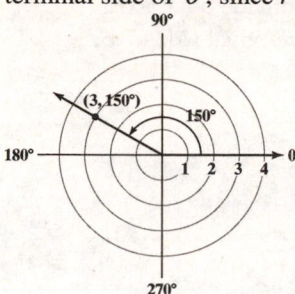

$$x = r\cos\theta = 3\cos 150° = 3\left(-\frac{\sqrt{3}}{2}\right) = -\frac{3\sqrt{3}}{2}$$

$$y = r\sin\theta = 3\sin 150° = 3\left(\frac{1}{2}\right) = \frac{3}{2}$$

The rectangular coordinates of $(3, 150°)$ are $\left(-\frac{3\sqrt{3}}{2}, \frac{3}{2}\right)$.

24. Draw $\theta = \frac{4\pi}{3} = 240°$ counterclockwise, since θ is positive, from the polar axis. Go 4 units out opposite the terminal side of θ, since $r < 0$.

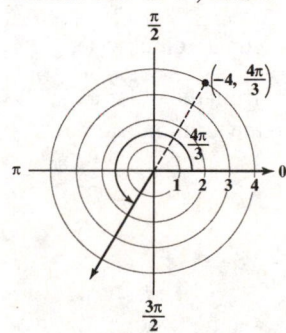

$$x = r\cos\theta = -4\cos\frac{4\pi}{3} = -4\left(-\frac{1}{2}\right) = 2$$

$$y = r\sin\theta = -4\sin\frac{4\pi}{3} = -4\left(-\frac{\sqrt{3}}{2}\right) = 2\sqrt{3}$$

The rectangular coordinates of $\left(-4, \frac{4\pi}{3}\right)$ are $\left(2, 2\sqrt{3}\right)$.

25. Draw $\theta = \frac{5\pi}{4} = 225°$ counterclockwise, since θ is positive from the polar axis. Go 2 units out opposite the terminal side of θ, since $r < 0$.

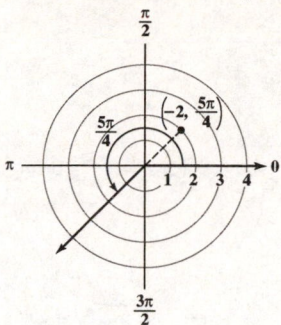

$$x = r\cos\theta = -2\cos\frac{5\pi}{4} = -2\left(-\frac{\sqrt{2}}{2}\right) = \sqrt{2}$$

$$y = r\sin\theta = -2\sin\frac{5\pi}{4} = -2\left(-\frac{\sqrt{2}}{2}\right) = \sqrt{2}$$

The rectangular coordinates of $\left(-2, \frac{5\pi}{4}\right)$ are $\left(\sqrt{2}, \sqrt{2}\right)$.

26. Draw $\theta = -\frac{\pi}{2} = -90°$ clockwise, since θ is negative, from the polar axis. Go 4 units out opposite the terminal side of θ, since $r < 0$.

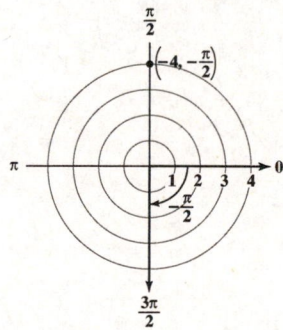

$$x = r\cos\theta = -4\cos\left(-\frac{\pi}{2}\right) = -4(0) = 0$$

$$y = r\sin\theta = -4\sin\left(-\frac{\pi}{2}\right) = -4(-1) = 4$$

The rectangular coordinates of $\left(-4, -\frac{\pi}{2}\right)$ are $(0, 4)$.

Additional Topics in Trigonometry

27. Draw $\theta = -\dfrac{\pi}{4} = -45°$ clockwise, since θ is negative, from the polar axis. Plot the point out 2 units opposite the terminal side of θ, since $r < 0$.

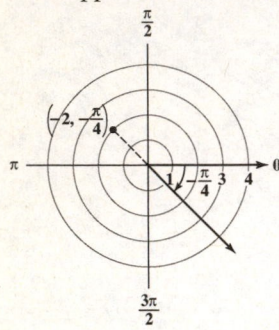

$$x = r\cos\theta = -2\cos\left(-\frac{\pi}{4}\right) = -2\left(\frac{\sqrt{2}}{2}\right) = -\sqrt{2}$$

$$y = r\sin\theta = -2\sin\left(-\frac{\pi}{4}\right) = -2\left(-\frac{\sqrt{2}}{2}\right) = \sqrt{2}$$

The rectangular coordinates of $\left(-2, -\dfrac{\pi}{4}\right)$ are $\left(-\sqrt{2}, \sqrt{2}\right)$.

28. Draw $\theta = \dfrac{\pi}{6} = 30°$ counterclockwise, since θ is positive, from the polar axis. Go out 3 units on the terminal side of θ, since $r > 0$.

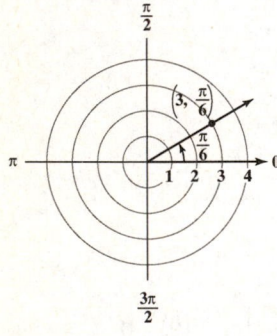

a. $\left(3, \dfrac{\pi}{6}\right) = \left(3, \dfrac{\pi}{6} + 2\pi\right) = \left(3, \dfrac{13\pi}{6}\right)$

b. $\left(3, \dfrac{\pi}{6}\right) = \left(-3, \dfrac{\pi}{6} + \pi\right) = \left(-3, \dfrac{7\pi}{6}\right)$

c. $\left(3, \dfrac{\pi}{6}\right) = \left(3, \dfrac{\pi}{6} - 2\pi\right) = \left(3, -\dfrac{11\pi}{6}\right)$

29. Draw $\theta = \dfrac{2\pi}{3} = 120°$ counterclockwise, since θ is positive, from the polar axis. Go out 3 units on the terminal side of θ, since $r > 0$.

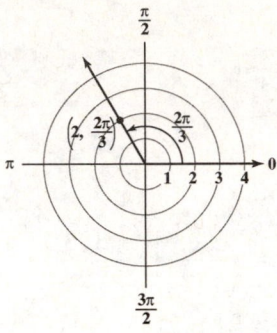

a. $\left(2, \dfrac{2\pi}{3}\right) = \left(2, \dfrac{2\pi}{3} + 2\pi\right) = \left(2, \dfrac{8\pi}{3}\right)$

b. $\left(2, \dfrac{2\pi}{3}\right) = \left(-2, \dfrac{2\pi}{3} + \pi\right) = \left(-2, \dfrac{5\pi}{3}\right)$

c. $\left(2, \dfrac{2\pi}{3}\right) = \left(2, \dfrac{2\pi}{3} - 2\pi\right) = \left(2, -\dfrac{4\pi}{3}\right)$

30. Draw $\theta = \dfrac{\pi}{2} = 90°$ counterclockwise, since θ is positive, from the polar axis. Go out 3.5 units on the terminal side of θ, since $r > 0$.

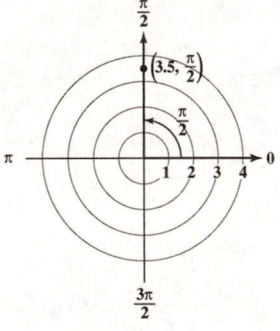

a. $\left(3, \dfrac{\pi}{2} + 2\pi\right) = \left(3, \dfrac{5\pi}{2}\right)$

b. $\left(-3.5, \dfrac{\pi}{2} + \pi\right) = \left(-3.5, \dfrac{3\pi}{2}\right)$

c. $\left(3.5, \dfrac{\pi}{2} - 2\pi\right) = \left(3.5, -\dfrac{3\pi}{2}\right)$

31. $(-4,4)$

$r = \sqrt{(-4)^2 + (-4)^2} = \sqrt{16+16} = \sqrt{32} = 4\sqrt{2}$

$\tan\theta = \dfrac{4}{-4} = -1$

Because $\tan\dfrac{\pi}{4} = 1$ and θ lies in quadrant II,

$\theta = \pi - \dfrac{\pi}{4} = \dfrac{3\pi}{4}$.

The polar coordinates of $(-4,4)$ are $\left(4\sqrt{2}, \dfrac{3\pi}{4}\right)$.

32. $(3,-3)$

$r = \sqrt{3^2 + (-3)^2} = \sqrt{9+9} = \sqrt{18} = 3\sqrt{2}$

$\tan\theta = \dfrac{-3}{3} = -1$

Because $\tan\dfrac{\pi}{4} = 1$, and θ lies in

quadrant IV, $\theta = 2\pi - \dfrac{\pi}{4} = \dfrac{7\pi}{4}$.

The polar coordinates of $(3, -3)$ are $\left(3\sqrt{2}, \dfrac{7\pi}{4}\right)$

33. $(5,12)$

$r = \sqrt{5^2 + 12^2} = \sqrt{25+144} = \sqrt{169} = 13$

$\tan\theta = \dfrac{12}{5}$

Because $\tan^{-1}\left(\dfrac{12}{5}\right) \approx 67°$ and θ lies in quadrant I,

$\theta \approx 67°$.

The polar coordinates of $(5,12)$ are approximately $(13, 67°)$.

34. $(-3,4)$

$r = \sqrt{(-3)^2 + 4^2} = \sqrt{9+16} = \sqrt{25} = 5$

$\tan\theta = \dfrac{4}{-3} = -\dfrac{4}{3}$

Because $\tan^{-1}\left(\dfrac{4}{3}\right) \approx 53°$ and θ lies in quadrant II,

$\theta \approx 180° - 53° = 127°$. The polar coordinates of $(-3, 4)$ are $(5, 127°)$.

35. $(0,-5)$

$r = \sqrt{0^2 + (-5)^2} = \sqrt{25} = 5$

$\tan\theta = \dfrac{-5}{0}$ is undefined

Because $\tan\dfrac{\pi}{2}$ is undefined and θ lies on the

negative y-axis, $\theta = \dfrac{\pi}{2} + \pi = \dfrac{3\pi}{2}$. The polar

coordinates of $(0,-5)$ are $\left(5, \dfrac{3\pi}{2}\right)$.

36. $(1,0)$

$r = \sqrt{1^2 + 0^2} = \sqrt{1} = 1$

$\tan\theta = \dfrac{0}{1} = 0$

Because $\tan 0 = 0$ and θ lies on the positive x-axis, $\theta = 0$.

The polar coordinates of $(1,0)$ are $(1,0)$.

37.
$$2x + 3y = 8$$
$$2r\cos\theta + 3r\sin\theta = 8$$
$$r(2\cos\theta + 3\sin\theta) = 8$$
$$r = \dfrac{8}{2\cos\theta + 3\sin\theta}$$

38. $x^2 + y^2 = 100$
$$r^2 = 100$$
$$r = 10$$

39.
$$(x-6)^2 + y^2 = 36$$
$$(r\cos\theta - 6)^2 + (r\sin\theta)^2 = 36$$
$$r^2\cos^2\theta - 12r\cos\theta + 36 + r^2\sin^2\theta = 36$$
$$r^2 - 12r\cos\theta = 0$$
$$r^2 = 12r\cos\theta$$
$$r = 12\cos\theta$$

40.
$$r = 3$$
$$r^2 = 3^2$$
$$x^2 + y^2 = 9$$

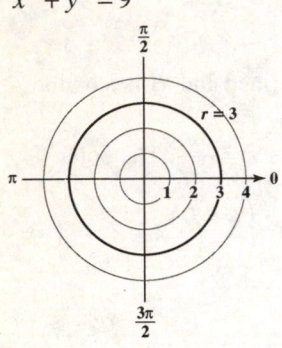

43.
$$r = 5\csc\theta$$
$$r = \frac{5}{\sin\theta}$$
$$r\sin\theta = 5$$
$$y = 5$$

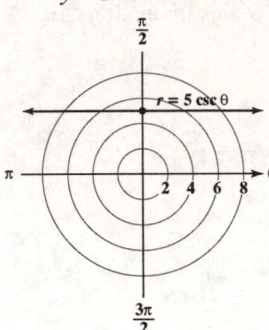

41.
$$\theta = \frac{3\pi}{4}$$
$$\tan\theta = \tan\frac{3\pi}{4}$$
$$\frac{x}{y} = -1$$
$$x = -y$$

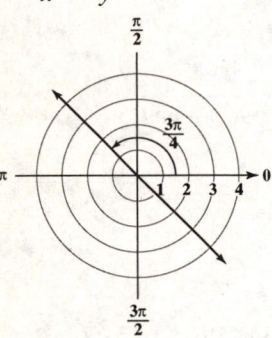

44.
$$r = 3\cos\theta$$
$$r\cdot r = r\cdot 3\cos\theta$$
$$r^2 = 3r\cos\theta$$
$$x^2 + y^2 = 3x$$
$$x^2 - 3x + y^2 = 0$$
$$x^2 - 3x + \frac{9}{4} + y^2 = \frac{9}{4}$$
$$\left(x - \frac{3}{2}\right)^2 + y^2 = \frac{9}{4}$$

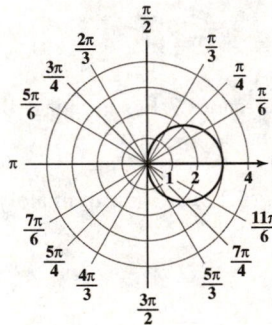

$r = 3\cos\theta$

42.
$$r\cos\theta = -1$$
$$x = -1$$

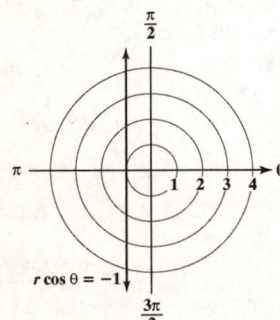

$r\cos\theta = -1$

45. $4r\cos\theta + r\sin\theta = 8$

$$4x + y = 8$$

$$y = -4x + 8$$

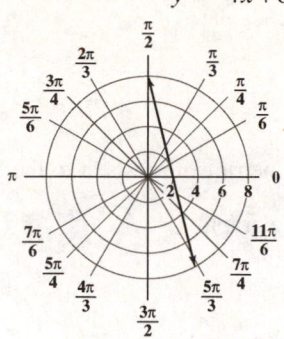

$4r\cos\theta + r\sin\theta = 8$

46. $$r^2\sin 2\theta = -2$$

$$r^2(2\sin\theta\cos\theta) = -2$$

$$2r\sin\theta r\cos\theta = -2$$

$$2yx = -2$$

$$y = \frac{-2}{2x}$$

$$y = -\frac{1}{x}$$

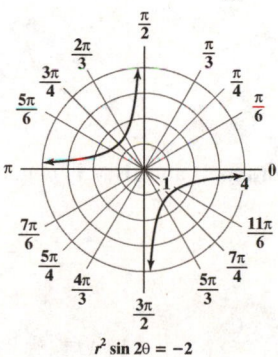

$r^2\sin 2\theta = -2$

47. $r = 5 + 3\cos\theta$

a. $r = 5 + 3\cos(-\theta)$

$r = 5 + 3\cos\theta$

The graph has symmetry about the polar axis.

b. $-r = 5 + 3\cos(-\theta)$

$-r = 5 + 3\cos\theta$

$r = -5 - 3\cos\theta$

The graph may or may not have symmetry with respect to the line $\theta = \dfrac{\pi}{2}$.

c. $-r = 5 + 3\cos\theta$

$r = -5 - 3\cos\theta$

The graph may or may not have symmetry with respect to the pole.

48. $r = 3\sin\theta$

a. $r = 3\sin(-\theta)$

$r = -3\sin\theta$

The graph may or may not have symmetry with respect to the polar axis.

b. $-r = 3\sin(-\theta)$

$-r = -3\sin\theta$

$r = 3\sin\theta$

The graph has symmetry with respect to line $\theta = \dfrac{\pi}{2}$.

c. $-r = 3\sin\theta$

$r = -3\sin\theta$

The graph may or may not have symmetry with respect to the pole.

49. $r^2 = 9\cos 2\theta$

a. $r^2 = 9\cos 2(-\theta)$

$r^2 = 9\cos(-2\theta)$

$r^2 = 9\cos 2\theta$

The graph has symmetry with respect to the polar axis.

b. $(-r)^2 = 9\cos 2(-\theta)$

$r^2 = 9\cos(-2\theta)$

$r^2 = 9\cos 2\theta$

The graph has symmetry with respect to the line $\theta = \dfrac{\pi}{2}$.

c. $(-r)^2 = 9\cos 2\theta$

$r^2 = 9\cos 2\theta$

The graph has symmetry with respect to the pole.

Additional Topics in Trigonometry

50. $r = 3\cos\theta$

Check for symmetry:

Polar Axis	The Line $\theta = \dfrac{\pi}{2}$	The Pole
$r = 3\cos(-\theta)$	$-r = 3\cos(-\theta)$	
	$-r = 3\cos\theta$	
$r = 3\cos\theta$	$r = -3\cos\theta$	

The graph has symmetry with respect to the polar axis. The graph may or may not be symmetric with respect to the line $\theta = \dfrac{\pi}{2}$ or the pole. Calculate values of r for θ from 0 to π and use symmetry to obtain the graph.

θ	0	$\dfrac{\pi}{6}$	$\dfrac{\pi}{3}$	$\dfrac{\pi}{2}$	$\dfrac{2\pi}{3}$	$\dfrac{5\pi}{6}$	π
r	3	2.6	1.5	0	−1.5	−2.6	−3

Use symmetry to obtain the graph.

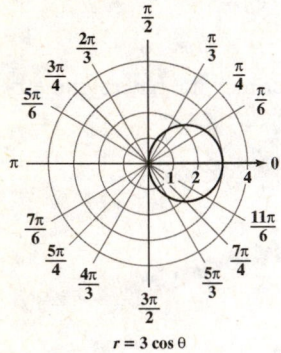

$r = 3\cos\theta$

Notice that there are no points in quadrants II or III. Because the cosine is negative in quadrants II and III, r is negative here. This places the points in quadrants IV and I respectively.

51. $r = 2 + 2\sin\theta$

Check for symmetry:

Polar Axis	The Line $\theta = \dfrac{\pi}{2}$	The Pole
$r = 2 + 2\sin(-\theta)$	$-r = 2 + 2\sin(-\theta)$	$-r = 2 + 2\sin\theta$
	$-r = 2 - 2\sin\theta$	
$r = 2 - 2\sin\theta$	$r = -2 + 2\sin\theta$	$r = -2 - 2\sin\theta$

There may be no symmetry, since each equation is not equivalent to $r = 2 + 2\sin\theta$. Calculate values of r for θ from 0 to 2π.

θ	0	$\frac{\pi}{6}$	$\frac{\pi}{3}$	$\frac{\pi}{2}$	$\frac{2\pi}{3}$	$\frac{5\pi}{6}$	π	$\frac{7\pi}{6}$	$\frac{4\pi}{3}$	$\frac{3\pi}{2}$	$\frac{5\pi}{3}$	$\frac{11\pi}{6}$	2π
r	2	3	3.73	4	3.73	3	2	1	0.27	0	0.27	1	2

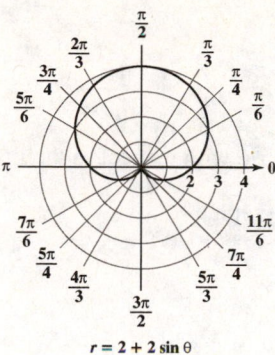

$r = 2 + 2\sin\theta$

52. $r = \sin 2\theta$

Check for symmetry:

Polar Axis	**The Line** $\theta = \dfrac{\pi}{2}$	**The Pole**
$r = \sin 2(-\theta)$	$-r = \sin 2(-\theta)$	$-r = \sin 2\theta$
$r = \sin(-2\theta)$	$-r = \sin(-2\theta)$	
	$-r = -\sin 2\theta$	
$r = -\sin 2\theta$	$r = \sin 2\theta$	$r = -\sin 2\theta$

The graph has symmetry with respect to the line $\theta = \dfrac{\pi}{2}$. The graph may or may not be symmetric with respect to the polar

axis or the pole. Calculate values of r for θ from 0 to $\dfrac{\pi}{2}$ and for θ from π to $\dfrac{3\pi}{2}$. Then, use symmetry to obtain the

graph.

θ	0	$\frac{\pi}{6}$	$\frac{\pi}{4}$	$\frac{\pi}{3}$	$\frac{\pi}{2}$	π	$\frac{7\pi}{6}$	$\frac{5\pi}{4}$	$\frac{4\pi}{3}$	$\frac{3\pi}{2}$
r	0	0.87	1	0.87	0	0	0.87	1	0.87	0

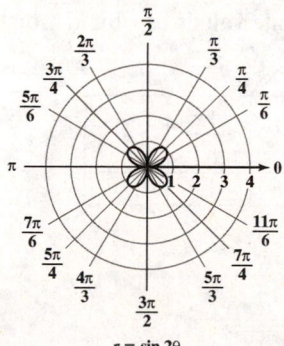

$r = \sin 2\theta$

53. $r = 2 + \cos\theta$

Check for symmetry:

Polar Axis	The Line $\theta = \dfrac{\pi}{2}$	The Pole
$r = 2 + \cos(-\theta)$	$-r = 2 + \cos(-\theta)$	$-r = 2 + \cos\theta$
	$-r = 2 + \cos\theta$	
$r = 2 + \cos\theta$	$r = -2 - \cos\theta$	$r = -2 - \cos\theta$

The graph is symmetric with respect to the polar axis. The graph may or may not be symmetric with respect to the line

$\theta = \dfrac{\pi}{2}$ or the pole. Calculate values of r for θ from 0 to π and use symmetry to obtain the graph.

θ	0	$\dfrac{\pi}{6}$	$\dfrac{\pi}{3}$	$\dfrac{\pi}{2}$	$\dfrac{2\pi}{3}$	$\dfrac{5\pi}{6}$	π
r	3	2.87	2.5	2	1.5	1.13	1

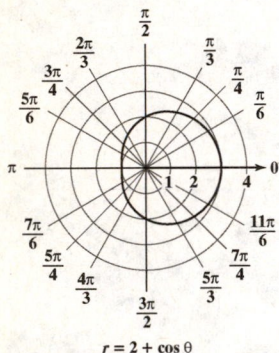

$r = 2 + \cos\theta$

54. $r = 1 + 3\sin\theta$

Check for symmetry:

Polar Axis	The Line $\theta = \dfrac{\pi}{2}$	The Pole
$r = 1 + 3\sin(-\theta)$	$-r = 1 + 3\sin(-\theta)$	$-r = 1 + 3\sin\theta$
	$-r = 1 - 3\sin\theta$	
$r = 1 - 3\sin\theta$	$r = -1 + 3\sin\theta$	$r = -1 - 3\sin\theta$

There may be no symmetry, since each equation is not equivalent to $r = 1 + 3\sin\theta$. Calculate values of r for θ from 0 to 2π.

θ	0	$\dfrac{\pi}{6}$	$\dfrac{\pi}{3}$	$\dfrac{\pi}{2}$	$\dfrac{2\pi}{3}$	$\dfrac{5\pi}{6}$	π	$\dfrac{7\pi}{6}$	$\dfrac{4\pi}{3}$	$\dfrac{3\pi}{2}$	$\dfrac{5\pi}{3}$	$\dfrac{11\pi}{6}$	2π
r	1	2.5	3.6	4	3.6	2.5	1	−0.5	−1.6	−2	−1.6	−0.5	1

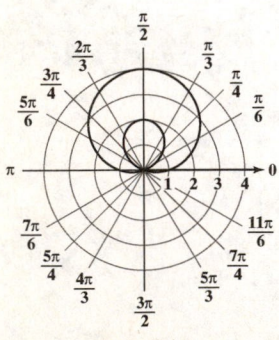

$r = 1 + 3\sin\theta$

55. $r = 1 - 2\cos\theta$

Check for symmetry:

Polar Axis	The Line $\theta = \dfrac{\pi}{2}$	The Pole
$r = 1 - 2\cos(-\theta)$	$-r = 1 - 2\cos(-\theta)$	$-r = 1 - 2\cos\theta$
	$-r = 1 - 2\cos\theta$	
$r = 1 - 2\cos\theta$	$r = -1 + 2\cos\theta$	$r = -1 + 2\cos\theta$

The graph is symmetric with respect to the polar axis. The graph may or may not be symmetric with respect to the line

$\theta = \dfrac{\pi}{2}$ or the pole. Calculate values of r for θ from 0 to π and use symmetry to

obtain the graph.

θ	0	$\dfrac{\pi}{6}$	$\dfrac{\pi}{3}$	$\dfrac{\pi}{2}$	$\dfrac{2\pi}{3}$	$\dfrac{5\pi}{6}$	π
r	-1	-0.73	0	1	2	2.73	3

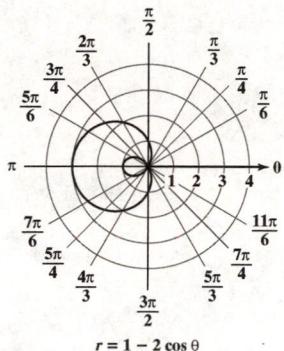

$r = 1 - 2\cos\theta$

56. $r^2 = \cos 2\theta$

Check for symmetry:

Polar Axis	The Line $\theta = \dfrac{\pi}{2}$	The Pole
$r^2 = \cos 2(-\theta)$	$(-r)^2 = \cos 2(-\theta)$	$(-r)^2 = \cos 2\theta$
$r^2 = \cos(-2\theta)$	$r^2 = \cos(-2\theta)$	
$r^2 = \cos 2\theta$	$r^2 = \cos 2\theta$	$r^2 = \cos 2\theta$

The graph has symmetry with respect to the polar axis, the line $\theta = \dfrac{\pi}{2}$, and the pole.

Calculate values of r for θ from 0 to $\dfrac{\pi}{2}$ and use symmetry to obtain the graph.

θ	0	$\dfrac{\pi}{6}$	$\dfrac{\pi}{4}$	$\dfrac{\pi}{3}$	$\dfrac{\pi}{2}$
r	± 1	± 0.71	0	undef	undef

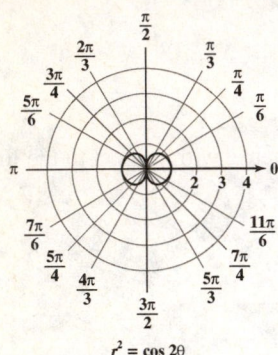

$r^2 = \cos 2\theta$

57. $z = 1 - i$ corresponds to the point $(1, -1)$.

Use $r = \sqrt{a^2 + b^2}$ and $\tan\theta = \dfrac{b}{a}$, with

$a = 1$ and $b = -1$, to find r and θ.

$r = \sqrt{1^2 + (-1)^2} = \sqrt{1+1} = \sqrt{2}$

$\tan\theta = \dfrac{-1}{1} = -1$

Because $\tan\dfrac{\pi}{4} = 1$ and θ lies in quadrant IV, $\theta = 2\pi - \dfrac{\pi}{4} = \dfrac{7\pi}{4}$.

$z = 1 - i = r(\cos\theta + i\sin\theta)$

$= \sqrt{2}\left(\cos\dfrac{7\pi}{4} + i\sin\dfrac{7\pi}{4}\right)$

or $\sqrt{2}(\cos 315° + i\sin 315°)$

58. $z = -2\sqrt{3} + 2i$ corresponds to the point

$(-2\sqrt{3},\ 2)$.

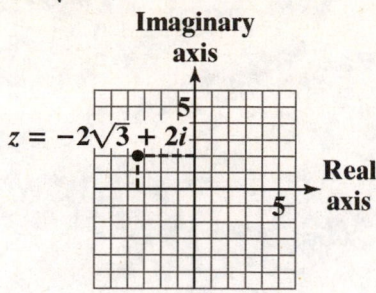

Use $r = \sqrt{a^2 + b^2}$ and $\tan\theta = \dfrac{b}{a}$, with $a = -2\sqrt{3}$ and $b = 2$, to find r and θ.

$r = \sqrt{\left(-2\sqrt{3}\right)^2 + 2^2} = \sqrt{12 + 4} = \sqrt{16} = 4$

$\tan\theta = \dfrac{2}{-2\sqrt{3}} = -\dfrac{1}{\sqrt{3}}$

Because $\tan 30° = \dfrac{1}{\sqrt{3}}$ and θ lies in quadrant II, $\theta = 180° - 30° = 150°$.

$z = -2\sqrt{3} + 2i$

$\quad = r(\cos\theta + i\sin\theta)$

$\quad = 4(\cos 150° + i\sin 150°)$

or $4\left(\cos\dfrac{5\pi}{6} + i\sin\dfrac{5\pi}{6}\right)$

59. $z = -3 - 4i$ corresponds to the point $(-3, -4)$.

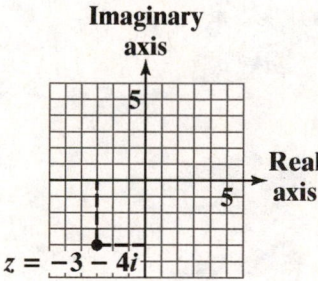

Use $r = \sqrt{a^2 + b^2}$ and $\tan\theta = \dfrac{b}{a}$, with

$a = -3$ and $b = -4$, to find r and θ.

$r = \sqrt{(-3)^2 + (-4)^2} = \sqrt{9 + 16} = \sqrt{25} = 5$

$\tan\theta = \dfrac{-4}{-3} = \dfrac{4}{3}$

Because $\tan^{-1}\left(\dfrac{4}{3}\right) \approx 53°$ and θ lies in quadrant III, $\theta \approx 180° + 53° = 233°$.

$z = -3 - 4i = r(\cos\theta + i\sin\theta)$

$\quad \approx 5(\cos 233° + i\sin 233°)$

60. $z = -5i = 0 - 5i$ corresponds to the point $(0, -5)$.

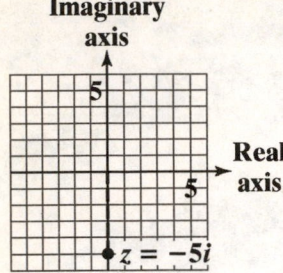

Use $r = \sqrt{a^2 + b^2}$ and $\tan\theta = \dfrac{b}{a}$, with $a = 0$ and $b = -5$, to find r and θ.

$r = \sqrt{0^2 + (-5)^2} = \sqrt{25} = 5$

$\tan\theta = \dfrac{-5}{0}$ is undefined

Because $\tan\dfrac{\pi}{2}$ is undefined and θ lies on the negative y-axis, $\theta = \dfrac{\pi}{2} + \pi = \dfrac{3\pi}{2}$.

$z = -5i = r(\cos\theta + i\sin\theta) = 5\left(\cos\dfrac{3\pi}{2} + i\sin\dfrac{3\pi}{2}\right)$ or $5(\cos 270° + i\sin 270°)$

61. $8(\cos 60° + i\sin 60°) = 8\left(\dfrac{1}{2} + i\dfrac{\sqrt{3}}{2}\right) = 4 + 4i\sqrt{3}$

The rectangular form of $z = 8(\cos 60° + i\sin 60°)$ is $z = 4 + 4i\sqrt{3}$.

62. $4(\cos 210° + i\sin 210°) = 4\left(-\dfrac{\sqrt{3}}{2} + i\left(-\dfrac{1}{2}\right)\right) = -2\sqrt{3} - 2i$

The rectangular form of $z = 4(\cos 210° + i\sin 210°)$ is $z = -2\sqrt{3} - 2i$.

63. $6\left(\cos\dfrac{2\pi}{3} + i\sin\dfrac{2\pi}{3}\right) = 6\left(-\dfrac{1}{2} + i\dfrac{\sqrt{3}}{2}\right) = -3 + 3i\sqrt{3}$

The rectangular form of $z = 6\left(\cos\dfrac{2\pi}{3} + i\sin\dfrac{2\pi}{3}\right)$ is $z = -3 + 3i\sqrt{3}$.

64. $0.6(\cos 100° + i\sin 100°) \approx 0.6(-0.17 + i(0.98)) \approx -0.1 + 0.6i$

The rectangular form of $z = 0.6(\cos 100° + i\sin 100°)$ is $z \approx -0.1 + 0.6i$.

65. $z_1 z_2 = \left[3(\cos 40° + i\sin 40°)\right]\left[5(\cos 70° + i\sin 70°)\right]$

$\qquad = (3 \cdot 5)\left[\cos(40° + 70°) + i\sin(40° + 70°)\right]$

$\qquad = 15(\cos 110° + i\sin 110°)$

66. $z_1 z_2 = \left[\cos 210° + i\sin 210°\right]\left[\cos 55° + i\sin 55°\right]$

$\qquad = \cos(210° + 55°) + i\sin(210° + 55°)$

$\qquad = \cos 265° + i\sin 265°$

67. $z_1 z_2 = \left[4\left(\cos\dfrac{3\pi}{7} + i\sin\dfrac{3\pi}{7}\right)\right]\left[10\left(\cos\dfrac{4\pi}{7} + i\dfrac{4\pi}{7}\right)\right]$

$\qquad = (4\cdot10)\left[\cos\left(\dfrac{3\pi}{7} + \dfrac{4\pi}{7}\right) + i\sin\left(\dfrac{3\pi}{7} + \dfrac{4\pi}{7}\right)\right]$

$\qquad = 40(\cos\pi + i\sin\pi)$

68. $\dfrac{z_1}{z_2} = \dfrac{10(\cos10° + i\sin10°)}{5(\cos5° + i\sin5°)} = \dfrac{10}{5}\left[\cos(10° - 5°) + i\sin(10° - 5°)\right] = 2(\cos5° + i\sin5°)$

69. $\dfrac{z_1}{z_2} = \dfrac{5\left(\cos\dfrac{4\pi}{3} + i\sin\dfrac{4\pi}{3}\right)}{10\left(\cos\dfrac{\pi}{3} + i\sin\dfrac{\pi}{3}\right)} = \dfrac{5}{10}\left[\cos\left(\dfrac{4\pi}{3} - \dfrac{\pi}{3}\right) + i\sin\left(\dfrac{4\pi}{3} - \dfrac{\pi}{3}\right)\right] = \dfrac{1}{2}(\cos\pi + i\sin\pi)$

70. $\dfrac{z_1}{z_2} = \dfrac{2\left(\cos\dfrac{5\pi}{3} + i\sin\dfrac{5\pi}{3}\right)}{\cos\dfrac{\pi}{2} + i\sin\dfrac{\pi}{2}} = 2\left[\cos\left(\dfrac{5\pi}{3} - \dfrac{\pi}{2}\right) + i\sin\left(\dfrac{5\pi}{3} - \dfrac{\pi}{2}\right)\right]$

$\qquad = 2\left[\cos\left(\dfrac{10\pi}{6} - \dfrac{3\pi}{6}\right) + i\sin\left(\dfrac{10\pi}{6} - \dfrac{3\pi}{6}\right)\right] = 2\left(\cos\dfrac{7\pi}{6} + i\sin\dfrac{7\pi}{6}\right)$

71. $\left[2(\cos20° + i\sin20°)\right]^3 = (2)^3\left[\cos(3\cdot20°) + i\sin(3\cdot20°)\right] = 8(\cos60° + i\sin60°)$

$\qquad\qquad\qquad = 8\left(\dfrac{1}{2} + i\dfrac{\sqrt{3}}{2}\right) = 4 + 4\sqrt{3}i$

72. $\left[4(\cos50° + i\sin50°)\right]^3 = (4)^3\left[\cos(3\cdot50°) + i\sin(3\cdot50°)\right] = 64(\cos150° + i\sin150°) = 64\left(-\dfrac{\sqrt{3}}{2} + i\dfrac{1}{2}\right)$

$\qquad\qquad\qquad = -32\sqrt{3} + 32i$

73. $\left[\dfrac{1}{2}\left(\cos\dfrac{\pi}{14} + i\sin\dfrac{\pi}{14}\right)\right]^7 = \left(\dfrac{1}{2}\right)^7\left[\cos\left(7\cdot\dfrac{\pi}{14}\right) + i\sin\left(7\cdot\dfrac{\pi}{14}\right)\right] = \dfrac{1}{128}\left(\cos\dfrac{\pi}{2} + i\sin\dfrac{\pi}{2}\right)$

$\qquad\qquad\qquad = \dfrac{1}{128}(0 + i1) = \dfrac{1}{128}i$

74. Write $1 - i\sqrt{3}$ in $r(\cos\theta + i\sin\theta)$ form.

$$r = \sqrt{a^2 + b^2} = \sqrt{1^2 + \left(-\sqrt{3}\right)^2} = \sqrt{1+3} = 2$$

$$\tan\theta = \frac{b}{a} = \frac{-\sqrt{3}}{1} = -\sqrt{3}$$

Because $\tan 60° = \sqrt{3}$ and θ lies in quadrant IV, $\theta = 360° - 60° = 300°$.

$1 - i\sqrt{3} = r(\cos\theta + i\sin\theta) = 2(\cos 300° + i\sin 300°)$

Use DeMoivre's Theorem to raise $1 - i\sqrt{3}$ to the seventh power.

$$\left(1 - i\sqrt{3}\right)^7 = \left[2(\cos 300° + i\sin 300°)\right]^7$$

$$= (2)^7 \left[\cos(7 \cdot 300°) + i\sin(7 \cdot 300°)\right]$$

$$= 128(\cos 2100° + i\sin 2100°)$$

$$= 128(\cos 300° + i\sin 300°)$$

$$= 128\left(\frac{1}{2} + i\left(-\frac{\sqrt{3}}{2}\right)\right)$$

$$= 64 - 64i\sqrt{3}$$

75. Write $-2 - 2i$ in $r(\cos\theta + i\sin\theta)$ form.

$$r = \sqrt{a^2 + b^2} = \sqrt{(-2)^2 + (-2)^2} = \sqrt{4+4} = 2\sqrt{2}$$

$$\tan\theta = \frac{b}{a} = \frac{-2}{-2} = 1$$

Because $\tan\frac{\pi}{4} = 1$ and θ lies in quadrant III, $\theta = \pi + \frac{\pi}{4} = \frac{5\pi}{4}$.

$$-2 - 2i = r(\cos\theta + i\sin\theta) = 2\sqrt{2}\left(\cos\frac{5\pi}{4} + i\sin\frac{5\pi}{4}\right)$$

Use DeMoivre's Theorem to raise $-2 - 2i$ to the fifth power.

$$(-2 - 2i)^5 = \left[2\sqrt{2}\left(\cos\frac{5\pi}{4} + i\sin\frac{5\pi}{4}\right)\right]^5$$

$$= \left(2\sqrt{2}\right)^5\left[\cos\left(5 \cdot \frac{5\pi}{4}\right) + i\sin\left(5 \cdot \frac{5\pi}{4}\right)\right]$$

$$= 128\sqrt{2}\left(\cos\frac{25\pi}{4} + i\sin\frac{25\pi}{4}\right)$$

$$= 128\sqrt{2}\left(\cos\frac{\pi}{4} + i\sin\frac{\pi}{4}\right)$$

$$= 128\sqrt{2}\left(\frac{\sqrt{2}}{2} + i\frac{\sqrt{2}}{2}\right)$$

$$= 128 + 128i$$

76. $49(\cos 50° + i \sin 50°)$

$$z_k = \sqrt[2]{49}\left[\cos\left(\frac{50° + 360°k}{2}\right) + i\sin\left(\frac{50° + 360°k}{2}\right)\right], \ k = 0,1$$

$$z_0 = \sqrt{49}\left[\cos\left(\frac{50° + 360° \cdot 0}{2}\right) + i\sin\left(\frac{50° + 360° \cdot 0}{2}\right)\right] = \sqrt{49}\left[\cos\left(\frac{50°}{2}\right) + i\sin\left(\frac{50°}{2}\right)\right]$$

$$= 7(\cos 25° + i \sin 25°)$$

$$z_1 = \sqrt{49}\left[\cos\left(\frac{50° + 360° \cdot 1}{2}\right) + i\sin\left(\frac{50° + 360° \cdot 1}{2}\right)\right] = \sqrt{49}\left[\cos\left(\frac{410°}{2}\right) + i\sin\left(\frac{410°}{2}\right)\right]$$

$$= 7(\cos 205° + i \sin 205°)$$

77. $125(\cos 165° + i \sin 165°)$

$$z_k = \sqrt[3]{125}\left[\cos\left(\frac{165° + 360°k}{3}\right) + i\sin\left(\frac{165° + 360°k}{3}\right)\right], \ k = 0,1,2$$

$$z_0 = \sqrt[3]{125}\left[\cos\left(\frac{165° + 360° \cdot 0}{3}\right) + i\sin\left(\frac{165° + 360° \cdot 0}{3}\right)\right] = \sqrt[3]{125}\left[\cos\left(\frac{165°}{3}\right) + i\sin\left(\frac{165°}{3}\right)\right]$$

$$= 5(\cos 55° + i \sin 55°)$$

$$z_1 = \sqrt[3]{125}\left[\cos\left(\frac{165° + 360° \cdot 1}{3}\right) + i\sin\left(\frac{165° + 360° \cdot 1}{3}\right)\right] = \sqrt[3]{125}\left[\cos\left(\frac{525°}{3}\right) + i\sin\left(\frac{525°}{3}\right)\right]$$

$$= 5(\cos 175° + i \sin 175°)$$

$$z_2 = \sqrt[3]{125}\left[\cos\left(\frac{165° + 360° \cdot 2}{3}\right) + i\sin\left(\frac{165° + 360° \cdot 2}{3}\right)\right] = \sqrt[3]{125}\left[\cos\left(\frac{885°}{3}\right) + i\sin\left(\frac{885°}{3}\right)\right]$$

$$= 5(\cos 295° + i \sin 295°)$$

78. $16\left(\cos\frac{2\pi}{3} + i\sin\frac{2\pi}{3}\right)$

$$z_k = \sqrt[4]{16}\left[\cos\left(\frac{\frac{2\pi}{3} + 2\pi k}{4}\right) + i\sin\left(\frac{\frac{2\pi}{3} + 2\pi k}{4}\right)\right], \ k = 0,1,2,3$$

$$z_0 = \sqrt[4]{16}\left[\cos\left(\frac{\frac{2\pi}{3} + 2\pi \cdot 0}{4}\right) + i\sin\left(\frac{\frac{2\pi}{3} + 2\pi \cdot 0}{4}\right)\right] = \sqrt[4]{16}\left[\cos\left(\frac{\pi}{6}\right) + i\sin\left(\frac{\pi}{6}\right)\right] = 2\left(\frac{\sqrt{3}}{2} + i\frac{1}{2}\right) = \sqrt{3} + i$$

$$z_1 = \sqrt[4]{16}\left[\cos\left(\frac{\frac{2\pi}{3} + 2\pi \cdot 1}{4}\right) + i\sin\left(\frac{\frac{2\pi}{3} + 2\pi \cdot 1}{4}\right)\right] = \sqrt[4]{16}\left(\cos\frac{2\pi}{3} + i\sin\frac{2\pi}{3}\right) = 2\left(-\frac{1}{2} + i\frac{\sqrt{3}}{2}\right) = -1 + i\sqrt{3}$$

$$z_2 = \sqrt[4]{16}\left[\cos\left(\frac{\frac{2\pi}{3} + 2\pi \cdot 2}{4}\right) + i\sin\left(\frac{\frac{2\pi}{3} + 2\pi \cdot 2}{4}\right)\right] = \sqrt[4]{16}\left(\cos\frac{7\pi}{6} + i\sin\frac{7\pi}{6}\right) = 2\left(-\frac{\sqrt{3}}{2} + i\left(-\frac{1}{2}\right)\right) = -\sqrt{3} - i$$

$$z_3 = \sqrt[4]{16}\left[\cos\left(\frac{\frac{2\pi}{3} + 2\pi \cdot 3}{4}\right) + i\sin\left(\frac{\frac{2\pi}{3} + 2\pi \cdot 3}{4}\right)\right] = \sqrt[4]{16}\left(\cos\frac{5\pi}{3} + i\sin\frac{5\pi}{3}\right) = 2\left(\frac{1}{2} + i\left(-\frac{\sqrt{3}}{2}\right)\right) = 1 - i\sqrt{3}$$

79. $8i = 8(\cos 90° + i \sin 90°)$

$$z_k = \sqrt[3]{8}\left[\cos\left(\frac{90° + 360°k}{3}\right) + i\sin\left(\frac{90° + 360°k}{3}\right)\right], \, k = 0, 1, 2$$

$$z_0 = \sqrt[3]{8}\left[\cos\left(\frac{90° + 360° \cdot 0}{3}\right) + i\sin\left(\frac{90° + 360° \cdot 0}{3}\right)\right] = \sqrt[3]{8}(\cos 30° + i\sin 30°) = 2\left(\frac{\sqrt{3}}{2} + i\frac{1}{2}\right) = \sqrt{3} + i$$

$$z_1 = \sqrt[3]{8}\left[\cos\left(\frac{90° + 360° \cdot 1}{3}\right) + i\sin\left(\frac{90° + 360° \cdot 1}{3}\right)\right] = \sqrt[3]{8}(\cos 150° + i\sin 150°) = 2\left(-\frac{\sqrt{3}}{2} + i\frac{1}{2}\right) = -\sqrt{3} + i$$

$$z_2 = \sqrt[3]{8}\left[\cos\left(\frac{90° + 360° \cdot 2}{3}\right) + i\sin\left(\frac{90° + 360° \cdot 2}{3}\right)\right] = \sqrt[3]{8}(\cos 270° + i\sin 270°) = 2(0 + i(-1)) = -2i$$

80. $-1 = \cos 180° + i \sin 180°$

$$z_k = \sqrt[3]{1}\left[\cos\left(\frac{180° + 360°k}{3}\right) + i\sin\left(\frac{180° + 360°k}{3}\right)\right], \, k = 0, 1, 2$$

$$z_0 = \sqrt[3]{1}\left[\cos\left(\frac{180° + 360° \cdot 0}{3}\right) + i\sin\left(\frac{180° + 360° \cdot 0}{3}\right)\right] = \sqrt[3]{1}(\cos 60° + i\sin 60°) = 1\left(\frac{1}{2} + i\frac{\sqrt{3}}{2}\right) = \frac{1}{2} + \frac{\sqrt{3}}{2}i$$

$$z_1 = \sqrt[3]{1}\left[\cos\left(\frac{180° + 360° \cdot 1}{3}\right) + i\sin\left(\frac{180° + 360° \cdot 1}{3}\right)\right] = \sqrt[3]{1}(\cos 180° + i\sin 180°) = 1(-1 + i0) = -1$$

$$z_2 = \sqrt[3]{1}\left[\cos\left(\frac{180° + 360° \cdot 2}{3}\right) + i\left(\frac{180° + 360° \cdot 2}{3}\right)\right] = \sqrt[3]{1}(\cos 300° + i\sin 300°) = 1\left(\frac{1}{2} + i\left(-\frac{\sqrt{3}}{2}\right)\right) = \frac{1}{2} - \frac{\sqrt{3}}{2}i$$

81. $-1 - i = \sqrt{2}(\cos 225° + i \sin 225°)$

$$z_k = \sqrt[5]{\sqrt{2}}\left[\cos\left(\frac{225° + 360°k}{5}\right) + i\sin\left(\frac{225° + 360°k}{5}\right)\right], \, k = 0, 1, 2, 3, 4$$

$$z_0 = \sqrt[5]{\sqrt{2}}\left[\cos\left(\frac{225° + 360° \cdot 0}{5}\right) + i\sin\left(\frac{225° + 360° \cdot 0}{5}\right)\right] = \sqrt[5]{\sqrt{2}}(\cos 45° + i\sin 45°) = \sqrt[5]{\sqrt{2}}\left(\frac{\sqrt{2}}{2} + i\frac{\sqrt{2}}{2}\right)$$

$$= \frac{\sqrt[5]{8}}{2} + \frac{\sqrt[5]{8}}{2}i$$

$$z_1 = \sqrt[5]{\sqrt{2}}\left[\cos\left(\frac{225° + 360° \cdot 1}{5}\right) + i\sin\left(\frac{225° + 360° \cdot 1}{5}\right)\right] = \sqrt[5]{\sqrt{2}}(\cos 117° + i\sin 117°) \approx -0.49 + 0.95i$$

$$z_2 = \sqrt[5]{\sqrt{2}}\left[\cos\left(\frac{225° + 360° \cdot 2}{5}\right) + i\sin\left(\frac{225° + 360° \cdot 2}{5}\right)\right] = \sqrt[5]{\sqrt{2}}(\cos 189° + i\sin 189°) \approx -1.06 - 0.17i$$

$$z_3 = \sqrt[5]{\sqrt{2}}\left[\cos\left(\frac{225° + 360° \cdot 3}{5}\right) + i\sin\left(\frac{225° + 360° \cdot 3}{5}\right)\right] = \sqrt[5]{\sqrt{2}}(\cos 261° + i\sin 261°) \approx -0.17 - 1.06i$$

$$z_4 = \sqrt[5]{\sqrt{2}}\left[\cos\left(\frac{225° + 360° \cdot 4}{5}\right) + i\sin\left(\frac{225° + 360° \cdot 4}{5}\right)\right] = \sqrt[5]{\sqrt{2}}(\cos 333° + i\sin 333°) \approx 0.95 - 0.49i$$

82.

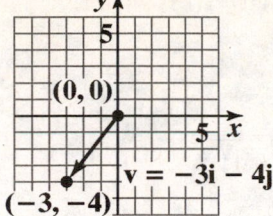

$$\|\mathbf{v}\| = \sqrt{a^2 + b^2}$$
$$= \sqrt{(-3)^2 + (-4)^2}$$
$$= \sqrt{9 + 16}$$
$$= \sqrt{25}$$
$$= 5$$

83.

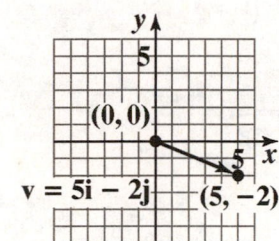

$$\|\mathbf{v}\| = \sqrt{a^2 + b^2}$$
$$= \sqrt{5^2 + (-2)^2}$$
$$= \sqrt{25 + 4}$$
$$= \sqrt{29}$$

84.

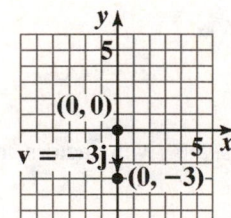

$$\|\mathbf{v}\| = \sqrt{a^2 + b^2}$$
$$= \sqrt{0^2 + (-3)^2}$$
$$= \sqrt{0 + 9}$$
$$= \sqrt{9}$$
$$= 3$$

85. $\mathbf{v} = (x_2 - x_1)\mathbf{i} + (y_2 - y_1)\mathbf{j}$
$$= (5 - 2)\mathbf{i} + [-3 - (-1)]\mathbf{j}$$
$$= 3\mathbf{i} - 2\mathbf{j}$$

86. $\mathbf{v} = (x_2 - x_1)\mathbf{i} + (y_2 - y_1)\mathbf{j}$
$$= [-2 - (-3)\mathbf{i}] + (-2 - 0)\mathbf{j}$$
$$= \mathbf{i} - 2\mathbf{j}$$

87. $\mathbf{v} + \mathbf{w} = (\mathbf{i} - 5\mathbf{j}) + (-2\mathbf{i} + 7\mathbf{j})$
$$= [1 + (-2)]\mathbf{i} + [-5 + 7]\mathbf{j}$$
$$= -\mathbf{i} + 2\mathbf{j}$$

88. $\mathbf{w} - \mathbf{v} = (-2\mathbf{i} + 7\mathbf{j}) - (\mathbf{i} - 5\mathbf{j})$
$$= (-2 - 1)\mathbf{i} + [7 - (-5)]\mathbf{j}$$
$$= -3\mathbf{i} + 12\mathbf{j}$$

89. $6\mathbf{v} - 3\mathbf{w} = 6(\mathbf{i} - 5\mathbf{j}) - 3(-2\mathbf{i} + 7\mathbf{j})$
$$= 6\mathbf{i} - 30\mathbf{j} + 6\mathbf{i} - 21\mathbf{j}$$
$$= 12\mathbf{i} - 51\mathbf{j}$$

90. $\|-2\mathbf{v}\| = |-2| \|\mathbf{v}\|$
$$= 2\|\mathbf{v}\|$$
$$= 2\sqrt{a^2 + b^2}$$
$$= 2\sqrt{1^2 + (-5)^2}$$
$$= 2\sqrt{1 + 25}$$
$$= 2\sqrt{26}$$

91. First, find the magnitude of **v**.
$$\|\mathbf{v}\| = \sqrt{a^2 + b^2}$$
$$= \sqrt{8^2 + (-6)^2}$$
$$= \sqrt{64 + 36}$$
$$= \sqrt{100}$$
$$= 10$$
A unit vector in the same direction as **v** is
$$\frac{\mathbf{v}}{\|\mathbf{v}\|} = \frac{8\mathbf{i} - 6\mathbf{j}}{10} = \frac{4}{5}\mathbf{i} - \frac{3}{5}\mathbf{j}.$$

92. First, find the magnitude of **v**.
$$\|\mathbf{v}\| = \sqrt{a^2 + b^2} = \sqrt{(-1)^2 + (2)^2} = \sqrt{1 + 4} = \sqrt{5}$$
A unit vector in the same direction as **v** is
$$\frac{\mathbf{v}}{\|\mathbf{v}\|} = \frac{-\mathbf{i} + 2\mathbf{j}}{\sqrt{5}} = -\frac{1}{\sqrt{5}}\mathbf{i} + \frac{2}{\sqrt{5}}\mathbf{j}.$$

93. $\mathbf{v} = \|\mathbf{v}\|\cos\theta\,\mathbf{i} + \|\mathbf{v}\|\sin\theta\,\mathbf{j}$

$\quad = 12\cos 60°\,\mathbf{i} + 12\sin 60°\,\mathbf{j}$

$\quad = 12\left(\dfrac{1}{2}\right)\mathbf{i} + 12\left(\dfrac{\sqrt{3}}{2}\right)\mathbf{j}$

$\quad = 6\mathbf{i} + 6\sqrt{3}\,\mathbf{j}$

94. $\quad \mathbf{F_1} = 100\cos 65°\,\mathbf{i} + 100\sin 65°\,\mathbf{j}$

$\qquad\quad = 42.3\mathbf{i} + 90.6\mathbf{j}$

$\quad \mathbf{F_2} = 200\cos 10°\,\mathbf{i} + 200\sin 10°\,\mathbf{j}$

$\qquad\quad = 197\mathbf{i} + 34.7\mathbf{j}$

$\quad \mathbf{F_1} + \mathbf{F_2} = (42.3 + 197)\mathbf{i} + (90.6 + 34.7)\mathbf{j}$

$\qquad\qquad\quad = 239.3\mathbf{i} + 125.3\mathbf{j}$

$\quad \sqrt{239.3^2 + 125.3^2} \approx 270$ pounds

$\qquad\qquad \cos\theta = \dfrac{239.3}{270}$

$\qquad\qquad\quad \theta = 27.6°$

95. $\mathbf{v} = 15\cos 25°\,\mathbf{i} + 15\sin 25°\,\mathbf{j} = 13.6\mathbf{i} + 6.3\mathbf{j}$

$\quad \mathbf{w} = 4\cos 270°\,\mathbf{i} + 4\sin 270°\,\mathbf{j} = -4\mathbf{j}$

a. $13.6\mathbf{i} + (6.3 - 4)\mathbf{j} = 13.6\mathbf{i} + 2.3\mathbf{j}$

b. $\sqrt{13.6^2 + 2.3^2} \approx 14$ mph

c. $\cos\theta = \dfrac{13.6}{14}; \theta = 13.7°$

96. $\mathbf{v}\cdot(\mathbf{v} + \mathbf{w}) = (5\mathbf{i} + 2\mathbf{j})[(\mathbf{i} - \mathbf{j}) + (3\mathbf{i} - 7\mathbf{j})]$

$\qquad\qquad\quad = (5\mathbf{i} + 2\mathbf{j})\cdot[4\mathbf{i} - 8\mathbf{j}]$

$\qquad\qquad\quad = 5(4) + 2(-8)$

$\qquad\qquad\quad = 20 - 16$

$\qquad\qquad\quad = 4$

97. $\mathbf{v}\cdot\mathbf{w} = (2\mathbf{i} + 3\mathbf{j})\cdot(7\mathbf{i} - 4\mathbf{j}) = 2(7) + 3(-4) = 2$

$\quad \cos\theta = \dfrac{2}{\sqrt{2^2 + 3^2}\,\sqrt{7^2 + (-4)^2}}$

$\qquad\quad = \dfrac{2}{\sqrt{13}\,\sqrt{65}}$

$\qquad\quad = \dfrac{2}{\sqrt{845}}$

The angle θ between the vectors is

$\theta = \cos^{-1}\left(\dfrac{2}{\sqrt{845}}\right) \approx 86.1°$.

98. $\mathbf{v}\cdot\mathbf{w} = (2\mathbf{i} + 4\mathbf{j})\cdot(6\mathbf{i} - 11\mathbf{j}) = 2(6) + 4(-11)$

$\qquad\quad = 12 - 44$

$\qquad\quad = -32$

$\quad \cos\theta = \dfrac{-32}{\sqrt{2^2 + 4^2}\,\sqrt{6^2 + (-11)^2}}$

$\qquad\quad = \dfrac{-32}{\sqrt{20}\,\sqrt{157}}$

$\qquad\quad = \dfrac{-32}{\sqrt{3140}}$

The angle θ between the vectors is

$\theta = \cos^{-1}\left(-\dfrac{32}{\sqrt{3140}}\right) \approx 124.8°$.

99. $\mathbf{v}\cdot\mathbf{w} = (2\mathbf{i} + \mathbf{j})\cdot(\mathbf{i} - \mathbf{j}) = 2(1) + 1(-1)$

$\qquad\quad = 2 - 1 = 1$

$\quad \cos\theta = \dfrac{1}{\sqrt{2^2 + 1^2}\,\sqrt{1^2 + (-1)^2}}$

$\qquad\quad = \dfrac{1}{\sqrt{5}\,\sqrt{2}}$

$\qquad\quad = \dfrac{1}{\sqrt{10}}$

The angle θ between the vectors is

$\theta = \cos^{-1}\left(\dfrac{1}{\sqrt{10}}\right) \approx 71.6°$.

100. $\mathbf{v}\cdot\mathbf{w} = (12\mathbf{i} - 8\mathbf{j})\cdot(2\mathbf{i} + 3\mathbf{j})$

$\qquad\quad = 12(2) + (-8)(3)$

$\qquad\quad = 24 - 24$

$\qquad\quad = 0$

The dot product is zero. Thus, the given vectors are orthogonal.

101. $\mathbf{v}\cdot\mathbf{w} = (\mathbf{i} + 3\mathbf{j})\cdot(-3\mathbf{i} - \mathbf{j})$

$\qquad\quad = 1(-3) + 3(-1)$

$\qquad\quad = -3 - 3$

$\qquad\quad = -6$

The dot product is not zero. Thus, the given vectors are not orthogonal.

102. $\text{proj}_{\mathbf{W}} \mathbf{v} = \dfrac{\mathbf{v} \cdot \mathbf{w}}{\|\mathbf{w}\|^2} \mathbf{w}$

$$= \dfrac{(-2\mathbf{i} + 5\mathbf{j}) \cdot (5\mathbf{i} + 4\mathbf{j})}{\left(\sqrt{5^2 + 4^2}\right)^2} \mathbf{w}$$

$$= \dfrac{-2(5) + 5(4)}{\left(\sqrt{41}\right)^2} \mathbf{w}$$

$$= \dfrac{10}{41}(5\mathbf{i} + 4\mathbf{j})$$

$$= \dfrac{50}{41}\mathbf{i} + \dfrac{40}{41}\mathbf{j}$$

$$\mathbf{v}_1 = \text{proj}_{\mathbf{W}} \mathbf{v} = \dfrac{50}{41}\mathbf{i} + \dfrac{40}{41}\mathbf{j}$$

$$\mathbf{v}_2 = \mathbf{v} - \mathbf{v}_1 = (-2\mathbf{i} + 5\mathbf{j}) - \left(\dfrac{50}{41}\mathbf{i} + \dfrac{40}{41}\mathbf{j}\right)$$

$$= -\dfrac{132}{41}\mathbf{i} + \dfrac{165}{41}\mathbf{j}$$

103. $\text{proj}_{\mathbf{W}} \mathbf{v} = \dfrac{\mathbf{v} \cdot \mathbf{w}}{\|\mathbf{w}\|^2} \mathbf{w}$

$$= \dfrac{(-\mathbf{i} + 2\mathbf{j}) \cdot (3\mathbf{i} - \mathbf{j})}{\left(\sqrt{3^2 + (-1)^2}\right)^2} \mathbf{w}$$

$$= \dfrac{-1(3) + 2(-1)}{\left(\sqrt{10}\right)^2} \mathbf{w}$$

$$= \dfrac{-5}{10} \mathbf{w}$$

$$= -\dfrac{1}{2}(3\mathbf{i} - 1\mathbf{j})$$

$$= \dfrac{-3}{2}\mathbf{i} + \dfrac{1}{2}\mathbf{j}$$

$$\mathbf{v}_1 = \text{proj}_{\mathbf{W}} \mathbf{v} = \dfrac{-3}{2}\mathbf{i} + \dfrac{1}{2}\mathbf{j}$$

$$\mathbf{v}_2 = \mathbf{v} - \mathbf{v}_1 = (-\mathbf{i} + 2\mathbf{j}) - \left(\dfrac{-3}{2}\mathbf{i} + \dfrac{1}{2}\mathbf{j}\right)$$

$$= \dfrac{1}{2}\mathbf{i} + \dfrac{3}{2}\mathbf{j}$$

104. $W = \|\mathbf{F}\| \|\overrightarrow{AB}\| \cos\theta$

$$= (30)(50)\cos 42°$$

$$\approx 1115$$

The work done is approximately 1115 foot-pounds.

Chapter 6 Test

1. The known ratio is $\dfrac{a}{\sin A}$, or $\dfrac{4.8}{\sin 34°}$. Because angle B is given, we use the Law of Sines to find side b

$$\dfrac{b}{\sin B} = \dfrac{a}{\sin A}$$

$$\dfrac{b}{\sin 68°} = \dfrac{4.8}{\sin 34°}$$

$$b = \dfrac{4.8 \sin 68°}{\sin 34°} \approx 8.0$$

2. Use the Law of Cosines to find c.

$$c^2 = a^2 + b^2 - 2ab\cos C$$

$$c^2 = 5^2 + 6^2 - 2(5)(6)\cos 68°$$

$$= 61 - 60\cos 68°$$

$$\approx 38.52$$

$$c \approx \sqrt{38.52} \approx 6.2$$

3. $s = \dfrac{1}{2}(a + b + c) = \dfrac{1}{2}(17 + 45 + 32) = 47$

$$\text{Area} = \sqrt{s(s-a)(s-b)(s-c)}$$

$$= \sqrt{47(47-17)(47-45)(47-32)}$$

$$= \sqrt{42,300} \approx 206$$

The area of the triangle is approximately 206 square inches.

4. Draw $\theta = \dfrac{5\pi}{4} = 225°$ counterclockwise, since θ is positive, from the polar axis. Go 4 units out on the terminal side of θ, since $r > 0$.

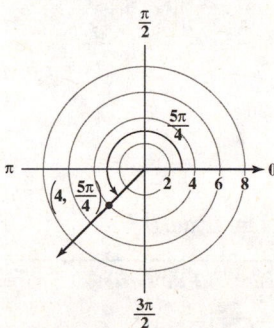

Ordered pairs may vary.

5. $(1, -1)$

$$r = \sqrt{1^2 + (-1)^2} = \sqrt{1+1} = \sqrt{2}$$

$$\tan\theta = \frac{-1}{1} = -1$$

Because $\tan\dfrac{\pi}{4} = 1$ and θ lies in quadrant IV, $\theta = 2\pi - \dfrac{\pi}{4} = \dfrac{7\pi}{4}$.

The polar coordinated of $(1, -1)$ are $(r, \theta) = \left(\sqrt{2}, \dfrac{7\pi}{4}\right)$.

6.
$$x^2 + (y+8)^2 = 64$$
$$(r\cos\theta)^2 + (r\sin\theta + 8)^2 = 64$$
$$r^2\cos^2\theta + r^2\sin^2\theta + 16r\sin\theta + 64 = 64$$
$$r^2 + 16r\sin\theta = 0$$
$$r^2 = -16r\sin\theta$$
$$r = -16\sin\theta$$

7.
$$r = -4\sec\theta$$
$$r = \frac{-4}{\cos\theta}$$
$$r\cos\theta = -4$$
$$x = -4$$

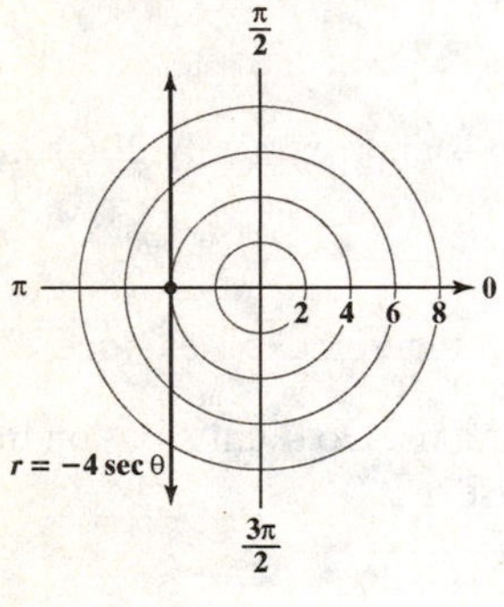

8. $r = 1 + \sin\theta$

Check for symmetry:

Polar Axis	The Line $\theta = \dfrac{\pi}{2}$	The Pole
$r = 1 + \sin(-\theta)$	$-r = 1 + \sin(-\theta)$	$-r = 1 + \sin\theta$
	$-r = 1 - \sin\theta$	
$r = 1 - \sin\theta$	$r = -1 + \sin\theta$	$r = -1 - \sin\theta$

There may be no symmetry, since each equation is not equivalent to $r = 1 + \sin\theta$.

Calculate values of r for θ from 0 to 2π .

θ	0	$\dfrac{\pi}{6}$	$\dfrac{\pi}{3}$	$\dfrac{\pi}{2}$	$\dfrac{2\pi}{3}$	$\dfrac{5\pi}{6}$	π	$\dfrac{7\pi}{6}$	$\dfrac{4\pi}{3}$	$\dfrac{3\pi}{2}$	$\dfrac{5\pi}{3}$	$\dfrac{11\pi}{6}$	2π
r	1	1.5	1.87	2	1.87	1.5	1	0.5	0.13	0	0.13	0.5	1

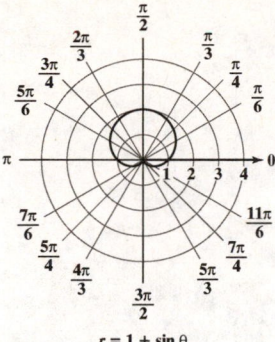

$r = 1 + \sin\theta$

9. $r = 1 + 3\cos\theta$

Check for symmetry:

Polar Axis	The Line $\theta = \dfrac{\pi}{2}$	The Pole
$r = 1 + 3\cos(-\theta)$	$-r = 1 + 3\cos(-\theta)$	$-r = 1 + 3\cos\theta$
	$-r = 1 + 3\cos\theta$	
$r = 1 + 3\cos\theta$	$r = -1 - 3\cos\theta$	$r = -1 - 3\cos\theta$

The graph has symmetry with respect to the polar axis. The graph may or may not be symmetric with respect to the line

$\theta = \dfrac{\pi}{2}$ or the pole. Calculate values of r for θ from 0 to π and use symmetry to complete the graph.

θ	0	$\dfrac{\pi}{6}$	$\dfrac{\pi}{3}$	$\dfrac{\pi}{2}$	$\dfrac{2\pi}{3}$	$\dfrac{5\pi}{6}$	π
r	4	3.6	2.5	1	−0.5	−1.6	−2

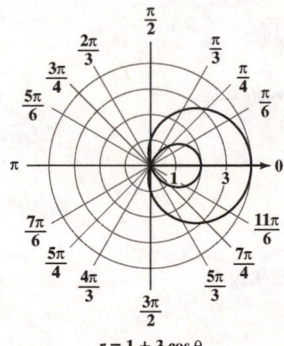

$r = 1 + 3\cos\theta$

10.   Use $r = \sqrt{a^2 + b^2}$ and $\tan\theta = \dfrac{b}{a}$, with $a = -\sqrt{3}$ and $b = 1$, to find r and θ.

$$r = \sqrt{\left(-\sqrt{3}\right)^2 + (1)^2} = \sqrt{3+1} = \sqrt{4} = 2$$

$$\tan\theta = \frac{1}{-\sqrt{3}} = -\frac{1}{\sqrt{3}}$$

Because $\tan 30° = \dfrac{1}{\sqrt{3}}$ and θ lies in quadrant II, $\theta = 180° - 30° = 150°$.

The polar form of $z = -\sqrt{3} + i$ is $z = r(\cos\theta + i\sin\theta) = 2(\cos 150° + i\sin 150°)$ or $2\left(\cos\dfrac{5\pi}{6} + i\sin\dfrac{5\pi}{6}\right)$.

11. $5(\cos 15° + i\sin 15°) \cdot 10(\cos 5° + i\sin 5°) = (5 \cdot 10)\left[\cos(15° + 5°) + i\sin(15° + 5°)\right]$

$$= 50(\cos 20° + i\sin 20°)$$

12. $\dfrac{2\left(\cos\dfrac{\pi}{2} + i\sin\dfrac{\pi}{2}\right)}{4\left(\cos\dfrac{\pi}{3} + i\sin\dfrac{\pi}{3}\right)} = \dfrac{2}{4}\left[\cos\left(\dfrac{\pi}{2} - \dfrac{\pi}{3}\right) + i\sin\left(\dfrac{\pi}{2} - \dfrac{\pi}{3}\right)\right] = \dfrac{2}{4}\left[\cos\left(\dfrac{3\pi}{6} - \dfrac{2\pi}{6}\right) + i\sin\left(\dfrac{3\pi}{6} - \dfrac{2\pi}{6}\right)\right]$

$$= \dfrac{1}{2}\left(\cos\dfrac{\pi}{6} + i\sin\dfrac{\pi}{6}\right)$$

13. $\left[2(\cos 10° + i\sin 10°)\right]^5 = (2)^5\left[\cos(5 \cdot 10°) + i\sin(5 \cdot 10°)\right] = 32(\cos 50° + i\sin 50°)$

14. $27 = 27(\cos 0° + i\sin 0°)$

$$z_k = \sqrt[3]{27}\left[\cos\left(\dfrac{0° + 360°k}{3}\right) + i\sin\left(\dfrac{0° + 360°k}{3}\right)\right], \; k = 0, 1, 2$$

$$z_0 = \sqrt[3]{27}\left[\cos\left(\dfrac{0° + 360° \cdot 0}{3}\right) + i\sin\left(\dfrac{0° + 360° \cdot 0}{3}\right)\right] = \sqrt[3]{27}(\cos 0° + i\sin 0°) = 3(1 + 0i) = 3$$

$$z_1 = \sqrt[3]{27}\left[\cos\left(\dfrac{0° + 360° \cdot 1}{3}\right) + i\sin\left(\dfrac{0° + 360° \cdot 1}{3}\right)\right] = \sqrt[3]{27}(\cos 120° + i\sin 120°) = 3\left(-\dfrac{1}{2} + i\dfrac{\sqrt{3}}{2}\right) = -\dfrac{3}{2} + \dfrac{3\sqrt{3}}{2}i$$

$$z_2 = \sqrt[3]{27}\left[\cos\left(\dfrac{0° + 360° \cdot 2}{3}\right) + i\sin\left(\dfrac{0° + 360° \cdot 2}{3}\right)\right] = \sqrt[3]{27}(\cos 240° + i\sin 240°) = 3\left(-\dfrac{1}{2} - \dfrac{\sqrt{3}}{2}i\right) = -\dfrac{3}{2} - \dfrac{3\sqrt{3}}{2}i$$

15. a. $\mathbf{v} = (x_2 - x_1)\mathbf{i} + (y_2 - y_1)\mathbf{j}$

$\mathbf{v} = [-1 - (-2)]\mathbf{i} + (5 - 3)\mathbf{j} = \mathbf{i} + 2\mathbf{j}$

b. $\|\mathbf{v}\| = \sqrt{a^2 + b^2} = \sqrt{1^2 + 2^2} = \sqrt{1 + 4} = \sqrt{5}$

16. $3\mathbf{v} - 4\mathbf{w} = 3(-5\mathbf{i} + 2\mathbf{j}) - 4(2\mathbf{i} - 4\mathbf{j}) = -15\mathbf{i} + 6\mathbf{j} - 8\mathbf{i} + 16\mathbf{j}$

$$= (-15 - 8)\mathbf{i} + (6 + 16)\mathbf{j} = -23\mathbf{i} + 22\mathbf{j}$$

17. $\mathbf{v} \cdot \mathbf{w} = (-5\mathbf{i} + 2\mathbf{j}) \cdot (2\mathbf{i} - 4\mathbf{j}) = -5(2) + 2(-4) = -10 - 8 = -18$

18. $\cos\theta = \dfrac{\mathbf{v} \cdot \mathbf{w}}{\|\mathbf{v}\|\|\mathbf{w}\|}\mathbf{w}$

$$= \dfrac{(-5\mathbf{i} + 2\mathbf{j}) \cdot (2\mathbf{i} - 4\mathbf{j})}{\sqrt{(-5)^2 + 2^2}\sqrt{2^2 + (-4)^2}}$$

$$= \dfrac{-5(2) + 2(-4)}{\sqrt{29}\sqrt{20}}$$

$$= -\dfrac{18}{\sqrt{580}}$$

The angle θ between the vectors is $\theta = \cos^{-1}\left(-\dfrac{18}{\sqrt{580}}\right) \approx 138°$.

19. $\text{proj}_{\mathbf{w}}\mathbf{v} = \dfrac{\mathbf{v}\cdot\mathbf{w}}{\|\mathbf{w}\|^2}\mathbf{w}$

$\qquad = \dfrac{(-5\mathbf{i}+2\mathbf{j})\cdot(2\mathbf{i}-4\mathbf{j})}{\left(\sqrt{2^2+(-4)^2}\right)^2}\mathbf{w}$

$\qquad = \dfrac{-5(2)+2(-4)}{\left(\sqrt{20}\right)^2}\mathbf{w}$

$\qquad = -\dfrac{18}{20}\mathbf{w}$

$\qquad = -\dfrac{9}{10}(2\mathbf{i}-4\mathbf{j})$

$\qquad = -\dfrac{9}{5}\mathbf{i}+\dfrac{18}{5}\mathbf{j}$

20.

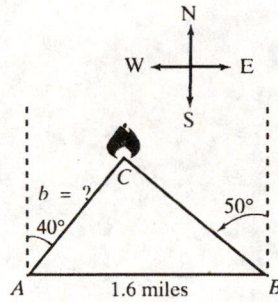

Using the figure,
$B = 90° - 50° = 40°$
$A = 90° - 40° = 50°$
$C = 180° - B - A = 180° - 40° - 50° = 90°$

Use the Law of Sines to find b. $\dfrac{b}{\sin B} = \dfrac{c}{\sin C}$

$\qquad\qquad\qquad\qquad\qquad \dfrac{b}{\sin 40°} = \dfrac{1.6}{\sin 90°}$

$\qquad\qquad\qquad\qquad\qquad\qquad b = \dfrac{1.6\sin 40°}{\sin 90°} \approx 1.0$

The fire is about 1.0 mile from the station.

21. $\qquad \mathbf{F}_1 = 250\cos 30°\,\mathbf{i} + 250\sin 30°\,\mathbf{j}$
$\qquad\qquad = 216.5\mathbf{i} + 125\mathbf{j}$

$\qquad \mathbf{F}_2 = 150\cos 315°\,\mathbf{i} + 150\sin 315°\,\mathbf{j}$
$\qquad\qquad = 106\mathbf{i} - 106\mathbf{j}$

$\mathbf{F}_1 + \mathbf{F}_2 = (216.5 + 106)\mathbf{i} + (125 - 106)\mathbf{j}$
$\qquad\qquad = 322.5\mathbf{i} + 19\mathbf{j}$

$\|\mathbf{F}_1 + \mathbf{F}_2\| = \sqrt{322.5^2 + 19^2} \approx 323$ pounds

$\cos\theta = \dfrac{322.5}{323} = 3.4°$

22. $W = \|\mathbf{F}\| \|\overline{AB}\| \cos\theta$

$= (40)(60)\cos 35° \approx 1966$

The work done is approximately 1966 foot-pounds.

Cumulative Review Exercises (Chapters 1–6)

1. $x^4 - x^3 - x^2 - x - 2 = 0$

$\dfrac{p}{q}: \pm\dfrac{2}{1}, \pm\dfrac{1}{1}$

$$
\begin{array}{r|rrrrr}
-1 & 1 & -1 & -1 & -1 & -2 \\
   &   & -1 & 2 & -1 & 2 \\
\hline
   & 1 & -2 & 1 & -2 & 0
\end{array}
$$

$x^4 - x^3 - x^2 - x - 2 = 0$

$(x+1)(x^3 - 2x^2 + x - 2) = 0$

$(x+1)[x^2(x-2) + 1(x-2)] = 0$

$(x+1)(x-2)(x^2+1) = 0$

$x+1=0 \quad x-2=0 \quad x^2+1=0$

$x=-1 \qquad x=2 \qquad x^2=-1$

$\qquad\qquad\qquad\qquad x=\pm i$

The solution set is $\{-1,\, 2\, i,\, -i\}$.

2. $2\sin^2\theta - 3\sin\theta + 1 = 0,\ 0 \le \theta < 2\pi$

$(2\sin\theta - 1)(\sin\theta - 1) = 0$

$2\sin\theta - 1 = 0 \quad \text{or} \quad \sin\theta - 1 = 0$

$\quad 2\sin\theta = 1 \qquad\qquad \sin\theta = 1$

$\quad\ \sin\theta = \dfrac{1}{2}$

The solutions in the interval $\left[0,\, 2\pi\right)$ are $\dfrac{\pi}{6}, \dfrac{5\pi}{6}$, and $\dfrac{\pi}{2}$.

3. Begin by solving the related quadratic equation. Thus, we will solve $x^2 + 2x + 3 = 11$.

$x^2 + 2x + 3 = 11$

$x^2 + 2x - 8 = 0$

$(x+4)(x-2) = 0$

$x+4=0 \quad \text{or} \quad x-2=0$

$\quad x=-4 \quad \text{or} \qquad x=2$

The boundary points are –4 and 2. The boundary points divide the number line into three test intervals, namely $(-\infty, -4)$, $(-4, 2)$, and $(2, \infty)$. Take one representative number within each test interval and substitute that number into the original inequality.

Test Interval	Representative Number	Substitute into $x^2 + 2x + 3 > 11$	Conclusion
$(-\infty, -4)$	-5	$(-5)^2 + 2(-5) + 3 > 11$ $18 > 11$ True	$(-\infty, -4)$ belongs to the solution set.
$(-4, -2)$	0	$0^2 + 2(0) + 3 > 11$ $3 > 11$ False	$(-4, -2)$ does not belong to the solution set.
$(2, \infty)$	3	$3^2 + 2(3) + 3 > 11$ $18 > 11$ True	$(2, \infty)$ belongs to the solution set.

The solution set is $\left\{ x \mid x < -4 \text{ or } x > 2 \right\}$.

4.

$$\sin\theta\cos\theta = -\frac{1}{2}$$

$$\frac{\sin 2\theta}{2} = -\frac{1}{2}$$

$$\sin 2\theta = -1$$

The period of the sine function is 2π. In the interval $[0, 2\pi)$, the only value for which the sine function is -1 is $\frac{3\pi}{2}$.

This means that $2\theta = \frac{3\pi}{2}$. Since the period is 2π, all the solutions to $\sin 2\theta = -1$ are given by $2\theta = \frac{3\pi}{2} + 2n\pi$

$$\theta = \frac{3\pi}{4} + n\pi$$

where n is any integer.

The solution in the interval $[0, 2\pi)$ is obtained by letting $n = 0$ and $n = 1$. The solutions are $\frac{3\pi}{4}$ and $\frac{7\pi}{4}$.

5. The equation $y = 3\sin(2x - \pi)$ is of the form $y = A\sin(Bx - C)$ with $A = 3$, $B = 2$, and $C = \pi$. The amplitude is $|A| = |3| = 3$. The period is $\frac{2\pi}{B} = \frac{2\pi}{2} = \pi$. The phase shift is $\frac{C}{B} = \frac{\pi}{2}$. The quarter-period is $\frac{\pi}{4}$. The cycle begins at $x = \frac{\pi}{2}$. Add quarter-periods to generate x-values for the key points.

$$x = \frac{\pi}{2}$$

$$x = \frac{\pi}{2} + \frac{\pi}{4} = \frac{3\pi}{4}$$

$$x = \frac{3\pi}{4} + \frac{\pi}{4} = \pi$$

$$x = \pi + \frac{\pi}{4} = \frac{5\pi}{4}$$

$$x = \frac{5\pi}{4} + \frac{\pi}{4} = \frac{3\pi}{2}$$

We evaluate the function at each value of x.

x	$y = 3\sin(2x - \pi)$	coordinates
$\dfrac{\pi}{2}$	$y = 3\sin\left(2 \cdot \dfrac{\pi}{2} - \pi\right)$ $= 3\sin(\pi - \pi)$ $= 3\sin 0 = 3 \cdot 0 = 0$	$\left(\dfrac{\pi}{2},\ 0\right)$
$\dfrac{3\pi}{4}$	$y = 3\sin\left(2 \cdot \dfrac{3\pi}{4} - \pi\right)$ $= 3\sin\left(\dfrac{3\pi}{2} - \pi\right)$ $= 3\sin\dfrac{\pi}{2} = 3 \cdot 1 = 3$	$\left(\dfrac{3\pi}{4},\ 3\right)$
π	$y = 3\sin(2 \cdot \pi - \pi)$ $= 3\sin(2\pi - \pi)$ $= 3\sin \pi = 3 \cdot 0 = 0$	$(\pi,\ 0)$
$\dfrac{5\pi}{4}$	$y = 3\sin\left(2 \cdot \dfrac{5\pi}{4} - \pi\right)$ $= 3\sin\left(\dfrac{5\pi}{2} - \pi\right)$ $= 3\sin\dfrac{3\pi}{2}$ $= 3(-1) = -3$	$\left(\dfrac{5\pi}{4},\ -3\right)$
$\dfrac{3\pi}{2}$	$y = 3\sin\left(2 \cdot \dfrac{3\pi}{2} - \pi\right)$ $= 3\sin(3\pi - \pi)$ $= 3\sin 2\pi = 3 \cdot 0 = 0$	$\left(\dfrac{3\pi}{2},\ 0\right)$

Connect the five points with a smooth curve and graph one complete cycle of the given function.

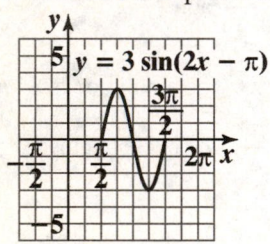

6. The equation $y = -4\cos\pi x$ is of the form $y = A\cos Bx$ with $A = -4$, and $B = \pi$. Thus, the amplitude is $|A| = |-4| = 4$.

The period is $\dfrac{2\pi}{B} = \dfrac{2\pi}{\pi} = 2$.

Find the x-values for the five key points by dividing the period, 2, by 4, $\dfrac{\text{period}}{4} = \dfrac{2}{4} = \dfrac{1}{2}$, then by adding quarter- periods to the value of x where the cycle begins, $x = 0$. The five x-values are

$x = 0$

$x = 0 + \dfrac{1}{2} = \dfrac{1}{2}$

$x = \dfrac{1}{2} + \dfrac{1}{2} = 1$

$x = 1 + \dfrac{1}{2} = \dfrac{3}{2}$

$x = \dfrac{3}{2} + \dfrac{1}{2} = 2$

We evaluate the function at each value of x.

x	$y = -4\cos\pi x$	coordinates
0	$y = -4\cos(\pi \cdot 0)$ $= -4\cos 0$ $= -4 \cdot 1 = -4$	$(0, -4)$
$\dfrac{1}{2}$	$y = -4\cos\left(\pi \cdot \dfrac{1}{2}\right)$ $= -4\cos\dfrac{\pi}{2}$ $= -4 \cdot 0 = 0$	$\left(\dfrac{1}{2}, 0\right)$
1	$y = -4\cos(\pi \cdot 1)$ $= -4\cos\pi$ $= -4 \cdot (-1) = 4$	$(1, 4)$
$\dfrac{3}{2}$	$y = -4\cos\left(\pi \cdot \dfrac{3}{2}\right)$ $= -4\cos\dfrac{3\pi}{2}$ $= -4 \cdot 0 = 0$	$\left(\dfrac{3}{2}, 0\right)$
2	$y = -4\cos(\pi \cdot 2)$ $= -4\cos 2\pi$ $= -4 \cdot 1 = -4$	$(2, -4)$

Connect the five key points with a smooth curve and graph one complete cycle of the given function.

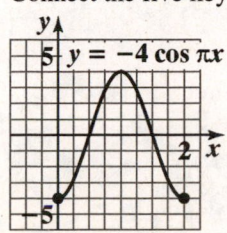

7. $\sin\theta\csc\theta - \cos^2\theta = \sin\theta\left(\dfrac{1}{\sin\theta}\right) - \cos^2\theta$

$\qquad\qquad\qquad\qquad = 1 - \cos^2\theta$

$\qquad\qquad\qquad\qquad = \sin^2\theta$

8. $\cos\left(\theta + \dfrac{3\pi}{2}\right) = \cos\theta\cos\dfrac{3\pi}{2} - \sin\theta\sin\dfrac{3\pi}{2}$

$\qquad\qquad\qquad = \cos\theta(0) - \sin\theta(-1)$

$\qquad\qquad\qquad = \sin\theta$

9. $2x + 4y - 8 = 0$

$\qquad 4y = -2x + 8$

$\qquad \dfrac{4y}{4} = \dfrac{-2x + 8}{4}$

$\qquad y = -\dfrac{1}{2}x + 2$

The slope is $-\dfrac{1}{2}$, and the y-intercept is 2.

10. $2\sin\dfrac{\pi}{3} - 3\tan\dfrac{\pi}{6} = 2\left(\dfrac{\sqrt{3}}{2}\right) - 3\left(\dfrac{1}{\sqrt{3}}\right)$

$\qquad\qquad\qquad = \sqrt{3} - \dfrac{3}{\sqrt{3}}$

$\qquad\qquad\qquad = \sqrt{3} - \sqrt{3}$

$\qquad\qquad\qquad = 0$

11. Let $\theta = \tan^{-1}\left(\dfrac{1}{2}\right)$, then $\tan\theta = \dfrac{1}{2}$. Because $\tan\theta$ is positive, θ is in the first quadrant.
Use the Pythagorean Theorem to find r.

$r = \sqrt{1^2 + 2^2} = \sqrt{1 + 4} = \sqrt{5}$

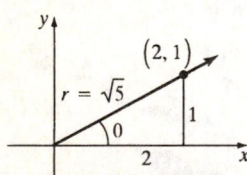

Use the right triangle to find the exact value.

$\sin\left(\tan^{-1}\dfrac{1}{2}\right) = \sin\theta = \dfrac{1}{\sqrt{5}} = \dfrac{\sqrt{5}}{5}$

12. $f(x) = \sqrt{5 - x}$

$\qquad 5 - x \geq 0$

$\qquad -x \geq -5$

$\qquad x \leq 5$

The domain of the function is $\{x \mid x \leq 5\}$.

13. $\qquad g(x) = \dfrac{x - 3}{x^2 - 9}$

$\qquad x^2 - 9 = 0$

$\qquad (x - 3)(x + 3) = 0$

$\qquad x - 3 = 0 \quad$ or $\quad x + 3 = 0$

$\qquad\quad x = 3 \qquad\qquad\quad x = -3$

The domain of the function is $\{x \mid x \neq 3 \text{ and } x \neq -3\}$.

14. $s(t) = -16t^2 + 48t + 8$

$\qquad = -16\left(t^2 - 3t - \dfrac{1}{2}\right)$

$\qquad = -16\left(t^2 - 3t + \dfrac{9}{4} - \dfrac{1}{2} - \dfrac{9}{4}\right)$

$\qquad = -16\left[\left(t - \dfrac{3}{2}\right)^2 - \dfrac{1}{2} - \dfrac{9}{4}\right]$

$\qquad = -16\left(t - \dfrac{3}{2}\right)^2 + 44$

The ball reaches its maximum height after the first 1.5 seconds. The maximum height is 44 feet.

15. $d = 4\sin 5t$ is of the form $d = a\sin\omega t$ with $a = 4$ and $\omega = 5$.

a. $|a| = |4| = 4$

The maximum displacement is 4 meters.

b. $f = \dfrac{\omega}{2\pi} = \dfrac{5}{2\pi}$

The frequency is $\dfrac{5}{2\pi}$ cycle per second.

c. period $= \dfrac{2\pi}{\omega} = \dfrac{2\pi}{5}$

$\dfrac{2\pi}{5}$ seconds are required for one cycle.

16. Because 22.5° lies in quadrant I, $\cos 22.5° > 0$.

$$\cos 22.5° = \cos \frac{45°}{2}$$

$$= \sqrt{\frac{1+\cos 45°}{2}}$$

$$= \sqrt{\frac{1+\frac{\sqrt{2}}{2}}{2}}$$

$$= \sqrt{\frac{2+\sqrt{2}}{4}}$$

$$= \frac{\sqrt{2+\sqrt{2}}}{2}$$

17. a. $3\mathbf{v} - \mathbf{w} = 3(2\mathbf{i} + 7\mathbf{j}) - (\mathbf{i} - 2\mathbf{j})$

$$= 6\mathbf{i} + 21\mathbf{j} - \mathbf{i} + 2\mathbf{j}$$

$$= 5\mathbf{i} + 23\mathbf{j}$$

b. $\mathbf{v} \cdot \mathbf{w} = (2\mathbf{i} + 7\mathbf{j}) \cdot (\mathbf{i} - 2\mathbf{j})$

$$= 2(1) + 7(-2) = 2 - 14$$

$$= -12$$

18. $\dfrac{1}{2} \log_b x - \log_b (x^2 + 1)$

$$= \log_b x^{1/2} - \log_b (x^2 + 1)$$

$$= \log_b \sqrt{x} - \log_b (x^2 + 1)$$

$$= \log_b \frac{\sqrt{x}}{x^2 + 1}$$

19. $(4, -1)$ and $(-8, 5)$

$$m = \frac{5 - (-1)}{-8 - 4} = \frac{6}{-12} = -\frac{1}{2}$$

$$y - (-1) = -\frac{1}{2}(x - 4)$$

$$y + 1 = -\frac{1}{2}x + 2$$

$$y = -\frac{1}{2}x + 1$$

20. $L = A\left(1 - e^{-kt}\right)$

a.
$$20 = 300\left(1 - e^{-k(5)}\right)$$

$$20 = 300 - 300e^{-5k}$$

$$300e^{-5k} = 280$$

$$e^{-5k} = \frac{14}{15}$$

$$\ln\left(e^{-5k}\right) = \ln\left(\frac{14}{15}\right)$$

$$-5k = \ln\left(\frac{14}{15}\right)$$

$$k = -\frac{\ln\left(\frac{14}{15}\right)}{5} \approx 0.014$$

b. $L = 300\left(1 - e^{-0.014(20)}\right) \approx 73$

After 20 minutes, the student will have learned approximately 73 words.

$$260 = 300\left(1 - e^{-0.014t}\right)$$

$$\frac{13}{15} = 1 - e^{-0.014t}$$

$$-\frac{2}{15} = -e^{-0.014t}$$

$$\frac{2}{15} = e^{-0.014t}$$

$$\ln\left(\frac{2}{15}\right) = \ln\left(e^{-0.014t}\right)$$

$$\ln\left(\frac{2}{15}\right) = -0.014t$$

$$t = -\frac{\ln\left(\frac{2}{15}\right)}{0.014} \approx 144$$

It will take about 144 minutes.

Chapter 7
Systems of Equations and Inequalities

Section 7.1

Check Point Exercises

1. a.
$$2x = 3y = -4$$
$$2(1) - 3(2) = -4$$
$$2 - 6 = -4$$
$$-4 = -4 \text{ true}$$
$$2x + y = 4$$
$$2(1) + 2 = 4$$
$$2 + 2 = 4$$
$$4 = 4 \text{ true}$$
$(1, 2)$ is a solution of the system.

b.
$$2x = 3y = -4$$
$$2(7) - 3(6) = -4$$
$$14 - 18 = -4$$
$$-4 = -4 \text{ true}$$
$$2x + y = 4$$
$$2(7) + 6 = 4$$
$$14 + 6 = 4$$
$$20 = 4 \text{ false}$$
$(7, 6)$ is not a solution of the system.

2.
$$3x + 2y = 4$$
$$2x + y = 1$$
Solve $2x + y = 1$ for y.
$$2x + y = 1$$
$$y = 1 - 2x$$
Substitute $1 - 2x$ for y in the other equation and solve.
$$3x + 2\overbrace{\left(1 - 2x\right)}^{y} = 4$$
$$3x + 2 - 4x = 4$$
$$-x = 2$$
$$x = -2$$
Back-substitute the obtained value:
$$3x + 2y = 4$$
$$3(-2) + 2y = 4$$
$$-6 + 2y = 4$$
$$2y = 10$$
$$y = 5$$
Checking these values in both equations shows that $(-2, 5)$ is the solution of the system.

3. Rewrite one or both equations:
$$4x + 5y = 3 \xrightarrow{\text{No change}} 4x + 5y = 3$$
$$2x - 3y = 7 \xrightarrow{\text{Mult. by } -2} \underline{-4x + 6y = -14}$$
$$11y = -11$$
$$y = -1$$
Back-substitute into either equation:
$$4x + 5y = 3$$
$$4x + 5(-1) = 3$$
$$4x - 5 = 3$$
$$4x = 8$$
$$x = 2$$
Checking confirms the solution set is $\{(2, -1)\}$.

4. Rewrite both equations in the form $Ax + By = C$:
$$2x = 9 + 3y \rightarrow 2x - 3y = 9$$
$$4y = 8 - 3x \rightarrow 3x + 4y = 8$$
Rewrite with opposite coefficients, then add and solve:
$$2x - 3y = 9 \xrightarrow{\text{Mult. by } 4} 8x - 12y = 36$$
$$3x + 4y = 8 \xrightarrow{\text{Mult. by } 3} \underline{9x + 12y = 24}$$
$$17x = 60$$
$$x = \frac{60}{17}$$
Back-substitute into either equation:
$$4y = 8 - 3x$$
$$4y = 8 - 3\left(\frac{60}{17}\right)$$
$$4y = -\frac{44}{17}$$
$$y = -\frac{11}{17}$$
Checking confirms the solution is $\left(\frac{60}{17}, -\frac{11}{17}\right)$.

5. Rewrite with a pair of opposite coefficients, then add:
$$5x - 2y = 4 \xrightarrow{\text{Mult. by } 2} 10x - 4y = 8$$
$$-10x + 4y = 7 \xrightarrow{\text{No change}} \underline{-10x + 4y = 7}$$
$$0 = 15$$
The statement $0 = 15$ is false which indicates that the system has no solution. The solution set is the empty set, $\varnothing$.

6. Substitute $4y - 8$ for x in the other equation:

$$5(\overset{x}{\overbrace{4y-8}}) - 20y = -40$$
$$20y - 40 - 20y = -40$$
$$-40 = -40$$

The statement $-40 = -40$ is true which indicates that the system has infinitely many solutions. The solution set is $\{(x, y) | x = 4y - 8\}$ or $\{(x, y) | 5x - 20y = -40\}$.

7. $x =$ liters of 18% acid
$y =$ liters of 45% acid

$$x + y = 12$$
$$0.18x + 0.45y = 0.36(12)$$

$$y = 12 - x$$

$$0.18x + 0.45(12 - x) = 4.32$$
$$0.18x + 5.4 - 0.45x = 4.32$$
$$-0.27x = -1.08$$
$$x = 4$$
$$y = 12 - 4 = 8$$

Mix 4 liters of 18% acid and 8 liters of 45% acid.

8. $x =$ velocity of the boat
$y =$ velocity of the current

Velocity	Time	Distance
$x + y$	2	$2(x+y)$
$x - y$	3	$3(x-y)$

$$2(x + y) = 84$$
$$3(x - y) = 84$$

$$x + y = 42$$
$$x - y = 29$$

$$2x = 70$$
$$x = 35$$

$$x + y = 42$$
$$35 + y = 42$$
$$y = 7$$

Velocity of the boat is 35 mph and the current is 7 mph.

9. **a.** $C(x) = 300,000 + 30x$

b. $R(x) = 80x$

c. $R(x) = C(x)$
$$80x = 300,000 + 30x$$
$$50x = 300,000$$
$$x = 6000$$
$$C(6000) = 300,000 + 30(6000) = 480,000$$
Break even point (6000, 480000)
The company will need to make 6000 pairs of shoes and earn \$480,000 to break even.

Exercise Set 7.1

1. $x + 3y = 11$
$$2 + 3(3) = 11$$
$$2 + 9 = 11$$
$$11 = 11 \text{ true}$$
$$x - 5y = -13$$
$$2 - 5(3) = -13$$
$$2 - 15 = -13$$
$$-13 = -13 \text{ true}$$
(2, 3) is a solution.

3. $2x + 3y = 17$
$$2(2) + 3(5) = 17$$
$$4 + 15 = 17$$
$$19 = 17 \text{ false}$$
(2, 5) is not a solution.

5. $x + y = 4$
$$y = 3x$$
Substitute the expression $3x$ for y in the first equation and solve for x.
$$x + 3x = 4$$
$$4x = 4$$
$$x = 1$$
Substitute 1 for x in the second equation.
$$y = 3(1) = 3$$
The solution set is $\{(1, 3)\}$.

7. $x + 3y = 8$
   $y = 2x - 9$
Substitute the expression $2x - 9$ for y in the first equation and solve for x.
$x + 3(2x - 9) = 8$
   $x + 6x - 27 = 8$
       $7x = 35$
        $x = 5$
Substitute 5 for x in the second equation.
$y = 2(5) - 9 = 10 - 9 = 1$
The solution set is $\{(5, 1)\}$.

9. $x = 4y - 2$
   $x = 6y + 8$
Substitute the expression $4y - 2$ for x in the second equation and solve for y.
$4y - 2 = 6y + 8$
   $-10 = 2y$
    $-5 = y$
Substitute -5 for y in the equation $x = 4y - 2$.
$x = 4(-5) - 2 = -22$
The solution set is $\{(-22, -5)\}$.

11. $5x + 2y = 0$
   $x - 3y = 0$
Solve the second equation for x.
$x = 3y$
Substitute the expression $3y$ for x in the first equation and solve for y.
$5(3y) + 2y = 0$
   $15y + 2y = 0$
       $17y = 0$
         $y = 0$
Substitute 0 for y in the equation $x = 3y$
$y = 3(0) = 0$
The solution set is $\{(0, 0)\}$.

13. $2x + 5y = -4$
    $3x - y = 11$
Solve the second equation for y.
$-y = -3x + 11$
   $y = 3x - 11$
Substitute the expression $3x - 11$ for y in the first equation and solve for x.
$2x + 5(3x - 11) = -4$
   $2x + 15x - 55 = -4$
           $17x = 51$
             $x = 3$
Substitute 3 for x in the equation $y = 3x - 11$.
$y = 3(3) - 11 = 9 - 11 = -2$
The solution set is $\{(3, -2)\}$.

15. $2x - 3y = 8 - 2x$
    $2x + 4y = x + 3y + 14$
Solve the second equation for y.
$y = -2x + 14$
Substitute the expression $-2x + 14$ for y in the first equation and solve for x.
$2x - 3(-2x + 14) = 8 - 2x$
   $2x + 6x - 42 = 8 - 2x$
       $8x - 42 = 8 - 2x$
         $10x = 50$
           $x = 5$
Substitute 5 for x in the equation $y = -2x + 14$.
$y = -2(5) + 14 = -10 + 14 = 4$
The solution set is $\{(5, 4)\}$.

17. $y = \dfrac{1}{3}x + \dfrac{2}{3}$

   $y = \dfrac{5}{7}x - 2$

Substitute the expression $y = \dfrac{1}{3}x + \dfrac{2}{3}$ for y in the second equation and solve for x.

$\dfrac{1}{3}x + \dfrac{2}{3} = \dfrac{5}{7}x - 2$

$7x + 14 = 15x - 42$

     $56 = 8x$

      $7 = x$

Substitute 7 for x in the equation $y = \dfrac{1}{3}x + \dfrac{2}{3}$ and

solve for y.

$y = \dfrac{1}{3}(7) + \dfrac{2}{3} = \dfrac{7}{3} + \dfrac{2}{3} = \dfrac{9}{3} = 3$

The solution set is $\{(7, 3)\}$.

19. Eliminate y by adding the equations.

$x + y = 1$

$\underline{x - y = 3}$

$2x = 4$

$x = 2$

Substitute 2 for x in the first equation.

$2 + y = 1$

$y = -1$

The solution set is $\{(2, -1)\}$.

21. Eliminate y by adding the equations.

$2x + 3y = 6$

$\underline{2x - 3y = 6}$

$4x = 12$

$x = 3$

Substitute 3 for x in the first equation.

$2(3) + 3y = 6$

$6 + 3y = 6$

$3y = 0$

$y = 0$

The solution set is $\{(3, 0)\}$.

23. $x + 2y = 2$

$-4x + 3y = 25$

Eliminate x by multiplying the first equation by 4 and adding the resulting equations.

$4x + 8y = 8$

$\underline{-4x + 3y = 25}$

$11y = 33$

$y = 3$

Substitute 3 for y in the first equation.

$x + 2(3) = 2$

$x + 6 = 2$

$x = -4$

The solution set is $\{(-4, 3)\}$.

25. $4x + 3y = 15$

$2x - 5y = 1$

Eliminate x by multiplying the second equation by -2 and adding the resulting equations.

$4x + 3y = 15$

$\underline{-4x + 10y = -2}$

$13y = 13$

$y = 1$

Substitute 1 for y in the second equation.

$2x - 5(1) = 1$

$2x = 6$

$x = 3$

The solution set is $\{(3, 1)\}$.

27. $3x - 4y = 11$

$2x + 3y = -4$

Eliminate x by multiplying the first equation by 2 and the second equation by -3. Add the resulting equations.

$6x - 8y = 22$

$\underline{-6x - 9y = 12}$

$-17y = 34$

$y = -2$

Substitute -2 for y in the second equation.

$2x + 3(-2) = -4$

$2x - 6 = -4$

$2x = 2$

$x = 1$

The solution set is $\{(1, -2)\}$.

29. $3x = 4y + 1$

$3y = 1 - 4x$

Arrange the system so that variable terms appear on the left and constants appear on the right.

$3x - 4y = 1$

$4x + 3y = 1$

Eliminate y by multiplying the first equation by 3 and the second equation by 4. Add the resulting equations.

$9x - 12y = 3$

$\underline{16x + 12y = 4}$

$25x = 7$

$x = \dfrac{7}{25}$

Substitute $\dfrac{7}{25}$ for x in the second equation.

$3y = 1 - 4\left(\dfrac{7}{25}\right)$

$3y = \dfrac{-3}{25}$

$y = \dfrac{-1}{25}$

The solution set is $\left\{\left(\dfrac{7}{25}, -\dfrac{1}{25}\right)\right\}$.

31. The substitution method is used here to solve the system.

$$x = 9 - 2y$$
$$x + 2y = 13$$

Substitute the expression $9 - 2y$ for x in the second equation and solve for y.

$$9 - 2y + 2y = 13$$
$$9 = 13$$

The false statement $9 = 13$ indicates that the system has no solution.

The solution set is the empty set, $\varnothing$.

33. The substitution method is used here to solve the system.

$$y = 3x - 5$$
$$21x - 35 = 7y$$

Substitute the expression $3x - 5$ for y in the second equation and solve for x.

$$21x - 35 = 7(3x - 5)$$
$$21x - 35 = 21x - 35$$
$$-35 = -35$$

This true statement indicates that the system has infinitely many solutions.

The solution set is $\left\{ (x, y) \mid y = 3x - 5 \right\}$

35. The elimination method is used here to solve the system.

$$3x - 2y = -5$$
$$4x + y = 8$$

Eliminate y by multiplying the second equation by 2 and adding the resulting equations.

$$3x - 2y = -5$$
$$\underline{8x + 2y = 16}$$
$$11x = 11$$
$$x = 1$$

Substitute 1 for x in the second equation.

$$4(1) + y = 8$$
$$y = 4$$

The solution set is $\{(1, 4)\}$.

37. The elimination method is used here to solve the system.

$$x + 3y = 2$$
$$3x + 9y = 6$$

Eliminate x by multiplying the first equation by -3 and adding the resulting equations.

$$-3x - 9y = -6$$
$$\underline{3x + 9y = 6}$$
$$0 = 0$$

This true statement indicates that the system has infinitely many solutions.

The solution set is $\left\{ (x, y) \mid x + 3y = 2 \right\}$.

39. First multiply each term in the first equation by 4 to eliminate the fractions.

$$\frac{x}{4} - \frac{y}{4} = -1$$
$$x - y = -4$$

Multiply the first equation by -1 and add to the second equation and solve for y.

$$-x + y = 4$$
$$x + 4y = -9$$
$$5y = -5$$
$$y = -1$$

Substitute -1 for y in the equation $x - y = -4$ and solve for x.

$$x - (-1) = -4$$
$$x + 1 = -4$$
$$x = -5$$

The solution set is $\{(-5, -1)\}$.

41. Rearrange the equations to get in the standard form.

$$2x - 3y = 4$$
$$4x + 5y = 3$$

Multiply the first equation by -2 and add to the second equation. Solve for y.

$$-4x + 6y = -8$$
$$4x + 5y = 3$$
$$11y = -5$$
$$y = -\frac{5}{11}$$

Multiply the first equation by 5 and the second equation by 3 and add the equations. Solve for x.

$$10x - 15y = 20$$
$$12x + 15y = 9$$
$$22x = 29$$
$$x = \frac{29}{22}$$

The solution set is $\left\{ \left(\frac{29}{22}, -\frac{5}{11} \right) \right\}$.

43. Add the equations to eliminate y.

$$x + y = 7$$
$$\underline{x - y = -1}$$
$$2x = 6$$
$$x = 3$$

Substitute 3 for x in the first equation.

$$3 + y = 7$$
$$y = 4$$

The numbers are 3 and 4.

45. $3x - y = 1$
$x + 2y = 12$
Eliminate y by multiplying the first equation by 2 and adding the resulting equations.
$6x - 2y = 2$
$\underline{x + 2y = 12}$
$7x = 14$
$x = 2$
Substitute 2 for x in the first equation.
$3(2) - y = 1$
$6 - y = 1$
$-y = -5$
$y = 5$
The numbers are 2 and 5.

47. $\dfrac{x+2}{2} - \dfrac{y+4}{3} = 3$

$\dfrac{x+y}{5} = \dfrac{x-y}{2} - \dfrac{5}{2}$

Start by multiplying each equation by its LCD and simplifying to clear the fractions.

$\dfrac{x+2}{2} - \dfrac{y+4}{3} = 3$

$\dfrac{x+y}{5} = \dfrac{x-y}{2} - \dfrac{5}{2}$

Start by multiplying each equation by its LCD and simplifying to clear the fractions.

$6\left(\dfrac{x+2}{2} - \dfrac{y+4}{3}\right) = 6(3)$

$3(x+2) - 2(y+4) = 18$

$3x + 6 - 2y - 8 = 18$

$3x - 2y = 20$

$10\left(\dfrac{x+y}{5}\right) = 10\left(\dfrac{x-y}{2} - \dfrac{5}{2}\right)$

$2(x+y) = 5(x-y) - 5(5)$

$2x + 2y = 5x - 5y - 25$

$3x - 7y = 25$

We now need to solve the equivalent system of equations:
$3x - 2y = 20$
$3x - 7y = 25$
Subtract the two equations:
$3x - 2y = 20$
$\underline{-(3x - 7y = 25)}$
$5y = -5$
$y = -1$

Back-substitute this value for y and solve for x.
$3x - 2y = 20$
$3x - 2(-1) = 20$
$3x + 2 = 20$
$3x = 18$
$x = 6$
The solution is $(6, -1)$.

49. $5ax + 4y = 17$
$ax + 7y = 22$
Multiply the second equation by -5 and add the equations.
$5ax + 4y = 17$
$\underline{-5ax - 35y = -110}$
$-31y = -93$
$y = 3$
Back-substitute into one of the original equations to solve for x.
$ax + 7y = 22$
$ax + 7(3) = 22$
$ax + 21 = 22$
$ax = 1$
$x = \dfrac{1}{a}$
The solution is $\left(\dfrac{1}{a}, 3\right)$.

51. $f(-2) = 11 \quad \rightarrow \quad -2m + b = 11$
$f(3) = -9 \quad \rightarrow \quad 3m + b = -9$

We need to solve the resulting system of equations:
$-2m + b = 11$
$3m + b = -9$

Subtract the two equations:
$-2m + b = 11$
$\underline{3m + b = -9}$
$-5m = 20$
$m = -4$

Back-substitute into one of the original equations to solve for b.
$-2m + b = 11$
$-2(-4) + b = 11$
$8 + b = 11$
$b = 3$
Therefore, $m = -4$ and $b = 3$.

53. The solution to a system of linear equations is the point of intersection of the graphs of the equations in the system. If $(6, 2)$ is a solution, then we need to find the lines that intersect at that point. Looking at the graph, we see that the graphs of $x + 3y = 12$ and $x - y = 4$ intersect at the point $(6, 2)$. Therefore, the desired system of equations is

$$x + 3y = 12 \quad \text{or} \quad y = -\frac{1}{3}x + 4$$
$$x - y = 4 \qquad\qquad y = x - 4$$

55. Let x = gallons of 5% wine. Let y = gallons of 9% wine.

$$x + y = 200$$
$$0.05x + 0.09y = 0.07(200)$$

or

$$x + y = 200$$
$$0.05x + 0.09y = 14$$

Solve the first equation for x.

$$x + y = 200$$
$$x = 200 - y$$

Substitute this expression for x in the second equation.

$$0.05(200 - y) + 0.09y = 14$$
$$10 - 0.05y + 0.09y = 14$$
$$0.04y = 4$$
$$y = 100$$

Back-substitute and solve for x.

$$x = 200 - y$$
$$= 200 - 100$$
$$= 100$$

The wine company should mix 100 gallons of the 5% California wine with 100 gallons of the 9% French wine.

57. Let x = grams of 18-karat gold. Let y = grams of 12-karat gold.

$$x + y = 300$$
$$0.75x + 0.5y = 0.58(300)$$

or

$$x + y = 300$$
$$0.75x + 0.5y = 174$$

Solve the first equation for x.

$$x = 300 - y$$

Substitute this result for x into the second equation and solve for y.

$$0.75(300 - y) + 0.5y = 174$$
$$225 - 0.75 + 0.5y = 174$$
$$-0.25y = -51$$
$$y = 204$$

Back-substitute to solve for x.

$$x = 300 - y = 300 - 204 = 96$$

You would need 96 grams of 18-karat gold and 204 grams of 12-karat gold.

59. Let x = pounds of cheaper candy. Let y = pounds of more expensive candy.

$$x + y = 75$$
$$1.6x + 2.1y = 1.9(75)$$

or

$$x + y = 75$$
$$1.6x + 2.1y = 142.5$$

Multiply the first equation by -1.6 and add the two equations.

$$-1.6x - 1.6y = -120$$
$$\underline{1.6x + 2.1y = 142.5}$$
$$0.5y = 22.5$$
$$y = 45$$

Back-substitute to solve for x.

$$x + 45 = 75$$
$$x = 30$$

The manager should mix 30 pounds of the cheaper candy and 45 pounds of the more expensive candy.

61. x = velocity of the plane
y = velocity of the wind

Velocity	Time	Distance
$x + y$	5	$5(x + y)$
$x - y$	8	$8(x - y)$

$5(x + y) = 800 \quad \rightarrow \quad x + y = 160$
$8(x - y) = 800 \quad \rightarrow \quad x - y = 100$
$$\overline{\qquad\qquad 2x \;\; = 260}$$
$$x = 130$$

$x + y = 160$
$130 + y = 160$
$y = 30$

Velocity of the plane is 130 mph and the wind is 30 mph.

63. x = velocity of the boat
y = velocity of the current

Velocity	Time	Distance
$x + y$	2	$2(x + y)$
$x - y$	4	$4(x - y)$

$2(x + y) = 16 \quad \rightarrow \quad x + y = 8$
$4(x - y) = 16 \quad \rightarrow \quad x - y = 4$
$$\overline{\qquad\qquad 2x \;\; = 12}$$
$$x = 6$$

$x + y = 8$
$6 + y = 8$
$y = 2$

Velocity of the boat is 6 mph and the current is 2 mph.

65. x = velocity of the boat
y = velocity of the current

Velocity	Time	Distance
$x + y$	4	$4(x + y)$
$x - y$	6	$6(x - y)$

$4(x + y) = 24 \qquad \rightarrow \quad x + y = 6$
$6(x - y) = \frac{3}{4}(24) \quad \rightarrow \quad x - y = 3$
$$\overline{\qquad\qquad 2x \;\; = 9}$$
$$x = 4.5$$

$x + y = 6$
$4.5 + y = 6$
$y = 1.5$

Velocity of the boat is 4.5 mph and the current is 1.5 mph.

67. At the break-even point, $R(x) = C(x)$.
$10000 + 30x = 50x$
$10000 = 20x$
$10000 = 20x$
$500 = x$
Five hundred radios must be produced and sold to break-even.

69. $R(x) = 50x$
$R(200) = 50(200) = 10000$
$C(x) = 10000 + 30x$
$C(200) = 10000 + 30(200)$
$\qquad = 10000 + 6000 = 16000$
$R(200) - C(200) = 10000 - 16000$
$$= -6000$$
This means that if 200 radios are produced and sold the company will lose $6,000.

71. a. $P(x) = R(x) - C(x)$
$\qquad = 50x - (10000 + 30x)$
$\qquad = 50x - 10000 - 30x$
$\qquad = 20x - 10000$
$P(x) = 20x - 10000$

b. $P(10000) = 20(10000) - 10000$
$\qquad = 200000 - 10000 = 190000$
If 10,0000 radios are produced and sold the profit will be $190,000.

73. a. The cost function is:
$C(x) = 18,000 + 20x$

b. The revenue function is:
$R(x) = 80x$

c. At the break-even point, $R(x) = C(x)$.
$80x = 18000 + 20x$
$60x = 18000$
$x = 300$
$R(x) = 80x$
$R(300) = 80(300)$
$\qquad = 24,000$
When approximately 300 canoes are produced the company will break-even with cost and revenue at $24,000.

75. **a.** The cost function is:
$$C(x) = 30000 + 2500x$$

b. The revenue function is:
$$R(x) = 3125x$$

c. At the break-even point, $R(x) = C(x)$.
$$3125x = 30000 + 2500x$$
$$625x = 30000$$
$$x = 48$$
After 48 sold out performances, the investor will break-even. ($150,000)

77. **a.** Substitute $0.375x + 3$ for p in the first equation.
$$p = -0.325x + 5.8$$

$$\overset{p}{\overbrace{0.375x + 3}} = -0.325x + 5.8$$
$$0.375x + 3 = -0.325x + 5.8$$
$$0.375x + 0.325x + 3 = -0.325x + 0.325x + 5.8$$
$$0.7x + 3 = 5.8$$
$$0.7x + 3 - 3 = 5.8 - 3$$
$$0.7x = 2.8$$
$$\frac{0.7x}{0.7} = \frac{2.8}{0.7}$$
$$x = 4$$
Back-substitute to find p.
$$p = -0.325x + 5.8$$
$$p = -0.325(4) + 5.8 = 4.5$$

The ordered pair is (4, 4.5).
Equilibrium number of workers: 4 million
Equilibrium hourly wage: $4.50

b. If workers are paid $4.50 per hour, there will be 4 million available workers and 4 million workers will be hired. In this state of market equilibrium, there is no unemployment.

c.
$$p = -0.325x + 5.8$$
$$5.15 = -0.325x + 5.8$$
$$0.65 = -0.325x$$
$$\frac{-0.65}{-0.325} = \frac{-0.325x}{-0.325}$$
$$2 = x$$
At $5.15 per hour, 2 million workers will be hired.

d.
$$p = 0.375x + 3$$
$$5.15 = 0.375x + 3$$
$$2.15 = 0.375x$$
$$\frac{2.15}{0.375} = \frac{0.375x}{0.375}$$
$$x \approx 5.7$$
At $5.15 per hour, there will be about 5.7 million available workers.

e. $5.7 - 2 = 3.7$
At $5.15 per hour, there will be about 3.7 million more people looking for work than employers are willing to hire.

79.
$$x + 2y = 50$$
$$-x + 2y = 24$$
Add the equations.
$$\begin{aligned} x + 2y &= 50 \\ \underline{-x + 2y} &= \underline{24} \\ 4y &= 74 \\ y &= 18.5 \end{aligned}$$
Back-substitute 18.5 for y and solve for x.
$$x + 2y = 50$$
$$x + 2(18.5) = 50$$
$$x + 37 = 50$$
$$x = 13$$
The solution is (13, 18.5). This means that 13 years after 1996, or 2009, the percentage who will be pro-life and pro-choice will be the same at 18.5%.

81. **a.** $y = 0.45x + 0.8$

b. $y = 0.15x + 2.6$

c. To find the week in the semester when both groups report the same number of symptoms, we set the two equations equal to each other and solve for x.
$$0.45x + 0.8 = 0.15x + 2.6$$
$$0.3x = 1.8$$
$$x = 6$$
The number of symptoms will be the same in week 6.
$$y = 0.15x + 2.6$$
$$y = 0.15(6) + 2.6$$
$$y = 3.5$$
The number of symptoms in week 6 will be 3.5 for both groups. This is shown in the graph by the intersection point (6, 3.5).

83. **a.** $m = \dfrac{27.3 - 38}{20 - 0} = \dfrac{-10.7}{20} \approx -0.54$

From the point $(0, 38)$ we have that the y-intercept is $b = 38$. Therefore, the equation of the line is $y = -0.54x + 38$.

b. $m = \dfrac{24.2 - 40}{20 - 0} = \dfrac{-15.8}{20} = -0.79$

From the point $(0, 40)$ we have that the y-intercept is $b = 40$. Therefore, the equation of the line is $y = -0.79x + 40$.

c. To find the year when cigarette use is the same, we set the two equations equal to each other and solve for x.

$-0.54x + 38 = -0.79x + 40$

$0.25x = 2$

$x = 8$

Cigarette use was the same for African Americans and Hispanics 8 years after 1985, or 1993.

$y = -0.54x + 38$

$y = -0.54(8) + 38$

$y = 33.68$

At that time about 33.68% of each group used cigarettes.

85. Let x = the number of calories in a Mr. Goodbar. Let y = the number of calories in a Mounds bar.

$x + 2y = 780$

$2x + y = 786$

Multiply the bottom equation by -2 and then add the equations to eliminate y.

$\begin{aligned} x + 2y &= 780 \\ -4x - 2y &= -1572 \\ \hline -3x &= -792 \end{aligned}$

$x = 264$

Back-substitute to find y.

$x + 2y = 780$

$264 + 2y = 780$

$2y = 516$

$y = 258$

There are 264 calories in a Mr. Goodbar and 258 calories in a Mounds bar.

87. Let x = the number of Mr. Goodbars. Let y = the number of Mounds bars.

$x + y = 5$

$16.3x + 14.1y - 70 = 7.1$

Solve the first equation for y in terms of x.

$x + y = 5$

$y = -x + 5$

Substitute $-x + 5$ for y in the second equation.

$16.3x + 14.1y - 70 = 7.1$

$16.3x + 14.1\overbrace{(-x + 5)}^{y} - 70 = 7.1$

$16.3x + 14.1(-x + 5) - 70 = 7.1$

$16.3x - 14.1x + 70.5 - 70 = 7.1$

$2.2x + 0.5 = 7.1$

$2.2x = 6.6$

$x = 3$

Back-substitute to find y.

$x + y = 5$

$3 + y = 5$

$y = 2$

There are 3 Mr. Goodbars and 2 Mounds bars.

89. $x + y = 200$

$100x + 80y = 17000$

Multiply the first equation by -100 and add to the second equation. Solve for y.

$\begin{aligned} -100x - 100y &= -20000 \\ 100x + 80y &= 17000 \\ \hline -20y &= -3000 \end{aligned}$

$y = 150$

Substitute 150 for y in the first equation and solve for x.

$x + 150 = 200$

$x = 50$

There are 50 rooms with kitchenettes and 150 rooms without.

91.
$$2x + 2y = 360$$
$$20x + 8(2y) = 3280$$
Multiply the first equation by -10 and add to the second equation. Solve for y.
$$-20x - 20y = -3600$$
$$20x + 16y = 3280$$
$$-4y = -320$$
$$y = 80$$
Substitute 80 for y in the first equation and solve for x.
$$2x + 2(80) = 360$$
$$2x + 160 = 360$$
$$2x = 200$$
$$x = 100$$
The lot is 100 feet long and 80 feet wide.

93.
$$x + 2y = 180$$
$$(2x - 30) + y = 180$$
Rewrite the second equation in standard form.
$$x + 2y = 180$$
$$2x + y = 210$$
Multiply the first equation by -2 and add the equations.
$$-2x - 4y = -360$$
$$\underline{2x + y = 210}$$
$$-3y = -150$$
$$y = 50$$
Back-substitute to solve for x.
$$x + 2y = 180$$
$$x + 2(50) = 180$$
$$x + 100 = 180$$
$$x = 80$$
The three interior angles measure $80°$, $50°$, and $50°$.

95. – 103. Answers may vary.

105. does not make sense; Explanations will vary. Sample explanation: Some linear systems have no solutions or one solution.

107. makes sense

109.
$$a_1 x + b_1 y = c_1$$
$$a_2 x + b_2 y = c_2$$
Solve the first equation for x.
$$x = \frac{c_1 - b_1 y}{a_1}$$
Substitute the expression $\dfrac{c_1 - b_1 y}{a_1}$ for x in the second equation and solve for y.
$$a_2 \left(\frac{c_1 - b_1 y}{a_1} \right) + b_2 y = c_2$$
$$a_2 \left(\frac{c_1 - b_1 y}{a_1} \right) + \frac{a_1 b_2 y}{a_1} = c_2$$
$$\frac{a_2 c_1 - a_2 b_1 y + a_1 b_2 y}{a_1} = c_2$$
$$a_2 c_1 - a_2 b_1 y + a_1 b_2 y = a_1 c_2$$
$$y(a_1 b_2 - a_2 b_1) = a_1 c_2 - a_2 c_1$$
$$y = \frac{a_1 c_2 - a_2 c_1}{a_1 b_2 - a_2 b_1}$$
Substitute the expression $\dfrac{a_1 c_2 - a_2 c_1}{a_1 b_2 - a_2 b_1}$ for y in the first equation and solve for x.
$$a_1 x + b_1 \left(\frac{a_1 c_2 - a_2 c_1}{a_1 b_2 - a_2 b_1} \right) = c_1$$
$$a_1 x + \frac{a_1 b_1 c_2 - a_2 b_1 c_1}{a_1 b_2 - a_2 b_1} = c_1$$
$$a_1 x = c_1 - \frac{a_1 b_1 c_2 - a_2 b_1 c_1}{a_1 b_2 - a_2 b_1}$$
$$= \frac{c_1(a_1 b_2 - a_2 b_1)}{a_1 b_2 - a_2 b_1} - \frac{a_1 b_1 c_2 - a_2 b_1 c_1}{a_1 b_2 - a_2 b_1}$$
$$= \frac{a_1 b_2 c_1 - a_1 b_1 c_2}{a_1 b_2 - a_2 b_1}$$
$$x = \frac{a_1 b_2 c_1 - a_1 b_1 c_2}{a_1 b_2 - a_2 b_1} \div a_1$$
$$= \frac{a_1 b_2 c_1 - a_1 b_1 c_2}{a_1(a_1 b_2 - a_2 b_1)} = \frac{a_1(b_2 c_1 - b_1 c_2)}{a_1(a_1 b_2 - a_2 b_1)}$$
$$x = \frac{b_2 c_1 - b_1 c_2}{a_1 b_2 - a_2 b_1}$$

111. x = number of hexagons formed

y = number of squares formed

$6x + y = 52$

$x + 4y = 24$

Eliminate x by multiplying the second equation by -6 and adding the resulting equations.

$6x + y = 52$

$\underline{-6x - 24y = -144}$

$-23y = -92$

$y = 4$

Substitute 4 for y in the second equation.

$x + 4(4) = 24$

$x + 16 = 24$

$x = 8$

Yes, they should make 8 hexagons and 4 squares.

113.
$$2x - y + 4z = -8$$
$$2(3) - (2) + 4(-3) = -8$$
$$-8 = -8, \text{ true}$$

Yes, the ordered triple satisfies the equation.

114. $5x - 2y - 4z = 3$

$3x + 3y + 2z = -3$

Multiply Equation 2 by 2.

$5x - 2y - 4z = 3$

$6x + 6y + 4z = -6$

Then add to eliminate z.

$5x - 2y - 4z = 3$

$\underline{6x + 6y + 4z = -6}$

$11x + 4y \qquad = -3$

115. $ax^2 + bx + c = y$

$a(4)^2 + b(4) + c = 1682$

$16a + 4b + c = 1682$

Check Point Exercises

1.
$$x - 2y + 3z = 22$$
$$-1 - 2(-4) + 3(5) = 22$$
$$-1 + 8 + 15 = 22$$
$$22 = 22 \text{ true}$$
$$2x - 3y - z = 5$$
$$2(-1) - 3(-4) - 5 = 5$$
$$-2 + 12 - 5 = 5$$
$$5 = 5 \text{ true}$$
$$3x + y - 5z = -32$$
$$3(-1) - 4 - 5(5) = -32$$
$$-3 - 4 - 25 = -32$$
$$-32 = -32 \text{ true}$$

$(-1, -4, 5)$ is a solution of the system.

2.
$$x + 4y - z = 20$$
$$3x + 2y + z = 8$$
$$2x - 3y + 2z = -16$$

Eliminate z from Equations 1 and 2 by adding Equation 1 and Equation 2.

$x + 4y - z = 20$

$\underline{3x + 2y + z = 8}$

$4x + 6y = 28$ Equation 4

Eliminate z from Equations 2 and 3 by multiplying Equation 2 by -2 and adding the resulting equation to Equation 3.

$-6x - 4y - 2z = -16$

$\underline{2x - 3y + 2z = -16}$

$-4x - 7y = -32$ Equation 5

Solve Equations 4 and 5 for x and y by adding Equation 4 and Equation 5.

$4x + 6y = 28$

$\underline{-4x - 7y = -32}$

$-y = -4$

$y = 4$

Substitute 4 for y in Equation 4 and solve for x.

$$4x + 6(4) = 28$$
$$4x + 24 = 28$$
$$4x = 4$$
$$x = 1$$

Substitute 1 for x and 4 for y in Equation 2 and solve for z.

$$3(1) + 2(4) + z = 8$$
$$3 + 8 + z = 8$$
$$11 + z = 8$$
$$z = -3$$

The solution set is $\{(1, 4, -3)\}$.

3.
$$2y - z = 7$$
$$x + 2y + z = 17$$
$$2x - 3y + 2z = -1$$

Eliminate x and z from Equations 2 and 3 by multiplying Equation 2 by -2 and adding the resulting equation to Equation 3.

$$-2x - 4y - 2z = -34$$
$$\underline{2x - 3y + 2z = -1}$$
$$-7y = -35$$
$$y = 5$$

Substitute 5 for y in Equation 1 and solve for z.

$$2(5) - z = 7$$
$$10 - z = 7$$
$$-z = -3$$
$$z = 3$$

Substitute 5 for y and 3 for z in Equation 2 and solve for x.

$$x + 2(5) + 3 = 17$$
$$x + 10 + 3 = 17$$
$$x + 13 = 17$$
$$x = 4$$

The solution set is $\{(4, 5, 3)\}$.

4. $(1, 4), (2, 1), (3, 4)$

$$y = ax^2 + bx + c$$

Substitute 1 for x and 4 for y in

$$y = ax^2 + bx + c.$$
$$4 = a(1)^2 + b(1) + c$$
$$4 = a + b + c \quad \text{Equation 1}$$

Substitute 2 for x and 1 for y in

$$y = ax^2 + bx + c.$$
$$1 = a(2)^2 + b(2) + c$$
$$1 = 4a + 2b + c \quad \text{Equation 2}$$

Substitute 3 for x and 4 for y in

$$y = ax^2 + bx + c.$$
$$4 = a(3)^2 + b(3) + c$$
$$4 = 9a + 3b + c \quad \text{Equation 3}$$

Eliminate c from Equations 1 and 2 by multiplying Equation 2 by -1 and adding the resulting equation to Equation 1.

$$4 = a + b + c$$
$$\underline{-1 = -4a - 2b - c}$$
$$3 = -3a - b \quad \text{Equation 4}$$

Eliminate c from Equation 2 and 3 by multiplying Equation 3 by -1 and adding the resulting equation to Equation 2.

$$1 = 4a + 2b + c$$
$$\underline{-4 = -9a - 3b - c}$$
$$-3 = -5a - b \quad \text{Equation 5}$$

Solve Equations 4 and 5 for a and b by multiplying Equation 5 by -1 and adding the resulting equation to Equation 4.

$$3 = -3a - b$$
$$\underline{3 = 5a + b}$$
$$6 = 2a$$
$$a = 3$$

Substitute 3 for a in Equation 4 and solve for b.

$$3 = -3(3) - b$$
$$3 = -9 - b$$
$$12 = -b$$
$$b = -12$$

Substitute 3 for a and -12 for b in Equation 1 and solve for c.

$$4 = 3 - 12 + c$$
$$4 = -9 + c$$
$$c = 13$$

Substituting 3 for a, -12 for b, and 13 for c in the quadratic equation $y = ax^2 + bx + c$ gives

$$y = 3x^2 - 12x + 13 .$$

Exercise Set 7.2

1. $x + y + z = 4$
$2 - 1 + 3 = 4$
$4 = 4$ true
$x - 2y - z = 1$
$2(2) - 2(-1) - 3 = 1$
$4 + 2 - 3 = 1$
$1 = 1$ true
$2x - y - 2z = -1$
$2(2) - (-1) - 2(3) = -1$
$4 + 1 - 6 = -1$
$-1 = -1$ false
$(2, -1, 3)$ is a solution.

3. $x - 2y = 2$
$4 - 2(1) = 2$
$4 - 2 = 2$
$2 = 2$ true
$2x + 3y = 11$
$2(4) + 3(1) = 11$
$8 + 3 = 11$
$11 = 11$ true
$y - 4z = -7$
$1 - 4(2) = -7$
$1 - 8 = -7$
$-7 = -7$ true
$(4, 1, 2)$ is a solution.

5. $x + y + 2z = 11$
$x + y + 3z = 14$
$x + 2y - z = 5$

Eliminate x and y from Equations 1 and 2 by multiplying Equation 2 by -1 and adding the resulting equation to Equation 1.
$-x - y - 3z = -14$
$\underline{x + y + 2z = 11}$
$-z = -3$
$z = 3$
Substitute 3 for z in Equations 1 and 3.
$x + y + 2(3) = 11$
$x + 2y - (3) = 5$
Simplify:
$x + y = 5$ Equation 4
$x + 2y = 8$ Equation 5

Solve Equations 4 and 5 for x and y by multiplying Equation 5 by -1 and adding the resulting equation to Equation 4.
$x + y = 5$
$\underline{-x - 2y = -8}$
$-y = -3$
$y = 3$
Substitute 3 for z and 3 for y in Equation 2 and solve for x.
$x + 3 + 3(3) = 14$
$x + 12 = 14$
$x = 2$
The solution set is $\{(2, 3, 3)\}$.

7. $4x - y + 2z = 11$
$x + 2y - z = -1$
$2x + 2y - 3z = -1$

Eliminate y from Equation 1 and 2 by multiplying Equation 1 by 2 and adding the resulting equation to Equation 2 and 3.
$8x - 2y + 4z = 22$
$\underline{x + 2y - z = -1}$
$9x + 3z = 21$ Equation 4

Eliminate y from Equations 1 and 3 by multiplying Equation 1 by 2 and adding the resulting equation to Equation 3.
$8x - 2y + 4z = 22$
$\underline{2x + 2y - 3z = -1}$
$10x + z = 21$ Equation 5

Solve Equations 4 and 5 for x and z by multiplying Equation 5 by -3 and adding the resulting equation to Equation 4.
$9x + 3z = 21$
$\underline{-30x - 3z = -63}$
$-21x = -42$
$x = 2$
Substitute 2 for x in Equation 5 and solve for z. $10(2) + z = 21$
$20 + z = 21$
$z = 1$
Substitute 2 for x and 1 for z in Equation 2 and solve for y.
$2 + 2y - 1 = -1$
$2y + 1 = -1$
$2y = -2$
$y = -1$
The solution set is $\{(2, -1, 1)\}$.

9. $3x + 2y - 3z = -2$
$2x - 5y + 2z = -2$
$4x - 3y + 4z = 10$
Eliminate z from Equations 1 and 2 by multiplying Equation 1 by 2 and Equation 2 by 3. Add the resulting equations.
$6x + 4y - 6z = -4$
$6x - 15y + 6z = -6$
———————————————
$12x - 11y = -10$ Equation 4
Eliminate z from Equations 2 and 3 by multiplying Equation 2 by –2.
$-4x + 10y - 4z = 4$
$4x - 3y + 4z = 10$
———————————————
$7y = 14$ Equation 5
Solve Equation 5 for y
$7y = 14$
$y = 2$
Solve for x by substituting 7 for y in Equation 4.
$12x - 11y = -10$
$12x - 11(2) = -10$
$12x - 22 = -10$
$12x = 12$
$x = 1$
Substitute 2 for y and 1 for x in Equation 2 and solve for z.
$2x - 5y + 2z = -2$
$2(1) - 5(2) + 2z = -2$
$2 - 10 + 2z = -2$
$2z = 6$
$z = 3$
The solution set is $\{(1, 2, 3)\}$.

11. $2x - 4y + 3z = 17$
$x + 2y - z = 0$
$4x - y - z = 6$
Eliminate z from Equations 1 and 2 by multiplying Equation 2 by 3 and adding the resulting equation to Equation 1.
$2x - 4y + 3z = 17$
$3x + 6y - 3z = 0$
———————————————
$5x + 2y = 17$ Equation 4
Eliminate z from Equations 2 and 3 by multiplying Equation 2 by –1 and adding the resulting equation to Equation 3.
$-x - 2y + z = 0$
$4x - y - z = 6$
———————————————
$3x - 3y = 6$ Equation 5
Solve Equations 4 and 5 for x and y by multiplying

Equation 5 by $\dfrac{2}{3}$ and adding the resulting equation to Equation 4.
$5x + 2y = 17$
$2x - 2y = 4$
———————————————
$7x = 21$
$x = 3$
Substitute 3 for x in Equation 4 and solve for y.
$5(3) + 2y = 17$
$15 + 2y = 17$
$2y = 2$
$y = 1$
Substitute 3 for x and 1 for y in Equation 2 and solve for z.
$3 + 2(1) - z = 0$
$3 + 2 - z = 0$
$5 - z = 0$
$5 = z$
The solution set is $\{(3, 1, 5)\}$.

13. $2x + y = 2$
$x + y - z = 4$
$3x + 2y + z = 0$
Eliminate z from Equations 2 and 3 by adding Equation 2 and Equation 3.
$x + y - z = 4$
$3x + 2y + z = 0$
———————————————
$4x + 3y = 4$ Equation 4
Solve Equations 1 and 4 for x and y by multiplying Equation 1 by –3 and adding the resulting equation to Equation 4.
$-6x - 3y = -6$
$4x + 3y = 4$
———————————————
$-2x = -2$
$x = 1$
Substitute 1 for x in Equation 1 and solve for y.
$2(1) + y = 2$
$2 + y = 2$
$y = 0$
Substitute 1 for x and 0 for y in Equation 2 and solve for z.
$1 + 0 - z = 4$
$1 - z = 4$
$-z = 3$
$z = -3$
The solution set is $\{(1, 0, -3)\}$.

15. $x + y = -4$
            $y - z = 1$
    $2x + y + 3z = -21$
Eliminate y from Equations 1 and 2 by multiplying
Equation 1 by -1 and adding
the resulting equation to Equation 2.
    $-x - y = 4$
    $\underline{y - z = 1}$
    $-x - z = 5$ Equation 4

Eliminate y from Equations 2 and 3 by multiplying
Equation 2 by -1 and adding
the resulting equation to Equation 3.
        $-y + z = -1$
    $\underline{2x + y + 3z = -21}$
        $2x + 4z = -22$ Equation 5

Solve Equations 4 and 5 for x and z by multiplying
Equation 4 by 2 and adding
the resulting equation to Equation 5.
    $-2x - 2z = 10$
    $\underline{2x + 4z = -22}$
        $2z = -12$
        $z = -6$
Substitute -6 for z in Equation 2 and solve for y.
$y - (-6) = 1$
    $y + 6 = 1$
        $y = -5$
Substitute -5 for y in Equation 1 and solve for x
$x + (-5) = -4$
        $x = 1$
The solution set is $\{(1, -5, -6)\}$.

17. $3(2x + y) + 5z = -1$
    $2(x - 3y + 4z) = -9$

    $4(1 + x) = -3(z - 3y)$

Simplify each equation.
$6x + 3y + 5z = -1$ Equation 4

$2x - 6y + 8z = -9$ Equation 5

$4 + 4x = -3z + 9y$

$4x - 9y + 3z = -4$ Equation 6

Eliminate x from Equations 4 and 5 by multiplying
Equation 5 by -3 and adding
the resulting equation to Equation 4.
    $-6x + 3y + 5z = -1$
    $\underline{-6x + 18y - 24z = 27}$
        $21y - 19z = 26$ Equation 7

Eliminate x from Equations 5 and 6 by multiplying
Equation 5 by -2 and adding
the resulting equation to Equation 6.

$-4x + 12y - 16z = 18$
$\underline{4x - 9y + 3z = -4}$
        $3y - 13z = 14$ Equation 8

Solve Equations 7 and 8 for y and z by multiplying
Equation 8 by -7 and adding
the resulting equation to Equation 7.
    $21y - 19z = 26$
    $\underline{-21y + 91z = -98}$
        $72z = -72$

        $z = -1$
Substitute -1 for z in Equation 8 and solve
for y.
$3y - 13(-1) = 14$
    $3y + 13 = 14$
        $3y = 1$

        $y = \dfrac{1}{3}$

Substitute $\dfrac{1}{3}$ for y and -1 for z in Equation 5 and

solve for x.

$2x - 6\left(\dfrac{1}{3}\right) + 8(-1) = -9$

        $2x - 2 - 8 = -9$

        $2x - 10 = -9$

        $2x = 1$

        $x = \dfrac{1}{2}$

The solution set is $\left\{\left(\dfrac{1}{2}, \dfrac{1}{3}, -1\right)\right\}$.

19. $(-1, 6), (1, 4), (2, 9)$

$y = ax^2 + bx + c$

Substitute -1 for x and 6 for y in $y = ax^2 + bx + c$.

$6 = a(-1)^2 + b(-1) + c$

$6 = a - b + c$ Equation 1

Substitute 1 for x and 4 for y in $y = ax^2 + bx + c$.

$4 = a(1)^2 + b(1) + c$

$4 = a + b + c$ Equation 2

Substitute 2 for x and 9 for y in
$y = ax^2 + bx + c$.

$9 = a(2)^2 + b(2) + c$

$9 = 4a + 2b + c$ Equation 3

Eliminate b from Equations 1 and 2 by adding
Equation 1 and Equation 2.

$6 = a - b + c$

$\underline{4 = a + b + c}$

$10 = 2a + 2c$ Equation 4

Eliminate b from Equations 1 and 3 by multiplying
Equation 1 by 2 and adding the resulting equation to
Equation 3.

$12 = 2a - 2b + 2c$

$\underline{9 = 4a + 2b + c}$

$21 = 6a + 3c$ Equation 5

Solve Equations 4 and 5 for a and c by multiplying
Equation 4 by -3 and adding the resulting equation to
Equation 5.

$-30 = -6a - 6c$

$\underline{21 = 6a + 3c}$

$-9 = -3c$

$c = 3$

Substitute 3 for c in Equation 4 and solve
for a.

$10 = 2a + 2(3)$

$10 = 2a + 6$

$4 = 2a$

$a = 2$

Substitute 2 for a and 3 for c in Equation 2 and solve
for b.

$4 = 2 + b + 3$

$4 = b + 5$

$b = -1$

Substituting 2 for a, -1 for b, and 3 for c in the
quadratic equation $y = ax^2 + bx + c$ gives

$y = 2x^2 - x + 3$.

21. $(-1, -4), (1, -2), (2, 5)$

Substitute -1 for x and -4 for y in $y = ax^2 + bx + c$.

$-4 = a(-1)^2 + b(-1) + c$

$-4 = a - b + c$ Equation 1

Substitute 1 for x and -2 for y in $y = ax^2 + bx + c$.

$-2 = a(1)^2 + b(1) + c$

$-2 = a + b + c$ Equation 2

Substitute 2 for x and 5 for y in $y = ax^2 + bx + c$.

$5 = a(2)^2 + b(2) + c$

$5 = 4a + 2b + c$ Equation 3

Eliminate a and b from Equations 1 and 2 by
multiplying Equation 1 by -1 and adding the
resulting equation to Equation 2.

$4 = -a + b - c$

$\underline{-2 = a + b + c}$

$2 = 2b$

$b = 1$

Eliminate c from Equations 1 and 3 by multiplying
Equation 1 by -1 and adding
the resulting equation to Equation 3.

$4 = -a + b - c$

$\underline{5 = 4a + 2b + c}$

$9 = 3a + 3b$ Equation 4

Substitute 1 for b in Equation 4 and solve
for a.

$9 = 3a + 3(1)$

$9 = 3a + 3$

$6 = 3a$

$a = 2$

Substitute 2 for a and 1 for b in Equation 2 and solve
for c.

$-2 = 2 + 1 + c$

$-2 = c + 3$

$c = -5$

Substituting 2 for a, 1 for b, and -5 for c in quadratic
equation $y = ax^2 + bx + c$ gives $y = 2x^2 + x - 5$.

23. $x + y + z = 16$
$2x + 3y + 4z = 46$
$5x - y = 31$

Eliminate z from Equations 1 and 2 by multiplying Equation 1 by -4 and adding the resulting equation to Equation 2.
$-4x - 4y - 4z = -64$
$\underline{2x + 3y + 4z = 46}$
$-2x - y = -18$ Equation 4

Solve Equations 3 and 4 for x and y by multiplying Equation 4 by -1 and adding the resulting equation to Equation 3.
$5x - y = 31$
$\underline{2x + y = 18}$
$7x = 49$
$x = 7$

Substitute 7 for x in Equation 3 and solve for y.
$5(7) - y = 31$
$35 - y = 31$
$-y = -4$
$y = 4$

Substitute 7 for x and 4 for y in Equation 1 and solve for z.
$7 + 4 + z = 16$
$z + 11 = 16$
$z = 5$

The numbers are 7, 4 and 5.

25.
$$\frac{x+2}{6} - \frac{y+4}{3} + \frac{z}{2} = 0$$
$$6\left(\frac{x+2}{6} - \frac{y+4}{3} + \frac{z}{2}\right) = 6(0)$$
$$(x+2) - 2(y+4) + 3z = 0$$
$$x + 2 - 2y - 8 + 3z = 0$$
$$x - 2y + 3z = 6$$

$$\frac{x+1}{2} + \frac{y-1}{2} - \frac{z}{4} = \frac{9}{2}$$
$$4\left(\frac{x+1}{2} + \frac{y-1}{2} - \frac{z}{4}\right) = 4\left(\frac{9}{2}\right)$$
$$2(x+1) + 2(y-1) - z = 18$$
$$2x + 2 + 2y - 2 - z = 18$$
$$2x + 2y - z = 18$$

$$\frac{x-5}{4} + \frac{y+1}{3} + \frac{z-2}{2} = \frac{19}{4}$$
$$12\left(\frac{x-5}{4} + \frac{y+1}{3} + \frac{z-2}{2}\right) = 12\left(\frac{19}{4}\right)$$
$$3(x-5) + 4(y+1) + 6(z-2) = 57$$
$$3x - 15 + 4y + 4 + 6z - 12 = 57$$
$$3x + 4y + 6z = 80$$

We need to solve the equivalent system:
$x - 2y + 3z = 6$
$2x + 2y - z = 18$
$3x + 4y + 6z = 80$

Add the first two equations together.
$x - 2y + 3z = 6$
$\underline{2x + 2y - z = 18}$
$3x + 2z = 24$

Multiply the second equation by -2 and add it to the third equation.

$-4x - 4y + 2z = -36$
$\underline{3x + 4y + 6z = 80}$
$-x + 8z = 44$

Using the two reduced equations, we solve the system
$3x + 2z = 24$
$-x + 8z = 44$

Multiply the second equation by 3 and add the equations.

$3x + 2z = 24$
$\underline{-3x + 24z = 132}$
$26z = 156$
$z = 6$

Back-substitute to find x.
$-x + 8(6) = 44$
$-x + 48 = 44$
$-x = -4$
$x = 4$

Back substitute to find y.
$x - 2y + 3z = 6$
$4 - 2y + 3(6) = 6$
$-2y = -16$
$y = 8$

The solution is $(4, 8, 6)$.

27. Selected points may vary, but the equation will be the same.

$$y = ax^2 + bx + c$$

Use the points $(2, -2)$, $(4, 1)$, and $(6, -2)$ to get the system

$$4a + 2b + c = -2$$
$$16a + 4b + c = 1$$
$$36a + 6b + c = -2$$

Multiply the first equation by -1 and add to the second equation.

$$-4a - 2b - c = 2$$
$$\underline{16a + 4b + c = 1}$$
$$12a + 2b = 3$$

Multiply the first equation by -1 and add to the third equation.

$$-4a - 2b - c = 2$$
$$\underline{36a + 6b + c = -2}$$
$$32a + 4b = 0$$

Using the two reduced equations, we get the system

$$12a + 2b = 3$$
$$32a + 4b = 0$$

Multiply the first equation by -2 and add to the second equation.

$$-24a - 4b = -6$$
$$\underline{32a + 4b = 0}$$
$$8a = -6$$
$$a = -\frac{3}{4}$$

Back-substitute to solve for b.

$$12a + 2b = 3$$
$$12\left(-\frac{3}{4}\right) + 2b = 3$$
$$-9 + 2b = 3$$
$$2b = 12$$
$$b = 6$$

Back-substitute to solve for c.

$$4a + 2b + c = -2$$
$$4\left(-\frac{3}{4}\right) + 2(6) + c = -2$$
$$-3 + 12 + c = -2$$
$$c = -11$$

The equation is:

$$y = -\frac{3}{4}x^2 + 6x - 11$$

29.

$$ax - by - 2cz = 21$$
$$ax + by + cz = 0$$
$$2ax - by + cz = 14$$

Add the first two equations.

$$ax - by - 2cz = 21$$
$$\underline{ax + by + cz = 0}$$
$$2ax - cz = 21$$

Multiply the first equation by -1 and add to the third equation.

$$-ax + by + 2cz = -21$$
$$\underline{2ax - by + cz = 14}$$
$$ax + 3cz = -7$$

Use the two reduced equations to get the following system:

$$2ax - cz = 21$$
$$ax + 3cz = -7$$

Multiply the second equation by -2 and add the equations.

$$2ax - cz = 21$$
$$\underline{-2ax - 6cz = 14}$$
$$-7cz = 35$$
$$z = -\frac{5}{c}$$

Back-substitute to solve for x.

$$ax + 3cz = -7$$
$$ax + 3c\left(-\frac{5}{c}\right) = -7$$
$$ax - 15 = -7$$
$$ax = 8$$
$$x = \frac{8}{a}$$

Back-substitute to solve for y.

$$ax + by + cz = 0$$
$$a\left(\frac{8}{a}\right) + by + c\left(-\frac{5}{c}\right) = 0$$
$$8 + by - 5 = 0$$
$$by = -3$$
$$y = -\frac{3}{b}$$

The solution is $\left(\dfrac{8}{a}, -\dfrac{3}{b}, -\dfrac{5}{c}\right)$.

31. a. Substitute the values for x and y into the quadratic form.
$$224 = a(1)^2 + b(1) + c$$
$$a + b + c = 224$$

$$176 = a(3)^2 + b(3) + c$$
$$9a + 3b + c = 176$$

$$104 = a(4)^2 + b(4) + c$$
$$16a + 4b + c = 104$$

Multiply the first equation by –1 and add to both the second and the third equations to obtain 2 new equations with 2 variables.
$$-a - b - c = -224$$
$$9a + 3b + c = 176$$
$$8a + 2b = -48$$

$$-a - b - c = -224$$
$$16a + 4b + c = 104$$
$$15a + 3b = -120$$

Use the two new equations to solve for a and b. Multiply the first equation by –3 and the second equation by 2 and add the results together. Solve for a. Substitute that value in $8a + 2b = -48$ and solve for b.
$$-24a - 6b = 144$$
$$30a + 6b = -240$$
$$6a = -96$$
$$a = -16$$

$$8(-16) + 2b = -48$$
$$-128 + 2b = -48$$
$$2b = 80$$
$$b = 40$$

Substitute –16 for a and 40 for b into the equation $a + b + c = 224$ and solve for c.
$$-16 + 40 + c = 224$$
$$c = 200$$
The equation is $y = -16x^2 + 40x + 200$.

b. $y = -16(5)^2 + 40(5) + 200 = 0$
The ball hit the ground after 5 seconds.

33. Let $w =$ the percent of body weight that consists of water.
Let $f =$ the percent of body weight that consists of fat.
Let $p =$ the percent of body weight that consists of protein.
The information is represented by the following system of equations.
$$w + f + p = 95$$
$$w - f = 35$$
$$f - p = 9$$
Solving $w - f = 35$ for w gives $w = f + 35$.
Solving $f - p = 9$ for p gives $p = f - 9$.
Use substitution to find f.
$$w + f + p = 95$$
$$\overbrace{f + 35}^{w} + f + \overbrace{f - 9}^{p} = 95$$
$$f + 35 + f + f - 9 = 95$$
$$3f + 26 = 95$$
$$3f = 69$$
$$f = 23$$
Find w.
$$w = f + 35$$
$$w = 23 + 35 = 58$$
Find p.
$$p = f - 9$$
$$p = 23 - 9 = 14$$
The total body weight consists of 58% water, 23% fat, and 14% protein.

35. x = number of $8 tickets sold
y = number of $10 tickets sold
z = number of $12 tickets sold
From the given conditions we have the following system of equations.

$$x + y + z = 400$$
$$8x + 10y + 12z = 3700$$
$$x + y = 7z \quad \text{or} \quad x + y - 7z = 0$$

Eliminate z from Equations 1 and 2 multiplying Equation 1 by -12 and adding the resulting equation to Equation 2.

$$-12x - 12y - 12z = -4800$$
$$\underline{8x + 10y + 12z = 3700}$$
$$-4x - 2y = -1100 \quad \text{Equation 4}$$

Eliminate z from Equations 1 and 3 by multiplying Equation 1 by 7 and adding the resulting equation to Equation 3.

$$7x + 7y + 7z = 2800$$
$$\underline{x + y - 7z = 0}$$
$$8x + 8y = 2800 \quad \text{Equation 5}$$

Solve Equations 4 and 5 for x and y by multiplying Equation 4 by 2 and adding the resulting equation to Equation 5.

$$-8x - 4y = -2200$$
$$\underline{8x + 8y = 2800}$$
$$4y = 600$$
$$y = 150$$

Substitute 150 for y in Equation 5 and solve for x.
$$8x + 8(150) = 2800$$
$$8x = 2800 - 1200$$
$$8x = 1600$$
$$x = 200$$

Substitute 200 for x and 150 for y in Equation 1 and solve for z.
$$200 + 150 + z = 400$$
$$350 + z = 400$$
$$z = 50$$

The number of $8 tickets sold was 200.
The number of $10 tickets sold was 150.
The number of $12 tickets sold was 50.

37. x = amount of money invested at 10%
y = amount of money invested at 12%
z = amount of money invested at 15%

$$x + y + z = 6700$$
$$0.08x + 0.10y + 0.12z = 716$$
$$z = x + y + 300$$

Arrange Equation 3 so that variable terms appear on the left and constants appear on the right.

$$-x - y + z = 300 \quad \text{Equation 4}$$

Eliminate x and y from Equations 1 and 4 by adding Equations 1 and 4.

$$x + y + z = 6700$$
$$\underline{-x - y + z = 300}$$
$$2z = 7000$$
$$z = 3500$$

Substitute 3500 for z in Equation 1 and Equation 2 and simplify.

$$x + y + 3500 = 6700$$
$$x + y = 3200 \quad \text{Equation 5}$$
$$0.08x + 0.10y + 0.12(3500) = 716$$
$$0.08x + 0.10y + 420 = 716 \qquad \text{Solve}$$
$$0.08x + 10y = 296 \quad \text{Equation 6}$$

Equations 5 and 6 for x and y by multiplying Equation 5 by -0.10 and adding the resulting equation to Equation 6.

$$-0.10x - 0.10y = -320$$
$$\underline{0.08x + 0.10y = 296}$$
$$-0.02x = 24$$
$$x = 1200$$

Substitute 1200 for x and 3,500 for z in Equation 1 and solve for y.
$$1200 + y + 3500 = 6700$$
$$y + 4700 = 6700$$
$$y = 2000$$

The person invested $1200 at 8%, $2000 at 10%, and $3500 at 12%.

39. $x + y + z = 180$

$2x - 5 + z = 180$

$2x + z = 185$

$2x + 5 + y = 180$

$2x + y = 175$

Multiply the second equation by -1 and add to the first equation. Use the new equation and the third equation to solve for x and z.

$-2x - z = -185$

$x + y + z = 180$

$-x + y = -5$

Multiply the new equation by -1.

$x - y = 5$

$2x + y = 175$

$3x = 180$

$x = 60$

$60 - y = 5$

$-y = -55$

$y = 55$

Substitute 60 for x and 55 for y in the first equation and solve for z.

$60 + 55 + z = 180$

$z = 65$

41. – 45. Answers may vary.

47. does not make sense; Explanations will vary. Sample explanation: A system of linear equations in three variables can contain an equation of the form $y = mx + b$. For this equation, the coefficient of z is 0.

49. makes sense

51. x = number of triangles
y = number of rectangles
z = number of pentagons

$x + y + z = 40$

$3x + 4y + 5z = 153$

$2y + 5z = 72$

Eliminate x from Equations 1 and 2 by multiplying Equation 1 by -3 and adding the resulting equation to Equation 2.

$-3x - 3y - 3z = -120$

$3x + 4y + 5z = 153$

$\overline{\qquad\qquad\qquad\qquad}$

$y + 2z = 33$ Equation 4

Solve for z by multiplying Equation 4 by -2 and adding the resulting equation to Equation 3.

$2y + 5z = 72$

$-2y - 4z = -66$

$\overline{\qquad\qquad}$

$z = 6$

Substitute 6 for z in Equation 4 and solve for y.

$y + 2(6) = 33$

$y + 12 = 33$

$y = 21$

Substitute 21 for y and 6 for z in Equation 1 and solve for x.

$x + 21 + 6 = 40$

$x + 27 = 40$

$x = 13$

The painting has 13 triangles, 21 rectangles, and 6 pentagons.

53.
$$\frac{3}{x-4} - \frac{2}{x+2} = \frac{3(x+2)}{(x-4)(x+2)} - \frac{2(x-4)}{(x-4)(x+2)}$$
$$= \frac{3x+6}{(x-4)(x+2)} - \frac{2x-8}{(x-4)(x+2)}$$
$$= \frac{3x+6-2x+8}{(x-4)(x+2)}$$
$$= \frac{x+14}{(x-4)(x+2)}$$

55. $A + B = 3$

$2A - 2B + C = 17$

$4A - 2C = 14$

Solving $A + B = 3$ for B gives $B = 3 - A$.
Solving $4A - 2C = 14$ for A gives $C = 2A - 7$.
Use substitution to find f.

$2A - 2B + C = 17$

$2A - 2\overbrace{(3-A)}^{B} + \overbrace{2A-7}^{C} = 17$

$2A - 2(3-A) + 2A - 7 = 17$

$2A - 6 + 2A + 2A - 7 = 17$

$6A - 13 = 17$

$6A = 30$

$A = 5$

Find B.

$B = 3 - A$

$B = 3 - 5 = -2$

Find C.

$C = 2A - 7$

$C = 2(5) - 7 = 3$

The solution set is $\{(5, -2, 3)\}$.

Section 7.3

Check Point Exercises

1. $\dfrac{5x-1}{(x-3)(x+4)} = \dfrac{A}{x-3} + \dfrac{B}{x+4}$

Multiply both sides of the equation by the least common denominator $(x-3)(x+4)$ and divide out common factors.

$5x-1 = A(x+4) + B(x-3)$

$5x-1 = Ax + 4A + Bx - 3B$

$5x-1 = (A+B)x + 4A - 3B$

Equate coefficients of like powers of x and equate constant terms.

$A + B = 5$

$4A - 3B = -1$

Solving the above system for A and B we find $A = 2$ and $B = 3$.

$\dfrac{5x-1}{(x-3)(x+4)} = \dfrac{2}{x-3} + \dfrac{3}{x+4}$

2. $\dfrac{x+2}{x(x-1)^2} = \dfrac{A}{x} + \dfrac{B}{x-1} + \dfrac{C}{(x-1)^2}$

Multiply both sides of the equation by the least common denominator $x(x-1)^2$ and divide out common factors.

$x+2 = A(x-1)^2 + Bx(x-1) + Cx$

$x+2 = A(x^2 - 2x + 1) + Bx^2 - Bx + Cx$

$x+2 = Ax^2 - 2Ax + A + Bx^2 - Bx + Cx$

$x+2 = Ax^2 + Bx^2 - 2Ax - Bx + Cx + A$

$x+2 = (A+B)x^2 + (-2A - B + C)x + A$

Equate coefficients of like powers of x and equate constant terms.

$A + B = 0$

$-2A - B + C = 1$

$A = 2$

Since $A = 2$, we find that $B = -2$ and $C = 3$ by substitution.

$\dfrac{x+2}{x(x-1)^2} = \dfrac{2}{x} - \dfrac{2}{x-1} + \dfrac{3}{(x-1)^2}$

3. $\dfrac{8x^2 + 12x - 20}{(x+3)(x^2 + x + 2)} = \dfrac{A}{x+3} + \dfrac{Bx+C}{x^2 + x + 2}$

Multiply both sides of the equation by the least common denominator $(x+3)(x^2 + x + 2)$ and divide out common factors.

$8x^2 + 12x - 20 = A(x^2 + x + 2) + (Bx + C)(x + 3)$

$8x^2 + 12x - 20 = Ax^2 + Ax + 2A + Bx^2 + 3Bx + Cx + 3C$

$8x^2 + 12x - 20 = Ax^2 + Bx^2 + Ax + 3Bx + Cx + 2A + 3C$

$8x^2 + 12x - 20 = (A+B)x^2 + (A + 3B + C)x + 2A + 3C$

Equate coefficients of like powers of x and equate constant terms.

$A + B = 8$

$A + 3B + C = 12$

$2A + 3C = -20$

Solving the above system for A, B, and C we find $A = 2$, $B = 6$, and $C = -8$.

$\dfrac{8x^2 + 12x - 20}{(x+3)(x^2 + x + 2)} = \dfrac{2}{x+3} + \dfrac{6x-8}{x^2 + x + 2}$

4. $\dfrac{2x^3 + x + 3}{(x^2 + 1)^2} = \dfrac{Ax + B}{x^2 + 1} + \dfrac{Cx + D}{(x^2 + 1)^2}$

Multiply both sides of the equation by the common denominator $(x^2 + 1)^2$ and divide out common factors.

$2x^3 + x + 3 = (Ax + B)(x^2 + 1) + Cx + D$

$2x^3 + x + 3 = Ax^3 + Bx^2 + Ax + B + Cx + D$

$2x^3 + x + 3 = Ax^3 + Bx^2 + Ax + Cx + B + D$

$2x^3 + x + 3 = Ax^3 + Bx^2 + (A + C)x + B + D$

Equate coefficients of like powers of x and equate constant terms.

$A = 2$

$B = 0$

$A + C = 1$

$B + D = 3$

Since $A = 2$ and $B = 0$ we find that $C = -1$ and $D = 3$ by substitution.

$\dfrac{2x^3 + x + 3}{(x^2 + 1)^2} = \dfrac{2x}{x^2 + 1} + \dfrac{-x+3}{(x^2 + 1)^2} = \dfrac{2x}{x^2 + 1} - \dfrac{x-3}{(x^2 + 1)^2}$

Exercise Set 7.3

1. $\dfrac{11x-10}{(x-2)(x+1)} = \dfrac{A}{x-2} + \dfrac{B}{x+1}$

3. $\dfrac{6x^2 - 14x - 27}{(x+2)(x-3)^2} = \dfrac{A}{x+2} + \dfrac{B}{x-3} + \dfrac{C}{(x-3)^2}$

5. $\dfrac{5x^2 - 6x + 7}{(x-1)(x^2+1)} = \dfrac{A}{x-1} + \dfrac{Bx+C}{x^2+1}$

7. $\dfrac{x^3 + x^2}{(x^2+4)^2} = \dfrac{Ax+B}{x^2+4} + \dfrac{Cx+D}{(x^2+4)^2}$

9. $\dfrac{x}{(x-3)(x-2)} = \dfrac{A}{x-3} + \dfrac{B}{x-2}$

Multiply both sides of the equation by the least common denominator $(x-3)(x-2)$ and divide out common factors.

$x = A(x-2) + B(x-3)$

$x = A\,x - 2A + Bx - 3B$

$x = Ax + Bx - 2A - 3B$

$x = (A+B)x - (2A+3B)$

Equate coefficients of like powers of *x*, and equate constant terms.

$A + B = 1$

$2A + 3B = 0$

Solving the above system for *A* and *B*, we find *A* = 3 and *B* = –2.

$\dfrac{x}{(x-3)(x-2)} = \dfrac{3}{x-3} - \dfrac{2}{x-2}$

11. $\dfrac{3x+50}{(x-9)(x+2)} = \dfrac{A}{x-9} + \dfrac{B}{x+2}$

Multiply both sides of the equation by the least common denominator $(x-9)(x+2)$ and divide out common factors.

$3x + 50 = A(x+2) + B(x-9)$

$3x + 50 = Ax + 2A + Bx - 9B$

$3x + 50 = Ax + Bx + 2A - 9B$

$3x + 50 = (A+B)x + (2A - 9B)$

Equate coefficients of like powers of *x*, and equate constant terms.

$A + B = 3$

$2A - 9B = 50$

Solving the above system for *A* and *B*, we find *A* = 7 and *B* = –4.

$\dfrac{3x+50}{(x-9)(x+2)} = \dfrac{7}{x-9} - \dfrac{4}{x+2}$

13. $\dfrac{7x-4}{x^2-x-12} = \dfrac{7x-4}{(x-4)(x+3)} = \dfrac{A}{x-4} + \dfrac{B}{x+3}$

Multiply both sides of the last equation by the least common denominator $(x-4)(x-3)$ and divide out common factors.

$7x - 4 = A(x+3) + B(x-4)$

$7x - 4 = A\,x + 3\,A + Bx - 4B$

$7x - 4 = Ax + Bx + 3A - 4B$

$7x - 4 = (A+B)x + (3A - 4B)$

Equate coefficients of like powers of *x*, and equate constant terms.

$A + B = 7$

$3A - 4B = -4$

Solving the above system for *A* and *B*, we find $A = \dfrac{24}{7}$ and $B = \dfrac{25}{7}$.

$\dfrac{7x-4}{x^2-x-12} = \dfrac{24}{7(x-4)} + \dfrac{25}{7(x+3)}$

15. $\dfrac{4}{(2x+1)(x-3)} = \dfrac{A}{2x+1} + \dfrac{B}{x-3}$

Multiply both sides of the equation by the least common denominator $(2x + 1)(x - 3)$ and divide out common factors.

$4 = A(x-3) + B(2x+1)$

$4 = Ax - A3 + B2x + B$

$4 = (A + 2B)x + (-3A + B)$

Equate coefficients of like powers of x and equate the constant terms. Solve for A and B.

$A + 2B = 0$

$-3A + B = 4$

$3A + 6B = 0$

$-3A + B = 4$

$7B = 4$

$B = \dfrac{4}{7}$

$A + 2B = 0$

$6A - 2B = -8$

$7A = -8$

$A = -\dfrac{8}{7}$

$\dfrac{4}{(2x+1)(x-3)} = \dfrac{-8}{7(2x+1)} + \dfrac{4}{7(x-3)}$

17. $\dfrac{4x^2 + 13x - 9}{x(x-1)(x+3)} = \dfrac{A}{x} + \dfrac{B}{x-3} + \dfrac{C}{x+3}$

Multiply both sides of the equation by the least common denominator $x(x-1)(x+3)$ and divide out common factors.

$4x^2 + 13x - 9 = A(x\text{-}1)(x+3) + Bx(x+3) + Cx(x-1)$

$4x^2 + 13x - 9 = A(x^2 + 2x - 3) + Bx^2 + 3Bx + Cx^2 - Cx$

$4x^2 + 13x - 9 = Ax^2 + 2Ax - 3A + Bx^2 + 3Bx + Cx^2 - Cx$

$4x^2 + 13x - 9 = Ax^2 + Bx^2 + Cx^2 + 2Ax + 3Bx - Cx - 3A$

$4x^2 + 13x - 9 = (A+B+C)x^2 + (2A+3B-C)x - 3A$

Equate coefficients of like powers of x, and equate constant terms.

$A + B + C = 4$

$2A + 3B - C = 13$

$-3A = -9$

Solving the above system for A, B, and C, we find $A = 3$ and $B = 2$, and $C = -1$.

$\dfrac{4x^2 + 13x - 9}{x(x-1)(x+3)} = \dfrac{3}{x} + \dfrac{2}{x-1} - \dfrac{1}{x+3}$

19. $\dfrac{4x^2 - 7x - 3}{x^3 - x} = \dfrac{4x^2 - 7x - 3}{x(x+1)(x-1)} = \dfrac{A}{x} + \dfrac{B}{x+1} + \dfrac{C}{x-1}$

Multiply both sides of the last equation by the least common denominator $x(x+1)(x-1)$ and divide out common factors.

$4x^2 - 7x - 3 = A(x+1)(x-1) + Bx(x-1) + Cx(x+1)$

$4x^2 - 7x - 3 = A(x^2 - 1) + Bx^2 - Bx + Cx^2 + Cx$

$4x^2 - 7x - 3 = Ax^2 - A + Bx^2 - Bx + Cx^2 + Cx$

$4x^2 - 7x - 3 = Ax^2 + Bx^2 + Cx^2 - Bx + Cx - A$

$4x^2 - 7x - 3 = (A+B+C)x^2 + (-B+C)x - A$

Equate coefficients of like powers of x, and equate constant terms.

$A + B + C = 4$

$-B + C = -7$

$-A = -3$

Solving the above system for A, B, and C, we find $A = 3$ and $B = 4$, and $C = -3$.

$\dfrac{4x^2 - 7x - 3}{x^3 - x} = \dfrac{3}{x} + \dfrac{4}{x+1} - \dfrac{3}{x-1}$

21. $\dfrac{6x - 11}{(x-1)^2} = \dfrac{A}{x-1} + \dfrac{B}{(x-1)^2}$

Multiply both sides of the equation by the least common denominator $(x-1)^2$ and divide out common factors.

$6x - 11 = A(x-1) + B$

$6x - 11 = Ax - A + B$

Equate coefficients of like powers of x, and equate constant terms.

$A = 6$

$-A + B = -11$

Since $A = 6$, we find that $B = -5$ by substitution.

$\dfrac{6x - 11}{(x-1)^2} = \dfrac{6}{x-1} - \dfrac{5}{(x-1)^2}$

23. $\dfrac{x^2 - 6x + 3}{(x-2)^3} = \dfrac{A}{x-2} + \dfrac{B}{(x-2)^2} + \dfrac{C}{(x-2)^3}$

Multiply both sides of the equation by the least common denominator $(x-2)^3$ and divide out common factors.

$x^2 - 6x + 3 = A(x-2)^2 + B(x-2) + C$

$x^2 - 6x + 3 = A(x^2 - 4x + 4) + Bx - 2B + C$

$x^2 - 6x + 3 = Ax^2 - 4Ax + 4A + Bx - 2B + C$

$x^2 - 6x + 3 = Ax^2 - 4Ax + Bx + 4A - 2B + C$

$x^2 - 6x + 3 = Ax^2 + (-4A + B)x + 4A - 2B + C$

Equate coefficients of like powers of x, and equate constant terms.

$A = 1$

$-4A + B = -6$

$4A - 2B + C = 3$

Since $A = 1$, we find that $B = -2$ and $C = -5$ by substitution. $\dfrac{x^2 - 6x + 3}{(x-2)^3} = \dfrac{1}{x-2} - \dfrac{2}{(x-2)^2} - \dfrac{5}{(x-2)^3}$

25. $\dfrac{x^2 + 2x + 7}{x(x-1)^2} = \dfrac{A}{x} + \dfrac{B}{x-1} + \dfrac{C}{(x-1)^2}$

Multiply both sides of the equation by the least common denominator $x(x-1)^2$ and divide out common factors.

$x^2 + 2x + 7 = A(x-1)^2 + Bx(x-1) + Cx$

$x^2 + 2x + 7 = A(x^2 - 2x + 1) + Bx^2 - Bx + Cx$

$x^2 + 2x + 7 = Ax^2 - 2Ax + A + Bx^2 - Bx + Cx$

$x^2 + 2x + 7 = Ax^2 + Bx^2 - 2Ax - Bx + Cx + A$

$x^2 + 2x + 7 = (A+B)x^2 + (-2A - B + C)x + A$

$A + B = 1$

$-2A - B + C = 2$

$A = 7$

Since $A = 7$, we find that $B = -6$ and $C = 10$ by substitution. $\dfrac{x^2 + 2x + 7}{x(x-1)^2} = \dfrac{7}{x} - \dfrac{6}{x-1} + \dfrac{10}{(x-1)^2}$

27. $\dfrac{x^2}{(x+1)(x-1)^2} = \dfrac{A}{x+1} + \dfrac{B}{x-1} + \dfrac{C}{(x-1)^2}$

Multiply both sides of the equation by the least common denominator $(x+1)(x-1)^2$ and divide out common factors.

$x^2 = A(x-1)^2 + B(x+1)(x-1) + C(x+1)$

$x^2 = x^2 A - 2xA + A + Bx^2 - B + Cx + C$

$x^2 = (A+B)x^2 + (-2A + C)x + (A - B + C)$

Equate coefficients of like powers of x, and equate constant terms.

$A + B = 1$

$-2A + C = 0$

$A - B + C = 0$

Solving the above system for A, B, and C, we find $A = \dfrac{1}{4}$, $B = \dfrac{3}{4}$, and $C = \dfrac{1}{2}$.

$\dfrac{x^2}{(x+1)(x-1)^2} = \dfrac{1}{4(x+1)} + \dfrac{3}{4(x-1)} + \dfrac{1}{2(x-1)^2}$

29. $\dfrac{5x^2-6x+7}{(x-1)(x^2+1)}=\dfrac{A}{x-1}+\dfrac{Bx+C}{x^2+1}$

Multiply both sides of the equation by the least common denominator $(x-1)(x^2+1)$ and divide out common factors.

$5x^2-6x+7 = A(x^2+1)+(Bx+C)(x-1)$

$5x^2-6x+7 = Ax^2+A+Bx^2-Bx+Cx-C$

$5x^2-6x+7 = Ax^2+Bx^2-Bx+Cx+A-C$

$5x^2-6x+7 = (A+B)x^2+(-B+C)x+A-C$

Equate coefficients of like powers of *x*, and equate constant terms.

$A+B=5$
$-B+C=-6$
$A-C=7$

Solving the above system for *A*, *B*, and *C*, we find $A=3$, $B=2$, and $C=-4$.

$\dfrac{5x^2-6x+7}{(x-1)(x^2+1)}=\dfrac{3}{x-1}+\dfrac{2x-4}{x^2+1}$

31. $\dfrac{5x^2+6x+3}{(x+1)(x^2+2x+2)}=\dfrac{A}{x+1}+\dfrac{Bx+C}{x^2+2x+2}$

Multiply both sides of the equation by the least common denominator $(x+1)(x^2+2x+2)$ and divide out common factors.

$5x^2+6x+3 = A(x^2+2x+2)+(Bx+C)(x+1)$

$5x^2+6x+3 = Ax^2+2Ax+2A+Bx^2+Bx+Cx+C$

$5x^2+6x+3 = Ax^2+Bx^2+2Ax+Bx+Cx+2A+C$

$5x^2+6x+3 = (A+B)x^2+(2A+B+C)x+2A+C$

Equate coefficients of like powers of *x*, and equate constant terms.

$A+B=5$
$2A+B+C=6$
$2A+C=3$

Solving the above system for *A*, *B*, and *C*, we find $A=2$, $B=3$, and $C=-1$.

$\dfrac{5x^2+6x+3}{(x+1)(x^2+2x+2)}=\dfrac{2}{x+1}+\dfrac{3x-1}{x^2+2x+2}$

33. $\dfrac{x+4}{x^2(x^2+4)}=\dfrac{A}{x}+\dfrac{B}{x^2}+\dfrac{Cx+D}{x^2+4}$

Multiply both sides of the equation by the least common denominator $x^2(x^2+4)$ and divide out common factors.

$x+4 = Ax(x^2+4)+B(x^2+4)+(Cx+D)x^2$

$x+4 = Ax^3+4Ax+Bx^2+4B+Cx^3+Dx^2$

$x+4 = (A+C)x^3+(B+D)x^2+4Ax+4B$

Equate coefficients of like powers of *x*, and equate constant terms

$A+C=0$

$B+D=0$

$4A=1$

$4B=4$

Solving the above system for *A*, *B*, and *C*, we find $A=\dfrac{1}{4}$, $B=1$, $C=-\dfrac{1}{4}$, and $D=-1$.

$\dfrac{x+4}{x^2(x^2+4)}=\dfrac{1}{4x}+\dfrac{1}{x^2}+\dfrac{-1x-4}{4\left(x^2+4\right)}$

35. $\dfrac{6x^2 - x + 1}{x^3 + x^2 + x + 1} = \dfrac{6x^2 - x + 1}{(x+1)(x^2+1)} = \dfrac{A}{x+1} + \dfrac{Bx+C}{x^2+1}$.

Multiply both sides of the last equation by the least common denominator $(x+1)(x^2+1)$ and divide out common factors.

$6x^2 - x + 1 = A(x^2 + 1) + (Bx + C)(x + 1)$

$6x^2 - x + 1 = Ax^2 + A + Bx^2 + Bx + Cx + C$

$6x^2 - x + 1 = Ax^2 + Bx^2 + Bx + Cx + A + C$

$6x^2 - x + 1 = (A + B)x^2 + (B + C)x + A + C$

Equate coefficients of like powers of x, and equate constant terms.

$A + B = 6$

$B + C = -1$

$A + C = 1$

Solving the above system for A, B, and C, we find $A = 4$, $B = 2$, and $C = -3$.

$\dfrac{6x^2 - x + 1}{x^3 + x^2 + x + 1} = \dfrac{4}{x+1} + \dfrac{2x-3}{x^2+1}$

37. $\dfrac{x^3 + x^2 + 2}{\left(x^2 + 2\right)^2} = \dfrac{Ax + B}{x^2 + 2} + \dfrac{Cx + D}{\left(x^2 + 2\right)^2}$

Multiply both sides of the last equation by the least common denominator $(x^2 + 2)^2$ and divide out common factors.

$x^3 + x^2 + 2 = \left(Ax + B\right)\left(x^2 + 2\right) + Cx + D$

$x^3 + x^2 + 2 = Ax^3 + Bx^2 + 2Ax + 2B + Cx + D$

$x^3 + x^2 + 2 = Ax^3 + Bx^2 + 2Ax + Cx + 2B + D$

$x^3 + x^2 + 2 = Ax^3 + Bx^2 + (2A + C)x + (2B + D)$

Equate coefficients of like powers of x, and equate constant terms.

$A = 1$

$B = 1$

$2A + C = 0$

$2B + D = 2$

Since $A = 1$ and $B = 1$, we find that $C = -2$ and $D = 0$ by substitution.

$\dfrac{x^3 + x^2 + 2}{\left(x^2 + 2\right)^2} = \dfrac{x+1}{x^2 + 2} - \dfrac{2x}{\left(x^2 + 2\right)^2}$

39. $\dfrac{x^3 - 4x^2 + 9x - 5}{(x^2 - 2x + 3)^2} = \dfrac{Ax + B}{x^2 - 2x + 3} + \dfrac{Cx + D}{(x^2 - 2x + 3)^2}$

Multiply both sides of the equation by the least common denominator $(x^2 - 2x + 3)^2$ and divide out common factors.

$x^3 - 4x^2 + 9x - 5 = (Ax + B)(x^2 - 2x + 3) + Cx + D$

$x^3 - 4x^2 + 9x - 5 = Ax^3 - 2Ax^2 + 3Ax + Bx^2 - 2Bx + 3B + Cx + D$

$x^3 - 4x^2 + 9x - 5 = Ax^3 - 2Ax^2 + Bx^2 + 3Ax - 2Bx + Cx + 3B + D$

$x^3 - 4x^2 + 9x - 5 = Ax^3 + (-2A + B)x^2 + (3A - 2B + C)x + 3B + D$

Equate coefficients of like powers of x, and equate constant terms.

$A = 1$

$-2A + B = -4$

$3A - 2B + C = 9$

$3B + D = -5$

Since $A = 1$, we find that $B = -2$, $C = 2$, and $D = 1$ by substitution.

$\dfrac{x^3 - 4x^2 + 9x - 5}{(x^2 - 2x + 3)^2} = \dfrac{x - 2}{x^2 - 2x + 3} + \dfrac{2x + 1}{(x^2 - 2x + 3)^2}$

41. $\dfrac{4x^2 + 3x + 14}{x^3 - 8} = \dfrac{4x^2 + 3x + 14}{(x - 2)(x^2 + 2x + 4)} = \dfrac{A}{x - 2} + \dfrac{Bx + C}{x^2 + 2x + 4}$

Multiply both sides of the last equation by the least common denominator $(x - 2)(x^2 + 2x + 4)$ and divide out common factors.

$4x^2 + 3x + 14 = A(x^2 + 2x + 4) + (Bx + C)(x - 2)$

$4x^2 + 3x + 14 = A^2 + 2Ax + 4A + Bx^2 - 2Bx + Cx - 2C$

$4x^2 + 3x + 14 = Ax^2 + Bx^2 + 2Ax - 2Bx + Cx + 4A - 2C$

$4x^2 + 3x + 14 = (A + B)x^2 + (2A - 2B + C)x + (4A - 2C)$

Equate coefficients of like powers of x, and equate constant terms.

$A + B = 4$

$2A - 2B + C = 3$

$4A - 2C = 14$

Solving the above system for A, B, and C, we find $A = 3$, $B = 1$, and $C = -1$.

$\dfrac{4x^2 + 3x + 4}{x^3 - 8} = \dfrac{3}{x - 2} + \dfrac{x - 1}{x^2 + 2x + 4}$

43. Divide $x^5 + 2$ by $x^2 - 1$.

$$
\begin{array}{r}
x^3 + x \\
x^2 - 1 \overline{)x^5 + 2} \\
\underline{x^5 - x^3} \\
x^3 \\
\underline{x^3 - x} \\
x + 2
\end{array}
$$

$\dfrac{x^5 + 2}{x^2 - 1} = x^3 + x + \dfrac{x + 2}{x^2 - 1}$

Decompose $\dfrac{x + 2}{x^2 - 1}$.

$$\frac{x+2}{x^2-1} = \frac{x+2}{(x+1)(x-1)} = \frac{A}{x+1} + \frac{B}{x-1}$$

$$(x+1)(x-1)\frac{x+2}{(x+1)(x-1)} = (x+1)(x-1)\left(\frac{A}{x+1} + \frac{B}{x-1}\right)$$

$$x+2 = (x-1)A + (x+1)B$$

$$x+2 = Ax - A + Bx + B$$

$$x+2 = (A+B)x + (-A+B)$$

Equate coefficients:

$$A+B = 1$$

$$-A+B = 2$$

Solving this system results in $A = -\dfrac{1}{2}$ and $B = \dfrac{3}{2}$.

$$\frac{x^5+2}{x^2-1} = x^3 + x + \frac{-\dfrac{1}{2}}{x+1} + \frac{\dfrac{3}{2}}{x-1} \quad \text{or} \quad x^3 + x + \frac{-1}{2(x+1)} + \frac{3}{2(x-1)}$$

45. Divide $x^4 - x^2 + 2$ by $x^3 - x^2$.

$$
\begin{array}{r}
x+1 \\
x^3-x^2\overline{\smash{\big)}\,x^4 \qquad\;\; -x^2+2} \\
\underline{x^4-x^3} \\
x^3-x^2 \\
\underline{x^3-x^2} \\
2
\end{array}
$$

$$\frac{x^4-x^2+2}{x^3-x^2} = x+1 + \frac{2}{x^3-x^2}$$

Decompose $\dfrac{2}{x^3-x^2}$.

$$\frac{2}{x^3-x^2} = \frac{2}{x^2(x-1)} = \frac{A}{x} + \frac{B}{x^2} + \frac{C}{x-1}$$

$$x^2(x-1)\frac{2}{x^2(x-1)} = x^2(x-1)\left(\frac{A}{x} + \frac{B}{x^2} + \frac{C}{x-1}\right)$$

$$2 = x(x-1)A + (x-1)B + x^2C$$

$$2 = Ax^2 - Ax + Bx - B + Cx^2$$

$$2 = (A+C)x^2 + (-A+B)x + (-B)$$

Equate coefficients:

$$A+C = 0$$

$$-A+B = 0$$

$$-B = 2$$

Solving this system results in $A = -2$, $B = -2$, and $C = 2$.

$$\frac{x^4-x^2+2}{x^3-x^2} = x+1 + \frac{-2}{x} + \frac{-2}{x^2} + \frac{2}{x-1}$$

47.
$$\frac{1}{x^2 - c^2} = \frac{1}{(x+c)(x-c)} = \frac{A}{x+c} + \frac{B}{x-c}$$

$$(x+c)(x-c)\frac{1}{(x+c)(x-c)} = (x+c)(x-c)\left(\frac{A}{x+c} + \frac{B}{x-c}\right)$$

$$1 = (x-c)A + (x+c)B$$

$$1 = Ax - Ac + Bx + Bc$$

$$1 = (A+B)x + (-Ac + Bc)$$

Equate coefficients:

$$A + B = 0$$

$$-Ac + Bc = 1$$

Solving this system results in $A = \dfrac{-1}{2c}$ and $B = \dfrac{1}{2c}$.

$$\frac{1}{x^2 - c^2} = \frac{\dfrac{-1}{2c}}{x+c} + \frac{\dfrac{1}{2c}}{x-c}$$

49.
$$\frac{ax+b}{(x-c)^2} = \frac{D}{x-c} + \frac{E}{(x-c)^2}$$

$$\frac{ax+b}{(x-c)^2} = \frac{D}{x-c} + \frac{E}{(x-c)^2}$$

$$(x-c)^2 \frac{ax+b}{(x-c)^2} = (x-c)^2\left(\frac{D}{x-c} + \frac{E}{(x-c)^2}\right)$$

$$ax+b = (x-c)D + E$$

$$ax+b = Dx - Dc + E$$

$$ax+b = (D)x + (-Dc + E)$$

Equate coefficients:

$$D = a$$

$$-Dc + E = b$$

Solving this system results in $D = a$ and $E = ac + b$.

$$\frac{ax+b}{(x-c)^2} = \frac{a}{x-c} + \frac{ac+b}{(x-c)^2}$$

51. $\dfrac{1}{x(x+1)} = \dfrac{A}{x} + \dfrac{B}{x+1}$

Multiply both sides of the equation by the least common denominator $x(x+1)$ and divide out common factors.

$1 = A(x+1) + Bx$

$1 = Ax + A + Bx$

$1 = Ax + Bx + A$

$1 = (A+B)x + A$

Equate coefficients of like powers of x, and equate constant terms.

$A + B = 0$

$\quad A = 1$

Since $A = 1$ we find that $B = -1$ by substitution.

$\dfrac{1}{x(x+1)} = \dfrac{1}{x} - \dfrac{1}{x+1}$

$\dfrac{1}{1\cdot 2} + \dfrac{1}{2\cdot 3} + \dfrac{1}{3\cdot 4} + \cdots \dfrac{1}{99\cdot 100} = \left(\dfrac{1}{1} - \dfrac{1}{2}\right) + \left(\dfrac{1}{2} - \dfrac{1}{3}\right) + \left(\dfrac{1}{3} - \dfrac{1}{4}\right) + \cdots \left(\dfrac{1}{99} - \dfrac{1}{100}\right)$

$\qquad\qquad\qquad\qquad\qquad = \dfrac{1}{1} - \dfrac{1}{100}$

$\qquad\qquad\qquad\qquad\qquad = \dfrac{99}{100}$

53. – 59. Answers may vary.

61. does not make sense; Explanations will vary. Sample explanation: To perform partial fraction decomposition, the degree in the numerator must be less than the degree in the denominator.

63. does not make sense; Explanations will vary. Sample explanation: The second denominator should be $(x+3)^2$.

65. $\dfrac{4x^2+5x-9}{x^3-6x-9} = \dfrac{4x^2+5x-9}{(x-3)(x^2+3x+3)} = \dfrac{A}{x-3} + \dfrac{Bx+C}{x^2+3x+3}$

Multiply both sides of the last equation by the common denominator $(x-3)(x^2+3x+3)$ and divide out common factors.

$4x^2+5x-9 = A(x^2+3x+3) + (Bx+C)(x-3)$

$4x^2+5x-9 = Ax^2+3Ax+3A+Bx^2-3Bx+Cx-3C$

$4x^2+5x-9 = Ax^2+Bx^2+3Ax-3Bx+Cx+3A-3C$

$4x^2+5x-9 = (A+B)x^2+(3A-3B+C)x+3A-3C$

Equate coefficients of like powers of x and equate constant terms.

$\qquad A + B = 4$

$3A - 3B + C = 5$

$\qquad 3A - 3C = -9$

Solving the above system for A, B, and C, we find $A = 2$, and $B = 2$, and $C = 5$.

$\dfrac{4x^2+5x-9}{x^3-6x-9} = \dfrac{2}{x-3} + \dfrac{2x+5}{x^2+3x+3}$

66. $4x + 3y = 4$

$y = 2x - 7$

Substitute.

$4x + 3y = 4$

$4x + 3\overbrace{(2x - 7)}^{y} = 4$

$4x + 3(2x - 7) = 4$

$4x + 6x - 21 = 4$

$10x = 25$

$x = 2.5$

Back-substitute.

$y = 2x - 7$

$y = 2(2.5) - 7 = -2$

The solution set is $\{(2.5, -2)\}$.

67. $2x + 4y = -4$

$3x + 5y = -3$

Multiply the first equation by -5 and the second equation by 4.

$-10x - 20y = 20$

$12x + 20y = -12$

Add the equations and solve for x.

$2x = 8$

$x = 4$

Back-substitute to find y.

$2x + 4y = -4$

$2(4) + 4y = -4$

$8 + 4y = -4$

$4y = -12$

$y = -3$

The solution set is $\{(4, -3)\}$.

68. The points of intersection are $(0, -3)$ and $(2, -1)$.

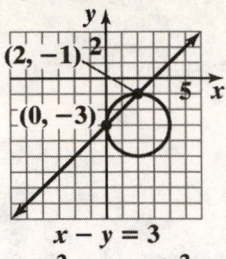

$x - y = 3$

$(x - 2)^2 + (y + 3)^2 = 4$

Point $(0, -3)$:

$x - y = 3$

$0 - (-3) = 3$

$3 = 3$, true

$(x - 2)^2 + (y + 3)^2 = 4$

$(0 - 2)^2 + (-3 + 3)^2 = 4$

$(-2)^2 + (0)^2 = 4$

$4 = 4$, true

Point $(2, -1)$:

$x - y = 3$

$2 - (-1) = 3$

$3 = 3$, true

$(x - 2)^2 + (y + 3)^2 = 4$

$(2 - 2)^2 + (-1 + 3)^2 = 4$

$(0)^2 + (2)^2 = 4$

$4 = 4$, true

Section 7.4

Check Point Exercises

1. $x^2 = y-1$

$4x - y = -1$

Solve the first equation for y.

$y = x^2 + 1$

Substitute the expression $x^2 + 1$ for y in the second equation and solve for x.

$4x - (x^2 + 1) = -1$

$4x - x^2 - 1 = -1$

$x^2 - 4x = 0$

$x(x-4) = 0$

$x = 0$ or $x - 4 = 0$

$x = 4$

If $x = 0$, $y = (0)^2 + 1 = 1$.

If $x = 4$, $y = (4)^2 + 1 = 17$.

The solution set is $\{(0, 1), (4, 17)\}$.

2. $x + 2y = 0$

$(x-1)^2 + (y-1)^2 = 5$

Solve the first equation for x.

$x = -2y$

Substitute the expression $-2y$ for x in the second equation and solve for y.

$(-2y-1)^2 + (y-1)^2 = 5$

$4y^2 + 4y + 1 + y^2 - 2y + 1 = 5$

$5y^2 + 2y - 3 = 0$

$(5y-3)(y+1) = 0$

$5y - 3 = 0$ or $y + 1 = 0$

$y = \dfrac{3}{5}$ or $y = -1$

If $y = \dfrac{3}{5}$, $x = -2\left(\dfrac{3}{5}\right) = -\dfrac{6}{5}$.

If $y = -1$, $x = -2(-1) = 2$.

The solution set is $\left\{\left(-\dfrac{6}{5}, \dfrac{3}{5}\right), (2, -1)\right\}$.

3. $3x^2 + 2y^2 = 35$

$4x^2 + 3y^2 = 48$

Eliminate the y^2-term by multiplying the first equation by –3 and the second equation by 2. Add the resulting equations.

$-9x^2 - 6y = -105$

$\underline{8x^2 + 6y^2 = 96}$

$-x^2 = -9$

$x^2 = 9$

$x = \pm 3$

If $x = 3$,

$3(3)^2 + 2y^2 = 35$

$y^2 = 4$

$y = \pm 2$

If $x = -3$,

$3(-3)^2 + 2y^2 = 35$

$y^2 = 4$

$y = \pm 2$

The solution set is $\{(3, 2), (3, -2), (-3, 2), (-3, -2)\}$.

4. $y = x^2 + 5$

$x^2 + y^2 = 25$

Arrange the first equation so that variable terms appear on the left, and constants appear on the right. Add the resulting equations to eliminate the x^2-terms and solve for y.

$-x^2 + y = 5$

$\underline{x^2 + y^2 = 25}$

$y^2 + y = 30$

$y^2 + y - 30 = 0$

$(y+6)(y-5) = 0$

$y + 6 = 0$ or $y - 5 = 0$

$y = -6$ or $y = 5$

If $y = -6$,

$x^2 + (-6)^2 = 25$

$x^2 = -11$

no real solution

If $y = 5$,

$x^2 + (5)^2 = 25$

$x^2 = 0$

$x = 0$

The solution set is $\{(0, 5)\}$.

5. $2x + 2y = 20$

$\quad xy = 21$

Solve the second equation for x.

$$x = \frac{21}{7}$$

Substitute the expression $\dfrac{21}{y}$ for x in the first

equation and solve for y.

$$2\left(\frac{21}{y}\right) + 2y = 20$$

$$\frac{42}{y} + 2y = 20$$

$$y^2 - 10y + 21 = 0$$

$$(y - 7)(y - 3) = 0$$

$\quad y - 7 = 0 \quad \text{or} \quad y - 3 = 0$

$\qquad y = 7 \quad \text{or} \qquad y = 3$

If $y = 7$, $x = \dfrac{21}{7} = 3$.

If $y = 3$, $x = \dfrac{21}{3} = 7$.

The dimensions are 7 feet by 3 feet.

Exercise Set 7.4

1. $x + y = 2$

$\quad y = x^2 - 4$

Solve the first equation for y. $y = 2 - x$.

Substitute the expression $2 - x$ for y in the second equation and solve for x.

$$2 - x = x^2 - 4$$

$$x^2 + x - 6 = 0$$

$$(x + 3)(x - 2) = 0$$

$x + 3 = 0 \quad \text{or} \quad x - 2 = 0$

$\quad x = -3 \quad \text{or} \qquad x = 2$

If $x = -3$, $y = 2 - (-3) = 5$.

If $x = 2$, $y = 2 - 2 = 0$.

The solution set is $\{(-3, 5), (2, 0)\}$.

3. $x + y = 2$

$\quad y = x^2 - 4x + 4$

Substitute the expression $x^2 - 4x + 4$ for y in the first equation and solve for x.

$$x + x^2 - 4x + 4 = 2$$

$$x^2 - 3x + 2 = 0$$

$$(x - 1)(x - 2) = 0$$

$\quad x - 1 = 0 \quad x - 2 = 0$

$\qquad x = 1 \qquad x = 2$

Substitute $x = 1$ and then $x = 2$ into the equation $x + y = 2$ and solve for each value of y.

$\quad 1 + y = 2 \quad 2 + y = 2$

$\qquad y = 1 \qquad\qquad y = 0$

The solution set is $\{(1, 1), (2, 0)\}$.

5. $y = x^2 - 4x - 10$

$\quad y = -x^2 - 2x + 14$

Substitute the expression $x^2 - 4x - 10$ for y in the second equation and solve for x.

$$x^2 - 4x - 10 = -x^2 - 2x + 14$$

$$2x^2 - 2x - 24 = 0$$

$$x^2 - x - 12 = 0$$

$$(x - 4)(x + 3) = 0$$

$x - 4 = 0 \quad \text{or} \quad x + 3 = 0$

$\quad x = 4 \qquad \text{or} \qquad x = -3$

If $x = 4$, $y = (4)^2 - 4(4) - 10 = -10$.

If $x = -3$, $y = (-3)^2 - 4(-3) - 10 = 11$.

The solution set is $\{(4, -10), (-3, 11)\}$.

7. $x^2 + y^2 = 25$

$\quad x - y = 1$

Solve the second equation for y. $y = x - 1$

Substitute the expression $x - 1$ for y in the first equation and solve for x.

$$x^2 + (x - 1)^2 = 25$$

$$x^2 + x^2 - 2x + 1 = 25$$

$$2x^2 - 2x - 24 = 0$$

$$x^2 - x - 12 = 0$$

$$(x - 4)(x + 3) = 0$$

$x - 4 = 0 \quad \text{or} \quad x + 3 = 0$

$\quad x = 4 \qquad \text{or} \qquad x = -3$

If $x = 4$, $y = 4 - 1 = 3$.

If $x = -3$, $y = -3 - 1 = -4$.

The solution set is $\{(4, 3), (-3, -4)\}$.

9. $xy = 6$
$2x - y = 1$
Solve the first equation for y.

$$y = \frac{6}{x}$$

Substitute the expression $\dfrac{6}{x}$ for y in the second equation and solve for x.

$$2x - \frac{6}{x} = 1$$
$$2x^2 - 6 = x$$
$$2x^2 - x - 6 = 0$$
$$(2x + 3)(x - 2) = 0$$
$$2x + 3 = 0 \quad \text{or} \quad x - 2 = 0$$
$$x = -\frac{3}{2} \quad \text{or} \quad x = 2$$

If $x = -\dfrac{3}{2}$, $y = \dfrac{6}{-\frac{3}{2}} = -4$.

If $x = 2$, $y = \dfrac{6}{2} = 3$.

The solution set is $\left\{ \left(-\dfrac{3}{2}, -4 \right), (2, 3) \right\}$.

11. $y^2 = x^2 - 9$
$2y = x - 3$
Solve the second equation for y.

$$y = \frac{x - 3}{2}$$

Substitute the expression $\dfrac{x-3}{2}$ for y in the first equation and solve for x.

$$\left(\frac{x-3}{2} \right)^2 = x^2 - 9$$

$$\frac{x^2 - 6x + 9}{4} = x^2 - 9$$

$$x^2 - 6x + 9 = 4x^2 - 36$$
$$3x^2 + 6x - 45 = 0$$
$$x^2 + 2x - 15 = 0$$
$$(x + 5)(x - 3) = 0$$
$$x + 5 = 0 \quad \text{or} \quad x - 3 = 0$$
$$x = -5 \quad \text{or} \quad x = 3$$

If $x = -5$, $y = \dfrac{-5 - 3}{2} = -4$.

If $x = 3$, $y = \dfrac{3 - 3}{2} = 0$.

The solution set is $\{(-5, -4), (3, 0)\}$.

13. $xy = 3$
$x^2 + y^2 = 10$
Solve the second equation for y.

$$y = \frac{3}{x}$$

Substitute the expression $\dfrac{3}{x}$ for y in the second equation and solve for x.

$$x^2 + \left(\frac{3}{x} \right)^2 = 10$$

$$x^2 + \frac{9}{x^2} - 10 = 0$$

$$x^4 - 10x^2 + 9 = 0$$
$$\left(x^2 - 9 \right)\left(x^2 - 1 \right) = 0$$
$$(x - 3)(x + 3)(x - 1)(x + 1) = 0$$
$$x - 3 = 0 \quad \text{or} \quad x + 3 = 0 \quad \text{or} \quad x - 1 = 0 \quad \text{or} \quad x + 1 = 0$$
$$x = 3 \quad \text{or} \quad x = -3 \quad \text{or} \quad x = 1 \quad \text{or} \quad x = -1$$

If $x = 3$, $y = \dfrac{3}{3} = 1$.

If $x = -3$, $y = \dfrac{3}{-3} = -1$.

If $x = 1$, $y = \dfrac{3}{1} = 3$.

If $x = -1$, $y = \dfrac{3}{-1} = -3$.

The solution set is $\{(3, 1), (-3, -1), (1, 3), (-1, -3)\}$.

15.
$$x + y = 1$$
$$x^2 + xy - y^2 = -5$$

Solve the first equation for y. $y = 1 - x$

Substitute the expression $1 - x$ for y in the second equation and solve for x.

$$x^2 + x(1-x) - (1-x)^2 = -5$$
$$x^2 + x - x^2 - (1 - 2x + x^2) = -5$$
$$x - 1 + 2x - x^2 = -5$$
$$x^2 - 3x - 4 = 0$$
$$(x-4)(x+1) = 0$$

$x - 4 = 0$ or $x + 1 = 0$

$x = 4$ or $x = -1$

If $x = 4$, $y = 1 - 4 = -3$.

If $x = -1$, $y = 1 - (-1) = 2$.

The solution set is $\{(4,-3),(-1,2)\}$.

17.
$$x + y = 1$$
$$(x-1)^2 + (y+2)^2 = 10$$

Solve the first equation for y.
$$y = 1 - x$$

Substitute the expression $1 - x$ for y in the second equation and solve for x.

$$(x-1)^2 + (1-x+2)^2 = 10$$
$$(x-1)^2 + (3-x)^2 = 10$$
$$x^2 - 2x + 1 + 9 - 6x + x^2 - 10 = 0$$
$$2x^2 - 8x = 0$$
$$x^2 - 4x = 0$$
$$x(x-4) = 0$$

$x = 0$ or $x - 4 = 0$

$x = 4$

If $x = 0$, $y = 1 - 0 = 1$.

If $x = 4$, $y = 1 - 4 = -3$.

The solution set is $\{(0,1),(4,-3)\}$.

19. Eliminate the y^2–terms by adding the equations.

$$\begin{array}{r} x^2 + y^2 = 13 \\ x^2 - y^2 = 5 \\ \hline 2x^2 = 18 \end{array}$$

$$x^2 = 9$$
$$x = \pm 3$$

If $x = 3$,
$$(3)^2 + y^2 = 13$$
$$y^2 = 4$$
$$y = \pm 2$$

If $x = -3$,
$$(-3)^2 + y^2 = 13$$
$$y^2 = 4$$
$$y = \pm 2$$

The solution set is
$\{(3, 2), (3, -2), (-3, 2), (-3, -2)\}$.

21.
$$x^2 - 4y^2 = -7$$
$$3x^2 + y^2 = 31$$

Eliminate the x^2–terms by multiplying the first equation by -3 and adding the resulting equations.

$$\begin{array}{r} -3x^2 + 12y^2 = 21 \\ 3x^2 + y^2 = 31 \\ \hline 13y^2 = 52 \end{array}$$

$$y^2 = 4$$
$$y = \pm 2$$

If $y = 2$,
$$x^2 - 4(2)^2 = -7$$
$$x^2 = 9$$
$$x = \pm 3$$

If $y = -2$,
$$x^2 - 4(-2)^2 = -7$$
$$x^2 = 9$$
$$x = \pm 3$$

The solution set is $\{(3,2),(3,-2),(-3,2),(-3,-2)\}$.

23. Arrange the equations so that variable terms appear on the left and constants appear on the right.

$3x^2 + 4y^2 = 16$

$2x^2 - 3y^2 = 5$

Eliminate the y^2–terms by multiplying the first equation by 3 and the second equation by 4. Add the resulting equations.

$9x^2 + 12y^2 = 48$

$\underline{8x^2 - 12y^2 = 20}$

$17x^2 = 68$

$x^2 = 4$

$x = \pm 2$

If $x = 2$,

$3(2)^2 + 4y^2 = 16$

$y^2 = 1$

$y = \pm 1$

If $x = -2$,

$3(-2)^2 + 4y^2 = 16$

$y = \pm 1$

The solution set is
{(2, 1), (2, –1), (–2, 1), (–2, –1)}.

25. $x^2 + y^2 = 25$

$(x-8)^2 + y^2 = 41$

Expand the second equation and eliminate x^2 and y^2–terms by multiplying the first equation by –1 and adding the resulting equations.

$x^2 - 16x + 64 + y^2 = 41$

$\underline{-x^2 - y^2 = -25}$

$-16x + 64 = 16$

$-16x = -48$

$x = 3$

If $x = 3$,

$(3)^2 + y^2 = 25$

$y^2 = 16$

$y = \pm 4$

The solution set is $\{(3,4), (3,-4)\}$.

27. $y^2 - x = 4$

$x^2 + y^2 = 4$

Eliminate the y^2–terms by multiplying the first equation by –1 and adding the resulting equations.

$x - y^2 = -4$

$\underline{x^2 + y^2 = 4}$

$x^2 + x = 0$

$x(x+1) = 0$

$x = 0$ or $x + 1 = 0$

$x = -1$

If $x = 0$,

$y^2 = 4$

$y = \pm 2$

If $x = -1$,

$y^2 - (-1) = 4$

$y^2 = 3$

$y = \pm\sqrt{3}$

The solution set is

$\left\{(0,2),(0,-2),\left(-1,\sqrt{3}\right),\left(-1,-\sqrt{3}\right)\right\}$.

29. The addition method is used here to solve the system.

$3x^2 + 4y^2 = 16$

$2x^2 - 3y^2 = 5$

Eliminate the y^2–terms by multiplying the first equation by 3 and the second equation by 4. Add the resulting equations.

$9x^2 + 12y^2 = 48$

$\underline{8x^2 - 12y^2 = 20}$

$17x^2 = 68$

$x^2 = 4$

$x = \pm 2$

If $x = 2$,

$3(2)^2 + 4y^2 = 16$

$y^2 = 1$

$y = \pm 1$

If $x = -2$,

$3(-2)^2 + 4y^2 = 16$

$y = \pm 1$

The solution set is
{(2, 1), (2, –1), (–2, 1), (–2, –1)}.

31. The substitution method is used here to solve the system.

$$2x^2 + y^2 = 18$$
$$xy = 4$$

Solve the second equation for y.

$$y = \frac{4}{x}$$

Substitute the expression $\frac{4}{x}$ for y in the first equation and solve for x.

$$2x^2 + \left(\frac{4}{x}\right)^2 = 18$$

$$2x^2 + \frac{16}{x^2} = 18$$

$$2x^4 + 16 = 18x^2$$

$$x^4 - 9x^2 + 8 = 0$$

$$(x^2 - 8)(x^2 - 1) = 0$$

$$x^2 - 8 = 0 \quad \text{or} \quad x^2 - 1 = 0$$

$$x^2 = 8 \quad \text{or} \quad x^2 = 1$$

$$x = \pm 2\sqrt{2} \quad \text{or} \quad x = \pm 1$$

If $x = 2\sqrt{2}$, $y = \dfrac{4}{2\sqrt{2}} = \sqrt{2}$.

If $x = -2\sqrt{2}$, $y = \dfrac{4}{-2\sqrt{2}} = -\sqrt{2}$.

If $x = 1$, $y = \dfrac{4}{1} = 4$.

If $x = -1$, $y = \dfrac{4}{-1} = -4$.

The solution set is

$$\left\{\left(2\sqrt{2}, \sqrt{2}\right), \left(-2\sqrt{2}, -\sqrt{2}\right), (1, 4), (-1, -4)\right\}.$$

33. The substitution method is used here to solve the system.

$$x^2 + 4y^2 = 20$$
$$x + 2y = 6$$

Solve the second equation for x.

$$x = 6 - 2y$$

Substitute the expression $6 - 2y$ for x in the first equation and solve for y.

$$(6 - 2y)^2 + 4y^2 = 20$$

$$36 - 24y + 4y^2 + 4y^2 - 20 = 0$$

$$8y^2 - 24y + 16 = 0$$

$$y^2 - 3y + 2 = 0$$

$$(y - 2)(y - 1) = 0$$

$$y - 2 = 0 \quad \text{or} \quad y - 1 = 0$$

$$y = 2 \quad \text{or} \quad y = 1$$

If $y = 2, x = 6 - 2(2) = 2$.

If $y = 1, x = 6 - 2(1) = 4$.

The solution set is $\{(2, 2), (4, 1)\}$.

35. Eliminate y by adding the equations.

$$x^3 + y = 0$$
$$\underline{x^2 - y = 0}$$
$$x^3 + x^2 = 0$$

$$x^2(x + 1) = 0$$

$$x^2 = 0 \quad \text{or} \quad x + 1 = 0$$

$$x = 0 \quad \text{or} \quad x = -1$$

If $x = 0$,

$$(0)^3 + y = 0$$
$$y = 0$$

If $x = -1$,

$$(-1)^3 + y = 0$$
$$y = 1$$

The solution set is $\{(0, 0), (-1, 1)\}$.

37. The substitution method is used here to solve the system.

$$x^2 + (y-2)^2 = 4$$
$$x^2 - 2y = 0$$

Solve the second equation for x^2.

$$x^2 = 2y$$

Substitute the expression $2y$ for x^2 in the first equation and solve for y.

$$2y + (y-2)^2 = 4$$
$$2y + y^2 - 4y + 4 = 4$$
$$y^2 - 2y = 0$$
$$y(y-2) = 0$$
$$y = 0 \quad \text{or} \quad y - 2 = 0$$
$$y = 2$$

If $y = 0$,
$$x^2 = 2(0)$$
$$x^2 = 0$$
$$x = 0$$
If $y = 2$,
$$x^2 = 2(2)$$
$$x^2 = 4$$
$$x = \pm 2$$

The solution set is $\{(0, 0), (-2, 2), (2, 2)\}$.

39. The substitution method is used here to solve the system.

$$y = (x+3)^2$$
$$x + 2y = -2$$

Solve the first equation for x.

$$x = -2y - 2$$

Substitute the expression $-2y-2$ for x in the first equation and solve for y.

$$y = (-2y - 2 + 3)^2 = (-2y + 1)^2$$
$$y = 4y^2 - 4y + 1$$
$$4y^2 - 5y + 1 = 0$$
$$(4y - 1)(y - 1) = 0$$
$$4y - 1 = 0 \quad \text{or} \quad y - 1 = 0$$
$$y = \frac{1}{4} \quad \text{or} \quad y = 1$$

If $y = \frac{1}{4}$, $x = -2\left(\frac{1}{4}\right) - 2 = -\frac{5}{2}$.

If $y = 1$, $x = -2(1) - 2 = -4$.

The solution set is $\left\{(-4, 1), \left(-\frac{5}{2}, \frac{1}{4}\right)\right\}$.

41. The substitution method is used here to solve the system.

$$x^2 + y^2 + 3y = 22$$
$$2x + y = -1$$

Solve the second equation for y.

$$y = -2x - 1$$

Substitute the expression $-2x-1$ for y in the first equation and solve for x.

$$x^2 + (-2x - 1)^2 + 3(-2x - 1) - 22 = 0$$
$$x^2 + 4x^2 + 4x + 1 - 6x - 3 - 22 = 0$$
$$5x^2 - 2x - 24 = 0$$
$$(5x - 12)(x + 2) = 0$$
$$5x - 12 = 0 \quad \text{or} \quad x + 2 = 0$$
$$x = \frac{12}{5} \quad \text{or} \quad x = -2$$

If $x = \frac{12}{5}$, $y = -2\left(\frac{12}{5}\right) - 1 = -\frac{29}{5}$.

If $x = -2$, $y = -2(-2) - 1 = 3$.

The solution set is $\left\{\left(\frac{12}{5}, -\frac{29}{5}\right), (-2, 3)\right\}$.

43. The substitution method is used here to solve the system.

$$x + y = 10$$
$$xy = 24$$

Solve the first equation for y.

$$y = 10 - x$$

Substitute the expression $10 - x$ for y in the second equation and solve for x.

$$x(10 - x) = 24$$
$$10x - x^2 = 24$$
$$x^2 - 10x + 24 = 0$$
$$(x - 4)(x - 6) = 0$$
$$x - 4 = 0 \quad \text{or} \quad x - 6 = 0$$
$$x = 4 \quad \text{or} \quad x = 6$$

If $x = 4$, $y = 10 - 4 = 6$.

If $x = 6$, $y = 10 - 6 = 4$.

The numbers are 4 and 6.

45. Eliminate the y^2-terms by adding the equations.

$$x^2 - y^2 = 3$$
$$\underline{2x^2 + y^2 = 9}$$
$$3x^2 = 12$$
$$x^2 = 4$$
$$x = \pm 2$$

If $x = 2$,
$$2(2)^2 + y^2 = 9$$
$$y^2 = 1$$
$$y = \pm 1$$

If $x = -2$,
$$2(-2)^2 + y^2 = 9$$
$$y^2 = 1$$
$$y = \pm 1$$

The numbers are 2 and 1, 2 and –1, –2 and 1, or –2 and –1.

47. $2x^2 + xy = 6$

$x^2 + 2xy = 0$

Multiply the first equation by -2 and add the two equations.

$$-4x^2 - 2xy = -12$$
$$\underline{x^2 + 2xy = 0}$$
$$-3x^2 = -12$$
$$x^2 = 4$$
$$x = \pm 2$$

Back-substitute these values for x in the second equation and solve for y.

For $x = -2$: $(-2)^2 + 2(-2)y = 0$
$$4 - 4y = 0$$
$$-4y = -4$$
$$y = 1$$

For $x = 2$: $(2)^2 + 2(2)y = 0$
$$4 + 4y = 0$$
$$4y = -4$$
$$y = -1$$

The solution set is $\{(-2,1),(2,-1)\}$.

49. $-4x + y = 12$

$y = x^3 + 3x^2$

Substitute $x^3 + 3x^2$ for y in the first equation and solve for x.

$$-4x + \left(x^3 + 3x^2\right) = 12$$
$$x^3 + 3x^2 - 4x - 12 = 0$$
$$x^2(x+3) - 4(x+3) = 0$$
$$(x+3)(x^2 - 4) = 0$$
$$(x+3)(x-2)(x+2) = 0$$
$$x = -3,\ x = 2,\ \text{or}\ x = -2$$

Substitute these values for x in the second equation and solve for y.

For $x = -3$: $y = (-3)^3 + 3(-3)^2$
$$= -27 + 27$$
$$= 0$$

For $x = 2$: $y = (2)^3 + 3(2)^2$
$$= 8 + 12$$
$$= 20$$

For $x = -2$: $y = (-2)^3 + 3(-2)^2$
$$= -8 + 12$$
$$= 4$$

The solution set is $\{(-3,0),(2,20),(-2,4)\}$.

51.
$$\frac{3}{x^2}+\frac{1}{y^2}=7$$
$$\frac{5}{x^2}-\frac{2}{y^2}=-3$$

Multiply the first equation by 2 and add the equations.

$$\frac{6}{x^2}+\frac{2}{y^2}=14$$
$$\frac{5}{x^2}-\frac{2}{y^2}=-3$$
$$\overline{}$$
$$\frac{11}{x^2}=11$$
$$x^2=1$$
$$x=\pm 1$$

Back-substitute these values for x in the first equation and solve for y.

For $x=-1$:
$$\frac{3}{(-1)^2}+\frac{1}{y^2}=7$$
$$3+\frac{1}{y^2}=7$$
$$\frac{1}{y^2}=4$$
$$y^2=\frac{1}{4}$$
$$y=\pm\frac{1}{2}$$

For $x=1$:
$$\frac{3}{(1)^2}+\frac{1}{y^2}=7$$
$$3+\frac{1}{y^2}=7$$
$$\frac{1}{y^2}=4$$
$$y^2=\frac{1}{4}$$
$$y=\pm\frac{1}{2}$$

The solution set is
$$\left\{\left(-1,-\frac{1}{2}\right),\left(-1,\frac{1}{2}\right),\left(1,-\frac{1}{2}\right),\left(1,\frac{1}{2}\right)\right\}.$$

53.

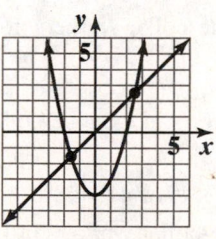

$$y=x^2-4$$
$$y=x$$

55.
$$16x^2+4y^2=64$$
$$y=x^2-4$$

Substitute the expression x^2-4 for y in the first equation and solve for x.

$$16x^2+4\left(x^2-4\right)^2=64$$
$$16x^2+4\left(x^4-8x^2+16\right)=64$$
$$16x^2+4x^4-32x^2+64=64$$
$$4x^4-16x^2=0$$
$$x^4-4x^2=0$$
$$x^2\left(x^2-4\right)=0$$
$$x^2=0 \quad\text{or}\quad x^2-4=0$$
$$x=0 \quad\text{or}\quad\quad x^2=4$$
$$x=\pm 2$$

If $x=0$, $y=(0)^2-4=-4$.

If $x=2$, $y=(2)^2-4=0$.

If $x=-2$, $y=(-2)^2-4=0$.

It is possible for the comet to intersect the orbiting body at $(0, -4)$, $(-2, 0)$, $(2, 0)$.

57.
$$2L+2W=36$$
$$LW=77$$

Divide each term in the first equation by 2 and solve L.
$$L+W=18$$
$$L=18-W$$

Substitute the expression $18-W$ for L in the second equation and solve for W.

$$(18-W)W=77$$
$$18W-W^2=77$$
$$W^2-18W+77=0$$
$$(W-11)(W-7)=0$$
$$W-11=0 \quad\text{or}\quad W-7=0$$
$$W=11 \quad\text{or}\quad\quad W=7$$

If $W=11$, $L=18-11=7$.

If $W=7$, $L=18-7=11$.

The dimensions are 11 feet by 7 feet.

59. $L^2 + W^2 = 10^2 = 100$

$LW = 48$

Solve the second equation for L. $L = \dfrac{48}{W}$

Substitute the expression $\dfrac{48}{W}$ for L in the first equation and solve for W.

$\left(\dfrac{48}{W}\right)^2 + W^2 = 100$

$\dfrac{2304}{W^2} + W^2 - 100 = 0$

$2304 + W^4 - 100W^2 = 0$

$W^4 - 100W^2 + 2304 = 0$

$\left(W^2 - 36\right)\left(W^2 - 64\right) = 0$

$W^2 - 36 = 0 \quad$ or $\quad W^2 - 64 = 0$

$W^2 = 36 \quad$ or $\quad W^2 = 64$

$W = \pm 6 \quad$ or $\quad W = \pm 8$

The width cannot be –6 or –8 inches.
If $W = 6$,

$L = \dfrac{48}{6} = 8$

If $W = 8$,

$L = \dfrac{48}{8} = 6$

The dimensions are 8 inches by 6 inches.

61. $x^2 - y^2 = 21$

$4x + 2y = 24$

Divide each term in the second equation by 2 and solve for y.

$2x + y = 12$

$y = 12 - 2x$

Substitute the expression $12 - 2x$ for y in the first equation and solve for x.

$x^2 - \left(12 - 2x\right)^2 = 21$

$x^2 - \left(144 - 48x + 4x^2\right) = 21$

$3x^2 - 48x + 165 = 0$

$x^2 - 16x + 55 = 0$

$(x - 5)(x - 11) = 0$

$x - 5 = 0 \quad$ or $\quad x - 11 = 0$

$x = 5 \quad$ or $\quad x = 11$

If $x = 11$, $y = 12 - 2(11) = -10$.

If $x = 5$, $y = 12 - 2(5) = 2$.

The dimensions of the floor are 5 meters by 5 meters and the dimensions of the square that will accommodate the pool are 2 meters by 2 meters.

63. **a.** It appears from the graphs that the percentage of white-collar workers was the same as blue-collar workers between the 1940s and 1960s.

b. $0.5x - y = -18$

$y = -0.004x^2 + 0.23x + 41$

Substitute $-0.004x^2 + 0.23x + 41$ in the first equation and solve for x.

$$0.5x - y = -18$$

$$0.5x - \overbrace{\left(-0.004x^2 + 0.23x + 41\right)}^{y} = -18$$

$0.5x + 0.004x^2 - 0.23x - 41 = -18$

$0.004x^2 + 0.27x - 23 = 0$

Use the quadratic formula.

$$x = \dfrac{-b \pm \sqrt{b^2 - 4ac}}{2a}$$

$$x = \dfrac{-(0.27) \pm \sqrt{0.27^2 - 4(0.004)(-23)}}{2(0.004)}$$

$x \approx 49 \quad$ or $\quad -117$

The percentage of white-collar workers was the same as blue-collar workers 49 years after 1900, or 1949.

Let $x = 49$ and solve for y in the white-collar model.

$0.5x - y = -18$

$0.5(49) - y = -18$

$24.5 - y = -18$

$-y = -42.5$

$y \approx 43$

The percentage of white-collar workers in 1949 was about 43%

Let $x = 49$ and solve for y in the blue-collar model.

$y = -0.004(49)^2 + 0.23(49) + 41 \approx 43\%$

The percentage of blue-collar workers in 1949 was about 43%.

c. According to the graph, the percentage of white-collar workers was the same as farmers in 1920.

The percentages of white-collar workers and farmers in 1920 were both 28%.

d. $0.5x - y = -18$
        $\underline{0.4x + y = \ \ 35}$
        $0.9x \ \ \ \ \ \ = 17$

$$x = \frac{17}{0.9}$$

$$x \approx 19$$

According to the models, the percentage of white-collar workers was the same as farmers 19 years after 1900, or 1919.

Let $x = 19$ and solve for y in the white-collar model.
    $0.5x - y = -18$
    $0.5(19) - y = -18$
    $9.5 - y = -18$
    $-y = -27.5$
    $y = 27.5$

The percentage of white-collar workers in 1919 was about 27.5%

Let $x = 19$ and solve for y in the farming model.
    $0.4x + y = 35$
    $0.4(19) + y = 35$
    $0.4(19) + y = 35$
    $7.6 + y = 35$
    $y = 27.4$

The percentage of farm workers in 1919 was about 27.4%

These answers model the actual data from part c (the graph) fairly well.

65. – 67. Answers may vary.

69.   makes sense

71.   makes sense

73.   false; Changes to make the statement true will vary. A sample change is: A circle and a line can intersect in at most two points, and therefore such a system has at most two real solutions.

75.   false; Changes to make the statement true will vary. A sample change is: It is possible that a system of two equations in two variables whose graphs

represent circles do not intersect, or intersect in a single point. This means that the system would have no solution, or a single solution, respectively.

77.   First determine the solution to the following system of equations.
    $$xy = 20$$
    $$x^2 + y^2 = 41$$
Solve the first equation for x.
$$x = \frac{20}{y}$$

Substitute the expression $\dfrac{20}{y}$ for x in the second equation and solve for in the second equation and solve for y.

$$\left(\frac{20}{y}\right)^2 + y^2 = 41$$

$$\frac{400}{y^2} + y^2 - 41 = 0$$

$$y^4 - 41y^2 + 400 = 0$$

$$\left(y^2 - 25\right)\left(y^2 - 16\right) = 0$$

$$y^2 - 25 = 0 \quad \text{or} \quad y^2 - 16 = 0$$
$$y^2 = 25 \quad \text{or} \quad y^2 = 16$$
$$y = \pm 5 \quad \text{or} \quad y = \pm 4$$

If $y = 5$, $x = \dfrac{20}{5} = 4$.

If $y = -5$, $x = \dfrac{20}{-5} = -4$.

If $y = 4$, $x = \dfrac{20}{4} = 5$.

If $y = -4$, $x = \dfrac{20}{-4} = -5$.

The rectangle formed by joining the points of intersection has sides a and b. The length of a is

$\sqrt{(5-4)^2 + (4-5)^2} = \sqrt{2}$. The length of b is

$\sqrt{(4-(-5))^2 + (5-(-4))^2} = 9\sqrt{2}$.

The area of the rectangle is $a \cdot b = \left(\sqrt{2}\right)\left(9\sqrt{2}\right) = 18$ square units.

79. $\log_y x = 3$

$\log_y (4x) = 5$

$x = y^3$

$4x = y^5$

Substitute y^3 for x in the equation $4x = y^5$ and solve for y.

$4(y^3) = y^5$

$4 = y^2$

$y = \pm 2$

Keep the positive base.

$x = 2^3 = 8$

The solution set is $\{(8, 2)\}$.

81.

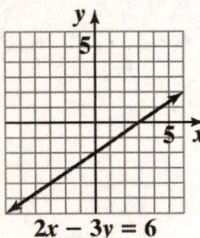

$2x - 3y = 6$

82.

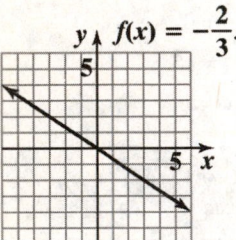

$f(x) = -\dfrac{2}{3}x$

83. $f(x) = -2$

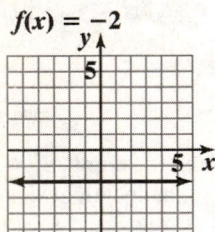

Mid-Chapter 7 Check Point

1.
$$x = 3y - 7$$
$$4x + 3y = 2$$

Since the first equation is solved for x already, we will use substitution.
Let $x = 3y - 7$ in the second equation and solve for y.

$4(3y - 7) + 3y = 2$

$12y - 28 + 3y = 2$

$15y = 30$

$y = 2$

Substitute this value for y in the first equation.

$x = 3(2) - 7 = 6 - 7 = -1$

The solution is $(-1, 2)$.

2.
$$3x + 4y = -5$$
$$2x - 3y = 8$$

Multiply the first equation by 3 and the second equation by 4, then add the equations.

$9x + 12y = -15$

$\underline{8x - 12y = 32}$

$17x = 17$

$x = 1$

Back-substitute to solve for y.

$3x + 4y = -5$

$3(1) + 4y = -5$

$3 + 4y = -5$

$4y = -8$

$y = -2$

The solution is $(1, -2)$.

3.
$$\frac{2x}{3} + \frac{y}{5} = 6$$
$$\frac{x}{6} - \frac{y}{2} = -4$$

Multiply the first equation by 15 and the second equation by 6 to eliminate the fractions.

$15\left(\dfrac{2x}{3} + \dfrac{y}{5}\right) = 15(6)$

$10x + 3y = 90$

$6\left(\dfrac{x}{6} - \dfrac{y}{2}\right) = 6(-4)$

$x - 3y = -24$

We now need to solve the equivalent system
$$10x + 3y = 90$$
$$x - 3y = -24$$

Add the two equations to eliminate y.
$$10x + 3y = 90$$
$$\underline{x - 3y = -24}$$
$$11x = 66$$
$$x = 6$$

Back-substitute to solve for y.
$$x - 3y = -24$$
$$6 - 3y = -24$$
$$-3y = -30$$
$$y = 10$$

The solution is $(6, 10)$.

4. $y = 4x - 5$

$8x - 2y = 10$

Since the first equation is already solved for y, we will use substitution.
Let $y = 4x - 5$ in the second equation and solve for x.

$$8x - 2(4x - 5) = 10$$
$$8x - 8x + 10 = 10$$
$$10 = 10$$

This statement is an identity. The system is dependent so there are an infinite number of solutions. The solution set is $\{(x, y) \mid y = 4x - 5\}$.

5. $2x + 5y = 3$

$3x - 2y = 1$

Multiply the first equation by 3 and the second equation by -2, then add the equations.

$$6x + 15y = 9$$
$$\underline{-6x + 4y = -2}$$
$$19y = 7$$
$$y = \frac{7}{19}$$

Back-substitute to solve for x.

$$2x + 5y = 3$$
$$2x + 5\left(\frac{7}{19}\right) = 3$$
$$2x + \frac{35}{19} = 3$$
$$2x = \frac{22}{19}$$
$$x = \frac{11}{19}$$

The solution is $\left(\frac{11}{19}, \frac{7}{19}\right)$.

6. $\dfrac{x}{12} - y = \dfrac{1}{4}$

$4x - 48y = 16$

Solve the first equation for y.

$$\frac{x}{12} - y = \frac{1}{4}$$
$$-y = -\frac{x}{12} + \frac{1}{4}$$
$$y = \frac{x}{12} - \frac{1}{4}$$

Let $y = \dfrac{x}{12} - \dfrac{1}{4}$ in the second equation and solve for x.

$$4x - 48\left(\frac{x}{12} - \frac{1}{4}\right) = 16$$
$$4x - 4x + 12 = 16$$
$$12 = 16$$

This statement is a contradiction. The system is inconsistent so there is no solution. The solution is $\{\ \}$ or $\varnothing$.

7.
$$2x - y + 2z = -8$$
$$x + 2y - 3z = 9$$
$$3x - y - 4z = 3$$

Multiply the first equation by 2 and add to the second equation.
$$4x - 2y + 4z = -16$$
$$\underline{x + 2y - 3z = 9}$$
$$5x + z = -7$$

Multiply the first equation by -1 and add to the third equation.

$$-2x + y - 2z = 8$$
$$\underline{3x - y - 4z = 3}$$
$$x - 6z = 11$$

Use the two reduced equations to get the following system:
$$5x + z = -7$$
$$x - 6z = 11$$

Multiply the first equation by 6 and add to the second equation.
$$30x + 6z = -42$$
$$\underline{x - 6z = 11}$$
$$31x = -31$$
$$x = -1$$

Back-substitute to solve for z.
$$5x + z = -7$$
$$5(-1) + z = -7$$
$$-5 + z = -7$$
$$z = -2$$

Back-substitute to solve for y.
$$2x - y + 2z = -8$$
$$2(-1) - y + 2(-2) = -8$$
$$-2 - y - 4 = -8$$
$$-y = -2$$
$$y = 2$$

The solution is $(-1, 2, -2)$.

8.
$$x \qquad - 3z = -5$$
$$2x - y + 2z = 16$$
$$7x - 3y - 5z = 19$$

Multiply the second equation by -3 and add to the third equation.
$$-6x + 3y - 6z = -48$$
$$\underline{7x - 3y - 5z = 19}$$
$$x - 11z = -29$$

Use this reduced equation and the original first equation to obtain the following system:
$$x - 3z = -5$$
$$x - 11z = -29$$

Multiply the second equation by -1 and add to the first equation.
$$x - 3z = -5$$
$$\underline{-x + 11z = 29}$$
$$8z = 24$$
$$z = 3$$

Back-substitute to solve for x.
$$x - 3z = -5$$
$$x - 3(3) = -5$$
$$x - 9 = -5$$
$$x = 4$$

Back-substitute to solve for y.
$$2x - y + 2z = 16$$
$$2(4) - y + 2(3) = 16$$
$$8 - y + 6 = 16$$
$$-y = 2$$
$$y = -2$$

The solution is $(4, -2, 3)$.

9. Solve $x + 2y - 3 = 0$ for x and substitute into the other equation.

$$\overset{x}{\overbrace{(-2y + 3)}}^2 + y^2 = 9$$
$$4y^2 - 12y + 9 + y^2 = 9$$
$$5y^2 - 12y = 0$$
$$y(5y - 12) = 0$$
$$y = 0 \text{ or } y = \frac{12}{5}$$

Back-substitute these values to find x.
When $y = 0$, $x = -2(0) + 3 = 3$.

When $y = \frac{12}{5}$, $x = -2\left(\frac{12}{5}\right) + 3 = -\frac{9}{5}$.

The solution set is $\left\{(3, 0), \left(-\frac{9}{5}, \frac{12}{5}\right)\right\}$

10. $3x^2 + 2y^2 = 14$

$2x^2 - y^2 = 7$

Multiply the second equation by 2 and add.

$3x^2 + 2y^2 = 14$

$\underline{4x^2 - 2y^2 = 14}$

$\qquad 7x^2 = 28$

$\qquad\quad x^2 = 4$

$\qquad\quad x = \pm 2$

Back-substitute these values to find y.

$3(2)^2 + 2y^2 = 14$

$\quad 12 + 2y^2 = 14$

$\qquad\quad 2y^2 = 2$

$\qquad\quad\ y^2 = 1$

$\qquad\quad\ \ y = \pm 1$

$3(-2)^2 + 2y^2 = 14$

$\quad 12 + 2y^2 = 14$

$\qquad\quad 2y^2 = 2$

$\qquad\quad\ y^2 = 1$

$\qquad\quad\ \ y = \pm 1$

The solution set is $\{(2,1),(2,-1),(-2,1),(-2,-1)\}$.

11. Use the first equation to substitute for y in the second equation.

$$x^2 + (\overbrace{x^2 - 6}^{y})^2 = 8$$

$$x^2 + (x^2 - 6)^2 = 8$$

$$x^2 + x^4 - 12x^2 + 36 = 8$$

$$x^4 - 11x^2 + 28 = 0$$

$$(x^2 - 4)(x^2 - 7) = 0$$

$x = \pm 2$ or $x = \pm\sqrt{7}$

Back-substitute these values to find x.

$y = (2)^2 - 6 \qquad y = (-2)^2 - 6$

$y = -2 \qquad\qquad y = -2$

$y = (\sqrt{7})^2 - 6 \quad y = (-\sqrt{7})^2 - 6$

$y = 1 \qquad\qquad y = 1$

The solution set is $\left\{(2,-2),(-2,-2),\left(\sqrt{7},1\right),\left(-\sqrt{7},1\right)\right\}$.

12. Use the first equation to substitute for x in the second equation.

$$2y^2 + \overbrace{(2y+4)}^{x}y = 8$$
$$2y^2 + (2y+4)y = 8$$
$$2y^2 + 2y^2 + 4y = 8$$
$$4y^2 + 4y - 8 = 0$$
$$y^2 + y - 2 = 0$$
$$(y+2)(y-1) = 0$$
$$y = -2 \ \text{ or } \ y = 1$$

Back-substitute these values to find x.

$$x = 2(-2)+4 \quad x = 2(1)+4$$
$$x = 0 \qquad\qquad y = 6$$

The solution set is $\{(0,-2),(6,1)\}$.

13. $\dfrac{x^2 - 6x + 3}{(x-2)^3} = \dfrac{A}{x-2} + \dfrac{B}{(x-2)^2} + \dfrac{C}{(x-2)^3}$

Multiply both sides of the equation by the common denominator $(x-2)^3$.

$$(x-2)^3 \frac{x^2-6x+3}{(x-2)^3} = (x-2)^3 \left(\frac{A}{x-2} + \frac{B}{(x-2)^2} + \frac{C}{(x-2)^3} \right)$$
$$x^2 - 6x + 3 = A(x-2)^2 + B(x-2) + C$$
$$x^2 - 6x + 3 = A(x^2 - 4x + 4) + Bx - 2B + C$$
$$x^2 - 6x + 3 = Ax^2 - 4Ax + 4A + Bx - 2B + C$$
$$x^2 - 6x + 3 = (A)x^2 + (-4A + B)x + (4A - 2B + C)$$

Equate coefficients of like powers of x and equate constant terms.

$$A = 1$$
$$-4A + B = -6$$
$$4A - 2B + C = 3$$

Since $A = 1$, we find that $B = -2$ and $C = -5$ by substitution.

$$\frac{2x^2 - 6x + 3}{(x-2)^3} = \frac{1}{x-2} + \frac{-2}{(x-2)^2} + \frac{-5}{(x-2)^3}$$

14. $\dfrac{10x^2+9x-7}{(x+2)(x^2-1)} = \dfrac{10x^2+9x-7}{(x+2)(x+1)(x-1)} = \dfrac{A}{x+2} + \dfrac{B}{x+1} + \dfrac{C}{x-1}$

Multiply both sides of the equation by the least common denominator.

$$(x+2)(x+1)(x-1)\dfrac{10x^2+9x-7}{(x+2)(x+1)(x-1)} = (x+2)(x+1)(x-1)\left(\dfrac{A}{x+2} + \dfrac{B}{x+1} + \dfrac{C}{x-1}\right)$$

$$10x^2+9x-7 = (x+1)(x-1)A + (x+2)(x-1)B + (x+2)(x+1)C$$

$$10x^2+9x-7 = (x^2-1)A + (x^2+x-2)B + (x^2+3x+2)C$$

$$10x^2+9x-7 = x^2A - A + x^2B + xB - 2B + x^2C + 3xC + 2C$$

$$10x^2+9x-7 = (A+B+C)x^2 + (B+3C)x + (-A-2B+2C)$$

Equate coefficients of like powers of x, and equate constant terms.

$$A+B+C = 10$$
$$B+3C = 9$$
$$-A-2B+2C = -7$$

Solving the above system for A, B, and C, we find $A=5$, $B=3$, and $C=2$.

$$\dfrac{10x^2+9x-7}{(x+2)(x^2-1)} = \dfrac{5}{x+2} + \dfrac{3}{x+1} + \dfrac{2}{x-1}$$

15. $\dfrac{x^2+4x-23}{(x+3)(x^2+4)} = \dfrac{A}{x+3} + \dfrac{Bx+C}{x^2+4}$

Multiply both sides of the equation by the least common denominator.

$$(x+3)(x^2+4)\dfrac{x^2+4x-23}{(x+3)(x^2+4)} = (x+3)(x^2+4)\left(\dfrac{A}{x+3} + \dfrac{Bx+C}{x^2+4}\right)$$

$$x^2+4x-23 = (x^2+4)A + (x+3)(Bx+C)$$

$$x^2+4x-23 = x^2A + 4A + x^2B + 3xB + xC + 3C$$

$$x^2+4x-23 = (A+B)x^2 + (3B+C)x + (4A+3C)$$

Equate coefficients of like powers of x, and equate constant terms.

$$A+B = 1$$
$$3B+C = 4$$
$$4A+3C = -23$$

Solving the above system for A, B, and C, we find $A=-2$, $B=3$, and $C=-5$.

$$\dfrac{x^2+4x-23}{(x+3)(x^2+4)} = \dfrac{-2}{x+3} + \dfrac{3x-5}{x^2+4}$$

16. $\dfrac{x^3}{\left(x^2+4\right)^2} = \dfrac{Ax+B}{x^2+4} + \dfrac{Cx+D}{\left(x^2+4\right)^2}$

Multiply both sides of the equation by the least common denominator.

$$\left(x^2+4\right)^2 \frac{x^3}{\left(x^2+4\right)^2} = \left(x^2+4\right)^2 \left(\frac{Ax+B}{x^2+4} + \frac{Cx+D}{\left(x^2+4\right)^2}\right)$$

$$x^3 = \left(x^2+4\right)(Ax+B) + (Cx+D)$$

$$x^3 = x^3A + x^2B + 4xA + 4B + Cx + D$$

$$x^3 = (A)x^3 + (B)x^2 + (4A+C)x + (4B+D)$$

Equate coefficients of like powers of x, and equate constant terms.

$A = 1$

$B = 0$

$4A + C = 0$

$4B + D = 0$

Since $A = 1$ and $B = 0$, we find that $C = -4$ and $D = 0$ by substitution.

$$\frac{x^3}{\left(x^2+4\right)^2} = \frac{1x+0}{x^2+4} + \frac{-4x+0}{\left(x^2+4\right)^2} = \frac{x}{x^2+4} + \frac{-4x}{\left(x^2+4\right)^2}$$

17.   **a.** $C(x) = 400,000 + 20x$

  b. $R(x) = 100x$

  c. $P(x) = R(x) - C(x)$

$$= 100x - (400,000 + 20x)$$

$$= 80x - 400,000$$

  d.   The break even point is the point where cost and revenue are the same. We need to solve the following system.

$y = 400,000 + 20x$

$y = 100x$

Let $y = 400,000 + 20x$ in the second equation and solve for x.

$400,000 + 20x = 100x$

$400,000 = 80x$

$5000 = x$

Back-substitute to solve for y.

$y = 100x$

$= 100(5000)$

$= 500,000$

Thus, the break-even point is $(5000,\ 500,000)$. The company will break even when it produces and sells 5000 PDAs. At this level, the cost and revenue will both be $500,000.

18. Let x = the number of roses.
Let y = the number of carnations.
$$x + y = 20$$
$$3x + 1.5y = 39$$
Solve the first equation for x.
$$x + y = 20$$
$$x = 20 - y$$
Substitute this expression for x in the second equation and solve for y.
$$3(20 - y) + 1.5y = 39$$
$$60 - 3y + 1.5y = 39$$
$$-1.5y = -21$$
$$y = 14$$
Back-substitute to solve for x.
$$x = 20 - y = 20 - 14 = 6$$
There are 6 roses and 14 carnations in the bouquet.

19. Let x = number of students at the north campus.
Let y = number of students at the south campus.
$$x + y = 1200$$
$$0.10x + 0.50y = 0.40(1200)$$
or
$$x + y = 1200$$
$$0.1x + 0.5y = 480$$
Solve the first equation for x.
$$x + y = 1200$$
$$x = 1200 - y$$
Substitute this expression for x in the second equation.
$$0.1(1200 - y) + 0.5y = 480$$
$$120 - 0.1y + 0.5y = 480$$
$$120 + 0.4y = 480$$
$$0.4y = 360$$
$$\frac{0.4y}{0.4} = \frac{360}{0.4}$$
$$y = 900$$
Back-substitute and solve for x.
$$x = 1200 - y$$
$$= 1200 - 900$$
$$= 300$$
There were 300 students at the north campus and 900 students at the south campus.

20. x = velocity of the boat
y = velocity of the current

Velocity	Time	Distance
$x + y$	2	$2(x + y)$
$x - y$	6	$6(x - y)$

$$2(x + y) = 9 \quad \rightarrow \quad x + y = 4.5$$
$$6(x - y) = 9 \quad \rightarrow \quad x - y = 1.5$$
$$\overline{\qquad\qquad 2x \ \ = 6}$$
$$x = 3$$
$$x + y = 4.5$$
$$3 + y = 4.5$$
$$y = 1.5$$
The rowing velocity in still water is 3 mph and the current is 1.5 mph.

21. Because the sum of the measures of the angles of any triangle is $180°$, or $x + y + 90 = 180$
$$x + y = 90.$$
Because the angle with measures x and $(3y + 20)$ are supplementary, $x + (3y + 20) = 180.$
Simplify this equation.
$$x + 3y + 20 = 180$$
$$x + 3y = 160$$
We now have the system
$$x + y = 90$$
$$x + 3y = 160.$$
To solve this system by the addition method, multiply the first equation by -1 and add to the second equation.
$$-x - \ y = -90$$
$$\underline{\ \ x + 3y = 160}$$
$$2y = 70$$
$$y = 35$$
Back-substitute.
$$x + y = 90$$
$$x + 35 = 90$$
$$x = 55$$
Thus, $x = 55°$, $y = 35°$, and $(3y + 20) = 125°$.

22. Using the points $(-1, 0)$, $(1, 4)$, and $(2, 3)$ in the equation $y = ax^2 + bx + c$, we get the following system of equations:
$$a - b + c = 0$$
$$a + b + c = 4$$
$$4a + 2b + c = 3$$
Add the first two equations.

$$a - b + c = 0$$
$$\underline{a + b + c = 4}$$
$$2a + 2c = 4$$

Multiply the first equation by 2 and add to the third equation.

$$2a - 2b + 2c = 0$$
$$\underline{4a + 2b + c = 3}$$
$$6a + 3c = 3$$

Using the two reduced equations, we get the following system of equations:

$$2a + 2c = 4$$
$$6a + 3c = 3$$

Multiply the first equation by -3 and add to the second equation.

$$-6a - 6c = -12$$
$$\underline{6a + 3c = 3}$$
$$-3c = -9$$
$$c = 3$$

Back-substitute to solve for a.

$$2a + 2c = 4$$
$$2a + 2(3) = 4$$
$$2a + 6 = 4$$
$$2a = -2$$
$$a = -1$$

Back-substitute to solve for b.

$$a + b + c = 4$$
$$-1 + b + 3 = 4$$
$$b = 2$$

The equation is $y = -x^2 + 2x + 3$.

23.
$$2l + 2w = 21$$
$$lw = 20$$

Solving $lw = 20$ for l gives $l = \dfrac{20}{w}$.

Substitute into the other equation:

$$2\left(\overset{l}{\overbrace{\dfrac{20}{w}}}\right) + 2w = 21$$

$$\dfrac{40}{w} + 2w = 21$$

$$w\left(\dfrac{40}{w} + 2w\right) = w(21)$$

$$40 + 2w^2 = 21w$$

$$2w^2 - 21w + 40 = 0$$

$$(2w - 5)(w - 8) = 0$$

$$w = \dfrac{5}{2} \text{ or } w = 8.$$

Back-substitute to find l.

When $w = \dfrac{5}{2}$, $l = \dfrac{20}{\frac{5}{2}} = 8$.

When $w = 8$, $l = \dfrac{20}{8} = \dfrac{5}{2}$.

The dimensions of the rectangle are $\dfrac{5}{2}$ m by 8 m.

Section 7.5

Check Point Exercises

1. $4x - 2y \geq 8$

Graph the equation $4x - 2y = 8$ as a solid line.

Choose a test point that is not on the line.

Test $(0,0)$

$4x - 2y \geq 8$

$4(0) - 2(0) \geq 8$

$0 \geq 8$, false

Since the statement is false, shade the other half-plane.

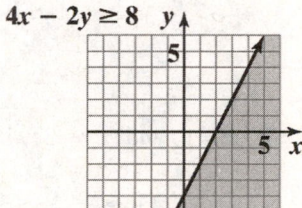

2. $y > -\dfrac{3}{4}x$

Graph the equation $y = -\dfrac{3}{4}x$ as a dashed line.

Choose a test point that is not on the line.

Test $(1,1)$

$y > -\dfrac{3}{4}x$

$1 > -\dfrac{3}{4}(1)$

$1 > -\dfrac{3}{4}$, true

Since the statement is true, shade the half-plane containing the point.

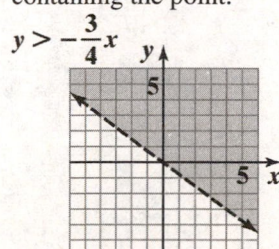

3. **a.** $y > 1$

Graph the equation $y = 1$ as a dashed line.

Choose a test point that is not on the line.

Test $(0,0)$

$y > 1$

$0 > 1$, false

Since the statement is false, shade the other half-plane.

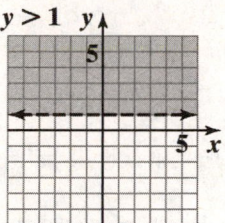

b. Graph the equation $x = -2$ as a solid line.

Choose a test point that is not on the line.

Test $(0,0)$

$x \leq -2$

$0 \leq -2$, false

Since the statement is false, shade the other half-plane.

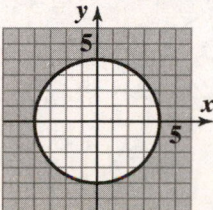

4. Graph $x^2 + y^2 = 16$ as a solid circle with radius 4 and center (0, 0).

Test $(0,0)$

$x^2 + y^2 \geq 16$

$(0)^2 + (0)^2 \geq 16$

$0 \geq 16$, false

Shade the region not containing (0, 0).

$x^2 + y^2 \geq 16$

5. Point $B = (66,130)$

$$4.9x - y \geq 165$$
$$4.9(66) - 130 \geq 165$$
$$193.4 \geq 165, \ \text{true}$$

$$3.7x - y \leq 125$$
$$3.7(66) - 130 \leq 125$$
$$114.2 \leq 125, \ \text{true}$$

Point B is a solution of the system.

6. Graph the equation $x - 3y = 6$ as a dashed line.

$\underline{\text{Test } (0,0)}$

$$x - 3y < 6$$
$$(0) - 3(0) < 6$$
$$0 < 6, \ \text{true}$$

Since the statement is true, shade the half-plane containing the point.
Graph the equation $2x + 3y = -6$ as a solid line.

$\underline{\text{Test } (0,0)}$

$$2x + 3y \geq -6$$
$$2(0) + 3(0) \geq -6$$
$$0 \geq -6, \ \text{true}$$

Since the statement is true, shade the half-plane containing the point.
For the solution graph, place an open circle at the point of intersection and shade the region that satisfies both inequalities.

$$x - 3y < 6$$
$$2x + 3y \geq -6$$

7. Begin by graphing $y = x^2 - 4$ as a solid parabola with vertex $(0, -4)$ and x-intercepts $(-2, 0)$ and $(2, 0)$. Since $(0, 0)$ makes the inequality $y \geq x^2 - 4$ true, shade the region containing $(0, 0)$. Graph $x + y = 2$ as a solid line by using its x-intercept, $(2, 0)$, and its y-intercept $(0, 2)$. Since $(0, 0)$ makes the inequality $x + y \leq 2$ true, shade the region containing $(0, 0)$.

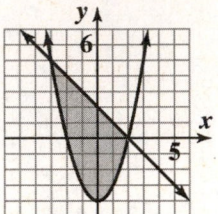

$$y \geq x^2 - 4$$
$$x + y \leq 2$$

8. Graph the lines $x + y = 2$, $x = 1$, and $y = -3$ with dashed lines.
Graph the line $x = -2$ with a solid line.
Test points indicate that the solution contains the region to the right of -2, to the left of 1, above -3, and below the line $x + y = 2$. The corner points are represented as open circles because none satisfy all three inequalities.

$$x + y < 2$$
$$-2 \leq x < 1$$
$$y > -3$$

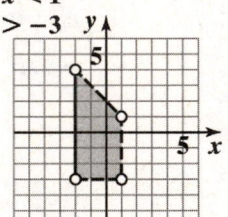

Exercise Set 7.5

1. Graph $x + 2y = 8$ as a solid line using its
   x-intercept, $(8, 0)$, and its y-intercept, $(0, 4)$.
   Test $(0, 0)$:
   $0 + 2(0) \leq 8?$
   $0 \leq 8$ true
   Shade the half-plane containing $(0, 0)$.

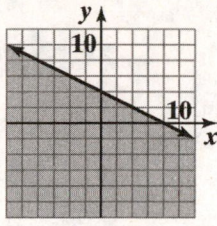

$$x + 2y \leq 8$$

3. Graph $x - 2y = 10$ as a dashed line using its x-
   intercept, $(10, 0)$, and its y-intercept,
   $(0, -5)$.
   Test $(0, 0)$:
   $0 - 2(0) > 10?$
   $0 > 10$ false
   Shade the half-plane not containing $(0, 0)$.

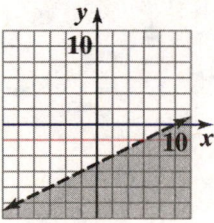

$$x - 2y > 10$$

5. Graph $y = \frac{1}{3}x$ as a solid line using its slope, $\frac{1}{3}$, and
   its y-intercept $(0, 0)$.
   Test $(1, 1)$:
   $1 \leq \frac{1}{3}(1)?$
   $1 \leq \frac{1}{3}$ false
   Shade the half-plane not containing $(1, 1)$.

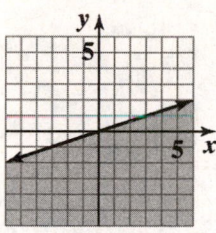

$$y \leq \frac{1}{3}x$$

7. Graph $y = 2x - 1$ as a dashed line using its
   x-intercept, $\left(\frac{1}{2}, 0\right)$ and its y-intercept,
   $(0, -1)$.
   Test $(0,0)$:
   $0 > 2(0) - 1?$
   $0 > -1$ true
   Shade the half-plane containing $(0, 0)$.

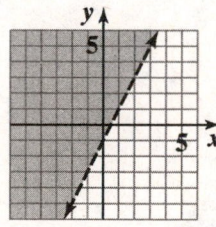

$$y > 2x - 1$$

9. Graph $x = 1$ as a solid vertical line.
Test $(0, 0)$:
$0 \leq 1$ true
Shade the half-plane containing $(0, 0)$.

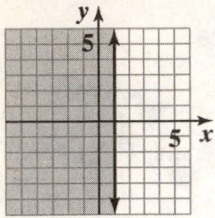

$x \leq 1$

11. Graph $y = 1$ as a dashed horizontal line.
Test $(0, 0)$:
$0 > 1$ false
Shade the half-plane not containing $(0, 0)$.

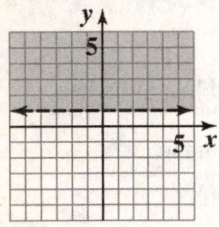

$y > 1$

13. Graph $x^2 + y^2 = 1$ as a solid circle with radius 1 and center $(0, 0)$.
Test $(0, 0)$:
$(0)^2 + (0)^2 \leq 1$?
$\qquad 0 \leq 1$ true
Shade the region containing $(0, 0)$.

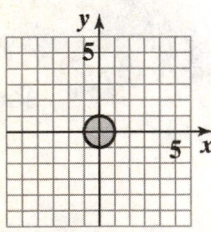

$x^2 + y^2 \leq 1$

15. Graph $x^2 + y^2 = 25$ as a dashed circle with radius 5 and center $(0, 0)$.
Test $(0, 0)$:
$(0)^2 + (0)^2 > 25$?
$\qquad 0 > 25$ false
Shade the region not containing $(0, 0)$.

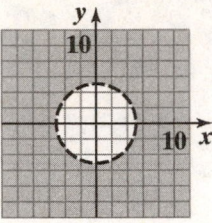

$x^2 + y^2 > 25$

17. Graph $(x - 2)^2 + (y + 1)^2 = 9$ as a dashed circle.
$\underline{\text{Test } (0, 0)}$
$(x - 2)^2 + (y + 1)^2 < 9$
$(0 - 2)^2 + (0 + 1)^2 < 9$
$\qquad 5 < 9, \quad$ true
Shade the region containing $(0, 0)$.

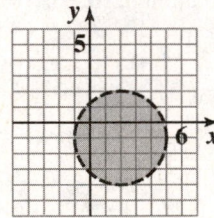

$(x - 2)^2 + (y + 1)^2 < 9$

19. Graph $y = x^2 - 1$ as a dashed parabola with vertex $(0, -1)$ and x-intercepts $(1, 0)$ and $(-1, 0)$.
Test $(0, 0)$:
$0 < (0)^2 - 1$?
$0 < -1$ false
Shade the region not containing $(0, 0)$.

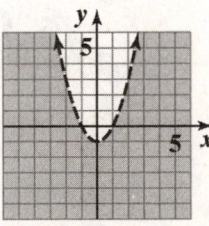

$y < x^2 - 1$

21. Graph $y = x^2 - 9$ as a solid parabola with vertex (0, –9) and x-intercepts (3, 0) and (–3, 0).
Test (0, 0):
$0 \geq (0)^2 - 9$?
$0 \geq -9$ true
Shade the region containing (0, 0).

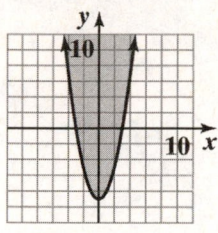

$y \geq x^2 - 9$

23. Graph $y = 2^x$ as a dashed exponential function with base 2 that passes through the point (0, 1).
Test (0, 0):
$0 > 2^0$?
$0 > 1$ false
Shade the region not containing (0, 0).

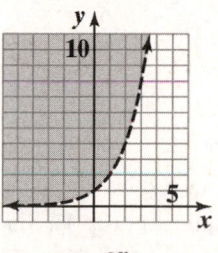

$y > 2^x$

25. Graph $y = \log_2(x+1)$ as a solid logarithmic function.
Test (0, 0):
$y \geq \log_2(x+1)$
$0 \geq \log_2(0+1)$
$0 \geq 0,$ true
Shade the region containing (0, 0).

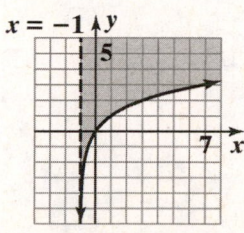

$y \geq \log_2(x+1)$

27. Begin by graphing $3x + 6y = 6$ as a solid line using its x-intercept, (2, 0), and its y-intercept, (0, 1). Since (0, 0) makes the inequality $3x + 6y \leq 6$ true, shade the half-plane containing (0, 0). Graph $2x + y = 8$ as a solid line using its x-intercept, (4, 0), and its y-intercept, (0, 8). Since (0, 0) makes the inequality $2x + y \leq 8$ true, shade the half-plane containing (0, 0). The solution set of the system is the intersection of the above shaded half-planes, and is shown as the shaded region in the following graph.

$3x + 6y \leq 6$
$2x + y \leq 8$

29. Begin by graphing $2x - 5y = 10$ as a solid line using its x-intercept, (5, 0), and its y-intercept, (0, –2). Since (0, 0) makes the inequality $2x - 5y \leq 10$ true, shade the half-plane containing (0, 0). Graph $3x - 2y = 6$ as a dashed line using its x-intercept, (2, 0), and its y-intercept, (0, –3). Since (0, 0) makes the inequality $3x - 2y > 6$ false, shade the half-plane containing (0, 0). The solution set of the system is the intersection of the above shaded half-planes, and is shown as the shaded region in the following graph.

$2x - 5y \leq 10$
$3x - 2y > 6$

31. Begin by graphing $y = 2x - 3$ as a dashed line using its slope, 2, and its y-intercept, $(0, -3)$. Since $(0, 0)$ makes the inequality $y > 2x - 3$ true, shade the half-plane containing $(0, 0)$. Graph $y = -x + 6$ as a dashed line using its slope, -1, and its y-intercept, $(0, 6)$. Since $(0, 0)$ makes the inequality $y < -x + 6$ true, shade the half-plane containing $(0, 0)$. The solution set of the system is the intersection of the above shaded half-planes, and is shown as the shaded region in the following graph.

$y > 2x - 3$
$y < -x + 6$
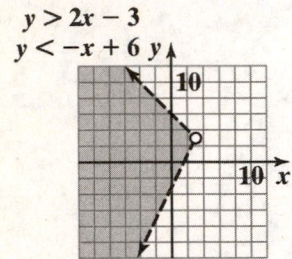

33. Begin by graphing $x + 2y = 4$ as a solid line using its x-intercept, $(4, 0)$, and its y-intercept, $(0, 2)$. Since $(0, 0)$ makes the inequality $x + 2y \le 4$ true, shade the half-plane containing $(0, 0)$. Graph $y = x - 3$ as a solid line using its slope, 1, and its y-intercept, $(0, -3)$. Since $(0, 0)$ makes the inequality $y \ge x - 3$ true, shade the half-plane containing $(0, 0)$.

The solution set of the system is the intersection of the above shaded half-planes, and is shown as the shaded region in the following graph.

$x + 2y \le 4$
$y \ge x - 3$
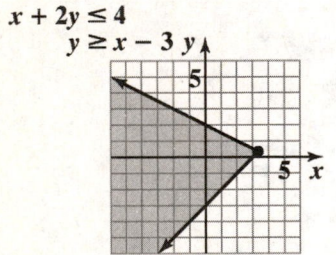

35. Begin by graphing $x = 2$ as a solid vertical line. Since $(0, 0)$ makes the inequality $x \le 2$ true, shade the half-plane containing $(0, 0)$. Graph $y = 1$ as a solid horizontal line. Since $(0, 0)$ makes the inequality $y \ge -1$ true, shade the half-plane containing $(0, 0)$. The solution set of the system is the intersection of the above shaded half-planes, and is shown as the shaded region in the following graph.

$x \le 2$
$y \ge -1$
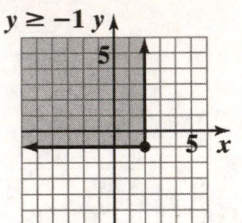

37. Graph $x = -2$ as a solid vertical line and $x = 5$ as a dashed vertical line. Since $(0, 0)$ makes the inequality $-2 \le x < 5$ true, shade the region between the two lines.

$-2 \le x < 5$
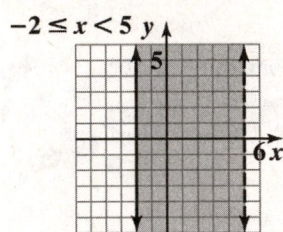

39. Begin by graphing $x - y = 1$ as a solid line using its x-intercept, $(1, 0)$, and its y-intercept $(0, -1)$. Since $(0, 0)$ makes the inequality $x - y \le 1$ true, shade the half-plane containing $(0, 0)$. Graph $x = 2$ as a solid horizontal line. Since $(0, 0)$ makes the inequality $x \ge 2$ false, shade the half-plane not containing $(0, 0)$. The solution set of the system is the intersection of the above shaded half-planes, and is shown as the shaded region in the following graph.

$x - y \le 1$
$x \ge 2$
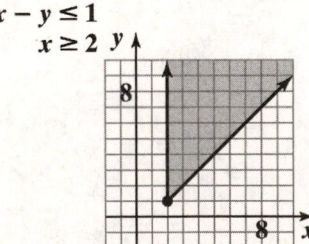

41. $x + y > 4$

$x + y < -1$

Begin by graphing $x + y = 4$ as a dashed line using its x-intercept, (4, 0), and its y-intercept (0, 4). Since (0, 0) makes the inequality $x + y > 4$ false, shade the half-plane not containing (0, 0). Graph $x + y = -1$ as a dashed line using its x-intercept, (–1, 0), and its y-intercept, (0, –1). Since (0, 0) makes the inequality $x + y < -1$ false, shade the half-plane not containing (0, 0). Since these half-planes do not intersect the system has no solution.

43. Begin by graphing $x + y = 4$ as a dashed line using its x-intercept, (4, 0), and its y-intercept, (0, 4). Since (0, 0) makes the inequality $x + y > 4$ false, shade the half-plane not containing (0, 0). Graph $x + y = -1$ as a dashed line using its x-intercept, (-1, 0), and its y-intercept, (0, –1). Since (0, 0) makes the inequality $x + y > -1$ true, shade the half-plane containing (0, 0). The solution set of the system is the intersection of the above half-planes, and is shown as the shaded region in the following graph.

$x + y > 4$

$x + y > -1$

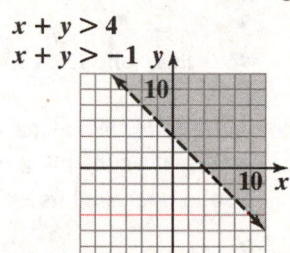

45. Begin by graphing $y = x^2 - 1$ as a solid parabola with vertex (0, –1) and x-intercepts, (–1, 0), and (1, 0). Since (0, 0) makes the inequality $y \geq x^2 - 1$ true, shade the half-plane containing (0, 0). Graph $x - y = -1$ as a solid line using its x-intercept, (–1, 0), and its y-intercept, (0, 1). Since (0, 0) makes the inequality $x - y \geq -1$ true, shade the half-plane containing (0, 0). The solution set of the system is the intersection of the above half-planes, and is shown as the shaded region in the following graph.

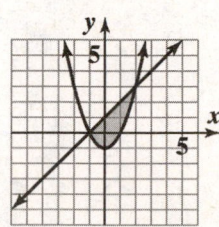

$y \geq x^2 - 1$

$x - y \geq -1$

47. Begin by graphing $x^2 + y^2 = 16$ as a solid circle with radius 4 and center, (0, 0). Since (0, 0) makes the inequality $x^2 + y^2 \leq 16$ true, shade the half-plane containing (0, 0). Graph $x + y = 2$ as a dashed line using its x-intercept, (2, 0), and its y-intercept, (0, 2). Since (0, 0) makes the inequality $x + y > 2$ false, shade the half-plane not containing (0, 0). The solution set of the system is the intersection of the above half-planes, and is shown as the shaded region in the following graph.

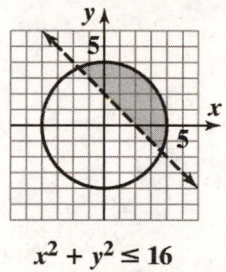

$x^2 + y^2 \leq 16$

$x + y > 2$

49. Begin by graphing $x^2 + y^2 = 1$ as a dashed circle with radius 1 and center, (0, 0). Since (0, 0) makes the inequality $x^2 + y^2 > 1$ false, shade the half-plane not containing (0, 0). Graph $x^2 + y^2 = 16$ as a dashed circle with radius 4 and center (0, 0). Since (0, 0) makes the inequality $x^2 + y^2 < 16$ true, shade the half-plane containing (0, 0). The solution set of the system is the intersection of the above half-planes, and is shown as the shaded region in the following graph.

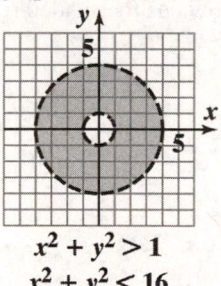

$x^2 + y^2 > 1$

$x^2 + y^2 < 16$

51.

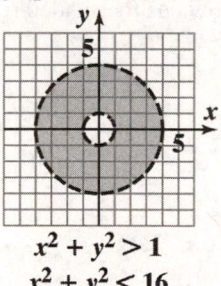

$(x - 1)^2 + (y + 1)^2 < 25$

$(x - 1) + (y + 1)^2 \geq 16$

53.

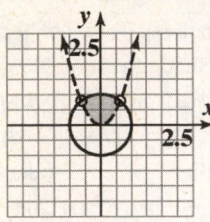

$$x^2 + y^2 \leq 1$$
$$y - x^2 > 0$$

55.

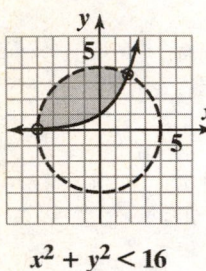

$$x^2 + y^2 < 16$$
$$y \geq 2^x$$

57. Begin by graphing $x - y = 2$ as a solid line using its x-intercept, $(2, 0)$, and its y-intercept, $(0, -2)$. Since $(0, 0)$ makes the inequality $x - y \leq 2$ true, shade the half-plane containing $(0, 0)$. Graph $x = -2$ as a solid vertical line. Since $(0, 0)$ makes the inequality $x \geq -2$ true, shade the half-plane containing $(0, 0)$. Graph $y = 3$ as a solid horizontal line. Since $(0, 0)$ makes the inequality $y \leq 3$ true, shade the half-plane containing $(0, 0)$. The solution set of the system is the intersection of the above half-planes, and is shown as the shaded region in the following graph.

$$x - y \leq 2$$
$$x > -2$$
$$y \leq 3$$

59. Since $x \geq 0$ and $y \geq 0$ the solution to the system lies in the first quadrant. Graph $2x + 5y = 10$ as a solid line using its x-intercept, $(5, 0)$, and its y-intercept, $(0, 2)$. Since $(0, 0)$ makes the inequality $2x + 5y \leq 10$ true, shade the half-plane containing $(0, 0)$. Graph $3x + 4y = 12$ as a solid line by using its x-intercept, $(4, 0)$, and its y-intercept, $(0, 3)$. Since $(0, 0)$ makes the inequality $3x + 4y \leq 12$ true, shade the half-plane containing $(0, 0)$. The solution set of the system is the intersection of the above half-planes which lies in the first quadrant, and is shown as the shaded region in the following graph.

$$x \geq 0$$
$$y \geq 0$$
$$2x + 5y < 10$$
$$3x + 4y \leq 12$$

61. Begin by graphing $3x + y = 6$ as a solid line using its x-intercept, $(2, 0)$, and its y-intercept, $(0, 6)$. Since $(0, 0)$ makes the inequality $3x + y \leq 6$ true, shade the half-plane containing $(0, 0)$. Graph $2x - y = -1$ as a solid line using its x-intercept, $\left(-\dfrac{1}{2}, 0\right)$, and its y-intercept, $(0, 1)$. Since $(0, 0)$ makes the inequality $2x - y \leq -1$ false, shade the half-plane not containing $(0, 0)$. Graph $x = -2$ as a solid vertical line. Since $(0, 0)$ makes the inequality $x \geq -2$ true, shade the half-plane containing $(0, 0)$. Graph $y = 4$ as a solid horizontal line. Since $(0, 0)$ makes the inequality $y \leq 4$ true, shade the half-plane containing $(0, 0)$. The solution set of the system is the intersection of the above half-planes, and is shown as the shaded region in the following graph.

$$3x + y \leq 6$$
$$2x - y \leq -1$$
$$x > -2$$
$$y < 4$$

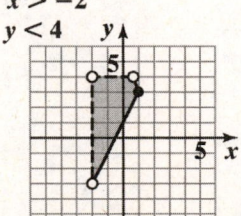

63. $y \geq -2x + 4$

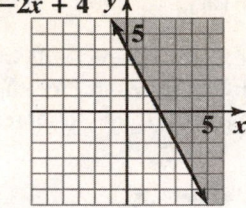

65. $x + y \leq 4$
$3x + y \leq 6$

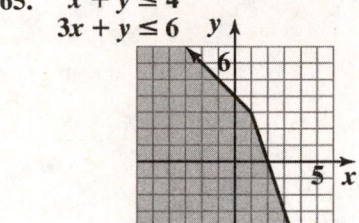

67.

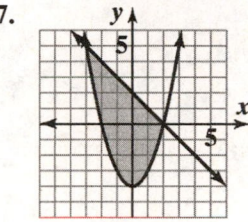

$x + y \leq 2$

$y \geq x^2 - 4$

69. $-2 \leq x \leq 2$
$-3 \leq y \leq 3$

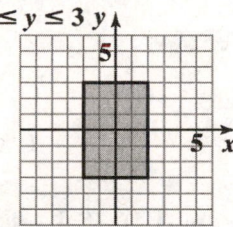

71. Find the union of solutions of

$y > \dfrac{3}{2}x - 2$ and $y < 4$.

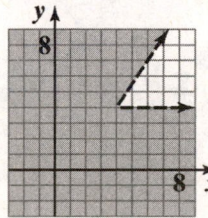

73. The system $\begin{matrix} 3x + 3y < 9 \\ 3x + 3y > 9 \end{matrix}$ has no solution.

The number $3x + 3y$ cannot both be less than 9 and greater than 9 at the same time.

75. The system has an infinite number of solutions. The solution is the set of points that make up the circle $(x + 4)^2 + (y - 3)^2 = 9$

77. Point $A = (66, 160)$
$5.3x - y \geq 180$
$5.3(66) - 160 \geq 180$
$189.8 \geq 180$, true

$4.1x - y \leq 14$
$4.1(66) - 160 \leq 140$
$110.6 \leq 140$, true

Point A is a solution of the system.

79. Point $= (72, 205)$
$5.3x - y \geq 180$
$5.3(72) - 205 \geq 180$
$176.6 \geq 180$, false

$4.1x - y \leq 14$
$4.1(72) - 205 \leq 140$
$90.2 \leq 140$, true

The data does not satisfy both inequalities. The person is not within the healthy weight region.

81. a. $50x + 150y > 2000$

b. Graph $50x + 150y$ as a dashed line using its x-intercept, $(40, 0)$, and its y-intercept, $\left(0, \dfrac{40}{3}\right)$.

Test $(0, 0)$:
$50(0) + 150(0) > 2000$?

$0 > 2000$ false
Shade the half-plane not containing $(0, 0)$.

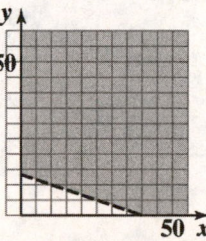

$50x + 150y > 2000$

c. Ordered pairs may vary.

83. **a.**
$$y \geq 0$$
$$x + y \geq 5$$
$$x \geq 1$$
$$200x + 100y \leq 700$$

b.
$$y \geq 0$$
$$x + y \geq 5$$
$$x \geq 1$$
$$200x + 100y \leq 700$$

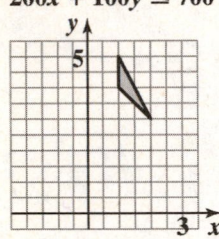

85. **a.** $\text{BMI} = \dfrac{703W}{H^2} = \dfrac{703(200)}{72^2} \approx 27.1$

b. A 20 year old man with a BMI of 27.1 is classified as overweight.

87. – 95. Answers may vary.

97.

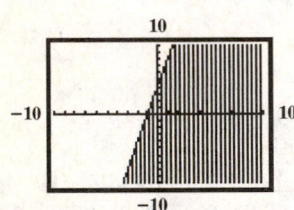

99. $y \geq x^2 - 4$

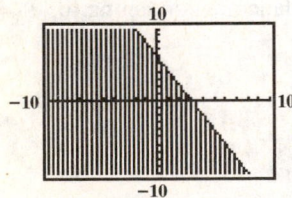

101. $2x + y \leq 6$

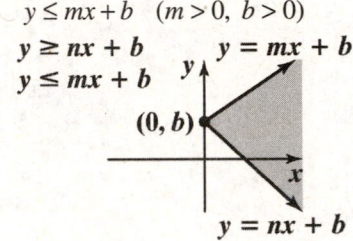

103. – 105. Answers may vary.

107. does not make sense; Explanations will vary. Sample explanation: It is necessary to graph the linear equation with a dashed line to represent its role as a borderline.

109. makes sense

111. $y > x - 3$
$$y \leq x$$

113. $x + 2y \leq 6$ or $2x + y \leq 6$

115. $y \geq nx + b \quad (n < 0,\ b > 0)$
$y \leq mx + b \quad (m > 0,\ b > 0)$

$y \geq nx + b$
$y \leq mx + b$

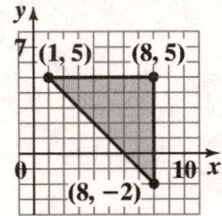

116. **a.**
$$x + y \geq 6$$
$$x \leq 8$$
$$y \leq 5$$

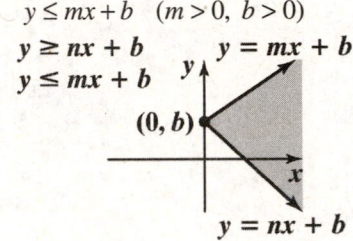

b. The corner points are $(1,5)$, $(8,5)$, and $(8,-2)$.

c. At $(1,5)$, $3x + 2y = 3(1) + 2(5) = 13$.
At $(8,5)$, $3x + 2y = 3(8) + 2(5) = 34$.
At $(8,-2)$, $3x + 2y = 3(8) + 2(-2) = 20$.

117. **a.**
$$x \geq 0$$
$$y \geq 0$$
$$3x - 2y \leq 6$$
$$y \leq -x + 7$$

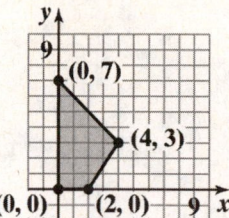

b. The corner points are $(0,0)$, $(2,0)$, $(4,3)$, and $(0,7)$.

c. At $(0,0)$, $2x + 5y = 2(0) + 5(0) = 0$.

At $(2,0)$, $2x + 5y = 2(2) + 5(0) = 4$.

At $(4,3)$, $2x + 5y = 2(4) + 5(3) = 23$.

At $(0,7)$, $2x + 5y = 2(0) + 5(7) = 35$.

118. $20x + 10y \leq 80,000$

Section 7.6

Check Point Exercises

1. The total profit is 25 times the number of bookshelves, x, plus 55 times the number of desks, y. The objective function is $z = 25x + 55y$

2. Not more than a total of 80 bookshelves and desks can be manufactured per day. This is represented by the inequality $x + y \leq 80$.

3. Objective function: $z = 25x + 55y$

Constraints: $x + y \leq 80$
$$30 \leq x \leq 80$$
$$10 \leq y \leq 30$$

4. Graph the constraints and find the corners, or vertices, of the region of intersection.

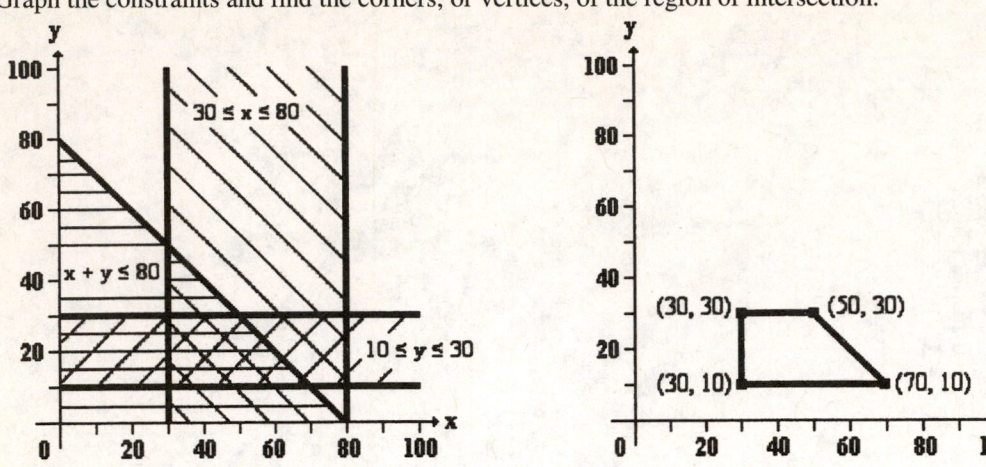

Find the value of the objective function at each corner of the graphed region.

Corner (x, y)	Objective Function $z = 25x + 55y$ z
(30, 10)	$z = 25(30) + 55(10)$ $= 750 + 550 = 1300$
(30, 30)	$z = 25(30) + 55(30)$ $= 750 + 1650 = 2400$
(50, 30)	$z = 25(50) + 55(30)$ $= 1250 + 1650 = 2900 \leftarrow$ Maximum
(70, 10)	$z = 25(70) + 55(10)$ $= 1750 + 550 = 2300$

The maximum value of z is 2900 and it occurs at the point (50, 30).
In order to maximize profit, 50 bookshelves and 30 desks must be produced each day for a profit of $2900.

5. objective function: $z = 3x + 5y$

constraints: $x \geq 0, \ y \geq 0$

$$x + y \geq 1$$
$$x + y \leq 6$$

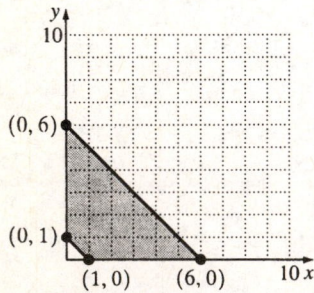

Evaluate the objective function at the four vertices of the region shown:

$(1, 0): 3(1) + 5(0) = 3$

$(0, 1): 3(0) + 5(1) = 5$

$(0, 6): 3(0) + 5(6) = 30$

$(6, 0): 3(6) + 5(0) = 18$

The maximum value of z is 30 and this occurs when $x = 0$ and $y = 6$.

Exercise Set 7.6

1. $z = 5x + 6y$

(1, 2): 5(1) + 6(2) = 5 + 12 = 17
(2, 10): 5(2) + 6(10) = 10 + 60 = 70
(7, 5): 5(7) + 6(5) = 35 + 30 = 65
(8, 3): 5(8) + 6(3) = 40 + 18 = 58
The maximum value is $z = 70$; the minimum value is $z = 17$.

3. $z = 40x + 50y$

(0, 0): 40(0) + 50(0) = 0 + 0 = 0
(0, 8): 40(0) + 50(8) = 0 + 400 = 400
(4, 9): 40(4) + 50(9) = 160 + 450 = 610
(8, 0): 40(8) + 50(0) = 320 + 0 = 320
The maximum value is $z = 610$; the minimum value is $z = 0$.

5. $z = 3x + 2y$

$x \geq 0, y \geq 0$

$2x + y \leq 8$

$x + y \geq 4$

a.

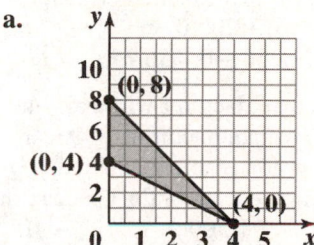

b. $(0,8): z = 3(0) + 2(8) = 16$

$(0,4): z = 3(0) + 2(4) = 8$

$(4,0): z = 3(4) + 2(0) = 12$

c. The maximum value is 16 at $x = 0$ and $y = 8$.

7. $z = 4x + y$

$x \geq 0, y \geq 0$

$2x + 3y \leq 12$

$x + y \geq 3$

a.

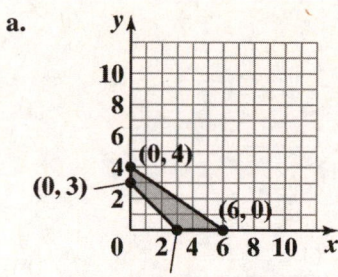

b. (0, 4): $z = 4(0) + 4 = 4$

(0, 3): $z = 4(0) + 3 = 3$

(3, 0): $z = 4(3) + 0 = 12$

(6, 0): $z = 4(6) + 0 = 24$

c. The maximum value is 24 at $x = 6$ and $y = 0$.

9. $z = 3x - 2y$

$1 \leq x \leq 5$

$y \geq 2$

$x - y \geq -3$

a.

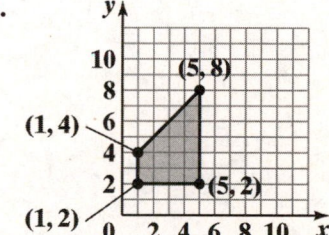

b. (1, 2): $z = 3(1) - 2(2) = -1$

(1, 4): $z = 3(1) - 2(4) = -5$

(5, 8): $z = 3(5) - 2(8) = -1$

(5, 2): $z = 3(5) - 2(2) = 11$

c. Maximum value is 11 at $x = 5$ and $y = 2$.

11. $z = 4x + 2y$

$x \geq 0, y \geq 0$

$2x + 3y \leq 12$

$3x + 2y \leq 12$

$x + y \geq 2$

a.

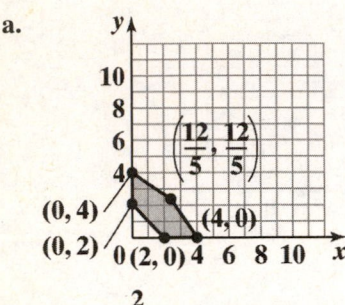

b. $(0, 4)$: $z = 4(0) + 2(4) = 8$

$(0, 2)$: $z = 4(0) + 2(2) = 4$

$(2, 0)$: $z = 4(2) + 2(0) = 8$

$(4, 0)$: $z = 4(4) + 2(0) = 16$

$\left(\dfrac{12}{5}, \dfrac{12}{5}\right)$: $z = 4\left(\dfrac{12}{5}\right) + 2\left(\dfrac{12}{5}\right)$

$= \dfrac{48}{5} + \dfrac{24}{5} = \dfrac{72}{5}$

c. The maximum value is 16 at $x = 4$ and $y = 0$.

13. $z = 10x + 12y$

$x \geq 0, y \geq 0$

$x + y \leq 7$

$2x + y \leq 10$

$2x + 3y \leq 18$

a.

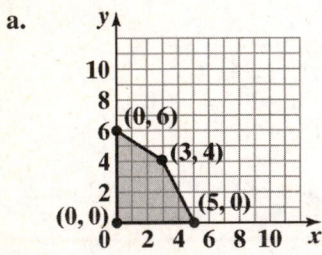

b. $(0, 6)$: $z = 10(0) + 12(6) = 72$

$(0, 0)$: $z = 10(0) + 12(0) = 0$

$(5, 0)$: $z = 10(5) + 12(0) = 50$

$(3, 4)$: $z = 10(3) + 12(4)$

$= 30 + 48 = 78$

c. The maximum value is 78 at $x = 3$ and $y = 4$.

15. a. $z = 125x + 200y$

b. $x \leq 450$

$y \leq 200$

$600x + 900y \leq 360{,}000$

c. Simplify the third inequality by dividing by 300 to get $2x + 3y \leq 1200$.

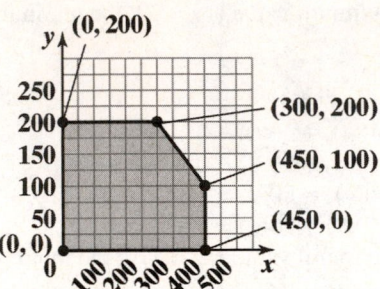

d. $(0, 0)$: $125(0) + 200(0) = 0 + 0 = 0$

$(0, 200)$: $125(0) + 200(200)$

$= 0 + 40{,}000 = 40{,}000$

$(300, 200)$: $125(300) + 200(200)$

$= 37{,}500 + 40{,}000 = 77{,}500$

$(450, 100)$: $125(450) + 200(100)$

$= 56{,}250 + 20{,}000 = 76{,}250$

$(450, 0)$: $125(450) + 200(0)$

$= 56{,}250 + 0 = 56{,}250$

e. The television manufacturer will make the greatest profit by manufacturing <u>300</u> rear-projection televisions each month and <u>200</u> plasma televisions each month. The maximum monthly profit is <u>$77,500</u>.

17. Let $x =$ number of model *A* bicycles and $y =$ number of model *B* bicycles.

The constraints are

$5x + 4y \leq 200$

$2x + 3y \leq 108$

Graph these inequalities in the first quadrant, since x and y cannot be negative.

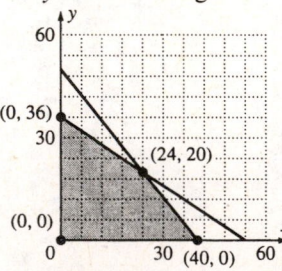

The quantity to be maximized is the profit, which is $25x + 15y$.

$(0, 0)$: $25(0) + 15(0) = 0 + 0 = 0$

$(0, 36)$: $25(0) + 15(36) = 0 + 540 = 540$

$(24, 20)$: $25(24) + 15(20) = 600 + 300 = 900$

$(40, 0)$: $25(40) + 15(0) = 1000 + 0 = 1000$

40 model *A* bicycles and no model *B* bicycles should be produced.

19. Let x = the number of cartons of food and y = the number of cartons of clothing.
The constraints are:
$20x + 10y \le 8,000$ or $2x + y \le 8000$
$50x + 20y \le 19,000$ or $5x + 2y \le 1900$
Graph these inequalities in the first quadrant, since x and y cannot be negative.

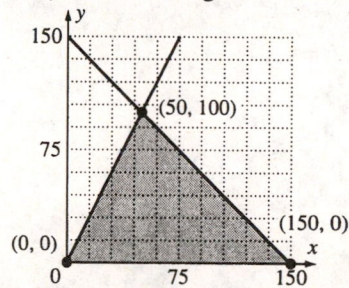

The quantity to be maximized is the number of people helped, which is $12x + 5y$.
(0, 0): $12(0) + 5(0) = 0 + 0 = 0$
(0, 800): $12(0) + 5(800) = 0 + 4000 = 4000$
(300, 200): $12(300) + 5(200) = 4600$
(380, 0): $12(380) + 5(0) = 4500$
300 cartons of food and 200 cartons of clothing should be shipped. This will help 4600 people.

21. Let x = number of students attending and y = number of parents attending.
The constraints are
$x + y \le 150$
$\quad 2x \ge y$
or
$\ x + y \le 150$
$2x - y \ge 0$
Graph these inequalities in the first quadrant, since x and y cannot be negative.

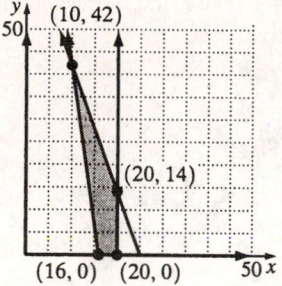

The quantity to be maximized is the amount of money raised, which is $x + 2y$.
(0, 0): $0 + 2(0) = 0 + 0 = 0$
(50, 100): $50 + 2(100) = 50 + 200 = 250$
(150, 0): $150 + 2(0) = 150 + 0 = 150$
50 students and 100 parents should attend.

23. Let x = number of Boeing 727s, y = number of Falcon 20s.

Maximize $z = x + y$ with the following constraints:
$1400x + 500y \le 35,000$ or $14x + 5y \le 350$
$42,000x + 6000y \ge 672,000$ or $7x + y \ge 112$
$x \le 20$
$x \ge 0, y \ge 0$

$(16, 0): z = 16$
$(20, 0): z = 20$
$(20, 14): z = 34$
$(10, 42): z = 52$

Federal Express should have purchased 10 Boeing 727s and 42 Falcon 20s.

25. – 27. Answers may vary.

29. does not make sense; Explanations will vary. Sample explanation: Solving a linear programming problem does not require graphing the objective function.

31. makes sense

33. Let x = amount invested in stocks and
y = amount invested in bonds.
The constraints are:
$$x + y \leq 10{,}000$$
$$y \geq 3000$$
$$x \geq 2000$$
$$y \geq x$$
Graph these inequalities in the first quadrant, since x and y cannot be negative.

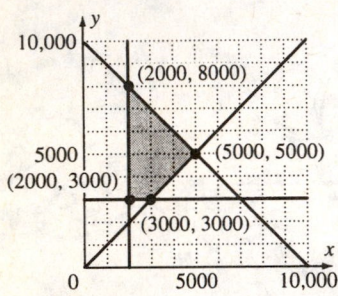

The quantity to be maximized is the return on the investment, which is $0.12x + 0.08y$.
(2000, 3000):
$0.12(2000) + 0.08(3000) = 240 + 240 = 480$
(2000, 8000):
$0.12(2000) + 0.08(8000) = 240 + 640 = 880$
(5000, 5000):
$0.12(5000) + 0.08(5000) = 600 + 400 = 1000$
(3000, 3000):
$0.12(3000) + 0.08(3000) = 360 + 240 = 600$
The greatest return occurs when $5000 is invested in stocks and $5000 is invested in bonds.

35. Answers may vary.

37. Back-substitute $z = 5$ to find y in the second equation.
$$y + 2z = 13$$
$$y + 2(5) = 13$$
$$y + 10 = 13$$
$$y = 3$$
Back-substitute to find x in the first equation.
$$x + y + 2z = 19$$
$$x + 3 + 2(5) = 19$$
$$x + 13 = 19$$
$$x = 6$$
The solution set is $\{(6, 3, 5)\}$.
Explanations may vary.

38. Back-substitute $z = 3$ to find y in the third equation.
$$y - z = 1$$
$$y - 3 = 1$$
$$y = 4$$
Back-substitute to find x in the second equation.
$$x - \frac{1}{3}y + z = \frac{8}{3}$$
$$x - \frac{1}{3}(4) + 3 = \frac{8}{3}$$
$$x - \frac{4}{3} + \frac{9}{3} = \frac{8}{3}$$
$$x + \frac{5}{3} = \frac{8}{3}$$
$$x = \frac{3}{3}$$
$$x = 1$$
Back-substitute to find w in the first equation.
$$w - x + 2y - 2z = -1$$
$$w - 1 + 2(4) - 2(3) = -1$$
$$w - 1 + 8 - 6 = -1$$
$$w + 1 = -1$$
$$w = -2$$
The solution set is $\{(-2, 1, 4, 3)\}$.
Explanations may vary.

39. $\begin{bmatrix} 1 & 2 & -1 \\ 1(-4)+4 & 2(-4)+(-3) & -1(-4)+(-15) \end{bmatrix}$
$= \begin{bmatrix} 1 & 2 & -1 \\ 0 & -11 & -11 \end{bmatrix}$

Chapter 7 Review Exercises

1.
$$y = 4x + 1$$
$$3x + 2y = 13$$
Substitute $4x + 1$ for y in the second equation:
$$3x + 2(4x + 1) = 13$$
$$3x + 8x + 2 = 13$$
$$11x = 11$$
$$x = 1$$
$$y = 4(1) + 1 = 5$$
The solution set is $\{(1, 5)\}$.

2.
$$x + 4y = 14$$
$$2x - y = 1$$
Multiply the second equation by 4 and add to the first equation.
$$x + 4y = 14$$
$$\underline{8x - 4y = 4}$$
$$9x = 18$$
$$x = 2$$
$$2(2) - y = 1$$
$$-y = -3$$
$$y = 3$$
The solution set is $\{(2, 3)\}$.

3.
$$5x + 3y = 1$$
$$3x + 4y = -6$$
Multiply the first equation by 4 and the second equation by -3.
Then add.
$$20x + 12y = 4$$
$$\underline{-9x - 12y = 18}$$
$$11x = 22$$
$$x = 2$$
$$5(2) + 3y = 1$$
$$3y = -9$$
$$y = -3$$
The solution set is $\{(2, -3)\}$.

4.
$$2y - 6x = 7$$
$$3x - y = 9$$
The second equation can be written as $y = 3x - 9$.
Substitute:
$$2(3x - 9) - 6x = 7$$
$$6x - 18 - 6x = 7$$
$$-18 = 7$$
Since this is false, the system has no solution.
The solution set is the empty set, $\varnothing$.

5.
$$4x - 8y = 16$$
$$3x - 6y = 12$$
Divide the first equation by 4 and the second equation by 3.
$$x - 2y = 4$$
$$x - 2y = 4$$
Since these equations are identical, the system has an infinite number of solutions.
The solution set is $\{(x, y) \mid 3x - 6y = 12\}$.

6. **a.** $C(x) = 60,000 + 200x$

    b. $R(x) = 450x$

    c. $450x = 60000 + 200x$
$$250x = 60000$$
$$x = 240$$
$$450(240) = 108,000$$
The company must make 240 desks at a cost of $108,000 to break even.

7. Let x = the selling price for Klint's work.
Let y = the selling price for Picasso's work.
$$x + y = 239$$
$$x - y = 31$$
Add the equations to eliminate y and solve for x.
$$x + y = 239$$
$$\underline{x - y = \ \ 31}$$
$$2x \quad\ \ = 270$$
$$x = 135$$
Back-substitute to find y.
$$x + y = 239$$
$$135 + y = 239$$
$$y = 104$$
Klint's work sold for $135 million and Picasso's work sold for $104 million.

8. **a.** Answers will vary. Approximate point is (2004,180). This means that in 2004 the number of cellphones and land-lines were both 180 million.

b. $y = 19.8x + 98$

c. Using substitution,
$$4.3x + y = 198$$

$$4.3x + \overbrace{(19.8x + 98)}^{y} = 198$$
$$4.3x + 19.8x + 98 = 198$$
$$24.1x + 98 = 198$$
$$24.1x = 100$$
$$x \approx 4$$

The number of cellphone and land-line customers will be they same 4 years after 2000, or 2004.
$$y = 19.8x + 98$$
$$y = 19.8(4) + 98$$
$$y = 177.2$$
$$y \approx 180$$

The number of customers that each will have in 2004 is about 180 million.

d. The models describe the point of intersection quite well.

9. Let x = the cost of the hotel
y = the cost of the car
$3x + 2y = 360$
$4x + 3y = 500$
Solve the system.
$$12x + 8y = 1440$$
$$-12x - 9y = -1500$$
$$-y = -60$$
$$y = 60$$

$$3x + 2(60) = 360$$
$$3x = 240$$
$$x = 80$$
The room costs $80 a day and the car rents for $60 a day.

10. x = ml of 34% solution
y = ml of 4% solution
$$x + y = 100$$
$$0.34x + 0.04y = 0.07(100)$$

$$y = 100 - x$$

$$0.34x + 0.04(100 - x) = 7$$
$$0.34x + 4 - 0.04x = 7$$
$$0.3x = 3$$
$$x = 10$$

$$y = 100 - 10 = 90$$
Mix 10 ml of 34% solution and 90 ml of 4% solution.

11. x = average velocity of the plane
y = average velocity of the wind

Velocity	Time	Distance
$x + y$	3	$3(x + y)$
$x - y$	4	$4(x - y)$

$$3(x + y) = 2160$$
$$4(x - y) = 2160$$

$$x + y = 720$$
$$x - y = 540$$

$$2x = 1260$$
$$x = 630$$

$$x + y = 720$$
$$630 + y = 720$$
$$y = 90$$
The average velocity of the plane is 630 mph and the wind is 90 mph.

12. $2x - y + z = 1$ (1)
$3x - 3y + 4z = 5$ (2)
$4x - 2y + 3z = 4$ (3)
Eliminate y from (1) and (2) by multiplying (1) by -3 and adding the result to (2).

$$-6x + 3y - 3z = -3$$
$$\underline{3x - 3y + 4z = 5}$$
$$-3x + z = 2 \quad (4)$$

Eliminate y from (1) and (3) by multiplying (1) by -2 and adding the result to (3).

$$-4x + 2y - 2z = -2$$
$$\underline{4x - 2y + 3z = 4}$$
$$z = 2$$

Substituting $z = 2$ into (4), we get:
$$-3x + 2 = 2$$
$$-3x = 0$$
$$x = 0$$
Substituting $x = 0$ and $z = 2$ into (1), we have:
$$2(0) - y + 2 = 1$$
$$-y = -1$$
$$y = 1$$
The solution set is $\{(0, 1, 2)\}$.

13. $x + 2y - z = 5$ (1)
$2x - y + 3z = 0$ (2)
$2y + z = 1$ (3)
Eliminate x from (1) and (2) by multiplying (1) by -2 and adding the result to (2).

$$-2x - 4y + 2z = -10$$
$$\underline{2x - y + 3z = 0}$$
$$-5y + 5z = -10$$
$$y - z = 2 \quad (4)$$

Adding (3) and (4), we get:
$$2y + z = 1$$
$$\underline{y - z = 2}$$
$$3y = 3$$
$$y = 1$$
Substituting $y = 1$ into (3), we have:
$$2(1) + z = 1$$
$$z = -1$$
Substituting $y = 1$ and $z = -1$ into (1), we obtain:
$$x + 2(1) - (-1) = 5$$
$$x + 3 = 5$$
$$x = 2$$
The solution set is $\{(2, 1, -1)\}$.

14. $y = ax^2 + bx + c$
$(1, 4): 4 = a + b + c$ (1)
$(3, 20): 20 = 9a + 3b + c$ (2)
$(-2, 25): 25 = 4a - 2b + c$ (3)
Multiply (1) by -1 and add to (2).

$$20 = 9a + 3b + c$$
$$\underline{-4 = -a - b - c}$$
$$16 = 8a + 2b$$
$$8 = 4a + b$$
$$8 = 4a + b \quad (4)$$

Multiply (1) by -1 and add to (3).

$$25 = 4a - 2b + c$$
$$\underline{-4 = -a - b - c}$$
$$21 = 3a - 3b$$
$$7 = a - b \quad (5)$$

Add (4) and (5).

$$8 = 4a + b$$
$$\underline{7 = a - b}$$
$$15 = 5a$$
$$a = 3$$

$$8 = 4(3) + b$$
$$b = -4$$

$$3 - 4 + c = 4$$
$$c = 5$$
Hence, the quadratic function is $y = 3x^2 - 4x + 5$.

15. Let x = average debt for the $18 - 29$ age group in the U.S.
Let y = average debt for the $30 - 39$ age group in the U.S.
Let z = average debt for the $40 - 49$ age group in the U.S.

$$x + y + z = 44,200$$
$$y - x = 8100$$
$$z - y = 3100$$

Solve the second equation for x.
$$y - x = 8100$$
$$-x = -y + 8100$$
$$x = y - 8100$$

Solve the third equation for z.
$$z - y = 3100$$
$$z = y + 3100$$

Substitute the expressions for x and z into the first equation and solve for y.

$$x + y + z = 44,200$$

$$(\overbrace{y-8100}^{x}) + y + (\overbrace{y+3100}^{z}) = 44,200$$

$$y-8100+y+y+3100 = 44,200$$

$$3y-5000 = 44,200$$

$$3y = 49,200$$

$$y = 16,400$$

Back-substitute to solve for x and z.

$$x = y-8100$$
$$= 16,400-8100$$
$$= 8300$$

$$z = y+3100$$
$$= 16,400+3100$$
$$= 19,500$$

The average debt for the $18-29$ age group in the U.S. is \$8300, for the $30-39$ age group is \$16,400, and for the $40-49$ age group is \$19,500.

16. $\dfrac{x}{(x-3)(x+2)} = \dfrac{A}{x-3} + \dfrac{B}{x+2}$

$$x = A(x+2)+B(x-3)$$
$$= (A+B)x+(2A-3B)$$

$$A+B=1$$
$$2A-3B=0$$

Multiply first equation by 3, then add to second equation.

$$3A+3B=3$$
$$\underline{2A-3B=0}$$
$$5A=3$$

$$A=\frac{3}{5},\ B=\frac{2}{5}$$

$$\frac{x}{(x-3)(x+2)} = \frac{3}{5(x-3)} + \frac{2}{5(x+2)}$$

17. $\dfrac{11x-2}{x^2-x-12} = \dfrac{11x-2}{(x-4)(x+3)} = \dfrac{A}{x-4} + \dfrac{B}{x+3}$

$$11x-2 = A(x+3)+B(x-4)$$
$$= Ax+3A+Bx-4B$$
$$= (A+B)x+(3A-4B)$$

$$A+B=11$$
$$3A-4B=-2$$

Multiply first equation by 4, then add to second equation.

$$3A-4B=-2$$
$$\underline{4A+4B=44}$$
$$7A=42$$

$$A=6,\ B=5$$

$$\frac{11x-2}{x^2-x-12} = \frac{6}{x-4} + \frac{5}{x+3}$$

18. $\dfrac{4x^2-3x-4}{x^3+x^2-2x} = \dfrac{4x^2-3x-4}{x(x+2)(x-1)}$

$$= \frac{A}{x} + \frac{B}{x+2} + \frac{C}{x-1}$$

$$4x^2-3x-4 = A(x+2)(x-1)+Bx(x-1)+Cx(x+2)$$
$$= A(x^2+x-2)+Bx^2-Bx+Cx^2+2Cx$$
$$= Ax^2+Ax-2A+Bx^2-Bx+Cx^2+2Cx$$
$$= (A+B+C)x^2+(A-B+2C)x-2A$$

$$A+B+C=4$$
$$A-B+2C=-3$$
$$-2A=-4$$
$$A=2$$
$$B+C=2$$
$$\underline{-B+2C=-5}$$
$$3C=-3$$
$$C=-1$$
$$B-1=2$$
$$B=3$$

$$\frac{4x^2-3x-4}{x^3+x^2-2x} = \frac{2}{x} + \frac{3}{x+2} - \frac{1}{x-1}$$

19.

$$\frac{2x+1}{(x-2)^2} = \frac{A}{x-2} + \frac{B}{(x-2)^2}$$

$$2x+1 = A(x-2) + B = Ax - 2A + B$$

$$A = 2$$

$$-2A + B = 1$$

$$-2(2) + B = 1$$

$$B = 5$$

$$\frac{2x+1}{(x-2)^2} = \frac{2}{x-2} + \frac{5}{(x-2)^2}$$

20.

$$\frac{2x-6}{(x-1)(x-2)^2} = \frac{A}{x-1} + \frac{B}{x-2} + \frac{C}{(x-2)^2}$$

$$2x-6 = A(x-2)^2 + B(x-1)(x-2) + C(x-1)$$

$$= A(x^2 - 4x + 4) + B(x^2 - 3x + 2) + C(x-1)$$

$$= Ax^2 - 4Ax + 4A + Bx^2 - 3Bx + 2B + Cx - C$$

$$= (A+B)x^2 + (-4A - 3B + C)x + (4A + 2B - C)$$

$$A + B = 0$$

$$-4A - 3B + C = 2$$

$$\underline{4A + 2B - C = -6}$$

$$-B = -4$$

$$B = 4$$

$$A = -4$$

$$4(-4) + 2(4) - C = -6$$

$$-16 + 8 - C = -6$$

$$-C - 8 = -6$$

$$-C = 2$$

$$C = -2$$

$$\frac{2x-6}{(x-1)(x-2)^2} = -\frac{4}{x-1} + \frac{4}{x-2} - \frac{2}{(x-2)^2}$$

21.

$$\frac{3x}{(x-2)(x^2+1)} = \frac{A}{x-2} + \frac{Bx+C}{x^2+1}$$

$$3x = A(x^2+1) + (Bx+C)(x-2)$$

$$= Ax^2 + A + Bx^2 - 2Bx + Cx - 2C$$

$$= (A+B)x^2 + (-2B+C)x - (2C-A)$$

$$A + B = 0$$

$$-2B + C = 3$$

$$2C - A = 0$$

$$A = 2C$$

$$B + 2C = 0$$

$$\underline{4B - 2C = -6}$$

$$5B = -6$$

$$B = -\frac{6}{5}$$

$$A = \frac{6}{5}$$

$$C = \frac{6}{10} = \frac{3}{5}$$

$$\frac{3x}{(x-2)(x^2+1)} = \frac{6}{5(x-2)} + \frac{-6x+3}{5(x^2+1)}$$

22.

$$\frac{7x^2 - 7x + 23}{(x-3)(x^2+4)} = \frac{A}{x-3} + \frac{Bx+C}{x^2+4}$$

$$7x^2 - 7x + 23 = A(x^2+4) + (Bx+C)(x-3)$$

$$= Ax^2 + 4A + Bx^2 - 3Bx + Cx - 3C$$

$$= (A+B)x^2 + (-3B+C)x + (4A-3C)$$

$$A + B = 7$$

$$-3B + C = -7$$

$$4A - 3C = 23$$

$$3A + 3B = 21$$

$$\underline{-3B + C = -7}$$

$$3A + C = 14$$

$$9A + 3C = 42$$

$$\underline{4A - 3C = 23}$$

$$13A = 65$$

$$A = 5$$

$$5 + B = 7$$

$$B = 7 - 5 = 2$$

$$-3(2) + C = -7$$

$$C = -7 + 6 = -1$$

$$\frac{7x^2 - 7x + 23}{(x-3)(x^2+4)} = \frac{5}{x-3} + \frac{2x-1}{x^2+4}$$

23. $\dfrac{x^3}{(x^2+4)^2} = \dfrac{Ax+B}{x^2+4} + \dfrac{Cx+D}{(x^2+4)^2}$

$x^3 = (Ax+B)(x^2+4)+Cx+D$

$\quad = Ax^3+4Ax+Bx^2+4B+Cx+D$

$\quad = Ax^3+Bx^2+(4A+C)x+(4B+D)$

$\qquad A=1$

$\qquad B=0$

$4A+C=0$

$4B+D=0$

$\qquad C=-4$

$0+D=0, D=0$

$\dfrac{x^2}{(x^2+4)^2} = \dfrac{x}{x^2+4} - \dfrac{4x}{(x^2+4)^2}$

24. $\dfrac{4x^3+5x^2+7x-1}{(x^2+x+1)^2} = \dfrac{Ax+B}{x^2+x+1} + \dfrac{Cx+D}{(x^2+x+1)^2}$

$4x^3+5x^2+7x-1$

$= (Ax+B)(x^2+x+1)+Cx+D$

$= Ax^3+Ax^2+Ax+Bx^2+Bx+B+Cx+D$

$= Ax^3+(A+B)x^2(A+B+C)x+(B+D)$

$\qquad A=4$

$\qquad A+B=5$

$\quad A+B+C=7$

$\qquad B+D=-1$

$\quad 4+B=5, B=1$

$4+1+C=7, C=2$

$\quad 1+D=-1, D=-2$

$\dfrac{4x^3+5x^2+7x-1}{(x^2+x+1)^2} = \dfrac{4x+1}{x^2+x+1} + \dfrac{2x-2}{(x^2+x+1)^2}$

25. $\qquad 5y=x^2-1$

$\qquad x-y=1$

$\qquad\quad y=x-1$

$\quad 5(x-1)=x^2-1$

$\quad 5x-5=x^2-1$

$x^2-5x+4=0$

$(x-4)(x-1)=0$

$\qquad\quad x=4,1$

If $x=4, y=4-1=3$.

If $x=1, y=1-1=0$.

The solution set is $\{(4,3),(1,0)\}$.

26. $\quad y=x^2+2x+1$

$\qquad x+y=1$

$\qquad\quad y=1-x$

$\quad 1-x=x^2+2x+1$

$\qquad x^2+3x=0$

$\qquad x(x+3)=0$

$\qquad\quad x=0,-3$

If $x=0, y=1-0=1$.

If $x=-3, y=1-(-3)=4$.

The solution set is $\{(0,1),(-3,4)\}$.

27. $\qquad x^2+y^2=2$

$\qquad\quad x+y=0$

$\qquad\qquad x=-y$

$\quad (-y)^2+y^2=2$

$\qquad\quad 2y^2=2$

$\qquad\quad y^2=1$

$\qquad\qquad y=1,-1$

If $y=1, x=-1$.

If $y=-1, x=1$.

The solution set is $\{(1,-1),(-1,1)\}$.

28. $\quad 2x^2+y^2=24$

$\qquad x^2+y^2=15$

$\quad 2x^2+y^2=24$

$\quad \underline{-x^2-y^2=-15}$

$\qquad\quad x^2=9$

$\qquad\quad x=3,-3$

If $x=3, 3^2+y^2=15, y^2=6$ and $y=\pm\sqrt{6}$.

If $x=-3, y=\pm\sqrt{6}$.

The solution set is

$\left\{\left(3,\sqrt{6}\right),\left(3,-\sqrt{6}\right),\left(-3,\sqrt{6}\right),\left(-3,-\sqrt{6}\right)\right\}$.

29. $\quad xy-4=0$

$\qquad y-x=0$

$\qquad\quad y=x$

$\qquad\quad xy=4$

$\qquad\quad x^2=4$

$\qquad\quad x=2,-2$

If $x=2, y=2$.

If $x=-2, y=-2$.

The solution set is $\{(2,2),(-2,-2)\}$.

694

30.

$$y^2 = 4x$$
$$x - 2y + 3 = 0$$
$$x = \frac{y^2}{4}$$
$$\frac{y^2}{4} - 2y + 3 = 0$$
$$y^2 - 8y + 12 = 0$$
$$(y - 6)(y - 2) = 0$$
$$y = 6, 2$$

If $y = 6, x = \frac{36}{4} = 9$.

If $y = 2, x = \frac{4}{4} = 1$.

The solution set is $\{(9, 6), (1, 2)\}$.

31.

$$x^2 + y^2 = 10$$
$$y = x + 2$$
$$x^2 + (x + 2)^2 = 10$$
$$x^2 + x^2 + 4x + 4 - 10 = 0$$
$$2x^2 + 4x - 6 = 0$$
$$x^2 + 2x - 3 = 0$$
$$(x + 3)(x - 1) = 0$$
$$x = -3, 1$$

If $x = -3, y = -3 + 2 = -1$.

If $x = 1, y = 1 + 2 = 3$.

The solution set is $\{(-3, -1), (1, 3)\}$.

32.

$$xy = 1$$
$$y = 2x + 1$$
$$x(2x + 1) = 1$$
$$2x^2 + x - 1 = 0$$
$$(2x - 1)(x + 1) = 0$$
$$x = \frac{1}{2}, -1$$

If $x = \frac{1}{2}, y = 2\left(\frac{1}{2}\right) + 1 = 2$.

If $x = -1, y = 2(-1) + 1 = -1$.

The solution set is $\left\{\left(\frac{1}{2}, 2\right), (-1, -1)\right\}$.

33.

$$x + y + 1 = 0$$
$$x^2 + y^2 + 6y - x = -5$$
$$x = -y - 1$$
$$(-y - 1)^2 + y^2 + 6y - (-y - 1) + 5 = 0$$
$$y^2 + 2y + 1 + y^2 + 6y + y + 1 + 5 = 0$$
$$2y^2 + 9y + 7 = 0$$
$$(2y + 7)(y + 1) = 0$$
$$y = -\frac{7}{2}, -1$$

If $y = -\frac{7}{2}, x = \frac{7}{2} - 1 = \frac{5}{2}$.

If $y = -1, x = 1 - 1 = 0$.

The solution set is $\left\{\left(\frac{5}{2}, -\frac{7}{2}\right), (0, -1)\right\}$.

34.

$$x^2 + y^2 = 13$$
$$x^2 - y = 7$$
$$x^2 + y^2 = 13$$
$$\underline{-x^2 + y = -7}$$
$$y^2 + y = 6$$
$$y^2 + y - 6 = 0$$
$$(y + 3)(y - 2) = 0$$
$$y = -3, 2$$

If $y = -3, x^2 + 3 = 7$

$$x^2 = 4, x = 2, -2$$

If $y = 2, x^2 - 2 = 7, x^2 = 9, x = 3, -3$.

The solution set is $\{(2, -3), (-2, -3), (3, 2), (-3, 2)\}$.

35.

$$2x^2 + 3y^2 = 21$$
$$3x^2 - 4y^2 = 23$$
$$8x^2 + 12y^2 = 84$$
$$\underline{9x^2 - 12y^2 = 69}$$
$$17x^2 = 153$$
$$x^2 = \frac{153}{17} = 9$$
$$x = 3, -3$$

If $x = 3, 2(3)^2 + 3y^2 = 21$.

$$3y^2 = 21 - 18 = 3$$
$$y^2 = 1, y = 1, -1$$

If $x = -3, y = 1, -1$.

The solution set is $\{(3, 1), (3, -1), (-3, 1), (-3, -1)\}$.

36.
$$2L + 2W = 26$$
$$LW = 40$$
$$L = \frac{40}{W}$$
$$2\left(\frac{40}{W}\right) + 2W = 26$$
$$\frac{80}{W} + 2W = 26$$
$$80 + 2W^2 = 26W$$
$$2W^2 - 26W + 80 = 0$$
$$W^2 - 13W + 40 = 0$$
$$(W - 8)(W - 5) = 0$$
$$W = 8, 5$$

If $W = 5, L = \dfrac{40}{5} = 8$

The dimensions are 8 m by 5 m.

37.
$$xy = 6$$
$$y = \frac{6}{x}$$
$$2x + y = 8$$
$$2x + \frac{6}{x} = 8$$
$$2x^2 + 6 = 8x$$
$$2x^2 - 8x + 6 = 0$$
$$x^2 - 4x + 3 = 0$$
$$(x - 1)(x - 3) = 0$$
$$x = 1, 3$$

If $x = 1$, $y = 6$.
If $x = 3$, $y = 2$.
The solution set is $\{(1, 6), (3, 2)\}$.

38.
$$x^2 + y^2 = 2900$$
$$4x + 2y = 240$$
$$2x + y = 120$$
$$y = 120 - 2x$$
$$x^2 + (120 - 2x)^2 = 2900$$
$$x^2 + 14,400 - 480x + 4x^2 - 2900 = 0$$
$$5x^2 - 480x + 11,500 = 0$$
$$x^2 - 96x + 2300 = 0$$
$$(x - 46)(x - 50) = 0$$
$$x = 46, 50$$

If $x = 46$, $y = 120 - 2(46) = 28$.
If $x = 50$, $y = 120 - 2(50) = 20$.
$x = 46$ ft and $y = 28$ ft or
$x = 50$ ft and $y = 20$ ft

39.

$3x - 4y > 12$

40.

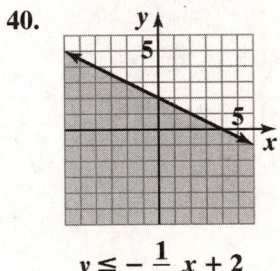

$y \leq -\dfrac{1}{2}x + 2$

41.

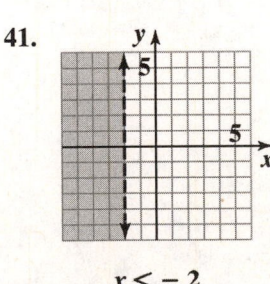

$x < -2$

42.

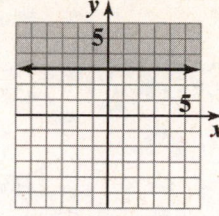

$y \geq 3$

43.

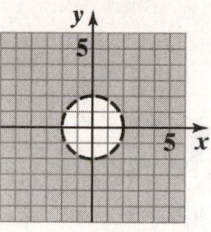

$x^2 + y^2 > 4$

44.

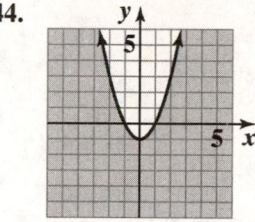

$y \leq x^2 - 1$

45.

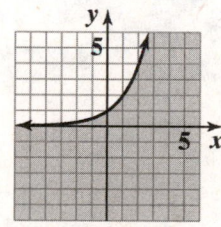

$y \leq 2^x$

46.

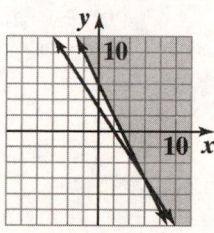

$3x + 2y \geq 6$
$2x + y \geq 6$

47.

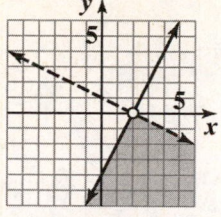

$2x - y \geq 4$
$x + 2y < 2$

48.

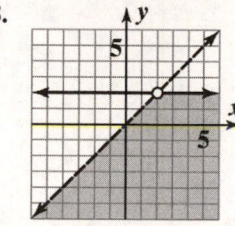

$y < x$
$y \leq 2$

49.

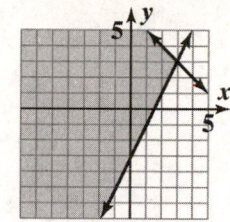

$x + y \leq 6$
$y \geq 2x - 3$

50.

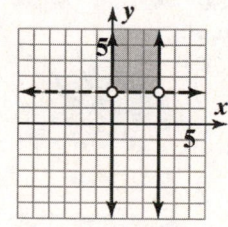

$0 \leq x \leq 3$
$y > 2$

51. No solution

52.

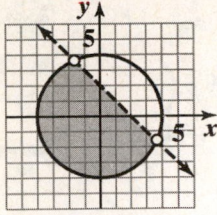

$x^2 + y^2 \le 16$
$x + y < 2$

53.

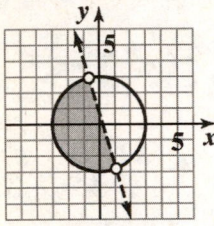

$x^2 + y^2 \le 9$
$y < -3x + 1$

54.

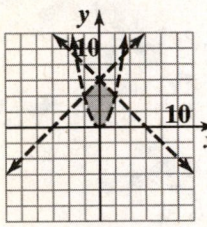

$y > x^2$
$x + y < 6$
$y < x + 6$

55.

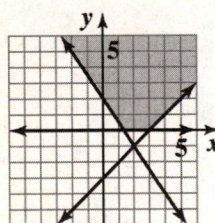

$y \ge 0$
$3x + 2y \ge 4$
$x - y \le 3$

56. $z = 2x + 3y$

$(2, 2) : z = 2(2) + 3(2) = 10$
$(4, 0) : z = 2(4) + 3(0) = 8$
$\left(\dfrac{1}{2}, \dfrac{1}{2}\right) : z = 2\left(\dfrac{1}{2}\right) + 3\left(\dfrac{1}{2}\right) = \dfrac{5}{2}$
$(1, 0) : z = 2(1) + 3(0) = 2$

The maximum value is 10 and the minimum value is 2.

57. $z = 2x + 3y$

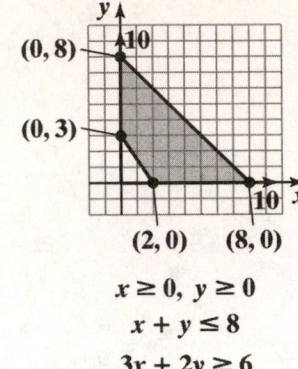

$x \ge 0,\ y \ge 0$
$x + y \le 8$
$3x + 2y \ge 6$

$(0, 8) : z = 2(0) + 3(8) = 24$
$(8, 0) : z = 2(8) + 3(0) = 16$
$(0, 3) : z = 2(0) + 3(3) = 9$
$(2, 0) : z = 2(2) + 3(0) = 6$
Maximum value is 24 at (0, 8).

58. $z = x + 4y$

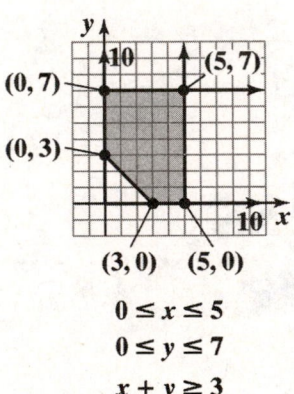

$0 \le x \le 5$
$0 \le y \le 7$
$x + y \ge 3$

$(0, 3) : z = 0 + 4(3) = 12$
$(3, 0) : z = 3 + 4(0) = 3$
$(0, 7) : z = 0 + 4(7) = 28$
$(5, 0) : z = 5 + 4(0) = 5$
$(5, 7) : z = 5 + 4(7) = 33$
Maximum value is 33 at (5, 7).

59. $z = 5x + 6y$

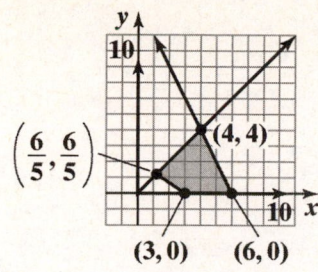

$$\left(\frac{6}{5}, \frac{6}{5}\right) \qquad (4, 4)$$

$$(3, 0) \qquad (6, 0)$$

$$x \geq 0, y \geq 0$$
$$y \leq x$$
$$2x + y \leq 12$$
$$2x + 3y \geq 6$$

$(3, 0): z = 5(3) + 6(0) = 15$

$(6, 0): z = 5(6) + 6(0) = 30$

$\left(\frac{6}{5}, \frac{6}{5}\right): z = 5\left(\frac{6}{5}\right) + 6\left(\frac{6}{5}\right) = \frac{66}{5} = 13.2$

$(4, 4): 5(4) + 6(4) = 44$

The maximum value is 44.

60. a. $z = 500x + 350y$

b. $x + y \leq 200$
$\quad\; x \geq 10$
$\quad\; y \geq 80$

c.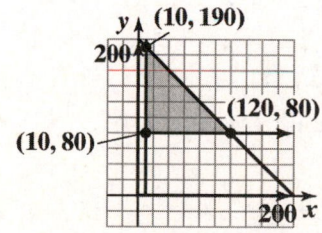

$$x + y \leq 200$$
$$x \geq 10, y \geq 80$$

d.

Vertex	Objective Function
	$z = 500x + 350y$
(10, 80)	$z = 500(10) + 350(80)$
	$= 33,000$
(10, 190)	$z = 500(10) + 350(190)$
	$= 71,500$
(120, 80)	$z = 500(120) + 350(80)$
	$= 88,000$

e. The company will make the greatest profit by producing 120 units of writing paper and 80 units of newsprint each day. The maximum daily profit is $88,000.

61. Let x = number of model A tents produced and y = number of model B tents produced. The constraints are:

$0.9x + 1.8y \leq 864$
$0.8x + 1.2y \leq 672$
$\quad\quad x \geq 0$
$\quad\quad y \geq 0$

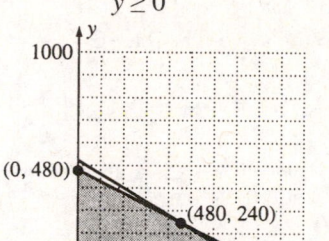

The vertices of the region are (0, 0), (0, 480), (480, 240), and (840, 0).

The objective is to maximize $25x + 40y$.

(0, 0): $25(0) + 40(0) = 0 + 0 = 0$

(0, 480): $25(0) + 40(480) = 0 + 19,200 = 19,200$

(480, 240): $25(480) + 40(240) = 12,000 + 9600$
$\quad\quad\quad\quad = 21,600$

(840, 0): $25(840) + 40(0) = 21,000 + 0 = 21,000$

The manufacturer should make 480 of model A and 240 of model B.

Chapter 7 Test

1. $x = y + 4$

$3x + 7y = -18$

Substitute $y + 4$ for x into second equation.

$3(y + 4) + 7y = -18$

$3y + 12 + 7y = -18$

$10y = -30$

$y = -3$

$x = -3 + 4 = 1$

The solution set to the system is $\{(1, -3)\}$.

2. $2x + 5y = -2$

$3x - 4y = 20$

Multiply the first equation by 3 and the second equation by -2 and add the result.

$6 + 15y = -6$

$\underline{-6x + 8y = -40}$

$23y = -46$

$y = -2$

Substitute $y = -2$ into the first equation:

$2x + 5(-2) = -2$

$2x - 10 = -2$

$2x = 8$

$x = 4$

The solution to the system is $\{(4, -2)\}$.

3. $x + y + z = 6 \,(1)$

$3x + 4y - 7z = 1 \,(2)$

$2x - y + 3z = 5 \,(3)$

Eliminate x by multiplying (1) by -3 and adding the result to (2) and by multiplying (1) by -2 and adding the result to (3).

$-3x - 3y - 3z = -18$

$\underline{3x + 4y - 7z = 1}$

$y - 10z = -17 \,(4)$

$-2x - 2y - 2z = -12$

$\underline{2x - y + 3z = 5}$

$-3y + z = -7 \,(5)$

Multiply (4) by 3 and add the result to (5) to eliminate y.

$3y - 30z = -51$

$\underline{-3y + z = -7}$

$-29z = -58$

$z = 2$

Substitute $z = 2$ into (5).

$-3y + 2 = -7$

$-3y = -9$

$y = 3$

Substitute $z = 2$ and $y = 3$ into (1).

$x + 3 + 2 = 6$

$x = 1$

The solution to the system is $\{(1, 3, 2)\}$.

4. $x^2 + y^2 = 25$

$x + y = 1$

$y = 1 - x$

Substitute $1 - x$ for y in the first equation.

$x^2 + (1 - x)^2 = 25$

$x^2 + 1 - 2x + x^2 = 25$

$2x^2 - 2x - 24 = 0$

$x^2 - x - 12 = 0$

$(x - 4)(x + 3) = 0$

$x = 4, -3$

If $x = 4$, $y = 1 - 4 = -3$.

If $x = -3$, $y = 1 - (-3) = 4$.

The solution set is $\{(4, -3), (-3, 4)\}$.

5. $2x^2 - 5y^2 = -2$

$3x^2 + 2y^2 = 35$

Multiply first equation by 2 and the second equation by 5. Then add.

$4x^2 - 10y^2 = -4$

$\underline{15x^2 + 10y^2 = 175}$

$19x^2 = 171$

$x^2 = 9$

$x = 3, -3$

If $x = 3$, $2(3)^2 - 5y^2 = -2$.

$18 - 5y^2 = -2$

$-5y^2 = -20$

$y^2 = 4$

$y = 2, -2$

If $x = -3$, $y = -2$.

The solution to the system is

$\{(3, 2), (3, -2), (-3, 2), (-3, -2)\}$.

6. $\dfrac{x}{(x+1)(x^2+9)} = \dfrac{A}{x+1} + \dfrac{Bx+C}{x^2+9}$

$x = A(x^2+9) + (Bx+C)(x+1)$

$\quad = Ax^2 + 9A + Bx^2 + Bx + Cx + C$

$\quad = (A+B)x^2 + (B+C)x + (9A+C)$

$A+B = 0 \rightarrow A = -B$

$\quad B+C = 1$

$\quad 9A+C = 0$

$\quad -9B+C = 0$

$\quad \underline{9B-C = 0}$

$\quad \underline{B+C = 1}$

$\quad 10B = 1$

$\qquad B = \dfrac{1}{10}$

$\qquad A = -\dfrac{1}{10}$

$\dfrac{1}{10} + C = 1, \ C = \dfrac{9}{10}$

$\dfrac{x}{(x+1)(x^2+9)} = \dfrac{-1}{10(x+1)} + \dfrac{x+9}{10(x^2+9)}$

7.

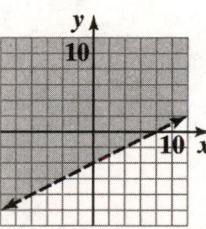

$x - 2y < 8$

8.

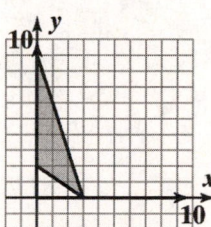

$x \geq 0, \ y \geq 0$

$3x + y \leq 9$

$2x + 3y \geq 6$

9.

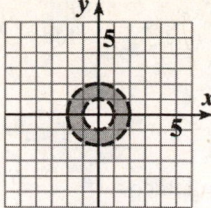

$x^2 + y^2 > 1$

$x^2 + y^2 < 4$

10.

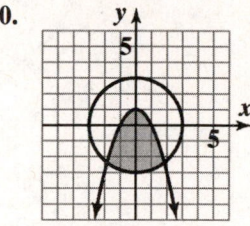

$y \leq 1 - x^2$

$x^2 + y^2 \leq 9$

11. $z = 3x + 5y$

$x \geq 0, y \geq 0$

$x + y \leq 6$

$x \geq 2$

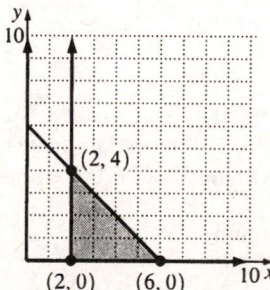

$(2,0): z = 3(2) + 5(0) = 6$

$(6,0): z = 3(6) + 5(0) = 18$

$(2,4): z = 3(2) + 5(4) = 26$

Maximum value is 26.

12. $x = $ mg of cholesterol in one ounce of shrimp
$y = $ mg of cholesterol in one ounce of scallops
$3x + 2y = 156$
$5x + 3y = 255$
Multiply the first equation by –3 and multiply the second equation by 2.
Add the resulting equations together.
$-9x - 6y = -468$

$$\underline{10x + 6y = 510}$$
$$x = 42$$

$3(42) + 2y = 156$
$126 + 2y = 156$
$2y = 30$
$y = 15$
$3(42) + 2y = 156$
$126 + 2y = 156$
$2y = 30$
$y = 15$

Shrimp: 42 mg of cholesterol per ounce
Scallops: 15 mg of cholesterol per ounce

13. a. $C(x) = 360,000 + 850x$

b. $R(x) = 1150x$

c. $1150x = 360000 + 850x$
$300x = 360000$
$x = 1200$
$1150(1200) = 1,380,000$
1200 computers need to be sold to make
$1,380,00 for the company to break even.

14. $x = $ ounces of 20% solution
$y = $ ounces of 50% solution
$x + y = 60$
$0.2x + 0.5y = 0.3(60)$

$y = 60 - x$

$0.2x + 0.5(60 - x) = 18$
$0.2x + 30 - 0.5x = 18$
$-0.3x = -12$
$x = 40$

$y = 60 - 40 = 20$
Mix 40 ounces of 20% solution and 20 ml of 50% solution.

15. $x = $ average velocity of the plane
$y = $ average velocity of the wind

Velocity	Time	Distance
$x + y$	2	$2(x + y)$
$x - y$	3	$3(x - y)$

$2(x + y) = 1600$
$3(x - y) = 1950$

$x + y = 800$
$x - y = 650$

$2x = 1450$
$x = 725$

$x + y = 800$
$725 + y = 75$
$y = 90$

The average velocity of the plane is 725 mph and the wind is 75 mph.

16. $y = ax^2 + bx + c$
$(-1, -2) : -2 = a - b + c$
$(2, 1) : 1 = 4a + 2b + c$
$(-2, 1) : 1 = 4a - 2b + c$

$$4a + 2b + c = 1$$
$$\underline{-4a + 2b - c = -1}$$
$$4b = 0$$
$$b = 0$$

$a + c = -2$
$4a + c = 1$

$$\underline{-a - c = 2}$$
$$3a = 3$$
$$a = 1$$
$a + c = -2$
$c = -3$

The quadratic function is $y = x^2 - 3$.

17. $2x + y = 39$

$xy = 180$

$y = 39 - 2x$

$x(39 - 2x) = 180$

$39x - 2x^2 = 180$

$2x^2 - 39x + 180 = 0$

$(2x - 15)(x - 12) = 0$

$$x = \frac{15}{2}, 12$$

If $x = \dfrac{15}{2}, \dfrac{15}{2}\, y = 180$ and $y = 24$.

If $x = 12, 12\, y = 180$ and $y = 15$.

The dimensions are 7.5 ft by 24 ft
or 12 ft by 15 ft

18. Let x = regular, y = deluxe.
objective function: $z = 200x + 250y$

constraints: $x \geq 50,\, y \geq 75$

$$x + y \leq 150$$

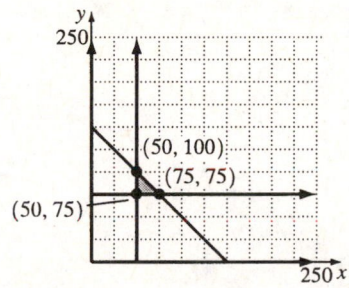

$(50, 75): z = 200(50) + 250(75) = 28,750$

$(50, 100): z = 200(50) + 250(100) = 35,000$

$(75, 75): z = 200(75) + 250(75) = 33,750$

For a maximum profit of $35,000 a week, the company should manufacture 50 regular and 100 deluxe jet skis.

Cumulative Review Exercises (Chapters P–7)

1. Domain: $(-\infty, \infty)$ Range: $[-\infty, 3]$

2. -1 and 1 are the zeros.

3. The relative maximum is $y = 3$ and it occurs at $x = 0$.

4. $f(x)$ is decreasing on the interval $(0, 2)$.

5. At $x = -0.7$, the curve is above the x-axis and thus $f(x)$ is positive.

6. $(f \circ f)(-1) = f\big(f(-1)\big) = f(0) = 3$

7. $f(x) \to -\infty$ as $x \to -2^{+}$ or as $x \to 2^{-}$.

8. $f(-x) = f(x)$ thus the function is even.

9. The graph of $g(x) = f(x + 2) - 1$ can be obtained by shifting $f(x)$ 2 units left and 1 unit down.

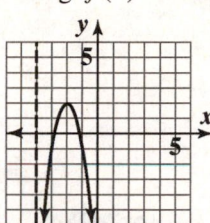

$g(x) = f(x + 2) - 1$

10. The graph of $h(x) = \frac{1}{2} f\left(\frac{1}{2} x\right)$ can be obtained by shrinking the graph of $f(x)$ horizontally by a factor of $\frac{1}{2}$ and vertically by a factor of $\frac{1}{2}$.

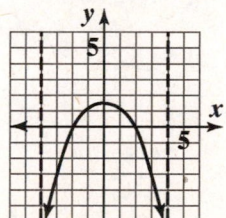

$h(x) = \frac{1}{2} f\left(\frac{1}{2} x\right)$

11.
$$\sqrt{x^2 - 3x} = 2x - 6$$
$$x^2 - 3x = 4x^2 - 24x + 36$$
$$3x^2 - 21x + 36 = 0$$
$$x^2 - 7x + 12 = 0$$
$$(x - 3)(x - 4) = 0$$
$$x = 3, 4$$

The solution set is $\{3, 4\}$.

12.
$$4x^2 = 8x - 7$$
$$4x^2 - 8x + 7 = 0$$
$$x = \frac{8 \pm \sqrt{64 - 112}}{8} = \frac{8 \pm \sqrt{-48}}{8}$$
$$= \frac{8 \pm 4\sqrt{3}i}{8} = \frac{2 \pm \sqrt{3}i}{2}$$

The solution set is $\left\{ \frac{2 + i\sqrt{3}}{2}, \frac{2 - i\sqrt{3}}{2} \right\}$.

13.
$$\left| \frac{x}{3} + 2 \right| < 4$$
$$-4 < \frac{x}{3} + 2 < 4$$
$$-6 < \frac{x}{3} < 2$$
$$-18 < x < 6$$

The solution is $\{x \mid -18 < x < 6\}$ or $(-18, 6)$.

14.
$$\frac{x + 5}{x - 1} > 2$$
$$\frac{x + 5}{x - 1} - 2 > 0$$
$$\frac{x + 5 - 2(x - 1)}{x - 1} > 0$$
$$\frac{x + 5 - 2x + 2}{x - 1} > 0$$
$$\frac{-x + 7}{x - 1} > 0$$

$\frac{-x + 7}{x - 1} = 0$ when $x = 7$ and is undefined when

$x = 1$.
Test $x = 0$:
$$\frac{0 + 5}{0 - 1} > 2?$$
$$\frac{5}{-1} > 2?$$
$$-5 \not> 2$$

Test $x = 2$:
$$\frac{2 + 5}{2 - 1} > 2?$$
$$\frac{7}{1} > 2?$$
$$7 > 2$$
Test $x = 8$:
$$\frac{8 + 5}{8 - 1} > 2?$$
$$\frac{13}{7} > 2?$$
$$\frac{13}{7} \geq \frac{14}{7}$$

The solution is $\{x \mid 1 < x < 7\}$ or $(1, 7)$.

15. $2x^3 + x^2 - 13x + 6 = 0$

$f(x) = 2x^3 + x^2 - 13x + 6$ has 2 sign changes: 2 or 0 positive real roots.

$f(-x) = -2x^3 + x^2 + 13x + 6$ has 1 sign change: 1 negative real root.

p: ±1, ±2, ±3, ±6
q: ±1, ±2

$\frac{p}{q}: \pm 1, \pm \frac{1}{2}, \pm 2, \pm 3, \pm \frac{3}{2}, \pm 6$

-3	2	1	-13	6
		-6	15	-6
	2	-5	2	0

$$2x^3 + x^2 - 13x + 6 = (x + 3)(2x^2 - 5x + 2)$$
$$= (x + 3)(2x - 1)(x - 2)$$

$$x = -3, \ x = \frac{1}{2}, \ x = 2$$

The solution set is $\left\{ -3, \frac{1}{2}, 2 \right\}$.

16. $6x - 3(5x + 2) = 4(1 - x)$

$6x - 15x - 6 = 4 - 4x$

$-9x - 6 = 4 - 4x$

$-5x = 10$

$x = -2$

The solution set is $\{-2\}$.

17. $\log (x + 3) + \log x = 1$

$\log x(x + 3) = 1$

$x(x + 3) = 10$

$x^2 + 3x - 10 = 0$

$(x + 5)(x - 2) = 0$

$x = -5 \text{ or } x = 2$

$x = -5 \text{ is extraneous.}$

$x = 2$

The solution set is $\{2\}$.

18. $3^{x+2} = 11$

$\log_3 3^{x+2} = \log_3 11$

$x + 2 = \log_3 11$

$x = -2 + \log_3 11$

$x = -2 + \dfrac{\log 11}{\log 3} \approx 0.18$

The solution set is $\left\{ -2 + \log_3 11 \right\}$.

19. $x^{\frac{1}{2}} - 2x^{\frac{1}{4}} - 15 = 0$

$\sqrt[4]{x}^{\,2} - 2\sqrt[4]{x} - 15 = 0$

$\left(\sqrt[4]{x} + 3 \right)\left(\sqrt[4]{x} - 5 \right) = 0$

$\sqrt[4]{x} + 3 = 0 \qquad\qquad \sqrt[4]{x} - 5 = 0$

$\sqrt[4]{x} = -3 \qquad \text{or} \qquad \sqrt[4]{x} = 5$

$x = (-3)^4 \qquad\qquad\quad x = 5^4$

$x = 81 \qquad\qquad\qquad x = 625$

81 does not check. The solution set is $\{625\}$.

20. $3x - y = -2$

$2x^2 - y = 0$

Solve the first equation for y.

$3x - y = -2$

$y = 3x + 2$

Use this equation to substitute into the other equation.

$2x^2 - (\overbrace{3x + 2}^{y}) = 0$

$2x^2 - 3x - 2 = 0$

$(2x + 1)(x - 2) = 0$

$x = -\dfrac{1}{2} \text{ or } x = 2.$

Back-substitute to find y.

$y = 3x + 2 \qquad\quad \text{or} \qquad y = 3x + 2$

$y = 3(2) + 2 \qquad\qquad\qquad y = 3\left(-\dfrac{1}{2}\right) + 2$

$y = 8$

$\qquad\qquad\qquad\qquad\qquad\quad y = \dfrac{1}{2}$

$y = 3x + 2$

The solution set is $\left\{ (2, 8), \left(-\dfrac{1}{2}, \dfrac{1}{2}\right) \right\}$.

21. $x + 2y + 3z = -2$

$3x + 3y + 10z = -2$

$2y - 5z = 6$

Multiply equation 1 by –3 and add to equation 2.

$-3x - 6y - 9z = 6$

$\underline{3x + 3y + 10z = -2}$

$\qquad -3y + z = 4 \quad \text{Equation 4}$

Multiply equation 4 by 5 and add to equation 3 and solve for y.

$-15y + 5z = 20$

$\underline{2y - 5z = 6}$

$-13y = 26$

$y = -2$

Back-substitute to find z.

$-3y + z = 4$

$-3(-2) + z = 4$

$z = -2$

Back-substitute to find x.

$x + 2y + 3z = -2$

$x + 2(-2) + 3(-2) = -2$

$x = 8$

The solution set is $\{8, -2, -2\}$.

22. vertex: $(-2, -4)$

y-intercept:

$$f(0) = (0+2)^2 - 4 = 0$$

x-intercepts:

$$(x+2)^2 - 4 = 0$$

$$x^2 + 4x + 4 - 4 = 0$$

$$x^2 + 4x = 0$$

$$x(x+4) = 0$$

$$x = 0, x - 4$$

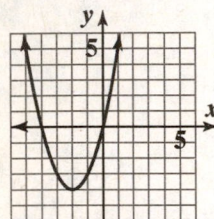

$$f(x) = (x+2)^2 - 4$$

23.

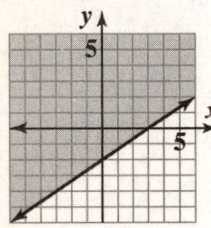

$$2x - 3y \leq 6$$

24.

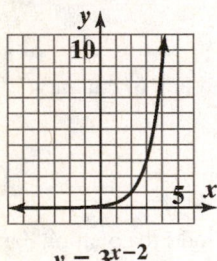

$$y = 3^{x-2}$$

25. vertical asymptote: $x = -1$

horizontal asymptote: $m > n$, none

x-intercepts:

$$x^2 - x - 6 = 0$$

$$(x-3)(x+2) = 0$$

$$x = 3, x = -2$$

y-intercept:

$$f(0) = \frac{0^2 - 0 - 6}{0 + 1} = -6$$

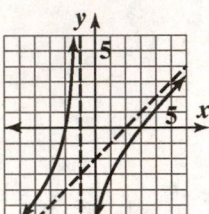

$$f(x) = \frac{x^2 - x - 6}{x + 1}$$

26.

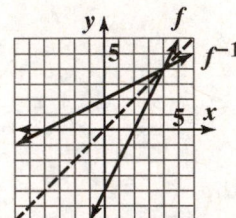

$$f(x) = 2x - 4$$

$$f^{-1}(x) = \frac{x+4}{2}$$

27.

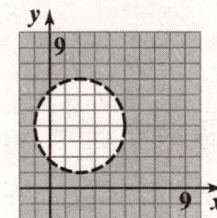

$$(x-2)^2 + (y-4)^2 > 9$$

28.

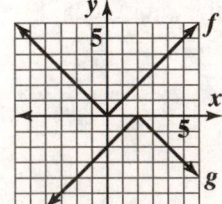

$$f(x) = |x|$$

$$g(x) = -|x-2|$$

706

29. $(f \circ g)(x) = f(g(x))$

$$= 2(1-x)^2 - (1-x) - 1$$
$$= 2x^2 - 3x$$

$(g \circ f) = g(f(x))$

$$= 1 - (2x^2 - x - 1)$$
$$= 1 - 2x^2 + x + 1$$
$$= -2x^2 + x + 2$$

30. $\dfrac{f(x+h) - f(x)}{h}$

$$= \frac{\left[2(x+h)^2 - (x+h) - 1\right] - \left[2x^2 - x - 1\right]}{h}$$

$$= \frac{2x^2 + 4hx + 2h^2 - x - h - 1 - 2x^2 + x + 1}{h}$$

$$= \frac{4hx + 2h^2 - h}{h}$$

$$= 4x + 2h - 1$$

31. Find slope: $m = \dfrac{4 - (-2)}{2 - 4} = \dfrac{6}{-2} = -3$

Use point slope form to find an equation.

$y - y_1 = m(x - x_1)$

$y - 4 = -3(x - 2)$

Put in slope-intercept form.

$y - 4 = -3(x - 2)$

$y - 4 = -3x + 6$

$y = -3x + 10$

32. Find the slope of the perpendicular line by putting in slope-intercept form.

$x + 3y - 6 = 0$

$3y = -x + 6$

$y = -\dfrac{1}{3}x + 2$

The slope of the perpendicular line is $-\dfrac{1}{3}$ so the slope of the desired line is the negative reciprocal, or 3.

Use point slope form to find an equation.

$y - y_1 = m(x - x_1)$

$y - 0 = 3(x + 1)$

Put in slope-intercept form.

$y - 0 = 3(x + 1)$

$y = 3x + 3$

33. Let x = the amount invested at 12%
Let $4000 - x$ = the amount invested at 14%

$0.12x + 0.14(4000 - x) = 508$

$0.12x + 560 - 0.14x = 508$

$-0.02x = -52$

$$x = \frac{-52}{-0.02}$$

$x = 2600$

$4000 - x = 4000 - 2600 = 1400$

Thus, \$2600 was invested at 12% and \$1400 was invested at 14%.

34.
$$L = 2W + 1$$
$$LW = 36$$
$$W(2W + 1) = 36$$
$$2W^2 + W - 36 = 0$$
$$(2W + 9)(W - 4) = 0$$
$$W = -\frac{9}{2} \text{ or } 4$$

Length cannot be negative. If $W = 4$, $4L = 36$, $L = 9$. The dimensions are 4 m by 9 m.

35. $A = Pe^{rt}$

$18{,}000 = 6000e^{10r}$

$3 = e^{10r}$

$\ln 3 = \ln e^{10r}$

$\ln 3 = 10r$

$r = \dfrac{\ln 3}{10} \approx 0.1099$

10.99%

36. $\sec\theta - \cos\theta = \dfrac{1}{\cos\theta} - \cos\theta$

$$= \frac{1 - \cos^2\theta}{\cos\theta}$$

$$= \frac{\sin^2\theta}{\cos\theta}$$

$$= \frac{\sin\theta}{\cos\theta} \cdot \sin\theta$$

$$= \tan\theta \cdot \sin\theta$$

37. $\tan x + \tan y = \dfrac{\sin x}{\cos x} + \dfrac{\sin y}{\cos y}$

$$= \frac{\sin x \cdot \cos y + \sin y \cdot \cos x}{\cos x \cdot \cos y}$$

$$= \frac{\sin(x + y)}{\cos x \cdot \cos y}$$

38.

$$\sin \theta = \tan \theta$$

$$\sin \theta = \frac{\sin \theta}{\cos \theta}$$

$$\sin \theta - \frac{\sin \theta}{\cos \theta} = 0$$

$$\sin \theta \left(1 - \frac{1}{\cos \theta} \right) = 0$$

$$\sin \theta = 0 \quad \text{or} \quad 1 - \frac{1}{\cos \theta} = 0$$

$$\theta = 0, \pi \quad \text{or} \quad 1 = \frac{1}{\cos \theta}$$

$$\cos \theta = 1$$

$$\theta = 0$$

The solutions in the interval $[0, 2\pi)$ are 0 and π.

39.

$$2 + \cos 2\theta = 3 \cos \theta$$

$$2 + (2 \cos^2 \theta - 1) = 3 \cos \theta$$

$$2 \cos^2 \theta + 1 = 3 \cos \theta$$

$$2 \cos^2 \theta - 3 \cos \theta + 1 = 0$$

$$(2 \cos \theta - 1)(\cos \theta - 1) = 0$$

$$2 \cos \theta - 1 = 0 \quad \text{or} \quad \cos \theta - 1 = 0$$

$$2 \cos \theta = 1 \qquad \cos \theta = 1$$

$$\cos \theta = \frac{1}{2}$$

$$\theta = \frac{\pi}{3}, \frac{5\pi}{3} \quad \text{or} \quad \theta = 0$$

The solutions in the interval $[0, 2\pi)$ are

$$0, \frac{\pi}{3}, \text{ and } \frac{5\pi}{3}.$$

40.

$$\frac{b}{\sin B} = \frac{a}{\sin A}$$

$$\frac{b}{\sin 75°} = \frac{20}{\sin 12°}$$

$$b = \frac{20 \sin 75°}{\sin 12°}$$

$$b \approx 92.9$$

Chapter 8
Matrices and Determinants

Check Point Exercises

1. **a.** The notation $R_1 \leftrightarrow R_2$ means to interchange the elements in row 1 and row 2. This results in the row-equivalent matrix
$$\begin{bmatrix} 1 & 6 & -3 & | & 7 \\ 4 & 12 & -20 & | & 8 \\ -3 & -2 & 1 & | & -9 \end{bmatrix}.$$

   b. The notation $\frac{1}{4}R_1$ means to multiply each element in row 1 by $\frac{1}{4}$. This results in the row-equivalent matrix
$$\begin{bmatrix} \frac{1}{4}(4) & \frac{1}{4}(12) & \frac{1}{4}(-20) & | & \frac{1}{4}(8) \\ 1 & 6 & -3 & | & 7 \\ -3 & -2 & 1 & | & -9 \end{bmatrix} = \begin{bmatrix} 1 & 3 & -5 & | & 2 \\ 1 & 6 & -3 & | & 7 \\ -3 & -2 & 1 & | & -9 \end{bmatrix}$$

   c. The notation $3R_2 + R_3$ means to add 3 times the elements in row 2 to the corresponding elements in row 3. Replace the elements in row 3 by these sums. First, we find 3 times the elements in row 2:
   $3(1) = 3, 3(6) = 18, 3(-3) = -9, 3(7) = 21$. Now we add these products to the corresponding elements in row 3. This results in the row equivalent matrix
$$\begin{bmatrix} 4 & 12 & -20 & | & 8 \\ 1 & 6 & -3 & | & 7 \\ -3+3=0 & -2+18=16 & 1-9=-8 & | & -9+21=12 \end{bmatrix} = \begin{bmatrix} 4 & 12 & -20 & | & 8 \\ 1 & 6 & -3 & | & 7 \\ 0 & 16 & -8 & | & 12 \end{bmatrix}.$$

2. $\begin{aligned} 2x + y + 2z &= 18 \\ x - y + 2z &= 9 \\ x + 2y - z &= 6 \end{aligned} \rightarrow \begin{bmatrix} 2 & 1 & 2 & | & 18 \\ 1 & -1 & 2 & | & 9 \\ 1 & 2 & -1 & | & 6 \end{bmatrix}$

   Interchange row 1 with row 2 to get 1 in the top position of the first column.
$$\begin{bmatrix} 1 & -1 & 2 & | & 9 \\ 2 & 1 & 2 & | & 18 \\ 1 & 2 & -1 & | & 6 \end{bmatrix}$$

   Multiply the first row by –2 and add these products to row 2.
$$\begin{bmatrix} 1 & -1 & 2 & | & 9 \\ 2+-2=0 & 1+2=3 & 2+-4=-2 & | & -18+18=0 \\ 1 & 2 & -1 & | & 6 \end{bmatrix} = \begin{bmatrix} 1 & -1 & 2 & | & 9 \\ 0 & 3 & -2 & | & 0 \\ 1 & 2 & -1 & | & 6 \end{bmatrix}$$

   Next, multiply the top row by –1 and add these products to row 3.
$$\begin{bmatrix} 1 & -1 & 2 & | & 9 \\ 0 & 3 & -2 & | & 0 \\ 1+-1=0 & 2+1=3 & -1-2=-3 & | & 6-9=-3 \end{bmatrix} = \begin{bmatrix} 1 & -1 & 2 & | & 9 \\ 0 & 3 & -2 & | & 0 \\ 0 & 3 & -3 & | & -3 \end{bmatrix}$$

Next, to obtain a 1 in the second row, second column, multiply 3 by its reciprocal, $\frac{1}{3}$. Therefore, we multiply all the

numbers in the second row by $\frac{1}{3}$ to get

$$\begin{bmatrix} 1 & -1 & 2 & | & 9 \\ 0 & 1 & -\frac{2}{3} & | & 0 \\ 0 & 3 & -3 & | & -3 \end{bmatrix}.$$

Next, to obtain a 0 in the third row, second column, multiply the second row by –3 and add the products to row three. The resulting matrix is

$$\begin{bmatrix} 1 & -1 & 2 & | & 9 \\ 0 & 1 & -\frac{2}{3} & | & 0 \\ 0 & 0 & -1 & | & -3 \end{bmatrix}.$$

To get 1 in the third row, third column, multiply –1 by its reciprocal, –1. Multiply all numbers in the third row by –1 to obtain the resulting matrix

$$\begin{bmatrix} 1 & -1 & 2 & | & 9 \\ 0 & 1 & -\frac{2}{3} & | & 0 \\ 0 & 0 & 1 & | & 3 \end{bmatrix}.$$

The system represented by this matrix is:

$x - y + 2z = 9$

$y - \dfrac{2}{3}z = 0$

$z = 3$

Use back substitution to find y and x.

$y - \dfrac{2}{3}(3) = 0 \qquad x - 2 + 6 = 9$

$y - 2 = 0 \qquad\qquad x + 4 = 9$

$y = 2 \qquad\qquad\quad x = 5$

The solution set for the original system is $\{(5, 2, 3)\}$.

3. $w - 3x - 2y + z = -3$

$2w - 7x - y + 2z = 1$

$3w - 7x - 3y + 3z = -5$

$5w + x + 4y - 2z = 18$

The augmented matrix is

$$\begin{bmatrix} 1 & -3 & -2 & 1 & | & -3 \\ 2 & -7 & -1 & 2 & | & 1 \\ 3 & -7 & -3 & 3 & | & -5 \\ 5 & 1 & 4 & -2 & | & 18 \end{bmatrix}.$$

Multiply the top row by –2 and add the products to the second row. Multiply the top row by –3 and add the pro-ducts to the third row. Multiply the top row by –5 and add the products to the fourth row. The resulting matrix is

$$\begin{bmatrix} 1 & -3 & -2 & 1 & | & -3 \\ 0 & -1 & 3 & 0 & | & 7 \\ 0 & 2 & 3 & 0 & | & 4 \\ 0 & 16 & 14 & -7 & | & 33 \end{bmatrix}.$$

Next, multiply the second row by –1 to obtain a 1 in the second row, second column.

$$\begin{bmatrix} 1 & -3 & -2 & 1 & | & -3 \\ 0 & 1 & -3 & 0 & | & -7 \\ 0 & 2 & 3 & 0 & | & 4 \\ 0 & 16 & 14 & -7 & | & 33 \end{bmatrix}$$

Next, multiply the second row by –2 and add the products to the third row. Multiply the second row by –16 and add the products to the fourth row. The resulting matrix is

$$\begin{bmatrix} 1 & -3 & -2 & 1 & | & -3 \\ 0 & 1 & -3 & 0 & | & -7 \\ 0 & 0 & 9 & 0 & | & 18 \\ 0 & 0 & 62 & -7 & | & 145 \end{bmatrix}.$$

Next, multiply the third row by $\frac{1}{9}$ to obtain a 1 in the third row, third column. The resulting matrix is

$$\begin{bmatrix} 1 & -3 & -2 & 1 & | & -3 \\ 0 & 1 & -3 & 0 & | & -7 \\ 0 & 0 & 1 & 0 & | & 2 \\ 0 & 0 & 62 & -7 & | & 145 \end{bmatrix}.$$

Multiply the third row by –62 and add the products to the fourth row to obtain the resulting matrix

$$\begin{bmatrix} 1 & -3 & -2 & 1 & | & -3 \\ 0 & 1 & -3 & 0 & | & -7 \\ 0 & 0 & 1 & 0 & | & 2 \\ 0 & 0 & 0 & -7 & | & 21 \end{bmatrix}.$$

Multiply the fourth row by $-\frac{1}{7}$, the reciprocal of –7. The resulting matrix is

$$\begin{bmatrix} 1 & -3 & -2 & 1 & | & -3 \\ 0 & 1 & -3 & 0 & | & -7 \\ 0 & 0 & 1 & 0 & | & 2 \\ 0 & 0 & 0 & 1 & | & -3 \end{bmatrix}.$$

The system of linear equations corresponding to the resulting matrix is

$$w - 3x - 2y + z = -3$$
$$x - 3y = -7$$
$$y = 2$$
$$z = -3$$

Using back-substitution solve for x and w.

$$x - 3(2) = -7$$
$$x = -1$$

$$w - 3(-1) - 2(2) - 3 = -3$$
$$w - 4 = -3$$
$$w = 1$$

The solution set is $\{(1, -1, 2, -3)\}$.

Matrices and Determinants

4. The matrix obtained in 3 will be the starting point.

$$\begin{bmatrix} 1 & -1 & 2 & | & 9 \\ 0 & 1 & -\frac{2}{3} & | & 0 \\ 0 & 0 & 1 & | & 3 \end{bmatrix}$$

Next, multiply the third row by $\dfrac{2}{3}$ and add the products to the second row. Multiply the third row by 2 and add the products to the first row. The resulting matrix is

$$\begin{bmatrix} 1 & -1 & 0 & 3 \\ 0 & 1 & 0 & 2 \\ 0 & 0 & 1 & 3 \end{bmatrix}.$$

Add the second row to the first row and replace the first row.

$$\begin{bmatrix} 1 & 0 & 0 & 5 \\ 0 & 1 & 0 & 2 \\ 0 & 0 & 1 & 3 \end{bmatrix}$$

This matrix corresponds to $x = 5$, $y = 2$ and $z = 3$. The solution set is $\{(5, 2, 3)\}$.

Exercise Set 8.1

1. $$\begin{bmatrix} 2 & 1 & 2 & | & 2 \\ 3 & -5 & -1 & | & 4 \\ 1 & -2 & -3 & | & -6 \end{bmatrix}$$

3. $$\begin{bmatrix} 1 & -1 & 1 & | & 8 \\ 0 & 1 & -12 & | & -15 \\ 0 & 0 & 1 & | & 1 \end{bmatrix}$$

5. $$\begin{bmatrix} 5 & -2 & -3 & | & 0 \\ 1 & 1 & 0 & | & 5 \\ 2 & 0 & -3 & | & 4 \end{bmatrix}$$

7. $$\begin{bmatrix} 2 & 5 & -3 & 1 & | & 2 \\ 0 & 3 & 1 & 0 & | & 4 \\ 1 & -1 & 5 & 0 & | & 9 \\ 5 & -5 & -2 & 0 & | & 1 \end{bmatrix}$$

9. $5x + 3z = -11$
$y - 4z = 12$
$7x + 2y = 3$

11. $w + x + 4y + z = 3$
$-w + x - y = 7$
$12w + 5z = 11$
$12y + 4z = 5$

13. $\begin{bmatrix} 2\left(\frac{1}{2}\right) & -6\left(\frac{1}{2}\right) & 4\left(\frac{1}{2}\right) & 10\left(\frac{1}{2}\right) \\ 1 & 5 & -5 & 0 \\ 3 & 0 & 4 & 7 \end{bmatrix} \frac{1}{2}R_1$

$\begin{bmatrix} 1 & -3 & 2 & 5 \\ 1 & 5 & -5 & 0 \\ 3 & 0 & 4 & 7 \end{bmatrix}$

15. $\begin{bmatrix} 1 & -3 & 2 & 0 \\ -3(1)+3 & -3(-3)+1 & -3(2)+-1 & -3(0)+7 \\ 2 & -2 & 1 & 3 \end{bmatrix} -3R_1 + R_2$

$\begin{bmatrix} 1 & -3 & 2 & 0 \\ 0 & 10 & -7 & 7 \\ 2 & -2 & 1 & 3 \end{bmatrix}$

17. $\begin{bmatrix} 1 & -1 & 1 & 1 & 3 \\ 0 & 1 & -2 & -1 & 0 \\ 2 & 0 & 3 & 4 & 11 \\ 5 & 1 & 2 & 4 & 6 \end{bmatrix} \begin{matrix} \\ \\ -2R_1 + R_3 \\ -5R_1 + R_4 \end{matrix}$

$\begin{bmatrix} 1 & -1 & 1 & 1 & 3 \\ 0 & 1 & -2 & -1 & 0 \\ -2(1)+2 & -2(-1)+0 & -2(1)+3 & -2(1)+4 & -2(3)+11 \\ -5(1)+5 & -5(-1)+1 & -5(1)+2 & -5(1)+4 & -5(3)+6 \end{bmatrix} = \begin{bmatrix} 1 & -1 & 1 & 1 & 3 \\ 0 & 1 & -2 & -1 & 0 \\ 0 & 2 & 1 & 2 & 5 \\ 0 & 6 & -3 & -1 & -9 \end{bmatrix}$

19. $\begin{bmatrix} 1 & -1 & 1 & 8 \\ 2 & 3 & -1 & -2 \\ 3 & -2 & -9 & 9 \end{bmatrix}$

$\begin{bmatrix} 1 & -1 & 1 & 8 \\ -2(1)+2 & -2(-1)+3 & -2(1)-1 & -2(8)-2 \\ -3(1)+3 & -3(-1)-2 & -3(1)-9 & -3(8)+9 \end{bmatrix}$

$\begin{bmatrix} 1 & -1 & 1 & 8 \\ 0 & 5 & \boxed{-3} & \boxed{-18} \\ 0 & 1 & \boxed{-12} & \boxed{-15} \end{bmatrix}$

$\begin{bmatrix} 1 & -1 & 1 & 8 \\ 0\left(\frac{1}{5}\right) & 1\left(\frac{1}{5}\right) & -3\left(\frac{1}{5}\right) & -18\left(\frac{1}{5}\right) \\ 0 & 1 & -12 & -15 \end{bmatrix}$

$\begin{bmatrix} 1 & -1 & 1 & 8 \\ 0 & 1 & \boxed{-\frac{3}{5}} & \boxed{-\frac{18}{5}} \\ 0 & 1 & \boxed{-12} & \boxed{-15} \end{bmatrix}$

21.
$$x + y - z = -2$$
$$2x - y + z = 5$$
$$-x + 2y + 2z = 1$$

$$\begin{bmatrix} 1 & 1 & -1 & -2 \\ 2 & -1 & 1 & 5 \\ -1 & 2 & 2 & 1 \end{bmatrix} \begin{matrix} \\ -2R_1 + R_2 \\ \end{matrix}$$

$$\begin{bmatrix} 1 & 1 & -1 & -2 \\ 0 & -3 & 3 & 9 \\ -1 & 2 & 2 & 1 \end{bmatrix} \begin{matrix} \\ 1R_1 + R_3 \\ \end{matrix}$$

$$\begin{bmatrix} 1 & 1 & -1 & -2 \\ 0 & -3 & 3 & 9 \\ 0 & 3 & 1 & -1 \end{bmatrix} \begin{matrix} \\ -\frac{1}{3}R_2 \\ \end{matrix}$$

$$\begin{bmatrix} 1 & 1 & -1 & -2 \\ 0 & 1 & -1 & -3 \\ 0 & 3 & 1 & -1 \end{bmatrix} \begin{matrix} \\ -3R_2 + R_3 \\ \end{matrix}$$

$$= \begin{bmatrix} 1 & 1 & -1 & -2 \\ 0 & 1 & -1 & -3 \\ 0 & 0 & 4 & 8 \end{bmatrix}$$

$$4z = 8$$
$$z = 2$$
$$y - z = -3$$
$$y - 2 = -3$$
$$y = -1$$
$$x + y - z = -2$$
$$x - 1 - 2 = -2$$
$$x - 3 = -2$$
$$x = 1$$

The solution set is $\{(1, -1, 2)\}$.

23.
$$x + 3y = 0$$
$$x + y + z = 1$$
$$3x - y - z = 11$$

$$\begin{bmatrix} 1 & 3 & 0 & 0 \\ 1 & 1 & 1 & 1 \\ 3 & -1 & -1 & 11 \end{bmatrix} \begin{matrix} \\ -1R_1 + R_2 \\ \end{matrix}$$

$$\begin{bmatrix} 1 & 3 & 0 & 0 \\ 0 & -2 & 1 & 1 \\ 3 & -1 & -1 & 11 \end{bmatrix} \begin{matrix} \\ -3R_1 + R_3 \\ \end{matrix}$$

$$\begin{bmatrix} 1 & 3 & 0 & 0 \\ 0 & -2 & 1 & 1 \\ 0 & -10 & -1 & 11 \end{bmatrix} \begin{matrix} \\ -\frac{1}{2}R_2 \\ \end{matrix}$$

$$\begin{bmatrix} 1 & 3 & 0 & 0 \\ 0 & 1 & -\frac{1}{2} & -\frac{1}{2} \\ 0 & -10 & -1 & 11 \end{bmatrix} \begin{matrix} \\ 10R_2 + R_3 \\ \end{matrix}$$

$$\begin{bmatrix} 1 & 3 & 0 & 0 \\ 0 & 1 & -\frac{1}{2} & -\frac{1}{2} \\ 0 & 0 & -6 & 6 \end{bmatrix} \begin{matrix} \\ -\frac{1}{6}R_3 \\ \end{matrix}$$

$$\begin{bmatrix} 1 & 3 & 0 & 0 \\ 0 & 1 & -\frac{1}{2} & -\frac{1}{2} \\ 0 & 0 & 1 & -1 \end{bmatrix}$$

$$z = -1$$
$$y - \frac{1}{2}z = -\frac{1}{2}$$
$$y - \frac{1}{2}(-1) = -\frac{1}{2}$$
$$y + \frac{1}{2} = -\frac{1}{2}$$
$$y = -1$$

Interchange row one and row two.
$$x + 3y = 0$$
$$x + 3(-1) = 0$$
$$x = 3$$

The solution set is $\{(3, -1, -1)\}$.

25. $2x - y - z = 4$
   $x + y - 5z = -4$
   $x - 2y = 4$

$$\begin{bmatrix} 2 & -1 & -1 & 4 \\ 1 & 1 & -5 & -4 \\ 1 & -2 & 0 & 4 \end{bmatrix}$$

Interchange rows one and two.

$$\begin{bmatrix} 1 & 1 & -5 & -4 \\ 2 & -1 & -1 & 4 \\ 1 & -2 & 0 & 4 \end{bmatrix}$$

Replace row two with $-2R_1 + R_2$.
Replace row three with $-R_1 + R_3$.

$$\begin{bmatrix} 1 & 1 & -5 & -4 \\ 0 & -3 & 9 & 12 \\ 0 & -3 & 5 & 8 \end{bmatrix}$$

Replace row two with $-\dfrac{1}{3}R_2$.

$$\begin{bmatrix} 1 & 1 & -5 & -4 \\ 0 & 1 & -3 & -4 \\ 0 & -3 & 5 & 8 \end{bmatrix}$$

Replace row three with $3R_2 + R_3$.

$$\begin{bmatrix} 1 & 1 & -5 & -4 \\ 0 & 1 & -3 & -4 \\ 0 & 0 & -4 & -4 \end{bmatrix}$$

Replace row three with $-\dfrac{1}{4}R_3$.

$$\begin{bmatrix} 1 & 1 & -5 & -4 \\ 0 & 1 & -3 & -4 \\ 0 & 0 & 1 & 1 \end{bmatrix}$$

$z = 1$
    $y - 3z = -4$
    $y - 3(1) = -4$
        $y = -1$
$x + y - 5z = -4$
$x - 1 - 5(1) = -4$
    $x - 6 = -4$
       $x = 2$

The solution set is $\{(2, -1, 1)\}$.

27. $x + y + z = 4$
   $x - y - z = 0$
   $x - y + z = 2$

$$\begin{bmatrix} 1 & 1 & 1 & 4 \\ 1 & -1 & -1 & 0 \\ 1 & -1 & 1 & 2 \end{bmatrix}$$

Replace row two with $-R_1 + R_2$.
Replace row three with $-R_1 + R_3$.

$$\begin{bmatrix} 1 & 1 & 1 & 4 \\ 0 & -2 & -2 & -4 \\ 0 & -2 & 0 & -2 \end{bmatrix}$$

Replace row two with $-\dfrac{1}{2}R_2$.

$$\begin{bmatrix} 1 & 1 & 1 & 4 \\ 0 & 1 & 1 & 2 \\ 0 & -2 & 0 & -2 \end{bmatrix}$$

Replace row 3 with $2R_2 + R_3$.

$$\begin{bmatrix} 1 & 1 & 1 & 4 \\ 0 & 1 & 1 & 2 \\ 0 & 0 & 2 & 2 \end{bmatrix}$$

Replace row 3 with $\dfrac{1}{2}R_3$.

$$\begin{bmatrix} 1 & 1 & 1 & 4 \\ 0 & 1 & 1 & 2 \\ 0 & 0 & 1 & 1 \end{bmatrix}$$

  $z = 1$
$y + 1 = 2$
    $y = 1$
$x + 1 + 1 = 4$
    $x = 2$
The solution set is $\{(2, 1, 1)\}$.

29. Write the equations in standard form.

$x + 2y - z = -1$

$x - y + z = 4$

$x + y - 3z = -2$

$$\begin{bmatrix} 1 & 2 & -1 & -1 \\ 1 & -1 & 1 & 4 \\ 1 & 1 & -3 & -2 \end{bmatrix}$$

Replace row two with $-R_1 + R_2$.

Replace row three with $-R_1 + R_3$.

$$\begin{bmatrix} 1 & 2 & -1 & -1 \\ 0 & -3 & 2 & 5 \\ 0 & -1 & -2 & -1 \end{bmatrix}$$

Replace row two with $-R_3$.

Replace row three with R_2.

$$\begin{bmatrix} 1 & 2 & -1 & -1 \\ 0 & 1 & 2 & 1 \\ 0 & -3 & 2 & 5 \end{bmatrix}$$

Replace row 3 with $3R_2 + R_3$.

$$\begin{bmatrix} 1 & 2 & -1 & -1 \\ 0 & 1 & 2 & 1 \\ 0 & 0 & 8 & 8 \end{bmatrix}$$

Replace row 3 with $\dfrac{1}{8}R_3$.

$$\begin{bmatrix} 1 & 2 & -1 & -1 \\ 0 & 1 & 2 & 1 \\ 0 & 0 & 1 & 1 \end{bmatrix}$$

$z = 1$

$y + 2(1) = 1$

$y = -1$

$x + 2(-1) - 1 = -1$

$x = 2$

The solution set is $\{(2, -1, 1)\}$.

31. $3a - b - 4c = 3$

$2a - b + 2c = -8$

$a + 2b - 3c = 9$

Interchange equations 1 and 3.

$$\begin{bmatrix} 1 & 2 & -3 & 9 \\ 2 & -1 & 2 & -8 \\ 3 & -1 & -4 & 3 \end{bmatrix}$$

Replace row two with $-2R_1 + R_2$.

Replace row three with $-3R_1 + R_3$.

$$\begin{bmatrix} 1 & 2 & -3 & 9 \\ 0 & -5 & 8 & -26 \\ 0 & -7 & 5 & -24 \end{bmatrix}$$

Replace row two with $-\dfrac{1}{5}R_2$

$$\begin{bmatrix} 1 & 2 & -3 & 9 \\ 0 & 1 & -\dfrac{8}{5} & \dfrac{26}{5} \\ 0 & -7 & 5 & -24 \end{bmatrix}$$

Replace row three with $7R_2 + R_3$.

$$\begin{bmatrix} 1 & 2 & -3 & 9 \\ 0 & 1 & -\dfrac{8}{5} & \dfrac{26}{5} \\ 0 & 0 & -\dfrac{31}{5} & \dfrac{62}{5} \end{bmatrix}$$

Replace row 3 with $-\dfrac{5}{31}R_3$.

$$\begin{bmatrix} 1 & 2 & -3 & 9 \\ 0 & 1 & -\dfrac{8}{5} & \dfrac{26}{5} \\ 0 & 0 & 1 & -2 \end{bmatrix}$$

$z = -2$

$y - \dfrac{8}{5}(-2) = \dfrac{26}{5}$

$y + \dfrac{16}{5} = \dfrac{26}{5}$

$y = 2$

$x + 2(2) - 3(-2) = 9$

$x + 4 + 6 = 9$

$x = -1$

The solution set is $\{(-1, 2, -2)\}$.

33.
$$2x + 2y + 7z = -1$$
$$2x + y + 2z = 2$$
$$4x + 6y + z = 15$$

$$\begin{bmatrix} 2 & 2 & 7 & | & -1 \\ 2 & 1 & 2 & | & 2 \\ 4 & 6 & 1 & | & 15 \end{bmatrix} \frac{1}{2}R_1$$

$$\begin{bmatrix} 1 & 1 & \frac{7}{2} & | & -\frac{1}{2} \\ 2 & 1 & 2 & | & 2 \\ 4 & 6 & 1 & | & 15 \end{bmatrix} -2R_1 + R_2$$

$$\begin{bmatrix} 1 & 1 & \frac{7}{2} & | & -\frac{1}{2} \\ 0 & -1 & -5 & | & 3 \\ 4 & 6 & 1 & | & 15 \end{bmatrix} -4R_1 + R_3$$

$$\begin{bmatrix} 1 & 1 & \frac{7}{2} & | & -\frac{1}{2} \\ 0 & -1 & -5 & | & 3 \\ 0 & 2 & -13 & | & 17 \end{bmatrix} -1R_2$$

$$\begin{bmatrix} 1 & 1 & \frac{7}{2} & | & -\frac{1}{2} \\ 0 & 1 & 5 & | & -3 \\ 0 & 2 & -13 & | & 17 \end{bmatrix} -2R_2 + R_3$$

$$\begin{bmatrix} 1 & 1 & \frac{7}{2} & | & -\frac{1}{2} \\ 0 & 1 & 5 & | & -3 \\ 0 & 0 & -23 & | & 23 \end{bmatrix} -\frac{1}{23}R_3$$

$$\begin{bmatrix} 1 & 1 & \frac{7}{2} & | & -\frac{1}{2} \\ 0 & 1 & 5 & | & -3 \\ 0 & 0 & 1 & | & -1 \end{bmatrix}$$

$$z = -1$$
$$y + 5z = -3$$
$$y + 5(-1) = -3$$
$$y - 5 = -3$$
$$y = 2$$

$$x + y + \frac{7}{2}z = -\frac{1}{2}$$
$$x + 2 + \frac{7}{2}(-1) = -\frac{1}{2}$$
$$x - \frac{3}{2} = -\frac{1}{2}$$
$$x = 1$$

The solution set is $\{(1, 2, -1)\}$.

35.
$$w + x + y + z = 4$$
$$2w + x - 2y - z = 0$$
$$w - 2x - y - 2z = -2$$
$$3w + 2x + y + 3z = 4$$

$$\begin{bmatrix} 1 & 1 & 1 & 1 & | & 4 \\ 2 & 1 & -2 & -1 & | & 0 \\ 1 & -2 & -1 & -2 & | & -2 \\ 3 & 2 & 1 & 3 & | & 4 \end{bmatrix} -2R_1 + R_2$$

$$\begin{bmatrix} 1 & 1 & 1 & 1 & | & 4 \\ 0 & -1 & -4 & -3 & | & -8 \\ 1 & -2 & -1 & -2 & | & -2 \\ 3 & 2 & 1 & 3 & | & 4 \end{bmatrix} -1R_1 + R_3$$

$$\begin{bmatrix} 1 & 1 & 1 & 1 & | & 4 \\ 0 & -1 & -4 & -3 & | & -8 \\ 0 & -3 & -2 & -3 & | & -6 \\ 3 & 2 & 1 & 3 & | & 4 \end{bmatrix} -3R_1 + R_4$$

$$\begin{bmatrix} 1 & 1 & 1 & 1 & | & 4 \\ 0 & -1 & -4 & -3 & | & -8 \\ 0 & -3 & -2 & -3 & | & -6 \\ 0 & -1 & -2 & 0 & | & -8 \end{bmatrix} -1R_2$$

$$\begin{bmatrix} 1 & 1 & 1 & 1 & | & 4 \\ 0 & 1 & 4 & 3 & | & 8 \\ 0 & -3 & -2 & -3 & | & -6 \\ 0 & -1 & -2 & 0 & | & -8 \end{bmatrix} 3R_2 + R_3$$

$$\begin{bmatrix} 1 & 1 & 1 & 1 & | & 4 \\ 0 & 1 & 4 & 3 & | & 8 \\ 0 & 0 & 10 & 6 & | & 18 \\ 0 & -1 & -2 & 0 & | & -8 \end{bmatrix} 1R_2 + R_4$$

$$\begin{bmatrix} 1 & 1 & 1 & 1 & | & 4 \\ 0 & 1 & 4 & 3 & | & 8 \\ 0 & 0 & 10 & 6 & | & 18 \\ 0 & 0 & 2 & 3 & | & 0 \end{bmatrix} \frac{1}{10}R_3$$

$$\begin{bmatrix} 1 & 1 & 1 & 1 & | & 4 \\ 0 & 1 & 4 & 3 & | & 8 \\ 0 & 0 & 1 & \frac{3}{5} & | & \frac{9}{5} \\ 0 & 0 & 2 & 3 & | & 0 \end{bmatrix} -2R_3 + R_4$$

$$\begin{bmatrix} 1 & 1 & 1 & 1 & | & 4 \\ 0 & 1 & 4 & 3 & | & 8 \\ 0 & 0 & 1 & \frac{3}{5} & | & \frac{9}{5} \\ 0 & 0 & 0 & \frac{9}{5} & | & -\frac{18}{5} \end{bmatrix} \frac{5}{9}R_4$$

$$\begin{bmatrix} 1 & 1 & 1 & 1 & | & 4 \\ 0 & 1 & 4 & 3 & | & 8 \\ 0 & 0 & 1 & \frac{3}{5} & | & \frac{9}{5} \\ 0 & 0 & 0 & 1 & | & -2 \end{bmatrix}$$

$z = -2$

$y + \frac{3}{5}z = \frac{9}{5}$

$y + \frac{3}{5}(-2) = \frac{9}{5}$

$y - \frac{6}{5} = \frac{9}{5}$

$y = 3$

$x + 4y + 3z = 8$

$x + 4(3) + 3(-2) = 8$

$x + 6 = 8$

$x = 2$

$w + x + y + z = 4$

$w + 2 + 3 - 2 = 4$

$w + 3 = 4$

$w = 1$

The solution set is $\{(1, 2, 3, -2)\}$.

37.
$$3w - 4x + y + z = 9$$
$$w + x - y - z = 0$$
$$2w + x + 4y - 2z = 3$$
$$-w + 2x + y - 3z = 3$$

$$\begin{bmatrix} 3 & -4 & 1 & 1 & | & 9 \\ 1 & 1 & -1 & -1 & | & 0 \\ 2 & 1 & 4 & -2 & | & 3 \\ -1 & 2 & 1 & -3 & | & 3 \end{bmatrix} R_1 \leftrightarrow R_2$$

$$\begin{bmatrix} 1 & 1 & -1 & -1 & | & 0 \\ 3 & -4 & 1 & 1 & | & 9 \\ 2 & 1 & 4 & -2 & | & 3 \\ -1 & 2 & 1 & -3 & | & 3 \end{bmatrix} -3R_1 + R_2$$

$$\begin{bmatrix} 1 & 1 & -1 & -1 & | & 0 \\ 0 & -7 & 4 & 4 & | & 9 \\ 2 & 1 & 4 & -2 & | & 3 \\ -1 & 2 & 1 & -3 & | & 3 \end{bmatrix} -2R_1 + R_3$$

$$\begin{bmatrix} 1 & 1 & -1 & -1 & | & 0 \\ 0 & -7 & 4 & 4 & | & 9 \\ 0 & -1 & 6 & 0 & | & 3 \\ -1 & 2 & 1 & -3 & | & 3 \end{bmatrix} 1R_1 + R_4$$

$$\begin{bmatrix} 1 & 1 & -1 & -1 & | & 0 \\ 0 & -7 & 4 & 4 & | & 9 \\ 0 & -1 & 6 & 0 & | & 3 \\ 0 & 3 & 0 & -4 & | & 3 \end{bmatrix} R_2 \leftrightarrow R_3$$

$$\begin{bmatrix} 1 & 1 & -1 & -1 & | & 0 \\ 0 & -1 & 6 & 0 & | & 3 \\ 0 & -7 & 4 & 4 & | & 9 \\ 0 & 3 & 0 & -4 & | & 3 \end{bmatrix} -R_2$$

$$\begin{bmatrix} 1 & 1 & -1 & -1 & | & 0 \\ 0 & 1 & -6 & 0 & | & -3 \\ 0 & -7 & 4 & 4 & | & 9 \\ 0 & 3 & 0 & -4 & | & 3 \end{bmatrix} 7R_2 + R_3$$

$$\begin{bmatrix} 1 & 1 & -1 & -1 & | & 0 \\ 0 & 1 & -6 & 0 & | & -3 \\ 0 & 0 & -38 & 4 & | & -12 \\ 0 & 3 & 0 & -4 & | & 3 \end{bmatrix} -3R_2 + R_4$$

$$\begin{bmatrix} 1 & 1 & -1 & -1 & | & 0 \\ 0 & 1 & -6 & 0 & | & -3 \\ 0 & 0 & -38 & 4 & | & -12 \\ 0 & 0 & 18 & -4 & | & 12 \end{bmatrix} -\frac{1}{38}R_3$$

$$\begin{bmatrix} 1 & 1 & -1 & -1 & | & 0 \\ 0 & 1 & -6 & 0 & | & -3 \\ 0 & 0 & 1 & -\frac{2}{19} & | & \frac{6}{19} \\ 0 & 0 & 18 & -4 & | & 12 \end{bmatrix} -18R_3 + R_4$$

$$\begin{bmatrix} 1 & 1 & -1 & -1 & | & 0 \\ 0 & 1 & -6 & 0 & | & -3 \\ 0 & 0 & 1 & -\frac{2}{19} & | & \frac{6}{19} \\ 0 & 0 & 0 & -\frac{40}{19} & | & \frac{120}{19} \end{bmatrix} -\frac{19}{40}R_4$$

$$\begin{bmatrix} 1 & 1 & -1 & -1 & | & 0 \\ 0 & 1 & -6 & 0 & | & -3 \\ 0 & 0 & 1 & -\frac{2}{19} & | & \frac{6}{19} \\ 0 & 0 & 0 & 1 & | & -3 \end{bmatrix}$$

$z = -3$

$$y - \frac{2}{19}z = \frac{6}{19}$$

$$y - \frac{2}{19}(-3) = \frac{6}{19}$$

$$y + \frac{6}{19} = \frac{6}{19}$$

$$y = 0$$

$$x - 6y = -3$$

$$x - 6(0) = -3$$

$$x = -3$$

$$w + x - y - z = 0$$

$$w - 3 + 0 + 3 = 0$$

$$w = 0$$

The solution set is $\{(0, -3, 0, -3)\}$.

39. $f(x) = ax^2 + bx + c$

Use the given function values to find three equations in terms of a, b, and c.

$$f(-2) = a(-2)^2 + b(-2) + c = -4$$

$$4a - 2b + c = -4$$

$$f(1) = a(1)^2 + b(1) + c = 2$$

$$a + b + c = 2$$

$$f(2) = a(2)^2 + b(2) + c = 0$$

$$4a + 2b + c = 0$$

System of equations:

$$4a - 2b + c = -4$$
$$a + b + c = 2$$
$$4a + 2b + c = 0$$

Matrix:

$$\begin{bmatrix} 4 & -2 & 1 & | & -4 \\ 1 & 1 & 1 & | & 2 \\ 4 & 2 & 1 & | & 0 \end{bmatrix}$$

This gives $a = -1$, $b = 1$, and $c = 2$.

Thus, $f(x) = -x^2 + x + 2$.

41. $f(x) = ax^3 + bx^2 + cx + d$

Use the given function values to find four equations in terms of a, b, c, and d.

$$f(-1) = a(-1)^3 + b(-1)^2 + c(-1) + d = 0$$

$$-a + b - c + d = 0$$

$$f(1) = a(1)^3 + b(1)^2 + c(1) + d = 2$$

$$a + b + c + d = 2$$

$$f(2) = a(2)^3 + b(2)^2 + c(2) + d = 3$$

$$8a + 4b + 2c + d = 3$$

$$f(3) = a(3)^3 + b(3)^2 + c(3) + d = 12$$

$$27a + 9b + 3c + d = 12$$

System of equations:

$$-a + b - c + d = 0$$
$$a + b + c + d = 2$$
$$8a + 4b + 2c + d = 3$$
$$27a + 9b + 3c + d = 12$$

Matrix:

$$\begin{bmatrix} -1 & 1 & -1 & 1 & | & 0 \\ 1 & 1 & 1 & 1 & | & 2 \\ 8 & 4 & 2 & 1 & | & 3 \\ 27 & 9 & 3 & 1 & | & 12 \end{bmatrix}$$

This gives $a = 1$, $b = -2$, $c = 0$, and $d = 3$.

Thus, $f(x) = x^3 - 2x^2 + 3$.

43. Let $A = \ln w$, $B = \ln x$, $C = \ln y$, and $D = \ln z$.

System of equations:

$$2A + B + 3C - 2D = -6$$
$$4A + 3B + C - D = -2$$
$$A + B + C + D = -5$$
$$A + B - C - D = 5$$

Matrix:

$$\begin{bmatrix} 2 & 1 & 3 & -2 & | & -6 \\ 4 & 3 & 1 & -1 & | & -2 \\ 1 & 1 & 1 & 1 & | & -5 \\ 1 & 1 & -1 & -1 & | & 5 \end{bmatrix}$$

This gives $A = -1$, $B = 1$, $C = -3$, and $D = -2$.

Substitute back to find w, x, y, and z.

$A = -1$	$B = 1$
$\ln w = -1$	$\ln x = 1$
$w = e^{-1}$	$x = e^{1}$
$w \approx 0.37$	$x \approx 2.72$
$C = -3$	$D = -2$
$\ln y = -3$	$\ln z = -2$
$y = e^{-3}$	$z = e^{-2}$
$y \approx 0.05$	$z \approx 0.14$

45. a. $s(t) = \dfrac{1}{2}at^2 + v_0 t + s_0$

Use the given function values to find three equations in terms of a, v_0, and s_0.

$$s(1) = \frac{1}{2}a(1)^2 + v_0(1) + s_0 = 40$$
$$\frac{1}{2}a + v_0 + s_0 = 40$$

$$s(2) = \frac{1}{2}a(2)^2 + v_0(2) + s_0 = 48$$
$$2a + 2v_0 + s_0 = 48$$

$$s(3) = \frac{1}{2}a(3)^2 + v_0(3) + s_0 = 24$$
$$\frac{9}{2}a + 3v_0 + s_0 = 24$$

System of equations:

$$\frac{1}{2}a + v_0 + s_0 = 40$$
$$2a + 2v_0 + s_0 = 48$$
$$\frac{9}{2}a + 3v_0 + s_0 = 24$$

Matrix:

$$\begin{bmatrix} \frac{1}{2} & 1 & 1 & | & 40 \\ 2 & 2 & 1 & | & 48 \\ \frac{9}{2} & 3 & 1 & | & 24 \end{bmatrix}$$

This gives $a = -32$, $v_0 = 56$, and $s_0 = 0$.

Thus, $s(t) = \dfrac{1}{2}(-32)t^2 + (56)t + (0)$

$$s(t) = -16t^2 + 56t$$

b. $s(t) = -16t^2 + 56t$

$$s(3.5) = -16(3.5)^2 + 56(3.5) = 0$$

This is the point $(3.5, 0)$.

The ball's height is 0 feet after 3.5 seconds.

This is the point $(3.5, 0)$.

c. The maximum occurs when $x = -\dfrac{b}{2a}$.

$$x = -\frac{b}{2a} = -\frac{v_0}{2a} = -\frac{56}{2(-16)} = 1.75$$

$$s(1.75) = -16(1.75)^2 + 56(1.75) = 49$$

At 1.75 seconds the ball will reach its maximum height of 49 feet.

47. Let $x = $ Food A

Let $y = $ Food B

Let $z = $ Food C

$$40x + 200y + 400z = 660$$
$$5x + 2y + 4z = 25$$
$$30x + 10y + 300z = 425$$
$$2x + 10y + 20z = 33$$
$$5x + 2y + 4z = 25$$
$$6x + 2y + 60z = 85$$

$$\begin{bmatrix} 2 & 10 & 20 & | & 33 \\ 5 & 2 & 4 & | & 25 \\ 6 & 2 & 60 & | & 85 \end{bmatrix} \frac{1}{2}R_1$$

$$\begin{bmatrix} 1 & 5 & 10 & | & \frac{33}{2} \\ 5 & 2 & 4 & | & 25 \\ 6 & 2 & 60 & | & 85 \end{bmatrix} -5R_1 + R_2$$

$$\begin{bmatrix} 1 & 5 & 10 & | & \frac{33}{2} \\ 0 & -23 & -46 & | & -\frac{115}{2} \\ 6 & 2 & 60 & | & 85 \end{bmatrix} -6R_1 + R_3$$

$$\begin{bmatrix} 1 & 5 & 10 & | & \frac{33}{2} \\ 0 & -23 & -46 & | & -\frac{115}{2} \\ 0 & -28 & 0 & | & -14 \end{bmatrix} -\frac{1}{23}R_2$$

$$\begin{bmatrix} 1 & 5 & 10 & | & \frac{33}{2} \\ 0 & 1 & 2 & | & \frac{5}{2} \\ 0 & -28 & 0 & | & -14 \end{bmatrix} 28R_2 + R_3$$

$$\begin{bmatrix} 1 & 5 & 10 & \Big| & \frac{33}{2} \\ 0 & 1 & 2 & \Big| & \frac{5}{2} \\ 0 & 0 & 56 & \Big| & 56 \end{bmatrix} \frac{1}{56}R_3$$

$$\begin{bmatrix} 1 & 5 & 10 & \Big| & \frac{33}{2} \\ 0 & 1 & 2 & \Big| & \frac{5}{2} \\ 0 & 0 & 1 & \Big| & 1 \end{bmatrix}$$

$$z = 1$$

$$y + 2z = \frac{5}{2}$$

$$y + 2 = \frac{5}{2}$$

$$2y + 4 = 5$$

$$2y = 1$$

$$y = \frac{1}{2}$$

$$x + 5y + 10z = \frac{33}{2}$$

$$x + \frac{5}{2} + 10 = \frac{33}{2}$$

$$2x + 5 + 20 = 33$$

$$2x + 25 = 33$$

$$2x = 8$$

$$x = 4$$

4 ounces of Food A

$\frac{1}{2}$ ounce of Food B

1 ounce of Food C

49. Let w = number of Asians
Let x = number of Africans
Let y = number of Europeans
Let z = number of Americans
Use the variables to model each sentence.

$$w + x + y + z = 183$$

$$w - x - y = 70$$

$$y - z = 15$$

$$2x - y - z = 23$$

$$\begin{bmatrix} 1 & 1 & 1 & 1 & \Big| & 183 \\ 1 & -1 & -1 & 0 & \Big| & 70 \\ 0 & 0 & 1 & -1 & \Big| & 15 \\ 0 & 2 & -1 & -1 & \Big| & 23 \end{bmatrix} -R_1 + R_2$$

$$\begin{bmatrix} 1 & 1 & 1 & 1 & \Big| & 183 \\ 0 & -2 & -2 & -1 & \Big| & -113 \\ 0 & 0 & 1 & -1 & \Big| & 15 \\ 0 & 2 & -1 & -1 & \Big| & 23 \end{bmatrix} R_4 + R_2$$

$$\begin{bmatrix} 1 & 1 & 1 & 1 & \Big| & 183 \\ 0 & -2 & -2 & -1 & \Big| & -113 \\ 0 & 0 & 1 & -1 & \Big| & 15 \\ 0 & 0 & -3 & -2 & \Big| & -90 \end{bmatrix} -\frac{1}{2}R_2$$

$$\begin{bmatrix} 1 & 1 & 1 & 1 & \Big| & 183 \\ 0 & 1 & 1 & \frac{1}{2} & \Big| & \frac{113}{2} \\ 0 & 0 & 1 & -1 & \Big| & 15 \\ 0 & 0 & -3 & -2 & \Big| & -90 \end{bmatrix} 3R_3 + R_4$$

$$\begin{bmatrix} 1 & 1 & 1 & 1 & \Big| & 183 \\ 0 & 1 & 1 & \frac{1}{2} & \Big| & \frac{113}{2} \\ 0 & 0 & 1 & -1 & \Big| & 15 \\ 0 & 0 & 0 & -5 & \Big| & -45 \end{bmatrix} -\frac{1}{5}R_4$$

$$\begin{bmatrix} 1 & 1 & 1 & 1 & \Big| & 183 \\ 0 & 1 & 1 & \frac{1}{2} & \Big| & \frac{113}{2} \\ 0 & 0 & 1 & -1 & \Big| & 15 \\ 0 & 0 & 0 & 1 & \Big| & 9 \end{bmatrix}$$

Back-substitute $z = 9$ to find y.

$$y - z = 15$$

$$y - 9 = 15$$

$$y = 24$$

Back-substitute to find x.

$$x + y + \frac{1}{2}z = \frac{113}{2}$$

$$x + 24 + \frac{1}{2}(9) = \frac{113}{2}$$

$$x + 24 + \frac{1}{2}(9) = \frac{113}{2}$$

$$x + 28\frac{1}{2} = 56\frac{1}{2}$$

$$x = 28$$

Back-substitute to find w.

$$w + x + y + z = 183$$

$$w + 28 + 24 + 9 = 183$$

$$w + 28 + 24 + 9 = 183$$

$$w + 61 = 183$$

$$w = 122$$

The number of Asians, Africans, Europeans, and Americans are 122, 28, 24, and 9, respectively.

51. – 57. Answers may vary.

59. makes sense

61. makes sense

63. false; Changes to make the statement true will vary. A sample change is: Multiplying a row by a negative fraction is permitted.

65. false; Changes to make the statement true will vary. A sample change is: When solving a system of three equations in three variables, we use row operations to obtain ones along the diagonal and zeros below the ones.

67.
$$y = ax^2 + bx + c$$
$$5900 = a(30)^2 + b(30) + c$$
$$5900 = 900a + 30b + c$$
$$7500 = a(50)^2 + b(50) + c$$
$$7500 = 2500a + 50b + c$$
$$4500 = a(100)^2 + b(100) + c$$
$$4500 = 10,000a + 100b + c$$

$$900a + 30b + c = 5900$$
$$2500a + 50b + c = 7500$$
$$10,000a + 100b + c = 4500$$

$$\begin{bmatrix} 900 & 30 & 1 & | & 5900 \\ 2500 & 50 & 1 & | & 7500 \\ 10,000 & 100 & 1 & | & 4500 \end{bmatrix} R_1 \leftrightarrow R_2$$

$$\begin{bmatrix} 2500 & 50 & 1 & | & 7500 \\ 900 & 30 & 1 & | & 5900 \\ 10,000 & 100 & 1 & | & 4500 \end{bmatrix} \frac{1}{2500} R_1$$

$$\begin{bmatrix} 1 & \frac{1}{50} & \frac{1}{2500} & | & 3 \\ 900 & 30 & 1 & | & 5900 \\ 10,000 & 100 & 1 & | & 4500 \end{bmatrix} \begin{matrix} -900R_1 + R_2 \\ -10,000R_1 + R_3 \end{matrix}$$

$$\begin{bmatrix} 1 & \frac{1}{50} & \frac{1}{2500} & | & 3 \\ 0 & 12 & \frac{16}{25} & | & 3200 \\ 0 & -100 & -3 & | & -25,500 \end{bmatrix} R_2 \leftrightarrow R_3$$

$$\begin{bmatrix} 1 & \frac{1}{50} & \frac{1}{2500} & | & 3 \\ 0 & -100 & -3 & | & -25,500 \\ 0 & 12 & \frac{16}{25} & | & 3200 \end{bmatrix} \frac{-1}{100} R_2$$

$$\begin{bmatrix} 1 & \frac{1}{50} & \frac{1}{2500} & | & 3 \\ 0 & 1 & \frac{3}{100} & | & 255 \\ 0 & 12 & \frac{16}{25} & | & 3200 \end{bmatrix} \begin{matrix} -12R_2 + R_3 \\ -\frac{1}{50}R_2 + R_1 \end{matrix}$$

$$\begin{bmatrix} 1 & 0 & \frac{-1}{5000} & | & \frac{-21}{10} \\ 0 & 1 & \frac{3}{100} & | & 255 \\ 0 & 0 & \frac{7}{25} & | & 140 \end{bmatrix} \frac{25}{7} R_3$$

$$\begin{bmatrix} 1 & 0 & \frac{-1}{5000} & | & \frac{-21}{10} \\ 0 & 1 & \frac{3}{100} & | & 255 \\ 0 & 0 & 1 & | & 500 \end{bmatrix} \begin{matrix} \frac{-3}{100}R_3 + R_2 \\ \frac{1}{5000}R_3 + R_1 \end{matrix}$$

$$\begin{bmatrix} 1 & 0 & 0 & | & -2 \\ 0 & 1 & 0 & | & 240 \\ 0 & 0 & 1 & | & 500 \end{bmatrix}$$

$$y = -2x^2 + 240x + 500$$
$$y = -2(x^2 - 120x) + 500$$
$$y = -2(x^2 - 120x + 3600) + 500 + 7200$$
$$y = -2(x - 60)^2 + 7700$$
60 units produce $7700.

68. When $z = 0$, $(12z + 1, 10z - 1, z)$ is equivalent to $(12(0) + 1, 10(0) - 1, 0)$ or $(1, -1, 0)$.

Check $(1, -1, 0)$ in each equation.
$$3x - 4y + 4z = 7$$
$$3(1) - 4(-1) + 4(0) = 7$$
$$7 = 7, \text{ true}$$
$$x - y - 2z = 2$$
$$1 - (-1) - 2(0) = 2$$
$$2 = 2, \text{ true}$$
$$2x - 3y - 2z = 5$$
$$2(1) - 3(-1) + 6(0) = 5$$
$$5 = 5, \text{ true}$$

$(1, -1, 0)$ satisfies each equation and, therefore, satisfies the system.

69. When $z = 1$, $(12z + 1, 10z - 1, z)$ is equivalent to $(12(1) + 1, 10(1) - 1, 1)$ or $(13, 9, 1)$.

Check $(13, 9, 1)$ in each equation.
$$3x - 4y + 4z = 7$$
$$3(13) - 4(9) + 4(1) = 7$$
$$7 = 7, \text{ true}$$
$$x - y - 2z = 2$$
$$13 - (9) - 2(1) = 2$$
$$2 = 2, \text{ true}$$
$$2x - 3y - 2z = 5$$
$$2(13) - 3(9) + 6(1) = 5$$
$$5 = 5, \text{ true}$$

$(13, 9, 1)$ satisfies each equation and, therefore, satisfies the system.

70. **a.** Answers may vary. A sample answer is given selecting $z = 10$.

When $z = 10$, $(12z + 1, 10z - 1, z)$ is equivalent to $(12(10) + 1, 10(10) - 1, 10)$ or $(121, 99, 10)$.

Check $(121, 99, 10)$ in each equation.

$$3x - 4y + 4z = 7$$
$$3(121) - 4(99) + 4(10) = 7$$
$$7 = 7, \text{ true}$$
$$x - y - 2z = 2$$
$$121 - (99) - 2(10) = 2$$
$$2 = 2, \text{ true}$$
$$2x - 3y - 2z = 5$$
$$2(121) - 3(99) + 6(10) = 5$$
$$5 = 5, \text{ true}$$

$(121, 99, 10)$ satisfies each equation and, therefore, satisfies the system.

b. This system has more than one solution.

Section 8.2

Check Point Exercises

1.
$$\begin{aligned}x - 2y - z &= 5 \\ 2x - 3y - z &= 0 \\ 3x - 4y - z &= 1\end{aligned} \rightarrow \begin{bmatrix} 1 & -2 & -1 & | & -5 \\ 2 & -3 & -1 & | & 0 \\ 3 & -4 & -1 & | & 1 \end{bmatrix}$$

$$\begin{bmatrix} 1 & -2 & -1 & | & -5 \\ 2 & -3 & -1 & | & 0 \\ 3 & -4 & -1 & | & 1 \end{bmatrix} \begin{aligned} -2R_1 + R_2 \\ -3R_1 + R_3 \end{aligned}$$

$$\begin{bmatrix} 1 & -2 & -1 & | & -5 \\ 0 & 1 & 1 & | & 10 \\ 0 & 2 & 2 & | & 16 \end{bmatrix} -2R_2 + R_3$$

$$\begin{bmatrix} 1 & -2 & -1 & | & -5 \\ 0 & 1 & -1 & | & -10 \\ 0 & 0 & 0 & | & -4 \end{bmatrix}$$

$0x + 0y + 0z = -4$ This equation can never be a true statement. Consequently, the system has no solution. The solution set is $\varnothing$, the empty set.

2.
$$\begin{aligned} x - 2y - z &= 5 \\ 2x - 5y + 3z &= 16 \\ x - 3y + 4z &= 1 \end{aligned} \rightarrow \begin{bmatrix} 1 & -2 & -1 & | & 5 \\ 2 & -5 & 3 & | & 6 \\ 1 & -3 & 4 & | & 1 \end{bmatrix}$$

$$\begin{bmatrix} 1 & -2 & -1 & | & 5 \\ 2 & -5 & 3 & | & 6 \\ 1 & -3 & 4 & | & 1 \end{bmatrix} \begin{aligned} -2R_1 + R_2 \\ -1R_1 + R_3 \end{aligned}$$

$$\begin{bmatrix} 1 & -2 & -1 & | & 5 \\ 0 & -1 & 5 & | & -4 \\ 0 & -1 & 5 & | & -4 \end{bmatrix} -1R_2$$

$$\begin{bmatrix} 1 & -2 & -1 & | & 5 \\ 0 & 1 & -5 & | & 4 \\ 0 & -1 & 5 & | & -4 \end{bmatrix} 1R_2 + R_3$$

$$\begin{bmatrix} 1 & -2 & -1 & | & 5 \\ 0 & 1 & -5 & | & 4 \\ 0 & 0 & 0 & | & 0 \end{bmatrix}$$

$0x + 0y + 0z = 0$ or $0 = 0$

This equation, $0x + 0y + 0z = 0$ is *dependent* on the other two equations. Thus, it can be dropped from the system which can now be expressed in the form

$$\begin{bmatrix} 1 & -2 & -1 & | & 5 \\ 0 & 1 & -5 & | & 4 \end{bmatrix}$$

The original system is equivalent to the system
$$x - 2y - z = 5$$
$$y - 5z = 4$$
Solve for x and y in terms of z.
$$y = 5z + 4$$
Use back-substitution for y in the previous equation.
$$x - 2(5z + 4) - z = 5$$
$$x - 10z - 8 - z = 5$$
$$x = 11z + 13$$
Finally, letting $z = t$ (or any letter of your choice), the solutions to the system are all of the form $x = 11t + 13$, $y = 5t + 4$, $z = t$, where t is a real number. The solution set of the system with dependent equations can be written as $\{(11t + 13, 5t + 4, t)\}$.

3. $\begin{aligned} x+2y+3z &= 70 \\ x+y+z &= 60 \end{aligned} \rightarrow \begin{bmatrix} 1 & 2 & 3 & | & 70 \\ 1 & 1 & 1 & | & 60 \end{bmatrix}$

$\begin{bmatrix} 1 & 2 & 3 & | & 70 \\ 1 & 1 & 1 & | & 60 \end{bmatrix} -1R_1 + R_2$

$\begin{bmatrix} 1 & 2 & 3 & | & 70 \\ 0 & -1 & -2 & | & -10 \end{bmatrix} -1R_2$

$\begin{bmatrix} 1 & 2 & 3 & | & 70 \\ 0 & 1 & 2 & | & 10 \end{bmatrix} \rightarrow \begin{aligned} x+2y+3z &= 70 \\ y+2z &= 10 \end{aligned}$

Express x and y in terms of z using back-substitution.

$y = -2z + 10$

$x + 2(-2z + 10) + 3z = 70$

$x - 4z + 20 + 3z = 70$

$x = z + 50$

With $z = t$, the ordered solution (x, y, z) enables us to express the system's solution set as $\{(t+50, -2t+10, t)\}$.

4. a. I_1: $10 + 5 = 15$ cars enter I_1, and $w + z$ cars leave I_1, then $w + z = 15$.

I_2: $20 + 10 = 30$ cars enter I_2 and $w + x$ cars leave I_2, then $w + x = 30$.

I_3: $15 + 30 = 45$ cars enter I_3 and $x + y$ cars leave I_3, then $x + y = 45$.

I_4: $10 + 20 = 30$ cars enter I_4 and $y + z$ cars leave I_4, then $y + z = 30$.

The system of equations that describes this situation is given by

$w + z = 15$

$w + x = 30$

$x + y = 45$

$y + z = 30$

b. $\begin{bmatrix} 1 & 0 & 0 & 1 & | & 15 \\ 1 & 1 & 0 & 0 & | & 30 \\ 0 & 1 & 1 & 0 & | & 45 \\ 0 & 0 & 1 & 1 & | & 30 \end{bmatrix} -1R_1 + R_2$

$\begin{bmatrix} 1 & 0 & 0 & 1 & | & 15 \\ 0 & 1 & 0 & -1 & | & 15 \\ 0 & 1 & 1 & 0 & | & 45 \\ 0 & 0 & 1 & 1 & | & 30 \end{bmatrix} -1R_2 + R_3$

$\begin{bmatrix} 1 & 0 & 0 & 1 & | & 15 \\ 0 & 1 & 0 & -1 & | & 15 \\ 0 & 0 & 1 & 1 & | & 30 \\ 0 & 0 & 1 & 1 & | & 30 \end{bmatrix} -1R_3 + R_4$

$\begin{bmatrix} 1 & 0 & 0 & 1 & | & 15 \\ 0 & 1 & 0 & -1 & | & 15 \\ 0 & 0 & 1 & 1 & | & 30 \\ 0 & 0 & 0 & 0 & | & 0 \end{bmatrix}$

$x + w = 15$

$y - w = 15$

$z + w = 30$

The last row of the matrix shows that the system has dependent equations and infinitely many solutions.

Let z be any real number.

Express w, x and y in terms of z:

$w = 15 - z$

$x = 15 + z$

$y = 30 - z$

With $w = t$, the ordered solution (w, x, y, z) enables us to express the system's solution set as $\{(-t+15, t+15, -t+30, t)\}$

Exercise Set 8.2

1. $\begin{bmatrix} 5 & 12 & 1 & | & 10 \\ 2 & 5 & 2 & | & -1 \\ 1 & 2 & -3 & | & 5 \end{bmatrix} R_1 \leftrightarrow R_3$

$\begin{bmatrix} 1 & 2 & -3 & | & 5 \\ 2 & 5 & 2 & | & -1 \\ 5 & 12 & 1 & | & 10 \end{bmatrix} \begin{aligned} -2R_1 + R_2 \\ -5R_1 + R_3 \end{aligned}$

$\begin{bmatrix} 1 & 2 & -3 & | & 5 \\ 0 & 1 & 8 & | & -11 \\ 0 & 2 & 16 & | & -15 \end{bmatrix} -2R_2 + R_3$

$\begin{bmatrix} 1 & 2 & 3 & | & 5 \\ 0 & 1 & 8 & | & -11 \\ 0 & 0 & 0 & | & 7 \end{bmatrix}$

From the last row, we see that the system has no solution. The solution set is $\varnothing$, the empty set.

3.
$$\begin{bmatrix} 5 & 8 & -6 & | & 14 \\ 3 & 4 & -2 & | & 8 \\ 1 & 2 & -2 & | & 3 \end{bmatrix} R_1 \leftrightarrow R_3$$

$$\begin{bmatrix} 1 & 2 & -2 & | & 3 \\ 3 & 4 & -2 & | & 8 \\ 5 & 8 & -6 & | & 14 \end{bmatrix} \begin{matrix} -3R_1 + R_2 \\ -5R_1 + R_3 \end{matrix}$$

$$\begin{bmatrix} 1 & 2 & -2 & | & 3 \\ 0 & -2 & 4 & | & -1 \\ 0 & -2 & 4 & | & -1 \end{bmatrix} -1R_2 + R_3$$

$$\begin{bmatrix} 1 & 2 & -2 & | & 3 \\ 0 & -2 & 4 & | & -1 \\ 0 & 0 & 0 & | & 0 \end{bmatrix} -\frac{1}{2}R_2$$

$$\begin{bmatrix} 1 & 2 & -2 & | & 3 \\ 0 & 1 & -2 & | & \frac{1}{2} \\ 0 & 0 & 0 & | & 0 \end{bmatrix}$$

The system $\begin{aligned} x + 2y - 2z &= 3 \\ y - 2z &= \frac{1}{2} \end{aligned}$ has no unique solution.

Express x and y in terms of z:

$$y = 2z + \frac{1}{2}$$

$$x + 2\left(2z + \frac{1}{2}\right) - 2z = 3$$

$$x + 4z + 1 - 2z = 3$$

$$x + 2z + 1 = 3$$

$$x = -2z + 2$$

With $z = t$, the complete solution to the system is

$$\left\{\left(-2t + 2, \; 2t + \frac{1}{2}, \; t\right)\right\}.$$

5.
$$\begin{bmatrix} 3 & 4 & 2 & | & 3 \\ 4 & -2 & -8 & | & -4 \\ 1 & 1 & -1 & | & 3 \end{bmatrix} R_1 \leftrightarrow R_3$$

$$\begin{bmatrix} 1 & 1 & -1 & | & 3 \\ 4 & -2 & -8 & | & -4 \\ 3 & 4 & 2 & | & 3 \end{bmatrix} \begin{matrix} -4R_1 + R_2 \\ -3R_1 + R_3 \end{matrix}$$

$$\begin{bmatrix} 1 & 1 & -1 & | & 3 \\ 0 & -6 & -4 & | & -16 \\ 0 & 1 & 5 & | & -6 \end{bmatrix} R_2 \leftrightarrow R_3$$

$$\begin{bmatrix} 1 & 1 & -1 & | & 3 \\ 0 & 1 & 5 & | & -6 \\ 0 & -6 & -4 & | & -16 \end{bmatrix} 6R_2 + R_3$$

$$\begin{bmatrix} 1 & 1 & 1 & | & 3 \\ 0 & 1 & 5 & | & -6 \\ 0 & 0 & 26 & | & -52 \end{bmatrix} \frac{1}{26}R_3$$

$$\begin{bmatrix} 1 & 1 & -1 & | & 3 \\ 0 & 1 & 5 & | & -6 \\ 0 & 0 & 1 & | & -2 \end{bmatrix}$$

This corresponds to the system
$$\begin{aligned} x + y - z &= 3 \\ y + 5z &= -6 \\ z &= -2 \end{aligned}$$

Use back-substitution to find the values of x and y:
$$\begin{aligned} y + 5(-2) &= -6 \\ y - 10 &= -6 \\ y &= 4 \\ x + 4 + 2 &= 3 \\ x + 6 &= 3 \\ x &= -3 \end{aligned}$$

The solution to the system is $\{(-3, 4, -2)\}$.

7.
$$\begin{bmatrix} 8 & 5 & 11 & | & 30 \\ -1 & -4 & 2 & | & 3 \\ 2 & -1 & 5 & | & 12 \end{bmatrix} R_1 \leftrightarrow R_2$$

$$\begin{bmatrix} -1 & -4 & 2 & | & 3 \\ 8 & 5 & 11 & | & 30 \\ 2 & -1 & 5 & | & 12 \end{bmatrix} -1R_1$$

$$\begin{bmatrix} 1 & 4 & -2 & | & -3 \\ 8 & 5 & 11 & | & 30 \\ 2 & -1 & 5 & | & 12 \end{bmatrix} \begin{matrix} -8R_1 + R_2 \\ \\ -2R_1 + R_3 \end{matrix}$$

$$\begin{bmatrix} 1 & 4 & -2 & | & -3 \\ 0 & -27 & 27 & | & 54 \\ 0 & -9 & 9 & | & 18 \end{bmatrix} -\frac{1}{27}R_2$$

$$\begin{bmatrix} 1 & 4 & -2 & | & -3 \\ 0 & 1 & -1 & | & -2 \\ 0 & -9 & 9 & | & 18 \end{bmatrix} 9R_2 + R_3$$

$$\begin{bmatrix} 1 & 4 & -2 & | & -3 \\ 0 & 1 & -1 & | & -2 \\ 0 & 0 & 0 & | & 0 \end{bmatrix}$$

The system $\begin{matrix} x + 4y - 2z = -3 \\ y - z = -2 \end{matrix}$ has no unique solution.

Express x and y in terms of z:

$y = -2 + z$

$x + 4(-2 + z) - 2z = -3$

$x - 8 + 4z - 2z = -3$

$x - 8 + 2z = -3$

$x = 5 - 2z$

With $z = t$, the complete solution to the system is $\{(5 - 2t, -2 + t, t)\}$.

9.
$$\begin{bmatrix} 1 & -2 & -1 & -3 & | & -9 \\ 1 & 1 & -1 & 0 & | & 0 \\ 3 & 4 & 0 & 1 & | & 6 \\ 0 & 2 & -2 & 1 & | & 3 \end{bmatrix} \begin{matrix} -1R_1 + R_2 \\ -3R_1 + R_3 \end{matrix}$$

$$\begin{bmatrix} 1 & -2 & -1 & -3 & | & -9 \\ 0 & 3 & 0 & 3 & | & 9 \\ 0 & 10 & 3 & 10 & | & 33 \\ 0 & 2 & -2 & 1 & | & 3 \end{bmatrix} \frac{1}{3}R_2$$

$$\begin{bmatrix} 1 & -2 & -1 & -3 & | & -9 \\ 0 & 1 & 0 & 1 & | & 3 \\ 0 & 10 & 3 & 10 & | & 33 \\ 0 & 2 & -2 & 1 & | & 3 \end{bmatrix} \begin{matrix} -10R_2 + R_3 \\ -2R_2 + R_4 \end{matrix}$$

$$\begin{bmatrix} 1 & -2 & -1 & -3 & | & -9 \\ 0 & 1 & 0 & 1 & | & 3 \\ 0 & 0 & 3 & 0 & | & 3 \\ 0 & 0 & -2 & -1 & | & -3 \end{bmatrix} \frac{1}{3}R_3$$

$$\begin{bmatrix} 1 & -2 & -1 & -3 & | & -9 \\ 0 & 1 & 0 & 1 & | & 3 \\ 0 & 0 & 1 & 0 & | & 1 \\ 0 & 0 & -2 & -1 & | & -3 \end{bmatrix} 2R_3 + R_4$$

$$\begin{bmatrix} 1 & -2 & -1 & -3 & | & -9 \\ 0 & 1 & 0 & 1 & | & 3 \\ 0 & 0 & 1 & 0 & | & 1 \\ 0 & 0 & 0 & -1 & | & -1 \end{bmatrix} -1R_4$$

$$\begin{bmatrix} 1 & -2 & -1 & -3 & | & -9 \\ 0 & 1 & 0 & 1 & | & 3 \\ 0 & 0 & 1 & 0 & | & 1 \\ 0 & 0 & 0 & 1 & | & 1 \end{bmatrix}$$

This corresponds to the system

$w - 2x - y - 3z = -9$

$x + z = 3$

$y = 1$

$z = 1$

Use back-substitution to find the values of w and x:

$x + 1 = 3$

$x = 2$

$w - 2(2) - 1 - 3(1) = -9$

$w - 4 - 1 - 3 = -9$

$w - 8 = -9$

$w = -1$

The solution to the system is $\{(-1, 2, 1, 1)\}$.

11. $\begin{bmatrix} 2 & 1 & -1 & 0 & | & 3 \\ 1 & -3 & 2 & 0 & | & -4 \\ 3 & 1 & -3 & 1 & | & 1 \\ 1 & 2 & -4 & -1 & | & -2 \end{bmatrix} R_1 \leftrightarrow R_2$

$\begin{bmatrix} 1 & -3 & 2 & 0 & | & -4 \\ 2 & 1 & -1 & 0 & | & 3 \\ 3 & 1 & -3 & 1 & | & 1 \\ 1 & 2 & -4 & -1 & | & -2 \end{bmatrix} \begin{matrix} -2R_1 + R_2 \\ -3R_1 + R_3 \\ -1R_1 + R_4 \end{matrix}$

$\begin{bmatrix} 1 & -3 & 2 & 0 & | & -4 \\ 0 & 7 & -5 & 0 & | & 11 \\ 0 & 10 & -9 & 1 & | & 13 \\ 0 & 5 & -6 & -1 & | & 2 \end{bmatrix} \frac{1}{7}R_2$

$\begin{bmatrix} 1 & -3 & 2 & 0 & | & -4 \\ 0 & 1 & -\frac{5}{7} & 0 & | & \frac{11}{7} \\ 0 & 10 & -9 & 1 & | & 13 \\ 0 & 5 & -6 & -1 & | & 2 \end{bmatrix} \begin{matrix} -10R_2 + R_3 \\ -5R_2 + R_4 \end{matrix}$

$\begin{bmatrix} 1 & -3 & 2 & 0 & | & -4 \\ 0 & 1 & -\frac{5}{7} & 0 & | & \frac{11}{7} \\ 0 & 0 & -\frac{13}{7} & 1 & | & -\frac{19}{7} \\ 0 & 0 & -\frac{17}{7} & -1 & | & -\frac{41}{7} \end{bmatrix} -\frac{7}{13}R_3$

$\begin{bmatrix} 1 & -3 & 2 & 0 & | & -4 \\ 0 & 1 & -\frac{5}{7} & 0 & | & \frac{11}{7} \\ 0 & 0 & 1 & -\frac{7}{13} & | & \frac{19}{13} \\ 0 & 0 & -\frac{17}{7} & -1 & | & -\frac{41}{7} \end{bmatrix} \frac{17}{7}R_3 + R_4$

$\begin{bmatrix} 1 & -3 & 2 & 0 & | & -4 \\ 0 & 1 & -\frac{5}{7} & 0 & | & \frac{11}{7} \\ 0 & 0 & 1 & -\frac{7}{13} & | & \frac{19}{13} \\ 0 & 0 & 0 & -\frac{30}{13} & | & -\frac{30}{13} \end{bmatrix} -\frac{13}{30}R_4$

$\begin{bmatrix} 1 & -3 & 2 & 0 & | & -4 \\ 0 & 1 & -\frac{5}{7} & 0 & | & \frac{11}{7} \\ 0 & 0 & 1 & -\frac{7}{13} & | & \frac{19}{13} \\ 0 & 0 & 0 & 1 & | & 1 \end{bmatrix}$

This corresponds to the system
$$w - 3x + 2y = -4$$
$$x - \frac{5}{7}y = \frac{11}{7}$$
$$y - \frac{7}{13}z = \frac{19}{13}$$
$$z = 1$$

Use back-substitution to find the values of w, x, and y:
$$y - \frac{7}{13}z = \frac{19}{13}$$
$$y - \frac{7}{13}(1) = \frac{19}{13}$$
$$y = 2$$
$$x - \frac{5}{7}(2) = \frac{11}{7}$$
$$x - \frac{10}{7} = \frac{11}{7}$$
$$x = 3$$
$$w - 3(3) + 2(2) = -4$$
$$w - 9 + 4 = -4$$
$$w - 5 = -4$$
$$w = 1$$

The solution to the system is $\{(1, 3, 2, 1)\}$.

13. $\begin{bmatrix} 1 & -3 & 1 & -4 & | & 4 \\ -2 & 1 & 2 & 0 & | & -2 \\ 3 & -2 & 1 & -6 & | & 2 \\ -1 & 3 & 2 & -1 & | & -6 \end{bmatrix} \begin{matrix} 2R_1 + R_2 \\ -3R_1 + R_3 \\ R_1 + R_4 \end{matrix}$

$\begin{bmatrix} 1 & -3 & 1 & -4 & | & 4 \\ 0 & -5 & 4 & -8 & | & 6 \\ 0 & 7 & -2 & 6 & | & -10 \\ 0 & 0 & 3 & -5 & | & -2 \end{bmatrix} -\frac{1}{5}R_2$

$\begin{bmatrix} 1 & -3 & 1 & -4 & | & 4 \\ 0 & 1 & -\frac{4}{5} & \frac{8}{5} & | & -\frac{6}{5} \\ 0 & 7 & -2 & 6 & | & -10 \\ 0 & 0 & 3 & -5 & | & -2 \end{bmatrix} -7R_2 + R_3$

$\begin{bmatrix} 1 & -3 & 1 & -4 & | & 4 \\ 0 & 1 & -\frac{4}{5} & \frac{8}{5} & | & -\frac{6}{5} \\ 0 & 0 & \frac{18}{5} & -\frac{26}{5} & | & -\frac{8}{5} \\ 0 & 0 & 3 & -5 & | & -2 \end{bmatrix} \frac{5}{18}R_3$

$\begin{bmatrix} 1 & -3 & 1 & -4 & | & 4 \\ 0 & 1 & -\frac{4}{5} & \frac{8}{5} & | & -\frac{6}{5} \\ 0 & 0 & 1 & -\frac{13}{9} & | & -\frac{4}{9} \\ 0 & 0 & 3 & -5 & | & -2 \end{bmatrix} -3R_3 + R_4$

$\begin{bmatrix} 1 & -3 & 1 & -4 & | & 4 \\ 0 & 1 & -\frac{4}{5} & \frac{8}{5} & | & -\frac{6}{5} \\ 0 & 0 & 1 & -\frac{13}{9} & | & -\frac{4}{9} \\ 0 & 0 & 0 & -\frac{2}{3} & | & -\frac{2}{3} \end{bmatrix} -\frac{3}{2}R_4$

$$\begin{bmatrix} 1 & -3 & 1 & -4 & | & 4 \\ 0 & 1 & -\frac{4}{5} & \frac{8}{5} & | & -\frac{6}{5} \\ 0 & 0 & 1 & -\frac{13}{9} & | & -\frac{4}{9} \\ 0 & 0 & 0 & 1 & | & 1 \end{bmatrix}$$

This corresponds to the system

$$w - 3x + y - 4z = 4$$

$$x - \frac{4}{5}y + \frac{8}{5}z = -\frac{6}{5}$$

$$y - \frac{13}{9}z = -\frac{4}{9}$$

$$z = 1$$

Use back-substitution to find the values of w, z, and y:

$$y - \frac{13}{9}(1) = -\frac{4}{9}$$

$$y = 1$$

$$x - \frac{4}{5}(1) + \frac{8}{5}(1) = -\frac{6}{5}$$

$$x + \frac{4}{5} = -\frac{6}{5}$$

$$x = -2$$

$$w - 3(-2) + 1 - 4 = 4$$

$$w + 6 - 3 = 4$$

$$w = 1$$

The solution to the system is $\{(1, -2, 1, 1)\}$.

15. $\begin{bmatrix} 2 & 1 & -1 & | & 2 \\ 3 & 3 & -2 & | & 3 \end{bmatrix} \frac{1}{2}R_1 \quad \begin{bmatrix} 1 & \frac{1}{2} & -\frac{1}{2} & | & 1 \\ 3 & 3 & -2 & | & 3 \end{bmatrix} -3R_1 + R_2$

$\begin{bmatrix} 1 & \frac{1}{2} & -\frac{1}{2} & | & 1 \\ 0 & \frac{3}{2} & -\frac{1}{2} & | & 0 \end{bmatrix} \frac{2}{3}R_2$

$\begin{bmatrix} 1 & \frac{1}{2} & -\frac{1}{2} & | & 1 \\ 0 & 1 & -\frac{1}{3} & | & 0 \end{bmatrix}$

The system $x + \frac{1}{2}y - \frac{1}{2}z = 1$ has no unique solution.

$$y - \frac{1}{3}z = 0$$

Express x and y in terms of z:

$$y = \frac{1}{3}z$$

$$x + \frac{1}{2}\left(\frac{1}{3}z\right) - \frac{1}{2}z = 1$$

$$x + \frac{1}{6}z - \frac{1}{2}z = 1$$

$$x - \frac{1}{3}z = 1$$

$$x = 1 + \frac{1}{3}z$$

With $z = t$, the complete solution to the system is

$$\left\{\left(1 + \frac{1}{3}t, \frac{1}{3}t, t\right)\right\}.$$

17. The system $\begin{array}{c} x + 2y + 3z = 5 \\ y - 5z = 0 \end{array}$ has no unique solution.

Express x and y in terms of z:

$y = 5z$

$x + 2(5z) + 3z = 5$

$x + 10z + 3z = 5$

$x = -13z + 5$

With $z = t$, the complete solution to the system is $\{(-13t + 5, 5t, t)\}$.

19. $\begin{bmatrix} 1 & 1 & -2 & | & 2 \\ 3 & -1 & -6 & | & -7 \end{bmatrix} -3R_1 + R_2$

$\begin{bmatrix} 1 & 1 & -2 & | & 2 \\ 0 & -4 & 0 & | & -13 \end{bmatrix} -\frac{1}{4}R_2$

$\begin{bmatrix} 1 & 1 & -2 & | & 2 \\ 0 & 1 & 0 & | & \frac{13}{4} \end{bmatrix}$

$$x + y - 2z = 2$$

The system $\begin{array}{c} \\ y = \dfrac{13}{4} \end{array}$ has no unique solution.

Express x in terms of z:

$$x + \frac{13}{4} - 2z = 2$$

$$x = 2z - \frac{5}{4}$$

With $z = t$, the complete solution to the system is

$$\left\{\left(2t - \frac{5}{4}, \frac{13}{4}, t\right)\right\}.$$

21. $\begin{bmatrix} 1 & 1 & -1 & 1 & | & -2 \\ 2 & -1 & 2 & -1 & | & 7 \\ -1 & 2 & 1 & 2 & | & -1 \end{bmatrix} \begin{matrix} -2R_1 + R_2 \\ 1R_1 + R_3 \end{matrix}$

$\begin{bmatrix} 1 & 1 & -1 & 1 & | & -2 \\ 0 & -3 & 4 & -3 & | & 11 \\ 0 & 3 & 0 & 3 & | & -3 \end{bmatrix} R_2 \leftrightarrow R_3$

$\begin{bmatrix} 1 & 1 & -1 & 1 & | & -2 \\ 0 & 3 & 0 & 3 & | & -3 \\ 0 & -3 & 4 & -3 & | & 11 \end{bmatrix} \frac{1}{3}R_2$

$\begin{bmatrix} 1 & 1 & -1 & 1 & | & -2 \\ 0 & 1 & 0 & 1 & | & -1 \\ 0 & -3 & 4 & -3 & | & 11 \end{bmatrix} 3R_2 + R_3$

$\begin{bmatrix} 1 & 1 & -1 & 1 & | & -2 \\ 0 & 1 & 0 & 1 & | & -1 \\ 0 & 0 & 4 & 0 & | & 8 \end{bmatrix} \frac{1}{4}R_3$

$\begin{bmatrix} 1 & 1 & -1 & 1 & | & -2 \\ 0 & 1 & 0 & 1 & | & -1 \\ 0 & 0 & 1 & 0 & | & 2 \end{bmatrix}$

$$x + y - z + w = -2$$

The system $\qquad x + z = -1 \quad$ has no unique

$$y = 2$$

solution. Let $z = t$ and use back substitution to find remaining variables. The complete solution to the system is $\{(1, -t - 1, 2, t)\}$.

23. $\begin{bmatrix} 1 & 2 & 3 & -1 & | & 7 \\ 0 & 2 & -3 & 1 & | & 4 \\ 1 & -4 & 1 & 0 & | & 3 \end{bmatrix} -1R_1 + R_3$

$\begin{bmatrix} 1 & 2 & 3 & -1 & | & 7 \\ 0 & 2 & -3 & 1 & | & 4 \\ 0 & -6 & -2 & 1 & | & -4 \end{bmatrix} \frac{1}{2}R_2$

$\begin{bmatrix} 1 & 2 & 3 & -1 & | & 7 \\ 0 & 1 & -\frac{3}{2} & \frac{1}{2} & | & 2 \\ 0 & -6 & -2 & 1 & | & -4 \end{bmatrix} 6R_2 + R_3$

$\begin{bmatrix} 1 & 2 & 3 & -1 & | & 7 \\ 0 & 1 & -\frac{3}{2} & \frac{1}{2} & | & 2 \\ 0 & 0 & -11 & 4 & | & 8 \end{bmatrix} -\frac{1}{11}R_3$

$\begin{bmatrix} 1 & 2 & 3 & -1 & | & 7 \\ 0 & 1 & -\frac{3}{2} & \frac{1}{2} & | & 2 \\ 0 & 0 & 1 & -\frac{4}{11} & | & -\frac{8}{11} \end{bmatrix}$

The system has no unique solution. Let $z = t$ and use back substitution to find remaining variables. The complete solution to the system is

$$\left\{ \left(-\frac{2}{11}t + \frac{81}{11}, \frac{1}{22}t + \frac{10}{11}, \frac{4}{11}t - \frac{8}{11}, t \right) \right\}.$$

25. a. System of equations:

$$4w - 2x + 2y - 3z = 0$$
$$7w - x - y - 3z = 0$$
$$w + x - y - z = 0$$

b. Reduced system:

$$w - 0.5z = 0$$
$$x = 0$$
$$y - 0.5z = 0$$

Find w and y in terms of z.

$$w - 0.5z = 0$$
$$w = 0.5z$$
$$y - 0.5z = 0$$
$$y = 0.5z$$

The complete solution to the system is

$$\{(0.5z,\ 0,\ 0.5z,\ z)\}.$$

27. a. System of equations:

$$w + 2x + 5y + 5z = 3$$
$$w + x + 3y + 4z = -1$$
$$w - x - y + 2z = 3$$

b. Reduced system:

$$w + y + 3z = 1$$
$$x + 2y + z = -2$$

Find w and x in terms of y and z.

$$w + y + 3z = 1$$
$$w = 1 - y - 3z$$
$$x + 2y + z = -2$$
$$x = -2 - 2y - z$$

The complete solution to the system is

$$\{(1 - y - 3z,\ -2 - 2y - z,\ y,\ z)\}.$$

29. $z + 12 = x + 6$

31. $x - y = 4$

$x - z = 6$

$y - z = 2$

$\begin{bmatrix} 1 & -1 & 0 & 4 \\ 1 & 0 & -1 & 6 \\ 0 & 1 & -1 & 2 \end{bmatrix} \begin{matrix} -1R_1 + R_2 \\ \\ \\ \end{matrix}$

$\begin{bmatrix} 1 & -1 & 0 & 4 \\ 0 & -1 & 1 & -2 \\ 0 & 1 & -1 & 2 \end{bmatrix} \begin{matrix} \\ -1R_2 \\ \\ \end{matrix}$

$\begin{bmatrix} 1 & -1 & 0 & 4 \\ 0 & 1 & -1 & 2 \\ 0 & 1 & -1 & 2 \end{bmatrix} \begin{matrix} \\ -1R_2 + R_3 \\ 1R_2 + R_1 \end{matrix}$

$\begin{bmatrix} 1 & 0 & -1 & 6 \\ 0 & 1 & -1 & 2 \\ 0 & 0 & 0 & 0 \end{bmatrix}$

The system has no unique solution. Express x and y in terms of z:

$x - z = 6$

$y - z = 2$

$x = z + 6$

$y = z + 2$

With $z = t$, the complete solution to the system is $\{(t + 6, t + 2, t)\}$.

33. a. From left to right along Palm Drive, then along Sunset Drive, we get the equations

$w + z = 200 + 180 = 380;$

$w + x = 400 + 200 = 600;$

$z + 70 = y + 20$ or $y - z = 50;$

$y + 200 = x + 30$ or $x - y = 170.$

The system is

$w + z = 380$

$w + x = 600$

$y - z = 50$

$x - y = 170$

b. $\begin{bmatrix} 1 & 0 & 0 & 1 & 380 \\ 0 & 1 & 0 & -1 & 220 \\ 0 & 0 & 1 & -1 & 50 \\ 0 & 1 & -1 & 0 & 170 \end{bmatrix} \begin{matrix} \\ -1R_2 + R_4 \\ \\ \end{matrix}$

$\begin{bmatrix} 1 & 0 & 0 & 1 & 380 \\ 0 & 1 & 0 & -1 & 220 \\ 0 & 0 & 1 & -1 & 50 \\ 0 & 0 & -1 & 1 & -50 \end{bmatrix} \begin{matrix} \\ 1R_3 + R_4 \\ \\ \end{matrix}$

$\begin{bmatrix} 1 & 0 & 0 & 1 & 380 \\ 0 & 1 & 0 & -1 & 220 \\ 0 & 0 & 1 & -1 & 50 \\ 0 & 0 & 0 & 0 & 0 \end{bmatrix}$

The system has no unique solution. Express x and y and z in terms of z:

$w = 380 - z$

$x = 220 + z$

$y = 50 + z$

With $z = t$, the complete solution to the system is $\{(380 - t, 220 + t, 50 + t, t)\}$.

c. Letting $z = 50$, the solution is

$w = 380 - 50 = 330$

$x = 220 + 50 = 270$

$y = 50 + 50 = 100$

35. Let $x =$ the number of ounces of Food 1, $y =$ the number of ounces of Food 2, and $z =$ the number of ounces of Food 3. The amount of vitamin A is $20x + 30y + 10z$; the amount of iron is $20x + 10y + 10z$; the amount of calcium is $10x + 10y + 30z$.

a. Not having Food 1 means that all x terms are left out. The vitamin A requirement can then be represented by $30y + 10z = 220$; the iron requirement is $10y + 10z = 180$; the calcium requirement is $10y + 30z = 340$. The corresponding system is

$30y + 10z = 220$

$10y + 10z = 180$

$10y + 30z = 340.$

730

Dividing all of the numbers by 10, the matrix for this system is

$$\begin{bmatrix} 3 & 1 & | & 22 \\ 1 & 1 & | & 18 \\ 1 & 3 & | & 34 \end{bmatrix} R_1 \leftrightarrow R_2 \quad \begin{bmatrix} 1 & 1 & | & 18 \\ 3 & 1 & | & 22 \\ 1 & 3 & | & 34 \end{bmatrix} \begin{matrix} \\ -3R_1 + R_2 \\ -1R_1 + R_3 \end{matrix}$$

$$\begin{bmatrix} 1 & 1 & | & 18 \\ 0 & -2 & | & -32 \\ 0 & 2 & | & 16 \end{bmatrix} 1R_2 + R_3$$

$$\begin{bmatrix} 1 & 1 & | & 18 \\ 0 & -2 & | & -32 \\ 0 & 0 & | & -16 \end{bmatrix}.$$

From the last row, we see that the system has no solution, so there is no way to satisfy these dietary requirements with no Food 1 available.

b. With Food 1 available, and dropping the vitamin A requirement, the system is
$20x + 10y + 10z = 180$
$10x + 10y + 30z = 340.$
Dividing all of the numbers by 10, the matrix for this system is

$$\begin{bmatrix} 2 & 1 & 1 & | & 18 \\ 1 & 1 & 3 & | & 34 \end{bmatrix} R_1 \leftrightarrow R_2$$

$$\begin{bmatrix} 1 & 1 & 3 & | & 34 \\ 2 & 1 & 1 & | & 18 \end{bmatrix} -2R_1 + R_2$$

$$\begin{bmatrix} 1 & 1 & 3 & | & 34 \\ 0 & -1 & -5 & | & -50 \end{bmatrix} -1R_2$$

$$\begin{bmatrix} 1 & 1 & 3 & | & 34 \\ 0 & 1 & 5 & | & 50 \end{bmatrix}.$$

The system $\begin{array}{l} x + y + 3z = 34 \\ y + 5z = 50 \end{array}$ has no unique solution. Express x and y in terms of z:
$y = -5z + 50$
$x + (-5z + 50) + 3z = 34$
$x - 2z + 50 = 34$
$x = 2z - 16$
Now we can choose a value for z, i.e., an amount of Food 3, and find the corresponding values of x and y. Note that negative amounts of food are not realistic, so $z \geq 0$, $y = -5z + 50 \geq 0$, and $x = 2z - 16 \geq 0$. These conditions are equivalent to
$8 \leq z \leq 10$.
Using $z = 8$ and $z = 10$, two possibilities are 0 ounces of Food 1, 10 ounces of Food 2, and 8 ounces of Food 3 or 4 ounces of Food 1, 0 ounces of Food 2, and 10 ounces of Food 3. (Other answers are possible.)

37. – 39. Answers may vary.

41. does not make sense; Explanations will vary. Sample explanation: Row 3 indicates that this system has no solution. If eliminated, it would falsely appear that the system has an infinite number of solutions.

43. does not make sense; Explanations will vary. Sample explanation: In a nonsquare system, the number of equations differs from the number of variables.

45.
$$\begin{bmatrix} 1 & 3 & 1 & | & a^2 \\ 2 & 5 & 2a & | & 0 \\ 1 & 1 & a^2 & | & -9 \end{bmatrix} \begin{matrix} \\ -2R_1 + R_2 \\ -1R_1 + R_3 \end{matrix}$$

$$\begin{bmatrix} 1 & 3 & 1 & | & a^2 \\ 0 & -1 & 2a-2 & | & -2a^2 \\ 0 & -2 & a^2-1 & | & -9-a^2 \end{bmatrix} -1R_2$$

$$\begin{bmatrix} 1 & 3 & 1 & | & a^2 \\ 0 & 1 & 2-2a & | & 2a^2 \\ 0 & -2 & a^2-1 & | & -9-a^2 \end{bmatrix} 2R_2 + R_3$$

$$\begin{bmatrix} 1 & 3 & 1 & | & a^2 \\ 0 & 1 & 2-2a & | & 2a^2 \\ 0 & 0 & a^2-4a+3 & | & -9+3a^2 \end{bmatrix}$$

The system will be inconsistent when $a^2 - 4a + 3 = 0$ but $-9 + 3a^2 \neq 0$.
$a^2 - 4a + 3 = (a-1)(a-3) = 0$ when $a = 1$ or $a = 3$. $-9 + 3a^2 = 0$ when $a = \pm\sqrt{3}$.
Thus, the system is inconsistent when $a = 1$ or $a = 3$.

47. $-6 - (-5) = -6 + 5 = -1$

48. $1(-4) + 2(5) + 3(-6) = -4 + 10 - 18 = -12$

49. $\frac{1}{2}[8 - (-8)] = \frac{1}{2}[8 + 8] = \frac{1}{2}[16] = 8$

Section 8.3

Check Point Exercises

1.　**a.**　The matrix $A = \begin{bmatrix} 5 & -2 \\ -3 & \pi \\ 1 & 6 \end{bmatrix}$ has 3 rows and 2

columns, so it is of order 3×2.

b.　The element a_{12} is in the first row and second column. Thus, $a_{12} = -2$. The element a_{31} is in the third row and first column. Thus, $a_{31} = 1$.

2.　**a.**　$\begin{bmatrix} -4 & 3 \\ 7 & -6 \end{bmatrix} + \begin{bmatrix} 6 & -3 \\ 2 & -4 \end{bmatrix}$

$= \begin{bmatrix} -4+6 & 3+(-3) \\ 7+2 & -6+(-4) \end{bmatrix} = \begin{bmatrix} 2 & 0 \\ 9 & -10 \end{bmatrix}$

b.　$\begin{bmatrix} 5 & 4 \\ -3 & 7 \\ 0 & 1 \end{bmatrix} - \begin{bmatrix} -4 & 8 \\ 6 & 0 \\ -5 & 3 \end{bmatrix}$

$= \begin{bmatrix} 5-(-4) & 4-8 \\ -3-6 & 7-0 \\ 0-(-5) & 1-3 \end{bmatrix}$

$= \begin{bmatrix} 9 & -4 \\ -9 & 7 \\ 5 & -2 \end{bmatrix}$

3.　**a.**　$-6B = -6 \begin{bmatrix} -1 & -2 \\ 8 & 5 \end{bmatrix}$

$= \begin{bmatrix} -6(-1) & -6(-2) \\ -6(8) & -6(5) \end{bmatrix}$

$= \begin{bmatrix} -6 & 12 \\ -48 & -30 \end{bmatrix}$

b.

$3A + 2B = \begin{bmatrix} -4 & 1 \\ 3 & 0 \end{bmatrix} + 2 \begin{bmatrix} -1 & -2 \\ 8 & 5 \end{bmatrix}$

$= \begin{bmatrix} 3(-4) & 3(1) \\ 3(3) & 3(0) \end{bmatrix} + \begin{bmatrix} 2(-1) & 2(-2) \\ 2(8) & 2(5) \end{bmatrix}$

$= \begin{bmatrix} -12 & 3 \\ 9 & 0 \end{bmatrix} + \begin{bmatrix} -2 & -4 \\ 16 & 10 \end{bmatrix}$

$= \begin{bmatrix} -12+(-2) & 3+(-4) \\ 9+16 & 0+10 \end{bmatrix}$

$= \begin{bmatrix} -14 & -1 \\ 25 & 10 \end{bmatrix}$

4.　$3X + A = B$
$3X = B - A$

$X = \frac{1}{3}(B - A)$

$X = \frac{1}{3}\left(\begin{bmatrix} -10 & 1 \\ -9 & 17 \end{bmatrix} - \begin{bmatrix} 2 & -8 \\ 0 & 4 \end{bmatrix} \right)$

$X = \frac{1}{3} \begin{bmatrix} -12 & 9 \\ -9 & 13 \end{bmatrix}$

$X = \begin{bmatrix} -4 & 3 \\ -3 & \frac{13}{3} \end{bmatrix}$

5.　Given $A = \begin{bmatrix} 1 & 3 \\ 2 & 5 \end{bmatrix}$ and $B = \begin{bmatrix} 4 & 6 \\ 1 & 0 \end{bmatrix}$,

$AB = \begin{bmatrix} 1 & 3 \\ 2 & 5 \end{bmatrix} \cdot \begin{bmatrix} 4 & 6 \\ 1 & 0 \end{bmatrix} = \begin{bmatrix} 1(4)+3(1) & 1(6)+3(0) \\ 2(4)+5(1) & 2(6)+5(0) \end{bmatrix}$

$= \begin{bmatrix} 7 & 6 \\ 13 & 12 \end{bmatrix}$

6. If $A = \begin{bmatrix} 2 & 0 & 4 \end{bmatrix}$ and $B = \begin{bmatrix} 1 \\ 3 \\ 7 \end{bmatrix}$, then

$$AB = \begin{bmatrix} 2 & 0 & 4 \end{bmatrix} \begin{bmatrix} 1 \\ 3 \\ 7 \end{bmatrix}$$

$$= \begin{bmatrix} 2(1) + 0(3) + 4(7) \end{bmatrix}$$

$$= \begin{bmatrix} 2 + 0 + 28 \end{bmatrix}$$

$$= \begin{bmatrix} 30 \end{bmatrix}$$

and $BA = \begin{bmatrix} 1 \\ 3 \\ 7 \end{bmatrix} \begin{bmatrix} 2 & 0 & 4 \end{bmatrix} = \begin{bmatrix} 1(2) & 1(0) & 1(4) \\ 3(2) & 3(0) & 3(4) \\ 7(2) & 7(0) & 7(4) \end{bmatrix} = \begin{bmatrix} 2 & 0 & 4 \\ 6 & 0 & 12 \\ 14 & 0 & 28 \end{bmatrix}$.

7. a.

$$\begin{bmatrix} 1 & 3 \\ 0 & 2 \end{bmatrix} \cdot \begin{bmatrix} 2 & 3 & -1 & 6 \\ 0 & 5 & 4 & 1 \end{bmatrix}$$

$$= \begin{bmatrix} 1(2) + 3(0) & 1(3) + 3(5) & 1(-1) + 3(4) & 1(6) + 3(1) \\ 0(2) + 2(0) & 0(3) + 2(5) & 0(-1) + 2(4) & 0(6) + 2(1) \end{bmatrix}$$

$$= \begin{bmatrix} 2 & 18 & 11 & 9 \\ 0 & 10 & 8 & 2 \end{bmatrix}$$

b.

$$\begin{bmatrix} 2 & 3 & -1 & 6 \\ 0 & 5 & 4 & 1 \end{bmatrix} \begin{bmatrix} 1 & 3 \\ 0 & 2 \end{bmatrix}$$

The number of columns in the first matrix does not equal the number of rows in the second matrix. Thus, the product of these two matrices is undefined.

8. Because the *L* is dark gray and the background is light gray, the digital photograph can be represented by the matrix
$$\begin{bmatrix} 2 & 1 & 1 \\ 2 & 1 & 1 \\ 2 & 2 & 1 \end{bmatrix}$$

We can make the *L* light gray by decreasing each 2 in the above matrix to 1. We can make the background black by increasing each 1 in the matrix to 3. This is accomplished using the following matrix addition.
$$\begin{bmatrix} 2 & 1 & 1 \\ 2 & 1 & 1 \\ 2 & 2 & 1 \end{bmatrix} + \begin{bmatrix} -1 & 2 & 2 \\ -1 & 2 & 2 \\ -1 & -1 & 2 \end{bmatrix} = \begin{bmatrix} 1 & 3 & 3 \\ 1 & 3 & 3 \\ 1 & 1 & 3 \end{bmatrix}$$

9. a.

$$\begin{bmatrix} 0 & 3 & 4 \\ 0 & 5 & 2 \end{bmatrix} + \begin{bmatrix} -3 & -3 & -3 \\ -1 & -1 & -1 \end{bmatrix} = \begin{bmatrix} -3 & 0 & 1 \\ -1 & 4 & 1 \end{bmatrix}$$

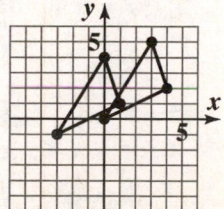

b. $2\begin{bmatrix} 0 & 3 & 4 \\ 0 & 5 & 2 \end{bmatrix} = \begin{bmatrix} 0 & 6 & 8 \\ 0 & 10 & 4 \end{bmatrix}$

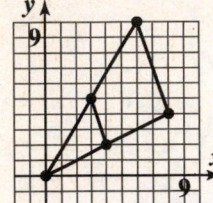

c. $\begin{bmatrix} 1 & 0 \\ 0 & -1 \end{bmatrix}\begin{bmatrix} 0 & 3 & 4 \\ 0 & 5 & 2 \end{bmatrix} = \begin{bmatrix} (-1)(0)+0(0) & (-1)(3)+0(5) & (-1)(4)+0(2) \\ 0(0)+1(0) & 0(3)+1(5) & 0(4)+1(2) \end{bmatrix}$

$= \begin{bmatrix} 0 & -3 & -4 \\ 0 & 5 & 2 \end{bmatrix}$

Multiplication with $\begin{bmatrix} 1 & 0 \\ 0 & -1 \end{bmatrix}$ reflects the triangle over the *x*-axis.

Exercise Set 8.3

1. **a.** 2×3

    b. a_{32} does not exist (*A* only has 2 rows).

       $a_{23} = -1$

3. **a.** 3×4

    b. $a_{32} = \dfrac{1}{2}; \; a_{23} = -6$

5. $\begin{bmatrix} x \\ 4 \end{bmatrix} = \begin{bmatrix} 6 \\ y \end{bmatrix}$

    $x = 6$

    $y = 4$

7. $\begin{bmatrix} x & 2y \\ z & 9 \end{bmatrix} = \begin{bmatrix} 4 & 12 \\ 3 & 9 \end{bmatrix}$

    $x = 4$

   $2y = 12$

    $y = 6$

    $z = 3$

9. **a.** $A + B = \begin{bmatrix} 4+5 & 1+9 \\ 3+0 & 2+7 \end{bmatrix} = \begin{bmatrix} 9 & 10 \\ 3 & 9 \end{bmatrix}$

   b. $A - B = \begin{bmatrix} 4-5 & 1-9 \\ 3-0 & 2-7 \end{bmatrix} = \begin{bmatrix} -1 & -8 \\ 3 & -5 \end{bmatrix}$

   c. $-4A = \begin{bmatrix} -16 & -4 \\ -12 & -8 \end{bmatrix}$

   d. $3A + 2B = \begin{bmatrix} 12+10 & 3+18 \\ 9+0 & 6+14 \end{bmatrix} = \begin{bmatrix} 22 & 21 \\ 9 & 20 \end{bmatrix}$

11. **a.** $A + B = \begin{bmatrix} 1+2 & 3+(-1) \\ 3+3 & 4+(-2) \\ 5+0 & 6+1 \end{bmatrix} = \begin{bmatrix} 3 & 2 \\ 6 & 2 \\ 5 & 7 \end{bmatrix}$

   b. $A - B = \begin{bmatrix} 1-2 & 3-(-1) \\ 3-3 & 4-(-2) \\ 5-0 & 6-1 \end{bmatrix} = \begin{bmatrix} -1 & 4 \\ 0 & 6 \\ 5 & 5 \end{bmatrix}$

   c. $-4A = \begin{bmatrix} -4 & -12 \\ -12 & -16 \\ -20 & -24 \end{bmatrix}$

   d. $3A + 2B = \begin{bmatrix} 3+4 & 9-2 \\ 9+6 & 12-4 \\ 15+0 & 18+2 \end{bmatrix} = \begin{bmatrix} 7 & 7 \\ 15 & 8 \\ 15 & 20 \end{bmatrix}$

13. **a.** $A + B = \begin{bmatrix} 2+(-5) \\ -4+3 \\ 1+(-1) \end{bmatrix} = \begin{bmatrix} -3 \\ -1 \\ 0 \end{bmatrix}$

   b. $A - B = \begin{bmatrix} 2-(-5) \\ -4-3 \\ 1-(-1) \end{bmatrix} = \begin{bmatrix} 7 \\ -7 \\ 2 \end{bmatrix}$

   c. $-4A = \begin{bmatrix} -8 \\ 16 \\ -4 \end{bmatrix}$

   d. $3A + 2B = \begin{bmatrix} 6-10 \\ -12+6 \\ 3-2 \end{bmatrix} = \begin{bmatrix} -4 \\ -6 \\ 1 \end{bmatrix}$

15. a. $A + B = \begin{bmatrix} 2+6 & -10+10 & -2+(-2) \\ 14+0 & 12+(-12) & 10+(-4) \\ 4+(-5) & -2+2 & 2+(-2) \end{bmatrix}$

$= \begin{bmatrix} 8 & 0 & -4 \\ 14 & 0 & 6 \\ -1 & 0 & 0 \end{bmatrix}$

b. $A - B = \begin{bmatrix} 2-6 & -10-10 & -2-(-2) \\ 14-0 & 12-(-12) & 10-(-4) \\ 4-(-5) & -2-2 & 2-(-2) \end{bmatrix}$

$= \begin{bmatrix} -4 & -20 & 0 \\ 14 & 24 & 14 \\ 9 & -4 & 4 \end{bmatrix}$

c. $-4A = \begin{bmatrix} -8 & 40 & 8 \\ -56 & -48 & -40 \\ -16 & 8 & -8 \end{bmatrix}$

d. $3A + 2B = \begin{bmatrix} 6+12 & -30+20 & -6-4 \\ 42+0 & 36-24 & 30-8 \\ 12-10 & -6+4 & 6-4 \end{bmatrix}$

$= \begin{bmatrix} 18 & -10 & -10 \\ 42 & 12 & 22 \\ 2 & -2 & 2 \end{bmatrix}$

17. $X - A = B$

$\quad X = A + B$

$X = \begin{bmatrix} -3 & -7 \\ 2 & -9 \\ 5 & 0 \end{bmatrix} + \begin{bmatrix} -5 & -1 \\ 0 & 0 \\ 3 & -4 \end{bmatrix} = \begin{bmatrix} -8 & -8 \\ 2 & -9 \\ 8 & -4 \end{bmatrix}$

19. $2X + A = B$

$\quad 2X = B - A$

$\quad X = \dfrac{1}{2}(B - A)$

$X = \dfrac{1}{2}\left(\begin{bmatrix} -5 & -1 \\ 0 & 0 \\ 3 & -4 \end{bmatrix} - \begin{bmatrix} -3 & -7 \\ 2 & -9 \\ 5 & 0 \end{bmatrix} \right) = \dfrac{1}{2}\begin{bmatrix} -2 & 6 \\ -2 & 9 \\ -2 & -4 \end{bmatrix} = \begin{bmatrix} -1 & 3 \\ -1 & \dfrac{9}{2} \\ -1 & -2 \end{bmatrix}$

21. $3X + 2A = B$

$\quad\;\; 3X = B - 2A$

$\quad\;\;\;\; X = \dfrac{1}{3}(B - 2A)$

$$X = \frac{1}{3}\left(\begin{bmatrix} -5 & -1 \\ 0 & 0 \\ 3 & -4 \end{bmatrix} - 2\begin{bmatrix} -3 & -7 \\ 2 & -9 \\ 5 & 0 \end{bmatrix}\right) = \frac{1}{3}\begin{bmatrix} 1 & 13 \\ -4 & 18 \\ -7 & -4 \end{bmatrix} = \begin{bmatrix} \dfrac{1}{3} & \dfrac{13}{3} \\ -\dfrac{4}{3} & 6 \\ -\dfrac{7}{3} & -\dfrac{4}{3} \end{bmatrix}$$

23. $\quad B - X = 4A$

$\quad\;\; B - 4A = X$

$$X = \begin{bmatrix} -5 & -1 \\ 0 & 0 \\ 3 & -4 \end{bmatrix} - 4\begin{bmatrix} -3 & -7 \\ 2 & -9 \\ 5 & 0 \end{bmatrix} = \begin{bmatrix} -5 & -1 \\ 0 & 0 \\ 3 & -4 \end{bmatrix} + \begin{bmatrix} 12 & 28 \\ -8 & 36 \\ -20 & 0 \end{bmatrix} = \begin{bmatrix} 7 & 27 \\ -8 & 36 \\ -17 & -4 \end{bmatrix}$$

25. $\quad\;\; 4A + 3B = -2X$

$\quad -\dfrac{1}{2}(4A + 3B) = X$

$$X = -\frac{1}{2}\left(4\begin{bmatrix} -3 & -7 \\ 2 & -9 \\ 5 & 0 \end{bmatrix} + 3\begin{bmatrix} -5 & -1 \\ 0 & 0 \\ 3 & -4 \end{bmatrix}\right) = -\frac{1}{2}\left(\begin{bmatrix} -12 & -28 \\ 8 & -36 \\ 20 & 0 \end{bmatrix} + \begin{bmatrix} -15 & -3 \\ 0 & 0 \\ 9 & -12 \end{bmatrix}\right) = -\frac{1}{2}\begin{bmatrix} -27 & -31 \\ 8 & -36 \\ 29 & -12 \end{bmatrix} = \begin{bmatrix} \dfrac{27}{2} & \dfrac{31}{2} \\ -4 & 18 \\ -\dfrac{29}{2} & 6 \end{bmatrix}$$

27. a. $AB = \begin{bmatrix} 1 & 3 \\ 5 & 3 \end{bmatrix}\begin{bmatrix} 3 & -2 \\ -1 & 6 \end{bmatrix} = \begin{bmatrix} (1)(3)+(3)(-1) & (1)(-2)+(3)(6) \\ (5)(3)+(3)(-1) & (5)(-2)+(3)(6) \end{bmatrix} = \begin{bmatrix} 3-3 & -2+18 \\ 15-3 & -10+18 \end{bmatrix} = \begin{bmatrix} 0 & 16 \\ 12 & 8 \end{bmatrix}$

b. $BA = \begin{bmatrix} 3 & -2 \\ -1 & 6 \end{bmatrix}\begin{bmatrix} 1 & 3 \\ 5 & 3 \end{bmatrix} = \begin{bmatrix} (3)(1)+(-2)(5) & (3)(3)+(-2)(3) \\ (-1)(1)+(6)(5) & (-1)(3)+(6)(3) \end{bmatrix} = \begin{bmatrix} 3-10 & 9-6 \\ -1+30 & -3+18 \end{bmatrix} = \begin{bmatrix} -7 & 3 \\ 29 & 15 \end{bmatrix}$

29. a. $AB = \begin{bmatrix} 1 & 2 & 3 & 4 \end{bmatrix}\begin{bmatrix} 1 \\ 2 \\ 3 \\ 4 \end{bmatrix} = [(1)(1)+(2)(2)+(3)(3)+(4)(4)] = [1+4+9+16] = [30]$

b. $BA = \begin{bmatrix} 1 \\ 2 \\ 3 \\ 4 \end{bmatrix}\begin{bmatrix} 1 & 2 & 3 & 4 \end{bmatrix} = \begin{bmatrix} (1)(1) & (1)(2) & (1)(3) & (1)(4) \\ (2)(1) & (2)(2) & (2)(3) & (2)(4) \\ (3)(1) & (3)(2) & (3)(3) & (3)(4) \\ (4)(1) & (4)(2) & (4)(3) & (4)(4) \end{bmatrix} = \begin{bmatrix} 1 & 2 & 3 & 4 \\ 2 & 4 & 6 & 8 \\ 3 & 6 & 9 & 12 \\ 4 & 8 & 12 & 16 \end{bmatrix}$

31. a. $AB = \begin{bmatrix} 1 & -1 & 4 \\ 4 & -1 & 3 \\ 2 & 0 & -2 \end{bmatrix} \begin{bmatrix} 1 & 1 & 0 \\ 1 & 2 & 4 \\ 1 & -1 & 3 \end{bmatrix}$

$= \begin{bmatrix} (1)(1)+(-1)(1)+(4)(1) & (1)(1)+(-1)(2)+(4)(-1) & (1)(0)+(-1)(4)+(4)(3) \\ (4)(1)+(-1)(1)+(3)(1) & (4)(1)+(-1)(2)+(3)(-1) & (4)(0)+(-1)(4)+(3)(3) \\ (2)(1)+(0)(1)+(-2)(1) & (2)(1)+(0)(2)+(-2)(-1) & (2)(0)+(0)(4)+(-2)(3) \end{bmatrix}$

$= \begin{bmatrix} 1-1+4 & 1-2-4 & 0-4+12 \\ 4-1+3 & 4-2-3 & 0-4+9 \\ 2+0-2 & 2+0+2 & 0+0-6 \end{bmatrix} = \begin{bmatrix} 4 & -5 & 8 \\ 6 & -1 & 5 \\ 0 & 4 & -6 \end{bmatrix}$

b. $BA = \begin{bmatrix} 1 & 1 & 0 \\ 1 & 2 & 4 \\ 1 & -1 & 3 \end{bmatrix} \begin{bmatrix} 1 & -1 & 4 \\ 4 & -1 & 3 \\ 2 & 0 & -2 \end{bmatrix}$

$= \begin{bmatrix} (1)(1)+(1)(4)+(0)(2) & (1)(-1)+(1)(-1)+(0)(0) & (1)(4)+(1)(3)+(0)(-2) \\ (1)(1)+(2)(4)+(4)(2) & (1)(-1)+(2)(-1)+(4)(0) & (1)(4)+(2)(3)+(4)(-2) \\ (1)(1)+(-1)(4)+(3)(2) & (1)(-1)+(-1)(-1)+(3)(0) & (1)(4)+(-1)(3)+(3)(-2) \end{bmatrix}$

$= \begin{bmatrix} 1+4+0 & -1-1+0 & 4+3+0 \\ 1+8+8 & -1-2+0 & 4+6-8 \\ 1-4+6 & -1+1+0 & 4-3-6 \end{bmatrix} = \begin{bmatrix} 5 & -2 & 7 \\ 17 & -3 & 2 \\ 3 & 0 & -5 \end{bmatrix}$

33. a. $AB = \begin{bmatrix} 4 & 2 \\ 6 & 1 \\ 3 & 5 \end{bmatrix} \begin{bmatrix} 2 & 3 & 4 \\ -1 & -2 & 0 \end{bmatrix} = \begin{bmatrix} (4)(2)+(2)(-1) & (4)(3)+(2)(-2) & (4)(4)+(2)(0) \\ (6)(2)+(1)(-1) & (6)(3)+(1)(-2) & (6)(4)+(1)(0) \\ (3)(2)+(5)(-1) & (3)(3)+(5)(-2) & (3)(4)+(5)(0) \end{bmatrix}$

$= \begin{bmatrix} 8-2 & 12-4 & 16+0 \\ 12-1 & 18-2 & 24+0 \\ 6-5 & 9-10 & 12+0 \end{bmatrix} = \begin{bmatrix} 6 & 8 & 16 \\ 11 & 16 & 24 \\ 1 & -1 & 12 \end{bmatrix}$

b. $BA = \begin{bmatrix} 2 & 3 & 4 \\ -1 & -2 & 0 \end{bmatrix} \begin{bmatrix} 4 & 2 \\ 6 & 1 \\ 3 & 5 \end{bmatrix} = \begin{bmatrix} (2)(4)+(3)(6)+(4)(3) & (2)(2)+(3)(1)+(4)(5) \\ (-1)(4)+(-2)(6)+(0)(3) & (-1)(2)+(-2)(1)+(0)(5) \end{bmatrix}$

$= \begin{bmatrix} 8+18+12 & 4+3+20 \\ -4-12+0 & -2-2+0 \end{bmatrix} = \begin{bmatrix} 38 & 27 \\ -16 & -4 \end{bmatrix}$

35. a. $AB = \begin{bmatrix} 2 & -3 & 1 & -1 \\ 1 & 1 & -2 & 1 \end{bmatrix} \begin{bmatrix} 1 & 2 \\ -1 & 1 \\ 5 & 4 \\ 10 & 5 \end{bmatrix} = \begin{bmatrix} (2)(1)+(-3)(-1)+(1)(5)+(-1)(10) & (2)(2)+(-3)(1)+(1)(4)+(-1)(5) \\ (1)(1)+(1)(-1)+(-2)(5)+(1)(10) & (1)(2)+(1)(1)+(-2)(4)+(1)(5) \end{bmatrix}$

$= \begin{bmatrix} 2+3+5-10 & 4-3+4-5 \\ 1-1-10+10 & 2+1-8+5 \end{bmatrix} = \begin{bmatrix} 0 & 0 \\ 0 & 0 \end{bmatrix}$

37. $4B - 3C = \begin{bmatrix} 20 & 4 \\ -8 & -8 \end{bmatrix} - \begin{bmatrix} 3 & -3 \\ -3 & 3 \end{bmatrix} = \begin{bmatrix} 20-3 & 4-(-3) \\ -8-(-3) & -8-3 \end{bmatrix} = \begin{bmatrix} 17 & 7 \\ -5 & -11 \end{bmatrix}$

39. $BC + CB = \begin{bmatrix} 5-1 & -5+1 \\ -2+2 & 2-2 \end{bmatrix} + \begin{bmatrix} 5+2 & 1+2 \\ -5-2 & -1-2 \end{bmatrix} = \begin{bmatrix} 4 & -4 \\ 0 & 0 \end{bmatrix} + \begin{bmatrix} 7 & 3 \\ -7 & -3 \end{bmatrix} = \begin{bmatrix} 11 & -1 \\ -7 & -3 \end{bmatrix}$

41. $A - C$ is not defined because A is 3 x 2 and C is 2 x 2.

43. $A(BC) = \begin{bmatrix} 4 & 0 \\ -3 & 5 \\ 0 & 1 \end{bmatrix} \begin{bmatrix} 5-1 & -5+1 \\ -2+2 & 2-2 \end{bmatrix} = \begin{bmatrix} 4 & 0 \\ -3 & 5 \\ 0 & 1 \end{bmatrix} \begin{bmatrix} 4 & -4 \\ 0 & 0 \end{bmatrix} = \begin{bmatrix} 16+0 & -16+0 \\ -12+0 & 12+0 \\ 0+0 & 0+0 \end{bmatrix} = \begin{bmatrix} 16 & -16 \\ -12 & 12 \\ 0 & 0 \end{bmatrix}$

45. $(A+B)(C-D) = \left(\begin{bmatrix} 1 & 0 \\ 0 & 1 \end{bmatrix} + \begin{bmatrix} 1 & 0 \\ 0 & -1 \end{bmatrix} \right) \left(\begin{bmatrix} -1 & 0 \\ 0 & 1 \end{bmatrix} - \begin{bmatrix} -1 & 0 \\ 0 & -1 \end{bmatrix} \right) = \left(\begin{bmatrix} 2 & 0 \\ 0 & 0 \end{bmatrix} \right) \left(\begin{bmatrix} 0 & 0 \\ 0 & 2 \end{bmatrix} \right) = \begin{bmatrix} 0 & 0 \\ 0 & 0 \end{bmatrix}$

47. Answers may vary.

49. $BZ = \begin{bmatrix} 1 & 0 \\ 0 & -1 \end{bmatrix} \begin{bmatrix} x \\ y \end{bmatrix} = \begin{bmatrix} x \\ -y \end{bmatrix}$ This reflects the graphic about the x-axis because all y-coordinates are negated.

51. a. $\begin{bmatrix} 1 & 3 & 1 \\ 3 & 3 & 3 \\ 1 & 3 & 1 \end{bmatrix}$

b. $\begin{bmatrix} 1 & 3 & 1 \\ 3 & 3 & 3 \\ 1 & 3 & 1 \end{bmatrix} + \begin{bmatrix} -1 & -1 & -1 \\ -1 & -1 & -1 \\ -1 & -1 & -1 \end{bmatrix} = \begin{bmatrix} 0 & 2 & 0 \\ 2 & 2 & 2 \\ 0 & 2 & 0 \end{bmatrix}$

c. $\begin{bmatrix} 1 & 3 & 1 \\ 3 & 3 & 3 \\ 1 & 3 & 1 \end{bmatrix} + \begin{bmatrix} 1 & -2 & 1 \\ -2 & -2 & -2 \\ 1 & -2 & 1 \end{bmatrix} = \begin{bmatrix} 2 & 1 & 2 \\ 1 & 1 & 1 \\ 2 & 1 & 2 \end{bmatrix}$

53. $\begin{bmatrix} 0 & 3 & 3 & 1 & 1 & 0 \\ 0 & 0 & 1 & 1 & 5 & 5 \end{bmatrix} + \begin{bmatrix} -2 & -2 & -2 & -2 & -2 & -2 \\ -3 & -3 & -3 & -3 & -3 & -3 \end{bmatrix} = \begin{bmatrix} -2 & 1 & 1 & -1 & -1 & -2 \\ -3 & -3 & -2 & -2 & 2 & 2 \end{bmatrix}$

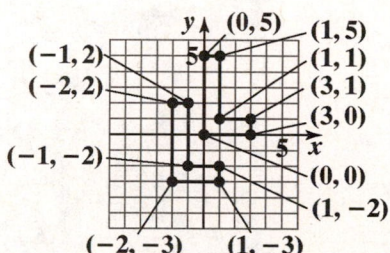

55. $0.5\begin{bmatrix} 0 & 3 & 3 & 1 & 1 & 0 \\ 0 & 0 & 1 & 1 & 5 & 5 \end{bmatrix} + \begin{bmatrix} 0 & 0 & 0 & 0 & 0 & 0 \\ 1 & 1 & 1 & 1 & 1 & 1 \end{bmatrix} = \begin{bmatrix} 0 & 1.5 & 1.5 & 0.5 & 0.5 & 0 \\ 0 & 0 & 0.5 & 0.5 & 2.5 & 2.5 \end{bmatrix} + \begin{bmatrix} 0 & 0 & 0 & 0 & 0 & 0 \\ 1 & 1 & 1 & 1 & 1 & 1 \end{bmatrix}$

$$= \begin{bmatrix} 0 & 1.5 & 1.5 & 0.5 & 0.5 & 0 \\ 1 & 1 & 1.5 & 1.5 & 3.5 & 3.5 \end{bmatrix}$$

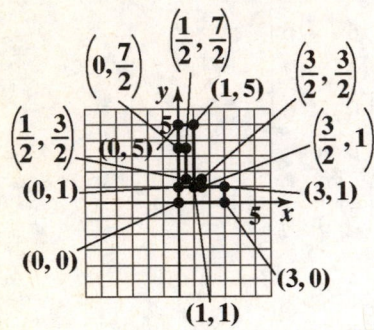

57. a. $AB = \begin{bmatrix} 1 & 0 \\ 0 & -1 \end{bmatrix} \cdot \begin{bmatrix} 0 & 3 & 3 & 1 & 1 & 0 \\ 0 & 0 & 1 & 1 & 5 & 5 \end{bmatrix} = \begin{bmatrix} 0 & 3 & 3 & 1 & 1 & 0 \\ 0 & 0 & -1 & -1 & -5 & -5 \end{bmatrix}$

b.

Rotated L about the *x*-axis.

59. a. $AB = \begin{bmatrix} 0 & -1 \\ 1 & 0 \end{bmatrix} \cdot \begin{bmatrix} 0 & 3 & 3 & 1 & 1 & 0 \\ 0 & 0 & 1 & 1 & 5 & 5 \end{bmatrix} = \begin{bmatrix} 0 & 0 & -1 & -1 & -5 & -5 \\ 0 & 3 & 3 & 1 & 1 & 0 \end{bmatrix}$

b.

Rotated L 90° counterclockwise about the origin.

61. a. $A = \begin{bmatrix} 2 & 6 \\ 31 & 46 \end{bmatrix}$

b. $B = \begin{bmatrix} 9 & 29 \\ 65 & 77 \end{bmatrix}$

c. $B - A = \begin{bmatrix} 9 & 29 \\ 65 & 77 \end{bmatrix} - \begin{bmatrix} 2 & 6 \\ 31 & 46 \end{bmatrix} = \begin{bmatrix} 7 & 23 \\ 34 & 31 \end{bmatrix}$

This matrix represents the percentages of people completing the transition to adulthood in 1960 and 2000 by age and gender.

63. a. System 1: The midterm and final both count for 50% of the course grade.
System 2: The midterm counts for 30% of the course grade and the final counts for 70%

b. $AB = \begin{bmatrix} 84 & 87.2 \\ 79 & 81 \\ 90 & 88.4 \\ 73 & 68.6 \\ 69 & 73.4 \end{bmatrix}$

System 1 grades are listed first (if different).
Student 1: B; Student 2: C or B; Student 3: A or B; Student 4: C or D; Student 5: D or C

65. – 75. Answers may vary.

77. makes sense

79. makes sense

81. Answers may vary.

83. $AB = \begin{bmatrix} 0 & -1 \\ 1 & 0 \end{bmatrix}\begin{bmatrix} 1 & 0 \\ 0 & -1 \end{bmatrix} = \begin{bmatrix} 0 & 1 \\ 1 & 0 \end{bmatrix}$

$-BA = \begin{bmatrix} 1 & 0 \\ 0 & -1 \end{bmatrix}\begin{bmatrix} 0 & -1 \\ 1 & 0 \end{bmatrix} = -\begin{bmatrix} 0 & -1 \\ -1 & 0 \end{bmatrix} = \begin{bmatrix} 0 & 1 \\ 1 & 0 \end{bmatrix}$

$AB = -BA$ so they are anticommutative.

85. $AB = \begin{bmatrix} a_{11} & a_{12} \\ a_{21} & a_{22} \end{bmatrix}\begin{bmatrix} 1 & 0 \\ 0 & 1 \end{bmatrix} = \begin{bmatrix} a_{11} & a_{12} \\ a_{21} & a_{22} \end{bmatrix}$

Nothing happens to the elements in the first matrix.

86. $\begin{bmatrix} -1 & -1 & -1 & | & 1 \\ 4 & 5 & 0 & | & 0 \\ 0 & 1 & -3 & | & 0 \end{bmatrix} 4R_1 + R_2$

$\begin{bmatrix} -1 & -1 & -1 & | & 1 \\ 0 & 1 & -4 & | & 4 \\ 0 & 1 & -3 & | & 0 \end{bmatrix} \begin{matrix} R_2 + R_1 \\ \\ -R_2 + R_3 \end{matrix}$

$\begin{bmatrix} -1 & 0 & -5 & | & 5 \\ 0 & 1 & -4 & | & 4 \\ 0 & 0 & 1 & | & -4 \end{bmatrix} -R_1$

$\begin{bmatrix} 1 & 0 & 5 & | & -5 \\ 0 & 1 & -4 & | & 4 \\ 0 & 0 & 1 & | & -4 \end{bmatrix} \begin{matrix} -5R_3 + R_1 \\ 4R_3 + R_2 \end{matrix}$

$\begin{bmatrix} 1 & 0 & 0 & | & 15 \\ 0 & 1 & 0 & | & -12 \\ 0 & 0 & 1 & | & -4 \end{bmatrix}$

The solution set is $\{(15, -12, -4)\}$.

87.
$$\begin{bmatrix} a_1 & b_1 & c_1 \\ a_2 & b_2 & c_2 \\ a_3 & b_3 & c_3 \end{bmatrix} \begin{bmatrix} x \\ y \\ x \end{bmatrix} = \begin{bmatrix} d_1 \\ d_2 \\ d_3 \end{bmatrix}$$

$$\begin{bmatrix} a_1x + b_1y + c_1x \\ a_2x + b_2y + c_2z \\ a_3x + b_3y + c_3z \end{bmatrix} = \begin{bmatrix} d_1 \\ d_2 \\ d_3 \end{bmatrix}$$

The linear system is written as follows.

$a_1x + b_1y + c_1x = d_1$
$a_2x + b_2y + c_2z = d_2$
$a_3x + b_3y + c_3z = d_3$

Mid-Chapter 8 Check Point

1.
$$\begin{bmatrix} 1 & 2 & -3 & | & -7 \\ 3 & -1 & 2 & | & 8 \\ 2 & -1 & 1 & | & 5 \end{bmatrix} \begin{matrix} \\ -3R_1 + R_2 \\ -2R_1 + R_3 \end{matrix}$$

$$\begin{bmatrix} 1 & 2 & -3 & | & -7 \\ 0 & -7 & 11 & | & 29 \\ 0 & -5 & 7 & | & 19 \end{bmatrix} \begin{matrix} \\ -\frac{1}{7}R_2 \\ -\frac{1}{5}R_3 \end{matrix}$$

$$\begin{bmatrix} 1 & 2 & -3 & | & -7 \\ 0 & 1 & -\frac{11}{7} & | & -\frac{29}{7} \\ 0 & 1 & -\frac{7}{5} & | & -\frac{19}{5} \end{bmatrix} \begin{matrix} \\ \\ -R_2 + R_3 \end{matrix}$$

$$\begin{bmatrix} 1 & 2 & -3 & | & -7 \\ 0 & 1 & -\frac{11}{7} & | & -\frac{29}{7} \\ 0 & 0 & \frac{6}{35} & | & \frac{12}{35} \end{bmatrix} \begin{matrix} \\ \\ \frac{35}{6}R_3 \end{matrix}$$

$$\begin{bmatrix} 1 & 2 & -3 & | & -7 \\ 0 & 1 & -\frac{11}{7} & | & -\frac{29}{7} \\ 0 & 0 & 1 & | & 2 \end{bmatrix}$$

Back-substitute to find y.

$$y - \frac{11}{7}z = -\frac{29}{7}$$
$$y - \frac{11}{7}(2) = -\frac{29}{7}$$
$$y - \frac{22}{7} = -\frac{29}{7}$$
$$y = -\frac{7}{7}$$
$$y = -1$$

Back-substitute to find x.

$x + 2y - 3z = -7$
$x + 2(-1) - 3(2) = -7$
$x - 2 - 6 = -7$
$x - 8 = -7$
$x = 1$

The solution is $\{(1, -1, 2)\}$.

2.
$$\begin{bmatrix} 2 & 4 & 5 & | & 2 \\ 1 & 1 & 2 & | & 1 \\ 3 & 5 & 7 & | & 4 \end{bmatrix} R_1 \leftrightarrow R_3$$

$$\begin{bmatrix} 1 & 1 & 2 & | & 1 \\ 2 & 4 & 5 & | & 2 \\ 3 & 5 & 7 & | & 4 \end{bmatrix} \begin{matrix} \\ -2R_1 + R_2 \\ -3R_1 + R_3 \end{matrix}$$

$$\begin{bmatrix} 1 & 1 & 2 & | & 1 \\ 0 & 2 & 1 & | & 0 \\ 0 & 2 & 1 & | & 1 \end{bmatrix} \begin{matrix} \\ \\ -R_2 + R_3 \end{matrix}$$

$$\begin{bmatrix} 1 & 1 & 2 & | & 1 \\ 0 & 2 & 1 & | & 0 \\ 0 & 0 & 0 & | & 1 \end{bmatrix}$$

The third row of the matrix is equivalent to
$0x + 0y + 0z = 1$ which is false.

The solution is $\varnothing$.

3.
$$\begin{bmatrix} 1 & -2 & 2 & | & -2 \\ 2 & 3 & -1 & | & 1 \end{bmatrix} -2R_1 + R_2$$

$$\begin{bmatrix} 1 & -2 & 2 & | & -2 \\ 0 & 7 & -5 & | & 5 \end{bmatrix} \frac{1}{7}R_2$$

$$\begin{bmatrix} 1 & -2 & 2 & | & -2 \\ 0 & 1 & -\frac{5}{7} & | & \frac{5}{7} \end{bmatrix}$$

Back-substitute to find y in terms of z.

$$y - \frac{5}{7}z = \frac{5}{7}$$
$$y = \frac{5}{7}z + \frac{5}{7}$$

Back-substitute to find x in terms of z.

$$x - 2y + 2z = -2$$
$$x - 2\left(\frac{5}{7}z + \frac{5}{7}\right) + 2z = -2$$
$$x - \frac{10}{7}z - \frac{10}{7} + 2z = -2$$
$$x + \frac{4}{7}z - \frac{10}{7} = -2$$
$$x = -\frac{4}{7}z - \frac{4}{7}$$

The solution is $\left\{\left(-\frac{4}{7}z - \frac{4}{7}, \frac{5}{7}z + \frac{5}{7}, z\right)\right\}$.

4.

$$\left[\begin{array}{cccc|c} 1 & 1 & 1 & 1 & 6 \\ 1 & -1 & 3 & 1 & -14 \\ 1 & 2 & 0 & -3 & 12 \\ 2 & 3 & 6 & 1 & 1 \end{array}\right] \begin{array}{l} \\ -R_1 + R_2 \\ -R_1 + R_3 \\ -2R_1 + R_4 \end{array}$$

$$\left[\begin{array}{cccc|c} 1 & 1 & 1 & 1 & 6 \\ 0 & -2 & 2 & 0 & -20 \\ 0 & 1 & -1 & -4 & 6 \\ 0 & 1 & 4 & -1 & -11 \end{array}\right] R_2 \leftrightarrow R_3$$

$$\left[\begin{array}{cccc|c} 1 & 1 & 1 & 1 & 6 \\ 0 & 1 & -1 & -4 & 6 \\ 0 & -2 & 2 & 0 & -20 \\ 0 & 1 & 4 & -1 & -11 \end{array}\right] \begin{array}{l} \\ \\ 2R_2 + R_3 \\ -R_2 + R_4 \end{array}$$

$$\left[\begin{array}{cccc|c} 1 & 1 & 1 & 1 & 6 \\ 0 & 1 & -1 & -4 & 6 \\ 0 & 0 & 0 & -8 & -8 \\ 0 & 0 & 5 & 3 & -17 \end{array}\right] R_3 \leftrightarrow R_4$$

$$\left[\begin{array}{cccc|c} 1 & 1 & 1 & 1 & 6 \\ 0 & 1 & -1 & -4 & 6 \\ 0 & 0 & 5 & 3 & -17 \\ 0 & 0 & 0 & -8 & -8 \end{array}\right] \begin{array}{l} \\ \\ \frac{1}{5}R_3 \\ -\frac{1}{8}R_4 \end{array}$$

$$\left[\begin{array}{cccc|c} 1 & 1 & 1 & 1 & 6 \\ 0 & 1 & -1 & -4 & 6 \\ 0 & 0 & 1 & \frac{3}{5} & -\frac{17}{5} \\ 0 & 0 & 0 & 1 & 1 \end{array}\right]$$

Back-substitute to find y in terms of z.

$$y + \frac{3}{5}z = -\frac{17}{5}$$
$$y + \frac{3}{5}(1) = -\frac{17}{5}$$
$$y + \frac{3}{5} = -\frac{17}{5}$$
$$y = -\frac{20}{5}$$
$$y = -4$$

Back-substitute to find x in terms of z.

$$x - y - 4z = 6$$
$$x - (-4) - 4(1) = 6$$
$$x + 4 - 4 = 6$$
$$x = 6$$

Back-substitute to find w in terms of z.

$$w + x + y + z = 6$$
$$w + (6) + (-4) + (1) = 6$$
$$w + 3 = 6$$
$$w = 3$$

The solution is $\{(3, 6, -4, 1)\}$.

5.

$$\left[\begin{array}{ccc|c} 2 & -2 & 2 & 5 \\ 1 & -1 & 1 & 2 \\ 2 & 1 & -1 & 1 \end{array}\right] \begin{array}{l} -R_2 + R_1 \\ \\ -R_1 + R_3 \end{array}$$

$$\left[\begin{array}{ccc|c} 1 & -1 & 1 & 3 \\ 1 & -1 & 1 & 2 \\ 0 & 3 & -3 & -4 \end{array}\right] -R_1 + R_2$$

$$\left[\begin{array}{ccc|c} 1 & -1 & 1 & 3 \\ 0 & 0 & 0 & -1 \\ 0 & 3 & -3 & -4 \end{array}\right]$$

The second row of the matrix is equivalent to
$0x + 0y + 0z = -1$ which is false.

The solution is $\varnothing$.

6. $2C - \frac{1}{2}B$

$$= 2\left[\begin{array}{cc} -1 & 0 \\ 0 & 1 \end{array}\right] - \frac{1}{2}\left[\begin{array}{cc} 4 & 1 \\ -6 & -2 \end{array}\right]$$

$$= \left[\begin{array}{cc} -2 & 0 \\ 0 & 2 \end{array}\right] - \left[\begin{array}{cc} 2 & \frac{1}{2} \\ -3 & -1 \end{array}\right]$$

$$= \left[\begin{array}{cc} -4 & -\frac{1}{2} \\ 3 & 3 \end{array}\right]$$

7. $A(B + C)$

$$= \left[\begin{array}{cc} 0 & 2 \\ -1 & 3 \\ 1 & 0 \end{array}\right]\left(\left[\begin{array}{cc} 4 & 1 \\ -6 & -2 \end{array}\right] + \left[\begin{array}{cc} -1 & 0 \\ 0 & 1 \end{array}\right]\right)$$

$$= \left[\begin{array}{cc} 0 & 2 \\ -1 & 3 \\ 1 & 0 \end{array}\right]\left(\left[\begin{array}{cc} 3 & 1 \\ -6 & -1 \end{array}\right]\right)$$

$$= \left[\begin{array}{cc} -12 & -2 \\ -21 & -4 \\ 3 & 1 \end{array}\right]$$

8. $A(BC)$

$$= \left[\begin{array}{cc} 0 & 2 \\ -1 & 3 \\ 1 & 0 \end{array}\right]\left(\left[\begin{array}{cc} 4 & 1 \\ -6 & -2 \end{array}\right] \cdot \left[\begin{array}{cc} -1 & 0 \\ 0 & 1 \end{array}\right]\right)$$

$$= \left[\begin{array}{cc} 0 & 2 \\ -1 & 3 \\ 1 & 0 \end{array}\right]\left(\left[\begin{array}{cc} -4 & 1 \\ 6 & -2 \end{array}\right]\right)$$

$$= \left[\begin{array}{cc} 12 & -4 \\ 22 & -7 \\ -4 & 1 \end{array}\right]$$

9. The operation is not defined. Matrices must have the same dimensions in order to be added.

10. $2X - 3C = B$

$$2X = B + 3C$$

$$X = \tfrac{1}{2}(B + 3C)$$

$$= \tfrac{1}{2}\left(\begin{bmatrix} 4 & 1 \\ -6 & -2 \end{bmatrix} + 3\begin{bmatrix} -1 & 0 \\ 0 & 1 \end{bmatrix} \right)$$

$$= \tfrac{1}{2}\left(\begin{bmatrix} 4 & 1 \\ -6 & -2 \end{bmatrix} + \begin{bmatrix} -3 & 0 \\ 0 & 3 \end{bmatrix} \right)$$

$$= \tfrac{1}{2}\left(\begin{bmatrix} 1 & 1 \\ -6 & 1 \end{bmatrix} \right)$$

$$= \begin{bmatrix} \tfrac{1}{2} & \tfrac{1}{2} \\ -3 & \tfrac{1}{2} \end{bmatrix}$$

Section 8.4

Check Point Exercises

1. We must show that: $AB = I_2 = \begin{bmatrix} 1 & 0 \\ 0 & 1 \end{bmatrix}$, and

$$BA = I_2 = \begin{bmatrix} 1 & 0 \\ 0 & 1 \end{bmatrix}.$$

$$AB = \begin{bmatrix} 2 & 1 \\ 1 & 1 \end{bmatrix}\begin{bmatrix} 1 & -1 \\ -1 & 2 \end{bmatrix}$$

$$= \begin{bmatrix} 2(1)+1(-1) & 2(-1)+1(2) \\ 1(1)+1(-1) & 1(-1)+1(2) \end{bmatrix}$$

$$= \begin{bmatrix} 1 & 0 \\ 0 & 1 \end{bmatrix}$$

$$BA = \begin{bmatrix} 1 & -1 \\ -1 & 2 \end{bmatrix}\begin{bmatrix} 2 & 1 \\ 1 & 1 \end{bmatrix}$$

$$= \begin{bmatrix} 1(2)+-1(1) & 1(1)+-1(1) \\ -1(2)+2(1) & -1(1)+2(1) \end{bmatrix}$$

$$= \begin{bmatrix} 1 & 0 \\ 0 & 1 \end{bmatrix}$$

Both products (AB and BA) give the multiplicative identity matrix, I_2. Thus, B is the multiplicative inverse of A.

2. Let us denote the multiplicative inverse of A by $A^{-1} = \begin{bmatrix} w & x \\ y & z \end{bmatrix}$. Because A is a 2×2 matrix, we use the equation $AA^{-1} = I_2$ to find values for w, x, y and z.

$$\begin{bmatrix} 5 & 7 \\ 2 & 3 \end{bmatrix}\begin{bmatrix} w & x \\ y & z \end{bmatrix} = \begin{bmatrix} 1 & 0 \\ 0 & 1 \end{bmatrix}$$

$$\begin{bmatrix} 5w+7y & 5x+7z \\ 2w+3y & 2x+3z \end{bmatrix} = \begin{bmatrix} 1 & 0 \\ 0 & 1 \end{bmatrix}$$

$$5w+7y = 1 \qquad 5x+7z = 0$$
$$2w+3y = 0 \qquad 2x+3z = 1$$

Each of these systems can be solved using the addition method.

Multiply by –2: $5w+7y = 1 \rightarrow -10w-14y = -2$

Multiply by 5: $2w+3y = 0 \rightarrow \quad 10w+15y = 0$

Use back substitution: $w = 3$, $y = -2$

Multiply by –2: $5x+7z = 0 \rightarrow -10x-14z = 0$

Multiply by 5: $2x+3z = 1 \rightarrow 10x+15z = 5$

Use back substitution: $x = -7$, $z = 5$

Using these values, we have

$$A^{-1} = \begin{bmatrix} 3 & -7 \\ -2 & 5 \end{bmatrix}.$$

3. $A^{-1} = \dfrac{1}{ad-bc}\begin{bmatrix} d & -b \\ -c & a \end{bmatrix}$

$$= \dfrac{1}{3(1)-(-2)(-1)}\begin{bmatrix} 1 & -(-2) \\ -(-1) & 3 \end{bmatrix}$$

$$= \dfrac{1}{3-2}\begin{bmatrix} 1 & 2 \\ 1 & 3 \end{bmatrix}$$

$$= \dfrac{1}{1}\begin{bmatrix} 1 & 2 \\ 1 & 3 \end{bmatrix}$$

$$= \begin{bmatrix} 1 & 2 \\ 1 & 3 \end{bmatrix}$$

4. The augmented matrix $\left[A \mid I_3 \right]$ is

$$\begin{bmatrix} 1 & 0 & 2 & | & 1 & 0 & 0 \\ -1 & 2 & 3 & | & 0 & 1 & 0 \\ 1 & -1 & 0 & | & 0 & 0 & 1 \end{bmatrix}.$$

Perform row transformations on $\left[A \mid I_3 \right]$ to obtain a matrix of the form $\left[I_3 \mid B \right]$.

$$\begin{bmatrix} 1 & 0 & 2 & | & 1 & 0 & 0 \\ -1 & 2 & 3 & | & 0 & 1 & 0 \\ 1 & -1 & 0 & | & 0 & 0 & 1 \end{bmatrix} 1R_1 / R_2$$

$$= \begin{bmatrix} 1 & 0 & 2 & | & 1 & 0 & 0 \\ 0 & 2 & 5 & | & 1 & 1 & 0 \\ 1 & -1 & 0 & | & 0 & 0 & 1 \end{bmatrix} -1R_3$$

$$= \begin{bmatrix} 1 & 0 & 2 & | & 1 & 0 & 0 \\ 0 & 2 & 5 & | & 1 & 1 & 0 \\ -1 & 1 & 0 & | & 0 & 0 & -1 \end{bmatrix} R_1 + R_3$$

$$= \begin{bmatrix} 1 & 0 & 2 & | & 1 & 0 & 0 \\ 0 & 2 & 5 & | & 1 & 1 & 0 \\ 0 & 1 & 2 & | & 1 & 0 & -1 \end{bmatrix} \frac{1}{2}R_2$$

$$= \begin{bmatrix} 1 & 0 & 2 & | & 1 & 0 & 0 \\ 0 & 1 & \frac{5}{2} & | & \frac{1}{2} & \frac{1}{2} & 0 \\ 0 & 1 & 2 & | & 1 & 0 & -1 \end{bmatrix} -1R_2 + R_3$$

$$= \begin{bmatrix} 1 & 0 & 2 & | & 1 & 0 & 0 \\ 0 & 1 & \frac{5}{2} & | & \frac{1}{2} & \frac{1}{2} & 0 \\ 0 & 0 & -\frac{1}{2} & | & \frac{1}{2} & -\frac{1}{2} & -1 \end{bmatrix} -2R_3$$

$$= \begin{bmatrix} 1 & 0 & 2 & | & 1 & 0 & 0 \\ 0 & 1 & \frac{5}{2} & | & \frac{1}{2} & \frac{1}{2} & 0 \\ 0 & 0 & 1 & | & -1 & 1 & 2 \end{bmatrix} \begin{matrix} -2R_3 + R_1 \\ -\frac{5}{2}R_3 + R_2 \end{matrix}$$

$$= \begin{bmatrix} 1 & 0 & 0 & | & 3 & -2 & -4 \\ 0 & 1 & 0 & | & 3 & -2 & -5 \\ 0 & 0 & 1 & | & 1 & 1 & 2 \end{bmatrix}$$

Thus, the multiplicative inverse of A is

$$A^{-1} = \begin{bmatrix} 3 & -2 & -4 \\ 3 & -2 & -5 \\ -1 & 1 & 2 \end{bmatrix}.$$

5. The linear system can be written as $AX = B$.

$$\begin{bmatrix} 1 & 0 & 2 \\ -1 & 2 & 3 \\ 1 & -1 & 0 \end{bmatrix} \begin{bmatrix} x \\ y \\ z \end{bmatrix} = \begin{bmatrix} 6 \\ -5 \\ 6 \end{bmatrix}.$$

$$X = A^{-1}B = \begin{bmatrix} 3 & -2 & -4 \\ 3 & -2 & -5 \\ -1 & 1 & 2 \end{bmatrix} \begin{bmatrix} 6 \\ -5 \\ 6 \end{bmatrix}$$

$$= \begin{bmatrix} 3(6) + -2(-5) + -4(6) \\ 3(6) + -2(-5) + -5(6) \\ -1(6) + 1(-5) + 2(6) \end{bmatrix}$$

$$= \begin{bmatrix} 18 + 10 - 24 \\ 18 + 10 - 30 \\ -6 - 5 + 12 \end{bmatrix} = \begin{bmatrix} 4 \\ -2 \\ 1 \end{bmatrix}$$

Thus, $x = 4$, $y = -2$, and $z = 1$. The solution set is $\{(4, -2, 1)\}$.

6. The numerical representation of the word BASE is 2, 1, 19, 5. The 2×2 matrix formed is $\begin{bmatrix} 2 & 19 \\ 1 & 5 \end{bmatrix}$.

$$\begin{bmatrix} -2 & -3 \\ 3 & 4 \end{bmatrix} \begin{bmatrix} 2 & 19 \\ 1 & 5 \end{bmatrix}$$

$$= \begin{bmatrix} -2(2) + -3(1) & -2(19) + -3(5) \\ 3(2) + 4(1) & 3(19) + 4(5) \end{bmatrix}$$

$$= \begin{bmatrix} -4 - 3 & -38 - 15 \\ 6 + 4 & 57 + 20 \end{bmatrix} = \begin{bmatrix} -7 & -53 \\ 10 & 77 \end{bmatrix}$$

The encoded message is -7, 10, -53, 77.

7. Use the multiplicative inverse of the coding matrix. It is $\begin{bmatrix} 4 & 3 \\ -3 & -2 \end{bmatrix}$.

$$\begin{bmatrix} 4 & 3 \\ -3 & -2 \end{bmatrix} \begin{bmatrix} -7 & -53 \\ 10 & 77 \end{bmatrix}$$

$$= \begin{bmatrix} 4(-7) + 3(10) & 4(-53) + 3(77) \\ -3(-7) + -2(10) & -3(-53) + -2(77) \end{bmatrix}$$

$$= \begin{bmatrix} -28 + 30 & -212 + 231 \\ 21 - 20 & 159 - 154 \end{bmatrix} = \begin{bmatrix} 2 & 19 \\ 1 & 5 \end{bmatrix}$$

The numbers are 2, 1, 19, and 5. Using letters, the decoded message is BASE.

Exercise Set 8.4

1. $A = \begin{bmatrix} 4 & -3 \\ -5 & 4 \end{bmatrix}$ $B = \begin{bmatrix} 4 & 3 \\ 5 & 4 \end{bmatrix}$

$$AB = \begin{bmatrix} 16 - 15 & 12 - 12 \\ -20 + 20 & -15 + 16 \end{bmatrix} = \begin{bmatrix} 1 & 0 \\ 0 & 1 \end{bmatrix}$$

$$BA = \begin{bmatrix} 16 - 15 & -12 + 12 \\ 20 - 20 & -15 + 16 \end{bmatrix} = \begin{bmatrix} 1 & 0 \\ 0 & 1 \end{bmatrix}$$

Since $AB = I_2$ $BA = I_2$, $B = A^{-1}$.

3. $AB = \begin{bmatrix} 8+0 & -16+0 \\ -2+0 & 4+3 \end{bmatrix} = \begin{bmatrix} 8 & -16 \\ -2 & 7 \end{bmatrix}$

$BA = \begin{bmatrix} 8+4 & 0+12 \\ 0+1 & 0+3 \end{bmatrix} = \begin{bmatrix} 12 & 12 \\ 1 & 3 \end{bmatrix}$

If B is the multiplicative inverse of A, both products (AB and BA) will be the multiplicative identity matrix, I_2. Therefore, B is not the multiplicative inverse of A. That is, $B \neq A^{-1}$.

5. $AB = \begin{bmatrix} -2+3 & -4+4 \\ \frac{3}{2}-\frac{3}{2} & 3-2 \end{bmatrix} = \begin{bmatrix} 1 & 0 \\ 0 & 1 \end{bmatrix}$

$BA = \begin{bmatrix} -2+3 & 1-1 \\ -6+6 & 3-2 \end{bmatrix} = \begin{bmatrix} 1 & 0 \\ 0 & 1 \end{bmatrix}$

Since $AB = I_2$ and $BA = I_2$, $B = A^{-1}$.

7. $A = \begin{bmatrix} 0 & 1 & 0 \\ 0 & 0 & 1 \\ 1 & 0 & 0 \end{bmatrix} \quad B = \begin{bmatrix} 0 & 0 & 1 \\ 1 & 0 & 0 \\ 0 & 1 & 0 \end{bmatrix}$

$AB = \begin{bmatrix} 0+1+0 & 0+0+0 & 0+0+0 \\ 0+0+0 & 0+0+1 & 0+0+0 \\ 0+0+0 & 0+0+0 & 1+0+0 \end{bmatrix} = \begin{bmatrix} 1 & 0 & 0 \\ 0 & 1 & 0 \\ 0 & 0 & 1 \end{bmatrix} \quad BA = \begin{bmatrix} 0+0+1 & 0+0+0 & 0+0+0 \\ 0+0+0 & 1+0+0 & 0+0+0 \\ 0+0+0 & 0+0+0 & 0+1+0 \end{bmatrix} = \begin{bmatrix} 1 & 0 & 0 \\ 0 & 1 & 0 \\ 0 & 0 & 1 \end{bmatrix}$

Since $AB = I_3$ and $BA = I_3$, $B = A^{-1}$.

9. $AB = \begin{bmatrix} \frac{7}{2}-1-\frac{3}{2} & -3+0+3 & \frac{1}{2}+1-\frac{3}{2} \\ \frac{7}{2}-\frac{3}{2}-2 & -3+0+4 & \frac{1}{2}+\frac{3}{2}-2 \\ \frac{7}{2}-2-\frac{3}{2} & -3+0+3 & \frac{1}{2}+2-\frac{3}{2} \end{bmatrix} = \begin{bmatrix} 1 & 0 & 0 \\ 0 & 1 & 0 \\ 0 & 0 & 1 \end{bmatrix}$

$BA = \begin{bmatrix} \frac{7}{2}-3+\frac{1}{2} & 7-9+2 & \frac{21}{2}-12+\frac{3}{2} \\ -\frac{1}{2}+0+\frac{1}{2} & -1+0+2 & -\frac{3}{2}+0+\frac{3}{2} \\ -\frac{1}{2}+1-\frac{1}{2} & -1+3-2 & -\frac{3}{2}+4-\frac{3}{2} \end{bmatrix} = \begin{bmatrix} 1 & 0 & 0 \\ 0 & 1 & 0 \\ 0 & 0 & 1 \end{bmatrix}$

Since $AB = I_3$ and $BA = I_3$, $B = A^{-1}$.

11. $AB = \begin{bmatrix} 0+0+0+1 & 0+0-2+2 & 0+0+0+0 & 0+0-2+2 \\ -1+0+0+1 & -2+0+1+2 & 0+0+0+0 & -3+0+1+2 \\ 0+0+0+0 & 0+1-1+0 & 0+1+0+0 & 0+1-1+0 \\ 1+0+0-1 & 2+0+0-2 & 0+0+0+0 & 3+0+0-2 \end{bmatrix} = \begin{bmatrix} 1 & 0 & 0 & 0 \\ 0 & 1 & 0 & 0 \\ 0 & 0 & 1 & 0 \\ 0 & 0 & 0 & 1 \end{bmatrix}$

$BA = \begin{bmatrix} 0-2+0+3 & 0+0+0+0 & -2+2+0+0 & 1+2+0-3 \\ 0-1+0+1 & 0+0+1+0 & 0+1-1+0 & 0+1+0-1 \\ 0-1+0+1 & 0+0+0+0 & 0+1+0+0 & 0+1+0-1 \\ 0-2+0+2 & 0+0+0+0 & -2+2+0+0 & 1+2+0-2 \end{bmatrix} = \begin{bmatrix} 1 & 0 & 0 & 0 \\ 0 & 1 & 0 & 0 \\ 0 & 0 & 1 & 0 \\ 0 & 0 & 0 & 1 \end{bmatrix}$

Since $AB = I_4$ and $BA = I_4$, $B = A^{-1}$.

13. $ad - bc = (2)(2) - (3)(-1) = 4 + 3 = 7$

$$A^{-1} = \frac{1}{7}\begin{bmatrix} 2 & -3 \\ 1 & 2 \end{bmatrix} = \begin{bmatrix} \frac{2}{7} & -\frac{3}{7} \\ \frac{1}{7} & \frac{2}{7} \end{bmatrix}$$

$$AA^{-1} = \begin{bmatrix} \frac{4}{7}+\frac{3}{7} & -\frac{6}{7}+\frac{6}{7} \\ -\frac{2}{7}+\frac{2}{7} & \frac{3}{7}+\frac{4}{7} \end{bmatrix} = \begin{bmatrix} 1 & 0 \\ 0 & 1 \end{bmatrix} \text{ and } A^{-1}A = \begin{bmatrix} \frac{4}{7}+\frac{3}{7} & \frac{6}{7}-\frac{6}{7} \\ \frac{2}{7}-\frac{2}{7} & \frac{3}{7}+\frac{4}{7} \end{bmatrix} = \begin{bmatrix} 1 & 0 \\ 0 & 1 \end{bmatrix}$$

15. $ad - bc = (3)(2) - (-1)(-4) = 6 - 4 = 2$

$$A^{-1} = \frac{1}{2}\begin{bmatrix} 2 & 1 \\ 4 & 3 \end{bmatrix} = \begin{bmatrix} 1 & \frac{1}{2} \\ 2 & \frac{3}{2} \end{bmatrix}$$

$$AA^{-1} = \begin{bmatrix} 3-2 & \frac{3}{2}-\frac{3}{2} \\ -4+4 & -\frac{4}{2}+\frac{6}{2} \end{bmatrix} = \begin{bmatrix} 1 & 0 \\ 0 & 1 \end{bmatrix} \text{ and } A^{-1}A = \begin{bmatrix} 3-\frac{4}{2} & -1+\frac{2}{2} \\ 6-\frac{12}{2} & -2+\frac{6}{2} \end{bmatrix} = \begin{bmatrix} 1 & 0 \\ 0 & 1 \end{bmatrix}$$

17. $ad - bc = (10)(1) - (-2)(-5) = 10 - 10 = 0$

Since division by zero is undefined, *A* does not have an inverse.

For Problems 19–24, verification that $AA^{-1} = I$ and $A^{-1}A = I$ is left to the student.

For Problems 19–23, verification that $AA^{-1} = I$ and $A^{-1}A = I$ is left to the student.

19. $\begin{bmatrix} 2 & 0 & 0 & 1 & 0 & 0 \\ 0 & 4 & 0 & 0 & 1 & 0 \\ 0 & 0 & 6 & 0 & 0 & 1 \end{bmatrix}$

Divide row 1 by 2, divide row 2 by 4 and divide row 4 by 6.

$$\begin{bmatrix} 1 & 0 & 0 & \frac{1}{2} & 0 & 0 \\ 0 & 1 & 0 & 0 & \frac{1}{4} & 0 \\ 0 & 0 & 1 & 0 & 0 & \frac{1}{6} \end{bmatrix}$$

$$A^{-1} = \begin{bmatrix} \frac{1}{2} & 0 & 0 \\ 0 & \frac{1}{4} & 0 \\ 0 & 0 & \frac{1}{6} \end{bmatrix}$$

21. $\begin{bmatrix} 1 & 2 & -1 & 1 & 0 & 0 \\ -2 & 0 & 1 & 0 & 1 & 0 \\ 1 & -1 & 0 & 0 & 0 & 1 \end{bmatrix}$

Replace row 2 with $2R_1 + R_2$.
Replace row 3 with $R_1 - R_3$.

$\begin{bmatrix} 1 & 2 & -1 & 1 & 0 & 0 \\ 0 & 4 & -1 & 2 & 1 & 0 \\ 0 & 3 & -1 & 1 & 0 & -1 \end{bmatrix}$

Replace row 1 with $R_2 - 2R_1$.
Replace row 3 with $-3R_2 + 4R_3$.

$\begin{bmatrix} -2 & 0 & 1 & 0 & 1 & 0 \\ 0 & 4 & -1 & 2 & 1 & 0 \\ 0 & 0 & -1 & -2 & -3 & -4 \end{bmatrix}$

Replace row 1 with $R_3 + R_1$.
Replace row 2 with $R_2 - R_3$.
Replace row 3 with $-R_3$.

$\begin{bmatrix} -2 & 0 & 0 & -2 & -2 & -4 \\ 0 & 4 & 0 & 4 & 4 & 4 \\ 0 & 0 & 1 & 2 & 3 & 4 \end{bmatrix}$

Divide row 1 by -2 and divide row 2 by 4.

$\begin{bmatrix} 1 & 0 & 0 & 1 & 1 & 2 \\ 0 & 1 & 0 & 1 & 1 & 1 \\ 0 & 0 & 1 & 2 & 3 & 4 \end{bmatrix}$

$A^{-1} = \begin{bmatrix} 1 & 1 & 2 \\ 1 & 1 & 1 \\ 2 & 3 & 4 \end{bmatrix}$

23. $\begin{bmatrix} 2 & 2 & -1 & 1 & 0 & 0 \\ 0 & 3 & -1 & 0 & 1 & 0 \\ -1 & -2 & 1 & 0 & 0 & 1 \end{bmatrix} R_1 \leftrightarrow R_3$

$\begin{bmatrix} -1 & -2 & 1 & 0 & 0 & 1 \\ 0 & 3 & -1 & 0 & 1 & 0 \\ 2 & 2 & -1 & 1 & 0 & 0 \end{bmatrix} -1R_1$

$\begin{bmatrix} 1 & 2 & -1 & 0 & 0 & -1 \\ 0 & 3 & -1 & 0 & 1 & 0 \\ 2 & 2 & -1 & 1 & 0 & 0 \end{bmatrix} -2R_1 + R_3$

$\begin{bmatrix} 1 & 2 & -1 & 0 & 0 & -1 \\ 0 & 3 & -1 & 0 & 1 & 0 \\ 0 & -2 & 1 & 1 & 0 & 2 \end{bmatrix} \frac{1}{3}R_2$

$\begin{bmatrix} 1 & 2 & -1 & 0 & 0 & -1 \\ 0 & 1 & -\frac{1}{3} & 0 & \frac{1}{3} & 0 \\ 0 & -2 & 1 & 1 & 0 & 2 \end{bmatrix} \begin{matrix} -2R_2 + R_1 \\ 2R_2 + R_3 \end{matrix}$

$\begin{bmatrix} 1 & 0 & -\frac{1}{3} & 0 & -\frac{2}{3} & -1 \\ 0 & 1 & -\frac{1}{3} & 0 & \frac{1}{3} & 0 \\ 0 & 0 & \frac{1}{3} & 1 & \frac{2}{3} & 2 \end{bmatrix} \begin{matrix} 1R_3 + R_1 \\ 1R_2 + R_1 \end{matrix}$

$\begin{bmatrix} 1 & 0 & 0 & 1 & 0 & 1 \\ 0 & 1 & 0 & 1 & 1 & 2 \\ 0 & 0 & \frac{1}{3} & 1 & \frac{2}{3} & 2 \end{bmatrix} 3R_3$

$\begin{bmatrix} 1 & 0 & 0 & 1 & 0 & 1 \\ 0 & 1 & 0 & 1 & 1 & 2 \\ 0 & 0 & 1 & 3 & 2 & 6 \end{bmatrix}$

$A^{-1} = \begin{bmatrix} 1 & 0 & 1 \\ 1 & 1 & 2 \\ 3 & 2 & 6 \end{bmatrix}$

25.
$$\left[\begin{array}{rrr|rrr} 5 & 0 & 2 & 1 & 0 & 0 \\ 2 & 2 & 1 & 0 & 1 & 0 \\ -3 & 1 & -1 & 0 & 0 & 1 \end{array}\right] \frac{1}{5}R_1$$

$$\left[\begin{array}{rrr|rrr} 1 & 0 & \frac{2}{5} & \frac{1}{5} & 0 & 0 \\ 2 & 2 & 1 & 0 & 1 & 0 \\ -3 & 1 & -1 & 0 & 0 & 1 \end{array}\right] \begin{array}{l} -2R_1 + R_2 \\ 3R_1 + R_3 \end{array}$$

$$\left[\begin{array}{rrr|rrr} 1 & 0 & \frac{2}{5} & \frac{1}{5} & 0 & 0 \\ 0 & 2 & \frac{1}{5} & -\frac{2}{5} & 1 & 0 \\ 0 & 1 & \frac{1}{5} & \frac{3}{5} & 0 & 1 \end{array}\right] R_2 \leftrightarrow R_3$$

$$\left[\begin{array}{rrr|rrr} 1 & 0 & \frac{2}{5} & \frac{1}{5} & 0 & 0 \\ 0 & 1 & \frac{1}{5} & \frac{3}{5} & 0 & 1 \\ 0 & 2 & \frac{1}{5} & -\frac{2}{5} & 1 & 0 \end{array}\right] -2R_2 + R_3$$

$$\left[\begin{array}{rrr|rrr} 1 & 0 & \frac{2}{5} & \frac{1}{5} & 0 & 0 \\ 0 & 1 & \frac{1}{5} & \frac{3}{5} & 0 & 1 \\ 0 & 0 & -\frac{1}{5} & -\frac{8}{5} & 1 & -2 \end{array}\right] \begin{array}{l} 2R_3 + R_1 \\ 1R_3 + R_2 \end{array}$$

$$\left[\begin{array}{rrr|rrr} 1 & 0 & 0 & -3 & 2 & -4 \\ 0 & 1 & 0 & -1 & 1 & -1 \\ 0 & 0 & -\frac{1}{5} & -\frac{8}{5} & 1 & -2 \end{array}\right] -5R_3$$

$$\left[\begin{array}{rrr|rrr} 1 & 0 & 0 & -3 & 2 & -4 \\ 0 & 1 & 0 & -1 & 1 & -1 \\ 0 & 0 & 1 & 8 & -5 & 10 \end{array}\right]$$

$$A^{-1} = \left[\begin{array}{rrr} -3 & 2 & -4 \\ -1 & 1 & -1 \\ 8 & -5 & 10 \end{array}\right]$$

27.
$$\left[\begin{array}{rrrr|rrrr} 1 & 0 & 0 & 0 & 1 & 0 & 0 & 0 \\ 0 & -1 & 0 & 0 & 0 & 1 & 0 & 0 \\ 0 & 0 & 3 & 0 & 0 & 0 & 1 & 0 \\ 1 & 0 & 0 & 1 & 0 & 0 & 0 & 1 \end{array}\right] -1R_1 + R_4$$

$$\left[\begin{array}{rrrr|rrrr} 1 & 0 & 0 & 0 & 1 & 0 & 0 & 0 \\ 0 & -1 & 0 & 0 & 0 & 1 & 0 & 0 \\ 0 & 0 & 3 & 0 & 0 & 0 & 1 & 0 \\ 0 & 0 & 0 & 1 & -1 & 0 & 0 & 1 \end{array}\right] -1R_2$$

$$\left[\begin{array}{rrrr|rrrr} 1 & 0 & 0 & 0 & 1 & 0 & 0 & 0 \\ 0 & 1 & 0 & 0 & 0 & -1 & 0 & 0 \\ 0 & 0 & 3 & 0 & 0 & 0 & 1 & 0 \\ 0 & 0 & 0 & 1 & -1 & 0 & 0 & 1 \end{array}\right] \frac{1}{3}R_3$$

$$\left[\begin{array}{rrrr|rrrr} 1 & 0 & 0 & 0 & 1 & 0 & 0 & 0 \\ 0 & 1 & 0 & 0 & 0 & -1 & 0 & 0 \\ 0 & 0 & 1 & 0 & 0 & 0 & \frac{1}{3} & 0 \\ 0 & 0 & 0 & 1 & -1 & 0 & 0 & 1 \end{array}\right]$$

$$A^{-1} = \left[\begin{array}{rrrr} 1 & 0 & 0 & 0 \\ 0 & -1 & 0 & 0 \\ 0 & 0 & \frac{1}{3} & 0 \\ -1 & 0 & 0 & 1 \end{array}\right]$$

29. $\begin{bmatrix} 6 & 5 \\ 5 & 4 \end{bmatrix}\begin{bmatrix} x \\ y \end{bmatrix} = \begin{bmatrix} 13 \\ 10 \end{bmatrix}$

31. $\begin{bmatrix} 1 & 3 & 4 \\ 1 & 2 & 3 \\ 1 & 4 & 3 \end{bmatrix}\begin{bmatrix} x \\ y \\ z \end{bmatrix} = \begin{bmatrix} -3 \\ -2 \\ -6 \end{bmatrix}$

33. $4x - 7y = -3$
$2x - 3y = 1$

35. $2x - z = 6$
$3y = 9$
$x + y = 5$

37. a.

$$\begin{bmatrix} 2 & 6 & 6 \\ 2 & 7 & 6 \\ 2 & 7 & 7 \end{bmatrix}\begin{bmatrix} x \\ y \\ z \end{bmatrix} = \begin{bmatrix} 8 \\ 10 \\ 9 \end{bmatrix}$$

$$\begin{bmatrix} \frac{7}{2} & 0 & -3 \\ -1 & 1 & 0 \\ 0 & -1 & 1 \end{bmatrix}\begin{bmatrix} 8 \\ 10 \\ 9 \end{bmatrix} = \begin{bmatrix} 28+0-27 \\ -8+10+0 \\ 0-10+9 \end{bmatrix} = \begin{bmatrix} 1 \\ 2 \\ -1 \end{bmatrix}$$

The solution to the system is $\{(1, 2, -1)\}$.

39. a.

$$\begin{bmatrix} 1 & -1 & 1 \\ 0 & 2 & -1 \\ 2 & 3 & 0 \end{bmatrix}\begin{bmatrix} x \\ y \\ z \end{bmatrix} = \begin{bmatrix} 8 \\ -7 \\ 1 \end{bmatrix}$$

b.

$$\begin{bmatrix} 3 & 3 & -1 \\ -2 & -2 & 1 \\ -4 & -5 & 2 \end{bmatrix}\begin{bmatrix} 8 \\ -7 \\ 1 \end{bmatrix}$$

$$= \begin{bmatrix} 24-21-1 \\ -16+14+1 \\ -32+35+2 \end{bmatrix} = \begin{bmatrix} 2 \\ -1 \\ 5 \end{bmatrix}$$

The solution to the system is
$\{(2, -1, 5)\}$.

41. a.

$$\begin{bmatrix} 1 & -1 & 2 & 0 \\ 0 & 1 & -1 & 1 \\ -1 & 1 & -1 & 2 \\ 0 & -1 & 1 & -2 \end{bmatrix}\begin{bmatrix} w \\ x \\ y \\ z \end{bmatrix} = \begin{bmatrix} -3 \\ 4 \\ 2 \\ -4 \end{bmatrix}$$

b.

$$\begin{bmatrix} 0 & 0 & -1 & -1 \\ 1 & 4 & 1 & 3 \\ 1 & 2 & 1 & 2 \\ 0 & -1 & 0 & -1 \end{bmatrix}\begin{bmatrix} -3 \\ 4 \\ 2 \\ -4 \end{bmatrix}$$

$$= \begin{bmatrix} 0+0-2+4 \\ -3+16+2-12 \\ -3+8+2-8 \\ 0-4+0+4 \end{bmatrix} = \begin{bmatrix} 2 \\ 3 \\ -1 \\ 0 \end{bmatrix}$$

The solution to the system is
$\{(2, 3, -1, 0)\}$.

43. $A = \begin{bmatrix} e^x & e^{3x} \\ -e^{3x} & e^{5x} \end{bmatrix}$

$$A^{-1} = \frac{1}{ad-bc}\begin{bmatrix} d & -b \\ -c & a \end{bmatrix}$$

$$A^{-1} = \frac{1}{(e^x)(e^{5x})-(e^{3x})(-e^{3x})}\begin{bmatrix} e^{5x} & -e^{3x} \\ -(-e^{3x}) & e^x \end{bmatrix}$$

$$A^{-1} = \frac{1}{e^{6x}+e^{6x}}\begin{bmatrix} e^{5x} & -e^{3x} \\ e^{3x} & e^x \end{bmatrix}$$

$$A^{-1} = \frac{1}{2e^{6x}}\begin{bmatrix} e^{5x} & -e^{3x} \\ e^{3x} & e^x \end{bmatrix}$$

$$A^{-1} = \begin{bmatrix} \frac{e^{5x}}{2e^{6x}} & \frac{-e^{3x}}{2e^{6x}} \\ \frac{e^{3x}}{2e^{6x}} & \frac{e^x}{2e^{6x}} \end{bmatrix}$$

$$A^{-1} = \begin{bmatrix} \frac{1}{2e^x} & -\frac{1}{2e^{3x}} \\ \frac{1}{2e^{3x}} & \frac{1}{2e^{5x}} \end{bmatrix} \text{ or } \begin{bmatrix} \frac{e^{-x}}{2} & -\frac{e^{-3x}}{2} \\ \frac{e^{-3x}}{2} & \frac{e^{-5x}}{2} \end{bmatrix}$$

Check:

$$\begin{bmatrix} e^x & e^{3x} \\ -e^{3x} & e^{5x} \end{bmatrix} \cdot \begin{bmatrix} \frac{1}{2e^x} & -\frac{1}{2e^{3x}} \\ \frac{1}{2e^{3x}} & \frac{1}{2e^{5x}} \end{bmatrix} = \begin{bmatrix} 1 & 0 \\ 0 & 1 \end{bmatrix}$$

45. $A = \begin{bmatrix} 8 & -5 \\ -3 & 2 \end{bmatrix}$

$$I - A = \begin{bmatrix} 1 & 0 \\ 0 & 1 \end{bmatrix} - \begin{bmatrix} 8 & -5 \\ -3 & 2 \end{bmatrix} = \begin{bmatrix} -7 & 5 \\ 3 & -1 \end{bmatrix}$$

$$(I-A)^{-1} = \frac{1}{(-7)(-1)-(5)(3)}\begin{bmatrix} -1 & -(5) \\ -(3) & -7 \end{bmatrix}$$

$$(I-A)^{-1} = \frac{1}{7-15}\begin{bmatrix} -1 & -5 \\ -3 & -7 \end{bmatrix}$$

$$(I-A)^{-1} = \frac{1}{-8}\begin{bmatrix} -1 & -5 \\ -3 & -7 \end{bmatrix}$$

$$(I-A)^{-1} = \begin{bmatrix} \frac{-1}{-8} & \frac{-5}{-8} \\ \frac{-3}{-8} & \frac{-7}{-8} \end{bmatrix}$$

$$(I-A)^{-1} = \begin{bmatrix} \frac{1}{8} & \frac{5}{8} \\ \frac{3}{8} & \frac{7}{8} \end{bmatrix}$$

47. $A = \begin{bmatrix} 2 & 1 \\ 3 & 1 \end{bmatrix} \quad B = \begin{bmatrix} 4 & 7 \\ 1 & 2 \end{bmatrix}$

$A^{-1} = \begin{bmatrix} -1 & 1 \\ 3 & -2 \end{bmatrix} \quad B^{-1} = \begin{bmatrix} 2 & -7 \\ -1 & 4 \end{bmatrix}$

$AB = \begin{bmatrix} 2 & 1 \\ 3 & 1 \end{bmatrix}\begin{bmatrix} 4 & 7 \\ 1 & 2 \end{bmatrix} = \begin{bmatrix} 9 & 16 \\ 13 & 23 \end{bmatrix}$

$(AB)^{-1} = \left(\begin{bmatrix} 9 & 16 \\ 13 & 23 \end{bmatrix} \right)^{-1} = \begin{bmatrix} -23 & 16 \\ 13 & -9 \end{bmatrix}$

$A^{-1}B^{-1} = \begin{bmatrix} -1 & 1 \\ 3 & -2 \end{bmatrix}\begin{bmatrix} 2 & -7 \\ -1 & 4 \end{bmatrix} = \begin{bmatrix} -3 & 11 \\ 8 & -29 \end{bmatrix}$

$B^{-1}A^{-1} = \begin{bmatrix} 2 & -7 \\ -1 & 4 \end{bmatrix}\begin{bmatrix} -1 & 1 \\ 3 & -2 \end{bmatrix} = \begin{bmatrix} -23 & 16 \\ 13 & -9 \end{bmatrix}$

Observe that $(AB)^{-1} = B^{-1}A^{-1}$.

49. $\begin{bmatrix} a & 0 & 0 \\ 0 & b & 0 \\ 0 & 0 & c \end{bmatrix}\begin{bmatrix} \frac{1}{a} & 0 & 0 \\ 0 & \frac{1}{b} & 0 \\ 0 & 0 & \frac{1}{c} \end{bmatrix}$

$= \begin{bmatrix} (a)(\frac{1}{a})+(0)(0)+(0)(0) & (a)(0)+(0)(\frac{1}{b})+(0)(0) & (a)(0)+(0)(0)+(0)(\frac{1}{c}) \\ (0)(\frac{1}{a})+(b)(0)+(0)(0) & (0)(0)+(b)(\frac{1}{b})+(0)(0) & (0)(0)+(b)(0)+(0)(\frac{1}{c}) \\ (0)(\frac{1}{a})+(0)(0)+(c)(0) & (0)(0)+(0)(\frac{1}{b})+(c)(0) & (0)(0)+(0)(0)+(c)(\frac{1}{c}) \end{bmatrix}$

$= \begin{bmatrix} \frac{a}{a}+0+0 & 0+0+0 & 0+0+0 \\ 0+0+0 & 0+\frac{b}{b}+0 & 0+0+0 \\ 0+0+0 & 0+0+0 & 0+0+\frac{c}{c} \end{bmatrix} = \begin{bmatrix} 1 & 0 & 0 \\ 0 & 1 & 0 \\ 0 & 0 & 1 \end{bmatrix}$

51. The numerical equivalent of HELP is
8, 5, 12, 16.

$\begin{bmatrix} 4 & -1 \\ -3 & 1 \end{bmatrix}\begin{bmatrix} 8 \\ 5 \end{bmatrix} = \begin{bmatrix} 27 \\ -19 \end{bmatrix}$,

$\begin{bmatrix} 4 & -1 \\ -3 & 1 \end{bmatrix}\begin{bmatrix} 12 \\ 16 \end{bmatrix} = \begin{bmatrix} 32 \\ -20 \end{bmatrix}$

The encoded message is 27, –19, 32, –20.

$\begin{bmatrix} 1 & 1 \\ 3 & 4 \end{bmatrix}\begin{bmatrix} 27 \\ -19 \end{bmatrix} = \begin{bmatrix} 8 \\ 5 \end{bmatrix}$, $\begin{bmatrix} 1 & 1 \\ 3 & 4 \end{bmatrix}\begin{bmatrix} 32 \\ -20 \end{bmatrix} = \begin{bmatrix} 12 \\ 16 \end{bmatrix}$

The decoded message is 8, 5, 12, 16 or HELP.

53.
$$\begin{bmatrix} 1 & -1 & 0 \\ 3 & 0 & 2 \\ -1 & 0 & -1 \end{bmatrix}\begin{bmatrix} 19 & 4 & 1 \\ 5 & 0 & 19 \\ 14 & 3 & 8 \end{bmatrix}$$

$$= \begin{bmatrix} 19-5+0 & 4+0+0 & 1-19+0 \\ 57+0+28 & 12+0+6 & 3+0+16 \\ -19+0-14 & -4+0-3 & -1+0-8 \end{bmatrix}$$

$$= \begin{bmatrix} 14 & 4 & -18 \\ 85 & 18 & 19 \\ -33 & -7 & -9 \end{bmatrix}$$

The encoded message is 14, 85, –33, 4, 18, –7, –18, 19, –9.

$$\begin{bmatrix} 0 & 1 & 2 \\ -1 & 1 & 2 \\ 0 & -1 & -3 \end{bmatrix}\begin{bmatrix} 14 & 4 & -18 \\ 85 & 18 & 19 \\ -33 & -7 & -9 \end{bmatrix}$$

$$= \begin{bmatrix} 0+85-66 & 0+18-14 & 0+19-18 \\ -14+85-66 & -4+18-14 & 18+19-18 \\ 0-85+99 & 0-18+21 & 0-19+27 \end{bmatrix}$$

$$= \begin{bmatrix} 19 & 4 & 1 \\ 5 & 0 & 19 \\ 14 & 3 & 8 \end{bmatrix}$$

The decoded message is 19, 5, 14, 4, 0, 3, 1, 19, 8 or SEND_CASH

55. – 63. Answers may vary.

65. Enter the matrix $\begin{bmatrix} 3 & -1 \\ -2 & 1 \end{bmatrix}$ as $[A]$, then use $[A]^{-1}$.

$[A]^{-1} = \begin{bmatrix} 1 & 1 \\ 2 & 3 \end{bmatrix}$

Verify this result by showing that $[A][A]^{-1} = I_2$ and $[A]^{-1}[A] = I_2$.

67. Enter the matrix $\begin{bmatrix} -2 & 1 & -1 \\ -5 & 2 & -1 \\ 3 & -1 & 1 \end{bmatrix}$ as $[A]$, then use $[A]^{-1}$.

$[A]^{-1} = \begin{bmatrix} 1 & 0 & 1 \\ 2 & 1 & 3 \\ -1 & 1 & 1 \end{bmatrix}$

Verify this result by showing that $[A][A]^{-1} = I_3$ and $[A]^{-1}[A] = I_3$.

69. Enter the matrix $\begin{bmatrix} 7 & -3 & 0 & 2 \\ -2 & 1 & 0 & -1 \\ 4 & 0 & 1 & -2 \\ -1 & 1 & 0 & -1 \end{bmatrix}$ as $[A]$, then use $[A]^{-1}$. $[A]^{-1} = \begin{bmatrix} 0 & -1 & 0 & 1 \\ -1 & -5 & 0 & 3 \\ -2 & -4 & 1 & -2 \\ -1 & -4 & 0 & 1 \end{bmatrix}$

Verify this result by showing that $[A][A]^{-1} = I_4$ and $[A]^{-1}[A] = I_4$.

71. The system is $AX = B$ where

$A = \begin{bmatrix} 1 & -1 & 1 \\ 4 & 2 & 1 \\ 4 & -2 & 1 \end{bmatrix}$, $X = \begin{bmatrix} x \\ y \\ z \end{bmatrix}$, and $B = \begin{bmatrix} -6 \\ 9 \\ -3 \end{bmatrix}$. $X = \begin{bmatrix} 2 \\ 3 \\ -5 \end{bmatrix}$, so the solution to the system is $\{(2, 3, -5)\}$.

73. The system is $AX = B$ where $A = \begin{bmatrix} 3 & -2 & 1 \\ 4 & -5 & 3 \\ 2 & -1 & 5 \end{bmatrix}$, $X = \begin{bmatrix} x \\ y \\ z \end{bmatrix}$, and $B = \begin{bmatrix} -2 \\ -9 \\ -5 \end{bmatrix}$.

$X = \begin{bmatrix} 1 \\ 2 \\ -1 \end{bmatrix}$ so the solution to the system is $\{(1, 2, -1)\}$.

75. The system is $AX = B$ where $A = \begin{bmatrix} 1 & 0 & -3 & 0 & 1 \\ 0 & 1 & 0 & 1 & 0 \\ 0 & 0 & 1 & 0 & 1 \\ 1 & 1 & -1 & 4 & 0 \\ 1 & 1 & 1 & 1 & 1 \end{bmatrix}$, $X = \begin{bmatrix} v \\ w \\ x \\ y \\ z \end{bmatrix}$ and $B = \begin{bmatrix} -3 \\ -1 \\ 7 \\ -8 \\ 8 \end{bmatrix}$. $X = \begin{bmatrix} 2 \\ 1 \\ 3 \\ -2 \\ 4 \end{bmatrix}$, so the solution to the system

is $\{(2, 1, 3, -2, 4)\}$.

77. Answers may vary.

79. does not make sense; Explanations will vary. Sample explanation: Only square matrices have inverses.

81. makes sense

83. false; Changes to make the statement true will vary. A sample change is: Not all square matrices have inverses.

85. false; Changes to make the statement true will vary. A sample change is: You need to multiply the inverse of A and B.

87. false; Changes to make the statement true will vary. A sample change is: $(A + B)^{-1} \neq A^{-1} + B^{-1}$

89. Answers may vary.

91. Using the statement before problems 9–14, we want to find values for a such that
$(1)(4) - (a + 1)(a - 2) = 0$.
$(1)(4) - (a+1)(a-2) = 4 - (a^2 - a - 2)$
$\quad = -a^2 + a + 6$
$0 = -a^2 + a + 6$
$0 = a^2 - a - 6$
$0 = (a - 3)(a + 2)$
$a = 3, -2$

93. $2(-5) - (-3)(4) = -10 + 12 = 2$

95. $2(-30 - (-3)) - 3(6 - 9) + (-1)(1 - 15)$
$= 2(-27) - 3(-3) + (-1)(-14)$
$= -54 + 9 + 14$
$= -31$

Matrices and Determinants

Section 8.5

Check Point Exercises

1. **a.** $\begin{vmatrix} 10 & 9 \\ 6 & 5 \end{vmatrix} = 10 \cdot 5 - 6 \cdot 9 = 50 - 54 = -4$

b. $\begin{vmatrix} 4 & 3 \\ -5 & -8 \end{vmatrix} = 4 \cdot (-8) - (-5) \cdot (3)$

$= -32 + 15 = -17$

2. $5x + 4y = 12$

$3x - 6y = 24$

$D = \begin{vmatrix} 5 & 4 \\ 3 & -6 \end{vmatrix} = 5 \cdot (-6) - 3 \cdot 4$

$= -30 - 12 = -42$

$D_x = \begin{vmatrix} 12 & 4 \\ 24 & -6 \end{vmatrix} = 12(-6) - 24(4)$

$= -72 - 96 = -168$

$D_y = \begin{vmatrix} 5 & 12 \\ 3 & 24 \end{vmatrix} = 5(24) - 3(12)$

$= 120 - 36 = 84$

Thus, $x = \dfrac{D_x}{D} = \dfrac{-168}{-42} = 4$

$y = \dfrac{D_y}{D} = \dfrac{84}{-42} = -2$

The solution set is $\{(4, -2)\}$.

3. $\begin{bmatrix} 2 & 1 & 7 \\ -5 & 6 & 0 \\ -4 & 3 & 1 \end{bmatrix}$

The minor for 2 is $\begin{vmatrix} 6 & 0 \\ 3 & 1 \end{vmatrix}$.

The minor for -5 is $\begin{vmatrix} 1 & 7 \\ 3 & 1 \end{vmatrix}$.

The minor for -4 is $\begin{vmatrix} 1 & 7 \\ 6 & 0 \end{vmatrix}$.

$\begin{bmatrix} 2 & 1 & 7 \\ -5 & 6 & 0 \\ -4 & 3 & 1 \end{bmatrix} = 2\begin{vmatrix} 6 & 0 \\ 3 & 1 \end{vmatrix} - (-5)\begin{vmatrix} 1 & 7 \\ 3 & 1 \end{vmatrix} - 4\begin{vmatrix} 1 & 7 \\ 6 & 0 \end{vmatrix}$

$= 2(6 \cdot 1 - 3 \cdot 0) + 5(1 \cdot 1 - 3 \cdot 7) - 4(1 \cdot 0 - 6 \cdot 7)$

$= 2(6 - 0) + 5(1 - 21) - 4(0 - 42)$

$= 12 - 100 + 168$

$= 80$

4.
$$\begin{vmatrix} 6 & 4 & 0 \\ -3 & -5 & 3 \\ 1 & 2 & 0 \end{vmatrix} = 0\begin{vmatrix} -3 & -5 \\ 1 & 2 \end{vmatrix} - 3\begin{vmatrix} 6 & 4 \\ 1 & 2 \end{vmatrix} + 0\begin{vmatrix} 6 & 4 \\ -3 & -5 \end{vmatrix}$$

$$= 0 - 3(6 \cdot 2 - 1 \cdot 4) + 0$$

$$= -3(12 - 4)$$

$$= -3(8)$$

$$= -24$$

5. $3x - 2y + z = 16$

$2x + 3y - z = -9$

$x + 4y + 3z = 2$

$$D = \begin{vmatrix} 3 & -2 & 1 \\ 2 & 3 & -1 \\ 1 & 4 & 3 \end{vmatrix}; \quad D_x = \begin{vmatrix} 16 & -2 & 1 \\ -9 & 3 & -1 \\ 2 & 4 & 3 \end{vmatrix}; \quad D_y = \begin{vmatrix} 3 & 16 & 1 \\ 2 & -9 & -1 \\ 1 & 2 & 3 \end{vmatrix}; \quad D_z = \begin{vmatrix} 3 & -2 & 16 \\ 2 & 3 & -9 \\ 1 & 4 & 2 \end{vmatrix}$$

$$D = \begin{vmatrix} 3 & -2 & 1 \\ 2 & 3 & -1 \\ 1 & 4 & 3 \end{vmatrix} = 3\begin{vmatrix} 3 & -1 \\ 4 & 3 \end{vmatrix} - 2\begin{vmatrix} -2 & 1 \\ 4 & 3 \end{vmatrix} + 1\begin{vmatrix} -2 & 1 \\ 3 & -1 \end{vmatrix}$$

$$= 3[(3) \cdot 3 - 4 \cdot (-1)] - 2[(-2) \cdot 3 - 4 \cdot 1] + 1[(-2) \cdot (-1) - (3) \cdot 1]$$

$$= 3(9 + 4) - 2(-6 - 4) + 1(2 - 3)$$

$$= 39 + 20 - 1$$

$$= 58$$

$$D_x = \begin{vmatrix} 16 & -2 & 1 \\ -9 & 3 & -1 \\ 2 & 4 & 3 \end{vmatrix} = 1\begin{vmatrix} -9 & 3 \\ 2 & 4 \end{vmatrix} - (-1)\begin{vmatrix} 16 & -2 \\ 2 & 4 \end{vmatrix} + 3\begin{vmatrix} 16 & -2 \\ -9 & 3 \end{vmatrix}$$

$$= 1[(-9) \cdot 4 - 2 \cdot (3)] + 1[16 \cdot 4 - 2(-2)] + 3[16 \cdot (3) - (-9) \cdot (-2)]$$

$$= 1(-36 - 6) + 1(64 + 4) + 3(48 - 18)$$

$$= -42 + 68 + 90$$

$$= 116$$

$$D_y = \begin{vmatrix} 3 & 16 & 1 \\ 2 & -9 & -1 \\ 1 & 2 & 3 \end{vmatrix} = 3\begin{vmatrix} -9 & -1 \\ 2 & 3 \end{vmatrix} - 2\begin{vmatrix} 16 & 1 \\ 2 & 3 \end{vmatrix} + 1\begin{vmatrix} 16 & 1 \\ -9 & -1 \end{vmatrix}$$

$$= 3[(-9) \cdot 3 - 2 \cdot (-1)] - 2[16 \cdot 3 - 2 \cdot 1] + 1[16(-1) - (-9) \cdot 1]$$

$$= 3(-27 + 2) - 2(48 - 2) + 1(-16 + 9)$$

$$= -75 - 92 - 7$$

$$= -174$$

$$D_z = \begin{vmatrix} 3 & -2 & 16 \\ 2 & 3 & -9 \\ 1 & 4 & 2 \end{vmatrix} = 3\begin{vmatrix} 3 & -9 \\ 4 & 2 \end{vmatrix} - 2\begin{vmatrix} -2 & 16 \\ 4 & 2 \end{vmatrix} + 1\begin{vmatrix} -2 & 16 \\ 3 & -9 \end{vmatrix}$$

$$= 3[(3)2 - 4(-9)] - 2[(-2)2 - 4 \cdot 16] + 1[(-2)(-9) - (3) \cdot 16]$$

$$= 3(6 + 36) - 2(-4 - 64) + 1(18 - 48)$$

$$= 126 + 136 - 30$$

$$= 232$$

$$x = \frac{D_x}{D} = \frac{116}{58} = 2$$

$$y = \frac{D_y}{D} = \frac{-174}{58} = -3$$

$$z = \frac{D_z}{D} = \frac{232}{58} = 4$$

The solution to the system is $\{(2, -3, 4)\}$.

6. $|A| = \begin{vmatrix} 0 & 4 & 0 & -3 \\ -1 & 1 & 5 & 2 \\ 1 & -2 & 0 & 6 \\ 3 & 0 & 0 & 1 \end{vmatrix} = (-1)^{2+3} 5 \begin{vmatrix} 0 & 4 & -3 \\ 1 & -2 & 6 \\ 3 & 0 & 1 \end{vmatrix} = -5 \begin{vmatrix} 0 & 4 & -3 \\ 1 & -2 & 6 \\ 3 & 0 & 1 \end{vmatrix}$

Evaluate the third-order determinant to get $|A| = -5(50) = -250$.

Exercise Set 8.5

1. $\begin{vmatrix} 5 & 7 \\ 2 & 3 \end{vmatrix} = 5 \cdot 3 - 2 \cdot 7 = 15 - 14 = 1$

3. $\begin{vmatrix} -4 & 1 \\ 5 & 6 \end{vmatrix} = (-4)6 - 5 \cdot 1 = -24 - 5 = -29$

5. $\begin{vmatrix} -7 & 14 \\ 2 & -4 \end{vmatrix} = (-7)(-4) - 2(14) = 28 - 28 = 0$

7. $\begin{vmatrix} -5 & -1 \\ -2 & -7 \end{vmatrix} = (-5)(-7) - (-2)(-1) = 35 - 2 = 33$

9. $\begin{vmatrix} \frac{1}{2} & \frac{1}{2} \\ \frac{1}{8} & -\frac{3}{4} \end{vmatrix} = \frac{1}{2}\left(-\frac{3}{4}\right) - \frac{1}{8} \cdot \frac{1}{2} = -\frac{3}{8} - \frac{1}{16} = -\frac{7}{16}$

11. $D = \begin{vmatrix} 1 & 1 \\ 1 & -1 \end{vmatrix} = -1 - 1 = -2$

$D_x = \begin{vmatrix} 7 & 1 \\ 3 & -1 \end{vmatrix} = -7 - 3 = -10$

$D_y = \begin{vmatrix} 1 & 7 \\ 1 & 3 \end{vmatrix} = 3 - 7 = -4$

$x = \dfrac{D_x}{D} = \dfrac{-10}{-2} = 5$

$y = \dfrac{D_y}{D} = \dfrac{-4}{-2} = 2$

The solution set is $\{(5, 2)\}$.

13. $D = \begin{vmatrix} 12 & 3 \\ 2 & -3 \end{vmatrix} = -36 - 6 = -42$

$D_x = \begin{vmatrix} 15 & 3 \\ 13 & -3 \end{vmatrix} = -45 - 39 = -84$

$D_y = \begin{vmatrix} 12 & 15 \\ 2 & 13 \end{vmatrix} = 156 - 30 = 126$

$x = \dfrac{D_x}{D} = \dfrac{-84}{-42} = 2$

$y = \dfrac{D_y}{D} = \dfrac{126}{-42} = -3$

The solution set is $\{(2, -3)\}$.

15. $D = \begin{vmatrix} 4 & -5 \\ 2 & 3 \end{vmatrix} = 12 - (-10) = 22$

$D_x = \begin{vmatrix} 17 & -5 \\ 3 & 3 \end{vmatrix} = 51 - (-15) = 66$

$D_y = \begin{vmatrix} 4 & 17 \\ 2 & 3 \end{vmatrix} = 12 - 34 = -22$

$x = \dfrac{D_x}{D} = \dfrac{66}{22} = 3$

$y = \dfrac{D_y}{D} = \dfrac{-22}{22} = -1$

The solution set is $\{(3, -1)\}$.

17. $D = \begin{vmatrix} 1 & 2 \\ 5 & 10 \end{vmatrix} = 10 - 10 = 0$

$D_x = \begin{vmatrix} 3 & 2 \\ 15 & 10 \end{vmatrix} = 30 - 30 = 0$

$D_y = \begin{vmatrix} 1 & 3 \\ 5 & 15 \end{vmatrix} = 15 - 15 = 0$

Because all 3 determinants equal zero, the system is dependent.

19. $D = \begin{vmatrix} 3 & -4 \\ 2 & 2 \end{vmatrix} = 6 - (-8) = 14$

$D_x = \begin{vmatrix} 4 & -4 \\ 12 & 2 \end{vmatrix} = 8 - (-48) = 56$

$D_y = \begin{vmatrix} 3 & 4 \\ 2 & 12 \end{vmatrix} = 36 - 8 = 28$

$x = \dfrac{D_x}{D} = \dfrac{56}{14} = 4$

$y = \dfrac{D_y}{D} = \dfrac{28}{14} = 2$

The solution set is $\{(4, 2)\}$.

21. $D = \begin{vmatrix} 2 & -3 \\ 5 & 4 \end{vmatrix} = 8 - (-15) = 23$

$D_x = \begin{vmatrix} 2 & -3 \\ 51 & 4 \end{vmatrix} = 8 - (-153) = 161$

$D_y = \begin{vmatrix} 2 & 2 \\ 5 & 51 \end{vmatrix} = 102 - 10 = 92$

$x = \dfrac{D_x}{D} = \dfrac{161}{23} = 7$

$y = \dfrac{D_y}{D} = \dfrac{92}{23} = 4$

The solution set is $\{(7, 4)\}$.

23. $D = \begin{vmatrix} 3 & 3 \\ 2 & 2 \end{vmatrix} = 6 - 6 = 0$

$D_x = \begin{vmatrix} 2 & 3 \\ 3 & 2 \end{vmatrix} = 4 - 9 = -5$

$D_y = \begin{vmatrix} 3 & 2 \\ 2 & 3 \end{vmatrix} = 9 - 4 = 5$

Because $D = 0$ but D_x or $D_y \neq 0$, the system is inconsistent.

25. Write the equations in standard form.

$3x + 4y = 16$

$6x + 8y = 32$

$D = \begin{bmatrix} 3 & 4 \\ 6 & 8 \end{bmatrix} = 24 - 24 = 0$

$D_x = \begin{bmatrix} 16 & 4 \\ 32 & 8 \end{bmatrix} = 128 - 128 = 0$

$D_y = \begin{bmatrix} 3 & 16 \\ 6 & 32 \end{bmatrix} = 96 - 69 = 0$

Since all determinants are zero, the system is dependent.

27. $\begin{vmatrix} 3 & 0 & 0 \\ 2 & 1 & -5 \\ -2 & 5 & -1 \end{vmatrix} = 3\begin{vmatrix} 1 & -5 \\ 5 & -1 \end{vmatrix} - 0\begin{vmatrix} 2 & -5 \\ -2 & -1 \end{vmatrix} + 0\begin{vmatrix} 2 & 1 \\ -2 & 5 \end{vmatrix}$

$= 3[(1)(-1) - (5)(-5)]$

$= 3(-1 + 25) = 3(24)$

$= 72$

29. $\begin{vmatrix} 3 & 1 & 0 \\ -3 & 4 & 0 \\ -1 & 3 & -5 \end{vmatrix} = 0\begin{vmatrix} -3 & 4 \\ -1 & 3 \end{vmatrix} - 0\begin{vmatrix} 3 & 1 \\ -1 & 3 \end{vmatrix} + (-5)\begin{vmatrix} 3 & 1 \\ -3 & 4 \end{vmatrix}$

$= -5[3 \cdot 4 - (-3)(1)]$

$= -5(12 + 3) = -5(15)$

$= -75$

31. $\begin{vmatrix} 1 & 1 & 1 \\ 2 & 2 & 2 \\ -3 & 4 & -5 \end{vmatrix} -2R_1 + R_2$

$\begin{vmatrix} 1 & 1 & 1 \\ 0 & 0 & 0 \\ -3 & 4 & -5 \end{vmatrix} = 0$

33. $D = \begin{vmatrix} 1 & 1 & 1 \\ 2 & -1 & 1 \\ -1 & 3 & -1 \end{vmatrix}$

$= \begin{vmatrix} -1 & 1 \\ 3 & -1 \end{vmatrix} - \begin{vmatrix} 2 & 1 \\ -1 & -1 \end{vmatrix} + \begin{vmatrix} 2 & -1 \\ -1 & 3 \end{vmatrix}$

$= (1 - 3) - [-2 - (-1)] + (6 - 1)$

$= -2 - (-1) + 5 = -2 + 1 + 5 = 4$

$D_x = \begin{vmatrix} 0 & 1 & 1 \\ -1 & -1 & 1 \\ -8 & 3 & -1 \end{vmatrix} = (-1)\begin{vmatrix} -1 & 1 \\ -8 & -1 \end{vmatrix} + \begin{vmatrix} -1 & -1 \\ -8 & 3 \end{vmatrix}$

$= (-1)[1 - (-8)] + (-3 - 8) = (-1)(9) - 11$

$= -20$

$D_y = \begin{vmatrix} 1 & 0 & 1 \\ 2 & -1 & 1 \\ -1 & -8 & -1 \end{vmatrix} = \begin{vmatrix} -1 & 1 \\ -8 & -1 \end{vmatrix} + \begin{vmatrix} 2 & -1 \\ -1 & -8 \end{vmatrix}$

$= 1 - (-8) + (-16 - 1) = 1 + 8 - 17 = -8$

$D_z = \begin{vmatrix} 1 & 1 & 0 \\ 2 & -1 & -1 \\ -1 & 3 & -8 \end{vmatrix} = 1\begin{vmatrix} -1 & -1 \\ 3 & -8 \end{vmatrix} - 1\begin{vmatrix} 2 & -1 \\ -1 & -8 \end{vmatrix}$

$= 8 - (-3) - 1(-16 - 1) = 11 + 17 = 28$

$x = \dfrac{D_x}{D} = \dfrac{-20}{4} = -5$

$y = \dfrac{D_y}{D} = \dfrac{-8}{4} = -2$

$z = \dfrac{D_z}{D} = \dfrac{28}{4} = 7$

The solution to the system is $\{(-5, -2, 7)\}$.

35. $D = \begin{vmatrix} 4 & -5 & -6 \\ 1 & -2 & -5 \\ 2 & -1 & 0 \end{vmatrix} = 2\begin{vmatrix} -5 & -6 \\ -2 & -5 \end{vmatrix} - (-1)\begin{vmatrix} 4 & -6 \\ 1 & -5 \end{vmatrix}$

$= 2(25 - 12) + [-20 - (-6)] = 2(13) + (-14)$

$= 26 - 14 = 12$

$D_x = \begin{vmatrix} -1 & -5 & -6 \\ -12 & -2 & -5 \\ 7 & -1 & 0 \end{vmatrix}$

$= 7\begin{vmatrix} -5 & -6 \\ -2 & -5 \end{vmatrix} - (-1)\begin{vmatrix} -1 & -6 \\ -12 & -5 \end{vmatrix}$

$= 7(25 - 12) + (5 - 72) = 7(13) - 67$

$= 91 - 67 = 24$

$D_y = \begin{vmatrix} 4 & -1 & -6 \\ 1 & -12 & -5 \\ 2 & 7 & 0 \end{vmatrix} = 2\begin{vmatrix} -1 & -6 \\ -12 & -5 \end{vmatrix} - 7\begin{vmatrix} 4 & -6 \\ 1 & -5 \end{vmatrix}$

$= 2(5 - 72) - 7[-20 - (-6)]$

$= 2(-67) - 7(-14) = -134 + 98 = -36$

$D_z = \begin{vmatrix} 4 & -5 & -1 \\ 1 & -2 & -12 \\ 2 & -1 & 7 \end{vmatrix}$

$= 4\begin{vmatrix} -2 & -12 \\ -1 & 7 \end{vmatrix} - (-5)\begin{vmatrix} 1 & -12 \\ 2 & 7 \end{vmatrix} + (-1)\begin{vmatrix} 1 & -2 \\ 2 & -1 \end{vmatrix}$

$= 4(-14 - 12) + 5[7 - (-24)] - [-1 - (-4)]$

$= 4(-26) + 5(31) - (3) = -104 + 155 - 3 = 48$

$x = \dfrac{D_x}{D} = \dfrac{24}{12} = 2, \ y = \dfrac{D_y}{D} = \dfrac{-36}{12} = -3,$

$z = \dfrac{D_z}{D} = \dfrac{48}{12} = 4$

The solution set is $\{(2, -3, 4)\}$.

37. $D = \begin{vmatrix} 1 & 1 & 1 \\ 1 & -2 & 1 \\ 1 & 3 & 2 \end{vmatrix} = 1\begin{vmatrix} -2 & 1 \\ 3 & 2 \end{vmatrix} - 1\begin{vmatrix} 1 & 1 \\ 1 & 2 \end{vmatrix} + 1\begin{vmatrix} 1 & -2 \\ 1 & 3 \end{vmatrix}$

$= -4 - 3 - (2 - 1) + [3 - (-2)]$

$= -7 - 1 + 5 = -3$

$D_x = \begin{vmatrix} 4 & 1 & 1 \\ 7 & -2 & 1 \\ 4 & 3 & 2 \end{vmatrix} = 4\begin{vmatrix} -2 & 1 \\ 3 & 2 \end{vmatrix} - 1\begin{vmatrix} 7 & 1 \\ 4 & 2 \end{vmatrix} + 1\begin{vmatrix} 7 & -2 \\ 4 & 3 \end{vmatrix}$

$= 4(-4 - 3) - (14 - 4) + [21 - (-8)]$

$= 4(-7) - 10 + 29 = -28 + 19 = -9$

$D_y = \begin{vmatrix} 1 & 4 & 1 \\ 1 & 7 & 1 \\ 1 & 4 & 2 \end{vmatrix} = 1\begin{vmatrix} 7 & 1 \\ 4 & 2 \end{vmatrix} - 1\begin{vmatrix} 4 & 1 \\ 4 & 2 \end{vmatrix} + 1\begin{vmatrix} 4 & 1 \\ 7 & 1 \end{vmatrix}$

$= 14 - 4 - (8 - 4) + (4 - 7) = 10 - 4 - 3 = 3$

$D_z = \begin{vmatrix} 1 & 1 & 4 \\ 1 & -2 & 7 \\ 1 & 3 & 4 \end{vmatrix} = 1\begin{vmatrix} -2 & 7 \\ 3 & 4 \end{vmatrix} - 1\begin{vmatrix} 1 & 4 \\ 3 & 4 \end{vmatrix} + 1\begin{vmatrix} 1 & 4 \\ -2 & 7 \end{vmatrix}$

$= -8 - 21 - (4 - 12) + [7 - (-8)]$

$= -29 + 8 + 15 = -6$

$x = \dfrac{D_x}{D} = \dfrac{-9}{-3} = 3, \ y = \dfrac{D_y}{D} = \dfrac{3}{-3} = -1,$

$z = \dfrac{D_z}{D} = \dfrac{-6}{-3} = 2$

The solution set is $\{3, -1, 2\}$.

39. $D = \begin{vmatrix} 1 & 0 & 2 \\ 0 & 2 & -1 \\ 2 & 3 & 0 \end{vmatrix} = \begin{vmatrix} 2 & -1 \\ 3 & 0 \end{vmatrix} + 2\begin{vmatrix} 0 & 2 \\ 2 & 3 \end{vmatrix}$

$= 0 - (-3) + 2(0 - 4) = 3 - 8 = -5$

$D_x = \begin{vmatrix} 4 & 0 & 2 \\ 5 & 2 & -1 \\ 13 & 3 & 0 \end{vmatrix} = 4\begin{vmatrix} 2 & -1 \\ 3 & 0 \end{vmatrix} + 2\begin{vmatrix} 5 & 2 \\ 13 & 3 \end{vmatrix}$

$= 4[0 - (-3)] + 2(15 - 26)$

$= 4(3) + 2(-11) = 12 - 22 = -10$

$D_y = \begin{vmatrix} 1 & 4 & 2 \\ 0 & 5 & -1 \\ 2 & 13 & 0 \end{vmatrix} = \begin{vmatrix} 5 & -1 \\ 13 & 0 \end{vmatrix} + 2\begin{vmatrix} 4 & 2 \\ 5 & -1 \end{vmatrix}$

$= 0 - (-13) + 2(-4 - 10)$

$= 13 + 2(-14) = 13 - 28 = -15$

$D_z = \begin{vmatrix} 1 & 0 & 4 \\ 0 & 2 & 5 \\ 2 & 3 & 13 \end{vmatrix} = \begin{vmatrix} 2 & 5 \\ 3 & 13 \end{vmatrix} + 4\begin{vmatrix} 0 & 2 \\ 2 & 3 \end{vmatrix}$

$= 26 - 15 + 4(0 - 4) = 11 + 4(-4)$

$= 11 - 16 = -5$

$x = \dfrac{D_x}{D} = \dfrac{-10}{-5} = 2, \ y = \dfrac{D_y}{D} = \dfrac{-15}{-5} = 3,$

$z = \dfrac{D_z}{D} = \dfrac{-5}{-5} = 1$

The solution set is $\{(2, 3, 1)\}$.

41. $\begin{vmatrix} 4 & 2 & 8 & -7 \\ -2 & 0 & 4 & 1 \\ 5 & 0 & 0 & 5 \\ 4 & 0 & 0 & -1 \end{vmatrix} = -2\begin{vmatrix} -2 & 4 & 1 \\ 5 & 0 & 5 \\ 4 & 0 & -1 \end{vmatrix} + 0\begin{vmatrix} 4 & 8 & -7 \\ 5 & 0 & 5 \\ 4 & 0 & -1 \end{vmatrix} - 0\begin{vmatrix} 4 & 8 & -7 \\ -2 & 4 & 1 \\ 4 & 0 & -1 \end{vmatrix} + 0\begin{vmatrix} 4 & 8 & -7 \\ -2 & 4 & 1 \\ 5 & 0 & 5 \end{vmatrix}$

$= (-2)\left[(-4)\begin{vmatrix} 5 & 5 \\ 4 & -1 \end{vmatrix} + 0\begin{vmatrix} -2 & 1 \\ 4 & -1 \end{vmatrix} - 0\begin{vmatrix} -2 & 1 \\ 5 & 5 \end{vmatrix} \right] = (-2)(-4)[5(-1) - 4 \cdot 5] = 8(-5 - 20) = 8(-25) = -200$

43. $\begin{vmatrix} -2 & -3 & 3 & 5 \\ 1 & -4 & 0 & 0 \\ 1 & 2 & 2 & -3 \\ 2 & 0 & 1 & 1 \end{vmatrix} = -1\begin{vmatrix} -3 & 3 & 5 \\ 2 & 2 & -3 \\ 0 & 1 & 1 \end{vmatrix} + (-4)\begin{vmatrix} -2 & 3 & 5 \\ 1 & 2 & -3 \\ 2 & 1 & 1 \end{vmatrix} - 0\begin{vmatrix} -2 & -3 & 5 \\ 1 & 2 & -3 \\ 2 & 0 & 1 \end{vmatrix} + 0\begin{vmatrix} -2 & -3 & 3 \\ 1 & 2 & 2 \\ 2 & 0 & 1 \end{vmatrix}$

$= (-1)\left[0\begin{vmatrix} 3 & 5 \\ 2 & -3 \end{vmatrix} - 1\begin{vmatrix} -3 & 5 \\ 2 & -3 \end{vmatrix} + 1\begin{vmatrix} -3 & 3 \\ 2 & 2 \end{vmatrix} \right] - 4\left[2\begin{vmatrix} 3 & 5 \\ 2 & -3 \end{vmatrix} - 1\begin{vmatrix} -2 & 5 \\ 1 & -3 \end{vmatrix} + 1\begin{vmatrix} -2 & 3 \\ 1 & 2 \end{vmatrix} \right]$

$= (-1)\{(-1)[(-3)(-3) - 2 \cdot 5] + [(-3)(2) - 2 \cdot 3]\} - 4\{2[3(-3) - 2 \cdot 5] - [(-2)(-3) - 1 \cdot 5] + [(-2)(2) - 1 \cdot 3]\} = 195$

45. $\begin{vmatrix} \begin{vmatrix} 3 & 1 \\ -2 & 3 \end{vmatrix} & \begin{vmatrix} 7 & 0 \\ 1 & 5 \end{vmatrix} \\ \begin{vmatrix} 3 & 0 \\ 0 & 7 \end{vmatrix} & \begin{vmatrix} 9 & -6 \\ 3 & 5 \end{vmatrix} \end{vmatrix} = \begin{vmatrix} 3(3) - (-2)(1) & 7(5) - 1(0) \\ 3(7) - 0(0) & 9(5) - 3(-6) \end{vmatrix} = \begin{vmatrix} 9+2 & 35-0 \\ 21-0 & 45+18 \end{vmatrix} = \begin{vmatrix} 11 & 35 \\ 21 & 63 \end{vmatrix}$

$= 11(63) - 21(35) = 693 - 735 = -42$

47. From $D = \begin{vmatrix} 2 & -4 \\ 3 & 5 \end{vmatrix}$ we obtain the coefficients of the variables in our equations:

$2x - 4y = c_1$

$3x + 5y = c_2$

From $D_x = \begin{vmatrix} 8 & -4 \\ -10 & 5 \end{vmatrix}$ we obtain the constant coefficients: 8 and -10

$2x - 4y = 8$

$3x + 5y = -10$

49. $\begin{vmatrix} -2 & x \\ 4 & 6 \end{vmatrix} = 32$

$-2(6) - 4(x) = 32$

$-12 - 4x = 32$

$-4x = 44$

$x = -11$

The solution is -11.

51.

$$\begin{vmatrix} 1 & x & -2 \\ 3 & 1 & 1 \\ 0 & -2 & 2 \end{vmatrix} = -8$$

$$0\begin{vmatrix} x & -2 \\ 1 & 1 \end{vmatrix} - (-2)\begin{vmatrix} 1 & -2 \\ 3 & 1 \end{vmatrix} + 2\begin{vmatrix} 1 & x \\ 3 & 1 \end{vmatrix} = -8$$

$$2\left[1(1) - 3(-2)\right] + 2\left[1(1) - 3(x)\right] = -8$$

$$2(1+6) + 2(1-3x) = -8$$

$$2(7) + 2(1-3x) = -8$$

$$14 + 2 - 6x = -8$$

$$-6x = -24$$

$$x = 4$$

The solution is 4.

53. Area $= \pm\dfrac{1}{2}\begin{vmatrix} 3 & -5 & 1 \\ 2 & 6 & 1 \\ -3 & 5 & 1 \end{vmatrix} = \pm\dfrac{1}{2}\begin{vmatrix} 3 & -5 & 1 \\ -1 & 11 & 0 \\ -6 & 10 & 0 \end{vmatrix} = \pm\dfrac{1}{2}\begin{vmatrix} -1 & 11 \\ -6 & 10 \end{vmatrix} = \pm\dfrac{1}{2}[-10 - (-66)] = \pm\dfrac{1}{2}(56) = 28$

The area is 28 square units.

The slope of the line through (3, –5) and (–3, 5) is $m = \dfrac{5 - (-5)}{-3 - 3} = \dfrac{10}{-6} = -\dfrac{5}{3}$.

The equation of the line is $y - (-5) = -\dfrac{5}{3}(x - 3)$ or $y = -\dfrac{5}{3}x$.

The line perpendicular to $y = -\dfrac{5}{3}x$ through (2, 6) has equation $y - 6 = \dfrac{3}{5}(x - 2)$ or $y = \dfrac{3}{5}x + \dfrac{24}{5}$.

These lines intersect where $-\dfrac{5}{3}x = \dfrac{3}{5}x + \dfrac{24}{5}$.

$-\dfrac{36}{17} = x$ and $-\dfrac{24}{5} = \dfrac{34}{15}x$ $y = -\dfrac{5}{3}\left(-\dfrac{36}{17}\right) = \dfrac{60}{17}$

Using the side connecting (3, –5) and

(–3, 5) as the base, the height is the distance from (2, 6) to $\left(-\dfrac{36}{17}, \dfrac{60}{17}\right)$.

$b = \sqrt{[3 - (-3)]^2 + (-5 - 5)^2}$

$= \sqrt{36 + 100} = \sqrt{136} = 2\sqrt{34}$

$h = \sqrt{\left[2 - \left(-\dfrac{36}{17}\right)\right]^2 + \left(6 - \dfrac{60}{17}\right)^2}$

$= \sqrt{\dfrac{4900}{289} + \dfrac{1764}{289}} = \dfrac{14\sqrt{34}}{17}$

$\dfrac{1}{2}bh = \dfrac{1}{2}\left(2\sqrt{34}\right)\left(\dfrac{14\sqrt{34}}{17}\right) = \dfrac{14(34)}{17}$

$= 14(2) = 28$ square units

55. $\begin{vmatrix} 3 & -1 & 1 \\ 0 & -3 & 1 \\ 12 & 5 & 1 \end{vmatrix} = \begin{vmatrix} 3 & -1 & 1 \\ -3 & -2 & 0 \\ 9 & 6 & 0 \end{vmatrix} = \begin{vmatrix} -3 & -2 \\ 9 & 6 \end{vmatrix}$

$= -18 - (-18) = 0$

Yes, the points are collinear.

57. $\begin{vmatrix} x & y & 1 \\ 3 & -5 & 1 \\ -2 & 6 & 1 \end{vmatrix} = x\begin{vmatrix} -5 & 1 \\ 6 & 1 \end{vmatrix} - y\begin{vmatrix} 3 & 1 \\ -2 & 1 \end{vmatrix} + \begin{vmatrix} 3 & -5 \\ -2 & 6 \end{vmatrix} = x(-5-6) - y[3-(-2)] + (18-10)$

$= -11x - 5y + 8$

The equation of the line is $-11x - 5y + 8 = 0$. The equation of the line in slope-intercept form is $y = -\dfrac{11}{5}x + \dfrac{8}{5}$.

59. – 67. Answers may vary.

69. Input the matrix as $[A]$, then use $\det[A]$ to find the determinant.

$\begin{vmatrix} 8 & 2 & 6 & -1 & 0 \\ 2 & 0 & -3 & 4 & 7 \\ 2 & 1 & -3 & 6 & -5 \\ -1 & 2 & 1 & 5 & -1 \\ 4 & 5 & -2 & 3 & -8 \end{vmatrix} = 13,200$

71. does not make sense; Explanations will vary. Sample explanation: Determinants must be square.

73. does not make sense; Explanations will vary. Sample explanation: The number of determinants needed is one greater than the number of variables.

75. In this exercise, expansions are all done about the first column of the matrix and the resulting products of 0 and a determinant are not shown.

a. $\begin{vmatrix} a & a \\ 0 & a \end{vmatrix} = a^2 - 0 = a^2$

b. $\begin{vmatrix} a & a & a \\ 0 & a & a \\ 0 & 0 & a \end{vmatrix} = a\begin{vmatrix} a & a \\ 0 & a \end{vmatrix} - 0 + 0$

$= a(a^2) = a^3$

c. $\begin{vmatrix} a & a & a & a \\ 0 & a & a & a \\ 0 & 0 & a & a \\ 0 & 0 & 0 & a \end{vmatrix} = a\begin{vmatrix} a & a & a \\ 0 & a & a \\ 0 & 0 & a \end{vmatrix} - 0 + 0 - 0$

$= a(a^3) = a^4$

d. Each determinant has zeros below the main diagonal and a's everywhere else.

e. Each determinant equals a raised to the power equal to the order of the determinant.

77. The sign of the value is changed when 2 columns are interchanged in a 2nd order determinant.

79. Evaluate the determinate and write the equation in slope intercept form.

$$\begin{vmatrix} x & y & 1 \\ x_1 & y_1 & 1 \\ x_2 & y_2 & 1 \end{vmatrix} = 0$$

$$x\begin{vmatrix} y_1 & 1 \\ y_2 & 1 \end{vmatrix} - y\begin{vmatrix} x_1 & 1 \\ x_2 & 1 \end{vmatrix} + 1\begin{vmatrix} x_1 & y_1 \\ x_2 & y_2 \end{vmatrix} = 0$$

$$x(y_1 - y_2) - y(x_1 - x_2) + x_1 y_2 - x_2 y_1 = 0$$

$$-y(x_1 - x_2) = -x(y_1 - y_2) + x_2 y_1 - x_1 y_2$$

$$y(x_2 - x_1) = x(y_2 - y_1) + x_2 y_1 - x_1 y_2$$

$$y = \frac{y_2 - y_1}{x_2 - x_1} x + \frac{x_2 y_1 - x_1 y_2}{x_2 - x_1}$$

$$m = \frac{y_2 - y_1}{x_2 - x_1} \qquad b = \frac{x_2 y_1 - x_1 y_2}{x_2 - x_1}$$

Write the slope-point equation of the line the in point slope form.

$$y - y_1 = \frac{y_2 - y_1}{x_2 - x_1}(x - x_1)$$

$$y - y_1 = \frac{y_2 - y_1}{x_2 - x_1} x + \frac{-x_1 y_2 + x_1 y_1}{x_2 - x_1}$$

$$y = \frac{y_2 - y_1}{x_2 - x_1} x + \frac{-x_1 y_2 + x_1 y_1}{x_2 - x_1} + y_1$$

$$y = \frac{y_2 - y_1}{x_2 - x_1} x + \frac{-x_1 y_2 + x_1 y_1}{x_2 - x_1} + \frac{x_2 y_1 - x_1 y_1}{x_2 - x_1}$$

$$y = \frac{y_2 - y_1}{x_2 - x_1} x + \frac{x_2 y_1 - x_1 y_2}{x_2 - x_1}$$

$$m = \frac{y_2 - y_1}{x_2 - x_1} \qquad b = \frac{x_2 y_1 - x_1 y_2}{x_2 - x_1}$$

Since both forms give the same slope and *y*-intercept, the determinant does give the equation of the line.

81. **a.** $\dfrac{x^2}{9} + \dfrac{y^2}{4} = 1$

$$\frac{x^2}{9} + \frac{0^2}{4} = 1$$

$$\frac{x^2}{9} = 1$$

$$x^2 = 9$$

$$x = \pm 3$$

b. $\dfrac{x^2}{9} + \dfrac{y^2}{4} = 1$

$$\frac{0^2}{9} + \frac{y^2}{4} = 1$$

$$\frac{y^2}{4} = 1$$

$$y^2 = 4$$

$$y = \pm 2$$

82. $25x^2 + 16y^2 = 400$

$$\frac{25x^2}{400} + \frac{16y^2}{400} = \frac{400}{400}$$

$$\frac{x^2}{16} + \frac{y^2}{25} = 1$$

83. $x^2 + y^2 - 2x + 4y = 4$

$$x^2 - 2x \quad + y^2 + 4y \quad = 4$$

$$x^2 - 2x + 1 + y^2 + 4y + 4 = 4 + 1 + 4$$

$$(x-1)^2 + (y+2)^2 = 9$$

$$(x-1)^2 + (y+2)^2 = 3^2$$

Center: (1, –2)
Radius: 3

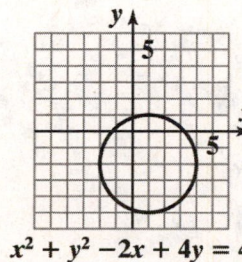

$x^2 + y^2 - 2x + 4y = 4$

Chapter 8 Review Exercises

1. $\begin{bmatrix} 1 & 2 & 2 & | & 2 \\ 0 & 1 & -1 & | & 2 \\ 0 & 5 & 4 & | & 1 \end{bmatrix} -5R_2 + R_3$

$\begin{bmatrix} 1 & 2 & 2 & | & 2 \\ 0 & 1 & -1 & | & 2 \\ 0 & 0 & 9 & | & -9 \end{bmatrix}$

2. $\begin{bmatrix} 2 & -2 & 1 & | & -1 \\ 1 & 2 & -1 & | & 2 \\ 6 & 4 & 3 & | & 5 \end{bmatrix} \frac{1}{2}R_1$

$\begin{bmatrix} 1 & -1 & \frac{1}{2} & | & -\frac{1}{2} \\ 1 & 2 & -1 & | & 2 \\ 6 & 4 & 3 & | & 5 \end{bmatrix}$

3. $\begin{bmatrix} 1 & 2 & 3 & | & -5 \\ 2 & 1 & 1 & | & 1 \\ 1 & 1 & -1 & | & 8 \end{bmatrix} \begin{matrix} -2R_1 + R_2 \\ -1R_1 + R_3 \end{matrix}$

$\begin{bmatrix} 1 & 2 & 3 & | & -5 \\ 0 & -3 & -5 & | & 11 \\ 0 & -1 & -4 & | & 13 \end{bmatrix} R_2 \leftrightarrow R_3$

$\begin{bmatrix} 1 & 2 & 3 & | & -5 \\ 0 & -1 & -4 & | & 13 \\ 0 & -3 & -5 & | & 11 \end{bmatrix} -1R_2$

$\begin{bmatrix} 1 & 2 & 3 & | & -5 \\ 0 & 1 & 4 & | & -13 \\ 0 & -3 & -5 & | & 11 \end{bmatrix} 3R_2 + R_3$

$\begin{bmatrix} 1 & 2 & 3 & | & -5 \\ 0 & 1 & 4 & | & -13 \\ 0 & 0 & 7 & | & -28 \end{bmatrix} \frac{1}{7}R_3$

$\begin{bmatrix} 1 & 2 & 3 & | & -5 \\ 0 & 1 & 4 & | & -13 \\ 0 & 0 & 1 & | & -4 \end{bmatrix} -2R_2 + R_1$

$\begin{bmatrix} 1 & 0 & -5 & | & 21 \\ 0 & 1 & 4 & | & -13 \\ 0 & 0 & 1 & | & -4 \end{bmatrix} \begin{matrix} 5R_3 + R_1 \\ -4R_3 + R_2 \end{matrix}$

$\begin{bmatrix} 1 & 0 & 0 & | & 1 \\ 0 & 1 & 0 & | & 3 \\ 0 & 0 & 1 & | & -4 \end{bmatrix}$

The solution set is $\{(1, 3, -4)\}$.

4. $\begin{bmatrix} 1 & -2 & 1 & | & 0 \\ 0 & 1 & -3 & | & -1 \\ 0 & 2 & 5 & | & -2 \end{bmatrix} -2R_2 + R_3$

$\begin{bmatrix} 1 & -2 & 1 & | & 0 \\ 0 & 1 & -3 & | & -1 \\ 0 & 0 & 11 & | & 0 \end{bmatrix} \frac{1}{11}R_3$

$\begin{bmatrix} 1 & -2 & 1 & | & 0 \\ 0 & 1 & -3 & | & -1 \\ 0 & 0 & 1 & | & 0 \end{bmatrix} 2R_2 + R_1$

$\begin{bmatrix} 1 & 0 & -5 & | & -2 \\ 0 & 1 & -3 & | & -1 \\ 0 & 0 & 1 & | & 0 \end{bmatrix} \begin{matrix} 3R_3 + R_2 \\ 5R_3 + R_1 \end{matrix}$

$\begin{bmatrix} 1 & 0 & 0 & | & -2 \\ 0 & 1 & 0 & | & -1 \\ 0 & 0 & 1 & | & 0 \end{bmatrix}$

$x = -2; y = -1; z = 0$
The solution set is $\{(-2, -1, 0)\}$.

5. $\begin{bmatrix} 3 & 5 & -8 & 5 & | & -8 \\ 1 & 2 & -3 & 1 & | & -7 \\ 2 & 3 & -7 & 3 & | & -11 \\ 4 & 8 & -10 & 7 & | & -10 \end{bmatrix} R_1 \leftrightarrow R_2$

$\begin{bmatrix} 1 & 2 & -3 & 1 & | & -7 \\ 3 & 5 & -8 & 5 & | & -8 \\ 2 & 3 & -7 & 3 & | & -11 \\ 4 & 8 & -10 & 7 & | & -10 \end{bmatrix} \begin{matrix} -3R_1 + R_2 \\ -2R_1 + R_3 \\ -4R_1 + R_4 \end{matrix}$

$\begin{bmatrix} 1 & 2 & -3 & 1 & | & -7 \\ 0 & -1 & 1 & 2 & | & 13 \\ 0 & -1 & -1 & 1 & | & 3 \\ 0 & 0 & 2 & 3 & | & 18 \end{bmatrix} -1R_2$

$\begin{bmatrix} 1 & 2 & -3 & 1 & | & -7 \\ 0 & 1 & -1 & -2 & | & -13 \\ 0 & -1 & -1 & 1 & | & 3 \\ 0 & 0 & 2 & 3 & | & 18 \end{bmatrix} \begin{matrix} -2R_2 + R_1 \\ 1R_2 + R_3 \end{matrix}$

$\begin{bmatrix} 1 & 0 & -1 & 5 & | & 19 \\ 0 & 1 & -1 & -2 & | & -13 \\ 0 & 0 & -2 & -1 & | & -10 \\ 0 & 0 & 2 & 3 & | & 18 \end{bmatrix} -\frac{1}{2}R_3$

$\begin{bmatrix} 1 & 0 & -1 & 5 & | & 19 \\ 0 & 1 & -1 & -2 & | & -13 \\ 0 & 0 & 1 & \frac{1}{2} & | & 5 \\ 0 & 0 & 2 & 3 & | & 18 \end{bmatrix} \begin{matrix} 1R_3 + R_1 \\ 1R_3 + R_2 \\ -2R_3 + R_4 \end{matrix}$

$$\begin{bmatrix} 1 & 0 & 0 & \frac{11}{2} & | & 24 \\ 0 & 1 & 0 & -\frac{3}{2} & | & -8 \\ 0 & 0 & 1 & \frac{1}{2} & | & 5 \\ 0 & 0 & 0 & 2 & | & 8 \end{bmatrix} \frac{1}{2}R_4$$

$$\begin{bmatrix} 1 & 0 & 0 & \frac{11}{2} & | & 24 \\ 0 & 1 & 0 & -\frac{3}{2} & | & -8 \\ 0 & 0 & 1 & \frac{1}{2} & | & 5 \\ 0 & 0 & 0 & 1 & | & 4 \end{bmatrix} \begin{matrix} -\frac{11}{2}R_4+R_1 \\ \frac{3}{2}R_4+R_2 \\ -\frac{1}{2}R_4+R_3 \end{matrix}$$

$$\begin{bmatrix} 1 & 0 & 0 & 0 & | & 2 \\ 0 & 1 & 0 & 0 & | & -2 \\ 0 & 0 & 1 & 0 & | & 3 \\ 0 & 0 & 0 & 1 & | & 4 \end{bmatrix}$$

The solution set is $\{(2, -2, 3, 4)\}$.

6. a. The function must satisfy:
$$98 = 4a = 2b + c$$
$$138 = 16a + 4b + c$$
$$162 = 100a + 10b + c.$$

$$\begin{bmatrix} 4 & 2 & 1 & | & 98 \\ 16 & 4 & 1 & | & 138 \\ 100 & 10 & 1 & | & 162 \end{bmatrix} \frac{1}{4}R_1$$

$$\begin{bmatrix} 1 & \frac{1}{2} & \frac{1}{4} & | & \frac{49}{2} \\ 16 & 4 & 1 & | & 138 \\ 100 & 10 & 1 & | & 162 \end{bmatrix} \begin{matrix} -16R_1+R_2 \\ -100R_1+R_3 \end{matrix}$$

$$\begin{bmatrix} 1 & \frac{1}{2} & \frac{1}{4} & | & \frac{49}{2} \\ 0 & -4 & -3 & | & -254 \\ 0 & -40 & -24 & | & -2288 \end{bmatrix} -\frac{1}{4}R_2$$

$$\begin{bmatrix} 1 & \frac{1}{2} & \frac{1}{4} & | & \frac{49}{2} \\ 0 & 1 & \frac{3}{4} & | & \frac{127}{2} \\ 0 & -40 & -24 & | & -2288 \end{bmatrix} 40R_2+R_3$$

$$\begin{bmatrix} 1 & \frac{1}{2} & \frac{1}{4} & | & \frac{49}{2} \\ 0 & 1 & \frac{3}{4} & | & \frac{127}{2} \\ 0 & 0 & 6 & | & 252 \end{bmatrix} \frac{1}{6}R_3$$

$$\begin{bmatrix} 1 & \frac{1}{2} & \frac{1}{4} & | & \frac{49}{2} \\ 0 & 1 & \frac{3}{4} & | & \frac{127}{2} \\ 0 & 0 & 1 & | & 42 \end{bmatrix} \begin{matrix} -\frac{1}{4}R_3+R_1 \\ -\frac{3}{4}R_3+R_2 \end{matrix}$$

$$\begin{bmatrix} 1 & \frac{1}{2} & 0 & | & 14 \\ 0 & 1 & 0 & | & 32 \\ 0 & 0 & 1 & | & 42 \end{bmatrix} -\frac{1}{2}R_3+R_1$$

$$\begin{bmatrix} 1 & 0 & 0 & | & -2 \\ 0 & 1 & 0 & | & 32 \\ 0 & 0 & 1 & | & 42 \end{bmatrix}$$

The function is $y = -2x^2 + 32x + 42$ and $a = -2$, $b = 32$ and $c = 42$.

b. $y = -2x^2 + 32x + 42$ is a parabola.

The maximum occurs when
$$x = \frac{-32}{2(-2)} = \frac{-32}{-4} = 8.$$

The air pollution level is a maximum 8 hours after 6 A.M., which is 2 P.M.
When $x = 8$, $y = -2(64) + 32(8) + 42$
$$= -128 + 256 + 42.$$
$$= 170.$$
The maximum level is 170 parts per million at 2 P.M.

7. Write the equations.
$$w + x + y + z = 80$$
$$y - w - x = 18$$
$$y - z = 4$$
$$3x - w - y = 10$$

Rewrite the equations with terms in consistent order.
$$w + x + y + z = 80$$
$$-w - x + y = 18$$
$$y - z = 4$$
$$-w + 3x - y = 10$$

Write as a matrix and solve.

$$\begin{bmatrix} 1 & 1 & 1 & 1 & | & 80 \\ -1 & -1 & 1 & 0 & | & 18 \\ 0 & 0 & 1 & -1 & | & 4 \\ -1 & 3 & -1 & 0 & | & 10 \end{bmatrix} \begin{matrix} R_1+R_2 \\ \\ R_1+R_4 \end{matrix}$$

$$\begin{bmatrix} 1 & 1 & 1 & 1 & | & 80 \\ 0 & 0 & 2 & 1 & | & 98 \\ 0 & 0 & 1 & -1 & | & 4 \\ 0 & 4 & 0 & 1 & | & 90 \end{bmatrix} R_2 \leftrightarrow R_4$$

$$\begin{bmatrix} 1 & 1 & 1 & 1 & | & 80 \\ 0 & 4 & 0 & 1 & | & 90 \\ 0 & 0 & 1 & -1 & | & 4 \\ 0 & 0 & 2 & 1 & | & 98 \end{bmatrix} \frac{1}{4}R_2$$

$$\begin{bmatrix} 1 & 1 & 1 & 1 & | & 80 \\ 0 & 1 & 0 & \frac{1}{4} & | & \frac{45}{2} \\ 0 & 0 & 1 & -1 & | & 4 \\ 0 & 0 & 2 & 1 & | & 98 \end{bmatrix} -2R_3+R_4$$

$$\begin{bmatrix} 1 & 1 & 1 & 1 & | & 80 \\ 0 & 1 & 0 & \frac{1}{4} & | & \frac{45}{2} \\ 0 & 0 & 1 & -1 & | & 4 \\ 0 & 0 & 0 & 3 & | & 90 \end{bmatrix} \frac{1}{3}R_4$$

$$\begin{bmatrix} 1 & 1 & 1 & 1 & | & 80 \\ 0 & 1 & 0 & \frac{1}{4} & | & \frac{45}{2} \\ 0 & 0 & 1 & -1 & | & 4 \\ 0 & 0 & 0 & 1 & | & 30 \end{bmatrix}$$

Back-substitute $z = 30$ to find y.

$y - z = 4$

$y - 30 = 4$

$\qquad y = 34$

Back-substitute to find x.

$$x + \frac{1}{4}z = \frac{45}{2}$$

$$x + \frac{1}{4}(30) = \frac{45}{2}$$

$$x + \frac{15}{2} = \frac{45}{2}$$

$\qquad x = 15$

Back-substitute to find x.

$w + x + y + z = 80$

$w + 15 + 34 + 30 = 80$

$\qquad w + 79 = 80$

$\qquad\qquad w = 1$

Capitalist: 1%

Upper Middle: 15%

Lower Middle: 34%

Working: 30%

8.

$$\begin{bmatrix} 2 & -3 & 1 & | & 1 \\ 1 & -2 & 3 & | & 2 \\ 3 & -4 & -1 & | & 1 \end{bmatrix} R_1 \leftrightarrow R_2$$

$$\begin{bmatrix} 1 & -2 & 3 & | & 2 \\ 2 & -3 & 1 & | & 1 \\ 3 & -4 & -1 & | & 1 \end{bmatrix} \begin{matrix} -2R_1 + R_2 \\ \\ -3R_1 + R_3 \end{matrix}$$

$$\begin{bmatrix} 1 & -2 & 3 & | & 2 \\ 0 & 1 & -5 & | & -3 \\ 0 & 2 & -10 & | & -5 \end{bmatrix} -2R_2 + R_3$$

$$\begin{bmatrix} 1 & -2 & 3 & | & 2 \\ 0 & 1 & -5 & | & -3 \\ 0 & 0 & 0 & | & 1 \end{bmatrix}$$

From the last line, we see that the system has no solution. Thus, the solution set is $\varnothing$.

9.

$$\begin{bmatrix} 1 & -3 & 1 & | & 1 \\ -2 & 1 & 3 & | & -7 \\ 1 & -4 & 2 & | & 0 \end{bmatrix} \begin{matrix} 2R_1 + R_2 \\ \\ -1R_1 + R_3 \end{matrix}$$

$$\begin{bmatrix} 1 & -3 & 1 & | & 1 \\ 0 & -5 & 5 & | & -5 \\ 0 & -1 & 1 & | & -1 \end{bmatrix} -\frac{1}{5}R_2$$

$$\begin{bmatrix} 1 & -3 & 1 & | & 1 \\ 0 & 1 & -1 & | & 1 \\ 0 & -1 & 1 & | & -1 \end{bmatrix} 1R_2 + R_3$$

$$\begin{bmatrix} 1 & -3 & 1 & | & 1 \\ 0 & 1 & -1 & | & 1 \\ 0 & 0 & 0 & | & 0 \end{bmatrix}$$

The system $\begin{matrix} x - 3y + z = 1 \\ y - z = 1 \end{matrix}$ has no unique solution.

Express x and y in terms of z:

$y = z + 1$

$x - 3(z + 1) + z = 1$

$\quad x - 3z - 3 + z = 1$

$\qquad\qquad x = 2z + 4$

With $z = t$, the complete solution to the system is $\{(2t + 4, t + 1, t)\}$.

10.
$$\begin{bmatrix} 1 & 4 & 3 & -6 & | & 5 \\ 1 & 3 & 1 & -4 & | & 3 \\ 2 & 8 & 7 & -5 & | & 11 \\ 2 & 5 & 0 & -6 & | & 4 \end{bmatrix} \begin{matrix} -1R_1 + R_2 \\ -2R_1 + R_3 \\ -2R_1 + R_4 \end{matrix}$$

$$\begin{bmatrix} 1 & 4 & 3 & -6 & | & 5 \\ 0 & -1 & -2 & 2 & | & -2 \\ 0 & 0 & 1 & 7 & | & 1 \\ 0 & -3 & -6 & 6 & | & -6 \end{bmatrix} -1R_2$$

$$\begin{bmatrix} 1 & 4 & 3 & -6 & | & 5 \\ 0 & 1 & 2 & -2 & | & 2 \\ 0 & 0 & 1 & 7 & | & 1 \\ 0 & -3 & -6 & 6 & | & -6 \end{bmatrix} 3R_2 + R_4$$

$$\begin{bmatrix} 1 & 4 & 3 & -6 & | & 5 \\ 0 & 1 & 2 & -2 & | & 2 \\ 0 & 0 & 1 & 7 & | & 1 \\ 0 & 0 & 0 & 0 & | & 0 \end{bmatrix}$$

The system $\quad \begin{matrix} x_1 + 4x_2 + 3x_3 - 6x_4 = 5 \\ x_2 + 2x_3 - 2x_4 = 2 \\ x_3 + 7x_4 = 1 \end{matrix}$

does not have a unique solution.
Express x_1, x_2, and x_3 in terms of x_4:
$x_3 = -7x_4 + 1$
$x_2 + 2(-7x_4 + 1) - 2x_4 = 2$
$\quad x_2 - 14x_4 + 2 - 2x_4 = 2$
$\quad\quad\quad\quad x_2 = 16x_4$
$x_1 + 4(16x_4) + 3(-7x_4 + 1) - 6x_4 = 5$
$\quad x_1 + 64x_4 - 21x_4 + 3 - 6x_4 = 5$
$\quad\quad\quad\quad\quad\quad x_1 = -37x_4 + 2$

With $x_4 = t$, the complete solution to the system is
$\{(-37t + 2, 16t, -7t + 1, t)\}$.

11.
$$\begin{bmatrix} 2 & 3 & -5 & | & 15 \\ 1 & 2 & -1 & | & 4 \end{bmatrix} R_1 \leftrightarrow R_2$$

$$\begin{bmatrix} 1 & 2 & -1 & | & 4 \\ 2 & 3 & -5 & | & 15 \end{bmatrix} -2R_1 + R_2$$

$$\begin{bmatrix} 1 & 2 & -1 & | & 4 \\ 0 & -1 & -3 & | & 7 \end{bmatrix} -1R_2$$

$$\begin{bmatrix} 1 & 2 & -1 & | & 4 \\ 0 & 1 & 3 & | & -7 \end{bmatrix}$$

The system $\quad \begin{matrix} x + 2y - z = 4 \\ y + 3z = -7 \end{matrix} \quad$ has no unique solution.

Express x and y in terms of z:
$y = -3z - 7$
$x + 2(-3z - 7) - z = 4$
$\quad x - 6z - 14 - z = 4$
$\quad\quad\quad\quad x = 7z + 18$
With $z = t$, the complete solution to the system is
$\{(7t + 18, -3t - 7, t)\}$.

12. a.
$350 + 400 = x + z$
$450 + z = y + 700$
$x + y = 300 + 200$
or
$x + z = 750$
$y - z = -250$
$x + y = 500$

b.
$$\begin{bmatrix} 1 & 0 & 1 & | & 750 \\ 0 & 1 & -1 & | & -250 \\ 1 & 1 & 0 & | & 500 \end{bmatrix} -1R_1 + R_3$$

$$\begin{bmatrix} 1 & 0 & 1 & | & 750 \\ 0 & 1 & -1 & | & -250 \\ 0 & 1 & -1 & | & -250 \end{bmatrix} -1R_2 + R_3$$

$$\begin{bmatrix} 1 & 0 & 1 & | & 750 \\ 0 & 1 & -1 & | & -250 \\ 0 & 0 & 0 & | & 0 \end{bmatrix}$$

The system $\quad \begin{matrix} x + z = 750 \\ y - z = -250 \end{matrix} \quad$ has no unique

solution.
Express x and y in terms of z:
$y = z - 250$
$x = -z + 750$
With $z = t$, the complete solution to the system
is $\{(-t + 750, t - 250, t)\}$.

c.
$x = -400 + 750 = 350$
$\quad y = 400 - 250 = 150$

13. $2x = -10$
$x = -5$
$y + 7 = 13$
$y = 6$
$z = 6$
$x = -5; y = 6; z = 6$

14. $A + D = \begin{bmatrix} 2-2 & -1+3 & 2+1 \\ 5+3 & 3-2 & -1+4 \end{bmatrix} = \begin{bmatrix} 0 & 2 & 3 \\ 8 & 1 & 3 \end{bmatrix}$

15. $2B = \begin{bmatrix} 2(0) & 2(-2) \\ 2(3) & 2(2) \\ 2(1) & 2(-5) \end{bmatrix} = \begin{bmatrix} 0 & -4 \\ 6 & 4 \\ 2 & -10 \end{bmatrix}$

16. $D - A = \begin{bmatrix} -2-2 & 3+1 & 1-2 \\ 3-5 & -2-3 & 4+1 \end{bmatrix}$
$= \begin{bmatrix} -4 & 4 & -1 \\ -2 & -5 & 5 \end{bmatrix}$

17. Not possible since B is 3×2 and C is 3×3.

18. $3A + 2D = \begin{bmatrix} 6 & -3 & 6 \\ 15 & 9 & -3 \end{bmatrix} + \begin{bmatrix} -4 & 6 & 2 \\ 6 & -4 & 8 \end{bmatrix}$
$= \begin{bmatrix} 2 & 3 & 8 \\ 21 & 5 & 5 \end{bmatrix}$

19.
$-2A + 4D = \begin{bmatrix} -4 & 2 & -4 \\ -10 & -6 & 2 \end{bmatrix} + \begin{bmatrix} -8 & 12 & 4 \\ 12 & -8 & 16 \end{bmatrix}$
$= \begin{bmatrix} -12 & 14 & 0 \\ 2 & -14 & 18 \end{bmatrix}$

20. $-5(A + D) = -5\left(\begin{bmatrix} 0 & 2 & 3 \\ 8 & 1 & 3 \end{bmatrix} \right) = \begin{bmatrix} 0 & -10 & -15 \\ -40 & -5 & -15 \end{bmatrix}$

21. $AB = \begin{bmatrix} 0-3+2 & -4-2-10 \\ 0+9-1 & -10+6+5 \end{bmatrix} = \begin{bmatrix} -1 & -16 \\ 8 & 1 \end{bmatrix}$

22. $BA = \begin{bmatrix} 0-10 & 0-6 & 0+2 \\ 6+10 & -3+6 & 6-2 \\ 2-25 & -1-15 & 2+5 \end{bmatrix} = \begin{bmatrix} -10 & -6 & 2 \\ 16 & 3 & 4 \\ -23 & -16 & 7 \end{bmatrix}$

23. $BD = \begin{bmatrix} 0-6 & 0+4 & 0-8 \\ -6+6 & 9-4 & 3+8 \\ -2-15 & 3+10 & 1-20 \end{bmatrix} = \begin{bmatrix} -6 & 4 & -8 \\ 0 & 5 & 11 \\ -17 & 13 & -19 \end{bmatrix}$

24. $DB = \begin{bmatrix} 0+9+1 & 4+6-5 \\ 0-6+4 & -6-4-20 \end{bmatrix} = \begin{bmatrix} 10 & 5 \\ -2 & -30 \end{bmatrix}$

25. Not possible since AB is 2 x 2 and BA is 3 x 3.

26.
$(A - D)C = \begin{bmatrix} 4 & -4 & 1 \\ 2 & 5 & -5 \end{bmatrix} \begin{bmatrix} 1 & 2 & 3 \\ -1 & 1 & 2 \\ -1 & 2 & 1 \end{bmatrix}$
$= \begin{bmatrix} 4+4-1 & 8-4+2 & 12-8+1 \\ 2-5+5 & 4+5-10 & 6+10-5 \end{bmatrix} = \begin{bmatrix} 7 & 6 & 5 \\ 2 & -1 & 11 \end{bmatrix}$

27. $B(AC) = \begin{bmatrix} 0 & -2 \\ 3 & 2 \\ 1 & -5 \end{bmatrix} \begin{bmatrix} 2+1-2 & 4-1+4 & 6-2+2 \\ 5-3+1 & 10+3-2 & 15+6-1 \end{bmatrix}$
$= \begin{bmatrix} 0 & -2 \\ 3 & 2 \\ 1 & -5 \end{bmatrix} \begin{bmatrix} 1 & 7 & 6 \\ 3 & 11 & 20 \end{bmatrix}$
$= \begin{bmatrix} 0-6 & 0-22 & 0-40 \\ 3+6 & 21+22 & 18+40 \\ 1-15 & 7-55 & 6-100 \end{bmatrix}$
$= \begin{bmatrix} -6 & -22 & -40 \\ 9 & 43 & 58 \\ -14 & -48 & -94 \end{bmatrix}$

28. $3X + A = B$
$3X = B - A$
$X = \frac{1}{3}(B - A)$
$X = \frac{1}{3}\left(\begin{bmatrix} -2 & -12 \\ 4 & 1 \end{bmatrix} - \begin{bmatrix} 4 & 6 \\ -5 & 0 \end{bmatrix} \right)$
$X = \frac{1}{3} \begin{bmatrix} -6 & -18 \\ 9 & 1 \end{bmatrix}$
$X = \begin{bmatrix} -2 & -6 \\ 3 & \frac{1}{3} \end{bmatrix}$

29. $\begin{bmatrix} 2 & 2 & 2 \\ 1 & 2 & 1 \\ 1 & 2 & 1 \end{bmatrix}$

30. $\begin{bmatrix} 2 & 2 & 2 \\ 1 & 2 & 1 \\ 1 & 2 & 1 \end{bmatrix} + \begin{bmatrix} 1 & 1 & 1 \\ -1 & 1 & -1 \\ -1 & 1 & -1 \end{bmatrix} = \begin{bmatrix} 3 & 3 & 3 \\ 0 & 3 & 0 \\ 0 & 3 & 0 \end{bmatrix}$

$B = \begin{bmatrix} 3 & 3 & 3 \\ 0 & 3 & 0 \\ 0 & 3 & 0 \end{bmatrix}$

31. $\begin{bmatrix} 0 & 2 & 2 \\ 0 & 0 & -4 \end{bmatrix} + \begin{bmatrix} -2 & -2 & -2 \\ 1 & 1 & 1 \end{bmatrix} = \begin{bmatrix} -2 & 0 & 0 \\ 1 & 1 & -3 \end{bmatrix}$

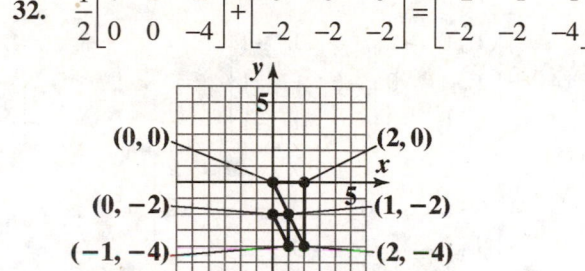

32. $\dfrac{1}{2}\begin{bmatrix} 0 & 2 & 2 \\ 0 & 0 & -4 \end{bmatrix} + \begin{bmatrix} 0 & 0 & 0 \\ -2 & -2 & -2 \end{bmatrix} = \begin{bmatrix} -2 & -1 & -1 \\ -2 & -2 & -4 \end{bmatrix}$

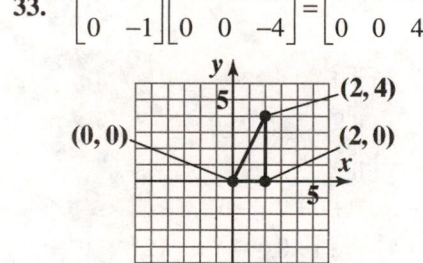

33. $\begin{bmatrix} 1 & 0 \\ 0 & -1 \end{bmatrix}\begin{bmatrix} 0 & 2 & 2 \\ 0 & 0 & -4 \end{bmatrix} = \begin{bmatrix} 0 & 2 & 2 \\ 0 & 0 & 4 \end{bmatrix}$

The triangle is reflected about the *x*-axis.

34. $\begin{bmatrix} -1 & 0 \\ 0 & 1 \end{bmatrix}\begin{bmatrix} 0 & 2 & 2 \\ 0 & 0 & -4 \end{bmatrix} = \begin{bmatrix} 0 & -2 & -2 \\ 0 & 0 & -4 \end{bmatrix}$

The triangle is reflected about the *y*-axis.

35. $\begin{bmatrix} 0 & -1 \\ 1 & 0 \end{bmatrix}\begin{bmatrix} 0 & 2 & 2 \\ 0 & 0 & -4 \end{bmatrix} = \begin{bmatrix} 0 & 0 & 4 \\ 0 & 2 & 2 \end{bmatrix}$

The triangle is rotated 90° counterclockwise about the origin.

36. $\begin{bmatrix} 2 & 0 \\ 0 & 1 \end{bmatrix}\begin{bmatrix} 0 & 2 & 2 \\ 0 & 0 & -4 \end{bmatrix} = \begin{bmatrix} 0 & 4 & 4 \\ 0 & 0 & -4 \end{bmatrix}$

The triangle is stretched by a factor of 2 horizontally.

37. $AB = \begin{bmatrix} 8-7 & -14+21 \\ 4-4 & -7+12 \end{bmatrix} = \begin{bmatrix} 1 & 7 \\ 0 & 5 \end{bmatrix}$

$BA = \begin{bmatrix} 8-7 & 28-28 \\ -2+3 & -7+12 \end{bmatrix} = \begin{bmatrix} 1 & 0 \\ 1 & 5 \end{bmatrix}$

If *B* is the multiplicative inverse of *A*, both products (*AB* and *BA*) will be the multiplicative identity matrix, I_2. Therefore, *B* is not the multiplicative inverse of *A*.

38. $AB = \begin{bmatrix} 1 & 0 & 0 \\ 0 & 2 & -7 \\ 0 & -1 & 4 \end{bmatrix}\begin{bmatrix} 1 & 0 & 0 \\ 0 & 4 & 7 \\ 0 & 1 & 2 \end{bmatrix} = \begin{bmatrix} 1 & 0 & 0 \\ 0 & 1 & 0 \\ 0 & 0 & 1 \end{bmatrix}$

$BA = \begin{bmatrix} 1 & 0 & 0 \\ 0 & 4 & 7 \\ 0 & 1 & 2 \end{bmatrix}\begin{bmatrix} 1 & 0 & 0 \\ 0 & 2 & -7 \\ 0 & -1 & 4 \end{bmatrix} = \begin{bmatrix} 1 & 0 & 0 \\ 0 & 1 & 0 \\ 0 & 0 & 1 \end{bmatrix}$

If *B* is the multiplicative inverse of *A*, both products (*AB* and *BA*) will be the multiplicative identity matrix, I_3. Therefore, *B* is the multiplicative inverse of *A*.

39. $A^{-1} = \dfrac{1}{3-2}\begin{bmatrix} 3 & 1 \\ 2 & 1 \end{bmatrix} = \begin{bmatrix} 3 & 1 \\ 2 & 1 \end{bmatrix}$

$AA^{-1} = \begin{bmatrix} 1 & -1 \\ -2 & 3 \end{bmatrix}\begin{bmatrix} 3 & 1 \\ 2 & 1 \end{bmatrix}$

$= \begin{bmatrix} 3-2 & 1-1 \\ -6+6 & -2+3 \end{bmatrix}$

$= \begin{bmatrix} 1 & 0 \\ 0 & 1 \end{bmatrix}$

$A^{-1}A = \begin{bmatrix} 3 & 1 \\ 2 & 1 \end{bmatrix}\begin{bmatrix} 1 & -1 \\ -2 & 3 \end{bmatrix}$

$= \begin{bmatrix} 3-2 & 3+3 \\ 2-2 & -2+3 \end{bmatrix}$

$= \begin{bmatrix} 1 & 0 \\ 0 & 1 \end{bmatrix}$

40. $A^{-1} = \dfrac{1}{0-5}\begin{bmatrix} 3 & -1 \\ -5 & 0 \end{bmatrix}$

$= \dfrac{-1}{5}\begin{bmatrix} 3 & -1 \\ -5 & 0 \end{bmatrix}$

$= \begin{bmatrix} -\frac{3}{5} & \frac{1}{5} \\ 1 & 0 \end{bmatrix}$

$AA^{-1} = \begin{bmatrix} 0 & 1 \\ 5 & 3 \end{bmatrix}\begin{bmatrix} -\frac{3}{5} & \frac{1}{5} \\ 1 & 0 \end{bmatrix}$

$= \begin{bmatrix} 0+1 & 0+0 \\ -3+3 & 1+0 \end{bmatrix}$

$= \begin{bmatrix} 1 & 0 \\ 0 & 1 \end{bmatrix}$

$A^{-1}A = \begin{bmatrix} -\frac{3}{5} & \frac{1}{5} \\ 1 & 0 \end{bmatrix}\begin{bmatrix} 0 & 1 \\ 5 & 3 \end{bmatrix}$

$= \begin{bmatrix} 0+1 & -\frac{3}{5}+\frac{3}{5} \\ 0+0 & 1+0 \end{bmatrix}$

$= \begin{bmatrix} 1 & 0 \\ 0 & 1 \end{bmatrix}$

41. $\left[\begin{array}{ccc|ccc} 1 & 0 & -2 & 1 & 0 & 0 \\ 2 & 1 & 0 & 0 & 1 & 0 \\ 1 & 0 & -3 & 0 & 0 & 1 \end{array}\right] \begin{array}{l} -2R_1+R_2 \\ -1R_1+R_3 \end{array}$

$\left[\begin{array}{ccc|ccc} 1 & 0 & -2 & 1 & 0 & 0 \\ 0 & 1 & 4 & -2 & 1 & 0 \\ 0 & 0 & -1 & -1 & 0 & 1 \end{array}\right] -1R_3$

$\left[\begin{array}{ccc|ccc} 1 & 0 & -2 & 1 & 0 & 0 \\ 0 & 1 & 4 & -2 & 1 & 0 \\ 0 & 0 & 1 & 1 & 0 & -1 \end{array}\right] \begin{array}{l} 2R_3+R_1 \\ -4R_3+R_2 \end{array}$

$\left[\begin{array}{ccc|ccc} 1 & 0 & 0 & 3 & 0 & -2 \\ 0 & 1 & 0 & -6 & 1 & 4 \\ 0 & 0 & 1 & 1 & 0 & -1 \end{array}\right]$

$A^{-1} = \begin{bmatrix} 3 & 0 & -2 \\ -6 & 1 & 4 \\ 1 & 0 & -1 \end{bmatrix}$

$AA^{-1} = \begin{bmatrix} 1 & 0 & -2 \\ 2 & 1 & 0 \\ 1 & 0 & -3 \end{bmatrix}\begin{bmatrix} 3 & 0 & -2 \\ -6 & 1 & 4 \\ 1 & 0 & -1 \end{bmatrix}$

$= \begin{bmatrix} 3+0-2 & 0+0+0 & -2+0+2 \\ 6-6+0 & 0+1+0 & -4+4+0 \\ 3+0-3 & 0+0+0 & -2+0+3 \end{bmatrix}$

$= \begin{bmatrix} 1 & 0 & 0 \\ 0 & 1 & 0 \\ 0 & 0 & 1 \end{bmatrix}$

$A^{-1}A = \begin{bmatrix} 3 & 0 & -2 \\ -6 & 1 & 4 \\ 1 & 0 & -1 \end{bmatrix}\begin{bmatrix} 1 & 0 & -2 \\ 2 & 1 & 0 \\ 1 & 0 & -3 \end{bmatrix}$

$= \begin{bmatrix} 3+0-2 & 0+0+0 & -6+0+6 \\ -6+2+4 & 0+1+0 & 12+0-12 \\ 1+0-1 & 0+0+0 & -2+0+3 \end{bmatrix}$

$= \begin{bmatrix} 1 & 0 & 0 \\ 0 & 1 & 0 \\ 0 & 0 & 1 \end{bmatrix}$

42. $\begin{bmatrix} 1 & 3 & -2 & | & 1 & 0 & 0 \\ 4 & 13 & -7 & | & 0 & 1 & 0 \\ 5 & 16 & -8 & | & 0 & 0 & 1 \end{bmatrix} \begin{matrix} -4R_1 + R_2 \\ -5R_1 + R_3 \end{matrix}$

$\begin{bmatrix} 1 & 3 & -2 & | & 1 & 0 & 0 \\ 0 & 1 & 1 & | & -4 & 1 & 0 \\ 0 & 1 & 2 & | & -5 & 0 & 1 \end{bmatrix} \begin{matrix} -1R_2 + R_3 \\ -3R_2 + R_1 \end{matrix}$

$\begin{bmatrix} 1 & 0 & -5 & | & 13 & -3 & 0 \\ 0 & 1 & 1 & | & -4 & 1 & 0 \\ 0 & 0 & 1 & | & -1 & -1 & 1 \end{bmatrix} \begin{matrix} -1R_3 + R_2 \\ 5R_3 + R_1 \end{matrix}$

$\begin{bmatrix} 1 & 0 & 0 & | & 8 & -8 & 5 \\ 0 & 1 & 0 & | & -3 & 2 & -1 \\ 0 & 0 & 1 & | & -1 & -1 & 1 \end{bmatrix}$

$A^{-1} = \begin{bmatrix} 8 & -8 & 5 \\ -3 & 2 & -1 \\ -1 & -1 & 1 \end{bmatrix}$

$AA^{-1} = \begin{bmatrix} 1 & 3 & -2 \\ 4 & 13 & -7 \\ 5 & 16 & -8 \end{bmatrix} \begin{bmatrix} 8 & -8 & 5 \\ -3 & 2 & -1 \\ -1 & -1 & 1 \end{bmatrix} = \begin{bmatrix} 8-9+2 & -8+6+2 & 5-3-2 \\ 32-39+7 & -32+26+7 & 20-13-7 \\ 40-48+8 & -40+32+8 & 25-16-8 \end{bmatrix} = \begin{bmatrix} 1 & 0 & 0 \\ 0 & 1 & 0 \\ 0 & 0 & 1 \end{bmatrix}$

$A^{-1}A = \begin{bmatrix} 8 & -8 & 5 \\ -3 & 2 & -1 \\ -1 & -1 & -1 \end{bmatrix} \begin{bmatrix} 1 & 3 & -2 \\ 4 & 13 & -7 \\ 5 & 16 & -8 \end{bmatrix} = \begin{bmatrix} 8-32+25 & 24-104+80 & -16+56-40 \\ -3+8-5 & -9+26-16 & 6-14+8 \\ -1-4+5 & -3-13+16 & 2+7-8 \end{bmatrix} = \begin{bmatrix} 1 & 0 & 0 \\ 0 & 1 & 0 \\ 0 & 0 & 1 \end{bmatrix}$

43. a. $\begin{bmatrix} 1 & 1 & 2 \\ 0 & 1 & 3 \\ 3 & 0 & -2 \end{bmatrix} \begin{bmatrix} x \\ y \\ z \end{bmatrix} = \begin{bmatrix} 7 \\ -2 \\ 0 \end{bmatrix}$.

b. $A^{-1}B = \begin{bmatrix} -2 & 2 & 1 \\ 9 & -8 & -3 \\ -3 & 3 & 1 \end{bmatrix} \begin{bmatrix} 7 \\ -2 \\ 0 \end{bmatrix} = \begin{bmatrix} -14-4+0 \\ 63+16+0 \\ -21-6+0 \end{bmatrix} = \begin{bmatrix} -18 \\ 79 \\ -27 \end{bmatrix}$

The solution to the system is $\{(-18, 79, -27)\}$.

44. a. $\begin{bmatrix} 1 & -1 & 2 \\ 0 & 1 & -1 \\ 1 & 0 & 2 \end{bmatrix} \begin{bmatrix} x \\ y \\ z \end{bmatrix} = \begin{bmatrix} 12 \\ -5 \\ 10 \end{bmatrix}$

b. $A^{-1}B = \begin{bmatrix} 2 & 2 & -1 \\ -1 & 0 & 1 \\ -1 & -1 & 1 \end{bmatrix} \begin{bmatrix} 12 \\ -5 \\ 10 \end{bmatrix} = \begin{bmatrix} 24-10-10 \\ -12+10 \\ -12+5+10 \end{bmatrix} = \begin{bmatrix} 4 \\ -2 \\ 3 \end{bmatrix}$

The solution to the system is $\{(4, -2, 3)\}$.

45. R U L E has a numerical equivalent of 18, 21, 12, 5.

$$\begin{bmatrix} 3 & 2 \\ 4 & 3 \end{bmatrix} \begin{bmatrix} 18 & 12 \\ 21 & 5 \end{bmatrix} = \begin{bmatrix} 54+42 & 36+10 \\ 72+63 & 48+15 \end{bmatrix} = \begin{bmatrix} 96 & 46 \\ 135 & 63 \end{bmatrix}$$

The encoded message is 96, 135, 46, 63.

$$\begin{bmatrix} 3 & -2 \\ -4 & 3 \end{bmatrix} \begin{bmatrix} 96 & 46 \\ 135 & 63 \end{bmatrix} = \begin{bmatrix} 288-270 & 138-126 \\ -384+405 & -184+189 \end{bmatrix} = \begin{bmatrix} 18 & 12 \\ 21 & 5 \end{bmatrix}$$

The decoded message is 18, 21, 12, 5 or RULE.

46. $\begin{vmatrix} 3 & 2 \\ -1 & 5 \end{vmatrix} = 15 - (-2) = 17$

47. $\begin{vmatrix} -2 & -3 \\ -4 & -8 \end{vmatrix} = 16 - 12 = 4$

48. $\begin{vmatrix} 2 & 4 & -3 \\ 1 & -1 & 5 \\ -2 & 4 & 0 \end{vmatrix} = -2 \begin{vmatrix} 4 & -3 \\ -1 & 5 \end{vmatrix} - 4 \begin{vmatrix} 2 & -3 \\ 1 & 5 \end{vmatrix} + 0 \begin{vmatrix} 2 & 4 \\ 1 & -1 \end{vmatrix}$

$$= -2(20 - 3) - 4[10 - (-3)] + 0$$
$$= -2(17) - 4(13)$$
$$= -34 - 52$$
$$= -86$$

49. $\begin{vmatrix} 4 & 7 & 0 \\ -5 & 6 & 0 \\ 3 & 2 & -4 \end{vmatrix} = 4 \begin{vmatrix} 6 & 0 \\ 2 & -4 \end{vmatrix} + 5 \begin{vmatrix} 7 & 0 \\ 2 & -4 \end{vmatrix} + 3 \begin{vmatrix} 7 & 0 \\ 6 & 0 \end{vmatrix}$

$$= 4(-24 - 0) + 5(-28 - 0) + 3(0 - 0)$$
$$= 4(-24) + 5(-28) + 0$$
$$= -236$$

50. $\begin{vmatrix} 1 & 1 & 0 & 2 \\ 0 & 3 & 2 & 1 \\ 0 & -2 & 4 & 0 \\ 0 & 3 & 0 & 1 \end{vmatrix} = \begin{vmatrix} 3 & 2 & 1 \\ -2 & 4 & 0 \\ 3 & 0 & 1 \end{vmatrix}$

$$= 3 \begin{vmatrix} 2 & 1 \\ 4 & 0 \end{vmatrix} + \begin{vmatrix} 3 & 2 \\ -2 & 4 \end{vmatrix}$$
$$= 3(0 - 4) + [12 - (-4)]$$
$$= 3(-4) + 16$$
$$= -12 + 16$$
$$= 4$$

51.

$$\begin{vmatrix} 2 & 2 & 2 & 2 \\ 0 & 2 & 2 & 2 \\ 0 & 0 & 2 & 2 \\ 0 & 0 & 0 & 2 \end{vmatrix} = 2\begin{vmatrix} 2 & 2 & 2 \\ 0 & 2 & 2 \\ 0 & 0 & 2 \end{vmatrix}$$

$$= 2(2)\begin{vmatrix} 2 & 2 \\ 0 & 2 \end{vmatrix}$$

$$= 2(2)(4)$$

$$= 16$$

52. $D = \begin{vmatrix} 1 & -2 \\ 3 & 2 \end{vmatrix} = 2 - (-6) = 2 + 6 = 8$

$$D_x = \begin{vmatrix} 8 & -2 \\ -1 & 2 \end{vmatrix} = 16 - 2 = 14$$

$$D_y = \begin{vmatrix} 1 & 8 \\ 3 & -1 \end{vmatrix} = -1 - 24 = -25$$

$$x = \frac{D_x}{D} = \frac{14}{8} = \frac{7}{4}, \ y = \frac{D_y}{D} = \frac{-25}{8} = -\frac{25}{8}$$

The solution to the system is $\left\{\left(\dfrac{7}{4}, -\dfrac{25}{8}\right)\right\}$.

53. $D = \begin{vmatrix} 7 & 2 \\ 2 & 1 \end{vmatrix} = 7 - 4 = 3$

$$D = \begin{vmatrix} 7 & 2 \\ 2 & 1 \end{vmatrix} = 7 - 4 = 3$$

$$D_x = \begin{vmatrix} 0 & 2 \\ -3 & 1 \end{vmatrix} = 0 - (-6) = 6$$

$$D_y = \begin{vmatrix} 7 & 0 \\ 2 & -3 \end{vmatrix} = -21 - 0 = -21$$

$$x = \frac{D_x}{D} = \frac{6}{3} = 2$$

$$y = \frac{D_y}{D} = \frac{-21}{3} = -7$$

The solution to the system is $\{(2, -7)\}$.

54. $D = \begin{vmatrix} 1 & 2 & 2 \\ 2 & 4 & 7 \\ -2 & -5 & -2 \end{vmatrix}$

$$= \begin{vmatrix} 1 & 2 & 2 \\ 0 & 0 & 3 \\ 0 & -1 & 2 \end{vmatrix}$$

$$= \begin{vmatrix} 0 & 3 \\ -1 & 2 \end{vmatrix}$$

$$= 0 - (-3)$$

$$= 3$$

$$D_x = \begin{vmatrix} 5 & 2 & 2 \\ 19 & 4 & 7 \\ 8 & -5 & -2 \end{vmatrix}$$

$$= 5\begin{vmatrix} 4 & 7 \\ -5 & -2 \end{vmatrix} - 2\begin{vmatrix} 19 & 7 \\ 8 & -2 \end{vmatrix} + 2\begin{vmatrix} 19 & 4 \\ 8 & -5 \end{vmatrix}$$

$$= 5[-8 - (-35)] - 2(-38 - 56) + 2(-95 - 32)$$

$$= 5(27) - 2(-94) - 2(127)$$

$$= 135 + 188 - 254$$

$$= 69$$

$$D_y = \begin{vmatrix} 1 & 5 & 2 \\ 2 & 19 & 7 \\ -2 & 8 & -2 \end{vmatrix}$$

$$= \begin{vmatrix} 1 & 5 & 2 \\ 0 & 9 & 3 \\ 0 & 18 & 2 \end{vmatrix}$$

$$= \begin{vmatrix} 9 & 3 \\ 18 & 2 \end{vmatrix}$$

$$= 18 - 54$$

$$= -36$$

$$D_z = \begin{vmatrix} 1 & 2 & 5 \\ 2 & 4 & 19 \\ -2 & -5 & 8 \end{vmatrix}$$

$$= \begin{vmatrix} 1 & 2 & 5 \\ 0 & 0 & 9 \\ 0 & -1 & 18 \end{vmatrix}$$

$$= \begin{vmatrix} 0 & 9 \\ -1 & 18 \end{vmatrix}$$

$$= 0 - (-9)$$

$$= 9$$

$$x = \frac{D_x}{D} = \frac{69}{3} = 23, \ y = \frac{D_y}{D} = \frac{-36}{3} = -12,$$

$$z = \frac{D_z}{D} = \frac{9}{3} = 3$$

The solution to the system is $\{(23, -12, 3)\}$.

55.

$$D = \begin{vmatrix} 2 & 1 & 0 \\ 0 & 1 & -2 \\ 3 & 0 & -2 \end{vmatrix}$$

$$= 2\begin{vmatrix} 1 & -2 \\ 0 & -2 \end{vmatrix} + 3\begin{vmatrix} 1 & 0 \\ 1 & -2 \end{vmatrix}$$

$$= 2(-2-0) + 3(-2-0)$$

$$= 2(-2) + 3(-2)$$

$$= -4 - 6$$

$$= -10$$

$$D_x = \begin{vmatrix} -4 & 1 & 0 \\ 0 & 1 & -2 \\ -11 & 0 & -2 \end{vmatrix}$$

$$= -1\begin{vmatrix} 0 & -2 \\ -11 & -2 \end{vmatrix} + 1\begin{vmatrix} -4 & 0 \\ -11 & -2 \end{vmatrix}$$

$$= -1(0-22) + 1(8-0)$$

$$= 22 + 8$$

$$= 30$$

$$D_y = \begin{vmatrix} 2 & -4 & 0 \\ 0 & 0 & -2 \\ 3 & -11 & -2 \end{vmatrix}$$

$$= 2\begin{vmatrix} 0 & -2 \\ -11 & -2 \end{vmatrix} + 3\begin{vmatrix} -4 & 0 \\ 0 & -2 \end{vmatrix}$$

$$= 2(0-22) + 3(8-0)$$

$$= 2(-22) + 3(8)$$

$$= -44 + 24$$

$$= -20$$

$$D_z = \begin{vmatrix} 2 & 1 & -4 \\ 0 & 1 & 0 \\ 3 & 0 & -11 \end{vmatrix}$$

$$= 2\begin{vmatrix} 1 & 0 \\ 0 & -11 \end{vmatrix} + 3\begin{vmatrix} 1 & -4 \\ 1 & 0 \end{vmatrix}$$

$$= 2(-11-0) + 3(0+4)$$

$$= 2(-11) + 3(+4) = -22 + 12$$

$$= -10$$

$$x = \frac{D_x}{D} = \frac{30}{-10} = -3$$

$$y = \frac{D_y}{D} = \frac{-20}{-10} = 2$$

$$z = \frac{D_z}{D} = \frac{-10}{-10} = 1$$

The solution to the system is $\{(-3, 2, 1)\}$.

56. The quadratic function must satisfy

$$f(20) = 400 = 400a + 20b + c$$
$$f(40) = 150 = 1600a + 40b + c$$
$$f(60) = 400 = 3600a + 60b + c$$

$$D = \begin{vmatrix} 400 & 20 & 1 \\ 1600 & 40 & 1 \\ 3600 & 60 & 1 \end{vmatrix}$$

$$= (400)(20)\begin{vmatrix} 1 & 1 & 1 \\ 4 & 2 & 1 \\ 9 & 3 & 1 \end{vmatrix}$$

$$= 8000\begin{vmatrix} 1 & 1 & 1 \\ 3 & 1 & 0 \\ 8 & 2 & 0 \end{vmatrix} = 8000\begin{vmatrix} 3 & 1 \\ 8 & 2 \end{vmatrix}$$

$$= 8000(6-8)$$

$$= 8000(-2)$$

$$= -16,000$$

$$D_a = \begin{vmatrix} 400 & 20 & 1 \\ 150 & 40 & 1 \\ 400 & 60 & 1 \end{vmatrix}$$

$$= (50)(20)\begin{vmatrix} 8 & 1 & 1 \\ 3 & 2 & 1 \\ 8 & 3 & 1 \end{vmatrix}$$

$$= 1000\begin{vmatrix} 8 & 1 & 1 \\ -5 & 1 & 0 \\ 0 & 2 & 0 \end{vmatrix}$$

$$= 1000\begin{vmatrix} -5 & 1 \\ 0 & 2 \end{vmatrix}$$

$$= 1000(-10-0)$$

$$= -10,000$$

$$D_b = \begin{vmatrix} 400 & 400 & 1 \\ 1600 & 150 & 1 \\ 3600 & 400 & 1 \end{vmatrix}$$

$$= (400)(50)\begin{vmatrix} 1 & 8 & 1 \\ 4 & 3 & 1 \\ 9 & 8 & 1 \end{vmatrix}$$

$$= 20,000\begin{vmatrix} 1 & 8 & 1 \\ 3 & -5 & 0 \\ 8 & 0 & 0 \end{vmatrix}$$

$$= 20,000\begin{vmatrix} 3 & -5 \\ 8 & 0 \end{vmatrix}$$

$$= 20,000[0 - (-40)]$$

$$= 20,000(40)$$

$$= 800,000$$

$D_c = \begin{vmatrix} 400 & 20 & 400 \\ 1600 & 40 & 150 \\ 3600 & 60 & 400 \end{vmatrix}$

$= (400)(20)(50)\begin{vmatrix} 1 & 1 & 8 \\ 4 & 2 & 3 \\ 9 & 3 & 8 \end{vmatrix}$

$= 400,000\begin{vmatrix} 1 & 0 & 0 \\ 4 & -2 & -29 \\ 2 & -6 & -64 \end{vmatrix}$

$= 400,000\begin{vmatrix} -2 & -29 \\ -6 & -64 \end{vmatrix}$

$= 400,000(128 - 174)$

$= 400,000(-46)$

$= -18,400,000$

$a = \dfrac{D_a}{D} = \dfrac{-10,000}{-16,000} = \dfrac{5}{8}$,

$b = \dfrac{D_b}{D} = \dfrac{800,000}{-16,000} = -50$,

$c = \dfrac{D_c}{D} = \dfrac{-18,400,000}{-16,000} = 1150$

The model is $f(x) = \dfrac{5}{8}x^2 - 50x + 1150$.

$f(30) = \dfrac{5}{8}(900) - 50(30) + 1150$

$= 562.5 - 1500 + 1150$

$= 212.5$

$f(50) = \dfrac{5}{8}(2500) - 50(50) + 1150$

$= 1562.8 - 2500 + 1150$

$= 212.5$

30- and 50-year-olds are involved in an average of 212.5 automobile accidents per day.

Chapter 8 Test

1. $\begin{bmatrix} 1 & 2 & -1 & | & -3 \\ 2 & -4 & 1 & | & -7 \\ -2 & 2 & -3 & | & 4 \end{bmatrix} \begin{matrix} \\ -2R_1 + R_2 \\ 2R_1 + R_3 \end{matrix}$

$\begin{bmatrix} 1 & 2 & -1 & | & -3 \\ 0 & -8 & 3 & | & -1 \\ 0 & 6 & -5 & | & -2 \end{bmatrix} -\frac{1}{8}R_2$

$\begin{bmatrix} 1 & 2 & -1 & | & -3 \\ 0 & 1 & -\frac{3}{8} & | & \frac{1}{8} \\ 0 & 6 & -5 & | & -2 \end{bmatrix} -6R_2 + R_3$

$\begin{bmatrix} 1 & 2 & -1 & | & -3 \\ 0 & 1 & -\frac{3}{8} & | & \frac{1}{8} \\ 0 & 0 & -\frac{11}{4} & | & -\frac{11}{4} \end{bmatrix} -\frac{4}{11}R_3$

$\begin{bmatrix} 1 & 2 & -1 & | & -3 \\ 0 & 1 & -\frac{3}{8} & | & \frac{1}{8} \\ 0 & 0 & 1 & | & 1 \end{bmatrix}$

$x + 2y - z = -3$

$y - \dfrac{3}{8}z = \dfrac{1}{8}$

$z = 1$

Using back substitution,

$y - \dfrac{3}{8}(1) = \dfrac{1}{8}$ and $x + 2\left(\dfrac{1}{2}\right) - 1 = -3$.

$y = \dfrac{1}{2}$ $x + 1 - 1 = -3$

                                        $x = -3$

The solution to the system is $\left\{\left(-3, \dfrac{1}{2}, 1\right)\right\}$.

2. $\begin{bmatrix} 1 & -2 & 1 & | & 2 \\ 2 & -1 & -1 & | & 1 \end{bmatrix} -2R_1 + R_2$

$\begin{bmatrix} 1 & -2 & 1 & | & 2 \\ 0 & 3 & -3 & | & -3 \end{bmatrix} \frac{1}{3}R_2$

$\begin{bmatrix} 1 & -2 & 1 & | & 2 \\ 0 & 1 & -1 & | & -1 \end{bmatrix}$

The system $\begin{matrix} x - 2y + z = 2 \\ y - z = -1 \end{matrix}$ has no unique solution.

Express x and y in terms of z:

$y = z - 1$

$x - 2(z - 1) + z = 2$

$x - 2z + 2 + z = 2$

$x = z$

With $z = t$, the complete solution to the system is $\{(t, t - 1, t)\}$.

Matrices and Determinants

3. $2B + 3C = \begin{bmatrix} 2 & -2 \\ 4 & 2 \end{bmatrix} + \begin{bmatrix} 3 & 6 \\ -3 & 9 \end{bmatrix} = \begin{bmatrix} 5 & 4 \\ 1 & 11 \end{bmatrix}$

4. $AB = \begin{bmatrix} 3+2 & -3+1 \\ 1+0 & -1+0 \\ 2+2 & -2+1 \end{bmatrix} = \begin{bmatrix} 5 & -2 \\ 1 & -1 \\ 4 & -1 \end{bmatrix}$

5. $C^{-1} = \dfrac{1}{(1)(3)-(2)(-1)}\begin{bmatrix} 3 & -2 \\ 1 & 1 \end{bmatrix}$

$= \dfrac{1}{3+2}\begin{bmatrix} 3 & -2 \\ 1 & 1 \end{bmatrix} = \begin{bmatrix} \frac{3}{5} & -\frac{2}{5} \\ \frac{1}{5} & \frac{1}{5} \end{bmatrix}$

6. $BC = \begin{bmatrix} 1+1 & 2-3 \\ 2-1 & 4+3 \end{bmatrix} = \begin{bmatrix} 2 & -1 \\ 1 & 7 \end{bmatrix}$

$BC - 3B = \begin{bmatrix} 2 & -1 \\ 1 & 7 \end{bmatrix} - \begin{bmatrix} 3 & -3 \\ 6 & 3 \end{bmatrix} = \begin{bmatrix} -1 & 2 \\ -5 & 4 \end{bmatrix}$

7. $AB = \begin{bmatrix} -3+14-10 & 2-8+6 & 0+2-2 \\ -6+21-15 & 4-12+9 & 0+3-3 \\ -3-7+10 & 2+4-6 & 0-1+2 \end{bmatrix}$

$= \begin{bmatrix} 1 & 0 & 0 \\ 0 & 1 & 0 \\ 0 & 0 & 1 \end{bmatrix} = I_3$

$BA = \begin{bmatrix} -3+4+0 & -6+6+0 & -6+6+0 \\ 7-8+1 & 14-12-1 & 14-12-2 \\ -5+6-1 & -10+9+1 & -10+9+2 \end{bmatrix}$

$= \begin{bmatrix} 1 & 0 & 0 \\ 0 & 1 & 0 \\ 0 & 0 & 1 \end{bmatrix} = I_3$

8. a. $\begin{bmatrix} 3 & 5 \\ 2 & -3 \end{bmatrix}\begin{bmatrix} x \\ y \end{bmatrix} = \begin{bmatrix} 9 \\ -13 \end{bmatrix}$

b. $A^{-1} = \dfrac{1}{(3)(-3)-(5)(2)}\begin{bmatrix} -3 & -5 \\ -2 & 3 \end{bmatrix}$

$= \dfrac{1}{-19}\begin{bmatrix} -3 & -5 \\ -2 & 3 \end{bmatrix} = \begin{bmatrix} \frac{3}{19} & \frac{5}{19} \\ \frac{2}{19} & -\frac{3}{19} \end{bmatrix}$

c. $A^{-1}B = \begin{bmatrix} \frac{3}{19} & \frac{5}{19} \\ \frac{2}{19} & -\frac{3}{19} \end{bmatrix}\begin{bmatrix} 9 \\ -13 \end{bmatrix} = \begin{bmatrix} \frac{27}{19}-\frac{65}{19} \\ \frac{18}{19}+\frac{39}{19} \end{bmatrix} = \begin{bmatrix} -2 \\ 3 \end{bmatrix}$

The solution to the system is $\{(-2, 3)\}$.

9. $\begin{vmatrix} 4 & -1 & 3 \\ 0 & 5 & -1 \\ 5 & 2 & 4 \end{vmatrix} = 4\begin{vmatrix} 5 & -1 \\ 2 & 4 \end{vmatrix} + 5\begin{vmatrix} -1 & 3 \\ 5 & -1 \end{vmatrix}$

$= 4[20-(-2)]+5(1-15)$

$= 4(22)+5(-14)$

$= 88-70$

$= 18$

10. $D = \begin{vmatrix} 3 & 1 & -2 \\ 2 & 7 & 3 \\ 4 & -3 & -1 \end{vmatrix} = 3\begin{vmatrix} 7 & 3 \\ -3 & -1 \end{vmatrix} - 1\begin{vmatrix} 2 & 3 \\ 4 & -1 \end{vmatrix} - 2\begin{vmatrix} 2 & 7 \\ 4 & -3 \end{vmatrix}$

$= 3[-7-(-9)]-1(-2-12)-2(-6-28)$

$= 3(2)-1(-14)-2(-34)$

$= 6+14+68$

$= 88$

$D_x = \begin{vmatrix} -3 & 1 & -2 \\ 9 & 7 & 3 \\ 7 & -3 & -1 \end{vmatrix} - 3\begin{vmatrix} 7 & 3 \\ -3 & -1 \end{vmatrix} - 1\begin{vmatrix} 9 & 3 \\ 7 & -1 \end{vmatrix} - 2\begin{vmatrix} 9 & 7 \\ 7 & -3 \end{vmatrix}$

$= -3[-7-(-9)]-1(-9-21)-2(-27-49)$

$= -3(2)-1(-30)-2(-76)$

$= -6+30+152$

$= 176$

$x = \dfrac{D_x}{D} = \dfrac{176}{88} = 2$

Cumulative Review Exercises (Chapters P–8)

1. $2x^2 = 4 - x$

$2x^2 + x - 4 = 0$

$x = \dfrac{-1 \pm \sqrt{1^2-4(2)(-4)}}{2(2)}$

$x = \dfrac{-1 \pm \sqrt{1-32}}{4}$

$x = \dfrac{-1 \pm \sqrt{33}}{4}$

The solution set is $\left\{\dfrac{-1+\sqrt{33}}{4}, \dfrac{-1-\sqrt{33}}{4}\right\}$.

776

2. $5x + 8 \le 7(1 + x)$

$5x + 8 \le 7 + 7x$

$-2x \le -1$

$x \ge \dfrac{1}{2}$

The solution set is $\left\{ x \mid x \ge \dfrac{1}{2} \right\}$ or $\left[\dfrac{1}{2}, \infty \right)$.

3. $x^3 + x^2 - 4x - 4 \ge 0$

Find boundary points by solving the related equation.

$x^3 + x^2 - 4x - 4 = 0$

$x^2(x + 1) - 4(x + 1) = 0$

$(x + 1)(x^2 - 4) = 0$

$(x + 1)(x + 2)(x - 2) = 0$

The boundary points are $x = -2, -1, 2$

Test values in each interval.

The solution set is $[-2, -1] \cup [2, \infty)$.

4. $3x^3 + 8x^2 - 15x + 4 = 0$

$p = \pm1, \pm2, \pm4$

$q = \pm1, \pm3$

$\dfrac{p}{q} = \pm1, \pm\dfrac{1}{3}, \pm2, \pm\dfrac{2}{3}, \pm4, \pm\dfrac{4}{3}$

$$
\begin{array}{r|rrrr}
-4 & 3 & 8 & -15 & 4 \\
   &   & -12 & 16 & -4 \\
\hline
   & 3 & -4 & 1 & 0
\end{array}
$$

$(x + 4)(3x^2 - 4x + 1) = 0$

$(x + 4)(3x - 1)(x - 1) = 0$

$x = -4, \quad x = \dfrac{1}{3}, \quad x = 1$

The solution set is $\left\{ -4, \dfrac{1}{3}, 1 \right\}$.

5. $e^{2x} - 14e^x + 45 = 0$ $\ let+t = e^x$

$t^2 - 14t + 45 = 0$

$(t - 5)(t - 9) = 0$

$t = 5 \quad t = 9$

$e^x = 5 \quad e^x = 9$

$\ln e^x = \ln 5 \quad \ln e^x = \ln 9$

$x = e^x = \ln 5 \quad x = \ln 9$

The solution set is $\{\ln 5, \ln 9\}$.

6. $\log_3 x + \log_3(x + 2) = 1$

$\log_3 x^2 + 2x = 1$

$3^1 = x^2 + 2x$

$x^2 + 2x - 3 = 0$

$(x - 1)(x + 3) = 0$

$x = 1, \ x = -3$

$x = -3$ does not check. The solution set is $\{1\}$.

7.

$$
\left[\begin{array}{rrr|r}
1 & -1 & 1 & 17 \\
2 & 3 & 1 & 8 \\
-4 & 1 & 5 & -2
\end{array} \right] \begin{array}{l} \\ -2R_1 + R_2 \\ 4R_1 + R_3 \end{array}
$$

$$
\left[\begin{array}{rrr|r}
1 & -1 & 1 & 17 \\
0 & 5 & -1 & -26 \\
0 & -3 & 9 & 66
\end{array} \right] -\tfrac{1}{3}R_3
$$

$$
\left[\begin{array}{rrr|r}
1 & -1 & 1 & 17 \\
0 & 1 & -3 & -22 \\
0 & 5 & -1 & -26
\end{array} \right] \begin{array}{l} \\ -5R_2 + R_3 \\ 1R_2 + R_1 \end{array}
$$

$$
\left[\begin{array}{rrr|r}
1 & 0 & -2 & -5 \\
0 & 1 & -3 & -22 \\
0 & 0 & 14 & 84
\end{array} \right] \tfrac{1}{14}R_3
$$

$$
\left[\begin{array}{rrr|r}
1 & 0 & -2 & -5 \\
0 & 1 & -3 & -22 \\
0 & 0 & 1 & 6
\end{array} \right] \begin{array}{l} \\ 3R_3 + R_2 \\ 2R_3 + R_1 \end{array}
$$

$$
\left[\begin{array}{rrr|r}
1 & 0 & 0 & 7 \\
0 & 1 & 0 & -4 \\
0 & 0 & 1 & 6
\end{array} \right]
$$

$x = 7 \ y = -4 \ z = 6$

The solution set is $\{(7, -4, 6)\}$.

8. $D = \begin{vmatrix} 1 & -2 & 1 \\ 2 & 1 & -1 \\ 3 & 2 & -2 \end{vmatrix}$

$= 1 \begin{vmatrix} 1 & -1 \\ 2 & -2 \end{vmatrix} - 2 \begin{vmatrix} -2 & 1 \\ 2 & -2 \end{vmatrix} + 3 \begin{vmatrix} -2 & 1 \\ 1 & 1 \end{vmatrix}$

$= 1(-2+2) - 2(4-2) + 3(2-1)$

$= 0 - 4 + 3$

$= -1$

$D_y = \begin{vmatrix} 1 & 7 & 1 \\ 2 & 0 & -1 \\ 3 & -2 & -2 \end{vmatrix} = 7 \begin{vmatrix} 2 & -1 \\ 3 & -2 \end{vmatrix} - 2 \begin{vmatrix} 1 & 1 \\ 2 & -1 \end{vmatrix}$

$= 7(-4+3) - 2(-1-2)$

$= -7 + 6 = 1$

$y = \dfrac{D_y}{D} = \dfrac{1}{-1} = -1$

$y = -1$

9. $y = \sqrt{4x-7}$

$x = \sqrt{4y-7}$

$x^2 = 4y - 7$

$x^2 + 7 - 4y$

$\dfrac{x^2 + 7}{4} = y$

$f^{-1}(x) = \dfrac{x^2 + 7}{4} \ (x \ge 0)$

10. $f(x) = \dfrac{x}{x^2 - 16}$

$f(0) = \dfrac{0}{-16} = 0$

y-intercept at 0

$0 = \dfrac{x}{x^2 - 16}$

$0 = x$

x-intercept at 0

$f(x) = \dfrac{x}{(x+4)(x-4)}$

vertical asymptotes at 4, –4
horizontal asymptote at 0

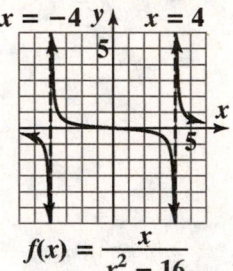

$f(x) = \dfrac{x}{x^2 - 16}$

11. $f(x) = 4x^4 - 4x^3 - 25x^2 + x + 6$

$\begin{array}{r|rrrrr} -2 & 4 & -4 & -25 & 1 & 6 \\ & & -8 & 24 & 2 & -6 \\ \hline 3 & 4 & -12 & -1 & 3 & 0 \\ & & 12 & 0 & -3 & \\ \hline & 4 & 0 & -1 & 0 & \end{array}$

$f(x) = (x+2)(x-3)(4x^2 - 1)$

$f(x) = (x+2)(x-3)(2x+1)(2x-1)$

12. $y = \log_2 x$

$2^y = x$

x	y
1	0
2	1
$\frac{1}{2}$	-1

$y = \log_2(x+1)$

Shift the graph of $y = \log_2{}^x$ left one unit.

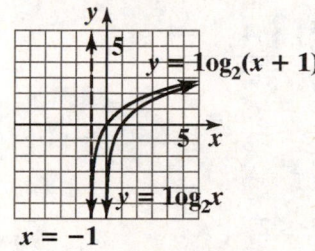

13. a. $A = A_0 e^{kt}$

$450 = 900 e^{k(40)}$

$\dfrac{1}{2} = e^{40k}$

$\ln\left(\dfrac{1}{2}\right) = \ln\left(e^{40k}\right)$

$\ln\left(\dfrac{1}{2}\right) = 40k$

$k = \dfrac{\ln\left(\dfrac{1}{2}\right)}{40} \approx -0.017$

$A = 900 e^{-0.017t}$

b. $A = 900 e^{-0.017(10)}$

$A = 900 e^{-0.17}$

$A \approx 759.30$ grams

14. $\begin{bmatrix} 1 & -1 & 0 \\ 2 & 1 & 3 \end{bmatrix} \begin{bmatrix} 4 & -1 \\ 2 & 0 \\ 1 & 1 \end{bmatrix} = \begin{bmatrix} 4-2+0 & -1+0+0 \\ 8+2+3 & -2+0+3 \end{bmatrix}$

$= \begin{bmatrix} 2 & -1 \\ 13 & 1 \end{bmatrix}$

15. $\dfrac{3x^2 + 17x - 38}{(x-3)(x-2)(x+2)} = \dfrac{A}{x-3} + \dfrac{B}{x-2} + \dfrac{C}{x+2}$

$3x^2 + 17x - 38$

$= A(x^2 - 4) + B(x^2 - x - 6) + C(x^2 - 5x + 6)$

$3x^2 + 17x - 38$

$= Ax^2 - 4A + Bx^2 - Bx - 6B + Cx^2 - 5Cx + 6c$

$3x^2 + 17x - 38$

$= (A+B+C)x^2 + (-B-5C)x - (4A+6B-6C)$

$A + B + C = 3$

$-B - 5C = 17$

$4A + 6B - 6C = 38$

$\begin{bmatrix} 1 & 1 & 1 & | & 3 \\ 0 & -1 & -5 & | & 17 \\ 4 & 6 & -6 & | & 38 \end{bmatrix} -4R_1 + R_3$

$\begin{bmatrix} 1 & 1 & 1 & | & 3 \\ 0 & -1 & -5 & | & 17 \\ 0 & 2 & -10 & | & 26 \end{bmatrix} -1R_2$

$\begin{bmatrix} 1 & 1 & 1 & | & 3 \\ 0 & 1 & 5 & | & -17 \\ 0 & 2 & -10 & | & 26 \end{bmatrix} \begin{matrix} -2R_2 + R_3 \\ -1R_2 + R_1 \end{matrix}$

$\begin{bmatrix} 1 & 0 & -4 & | & 20 \\ 0 & 1 & 5 & | & -17 \\ 0 & 0 & -20 & | & 60 \end{bmatrix} -\frac{1}{20}R_3$

$\begin{bmatrix} 1 & 0 & -4 & | & 20 \\ 0 & 1 & 5 & | & -17 \\ 0 & 0 & 1 & | & -3 \end{bmatrix} \begin{matrix} -5R_3 + R_2 \\ 4R_3 + R_1 \end{matrix}$

$\begin{bmatrix} 1 & 0 & 0 & | & 8 \\ 0 & 1 & 0 & | & -2 \\ 0 & 0 & 1 & | & -3 \end{bmatrix}$

$A = 8, \quad B = -2, \quad C = -3$

$\dfrac{3x^2 + 17x - 38}{(x-3)(x-2)(x+2)} = \dfrac{8}{x-3} + \dfrac{-2}{x-2} + \dfrac{-3}{x+2}$

16. $y = -\dfrac{2}{3}x - 1$

x	y
0	−1
3	−3
−3	1

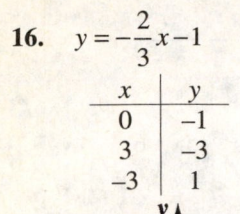

$$y = -\dfrac{2}{3}x - 1$$

17. $3x - 5y < 15$

$-5y < -3x + 15$

$y > \dfrac{3}{5}x - 3$

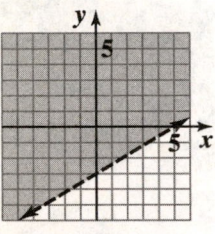

$3x - 5y < 15$

18. $f(x) = x^2 - 2x - 3$

$f(x) = (x^2 - 2x + 1) - 3 - 1$

$f(x) = (x - 1)^2 - 4$

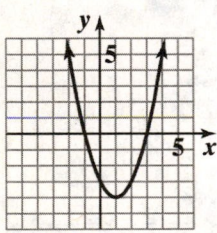

$f(x) = x^2 - 2x - 3$

19. $(x - 1)^2 + (y + 1)^2 = 9$

center $(1, -1)$

radius = 3

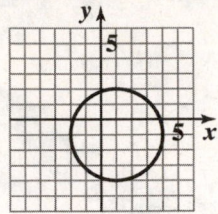

$(x - 1)^2 + (y + 1)^2 = 9$

20.

$$\begin{array}{r|rrrr} 2 & 1 & 0 & -6 & 4 \\ & & 2 & 4 & -4 \\ \hline & 1 & 2 & -2 & 0 \end{array}$$

$$\dfrac{x^3 - 6x + 4}{x - 2} = x^2 + 2x - 2$$

21. $y = 2\sin 2\pi x,\ 0 \le x \le 2$

Amplitude: $|A| = |2| = 2$

Period: $\dfrac{2\pi}{B} = \dfrac{2\pi}{2\pi} = 1$

x-intercepts:

$(0, 0), \left(\dfrac{1}{2}, 0\right), (1, 0), \left(\dfrac{3}{2}, 0\right), (2, 0)$

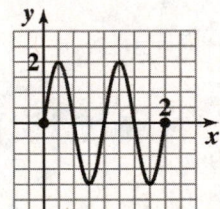

$y = 2\sin 2\pi x, 0 \le x \le 2$

22. $\cos\left[\tan^{-1}\left(-\dfrac{4}{3}\right)\right]$

If $\tan\theta = -\dfrac{4}{3}$, θ lies in quadrant IV.

$\tan\theta = -\dfrac{4}{3} = \dfrac{y}{x} = \dfrac{-4}{3}$

$r = \sqrt{(3)^2 + (-4)^2} = \sqrt{9 + 16} = \sqrt{25} = 5$

$\cos\left[\tan^{-1}\left(-\dfrac{4}{3}\right)\right] = \dfrac{x}{r} = \dfrac{3}{5}$

23. $\dfrac{\cos 2x}{\cos x - \sin x} = \dfrac{\cos^2 x - \sin^2 x}{\cos x - \sin x}$

$\qquad\qquad\quad = \dfrac{(\cos x - \sin x)(\cos x + \sin x)}{\cos x - \sin x}$

$\qquad\qquad\quad = \cos x + \sin x$

24. $\cos^2 x + \sin x + 1 = 0$

$\qquad (1 - \sin^2 x) + \sin x + 1 = 0$

$\qquad\quad -\sin^2 x + \sin x + 2 = 0$

$\qquad\quad\;\; \sin^2 x - \sin x - 2 = 0$

$\qquad\quad (\sin x - 2)(\sin x + 1) = 0$

$\quad \sin x - 2 = 0 \quad \text{or} \quad \sin x + 1 = 0$

$\qquad \sin x = 2 \qquad\qquad \sin x = -1$

$\quad \text{no solution} \quad \text{or} \qquad x = \dfrac{3\pi}{2}$

The solution in the interval $[0,\, 2\pi)$ is $\dfrac{3\pi}{2}$.

25. $4\mathbf{w} - 5\mathbf{v} = 4(-7\mathbf{i} + 3\mathbf{j}) - 5(-6\mathbf{i} + 5\mathbf{j})$

$\qquad\qquad\quad = -28\mathbf{i} + 12\mathbf{j} + 30\mathbf{i} - 25\mathbf{j}$

$\qquad\qquad\quad = 2\mathbf{i} - 13\mathbf{j}$

Section 9.1

Check Point Exercises

1. $a^2 = 36, a = 6$

$b^2 = 9, b = 3$

$c^2 = a^2 - b^2 = 36 - 9 = 27$

$c = \sqrt{27} = 3\sqrt{3}$

The foci are located at $(-3\sqrt{3}, 0)$ and $(3\sqrt{3}, 0)$.

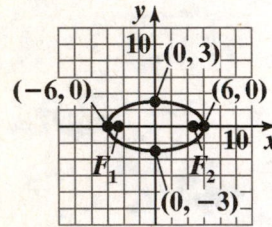

$$\frac{x^2}{36} + \frac{y^2}{9} = 1$$

2. $\dfrac{16x^2}{144} + \dfrac{9y^2}{144} = \dfrac{144}{144}$

$\dfrac{x^2}{9} + \dfrac{y^2}{16} = 1$

$a^2 = 16, a = 4$

$b^2 = 9, b = 3$

$c^2 = a^2 - b^2 = 16 - 9 = 7$

$c = \sqrt{7}$

The foci are located at $(0, -\sqrt{7})$ and $(0, \sqrt{7})$.

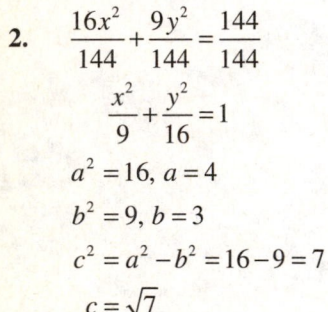

$16x^2 + 9y^2 = 144$

3. $c^2 = 4, a^2 = 9$

$b^2 = a^2 - c^2 = 9 - 4 = 5$

$\dfrac{x^2}{9} + \dfrac{y^2}{5} = 1$

4. $a^2 = 9, a = 3$

$b^2 = 4, b = 2$

center at $(-1, 2)$

$c^2 = a^2 - b^2$

$c^2 = 9 - 4$

$c^2 = 5$

$c = \sqrt{5}$

The foci are located at $(-1 - \sqrt{5}, 2)$ and $(-1 + \sqrt{5}, 2)$.

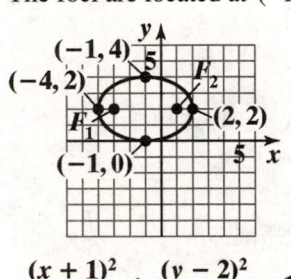

$$\frac{(x+1)^2}{9} + \frac{(y-2)^2}{4} = 1$$

5. $\dfrac{x^2}{20^2} + \dfrac{y^2}{10^2} = 1$

$\dfrac{x^2}{400} + \dfrac{y^2}{100} = 1$

Since the truck is 12 feet wide, substitute $x = 6$ into the equation to find y.

$$\frac{6^2}{400} + \frac{y^2}{100} = 1$$

$$400\left(\frac{36}{400} + \frac{y^2}{100}\right) = 400(1)$$

$$36 + 4y^2 = 400$$

$$4y^2 = 364$$

$$y^2 = 91$$

$$y = \sqrt{91}$$

$$y \approx 9.54$$

6 feet from the center, the height of the archway is 9.54 feet. Since the truck's height is 9 feet, it will fit under the archway.

Exercise Set 9.1

1. $\dfrac{x^2}{16}+\dfrac{y^2}{4}=1$

$a^2=16,\ a=4$

$b^2=4,\ b=2$

$c^2=a^2-b^2=16-4=12$

$c=\sqrt{12}=2\sqrt{3}$

The foci are located at $(-2\sqrt{3},\,0)$ and $(2\sqrt{3},\,0)$.

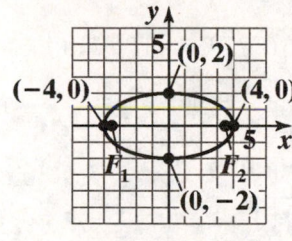

$$\dfrac{x^2}{16}+\dfrac{y^2}{4}=1$$

3. $a^2=36,\ a=6$

$b^2=9,\ b=3$

$c^2=a^2-b^2=36-9=27$

$c=\sqrt{27}=3\sqrt{3}$

The foci are located at $(0,\,-3\sqrt{3})$ and $(0,\,3\sqrt{3})$.

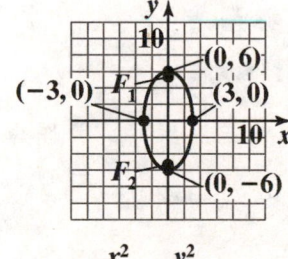

$$\dfrac{x^2}{9}+\dfrac{y^2}{36}=1$$

5. $a^2=64,\ a=8$

$b^2=25,\ b=5$

$c^2=a^2-b^2=64-25=39$

$c=\sqrt{39}$

The foci are located at $(0,\,-\sqrt{39})$ and $(0,\,\sqrt{39})$.

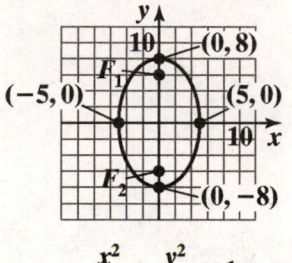

$$\dfrac{x^2}{25}+\dfrac{y^2}{64}=1$$

7. $a^2=81,\ a=9$

$b^2=49,\ b=7$

$c^2=a^2-b^2=81-49=32$

$c=\sqrt{32}=4\sqrt{2}$

The foci are located at $(0,\,-4\sqrt{2})$ and $(0,\,4\sqrt{2})$.

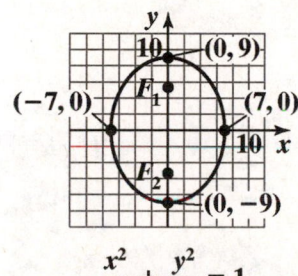

$$\dfrac{x^2}{49}+\dfrac{y^2}{81}=1$$

9. $\dfrac{x^2}{\frac{9}{4}}+\dfrac{y^2}{\frac{25}{4}}=1$

$c^2=\dfrac{25}{4}-\dfrac{9}{4}$

$c^2=\dfrac{16}{4}$

$c^2=4$

$c=2$

The foci are located at $(0,\,2)$ and $(0,\,-2)$.

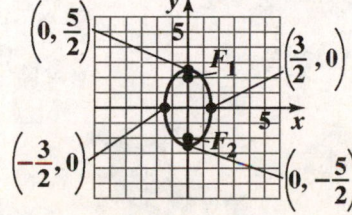

11.
$$x^2 = 1 - 4y^2$$
$$x^2 + 4y^2 = 1$$
$$x^2 + \frac{y^2}{\frac{1}{4}} = 1$$
$$c^2 = 1 - \frac{1}{4}$$
$$c^2 = \frac{3}{4}$$
$$c = \pm\frac{\sqrt{3}}{2}$$
$$c \approx \pm 0.9$$

The foci are located at $\left(\frac{\sqrt{3}}{2}, 0\right)$ and $\left(-\frac{\sqrt{3}}{2}, 0\right)$.

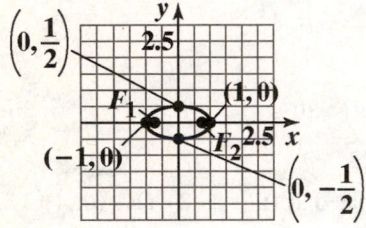

13.
$$25x^2 + 4y^2 = 100$$
$$\frac{25x^2}{100} + \frac{4y^2}{100} = \frac{100}{100}$$
$$\frac{x^2}{4} + \frac{y^2}{25} = 1$$
$$a^2 = 25, \ a = 5$$
$$b^2 = 4, \ b = 2$$
$$c^2 = a^2 = b^2 = 25 - 4 = 21$$

The foci are located at $(0, -\sqrt{21})$ and $(0, \sqrt{21})$.

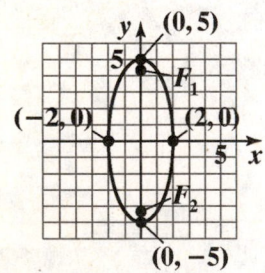

15. $4x^2 + 16y^2 = 64$

$$\frac{x^2}{16} + \frac{y^2}{4} = 1$$
$$a^2 = 16, a = 4$$
$$b^2 = 4, b = 2$$
$$c^2 = 16 - 4$$
$$c^2 = 12$$
$$c = \pm\sqrt{12}$$
$$c = \pm 2\sqrt{3}$$
$$c \approx \pm 3.5$$

The foci are located at $(2\sqrt{3}, 0)$ and $(-2\sqrt{3}, 0)$.

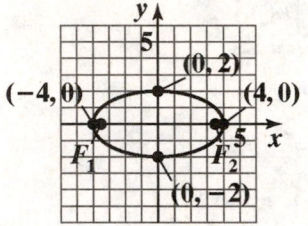

17.
$$7x^2 = 35 - 5y^2$$
$$7x^2 + 5y^2 = 35$$
$$\frac{x^2}{5} + \frac{y^2}{7} = 1$$
$$a^2 = 7, a = \sqrt{7}$$
$$b^2 = 5, b = \sqrt{5}$$
$$c^2 = 7 - 5$$
$$c^2 = 2$$
$$c = \pm\sqrt{2}$$
$$c \approx \pm 1.4$$

The foci are located at $(0, \sqrt{2})$ and $(0, -\sqrt{2})$.

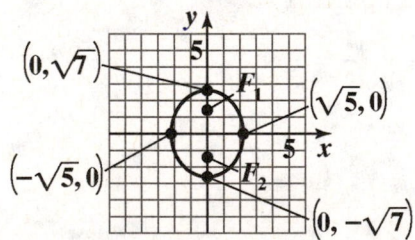

19. $a^2 = 4, b^2 = 1$, center at $(0, 0)$

$$\frac{x^2}{4} + \frac{y^2}{1} = 1$$

$$c^2 = a^2 - b^2 = 4 - 1 = 3$$

$$c = \sqrt{3}$$

The foci are at $(-\sqrt{3}, 0)$ and $(\sqrt{3}, 0)$.

21. $a^2 = 4, b^2 = 1$,

center: $(0, 0)$

$$\frac{x^2}{1} + \frac{y^2}{4} = 1$$

$$c^2 = a^2 - b^2 = 4 - 1 = 3$$

$$c = \sqrt{3}$$

The foci are at $(0, \sqrt{3})$ and $(0, -\sqrt{3})$.

23. $\dfrac{(x+1)^2}{4} + \dfrac{(y-1)^2}{1} = 1$

$$a^2 = 4, \ b^2 = 1$$

$$c^2 = 4 - 1$$

$$c^2 = 3$$

$$c = \pm\sqrt{3}$$

The foci are located at $(-1+\sqrt{3}, 1)$ and $(-1-\sqrt{3}, 1)$.

25. $c^2 = 25, a^2 = 64$

$$b^2 = a^2 - c^2 = 64 - 25 = 39$$

$$\frac{x^2}{64} + \frac{y^2}{39} = 1$$

27. $c^2 = 16, a^2 = 49$

$$b^2 = a^2 - c^2 = 49 - 16 = 33$$

$$\frac{x^2}{33} + \frac{y^2}{49} = 1$$

29. $c^2 = 4, b^2 = 9$

$$a^2 = b^2 + c^2 = 9 + 4 = 13$$

$$\frac{x^2}{13} + \frac{y^2}{9} = 1$$

31. $2a = 8, a = 4, a^2 = 16$

$$2b = 4, b = 2, b^2 = 4$$

$$\frac{x^2}{16} + \frac{y^2}{4} = 1$$

33. $2a = 10, a = 5, a^2 = 25$

$2b = 4, b = 2, b^2 = 4$

$$\frac{(x+2)^2}{4} + \frac{(y-3)^2}{25} = 1$$

35. length of the major axis $= 9 - 3 = 6$

$2a = 6, a = 3$ major axis is vertical

length of the minor axis $= 9 - 5 = 4$

$2b = 4, b = 2$

Center is at $(7, 6)$.

$$\frac{(x-7)^2}{4} + \frac{(y-6)^2}{9} = 1$$

37. $a^2 = 9, a = 3$

$b^2 = 4, b = 2$

center: $(2, 1)$

$$c^2 = a^2 - b^2 = 9 - 4 = 5$$

$$c = \sqrt{5}$$

The foci are at $(2 - \sqrt{5}, 1)$ and $(2 + \sqrt{5}, 1)$.

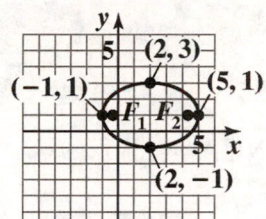

$$\frac{(x-2)^2}{9} + \frac{(y-1)^2}{4} = 1$$

39. $\dfrac{(x+3)^2}{16} + \dfrac{4(y-2)^2}{16} = \dfrac{16}{16}$

$$\frac{(x+3)^2}{16} + \frac{(y-2)^2}{4} = 1$$

$$a^2 = 16, a = 4$$

$$b^2 = 4, b = 2$$

center: $(-3, 2)$

$$c^2 = a^2 - b^2 = 16 - 4 = 12$$

$$c = \sqrt{12} = 2\sqrt{3}$$

The foci are at $(-3 - 2\sqrt{3}, 2)$ and $(-3 + 2\sqrt{3}, 2)$.

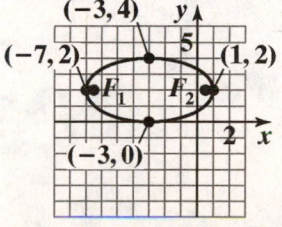

$$(x + 3)^2 + 4(y - 2)^2 = 16$$

41. $a^2 = 25, a = 5$

$b^2 = 9, b = 3$

center: $(4, -2)$

$c^2 = a^2 - b^2 = 25 - 9 = 16$

$c = 4$

The foci are at $(4, 2)$ and $(4, -6)$.

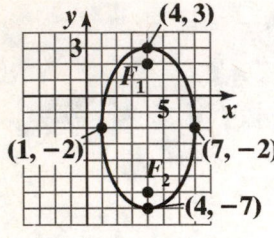

$$\frac{(x-4)^2}{9} + \frac{(y+2)^2}{25} = 1$$

43. $a^2 = 36, a = 6$

$b^2 = 25, b = 5$

center: $(0, 2)$

$c^2 = a^2 - b^2 = 36 - 25 = 11$

$c = \sqrt{11}$

The foci are at $(0, 2 + \sqrt{11})$ and $(0, 2 - \sqrt{11})$.

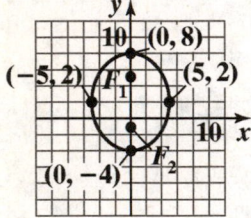

$$\frac{x^2}{25} + \frac{(y-2)^2}{36} = 1$$

45. $a^2 = 9, a = 3$

$b^2 = 1, b = 1$

center: $(-3, 2)$

$c^2 = a^2 - b^2 = 9 - 1 = 8$

$c = \sqrt{8} = 2\sqrt{2}$

The foci are at $(-3 - 2\sqrt{2}, \ 2)$ and

$(-3 + 2\sqrt{2}, \ 2)$.

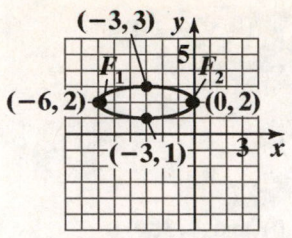

$$\frac{(x+3)^2}{9} + (y-2)^2 = 1$$

47. $c^2 = 5 - 2$

$c^2 = 3$

$c = \pm\sqrt{3}$

$c \approx \pm 1.7$

The foci are located at $(1, -3 + \sqrt{3})$ and $(1, -3 - \sqrt{3})$.

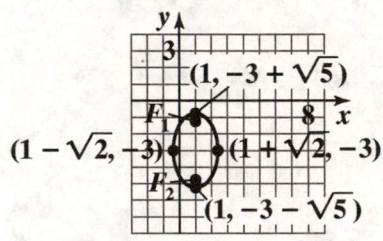

$$\frac{(x-1)^2}{2} + \frac{(y+3)^2}{5} = 1$$

49. $\dfrac{9(x-1)^2}{36} + \dfrac{4(y+3)^2}{36} = \dfrac{36}{36}$

$$\frac{(x-1)^2}{4} + \frac{(y+3)^2}{9} = 1$$

$a^2 = 9, a = 3$

$b^2 = 4, b = 2$

center: $(1, -3)$

$c^2 = a^2 - b^2 = 9 - 4 = 5$

$c = \sqrt{5}$

The foci are at $(1, -3 + \sqrt{5})$ and $(1, -3 - \sqrt{5})$.

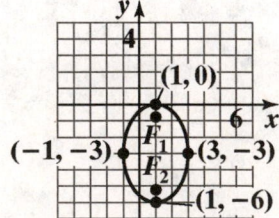

$$9(x-1)^2 + 4(y+3)^2 = 36$$

51. $(9x^2 - 36x) + (25y^2 + 50y) = 164$

$9(x^2 - 4x) + 25(y^2 + 2y) = 164$

$9(x^2 - 4x + 4) + 25(y^2 + 2y + 1)$

$\qquad = 164 + 36 + 25$

$9(x - 2)^2 + 25(y + 1)^2 = 225$

$\dfrac{9(x - 2)^2}{225} + \dfrac{25(y + 1)^2}{225} = \dfrac{225}{225}$

$\dfrac{(x - 2)^2}{25} + \dfrac{(y + 1)^2}{9} = 1$

center: (2, –1)

$a^2 = 25, a = 5$

$b^2 = 9, b = 3$

$c^2 = a^2 - b^2 = 25 - 9 = 16$

$c = 4$

The foci are at (–2, –1) and (6, –1).

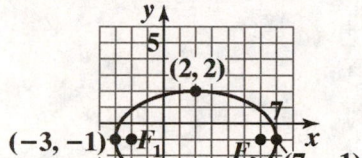

$9x^2 + 25y^2 - 36x + 50y - 164 = 0$

53. $(9x^2 - 18x) + (16y^2 + 64y) = 71$

$9(x^2 - 2x) + 16(y^2 + 4y) = 71$

$9(x^2 - 2x + 1) + 16(y^2 + 4y + 4)$

$\qquad = 71 + 9 + 64$

$9(x - 1)^2 + 16(y + 2)^2 = 144$

$\dfrac{9(x - 1)^2}{144} + \dfrac{16(y + 2)^2}{144} = \dfrac{144}{144}$

$\dfrac{(x - 1)^2}{16} + \dfrac{(y + 2)^2}{9} = 1$

center: (1, –2)

$a^2 = 16, a = 4$

$b^2 = 9, b = 3$

$c^2 = a^2 - b^2 = 16 - 9 = 7$

$c = \sqrt{7}$

The foci are at

$(1 - \sqrt{7}, -2)$ and $(1 + \sqrt{7}, -2)$.

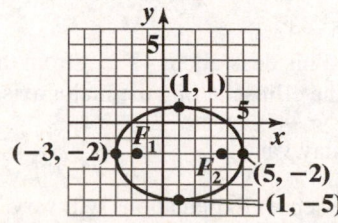

$9x^2 + 16y^2 - 18x + 64y - 71 = 0$

55. $(4x^2 + 16x) + (y^2 - 6y) = 39$

$4(x^2 + 4x) + (y^2 - 6y) = 39$

$4(x^2 + 4x + 4) + (y^2 - 6y + 9) = 39 + 16 + 9$

$4(x + 2)^2 + (y - 3)^2 = 64$

$\dfrac{4(x + 2)^2}{64} + \dfrac{(y - 3)^2}{64} = \dfrac{64}{64}$

$\dfrac{(x + 2)^2}{16} + \dfrac{(y - 3)^2}{64} = 1$

center: (–2, 3)

$a^2 = 64, a = 8$

$b^2 = 16, b = 4$

$c^2 = a^2 - b^2 = 64 - 16 = 48$

$c = \sqrt{48} = 4\sqrt{3}$

The foci are at $(-2, 3 + 4\sqrt{3})$ and

$(-2, 3 - 4\sqrt{3})$.

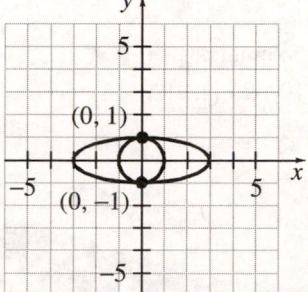

$4x^2 + y^2 + 16x - 6y - 39 = 0$

57. $x^2 + y^2 = 1 \qquad x^2 + 9y^2 = 9$

$\qquad\qquad\qquad \dfrac{x^2}{9} + \dfrac{9y^2}{9} = \dfrac{9}{9}$

$\qquad\qquad\qquad \dfrac{x^2}{9} + \dfrac{y^2}{1} = 1$

The first equation is that of a circle with center at the origin and $r = 1$. The second equation is that of an ellipse with center at the origin, horizontal major axis of length 6 units $(a = 3)$, and vertical minor axis of

length 2 units $(b = 1)$.

Check each intersection point.

The solution set is $\{(0, -1), (0, 1)\}$.

59. $\dfrac{x^2}{25} + \dfrac{y^2}{9} = 1$ $\qquad y = 3$

The first equation is for an ellipse centered at the origin with horizontal major axis of length 10 units and vertical minor axis of length 6 units. The second equation is for a horizontal line with a *y*-intercept of 3.

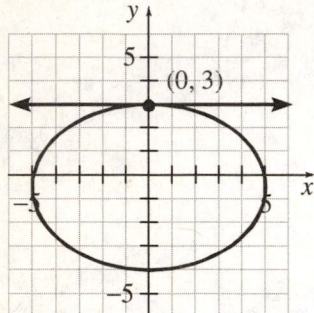

Check the intersection point.
The solution set is $\{(0,3)\}$.

61. $4x^2 + y^2 = 4$ $\qquad 2x - y = 2$

$\dfrac{4x^2}{4} + \dfrac{y^2}{4} = \dfrac{4}{4}$ $\qquad -y = -2x + 2$

$\dfrac{x^2}{1} + \dfrac{y^2}{4} = 1$ $\qquad y = 2x - 2$

The first equation is for an ellipse centered at the origin with vertical major axis of length 4 units ($b = 2$) and horizontal minor axis of length 2 units ($a = 1$). The second equation is for a line with slope 2 and *y*-intercept -2.

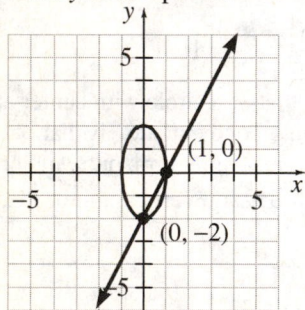

Check the intersection points.
The solution set is $\{(0,-2),(1,0)\}$.

63. $\qquad y^2 = \left(-\sqrt{16-4x^2}\right)^2$

$\qquad y^2 = 16 - 4x^2$

$\qquad 4x^2 + y^2 = 16$

$\qquad \dfrac{x^2}{4} + \dfrac{y^2}{16} = 1$

We want to graph the bottom half of an ellipse centered at the origin with a vertical major axis of length 8 units ($b = 4$) and horizontal minor axis of length 4 units ($a = 2$).

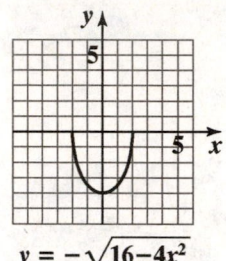

$y = -\sqrt{16-4x^2}$

65. $a = 15, b = 10$

$\dfrac{x^2}{225} + \dfrac{y^2}{100} = 1$

Let $x = 4$

$\qquad \dfrac{4^2}{225} + \dfrac{y^2}{100} = 1$

$\quad 900\left(\dfrac{16}{225} + \dfrac{y^2}{100}\right) = 900(1)$

$\qquad 64 + 9y^2 = 900$

$\qquad 9y^2 = 836$

$\qquad y = \sqrt{\dfrac{836}{9}} \approx 9.64$

Yes, the truck only needs 7 feet so it will clear.

67. a. $a = 48, a^2 = 2304$
$\qquad b = 23, b^2 = 529$

$\qquad \dfrac{x^2}{2304} + \dfrac{y^2}{529} = 1$

b. $c^2 = a^2 - b^2 = 2304 - 529 = 1775$
$\quad c = \sqrt{1775} \approx 42.13$
He situated his desk about 42 feet from the center of the ellipse, along the major axis.

69. – 77. Answers may vary.

79. does not make sense; Explanations will vary.
Sample explanation: The foci are on the major axis.

81. does not make sense; Explanations will vary. Sample explanation: We must also know the other vertices.

83. $a = 6$, $a^2 = 36$

$$\frac{x^2}{b^2} + \frac{y^2}{36} = 1$$

When $x = 2$ and $y = -4$,

$$\frac{2^2}{b^2} + \frac{(-4)^2}{36} = 1$$

$$\frac{4}{b^2} + \frac{16}{36} = 1$$

$$\frac{4}{b^2} = \frac{5}{9}$$

$$36 = 5b^2$$

$$b^2 = \frac{36}{5}$$

$$\frac{x^2}{\frac{36}{5}} + \frac{y^2}{36} = 1$$

b. The apogee is at the point $(-5000, 0)$. The left endpoint of the earth along the major axis is $(-3984, 0)$. The apogee is $|-5000 - (-3984)| = 1016$ miles above the earth's surface.

85. The large circle has radius 5 with center $(0, 0)$. Its equation is $x^2 + y^2 = 25$. The small circle has radius 3 with center $(0, 0)$. Its equation is $x^2 + y^2 = 9$.

87. $4x^2 - 9y^2 = 36$

$$\frac{4x^2}{36} - \frac{9y^2}{36} = \frac{36}{36}$$

$$\frac{x^2}{9} - \frac{y^2}{4} = 1$$

The terms are separated by subtraction rather than by addition.

88. $\dfrac{x^2}{16} - \dfrac{y^2}{9} = 1$

a. Substitute 0 for y.

$$\frac{x^2}{16} - \frac{0^2}{9} = 1$$

$$\frac{x^2}{16} = 1$$

$$x^2 = 16$$

$$x = \pm 4$$

The x-intercepts are -4 and 4.

b.
$$\frac{0^2}{16} - \frac{y^2}{9} = 1$$

$$-\frac{y^2}{9} = 1$$

$$y^2 = -9$$

The equation $y^2 = -9$ has no real solutions.

89. $\dfrac{y^2}{9} - \dfrac{x^2}{16} = 1$

a. Substitute 0 for x.

$$\frac{y^2}{9} - \frac{0^2}{16} = 1$$

$$\frac{y^2}{9} = 1$$

$$y^2 = 9$$

$$y = \pm 3$$

The y-intercepts are -3 and 3.

b.
$$\frac{0^2}{9} - \frac{x^2}{16} = 1$$

$$-\frac{x^2}{16} = 1$$

$$x^2 = -16$$

The equation $x^2 = -16$ has no real solutions.

Section 9.2

Check Point Exercises

1. a. $a^2 = 25$, $a = 5$

vertices: $(5, 0)$ and $(-5, 0)$

$b^2 = 16$

$c^2 = a^2 + b^2 = 25 + 16 = 41$

$c = \sqrt{41}$

The foci are at $(\sqrt{41},\, 0)$ and $(-\sqrt{41},\, 0)$.

b. $a^2 = 25$, $a = 5$

vertices: $(0, 5)$ and $(0, -5)$

$b^2 = 16$

$c^2 = a^2 + b^2 = 25 + 16 = 41$

$c = \sqrt{41}$

The foci are at $(0,\, \sqrt{41})$ and $(0,\, -\sqrt{41})$.

2. $a = 3$, $c = 5$

$b^2 = c^2 - a^2 = 25 - 9 = 16$

$\dfrac{y^2}{9} - \dfrac{x^2}{16} = 1$

3. $a^2 = 36$, $a = 6$

The vertices are $(6, 0)$ and $(-6, 0)$.

$b^2 = 9$, $b = 3$

asymptotes: $y = \pm \dfrac{b}{a}x = \pm \dfrac{3}{6}x = \pm \dfrac{1}{2}x$

$c^2 = a^2 + b^2 = 36 + 9 = 45$

$c = \sqrt{45} = 3\sqrt{5}$

The foci are at $(-3\sqrt{5},\, 0)$ and $(3\sqrt{5},\, 0)$.

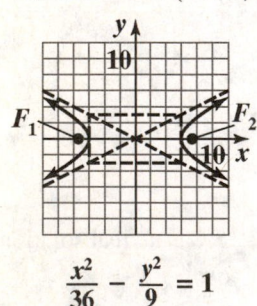

$$\frac{x^2}{36} - \frac{y^2}{9} = 1$$

4. $\dfrac{y^2}{4} - \dfrac{4x^2}{4} = \dfrac{4}{4}$

$\dfrac{y^2}{4} - x^2 = 1$

$a^2 = 4$, $a = 2$

The vertices are $(0, 2)$ and $(0, -2)$.

$b^2 = 1$, $b = 1$

asymptotes: $y = \pm \dfrac{a}{b}x = \pm 2x$

$c^2 = a^2 + b^2 = 4 + 1 = 5$

$c = \sqrt{5}$

The foci are at $(0,\, \sqrt{5})$ and $(0,\, -\sqrt{5})$.

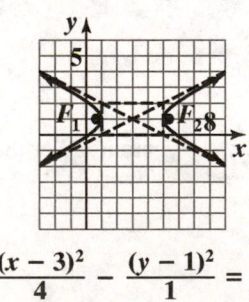

$$y^2 - 4x^2 = 4$$

5. center at $(3, 1)$

$a^2 = 4$, $a = 2$

$b^2 = 1$, $b = 1$

The vertices are $(1, 1)$ and $(5, 1)$.

asymptotes: $y - 1 = \pm \dfrac{1}{2}(x - 3)$

$c^2 = a^2 + b^2 = 4 + 1 = 5$

$c = \sqrt{5}$

The foci are at $(3 - \sqrt{5},\, 1)$ and $(3 + \sqrt{5},\, 1)$.

$$\frac{(x-3)^2}{4} - \frac{(y-1)^2}{1} = 1$$

6. $4\left(x^2 - 6x \quad\right) - 9\left(y^2 + 10y \quad\right) = 153$

$4\left(x^2 - 6x + 9\right) - 9\left(y^2 + 10y + 25\right) = 153 + 36 + (-225)$

$4(x-3)^2 - 9(y+5)^2 = -36$

$\dfrac{4(x-3)^2}{-36} - \dfrac{9(y+5)^2}{-36} = \dfrac{-36}{-36}$

$-\dfrac{(x-3)^2}{9} + \dfrac{(y+5)^2}{4} = 1$

$\dfrac{(y+5)^2}{4} - \dfrac{(x-3)^2}{9} = 1$

center at $(3, -5)$

$a^2 = 4,\, a = 2$

$b^2 = 9,\, b = 3$

The vertices are $(3, -3)$ and $(3, -7)$.

asymptotes: $y + 5 = \pm\dfrac{2}{3}(x-3)$

$c^2 = a^2 + b^2 = 4 + 9 = 13$

$c = \sqrt{13}$

The foci are at $(3, -5 - \sqrt{13})$ and $(3, -5 + \sqrt{13})$.

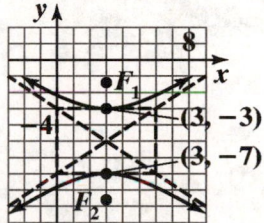

$4x^2 - 24x - 9y^2 - 90y - 153 = 0$

7. $c = 5280$

$2a = 3300,\, a = 1650$

$b^2 = c^2 - a^2 = 5280^2 - 1650^2 = 25{,}155{,}900$

The explosion occurred somewhere at the right branch of the hyperbola given by

$\dfrac{x^2}{2{,}722{,}500} - \dfrac{y^2}{25{,}155{,}900} = 1.$

Exercise Set 9.2

1. $a^2 = 4,\, a = 2$

The vertices are $(2, 0)$ and $(-2, 0)$.

$b^2 = 1$

$c^2 = a^2 + b^2 = 4 + 1 = 5$

$c = \sqrt{5}$

The foci are located at $(\sqrt{5}, 0)$ and $(-\sqrt{5}, 0)$.

graph (b)

3. $a^2 = 4,\, a = 2$

The vertices are $(0, 2)$ and $(0, -2)$.

$b^2 = 1$

$c^2 = a^2 + b^2 = 4 + 1 = 5$

$c = \sqrt{5}$

The foci are located at

$(0, \sqrt{5})$ and $(0, -\sqrt{5})$.

graph (a)

5. $a = 1,\, c = 3$

$b^2 = c^2 - a^2 = 9 - 1 = 8$

$y^2 - \dfrac{x^2}{8} = 1$

7. $a = 3,\, c = 4$

$b^2 = c^2 - a^2 = 16 - 9 = 7$

$\dfrac{x^2}{9} - \dfrac{y^2}{7} = 1$

9. $2a = 6 - (-6)$

$2a = 12$

$a = 6$

$\dfrac{a}{b} = 2$

$\dfrac{6}{b} = 2$

$6 = 2b$

$3 = b$

Transverse axis is vertical.

$\dfrac{y^2}{36} - \dfrac{x^2}{9} = 1$

11. $a = 2,\, c = 7 - 4 = 3$

$2^2 + b^2 = 3^2$

$4 + b^2 = 9$

$b^2 = 5$

Transverse axis is horizontal.

$\dfrac{(x-4)^2}{4} - \dfrac{(y+2)^2}{5} = 1$

13. $a^2 = 9, a = 3$
$b^2 = 25, b = 5$
vertices: (3, 0) and (–3, 0)

asymptotes: $y = \pm \dfrac{b}{a} x = \pm \dfrac{5}{3} x$

$c^2 = a^2 + b^2 = 9 + 25 = 34$

$c = \sqrt{34}$ on x-axis

The foci are at $(\sqrt{34}, 0)$ and $(-\sqrt{34}, 0)$.

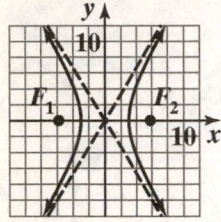

$$\dfrac{x^2}{9} - \dfrac{y^2}{25} = 1$$

15. $a^2 = 100, a = 10$
$b^2 = 64, b = 8$
vertices: (10, 0) and (–10, 0)

asymptotes: $y = \pm \dfrac{b}{a} x = \pm \dfrac{8}{10} x$

or $y = \pm \dfrac{4}{5} x$

$c^2 = a^2 + b^2 = 100 + 64 = 164$

$c = \sqrt{164} = 2\sqrt{41}$ on x-axis

The foci are at $(2\sqrt{41}, 0)$ and $(-2\sqrt{41}, 0)$.

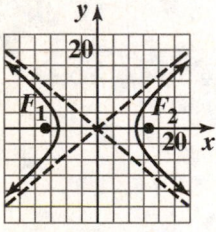

$$\dfrac{x^2}{100} - \dfrac{y^2}{64} = 1$$

17. $a^2 = 16, a = 4$
$b^2 = 36, b = 6$
vertices: (0, 4) and (0, –4)

asymptotes: $y = \pm \dfrac{a}{b} x = \pm \dfrac{4}{6} x = \pm \dfrac{2}{3} x$

or $y = \pm \dfrac{2}{3} x$

$c^2 = a^2 + b^2 = 16 + 36 = 52$

$c = \sqrt{52} = 2\sqrt{13}$ on y-axis

The foci are at $(0, 2\sqrt{13})$ and $(0, -2\sqrt{13})$.

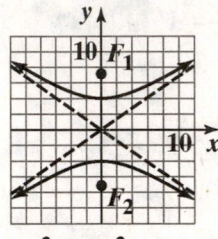

$$\dfrac{y^2}{16} - \dfrac{x^2}{36} = 1$$

19. $\dfrac{y^2}{\frac{1}{4}} - x^2 = 1$

$a^2 = \dfrac{1}{4}, a = \dfrac{1}{2}$

$b^2 = 1, b = 1$

$c^2 = a^2 + b^2$

$c^2 = \dfrac{1}{4} + 1$

$c^2 = \dfrac{5}{4}$

$c = \pm \dfrac{\sqrt{5}}{2}$

$c \approx \pm 1.1$

The foci are located at $\left(0, \dfrac{\sqrt{5}}{2}\right)$ and $\left(0, -\dfrac{\sqrt{5}}{2}\right)$.

asymptotes: $y = \pm \dfrac{\frac{1}{2}}{1} x$

$y = \pm \dfrac{1}{2} x$

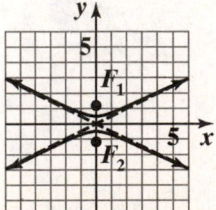

$$4y^2 - x^2 = 1$$

21. $\dfrac{9x^2}{36} - \dfrac{4y^2}{36} = \dfrac{36}{36}$

$\dfrac{x^2}{4} - \dfrac{y^2}{9} = 1$

$a^2 = 4,\ a = 2$
$b^2 = 9,\ b = 3$
vertices: $(2, 0)$ and $(-2, 0)$

asymptotes: $y = \pm\dfrac{b}{a}x = \pm\dfrac{3}{2}x$

$c^2 = a^2 + b^2 = 4 + 9 = 13$

$c = \sqrt{13}$ on x-axis

The foci are at $(\sqrt{13}, 0)$ and $(-\sqrt{13}, 0)$.

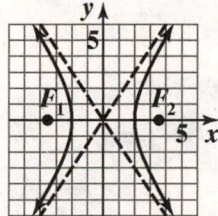

$9x^2 - 4y^2 = 36$

23. $\dfrac{9y^2}{225} - \dfrac{25x^2}{225} = \dfrac{225}{225}$

$\dfrac{y^2}{25} - \dfrac{x^2}{9} = 1$

$a^2 = 25,\ a = 5$
$b^2 = 9,\ b = 3$
vertices: $(0, 5)$ and $(0, -5)$

asymptotes: $y = \pm\dfrac{a}{b}x = \pm\dfrac{5}{3}x$

$c^2 = a^2 + b^2 = 25 + 9 = 34$

$c = \sqrt{34}$ on y-axis

The foci are at $(0, \sqrt{34})$ and $(0, -\sqrt{34})$.

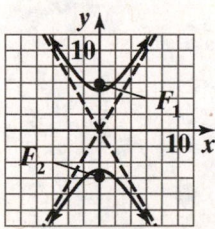

$9y^2 - 25x^2 = 225$

25. $y^2 = x^2 - 2$

$2 = x^2 - y^2$

$1 = \dfrac{x^2}{2} - \dfrac{y^2}{2}$

$a^2 = 2,\ a = \sqrt{2}$

$b^2 = 2,\ b = \sqrt{2}$

$c^2 = 2 + 2$

$c^2 = 4$

$c = 2$

The foci are located at $(2, 0)$ and $(-2, 0)$.

asymptotes: $y = \pm\dfrac{\sqrt{2}}{\sqrt{2}}x$

$y = \pm x$

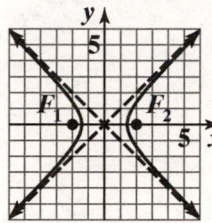

$y = \pm\sqrt{x^2 - 2}$

27. $a = 3,\ b = 5$

$\dfrac{x^2}{9} - \dfrac{y^2}{25} = 1$

29. $a = 2,\ b = 3$

$\dfrac{y^2}{4} - \dfrac{x^2}{9} = 1$

31. Center $(2, -3)$, $a = 2$, $b = 3$

$\dfrac{(x-2)^2}{4} - \dfrac{(y+3)^2}{9} = 1$

33. center: $(-4, -3)$

$a^2 = 9, a = 3$

$b^2 = 16, b = 4$

vertices: $(-7, -3)$ and $(-1, -3)$

asymptotes: $y + 3 = \pm\dfrac{4}{3}(x+4)$

$c^2 = a^2 + b^2 = 9 + 16 = 25$

$c = \pm 5$ parallel to x-axis

The foci are at $(-9, -3)$ and $(1, -3)$.

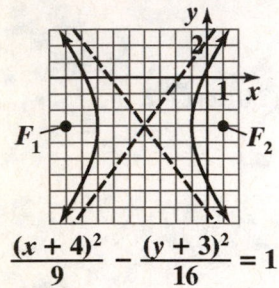

$$\dfrac{(x+4)^2}{9} - \dfrac{(y+3)^2}{16} = 1$$

35. center: $(-3, 0)$

$a^2 = 25, a = 5$

$b^2 = 16, b = 4$

vertices: $(2, 0)$ and $(-8, 0)$

asymptotes: $y = \pm\dfrac{4}{5}(x+3)$

$c^2 = a^2 + b^2 = 25 + 16 = 41$

$c = \sqrt{41}$

The foci are at $(-3+\sqrt{41}, 0)$ and $(-3-\sqrt{41}, 0)$.

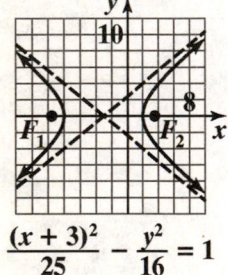

$$\dfrac{(x+3)^2}{25} - \dfrac{y^2}{16} = 1$$

37. center: $(1, -2)$

$a^2 = 4, a = 2$

$b^2 = 16, b = 4$

vertices: $(1, 0)$ and $(1, -4)$

asymptotes: $y + 2 = \pm\dfrac{1}{2}(x-1)$

$c^2 = a^2 + b^2 = 4 + 16 = 20$

$c = \sqrt{20} = 2\sqrt{5}$ parallel to y-axis

The foci are at $(1, -2+2\sqrt{5})$ and $(1, -2-2\sqrt{5})$.

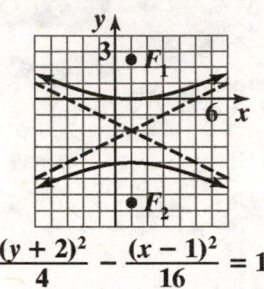

$$\dfrac{(y+2)^2}{4} - \dfrac{(x-1)^2}{16} = 1$$

39. $\dfrac{(x-3)^2}{4} - \dfrac{4(y+3)^2}{4} = \dfrac{4}{4}$

$\dfrac{(x-3)^2}{4} - (y+3)^2 = 1$

center: $(3, -3)$

$a^2 = 4, a = 2$

$b^2 = 1, b = 1$

vertices: $(1, -3)$ and $(5, -3)$

asymptotes: $y + 3 = \pm\dfrac{1}{2}(x-3)$

$c^2 = a^2 + b^2 = 4 + 1 = 5$

$c = \sqrt{5}$

The foci are at $(3+\sqrt{5}, -3)$ and $(3-\sqrt{5}, -3)$.

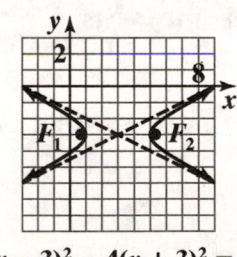

$$(x-3)^2 - 4(y+3)^2 = 4$$

41. $\dfrac{(x-1)^2}{3} - \dfrac{(y-2)^2}{3} = 1$

center: (1, 2)

$a^2 = 3, a = \sqrt{3}$

$b^2 = \;, b = \sqrt{3}$

vertices: (−1, 2) and (3, 2)

asymptotes: $y - 2 = \pm(x-1)$

$c^2 = a^2 + b^2 = 3 + 3 = 6$

$c = \sqrt{6}$ parallel to y-axis

The foci are at $(1+\sqrt{6}, 2)$ and $(1-\sqrt{6}, 2)$.

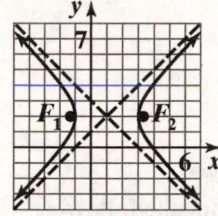

$(x-1)^2 - (y-2)^2 = 3$

43. $(x^2 - 2x) - (y^2 + 4y) = 4$

$(x^2 - 2x + 1) - (y^2 + 4y + 4) = 4 + 1 - 4$

$(x-1)^2 - (y+2)^2 = 1$

center: (1, −2)

$a^2 = 1, a = 1$

$b^2 = 1, b = 1$

$c^2 = a^2 + b^2 = 1 + 1 = 2$

$c = \sqrt{2}$

asymptotes: $y + 2 = \pm(x-1)$

The foci are at $(1+\sqrt{2}, -2)$ and $(1-\sqrt{2}, -2)$.

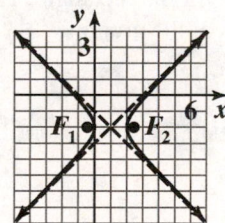

$x^2 - y^2 - 2x - 4y - 4 = 0$

45. $(16x^2 + 64x) - (y^2 + 2y) = -67$

$16(x^2 + 4x + 4) - (y^2 + 2y + 1) = -67 + 64 - 1$

$16(x+2)^2 - (y+1)^2 = -4$

$\dfrac{16(x+2)^2}{-4} - \dfrac{(y+1)^2}{-4} = \dfrac{-4}{-4}$

$\dfrac{(y+1)^2}{4} - \dfrac{(x+2)^2}{\frac{1}{4}} = 1$

center: (−2, −1)

$a^2 = 4, a = 2$

$b^2 = \dfrac{1}{4}, b = \dfrac{1}{2}$

$c^2 = a^2 + b^2 = 4 + \dfrac{1}{4} = \dfrac{17}{4}$

$c = \sqrt{\dfrac{17}{4}} = \sqrt{4.25}$

$$(y+1) = \pm \dfrac{2}{\frac{1}{2}}(x+2)$$

asymptotes:

$$y + 1 = \pm 4(x+2)$$

The foci are at $\left(-2, -1+\sqrt{4.25}\right)$ and $\left(-2, -1-\sqrt{4.25}\right)$.

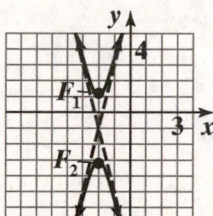

$16x^2 - y^2 + 64x - 2y + 67 = 0$

47. $(4x^2 - 16x) - (9y^2 - 54y) = 101$

$4(x^2 - 4x + 4) - 9(y^2 - 6y + 9) = 101 + 16 - 81$

$4(x-2)^2 - 9(y-3)^2 = 36$

$\dfrac{(x-2)^2}{9} - \dfrac{(y-3)^2}{4} = 1$

center: (2, 3)

$a^2 = 9, a = 3$

$b^2 = 4, b = 2$

$c^2 = a^2 + b^2 = 9 + 4 = 13$

$c = \sqrt{13}$

asymptotes: $y - 3 = \pm\dfrac{2}{3}(x-2)$

The foci are at $(2+\sqrt{13},\, 3)$ and $(2-\sqrt{13},\, 3)$.

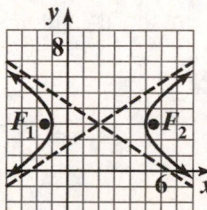

$4x^2 - 9y^2 - 16x + 5y - 101 = 0$

49.
$$(4x^2 - 32x) - 25y^2 = -164$$
$$4(x^2 - 8x + 16) - 25y^2 = -164 + 64$$
$$4(x - 4)^2 - 25y^2 = -100$$
$$\frac{4(x-4)^2}{-100} - \frac{25y^2}{-100} = \frac{-100}{-100}$$
$$\frac{y^2}{4} - \frac{(x-4)^2}{25} = 1$$
center: (4, 0)
$a^2 = 4, a = 2$
$b^2 = 25, b = 5$
$c^2 = a^2 + b^2 = 4 + 25 = 29$
$c = \sqrt{29}$

asymptotes: $y = \pm\dfrac{2}{5}(x-4)$

The foci are at $(4, \sqrt{29})$ and $(4, -\sqrt{29})$.

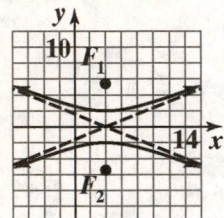

$4x^2 - 25y^2 - 32x + 164 = 0$

51.
$$\frac{x^2}{9} - \frac{y^2}{16} = 1$$
The equation is for a hyperbola in standard form with the transverse axis on the *x*-axis. We have $a^2 = 9$ and $b^2 = 16$, so $a = 3$ and $b = 4$.
Therefore, the vertices are at $(\pm a, 0)$ or $(\pm 3, 0)$.

Using a dashed line, we construct a rectangle using the ±3 on the *x*-axis and ±4 on the *y*-axis. Then use dashed lines to draw extended diagonals for the rectangle. These represent the asymptotes of the graph.

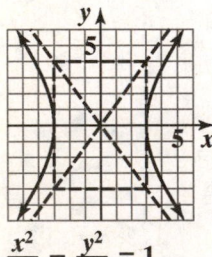

$\dfrac{x^2}{9} - \dfrac{y^2}{16} = 1$

From the graph we determine the following:
Domain: $\{x \mid x \le -3 \text{ or } x \ge 3\}$ or $(-\infty, -3] \cup [3, \infty)$

Range: $\{y \mid y \text{ is a real number}\}$ or $(-\infty, \infty)$

53.
$$\frac{x^2}{9} + \frac{y^2}{16} = 1$$
The equation is for an ellipse in standard form with major axis along the y-axis. We have $a^2 = 16$ and $b^2 = 9$, so $a = 4$ and $b = 3$. Therefore, the vertices are $(0, \pm a)$ or $(0, \pm 4)$. The endpoints of the minor axis are $(\pm b, 0)$ or $(\pm 3, 0)$.

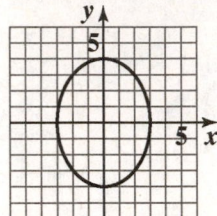

$\dfrac{x^2}{9} + \dfrac{y^2}{16} = 1$

From the graph we determine the following:
Domain: $\{x \mid -3 \le x \le 3\}$ or $[-3, 3]$

Range: $\{y \mid -4 \le y \le 4\}$ or $[-4, 4]$.

55.
$$\frac{y^2}{16} - \frac{x^2}{9} = 1$$
The equation is in standard form with the transverse axis on the y-axis. We have $a^2 = 16$ and $b^2 = 9$, so $a = 4$ and $b = 3$. Therefore, the vertices are at $(0, \pm a)$ or $(0, \pm 4)$. Using a dashed line, we construct a rectangle using the ±4 on the y-axis and ±3 on the x-axis. Then use dashed lines to draw extended diagonals for the rectangle. These represent the asymptotes of the graph.

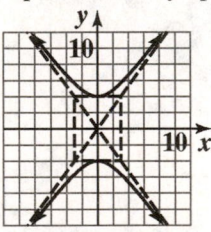

$\dfrac{y^2}{16} - \dfrac{x^2}{9} = 1$

From the graph we determine the following:
Domain: $\{x \mid x \text{ is a real number}\}$ or $(-\infty, \infty)$

Range: $\{y \mid y \le -4 \text{ or } y \ge 4\}$ or $(-\infty, -4] \cup [4, \infty)$

57. $x^2 - y^2 = 4$

$x^2 + y^2 = 4$

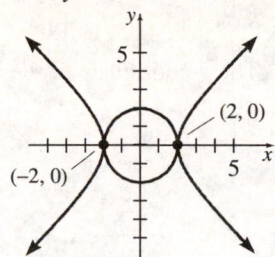

Check $(-2, 0)$:

$(-2)^2 - 0^2 = 4 \qquad (-2)^2 + 0^2 = 4$

$\quad 4 - 0 = 4 \qquad\qquad 4 + 0 = 4$

$\qquad 4 = 4 \ \text{true} \qquad\qquad 4 = 4 \ \text{true}$

Check $(2, 0)$:

$(2)^2 - 0^2 = 4 \qquad (2)^2 + 0^2 = 4$

$\quad 4 - 0 = 4 \qquad\qquad 4 + 0 = 4$

$\qquad 4 = 4 \ \text{true} \qquad\qquad 4 = 4 \ \text{true}$

The solution set is $\{(-2, 0), (2, 0)\}$.

59.

$9x^2 + y^2 = 9 \quad \text{or} \quad \dfrac{x^2}{1} + \dfrac{y^2}{9} = 1$

$y^2 - 9x^2 = 9 \qquad\qquad \dfrac{y^2}{9} - \dfrac{x^2}{1} = 1$

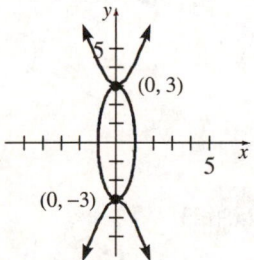

Check $(0, -3)$:

$9(0)^2 + (-3)^2 = 9 \qquad (-3)^2 - 9(0)^2 = 9$

$\quad 0 + 9 = 9 \qquad\qquad\quad 9 - 0 = 9$

$\qquad 9 = 9 \qquad\qquad\qquad\quad 9 = 9$

$\qquad\quad \text{true} \qquad\qquad\qquad\quad \text{true}$

Check $(0, 3)$:

$9(0)^2 + (3)^2 = 9 \qquad (3)^2 - 9(0)^2 = 9$

$\quad 0 + 9 = 9 \qquad\qquad\quad 9 - 0 = 9$

$\qquad 9 = 9 \qquad\qquad\qquad\quad 9 = 9$

$\qquad\quad \text{true} \qquad\qquad\qquad\quad \text{true}$

The solution set is $\{(0, -3), (0, 3)\}$.

61. $|d_2 - d_1| = 2a = (2 \text{ s})(1100 \text{ ft / s}) = 2200 \text{ ft}$

$a = 1100 \text{ ft}$

$2c = 5280 \text{ ft}, c = 2640 \text{ ft}$

$b^2 = c^2 - a^2 = (2640)^2 - (1100)^2$

$\quad = 5,759,600$

$\dfrac{x^2}{(1100)^2} - \dfrac{y^2}{5,759,600} = 1$

$\dfrac{x^2}{1,210,000} - \dfrac{y^2}{5,759,600} = 1$

If M_1 is located 2640 feet to the right of the origin on the x-axis, the explosion is located on the right branch of the hyperbola given by the equation above.

63. $\qquad 625y^2 - 400x^2 = 250,000$

$\dfrac{625y^2}{250,000} - \dfrac{400x^2}{250,000} = \dfrac{250,000}{250,000}$

$\dfrac{y^2}{400} - \dfrac{x^2}{625} = 1$

$a^2 = 400, a = \sqrt{400} = 20$

$2a = 40$

The houses are 40 yards apart at their closest point.

65. **a.** ellipse

b. $x^2 + 4y^2 = 4$

67. – 75. Answers may vary.

77. $\dfrac{x^2}{4} - \dfrac{y^2}{9} = 0$

$y^2 = \dfrac{9}{4}x^2$

$y = \pm\dfrac{3}{2}x$

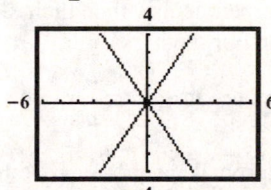

No; in general, the graph is two intersecting lines.

79. $4x^2 - 6xy + 2y^2 - 3x + 10y - 6 = 0$

$2y^2 + (10 - 6x)y + (4x^2 - 3x - 6) = 0$

$y = \dfrac{6x - 10 \pm \sqrt{(10 - 6x)^2 - 8(4x^2 - 3x - 6)}}{4}$

$y = \dfrac{6x - 10 \pm \sqrt{4(x^2 - 24x + 37)}}{4}$

$y = \dfrac{3x - 5 \pm \sqrt{x^2 - 24x + 37}}{2}$

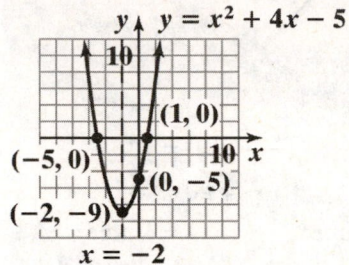

The *xy*-term rotates the hyperbola. Separation of terms into ones containing only *x* or only *y* would not be possible.

81. does not make sense; Explanations will vary. Sample explanation: This would change the ellipse to a hyperbola.

83. makes sense

85. false; Changes to make the statement true will vary. A sample change is: If a hyperbola has a transverse axis along the *x*–axis and one of the branches is removed, the remaining branch does not define a function of *x*.

87. true

89. $\dfrac{c}{a}$ will be large when *a* is small. When this happens, the asymptotes will be nearly vertical.

91. If the asymptotes are perpendicular, then their slopes are negative reciprocals. For the hyperbola $\dfrac{x^2}{a^2} - \dfrac{y^2}{b^2} = 1$, the asymptotes are $y = \pm \dfrac{b}{a}x$. The slopes are negative reciprocals when $\dfrac{b}{a} = \dfrac{a}{b}$ (since one is already the negative of the other). This happens when $b^2 = a^2$, so $a = b$. Any hyperbola where $a = b$, such as $\dfrac{x^2}{4} - \dfrac{y^2}{4} = 1$, has perpendicular asymptotes.

92. $y = x^2 + 4x - 5$

Since $a = 1$ is positive, the parabola opens upward. The *x*-coordinate of the vertex is

$x = -\dfrac{b}{2a} = -\dfrac{4}{2(1)} = -2$. The *y*-coordinate of the

vertex is $y = (-2)^2 + 4(-2) - 5 = -9$.
Vertex: $(-2, -9)$.

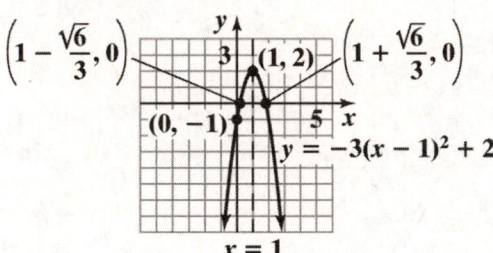

93. $y = -3(x - 1)^2 + 2$

Since $a = -3$ is negative, the parabola opens downward. The vertex of the parabola is $(h, k) = (1, 2)$.

The *y*–intercept is -1.

$\left(1 - \dfrac{\sqrt{6}}{3}, 0\right)$ $\left(1 + \dfrac{\sqrt{6}}{3}, 0\right)$

$(0, -1)$

$y = -3(x - 1)^2 + 2$

$x = 1$

94. $y^2 + 2y + 12x - 23 = 0$

$y^2 + 2y = -12x + 23$

$y^2 + 2y + 1 = -12x + 23 + 1$

$(y + 1)^2 = -12x + 24$

Section 9.3

Check Point Exercises

1. $4p = 8$, $p = 2$
focus: $(2, 0)$
directrix: $x = -2$

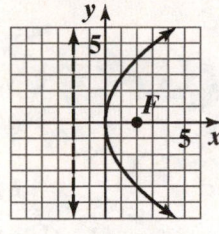

$$y^2 = 8x$$

2. $x^2 = -12y$
$4p = -12$, $p = 3$
focus: $(0, -3)$
directrix: $y = 3$

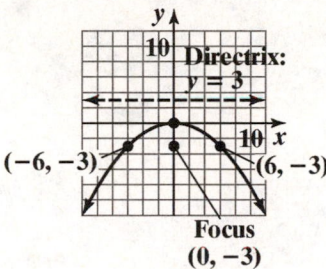

3. $p = 8$
$y^2 = 4 \cdot 8x$
$y^2 = 32x$

4. $4p = 4$, $p = 1$
vertex: $(2, -1)$
focus: $(2, 0)$
directrix: $y = -2$

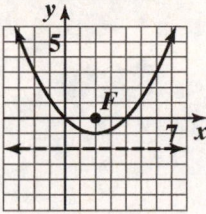

$$(x - 2)^2 = 4(y + 1)$$

5. $y^2 + 2y = -4x + 7$
$y^2 + 2y + 1 = -4x + 7 + 1$
$(y + 1)^2 = -4(x - 2)$
$4p = -4$, $p = -1$
vertex: $(2, -1)$
focus: $(1, -1)$
directrix: $x = 3$

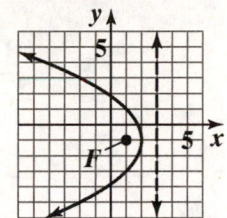

$$y^2 + 2y + 4x - 7 = 0$$

6. $x^2 = 4py$
Let $x = 3$ and $y = 4$.
$3^2 = 4p \cdot 4$
$9 = 16p$
$$p = \frac{9}{16}$$
$$x^2 = \frac{9}{4}y$$

The light should be placed at $\left(0, \dfrac{9}{16}\right)$ or $\dfrac{9}{16}$ inch
above the vertex.

Exercise Set 9.3

1. $y^2 = 4x$
$4p = 4,\ p = 1$
vertex: $(0, 0)$
focus: $(1, 0)$
directrix: $x = -1$
graph (c)

3. $x^2 = -4y$
$4p = -4,\ p = -1$
vertex: $(0, 0)$
focus: $(0, -1)$
directrix: $y = 1$
graph (b)

5. $4p = 16,\ p = 4$
vertex: $(0, 0)$
focus: $(4, 0)$
directrix: $x = -4$

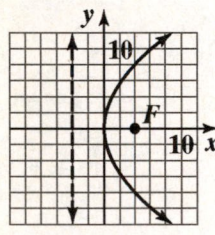

$y^2 = 16x$

7. $4p = -8,\ p = -2$
vertex: $(0, 0)$
focus: $(-2, 0)$
directrix: $x = 2$

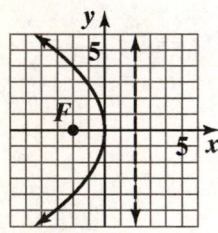

$y^2 = -8x$

9. $4p = 12,\ p = 3$
vertex: $(0, 0)$
focus: $(0, 3)$
directrix: $y = -3$

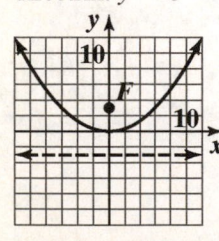

$x^2 = 12y$

11. $4p = -16,\ p = -4$
vertex: $(0, 0)$
focus: $(0, -4)$
directrix: $y = 4$

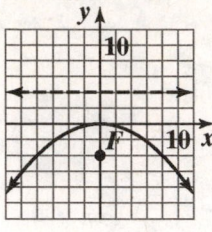

$x^2 = -16y$

13. $y^2 = 6x$
$4p = 6,\ p = \dfrac{6}{4} = \dfrac{3}{2}$
vertex: $(0, 0)$
focus: $\left(\dfrac{3}{2}, 0\right)$
directrix: $x = -\dfrac{3}{2}$

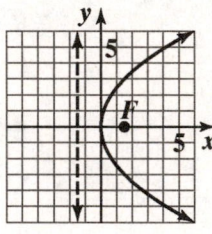

$y^2 - 6x = 0$

15. $8x^2 = -4y$
$x^2 = -\dfrac{1}{2}y$
$4p = -\dfrac{1}{2}$
$p = -\dfrac{1}{8}$
focus: $\left(0, -\dfrac{1}{8}\right)$
directrix: $y = \dfrac{1}{8}$

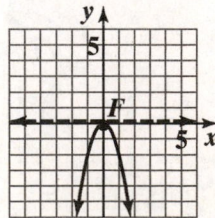

$8x^2 + 4y = 0$

17. $p = 7, 4p = 28$
$y^2 = 28x$

19. $p = -5, 4p = -20$
$y^2 = -20x$

21. $p = 15, 4p = 60$
$x^2 = 60y$

23. $p = -25, 4p = -100$
$x^2 = -100y$

25. $p = -5 - (-3) = -2$ Vertex, $(2, -3)$
$(x-2)^2 = -8(y+3)$

27. vertex: $(1, 2)$ $p = 2$
$(y-2)^2 = 8(x-1)$

29. vertex: $(-3, 3), p = 1$
$(x+3)^2 = 4(y-3)$

31. $(y-1)^2 = 4(x-1)$
$4p = 4, p = 1$
vertex: $(1, 1)$
focus: $(2, 1)$
directrix: $x = 0$
graph (c)

33. $(x+1)^2 = -4(y+1)$
$4p = -4, p = -1$
vertex: $(-1, -1)$
focus: $(-1, -2)$
directrix: $y = 0$
graph (d)

35. $4p = 8, p = 2$
vertex: $(2, 1)$
focus: $(2, 3)$
directrix: $y = -1$

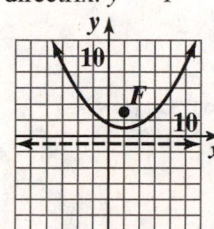

$(x - 2)^2 = 8(y - 1)$

37. $4p = -8, p = -2$
vertex: $(-1, -1)$
focus: $(-1, -3)$
directrix: $y = 1$

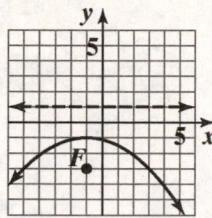

$(x + 1)^2 = -8(y + 1)$

39. $4p = 12, p = 3$
vertex: $(-1, -3)$
focus: $(2, -3)$
directrix: $x = -4$

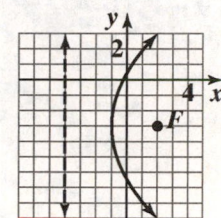

$(y + 3)^2 = 12(x + 1)$

41. $(y + 1)^2 = -8(x - 0)$
$4p = -8, p = -2$
vertex: $(0, -1)$
focus: $(-2, -1)$
directrix: $x = 2$

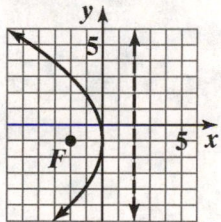

$(y + 1)^2 = -8x$

43.
$$x^2 - 2x + 1 = 4y - 9 + 1$$
$$(x-1)^2 = 4y - 8$$
$$(x-1)^2 = 4(y-2)$$
$4p = 4, p = 1$
vertex: (1, 2)
focus: (1, 3)
directrix: $y = 1$

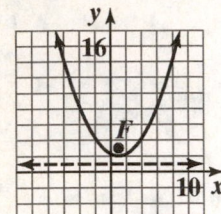

$x^2 - 2x - 4y + 9 = 0$

45.
$$y^2 - 2y + 1 = -12x + 35 + 1$$
$$(y-1)^2 = -12x + 36$$
$$(y-1)^2 = -12(x-3)$$
$4p = -12, p = -3$
vertex: (3, 1)
focus: (0, 1)
directrix: $x = 6$

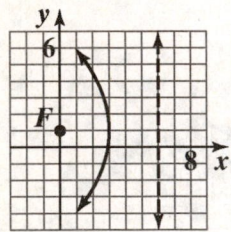

$y^2 - 2y + 12x - 35 = 0$

47.
$$x^2 + 6x = 4y - 1$$
$$x^2 + 6x + 9 = 4y - 1 + 9$$
$$(x+3)^2 = 4(y+2)$$
$4p = 4, p = 1$
vertex: (−3, −2)
focus: (−3, −1)
directrix: $y = -3$

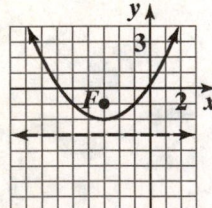

$x^2 + 6x - 4y + 1 = 0$

49. The y-coordinate of the vertex is
$$y = -\frac{b}{2a} = -\frac{6}{2(1)} = -3$$
The x-coordinate of the vertex is
$$x = (-3)^2 + 6(-3) + 5$$
$$= 9 - 18 + 5$$
$$= -4$$
The vertex is $(-4, -3)$.

Since the squared term is y and $a > 0$, the graph opens to the right.

Domain: $\{x \mid x \geq -4\}$ or $[-4, \infty)$

Range: $\{y \mid y$ is a real number$\}$ or $(-\infty, \infty)$

The relation is not a function.

51. The x-coordinate of the vertex is
$$x = -\frac{b}{2a} = -\frac{(4)}{2(-1)} = 2$$
The y-coordinate of the vertex is
$$y = -(2)^2 + 4(2) - 3$$
$$= -4 + 8 - 3$$
$$= 1$$
The vertex is $(2, 1)$.

Since the squared term is x and $a < 0$, the graph opens down.

Domain: $\{x \mid x$ is a real number$\}$ or $(-\infty, \infty)$

Range: $\{y \mid y \leq 1\}$ or $(-\infty, 1]$

The relation is a function.

53. The equation is in the form $x = a(y-k)^2 + h$

From the equation, we can see that the vertex is $(3, 1)$.

Since the squared term is y and $a < 0$, the graph opens to the left.

Domain: $\{x \mid x \leq 3\}$ or $(-\infty, 3]$

Range: $\{y \mid y$ is a real number$\}$ or $(-\infty, \infty)$

The relation is not a function.

55.

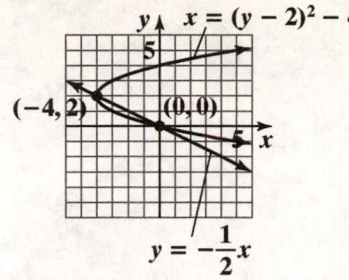

Check $(-4,2)$:

$-4 = (2-2)^2 - 4 \qquad 2 = -\dfrac{1}{2}(-4)$

$-4 = 0 - 4 \qquad\qquad 2 = 2$

$-4 = -4 \qquad\qquad\qquad$ true

$\qquad$ true

Check $(0,0)$:

$0 = (0-2)^2 - 4 \qquad 0 = -\dfrac{1}{2}(0)$

$0 = 4 - 4 \qquad\qquad 0 = 0$

$0 = 0 \qquad\qquad\qquad$ true

$\quad$ true

The solution set is $\{(-4,2),(0,0)\}$.

57.

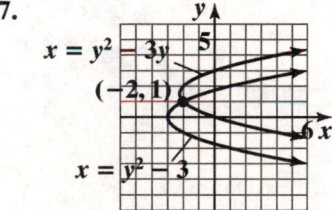

Check $(-2,1)$:

$-2 = (1)^2 - 3 \qquad -2 = (1)^2 - 3(1)$

$-2 = 1 - 3 \qquad\quad -2 = 1 - 3$

$-2 = -2$ true $\qquad -2 = -2$ true

The solution set is $\{(-2,1)\}$.

59.

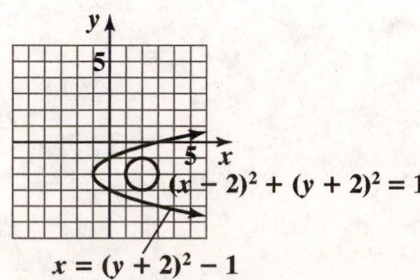

The two graphs do not cross. Therefore, the solution set is the empty set, $\{\ \ \}$ or $\varnothing$.

61. $x^2 = 4py$

$2^2 = 4p(1)$

$4 = 4$

$p = 1$

The light bulb should be placed 1 inch above the vertex.

63. $x^2 = 4py$

$6^2 = 4p(2)$

$36 = 8p$

$p = \dfrac{36}{8} = \dfrac{9}{2} = 4.5$

The receiver should be located 4.5 feet from the base of the dish.

65. $\qquad x^2 = 4py$

$(640)^2 = 4p(160)$

$p = \dfrac{(640)^2}{640} = 640$

$x = 640 - 200 = 440$

$(440)^2 = 4(640)y$

$y = \dfrac{(440)^2}{4(640)} = 75.625$

The height is 76 meters.

67. $\qquad x^2 = 4py$

$\left(\dfrac{200}{2}\right)^2 = 4p(-50)$

$\dfrac{10,000}{-50} = 4p$

$4p = -200$

$x^2 = -200y$

$(30)^2 = -200y$

$y = \dfrac{900}{-200} = -4.5$

(height of bridge) $= 50 - 4.5 = 45.5$ feet.

Yes, the boat will clear the arch.

69. – 75. Answers may vary.

77. $y^2 + 2y - 6x + 13 = 0$

$y^2 + 2y + (-6x + 13) = 0$

$y = \dfrac{-2 \pm \sqrt{2^2 - 4(-6x + 13)}}{2}$

$y = \dfrac{-2 \pm \sqrt{24x - 48}}{2}$

$y = -1 \pm \sqrt{6x - 12}$

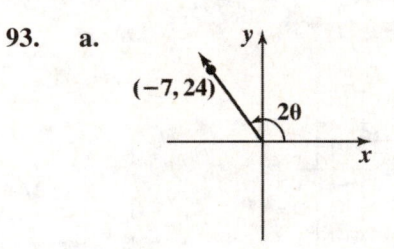

79. $16x^2 - 24xy + 9y^2 - 60x - 80y + 100 = 0$

$9y^2 - (24x + 80)y + (16x^2 - 60x + 100) = 0$

$y = \dfrac{24x + 80 \pm \sqrt{(24x + 80)^2 - 36(16x^2 - 60x + 100)}}{18}$

$y = \dfrac{24x + 80 \pm \sqrt{6000x + 2800}}{18}$

$y = \dfrac{24x + 80 \pm 20\sqrt{15x + 7}}{18}$

$y = \dfrac{12x + 40 \pm 10\sqrt{15x + 7}}{9}$

81. does not make sense; Explanations will vary. Sample explanation: Horizontal parabolas will rise without limit.

83. makes sense

85. false; Changes to make the statement true will vary. A sample change is: Because $a = -1$, the parabola will open to the left.

87. false; Changes to make the statement true will vary. A sample change is: If a parabola defines y as a function of x, it will open up or down.

89. $Ax^2 + Ey = 0$

$Ax^2 = -Ey \qquad\qquad 4p = -\dfrac{E}{A}y$

$x^2 = -\dfrac{E}{A}y \qquad\qquad p = -\dfrac{E}{4A}y$

focus: $\left(0, -\dfrac{E}{4A}\right),$

directrix: $y = \dfrac{E}{4A}$

91. Answers may vary.

92. $\left[\dfrac{\sqrt{2}}{2}(x' - y')\right]\left[\dfrac{\sqrt{2}}{2}(x' + y')\right] = 1$

$\dfrac{\sqrt{2}}{2} \dfrac{\sqrt{2}}{2}(x' - y')(x' + y') = 1$

$\dfrac{2}{4}\left((x')^2 - (y')^2\right) = 1$

$\dfrac{x'^2}{2} - \dfrac{y'^2}{2} = 1$

$x'^2 - y'^2 = 2$

93. **a.**

b. $\cos 2\theta = -\dfrac{7}{25}$

c. $\sin\theta = \sqrt{\dfrac{1-\cos 2\theta}{2}}$

$$\sin\theta = \sqrt{\dfrac{1-\left(-\dfrac{7}{25}\right)}{2}}$$

$$\sin\theta = \sqrt{\dfrac{16}{25}}$$

$$\sin\theta = \dfrac{4}{5}$$

$$\cos\theta = \sqrt{\dfrac{1+\cos 2\theta}{2}}$$

$$\cos\theta = \sqrt{\dfrac{1+\left(-\dfrac{7}{25}\right)}{2}}$$

$$\cos\theta = \sqrt{\dfrac{9}{25}}$$

$$\cos\theta = \dfrac{3}{5}$$

d.   Since $90° < 2\theta < 180°$, we have $45° < \theta < 90°$.
Both $\sin\theta$ and $\cos\theta$ are positive when
$45° < \theta < 90°$.

94. $B^2 - 4AC = (-2\sqrt{3})^2 - 4(3)(1)$
$$= 12 - 12$$
$$= 0$$

Mid-Chapter 9 Check Point

1.   Center: $(0,0)$

Because the denominator of the x^2 – term is
greater than the denominator of the y^2 – term,
the major axis is horizontal. Since $a^2 = 25$,
$a = 5$ and the vertices are $(-5,0)$ and $(5,0)$.

Since $b^2 = 4$, $b = 2$ and endpoints of the minor
axis are $(0,-2)$ and $(0,2)$.

Foci: $\left(\pm\sqrt{21}, 0\right)$

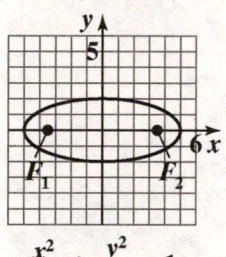

$$\frac{x^2}{25} + \frac{y^2}{4} = 1$$

2.   Divide both sides by 36 to get the standard form:
$$\frac{x^2}{4} + \frac{y^2}{9} = 1$$

Center: $(0,0)$

Because the denominator of the y^2 – term is
greater than the denominator of the x^2 – term,
the major axis is vertical. Since $a^2 = 9$, $a = 3$
and the vertices are $(0,-3)$ and $(0,3)$. Since
$b^2 = 4$, $b = 2$ and endpoints of the minor axis
are $(-2,0)$ and $(2,0)$.

Foci: $\left(0, \pm\sqrt{5}\right)$

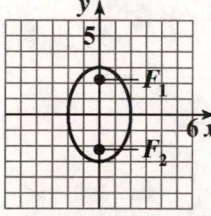

$$9x^2 + 4y^2 = 36$$

3. Center: $(2,-1)$

Because the denominator of the y^2-term is greater than the denominator of the x^2-term, the major axis is vertical. We have $a^2=25$ and $b^2=16$, so $a=5$ and $b=4$. The vertices lie 5 units above and below the center. The endpoints of the minor axis lie 4 units to the left and right of the center.

Vertices: $(2,4)$ and $(2,-6)$

Minor endpoints: $(-2,-1)$ and $(6,-1)$

Foci: $(2,2)$, $(2,-4)$

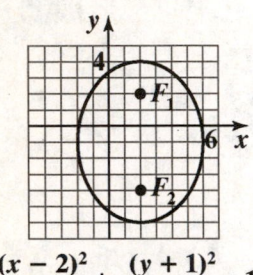

$$\frac{(x-2)^2}{16}+\frac{(y+1)^2}{25}=1$$

4. Center: $(-2,1)$

Because the denominator of the x^2-term is greater than the denominator of the y^2-term, the major axis is horizontal. We have $a^2=25$ and $b^2=16$, so $a=5$ and $b=4$. The vertices lie 5 units to the left and right of the center. The endpoints of the minor axis lie 4 units above and below the center.

Vertices: $(-7,1)$ and $(3,1)$

Minor endpoints: $(-2,5)$ and $(-2,-3)$

Foci: $(-5,1)$, $(1,1)$

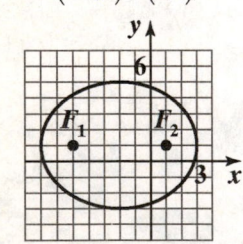

$$\frac{(x+2)^2}{25}+\frac{(y-1)^2}{16}=1$$

5.
$$x^2-4x\ +9y^2+54y\ =-49$$
$$\left(x^2-4x\ \right)+9\left(y^2+6y\ \right)=-49$$
$$\left(x^2-4x+4\right)+9\left(y^2+6y+9\right)=-49+4+81$$
$$(x-2)^2+9(y+3)^2=36$$
$$\frac{(x-2)^2}{36}+\frac{9(y+3)^2}{36}=\frac{36}{36}$$
$$\frac{(x-2)^2}{36}+\frac{(y+3)^2}{4}=1$$

Center: $(2,-3)$

Foci: $\left(2\pm4\sqrt{2},-3\right)$

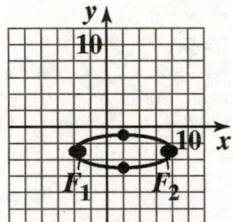

$$x^2+9y^2-4x+54y+49=0$$

6. The equation is for a hyperbola in standard form with the transverse axis on the x-axis. We have $a^2=9$ and $b^2=1$, so $a=3$ and $b=1$.

Therefore, the vertices are at $(\pm a,0)$ or $(\pm3,0)$.

Using a dashed line, we construct a rectangle using the ±3 on the x-axis and ±1 on the y-axis. Then use dashed lines to draw extended diagonals for the rectangle. These represent the asymptotes of the graph.

Foci: $\left(\pm\sqrt{10},0\right)$

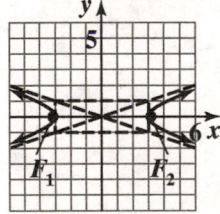

$$\frac{x^2}{9}-y^2=1$$

7.

The equation is in the form $\dfrac{y^2}{a^2} - \dfrac{x^2}{b^2} = 1$ with

$a^2 = 9$, and $b^2 = 1$. We know the transverse axis lies on the y-axis and the vertices are $(0, -3)$ and $(0, 3)$. Because $a^2 = 9$ and $b^2 = 1$, $a = 3$ and $b = 1$. Construct a rectangle using -1 and 1 on the x–axis, and -3 and 3 on the y–axis. Draw extended diagonals to obtain the asymptotes.

Foci: $\left(0, \pm\sqrt{10}\right)$

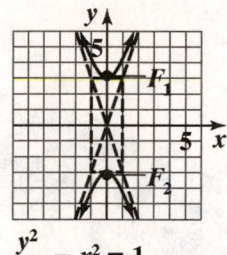

$$\dfrac{y^2}{9} - x^2 = 1$$

8. $\dfrac{y^2}{16} - \dfrac{x^2}{4} = 1$

The equation is in the form $\dfrac{y^2}{a^2} - \dfrac{x^2}{b^2} = 1$ with

$a^2 = 16$, and $b^2 = 4$. We know the transverse axis lies on the y-axis and the vertices are $(0, -4)$ and $(0, 4)$. Because

$a^2 = 16$ and $b^2 = 4$, $a = 4$ and $b = 2$. Construct a rectangle using -2 and 2 on the x–axis, and -4 and 4 on the y–axis. Draw extended diagonals to obtain the asymptotes.

Foci: $\left(0, \pm 2\sqrt{5}\right)$

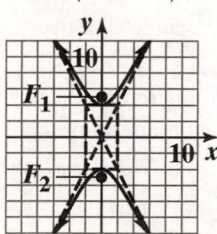

$$y^2 - 4x^2 = 16$$

9. $\dfrac{x^2}{49} - \dfrac{y^2}{4} = 1$

The equation is for a hyperbola in standard form with the transverse axis on the x-axis. We have $a^2 = 49$ and $b^2 = 4$, so $a = 7$ and $b = 2$. Therefore, the vertices are at $(\pm a, 0)$ or $(\pm 7, 0)$. Using a dashed line, we construct a rectangle using the ± 7 on the x-axis and ± 2 on the y-axis. Then use dashed lines to draw extended diagonals for the rectangle. These represent the asymptotes of the graph.

Foci: $\left(\pm\sqrt{53}, 0\right)$

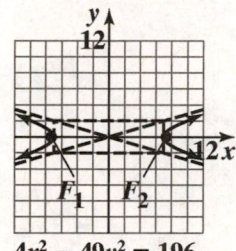

$$4x^2 - 49y^2 = 196$$

10. The equation is for a hyperbola in standard form with center $(2, -2)$. We have $a^2 = 9$ and $b^2 = 16$, so $a = 3$ and $b = 4$.

Asymptotes: $y + 2 = \pm\dfrac{4}{3}(x - 2)$

Foci: $(-3, -2)$, $(7, -2)$

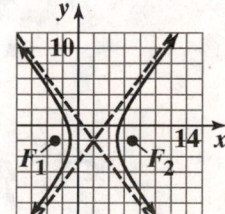

$$\dfrac{(x-2)^2}{9} - \dfrac{(y+2)^2}{16} = 1$$

11. Write the equation for the hyperbola in standard form:

$$4x^2 - y^2 + 8x + 6y + 11 = 0$$

$$4x^2 + 8x \quad - y^2 + 6y \quad = -11$$

$$4\left(x^2 + 2x \quad\right) - \left(y^2 - 6y \quad\right) = -11$$

$$4\left(x^2 + 2x + 1\right) - \left(y^2 - 6y + 9\right) = -11 + 4 - 9$$

$$4(x+1)^2 - (y-3)^2 = -16$$

$$\frac{4(x+1)^2}{-16} - \frac{(y-3)^2}{-16} = \frac{-16}{-16}$$

$$\frac{(y-3)^2}{16} - \frac{(x+1)^2}{4} = 1$$

Center $(-1, 3)$.

Asymptotes: $y - 3 = \pm 2(x+1)$

Foci: $\left(-1, 3 \pm 2\sqrt{5}\right),\ (7, -2)$

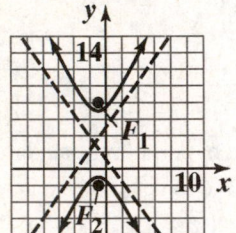

$4x^2 - y^2 + 8x + 6y + 11 = 0$

12. $(x-2)^2 = -12(y+1)$

$h = 2$

$k = -1$

$4p = -12$

$p = -3$

Vertex: (h, k)

$(2, -1)$

Focus: $(h, k + p)$

$(2, -1 - 3)$

$(2, -4)$

Directrix: $y = k - p$

$y = -1 - (-3)$

$y = 2$

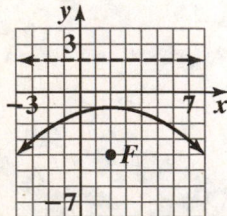

$(x - 2)^2 = -12(y + 1)$

13. $y^2 - 2x - 2y - 5 = 0$

$$y^2 - 2y \quad = 2x + 5$$

$$y^2 - 2y + 1 = 2x + 5 + 1$$

$$(y-1)^2 = 2x + 6$$

$$(y-1)^2 = 2(x+3)$$

$h = -3$

$k = 1$

$4p = 2$

$p = \dfrac{1}{2}$

Vertex: (h, k)

$(-3, 1)$

Focus: $(h + p, k)$

$$\left(-3 + \frac{1}{2}, 1\right)$$

$$\left(-\frac{5}{2}, 1\right)$$

Directrix: $y = h - p$

$$y = -3 - \left(\frac{1}{2}\right)$$

$$y = -\frac{7}{2}$$

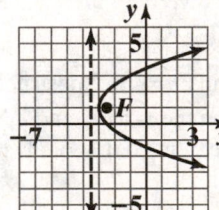

14. This is the equation of a circle centered at the origin with radius $r = \sqrt{4} = 2$.

We can plot points that are 2 units to the left, right, above, and below the origin and then graph the circle. The points are $(-2, 0)$, $(2, 0)$, $(0, 2)$, and $(0, -2)$.

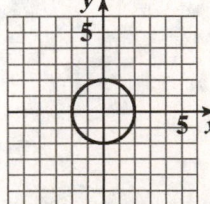

$x^2 + y^2 = 4$

15. $x + y = 4$

$y = -x + 4$

This is the equation of a line with slope $m = -1$ and a y-intercept of 4. We can plot the point $(0,4)$, use the slope to get an additional point, connect the points with a straight line and then extend the line to represent the graph of the equation.

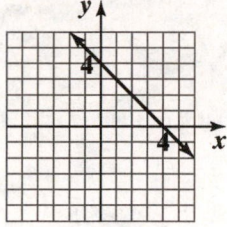

$x + y = 4$

16. $x^2 - y^2 = 4$

$$\frac{x^2}{4} - \frac{y^2}{4} = 1$$

The equation is for a hyperbola in standard form with the transverse axis on the x-axis. We have $a^2 = 4$ and $b^2 = 4$, so $a = 2$ and $b = 2$.

Therefore, the vertices are at $(\pm a, 0)$ or $(\pm 2, 0)$.

Using a dashed line, we construct a rectangle using the ± 2 on the x-axis and ± 2 on the y-axis. Then use dashed lines to draw extended diagonals for the rectangle. These represent the asymptotes of the graph.

Graph the hyperbola.

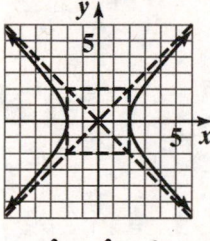

$x^2 - y^2 = 4$

17. $x^2 + 4y^2 = 4$

$$\frac{x^2}{4} + \frac{y^2}{1} = 1$$

Center: $(0,0)$

Because the denominator of the x^2 – term is greater than the denominator of the y^2 – term, the major axis is horizontal. We have $a^2 = 4$ and $b^2 = 1$, so $a = 2$ and $b = 1$. The vertices lie 2 units to the left and right of the center. The endpoints of the minor axis lie 1 unit above and below the center.

Vertices: $(-2,0)$ and $(2,0)$

Minor endpoints: $(0,-1)$ and $(0,1)$

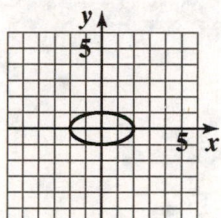

$x^2 + 4y^2 = 4$

18. Center: $(-1,1)$

Radius: $r = \sqrt{4} = 2$

We plot the points that are 2 units to the left, right, above and below the center.

These points are $(-3,1)$, $(1,1)$, $(-1,3)$, and $(-1,-1)$.

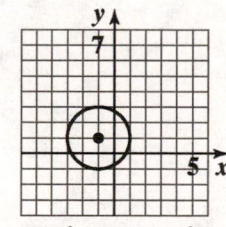

$(x + 1)^2 + (y - 1)^2 = 4$

19. $x^2 + 4(y-1)^2 = 4$

$$\frac{x^2}{4} + \frac{(y-1)^2}{1} = 1$$

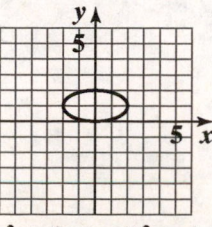

$x^2 + 4(y - 1)^2 = 4$

20. $(x-1)^2 - (y-1)^2 = 4$

$$\frac{(x-1)^2}{4} - \frac{(y-1)^2}{4} = 1$$

The equation is for a hyperbola in standard form centered at (1, 1). We have $a^2 = 4$ and $b^2 = 4$, so $a = 2$ and $b = 2$.

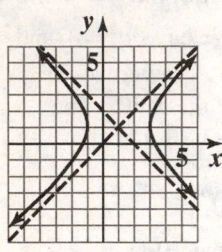

$(x-1)^2 - (y-1)^2 = 4$

21. $(y+1)^2 = 4(x-1)$

$h = 1$

$k = -1$

$4p = 4$

$p = 1$

Vertex: (h, k)

$\qquad (1, -1)$

Focus: $(h+p, k)$

$\qquad (1+1, -1)$

$\qquad (2, -1)$

Directrix: $y = h - p$

$\qquad y = 1 - 1$

$\qquad y = 0$

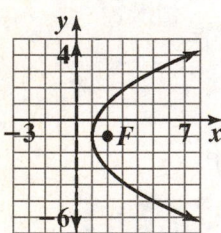

22. The foci and vertices show that c is 4 and a is 5.

$c^2 = a^2 - b^2$

$4^2 = 5^2 - b^2$

$b^2 = 25 - 16$

$b^2 = 9$

$$\frac{x^2}{25} + \frac{y^2}{9} = 1$$

23. The endpoints show that the center is (1, 2).

Since $2a = 18$, $a = 9$ and $a^2 = 81$.

Since $2c = 10$, $c = 5$ and $c^2 = 25$.

$c^2 = a^2 - b^2$

$25 = 81 - b^2$

$b^2 = 81 - 25$

$b^2 = 56$

$$\frac{(x-1)^2}{81} + \frac{(y-2)^2}{56} = 1$$

24. The foci and vertices show that c is 3 and a is 2.

$b^2 = c^2 - a^2$

$b^2 = 3^2 - 2^2$

$b^2 = 9 - 4$

$b^2 = 5$

$$\frac{x^2}{4} - \frac{y^2}{5} = 1$$

25. The endpoints show that the center is (–1, 5).

Since $2a = 4$, $a = 2$ and $a^2 = 4$.

Since $2c = 6$, $c = 3$ and $c^2 = 9$.

$b^2 = c^2 - a^2$

$b^2 = 9 - 4$

$b^2 = 5$

$$\frac{(x+1)^2}{4} - \frac{(y-5)^2}{5} = 1$$

26. Focus: $(h, \overset{4}{k} + \overset{5}{p})$

Directrix: $y = \overset{-1}{k - p}$

$(k+p) + (k-p) = (5) + (-1)$

$\qquad\qquad\qquad 2k = 4$

$\qquad\qquad\qquad\quad k = 2$

$k + p = 5$

$\qquad p = 5 - k$

$\qquad p = 5 - 2$

$\qquad p = 3$

$(x-h)^2 = 4p(y-k)$

$(x-4)^2 = 4(3)(y-2)$

$(x-4)^2 = 12(y-2)$

27. Focus: $(\overbrace{h+p}^{-2}, \overbrace{k}^{6})$

Directrix: $x = \overbrace{h-p}^{8}$

$(h+p)+(h-p) = (-2)+(8)$

$$2h = 6$$
$$h = 3$$

$h+p = -2$

$p = -2-h$

$p = -2-3$

$p = -5$

$(y-k)^2 = 4p(x-h)$

$(y-6)^2 = 4(-5)(x-3)$

$(y-6)^2 = -20(x-3)$

28. $a = 15, b = 10$

$$\frac{x^2}{15^2} + \frac{y^2}{10^2} = 1$$

$$\frac{x^2}{225} + \frac{y^2}{100} = 1$$

Since the truck is 10 feet wide, substitute $x = 5$ into the equation to find y.

$$\frac{5^2}{225} + \frac{y^2}{100} = 1$$

$$\frac{25}{225} + \frac{y^2}{100} = 1$$

$$\frac{1}{9} + \frac{y^2}{100} = 1$$

$$900\left(\frac{1}{9} + \frac{y^2}{100}\right) = 900(1)$$

$$100 + 9y^2 = 900$$

$$9y^2 = 800$$

$$y^2 = 88.8889$$

$$y = \sqrt{88.8889}$$

$$y \approx 9.43$$

5 feet from the center, the height of the archway is 9.43 feet. Since the truck's height is 9.5 feet, it will not fit under the archway.

29. Find the distance between the foci.

Since $2a = 40$, $a = 20$ and $a^2 = 400$.

Since $2b = 20$, $b = 10$ and $b^2 = 100$.

$c^2 = a^2 - b^2$

$c^2 = 400 - 100$

$c^2 = 300$

$c = \sqrt{300}$

$ = 10\sqrt{3}$

$2c = 20\sqrt{3}$

$2c \approx 34.64$

The kidney stone should be 34.64 cm from the electrode that sends the ultrasound waves.

30. a. Since $2c = 6$, $c = 3$ and $c^2 = 9$.

The ranger at the primary station heard the explosion 6 seconds before the other ranger. This means that the explosion occurred $6 \times 0.35 = 2.1$ miles closer to the primary station.

Since $2a = 2.1$, $a = 1.05$ and $a^2 = 1.1025$.

$b^2 = c^2 - a^2$

$b^2 = 9 - 1.1025$

$b^2 = 7.8975$

$$\frac{x^2}{1.1025} - \frac{y^2}{7.8975} = 1$$

b.

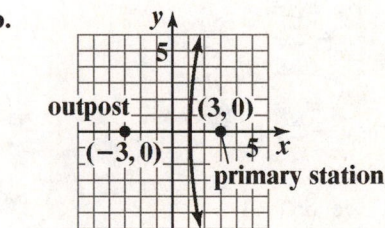

31. Consider the peak of the dome as the point $(0,10)$.

Since the width is 15 meters, the points $(\pm 7.5, 0)$ are on the parabola.

Use standard form to find p.

$(x-h)^2 = 4p(y-k)$

$(7.5-0)^2 = 4p(0-10)$

$$7.5^2 = -40p$$

$$\frac{56.25}{-40} = p$$

$$p \approx -1.4$$

The light should be about 1.4 meters below the peak of the ceiling.

Section 9.4

Check Point Exercises

1. **a.** $A = 3$ and $C = 2$.
        $AC = 3(2) = 6$. Since $A \neq C$ and $AC > 0$, the graph is an ellipse.

    b. $A = 1$ and $C = 1$. Since $A = C$, the graph is a circle.

    c. $A = 0$ and $C = 1$. Since $AC = 0$, the graph is a parabola.

    d. $A = 9$ and $C = -16$. Since $AC < 0$, the graph is a hyperbola.

2.

$$x = x' \cos 45° - y' \sin 45° = x'\left(\frac{\sqrt{2}}{2}\right) - y'\left(\frac{\sqrt{2}}{2}\right) = \frac{\sqrt{2}}{2}(x' - y')$$

$$y = x' \sin 45° + y' \cos 45° = x'\left(\frac{\sqrt{2}}{2}\right) + y'\left(\frac{\sqrt{2}}{2}\right) = \frac{\sqrt{2}}{2}(x' + y')$$

Substitute into the equation: $xy = 2$

$$\left[\frac{\sqrt{2}}{2}(x' - y')\right]\left[\frac{\sqrt{2}}{2}(x' + y')\right] = 2$$

$$\frac{1}{2}\left(x'^2 - y'^2\right) = 2$$

$$\frac{x'^2}{4} - \frac{y'^2}{4} = 1$$

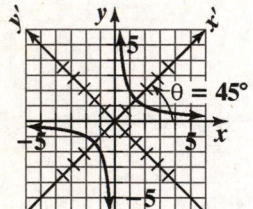

3. $2x^2 + \sqrt{3}xy + y^2 - 2 = 0$
    Step 1
    $A = 2, B = \sqrt{3},$ and $C = 1$.

    $$\cot 2\theta = \frac{A - C}{B} = \frac{2 - 1}{\sqrt{3}} = \frac{1}{\sqrt{3}} = \frac{\sqrt{3}}{3}$$

    Step 2
    Since $\cot 2\theta = \frac{\sqrt{3}}{3}, 2\theta = 60°$. Thus, $\theta = 30°$.

    Step 3

    $$x = x' \cos 30° - y' \sin 30° = x'\left(\frac{\sqrt{3}}{2}\right) - y'\left(\frac{1}{2}\right) = \frac{\sqrt{3}x' - y'}{2}$$

    $$y = x' \sin 30° + y' \cos 30° = x'\left(\frac{1}{2}\right) + y'\left(\frac{\sqrt{3}}{2}\right) = \frac{x' + \sqrt{3}y'}{2}$$

Step 4

Substitute into the equation:

$2x^2 + \sqrt{3}xy + y^2 - 2 = 0$

$2\left(\dfrac{\sqrt{3}x' - y'}{2}\right)^2 + \sqrt{3}\left(\dfrac{\sqrt{3}x' - y'}{2}\right)\left(\dfrac{x' + \sqrt{3}y'}{2}\right) + \left(\dfrac{x' + \sqrt{3}y'}{2}\right)^2 - 2 = 0$

$2\left(\dfrac{3x'^2 - 2\sqrt{3}x'y' + y'^2}{4}\right) + \sqrt{3}\left(\dfrac{\sqrt{3}x'^2 + 2x'y' - \sqrt{3}y'^2}{4}\right) + \dfrac{x'^2 + 2\sqrt{3}x'y' + 3y'^2}{4} = 2$

$6x'^2 - 4\sqrt{3}x'y' + 2y'^2 + 3x'^2 + 2\sqrt{3}x'y' - 3y'^2 + x'^2 + 2\sqrt{3}x'y' + 3y'^2 = 8$

$10x'^2 + 2y'^2 = 8$

$\dfrac{10x'^2}{8} + \dfrac{2y'^2}{8} = \dfrac{8}{8}$

$\dfrac{x'^2}{\frac{4}{5}} + \dfrac{y^2}{4} = 1$

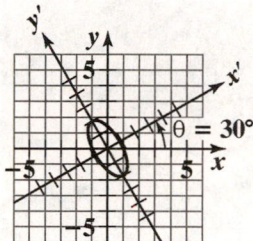

4. $4x^2 - 4xy + y^2 - 8\sqrt{5}x - 16\sqrt{5}y = 0$

Step 1:

$A = 4, B = -4,$ and $C = 1$.

$\cot 2\theta = \dfrac{A - C}{B} = \dfrac{4 - 1}{-4} = \dfrac{-3}{4}$

Step 2: Since θ is always acute, and $\cot 2\theta$ is negative, 2θ is in quadrant II.

The third side of the right triangle is found using the Pythagorean theorem:

$(-3)^2 + 4^2 = r^2$

$\qquad 25 = r^2$

$\qquad r = 5$

So, $\cos 2\theta = \dfrac{\text{adjacent}}{\text{hypotenuse}} = \dfrac{-3}{5}$.

$\sin\theta = \sqrt{\dfrac{1 - \cos 2\theta}{2}} = \sqrt{\dfrac{1 - \left(-\frac{3}{5}\right)}{2}} = \sqrt{\dfrac{8}{10}} = \sqrt{\dfrac{4}{5}} = \dfrac{2\sqrt{5}}{5}$ and

$\cos\theta = \sqrt{\dfrac{1 + \cos 2\theta}{2}} = \sqrt{\dfrac{1 + \left(-\frac{3}{5}\right)}{2}} = \sqrt{\dfrac{1}{5}} = \dfrac{\sqrt{5}}{5}$

Step 3:

$x = x'\cos\theta - y'\sin\theta = x'\left(\dfrac{\sqrt{5}}{5}\right) - y'\left(\dfrac{2\sqrt{5}}{5}\right) = \sqrt{5}\left(\dfrac{x' - 2y'}{5}\right)$

$y = x'\sin\theta + y'\cos\theta = x'\left(\dfrac{2\sqrt{5}}{5}\right) + y'\left(\dfrac{\sqrt{5}}{5}\right) = \sqrt{5}\left(\dfrac{2x' + y'}{5}\right)$

Step 4: Substitute into the equation: $4x^2 - 4xy + y^2 - 8\sqrt{5}x - 16\sqrt{5}y = 0$

$$4\left[\sqrt{5}\left(\frac{x'-2y'}{5}\right)\right]^2 - 4\left[\sqrt{5}\left(\frac{x'-2y'}{5}\right)\right]\left[\sqrt{5}\left(\frac{2x'+y'}{5}\right)\right] + \left[\sqrt{5}\left(\frac{2x'+y'}{5}\right)\right]^2 - 8\sqrt{5}\left[\sqrt{5}\left(\frac{x'-2y'}{5}\right)\right]$$

$$-16\sqrt{5}\left[\sqrt{5}\left(\frac{2x'+y'}{5}\right)\right] = 0$$

$$20\left(\frac{x'^2 - 4x'y' + 4y'^2}{25}\right) - 20\left(\frac{2x'^2 - 3x'y' - 2y'^2}{25}\right) + 5\left(\frac{4x'^2 + 4x'y' + y'^2}{25}\right) - 40\left(\frac{x'-2y'}{5}\right)$$

$$-80\left(\frac{2x'+y'}{5}\right) = 0$$

Multiply both sides by 25:

$20x'^2 - 80x'y' + 80y'^2 - 40x'^2 + 60x'y' + 40y'^2 + 20x'^2 + 20x'y' + 5y'^2 - 200x' + 400y'$

$-800x' - 400y' = 0$

$125y'^2 - 1000x' = 0$

$\quad\quad y'^2 - 8x' = 0$

Step 5: This is a parabola, since it has only the y' squared.

$y'^2 - 8x' = 0$

$\quad\quad y'^2 = 8x'$

The vertex of the parabola, relative to the $x'y'$ system, is (0, 0). Using a calculator to solve

$\sin\theta = \dfrac{2\sqrt{5}}{5}$, we find $\theta = \sin^{-1}\left(\dfrac{2\sqrt{5}}{5}\right) \approx 63°$. Rotate the axes through approximately 63°.

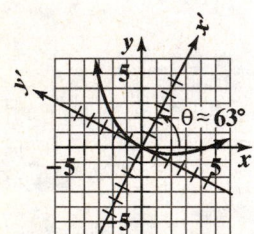

5. $\quad 3x^2 - 2\sqrt{3}xy + y^2 + 2x + 2\sqrt{3}y = 0$

$A = 3, B = -2\sqrt{3},$ and $C = 1$.

$B^2 - 4AC = \left(-2\sqrt{3}\right)^2 - 4(3)(1) = 12 - 12 = 0$

Because $B^2 - 4AC = 0$, the graph of the equation is a parabola.

Exercise Set 9.4

1. $\quad A = 0$ and $C = 1$. Since $AC = 0$, the graph is a parabola.

3. $\quad A = 4$ and $C = -9$. Since $AC = -36 < 0$, the graph is a hyperbola.

5. $\quad A = 4$ and $C = 4$. Since $A = C$, the graph is a circle.

7. $\quad A = 100$ and $C = -7$. Since $AC = -700 < 0$, the graph is a hyperbola.

9. $x = x' \cos 45° - y' \sin 45° = x'\left(\dfrac{\sqrt{2}}{2}\right) - y'\left(\dfrac{\sqrt{2}}{2}\right) = \dfrac{\sqrt{2}}{2}(x' - y')$

$y = x' \sin 45° + y' \cos 45° = x'\left(\dfrac{\sqrt{2}}{2}\right) + y'\left(\dfrac{\sqrt{2}}{2}\right) = \dfrac{\sqrt{2}}{2}(x' + y')$

Substitute into the equation: $xy = -1$

$\left[\dfrac{\sqrt{2}}{2}(x' - y')\right]\left[\dfrac{\sqrt{2}}{2}(x' + y')\right] = -1$

$\dfrac{1}{2}\left(x'^2 - y'^2\right) = -1$

$\dfrac{y'^2}{2} - \dfrac{x'^2}{2} = 1$

11. $x = x' \cos 45° - y' \sin 45° = x'\left(\dfrac{\sqrt{2}}{2}\right) - y'\left(\dfrac{\sqrt{2}}{2}\right) = \dfrac{\sqrt{2}}{2}(x' - y')$

$y = x' \sin 45° + y' \cos 45° = x'\left(\dfrac{\sqrt{2}}{2}\right) + y'\left(\dfrac{\sqrt{2}}{2}\right) = \dfrac{\sqrt{2}}{2}(x' + y')$

Substitute into the equation: $x^2 - 4xy + y^2 - 3 = 0$

$\left[\dfrac{\sqrt{2}}{2}(x' - y')\right]^2 - 4\left[\dfrac{\sqrt{2}}{2}(x' - y')\right]\left[\dfrac{\sqrt{2}}{2}(x' + y')\right] + \left[\dfrac{\sqrt{2}}{2}(x' + y')\right]^2 - 3 = 0$

$\dfrac{1}{2}(x' - y')^2 - 4\left[\dfrac{1}{2}\left(x'^2 - y'^2\right)\right] + \dfrac{1}{2}(x' + y')^2 = 3$

$\dfrac{1}{2}\left(x'^2 - 2x'y' + y'^2\right) - 2x'^2 + 2y'^2 + \dfrac{1}{2}\left(x'^2 + 2x'y' + y'^2\right) = 3$

$\dfrac{1}{2}x'^2 - x'y' + \dfrac{1}{2}y'^2 - 2x'^2 + 2y'^2 + \dfrac{1}{2}x'^2 + x'y' + \dfrac{1}{2}y'^2 = 3$

$-x'^2 + 3y'^2 = 3$

$\dfrac{-x'^2}{3} + \dfrac{3y'^2}{3} = \dfrac{3}{3}$

$\dfrac{y'^2}{1} - \dfrac{x'^2}{3} = 1$

13. $x = x' \cos 30° - y' \sin 30° = x'\left(\dfrac{\sqrt{3}}{2}\right) - y'\left(\dfrac{1}{2}\right) = \dfrac{\sqrt{3}x' - y'}{2}$

$y = x' \sin 30° + y' \cos 30° = x'\left(\dfrac{1}{2}\right) + y'\left(\dfrac{\sqrt{3}}{2}\right) = \dfrac{x' + \sqrt{3}y'}{2}$

Substitute into the equation: $23x^2 + 26\sqrt{3}xy - 3y^2 - 144 = 0$

$$23\left(\dfrac{\sqrt{3}x' - y'}{2}\right)^2 + 26\sqrt{3}\left(\dfrac{\sqrt{3}x' - y'}{2}\right)\left(\dfrac{x' + \sqrt{3}y'}{2}\right) - 3\left(\dfrac{x' + \sqrt{3}y'}{2}\right)^2 = 144$$

$$23\left(\dfrac{3x'^2 - 2\sqrt{3}x'y' + y'^2}{4}\right) + 26\sqrt{3}\left(\dfrac{\sqrt{3}x'^2 + 2x'y' - \sqrt{3}y'^2}{4}\right) - 3\left(\dfrac{x'^2 + 2\sqrt{3}x'y' + 3y'^2}{4}\right) = 144$$

$$69x'^2 - 46\sqrt{3}x'y' + 23y'^2 + 78x'^2 + 52\sqrt{3}x'y' - 78y'^2 - 3x'^2 - 6\sqrt{3}x'y' - 9y'^2 = 576$$

$$144x'^2 - 64y'^2 = 576$$

$$\dfrac{144x'^2}{576} - \dfrac{64y'^2}{576} = \dfrac{576}{576}$$

$$\dfrac{x'^2}{4} - \dfrac{y'^2}{9} = 1$$

15. $x^2 + xy + y^2 - 10 = 0$

$A = 1, B = 1,$ and $C = 1.$

$\cot 2\theta = \dfrac{A - C}{B} = \dfrac{1 - 1}{1} = 0$

$\quad 2\theta = 90°$

$\quad\ \theta = 45°$

$x = x' \cos 45° - y' \sin 45° = x'\left(\dfrac{\sqrt{2}}{2}\right) - y'\left(\dfrac{\sqrt{2}}{2}\right) = \dfrac{\sqrt{2}}{2}(x' - y')$

$y = x' \sin 45° + y' \cos 45° = x'\left(\dfrac{\sqrt{2}}{2}\right) + y'\left(\dfrac{\sqrt{2}}{2}\right) = \dfrac{\sqrt{2}}{2}(x' + y')$

17. $3x^2 - 10xy + 3y^2 - 32 = 0$

$A = 3, B = -10,$ and $C = 3.$

$\cot 2\theta = \dfrac{A - C}{B} = \dfrac{3 - 3}{-10} = 0$

$\quad 2\theta = 90°$

$\quad\ \theta = 45°$

$x = x' \cos 45° - y' \sin 45° = x'\left(\dfrac{\sqrt{2}}{2}\right) - y'\left(\dfrac{\sqrt{2}}{2}\right) = \dfrac{\sqrt{2}}{2}(x' - y')$

$y = x' \sin 45° + y' \cos 45° = x'\left(\dfrac{\sqrt{2}}{2}\right) + y'\left(\dfrac{\sqrt{2}}{2}\right) = \dfrac{\sqrt{2}}{2}(x' + y')$

19. $11x^2 + 10\sqrt{3}xy + y^2 - 4 = 0$

$A = 11, B = 10\sqrt{3},$ and $C = 1.$

$$x = x' \cos 30° - y' \sin 30°$$

$$x = x'\left(\frac{\sqrt{3}}{2}\right) - y'\left(\frac{1}{2}\right)$$

$$\cot 2\theta = \frac{A-C}{B} = \frac{11-1}{10\sqrt{3}} = \frac{10}{10\sqrt{3}} = \frac{1}{\sqrt{3}} = \frac{\sqrt{3}}{3} \qquad x = \frac{\sqrt{3}x' - y'}{2}$$

$$2\theta = 60° \qquad\qquad y = x' \sin 30° + y' \cos 30°$$

$$\theta = 30° \qquad\qquad y = x'\left(\frac{1}{2}\right) + y'\left(\frac{\sqrt{3}}{2}\right)$$

$$y = \frac{x' + \sqrt{3}y'}{2}$$

21. $10x^2 + 24xy + 17y^2 - 9 = 0$

$A = 10, B = 24,$ and $C = 17.$

$$\cot 2\theta = \frac{A-C}{B} = \frac{10-17}{24} = \frac{-7}{24}$$

Since θ is always acute, and $\cot 2\theta$ is negative, 2θ is in quadrant II.
The third side of the right triangle is found by using the Pythagorean theorem:

$$(-7)^2 + 24^2 = r^2$$

$$625 = r^2$$

$$r = 25$$

So, $\cos 2\theta = \dfrac{-7}{25}.$

$$\sin \theta = \sqrt{\frac{1 - \cos 2\theta}{2}} = \sqrt{\frac{1 - \left(\frac{-7}{25}\right)}{2}} = \frac{4}{5} \text{ and}$$

$$\cos \theta = \sqrt{\frac{1 + \cos 2\theta}{2}} = \sqrt{\frac{1 + \left(\frac{-7}{25}\right)}{2}} = \frac{3}{5}$$

$$x = x' \cos\theta - y' \sin\theta$$

$$x = x'\left(\frac{3}{5}\right) - y'\left(\frac{4}{5}\right)$$

$$x = \frac{3x' - 4y'}{5}$$

$$y = x' \sin\theta + y' \cos\theta$$

$$y = x'\left(\frac{4}{5}\right) + y'\left(\frac{3}{5}\right)$$

$$y = \frac{4x' + 3y'}{5}$$

23. $x^2 + 4xy - 2y^2 - 1 = 0$

$A = 1, B = 4,$ and $C = -2.$

$\cot 2\theta = \dfrac{A-C}{B} = \dfrac{1-(-2)}{4} = \dfrac{3}{4}$

Since θ is always acute, and $\cot 2\theta$ is positive, 2θ is in quadrant I.
The third side of the right triangle is found using the Pythagorean theorem:

$3^2 + 4^2 = r^2$

$\qquad 25 = r^2$

$\qquad r = 5$

So, $\cos 2\theta = \dfrac{3}{5}.$

$\sin \theta = \sqrt{\dfrac{1 - \cos 2\theta}{2}} = \sqrt{\dfrac{1 - \frac{3}{5}}{2}} = \dfrac{\sqrt{5}}{5}$ and

$\cos \theta = \sqrt{\dfrac{1 + \cos 2\theta}{2}} = \sqrt{\dfrac{1 + \frac{3}{5}}{2}} = \dfrac{2\sqrt{5}}{5}$

$x = x' \cos \theta - y' \sin \theta$

$x = x'\left(\dfrac{2\sqrt{5}}{5}\right) - y'\left(\dfrac{\sqrt{5}}{5}\right)$

$x = \sqrt{5}\left(\dfrac{2x' - y'}{5}\right)$

$y = x' \sin \theta + y' \cos \theta$

$y = x'\left(\dfrac{\sqrt{5}}{5}\right) + y'\left(\dfrac{2\sqrt{5}}{5}\right)$

$y = \sqrt{5}\left(\dfrac{x' + 2y'}{5}\right)$

25. $34x^2 - 24xy + 41y^2 - 25 = 0$

$A = 34, B = -24,$ and $C = 41.$

$\cot 2\theta = \dfrac{A-C}{B} = \dfrac{34-41}{-24} = \dfrac{-7}{-24} = \dfrac{7}{24}$

Since θ is always acute, and $\cot 2\theta$ is positive, 2θ is in quadrant I.
The third side of the right triangle is found using the Pythagorean theorem:

$7^2 + 24^2 = r^2$

$\qquad 625 = r^2$

$\qquad r = 25$

So, $\cos 2\theta = \dfrac{7}{25}.$

$\sin \theta = \sqrt{\dfrac{1 - \cos 2\theta}{2}} = \sqrt{\dfrac{1 - \frac{7}{25}}{2}} = \dfrac{3}{5}$ and $\cos \theta = \sqrt{\dfrac{1 + \cos 2\theta}{2}} = \sqrt{\dfrac{1 + \frac{7}{25}}{2}} = \dfrac{4}{5}$

$x = x' \cos \theta - y' \sin \theta = x'\left(\dfrac{4}{5}\right) - y'\left(\dfrac{3}{5}\right) = \dfrac{4x' - 3y'}{5}$

$y = x' \sin \theta + y' \cos \theta = x'\left(\dfrac{3}{5}\right) + y'\left(\dfrac{4}{5}\right) = \dfrac{3x' + 4y'}{5}$

27. **a.** From Exercise 15,

$$x = \frac{\sqrt{2}}{2}(x' - y') \text{ and } y = \frac{\sqrt{2}}{2}(x' + y').$$

Substitute into the equation: $x^2 + xy + y^2 - 10 = 0$

$$\left[\frac{\sqrt{2}}{2}(x' - y')\right]^2 + \left[\frac{\sqrt{2}}{2}(x' - y')\right]\left[\frac{\sqrt{2}}{2}(x' + y')\right] + \left[\frac{\sqrt{2}}{2}(x' + y')\right]^2 - 10 = 0$$

$$\frac{1}{2}\left(x'^2 - 2x'y' + y'^2\right) + \frac{1}{2}\left(x'^2 - y'^2\right) + \frac{1}{2}\left(x'^2 + 2x'y' + y'^2\right) = 10$$

Multiply both sides by 2.

$$x'^2 - 2x'y' + y'^2 + x'^2 - y'^2 + x'^2 + 2x'y' + y'^2 = 20$$

$$3x'^2 + y'^2 = 20$$

b. $\dfrac{x'^2}{\frac{20}{3}} + \dfrac{y'^2}{20} = 1$

c.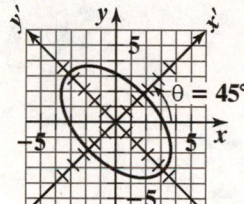

29. **a.** From Exercise 17, $x = \dfrac{\sqrt{2}}{2}(x' - y')$ and $y = \dfrac{\sqrt{2}}{2}(x' + y')$.

Substitute into the equation: $3x^2 - 10xy + 3y^2 - 32 = 0$

$$3\left[\frac{\sqrt{2}}{2}(x' - y')\right]^2 - 10\left[\frac{\sqrt{2}}{2}(x' - y')\right]\left[\frac{\sqrt{2}}{2}(x' + y')\right] + 3\left[\frac{\sqrt{2}}{2}(x' + y')\right]^2 - 32 = 0$$

$$3\left[\frac{1}{2}\left(x'^2 - 2x'y' + y'^2\right)\right] - 10\left[\frac{1}{2}\left(x'^2 - y'^2\right)\right] + 3\left[\frac{1}{2}\left(x'^2 + 2x'y' + y'^2\right)\right] = 32$$

Multiply both sides by 2.

$$3x'^2 - 6x'y' + 3y'^2 - 10x'^2 + 10y'^2 + 3x'^2 + 6x'y' + 3y'^2 = 64$$

$$-4x'^2 + 16y'^2 = 64$$

b. $\dfrac{y'^2}{4} - \dfrac{x'^2}{16} = 1$

c.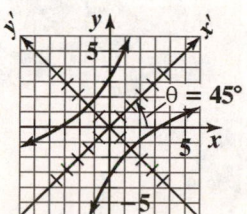

31. a. From Exercise 19,

$$x = \frac{\sqrt{3}x' - y'}{2} \text{ and } y = \frac{x' + \sqrt{3}y'}{2}.$$

Substitute into the equation: $11x^2 + 10\sqrt{3}xy + y^2 - 4 = 0$

$$11\left(\frac{\sqrt{3}x' - y'}{2}\right)^2 + 10\sqrt{3}\left(\frac{\sqrt{3}x' - y'}{2}\right)\left(\frac{x' + \sqrt{3}y'}{2}\right) + \left(\frac{x' + \sqrt{3}y'}{2}\right)^2 - 4 = 0$$

$$11\left(\frac{3x'^2 - 2\sqrt{3}x'y' + y'^2}{4}\right) + 10\sqrt{3}\left(\frac{\sqrt{3}x'^2 + 2x'y' - \sqrt{3}y'^2}{4}\right) + \frac{x'^2 + 2\sqrt{3}x'y' + 3y'^2}{4} = 4$$

Multiply both sides by 4:

$$33x'^2 - 22\sqrt{3}x'y' + 11y'^2 + 30x'^2 + 20\sqrt{3}x'y' - 30y'^2 + x'^2 + 2\sqrt{3}x'y' + 3y'^2 = 16$$

$$64x'^2 - 16y'^2 = 16$$

b. $\dfrac{x'^2}{\frac{1}{4}} - \dfrac{y'^2}{1} = 1$

c.

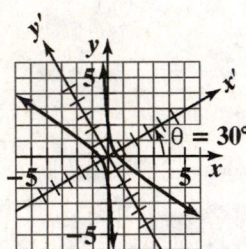

$\theta = 30°$

33. a. From Exercise 21, $x = \dfrac{3x' - 4y'}{5}$ and $y = \dfrac{4x' + 3y'}{5}$.

Substitute into the equation: $10x^2 + 24xy + 17y^2 - 9 = 0$

$$10\left(\frac{3x' - 4y'}{5}\right)^2 + 24\left(\frac{3x' - 4y'}{5}\right)\left(\frac{4x' + 3y'}{5}\right) + 17\left(\frac{4x' + 3y'}{5}\right)^2 - 9 = 0$$

$$10\left(\frac{9x'^2 - 24x'y' + 16y'^2}{25}\right) + 24\left(\frac{12x'^2 - 7x'y' - 12y'^2}{25}\right) + 17\left(\frac{16x'^2 + 24x'y' + 9y'^2}{25}\right) = 9$$

Multiply both sides by 25:

$$90x'^2 - 240x'y' + 160y'^2 + 288x'^2 - 168x'y' - 288y'^2 + 272x'^2 + 408x'y' + 153y'^2 = 225$$

$$650x'^2 + 25y'^2 = 225$$

b. $\dfrac{x'^2}{\frac{9}{26}} + \dfrac{y'^2}{9} = 1$

c.

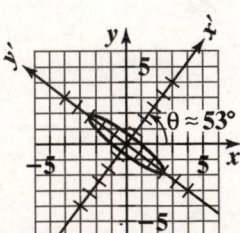

$\theta \approx 53°$

The axes are rotated by $\theta = \sin^{-1}\left(\dfrac{4}{5}\right) \approx 53°$.

35. a. From Exercise 23,

$$x = \sqrt{5}\left(\frac{2x' - y'}{5}\right) \text{ and } y = \sqrt{5}\left(\frac{x' + 2y'}{5}\right).$$

Substitute into the equation: $x^2 + 4xy - 2y^2 - 1 = 0$

$$\left[\sqrt{5}\left(\frac{2x' - y'}{5}\right)\right]^2 + 4\left[\sqrt{5}\left(\frac{2x' - y'}{5}\right)\right]\left[\sqrt{5}\left(\frac{x' + 2y'}{5}\right)\right] - 2\left[\sqrt{5}\left(\frac{x' + 2y'}{5}\right)\right]^2 = 1$$

$$5\left(\frac{4x'^2 - 4x'y' + y'^2}{25}\right) + 20\left(\frac{2x'^2 + 3x'y' - 2y'^2}{25}\right) - 10\left(\frac{x'^2 + 4x'y' + 4y'^2}{25}\right) = 1$$

Multiply both sides by 25:

$$20x'^2 - 20x'y' + 5y'^2 + 40x'^2 + 60x'y' - 40y'^2 - 10x'^2 - 40x'y' - 40y'^2 = 25$$

$$50x'^2 - 75y'^2 = 25$$

b. $\dfrac{x'^2}{\frac{1}{2}} - \dfrac{y'^2}{\frac{1}{3}} = 1$

c.

The axes are rotated by $\theta = \sin^{-1}\left(\dfrac{\sqrt{5}}{5}\right) \approx 27°$.

37. a. From Exercise 25,

$$x = \frac{4x' - 3y'}{5} \text{ and } y = \frac{3x' + 4y'}{5}.$$

Substitute into the equation: $34x^2 - 24xy + 41y^2 - 25 = 0$

$$34\left(\frac{4x' - 3y'}{5}\right)^2 - 24\left(\frac{4x' - 3y'}{5}\right)\left(\frac{3x' + 4y'}{5}\right) + 41\left(\frac{3x' + 4y'}{5}\right)^2 = 25$$

$$34\left(\frac{16x'^2 - 24x'y' + 9y'^2}{25}\right) - 24\left(\frac{12x'^2 + 7x'y' - 12y'^2}{25}\right) + 41\left(\frac{9x'^2 + 24x'y' + 16y'^2}{25}\right) = 25$$

Multiply both sides by 25:

$$544x'^2 - 816x'y' + 306y'^2 - 288x'^2 - 168x'y' + 288y'^2 + 369x'^2 + 984x'y' + 656y'^2 = 625$$

$$625x'^2 + 1250y'^2 = 625$$

b. $\dfrac{625x'^2 + 1250y'^2}{625} = \dfrac{625}{625}$

$$\frac{x'^2}{1} + \frac{y'^2}{\frac{1}{2}} = 1$$

c.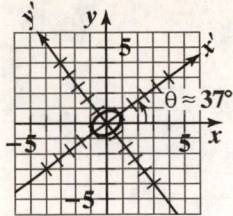

The axes are rotated by $\theta = \sin^{-1}\left(\dfrac{3}{5}\right) \approx 37°$.

39. $5x^2 - 2xy + 5y^2 - 12 = 0$

$A = 5, B = -2,$ and $C = 5.$

$B^2 - 4AC = (-2)^2 - 4(5)(5) = -96.$

Since $B^2 - 4AC < 0$, the graph is an ellipse or a circle.

41. $24x^2 + 16\sqrt{3}xy + 8y^2 - x + \sqrt{3}y - 8 = 0$

$A = 24, B = 16\sqrt{3},$ and $C = 8.$

$B^2 - 4AC = \left(16\sqrt{3}\right)^2 - 4(24)(8) = 768 - 768 = 0$

Since $B^2 - 4AC = 0$, the graph is a parabola.

43. $23x^2 + 26\sqrt{3}xy - 3y^2 - 144 = 0$

$A = 23, B = 26\sqrt{3},$ and $C = -3.$

$B^2 - 4AC = \left(26\sqrt{3}\right)^2 - 4(23)(-3) = 2028 + 276 = 2304$

Since $B^2 - 4AC > 0$, the graph is a hyperbola.

45. Find θ :

$$\cot 2\theta = \frac{A-C}{B} = \frac{5-5}{-6} = 0$$

$$2\theta = 90°$$

$$\theta = 45°$$

Substitute θ into rotation formulas:

$$x = x'\cos\theta - y'\sin\theta = x'\cos 45° - y'\sin 45° = x'\frac{\sqrt{2}}{2} - y'\frac{\sqrt{2}}{2} = \frac{x'\sqrt{2} - y'\sqrt{2}}{2}$$

$$y = x'\sin\theta + y'\cos\theta = x'\sin 45° + y'\cos 45° = x'\frac{\sqrt{2}}{2} + y'\frac{\sqrt{2}}{2} = \frac{x'\sqrt{2} + y'\sqrt{2}}{2}$$

Substitute into equation:

$$5x^2 - 6xy + 5y^2 - 8 = 0$$

$$5\left(\frac{x'\sqrt{2} - y'\sqrt{2}}{2}\right)^2 - 6\left(\frac{x'\sqrt{2} - y'\sqrt{2}}{2}\right)\left(\frac{x'\sqrt{2} + y'\sqrt{2}}{2}\right) + 5\left(\frac{x'\sqrt{2} + y'\sqrt{2}}{2}\right)^2 - 8 = 0$$

$$\frac{x'^2}{4} + \frac{y'^2}{1} = 1$$

$\dfrac{x'^2}{4} + \dfrac{y'^2}{1} = 1$ is an ellipse with minor axis vertices of $(0, -1)$ and $(0, 1)$.

47. Find $\sin\theta$ and $\cos\theta$:

$$\cot 2\theta = \frac{A-C}{B} = \frac{1-4}{-4} = \frac{3}{4}$$

$$\cot 2\theta = \frac{x}{y} = \frac{3}{4}$$

$$r = \sqrt{x^2+y^2} = \sqrt{3^2+4^2} = 5$$

$$\sin\theta = \sqrt{\frac{1-\cos 2\theta}{2}} = \sqrt{\frac{1-\left(\frac{3}{5}\right)}{2}} = \frac{\sqrt{5}}{5} \quad \text{and} \quad \cos\theta = \sqrt{\frac{1+\cos 2\theta}{2}} = \sqrt{\frac{1+\left(\frac{3}{5}\right)}{2}} = \frac{2\sqrt{5}}{5}$$

Substitute into rotation formulas:

$$x = x'\cos\theta - y'\sin\theta = x'\frac{2\sqrt{5}}{5} - y'\frac{\sqrt{5}}{5} = \frac{x'2\sqrt{5}-y'\sqrt{5}}{5}$$

$$y = x'\sin\theta + y'\cos\theta = x'\frac{\sqrt{5}}{5} + y'\frac{2\sqrt{5}}{5} = \frac{x'\sqrt{5}-y'2\sqrt{5}}{5}$$

Substitute into equation:

$$x^2 - 4xy + 4y^2 + 5\sqrt{5}\,y - 10 = 0$$

$$\left(\frac{x'2\sqrt{5}-y'\sqrt{5}}{5}\right)^2 - 4\left(\frac{x'2\sqrt{5}-y'\sqrt{5}}{5}\right)\left(\frac{x'\sqrt{5}+y'2\sqrt{5}}{5}\right) + 4\left(\frac{x'\sqrt{5}+y'2\sqrt{5}}{5}\right)^2 + 5\sqrt{5}\left(\frac{x'\sqrt{5}+y'2\sqrt{5}}{5}\right) - 10 = 0$$

$$x' + y'^2 + 2y' - 2 = 0$$

$$x' = -y'^2 - 2y' + 2$$

$$x' = -\left(y'^2 + 2y'\right) + 2$$

$$x' = -\left(y'^2 + 2y' + 1\right) + 2 + 1$$

$$x' = -\left(y'+1\right)^2 + 3$$

$x' = -\left(y'+1\right)^2 + 3$ is a parabola with vertex $(3,-1.)$

49. – 53. Answers may vary.

$$y_1 = \frac{-(Bx+E)+\sqrt{(Bx+E)^2 - 4C\left(Ax^2+Dx+F\right)}}{2C} \quad \text{and} \quad y_2 = \frac{-(Bx+E)-\sqrt{(Bx+E)^2 - 4C\left(Ax^2+Dx+F\right)}}{2C}$$

for equations of the form $Ax^2 + Bxy + Cy^2 + Dx + Ey + F = 0$.

55. $y_1 = \dfrac{-(Bx+E)+\sqrt{(Bx+E)^2 - 4C\left(Ax^2+Dx+F\right)}}{2C}$ and $y_2 = \dfrac{-(Bx+E)-\sqrt{(Bx+E)^2 - 4C\left(Ax^2+Dx+F\right)}}{2C}$

for equations of the form $Ax^2 + Bxy + Cy^2 + Dx + Ey + F = 0$.

$A = 7$, $B = 8$, $C = 1$, $D = 0$, $E = 0$, and $F = -1$.

Graph $y_1 = \dfrac{-8x+\sqrt{64x^2 - 4\left(7x^2-1\right)}}{2}$ and $y_2 = \dfrac{-8x-\sqrt{64x^2 - 4\left(7x^2-1\right)}}{2}$.

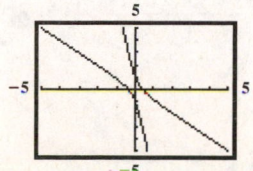

57. $y_1 = \dfrac{-(Bx+E)+\sqrt{(Bx+E)^2-4C(Ax^2+Dx+F)}}{2C}$ and $y_2 = \dfrac{-(Bx+E)-\sqrt{(Bx+E)^2-4C(Ax^2+Dx+F)}}{2C}$

for equations of the form $Ax^2+Bxy+Cy^2+Dx+Ey+F=0$.

$A=3, B=-6, C=3, D=10, E=-8,$ and $F=-2$.

Graph $y_1 = \dfrac{-(-6x-8)+\sqrt{(-6x-8)^2-12(3x^2+10x-2)}}{6}$ and $y_2 = \dfrac{-(-6x-8)-\sqrt{(-6x-8)^2-12(3x^2+10x-2)}}{6}$.

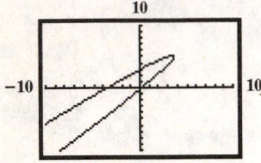

59. $y_1 = \dfrac{-(Bx+E)+\sqrt{(Bx+E)^2-4C(Ax^2+Dx+F)}}{2C}$ and $y_2 = \dfrac{-(Bx+E)-\sqrt{(Bx+E)^2-4C(Ax^2+Dx+F)}}{2C}$

for equations of the form $Ax^2+Bxy+Cy^2+Dx+Ey+F=0$.

$A=1, B=4, C=4, D=10\sqrt{5}, E=0,$ and $F=-9$.

Graph $y_1 = \dfrac{-4x+\sqrt{16x^2-16(x^2+10\sqrt{5}x-9)}}{8}$ and $y_2 = \dfrac{-4x-\sqrt{16x^2-16(x^2+10\sqrt{5}x-9)}}{8}$.

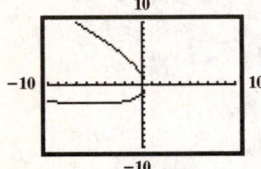

61. does not make sense; Explanations will vary. Sample explanation: This is not necessary because there is no xy term.

63. makes sense

65. $A=3, B=2,$ and $C=3$.

$\cot 2\theta = \dfrac{A-C}{B} = \dfrac{3-3}{2} = 0$

$2\theta = 90°$

$\theta = 45°$

$x = x'\cos 45° - y'\sin 45° = x'\left(\dfrac{\sqrt{2}}{2}\right) - y'\left(\dfrac{\sqrt{2}}{2}\right) = \dfrac{\sqrt{2}}{2}(x'-y')$

$y = x'\sin 45° + y'\cos 45° = x'\left(\dfrac{\sqrt{2}}{2}\right) + y'\left(\dfrac{\sqrt{2}}{2}\right) = \dfrac{\sqrt{2}}{2}(x'+y')$

Substitute into the equation: $3x^2 - 2xy + 3y^2 + 2 = 0$

$3\left[\dfrac{\sqrt{2}}{2}(x'-y')\right]^2 - 2\left[\dfrac{\sqrt{2}}{2}(x'-y')\right]\left[\dfrac{\sqrt{2}}{2}(x'+y')\right] + 3\left[\dfrac{\sqrt{2}}{2}(x'+y')\right]^2 + 2 = 0$

$\dfrac{3}{2}\left(x'^2 - 2x'y' + y'^2\right) - \left(x'^2 - y'^2\right) + \dfrac{3}{2}\left(x'^2 + 2x'y' + y'^2\right) = -2$

Multiply both sides by 2:

$3x'^2 - 6x'y' + 3y'^2 - 2x'^2 + 2y'^2 + 3x'^2 + 6x'y' + 3y'^2 = -4$

$4x'^2 + 8y'^2 = -4$

$\dfrac{4x'^2 + 8y'^2}{-4} = \dfrac{-4}{-4}$

$-x'^2 - 2y'^2 = 1$

There are no solutions to this equation since the left side of the equation is negative or 0 for all values of x' and y'. Thus, there are no points on the graph of this equation, just as one hand clapping makes no sound.

67. $A' = A\cos^2\theta + B\sin\theta\cos\theta + C\sin^2\theta$

$C' = A\sin^2\theta - B\sin\theta\cos\theta + C\cos^2\theta$

$A' + C' = A\cos^2\theta + B\sin\theta\cos\theta + C\sin^2\theta + A\sin^2\theta - B\sin\theta\cos\theta + C\cos^2\theta$

$\qquad = A\left(\cos^2\theta + \sin^2\theta\right) + B\left(\sin\theta\cos\theta - \sin\theta\cos\theta\right) + C\left(\sin^2\theta + \cos^2\theta\right)$

$\qquad = A(1) + B(0) + C(1)$

$\qquad = A + C$

69.   Answers may vary.

70.   parabola

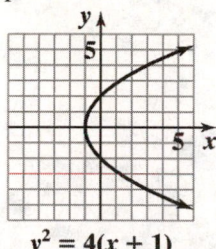

$y^2 = 4(x + 1)$

71.   parabola (partial)

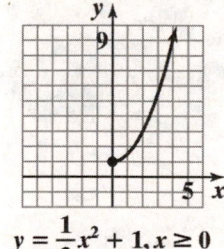

$y = \dfrac{1}{2}x^2 + 1, x \geq 0$

72.   ellipse

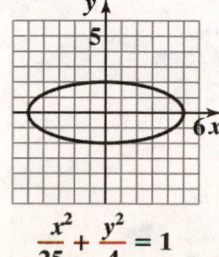

$\dfrac{x^2}{25} + \dfrac{y^2}{4} = 1$

Section 9.5

Check Point Exercises

1.

t	$x = t^2 + 1$	$y = 3t$	(x, y)
-2	$(-2)^2 + 1 = 5$	$3(-2) = -6$	$(5, -6)$
-1	$(-1)^2 + 1 = 2$	$3(-1) = -3$	$(2, -3)$
0	$0^2 + 1 = 1$	$3(0) = 0$	$(1, 0)$
1	$1^2 + 1 = 2$	$3(1) = 3$	$(2, 3)$
2	$2^2 + 1 = 5$	$3(2) = 6$	$(5, 6)$

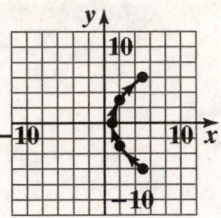

2. $x = \sqrt{t} \implies t = x^2$

Using $t = x^2$ and $y = 2t - 1$, $y = 2x^2 - 1$.

Since $t \geq 0$, x is nonnegative.

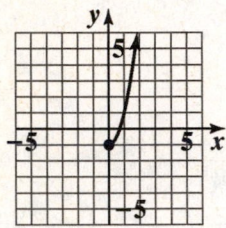

3. $x = 6\cos t$ and $y = 4\sin t$, $\pi \leq t \leq 2\pi$

$\dfrac{x}{6} = \cos t$ $\qquad \dfrac{y}{4} = \sin t$

Square and add the equations:

$$\dfrac{x^2}{36} = \cos^2 t$$

$$+ \quad \dfrac{y^2}{16} = \sin^2 t$$

$$\dfrac{x^2}{36} + \dfrac{y^2}{16} = \cos^2 t + \sin^2 t$$

$$\dfrac{x^2}{36} + \dfrac{y^2}{16} = 1$$

Since t is in the interval $[\pi, 2\pi]$, we use $t = \pi$, $t = \dfrac{3\pi}{2}$, and $t = 2\pi$:

$t = \pi$: $x = 6\cos\pi = -6$

$\qquad\qquad y = 4\sin\pi = 0$

$t = \dfrac{3\pi}{2}$: $x = 6\cos\dfrac{3\pi}{2} = 0$

$\qquad\qquad y = 4\sin\dfrac{3\pi}{2} = -4$

$t = 2\pi$: $x = 6\cos 2\pi = 6$

$\qquad\qquad y = 4\cos 2\pi = 0$

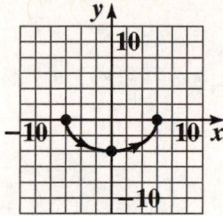

4. $y = x^2 - 25$

Let $x = t$. Then $y = t^2 - 25$.

The parametric equations are $x = t$ and $y = t^2 - 25$.

Exercise Set 9.5

1. $x = 3 - 5(1) = -2$

$y = 4 + 2(1) = 6$;

$\quad (-2, 6)$

3. $x = 2^2 + 1 = 5$

$y = 5 - 2^3 = 5 - 8 = -3$;

$\quad (5, -3)$

5. $x = 4 + 2\cos\dfrac{\pi}{2} = 4 + 2(0) = 4$

$y = 3 + 5\sin\dfrac{\pi}{2} = 3 + 5(1) = 8$;

$\quad (4, 8)$

7. $x = (60\cos 30°)(2) = \left(60 \cdot \dfrac{\sqrt{3}}{2}\right)(2) = 60\sqrt{3}$

$y = 5 + (60\sin 30°)(2) - 16(2)^2$

$\quad = 5 + \left(60 \cdot \dfrac{1}{2}\right)(2) - 16 \cdot 4$

$\quad = 5 + 60 - 64 = 1$;

$\left(60\sqrt{3}, 1\right)$

9.

t	$x = t + 2$	$y = t^2$	(x, y)
-2	$-2 + 2 = 0$	$(-2)^2 = 4$	$(0, 4)$
-1	$-1 + 2 = 1$	$(-1)^2 = 1$	$(1, 1)$
0	$0 + 2 = 2$	$0^2 = 0$	$(2, 0)$
1	$1 + 2 = 3$	$1^2 = 1$	$(3, 1)$
2	$2 + 2 = 4$	$2^2 = 4$	$(4, 4)$

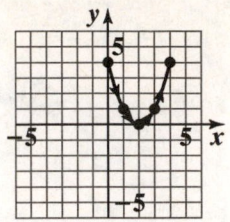

11.

t	$x = t - 2$	$y = 2t + 1$	(x, y)
-2	$-2 - 2 = -4$	$2(-2) + 1 = -3$	$(-4, -3)$
-1	$-1 - 2 = -3$	$2(-1) + 1 = -1$	$(-3, -1)$
0	$0 - 2 = -2$	$2(0) + 1 = 1$	$(-2, 1)$
1	$1 - 2 = -1$	$2(1) + 1 = 3$	$(-1, 3)$
2	$2 - 2 = 0$	$2(2) + 1 = 5$	$(0, 5)$
3	$3 - 2 = 1$	$2(3) + 1 = 7$	$(1, 7)$

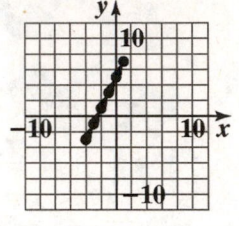

13.

t	$x = t + 1$	$y = \sqrt{t}$	(x, y)
0	$0 + 1 = 1$	$\sqrt{0} = 0$	$(1, 0)$
1	$1 + 1 = 2$	$\sqrt{1} = 1$	$(2, 1)$
4	$4 + 1 = 5$	$\sqrt{4} = 2$	$(5, 2)$
9	$9 + 1 = 10$	$\sqrt{9} = 3$	$(10, 3)$

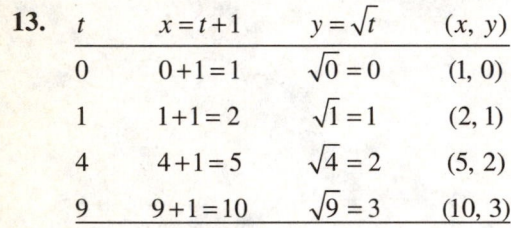
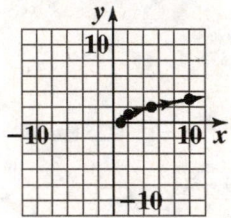

15.

t	$x = \cos t$	$y = \sin t$	(x, y)
0	$\cos 0 = 1$	$\sin 0 = 0$	$(1, 0)$
$\dfrac{\pi}{2}$	$\cos \dfrac{\pi}{2} = 0$	$\sin \dfrac{\pi}{2} = 1$	$(0, 1)$
π	$\cos \pi = -1$	$\sin \pi = 0$	$(-1, 0)$
$\dfrac{3\pi}{2}$	$\cos \dfrac{3\pi}{2} = 0$	$\sin \dfrac{3\pi}{2} = -1$	$(0, -1)$
2π	$\cos 2\pi = 1$	$\sin 2\pi = 0$	$(1, 0)$

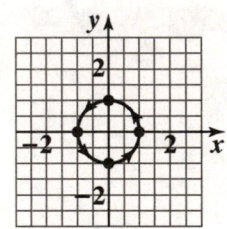

17.

t	$x = t^2$	$y = t^3$	(x, y)
-2	$(-2)^2 = 4$	$(-2)^3 = -8$	$(4, -8)$
-1	$(-1)^2 = 1$	$(-1)^3 = -1$	$(1, -1)$
0	$0^2 = 0$	$0^3 = 0$	$(0, 0)$
1	$1^2 = 1$	$1^3 = 1$	$(1, 1)$
2	$2^2 = 4$	$2^3 = 8$	$(4, 8)$

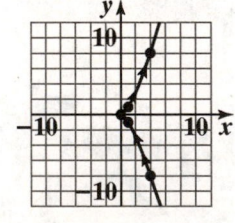

19.

t	$x = 2t$	$y = |t - 1|$	(x, y)
-2	$2(-2) = -4$	$|-2 - 1| = 3$	$(-4, 3)$
-1	$2(-1) = -2$	$|-1 - 1| = 2$	$(-2, 2)$
0	$2(0) = 0$	$|0 - 1| = 1$	$(0, 1)$
1	$2(1) = 2$	$|1 - 1| = 0$	$(2, 0)$
2	$2(2) = 4$	$|2 - 1| = 1$	$(4, 1)$

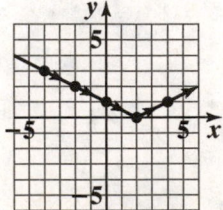

21. $x = t \implies y = 2x$

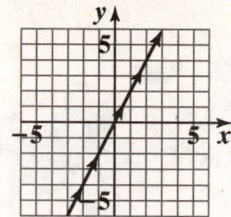

23. $x = 2t - 4$

$\dfrac{x+4}{2} = t$

Substitute into y:

$y = 4\left(\dfrac{x+4}{2}\right)^2 = (x+4)^2$

$y = (x+4)^2$

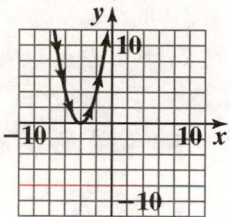

25. $x = \sqrt{t}$

$x^2 = t$

Substitute into y:

$y = x^2 - 1$

Since $t \geq 0$ in $x = \sqrt{t}$, $x \geq 0$.

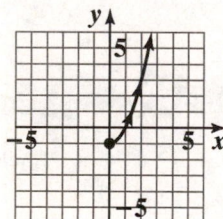

27. $x = 2\sin t$ and $y = 2\cos t$

$\dfrac{x}{2} = \sin t \qquad \dfrac{y}{2} = \cos t$

Square and add the equations:

$\dfrac{x^2}{4} = \sin^2 t$

$+ \qquad \dfrac{y^2}{4} = \cos^2 t$

$\rule{4cm}{0.4pt}$

$\dfrac{x^2}{4} + \dfrac{y^2}{4} = \sin^2 t + \cos^2 t$

$\dfrac{x^2}{4} + \dfrac{y^2}{4} = 1$

$x^2 + y^2 = 4$

The circle centered at $(0, 0)$ with radius 2.

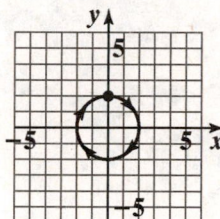

29. $x = 1 + 3\cos t$ and $y = 2 + 3\sin t$

$\dfrac{x-1}{3} = \cos t \qquad \dfrac{y-2}{3} = \sin t$

Square and add the equations:

$\dfrac{(x-1)^2}{9} = \cos^2 t$

$+ \qquad \dfrac{(y-2)^2}{9} = \sin^2 t$

$\rule{4cm}{0.4pt}$

$\dfrac{(x-1)^2}{9} + \dfrac{(y-2)^2}{9} = \cos^2 t + \sin^2 t$

$\dfrac{(x-1)^2}{9} + \dfrac{(y-2)^2}{9} = 1$

$(x-1)^2 + (y-2)^2 = 9$

This is a circle centered at $(1, 2)$ with radius 3.

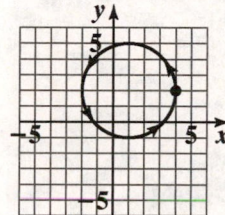

31. $x = 2\cos t$ and $y = 3\sin t$

$\dfrac{x}{2} = \cos t \qquad \dfrac{y}{3} = \sin t$

Square and add the equations:

$\dfrac{x^2}{4} = \cos^2 t$

$+ \quad \dfrac{y^2}{9} = \sin^2 t$

$\overline{\qquad\qquad\qquad\qquad}$

$\dfrac{x^2}{4} + \dfrac{y^2}{9} = \cos^2 t + \sin^2 t$

$\dfrac{x^2}{4} + \dfrac{y^2}{9} = 1$

This is an ellipse centered at (0, 0).

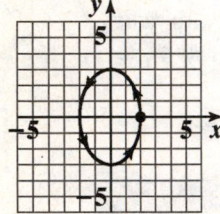

33. $x = 1 + 3\cos t$ and $y = -1 + 2\sin t$

$\dfrac{x-1}{3} = \cos t \qquad \dfrac{y+1}{2} = \sin t$

Square and add the equations:

$\dfrac{(x-1)^2}{9} = \cos^2 t$

$+ \quad \dfrac{(y+1)^2}{4} = \sin^2 t$

$\overline{\qquad\qquad\qquad\qquad}$

$\dfrac{(x-1)^2}{9} + \dfrac{(y+1)^2}{4} = \cos^2 t + \sin^2 t$

$\dfrac{(x-1)^2}{9} + \dfrac{(y+1)^2}{4} = 1$

Since $0 \le t \le \pi$, $-1 \le \cos t \le 1$ and $0 \le \sin t \le 1$.

Thus, $-1 \le \dfrac{x-1}{3} \le 1$ and $0 \le \dfrac{y+1}{2} \le 1$. Hence,

$-2 \le x \le 4$ and $-1 \le y \le 1$. This is the upper half of an ellipse centered at (–1, 1).

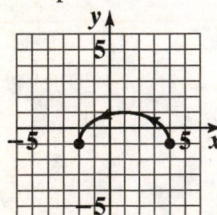

35. $x = \sec t$ and $y = \tan t$

Square and subtract the equations:

$x^2 = \sec^2 t$

$- \quad (y^2 = \tan^2 t)$

$\overline{\qquad\qquad\qquad\qquad}$

$x^2 - y^2 = \sec^2 t - \tan^2 t$

$x^2 - y^2 = 1$

This is a hyperbola centered at (0, 0).

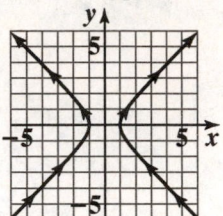

37. $x = t^2 + 2$ and $y = t^2 - 2$

$x - 2 = t^2$

Substitute $x - 2$ into $y = t^2 - 2$ for t^2:

$y = (x - 2) - 2$

$y = x - 4$

Since $t^2 \ge 0$ for all t, $x \ge 2$ and $y \ge -2$.

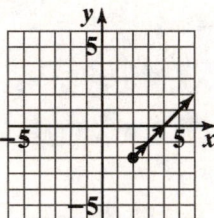

39. $x = 2^t$ and $y = 2^{-t}$

$y = \left(2^t\right)^{-1}$

Substitute x in $y = \left(2^t\right)^{-1}$ for 2^t:

$y = (x)^{-1}$

$y = \dfrac{1}{x}$

Since $t \ge 0$, $x \ge 1$ and $y \ge 0$.

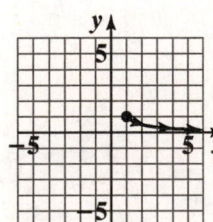

41. $x = h + r\cos t$ and $y = k + r\sin t$

$\dfrac{x-h}{r} = \cos t$ $\dfrac{y-k}{r} = \sin t$

Square and add the equations:

$$\dfrac{(x-h)^2}{r^2} = \cos^2 t$$

$$+ \quad \dfrac{(y-k)^2}{r^2} = \sin^2 t$$

$$\rule{6cm}{0.4pt}$$

$$\dfrac{(x-h)^2}{r^2} + \dfrac{(y-k)^2}{r^2} = \cos^2 t + \sin^2 t$$

$$\dfrac{(x-h)^2}{r^2} + \dfrac{(y-k)^2}{r^2} = 1$$

$$(x-h)^2 + (y-k)^2 = r^2$$

43. $x = h + a\sec t$ and $y = k + b\tan t$

$\dfrac{x-h}{a} = \sec t$ $\dfrac{y-k}{b} = \tan t$

Square and subtract the equations:

$$\dfrac{(x-h)^2}{a^2} = \sec^2 t$$

$$- \quad \left(\dfrac{(y-k)^2}{b^2} = \tan^2 t \right)$$

$$\rule{6cm}{0.4pt}$$

$$\dfrac{(x-h)^2}{a^2} - \dfrac{(y-k)^2}{b^2} = \sec^2 t - \tan^2 t$$

$$\dfrac{(x-h)^2}{a^2} - \dfrac{(y-k)^2}{b^2} = 1$$

45. $h = 3,\ k = 5,$ and $r = 6$

$x = h + r\cos t$ and $y = k + r\sin t$

$x = 3 + 6\cos t$ $y = 5 + 6\sin t$

47. $h = -2,\ k = 3,\ a = 5,\ b = 2$

$x = h + a\cos t$ and $y = k + b\sin t$

$x = -2 + 5\cos t$ $y = 3 + 2\sin t$

49. $h = 0,\ k = 0,\ a = 4,\ c = 6$

$$c^2 = a^2 + b^2$$

$$6^2 = 4^2 + b^2$$

$$36 - 16 = b^2$$

$$b = \sqrt{20} = 2\sqrt{5}$$

$x = h + a\sec t$ and $y = k + b\tan t$

$x = 0 + 4\sec t$ $y = 0 + 2\sqrt{5}\tan t$

$x = 4\sec t$ $y = 2\sqrt{5}\tan t$

51. $x_1 = -2,\ y_1 = 4,\ x_2 = 1,\ y_2 = 7$

53.   Answers may vary.
Sample answer:
$x = t$ and $y = 4t - 3$; $x = t + 1$ and $y = 4t + 1$

55.   Answers may vary.
Sample answer:
$x = t$ and $y = t^2 + 4$; $x = t + 1$ and $y = t^2 + 2t + 5$

57. a.

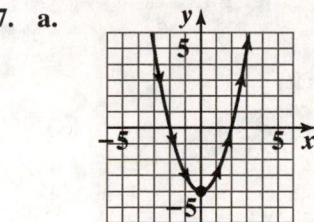

b.

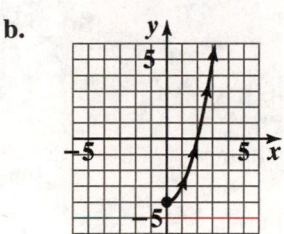

c.

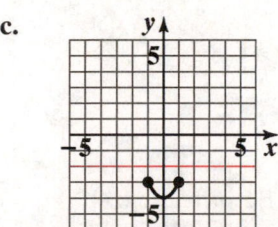

d.

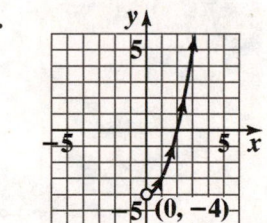

Explanations of how the curves differ from each other may vary.

59.

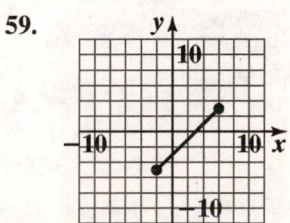

domain: $[-2, 6]$;
range: $[-5, 3]$

61.

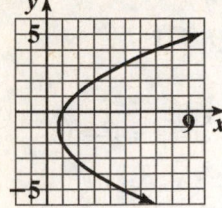

domain: $\left[\dfrac{3}{4}, \infty\right)$;

range: $[-\infty, \infty]$

63.

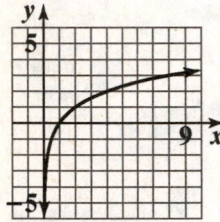

a. increasing: $[-\infty, \infty]$

b. no maximum or minimum

65.

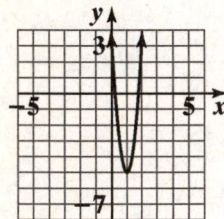

a. decreasing: $[-\infty, 1]$; increasing: $(1, \infty)$

b. minimum of -5 at $x = 1$

67.

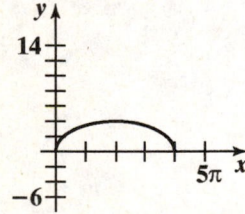

a. increasing: $(1, 2\pi)$; decreasing: $(2\pi, 4\pi)$

b. maximum of 4 at $x = 2\pi$,
minimum of 0 at $x = 0$ and $x = 4\pi$

69. a. $x = (180\cos 40°)t$

$y = 3 + (180\sin 40°)t - 16t^2$

b. After 1 second:
$x = (180\cos 40°) \cdot 1$

≈ 137.9 feet in distance

$y = 3 + (180\sin 40°)1 - 16 \cdot 1^2$

≈ 102.7 feet in height

After 2 seconds:
$x = (180\cos 40°) \cdot 2$

≈ 275.8 feet in distance

$y = 3 + (180\sin 40°) \cdot 2 - 16 \cdot 2^2$

≈ 170.4 feet in height

After 3 seconds:
$x = (180\cos 40°) \cdot 3$

≈ 413.7 feet in distance

$y = 3 + (180\sin 40°) \cdot 3 - 16 \cdot 3^2$

≈ 206.1 feet in height

The points on the curve are (137.9, 102.7), (275.8, 170.4), (413.7, 206.1).

c. The ball is no longer in flight when its height above ground is zero:

$0 = 3 + (180\sin 40°)t - 16t^2$

$0 = -16t^2 + (180\sin 40°)t + 3$

$t = \dfrac{-(180\sin 40°) \pm \sqrt{(180\sin 40°)^2 - 4 \cdot (-16)(3)}}{2(-16)}$

$t \approx -.03$ or $t \approx 7.3$

Since we cannot use the negative time, the ball hits the ground at $t \approx 7.3$ seconds.
The total horizontal distance is:
$x = (180\cos 40°) \cdot (7.3) \approx 1006.6$ feet

d. Answers may vary.

71. – 77. Answers may vary.

79.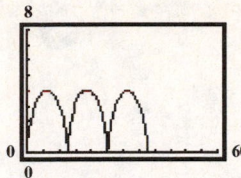

81.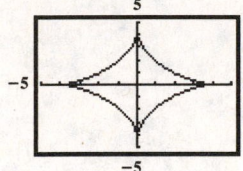

83. $x = (v_0 \cos\theta)t$ and $y = h + (v_0 \sin\theta)t - 16t^2$ where v_0 is the initial velocity, θ is the angle from horizontal, h is the height above the ground, and t is the time, in seconds.

$x = (200\cos 55°)t$ and $y = (200\sin 55°)t - 16t^2$

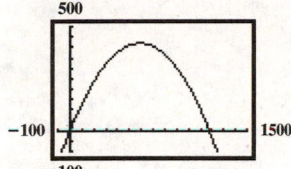

Window: $[-100, 1500] \times [-100, 500]$

The maximum height is 419.4 feet at a time of 5.1 seconds. The range of the projective is 1174.6 feet horizontally. It hits the ground at 10.2 seconds.

85. a. $x = (v_0 \cos\theta)t$ and $y = h + (v_0 \sin\theta)t - 16t^2$ where v_0 is the initial velocity, θ is the angle from horizontal, h is the height above the ground, and t is the time, in seconds.

$x = (140\cos 22°)t$ and $y = 5 + (140\sin 22°)t - 16t^2$

b.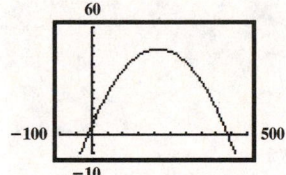

Window: $[-100, 500] \times [-10, 60]$

c. The maximum height is 48.0 feet. It occurs at 1.6 seconds.

d. The ball is in the air for 3.4 seconds.

e. The ball travels 437.5 feet.

87. makes sense

89. makes sense

91. $x = 3\sin t$ and $y = 3\cos t$

93. $r = \dfrac{2}{1 + \frac{1}{2}\cos\theta}$

94.

θ	0	$\dfrac{\pi}{2}$	$\dfrac{2\pi}{3}$	$\dfrac{3\pi}{4}$	$\dfrac{5\pi}{6}$	π
$r = \dfrac{4}{2+\cos\theta}$	$\dfrac{4}{3}$ ≈ 1.33	2	$\dfrac{8}{3}$ ≈ 2.66	$\dfrac{16+4\sqrt{2}}{7}$ ≈ 3.09	$\dfrac{32+8\sqrt{3}}{13}$ ≈ 3.53	4

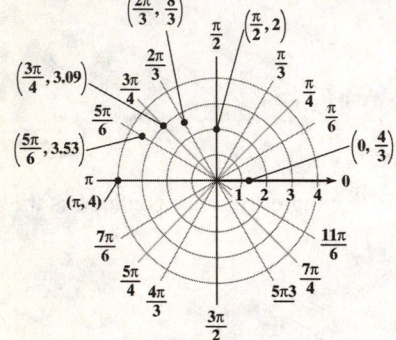

95. **a.**
$$r = \frac{1}{3 - 3\cos\theta}$$
$$r(3 - 3\cos\theta) = 1$$
$$3r - 3r\cos\theta = 1$$
$$3r = 1 + 3r\cos\theta$$
$$(3r)^2 = (1 + 3r\cos\theta)^2$$
$$9r^2 = (1 + 3r\cos\theta)^2$$

b.
$$9r^2 = (1 + 3r\cos\theta)^2$$
$$9(x^2 + y^2) = (1 + 3x)^2$$
$$9x^2 + 9y^2 = 1 + 6x + 9x^2$$
$$9y^2 = 1 + 6x$$

This is the equation of a parabola.

Section 9.6

Check Point Exercises

1. Graph $r = \dfrac{4}{2 - \cos\theta}$.

    Step 1: Divide numerator and denominator by 2 to write the equation in standard form:
    $$r = \frac{2}{1 - \frac{1}{2}\cos\theta}$$

    Step 2: $e = \dfrac{1}{2}$ and $ep = \dfrac{1}{2}p = 2$, so $p = 4$. Since $e < 1$, the graph is an ellipse.

    Step 3: The graph has symmetry with respect to the polar axis. One focus is at the pole and the directrix is $x = -4$.

    Find the vertices by selecting $\theta = 0$ and $\theta = \pi$: $(4, 0)$ and $\left(\dfrac{4}{3}, \pi\right)$.

    Sketch the upper half by plotting some points, then use the symmetry of the graph to sketch the lower half.

    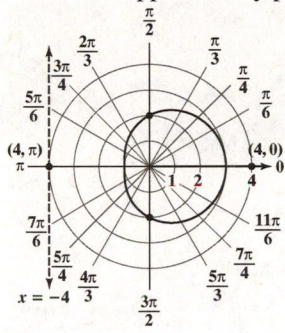

2. Graph $r = \dfrac{8}{4 + 4\sin\theta}$.

    Step 1: Divide numerator and denominator by 4 to write the equation in standard form:
    $$r = \frac{2}{1 + \sin\theta}$$
    Step 2: $e = 1$ and $ep = 1$ $p = 2$, so $p = 2$. Since $e = 1$, the graph is parabola.

    Step 3: The graph has symmetry with respect to $\theta = \dfrac{\pi}{2}$. The focus is at the pole and the directrix is $y = 2$. Since the

    vertex is on the line $\theta = \dfrac{\pi}{2}$ (y-axis) the vertex is at $\left(1, \dfrac{\pi}{2}\right)$. To find the intercepts on the polar axis, select $\theta = 0$ and

    $\theta = \pi$: $(2, 0)$ and $(2, \pi)$.

    Sketch the right half by plotting some points, then use symmetry to sketch the left half.

    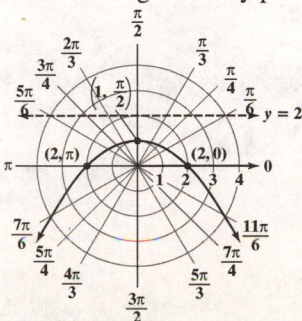

3. Graph $r = \dfrac{9}{3 - 9\cos\theta}$.

Step 1: Divide numerator and denominator by 3 to write the equation in standard form:

$$r = \frac{3}{1 - 3\cos\theta}$$

Step 2: $e = 3$ and $ep = 3p = 3$, so $p = 1$. Since $e > 1$, the graph is a hyperbola.

Step 3: The graph is symmetric with respect to the polar axis. One focus is at the pole and the directrix is at $x = -1$. The transverse axis is horizontal and the vertices lie on the polar axis. Find them by selecting $\theta = 0$ and $0 = \pi$:

$$\left(-\frac{3}{2}, 0\right) \text{ and } \left(\frac{3}{4}, \pi\right).$$

Sketch the upper half of the hyperbola by plotting some points, then use symmetry to sketch the lower half.

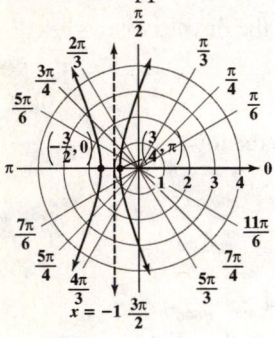

Exercise Set 9.6

1. $r = \dfrac{3}{1 + \sin\theta}$

$e = 1$ and $ep = 3$, so $p = 3$.

    a. The graph is a parabola.

    b. The directrix is 3 units above the pole, at $y = 3$.

3. $r = \dfrac{6}{3 - 2\cos\theta}$

Divide numerator and denominator by 3: $r = \dfrac{2}{1 - \frac{2}{3}\cos\theta}$.

$e = \dfrac{2}{3}$ and $ep = 2$, so $p = 3$.

    a. The graph is an ellipse.

    b. The directrix is 3 units to the left of the pole, at $x = -3$.

5. $r = \dfrac{8}{2+2\sin\theta}$

Divide numerator and denominator

by 2: $r = \dfrac{4}{1+\sin\theta}$.

$e = 1$ and $ep = 4$, so $p = 4$.

a. The graph is a parabola.

b. The directrix is 4 units above the pole, at $y = 4$.

7. $r = \dfrac{12}{2-4\cos\theta}$

Divide numerator and denominator

by 2: $r = \dfrac{6}{1-2\cos\theta}$.

$e = 2$ and $ep = 6$, so $p = 3$.

a. The graph is a hyperbola.

b. The directrix is 3 units to the left of the pole, at $x = -3$.

9. $r = \dfrac{1}{1+\sin\theta}$

$e = 1$ and $ep = 1$, so $p = 1$.

Since $e = 1$, the graph is a parabola. It is symmetric with respect to the y-axis and has a directrix at $y = 1$.

The vertex is at $\left(\dfrac{1}{2}, \dfrac{\pi}{2}\right)$.

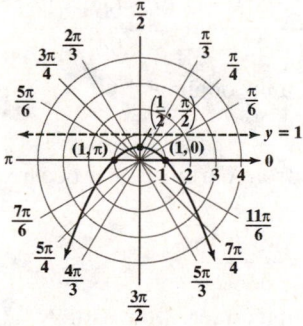

11. $r = \dfrac{2}{1-\cos\theta}$

$e = 1$ and $ep = 2$, so $p = 2$.

Since $e = 1$, the graph is a parabola. It is symmetric with respect to the polar axis and has a directrix at $x = -2$. The vertex is at $(1, \pi)$.

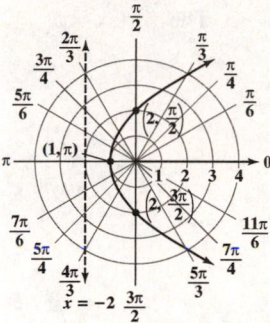

13. $r = \dfrac{12}{5+3\cos\theta}$

Write in standard form:

$r = \dfrac{\frac{12}{5}}{1+\frac{3}{5}\cos\theta}$

$e = \dfrac{3}{5}$ and $ep = \dfrac{12}{5}$, so $p = 4$. Since $e < 1$, the graph

is an ellipse. It is symmetric with respect to the polar axis and has a directrix at $x = 4$.

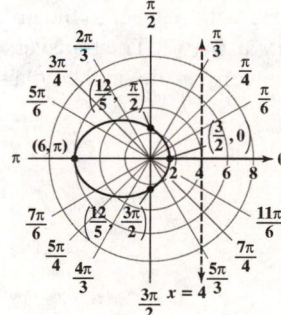

15. $r = \dfrac{6}{2 - 2\sin\theta}$

Write in standard form: $r = \dfrac{3}{1 - \sin\theta}$

$e = 1$ and $ep = 3$, so $p = 3$. Since $e = 1$, the graph is a parabola. It is symmetric with respect to the y-axis and has a directrix at $y = -3$. The vertex is at $\left(\dfrac{3}{2}, \dfrac{3\pi}{2}\right)$.

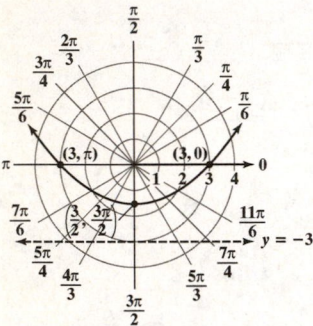

17. $r = \dfrac{8}{2 - 4\cos\theta}$

Write in standard form:

$r = \dfrac{4}{1 - 2\cos\theta}$

$e = 2$ and $ep = 4$, so $p = 2$. Since $e > 1$, the graph is a hyperbola. It is symmetric with respect to the polar axis and it has a directrix at $x = -2$. The transverse axis is horizontal and the vertices lie on the polar axis.

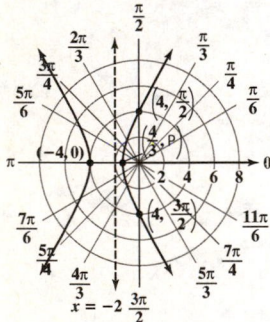

19. $r = \dfrac{12}{3 - 6\cos\theta}$

Write in standard form:

$r = \dfrac{4}{1 - 2\cos\theta}$

$e = 2$ and $ep = 4$, so $p = 2$. Since $e > 2$, the graph is a hyperbola. It is symmetric with respect to the polar axis and has a directrix at $x = -2$. The transverse axis is horizontal and the vertices lie on the polar axis.

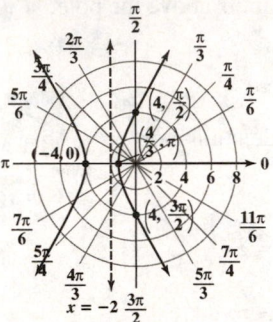

21. $[-3, 15, 1]$ by $[-7, 7, 1]$

23. $[-4, 2, 1]$ by $[-10, 10, 1]$

25. $[-2, 5, 1]$ by $[-10, 10, 1]$

27. $[-4, 4, 1]$ by $[-10, 0.4, 1]$

29. The shortest distance from the sun occurs on the positive y-axis, at $\theta = \dfrac{\pi}{2}$.

When $\theta = \dfrac{\pi}{2}$, $r = \dfrac{1.069}{1 + 0.967\sin\frac{\pi}{2}} = \dfrac{1.069}{1.967}$

≈ 0.54 astronomical units or about 51 million miles.

31. His greatest distance from Earth's center occurred when $\theta = 0$:

$r = \dfrac{4090.76}{1 - 0.0076\cos 0} = \dfrac{4090.76}{0.9924}$

≈ 4122 miles from the center of the earth. Assuming the earth to be perfectly spherical, he was $4122 - 3960 = 162$ miles from the surface of the earth.

33. – 39. Answers may vary.

41. Write the equation in standard form:

$$r = \frac{3}{1 + \frac{5}{4}\sin\theta}$$

Since $e = \dfrac{5}{4} > 1$, the graph is a hyperbola.

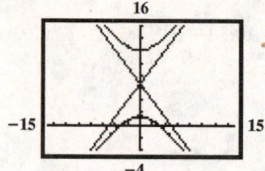

43.

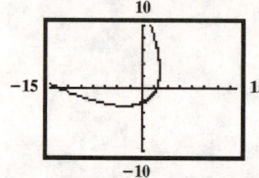

The graph appears to be rotated counter-clockwise through an angle of $\dfrac{\pi}{4}$ radians.

45. Mercury: $r = \dfrac{\left(1 - 0.2056^2\right)\left(36.0 \times 10^6\right)}{1 - 0.2056\cos\theta}$

Earth: $r = \dfrac{\left(1 - 0.0167^2\right)\left(92.96 \times 10^6\right)}{1 - 0.0167\cos\theta}$

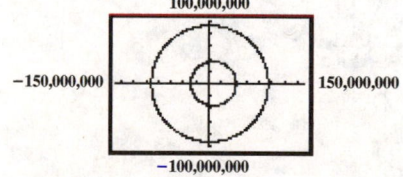

$r : -150,000,000 - 150,000,000$
$y : -100,000,000, -100,000,000$
Observations may vary.

47. does not make sense; Explanations will vary.
Sample explanation: This form is not symmetrical with respect to the y-axis.

49. does not make sense; Explanations will vary.
Sample explanation: A knowledge of conic sections is necessary to graph such equations.

51. Since the equation is an ellipse with a vertex at (4, 0), the polar axis is the major axis. Since $\dfrac{PF}{PD} = e$, then at the point (4, 0),

$\dfrac{4}{x} = \dfrac{1}{2}$, where x is the distance between the point (4, 0) and the directrix. Thus, $x = 8$ and the distance between the point (4, 0) and the directrix is 8. The directrix is either
$x = -4$ or $x = 12$. There are two polar equations that meet the given conditions. If the directrix is $x = -4$,

$$r = \frac{\frac{1}{2}(4)}{1 - \frac{1}{2}\cos\theta} = \frac{2}{1 - \frac{1}{2}\cos\theta}.$$

If the directrix is $x = 12$, $r = \dfrac{\frac{1}{2}(12)}{1 + \frac{1}{2}\cos\theta} = \dfrac{6}{1 + \frac{1}{2}\cos\theta}.$

53. $r = \dfrac{1}{2 - 2\cos\theta}$

Write the equation in standard form: $r = \dfrac{\frac{1}{2}}{1 - \cos\theta}$

Since $e = 1$, the graph is a parabola.
Write in rectangular coordinates:

$$r = \frac{\frac{1}{2}}{1 - \cos\theta}$$

$$r(1 - \cos\theta) = \frac{1}{2}$$

$$r - r\cos\theta = \frac{1}{2}$$

$$r = r\cos\theta + \frac{1}{2}$$

$$r = x + \frac{1}{2}$$

Substitution: $x = r\cos\theta$

$$r^2 = \left(x + \frac{1}{2}\right)^2$$

Square both sides

$$x^2 + y^2 = \left(x + \frac{1}{2}\right)^2$$

Substitution: $r^2 = x^2 + y^2$

$$x^2 + y^2 = x^2 + x + \frac{1}{4}$$

$$y^2 = x + \frac{1}{4}$$

55. For $n = 1$; $\dfrac{(-1)^n}{3^n - 1} = \dfrac{(-1)^1}{3^1 - 1} = \dfrac{-1}{3 - 1} = -\dfrac{1}{2}$

For $n = 2$; $\dfrac{(-1)^n}{3^n - 1} = \dfrac{(-1)^2}{3^2 - 1} = \dfrac{1}{9 - 1} = \dfrac{1}{8}$

For $n = 3$; $\dfrac{(-1)^n}{3^n - 1} = \dfrac{(-1)^3}{3^3 - 1} = \dfrac{-1}{27 - 1} = -\dfrac{1}{26}$

For $n = 4$; $\dfrac{(-1)^n}{3^n - 1} = \dfrac{(-1)^4}{3^4 - 1} = \dfrac{1}{81 - 1} = \dfrac{1}{80}$

56. $5 \cdot 4 \cdot 3 \cdot 2 \cdot 1 = 120$

57. For $i = 1$; $i^2 + 1 = 1^2 + 1 = 1 + 1 = 2$

For $i = 2$; $i^2 + 1 = 2^2 + 1 = 4 + 1 = 5$

For $i = 3$; $i^2 + 1 = 3^2 + 1 = 9 + 1 = 10$

For $i = 4$; $i^2 + 1 = 4^2 + 1 = 16 + 1 = 17$

For $i = 5$; $i^2 + 1 = 5^2 + 1 = 25 + 1 = 26$

For $i = 6$; $i^2 + 1 = 6^2 + 1 = 36 + 1 = 37$

$2 + 5 + 10 + 17 + 26 + 37 = 97$

Chapter 9 Review Exercises

1. $a^2 = 36, a = 6$
$b^2 = 25, b = 5$
$c^2 = a^2 - b^2 = 36 - 25 = 11$
$c = \sqrt{11}$
The foci are at ($\sqrt{11}$, 0) and ($-\sqrt{11}$, 0)

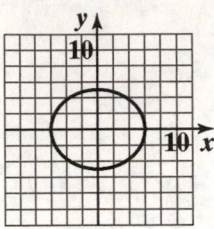

$$\frac{x^2}{36} + \frac{y^2}{25} = 1$$

2. $a^2 = 25, a = 5$
$b^2 = 16, b = 4$
$c^2 = a^2 - b^2$
$c^2 = 25 - 16$
$c^2 = 9$
$c = 3$
The foci are (0, 3) and (0, −3).

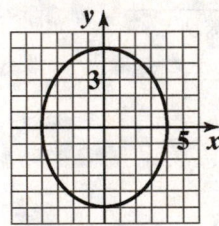

$$\frac{y^2}{25} + \frac{x^2}{16} = 1$$

3. $\dfrac{4x^2}{16} + \dfrac{y^2}{16} = \dfrac{16}{16}$

$\dfrac{x^2}{4} + \dfrac{y^2}{16} = 1$

$b^2 = 4,\ b = 2$

$a^2 = 16,\ a = 4$

$c^2 = a^2 - b^2 = 16 - 4 = 12$

$c = \sqrt{12} = 2\sqrt{3}$

The foci are at $(0,\ 2\sqrt{3})$ and $(0,\ -2\sqrt{3})$.

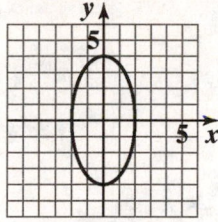

$4x^2 + y^2 = 16$

4. $\dfrac{4x^2}{36} + \dfrac{9y^2}{36} = \dfrac{36}{36}$

$\dfrac{x^2}{9} + \dfrac{y^2}{4} = 1$

$a^2 = 9,\ a = 3$

$b^2 = 4,\ b = 2$

$c^2 = a^2 - b^2 = 9 - 4 = 5,\ c = \sqrt{5}$

The foci are at $(\sqrt{5},\ 0)$ and $(-\sqrt{5},\ 0)$.

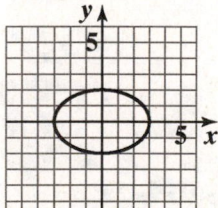

$4x^2 + 9y^2 = 36$

5. $a^2 = 16\ \ a = 4$

$b^2 = 9\ \ b = 3$

$c^2 = 16 - 9 = 7,\ c = \sqrt{7}$

center: $(1, -2)$

The foci are at $(1+\sqrt{7},\ -2)$ and $(1-\sqrt{7},\ -2)$.

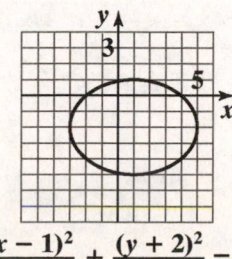

$\dfrac{(x-1)^2}{16} + \dfrac{(y+2)^2}{9} = 1$

6. $a^2 = 16,\ a = 4$

$b^2 = 9,\ b = 3$

$c^2 = a^2 - b^2 = 16 - 9 = 7,\ c = \sqrt{7}$

center: $(-1, 2)$

The foci are at $(-1,\ 2+\sqrt{7})$ and $(-1,\ 2-\sqrt{7})$.

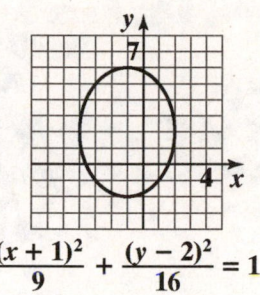

$\dfrac{(x+1)^2}{9} + \dfrac{(y-2)^2}{16} = 1$

7. $4x^2 + 24x + 9y^2 - 36y = -36$

$4(x^2 + 6x + 9) + 9(y^2 - 4y + 4)$

$= -36 + 36 + 36$

$= 4(x+3)^2 + 9(y-2)^2 = 36$

$\dfrac{(x+3)^2}{9} + \dfrac{(y-2)^2}{4} = 1$

$c^2 = a^2 - b^2 = 5,\ c = \sqrt{5}$

center: $(-3, 2)$

The foci are at $(-3+\sqrt{5},\ 2)$ and $(-3-\sqrt{5},\ 2)$.

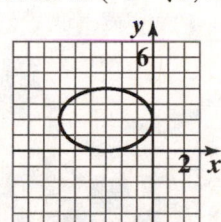

$4x^2 + 9y^2 + 24x - 36y + 36 = 0$

8. $9x^2 - 18x + 4y^2 + 8y = 23$

$9(x^2 - 2x + 1) + 4(y^2 + 2y + 1) = 23 + 9 + 4$

$9(x-1)^2 + 4(y+1)^2 = 36$

$\dfrac{(x-1)^2}{4} + \dfrac{(y+1)^2}{9} = 1$

$c^2 = a^2 - b^2 = 9 - 4 = 5,\ c = \sqrt{5}$

center: $(1, -1)$

The foci are at $(1,\ -1+\sqrt{5})$ and $(1,\ -1-\sqrt{5})$.

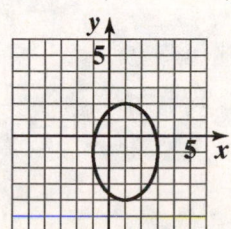

$9x^2 + 4y^2 - 18x + 8y - 23 = 0$

9. $c = 4, c^2 = 16$
$a = 5, a^2 = 25$
$b^2 = a^2 - c^2 = 25 - 16 = 9$
$$\frac{x^2}{25} + \frac{y^2}{9} = 1$$

10. $c = 3, c^2 = 9$
$a = 6, a^2 = 36$
$b^2 = a^2 - c^2 = 36 - 9 = 27$
$$\frac{x^2}{27} + \frac{y^2}{36} = 1$$

11. $2a = 12, a = 6, a^2 = 36$
$2b = 4, b = 2, b^2 = 4$
$$\frac{(x+3)^2}{36} + \frac{(y-5)^2}{4} = 1$$

12. $2a = 20, a = 10, a^2 = 100$
$b = 6, b^2 = 36$
$$\frac{x^2}{100} + \frac{y^2}{36} = 1$$

13. $2a = 50, a = 25$
$b = 15$
$$\frac{x^2}{625} + \frac{y^2}{225} = 1$$
Let $x = 14$.
$$\frac{(14)^2}{625} + \frac{y^2}{225} = 1$$
$$y^2 = 225\left(1 - \frac{196}{625}\right)$$
$y \approx 15(0.8285) \approx 12.4 > 12$
Yes, the truck can drive under the archway.

14. The hit ball will collide with the other ball.

15. $c^2 = a^2 + b^2 = 16 + 1 = 17, c = \sqrt{17}$
The foci are at $(\sqrt{17}, 0)$ and $(-\sqrt{17}, 0)$.

Asymptotes: $y = \pm\frac{1}{4}x$

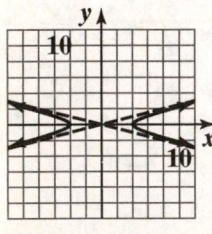

$$\frac{x^2}{16} - y^2 = 1$$

16. $c^2 = a^2 + b^2 = 16 + 1 = 17$
$c = \sqrt{17}$
The foci are at $(0, \sqrt{17})$ and $(0, -\sqrt{17})$.
Asymptotes: $y = \pm 4x$

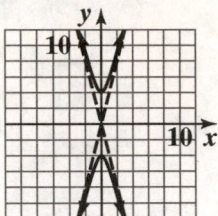

$$\frac{y^2}{16} - x^2 = 1$$

17. $\dfrac{x^2}{16} - \dfrac{y^2}{9} = 1$
$c^2 = a^2 + b^2 = 16 + 9 = 25, c = 5$
The foci are at $(5, 0)$ and $(-5, 0)$.

Asymptotes: $y = \pm\frac{3}{4}x$

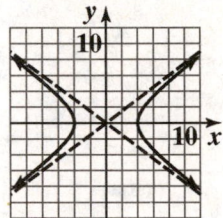

$$9x^2 - 16y^2 = 144$$

18. $\dfrac{y^2}{4} - \dfrac{x^2}{16} = 1$
$c^2 = a^2 + b^2 = 4 + 16 = 20$
$c = \sqrt{20} = 2\sqrt{5}$

The foci are at $(0, 2\sqrt{5})$ and $(0, -2\sqrt{5})$.

Asymptotes: $y = \pm\frac{1}{2}x$

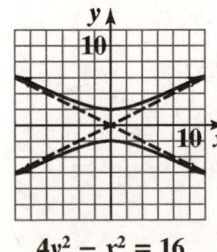

$$4y^2 - x^2 = 16$$

19. $c^2 = a^2 + b^2 = 25 + 16 = 41, c = \sqrt{41}$

center: $(2, -3)$

The foci are at $(2+\sqrt{41}, -3)$ and $(2-\sqrt{41}, -3)$.

Asymptotes: $y + 3 = \pm\dfrac{4}{5}(x-2)$

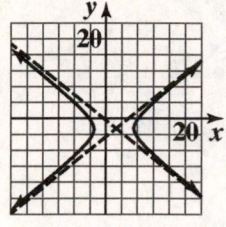

$$\dfrac{(x-2)^2}{25} - \dfrac{(y+3)^2}{16} = 1$$

20. $c^2 = a^2 + b^2 = 25 + 16 = 41, c = \sqrt{41}$

center: $(3, -2)$

The foci are at $(3, -2+\sqrt{41})$ and $(3, -2-\sqrt{41})$.

Asymptotes: $y + 2 = \pm\dfrac{5}{4}(x-3)$

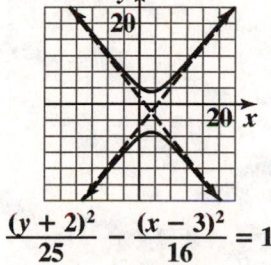

$$\dfrac{(y+2)^2}{25} - \dfrac{(x-3)^2}{16} = 1$$

21. $\qquad\quad y^2 - 4y - 4x^2 + 8x - 4 = 0$

$\qquad (y^2 - 4y + 4) - 4(x^2 - 2x + 1) = 4 + 4 - 4$

$\qquad\qquad (y-2)^2 - 4(x-1)^2 = 4$

$\dfrac{(y-2)^2}{4} - (x-1)^2 = 1$

$c^2 = a^2 + b^2 = 4 + 1 = 5, c = \sqrt{5}$

center: $(1, 2)$

The foci are at $(1, 2+\sqrt{5})$ and $(1, 2-\sqrt{5})$.

Asymptotes: $y - 2 = \pm 2(x-1)$

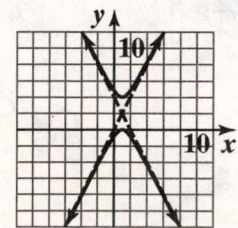

$$y^2 - 4y - 4x^2 + 8x - 4 = 0$$

22. $x^2 - 2x - y^2 - 2y = 1$

$(x^2 - 2x + 1) - (y^2 + 2y + 1) = 1 + 1 - 1$

$(x-1)^2 - (y+1)^2 = 1$

$c^2 = a^2 + b^2 = 1 + 1 = 2, c = \sqrt{2}$

center: $(1, -1)$

The foci are at $(1+\sqrt{2}, -1)$ and $(1-\sqrt{2}, -1)$.

asymptotes: $y + 1 = \pm(x-1)$

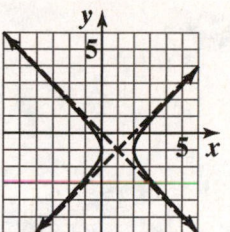

$$x^2 - y^2 - 2x - 2y - 1 = 0$$

23. $c = 4, c^2 = 16$

$a = 2, a^2 = 4$

$b^2 = c^2 - a^2 = 16 - 4 = 12$

$$\dfrac{y^2}{4} - \dfrac{x^2}{12} = 1$$

24. $c = 8, c^2 = 64$

$a = 3, a^2 = 9$

$b^2 = c^2 - a^2 = 64 - 9 = 55$

$$\dfrac{x^2}{9} - \dfrac{y^2}{55} = 1$$

25. If the foci are at $(0, -2)$ and $(0, 2)$, then $c = 2$. If the vertices are at $(0, -3)$ and $(0, 3)$ then $a = 3$. This is not possible since c must be greater than a.

26. foci: $(\pm100, 0)$, $c = 100$

$$|d_1 - d_2| = \left(0.186\,\dfrac{\text{mi}}{\mu\text{s}}\right)(500\mu\text{s}) = 93 \text{ mi} = 2a$$

$$a = \dfrac{93}{2}$$

$$b^2 = c^2 - a^2 = (100)^2 - \left(\dfrac{93}{2}\right)^2 = 7837.75$$

$$\dfrac{x^2}{\left(\frac{93}{2}\right)^2} - \dfrac{y^2}{7837.75} = 1$$

$$\dfrac{x^2}{2162.25} - \dfrac{y^2}{7837.75} = 1$$

27. $4p = 8, p = 2$
vertex: $(0, 0)$
focus: $(2, 0)$
directrix: $x = -2$

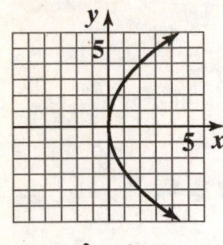

$y^2 = 8x$

28. $x^2 + 16y = 0$

$\qquad x^2 = -16y$

$4p = -16$
$\ p = -4$
vertex: $(0, 0)$
focus: $(0, -4)$
directrix: $y = 4$

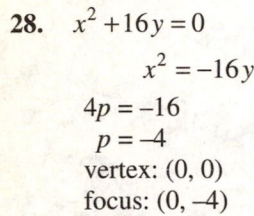

$x^2 + 16y = 0$

29. $4p = -16$
$p = -4$
vertex: $(0, 2)$
focus: $(-4, 2)$
directrix: $x = 4$

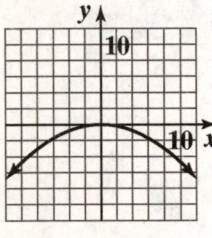

$(y - 2)^2 = -16x$

30. $4p = 4, p = 1$
vertex: $(4, -1)$
focus: $(4, 0)$
directrix: $y = -2$

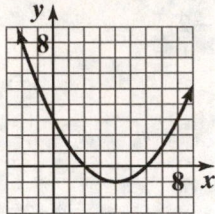

$(x - 4)^2 = 4(y + 1)$

31. $x^2 = -4y + 4$
$x^2 = -4(y - 1)$
$4p = -4, p = -1$
vertex: $(0, 1)$
focus: $(0, 0)$
directrix: $y = 2$

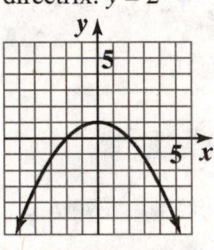

$x^2 + 4y = 4$

32. $\qquad y^2 - 10y = 4x - 21$

$y^2 - 10y + 25 = 4x - 21 + 25$

$\qquad (y - 5)^2 = 4(x + 1)$

$4p = 4, p = 1$
vertex: $(-1, 5)$
focus: $(0, 5)$
directrix: $x = -2$

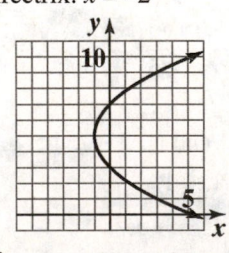

$y^2 - 4x - 10y + 21 = 0$

33.
$$x^2 - 4x - 2y = 0$$
$$x^2 - 4x = 2y$$
$$(x^2 - 4x + 4) = 2y + 4$$
$$(x - 2)^2 = 2(y + 2)$$
$$4p = 2, p = \frac{1}{2}$$
vertex: $(2, -2)$

focus: $\left(2, -\frac{3}{2}\right)$

directrix: $y = -\frac{5}{2}$

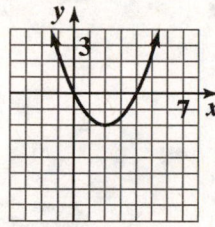

$$x^2 - 4x - 2y = 0$$

34.
$$p = 12$$
$$y^2 = 48x$$

35.
$$p = -11$$
$$x^2 = -44y$$

36.
$$x^2 = 4py$$
$$(6)^2 = 4p(3)$$
$$p = 3$$
$$x^2 = 12y$$
Place the light 3 inches from the vertex at $(0, 3)$.

37.
$$x^2 = 4py$$
$$(1750)^2 = 4p(316)$$
$$4p \approx 9691$$
$$x^2 = 9691y$$
Let $x = 1750 - 1000 = 750$.
$$y = \frac{x^2}{9691} = \frac{(750)^2}{9691} \approx 58$$
The height is approximately 58 feet.

38.
$$x^2 = 4py$$
$$(150)^2 = 4p(44)$$
$$22{,}500 = 176p$$
$$p \approx 128$$
The receiver should be placed approximately 128 feet from the base of the dish.

39. $A = 0$, $C = 1$.
$AC = 0$, so the graph is a parabola.

40. $A = 1$, $C = 16$.
$AC = 16 > 0$ and $A \neq C$, so the graph is an ellipse.

41. $A = 16$, $C = 9$.
$AC = 16 \cdot 9 = 144 > 0$ and $A \neq C$, so the graph is an ellipse.

42. $A = 4$, $C = -9$.
$AC = 4(-9) = 36 < 0$, so the graph is a hyperbola.

43. $A = 5$, $B = 2\sqrt{3}$, $C = 3$.
$B^2 - 4AC = \left(2\sqrt{3}\right)^2 - 4(5)(3) = 12 - 60 = -48$ Since
$B^2 - 4AC < 0$, the graph is an ellipse or a circle.

44. $A = 5$, $B = -8$, $C = 7$.
$B^2 - 4AC = (-8)^2 - 4(5)(7) = 64 - 140 = -76$. Since
$B^2 - 4AC < 0$, the graph is an ellipse or a circle.

45. $A = 1$, $B = 6$, $C = 9$.
$B^2 - 4AC = 6^2 - 4(1)(9) = 36 - 36 = 0$. Since
$B^2 - 4AC = 0$, the graph is a parabola.

46. $A = 1$, $B = -2$, $C = 3$.
$B^2 - 4AC = (-2)^2 - 4(1)(3) = 4 - 12 = -8$ Since
$B^2 - 4AC < 0$, the graph is an ellipse or a circle.

47. $xy - 4 = 0$

    a. $A = 0, B = 1, C = 0$.

$$\cot 2\theta = \frac{A - C}{B} = \frac{0 - 0}{1} = 0$$

$$2\theta = 90°$$

$$\theta = 45°$$

$$x = x'\cos 45° - y'\sin 45° = x'\left(\frac{\sqrt{2}}{2}\right) - y'\left(\frac{\sqrt{2}}{2}\right) = \frac{\sqrt{2}}{2}(x' - y')$$

$$y = x'\sin 45° + y'\cos 45° = x'\left(\frac{\sqrt{2}}{2}\right) + y'\left(\frac{\sqrt{2}}{2}\right) = \frac{\sqrt{2}}{2}(x' + y')$$

Substitute into the equation: $xy - 4 = 0$

$$\left[\frac{\sqrt{2}}{2}(x' - y')\right]\left[\frac{\sqrt{2}}{2}(x' - y')\right] - 4 = 0$$

$$\frac{1}{2}\left(x'^2 - y'^2\right) - 4 = 0$$

$$x'^2 - y'^2 = 8$$

    b. $\dfrac{x'^2}{8} - \dfrac{y'^2}{8} = 1$

    c.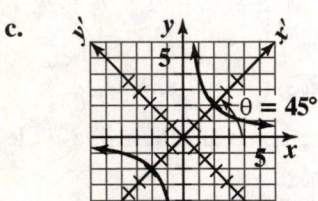

48. $x^2 + xy + y^2 - 1 = 0$

    a. $A = 1, B = 1, C = 1$.

$$\cot 2\theta = \frac{A - C}{B} = \frac{1 - 1}{1} = \frac{0}{1} = 0$$

$$2\theta = 90°$$

$$\theta = 45°$$

$$x = x'\cos 45° - y'\sin 45° = \frac{\sqrt{2}}{2}(x' - y')$$

$$y = x'\sin 45° + y'\cos 45° = \frac{\sqrt{2}}{2}(x' + y')$$

Substitute into the equation: $x^2 + xy + y^2 - 1 = 0$

$$\left[\frac{\sqrt{2}}{2}(x' - y')\right]^2 + \left[\frac{\sqrt{2}}{2}(x' - y')\right]\left[\frac{\sqrt{2}}{2}(x' + y')\right] + \left[\frac{\sqrt{2}}{2}(x' + y')\right]^2 - 1 = 0$$

$$\frac{1}{2}\left(x'^2 - 2x'y' + y'^2\right) + \frac{1}{2}\left(x'^2 - y'^2\right) + \frac{1}{2}\left(x'^2 + 2x'y' + y'^2\right) = 1$$

Multiply both sides by 2 and simplify: $3x'^2 + y'^2 = 2$

b. $\dfrac{x'^2}{\frac{2}{3}} + \dfrac{y'^2}{2} = 1$

c.

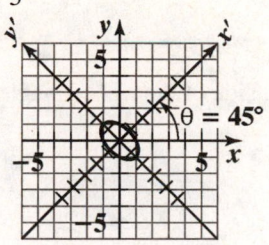

$\theta = 45°$

49. $4x^2 + 10xy + 4y^2 - 9 = 0$

a. $A = 4$, $B = 10$, $C = 4$.

$$\cot 2\theta = \frac{A-C}{B} = \frac{4-4}{10} = \frac{0}{10} = 0$$

$$2\theta = 90°$$

$$\theta = 45°$$

$$x = x'\cos 45° - y'\sin 45° = \frac{\sqrt{2}}{2}\left(x' - y'\right)$$

$$y = x'\sin 45° + y'\cos 45° = \frac{\sqrt{2}}{2}\left(x' + y'\right)$$

Substitute into the equation: $4x^2 + 10xy + 4y^2 - 9 = 0$

$$4\left[\frac{\sqrt{2}}{2}\left(x' - y'\right)\right]^2 + 10\left[\frac{\sqrt{2}}{2}\left(x' - y'\right)\right]\left[\frac{\sqrt{2}}{2}\left(x' + y'\right)\right] + 4\left[\frac{\sqrt{2}}{2}\left(x' + y'\right)\right]^2 - 9 = 0$$

$$4\cdot\frac{1}{2}\left(x'^2 - 2x'y' + y'^2\right) + 10\cdot\frac{1}{2}\left(x'^2 - y'^2\right) + 4\cdot\frac{1}{2}\left(x'^2 + 2x'y' + y'^2\right) = 9$$

Multiply both sides by 2 and simplify: $18x'^2 - 2y'^2 = 18$

b. $\dfrac{x'^2}{1} - \dfrac{y'^2}{9} = 1$

c.

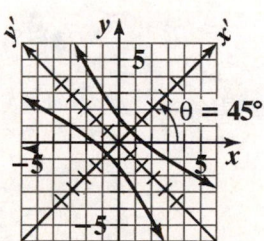

$\theta = 45°$

50. $6x^2 - 6xy + 14y^2 - 45 = 0$

a. $A = 6, B = -6, C = 14$

$$\cot 2\theta = \frac{A-C}{B} = \frac{6-14}{-6} = \frac{-8}{-6} = \frac{4}{3}$$

Since θ is always acute, and $\cot 2\theta$ is positive, 2θ lies in quadrant I. The third side of the right triangle is found using the Pythagorean Theorem.

$$4^2 + 3^2 = r^2$$
$$r = 5$$

So, $\cos 2\theta = \dfrac{4}{5}$.

$$\sin\theta = \sqrt{\frac{1-\cos 2\theta}{2}} = \sqrt{\frac{1-\frac{4}{5}}{2}} = \frac{\sqrt{10}}{10} \quad \text{and} \quad \cos\theta = \sqrt{\frac{1+\cos 2\theta}{2}} = \sqrt{\frac{1+\frac{4}{5}}{2}} = \frac{3\sqrt{10}}{10}.$$

So, $x = x'\cos\theta - y'\sin\theta = x'\left(\dfrac{3\sqrt{10}}{10}\right) - y'\left(\dfrac{\sqrt{10}}{10}\right) = \dfrac{\sqrt{10}}{10}(3x' - y')$

and $y = x'\sin\theta - y'\cos\theta = x'\left(\dfrac{\sqrt{10}}{10}\right) + y'\left(\dfrac{3\sqrt{10}}{10}\right) = \dfrac{\sqrt{10}}{10}(x' + 3y')$.

Substitute into the equation: $6x^2 - 6xy + 14y^2 - 45 = 0$

$$6\left[\frac{\sqrt{10}}{10}(3x' - y')\right]^2 - 6\left[\frac{\sqrt{10}}{10}(3x' - y')\right]\left[\frac{\sqrt{10}}{10}(x' + 3y')\right] + 14\left[\frac{\sqrt{10}}{10}(x' + 3y')\right]^2 - 45 = 0$$

$$6\left[\frac{1}{10}\left(9x'^2 - 6x'y' + y'^2\right)\right] - 6\left[\frac{1}{10}\left(3x'^2 + 8x'y' - 3y'^2\right)\right] + 14\left[\frac{1}{10}\left(x'^2 + 6x'y' + 9y'^2\right)\right] = 45$$

Multiply both sides by 10 and simplify: $50x'^2 + 150y'^2 = 450$

b. $\dfrac{x'^2}{9} + \dfrac{y'^2}{3} = 1$

c.

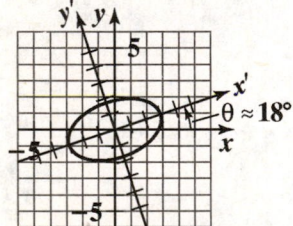

The axes are rotated by $\theta = \sin^{-1}\left(\dfrac{\sqrt{10}}{10}\right) \approx 18°$.

51. $x^2 + 2\sqrt{3}xy + 3y^2 - 12\sqrt{3}x + 12y = 0$

$A = 1,\ B = 2\sqrt{3},\ C = 3$

$\cot 2\theta = \dfrac{A - C}{B} = \dfrac{1 - 3}{2\sqrt{3}} = \dfrac{-2}{2\sqrt{3}} = \dfrac{-\sqrt{3}}{3}$

$2\theta = 120°$

$\theta = 60°$

$x = x'\cos 60° - y'\sin 60° = x'\left(\dfrac{1}{2}\right) - y'\left(\dfrac{\sqrt{3}}{2}\right) = \dfrac{1}{2}\left(x' - \sqrt{3}y'\right)$

and $y = x'\sin 60° + y'\cos 60° = x'\left(\dfrac{\sqrt{3}}{2}\right) + y'\left(\dfrac{1}{2}\right) = \dfrac{1}{2}\left(\sqrt{3}x' + y'\right)$

Substitute into the equation: $x^2 + 2\sqrt{3}xy + 3y^2 - 12\sqrt{3}x + 12y = 0$

$\left[\dfrac{1}{2}\left(x' - \sqrt{3}y'\right)\right]^2 + 2\sqrt{3}\left[\dfrac{1}{2}\left(x' - \sqrt{3}y'\right)\right]\left[\dfrac{1}{2}\left(\sqrt{3}x' + y'\right)\right] + 3\left[\dfrac{1}{2}\left(\sqrt{3}x' + y'\right)\right]^2 - 12\sqrt{3}\left[\dfrac{1}{2}\left(x' - \sqrt{3}y'\right)\right]$

$+12\left[\dfrac{1}{2}\left(\sqrt{3}x' + y'\right)\right] = 0$

$\dfrac{1}{4}\left(x'^2 - 2\sqrt{3}x'y' + 3y'^2\right) + 2\sqrt{3}\cdot\dfrac{1}{4}\left(\sqrt{3}x'^2 - 2x'y' - \sqrt{3}y'^2\right) + 3\cdot\dfrac{1}{4}\left(3x'^2 + 2\sqrt{3}x'y' + y'^2\right) - 6\sqrt{3}x' + 18y'$

$+6\sqrt{3}x' + 6y' = 0$

Multiply both sides by 4 and simplify: $16x'^2 + 96y' = 0$

b. $16x'^2 = -96y'$

$x'^2 = -6y'$

c.

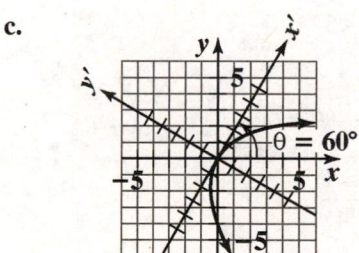

52. $x = 2t - 1$ and $y = 1 - t$; $-\infty < t < \infty$

$$\frac{x+1}{2} = t$$

Substitute into y: $y = 1 - \left(\frac{x+1}{2}\right)$

$$y = -\frac{1}{2}x + \frac{1}{2}$$

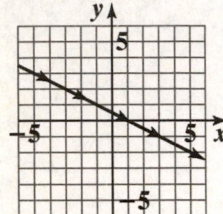

53. $x = t^2$ and $y = t - 1$; $-1 \leq t \leq 3$

$$y + 1 = t$$

Substitute into x:

$$x = (y+1)^2$$

$$(y+1)^2 = x$$

$$0 \leq x \leq 9, \; -2 \leq y \leq 2$$

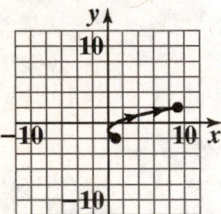

54. $x = 4t^2$ and $y = t + 1$; $-\infty < t < \infty$

$$y - 1 = t$$

Substitute into x:

$$x = 4(y-1)^2$$

$$\frac{1}{4}x = (y-1)^2$$

$$(y-1)^2 = \frac{1}{4}x$$

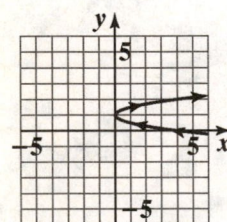

55. $x = 4\sin t$, $y = 3\cos t$; $0 \leq t < \pi$

$$\frac{x}{4} = \sin t \qquad \frac{y}{3} = \cos t$$

Square and add the equations:

$$\frac{x^2}{16} = \sin^2 t$$

$$+ \quad \frac{y^2}{9} = \cos^2 t$$

$$\overline{\frac{x^2}{16} + \frac{y^2}{9} = \sin^2 t + \cos^2 t}$$

$$\frac{x^2}{16} + \frac{y^2}{9} = \sin^2 t + \cos^2 t$$

$$\frac{x^2}{16} + \frac{y^2}{9} = 1$$

$$0 \leq x \leq 4, \; -3 \leq y \leq 3$$

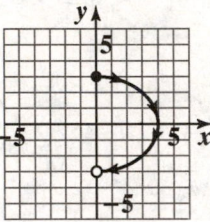

56. $x = 3 + 2\cos t$, $y = 1 + 2\sin t$; $0 \leq t < 2\pi$

$$\frac{x-3}{2} = \cos t \qquad \frac{y-1}{2} = \sin t$$

Square and add the equations:

$$\frac{(x-3)^2}{4} = \cos^2 t$$

$$+ \quad \frac{(y-1)^2}{4} = \sin^2 t$$

$$\overline{\frac{(x-3)^2}{4} + \frac{(y-1)^2}{4} = \cos^2 t + \sin^2 t}$$

$$\frac{(x-3)^2}{4} + \frac{(y-1)^2}{4} = 1$$

or $(x-3)^2 + (y-1)^2 = 4$

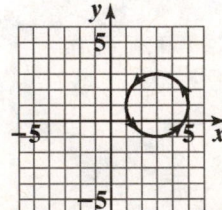

57. $x = 3\sec t,\ y = 3\tan t;\ 0 \le t \le \dfrac{\pi}{4}$

$\dfrac{x}{3} = \sec t \qquad \dfrac{y}{3} = \tan t$

Square and subtract the equations:

$\dfrac{x^2}{9} = \sec^2 t$

$-\left(\dfrac{y^2}{9} = \tan^2 t\right)$

$\overline{\dfrac{x^2}{9} - \dfrac{y^2}{9} = \sec^2 t - \tan^2 t}$

$\dfrac{x^2}{9} - \dfrac{y^2}{9} = 1$

$3 \le x \le 3\sqrt{2},\ 0 \le y \le 3$

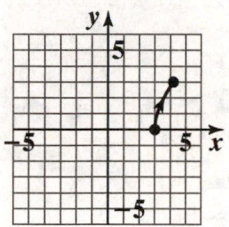

58. Answers may vary. Sample answer:
$x = t$ and $y = t^2 + 6$; $x = t+1$ and $y = t^2 + 2t + 7$

59. a. $x = (100\cos 40°)t$

$y = 6 + (100\sin 40°)t - 16t^2$

b. After 1 second:
$x = (100\cos 40°) \cdot 1$

≈ 76.6 feet in distance

$y = 6 + (100\sin 40°) \cdot 1 - 16(1)^2$

≈ 54.3 feet in height

After 2 seconds:
$x = (100\cos 40°) \cdot 2$

≈ 153.2 feet in distance

$y = 6 + (100\sin 40°) \cdot 2 - 16(2)^2$

≈ 70.6 feet in height

After 3 seconds:
$x = (100\cos 40°) \cdot 3$

≈ 229.8 feet in distance

$y = 6 + (100\sin 40°) \cdot 3 - 16(3)^2$

≈ 54.8 feet in height

c. $0 = 6t(100\sin 40°)t - 16t^2$

Using the quadratic formula with
$a = -16$, $b = 100\sin 40°$, and $c = 6$,

$t = \dfrac{-100\sin 40° \pm \sqrt{(100\sin 40°)^2 - 4(-16)(6)}}{2(-16)}$

$t \approx -0.1$ or $t \approx 4.1$

Since t cannot be negative, we discard
$t \approx -0.1$.

At $t \approx 4.1$, $x = (100\cos 40°)(4.1) \approx 314.1$

The ball is in flight for 4.1 seconds. It travels a total horizontal distance of 314.1 feet.

d.

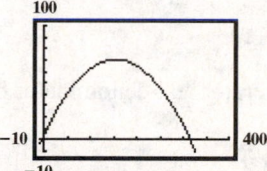

The ball is at its maximum height at 2.0 seconds. The maximum height is 70.6 feet.

60. a. $r = \dfrac{4}{1 - \sin\theta}$

b. $e = 1$ and $ep = 4$, so $p = 4$. Since $e = 1$, the graph is a parabola.

c.

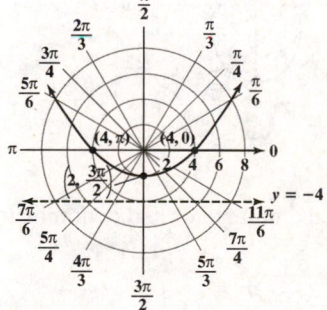

61. **a.** $r = \dfrac{6}{1 + \cos\theta}$

b. $e = 1$ and $ep = 6$, so $p = 6$. Since $e = 1$, the graph is a parabola.

c.

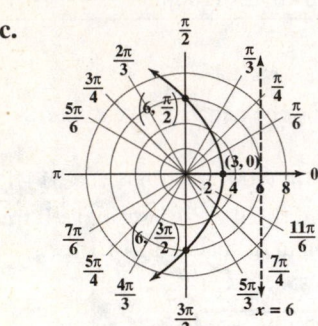

62. **a.** Divide numerator and denominator by 2:

$$r = \dfrac{3}{1 + \frac{1}{2}\sin\theta}$$

b. $e = \dfrac{1}{2}$ and $ep = 3$, so $p = 6$. Since $e < 1$, the graph is an ellipse.

c.

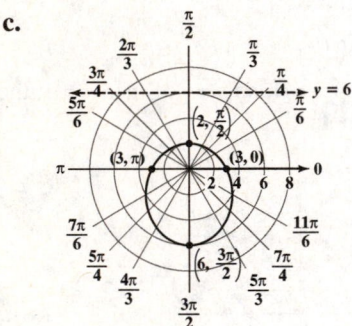

63. **a.** Divide the numerator and denominator by 3:

$$r = \dfrac{\frac{2}{3}}{1 - \frac{2}{3}\cos\theta}$$

b. $e = \dfrac{2}{3}$ and $ep = \dfrac{2}{3}$, so $p = 1$. Since $e < 1$, the graph is an ellipse.

c.

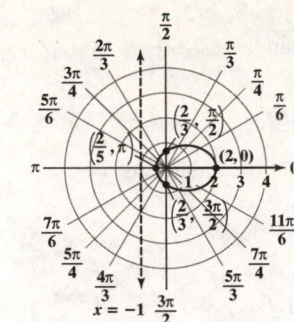

64. **a.** Divide the numerator and denominator by 3:

$$r = \dfrac{2}{1 + 2\sin\theta}$$

b. $e = 2$ and $ep = 2$, so $p = 1$. Since $e > 1$, the graph is a hyperbola.

c.

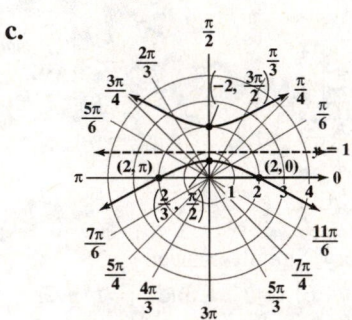

65. **a.** Divide the numerator and denominator by 4:

$$r = \dfrac{2}{1 + 4\cos\theta}$$

b. $e = 4$ and $ep = 2$, so $p = \dfrac{1}{2}$. Since $e > 1$, the graph is a hyperbola.

c.

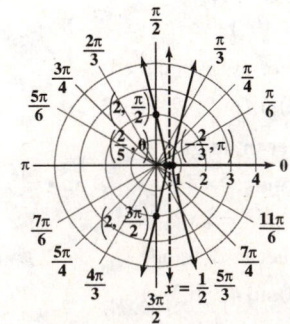

Chapter 9 Test

1. $\dfrac{x^2}{4} - \dfrac{y^2}{9} = 1$

$c^2 = a^2 + b^2 = 4 + 9 = 13$, $c = \sqrt{13}$

hyperbola

The foci are at $\left(\sqrt{13}, 0\right)$ and $\left(-\sqrt{13}, 0\right)$.

Asymptotes: $y = \pm \dfrac{3}{2} x$

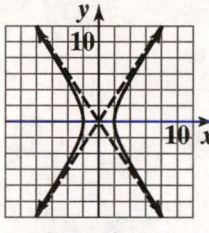

$9x^2 - 4y^2 = 36$

2. $4p = -8, \, p = -2$
parabola
vertex: $(0, 0)$
focus: $(0, -2)$
directrix: $y = 2$

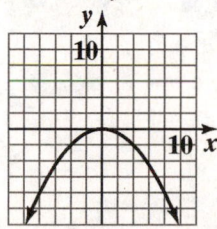

$x^2 = -8y$

3.   The center is at $(-2, 5)$.
$c^2 = a^2 - b^2 = 25 - 9 = 16$, $c = 4$
ellipse
The foci are at $(-6, 5)$ and $(2, 5)$.

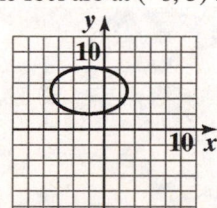

$\dfrac{(x + 2)^2}{25} + \dfrac{(y - 5)^2}{9} = 1$

4. $4x^2 - y^2 + 8x + 2y + 7 = 0$

$\left(4x^2 + 8x\right) - \left(y^2 - 2y\right) = -7$

$4\left(x^2 + 2x + 1\right) - \left(y^2 - 2y + 1\right) = -7 + 4 - 1$

$4\left(x + 1\right)^2 - \left(y - 1\right)^2 = -4$

$\left(y - 1\right)^2 - 4\left(x + 1\right)^2 = 4$

$\dfrac{\left(y - 1\right)^2}{4} - \left(x + 1\right)^2 = 1$

$c^2 = a^2 + b^2 = 4 + 1 = 5$, $c = \sqrt{5}$

The center is at $(-1, 1)$.

Asymptotes: $y - 1 = \pm 2(x + 1)$

hyperbola

The foci are at $\left(-1, 1 + \sqrt{5}\right)$ and $\left(-1, 1 - \sqrt{5}\right)$.

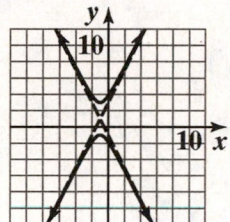

$4x^2 - y^2 + 8x + 2y + 7 = 0$

5. $4p = 8, \, p = 2$
parabola
vertex: $(-5, 1)$
focus: $(-5, 3)$
directrix: $y = -1$

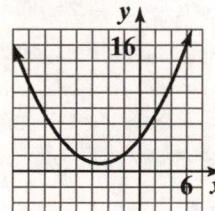

$(x + 5)^2 = 8(y - 1)$

6. $c = 7, \, c^2 = 49$
$a = 10, \, a^2 = 100$
$b^2 = a^2 - c^2 = 100 - 49 = 51$
$\dfrac{x^2}{100} + \dfrac{y^2}{51} = 1$

7. $c = 10, \, c^2 = 100$
$a = 7, \, a^2 = 49$
$b^2 = c^2 - a^2 = 100 - 49 = 51$
$\dfrac{y^2}{49} - \dfrac{x^2}{51} = 1$

8. $p = 50$

$y^2 = 4px$

$y^2 = 200x$

9. $b = 24, b^2 = 576$

$2a = 80, a = 40, a^2 = 1600$

$c^2 = a^2 - b^2 = 1600 - 576 = 1024$

$c = \sqrt{1024} = 32$

The two people should each stand 32 feet from the center of the room, along the major axis.

10. a. $x^2 = 4py$

when $x = \pm 3, y = 3$

$9 = 4p(3)$

$3 = 4p$

$\dfrac{3}{4} = p$

$x^2 = 3y$

b. focus: $\left(0, \dfrac{3}{4}\right)$

The light is placed $\dfrac{3}{4}$ inch above the vertex.

11. $A = 1, C = 9$

$AC = 1 \cdot 9 = 9 > 0$, so the graph is an ellipse.

12. $A = 1, B = 1, C = 1$

$B^2 - 4AC = 1^2 - 4(1)(1) = -3$.

Since $B^2 - 4AC < 0$, the graph is an ellipse or circle.

13. $7x^2 - 6\sqrt{3}xy + 13y^2 - 16 = 0$

$A = 7, B = -6\sqrt{3}, C = 13$

$\cot 2\theta = \dfrac{A - C}{B} = \dfrac{7 - 13}{-6\sqrt{3}}$

$= \dfrac{-6}{-6\sqrt{3}} = \dfrac{1}{\sqrt{3}} = \dfrac{\sqrt{3}}{3}$

$2\theta = 60°$

$\theta = 30°$

14. $x = t^2, \quad y = t - 1; -\infty < t < \infty$

$y + 1 = t$

Substitute into x:

$x = (y + 1)^2$

$(y + 1)^2 = x$

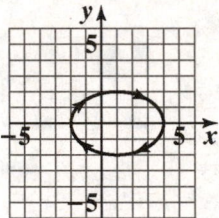

15. $x = 1 + 3\sin t, \quad y = 2\cos t; \ 0 \le t < 2\pi$

$\dfrac{x - 1}{3} = \sin t \qquad \dfrac{y}{2} = \cos t$

Square and add the equations:

$\dfrac{(x - 1)^2}{9} = \sin^2 t$

$+ \quad \dfrac{y^2}{4} = \cos^2 t$

$\overline{\dfrac{(x - 1)^2}{9} + \dfrac{y^2}{4} = \sin^2 t + \cos^2 t}$

$\dfrac{(x - 1)^2}{9} + \dfrac{y^2}{4} = 1$

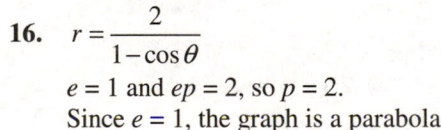

16. $r = \dfrac{2}{1 - \cos\theta}$

$e = 1$ and $ep = 2$, so $p = 2$.

Since $e = 1$, the graph is a parabola.

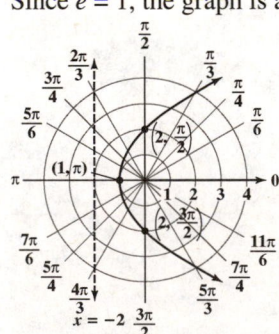

17. $r = \dfrac{4}{2 + \sin\theta}$

Divide the numerator and denominator by 2:

$r = \dfrac{2}{1 + \dfrac{1}{2}\sin\theta}$

$e = \dfrac{1}{2}$ and $ep = 2$, so $p = 4$. Since $e < 1$, the graph is an ellipse.

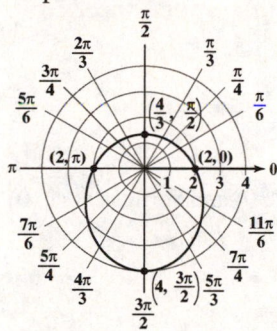

Cumulative Review Exercises (Chapters 1–9)

1. $2(x-3) + 5x = 8(x-1)$
$2x - 6 + 5x = 8x - 8$
$7x - 6 = 8x - 8$
$-x = -2$
$x = 2$
The solution set is $\{2\}$.

2. $-3(2x-4) > 2(6x-12)$
$-6x + 12 > 12x - 24$
$-18x > -36$
$x < 2$
The solution set is $\{x \mid x < 2\}$.

3. $x - 5 = \sqrt{x+7}$
$(x-5)^2 = x + 7$
$x^2 - 10x + 25 = x + 7$
$x^2 - 11x + 18 = 0$
$(x-2)(x-9) = 0$
$x = 2$ or $x = 9$
The solution $x = 2$ is extraneous, so the only solution is $x = 9$.
The solution set is $\{9\}$.

4. $(x-2)^2 = 20$
$x - 2 = \pm\sqrt{20}$
$x - 2 = \pm 2\sqrt{5}$

$x = 2 \pm 2\sqrt{5}$
The solution set is $\{2 + 2\sqrt{5},\ 2 - 2\sqrt{5}\}$.

5. $|2x - 1| \geq 7$
$2x - 1 \geq 7$ or $2x - 1 \leq -7$
$2x \geq 8$ $2x \leq -6$
$x \geq 4$ or $x \leq -3$
The solution set is $\{x \mid x \leq -3 \text{ or } x \geq 4\}$

6. $3x^3 + 4x^2 - 7x + 2 = 0$
$p: \pm 1, \pm 2$
$q: \pm 1, \pm 3$

$\dfrac{p}{q}: \pm 1,\ \pm 2,\ \pm\dfrac{1}{3},\ \pm\dfrac{2}{3}$

Let $f(x) = 3x^3 + 4x^2 - 7x + 2$.
Evaluate f at the possible rational zeros to find $f\left(\dfrac{2}{3}\right) = 0$.

$$
\begin{array}{c|cccc}
\dfrac{2}{3} & 3 & 4 & -7 & 2 \\[2mm]
 & & 2 & 4 & -2 \\
\hline
 & 3 & 6 & -3 & 0
\end{array}
$$

$\left(x - \dfrac{2}{3}\right)(3x^2 + 6x - 3) = 0$

$(3x - 2)(x^2 + 2x - 1) = 0$

$x = \dfrac{2}{3}$ or $x = \dfrac{-2 \pm \sqrt{(2)^2 - 4(1)(-1)}}{2}$

$x = \dfrac{-2 \pm \sqrt{8}}{2}$

$x = -1 \pm \sqrt{2}$

The solution set is $\left\{\dfrac{2}{3},\ -1 + \sqrt{2},\ -1 - \sqrt{2}\right\}$.

7. $\log_2(x+1) + \log_2(x-1) = 3$
$\log_2(x^2 - 1) = 3$
$x^2 - 1 = 2^3$
$x^2 = 9$
$x = \pm 3$
$x = -3$ is not a solution of the original equation. The solution set is $\{3\}$.

8. $3x + 4y = 2$

$2x + 5y = -1$

$6x + 8y = 4$

$-6x - 15y = 3$

—————————

$-7y = 7$

$y = -1$

$3x + 4(-1) = 2$

$3x = 6$

$x = 2$

The solution set is $\{(2, -1)\}$.

9. $2x^2 - y^2 = -8$

$x - y = 6$

$x - y = 6$

$x = y + 6$

$x^2 = (y + 6)^2 = y^2 + 12y + 36$

Substitute into first equation.

$2(y^2 + 12y + 36) - y^2 = -8$

$2y^2 + 24y + 72 - y^2 = -8$

$y^2 + 24y + 80 = 0$

$(y + 4)(y + 20) = 0$

$y = -4$ or $y = -20$

$x = 2 \qquad x = -14$

The solution set is $\{(2, -4), (-14, -20)\}$.

10. Set up the augmented matrix and use Gauss-Jordan reduction.

$$\begin{bmatrix} 1 & -1 & 1 & | & 17 \\ -4 & 1 & 5 & | & -2 \\ 2 & 3 & 1 & | & 8 \end{bmatrix}$$

$$\begin{bmatrix} 1 & -1 & 1 & | & 17 \\ 0 & -3 & 9 & | & 66 \\ 0 & 5 & -1 & | & -26 \end{bmatrix} \begin{matrix} \\ 4R_1 + R_2 \\ -2R_1 + R_3 \end{matrix}$$

$$\begin{bmatrix} 1 & -1 & 1 & | & 17 \\ 0 & 1 & -3 & | & -22 \\ 0 & 5 & -1 & | & -26 \end{bmatrix} -\frac{1}{3}R_2$$

$$\begin{bmatrix} 1 & 0 & -2 & | & -5 \\ 0 & 1 & -3 & | & -22 \\ 0 & 0 & 14 & | & 84 \end{bmatrix} \begin{matrix} R_2 + R_1 \\ \\ -5R_2 + R_3 \end{matrix}$$

$$\begin{bmatrix} 1 & 0 & -2 & | & -5 \\ 0 & 1 & -3 & | & -22 \\ 0 & 0 & 1 & | & 6 \end{bmatrix} \frac{1}{14}R_3$$

$$\begin{bmatrix} 1 & 0 & 0 & | & 7 \\ 0 & 1 & 0 & | & -4 \\ 0 & 0 & 1 & | & 6 \end{bmatrix} \begin{matrix} 2R_3 + R_1 \\ 3R_3 + R_2 \\ \end{matrix}$$

$x = 7, y = -4, z = 6$

The solution set is $\{(7, -4, 6)\}$.

11. Parabola with vertex at $(1, -4)$.

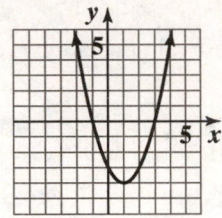

$f(x) = (x - 1)^2 - 4$

12. Ellipse with center at $(0, 0)$ and vertices at $(3, 0)$ and $(-3, 0)$.

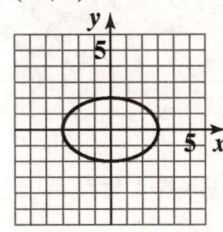

$$\frac{x^2}{9} + \frac{y^2}{4} = 1$$

13. $5x + y \le 10$ $\qquad$ $y \ge \frac{1}{4}x + 2$

$y \le -5x + 10$

Graph with solid line $y = -5x + 10$ and $y = \frac{1}{4}x + 2$.

Shade the region that is below the line $y = -5x + 10$

and above the line $y = \frac{1}{4}x + 2$. Then dash the solid

lines that do not contain the solution set.

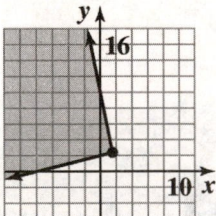

14. a. $p : \pm 1, \pm 3$

$q : \pm 1, \pm 2, \pm 4, \pm 8, \pm 16, \pm 32$

$\dfrac{p}{q}: \pm 1, \pm 3, \pm \dfrac{1}{2}, \pm \dfrac{3}{2}, \pm \dfrac{1}{4}, \pm \dfrac{3}{4}, \pm \dfrac{1}{8},$

$\pm \dfrac{3}{8}, \pm \dfrac{1}{16}, \pm \dfrac{3}{16}, \pm \dfrac{1}{32}, \pm \dfrac{3}{32}$

b. $x = 1$ appears to be a root.

$$\begin{array}{r|rrrr}
1 & 32 & -52 & 17 & 3 \\
  &   & 32 & -20 & -3 \\
\hline
  & 32 & -20 & -3 & 0
\end{array}$$

$32x^3 - 52x^2 + 17x + 3 = 0$

$(x-1)(32x^2 - 20x - 3) = 0$

$(x-1)(4x-3)(8x+1) = 0$

$x = 1$ or $x = \dfrac{3}{4}$ or $x = -\dfrac{1}{8}$

The solution set is $\left\{ -\dfrac{1}{8}, \dfrac{3}{4}, 1 \right\}$.

15. a.   domain: $(-2, 2)$

range: $[-3, \infty)$

b.   the relative minimum of -3 occurs at $x = 0$.

c.   increasing: $(0, 2)$

d. $f(-1) - f(0) = 0 - (-3) = 3$

e. $(f \circ f)(1) = f(f(1)) = f(0) = -3$

f. $f(x) \to \infty$ as $x \to -2^+$ or as $x \to 2^-$

g.

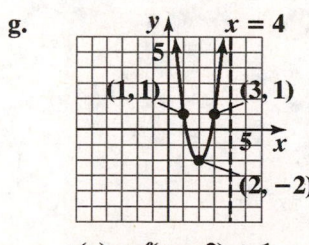

$g(x) = f(x-2) + 1$

h.

$h(x) = -f(2x)$

16. $f(x) = x^2 - 4, g(x) = x + 2$

$(g \circ f)(x) = g(x^2 - 4) = (x^2 - 4) + 2 = x^2 - 2$

17. $\log_5 \dfrac{x^3 \sqrt{y}}{125} = \log_5 x^3 \sqrt{y} - \log_5 125$

$= \log_5 x^3 + \log_5 \sqrt{y} - 3$

$= 3\log_5 x + \dfrac{1}{2}\log_5 y - 3$

18. $m = \dfrac{y_2 - y_1}{x_2 - x_1} = \dfrac{8 - (-4)}{-5 - 1} = \dfrac{12}{-6} = -2$

$y - y_1 = m(x - x_1)$

$y + 4 = -2(x - 1)$

$y = -2x - 2$

19.   Let R = the cost of a rental at Rent-a-Truck and let A = the cost of a rental at Ace Truck Rentals.

$R = 39 + 0.16m$

$A = 25 + 0.24m$

where m is the number of miles.

$39 + 0.16m = 25 + 0.24m$

$14 = 0.08m$

$m = 175$

$R = 39 + 0.16(175) = 67$

The cost will be the same when the number of miles driven is 175 miles. The cost will be $67.

20.   Let x = cost of basic cable,

Let y = cost of movie channel.

$x + y = 35$

$x + 2y = 45$

Multiply the first equation by -1 and then add the two equations.

$-x - y = -35$

$\underline{x + 2y = 45}$

$y = 10$

Use back-substitution to find x.

$x + 10 = 35$

$x = 25$

Basic cable costs $25 and one movie channel costs $10.

21.
$$\frac{\csc\theta - \sin\theta}{\sin\theta} = \frac{\dfrac{1}{\sin\theta} - \sin\theta}{\sin\theta} \cdot \frac{\sin\theta}{\sin\theta}$$

$$= \frac{1 - \sin^2\theta}{\sin^2\theta}$$

$$= \frac{\cos^2\theta}{\sin^2\theta}$$

$$= \left(\frac{\cos\theta}{\sin\theta}\right)^2$$

$$= \cot^2\theta$$

22. $y = 2\cos(2x + \pi)$

$A = 2,\ B = 2,\ C = -\pi$

Amplitude: $|A| = |2| = 2$

Period: $\dfrac{2\pi}{B} = \dfrac{2\pi}{2} = \pi$

Phase Shift: $\dfrac{C}{B} = \dfrac{-\pi}{2} = -\dfrac{\pi}{2}$

$(0, -2), \left(\dfrac{\pi}{4}, 0\right), \left(\dfrac{\pi}{2}, 2\right), \left(\dfrac{3\pi}{4}, 0\right), (\pi, -2)$

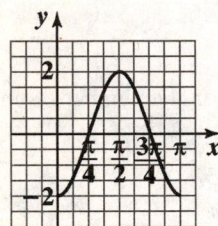

23.
$$(\mathbf{v}\cdot\mathbf{w})\mathbf{w} = \big[(3\mathbf{i} - 6\mathbf{j})\cdot(\mathbf{i} + \mathbf{j})\big](\mathbf{i} + \mathbf{j})$$

$$= \big[3(1) - 6(1)\big](\mathbf{i} + \mathbf{j})$$

$$= (3 - 6)(\mathbf{i} + \mathbf{j})$$

$$= -3(\mathbf{i} + \mathbf{j})$$

$$= -3\mathbf{i} - 3\mathbf{j}$$

24.
$$\sin 2\theta = \sin\theta,\ 0 \le \theta < 2\pi$$

$$2\sin\theta\cos\theta = \sin\theta$$

$$2\sin\theta\cos\theta - \sin\theta = 0$$

$$\sin\theta(2\cos\theta - 1) = 0$$

$\sin\theta = 0 \qquad$ or $\quad 2\cos\theta - 1 = 0$

$\theta = 0,\ \pi \qquad\qquad 2\cos\theta = 1$

$$\cos\theta = \frac{1}{2}$$

$$\theta = \frac{\pi}{3},\ \frac{5\pi}{3}$$

The solutions in the interval $[0, 2\pi)$ are $0,\ \pi,\ \dfrac{\pi}{3}$,

and $\dfrac{5\pi}{3}$.

25.
$$A + B + C = 180°$$

$$64° + 72° + C = 180°$$

$$136° + C = 180°$$

$$C = 44°$$

$$\frac{b}{\sin B} = \frac{a}{\sin A}$$

$$\frac{b}{\sin 72°} = \frac{13.6}{\sin 64°}$$

$$b = \frac{13.6\sin 72°}{\sin 64°} \approx 14.4$$

$$\frac{c}{\sin C} = \frac{a}{\sin A}$$

$$\frac{c}{\sin 44°} = \frac{13.6}{\sin 64°}$$

$$c = \frac{13.6\sin 44°}{\sin 64°} \approx 10.5$$

The solution is $C = 44°,\ b \approx 14.4$, and $c \approx 10.5$.

Chapter 10
Sequences, Induction, and Probability

Section 10.1
Check Point Exercises

1. **a.** $a_n = 2n + 5$

$a_1 = 2(1) + 5 = 7$

$a_2 = 2(2) + 5 = 9$

$a_3 = 2(3) + 5 = 11$

$a_4 = 2(4) + 5 = 13$

The first four terms are 7, 9, 11, and 13.

b. $a_n = \dfrac{(-1)^n}{2^n + 1}$

$a_1 = \dfrac{(-1)^1}{2^1 + 1} = \dfrac{-1}{3} = -\dfrac{1}{3}$

$a_2 = \dfrac{(-1)^2}{2^2 + 1} = \dfrac{1}{5}$

$a_3 = \dfrac{(-1)^3}{2^3 + 1} = \dfrac{-1}{9} = -\dfrac{1}{9}$

$a_4 = \dfrac{(-1)^4}{2^4 + 1} = \dfrac{1}{17}$

The first four terms are $-\frac{1}{3}, \frac{1}{5}, -\frac{1}{9},$ and $\frac{1}{17}$.

2. $a_1 = 3$ and $a_n = 2a_{n-1} + 5$ for $n \geq 2$

$a_2 = 2a_1 + 5$

$\quad = 2(3) + 5 = 11$

$a_3 = 2a_2 + 5$

$\quad = 2(11) + 5 = 27$

$a_4 = 2a_3 + 5$

$\quad = 2(27) + 5 = 59$

The first four terms are 3, 11, 27, and 59.

3. $a_n = \dfrac{20}{(n+1)!}$

$a_1 = \dfrac{20}{(1+1)!} = \dfrac{20}{2!} = 10$

$a_2 = \dfrac{20}{(2+1)!} = \dfrac{20}{3!} = \dfrac{20}{6} = \dfrac{10}{3}$

$a_3 = \dfrac{20}{(3+1)!} = \dfrac{20}{4!} = \dfrac{20}{24} = \dfrac{5}{6}$

$a_4 = \dfrac{20}{(4+1)!} = \dfrac{20}{5!} = \dfrac{20}{120} = \dfrac{1}{6}$

The first four terms are $10, \frac{10}{3}, \frac{5}{6},$ and $\frac{1}{6}$.

4. **a.** $\dfrac{14!}{2!\,12!} = \dfrac{14 \cdot 13 \cdot 12!}{2!\,12!} = \dfrac{14 \cdot 13}{2 \cdot 1} = 91$

b. $\dfrac{n!}{(n-1)!} = \dfrac{n \cdot (n-1)!}{(n-1)!} = n$

5. **a.** $\displaystyle\sum_{i=1}^{6} 2i^2$

$= 2(1)^2 + 2(2)^2 + 2(3)^2$

$\quad + 2(4)^2 + 2(5)^2 + 2(6)^2$

$= 2 + 8 + 18 + 32 + 50 + 72$

$= 182$

b. $\displaystyle\sum_{k=3}^{5} \left(2^k - 3\right)$

$= \left(2^3 - 3\right) + \left(2^4 - 3\right) + \left(2^5 - 3\right)$

$= (8 - 3) + (16 - 3) + (32 - 3)$

$= 5 + 13 + 29$

$= 47$

c. $\displaystyle\sum_{i=1}^{5} 4 = 4 + 4 + 4 + 4 + 4 = 20$

6. **a.** The sum has nine terms, each of the form i^2, starting at $i = 1$ and ending at $i = 9$.

$1^2 + 2^2 + 3^2 + \cdots + 9^2 = \displaystyle\sum_{i=1}^{9} i^2$

b. The sum has n terms, each of the form $\frac{1}{2^{i-1}}$, starting at $i = 1$ and ending at $i = n$.

$1 + \dfrac{1}{2} + \dfrac{1}{4} + \dfrac{1}{8} + \cdots + \dfrac{1}{2^{n-1}} = \displaystyle\sum_{i=1}^{n} \dfrac{1}{2^{i-1}}$

Exercise Set 10.1

1. $a_n = 3n + 2$

$a_1 = 3(1) + 2 = 5$

$a_2 = 3(2) + 2 = 8$

$a_3 = 3(3) + 2 = 11$

$a_4 = 3(4) + 2 = 14$

The first four terms are 5, 8, 11, and 14.

3. $a_n = 3^n$

$a_1 = 3^1 = 3$

$a_2 = 3^2 = 9$

$a_3 = 3^3 = 27$

$a_4 = 3^4 = 81$

The first four terms are 3, 9, 27, and 81.

5. $a_n = (-3)^n$

$a_1 = (-3)^1 = -3$

$a_2 = (-3)^2 = 9$

$a_3 = (-3)^3 = -27$

$a_4 = (-3)^4 = 81$

The first four terms are –3, 9, –27, and 81.

7. $a_n = (-1)^n (n+3)$

$a_1 = (-1)^1 (1+3) = -4$

$a_2 = (-1)^2 (2+3) = 5$

$a_3 = (-1)^3 (3+3) = -6$

$a_4 = (-1)^4 (4+3) = 7$

The first four terms are –4, 5, –6, and 7.

9. $a_n = \dfrac{2n}{n+4}$

$a_1 = \dfrac{2(1)}{1+4} = \dfrac{2}{5}$

$a_2 = \dfrac{2(2)}{2+4} = \dfrac{4}{6} = \dfrac{2}{3}$

$a_3 = \dfrac{2(3)}{3+4} = \dfrac{6}{7}$

$a_4 = \dfrac{2(4)}{4+4} = \dfrac{8}{8} = 1$

The first four terms are $\frac{2}{5}, \frac{2}{3}, \frac{6}{7},$ and 1.

11. $a_n = \dfrac{(-1)^{n+1}}{2^n - 1}$

$a_1 = \dfrac{(-1)^{1+1}}{2^1 - 1} = \dfrac{1}{1} \quad n = 1$

$a_2 = \dfrac{(-1)^{2+1}}{2^2 - 1} = -\dfrac{1}{3}$

$a_3 = \dfrac{(-1)^{3+1}}{2^3 - 1} = \dfrac{1}{7}$

$a_4 = \dfrac{(-1)^{4+1}}{2^4 - 1} = -\dfrac{1}{15}$

The first four terms are $1, -\frac{1}{3}, \frac{1}{7},$ and $-\frac{1}{15}$.

13. $a_1 = 7$ and $a_n = a_{n-1} + 5$ for $n \geq 2$

$a_2 = a_1 + 5 = 7 + 5 = 12$

$a_3 = a_2 + 5 = 12 + 5 = 17$

$a_4 = a_3 + 5 = 17 + 5 = 22$

The first four terms are 7, 12, 17, and 22.

15. $a_1 = 3$ and $a_n = 4a_{n-1}$ for $n \geq 2$

$a_2 = 4a_1 = 4(3) = 12$

$a_3 = 4a_2 = 4(12) = 48$

$a_4 = 4a_3 = 4(48) = 192$

The first four terms are 3, 12, 48, and 192.

17. $a_1 = 4$ and $a_n = 2a_{n-1} + 3$

$a_2 = 2(4) + 3 = 11$

$a_3 = 2(11) + 3 = 25$

$a_4 = 2(25) + 3 = 53$

The first four terms are 4, 11, 25, and 53.

19. $a_n = \dfrac{n^2}{n!}$

$a_1 = \dfrac{1^2}{1!} = 1$

$a_2 = \dfrac{2^2}{2!} = 2$

$a_3 = \dfrac{3^2}{3!} = \dfrac{9}{6} = \dfrac{3}{2}$

$a_4 = \dfrac{4^2}{4!} = \dfrac{16}{24} = \dfrac{2}{3}$

The first four terms are $1, \ 2, \ \frac{3}{2},$ and $\frac{2}{3}$.

21. $a_n = 2(n+1)!$

$a_1 = 2(1+1)! = 2(2) = 4$

$a_2 = 2(2+1)! = 2(6) = 12$

$a_3 = 2(3+1)! = 2(24) = 48$

$a_4 = 2(4+1)! = 2(120) = 240$

The first four terms are 4, 12, 48, and 240.

23. $\dfrac{17!}{15!} = \dfrac{17 \cdot 16 \cdot 15!}{15!} = 17 \cdot 16 = 272$

25. $\dfrac{16!}{2! \, 14!} = \dfrac{16 \cdot 15 \cdot 14!}{2! \, 14!} = \dfrac{16 \cdot 15}{2 \cdot 1} = \dfrac{8 \cdot 15}{1} = 120$

27. $\dfrac{(n+2)!}{n!} = \dfrac{(n+2)(n+1)n!}{n!} = (n+2)(n+1)$

29. $\displaystyle\sum_{i=1}^{6} 5i = 5 \cdot 1 + 5 \cdot 2 + 5 \cdot 3 + 5 \cdot 4 + 5 \cdot 5 + 5 \cdot 6$

$= 5 + 10 + 15 + 20 + 25 + 30$

$= 105$

31. $\displaystyle\sum_{i=1}^{4} 2i^2 = 2 \cdot 1^2 + 2 \cdot 2^2 + 2 \cdot 3^2 + 2 \cdot 4^2$

$= 2 + 8 + 18 + 32$

$= 60$

33. $\displaystyle\sum_{k=1}^{5} k(k+4) = 1(5) + 2(6) + 3(7) + 4(8) + 5(9)$

$= 5 + 12 + 21 + 32 + 45$

$= 115$

35. $\displaystyle\sum_{i=1}^{4}\left(\dfrac{-1}{2}\right)^i = \left(-\dfrac{1}{2}\right)^1 + \left(-\dfrac{1}{2}\right)^2 + \left(-\dfrac{1}{2}\right)^3 + \left(-\dfrac{1}{2}\right)^4$

$= -\dfrac{1}{2} + \dfrac{1}{4} + -\dfrac{1}{8} + \dfrac{1}{16}$

$= -\dfrac{5}{16}$

37. $\displaystyle\sum_{i=5}^{9} 11 = 11 + 11 + 11 + 11 + 11 = 55$

39. $\displaystyle\sum_{i=0}^{4} \dfrac{(-1)^i}{i!}$

$= \dfrac{(-1)^0}{0!} + \dfrac{(-1)^1}{1!} + \dfrac{(-1)^2}{2!} + \dfrac{(-1)^3}{3!} + \dfrac{(-1)^4}{4!}$

$= 1 - 1 + \dfrac{1}{2} - \dfrac{1}{6} + \dfrac{1}{24}$

$= \dfrac{9}{24} = \dfrac{3}{8}$

41. $\displaystyle\sum_{i=1}^{5} \dfrac{i!}{(i-1)!} = \dfrac{1!}{0!} + \dfrac{2!}{1!} + \dfrac{3!}{2!} + \dfrac{4!}{3!} + \dfrac{5!}{4!}$

$= 1 + 2 + 3 + 4 + 5 = 15$

43. $1^2 + 2^2 + 3^2 + \cdots + 15^2 = \displaystyle\sum_{i=1}^{15} i^2$

45. $2 + 2^2 + 2^3 + 2^4 + \cdots + 2^{11} = \displaystyle\sum_{i=1}^{11} 2^i$

47. $1 + 2 + 3 + \cdots + 30 = \displaystyle\sum_{i=1}^{30} i$

49. $\dfrac{1}{2} + \dfrac{2}{3} + \dfrac{3}{4} + \cdots + \dfrac{14}{14+1} = \displaystyle\sum_{i=1}^{14} \dfrac{i}{i+1}$

51. $4 + \dfrac{4^2}{2} + \dfrac{4^3}{3} + \cdots + \dfrac{4^n}{n} = \displaystyle\sum_{i=1}^{n} \dfrac{4^i}{i}$

53. $1 + 3 + 5 + \cdots + (2n-1) = \displaystyle\sum_{i=1}^{n} (2i-1)$

55. $5 + 7 + 9 + \cdots + 31$

Possible answer: $\displaystyle\sum_{k=1}^{14} (2k+3)$

57. $a + ar + ar^2 + \cdots + ar^{12}$

Possible answer: $\displaystyle\sum_{k=0}^{12} ar^k$

59. $a + (a+d) + (a+2d) + \cdots + (a+nd)$

Possible answer: $\displaystyle\sum_{k=0}^{n} (a+kd)$

61. $\displaystyle\sum_{i=1}^{5}(a_i^2+1)=\left((-4)^2+1\right)+\left((-2)^2+1\right)+\left((0)^2+1\right)+\left((2)^2+1\right)+\left((4)^2+1\right)$

$$=17+5+1+5+17$$
$$=45$$

63. $\displaystyle\sum_{i=1}^{5}(2a_i+b_i)=\left(2(-4)+4\right)+\left(2(-2)+2\right)+\left(2(0)+0\right)+\left(2(2)+(-2)\right)+\left(2(4)+(-4)\right)$

$$=-4+(-2)+0+2+4=0$$

65. $\displaystyle\sum_{i=4}^{5}\left(\frac{a_i}{b_i}\right)^2=\left(\frac{2}{-2}\right)^2+\left(\frac{4}{-4}\right)^2=(-1)^2+(-1)^2=1+1=2$

67. $\displaystyle\sum_{i=1}^{5}a_i^2+\sum_{i=1}^{5}b_i^2=\left((-4)^2+(-2)^2+0^2+2^2+4^2\right)+\left(4^2+2^2+0^2+(-2)^2+(-4)^2\right)$

$$=(16+4+0+4+16)+(16+4+0+4+16)=80$$

69. a. $\displaystyle\frac{1}{7}\sum_{i=1}^{7}a_i=\frac{1}{7}(8.1+7.2+6.1+8.1+10.0+13.1+16.7)=\frac{1}{7}(69.3)=9.9$

From 2000 through 2006, Online ad spending averaged \$9.9 billion per year.

b. $a_n=0.5n^2-1.5n+8 \qquad a_4=0.5(4)^2-1.5(4)+8=10$

$a_1=0.5(1)^2-1.5(2)+8=7 \qquad a_5=0.5(5)^2-1.5(5)+8=13$

$a_2=0.5(2)^2-1.5(2)+8=7 \qquad a_6=0.5(6)^2-1.5(6)+8=17$

$a_3=0.5(3)^2-1.5(3)+8=8 \qquad a_7=0.5(7)^2-1.5(7)+8=22$

$\displaystyle\frac{1}{7}\sum_{i=1}^{7}a_i=\frac{1}{7}(7+7+8+10+13+17+22)=\frac{1}{7}(84)=12$

This overestimates the actual sum by \$2.1 billion.

71. $\displaystyle a_n=6000\left(1+\frac{0.06}{4}\right)^n,n=1,2,3,\cdots$

$$a_{20}=6000\left(1+\frac{0.06}{4}\right)^{20}\approx8081.13$$

After five years, the balance is \$8081.13.

73. – 79. Answers may vary.

81. Most calculators give error message if the expression is entered directly.

However, $\displaystyle\frac{200!}{198!}=\frac{200\cdot199\cdot198!}{198!}=200\cdot199=39,800$

83. $\displaystyle\frac{20!}{300}=8,109,673,360,588,800$

However, most calculators give a rounded answer in scientific notation.

85. $\displaystyle\frac{54!}{(54-3)3!}=24,804$

862

87. Answers may vary.

89. $a_n = \dfrac{n}{n+1}$

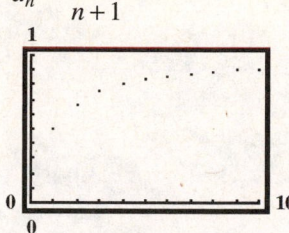

As n gets larger, a_n approaches 1.

91. $a_n = \dfrac{2n^2 + 5n - 7}{n^3}$

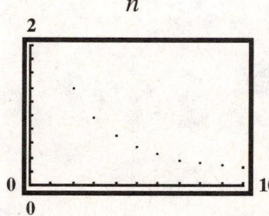

As n gets larger, a_n approaches 0.

93. does not make sense; Explanations will vary. Sample explanation: There is nothing that implies that there is a negative number of sheep.

95. makes sense

97. false; Changes to make the statement true will vary. A sample change is: $\dfrac{n!}{(n-1)!} = \dfrac{n \cdot (n-1)!}{(n-1)!} = n$

99. false; Changes to make the statement true will vary. A sample change is:

$$\sum_{i=1}^{2}(-1)^i 2^i = (-1)^1 2^1 + (-1)^2 2^2 = -1(2) + 1(4) = -2 + 4 = 2$$

101. $a_n = \begin{cases} \dfrac{a_{n-1}}{2} & \text{if } a_{n-1} \text{ is even.} \\ 3a_n + 5 & \text{if } a_{n-1} \text{ is odd} \end{cases}$

     for $n \geq 2$.

$a_1 = 9$

Since 9 is odd, $a_2 = 3(9) + 5 = 32$.

Since 32 is even, $a_3 = \dfrac{32}{2} = 16$.

Similarly, $a_4 = \dfrac{16}{2} = 8$, $a_5 = \dfrac{8}{2} = 4$.

The first five terms of the sequence are 9, 32, 16, 8, and 4.

103. $a_2 - a_1 = 3 - 8 = -5$

$a_3 - a_2 = -2 - 3 = -5$

$a_4 - a_3 = -7 - (-2) = -5$

$a_5 - a_4 = -12 - (-7) = -5$

The difference between consecutive terms is always -5.

104. $a_2 - a_1 = (4(2) - 3) - (4(1) - 3) = 4$

$a_3 - a_2 = (4(3) - 3) - (4(2) - 3) = 4$

$a_4 - a_3 = (4(4) - 3) - (4(3) - 3) = 4$

$a_5 - a_4 = (4(5) - 3) - (4(4) - 3) = 4$

The difference between consecutive terms is always 4.

105. $a_n = 4 + (n - 1)(-7)$

$a_8 = 4 + (8 - 1)(-7) = 4 + (7)(-7) = 4 - 49 = -45$

Section 10.2

Check Point Exercises

1. $a_1 = 100$

$a_2 = a_1 - 30 = 100 - 30 = 70$

$a_3 = a_2 - 30 = 70 - 30 = 40$

$a_4 = a_3 - 30 = 40 - 30 = 10$

$a_5 = a_4 - 30 = 10 - 30 = -20$

$a_6 = a_5 - 30 = -20 - 30 = -50$

The first five terms are 100, 70, 40, 10, -20, -50.

2. $a_1 = 6, d = -5$

To find the ninth term, a_9, replace n in the formula with 9, a_1 with 6, and d with -5.

$a_n = a_1 + (n - 1)d$

$a_9 = 6 + (9 - 1)(-5)$

$\quad = 6 + 8(-5)$

$\quad = 6 + (-40)$

$\quad = -34$

3. **a.** $a_n = a_1 + (n - 1)d$

$\quad = 32 + (n - 1)0.7$

$\quad = 0.7n + 31.3$

b. $a_n = 0.7n + 31.3$

$a_{11} = 0.7(11) + 31.3 = 39$

In 2014 Americans will average 39 car meals.

4. 3, 6, 9, 12, ...

To find the sum of the first 15 terms, S_{15}, replace n in the formula with 15.

$$S_n = \frac{n}{2}(a_1 + a_n)$$

$$S_{15} = \frac{15}{2}(a_1 + a_{15})$$

Use the formula for the general term of a sequence to find a_{15}. The common difference, d, is 3, and the first term, a_1, is 3.

$$a_n = a_1 + (n-1)d$$

$$a_{15} = 3 + (15-1)(3)$$

$$= 3 + 14(3)$$

$$= 3 + 42$$

$$= 45$$

Thus, $S_{15} = \frac{15}{2}(3+45) = \frac{15}{2}(48) = 360$.

5. $\sum\limits_{i=1}^{30}(6i-11) = (6\cdot 1 - 11) + (6\cdot 2 - 11) + (6\cdot 3 - 11) + \ldots + (6\cdot 30 - 11)$

$$= -5 + 1 + 7 + \ldots + 169$$

So the first term, a_1, is -5; the common difference, d, is $1 - (-5) = 6$; the last term, a_{30}, is 169.

Substitute $n = 30$, $a_1 = -5$, and $a_{30} = 169$ in the formula $S_n = \frac{n}{2}(a_1 + a_n)$.

$$S_{30} = \frac{30}{2}(-5 + 169) = 15(164) = 2460$$

Thus, $\sum\limits_{i=1}^{30}(6i-11) = 2460$

6. $a_n = 1800n + 64{,}130$

$a_1 = 1800(1) + 64{,}130 = 65{,}930$

$a_{10} = 1800(10) + 64{,}130 = 82{,}130$

$$S_n = \frac{n}{2}(a_1 + a_n)$$

$$S_{10} = \frac{10}{2}(a_1 + a_{10})$$

$$= 5(65{,}930 + 82{,}130)$$

$$= 5(148{,}060)$$

$$= \$740{,}300$$

It would cost \$740,300 for the ten-year period beginning in 2009.

Exercise Set 10.2

1. $a_1 = 200,\ d = 20$

The first six terms are 200, 220, 240, 260, 280, and 300.

3. $a_1 = -7,\ d = 4$

The first six terms are $-7, -3, 1, 5, 9$, and 13.

5. $a_1 = 300, \ d = -90$
The first six terms are 300, 210, 120, 30, –60, and –150.

7. $a_1 = \dfrac{5}{2}, \ d = -\dfrac{1}{2}$

The first six terms are $\dfrac{5}{2}, 2, \dfrac{3}{2}, 1, \dfrac{1}{2}$, and 0.

9. $a_n = a_{n-1} + 6, \ a_1 = -9$
The first six terms are –9, –3, 3, 9, 15, and 21.

11. $a_n = a_{n-1} - 10, \ a_1 = 30$
The first six terms are 30, 20, 10, 0, –10, and –20.

13. $a_n = a_{n-1} - 0.4, \ a_1 = 1.6$
The first six terms are 1.6, 1.2, 0.8, 0.4, 0, and –0.4.

15. $a_1 = 13, \ d = -4$
$a_n = 13 + (n-1)4$
$a_6 = 13 + 5(4) = 13 + 20 = 33$

17. $a_1 = 7, \ d = 5$
$a_n = 7 + (n-1)2$
$a_{50} = 7 + 49(5) = 252$

19. $a_1 = -40, \ d = 5$
$a_n = -40 + (n-1)5$
$a_{200} = -40 + (199)5 = 955$

21. $a_1 = 35, \ d = -3$
$a_n = 35 - 3(n-1)$
$a_{60} = 35 - 3(59) = -142$

23. 1, 5, 9, 13, ...
$d = 5 - 1, \ = 4$
$a_n = 1 + (n-1)4 = 1 + 4n - 4$
$a_n = 4n - 3$
$a_{20} = 4(20) - 3 = 77$

25. 7, 3, –1, –5, ...
$d = 3 - 7 = -4$
$a_n = 7 + (n-1)(-4) = 7 - 4n + 4$
$a_n = 11 - 4n$
$a_{20} = 11 - 4(20) = -69$

27. $a_1 = 9, \ d = 2$

$a_n = 9 + (n-1)(2)$

$a_n = 7 + 2n$

$a_{20} = 7 + 2(20) = 47$

29. $a_1 = -20, \ d = -4$

$a_n = -20 + (n-1)(-4)$

$a_n = -20 - 4n + 4$

$a_n = -16 - 4n$

$a_{20} = -16 - 4(20) = -96$

31. $a_n = a_{n-1} + 3, \ a_1 = 4$

$d = 3$

$a_n = 4 + (n-1)(3)$

$a_n = 1 + 3n$

$a_{20} = 1 + 3(20) = 61$

33. $a_n = a_{n-1} - 10, \ a_1 = 30, \ d = -10$

$a_n = 30 - 10(n-1) = 30 - 10n + 10$

$a_n = 40 - 10n$

$a_{20} = 40 - 10(20) = -160$

35. $4, 10, 16, 22, \ldots$

$d = 10 - 4 = 6$

$a_n = 4 + (n-1)(6)$

$a_{20} = 4 + (19)(6) = 118$

$S_{20} = \dfrac{20}{2}(4 + 118) = 1220$

37. $-10, -6, -2, 2, \ldots$

$d = -6 - (-10) = -6 + 10 = 4$

$a_n = -10 + (n-1)4$

$a_{50} = -10 + (49)4 = 186$

$S_{50} = \dfrac{50}{2}(-10 + 186) = 4400$

39. $1 + 2 + 3 + 4 + \cdots + 100$

$S_{100} = \dfrac{100}{2}(1 + 100) = 5050$

41. $2 + 4 + 6 + \cdots + 120$

$S_{60} = \dfrac{60}{2}(2 + 120) = 3660$

43. even integers between 21 and 45;

$22 + 24 + 26 + \cdots + 44$

$$S_{12} = \frac{12}{2}(22 + 44) = 396$$

45. $\displaystyle\sum_{i=1}^{17}(5i + 3) = (5 + 3) + (10 + 3) + (15 + 3) + \cdots + (85 + 3) = 8 + 13 + 18 + \cdots + 88$

$$S_{17} = \frac{17}{2}(8 + 88) = 816$$

47. $\displaystyle\sum_{i=1}^{30}(-3i + 5) = (-3 + 5) + (-6 + 5) + (-9 + 5) + \cdots + (-90 + 5) = 2 - 1 - 4 - \cdots - 85$

$$S_{30} = \frac{30}{2}(2 - 85) = -1245$$

49. $\displaystyle\sum_{i=1}^{100}4i = 4 + 8 + 12 + \cdots + 400$

$$S_{100} = \frac{100}{2}\left(4 + 400\right) = 20,200$$

51. First find a_{14} and b_{12}:

$a_{14} = a_1 + (n-1)d$

$\quad = 1 + (14-1)(-3-1) = -51$

$b_{12} = b_1 + (n-1)d$

$\quad = 3 + (12-1)(8-3) = 58$

So, $a_{14} + b_{12} = -51 + 58 = 7$.

53. $a_n = a_1 + (n-1)d$

$-83 = 1 + (n-1)(-3-1)$

$-83 = 1 + -4(n-1)$

$-84 = -4n + 4$

$-88 = -4n$

$\quad n = 22$

There are 22 terms.

55. $S_n = \dfrac{n}{2}(a_1 + a_n)$

For $\{a_n\}$: $S_{14} = \dfrac{14}{2}(a_1 + a_{14}) = 7(1 + (-51)) = -350$ For $\{b_n\}$: $S_{14} = \dfrac{14}{2}(b_1 + b_{14}) = 7(3 + 68) = 497$

So $\displaystyle\sum_{n=1}^{14}b_n - \sum_{n=1}^{14}a_n = 497 - (-350) = 847$

57. Two points on the graph are $(1, 1)$ and $(2, -3)$.
Finding the slope of the line;

$$m = \frac{y_2 - y_1}{x_2 - x_2} = \frac{-3 - 1}{2 - 1} = \frac{-4}{1} = -4$$

Using the point-slope form of an equation of a line;

$$y - y_2 = m(x - x_2)$$
$$y - 1 = -4(x - 1)$$
$$y - 1 = -4x + 4$$
$$y = -4x + 5$$

Thus, $f(x) = -4x + 5$.

59. Using $a_n = a_1 + (n - 1)d$ and $a_2 = 4$:

$$a_2 = a_1 + (2 - 1)d$$
$$4 = a_1 + d$$

And since $a_6 = 16$:

$$a_6 = a_1 + (6 - 1)d$$
$$16 = a_1 + 5d$$

The system of equations is

$$4 = a_1 + d$$
$$16 = a_1 + 5d$$

Solving the first equation for a_1:

$$a_1 = 4 - d$$

Substituting the value into the second equation and solving for d:

$$16 = (4 - d) + 5d$$
$$16 = 4 + 4d$$
$$12 = 4d$$
$$3 = d$$

Back-substitute:

$$a_1 = 4 - d$$
$$a_1 = 4 - 3)$$
$$a_1 = 1$$

Then $a_n = a_1 + (n - 1)d$

$$a_n = 1 + (n - 1)3$$
$$a_n = 1 + 3n - 3$$
$$a_n = 3n - 2$$

61. **a.**
$$a_n = a_1 + (n - 1)d$$
$$a_n = 10 + (n - 1)(0.77)$$
$$a_n = 10 + 0.77n - 0.77$$
$$a_n = 0.77n + 9.23$$

b. 2011 is 27 years after 1984.

$$a_{27} = 0.77(27) + 9.23 \approx 30.0$$

If trends continue, in 2011 the percentage of Americans with no close friends will be 30.0%.

63. Company *A*:

$$a_n = 24000 + (n - 1)1600$$
$$= 24000 + 1600n - 1600$$
$$= 1600n + 22400$$
$$a_{10} = 1600(10) + 22400$$
$$= 16000 + 22400 = 38400$$

Company *B*:

$$a_n = 28000 + (n - 1)1000$$
$$= 28000 + 1000n - 1000$$
$$= 1000n + 27000$$
$$a_{10} = 1000(10) + 27000$$
$$= 10000 + 27000 = 37000$$

Company *A* will pay $1400 more in year 10.

65. **a.** Total cost:

$4694 + \$5132 + \$5491 + \$5836 = \$21,153$

b.
$$a_1 = 379(1) + 4342 = 4721$$
$$a_4 = 379(4) + 4342 = 5858$$

$$S_n = \frac{n}{2}(a_1 + a_n)$$

$$S_4 = \frac{4}{2}(4721 + 5858) = 2(10,579) = \$21,158$$

The model overestimates the actual sum by $5.

67. Answers will vary.

69. Company A:

$a_n = 19,000 + (n-1)2600$

$a_{10} = 19,000 + (9)2600 = \$42,200$

$S_{10} = \dfrac{10}{2}(19000 + 42400) = \$307,000$

Company B:

$a_n = 27,000 + (n-1)1200$

$a_{10} = 27,000 + (9)1200 = \$37,800$

$S_{10} = \dfrac{10}{2}(27,000 + 37,800) = \$324,000$

Company B pays the greater total amount.

71. $a_n = 20 + (n+1)3$

$a_{38} = 20 + (37)3 = 131$

$S_{38} = \dfrac{38}{2}(20 + 131) = 2869$

The theater has 2869 seats.

73. – 77. Answers may vary.

79. makes sense

81. makes sense

83. Degree days: 23, 25, 27, …

$a_1 = 23, d = 2$

$a_{10} = 23 + 9(2) = 41$

$S_{10} = \dfrac{10}{2}(a_1 + a_{10})$

$S_{10} = \dfrac{10}{2}(23 + 41) = 320$

There are 320 degree-days.

85. $\dfrac{a_2}{a_1} = \dfrac{-2}{1} = -2$

$\dfrac{a_3}{a_2} = \dfrac{4}{-2} = -2$

$\dfrac{a_4}{a_3} = \dfrac{-8}{4} = -2$

$\dfrac{a_5}{a_4} = \dfrac{16}{-8} = -2$

The ratio of a term to the term that directly precedes it is always -2.

86. $\dfrac{a_2}{a_1} = \dfrac{3 \cdot 5^2}{3 \cdot 5^1} = 5$

$\dfrac{a_3}{a_2} = \dfrac{3 \cdot 5^3}{3 \cdot 5^2} = 5$

$\dfrac{a_4}{a_3} = \dfrac{3 \cdot 5^4}{3 \cdot 5^3} = 5$

$\dfrac{a_5}{a_4} = \dfrac{3 \cdot 5^5}{3 \cdot 5^4} = 5$

The ratio of a term to the term that directly precedes it is always 5.

87. $a_n = a_1 3^{n-1}$

$a_7 = 11 \cdot 3^{7-1} = 11 \cdot 3^6 = 11 \cdot 729 = 8019$

Section 10.3

Check Point Exercises

1. $a_1 = 12, r = \dfrac{1}{2}$

$a_2 = 12\left(\dfrac{1}{2}\right)^1 = 6$

$a_3 = 12\left(\dfrac{1}{2}\right)^2 = \dfrac{12}{4} = 3$

$a_4 = 12\left(\dfrac{1}{2}\right)^3 = \dfrac{12}{8} = \dfrac{3}{2}$

$a_5 = 12\left(\dfrac{1}{2}\right)^4 = \dfrac{12}{16} = \dfrac{3}{4}$

$a_6 = 12\left(\dfrac{1}{2}\right)^5 = \dfrac{12}{32} = \dfrac{3}{8}$

The first six terms are $12, 6, 3, \dfrac{3}{2}, \dfrac{3}{4},$ and $\dfrac{3}{8}$.

2. $a_1 = 5, r = -3$

$a_n = 5r^{n-1}$

$a_7 = 5(-3)^{7-1} = 5(-3)^6 = 5(729) = 3645$

The seventh term is 3645.

3. 3, 6, 12, 24, 48, …

$r = \dfrac{6}{3} = 2, a_1 = 3$

$a_n = 3(2)^{n-1}$

$a_8 = 3(2)^{8-1} = 3(2)^7 = 3(128) = 384$

The eighth term is 384.

4. $a_1 = 2,\ r = \dfrac{-6}{2} = -3$

$S_n = \dfrac{a_1(1 - r^r)}{1 - r}$

$S_9 = \dfrac{2\left(1 - (-3)^9\right)}{1 - (-3)} = \dfrac{2(19,684)}{4} = 9842$

The sum of the first nine terms is 9842.

5. $\displaystyle\sum_{i=1}^{8} 2 \cdot 3^i$

$a_1 = 2 \cdot (3)^1 = 6,\ r = 3$

$S_n = \dfrac{a_1(1 - r^n)}{1 - r}$

$S_8 = \dfrac{6\left(1 - 3^8\right)}{1 - 3} = \dfrac{6(-6560)}{-2} = 19,680$

Thus, $\displaystyle\sum_{i=1}^{8} 2 \cdot 3^i = 19,680.$

6. $a_1 = 30,000,\ r = 1.06$

$S_n = \dfrac{a_1(1 - r^n)}{1 - r}$

$S_{30} = \dfrac{30,000\left(1 - (1.06)^{30}\right)}{1 - 1.06} \approx 2,371,746$

The total lifetime salary is \$2,371,746.

7.   **a.** $A = \dfrac{P\left[\left(1 + \frac{r}{n}\right)^{nt} - 1\right]}{\frac{r}{n}}$

$P = 100,\ r = 0.095,\ n = 12,\ t = 35$

$A = \dfrac{100\left[\left(1 + \dfrac{0.095}{12}\right)^{12 \cdot 35} - 1\right]}{\dfrac{0.095}{12}} \approx 333,946$

The value of the IRA will be \$333,946.

   b.   Interest = Value of IRA − Total deposits

$\approx \$333,946 - \$100 \cdot 12 \cdot 35$

$\approx \$333,946 - \$42,000$

$\approx \$291,946$

8. $3 + 2 + \dfrac{4}{3} + \dfrac{8}{9} + \cdots$

$a_1 = 3,\ r = \dfrac{2}{3}$

$S = \dfrac{a_1}{1 - r}$

$S = \dfrac{3}{1 - \frac{2}{3}} = \dfrac{3}{\frac{1}{3}} = 9$

The sum of this infinite geometric series is 9.

9. $0.\overline{9} = 0.9999\cdots = \dfrac{9}{10} + \dfrac{9}{100} + \dfrac{9}{1000} + \cdots$

$a_1 = \dfrac{9}{10},\ r = \dfrac{1}{10}$

$S = \dfrac{\frac{9}{10}}{1 - \frac{1}{10}} = \dfrac{\frac{9}{10}}{\frac{9}{10}} = 1$

An equivalent fraction for $0.\overline{9}$ is 1.

10. $a_1 = 1000(0.8) = 800,\ r = 0.8$

$S = \dfrac{800}{1 - 0.8} = 4000$

The total amount spent is \$4000.

Exercise Set 10.3

1. $a_1 = 5,\ r = 3$

First five terms: 5, 15, 45, 135, 405.

3. $a_1 = 20,\ r = \dfrac{1}{2}$

First five terms: $20,\ 10,\ 5,\ \frac{5}{2},\ \frac{5}{4}\ \frac{5}{4}.$

5. $a_n = -4a_{n-1},\ a_1 = 10$

First five terms: 10, −40, 160, −640, 2560.

7. $a_n = -5a_{n-1},\ a_1 = -6$

First five terms: −6, 30, −150, 750, −3750.

9. $a_1 = 6,\ r = 2$

$a_n = 6 \cdot 2^{n-1}$

$a_8 = 6 \cdot 2^7 = 768$

11. $a_1 = 5,\ r = -2$

$a_n = 5 \cdot (-2)^{n-1}$

$a_{12} = 5 \cdot (-2)^{11} = -10,240$

13. $a_1 = 1000, \ r = -\dfrac{1}{2}$

$a_n = 1000\left(-\dfrac{1}{2}\right)^{n-1}$

$a_{40} = 1000\left(-\dfrac{1}{2}\right)^{39}$

≈ 0.000000002

15. $a_1 = 1,000,000, \ r = 0.1$

$a_n = 1,000,000(0.1)^{n-1}$

$a_8 = 1,000,000(0.1)^7 = 0.1$

17. $3, 12, 48, 192, \ldots$

$r = \dfrac{12}{3} = 4$

$a_n = 3(4)^{n-1}$

$a_7 = 3(4)^6 = 12,288$

19. $19, 6, 2, \dfrac{2}{3}, \cdots \qquad r = \dfrac{6}{18} = \dfrac{1}{3}$

$a_n = 18\left(\dfrac{1}{3}\right)^{n-1}$

$a_7 = 18\left(\dfrac{1}{3}\right)^6 = \dfrac{2}{81}$

21. $1.5, -3, 6, -12, \ldots$

$r = \dfrac{6}{-3} = -2$

$a_n = 1.5(-2)^{n-1}$

$a_7 = 1.5(-2)^6 = 96$

23. $0.0004, -0.004, 0.04, -0.4, \ldots$

$r = \dfrac{-0.004}{0.0004} = -10$

$a_n = 0.0004(-10)^{n-1}$

$a_7 = 0.0004(-10)^6 = 400$

25. $2, 6, 18, 54, \ldots$

$r = \dfrac{6}{2} = 3$

$S_{12} = \dfrac{2\left(1-3^{12}\right)}{1-3} = \dfrac{2(-531,440)}{-2} = 531,440$

27. $3, -6, 12, -24, \ldots$

$r = \dfrac{-6}{3} = -2$

$S_{11} = \dfrac{3\left[1-(-2)^{11}\right]}{1-(-2)} = \dfrac{3(2049)}{3} = 2049$

29. $-\dfrac{3}{2}, 3, -6, 12, \cdots$

$r = \dfrac{3}{\dfrac{-3}{2}} = -2$

$S_{14} = \dfrac{-\dfrac{3}{2}\left[1-(-2)^{14}\right]}{1-(-2)} = \dfrac{-\dfrac{3}{2}(-16,383)}{3} = \dfrac{16,383}{2}$

31. $\displaystyle\sum_{i=1}^{8} 3^i$

$r = 3, \ a_1 = 3$

$S_8 = \dfrac{3\left(1-3^8\right)}{1-3} = \dfrac{3(-6560)}{-2} = 9840$

33. $\displaystyle\sum_{i=1}^{10} 5 \cdot 2^i$

$r = 2, \ a_1 = 10$

$S_{10} = \dfrac{10\left(1-2^{10}\right)}{1-2} = \dfrac{10(-1023)}{-1} = 10,230$

35. $\displaystyle\sum_{i=1}^{6}\left(\dfrac{1}{2}\right)^{i+1}$

$r = \dfrac{1}{2}, \ a_1 = \dfrac{1}{4}$

$S_6 = \dfrac{\dfrac{1}{4}\left(1-\left(\dfrac{1}{2}\right)^6\right)}{1-\dfrac{1}{2}} = \dfrac{\dfrac{1}{4}\left(\dfrac{63}{64}\right)}{\dfrac{1}{2}} = \dfrac{63}{128}$

37. $r = \dfrac{1}{3}$

$S_\infty = \dfrac{1}{1-\dfrac{1}{3}} = \dfrac{1}{\dfrac{2}{3}} = \dfrac{3}{2}$

39. $r = \dfrac{1}{4}$

$S_\infty = \dfrac{3}{1-\dfrac{1}{4}} = \dfrac{3}{\dfrac{3}{4}} = 4$

41. $r = -\dfrac{1}{2}$

$$S_\infty = \frac{1}{1-\left(-\dfrac{1}{2}\right)} = \frac{1}{\dfrac{3}{2}} = \frac{2}{3}$$

43. $r = -0.3$

$$S_\infty = \frac{8}{1-(-0.3)} = \frac{8}{1.3} \approx 6.15385$$

45. $r = \dfrac{1}{10}$

$$S_\infty = \frac{\dfrac{5}{10}}{1-\dfrac{1}{10}} = \frac{\dfrac{5}{10}}{\dfrac{9}{10}} = \frac{5}{9}$$

47. $r = \dfrac{1}{100}$

$$S_\infty = \frac{\dfrac{47}{100}}{1-\dfrac{1}{100}} = \frac{\dfrac{47}{100}}{\dfrac{99}{100}} = \frac{47}{99}$$

49. $0.\overline{257} = \dfrac{257}{1000} + \dfrac{257}{10^6} + \dfrac{257}{10^9} + \cdots$

$r = \dfrac{1}{1000}$

$$S_\infty = \frac{\dfrac{257}{1000}}{1-\dfrac{1}{1000}} = \frac{\dfrac{257}{1000}}{\dfrac{999}{1000}} = \frac{257}{999}$$

51. $a_n = n+5$

arithmetic, $d = 1$

53. $a_n = 2^n$

geometric, $r = 2$

55. $a_n = n^2 + 5$

neither

57. First find a_{10} and b_{10}:

$$a_{10} = a_1 r^{n-1}$$
$$= (-5)\left(\frac{10}{-5}\right)^{10-1} = (-5)(-2)^9$$
$$= 2560$$
$$b_{10} = b_1 + (n-1)d$$
$$= 10 + (10-1)(-5-10)$$
$$= 10 + (9)(-15) = -125$$

So, $a_{10} + b_{10} = 2560 + (-125) = 2435$.

59. For $\{a_n\}$, $r = \dfrac{10}{-5} = -2$ and

$$S_{10} = \frac{a_1(1-r^n)}{1-r} = \frac{(-5)\left(1-(-2)^{10}\right)}{1-(-2)}$$
$$= \frac{(-5)(-1023)}{3} = 1705$$

For $\{b_n\}$,

$$b_{10} = b_1 + (n-1)d$$
$$= 10 + (10-1)(-5-10)$$
$$= 10 + (9)(-15) = -125$$
$$S_n = \frac{n}{2}\left(b_1 + b_n\right)$$
$$S_{10} = \frac{n}{2}\left(b_1 + b_{10}\right)$$
$$= \frac{10}{2}(10+(-125))$$
$$= 5(-115) = -575$$

61. For $\{a_n\}$,

$$S_6 = \frac{a_1(1-r^n)}{1-r} = \frac{(-5)\left(1-(-2)^6\right)}{1-(-2)}$$
$$= \frac{(-5)(-63)}{3} = 105$$

For $\{c_n\}$,

$$S = \frac{a_1}{1-r} = \frac{-2}{1-\dfrac{1}{-2}} = \frac{-2}{\dfrac{3}{2}} = -\frac{4}{3}$$

So, $S_6 \cdot S = 105\left(-\dfrac{4}{3}\right) = -140$

63. It is given that $a_4 = 27$. Using the formula $a_n = a_1 r^{n-1}$ when $n = 4$ we have

$$27 = 8r^{4-1}$$

$$\frac{27}{8} = r^3$$

$$r = \sqrt[3]{\frac{27}{8}} = \frac{3}{2}$$

Thus,

$$a_n = a_1 r^{n-1}$$

$$a_2 = 8\left(\frac{3}{2}\right)^{2-1} = 8\left(\frac{3}{2}\right) = 12$$

$$a_3 = 8\left(\frac{3}{2}\right)^{3-1} = 8\left(\frac{3}{2}\right)^2 = 8\left(\frac{9}{4}\right) = 18$$

65. $1, 2, 4, 8, \ldots$

$$r = 2$$

$$a_n = 2^{n-1}$$

$$a_{15} = 2^{14} = \$16,384$$

67. $a_1 = 3,000,000$

$$r = 1.04$$

$$a_n = 3,000,000(1.04)^{n-1}$$

$$a_7 = 3,000,000(1.04)^6 = \$3,795,957$$

69. **a.** $r_{2003 \text{ to } 2004} = \dfrac{35.89}{35.48} \approx 1.01$

$$r_{2004 \text{ to } 2005} = \dfrac{36.13}{35.89} \approx 1.01$$

$$r_{2005 \text{ to } 2006} = \dfrac{36.46}{36.13} \approx 1.01$$

r is approximately 1.01 for each division.

b. $a_n = a_1 r^{n-1}$

$$a_n = 35.48(1.01)^{n-1}$$

c. Since year 2010 is the 8th term, find a_8.

$$a_n = 35.48(1.01)^{n-1}$$

$$a_8 = 35.48(1.01)^{8-1} \approx 38.04$$

The population of California will be approximately 38.04 million in 2010.

71. $1, 2, 4, 8, \ldots$

$r = 2$

$$S_{15} = \frac{1(1 - 2^{15})}{1 - 2} = 32,767$$

The total savings is $32,767.

73. $a_1 = 24,000, \ r = 1.05$

$$S_{20} = \frac{24,000\left[1 - (1.05)^{20}\right]}{1 - 1.05} = 793,582.90$$

The total salary is $793,583.

75. $r = 0.9$

$$S_{10} = \frac{20(1 - 0.9^{10})}{1 - 0.9} \approx 130.26$$

The total length is 130.26 inches.

77. **a.** $A = \dfrac{P\left[\left(1 + \dfrac{r}{n}\right)^{nt} - 1\right]}{\dfrac{r}{n}} = \dfrac{2000\left[\left(1 + \dfrac{0.075}{1}\right)^{5} - 1\right]}{\dfrac{0.075}{1}} \approx \$11,617$

  b. $\$11,617 - 5 \times \$2000 = \$1617$

79. **a.** $A = \dfrac{P\left[\left(1 + \dfrac{r}{n}\right)^{nt} - 1\right]}{\dfrac{r}{n}} = \dfrac{50\left[\left(1 + \dfrac{0.055}{12}\right)^{12 \times 40} - 1\right]}{\dfrac{0.055}{12}} \approx \$87,052$

  b. $\$87,052 - \$50 \cdot 12 \cdot 40 = \$63,052$

81. **a.** $A = \dfrac{P\left[\left(1 + \dfrac{r}{n}\right)^{nt} - 1\right]}{\dfrac{r}{n}} = \dfrac{10,000\left[\left(1 + \dfrac{0.105}{4}\right)^{4 \times 10} - 1\right]}{\dfrac{0.105}{4}} \approx \$693,031$

  b. $\$693,031 - \$10,000 \cdot 4 \cdot 10 = \$293,031$

83. Find the total value of the lump-sum investment.

$$A = P(1 + r)^t = 30,000(1 + 0.05)^{20} \approx 79,599$$

Find the total value of the annuity.

$$A = \frac{P\left[\left(1 + \dfrac{r}{n}\right)^{nt} - 1\right]}{\dfrac{r}{n}} = \frac{1500\left[\left(1 + \dfrac{0.05}{1}\right)^{20} - 1\right]}{\dfrac{0.05}{1}} \approx 49,599$$

$\$79,599 - \$49,599 = \$30,000$

You will have $30,000 more from the lump-sum investment.

85. $r = 0.6$

$$S_\infty = \frac{6(0.6)}{1-0.6} = 9$$

The total economic impact is $9 million.

87. $r = \dfrac{1}{4}$

$$S_\infty = \frac{\frac{1}{4}}{1-\frac{1}{4}} = \frac{1}{4} \cdot \frac{4}{3} = \frac{1}{3}$$

89. – 97. Answers may vary.

99. $f(x) = \dfrac{2\left[1-\left(\frac{1}{3}\right)^x\right]}{1-\frac{1}{3}}$

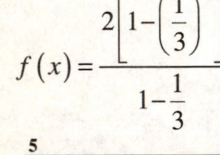

$$S = \frac{2}{1-\frac{1}{3}} = \frac{2}{\frac{2}{3}} = 2 \div \frac{2}{3} = 2 \cdot \frac{3}{2} = 3$$

The sum of the series is 3 and the asymptote of the function is $y = 3$.

101. makes sense

103. makes sense

105. false; Changes to make the statement true will vary. A sample change is: The sequence is not geometric. There is not a common ratio.

107. false; Changes to make the statement true will vary. A sample change is: The sum of the sequence is $\dfrac{10}{1-\left(-\frac{1}{2}\right)}$.

109. Let a_1 equal the number of flies released each day. On any day, the total number of flies is the number released that day, plus 90% of those released the day before, plus 90% of 90% of those released two days before, etc.:

$$S = \frac{a_1}{1-r}$$

$$20,000 = \frac{a_1}{1-.9}$$

$$20,000 = \frac{a_1}{.1}$$

$$2000 = a_1$$

2000 flies to be released each day.

111. Answers may vary.

112. $1+2+3 = \dfrac{3(3+1)}{2}$

$6 = \dfrac{3(4)}{2}$

$6 = 6$

113. $1+2+3+4+5 = \dfrac{5(5+1)}{2}$

$15 = \dfrac{5(6)}{2}$

$15 = 15$

114. $\dfrac{k(k+1)(2k+1)}{6} + (k+1)^2$

$= \dfrac{k(k+1)(2k+1)}{6} + \dfrac{6(k+1)^2}{6}$

$= \dfrac{k(k+1)(2k+1) + 6(k+1)^2}{6}$

$= \dfrac{(k+1)[k(2k+1) + 6(k+1)]}{6}$

$= \dfrac{(k+1)[2k^2 + k + 6k + 6]}{6}$

$= \dfrac{(k+1)[2k^2 + 7k + 6]}{6}$

$= \dfrac{(k+1)(k+2)(2k+3)}{6}$

Mid-Chapter 10 Check Point

1. $a_n = (-1)^{n+1} \dfrac{n}{(n-1)!}$

$a_1 = (-1)^{1+1} \dfrac{1}{(1-1)!} = (-1)^2 \dfrac{1}{0!} = 1 \cdot 1 = 1$

$a_2 = (-1)^{2+1} \dfrac{2}{(2-1)!} = (-1)^3 \dfrac{2}{1!} = (-1)(2) = -2$

$a_3 = (-1)^{3+1} \dfrac{3}{(3-1)!} = (-1)^4 \dfrac{3}{2!} = 1 \cdot \dfrac{3}{2} = \dfrac{3}{2}$

$a_4 = (-1)^{4+1} \dfrac{4}{(4-1)!} = (-1)^5 \dfrac{4}{3!} = (-1)\dfrac{4}{6} = -\dfrac{2}{3}$

$a_5 = (-1)^{5+1} \dfrac{5}{(5-1)!} = (-1)^6 \dfrac{5}{4!} = 1 \cdot \dfrac{5}{24} = \dfrac{5}{24}$

2. Using $a_n = a_1 + (n-1)d$;

$a_1 = 5$

$a_2 = 5 + (2-1)(-3) = 5 + 1(-3) = 5 - 3 = 2$

$a_3 = 5 + (3-1)(-3) = 5 + 2(-3) = 5 - 6 = -1$

$a_4 = 5 + (4-1)(-3) = 5 + 3(-3) = 5 - 9 = -4$

$a_5 = 5 + (5-1)(-3) = 5 + 4(-3) = 5 - 12 = -7$

3. Using $a_n = a_1 r^{n-1}$;

$a_1 = 5$

$a_2 = 5(-3)^{2-1} = 5(-3)^1 = 5(-3) = -15$

$a_3 = 5(-3)^{3-1} = 5(-3)^2 = 5(9) = 45$

$a_4 = 5(-3)^{4-1} = 5(-3)^3 = 5(-27) = -135$

$a_5 = 5(-3)^{5-1} = 5(-3)^4 = 5(81) = 405$

4. $a_n = -a_{n-1} + 4$

$a_1 = 3$

$a_2 = -a_1 + 4 = -3 + 4 = 1$

$a_3 = -a_2 + 4 = -1 + 4 = 3$

$a_4 = -a_3 + 4 = -3 + 4 = 1$

$a_5 = -a_4 + 4 = -1 + 4 = 3$

5. $d = a_2 - a_1 = 6 - 2 = 4$

$a_n = a_1 + (n-1)d$

$= 2 + (n-1)4$

$= 2 + 4n - 4$

$= 4n - 2$

$a_{20} = 4(20) - 2 = 78$

6. $r = \dfrac{a_2}{a_1} = \dfrac{6}{3} = 2$

$a_n = a_1 r^{n-1}$

$= 3(2)^{n-1}$

$a_{10} = 3(2)^{10-1}$

$= 3(2)^9$

$= 1536$

7. $d = a_2 - a_1 = 1 - \dfrac{3}{2} = -\dfrac{1}{2}$

$a_n = a_1 + (n-1)d$

$\quad = \dfrac{3}{2} + (n-1)\left(-\dfrac{1}{2}\right)$

$\quad = \dfrac{3}{2} - \dfrac{1}{2}n + \dfrac{1}{2}$

$\quad = -\dfrac{1}{2}n + 2$

$a_{30} = -\dfrac{1}{2}(30) + 2$

$\quad = -15 + 2$

$\quad = -13$

8. $S_n = \dfrac{a_1(1-r^n)}{1-r}; \; r = \dfrac{a_2}{a_1} = \dfrac{10}{5} = 2$

$S_{10} = \dfrac{5(1-2^{10})}{1-2} = \dfrac{5(-1023)}{-1} = 5115$

9.   First find a_{10};

$d = a_2 - a_1 = 0 - (-2) = 2$

$a_{50} = a_1 + (n-1)d = -2 + (50-1)(2) = -2 + 49(2) = 96$

$S_{50} = \dfrac{n}{2}(a_1 + a_n) = \dfrac{50}{2}(-2 + 96) = 25(94) = 2350$

10. $r = \dfrac{a_2}{a_1} = \dfrac{40}{-20} = -2$

$S_{10} = \dfrac{a_1(1-r^n)}{1-r} = \dfrac{-20(-1-(-2)^{10})}{1-(-2)} = \dfrac{-20(-1023)}{3} = \dfrac{20460}{3} = 6820$

11.   First find a_{100};

$d = a_2 - a_1 = -2 - 4 = -6$

$a_{100} = a_1 + (n-1)d = 4 + (100-1)(-6) = 4 + 99(-6) = -590$

$S_{100} = \dfrac{n}{2}(a_1 + a_n) = \dfrac{100}{2}(4 - 590) = 50(-586) = -29,300$

12. $\displaystyle\sum_{i=1}^{4}(i+4)(i-1) = (1+4)(1-1) + (2+4)(2-1) + (3+4)(3-1) + (4+4)(4-1)$

$\qquad\qquad = 5(0) + 6(1) + 7(2) + 8(3) = 0 + 6 + 14 + 24 = 44$

13. $\displaystyle\sum_{i=1}^{50}(3i-2) = (3\cdot1-2)+(3\cdot2-2)+(3\cdot3-3)+\ldots+(3\cdot50-2)$

$$= (3-2)+(6-2)+(9-3)+\ldots+(150-2)$$
$$= 1+4+6+\ldots+148$$

The sum of this arithmetic sequence is given by $S_n = \dfrac{n}{2}(a_1+a_n)$;

$$S_{50} = \frac{50}{2}(1+148) = 25(149) = 3725$$

14. $\displaystyle\sum_{i=1}^{6}\left(\frac{3}{2}\right)^i = \left(\frac{3}{2}\right)^1 + \left(\frac{3}{2}\right)^2 + \left(\frac{3}{2}\right)^3 + \left(\frac{3}{2}\right)^4 + \left(\frac{3}{2}\right)^5 + \left(\frac{3}{2}\right)^6$

$$= \frac{3}{2} + \frac{9}{4} + \frac{27}{8} + \frac{81}{16} + \frac{243}{32} + \frac{729}{64} = \frac{1995}{64}$$

15. $\displaystyle\sum_{i=1}^{\infty}\left(-\frac{2}{5}\right)^{i-1} = \left(-\frac{2}{5}\right)^{1-1} + \left(-\frac{2}{5}\right)^{2-1} + \left(-\frac{2}{5}\right)^{3-1} + \ldots$

$$= \left(-\frac{2}{5}\right)^0 + \left(-\frac{2}{5}\right)^1 + \left(-\frac{2}{5}\right)^2 + \ldots$$

$$= 1 + \left(-\frac{2}{5}\right) + \frac{4}{25} + \ldots$$

This is an infinite geometric sequence with $r = \dfrac{a_2}{a_1} = \dfrac{-\frac{2}{5}}{1} = -\dfrac{2}{5}$.

$$S = \frac{a_1}{1-r} = \frac{1}{1-\left(-\frac{2}{5}\right)} = \frac{1}{\frac{7}{5}} = \frac{5}{7}$$

16. $0.\overline{45} = \dfrac{a_1}{1-r} = \dfrac{\frac{45}{100}}{1-\frac{1}{100}} = \dfrac{\frac{45}{100}}{\frac{99}{100}}$

$$= \frac{45}{100} \div \frac{99}{100} = \frac{45}{100} \cdot \frac{100}{99} = \frac{45}{99} = \frac{5}{11}$$

17. Answers may vary. An example is $\displaystyle\sum_{i=1}^{18}\frac{i}{i+2}$.

18. The arithmetic sequence is 16, 48, 80, 112, ….
First find a_{15} where $d = a_2 - a_1 = 48-16 = 32$.
$$a_{15} = a_1 + (n-1)d = 16 + (15-1)(32) = 16 + 14(32) = 16 + 448 = 464$$
The distance the skydiver falls during the 15th second is 464 feet.
$$S_{15} = \frac{n}{2}(a_1+a_n) = \frac{15}{2}(16+464) = 7.5(480) = 3600$$
The total distance the skydiver falls in 15 seconds is 3600 feet.

19. $r = 0.10$
$$A = P(1+r)^t$$
$$= 120000(1+0.10)^{10}$$
$$\approx 311249$$
The value of the house after 10 years is \$311,249.

Section 10.4

Check Point Exercises

1. **a.** $S_1 : 2 = 1(1+1)$

   $S_k : 2+4+6+\cdots 2k = k(k+1)$

   $S_{k+1} : 2+4+6+\cdots+2(k+1) = (k+1)(k+2)$

   b. $S_1 : 1^3 = \dfrac{1^2(1+1)^2}{4}$

   $S_k = 1^3 + 2^3 + 3^3 + \cdots + k^3 = \dfrac{k^2(k+1)^2}{4}$

   $S_{k+1} = 1^3 + 2^3 + 3^3 + \cdots + (k+1)^3 = \dfrac{(k+1)^2(k+2)^2}{4}$

2. $S_1 : 2 = 1(1+1)$

   $2 = 2$ is true.

   $S_k : 2+4+6+\cdots+2k = k(k+1)$

   $S_{k+1} : 2+4+6+\cdots+2k+2(k+1) = (k+1)(k+2)$

   Add $2(k+1)$ to both sides of S_k :

   $2+4+6+\cdots+2k+2(k+1) = k(k+1)+2(k+1)$

   Simplify the right-hand side:

   $k(k+1)+2(k+1) = (k+1)(k+2)$

   If S_k is true, then S_{k+1} is true. The statement is true for all n.

3. $S_1 : 1^3 = \dfrac{1^2(1+1)^2}{4}$

   $1 = \dfrac{4}{4}$

   $1 = 1$ is true.

   $S_k : 1^3 + 2^3 + 3^3 + \cdots + k^3 = \dfrac{k^2(k+1)^2}{4}$

   $S_{k+1} : 1^3 + 2^3 + 3^3 + \cdots + k^3 + (k+1)^3 = \dfrac{(k+1)^2(k+2)^2}{4}$

   Add $(k + 1)^3$ to both sides of S_k :

   $1^3 + 2^3 + 3^3 + \cdots + k^3 + (k+1)^3 = \dfrac{k^2(k+1)^2}{4} + (k+1)^3$

   Simplify the right hand side:

   $\dfrac{k^2(k+1)^2}{4} + (k+1)^3 = \dfrac{k^2(k+1)^2 + 4(k+1)^3}{4} = \dfrac{(k+1)^2\left[k^2+4(k+1)\right]}{4} = \dfrac{(k+1)^2(k^2+4k+4)}{4}$

   $\qquad = \dfrac{(k+1)^2(k+2)^2}{4}$

   If S_k is true, then S_{k+1} is true. The statement is true for all n.

4. S_1: 2 is a factor of $1^2 + 1 = 2$, since $2 = 2 \cdot 1$.

        S_k: 2 is a factor of $k^2 + k$

S_{k+1}: 2 is a factor of $(k+1)^2 + (k+1)$

Simplify:

$$(k+1)^2 + (k+1) = k^2 + 2k + 1 + k + 1$$
$$= k^2 + 3k + 2$$
$$= k^2 + k + 2k + 2$$
$$= (k^2 + k) + 2(k+1)$$

Because we assume S_k is true, we know 2 is a factor of $k^2 + k$. Since 2 is a factor of $2(k+1)$, we conclude 2 is a factor of the sum $(k^2 + k) + 2(k+1)$. If S_k is true, then S_{k+1} is true. The statement is true for all n.

Exercise Set 10.4

1. $S_n = 1 + 3 + 5 + \cdots + (n-1) = n^2$

    $S_1 : 1 = 1^2$

    $1 = 1$ true

    $S_2 : 1 + 3 = 2^2$

           $4 = 4$ true

    $S_3 : 1 + 3 + 5 = 3^2$

              $9 = 9$ true

3. S_n : 2 is a factor of $n^2 - n$

    S_1 : 2 is a factor of $1^2 - 1 = 0$

       $0 = 0 \cdot 2$ so 2 is a factor of 0 is true.

    S_2 : 2 is a factor of $2^2 - 2 = 2$

       $2 = 1 \cdot 2$ so 2 is a factor of 2 is true.

    S_3 : 2 is a factor of $3^2 - 3 = 6$

       $6 = 3 \cdot 2$ so 2 is a factor of 6 is true.

5. $S_n : 4 + 8 + 12 + \cdots + 4n = 2n(n+1)$

      $S_k : 4 + 8 + 12 + \cdots + 4k = 2k(k+1)$

$S_{k+1} : 4 + 8 + 12 + \cdots + 4(k+1) = 2(k+1)(k+1+1)$

      $4 + 8 + 12 + \cdots + 4(k+1) = 2(k+1)(k+2)$

7. $S_n : 3 + 7 + 11 + \cdots + (4n-1) = n(2n+1)$

       $S_k : 3 + 7 + 11 + \cdots + (4k-1) = k(2k+1)$

$S_{k+1} : 3 + 7 + 11 + \cdots + [4(k+1) - 1] = (k+1)[2(k+1)+1]$

        $3 + 7 + 11 + \cdots + (4k+3) = (k+1)(2k+3)$

9. $S_n : 2$ is a factor of $n^2 - n + 2$

   $S_k : 2$ is a factor of $k^2 - k + 2$

   $S_{k+1} : 2$ is a factor of $(k+1)^2 - (k+1) + 2$

   $\quad\quad k^2 + 2k + 1 - k - 1 + 2 = k^2 + k + 2$

   $S_{k+1} : 2$ is a factor of $k^2 + k + 2$.

11. $S_1 : 4 = 2(1)(1+1)$

   $4 = 2(2)$

   $4 = 4$ is true.

   $\quad S_k : 4 + 8 + 12 + \cdots + 4k = 2k(k+1)$

   $S_{k+1} : 4 + 8 + 12 + \cdots 4(k+1) = 2(k+1)(k+1+1)$

   Add $4(k + 1)$ to both sides of S_k:

   $4 + 8 + 12 + \cdots + 4(k+1) = 2k(k+1) + 4(k+1)$

   Simplify the right-hand side:

   $= 2k(k+1) + 4(k+1) = (2k+4)(k+1)$

   $= 2(k+2)(k+1)$

   $= 2(k+1)(k+1+1)$

   If S_k is true, then S_{k+1} is true. The statement is true for all *n*.

13. $S_1 : 1 = 1^2$

   $1 = 1$ is true.

   $\quad S_k : 1 + 3 + 5 + \cdots + (2k-1) = k^2$

   $S_{k+1} : 1 + 3 + 5 + \cdots + (2k-1) + [2(k+1)-1] = (k+1)^2$

   $\quad\quad\quad 1 + 3 + 5 + \cdots + (2k-1) + (2k+1) = (k+1)^2$

   Add $(2k + 1)$ to both sides of S_k:

   $1 + 3 + 5 + \cdots + (2k-1) + (2k+1) = k^2 + (2k+1)$

   Simplify the right-hand side:

   $= k^2 + (2k+1)$

   $= (k+1)^2$

   If S_k is true, then S_{k+1} is true. The statement is true for all *n*.

15. $S_1 : 3 = 1[2(1)+1)]$

   $3 = 3$ is true.

   $\quad S_k : 3 + 7 + 11 + \cdots + (4k-1) = k(2k+1)$

   $S_{k+1} : 3 + 7 + 11 + \cdots + (4k-1) + [4(k+1)-1] = (k+1)[2(k+1)+1]$

   $\quad\quad\quad 3 + 7 + 11 + \cdots + (4k-1) + (4k+3) = (k+1)(2k+3)$

   Add $(4k + 3)$ to both sides of S_k:

   $3 + 7 + 11 + \ldots + (4k - 1) + (4k + 3) = k(2k+1) + 4(k+3)$

   Simplify the right-hand side:

   $= k(2k+1) + (4k+3) = 2k^2 + k + 4k + 3$

   $= 2k^2 + 5k + 3$

   $= (k+1)(2k+3)$

   If S_k is true, then S_{k+1} is true. The statement is true for all *n*.

17. $S_1: 1 = 2^1 - 1$

$\qquad 1 = 1$ is true.

$S_k: 1 + 2 + 2^2 + \cdots + 2^{k-1} = 2^k - 1$

$S_{k+1}: 1 + 2 + 2^2 + \cdots + 2^{k-1} + 2^{k+1-1} = 2^{k+1} - 1$

$\qquad\qquad 1 + 2 + 2^2 + \cdots + 2^{k-1} + 2^k = 2^{k+1} - 1$

Add 2^k to both sides of S_k:

$1 + 2 + 2^2 + \cdots + 2^{k-1} + 2^k = 2^k + 2^k - 1$

Simplify the right-hand side:

$= 2^k + 2^k - 1 = 2\left(2^k\right) - 1$

$= 2^{k+1} - 1$

If S_k is true, then S_{k+1} is true. The statement is true for all n.

19. $S_1: 2 = 2^{1+1} - 2$

$\qquad 2 = 4 - 2$

$\qquad 2 = 2$ is true.

$\qquad S_k: 2 + 4 + 8 + \cdots + 2^k = 2^{k+1} - 2$

$S_{k+1}: 2 + 4 + 8 + \cdots + 2^k + 2^{k+1} = 2^{k+2} - 2$

Add 2^{k+1} to both sides of S_k:

$2 + 4 + 8 + \cdots + 2^k + 2^{k+1} = 2^{k+1} + 2^{k+1} - 2$

Simplify the right-hand side:

$= 2^{k+1} + 2^{k+1} - 1 = 2\left(2^{k+1}\right) - 2$

$= 2^{k+2} - 2$

If S_k is true, then S_{k+1} is true. The statement is true for all n.

21. $S_1: 1 \cdot 2 = \dfrac{1(1+1)(1+2)}{3}$

$2 = \dfrac{6}{3}$

$2 = 2$ is true.

$\qquad S_k: 1 \cdot 2 + 2 \cdot 3 + 3 \cdot 4 + \cdots + k(k+1) = \dfrac{k(k+1)(k+2)}{3}$

$S_{k+1}: 1 \cdot 2 + 2 \cdot 3 + 3 \cdot 4 + \cdots + k(k+1) + (k+1)(k+2) = \dfrac{(k+1)(k+2)(k+3)}{3}$

Add $(k+1)(k+2)$ to both sided of S_k:

$1 \cdot 2 + 2 \cdot 3 + 3 \cdot 4 + \cdots + k(k+1) + (k+1)(k+2) = \dfrac{k(k+1)(k+2)}{3} + (k+1)(k+2)$

Simplify the right-hand side:

$= \dfrac{k(k+1)(k+2)}{3} + (k+1)(k+2) = \dfrac{k(k+1)(k+2) + 3(k+1)(k+2)}{3}$

$= \dfrac{(k+1)(k+2)(k+3)}{3}$

If S_k is true, then S_{k+1} is true. The statement is true for all n.

23. $S_1 : \dfrac{1}{1 \cdot 2} = \dfrac{1}{1+1}$

$\dfrac{1}{2} = \dfrac{1}{2}$ is true.

$S_k : \dfrac{1}{1 \cdot 2} + \dfrac{1}{2 \cdot 3} + \dfrac{1}{3 \cdot 4} + \cdots + \dfrac{1}{k(k+1)} = \dfrac{k}{k+1}$

$S_{k+1} : \dfrac{1}{1 \cdot 2} + \dfrac{1}{2 \cdot 3} + \dfrac{1}{3 \cdot 4} + \cdots + \dfrac{1}{k(k+1)} + \dfrac{1}{(k+1)(k+2)} = \dfrac{k+1}{k+2}$

Add $\dfrac{1}{(k+1)(k+2)}$ to both sides of S_k:

$\dfrac{1}{1 \cdot 2} + \dfrac{1}{2 \cdot 3} + \dfrac{1}{3 \cdot 4} + \cdots + \dfrac{1}{k(k+1)} + \dfrac{1}{(k+1)(k+2)} = \dfrac{k}{k+1} + \dfrac{1}{(k+1)(k+2)}$

Simplify the right-hand side:

$\dfrac{k}{(k+1)} + \dfrac{1}{(k+1)(k+2)} = \dfrac{k(k+2)+1}{(k+1)(k+2)}$

$= \dfrac{k^2 + 2k + 1}{(k+1)(k+2)}$

$= \dfrac{(k+1)(k+1)}{(k+1)(k+2)}$

$= \dfrac{k+1}{k+2}$

If S_k is true, then S_{k+1} is true. The statement is true for all n.

25. S_1 : 2 is a factor of $1^2 - 1 = 0$, since $0 = 2 \cdot 0$.

S_k : 2 is a factor of $k^2 - k$

S_{k+1} : 2 is a factor of $(k+1)^2 - (k+1)$

$(k+1)^2 - (k-1) = k^2 + 2k + 1 - k - 1$

$\qquad\qquad\qquad = k^2 + k$

$\qquad\qquad\qquad = k^2 - k + 2k$

$\qquad\qquad\qquad = (k^2 - k) + 2k$

Because we assume S_k is true, we know 2 as a factor of $k^2 - k$. Since 2 is a factor of $2k$, we conclude 2 is factor of the sum $(k^2 + k) + 2k$. If S_k is true, then S_{k+1} is true. The statement is true for all n.

27. S_1: 6 is a factor of $1(1+1)(1+2) = 6$, since $6 = 6 \cdot 1$.

S_k: 6 is a factor of $k(k+1)(k+2)$

S_{k+1}: 6 is a factor of $(k+1)(k+2)(k+3)$

$(k+1)(k+2)(k+3) = k(k+1)(k+2) + 3(k+1)(k+2)$

Because we assume S_k is true, we know 6 as a factor of $k(k+1)(k+2)$. Since either $k+1$ or $k+2$ must be even, the product $(k+1)(k+2)$ is even. Thus 2 is a factor of $(k+1)(k+2)$, and we can conclude that 6 is factor of $3(k+1)(k+2)$ If S_k is true, then S_{k+1} is true.

The statement is true for all n.

29. $\displaystyle\sum_{i=1}^{n} 5 \cdot 6^i = 6\left(6^n - 1\right)$

Show that S_1 is true: $\displaystyle\sum_{i=1}^{1} 5 \cdot 6^i = 6\left(6^1 - 1\right)$

$$5 \cdot 6^1 = 6(6 - 1)$$
$$5 \cdot 6 = 6 \cdot 5, \ \text{True}$$

Show that if S_k is true, then S_{k+1} is true:

Assume $S_k : \displaystyle\sum_{i=1}^{k} 5 \cdot 6^i = 6\left(6^k - 1\right)$ is true. Then,

$$\sum_{i=1}^{k} 5 \cdot 6^i + 5 \cdot 6^{k+1} = 6\left(6^k - 1\right) + 5 \cdot 6^{k+1}$$

$$\sum_{i=1}^{k+1} 5 \cdot 6^i = 6^{k+1} - 6 + 5 \cdot 6^{k+1}$$

$$\sum_{i=1}^{k+1} 5 \cdot 6^i = 6 \cdot 6^{k+1} - 6$$

$$\sum_{i=1}^{k+1} 5 \cdot 6^i = 6\left(6^{k+1} - 1\right)$$

The final statement is S_{k+1}. Thus, by mathematical induction, we have proven that $\displaystyle\sum_{i=1}^{n} 5 \cdot 6^i = 6\left(6^n - 1\right)$.

31. $n + 2 > n$

Show that S_1 is true: $1 + 2 > 1$

$$3 > 1, \ \text{True}$$

Show that if S_k is true, then S_{k+1} is true:

Assume $S_k : k + 2 > k$ is true. Then,

$k + 2 + 1 > k + 1$

$(k+1) + 2 > k + 1$

The final statement is S_{k+1}. Thus, by mathematical induction, we have proven that $n + 2 > n$.

33. $S_1 : (ab)^1 = a^1 b^1$

$ab = ab$ is true.

$S_k : (ab)^k = a^k b^k$

$S_{k+1} : (ab)^{k+1} = a^{k+1} b^{k+1}$

Multiply both sides of S_k by ab:

$(ab)^k (ab) = a^k b^k (ab)$

$(ab)^{k+1} = a^{k+1} b^{k+1}$

If S_k is true, then S_{k+1} is true.

The statement is true for all n.

35. Answers may vary.

37. does not make sense; Explanations will vary. Sample explanation: We use mathematical induction to prove statements involving positive integers.

39. does not make sense; Explanations will vary. Sample explanation: It is necessary for all the dominoes to topple.

41. $n^2 > 2n + 1$ for $n \geq 3$

$S_3 : 3^2 > 2 \cdot 3 + 1$

$9 > 7$

$S_k : k^2 > 2k + 1$ for $k \geq 3$

$S_{k+1} : (k+1)^2 > 2k + 3$.

Add $2k + 1$ to both sides of S_k.

$k^2 + (2k+1) > 2k + 1 + (2k+1)$

Write the left side of the inequalities as the square of a binomial and simplify the right side. $(k+1)^2 > 4k + 2$

Since $4k + 2 > 2k + 3$ for $k \geq 3$, we can conclude that $(k+1)^2 > 4k + 2 > 2k + 3$.

By the transitive property,

$(k+1)^2 > 2k + 3$

$(k+1)^2 > 2(k+1) + 1$

If S_k is true, then S_{k+1} is true.

The statement is true for all n.

43. $S_1 = \dfrac{1}{4} = \dfrac{1}{4}$

$S_2 = \dfrac{1}{4} + \dfrac{1}{12} = \dfrac{1}{3}$

$S_3 = \dfrac{1}{4} + \dfrac{1}{12} + \dfrac{1}{24} = \dfrac{3}{8}$

$S_4 = \dfrac{1}{4} + \dfrac{1}{12} + \dfrac{1}{24} + \dfrac{1}{40} = \dfrac{2}{5}$

$S_5 = \dfrac{1}{4} + \dfrac{1}{12} + \dfrac{1}{24} + \dfrac{1}{40} + \dfrac{1}{60} = \dfrac{5}{12}$

$S_n = \dfrac{1}{4} + \dfrac{1}{12} + \dfrac{1}{24} + \cdots + \dfrac{1}{2n(n+1)} = \dfrac{n}{2n+2}$

$S_k = \dfrac{1}{4} + \dfrac{1}{12} + \dfrac{1}{24} + \cdots + \dfrac{1}{2k(k+1)} = \dfrac{k}{2k+2}$

$S_{k+1} = \dfrac{1}{4} + \dfrac{1}{12} + \dfrac{1}{24} + \cdots + \dfrac{1}{2k(k+1)} + \dfrac{1}{2(k+1)(k+2)} = \dfrac{k+1}{2k+4}$

Add $\dfrac{1}{2(k+1)(k+2)}$ to both sides of S_k:

$\dfrac{1}{4} + \dfrac{1}{12} + \dfrac{1}{24} + \cdots + \dfrac{1}{2k(k+1)} + \dfrac{1}{2(k+1)(k+2)} = \dfrac{k}{2k+2} + \dfrac{1}{2(k+1)(k+2)}$

Simplify the right-hand side:

$\dfrac{k}{2k+2} + \dfrac{1}{2(k+1)(k+2)} = \dfrac{k(k+2)+1}{2(k+1)(k+2)} = \dfrac{k^2+2k+1}{2(k+1)(k+2)} = \dfrac{(k+1)^2}{2(k+1)(k+2)} = \dfrac{k+1}{2k+4}$

If S_k is true, then S_{k+1} is true. The conjecture is proven.

45. Answers may vary.

46. The exponents begin with the exponent on $a+b$ and decrease by 1 in each successive term.

47. The exponents begin with 0, increase by 1 in each successive term, and end with the exponent on $a+b$.

48. The sum of the exponents is the exponent on $a+b$.

Section 10.5

Check Point Exercises

1. **a.** $\dbinom{6}{3} = \dfrac{6!}{3!(6-3)!} = \dfrac{6!}{3!3!} = \dfrac{5 \cdot 4}{1} = 20$

b. $\dbinom{6}{0} = \dfrac{6!}{0!(6-0)!} = \dfrac{6!}{6!} = 1$

c. $\dbinom{8}{2} = \dfrac{8!}{2!(8-2)!} = \dfrac{8!}{2!6!} = \dfrac{8 \cdot 7}{2} = 28$

d. $\dbinom{3}{3} = \dfrac{3!}{3!(3-3)!} = \dfrac{3!}{3!0!} = \dfrac{3!}{3!} = 1$

2. $(x+1)^4 = \binom{4}{0}x^4 + \binom{4}{1}x^3 + \binom{4}{2}x^2 + \binom{4}{1}x + \binom{4}{0} = x^4 + 4x^3 + 6x^2 + 4x + 1$

3. $(x-2y)^5 = \binom{5}{0}x^5(-2y)^0 + \binom{5}{1}x^4(-2y)^1 + \binom{5}{2}x^3(-2y)^2 + \binom{5}{3}x^2(-2y)^3 + \binom{5}{4}x(-2y)^4 + \binom{5}{5}x^0(-2y)^5$

$$= x^5 - 5x^4(2y) + 10x^3(4y^2) - 10x^2(8y^3) + 5x(16y^4) - 32y^5$$

$$= x^5 - 10x^4y + 40x^3y^2 - 80x^2y^3 + 80xy^4 - 32y^5$$

4. $(2x+y)^9$

fifth term $= \binom{9}{4}(2x)^5 y^4 = \frac{9!}{4!5!}(32x^5)y^4 = 4032x^5 y^4$

Exercise Set 10.5

1. $\binom{8}{3} = \frac{8!}{3!(8-3)!} = \frac{8 \cdot 7 \cdot 6}{3 \cdot 2 \cdot 1} = 56$

3. $\binom{12}{1} = \frac{12!}{1!11!} = 12$

5. $\binom{6}{6} = \frac{6!}{0!6!} = 1$

7. $\binom{100}{2} = \frac{100!}{2!98!} = \frac{100 \cdot 99}{2} = 4950$

9. $(x+2)^3 = \binom{3}{0}x^3 + \binom{3}{1}2x^2 + \binom{3}{2}4x + \binom{3}{3}8 = x^3 + 3x^2 \cdot 2 + 3x \cdot 4 + 8 = x^3 + 6x^2 + 12x + 8$

11. $(3x+y)^3 = \binom{3}{0}27x^3 + \binom{3}{1}9x^2y + \binom{3}{2}3xy^2 + \binom{3}{3}y^3 = 27x^3 + 27x^2y + 9xy^2 + y^3$

13. $(5x-1)^3 = \binom{3}{0}125x^3 - \binom{3}{1}25x^2 + \binom{3}{2}5x - \binom{3}{3} = 125x^3 - 75x^2 + 15x - 1$

15. $(2x+1)^4 = \binom{4}{0}16x^4 - \binom{4}{1}8x^3 + \binom{4}{2}4x^2 + \binom{4}{3}2x + \binom{4}{4} = 16x^4 + 32x^3 + 24x^2 + 8x + 1$

17. $(x^2+2y)^4 = \binom{4}{0}(x^2)^4 + \binom{4}{1}(x^2)^3(2y) + \binom{4}{2}(x^2)^2(2y)^2 + \binom{4}{3}(x^2)^1(2y)^3 + \binom{4}{4}(2y)^4$

$$= 1(x^8) + 4(x^6)(2y) + 6(x^4)(4y^2) + 4x^2(8y^3) + 1(16y^4)$$

$$= x^8 + 8x^6y + 24x^4y^2 + 32x^2y^3 + 16y^4$$

19. $(y-3)^4 = \binom{4}{0}y^4 + \binom{4}{1}y^3(-3) + \binom{4}{2}y^2(-3)^2 + \binom{4}{3}y(-3)^3 + \binom{4}{4}(-3)^4$

$\qquad = y^4 + 4(y^3)(-3) + 6(y^2)(9) + 4(y)(-27) + 81$

$\qquad = y^4 - 12y^3 + 54y^2 - 108y + 81$

21. $(2x^3-1)^4 = \binom{4}{0}(2x^3)^4 + \binom{4}{1}(2x^3)^3(-1) + \binom{4}{2}(2x^3)^2(-1)^2 + \binom{4}{3}(2x^3)(-1)^3 + \binom{4}{4}(-1)^4$

$\qquad = 16x^{12} - 4(8x^9) + 6(4x^6) - 4(2x^3) + 1$

$\qquad = 16x^{12} - 32x^9 + 24x^6 - 8x^3 + 1$

23. $(c+2)^5 = \binom{5}{0}c^5 + \binom{5}{1}c^4(2) + \binom{5}{2}c^3(2^2) + \binom{5}{3}c^2(2^3) + \binom{5}{4}c(2^4) + \binom{5}{5}(2^5)$

$\qquad = c^5 + 5c^4(2) + 10c^3(4) + 10c^2(8) + 5c(16) + 32$

$\qquad = c^5 + 10c^4 + 40c^3 + 80c^2 + 80c + 32$

25. $(x-1)^5 = \binom{5}{0}x^5 - \binom{5}{1}x^4 + \binom{5}{2}x^3 - \binom{5}{3}x^2 + \binom{5}{4}x - \binom{5}{5} = x^5 - 5x^4 + 10x^3 - 10x^2 + 5x - 1$

27. $(3x-y)^5 = \binom{5}{0}(3x)^5 - \binom{5}{1}(3x)^4 y + \binom{5}{2}(3x)^3 y^2 - \binom{5}{3}(3x)^2 y^3 + \binom{5}{4}3xy^4 - \binom{5}{5}y^5$

$\qquad = (1)243x^5 - 5(81x^4)y + 10(27x^3)y^2 - 10(9x^2)y^3 + 5(3x)y^4 - (1)y^5$

$\qquad = 243x^5 - 405x^4 y + 270x^3 y^2 - 90x^2 y^3 + 15xy^4 - y^5$

29. $(2a+b)^6 = \binom{6}{0}(2a)^6 + \binom{6}{1}(2a)^5 b + \binom{6}{2}(2a)^4 b^2 + \binom{6}{3}(2a)^3 b^3 + \binom{6}{4}(2a)^2 b^4 + \binom{6}{5}(2a)b^5 + \binom{6}{6}b^6$

$\qquad = 64a^6 + 6(32a^5)b + 15(16a^4)b^2 + 20(8a^3)b^3 + 15(4a^2)b^4 + 6(2a)b^5 + b^6$

$\qquad = 64a^6 + 192a^5 b + 240a^4 b^2 + 160a^3 b^3 + 60a^2 b^4 + 12ab^5 + b^6$

31. $(x+2)^8 = \binom{8}{0}x^8 + \binom{8}{1}x^7 2 + \binom{8}{3}x^6(2)^2 + \cdots$

$\qquad = x^8 + 16x^7 + 112x^6 + \cdots$

33. $(x-2y)^{10} = \binom{10}{0}x^{10} - \binom{10}{1}x^9(2y) + \binom{10}{2}x^8(2y)^2 - \cdots$

$\qquad = x^{10} - 20x^9 y + 180x^8 y^2 - \cdots$

35. $(x^2+1)^{16} = \binom{16}{0}(x^2)^{16} + \binom{16}{1}(x^2)^{15} + \binom{16}{2}(x^2)^{14} + \cdots$

$\qquad = x^{32} + 16x^{30} + 120x^{28} + \cdots$

37. $(y^3-1)^{20} = \binom{20}{0}(y^3)^{20} - \binom{20}{1}(y^3)^{19} + \binom{20}{2}(y^3)^{18} - \cdots$

$\qquad = y^{60} - 20y^{57} + 190y^{54} - \cdots$

39. $(2x+y)^6$; third term $= \binom{6}{2}(2x)^4(y)^2 = 15\left(16x^4y^2\right) = 240x^4y^2$

41. $(x-1)^9$; fifth term $= \binom{9}{4}x^5(-1)^4 = 126x^5$

43. $\left(x^2+y^3\right)^8$; sixth term $= \binom{8}{5}\left(x^2\right)^3\left(y^3\right)^5 = 56x^6y^{15}$

45. $\left(x-\tfrac{1}{2}\right)^9$; fourth term $= \binom{9}{3}x^6\left(-\tfrac{1}{2}\right)^3 = 84x^6\left(-\tfrac{1}{8}\right) = -\tfrac{21}{2}x^6$

47. $\binom{22}{14}(x^2)^8 y^{14} = 319{,}770x^{16}y^{14}$

49.

$$\left(x^3+x^{-2}\right)^4 = \binom{4}{0}\left(x^3\right)^4 + \binom{4}{1}\left(x^3\right)^3\left(x^{-2}\right) + \binom{4}{2}\left(x^3\right)^2\left(x^{-2}\right)^2 + \binom{4}{3}\left(x^3\right)^1\left(x^{-2}\right)^3 + \binom{4}{4}\left(x^{-2}\right)^4$$

$$= \frac{4!}{0!(4-0)!}x^{12} + \frac{4!}{1!(4-1)!}x^9x^{-2} + \frac{4!}{2!(4-2)!}x^6x^{-4} + \frac{4!}{3!(4-3)!}x^3x^{-6} + \frac{4!}{4!(4-4)!}x^{-8}$$

$$= \frac{\cancel{4!}}{0!\,\cancel{4!}}x^{12} + \frac{4\cdot\cancel{3!}}{1!\,\cancel{3!}}x^7 + \frac{4\cdot3\cdot\cancel{2!}}{2\cdot1\cdot\cancel{2!}}x^2 + \frac{4\cdot\cancel{3!}}{\cancel{3!}\cdot1!}x^{-3} + \frac{\cancel{4!}}{\cancel{4!}\cdot0!}x^{-8}$$

$$= x^{12} + 4x^7 + 6x^2 + \frac{4}{x^3} + \frac{1}{x^8}$$

51.

$$\left(x^{\frac{1}{3}}-x^{-\frac{1}{3}}\right)^3 = \left(x^{\frac{1}{3}}+\left(-x^{-\frac{1}{3}}\right)\right)^3 = \binom{3}{0}\left(x^{\frac{1}{3}}\right)^3 + \binom{3}{1}\left(x^{\frac{1}{3}}\right)^2\left(-x^{-\frac{1}{3}}\right) +$$

$$+ \binom{3}{2}\left(x^{\frac{1}{3}}\right)^1\left(-x^{-\frac{1}{3}}\right)^2 + \binom{3}{3}\left(-x^{-\frac{1}{3}}\right)^3$$

$$= \frac{3!}{0!(3-0)!}x^1 + \frac{3!}{1!(3-1)!}x^{\frac{2}{3}}\cdot-x^{-\frac{1}{3}} + \frac{3!}{2!(3-2)!}x^{\frac{1}{3}}x^{-\frac{2}{3}} + \frac{3!}{3!(3-3)!}\cdot-x^{-1}$$

$$= \frac{\cancel{3!}}{0!\,\cancel{3!}}x + \frac{3\cdot\cancel{2!}}{1!\,\cancel{2!}}\cdot-x^{\frac{1}{3}} + \frac{\cdot3\cdot\cancel{2!}}{\cancel{2!}\cdot1!}x^{-\frac{1}{3}} + \frac{\cancel{3!}}{\cancel{3!}\cdot0!}\cdot-x^{-1}$$

$$= x - 3x^{\frac{1}{3}} + \frac{3}{x^{\frac{1}{3}}} - \frac{1}{x}$$

53. $f(x) = x^4 + 7$;

$\dfrac{f(x+h) - f(x)}{h}$

$= \dfrac{(x+h)^4 + 7 - (x^4 + 7)}{h}$

$= \dfrac{\binom{4}{0}x^4 + \binom{4}{1}x^3 h + \binom{4}{2}x^2 h^2 + \binom{4}{3}xh^3 + \binom{4}{4}h^4 + 7 - x^4 - 7}{h}$

$= \dfrac{\dfrac{4!}{0!(4-0)!}x^4 + \dfrac{4!}{1!(4-1)!}x^3 h + \dfrac{4!}{2!(4-2)!}x^2 h^2 + \dfrac{4!}{3!(4-3)!}xh^3 + \dfrac{4!}{4!(4-4)!}h^4 - x^4}{h}$

$= \dfrac{\dfrac{\cancel{4!}}{0!\cancel{4!}}x^4 + \dfrac{4 \cdot \cancel{3!}}{1!\cancel{3!}}x^3 h + \dfrac{4 \cdot 3 \cdot \cancel{2!}}{2! \cdot 2 \cdot 1}x^2 h^2 + \dfrac{4 \cdot \cancel{3!}}{\cancel{3!}1!}xh^3 + \dfrac{\cancel{4!}}{\cancel{4!}0!}h^4 - x^4}{h}$

$= \dfrac{\cancel{x^4} + 4x^3 h + 6x^2 h^2 + 4xh^3 + h^4 - \cancel{x^4}}{h}$

$= \dfrac{h(4x^3 + 6x^2 h + 4xh^2 + h^3)}{h}$

$= 4x^3 + 6x^2 h + 4xh^2 + h^3$

55.   Find the $(5+1) = 6^{th}$ term.

$\binom{n}{r}a^{n-r}b^r = \binom{10}{5}\left(\dfrac{3}{x}\right)^{10-5}\left(\dfrac{x}{3}\right)^5 = \dfrac{10!}{5!(10-5)!}\left(\dfrac{3}{x}\right)^5\left(\dfrac{x}{3}\right)^5$

$= \dfrac{10 \cdot 9 \cdot 8 \cdot 7 \cdot 6 \cdot \cancel{5!}}{\cancel{5!} \cdot 5 \cdot 4 \cdot 3 \cdot 2 \cdot 1}\left(\dfrac{3}{x}\right)^5\left(\dfrac{x}{3}\right)^5 = 252 \cdot \dfrac{3^5}{x^5} \cdot \dfrac{x^5}{3^5} = 252$

57. $(0.28 + 0.72)^5$

Third Term $(r = 2)$: $\binom{n}{r}a^{n-r}b^r = \binom{5}{2}0.28^{5-2}0.72^2 = \dfrac{5!}{2!(5-2)!}0.28^{5-2}0.72^2 = \dfrac{5!}{2!3!}0.28^3 0.72^2 \approx 0.1138$

59. – 67.   Answers may vary.

69. $f_1(x) = (x+1)^4$

$f_2(x) = x^4$

$f_3(x) = x^4 + 4x^3$

$f_4(x) = x^4 + 4x^3 + 6x^2$

$f_5(x) = x^4 + 4x^3 + 6x^2 + 4x$

$f_6(x) = x^4 + 4x^3 + 6x^2 + 4x + 1$

$f_2, f_3, f_4,$ and f_5 are approaching $f_1 = f_6$.

71. $f_1(x) = (x-2)^4$

$$= \binom{4}{0}x^4 + \binom{4}{1}x^3(-2) + \binom{4}{2}x^2(-2)^2 + \binom{4}{3}x(-2)^3 + \binom{4}{4}(-2)^4$$

$$= x^4 + 4x^3(-2) + 6x^2(4) + 4x(-8) + 16$$

$$= x^4 - 8x^3 + 24x^2 - 32x + 16$$

73. makes sense

75. does not make sense; Explanations will vary. Sample explanation: $\binom{n}{0}$ and $\binom{n}{1}$ are the coefficients of the first and second term.

77. false; Changes to make the statement true will vary. A sample change is: The binomial expansion for $(a+b)^n$ contains $n+1$ terms.

79. false; Changes to make the statement true will vary. A sample change is: The sum of the binomial coefficients in $(a+b)^n$ is 2^n.

81. $(x^2 + x + 1)^3 = \left[x^2 + (x+1)\right]^3$

$$= \binom{3}{0}(x^2)^3 + \binom{3}{1}(x^2)^2(x+1) + \binom{3}{2}x^2(x+1)^2 + \binom{3}{3}(x+1)^3$$

$$= x^6 + 3x^4(x+1) + 3x^2(x^2 + 2x + 1) + x^3 + 3x^2 + 3x + 1$$

$$= x^6 + 3x^5 + 3x^4 + 3x^4 + 6x^3 + 3x^2 + x^3 + 3x^2 + 3x + 1$$

$$= x^6 + 3x^5 + 6x^4 + 7x^3 + 6x^2 + 3x + 1$$

83. $\binom{n}{r} = \dfrac{n!}{r!(n-r)!} = \dfrac{n!}{(n-r)!r!} = \dfrac{n!}{(n-r)!\left[n-(n-r)\right]!} = \binom{n}{n-r}$

85. a. $S_1 : (a+b)^1 = \binom{1}{0}a^1 + \binom{1}{1}a^{1-1}b = a + b$

b. $S_k : (a+b)^k = \binom{k}{0}a^k + \binom{k}{1}a^{k-1}b + \binom{k}{2}a^{k-2}b^2 + \cdots + \binom{k}{k-1}ab^{k-1} + \binom{k}{k}b^k$

$S_{k+1} : (a+b)^{k+1} = \binom{k+1}{0}a^{k+1} + \binom{k+1}{1}a^k b + \binom{k+1}{2}a^{k-1}b^2 + \cdots + \binom{k+1}{k}ab^k + \binom{k+1}{k+1}b^{k+1}$

c. $(a+b)(a+b)^k$

$$(a+b)^{k+1} = \binom{k}{0}a^{k+1} + \binom{k}{0}a^k b + \binom{k}{1}a^k b + \binom{k}{1}a^{k-1}b^2 + \binom{k}{2}a^{k-1}b^2 + \binom{k}{2}a^{k-2}b^3 + \cdots$$

$$= \binom{k}{k-1}a^2 b^{k-1} + \binom{k}{k-1}ab^k + \binom{k}{k}ab^k + \binom{k}{k}b^{k+1}$$

d. $(a+b)^{k+1} = \binom{k}{0}a^{k+1} + \left[\binom{k}{0}+\binom{k}{1}\right]a^k b + \left[\binom{k}{1}+\binom{k}{2}\right]a^{k-1}b^2 + \left[\binom{k}{2}+\binom{k}{3}\right]a^{k-2}b^3 + \cdots$

$$+ \left[\binom{k}{k-1}+\binom{k}{k}\right]ab^k + \binom{k}{k}b^{k+1}$$

e. $(a+b)^{k+1} = \binom{k}{0}a^{k+1} + \binom{k+1}{1}a^k b + \binom{k+1}{2}a^{k-1}b^2 + \binom{k+1}{3}a^{k-2}b^3 + \cdots + \binom{k+1}{k}ab^k + \binom{k}{k}b^{k+1}$

f. $\binom{k}{0} = \binom{k+1}{0}$ because both equal 1. $\binom{k}{k} = \binom{k+1}{k+1}$ also because both equal 1.

$$S_{k+1} : (a+b)^{k+1} = \binom{k+1}{0}a^{k+1} + \binom{k+1}{1}a^k b + \binom{k+1}{2}a^{k-1}b^2 + \cdots + \binom{k+1}{k}ab^k + \binom{k+1}{k+1}b^{k+1}$$

86. $\dfrac{n!}{(n-r)!} = \dfrac{20!}{(20-3)!} = \dfrac{20!}{17!} = \dfrac{20 \cdot 19 \cdot 18 \cdot 17!}{17!} = 20 \cdot 19 \cdot 18 = 6840$

87. $\dfrac{n!}{(n-r)!r!} = \dfrac{8!}{(8-3)!3!} = \dfrac{8!}{5!3!} = \dfrac{8 \cdot 7 \cdot 6 \cdot 5!}{3 \cdot 2 \cdot 1 \cdot 5!} = \dfrac{8 \cdot 7 \cdot 6}{3 \cdot 2 \cdot 1} = 56$

88. true

Section 10.6

Check Point Exercises

1. We use the Fundamental Counting Principal to find the number of ways a one-topping pizza can be ordered.
Size : Crust : Topping:

$\quad$ 3 $\times$ 4 $\times$ 6 $\quad = 72$

There are 72 different ways of ordering a one-topping pizza.

2. We use the Fundamental Counting Principal to find the number of ways we can answer the questions.
Question #1: Question #2: Question #3: Question #4: Question #5: Question #6:

$\quad$ 3 $\times$ 3 $\times$ 3 $\times$ 3 $\times$ 3 $\times$ 3 $= 3^6 = 729$

There are 729 ways of answering the questions.

3. We use the Fundamental Counting Principal to find the number of different license plates that can be manufactured. Multiply the number of different letters, 26, for the first two places and the number of different digits, 10, for the next three places. $26 \cdot 26 \cdot 10 \cdot 10 \cdot 10 = 26^2 \cdot 1000 = 676{,}000$ plates There are 676,000 different license plates possible.

4. Your group is choosing $r = 4$ officers from a group of $n = 7$ people. The order in which the officers are chosen matters because the four officers to be chosen have different responsibilities. Thus, we are looking for the number of permutations of 7 things taken 4 at a time.

We use the formula $_nP_r = \dfrac{n!}{(n-r)!}$ with $n = 7$ and $r = 4$. $_7P_4 = \dfrac{7!}{(7-4)!} = \dfrac{7!}{3!} = 840.$

Thus, there are 840 different ways of filling the four offices.

5. Because you are using all six of your books in every possible arrangement, you are arranging $r = 6$ books from a group of $n = 6$ books. Thus, we are looking for the number of permutations of 6 things taken 6 at a time. We use the formula

$$_nP_r = \frac{n!}{(n-r)!} \text{ with } n = 6 \text{ and } r = 6.$$

$$_6P_6 = \frac{6!}{(6-6)!} = \frac{6!}{0!} = 6! = 720.$$

There are 720 different possible permutations. Thus, you can arrange the books in 720 ways.

6. **a.** The order does not matter; this is a combination.

b. Since what place each runner finishes matters, this is a permutation.

7. The order in which the four people are selected does not matter. This is a problem of selecting $r = 4$ people from a group of $n = 10$ people. We are looking for the number of combinations of 10 things taken 4 at a time. We use the formula

$$_nC_r = \frac{n!}{(n-r)! \, r!} \text{ with } n = 10 \text{ and } r = 4.$$

$$_{10}C_4 = \frac{10!}{(10-4)!4!} = \frac{10!}{6!4!} = \frac{10 \cdot 9 \cdot 8 \cdot 7 \cdot 6!}{6! \cdot 4 \cdot 3 \cdot 2 \cdot 1} = \frac{10 \cdot 9 \cdot 8 \cdot 7}{4 \cdot 3 \cdot 2 \cdot 1} = 210$$

Thus, 210 committees of 4 people each can be found from 10 people at the conference on acupuncture.

8. Because the order in which the 4 cards are dealt does not matter, this is a problem involving combinations. We are looking for the number of combinations of $n = 16$ cards drawn $r = 4$ at a time. We use the formula $_nC_r = \frac{n!}{(n-r)! \, r!}$ with

$n = 16$ and $r = 4$.

$$_{16}C_4 = \frac{16!}{(16-4)!4!} = \frac{16!}{12!4!} = \frac{16 \cdot 15 \cdot 14 \cdot 13 \cdot 12!}{12! \cdot 4 \cdot 3 \cdot 2 \cdot 1} = 1820$$

Thus, there are 1820 different 4-card hands possible.

Exercise Set 10.6

1. $\quad _9P_4 = \frac{9!}{5!} = 3024$

3. $\quad _8P_5 = \frac{8!}{3!} = 8 \cdot 7 \cdot 6 \cdot 5 \cdot 4 = 6720$

5. $\quad _6P_6 = \frac{6!}{0!} = 720$

7. $\quad _8P_0 = \frac{8!}{8!} = 1$

9. $\quad _9C_5 = \frac{9!}{4!5!} = \frac{9 \cdot 8 \cdot 7 \cdot 6}{4 \cdot 3 \cdot 2 \cdot 1} = \frac{3 \cdot 7 \cdot 6}{1} = 126$

11. $\quad _{11}C_4 = \frac{11!}{7!4!} = \frac{11 \cdot 10 \cdot 9 \cdot 8}{4 \cdot 3 \cdot 2 \cdot 1} = \frac{11 \cdot 10 \cdot 3}{1} = 330$

13. $\quad _7C_7 = \frac{7!}{0!7!} = 1$

15. $_5C_0 = \dfrac{5!}{5!0!} = 1$

17. combinations; The order in which the volunteers are chosen does not matter.

19. permutations; The order of the letters matters because ABCD is not the same as BADC.

21. $\dfrac{_7P_3}{3!} - {_7C_3} = \dfrac{\dfrac{7!}{(7-3)!}}{3!} - \dfrac{7!}{(7-3)!3!} = \dfrac{\dfrac{7!}{4!}}{3!} - \dfrac{7!}{4!3!} = \dfrac{7!}{4!3!} - \dfrac{7!}{4!3!} = 0$

23. $1 - \dfrac{_3P_2}{_4P_3} = 1 - \dfrac{\dfrac{3!}{(3-2)!}}{\dfrac{4!}{(4-3)!}} = 1 - \dfrac{\dfrac{3!}{1!}}{\dfrac{4!}{1!}} = 1 - \dfrac{3!}{4!} = 1 - \dfrac{3!}{4\cdot3!} = 1 - \dfrac{1}{4} = \dfrac{3}{4}$

25. $\dfrac{_7C_3}{_5C_4} - \dfrac{98!}{96!} = \dfrac{\dfrac{7!}{(7-3)!3!}}{\dfrac{5!}{(5-4)!4!}} - \dfrac{98\cdot97\cdot96!}{96!} = \dfrac{\dfrac{7!}{4!3!}}{\dfrac{5!}{1!4!}} - 95067 = \dfrac{\dfrac{7\cdot6\cdot5\cdot4!}{4!3\cdot2\cdot1}}{\dfrac{5\cdot4!}{1!4!}} - 9506 = \dfrac{35}{5} - 9506 = 7 - 9506 = -9499$

27. $\dfrac{_4C_2\cdot{_6C_1}}{_{18}C_3} = \dfrac{\dfrac{4!}{(4-2)!2!}\cdot\dfrac{6!}{(6-1)!1!}}{\dfrac{18!}{(18-3)!3!}} = \dfrac{\dfrac{4!}{2!2!}\cdot\dfrac{6!}{5!1!}}{\dfrac{18!}{15!3!}} = \dfrac{\dfrac{4\cdot3\cdot2!}{2!2\cdot1}\cdot\dfrac{6\cdot5!}{5!1!}}{\dfrac{18\cdot17\cdot16\cdot15!}{15!3\cdot2\cdot1}} = \dfrac{36}{816} = \dfrac{3}{68}$

29. $9\cdot3 = 27$ ways

31. $2\cdot4\cdot5 = 40$ ways

33. $3^5 = 243$ ways

35. $8\cdot2\cdot9 = 144$ area codes

37. $5\cdot4\cdot3\cdot2\cdot1\cdot1 = 120$ ways

39. $1\cdot3\cdot2\cdot1\cdot1 = 6$ paragraphs

41. $_{10}P_3 = \dfrac{10!}{7!3!} = 10\cdot9\cdot8 = 720$ ways

43. $_{13}P_7 = \dfrac{13!}{6!} = 13\cdot12\cdot11\cdot10\cdot9\cdot8\cdot7$
$= 8,648,640$ ways

45. $_6P_3 = \dfrac{6!}{3!} = 6\cdot5\cdot4 = 120$ ways

47. $_9P_5 = \dfrac{9!}{4!} = 9\cdot8\cdot7\cdot6\cdot5 = 15,120$ lineups

49. $_6C_3 = \dfrac{6!}{3!3!} = \dfrac{6 \cdot 5 \cdot 4}{3 \cdot 2 \cdot 1} = 20$ ways

51. $_{12}C_4 = \dfrac{12!}{8!4!} = \dfrac{12 \cdot 11 \cdot 10 \cdot 9}{4 \cdot 3 \cdot 2 \cdot 1}$
=495 collections

53. $_{17}C_8 = \dfrac{17!}{9!8!} = \dfrac{17 \cdot 16 \cdot 15 \cdot 14 \cdot 13 \cdot 12 \cdot 11 \cdot 10}{8 \cdot 7 \cdot 6 \cdot 5 \cdot 4 \cdot 3 \cdot 2 \cdot 1}$
=24,310 groups

55. $_{53}C_6 = \dfrac{53!}{47!6!} = 22,957,480$ selections

57. $_6P_4 = \dfrac{6!}{2!} = 6 \cdot 5 \cdot 4 \cdot 3 = 360$ ways

59. $_{13}C_6 = \dfrac{13!}{7!6!} = \dfrac{13 \cdot 12 \cdot 11 \cdot 10 \cdot 9 \cdot 8}{6 \cdot 5 \cdot 4 \cdot 3 \cdot 2 \cdot 1}$
=1716 ways

61. $_{20}C_3 = \dfrac{20!}{17!3!} = \dfrac{20 \cdot 19 \cdot 18}{3 \cdot 2 \cdot 1} = 1140$ ways

63. $_7P_4 = \dfrac{7!}{3!} = 840$ passwords

65. $_{15}P_3 = \dfrac{15!}{12!} = 15 \cdot 14 \cdot 13 = 2730$ cones

67. $_6P_6 = \dfrac{6!}{0!} = 6! = 720$ rankings

69. $_6C_3 = \dfrac{6!}{3!3!} = \dfrac{6 \cdot 5 \cdot 4}{3 \cdot 2 \cdot 1} = 20$ ways

71. $_4P_4 = \dfrac{4!}{0!} = 4! = 24$ ways

73. – 81. Answers may vary.

83. makes sense

85. does not make sense; Explanations will vary. Sample explanation: Since order matters use permutations.

87. false; Changes to make the statement true will vary. A sample change is: The number of ways to choose four questions out of ten questions is $_{10}C_4$.

89. true

91. $5 \cdot 5 \cdot 4 \cdot 4 \cdot 3 \cdot 3 \cdot 2 \cdot 2 \cdot 1 \cdot 1 = 14,400$ ways

93. $_{10}C_8 \cdot {}_5 C_3 = \dfrac{10!}{(10-8)!8!} \cdot \dfrac{5!}{(5-3)!3!}$

$= 45 \cdot 10 = 450$

They can be chosen in 450 ways.

95. $\dfrac{4}{6}$ or $\dfrac{2}{3}$ are less than 5.

96. $\dfrac{2}{6}$ or $\dfrac{1}{3}$ are not less than 5.

97. 2, 4, 5, and 6 are even or greater than three. This is a fraction of $\dfrac{4}{6}$ or $\dfrac{2}{3}$.

Section 10.7

Check Point Exercises

1. **a.** $P(\text{positive test}) = \dfrac{\#\text{ women w/positive test}}{\text{total number of women}} = \dfrac{7664}{100{,}000} = \dfrac{479}{6250} \approx 0.077$

b. $P(\text{positive test}) = \dfrac{\#\text{ women w/breast cancer and positive test}}{\text{total number of women w/breast cancer}} = \dfrac{720}{800} = \dfrac{9}{10} = 0.9$

c. $P(\text{breast cancer}) = \dfrac{\#\text{ women w/breast cancer and positive test}}{\text{total number of women w/positive test}} = \dfrac{720}{7664} = \dfrac{45}{479} = 0.094$

2. The sample space of equally likely outcomes is $S = \{1, 2, 3, 4, 5, 6\}$. There are six outcomes in the sample space, so $n(S) = 6$. The event of getting a number greater than 4 can be represented by $E = \{5, 6\}$. There are two outcomes in this event, so $n(E) = 2$.

The probability of rolling a number greater than 4 is $P(E) = \dfrac{n(E)}{n(S)} = \dfrac{2}{6} = \dfrac{1}{3}$.

3. We have $n(S) = 36$. The phrase "getting a sum of 5" describes the event $E = \{(1,4),(2,3),(3,2),(4,1)\}$. This event has 4 outcomes, so $n(E) = 4$. Thus, the probability of getting a sum of 5 is $P(E) = \dfrac{n(E)}{n(S)} = \dfrac{4}{36} = \dfrac{1}{9}$.

4. Let E be the event of being dealt a king. Because there are 4 kings in the deck, the event of being dealt a king can occur in 4 ways, i.e., $n(E) = 4$. With 52 cards in the deck, $n(S) = 52$.

The probability of being dealt a king is $P(E) = \dfrac{n(E)}{n(S)} = \dfrac{4}{52} = \dfrac{1}{13}$.

5. Because the order of the six numbers does not matter, this is a situation involving combinations. With one lottery ticket, there is only one way of winning so $n(E) = 1$. Using the combinations formula $_nC_r = \dfrac{n!}{(n-r)!r!}$ to find the number of outcomes in the sample space, we are selecting $r = 6$ numbers from a collection of $n = 49$ numbers.

$_{49}C_6 = \dfrac{49!}{43! \cdot 6!} = 13,983,816$ So $n(S) = 13,983,816$. If a person bought one lottery ticket, the probability of winning was

$P(E) = \dfrac{n(E)}{n(S)} = \dfrac{1}{13,983,816}$

The probability of winning the state lottery was 0.0000000715.

6. $P(\text{not } 50-59) = 1 - P(50-59) = 1 - \dfrac{31}{191} = \dfrac{160}{191}$

7. We find the probability that either of these mutually exclusive events will occur by adding their individual probabilities.

$P(4 \text{ or } 5) = P(4) + P(5) = \dfrac{1}{6} + \dfrac{1}{6} = \dfrac{2}{6} = \dfrac{1}{3}$

The probability of selecting a 4 or a 5 is $\dfrac{1}{3}$.

8. It is possible for the pointer to land on a number that is odd and less than 5. Two of the numbers , 1 and 3, are odd and less than 5. These events are not mutually exclusive. The probability of landing on a number that is odd and less than 5 is P (odd or less than 5)

$= P \text{ (odd)} + P \text{ (less than 5)} - P \text{ (odd and less than 5)} = \dfrac{4}{8} + \dfrac{4}{8} - \dfrac{2}{8} = \dfrac{6}{8} = \dfrac{3}{4}$

The probability that the pointer will stop on an odd number or a number less than 5 is $\dfrac{3}{4}$.

9. **a.** These events are not mutually exclusive.
   $P(\text{at least } \$100,000 \text{ or was not audited})$

   $= P(\text{at least } \$100,000) + P(\text{was not audited}) - P(\text{at least } \$100,000 \text{ and was not audited})$

   $= \dfrac{10,927,511}{120,851,273} + \dfrac{120,035,962}{120,851,273} - \dfrac{10,775,542}{120,851,273} = \dfrac{120,187,931}{120,851,273} \approx 0.99$

   b. These events are mutually exclusive.
   $P(\text{less than } \$25,000 \text{ or between } \$50,000 \text{ and } \$99,999, \text{ inclusive})$

   $= P(\text{less than } \$25,000) + P(\text{between } \$50,000 \text{ and } \$99,999, \text{ inclusive})$

   $= \dfrac{53,207,268}{120,851,273} + \dfrac{25,616,486}{120,851,273} = \dfrac{78,823,754}{120,851,273} \approx 0.65$

10. The wheel has 38 equally likely outcomes and 2 are green. Thus, the probability of a green occurring on a play is $\frac{2}{38}$, or

    $\frac{1}{19}$. The result that occurs on each play is independent of all previous results. Thus,

    $$P \text{ (green and green)} = P \text{ (green)} \cdot P \text{ (green)} = \frac{1}{19} \cdot \frac{1}{19} = \frac{1}{361} \approx 0.00277.$$

    The probability of green occurring on two consecutive plays is $\frac{1}{361}$.

11. If two or more events are independent, we can find the probability of them all occurring by multiplying the probabilities.

    The probability of a baby boy is $\frac{1}{2}$, so the probability of having four boys in a row is P (4 boys in a row)

    $$= \frac{1}{2} \cdot \frac{1}{2} \cdot \frac{1}{2} \cdot \frac{1}{2} = \frac{1}{16}.$$

Exercise Set 10.7

1. $P(\text{divorced}) = \dfrac{\text{number of persons divorced}}{\text{total number of U.S. adults}} = \dfrac{21.7}{212.5} \approx 0.10$

3. $P(\text{female}) = \dfrac{\text{number of females}}{\text{total number of U.S. adults}} = \dfrac{110.1}{212.5} \approx 0.52$

5. $P(\text{widowed male}) = \dfrac{\text{number of widowed males}}{\text{total number of U.S. adults}} = \dfrac{2.7}{212.5} \approx 0.01$

7. $P(\text{selecting a woman from the divorced population}) = \dfrac{\text{number of divorced women}}{\text{number of persons divorced}} = \dfrac{12.7}{21.7} \approx 0.59$

9. $P(\text{selecting a married man from the adult male population}) = \dfrac{\text{number of married men}}{\text{number of males}} = \dfrac{62.1}{102.4} \approx 0.61$

11. $P(R) = \dfrac{n(E)}{n(S)} = \dfrac{1}{6}$

13. $P(E) = \dfrac{n(E)}{n(S)} = \dfrac{3}{6} = \dfrac{1}{2}$

15. $P(E) = \dfrac{n(E)}{n(S)} = \dfrac{2}{6} = \dfrac{1}{3}$

17. $P(E) = \dfrac{n(E)}{n(S)} = \dfrac{4}{52} = \dfrac{1}{13}$

19. $P(E) = \dfrac{n(E)}{n(S)} = \dfrac{12}{52} = \dfrac{3}{13}$

21. $P(E) = \dfrac{n(E)}{n(S)} = \dfrac{1}{4}$

23. $P(E) = \dfrac{n(E)}{n(S)} = \dfrac{7}{8}$

25. $P(E) = \dfrac{n(E)}{n(S)} = \dfrac{3}{36} = \dfrac{1}{12}$

27. Buying 1 ticket:

$$P(E) = \frac{n(E)}{n(S)} = \frac{1}{{}_{51}C_6} = \frac{1}{18,009,460}$$

Buying 100 tickets:

$$P(E) = \frac{100}{18,009,460} = \frac{5}{900,473}$$

29. a. $\quad {}_{52}C_5 = \dfrac{52!}{47!5!}$

$$= \frac{52 \cdot 51 \cdot 50 \cdot 49 \cdot 48}{5 \cdot 4 \cdot 3 \cdot 2 \cdot 1} = 2,598,960$$

b. $\quad {}_{13}C_5 = \dfrac{13!}{8!5!} = \dfrac{13 \cdot 12 \cdot 11 \cdot 10 \cdot 9}{5 \cdot 4 \cdot 3 \cdot 2 \cdot 1} = 1287$

c. $\quad P(E) = \dfrac{n(E)}{n(S)} = \dfrac{1287}{2,598,960} \approx 0.0005$

31. $P(\text{not completed 4 years of college}) = 1 - P(\text{completed 4 years of college}) = 1 - \dfrac{45}{174} = \dfrac{43}{58}$

33. $P(\text{completed H.S. or less than 4 yrs college}) = P(\text{completed H.S}) + P(\text{less than 4 yrs college})$

$$= \frac{56}{174} + \frac{44}{174} = \frac{100}{174} = \frac{50}{87}$$

35. $P(\text{completed 4 yrs H.S. or man}) = P(\text{completed 4 yrs H.S.}) + P(\text{man}) - P(\text{man who completed 4 yrs H.S})$

$$= \frac{56}{174} + \frac{82}{174} - \frac{25}{174} = \frac{113}{174}$$

37. $P(\text{not king}) = 1 - P(\text{king}) = 1 - \dfrac{4}{52} = 1 - \dfrac{1}{13} = \dfrac{12}{13}$

39. $P(2 \text{ or } 3) = P(2) + P(3) = \dfrac{4}{52} + \dfrac{4}{52} = \dfrac{8}{52} = \dfrac{2}{13}$

41. $P(7 \text{ or red card}) = P(7) + P(\text{red card}) - P(7 \text{ and red}) = \dfrac{4}{52} + \dfrac{26}{52} - \dfrac{2}{52} = \dfrac{28}{52} = \dfrac{7}{13}$

43. $P(\text{odd or less than 6}) = P(\text{odd}) + P(\text{less than 6}) - P(\text{odd \# less than 6}) = \dfrac{4}{8} + \dfrac{5}{8} - \dfrac{3}{8} = \dfrac{6}{8} = \dfrac{3}{4}$

45. $P(\text{professor or male}) = P(\text{professor}) + P(\text{male}) - P(\text{male professor}) = \dfrac{19}{40} + \dfrac{22}{40} - \dfrac{8}{40} = \dfrac{33}{40}$

47. $P(2 \text{ and } 3) = P(2) \cdot P(3) = \dfrac{1}{6} \cdot \dfrac{1}{6} = \dfrac{1}{36}$

49. $P(\text{even and greater than } 2) = P(\text{even}) \cdot P(\text{greater than } 2) = \dfrac{3}{6} \cdot \dfrac{4}{6} = \dfrac{1}{3}$

51. $P(\text{all heads}) = \dfrac{1}{2} \cdot \dfrac{1}{2} \cdot \dfrac{1}{2} \cdot \dfrac{1}{2} \cdot \dfrac{1}{2} \cdot \dfrac{1}{2} = \left(\dfrac{1}{2}\right)^6 = \dfrac{1}{64}$

53. **a.** $P(\text{hit 2 yrs in a row}) = P(\text{hit in 1}^{\text{st}} \text{ year and 2}^{\text{nd}} \text{ year}) = P(\text{hit 1}^{\text{st}} \text{ year}) \cdot P(\text{hit 2}^{\text{nd}} \text{ year}) = \dfrac{1}{16} \cdot \dfrac{1}{16} = \dfrac{1}{256}$

   b. $P(\text{hit 3 yrs in a row}) = \dfrac{1}{16} \cdot \dfrac{1}{16} \cdot \dfrac{1}{16} = \dfrac{1}{4096}$

   c. First find the probability that South Florida will not be hit by a major hurricane in a single year:

   $P(\text{not hit}) = 1 - \dfrac{1}{16} = \dfrac{15}{16}$

   $P(\text{not hit in next 10 years}) = \left(\dfrac{15}{16}\right)^{10} \approx 0.524$

   d. $P(\text{hit at least once}) = 1 - P(\text{hit none}) = 1 - \left(\dfrac{15}{16}\right)^{10} \approx 0.476$

55. – 63. Answers may vary.

65. does not make sense; Explanations will vary. Sample explanation: The probability of a Democrat winning and the probability of a Republican winning are not necessarily equal. For instance, it is possible that the probability of a Democrat winning is 0.6 and the probability of a Republican winning is 0.4.

67. makes sense

69. Answers may vary.

71. **a.** $P(\text{Democrat who is not a business major})$

   $= \dfrac{\text{\# of students who are Democrats but not business majors}}{\text{\# of students}}$

   $= \dfrac{29 - 5}{50} = \dfrac{24}{50} = \dfrac{12}{25}$

   b. $P(\text{neither Democrat nor business major})$

   $= 1 - P(\text{Democrat or business major})$

   $= 1 - \big(P(\text{Democrat}) + P(\text{business major}) - P(\text{Democrat and business major})\big)$

   $= 1 - \left(\dfrac{29}{50} + \dfrac{11}{50} - \dfrac{5}{50}\right) = 1 - \dfrac{35}{50} = \dfrac{15}{50} = \dfrac{3}{10}$

73. **a.** The first person can have any birthday in the year. The second person can have all but one birthday.

   b. $\dfrac{365}{365} \cdot \dfrac{364}{365} \cdot \dfrac{363}{365} \approx 0.99$

   c. $100\% - 99\% = 0.01$

d.
$$\frac{365}{365} \cdot \frac{364}{365} \cdot \frac{363}{365} \cdots \frac{346}{365} \approx 0.59$$
$$1 - 0.59 = 0.41$$

e. With 23 people, the probability that at least two people have the same birthday is
$$P(E) = 1 - \frac{365}{365} \cdot \frac{364}{365} \cdot \frac{363}{365} \cdots + \frac{342}{365}$$
$$\approx 1 - 0.4927 \approx 0.5073$$

75. a. The denominator is zero when $x = 4$.

    b. The values are getting closer to the integer 2.

    c. The values are getting closer to the integer 2.

76. $f(x) = \dfrac{x^2 - 6x + 8}{x - 4} = \dfrac{(x+2)(x-4)}{x-4} = x + 2$

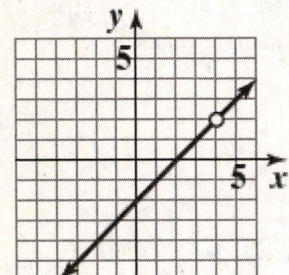

$$f(x) = \frac{x^2 - 6x + 8}{x - 4}$$

The two pieces of the graph approach the point (4, 2).

77. Graph of the compound function.

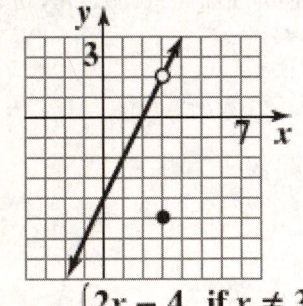

$$f(x) = \begin{cases} 2x - 4 & \text{if } x \neq 3 \\ -5 & \text{if } x = 3 \end{cases}$$

Chapter 10 Review Exercises

1. $a_n = 7n - 4$

$a_1 = 7 - 4 = 3$

$a_2 = 14 - 4 = 10$

$a_3 = 21 - 4 = 17$

$a_4 = 28 - 4 = 24$

First four terms: 3, 10, 17, 24.

2. $a_n = (-1)^n \dfrac{n+2}{n+1}$

$a_1 = (-1)^1 \dfrac{1+2}{1+1} = -\dfrac{3}{2}$

$a_2 = (-1)^2 \dfrac{2+2}{2+1} = \dfrac{4}{3}$

$a_3 = (-1)^3 \dfrac{3+2}{3+1} = -\dfrac{5}{4}$

$a_4 = (-1)^4 \dfrac{4+2}{4+1} = \dfrac{6}{5}$

First four terms: $-\dfrac{3}{2}, \dfrac{4}{3}, -\dfrac{5}{4}, \dfrac{6}{5}$.

3. $a_n = \dfrac{1}{(n-1)!}$

$a_1 = \dfrac{1}{0!} = 1$

$a_2 = \dfrac{1}{1!} = 1$

$a_3 = \dfrac{1}{2!} = \dfrac{1}{2}$

$a_4 = \dfrac{1}{3!} = \dfrac{1}{6}$

First four terms: $1, 1, \dfrac{1}{2}, \dfrac{1}{6}$.

4. $a_n = \dfrac{(-1)^{n+1}}{2^n}$

$a_1 = \dfrac{(-1)^2}{2^1} = \dfrac{1}{2}$

$a_2 = \dfrac{(-1)^3}{2^2} = -\dfrac{1}{4}$

$a_3 = \dfrac{(-1)^4}{2^3} = \dfrac{1}{8}$

$a_4 = \dfrac{(-1)^5}{2^4} = -\dfrac{1}{16}$

First four terms: $\dfrac{1}{2}, -\dfrac{1}{4}, \dfrac{1}{8}, -\dfrac{1}{16}$.

5. $a_1 = 9$ and $a_n = \dfrac{2}{3a_{n-1}}$

$a_1 = 9$

$a_2 = \dfrac{2}{3 \cdot 9} = \dfrac{2}{27}$

$a_3 = \dfrac{2}{3} \cdot \dfrac{27}{2} = \dfrac{54}{6} = 9$

$a_4 = \dfrac{2}{3 \cdot 9} = \dfrac{2}{27}$

First four terms: $9, \dfrac{2}{27}, 9, \dfrac{2}{27}$.

6. $a_1 = 4$ and $a_n = 2a_{n-1} + 3$

$a_1 = 4$

$a_2 = 2 \cdot 4 + 3 = 8 + 3 = 11$

$a_3 = 2 \cdot 11 + 3 = 22 + 3 = 25$

$a_4 = 2 \cdot 25 + 3 = 50 + 3 = 53$

First four terms: 4, 11, 25, and 53.

7. $\dfrac{40!}{4! \, 38!} = \dfrac{40 \cdot 39 \cdot 38!}{4 \cdot 3 \cdot 2 \cdot 1 \cdot 38!} = 65$

8.

$$\sum_{i=1}^{5} (2i^2 - 3) = (2 - 3) + (2 \cdot 2^2 - 3) + (2 \cdot 3^2 - 3) + (2 \cdot 4^2 - 3) + ($$

$$= -1 + 5 + 15 + 29 + 47$$
$$= 95$$

9. $\displaystyle\sum_{i=0}^{4} (-1)^{i+1} i! = (-1)^1 0! + (-1)^2 1! + (-1)^3 3! + (-1)^4 4!$

$$= -1 + 1 - 2 + 6 - 24$$
$$= -20$$

10. $\dfrac{1}{3}+\dfrac{2}{4}+\dfrac{3}{5}+\cdots+\dfrac{15}{17}=\displaystyle\sum_{i=1}^{15}\dfrac{i}{i+2}$

11. $4^3+5^3+6^3+\cdots+13^3=\displaystyle\sum_{i=1}^{10}(i+3)^3$

12. $a_1=7,\,d=4$
First six terms: 7, 11, 15, 19, 23, 27.

13. $a_1=-4,\,d=-5$
First six terms: $-4,-9,-14,-19,-24,-29$.

14. $a_1=\dfrac{3}{2},\,d=-\dfrac{1}{2}$

First six terms: $\dfrac{3}{2},1,\dfrac{1}{2},0,-\dfrac{1}{2},-1$.

15. $a_{n+1}=a_n+5,\,a_1=-2$
First six terms: $-2, 3, 8, 13, 18,\ 23$.

16. $a_1=5,\,d=3$
$a_n=5+(n-1)3$
$a_6=5+(5)3=20$

17. $a_1=-8,\,d=-2$
$a_n=-8+(n-1)(-2)$
$a_{12}=-8+11(-2)=-30$

18. $a_1=14,\,d=-4$
$a_n=14+(n-1)(-4)$
$a_{14}=14+(13)(-4)=-38$

19. $-7,-3,1,5,\ldots$
$d=-3-(-7)=4$
$a_n=-7+(n-1)(4)$
$a_n=4n-11$
$a_{20}=4(20)-11$
$a_{20}=69$

20. $a_1=200,\,d=-20$
$a_n=200+(n-1)(-20)$
$a_n=220-20n$
$a_{20}=220-20(20)$
$a_{20}=-180$

21. $a_n=a_{n-1}-5,\,a_1=3$
$d=-5$
$a_n=3+(n-1)(-5)=3-5n+5$
$a_n=8-5n$
$a_{20}=8-5(20)=-92$

22. $5, 12, 19, 26, \ldots$
$d=7$
$a_n=5+(n-1)(7)$
$a_{22}=5+21(7)=152$
$S_{22}=\dfrac{22}{2}(5+152)=1727$

23. $-6,-3,0,3,\ldots$
$d=3$
$a_n=-6+(n-1)3$
$a_{15}=-6+(14)3=36$
$S_{15}=\dfrac{15}{2}(-6+36)=225$

24. $3+6+9+\ldots+300$
$S_{100}=\dfrac{100}{2}(3+300)=15{,}150$

25. $\displaystyle\sum_{i=1}^{16}(3i+2)$
$a_1=3+2=5$
$a_{16}=3(16)+2=50$
$S_{16}=\dfrac{16}{2}(5+50)=440$

26. $\displaystyle\sum_{i=1}^{25}(-2i+6)$
$a_1=-2+6=4$
$a_{25}=-2(25)+6=-44$
$S_{25}=\dfrac{25}{2}(4-44)=-500$

27. $\displaystyle\sum_{i=1}^{30}-5i$
$a_1=-5$
$a_{30}=-5(30)=-150$
$S_{30}=\dfrac{30}{2}(-5-150)=-2325$

28. **a.** $a_n = 39 + (n-1)(4.75)$

$$= 39 + 4.75n - 4.75$$

$$= 4.75n + 34.25$$

b. $a_n = 4.75n + 34.25$

$a_{12} = 4.75(13) + 34.25$

$$= 96$$

The percentage of students ages 12 – 18 who will report seeing security cameras at school in the year 2013 will be approximately 96%.

29. $a_n = 31,500 + (n-1)2300$

$a_{10} = 31,500 + (9)2300 = 52,200$

$S_{10} = \dfrac{10}{2}(31,500 + 52,200) = 418,500$

The total salary is $418,500.

30. $a_n = 25 + (n-1)$

$a_{35} = 25 + 34 = 59$

$S_{35} = \dfrac{35}{2}(25 + 59) = 1470$

There are 1470 seats.

31. $a_1 = 3$, $r = 2$

First five terms: 3, 6, 12, 24, 48.

32. $a_1 = \dfrac{1}{2}$, $r = \dfrac{1}{2}$

First five terms: $\dfrac{1}{2}, \dfrac{1}{4}, \dfrac{1}{8}, \dfrac{1}{16}, \dfrac{1}{32}$.

33. $a_1 = 16$, $r = -\dfrac{1}{2}$

First five terms: 16, −8, 4, −2, 1.

34. $a_n = -5a_{n-1}$, $a_1 = -1$

First five terms: −1, 5, −25, 125, −625.

35. $a_1 = 2$, $r = 3$

$a_n = 2 \cdot 3^{n-1}$

$a_7 = 2 \cdot 3^6 = 1458$

36. $a_1 = 16$, $r = \dfrac{1}{2}$

$a_n = 16\left(\dfrac{1}{2}\right)^{n-1}$

$a_6 = 16\left(\dfrac{1}{2}\right)^5 = \dfrac{16}{32} = \dfrac{1}{2}$

37. $a_1 = -3$, $r = 2$

$a_n = -3 \cdot 2^{n-1}$

$a_5 = -3 \cdot 2^4 = -48$

38. 1, 2, 4, 8, ...

$a_1 = 1$, $r = \dfrac{2}{1} = 2$

$a_n = 2^{n-1}$

$a_8 = 2^7 = 128$

39. $100, 10, 1, \dfrac{1}{10}, \ldots$

$a_1 = 100$, $r = \dfrac{10}{100} = \dfrac{1}{10}$

$a_n = 100\left(\dfrac{1}{10}\right)^{n-1}$

$a_8 = 100\left(\dfrac{1}{10}\right)^7 = \dfrac{1}{100,000}$

40. $12, -4, \dfrac{4}{3}, -\dfrac{4}{9} \cdots$

$a_1 = 12$, $r = -\dfrac{4}{12} = -\dfrac{1}{3}$

$a_n = 12\left(-\dfrac{1}{3}\right)^{n-1}$

$a_8 = 12\left(-\dfrac{1}{3}\right)^7 = -\dfrac{4}{729}$

41. 5, −15, 45, −135, ...

$r = \dfrac{-15}{5} = -3$

$S_{15} = \dfrac{5\left[1 - (-3)^{15}\right]}{1 - (-3)} = 17,936,135$

42. $r = \dfrac{1}{2}$, $a_1 = 8$

$S_{78} = \dfrac{8\left[1 - \left(\dfrac{1}{2}\right)^7\right]}{1 - \dfrac{1}{2}} = -16\left(1 - \dfrac{1}{128}\right)$

$$= -16\left(-\dfrac{127}{128}\right) = \dfrac{127}{8}$$

43. $S_6 = \dfrac{5(1 - 5^6)}{1 - 5} = \dfrac{5(-15624)}{-4} = 19,530$

44. $\displaystyle\sum_{i=1}^{7}3(-2)^i$

$a_1=-6,\ r=-2$

$S_7=\dfrac{-6\left[1-(-2)^7\right]}{1-(-2)}=\dfrac{-6(129)}{3}=-258$

45. $\displaystyle\sum_{i=1}^{5}2\left(\tfrac{1}{4}\right)^{i-1}$

$a_1=2,\ r=\dfrac{1}{4}$

$S_5=\dfrac{2\left[1-\left(\tfrac14\right)^5\right]}{1-\tfrac14}=\dfrac{2\left(\tfrac{1023}{1024}\right)}{\tfrac34}=\dfrac{341}{128}$

46. $a_1=9,\ r=\dfrac{1}{3}$

$S_\infty=\dfrac{9}{1-\tfrac13}=\dfrac{9}{\tfrac23}=9\cdot\dfrac{3}{2}=\dfrac{27}{2}$

47. $a_1=2,\ r=-\dfrac{1}{2}$

$S_\infty=\dfrac{2}{1-\left(-\tfrac12\right)}=\dfrac{2}{\tfrac32}=\dfrac{4}{3}$

48. $a_1=-6,\ r=-\dfrac{2}{3}$

$S_\infty=\dfrac{-6}{1-\left(-\tfrac23\right)}=\dfrac{-6}{\tfrac53}=-\dfrac{18}{5}$

49. $r=0.8$

$S_\infty=\dfrac{4}{1-0.8}=20$

50. $0.\overline{6}=0.6+0.06+0.006+\cdots$

$a_1=\dfrac{6}{10},\ r=\dfrac{1}{10}$

$S_\infty=\dfrac{\tfrac{6}{10}}{1-\tfrac{1}{10}}=\dfrac{\tfrac{6}{10}}{\tfrac{9}{10}}=\dfrac{6}{9}=\dfrac{2}{3}$

51. $0.\overline{47}=0.47+0.0047+0.000047+\cdots$

$a_1=\dfrac{47}{100},\ r=\dfrac{1}{100}$

$S_\infty=\dfrac{\tfrac{47}{100}}{1-\tfrac{1}{100}}=\dfrac{\tfrac{47}{100}}{\tfrac{99}{100}}=\dfrac{47}{99}$

52. a. Divide each value by the previous value:

$\dfrac{5.9}{4.2}=1.405$

$\dfrac{8.3}{5.9}=1.407$

$\dfrac{11.6}{8.3}=1.398$

$\dfrac{16.2}{11.6}=1.397$

$\dfrac{22.7}{16.2}=1.401$

The population is increasing geometrically with $r=1.4$.

b. $a_n=4.2\cdot1.4^n$

c. 2080 is 8 decades after 2000 so $n=8$.

$a_n=4.2\cdot1.4^n$

$a_8=4.2\cdot1.4^8\approx62.0$

In 2080, the model predicts the U.S. population, ages 85 and older, will be 62.0 million

53. $a_1=32{,}000,\ r=1.06$

$a_6=32{,}000(1.06)^5\approx\$42{,}823$

The sixth year salary is \$42,823.

$S_6=\dfrac{32{,}000\left(1-1.06^6\right)}{1-1.06}$

$=\dfrac{32{,}000\left(1-1.06^6\right)}{-0.06}$

$\approx223{,}210$

The total salary paid is \$223,210.

54.   **a.** $A = \dfrac{P\left[\left(1+\frac{r}{n}\right)^{nt}-1\right]}{\frac{r}{n}}$

$P = \$520, \quad r = 0.06, \quad n = 1, \quad t = 20$

$$A = \dfrac{\$520\left[\left(1+\dfrac{0.06}{1}\right)^{1\cdot 20}-1\right]}{\dfrac{0.06}{1}} = \dfrac{\$520\left[(1.06)^{20}-1\right]}{0.06} \approx \$19{,}129$$

The value of the annuity will be \$19,129.

   b.   Interest = Value of annuity − Total deposits

$\approx \quad \$19{,}129 \ - \ \$520\cdot 20$

$\approx \$8729$

55.   **a.** $A = \dfrac{P\left[\left(1+\frac{r}{n}\right)^{nt}-1\right]}{\frac{r}{n}}$

$P = 100, \quad r = 0.055, \quad n = 12, \quad t = 30$

$$A = \dfrac{\$100\left[\left(1+\dfrac{0.055}{12}\right)^{12\cdot 30}-1\right]}{\dfrac{0.055}{12}} \approx \$91{,}361$$

The value of the IRA will be \$91,361.

   b.   Interest = Value of IRA − Total deposits

$\approx \quad \$91{,}361 \ - \ \$100\cdot 12\cdot 30$

$\approx \$55{,}361$

56. $4(0.7) + 4(0.7)^2 + \cdots; \quad r = 0.7$

$S_{\infty} = \dfrac{4(0.7)}{1-0.7} = 9.\overline{3}$

The total spending is $\$9\dfrac{1}{3}$ million.

57. $S_1 : 5 = \dfrac{5(1)(1+1)}{2}$

$5 = \dfrac{5(2)}{2}$

$5 = 5$ is true.

$S_k : 5 + 10 + 15 + \cdots + 5k = \dfrac{5k(k+1)}{2}$

$S_{k+1} : 5 + 10 + 15 + \cdots + 5k + 5(k+1)$

$\qquad = \dfrac{5(k+1)(k+2)}{2}$

Add $5(k+1)$ to both sides of S_k:

$5 + 10 + 15 + \cdots + 5k + 5(k+1)$

$\qquad = \dfrac{5k(k+1)}{2} + 5k(k+1)$

Simplify the right-hand side:

$\dfrac{5k(k+1)}{2} + 5(k+1) = \dfrac{5k(k+1) + 10(k+1)}{2}$

$\qquad\qquad = \dfrac{(5k+10)(k+1)}{2}$

$\qquad\qquad = \dfrac{5(k+1)(k+2)}{2}$

If S_k is true, then S_{k+1} is true.

The statement is true for all n.

58. $S_1 : 1 = \dfrac{4^1 - 1}{3}$

$1 = \dfrac{3}{3}$

$1 = 1$ is true.

$S_k : 1 + 4 + 4^2 + \cdots + 4^{k-1} = \dfrac{4^k - 1}{3}$

$S_{k+1} : 1 + 4 + 4^2 + \cdots + 4^{k-1} + 4^k = \dfrac{4^{k+1} - 1}{3}$ Add 4^k to both sides of S_k:

$S_k : 1 + 4 + 4^2 + \cdots + 4^{k-1} = \dfrac{4^k - 1}{3}$

$1 + 4 + 4^2 + \cdots + 4^{k-1} + 4^k = \dfrac{4^k - 1}{3} + 4^k$

Simplify the right-hand side:

$\dfrac{4^k - 1}{3} + 4^k = \dfrac{4^k - 1 + 3 \cdot 4^k}{3}$

$\qquad\qquad = \dfrac{4 \cdot 4^k - 1}{3}$

$\qquad\qquad = \dfrac{4^{k+1} - 1}{3}$

If S_k is true, then S_{k+1} is true.

The statement is true for all n.

59. $S_1 : 2 = 2(1)^2$

       $2 = 2$ is true.

       $S_k : 2 + 6 + 10 + \cdots + (4k - 2) = 2k^2$

       $S_{k+1} : 2 + 6 + 10 + \cdots + (4k - 2) + (4k + 2) = 2(k + 1)^2$

Add $(4k + 2)$ to both sides of S_k :

$2 + 6 + 10 + \cdots + (4k - 2) + (4k + 2) = 2k^2 + (4k + 2)$

Simplify the right-hand side:

   $2k^2 + 4k + 2 = 2(k^2 + 2k + 1)$

                 $= 2(k + 1)^2$

If S_k is true, then S_{k+1} is true. The statement is true for all n.

60. $S_1 : 1 \cdot 3 = \dfrac{1(1+1)[2(1) + 7]}{6}$

      $3 = \dfrac{2 \cdot 9}{6}$

      $3 = \dfrac{18}{6}$

      $3 = 3$ is true.

   $S_k : 1 \cdot 3 + 2 \cdot 4 + 3 \cdot 5 + \cdots + k(k+2) = \dfrac{k(k+1)(2k+7)}{6}$

   $S_{k+1} : 1 \cdot 3 + 2 \cdot 4 + 3 \cdot 5 + \cdots + k(k+2) + (k+1)(k+3) = \dfrac{(k+1)(k+2)(2k+9)}{6}$

Add $(k+1)(k+3)$ to both sides of S_k :

$1 \cdot 3 + 2 \cdot 4 + 3 \cdot 5 + \cdots + k(k+2) + (k+1)(k+3) = \dfrac{k(k+1)(2k+7)}{6} + (k+1)(k+3)$

Simplify the right-hand side:

$= \dfrac{k(k+1)(2k+7)}{6} + (k+1)(k+3)$

$= \dfrac{k(k+1)(2k+7) + 6(k+1)(k+3)}{6}$

$= \dfrac{(k+1)[k(2k+7) + 6(k+3)]}{6}$

$= \dfrac{(k+1)(2k^2 + 13k + 18)}{6}$

$= \dfrac{(k+1)(k+2)(2k+9)}{6}$

If S_k is true, then S_{k+1} is true. The statement is true for all n.

61. $S_1 : 2$ is a factor of $1^2 + 5(1) = 6$ since $6 = 2 \cdot 3$.

$S_k : 2$ is a factor of $k^2 + 5k$.

$S_{k+1} : 2$ is a factor of $(k+1)^2 + 5(k+1)$.

$$(k+1)^2 + 5(k+1) = k^2 + 2k + 1 + 5k + 5$$
$$= k^2 + 7k + 6$$
$$= k^2 + 5k + 2(k+3)$$
$$= (k^2 + 5k) + 2(k+3)$$

Because we assume S_k is true, we know 2 is a factor of $k^2 + 5k$. Since 2 is a factor of $2(k+3)$, we conclude 2 is a factor of the sum $(k^2 + 5k) + 2(k+3)$. If S_k is true, then S_{k+1} is true. The statement is true for all n.

62. $\dbinom{11}{8} = \dfrac{11!}{3!8!} = \dfrac{11 \cdot 10 \cdot 9}{3 \cdot 2 \cdot 1} = 165$

63. $\dbinom{90}{2} = \dfrac{90!}{88!2!} = \dfrac{90 \cdot 89}{2 \cdot 1} = 4005$

64. $(2x+1)^3 = \dbinom{3}{0}(2x)^3 + \dbinom{3}{1}(2x)^2 \cdot 1 + \dbinom{3}{2}(2x)1^2 + \dbinom{3}{3}1^3$

$\qquad = 8x^3 + 3(4x^2) + 3(2x) + 1$

$\qquad = 8x^3 + 12x^2 + 6x + 1$

65. $(x^2-1)^4 = \dbinom{4}{0}(x^2)^4 + \dbinom{4}{1}(x^2)^3(-1) + \dbinom{4}{2}(x^2)^2(-1)^2 + \dbinom{4}{3}x^2(-1)^3 + \dbinom{4}{4}(-1)^4$

$\qquad = x^8 - 4x^6 + 6x^4 - 4x^2 + 1$

66. $(x+2y)^5 = \dbinom{5}{0}x^5 + \dbinom{5}{1}x^4(2y) + \dbinom{5}{2}x^3(2y)^2 + \dbinom{5}{3}x^2(2y)^3 + \dbinom{5}{4}x(2y)^4 + \dbinom{5}{5}(2y)^5$

$\qquad = x^5 + 5(2)x^4y + 10(4)x^3y^2 + 10(8)x^2y^3 + 5(16)xy^4 + 32y^5$

$\qquad = x^5 + 10x^4y + 40x^3y^2 + 80x^2y^3 + 80xy^4 + 32y^5$

67. $(x-2)^6 = \dbinom{6}{0}x^6 + \dbinom{6}{1}x^5(-2) + \dbinom{6}{2}x^4(-2)^2 + \dbinom{6}{3}x^3(-2)^3 + \dbinom{6}{4}x^2(-2)^4 + \dbinom{6}{5}x(-2)^5 \dbinom{6}{6}(-2)^6$

$\qquad = x^6 + 6x^5(-2) + 15x^4(4) + 20x^3(-8) + 15x^2(16) + 6x(-32) + 64$

$\qquad = x^6 - 12x^5 + 60x^4 - 160x^3 + 240x^2 - 192x + 64$

68. $(x^2+3)^8 = \dbinom{8}{0}(x^2)^8 + \dbinom{8}{1}(x^2)^7 3 + \dbinom{8}{2}(x^2)^6 3^2 + \cdots$

$\qquad = x^{16} + 8x^{14}3 + 28x^{12}9 + \cdots$

$\qquad = x^{16} + 24x^{14} + 252x^{12} + \cdots$

69. $(x-3)^9 = \binom{9}{0}x^9 + \binom{9}{1}x^8(-3) + \binom{9}{2}x^7(-3)^2 - \cdots$

$$= x^9 + 9(-3)x^8 + 36(9)x^7 - \cdots$$

$$= x^9 - 27x^8 + 324x^7 - \cdots$$

70. $(x+2)^5$

fourth term $= \binom{5}{3}x^2(2)^3$

$$= 10(8)x^2 = 80x^2$$

71. $(2x-3)^6$

fifth term $= \binom{6}{4}(2x)^2(-3)^4$

$$= 15(4x^2)(81) = 4860x^2$$

72. $_8P_3 = \dfrac{8!}{5!} = 8 \cdot 7 \cdot 6 = 336$

73. $_9P_5 = \dfrac{9!}{4!} = 9 \cdot 8 \cdot 7 \cdot 6 \cdot 5 = 15,120$

74. $_8C_3 = \dfrac{8!}{5!3!} = \dfrac{8 \cdot 7 \cdot 6}{3 \cdot 2 \cdot 1} = 56$

75. $_{13}C_{11} = \dfrac{13!}{2!11!} = \dfrac{13 \cdot 12}{2 \cdot 1} = 78$

76. $4 \cdot 5 = 20$ choices

77. $3^5 = 243$ possibilities

78. $_{15}P_4 = \dfrac{15!}{11!} = 15 \cdot 14 \cdot 13 \cdot 12 = 32,760$ ways

79. $_{20}C_4 = \dfrac{20!}{16!4!} = \dfrac{20 \cdot 19 \cdot 18 \cdot 17}{4 \cdot 3 \cdot 2 \cdot 1} = 4845$ ways

80. $_{20}C_3 = \dfrac{20!}{17!3!} = \dfrac{20 \cdot 19 \cdot 18}{3 \cdot 2 \cdot 1} = 1140$ sets

81. $_{20}P_4 = \dfrac{20!}{16!}$

$$= 20 \cdot 19 \cdot 18 \cdot 17$$

$$= 116,280 \text{ ways}$$

82. $5! = 120$ ways

83. $P(\text{public college}) = \dfrac{252}{350} = \dfrac{18}{25}$

84. $P(\text{not from high-income family}) = 1 - P(\text{from high-income family}) = 1 - \dfrac{50}{350} = \dfrac{350}{350} - \dfrac{50}{350} = \dfrac{300}{350} = \dfrac{6}{7}$

85. $P(\text{from middle-income family or high-income family}) = \dfrac{160 + 50}{350} = \dfrac{210}{350} = \dfrac{3}{5}$

86. $P(\text{attended private college or is from a high income family})$

$\quad = P(\text{private college}) + P(\text{high income family}) - P(\text{attended private college and is from a high income family})$

$\quad = \dfrac{98}{350} + \dfrac{50}{350} - \dfrac{28}{350} = \dfrac{120}{350} = \dfrac{12}{35}$

87. $P(\text{selecting a student from a low-income family from among those attending public college}) = \dfrac{120}{252} = \dfrac{10}{21}$

88. $P(\text{selecting a student that attends a private college from among those in middle-income families}) = \dfrac{50}{160} = \dfrac{5}{16}$

89. $P(\text{less than 5}) = P(\text{rolling a 1, 2, 3, 4})$

$\qquad = \dfrac{4}{6} = \dfrac{2}{3}$

90. $P(\text{less than 3 or greater than 4})$

$\quad = P(\text{less than 3}) + P(\text{greater than 4})$

$\quad = P(\text{rolling a 1, 2}) + P(\text{rolling a 5, 6})$

$\quad = \dfrac{2}{6} + \dfrac{2}{6} = \dfrac{4}{6} = \dfrac{2}{3}$

91. $P(E) = \dfrac{4}{52} + \dfrac{4}{52} = \dfrac{8}{52} = \dfrac{2}{13}$

92. $P(E) = \dfrac{4}{52} + \dfrac{26}{52} - \dfrac{2}{52} = \dfrac{28}{52} = \dfrac{7}{13}$

93. $P(\text{not yellow}) = 1 - P(\text{yellow}) = 1 - \dfrac{1}{6} = \dfrac{5}{6}$

94. $P(\text{red or greater than 3}) = \dfrac{3}{6} + \dfrac{3}{6} - \dfrac{1}{6} = \dfrac{5}{6}$

95. $P(\text{green, then less than 4}) = \dfrac{2}{6} \cdot \dfrac{3}{6} = \dfrac{6}{36} = \dfrac{1}{6}$

96. **a.** $P(E) = \dfrac{n(E)}{n(S)} = \dfrac{1}{{}_{20}C_5} = \dfrac{1}{15,504}$

     b. $P(E) = \dfrac{100}{15,504} = \dfrac{25}{3876}$

97. $P(E) = \left(\dfrac{1}{2}\right)^5 = \dfrac{1}{32}$

98. a. $(0.2)^2 = 0.04$

b. $(0.2)^3 = 0.008$

c. $(1-0.2)^4 = (0.8)^4 = 0.4096$

Chapter 10 Test

1. $a_n = \dfrac{(-1)^{n+1}}{n^2}$

$a_1 = \dfrac{(-1)^2}{1^2} = 1$

$a_2 = \dfrac{(-1)^3}{2^2} = -\dfrac{1}{4}$

$a_3 = \dfrac{(-1)^4}{3^2} = \dfrac{1}{9}$

$a_4 = \dfrac{(-1)^5}{4^2} = -\dfrac{1}{16}$

$a_5 = \dfrac{(-1)^6}{5^2} = \dfrac{1}{25}$

First five terms: $1, -\dfrac{1}{4}, \dfrac{1}{9}, -\dfrac{1}{16}, \dfrac{1}{25}$.

2. $\displaystyle\sum_{i=1}^{5} \left(i^2 + 10\right) = 11 + 14 + 19 + 26 + 35 = 105$

3. $\displaystyle\sum_{i=1}^{20} (3i - 4)$

$a_1 = 3 - 4 = -1$

$d = 3$

$a_n = -1 + (n-1)3$

$a_{20} = -1 + (19)3 = 56$

$S_{20} = \dfrac{20}{2}(-1 + 56) = 550$

4. $\displaystyle\sum_{i=1}^{15} (-2)^i$

$a_1 = -2,\ r = -2$

$S_{15} = \dfrac{-2\left[1 - (-2)^{15}\right]}{1 - (-2)} = -21,846$

5. $\dbinom{9}{2} = \dfrac{9!}{7!2!} = \dfrac{9 \cdot 8}{2 \cdot 1} = 36$

6. $_{10}P_3 = \dfrac{10!}{7!} = 10 \cdot 9 \cdot 8 = 720$

7. $_{10}C_3 = \dfrac{10!}{7!3!} = \dfrac{10 \cdot 9 \cdot 8}{3 \cdot 2 \cdot 1} = 120$

8. $\dfrac{2}{3} + \dfrac{3}{4} + \dfrac{4}{5} + \cdots + \dfrac{21}{22} = \displaystyle\sum_{i=1}^{20} \dfrac{i+1}{i+2}$

9. $4, 9, 14, 19, \ldots$

$a_1 = 4,\ d = 5$

$a_n = 4 + (n-1) \cdot 5 = 4 + 5n - 1$

$a_n = 5n - 1$

$a_{12} = 5(12) - 1 = 59$

10. $16, 4, 1, \dfrac{1}{4}, \cdots$

$a_1 = 16,\ r = \dfrac{1}{4}$

$a_n = 16\left(\dfrac{1}{4}\right)^{n-1}$

$a_{12} = 16\left(\dfrac{1}{4}\right)^{11} = \dfrac{1}{262,144}$

11. $7, -14, 28, -56, \ldots$

$a_1 = 7,\ r = -2$

$S_{10} = \dfrac{7\left[1 - (-2)^{10}\right]}{1 - (-2)} = \dfrac{7(-1023)}{3} = -2387$

12. $-7, -14, -21, -28, \ldots$

$a_1 = -7,\ d = -7$

$a_n = -7 + (n-1)(-7)$

$a_{10} = -7 + 9(-7) = -70$

$S_{10} = \dfrac{10}{2}(-7 - 70) = -385$

13. $4 + \dfrac{4}{2} + \dfrac{4}{2^2} + \dfrac{4}{2^3} + \cdots$

$r = \dfrac{1}{2}$

$S_\infty = \dfrac{4}{1 - \dfrac{1}{2}} = 8$

14. $0.\overline{73} = 0.73 + 0.0073 + 0.000073 + \cdots$

$$a_1 = \frac{73}{100}, \ r = \frac{1}{100}$$

$$S_\infty = \frac{\dfrac{73}{100}}{1 - \dfrac{1}{100}} = \frac{\dfrac{73}{100}}{\dfrac{99}{100}} = \frac{73}{99}$$

15. $a_1 = 30,000, \ r = 1.04$

$$S_8 = \frac{30,000\left[1 - (1.04)^8\right]}{1 - 1.04} \approx 276,426.79$$

The total salary is \$276,427.

16. $S_1 : 1 = \dfrac{1[3(1) - 1]}{2}$

$$1 = \frac{2}{2}$$

$1 = 1$ is true.

$$S_k : 1 + 4 + 7 + \cdots + (3k - 2) = \frac{k(3k - 1)}{2}$$

$$S_{k+1} : 1 + 4 + 7 + \cdots + (3k - 2) + (3k + 1) = \frac{(k + 1)(3k + 2)}{2}$$

Add $(3k + 1)$ to both sides of S_k:

$$1 + 4 + 7 + \cdots + (3k - 2) + (3k + 1) = \frac{k(3k - 1)}{2} + (3k + 1)$$

Simplify the right-hand side:

$$\frac{k(3k - 1)}{2} + (3k + 1) = \frac{k(3k - 1) + 2(3k + 1)}{2}$$

$$= \frac{3k^2 + 5k + 2}{2}$$

$$= \frac{(k + 1)(3k + 2)}{2}$$

If S_k is true, then S_{k+1} is true. The statement is true for all n.

17. $\left(x^2 - 1\right)^5 = \binom{5}{0}\left(x^2\right)^5 + \binom{5}{1}\left(x^2\right)^4 (-1) + \binom{5}{2}\left(x^2\right)^3 (-1)^2 + \binom{5}{3}\left(x^2\right)^2 (-1)^3 + \binom{5}{4}x^2 (-1)^4 + \binom{5}{5}(-1)^5$

$$= x^{10} - 5x^8 + 10x^6 - 10x^4 + 5x^2 - 1$$

18. $\left(x+y^2\right)^8$

First Term $\binom{n}{r-1}a^{n-r+1}b^{r-1}=\binom{8}{1-1}x^{8-1+1}\left(y^2\right)^{1-1}=\binom{8}{0}x^8\left(y^2\right)^0=\dfrac{8!}{0!(8-0)!}x^8\cdot 1=\dfrac{\cancel{8!}}{0!\cancel{8!}}x^8$

$\qquad\qquad\qquad = x^8$

Second Term $\binom{n}{r-1}a^{n-r+1}b^{r-1}=\binom{8}{2-1}x^{8-2+1}\left(y^2\right)^{2-1}=\binom{8}{1}x^7\left(y^2\right)^1=\dfrac{8!}{1!(8-1)!}x^7y^2=\dfrac{8\cdot\cancel{7!}}{1\cdot\cancel{7!}}x^7y^2$

$\qquad\qquad\qquad = 8x^7y^2$

Third Term $\binom{n}{r-1}a^{n-r+1}b^{r-1}=\binom{8}{3-1}x^{8-3+1}\left(y^2\right)^{3-1}=\binom{8}{2}x^6\left(y^2\right)^2=\dfrac{8!}{2!(8-2)!}x^6y^4=\dfrac{8\cdot7\cdot\cancel{6!}}{2\cdot1\cdot\cancel{6!}}x^6y^4$

$\qquad\qquad\qquad = 28x^6y^4$

$x^8+8x^7y^2+28x^6y^4+\cdots$

19. $_{11}P_3=\dfrac{11!}{8!}=11\cdot10\cdot9=990\ \text{ways}$

20. $_{10}C_4=\dfrac{10!}{6!4!}=\dfrac{10\cdot9\cdot8\cdot7}{4\cdot3\cdot2\cdot1}=210\ \text{sets}$

21. Four digits are open: $10^4=10{,}000$

22. $P(\text{not brown eyes})=1-P(\text{brown eyes})$

$\qquad\qquad = 1-\dfrac{40}{50}$

$\qquad\qquad = \dfrac{60}{100}=\dfrac{3}{5}$

23. $P(\text{brown eyes or blue eyes})$

$= P(\text{brown eyes})+P(\text{blue eyes})$

$= \dfrac{40}{100}+\dfrac{38}{100}=\dfrac{78}{100}=\dfrac{39}{50}$

24. $P(\text{female or green eyes})$

$= P(\text{female})+P(\text{green eyes})$

$\qquad -P(\text{female with green eyes})$

$= \dfrac{50}{100}+\dfrac{22}{100}-\dfrac{12}{100}=\dfrac{60}{100}=\dfrac{3}{5}$

25. $P(\text{male, given blue eyes})=\dfrac{18}{38}=\dfrac{9}{19}$

26. $_{15}C_6=\dfrac{15!}{9!6!}=\dfrac{15\cdot14\cdot13\cdot12\cdot11\cdot10}{6\cdot5\cdot4\cdot3\cdot2}=5005$

$P(E)=\dfrac{50}{5005}=\dfrac{10}{1001}$

27. $P(E) = \dfrac{26}{52} + \dfrac{12}{52} - \dfrac{6}{52} = \dfrac{32}{52} = \dfrac{8}{13}$

28. $P(E) = \dfrac{25}{50} + \dfrac{20}{50} - \dfrac{15}{50} = \dfrac{30}{50} = \dfrac{3}{5}$

29. $P(E) = \left(\dfrac{1}{4}\right)^4 = \dfrac{1}{256}$

30. $P(E) = \dfrac{2}{8} \cdot \dfrac{2}{8} = \dfrac{1}{16}$

Cumulative Review Exercises (Chapters P–10)

1. domain: $[-4, 1)$; range: $(-\infty, 2]$

2. maximum of 2 at $x = -2$

3. decreasing interval: $(-2, 1)$

4. neither

5. $f(-3) = 1$ and $f(-1) = 1$

6. $(f \circ f)(-4) = f(f(-4)) = f(0) = 0$

7. $f(x) \to -\infty$ as $x \to 1^-$

8.

$g(x) = f(x - 2) + 1$

9.

$h(x) = -f(2x)$

10.
$$-2(x - 5) + 10 = 3(x + 2)$$
$$-2x + 10 + 10 = 3x + 6$$
$$14 = 5x$$
$$x = \dfrac{14}{5}$$

The solution set is $\left\{\dfrac{14}{5}\right\}$.

11. $3x^2 - 6x + 2 = 0$
$$x = \dfrac{6 \pm \sqrt{36 - 24}}{6}$$
$$= \dfrac{6 \pm \sqrt{12}}{6}$$
$$= \dfrac{6 \pm 2\sqrt{3}}{6}$$
$$= \dfrac{3 \pm \sqrt{3}}{3}$$

The solution set is $\left\{\dfrac{3 + \sqrt{3}}{3}, \dfrac{3 - \sqrt{3}}{3}\right\}$.

12.
$$\log_2 x + \log_2 (2x - 3) = 1$$
$$\log_2 x(2x - 3) = 1$$
$$x(2x - 3) = 2$$
$$2x^2 - 3x - 2 = 0$$
$$(2x + 1)(x - 2) = 0$$
$$2x + 1 = 0 \quad \text{or} \quad x - 2 = 0$$
$$x = -\dfrac{1}{2} \qquad\qquad x = 2$$

$x = -\dfrac{1}{2}$ does not check since $\log_2\left(-\dfrac{1}{2}\right)$ does not exist.
The solution set is {2}.

13. $x^{1/2} - 6x^{1/4} + 8 = 0$
Let $t = x^{1/4}$.
$$t^2 - 6t + 8 = 0$$
$$(t - 2)(t - 4) = 0$$
$$t - 2 = 0 \quad \text{or} \quad t - 4 = 0$$
$$t = 2 \qquad\qquad t = 4$$
$$x^{1/4} = 2 \qquad\quad x^{1/4} = 4$$
$$x = 16 \qquad\qquad x = 256$$
The solution set is {16, 256}.

14. $e^{2x} - 6e^x + 8 = 0$

$(e^x - 2)(e^x - 4) = 0$

$e^x - 2 = 0$ or $e^x - 4 = 0$

$e^x = 2$ $e^x = 4$

$x = \ln 2$ $x = \ln 4$

The solution set is $\{\ln 2, \ln 4\}$.

15. $|2x + 1| \le 1$

$-1 \le 2x + 1 \le 1$

$-2 \le 2x \le 0$

$-1 \le x \le 0$ or $[-1, 0]$

The solution set is $\{x | -1 \le x \le 0\}$ or $[-1, 0]$.

16. $6x^2 - 6 < 5x$

$6x^2 - 5x - 6 < 0$

$6x^2 - 5x - 6 = 0$

$(3x + 2)(2x - 3) = 0$

$3x + 2 = 0$ or $2x - 3 = 0$

$x = -\dfrac{2}{3}$ $x = \dfrac{3}{2}$

The test intervals are $\left(-\infty, -\frac{2}{3}\right)$, $\left(-\frac{2}{3}, \frac{3}{2}\right)$, and

$\left(\frac{3}{2}, \infty\right)$. Testing a point in each interval shows that

the solution is $\left(-\frac{2}{3}, \frac{3}{2}\right)$.

17. $\dfrac{x-1}{x+3} \le 0$

The test intervals are $(-\infty, -3)$, $(-3, 1)$
and $(1, \infty)$.
Testing a point in each interval shows that the
solution is $(-3, 1]$.

18. $30e^{0.7x} = 240$

$e^{0.7x} = 8$

$\ln e^{0.7x} = \ln 8$

$0.7x = \ln 8$

$x = \dfrac{\ln 8}{0.7} \approx 2.9706$

The solution set is $\left\{\dfrac{\ln 8}{0.7}\right\}$ or $\{2.9706\}$.

19. $2x^3 + 3x^2 - 8x + 3 = 0$

$p:\ \pm 1,\ \pm 3$

$q:\ \pm 1,\ \pm 2$

$\dfrac{p}{q}:\ \pm 1, \pm 3, \pm\dfrac{1}{2}, \pm\dfrac{3}{2}$

$$
\begin{array}{r|rrrr}
1 & 2 & 3 & -8 & 3 \\
  &   & 2 & 5 & -3 \\
\hline
  & 2 & 5 & -3 & 0
\end{array}
$$

$(x - 1)(2x^2 + 5x - 3) = 0$

$(x - 1)(2x - 1)(x + 3) = 0$

$x = 1$ or $x = \dfrac{1}{2}$ or $x = -3$

The solution set is $\left\{-3, \dfrac{1}{2}, 1\right\}$.

20. $4x^2 + 3y^2 = 48$

$3x^2 + 2y^2 = 35$

Multiply equation 1 by -2.

Multiply equation 2 by 3.

$-8x^2 - 6y^2 = -96$

$\underline{9x^2 + 6y^2 = 105}$

Add: $x^2 = 9$

$x = \pm 3$

Let $x = -3$:

$4(-3)^2 + 3y^2 = 48$

$36 + 3y^2 = 48$

$3y^2 = 12$

$y^2 = 4$

$y = \pm 2$

Let $x = 3$:

$4(3)^2 + 3y^2 = 48$

$36 + 3y^2 = 48$

$3y^2 = 12$

$y^2 = 4$

$y = \pm 2$

The solution set is
$\{(3, 2), (3, -2), (-3, 2), (-3, -2)\}$.

21.

$$x - 2y + z = 16$$
$$2x - y - z = 14$$
$$3x + 5y - 4z = -10$$

$$\begin{bmatrix} 1 & -2 & 1 & | & 16 \\ 0 & -1 & -1 & | & 14 \\ 3 & 5 & -4 & | & -10 \end{bmatrix} \begin{matrix} \\ -2R_1 + R_2 \\ -3R_1 + R_3 \end{matrix}$$

$$\begin{bmatrix} 1 & -2 & 1 & | & 16 \\ 0 & 3 & -3 & | & -18 \\ 0 & 11 & -7 & | & -58 \end{bmatrix} \begin{matrix} \\ \frac{1}{3}R_2 \\ \\ \end{matrix}$$

$$\begin{bmatrix} 1 & -2 & 1 & | & 16 \\ 0 & 1 & -1 & | & -6 \\ 0 & 11 & -7 & | & -58 \end{bmatrix} \begin{matrix} \\ 2R_2 + R_1 \\ -11R_2 + R_3 \end{matrix}$$

$$\begin{bmatrix} 1 & 0 & -1 & | & 4 \\ 0 & 1 & -1 & | & -6 \\ 0 & 0 & 4 & | & 8 \end{bmatrix} \begin{matrix} \\ \\ \frac{1}{4}R_3 \end{matrix}$$

$$\begin{bmatrix} 1 & 0 & -1 & | & 4 \\ 0 & 1 & -1 & | & -6 \\ 0 & 0 & 1 & | & 2 \end{bmatrix} \begin{matrix} R_3 + R_1 \\ R_2 + R_1 \\ \\ \end{matrix}$$

$$\begin{bmatrix} 1 & 0 & 0 & | & 6 \\ 0 & 1 & 0 & | & -4 \\ 0 & 0 & 1 & | & 2 \end{bmatrix} \begin{matrix} R_3 + R_1 \\ R_2 + R_1 \\ \\ \end{matrix}$$

The solution set is $\{(6, -4, 2)\}$.

22.

$$\begin{cases} x - y = 1 \\ x^2 - x - y = 1 \end{cases}$$

Solving $x - y = 1$ for y gives $y = x - 1$.

Substitute:

$$x^2 - x - y = 1$$

$$x^2 - x - \overbrace{(x-1)}^{y} = 1$$

$$x^2 - x - x + 1 = 1$$

$$x^2 - 2x + 1 = 1$$

$$x^2 - 2x = 0$$

$$x(x - 2) = 0$$

$x = 0$ or $x = 2$

If $x = 0$, $0 - y = 1$ so $y = -1$.

If $x = 2$, $2 - y = 1$ so $y = 1$.

The solution set is $\{(0, -1), (2, 1)\}$.

23. $100x^2 + y^2 = 25$

$$4x^2 + \frac{y^2}{25} = 1$$

$$\frac{x^2}{\left(\frac{1}{4}\right)} + \frac{y^2}{25} = 1$$

Ellipse

Foci on the y-axis

$a^2 = 25$ and $b^2 = \frac{1}{4}$, so $\frac{1}{4} = 25 - c^2$.

$$c^2 = \frac{99}{4}$$

$$c = \frac{3\sqrt{11}}{2}$$

Foci: $\left(0, -\frac{3\sqrt{11}}{2}\right), \left(0, \frac{3\sqrt{11}}{2}\right)$

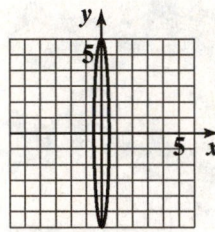

$100x^2 + y^2 = 25$

24. $4x^2 - 9y^2 - 16x + 54y - 29 = 0$

$$4(x^2 - 4x) - 9(y^2 - 6y) = 29$$

$$4(x^2 - 4x + 4) - 9(y^2 - 6y + 9) = 16 - 81 + 29$$

$$4(x - 2)^2 - 9(y - 3)^2 = -36$$

$$\frac{(y-3)^2}{4} - \frac{(x-2)^2}{9} = 1$$

Hyperbola with center at (2, 3)

Transverse axis vertical

$a^2 = 4$ and $b^2 = 9$, so $9 = c^2 - 4$.

$$c^2 = 13$$

$$c = \sqrt{13}$$

Foci: $\left(2, 3 - \sqrt{13}\right), \left(2, 3 + \sqrt{13}\right)$

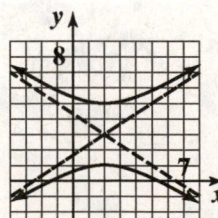

$4x^2 - 9y^2 - 16x + 54y - 29 = 0$

25. Symmetry:

$$f(-x) = \frac{x^2 - 1}{-x - 2}$$

No symmetry since $f(-x) \neq f(x)$ and $f(-x) \neq -f(-x)$.

x-intercepts:

$$x^2 - 1 = 0$$
$$x = \pm 1$$

y-intercept:

$$f(0) = \frac{1}{2}$$
$$y = \frac{1}{2}$$

Vertical asymptote:

$$x - 2 = 0$$
$$x = 2$$

Horizontal asymptote:

$n > m$, so no horizontal asymptote.

Slant asymptote: $n = m + 1$

$$f(x) = x + 2 + \frac{3}{x - 2}$$
$$y = x + 2$$

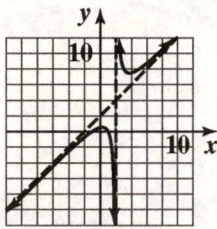

$$f(x) = \frac{x^2 - 1}{x - 2}$$

26.

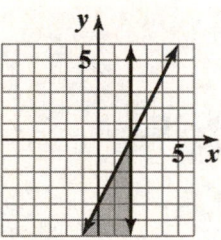

$$2x - y \geq 4$$
$$x \leq 2$$

27. $f(x) = x^2 - 4x - 5$

$$x = \frac{-b}{2a} = \frac{4}{2} = 2$$
$$f(2) = 2^2 - 8 - 5 = -9$$

vertex: $(2, -9)$

x-intercepts:

$$x^2 - 4x - 5 = 0$$
$$(x - 5)(x + 1) = 0$$
$$x = 5, -1$$

y-intercept: $f(0) = -5$

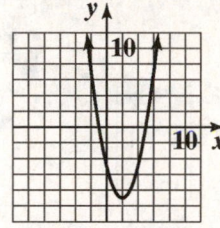

$$f(x) = x^2 - 4x - 5$$

28.

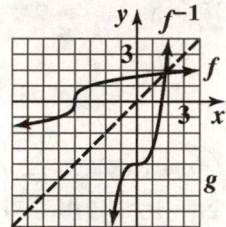

29.

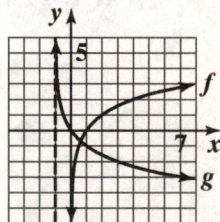

30. $f(x) = -x^2 - 2x + 1, \quad g(x) = x - 1$

$$(f \circ g)(x) = f(g(x))$$
$$= -(x-1)^2 - 2(x-1) + 1$$
$$= -x^2 + 2x - 1 - 2x + 2 + 1$$
$$= -x^2 + 2$$
$$(g \circ f)(x) = g(f(x))$$
$$= (-x^2 - 2x + 1) - 1$$
$$= -x^2 - 2x$$

31. $f(x) = -x^2 - 2x + 1$

$$\frac{f(x+h) - f(x)}{h} = \frac{\left(-(x+h)^2 - 2(x+h) + 1\right) - \left(-x^2 - 2x + 1\right)}{h}$$

$$= \frac{-x^2 - 2xh - h^2 - 2x - 2h + 1 + x^2 + 2x - 1}{h}$$

$$= \frac{-2xh - h^2 - 2h}{h}$$

$$= \frac{h(-2x - h - 2)}{h}$$

$$= -2x - h - 2$$

32. $AB - 4A = \begin{bmatrix} 4 & 2 \\ 1 & -1 \\ 0 & 5 \end{bmatrix} \begin{bmatrix} 2 & 4 \\ 3 & 1 \end{bmatrix} - 4 \begin{bmatrix} 4 & 2 \\ 1 & 1 \\ 0 & 5 \end{bmatrix}$

$$= \begin{bmatrix} 14 & 18 \\ -1 & 3 \\ 15 & 5 \end{bmatrix} - \begin{bmatrix} 16 & 8 \\ 4 & -4 \\ 0 & 20 \end{bmatrix} = \begin{bmatrix} -2 & 10 \\ -5 & 7 \\ 15 & -15 \end{bmatrix}$$

33. $\dfrac{2x^2 - 10x + 2}{(x-2)(x^2 + 2x + 2)} = \dfrac{A}{x-2} + \dfrac{Bx + C}{x^2 + 2x + 2}$

$2x^2 - 10x + 2 = A(x^2 + 2x + 2) + (Bx + C)(x - 2)$

$\qquad\qquad\qquad = Ax^2 + 2Ax + 2A + Bx^2 - 2Bx + Cx - 2C$

$\qquad\qquad\qquad = (A + B)x^2 + (2A - 2B + C)x + 2A - 2C$

Thus we have the following system of equations.

$\qquad A + B = 2$

$2A - 2B + C = -10$

$\qquad 2A - 2C = 2$

Add twice the first equation to the second equation.

$\qquad 2A + 2B = 4$

$\underline{2A - 2B + C = -10}$

$\qquad 4A + C = -6$

Add twice the resulting equation to the third equation.

$8A + 2C = -12$

$\underline{2A - 2C = 2}$

$\quad 10A = -10$

$\qquad A = -1$

Back-substitute to find B and C.

$2(-1) - 2C = 2$

$\quad -2 - 2C = 2$

$\qquad -2C = 4$

$\qquad\quad C = -2$

$-1 + B = 2$

$\qquad B = 3$

$\dfrac{-1}{x-2} + \dfrac{3x - 2}{x^2 + 2x + 2}$

34.

$$(x^3 + 2y)^5 = \binom{5}{0}(x^3)^5 + \binom{5}{1}(x^3)^4(2y) + \binom{5}{2}(x^3)^3(2y)^2 + \binom{5}{3}(x^3)^2(2y)^3 + \binom{5}{4}(x^3)(2y)^4 + \binom{5}{5}(2y)^5$$

$$= x^{15} + 5x^{12}(2y) + 10x^9(4y^2) + 10x^6(8y^3) + 5x^3(16y^4) + 32y^5$$

$$= x^{15} + 10x^{12}y + 40x^9y^2 + 80x^6y^3 + 80x^3y^4 + 32y^5$$

35. $\displaystyle\sum_{i=1}^{50}(4i - 25)$

$$a_1 = 4(1) - 25 = -21$$

$$a_{50} = 4(50) - 25 = 175$$

$$S_{50} = \frac{50}{2}(-21 + 175) = 3850$$

36. Find slope:

$$m = \frac{y_2 - y_1}{x_2 - x_1} = \frac{1 - 3}{-2 - 6} = \frac{-2}{-8} = \frac{1}{4}$$

Find equation:

$$y - y_1 = m(x - x_1)$$

$$y - 3 = \frac{1}{4}(x - 6)$$

$$y - 3 = \frac{1}{4}x - \frac{3}{2}$$

$$y = \frac{1}{4}x + \frac{3}{2}$$

37. Find the slope:

$$x - 5y - 20 = 0$$

$$-5y = -x + 20$$

$$\frac{-5y}{-5} = \frac{-x}{-5} + \frac{20}{-5}$$

$$y = \frac{1}{5}x - 4$$

$$m = -\frac{1}{\frac{1}{5}} = -5$$

Find the equation:

$$y - y_1 = m(x - x_1)$$

$$y - (-2) = -5(x - 0)$$

$$y + 2 = -5x$$

$$y = -5x - 2$$

38.

$$200 + 0.05x = 0.15x$$

$$200 = 0.1x$$

$$\frac{200}{0.1} = \frac{0.1x}{0.1}$$

$$2000 = x$$

At \$2000 in sales, the two earnings will be the same.

39. $2L + 2W = 300$
$\quad\quad L = W + 50$
Rearrange the equations and add:
$L + W = 150$
$\underline{L - W = 50}$
$\quad 2L = 200$
$\quad\quad L = 100$
$\quad\quad W = 50$
length: 100 yards, width 50 yards

40. $10x + 12y = 42$
$\quad 5x + 10y = 29$
Multiply second equation by -2 and add:
$\quad 10x + 12y = 42$
$\underline{-10x - 20y = -58}$
$\quad\quad\quad -8y = -16$
$\quad\quad\quad\quad y = 2$
Back substitute:
$5x + 10(2) = 29$
$\quad\quad 5x = 9$
$\quad\quad\; x = 1.8$
pen: \$1.80, pad: \$2

41. $s(t) = -16t^2 + 80t + 96$

a. $-16t^2 + 80t + 96 = 0$
$\quad t^2 - 5t - 6 = 0$
$\quad (t + 1)(t - 6) = 0$
$\quad\quad\quad t = -1 \text{ or } t = 6$
The ball will strike the ground after 6 seconds.

b. $t = \dfrac{-b}{2a} = \dfrac{-80}{-32} = \dfrac{5}{2} \text{ or } 2.5$

$S(2.5) = -16(2.5)^2 + 80(2.5) + 96 = 196$
The ball reaches a maximum height of 196 feet, 2.5 seconds after it is thrown.

42. $I = \dfrac{k}{R}$

$5 = \dfrac{k}{22}$

$k = 110$

$I = \dfrac{110}{10} = 11$

11 amperes

43. Let x represent the number of years after 1980. The data from 1980 and 2004 are represented as $(0, 33.2)$ and $(24, 20.9)$

Find slope: $m = \dfrac{20.9 - 33.2}{24 - 0} \approx -0.51$

Thus, $y = -0.51x + 33.2$

At this rate there may eventually be no smokers among U.S. adults.

44. $d = 10 \sin \dfrac{3\pi}{4} t$

a. $\mid a \mid \; = \; \mid 10 \mid \; = 10 \quad 2a = 20$
The maximum displacement is 20 inches.

b. $f = \dfrac{\omega}{2\pi} = \dfrac{\frac{3\pi}{4}}{2\pi} = \dfrac{3}{8}$

The frequency is $\dfrac{3}{8}$ cycle per second.

c. period $= \dfrac{2\pi}{\omega} = \dfrac{2\pi}{\frac{3\pi}{4}} = \dfrac{8}{3}$

The time required for one oscillation is $\dfrac{8}{3}$ seconds.

45. $\tan x + \dfrac{1}{\tan x} = \dfrac{\sin x}{\cos x} + \dfrac{1}{\frac{\sin x}{\cos x}} = \dfrac{\sin x}{\cos x} + \dfrac{\cos x}{\sin x}$

$\quad = \dfrac{\sin^2 x + \cos^2 x}{\cos x \cdot \sin x}$

$\quad = \dfrac{1}{\cos x \cdot \sin x}$

46. $\dfrac{1 - \tan^2 x}{1 + \tan^2 x} = \dfrac{1 - \frac{\sin^2 x}{\cos^2 x}}{1 + \frac{\sin^2 x}{\cos^2 x}} \cdot \dfrac{\cos^2 x}{\cos^2 x}$

$\quad = \dfrac{\cos^2 x - \sin^2 x}{\cos^2 x + \sin^2 x}$

$\quad = \dfrac{\cos 2x}{1} = \cos 2x$

47. $y = -2\cos(3x - \pi)$

Amplitude: $|A| = |-2| = 2$

Period: $\dfrac{2\pi}{B} = \dfrac{2\pi}{3}$

Phase shift: $\dfrac{C}{B} = \dfrac{\pi}{3}$

$\dfrac{\pi}{3}, -2$, $\dfrac{\pi}{2}, 0$, $\dfrac{2\pi}{3}, 2$, $\dfrac{5\pi}{6}, 0$,

$(\pi, -2)$

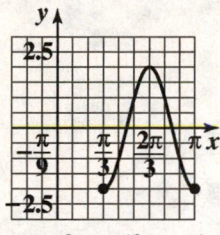

$y = -2\cos(3x - \pi)$

48. $4\cos^2 x = 3$

$\cos^2 x = \dfrac{3}{4}$

$\cos x = \pm\sqrt{\dfrac{3}{4}} = \pm\dfrac{\sqrt{3}}{2}$

$x = \dfrac{\pi}{6}, \dfrac{5\pi}{6}, \dfrac{7\pi}{6}, \dfrac{11\pi}{6}$

The solutions in the interval $[0, 2\pi)$ are

$\dfrac{\pi}{6}, \dfrac{5\pi}{6}, \dfrac{7\pi}{6}$, and $\dfrac{11\pi}{6}$.

49. $2\sin^2 x + 3\cos x - 3 = 0$

$2(1 - \cos^2 x) + 3\cos x - 3 = 0$

$2 - 2\cos^2 x + 3\cos x - 3 = 0$

$2\cos^2 x - 3\cos x + 1 = 0$

$(2\cos x - 1)(\cos x - 1) = 0$

$2\cos x - 1 = 0$ or $\cos x - 1 = 0$

$2\cos x = 1$ $\cos x = 1$

$\cos x = \dfrac{1}{2}$

$x = \dfrac{\pi}{3}, \dfrac{5\pi}{3}$ or $x = 0$

The solutions in the interval $[0, 2\pi)$ are

$0, \dfrac{\pi}{3}$, and $\dfrac{5\pi}{3}$.

50. $\cot\,\cos^{-1}\left(-\dfrac{5}{6}\right)$

If $\cos\theta = -\dfrac{5}{6}$, θ lies in quadrant II.

$\cos\theta = -\dfrac{5}{6} = \dfrac{x}{r} = \dfrac{-5}{6}$

$x^2 + y^2 = r^2$

$(-5)^2 + y^2 = 6^2$

$25 + y^2 = 36$

$y^2 = 11$

$\cot\,\cos^{-1}\left(-\dfrac{5}{6}\right) = \dfrac{x}{y} = \dfrac{-5}{\sqrt{11}} = -\dfrac{5\sqrt{11}}{11}$

51. $r = 1 + 2\cos\theta$
Check for symmetry:

Polar Axis	The Line $\theta = \dfrac{\pi}{2}$	The Pole
$r = 1 + 2\cos(-\theta)$	$-r = 1 + 2\cos(-\theta)$	$-r = 1 + 2\cos\theta$
$r = 1 + 2\cos\theta$	$r = -1 - 2\cos\theta$	$r = -1 - 2\cos\theta$

Graph is symmetric with respect to the polar axis.

θ	0	$\dfrac{\pi}{6}$	$\dfrac{\pi}{3}$	$\dfrac{\pi}{2}$	$\dfrac{2\pi}{3}$	$\dfrac{5\pi}{6}$	π
r	3	2.73	2	1	0	-0.73	-1

Use symmetry to obtain the graph.

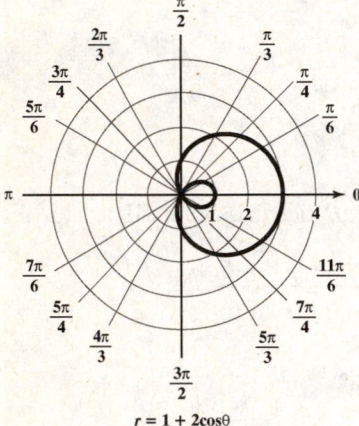

$r = 1 + 2\cos\theta$

Chapter 11
Introduction to Calculus

Section 11.1

Check Point Exercises

1. $\lim\limits_{x \to 3} 4x^2$

x	2.99	2.999	$2.9999 \to \leftarrow 3.0001$	3.001	3.01
$f(x) = 4x^2$	35.7604	35.9760	$35.9976 \to \leftarrow 36.0024$	36.0240	36.2404

The limit of $4x^2$ as x approaches 3 equals the number 36.

2. $\lim\limits_{x \to 0} \dfrac{\cos x - 1}{x}$

x	-0.1	-0.01	$-0.001 \to \leftarrow 0.001$	0.01	0.1
$f(x) = \dfrac{\cos x - 1}{x}$	0.0500	0.0050	$0.0005 \to \leftarrow -0.0005$	-0.0050	-0.0500

The limit of $\dfrac{\cos x - 1}{x}$ as x approaches 0 equals the number 0.

3. **a.** Figure 12.4 shows that as x approaches -2, $f(x)$ approaches 5. Thus, $\lim\limits_{x \to 2} f(x) = 5$.

   b. In figure 12.4, the graph of $f(x)$ at $x = -2$ is shown by the closed dot with coordinates $(-2, 3)$. Thus, $f(-2) = 3$.

4. Graph the piece defined by the linear function, $f(x) = 3x - 2$, using the y-intercept, -2, and the slope, 3. Because $x = 2$ is not included, show an open dot on the line corresponding to $x = 2$. Complete the graph using $f(x) = 1$ if $x = 2$. This part of the graph is the point $(2, 1)$, shown as a dot.

   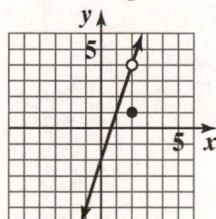

   As x gets closer to 2, the values of $f(x)$ get closer to the y-coordinate of the open dot or 2. We conclude that $\lim\limits_{x \to 2} f(x) = 4$.

5. **a.** As x approaches 0 from the left, the values of $f(x)$ get close to the y-coordinate of 2. Thus $\lim\limits_{x \to 0^-} f(x) = 2$.

   b. As x approaches 0 from the right, the values of $f(x)$ get close to the y-coordinate of 1. Thus $\lim\limits_{x \to 0^+} f(x) = 1$.

   c. The $\lim\limits_{x \to 0^-} f(x) = 2$ and $\lim\limits_{x \to 0^+} f(x) = 1$. Because the left and right-hand limits are unequal, $\lim\limits_{x \to 0} f(x)$ does not exist.

   d. $f(0) = 1$ because this point is shown as a solid dot.

Introduction to Calculus

Exercise Set 11.1

1. The limit of $2x^2$ as x approaches 2 equals the number 8.

3. The limit of $\dfrac{\sin 3x}{x}$ as x approaches 0 equals the number 3.

5. $\lim\limits_{x \to 2} 5x^2$

x	1.99	1.999	1.9999 $\to$ $\gets$ 2.0001	2.001	2.01
$f(x) = 5x^2$	19.801	19.980	19.998 $\to$ $\gets$ 20.002	20.020	20.201

The limit of $5x^2$ as x approaches 2 equals the number 20.

7. $\lim\limits_{x \to 3} \dfrac{1}{x-2}$

x	2.99	2.999	2.9999 $\to$ $\gets$ 3.0001	3.001	3.01
$f(x) = \dfrac{1}{x-2}$	1.0101	1.0010	1.0001 $\to$ $\gets$ 0.9999	0.9990	0.9901

The limit of $\dfrac{1}{x-2}$ as x approaches 3 equals the number 1.

9. $\lim\limits_{x \to 0} \dfrac{x}{x^2+1}$

x	−0.01	−0.001	−0.0001 $\to$ $\gets$ 0.0001	0.001	0.01
$f(x) = \dfrac{x}{x^2+1}$	−0.0100	−0.0010	−0.0001 $\to$ $\gets$ 0.0001	0.0010	0.0100

The limit of $\dfrac{x}{x^2+1}$ as x approaches 0 equals the number 0.

11. $\lim\limits_{x \to -2} \dfrac{x^3+8}{x+2}$

x	-2.01	-2.001	$-2.0001 \to \leftarrow -1.9999$	-1.999	-1.99
$f(x) = \dfrac{x^3+8}{x+2}$	12.0601	12.0060	$12.0006 \to \leftarrow 11.9994$	11.9940	11.9401

The limit of $\dfrac{x^3+8}{x+2}$ as x approaches -2 equals the number 12.

13. $\lim\limits_{x \to 0} \dfrac{2x^2+x}{\sin x}$

x	-0.01	-0.001	$-0.0001 \to \leftarrow 0.0001$	0.001	0.01
$f(x) = \dfrac{2x^2+x}{\sin x}$	0.9800	0.9980	$0.9998 \to \leftarrow 1.0002$	1.0020	1.0200

The limit of $\dfrac{2x^2+x}{\sin x}$ as x approaches 0 equals the number 1.

15. $\lim\limits_{x \to 0} \dfrac{\tan x}{x}$

x	-1	-0.1	$-0.01 \to \leftarrow 0.01$	0.1	1
$f(x) = \dfrac{\tan x}{x}$	1.5574	1.0033	$1.00003 \to \leftarrow 1.00003$	1.0033	1.5574

The limit of $\dfrac{\tan x}{x}$ as x approaches 0 equals the number 1.

17. $\lim\limits_{x \to 0} f(x)$, where $f(x) = \begin{cases} x+1 \text{ if } x < 0 \\ 2x+1 \text{ if } x \geq 0 \end{cases}$

x	-0.01	-0.001	$-0.0001 \to \leftarrow 0.0001$	0.001	0.01
$f(x)$	0.99	0.999	$0.9999 \to \leftarrow 1.0002$	1.002	1.02

The limit of $f(x)$ as x approaches 0 equals the number 1.

Introduction to Calculus

19. a. The graph shows that as x approaches 3, $f(x)$ approaches -1. Thus, $\lim_{x \to 3} f(x) = -1$.

   b. The graph of $f(x)$ at $x = 3$ is the point with coordinates $(3, -1)$. Thus, $f(3) = -1$.

21. a. The graph shows that as x approaches 2, $f(x)$ approaches 2. Thus, $\lim_{x \to 2} f(x) = 2$.

   b. The graph of $f(x)$ at $x = 2$ is the point with coordinates $(2, 1)$. Thus, $f(2) = 1$.

23. The graph shows that as x approaches 2, $f(x)$ approaches -3. Thus, $\lim_{x \to 2}(1 - x^2) = -3$.

25. The graph shows that as x approaches $-\dfrac{\pi}{2}$, $f(x)$ approaches -1. Thus, $\lim_{x \to -\frac{\pi}{2}} \sin x = -1$.

27. a. As x approaches 2 from the left, the values of $f(x)$ get close to the y-coordinate of 4. Thus $\lim_{x \to 2^-} f(x) = 4$.

   b. As x approaches 2 from the right, the values of $f(x)$ get close to the y-coordinate of 2. Thus $\lim_{x \to 2^+} f(x) = 2$.

   c. The $\lim_{x \to 2^-} f(x) = 4$ and $\lim_{x \to 2^+} f(x) = 2$. Because the left and right-hand limits are unequal, $\lim_{x \to 2} f(x)$ does not exist.

   d. $f(2) = 4$ because this point is shown as a solid dot.

29. a. As x approaches -3 from the left, the values of $f(x)$ get close to the y-coordinate of 2. Thus $\lim_{x \to -3^-} f(x) = 2$.

   b. As x approaches -3 from the right, the values of $f(x)$ get close to the y-coordinate of 2. Thus $\lim_{x \to -3^+} f(x) = 2$.

   c. The $\lim_{x \to -3^-} f(x) = 2$ and $\lim_{x \to -3^+} f(x) = 2$, thus $\lim_{x \to -3} f(x) = 2$.

d. $f(-3) = 2$ because this point is shown as a solid dot.

e. As x approaches -1 from the left, the values of $f(x)$ get close to the y-coordinate of 4. Thus $\lim_{x \to -1^-} f(x) = 4$.

f. As x approaches -3 from the right, the values of $f(x)$ get close to the y-coordinate of 3. Thus $\lim_{x \to -1^+} f(x) = 3$.

g. The $\lim_{x \to -1^-} f(x) = 4$ and $\lim_{x \to -1^+} f(x) = 3$. Because the left and right-hand limits are unequal, $\lim_{x \to -1} f(x)$ does not exist.

h. $f(-1)$ does not exist because these points are shown as open dots.

i. As x approaches 3 from the left, the values of $f(x)$ get close to the y-coordinate of 2. Thus $\lim_{x \to 3^-} f(x) = 2$.

j. As x approaches 3 from the right, the values of $f(x)$ get close to the y-coordinate of 2. Thus $\lim_{x \to 3^+} f(x) = 2$.

k. The $\lim_{x \to 3^-} f(x) = 2$ and $\lim_{x \to 3^+} f(x) = 2$, thus $\lim_{x \to 3} f(x) = 2$.

l. $f(3) = 1$ because this point is shown as a solid dot.

31. a. As x approaches 2 from the left, the values of $f(x)$ get close to the y-coordinate of 1. Thus $\lim_{x \to 2^-} f(x) = 1$.

   b. As x approaches 2 from the right, the values of $f(x)$ get close to the y-coordinate of 2. Thus $\lim_{x \to 2^+} f(x) = 2$.

   c. The $\lim_{x \to 2^-} f(x) = 1$ and $\lim_{x \to 2^+} f(x) = 2$. Because the left and right-hand limits are unequal, $\lim_{x \to 2} f(x)$ does not exist.

d. $f(2) = 2$ because this point is shown as a solid dot.

e. As x approaches 2.5 from the left, the values of $f(x)$ get close to the y-coordinate of 2. Thus

$$\lim_{x \to 2.5^-} f(x) = 2 \,.$$

f. As x approaches 2.5 from the right, the values of $f(x)$ get close to the y-coordinate of 2. Thus

$$\lim_{x \to 2.5^+} f(x) = 2 \,.$$

g. The $\lim_{x \to 2.5^-} f(x) = 2$ and $\lim_{x \to 2.5^+} f(x) = 2$, thus

$$\lim_{x \to 2.5} f(x) = 2 \,.$$

h. $f(2.5) = 2$ because this point is shown as a solid dot.

33. Graph using the y-intercept, 1, and the slope 2.

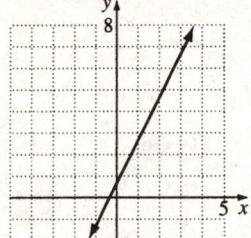

As x gets closer to 3, the values of $f(x)$ get closer to 7. We conclude that $\lim_{x \to 3}(2x + 1) = 7$.

35. Graph by reflecting the graph of $f(x) = x^2$ across the x-axis, then shifting the graph up 4 units.

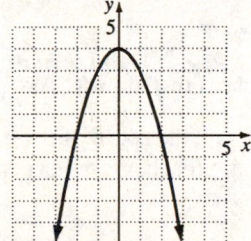

As x gets closer to -3, the values of $f(x)$ get closer to -5. We conclude that $\lim_{x \to -3}(4 - x^2) = -5$.

37. Graph by shifting the graph of $f(x) = |x|$ left 1 unit.

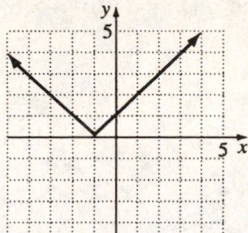

As x gets closer to -1, the values of $f(x)$ get closer to 0. We conclude that $\lim_{x \to -1} |x + 1| = 0$.

39. Graph by using the vertical asymptote at $x = 0$, and the horizontal asymptote at $y = 0$.

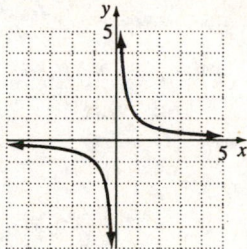

As x gets closer to -1, the values of $f(x)$ get closer to -1. We conclude that $\lim_{x \to -1} \dfrac{1}{x} = -1$.

41. Graph by first factoring the numerator and then dividing out the factor $x - 1$. Then graph the equivalent function, $f(x) = x + 1$ where $x \ne 1$. Use the y-intercept, 1, and the slope 1. Because $x = 1$ is not included, show an open dot on the line corresponding to $x - 1$.

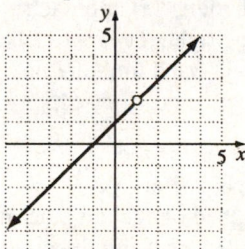

As x gets closer to 1, the values of $f(x)$ get closer to 2. We conclude that $\lim_{x \to 1} \dfrac{x^2 - 1}{x - 1} = 2$.

43. Graph using the *y*-intercept, (0, 1), and the properties of the exponential function $f(x) = a^x$ for $a > 1$.

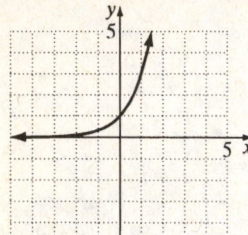

As *x* gets closer to 0, the values of *f(x)* get closer to 1. We conclude that $\lim_{x \to 0} e^x = 1$.

45. Graph using the period = 2π, amplitude = 1, and no phase shift. Use the key points,

$(0, 0)$, $\left(\dfrac{\pi}{2}, 1\right)$, $(\pi, 0)$, $\left(\dfrac{3\pi}{2}, -1\right)$, and $(2\pi, 0)$.

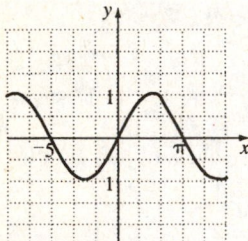

As *x* gets closer to π, the values of *f(x)* get closer to 0. We conclude that $\lim_{x \to \pi} \sin x = 0$.

47. As *x* approaches 2 from the left and the right, *f(x)* approaches 3, thus $\lim_{x \to 2} f(x) = 3$.

49. As *x* approaches 0 from the left, *f(x)* approaches 3. As *x* approaches 0 from the right, *f(x)* approaches 4. The left-hand limit and right-hand limit are not equal, thus the limit does not exist.

51. Graph $f(x) = 2x$ for $x < 1$ by using the *y*-intercept, 0, and the slope, 2. Because $x < 1$, this line is only graphed on the left side of $x = 1$. Graph $f(x) = x + 1$ for $x \geq 1$ by using the point (1, 2) and the slope, 1. Graph this line on the right side of $x = 1$.

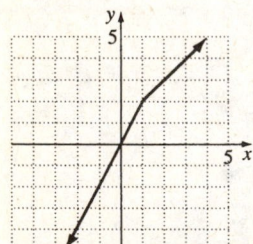

As *x* gets closer to 1, the values of *f(x)* get closer to 2. We conclude that $\lim_{x \to 1} f(x) = 2$.

53. As *x* approaches 0 from the left, *f(x)* approaches 1. As *x* approaches 0 from the right, *f(x)* approaches 0. The left-hand limit and right-hand limit are not equal, thus the limit does not exist.

55. $(f \circ g)(x) = \left(\sqrt{x}\right)^2 - 5 = x - 5$

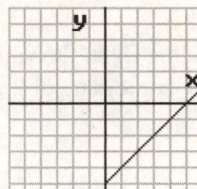

$\lim_{x \to 2}(f \circ g)(x) = -3$

57. $f^{-1}(x) = \sqrt[3]{x + 2}$

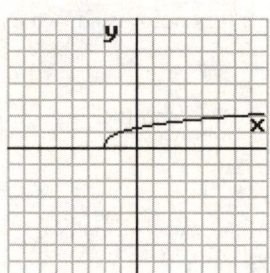

$\lim_{x \to 6} f^{-1}(x) = 2$

59. $\lim_{x \to 3} g(x) = 3$

61. $\lim_{x \to 1^-} g(x) = 0$

63. $\lim_{x \to -3^+} g(x) = 2$

65. $\lim_{x \to 1} g(x) = 0$

67. a. According to the table, $\lim_{x \to 3} f(x) = 8$. As your nose approaches the fan, the speed of the breeze that your nose feels approaches 8 miles per hour.

b. Answers may vary.

69. $\lim_{x \to 67} f(x) = 45$

The mean score in spatial orientation for someone 67 is 45.

71. a. $\lim\limits_{x \to 100} f(x) = 30$

The cost to rent the car one day and drive it 100 miles is $30.

b.   The rental cost on the first day approaches $40 as the mileage approaches 200.

c.   At the start of the second day, the rental cost is $60.

73. – 83.   Answers may vary.

85.

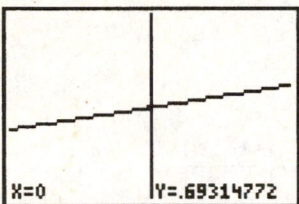

$\lim\limits_{x \to 0} \dfrac{2^x - 1}{x} \approx 0.69315$

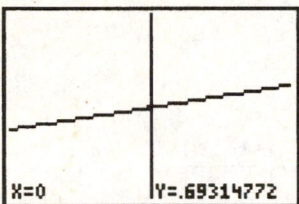

$\lim\limits_{x \to 0} \dfrac{2^x - 1}{x} \approx 0.693147$

87.

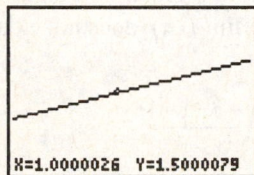

$\lim\limits_{x \to 1} \dfrac{x^{3/2} - 1}{x - 1} \approx 1.5000$

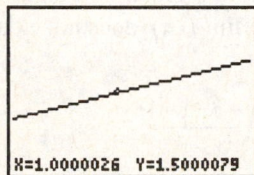

$\lim\limits_{x \to 1} \dfrac{x^{3/2} - 1}{x - 1} \approx 1.50000$

89.   makes sense

91.   makes sense

93.   Answers may vary.

95.   Let $x_1 = \dfrac{314}{100}$, $x_2 = \dfrac{31{,}415}{10{,}000}$,

$x_3 = \dfrac{3{,}141{,}592}{1{,}000{,}000}$, and $x_4 = \dfrac{314{,}159{,}265}{100{,}000{,}000}$.

$3^{x_1} \approx 31.5$

$3^{x_2} \approx 31.541$

$3^{x_3} \approx 31.54426$

$3^{x_4} \approx 31.544281$

3^{π} is approximately equal to 31.544281.

96. a.   Graph:

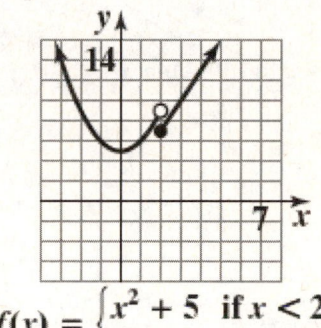

$$f(x) = \begin{cases} x^2 + 5 & \text{if } x < 2 \\ 3x + 1 & \text{if } x \ge 2 \end{cases}$$

b. $\lim\limits_{x \to 2^-} f(x) = 9$

$\lim\limits_{x \to 2^+} f(x) = 7$

$\lim\limits_{x \to 2} f(x)$ does not exist

97. $\dfrac{x^2 - x - 6}{x - 3} = \dfrac{(x+2)(x-3)}{x-3} = x + 2$

98. $\dfrac{\sqrt{4+x} - 2}{x} \cdot \dfrac{\sqrt{4+x} + 2}{\sqrt{4+x} + 2} = \dfrac{(\sqrt{4+x})^2 - 4}{x(\sqrt{4+x} + 2)}$

$$= \dfrac{4 + x - 4}{x(\sqrt{4+x} + 2)}$$

$$= \dfrac{x}{x(\sqrt{4+x} + 2)}$$

$$= \dfrac{1}{\sqrt{4+x} + 2}$$

Introduction to Calculus

Section 11.2

Check Point Exercises

1. **a.** $\lim_{x \to 8} 11 = 11$

    b. $\lim_{x \to 0} (-9) = -9$

2. **a.** $\lim_{x \to 19} x = 19$

    b. $\lim_{x \to -\sqrt{2}} x = -\sqrt{2}$

3. $\lim_{x \to -3} (x + 16) = \lim_{x \to -3} x + \lim_{x \to -3} 16$

             $= -3 + 16$

             $= 13$

4. $\lim_{x \to 14} (19 - x) = \lim_{x \to 14} 19 - \lim_{x \to 14} x$

             $= 19 - 14$

             $= 5$

5. $\lim_{x \to 7} (-10x) = \lim_{x \to 7} (-10) \cdot \lim_{x \to 7} x$

             $= -10 \cdot 7$

             $= -70$

6. **a.** $\lim_{x \to -5} (3x - 7)$

         $= \lim_{x \to -5} 3 \cdot \lim_{x \to -5} x - \lim_{x \to -5} 7$

         $= 3(-5) - 7$

         $= -15 - 7$

         $= -22$

    b. $\lim_{x \to 3} 8x^2 = \lim_{x \to 3} 8 \cdot \lim_{x \to 3} x^2$

               $= 8 \cdot \lim_{x \to 3} (x \cdot x)$

               $= 8 \cdot \lim_{x \to 3} x \cdot \lim_{x \to 3} x$

               $= 8 \cdot 3 \cdot 3$

               $= 72$

7. $\lim_{x \to 2} (-7x^3) = -7 \cdot 2^3$

            $= -7 \cdot 8$

            $= -56$

8. $\lim_{x \to 2} (7x^3 + 3x^2 - 5x + 3)$

     $= 7 \cdot 2^3 + 3 \cdot 2^2 - 5 \cdot 2 + 3$

     $= 7 \cdot 8 + 3 \cdot 4 - 5 \cdot 2 + 3$

     $= 56 + 12 - 10 + 3$

     $= 61$

9. $\lim_{x \to 4} (3x - 5)^3 = \left[\lim_{x \to 4} (3x - 5) \right]^3$

                 $= (3 \cdot 4 - 5)^3$

                 $= 7^3$

                 $= 343$

10. $\lim_{x \to -1} \sqrt{6x^2 - 4} = \sqrt{\lim_{x \to -1} (6x^2 - 4)}$

               $= \sqrt{6(-1)^2 - 4}$

               $= \sqrt{6 - 4}$

               $= \sqrt{2}$

11. $\lim_{x \to 2} \dfrac{x^2 - 4x + 1}{3x - 5} = \dfrac{\lim_{x \to 2} (x^2 - 4x + 1)}{\lim_{x \to 2} (3x - 5)}$

                   $= \dfrac{2^2 - 4 \cdot 2 + 1}{3 \cdot 2 - 5}$

                   $= \dfrac{-3}{1}$

                   $= -3$

12. **a.** Because x is less than 1, use the first line of the piecewise function.

         $\lim_{x \to 1^-} f(x) = -1$

    b. Because x is greater than 1, use the second line of the piecewise function.

         $\lim_{x \to 1^+} f(x) = \sqrt[3]{2(1) - 1} = 1$

    c. Because the left-hand limit and the right-hand limit are unequal, $\lim_{x \to 1} f(x)$ does not exist.

13. $\lim_{x \to 1} \dfrac{x^2 + 2x - 3}{x - 1} = \lim_{x \to 1} \dfrac{(x + 3)(x - 1)}{x - 1}$

                   $= \lim_{x \to 1} (x + 3)$

                   $= 1 + 3$

                   $= 4$

14. $\lim\limits_{x \to 0} \dfrac{\sqrt{9+x}-3}{x} = \lim\limits_{x \to 0} \dfrac{\sqrt{9+x}-3}{x} \cdot \dfrac{\sqrt{9+x}+3}{\sqrt{9+x}+3}$

$\qquad = \lim\limits_{x \to 0} \dfrac{\left(\sqrt{9+x}\right)^2 - 3^2}{x\left(\sqrt{9+x}+3\right)}$

$\qquad = \lim\limits_{x \to 0} \dfrac{9+x-9}{x\left(\sqrt{9+x}+3\right)}$

$\qquad = \lim\limits_{x \to 0} \dfrac{x}{x\left(\sqrt{9+x}+3\right)}$

$\qquad = \lim\limits_{x \to 0} \dfrac{1}{\sqrt{9+x}+3}$

$\qquad = \dfrac{\lim\limits_{x \to 0} 1}{\sqrt{\lim\limits_{x \to 0}(9+x)} + \lim\limits_{x \to 0} 3}$

$\qquad = \dfrac{1}{\sqrt{9+0}+3}$

$\qquad = \dfrac{1}{3+3}$

$\qquad = \dfrac{1}{6}$

Exercise Set 11.2

1. $\lim\limits_{x \to 2} 8 = 8$

3. $\lim\limits_{x \to 2} x = 2$

5. $\lim\limits_{x \to 6}(3x-4) = \lim\limits_{x \to 6} 3 \cdot \lim\limits_{x \to 6} x - \lim\limits_{x \to 6} 4$

$\qquad = 3 \cdot 6 - 4$

$\qquad = 18 - 4$

$\qquad = 14$

7. $\lim\limits_{x \to -2} 7x^2 = \lim\limits_{x \to -2} 7 \cdot \lim\limits_{x \to -2} x^2$

$\qquad = 7 \cdot \lim\limits_{x \to -2}(x \cdot x)$

$\qquad = 7 \cdot \lim\limits_{x \to -2} x \cdot \lim\limits_{x \to -2} x$

$\qquad = 7(-2)(-2)$

$\qquad = 28$

9. $\lim\limits_{x \to 5}(x^2 - 3x - 4) = 5^2 - 3 \cdot 5 - 4$

$\qquad = 25 - 15 - 4$

$\qquad = 6$

11. $\lim\limits_{x \to 2}(5x-8)^3 = \left[\lim\limits_{x \to 2}(5x-8)\right]^3$

$\qquad = (5 \cdot 2 - 8)^3$

$\qquad = 2^3$

$\qquad = 8$

13. $\lim\limits_{x \to 1}(2x^2 - 3x + 5)^2 = \left[\lim\limits_{x \to 1}(2x^2 - 3x + 5)\right]^2$

$\qquad = (2 \cdot 1^2 - 3 \cdot 1 + 5)^2$

$\qquad = (2 \cdot 1 - 3 \cdot 1 + 5)^2$

$\qquad = (2 - 3 + 5)^2$

$\qquad = 4^2$

$\qquad = 16$

15. $\lim\limits_{x \to -4} \sqrt{x^2 + 9} = \sqrt{\lim\limits_{x \to -4}(x^2 + 9)}$

$\qquad = \sqrt{(-4)^2 + 9}$

$\qquad = \sqrt{16 + 9}$

$\qquad = \sqrt{25}$

$\qquad = 5$

17. $\lim\limits_{x \to 5} \dfrac{x}{x+1} = \dfrac{\lim\limits_{x \to 5} x}{\lim\limits_{x \to 5}(x+1)}$

$\qquad = \dfrac{5}{5+1}$

$\qquad = \dfrac{5}{6}$

19. $\lim\limits_{x \to 2} \dfrac{x^2 - 1}{x - 1} = \lim\limits_{x \to 2} \dfrac{(x+1)(x-1)}{x-1}$

$\qquad = \lim\limits_{x \to 2}(x+1)$

$\qquad = 2 + 1$

$\qquad = 3$

21. $\lim\limits_{x \to 1} \dfrac{x^2 - 1}{x - 1} = \lim\limits_{x \to 1} \dfrac{(x+1)(x-1)}{x-1}$

$\qquad = \lim\limits_{x \to 1}(x+1)$

$\qquad = 1 + 1$

$\qquad = 2$

23. $\lim\limits_{x \to 2} \dfrac{2x - 4}{x - 2} = \lim\limits_{x \to 2} \dfrac{2(x-2)}{x-2}$

$\qquad = \lim\limits_{x \to 2} 2$

$\qquad = 2$

25. $\lim\limits_{x\to 1}\dfrac{x^2+2x-3}{x^2-1}=\lim\limits_{x\to 1}\dfrac{(x+3)(x-1)}{(x+1)(x-1)}=\lim\limits_{x\to 1}\dfrac{x+3}{x+1}=\dfrac{\lim\limits_{x\to 1}(x+3)}{\lim\limits_{x\to 1}(x+1)}=\dfrac{1+3}{1+1}=\dfrac{4}{2}=2$

27. $\lim\limits_{x\to 2}\dfrac{x^3-2x^2+4x-8}{x^4-2x^3+x-2}=\lim\limits_{x\to 2}\dfrac{x^2(x-2)+4(x-2)}{x^3(x-2)+(x-2)}=\lim\limits_{x\to 2}\dfrac{(x^2+4)(x-2)}{(x^3+1)(x-2)}=\lim\limits_{x\to 2}\dfrac{x^2+4}{x^3+1}$

$=\dfrac{\lim\limits_{x\to 2}(x^2+4)}{\lim\limits_{x\to 2}(x^3+1)}=\dfrac{2^2+4}{2^3+1}=\dfrac{8}{9}$

29. $\lim\limits_{x\to 0}\dfrac{\sqrt{1+x}-1}{x}=\lim\limits_{x\to 0}\dfrac{\sqrt{1+x}-1}{x}\cdot\dfrac{\sqrt{1+x}+1}{\sqrt{1+x}+1}=\lim\limits_{x\to 0}\dfrac{\left(\sqrt{1+x}\right)^2-1^2}{x\left(\sqrt{1+x}+1\right)}=\lim\limits_{x\to 0}\dfrac{1+x-1}{x\left(\sqrt{1+x}+1\right)}=\lim\limits_{x\to 0}\dfrac{x}{x\left(\sqrt{1+x}+1\right)}$

$=\lim\limits_{x\to 0}\dfrac{1}{\sqrt{1+x}+1}=\dfrac{\lim\limits_{x\to 0}1}{\sqrt{\lim\limits_{x\to 0}(1+x)}+\lim\limits_{x\to 0}1}=\dfrac{1}{\sqrt{1+0}+1}=\dfrac{1}{1+1}=\dfrac{1}{2}$

31. $\lim\limits_{x\to 2}\left[(x+1)^2(3x-1)^3\right]=\left[\lim\limits_{x\to 2}(x+1)\right]^2\left[\lim\limits_{x\to 2}(3x-1)\right]^3=(2+1)^2\cdot(3\cdot 2-1)^3=3^2\cdot 5^3=1125$

33. $\lim\limits_{x\to 4}\dfrac{\sqrt{x}-2}{x-4}=\lim\limits_{x\to 4}\dfrac{\sqrt{x}-2}{x-4}\cdot\dfrac{\sqrt{x}+2}{\sqrt{x}+2}=\lim\limits_{x\to 4}\dfrac{\left(\sqrt{x}\right)^2-2^2}{(x-4)\left(\sqrt{x}+2\right)}=\lim\limits_{x\to 4}\dfrac{x-4}{(x-4)\left(\sqrt{x}+2\right)}$

$=\lim\limits_{x\to 4}\dfrac{1}{\sqrt{x}+2}=\dfrac{\lim\limits_{x\to 4}1}{\sqrt{\lim\limits_{x\to 4}x}+\lim\limits_{x\to 4}2}=\dfrac{1}{\sqrt{4}+2}=\dfrac{1}{2+2}=\dfrac{1}{4}$

35. $\lim\limits_{x\to 2}\dfrac{\frac{1}{x}-\frac{1}{2}}{x-2}=\lim\limits_{x\to 2}\dfrac{\frac{2-x}{2x}}{x-2}=\lim\limits_{x\to 2}\dfrac{2-x}{2x(x-2)}=\lim\limits_{x\to 2}\dfrac{-(x-2)}{2x(x-2)}=\lim\limits_{x\to 2}\left(-\dfrac{1}{2x}\right)=-\dfrac{1}{2\cdot 2}=-\dfrac{1}{4}$

37. $\lim\limits_{x\to 4}\dfrac{\sqrt{x}+5}{x-5}=\dfrac{\sqrt{\lim\limits_{x\to 4}x+\lim\limits_{x\to 4}5}}{\lim\limits_{x\to 4}x-\lim\limits_{x\to 4}5}=\dfrac{\sqrt{4}+5}{4-5}=\dfrac{7}{-1}=-7$

39. $\lim\limits_{x\to 0}\dfrac{\frac{1}{x+3}-\frac{1}{3}}{x}=\lim\limits_{x\to 0}\dfrac{\frac{3-(x+3)}{3(x+3)}}{x}=\lim\limits_{x\to 0}\dfrac{-x}{3x(x+3)}=\lim\limits_{x\to 0}\dfrac{-1}{3(x+3)}=\dfrac{\lim\limits_{x\to 0}(-1)}{3\lim\limits_{x\to 0}(x+3)}=\dfrac{-1}{3(0+3)}=-\dfrac{1}{9}$

41. $\lim\limits_{x\to 2}\dfrac{x^2-4}{x^3-8}=\lim\limits_{x\to 2}\dfrac{(x+2)(x-2)}{(x-2)(x^2+2x+4)}=\lim\limits_{x\to 2}\dfrac{x+2}{x^2+2x+4}=\dfrac{\lim\limits_{x\to 2}(x+2)}{\lim\limits_{x\to 2}(x^2+2x+4)}=\dfrac{2+2}{2^2+2\cdot 2+4}=\dfrac{4}{12}=\dfrac{1}{3}$

43. $f(x) = \begin{cases} x+5 \text{ if } x < 1 \\ x+7 \text{ if } x \geq 1 \end{cases}$

   a.　$\lim\limits_{x \to 1^-} f(x) = \lim\limits_{x \to 1^-}(x+5) = 1+5 = 6$

   b.　$\lim\limits_{x \to 1^+} f(x) = \lim\limits_{x \to 1^+}(x+7) = 1+7 = 8$

   c.　$\lim\limits_{x \to 1^-} f(x) \neq \lim\limits_{x \to 1^+} f(x)$, therefore, $\lim\limits_{x \to 1} f(x)$
     does not exist.

45. $f(x) = \begin{cases} x^2+5 \text{ if } x < 2 \\ x^3+1 \text{ if } x \geq 2 \end{cases}$

   a.　$\lim\limits_{x \to 2^-} f(x) = \lim\limits_{x \to 2^-}(x^2+5) = 2^2+5 = 9$

   b.　$\lim\limits_{x \to 2^+} f(x) = \lim\limits_{x \to 2^+}(x^3+1) = 2^3+1 = 9$

   c.　$\lim\limits_{x \to 2^-} f(x) = \lim\limits_{x \to 2^+} f(x)$, therefore,
     $\lim\limits_{x \to 2} f(x)$ exists, and is 9

47. $f(x) = \begin{cases} \dfrac{x^2-9}{x-3} & \text{if } x \neq 3 \\ 5 & \text{if } x = 3 \end{cases}$

   a.　$\lim\limits_{x \to 3^-} f(x) = \lim\limits_{x \to 3^-} \dfrac{x^2-9}{x-3}$

     $= \lim\limits_{x \to 3^-} \dfrac{(x+3)(x-3)}{x-3}$

     $= \lim\limits_{x \to 3^-}(x+3)$

     $= 3+3$

     $= 6$

   b.　$\lim\limits_{x \to 3^+} f(x) = \lim\limits_{x \to 3^+} \dfrac{x^2-9}{x-3}$

     $= \lim\limits_{x \to 3^+} \dfrac{(x+3)(x-3)}{x-3}$

     $= \lim\limits_{x \to 3^+}(x+3)$

     $= 3+3$

     $= 6$

   c.　$\lim\limits_{x \to 3^-} f(x) = \lim\limits_{x \to 3^+} f(x)$, therefore, $\lim\limits_{x \to 3} f(x)$
     exists, and is 6

49. $f(x) + \begin{cases} 1-x & \text{if } x < 1 \\ 2 & \text{if } x = 1 \\ x^2-1 & \text{if } x > 1 \end{cases}$

   a.　$\lim\limits_{x \to 1^-} f(x) = \lim\limits_{x \to 1^-}(1-x) = 1-1 = 0$

   b.　$\lim\limits_{x \to 1^+} f(x) = \lim\limits_{x \to 1^+}(x^2-1) = 1^2-1 = 0$

   c.　$\lim\limits_{x \to 1^-} f(x) = \lim\limits_{x \to 1^+} f(x)$, therefore, $\lim\limits_{x \to 1} f(x)$
     exists, and is 0

51. $\lim\limits_{x \to 3}(f \circ g)(x) = \lim\limits_{x \to 3}\left[2^3 - 2^2 + 5(2) - 1\right] = \lim\limits_{x \to 3} 13 = 13$
   $\lim\limits_{x \to 3}(g \circ f)(x) = \lim\limits_{x \to 3} 2 = 2$

53. $\lim\limits_{x \to 1}(f \circ g)(x) = \lim\limits_{x \to 1} \dfrac{2}{\frac{3}{x-1}} = \lim\limits_{x \to 1} \dfrac{2x-2}{3} = \dfrac{2(1)-2}{3} = 0$

   $\lim\limits_{x \to 1}(g \circ f)(x) = \lim\limits_{x \to 1} \dfrac{3}{\frac{2}{x}-1} = \lim\limits_{x \to 1} \dfrac{3}{\frac{2}{x}-1} = \dfrac{3}{\frac{2}{1}-1} = 3$

55. $f^{-1}(x) = \sqrt{x-4}$

   $\lim\limits_{x \to 8} f^{-1}(x) = \lim\limits_{x \to 8} \sqrt{x-4} = \sqrt{8-4} = 2$

57. $f^{-1}(x) = \dfrac{x+1}{x-2}$

   $\lim\limits_{x \to 4} f^{-1}(x) = \lim\limits_{x \to 4} \dfrac{x+1}{x-2} = \dfrac{4+1}{4-2} = \dfrac{5}{2}$

59.　a.　$\lim\limits_{v \to c^-} L = \lim\limits_{v \to c^-} L_0\sqrt{1 - \dfrac{v^2}{c^2}}$

      $= \lim\limits_{v \to c^-} L_0\sqrt{1-1}$

      $= \lim\limits_{v \to c^-} 0 = 0$

   b.　The length of the starship appears to approach 0.

   c.　It is not possible to exceed the speed of light.

61. – 73.　Answers may vary.

75.　makes sense

77.　makes sense

79. $\lim\limits_{x \to 4}\left(\dfrac{1}{x} - \dfrac{1}{4}\right)\left(\dfrac{1}{x-4}\right) = \lim\limits_{x \to 4}\left(\dfrac{4-x}{4x}\right)\left(\dfrac{1}{x-4}\right)$

$$= \lim\limits_{x \to 4}\dfrac{-(x-4)}{4x(x-4)}$$

$$= \lim\limits_{x \to 4}\left(-\dfrac{1}{4x}\right)$$

$$= -\dfrac{1}{4\cdot 4}$$

$$= -\dfrac{1}{16}$$

81. $\lim\limits_{h \to 0}\dfrac{f(a+h)-f(a)}{h} = \lim\limits_{h \to 0}\dfrac{\sqrt{1+h}-\sqrt{1}}{h}$

$$= \lim\limits_{h \to 0}\dfrac{\sqrt{1+h}-1}{h}\cdot\dfrac{\sqrt{1+h}+1}{\sqrt{1+h}+1}$$

$$= \lim\limits_{h \to 0}\dfrac{\left(\sqrt{1+h}\right)^2 - 1^2}{h\left(\sqrt{1+h}+1\right)}$$

$$= \lim\limits_{h \to 0}\dfrac{1+h-1}{h\left(\sqrt{1+h}+1\right)}$$

$$= \lim\limits_{h \to 0}\dfrac{h}{h\left(\sqrt{1+h}+1\right)}$$

$$= \lim\limits_{h \to 0}\dfrac{1}{\sqrt{1+h}+1}$$

$$= \dfrac{\lim\limits_{h \to 0}1}{\sqrt{\lim\limits_{h \to 0}(1+h)} + \lim\limits_{h \to 0}1}$$

$$= \dfrac{1}{\sqrt{1+0}+1}$$

$$= \dfrac{1}{1+1}$$

$$= \dfrac{1}{2}$$

83. $\lim\limits_{x \to 0}\dfrac{2\sin x + \cos x - 1}{3x}$

$$= \lim\limits_{x \to 0}\left(\dfrac{2\sin x}{3x} + \dfrac{\cos x - 1}{3x}\right)$$

$$= \dfrac{2}{3}\lim\limits_{x \to 0}\dfrac{\sin x}{x} + \dfrac{1}{3}\lim\limits_{x \to 0}\dfrac{\cos x - 1}{x}$$

$$= \dfrac{2}{3}(1) + \dfrac{1}{3}(0)$$

$$= \dfrac{2}{3}$$

85. Answers may vary.

86. No, it is not necessary to lift your pencil off the paper.

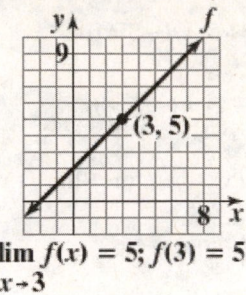

$\lim\limits_{x \to 3} f(x) = 5;\ f(3) = 5$

87. Yes, it is necessary to lift your pencil off the paper.

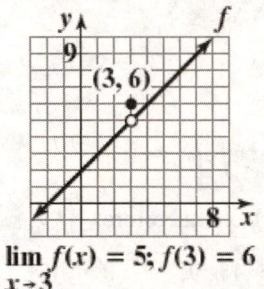

$\lim\limits_{x \to 3} f(x) = 5;\ f(3) = 6$

88. Yes, it is necessary to lift your pencil off the paper.

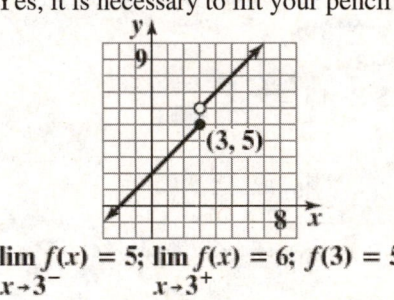

$\lim\limits_{x \to 3^-} f(x) = 5;\ \lim\limits_{x \to 3^+} f(x) = 6;\ f(3) = 5$

Section 11.3
Check Point Exercises

1. $f(x) = \dfrac{x-2}{x^2-4}$

    a. Is $f(1)$ defined?

    $f(1) = \dfrac{1-2}{1^2-4} = \dfrac{1-2}{1-4} = \dfrac{-1}{-3} = \dfrac{1}{3}$

    Because $f(1)$ is a real number, $f(1)$ is defined.
    Does $\lim\limits_{x \to 1} f(x)$ exist?

    $\lim\limits_{x \to 1} f(x) = \lim\limits_{x \to 1} \dfrac{x-2}{x^2-4}$

    $= \dfrac{\lim\limits_{x \to 1}(x-2)}{\lim\limits_{x \to 1}(x^2-4)}$

    $= \dfrac{1-2}{1^2-4}$

    $= \dfrac{1-2}{1-4}$

    $= \dfrac{-1}{-3}$

    $= \dfrac{1}{3}$

    Using the properties of limits, we see that
    $\lim\limits_{x \to 1} f(x)$ exists.
    Does $\lim\limits_{x \to 1} f(x) = f(1)$? Yes.

    Because the three conditions for continuity are
    satisfied, $f(x)$ is continuous at 1.

    b. Is $f(2)$ defined?

    $f(x) = \dfrac{x-2}{(x+2)(x-2)}$

    Because division by zero is undefined, the
    domain of f is $\{x \mid x \neq 2, x \neq -2\}$.

    Thus, f is not defined at 2. Because one of the
    three conditions is not satisfied, we conclude
    that f is discontinuous at 2.

2. $f(x) = \begin{cases} 2x & \text{if } x \leq 0 \\ x^2+1 & \text{if } 0 < x \leq 2 \\ 7-x & \text{if } x > 2 \end{cases}$

    Because $2x$ and $7-x$ are linear functions and x^2+1
    is a polynomial function, each is continuous for every
    number x. However, since the functional form of f
    changes at
    $x = 0$ and $x = 2$, we must check for continuity at these
    two points.
    Is $f(0)$ defined?

$f(0) = 2(0) = 0$
Because $f(0)$ is a real number, $f(0)$ is defined.
Does $\lim\limits_{x \to 0} f(x)$ exist?

$\lim\limits_{x \to 0^-} f(x) = \lim\limits_{x \to 0^-} 2x = 2 \cdot 0 = 0$

$\lim\limits_{x \to 0^+} f(x) = \lim\limits_{x \to 0^+} (x^2+1) = 0^2+1 = 1$

Because the left-and right-hand limits are
unequal, $\lim\limits_{x \to 0} f(x)$ does not exist.

Because one of the three conditions is not satisfied,
we conclude that f is not continuous at 0.
Is $f(2)$ defined?

$f(2) = 2^2+1 = 5$

Because $f(2)$ is a real number, $f(2)$ is defined.
Does $\lim\limits_{x \to 2} f(x)$ exist?

$\lim\limits_{x \to 2^-} f(x) = \lim\limits_{x \to 2^-} (x^2+1) = 2^2+1 = 5$

$\lim\limits_{x \to 2^+} f(x) = \lim\limits_{x \to 2^+} (7-x) = 7-2 = 5$

Because the left- and right-hand limits are
equal, $\lim\limits_{x \to 2} f(x) = 5$.
Does $\lim\limits_{x \to 2} f(x) = f(2)$?

$\lim\limits_{x \to 2} f(x) = 5 = f(2)$

Because the three conditions for continuity are
satisfied, f is continuous at 2.
In summary, the given function is discontinuous at 0
only.

Exercise Set 11.3

1. $f(1) = 2 \cdot 1 + 5 = 7$
    f is defined at 1.
    $\lim\limits_{x \to 1} f(x) = \lim\limits_{x \to 1}(2x+5) = 2 \cdot 1 + 5 = 7$
    $\lim\limits_{x \to 1} f(x)$ exists.
    $\lim\limits_{x \to 1} f(x) = 7 = f(1)$, therefore, f is

    continuous at 1.

3. $f(4) = 4^2 - 3 \cdot 4 + 7 = 11$
    f is defined at 4.
    $\lim\limits_{x \to 4} f(x) = \lim\limits_{x \to 4}(x^2 - 3x + 7) = 4^2 - 3 \cdot 4 + 7 = 11$
    $\lim\limits_{x \to 4} f(x)$ exists.
    $\lim\limits_{x \to 1} f(x) = f(4) = 11$, therefore, f is

    continuous at 4.

5. $f(3) = \dfrac{3^2+4}{3-2} = 13$

f is defined at 3.

$\lim\limits_{x\to 3} f(x) = \lim\limits_{x\to 3} \dfrac{x^2+4}{x-2} = \dfrac{3^2+4}{3-2} = 13$

$\lim\limits_{x\to 3} f(x)$ exists.

$\lim\limits_{x\to 3} f(x) = 13 = f(3),$ therefore, f is

continuous at 3.

7. $f(5)$ results in division by zero, so $f(5)$ is undefined and, therefore, discontinuous at 5.

9. $f(5) = \dfrac{5-5}{5+5} = 0$

f is defined at 5.

$\lim\limits_{x\to 5} f(x) = \lim\limits_{x\to 5} \dfrac{x-5}{x+5} = \dfrac{5-5}{5+5} = 0$

$\lim\limits_{x\to 5} f(x)$ exists.

$\lim\limits_{x\to 5} f(x) = 0 = f(5),$ therefore, f is

continuous at 5.

11. $f(0)$ results in division by zero, so $f(0)$ is undefined and, therefore, discontinuous at 0.

13. $f(x) = \begin{cases} \dfrac{x^2-4}{x-2} & \text{if } x \neq 2 \\ 5 & \text{if } x = 2 \end{cases}$

$f(2) = 5,$ therefore, f is defined at 2.

$\lim\limits_{x\to 2} f(x) = \lim\limits_{x\to 2} \dfrac{x^2-4}{x-2} = \lim\limits_{x\to 2} \dfrac{(x+2)(x-2)}{x-2}$

$\qquad\qquad = \lim\limits_{x\to 2}(x+2)$

$\qquad\qquad = 2+2$

$\qquad\qquad = 4$

$\lim\limits_{x\to 2} f(x) \neq f(2),$ therefore, f is

discontinuous at 2.

15. $f(x) = \begin{cases} x-5 & \text{if } x \leq 0 \\ x^2+x-5 & \text{if } x > 0 \end{cases}$

$f(0) = 0 - 5 = -5$

f is defined at 0.

$\lim\limits_{x\to 0^-} f(x) = \lim\limits_{x\to 0^-}(x-5) = 0 - 5 = -5$

$\lim\limits_{x\to 0^+} f(x) = \lim\limits_{x\to 0^+}(x^2+x-5) = 0^2 + 0 - 5 = -5$

$\lim\limits_{x\to 0^-} f(x) = \lim\limits_{x\to 0^+} f(x),$ therefore,

$\lim\limits_{x\to 0} f(x)$ exists and is -5.

$\lim\limits_{x\to 0} f(x) = -5 = f(0),$ therefore, f is continuous at 0.

17. $f(x) = \begin{cases} 1-x & \text{if } x < 1 \\ 0 & \text{if } x = 1 \\ x^2-1 & \text{if } x > 1 \end{cases}$

$f(1) = 0,$ therefore, f is defined at 1.

$\lim\limits_{x\to 1^-} f(x) = \lim\limits_{x\to 1^-}(1-x) = 1 - 1 = 0$

$\lim\limits_{x\to 1^+} f(x) = \lim\limits_{x\to 1^+}(x^2-1) = 1^2 - 1 = 0$

$\lim\limits_{x\to 1^-} f(x) = \lim\limits_{x\to 1^+} f(x),$ therefore,

$\lim\limits_{x\to 1} f(x)$ exists and is 0.

$\lim\limits_{x\to 1} f(x) = 0 = f(1),$ therefore, f is continuous at 1.

19. $f(x) = x^2 + 4x - 6$ is a polynomial function, thus, f is continuous for every number x.

21. $f(x) = \dfrac{x+1}{(x+1)(x-4)}$

f is not defined at $x = -1$ and $x = 4$, therefore, f is discontinuous at -1 and 4.

23. $f(x) = \dfrac{\sin x}{x}$

f is not defined at $x = 0$, therefore, f is discontinuous at 0.

25. $f(x) = \pi$

f is a constant function, thus, f is continuous for every number x.

27. $f(x) = \begin{cases} x-1 & \text{if } x \leq 1 \\ x^2 & \text{if } x > 1 \end{cases}$

$x - 1$ and x^2 are continuous for every number x, but because $f(x)$ changes its functional form at $x = 1$, we must investigate continuity at $x = 1$.

$f(1) = 1 - 1 = 0$

f is defined at 1.

$\lim\limits_{x\to 1^-} f(x) = \lim\limits_{x\to 1^-}(x-1) = 1 - 1 = 0$

$\lim\limits_{x\to 1^+} f(x) = \lim\limits_{x\to 1^+} x^2 = 1^2 = 1$

The left- and right-hand limits are not equal, so $\lim\limits_{x\to 1} f(x)$ does not exist. Therefore, f is discontinuous at 1.

29. $f(x) = \begin{cases} \dfrac{x^2-1}{x-1} & \text{if } x \neq 1 \\ 2 & \text{if } x = 1 \end{cases}$

$\dfrac{x^2-1}{x-1}$ is continuous for every number x except at $x = 1$.

We must investigate continuity at $x = 1$.

$f(1) = 2$

f is defined at 1.

$\displaystyle\lim_{x \to 1} f(x) = \lim_{x \to 1} \frac{x^2-1}{x-1}$

$\displaystyle = \lim_{x \to 1} \frac{(x+1)(x-1)}{x-1}$

$\displaystyle = \lim_{x \to 1}(x+1)$

$= 1 + 1$

$= 2$

$\displaystyle\lim_{x \to 1} f(x)$ exists.

$\displaystyle\lim_{x \to 1} f(x) = 2 = f(1)$, therefore, f is continuous at 1.

In summary, $f(x)$ is continuous for every number x.

31. $f(x) = \begin{cases} x+6 & \text{if } x \leq 0 \\ 6 & \text{if } 0 < x \leq 2 \\ x^2+1 & \text{if } x > 2 \end{cases}$

$x + 6$, 6, and $x^2 + 1$ are continuous for every number x. We must investigate continuity at $x = 0$ and $x = 2$.

$f(0) = 0 + 6 = 6$

f is defined at 0.

$\displaystyle\lim_{x \to 0^-} f(x) = \lim_{x \to 0^-}(x+6) = 0 + 6 = 6$

$\displaystyle\lim_{x \to 0^+} f(x) = \lim_{x \to 0^+} 6 = 6$

$\displaystyle\lim_{x \to 0^-} f(x) = \lim_{x \to 0^+} f(x)$, therefore,

$\displaystyle\lim_{x \to 0} f(x)$ exists and is 6.

$\displaystyle\lim_{x \to 0} f(x) = 6 = f(0)$, therefore, f is continuous at 0.

$f(2) = 6$

f is defined at 2.

$\displaystyle\lim_{x \to 2^-} f(x) = \lim_{x \to 2^-} 6 = 6$

$\displaystyle\lim_{x \to 2^+} f(x) = \lim_{x \to 2^+}(x^2+1) = 2^2 + 1 = 5$

The left- and right-hand limits are not equal, therefore, $\displaystyle\lim_{x \to 2} f(x)$ does not exist and $f(x)$ is discontinuous at 2.

In summary, the given function is discontinuous at 2 only.

33. $f(x) = \begin{cases} 5x & \text{if } x < 4 \\ 21 & \text{if } x = 4 \\ x^2 + 4 & \text{if } x > 4 \end{cases}$

$5x$ and $x^2 + 4$ are continuous for every number x. We must investigate continuity at $x = 4$.

$f(4) = 21$

f is defined at 4.

$\displaystyle\lim_{x \to 4^-} f(x) = \lim_{x \to 4^-} 5x = 5 \cdot 4 = 20$

$\displaystyle\lim_{x \to 4^+} f(x) = \lim_{x \to 4^+}(x^2+4) = 4^2 + 4 = 20$

$\displaystyle\lim_{x \to 4^-} f(x) = \lim_{x \to 4^+} f(x)$, therefore,

$\displaystyle\lim_{x \to 4} f(x)$ exists and is 20.

$\displaystyle\lim_{x \to 4} f(x) \neq f(4)$, therefore, f is discontinuous at 4.

35. $f(x)$ is discontinuous at π.

37. $f(x)$ is discontinuous at each integer.

39.

x	-0.1	-0.01	0.01	0.1
$f(x)$	1.9867	1.9999	1.9999	1.9867

$f(x)$ is continuous for every number x.

41.

x	$\dfrac{\pi}{2}-0.1$	$\dfrac{\pi}{2}-0.01$	$\dfrac{\pi}{2}+0.01$	$\dfrac{\pi}{2}+0.1$
$f(x)$	-0.9983	-0.9999	-0.9999	-0.9983

$f(x)$ is discontinuous at $\dfrac{\pi}{2}$.

43. a. $\displaystyle\lim_{t \to t_1} p(t) = 20$ and $p(t_1) = 20$

b. Yes, p is continuous at t_1.

1. p is defined at t_1.

2. the limit exists at t_1.

3. $\displaystyle\lim_{t \to t_1} p(t) = p(t_1)$

c. $\displaystyle\lim_{t \to t_2} p(t) = 40$ and $p(t_2) = 10$

d. No, p is discontinuous at t_2 because $\displaystyle\lim_{t \to t_2} p(t) \neq p(t_2)$.

e. $\displaystyle\lim_{t \to t_3} p(t)$ does not exist because the left-hand and right-hand limits are not the same.

$p(t_3) = 70$

f. No, p is discontinuous at t_3 because the limit does not exist at t_3.

g. $\lim\limits_{t \to t_4^-} p(t) = 100$ and $p(t_4) = 100$

h. As the end of the course approached, the percentage of topics learned by the student approached, and then reached, 100%.

45. a. Yes, T is continuous at 7825.
 1. T is defined at 7825.
 2. The limit exists at 7825.
 3. $\lim\limits_{x \to 7825} T(x) = T(7825)$

b. Yes, T is continuous at 31,850.
 1. T is defined at 31,850.
 2. The limit exists at 31,850.
 3. $\lim\limits_{x \to 7825} T(x) = T(31,850)$

c. Answers may vary.

47. – 51. Answers may vary.

53.

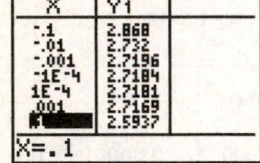

$$\lim_{x \to 0^+} (1+x)^{1/x} \approx 2.7183$$

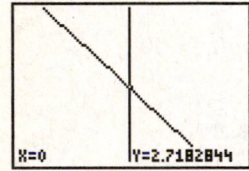

$$\lim_{x \to 0^+} (1+x)^{1/x} \approx 2.71828$$

55. does not make sense; Explanations will vary. Sample explanation: There is a jump in the graph at $x = a$, thus redefining $f(a)$ will not make f continuous.

57. makes sense

59. No, it is not possible to define $f(x) = 1/(x-9)$ at $x-9$ such that the function is continuous at 9, since $\lim\limits_{x \to 9} \dfrac{1}{x-9}$ does not exist. The graph of the function in Exercise 54 is a line with a hole at $x = 9$. In contrast, this function approaches infinity as x approaches 9 from the right-hand side, and negative infinity as x approaches 9 from the left-hand side.

61. Answers may vary.

62. $f(x) = x^2 + x$

$$\frac{f(2+h) - f(2)}{h}$$

$$= \frac{(2+h)^2 + (2+h) \ - \ (2)^2 + (2)}{h}$$

$$= \frac{4 + 4h + h^2 + 2 + h \ - [6]}{h}$$

$$= \frac{6 + 5h + h^2 \ - [6]}{h}$$

$$= \frac{5h + h^2}{h}$$

$$= 5 + h$$

63. $f(x) = x^3$

$$\frac{f(x+h) - f(x)}{h}$$

$$= \frac{(x+h)^3 \ - \ x^3}{h}$$

$$= \frac{x^3 + 3x^2h + 3xh^2 + h^3 \ - \ x^3}{h}$$

$$= \frac{3x^2h + 3xh^2 + h^3}{h}$$

$$= 3x^2 + 3xh + h^2$$

64. $s(t) = -16t^2 + 48t + 160$

$\dfrac{s(a+h)-s(a)}{h}$

$= \dfrac{\left[-16(a+h)^2 + 48(a+h) + 160\right] - \left[-16a^2 + 48a + 160\right]}{h}$

$= \dfrac{\left[-16a^2 - 32ah - 16h^2 + 48a + 48h + 160\right] - \left[-16a^2 + 48a + 160\right]}{h}$

$= \dfrac{-32ah - 16h^2 + 48h}{h}$

$= -32a - 16h + 48$

Mid-Chapter 11 Check Point

1. $\displaystyle\lim_{x \to -1^-} f(x) = 2$

2. $\displaystyle\lim_{x \to -1^+} f(x) = 1$

3. $\displaystyle\lim_{x \to -1} f(x)$ does not exist.

4. $\displaystyle\lim_{x \to 1}\left[f(x) + g(x)\right] = -2$

5. $\displaystyle\lim_{x \to 0}\left[f(x) - g(x)\right] = -1$

6. $(f-g)(0) = f(0) - g(0) = -2 - 1 = -3$

7. $\displaystyle\lim_{x \to 1}\sqrt{10 + f(x)} = \sqrt{10 + (-1)} = \sqrt{9} = 3$

8. $f(x)$ is discontinuous at -1 because the limit does not exist at -1.
$f(x)$ is discontinuous at 0 because $\displaystyle\lim_{x \to 0} f(x)$ does not equal $f(0)$.

9. $\displaystyle\lim_{x \to 0} f(x) = -1$

10. $\displaystyle\lim_{x \to 0} g(x) = 1$

11. $\displaystyle\lim_{x \to 0}\dfrac{4g(x)}{\left[f(x)\right]^2} = \dfrac{4(1)}{(-1)^2} = 4$

12. $\displaystyle\lim_{x \to -2}\left(x^3 - x + 5\right) = (-2)^3 - (-2) + 5 = -1$

13. $\displaystyle\lim_{x \to 3}\sqrt{x^2 - 3x + 4} = \sqrt{3^2 - 3(3) + 4} = \sqrt{4} = 2$

Introduction to Calculus

14. $\lim_{x \to 5} \dfrac{2x^2 - x + 4}{x - 1} = \dfrac{2(5)^2 - (5) + 4}{(5) - 1} = \dfrac{49}{4}$

15. $\lim_{x \to 5} \dfrac{2x^2 - 7x - 15}{x - 5} = \lim_{x \to 5} \dfrac{(2x+3)(x-5)}{x-5}$

$= \lim_{x \to 5}(2x + 3)$

$= 2(5) + 3$

$= 13$

16. $\lim_{x \to 0} \dfrac{\sqrt{x^2 + 9} - 3}{x^2}$

$= \lim_{x \to 0}\left(\dfrac{\sqrt{x^2+9}-3}{x^2} \cdot \dfrac{\sqrt{x^2+9}+3}{\sqrt{x^2+9}+3}\right)$

$= \lim_{x \to 0} \dfrac{x^2}{x^2\left(\sqrt{x^2+9}+3\right)}$

$= \lim_{x \to 0} \dfrac{1}{\sqrt{x^2+9}+3}$

$= \dfrac{1}{\sqrt{0^2+9}+3}$

$= \dfrac{1}{6}$

17. The limit does not exist.

$\lim_{x \to 0} \dfrac{\dfrac{1}{x+10} - \dfrac{1}{x}}{x} = \lim_{x \to 0} \dfrac{\dfrac{x}{x(x+10)} - \dfrac{(x+10)}{x(x+10)}}{x}$

$= \lim_{x \to 0} \dfrac{\dfrac{x-(x+10)}{x(x+10)}}{x}$

$= \lim_{x \to 0} \dfrac{-10}{x^2(x+10)}$

$= \lim_{x \to 0} \dfrac{-10}{x^3 + 10x^2}$

18. a. $\lim_{x \to 4^-} f(x) = 9 - 2(4) = 1$

b. $\lim_{x \to 4^+} f(x) = \sqrt{4 - 4} = 0$

c. The limit does not exist because the left-hand and right-hand limits are not the same.

19. a. $\lim_{x \to 2^-} f(x)$

$= \lim_{x \to 2^-} \dfrac{x^4 - 16}{x - 2}$

$= \lim_{x \to 2^-} \dfrac{\left(x^2+4\right)(x+2)(x-2)}{x-2}$

$= \lim_{x \to 2^-}\left[\left(x^2+4\right)(x+2)\right]$

$= \left(2^2 + 4\right)(2 + 2)$

$= 32$

b. $\lim_{x \to 2^+} f(x)$

$= \lim_{x \to 2^+} \dfrac{x^4 - 16}{x - 2}$

$= \lim_{x \to 2^+} \dfrac{\left(x^2+4\right)(x+2)(x-2)}{x-2}$

$= \lim_{x \to 2^+}\left[\left(x^2+4\right)(x+2)\right]$

$= \left(2^2 + 4\right)(2 + 2)$

$= 32$

c. $\lim_{x \to 2} f(x) = 32$ because the left-hand and right-hand limits are both 32.

20. $f(x)$ is continuous at 3.
1. $f(x)$ is defined at 3, $f(3) = 0$.
2. $\lim_{x \to 3} f(x) = 0$
3. $\lim_{x \to 3} f(x) = f(3) = 0$

21. $f(x)$ is continuous at 0.
1. $f(x)$ is defined at 0, $f(0) = 6$.
2. $\lim_{x \to 0} f(x)$ exists:

$\lim_{x \to 0} \dfrac{(x+3)^2 - 9}{x}$

$= \lim_{x \to 0} \dfrac{x^2 + 6x + 9 - 9}{x}$

$= \lim_{x \to 0} \dfrac{x^2 + 6x}{x}$

$= \lim_{x \to 0}(x + 6)$

$= 0 + 6$

$= 6$

3. $\lim_{x \to 0} f(x) = f(0) = 6$

22. Each of the three pieces are continuous. The pieces change at –1 and 5 so we must check for continuity at these two values.

Conditions of continuity at –1.

1. $f(-1) = 2(-1) = -2$ exists.

2. $\lim\limits_{x \to -1} f(x)$ exists:

$$\lim_{x \to -1^-} \frac{x^2 - 1}{x + 1} \quad \text{and} \quad \lim_{x \to -1^+} 2x$$

$$= \lim_{x \to -1^-} (x - 1) \qquad\qquad = 2(-1)$$

$$= -1 - 1 \qquad\qquad\qquad = -2$$

$$= -2$$

3. $\lim\limits_{x \to -1} f(x) = f(-1) = -2$

Conditions of continuity at 5.

1. $f(5) = 2(5) = 10$ exists.

2. $\lim\limits_{x \to 5} f(x)$ does not exist:

$$\lim_{x \to 5^-} 2x \quad \text{but} \quad \lim_{x \to 5^+} (3x - 4)$$

$$= 2(5) \qquad\qquad = 3(5) - 4$$

$$= 10 \qquad\qquad\ \ = 11$$

Thus $f(x)$ is discontinuous at 5.

Section 11.4

Check Point Exercises

1. $f(x) = x^2 - x$

$(a, f(a)) = (4, 12)$

$$m_{\tan} = \lim_{h \to 0} \frac{f(4 + h) - f(4)}{h}$$

$$= \lim_{h \to 0} \frac{(4 + h)^2 - (4 + h) - (4^2 - 4)}{h}$$

$$= \lim_{h \to 0} \frac{(16 + 8h + h^2 - 4 - h) - 12}{h}$$

$$= \lim_{h \to 0} \frac{h^2 + 7h}{h}$$

$$= \lim_{h \to 0} \frac{h(h + 7)}{h}$$

$$= \lim_{h \to 0} (h + 7)$$

$$= 0 + 7$$

$$= 7$$

2. $f(x) = \sqrt{x}$

$(a, f(a)) = (1, 1)$

$$m_{\tan} = \lim_{h \to 0} \frac{f(1 + h) - f(1)}{h}$$

$$= \lim_{h \to 0} \frac{\sqrt{1 + h} - \sqrt{1}}{h}$$

$$= \lim_{h \to 0} \frac{\sqrt{1 + h} - 1}{h} \cdot \frac{\sqrt{1 + h} + 1}{\sqrt{1 + h} + 1}$$

$$= \lim_{h \to 0} \frac{\left(\sqrt{1 + h}\right)^2 - 1}{h\left(\sqrt{1 + h} + 1\right)}$$

$$= \lim_{h \to 0} \frac{1 + h - 1}{h\left(\sqrt{1 + h} + 1\right)}$$

$$= \lim_{h \to 0} \frac{1}{\sqrt{1 + h} + 1}$$

$$= \frac{1}{\sqrt{1 + 0} + 1}$$

$$= \frac{1}{2}$$

$$y - y_1 = m(x - x_1)$$

$$y - 1 = \frac{1}{2}(x - 1)$$

$$y - 1 = \frac{1}{2}x - \frac{1}{2}$$

$$y = \frac{1}{2}x + \frac{1}{2}$$

3. a. $f(x) = x^2 - 5x$

$$f'(x) = \lim_{h \to 0} \frac{f(x + h) - f(x)}{h}$$

$$= \lim_{h \to 0} \frac{(x + h)^2 - 5(x + h) - (x^2 - 5x)}{h}$$

$$= \lim_{h \to 0} \frac{x^2 + 2xh + h^2 - 5x - 5h - x^2 + 5x}{h}$$

$$= \lim_{h \to 0} \frac{h^2 + 2xh - 5h}{h}$$

$$= \lim_{h \to 0} \frac{h(h + 2x - 5)}{h}$$

$$= \lim_{h \to 0} (h + 2x - 5)$$

$$= 0 + 2x - 5$$

$$= 2x - 5$$

b. $f'(-1) = 2(-1) - 5 = -2 - 5 = -7$

$f'(3) = 2 \cdot 3 - 5 = 6 - 5 = 1$

4. $f(x) = x^3$

a.
$$\frac{f(a+h) - f(a)}{h} = \frac{f(4+0.1) - f(4)}{0.1}$$
$$= \frac{f(4.1) - f(4)}{0.1}$$
$$= \frac{(4.1)^3 - 4^3}{0.1}$$
$$= 49.21$$

The average rate of change of the volume is 49.21 cubic inches per inch as x changes from 4 to 4.1 inches.

$$\frac{f(a+h) - f(a)}{h} = \frac{f(4+0.01) - f(4)}{0.01}$$
$$= \frac{f(4.01) - f(4)}{0.01}$$
$$= \frac{4.01^3 - 4^3}{0.01}$$
$$= 48.1201$$

The average rate of change of the volume is 48.1201 cubic inches per inch as x changes from 4 to 4.01 inches.

b.
$$f'(x) = \lim_{h \to 0} \frac{f(x+h) - f(x)}{h}$$
$$= \lim_{h \to 0} \frac{(x+h)^3 - x^3}{h}$$
$$= \lim_{h \to 0} \frac{x^3 + 3hx^2 + 3h^2x + h^3 - x^3}{h}$$
$$= \lim_{h \to 0} \frac{3hx^2 + 3h^2x + h^3}{h}$$
$$= \lim_{h \to 0} \frac{h(3x^2 + 3hx + h^2)}{h}$$
$$= \lim_{h \to 0} (3x^2 + 3hx + h^2)$$
$$= 3x^2 + 3 \cdot 0 \cdot x + 0^2$$
$$= 3x^2$$

$$f'(4) = 3 \cdot 4^2 = 48$$

The instantaneous rate of change of the volume with respect to x at the moment when $x = 4$ inches is 48 cubic inches per inch.

5. $s(t) = -16t^2 + 96t$

a.
$$s'(a) = \lim_{h \to 0} \frac{s(a+h) - s(a)}{h} = \lim_{h \to 0} \frac{-16(a+h)^2 + 96(a+h) - (-16a^2 + 96a)}{h}$$
$$= \lim_{h \to 0} \frac{-16a^2 - 32ah - 16h^2 + 96a + 96h + 16a^2 - 96a}{h} = \lim_{h \to 0} \frac{-16h^2 - 32ah + 96h}{h}$$
$$= \lim_{h \to 0} (-16h - 32a + 96) = -32a + 96$$

$$s'(4) = -32 \cdot 4 + 96 = -32$$

After 4 seconds, the ball's instantaneous velocity is −32 feet per second.

b. Set $s(t) = 0$.

$$-16t^2 + 96t = 0$$

$$t^2 - 6t = 0$$

$$t(t-6) = 0$$

$$t = 0 \quad t - 6 = 0$$

$$t = 6$$

The ball hits the ground at $t = 6$ seconds.

$$s'(6) = -32 \cdot 6 + 96 = -96$$

The instantaneous velocity of the ball when it hits the ground is -96 feet per second.

Exercise Set 11.4

1. $f(x) = 2x + 3$ at $(1, 5)$

a.. $m_{\tan} = \lim\limits_{h \to 0} \dfrac{f(1+h) - f(1)}{h} = \lim\limits_{h \to 0} \dfrac{2(1+h) + 3 - [2 \cdot 1 + 3]}{h} = \lim\limits_{h \to 0} \dfrac{2 + 2h + 3 - 5}{h} = \lim\limits_{h \to 0} \dfrac{2h}{h} = \lim\limits_{h \to 0} 2 = 2$

b. $y - y_1 = m(x - x_1)$

$$y - 5 = 2(x - 1)$$

$$y - 5 = 2x - 2$$

$$y = 2x + 3$$

3. $f(x) = x^2 + 4$ at $(-1, 5)$

a. $m_{\tan} = \lim\limits_{h \to 0} \dfrac{f(-1+h) - f(-1)}{h} = \lim\limits_{h \to 0} \dfrac{(-1+h)^2 + 4 - [(-1)^2 + 4]}{h}$

$\qquad = \lim\limits_{h \to 0} \dfrac{1 - 2h + h^2 + 4 - 5}{h} = \lim\limits_{h \to 0} \dfrac{-2h + h^2}{h} = \lim\limits_{h \to 0}(h - 2) = 0 - 2 = -2$

b. $y - y_1 = m(x - x_1)$

$$y - 5 = -2(x + 1)$$

$$y - 5 = -2x - 2$$

$$y = -2x + 3$$

5. $f(x) = 5x^2$ at $(-2, 20)$

a. $m_{\tan} = \lim\limits_{h \to 0} \dfrac{f(-2+h) - f(-2)}{h} = \lim\limits_{h \to 0} \dfrac{5(-2+h)^2 - 5(-2)^2}{h} = \lim\limits_{h \to 0} \dfrac{5(4 - 4h + h^2) - 5 \cdot 4}{h} = \lim\limits_{h \to 0}(5h - 20) = -20$

b. $y - y_1 = m(x - x_1)$

$$y - 20 = -20(x + 2)$$

$$y - 20 = -20x - 40$$

$$y = -20x - 20$$

7. $f(x) = 2x^2 - x$ at $(2, 6)$

a. $m_{\tan} = \lim\limits_{h \to 0} \dfrac{f(2+h) - f(2)}{h} = \lim\limits_{h \to 0} \dfrac{2(2+h)^2 - (2+h) - [2 \cdot 2^2 - 2]}{h} = \lim\limits_{h \to 0} \dfrac{2(4 + 4h + h^2) - 2 - h - 6}{h}$

$= \lim\limits_{h \to 0} \dfrac{8 + 8h + 2h^2 - 2 - h - 6}{h} = \lim\limits_{h \to 0} \dfrac{7h + 2h^2}{h} = \lim\limits_{h \to 0} (2h + 7) = 2 \cdot 0 + 7 = 7$

b. $y - y_1 = m(x - x_1)$

$y - 6 = 7(x - 2)$

$y - 6 = 7x - 14$

$y = 7x - 8$

9. $f(x) = 2x^2 + x - 3$ at $(0, -3)$

a. $m_{\tan} = \lim\limits_{h \to 0} \dfrac{f(0+h) - f(0)}{h} = \lim\limits_{h \to 0} \dfrac{f(h) - f(0)}{h} = \lim\limits_{h \to 0} \dfrac{2h^2 + h - 3 - (0 + 0 - 3)}{h}$

$= \lim\limits_{h \to 0} \dfrac{2h^2 + h}{h} = \lim\limits_{h \to 0} (2h + 1) = 2 \cdot 0 + 1 = 1$

b. $y - y_1 = m(x - x_1)$

$y + 3 = 1(x - 0)$

$y + 3 = x - 0$

$y = x - 3$

11. $f(x) = \sqrt{x}$ at $(9, 3)$

a. $m_{\tan} = \lim\limits_{h \to 0} \dfrac{f(9+h) - f(9)}{h} = \lim\limits_{h \to 0} \dfrac{\sqrt{9+h} - \sqrt{9}}{h} = \lim\limits_{h \to 0} \dfrac{\sqrt{9+h} - \sqrt{9}}{h} \cdot \dfrac{\sqrt{9+h} + \sqrt{9}}{\sqrt{9+h} + \sqrt{9}}$

$= \lim\limits_{h \to 0} \dfrac{9 + h - 9}{h\left(\sqrt{9+h} + 3\right)} = \lim\limits_{h \to 0} \dfrac{1}{\sqrt{9+h} + 3} = \dfrac{1}{\sqrt{9+0} + 3} = \dfrac{1}{6}$

b. $y - y_1 = m(x - x_1)$

$y - 3 = \dfrac{1}{6}(x - 9)$

$y - 3 = \dfrac{1}{6}x - \dfrac{9}{6}$

$y = \dfrac{1}{6}x + \dfrac{3}{2}$

13. $f(x) = \dfrac{1}{x}$ at $(1, 1)$

     a. $m_{\tan} = \lim\limits_{h \to 0} \dfrac{f(1+h) - f(1)}{h} = \lim\limits_{h \to 0} \dfrac{\frac{1}{1+h} - \frac{1}{1}}{h} = \lim\limits_{h \to 0} \dfrac{\frac{1-(1+h)}{1+h}}{h} = \lim\limits_{h \to 0} \dfrac{1-1-h}{h(1+h)} = \lim\limits_{h \to 0} \dfrac{-1}{1+h} = \dfrac{-1}{1+0} = -1$

     b. $y - y_1 = m(x - x_1)$

           $y - 1 = -1(x - 1)$

           $y - 1 = -x + 1$

               $y = -x + 2$

15. $f(x) = -3x + 7$; $x = 1, x = 4$

     a. $f'(x) = \lim\limits_{h \to 0} \dfrac{f(x+h) - f(x)}{h} = \lim\limits_{h \to 0} \dfrac{-3(x+h) + 7 - (-3x+7)}{h} = \lim\limits_{h \to 0} \dfrac{-3x - 3h + 7 + 3x - 7}{h}$

               $= \lim\limits_{h \to 0} \dfrac{-3h}{h} = \lim\limits_{h \to 0}(-3) = -3$

     b. $f'(1) = -3$

           $f'(4) = -3$

17. $f(x) = x^2 - 6$; $x = -1, x = 3$

     a. $f'(x) = \lim\limits_{h \to 0} \dfrac{f(x+h) - f(x)}{h} = \lim\limits_{h \to 0} \dfrac{(x+h)^2 - 6 - (x^2 - 6)}{h} = \lim\limits_{h \to 0} \dfrac{x^2 + 2xh + h^2 - 6 - x^2 + 6}{h}$

               $= \lim\limits_{h \to 0} \dfrac{2xh + h^2}{h} = \lim\limits_{h \to 0}(2x + h) = 2x + 0 = 2x$

     b. $f'(-1) = 2(-1) = -2$

           $f'(3) = 2 \cdot 3 = 6$

19. $f(x) = x^2 - 3x + 5$; $x = \dfrac{3}{2}, x = 2$

     a. $f'(x) = \lim\limits_{h \to 0} \dfrac{f(x+h) - f(x)}{h} = \lim\limits_{h \to 0} \dfrac{(x+h)^2 - 3(x+h) + 5 - (x^2 - 3x + 5)}{h}$

               $= \lim\limits_{h \to 0} \dfrac{x^2 + 2xh + h^2 - 3x - 3h + 5 - x^2 + 3x - 5}{h} = \lim\limits_{h \to 0} \dfrac{2xh - 3h + h^2}{h}$

               $= \lim\limits_{h \to 0}(2x - 3 + h) = 2x - 3 + 0 = 2x - 3$

     b. $f'\left(\dfrac{3}{2}\right) = 2 \cdot \dfrac{3}{2} - 3 = 3 - 3 = 0$

           $f'(2) = 2 \cdot 2 - 3 = 4 - 3 = 1$

21. $f(x) = x^3 + 2$; $x = -1$, $x = 1$

a. $f'(x) = \lim\limits_{h \to 0} \dfrac{f(x+h) - f(x)}{h} = \lim\limits_{h \to 0} \dfrac{(x+h)^3 + 2 - (x^3 + 2)}{h} = \lim\limits_{h \to 0} \dfrac{x^3 + 3hx^2 + 3h^2x + h^3 + 2 - x^3 - 2}{h}$

$= \lim\limits_{h \to 0} \dfrac{3hx^2 + 3h^2x + h^3}{h} = \lim\limits_{h \to 0} (3x^2 + 3hx + h^2) = 3x^2 + 3 \cdot 0 \cdot x + 0^2 = 3x^2$

b. $f'(-1) = 3(-1)^2 = 3 \cdot 1 = 3$

$f'(1) = 3 \cdot 1^2 = 3 \cdot 1 = 3$

23. $f(x) = \sqrt{x}$; $x = 1$, $x = 4$

a. $f'(x) = \lim\limits_{h \to 0} \dfrac{f(x+h) - f(x)}{h} = \lim\limits_{h \to 0} \dfrac{\sqrt{x+h} - \sqrt{x}}{h} = \lim\limits_{h \to 0} \dfrac{\sqrt{x+h} - \sqrt{x}}{h} \cdot \dfrac{\sqrt{x+h} + \sqrt{x}}{\sqrt{x+h} + \sqrt{x}}$

$= \lim\limits_{h \to 0} \dfrac{x+h-x}{h\left(\sqrt{x+h} + \sqrt{x}\right)} = \lim\limits_{h \to 0} \dfrac{1}{\sqrt{x+h} + \sqrt{x}} = \dfrac{1}{\sqrt{x+0} + \sqrt{x}} = \dfrac{1}{2\sqrt{x}}$

b. $f'(1) = \dfrac{1}{2\sqrt{1}} = \dfrac{1}{2 \cdot 1} = \dfrac{1}{2}$

$f'(4) = \dfrac{1}{2\sqrt{4}} = \dfrac{1}{2 \cdot 2} = \dfrac{1}{4}$

25. $f(x) = \dfrac{4}{x}$; $x = -2$, $x = 1$

a. $f'(x) = \lim\limits_{h \to 0} \dfrac{f(x+h) - f(x)}{h} = \lim\limits_{h \to 0} \dfrac{\frac{4}{x+h} - \frac{4}{x}}{h} = \lim\limits_{h \to 0} \dfrac{\frac{4x - 4(x+h)}{x(x+h)}}{h} = \lim\limits_{h \to 0} \dfrac{4x - 4x - 4h}{hx(x+h)}$

$= \lim\limits_{h \to 0} \dfrac{-4}{x(x+h)} = \dfrac{-4}{x(x+0)} = -\dfrac{4}{x^2}$

b. $f'(-2) = -\dfrac{4}{(-2)^2} = -\dfrac{4}{4} = -1$

$f'(1) = -\dfrac{4}{1^2} = -\dfrac{4}{1} = -4$

27. $f(x) = 3.2x^2 + 2.1x$; $x = 0$, $x = 4$

a. $f'(x) = \lim\limits_{h \to 0} \dfrac{f(x+h) - f(x)}{h} = \lim\limits_{h \to 0} \dfrac{3.2(x+h)^2 + 2.1(x+h) - (3.2x^2 + 2.1x)}{h}$

$= \lim\limits_{h \to 0} \dfrac{3.2(x^2 + 2xh + h^2) + 2.1x + 2.1h - 3.2x^2 - 2.1x}{h}$

$= \lim\limits_{h \to 0} \dfrac{3.2x^2 + 6.4xh + 3.2h^2 + 2.1x + 2.1h - 3.2x^2 - 2.1x}{h} = \lim\limits_{h \to 0} \dfrac{6.4xh + 3.2h^2 + 2.1h}{h}$

$= \lim\limits_{h \to 0} (6.4x + 3.2h + 2.1) = 6.4x + 3.2 \cdot 0 + 2.1 = 6.4x + 2.1$

b. $f'(0) = 6.4 \cdot 0 + 2.1 = 0 + 2.1 = 2.1$

$f'(4) = 6.4 \cdot 4 + 2.1 = 25.6 + 2.1 = 27.7$

29. a. & c.

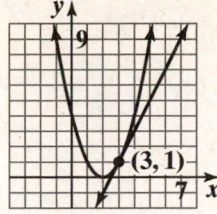

b. $f(3) = 1$, so the tangent line passes through $(3,1)$.

$$m_{\tan} = \lim_{h \to 0} \frac{f(3+h) - f(3)}{h} = \lim_{h \to 0}(h+2) = 2$$

$$y - y_1 = m(x - x_1)$$

$$y - 1 = 2(x - 3)$$

$$y - 1 = 2x - 6$$

$$y = 2x - 5$$

31. a. & c.

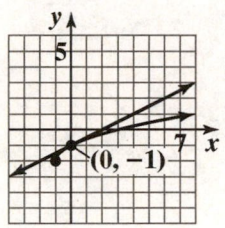

b. $f(0) = -1$, so the tangent line passes through $(0, -1)$.

$$m_{\tan} = \lim_{h \to 0} \frac{f(0+h) - f(0)}{h} = \lim_{h \to 0}\frac{\sqrt{h+1} - 1}{h} = \lim_{h \to 0}\frac{1}{\sqrt{h+1} + 1} = \frac{1}{2}$$

$$y - y_1 = m(x - x_1)$$

$$y - (-1) = \frac{1}{2}\left[x - 0\right]$$

$$y + 1 = \frac{1}{2}x$$

$$y = \frac{1}{2}x - 1$$

33. a. & c.

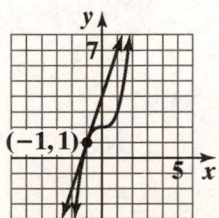

b. $f(-1) = 1$, so the tangent line passes through $(-1, 1)$.

$$m_{\tan} = \lim_{h \to 0} \frac{f(-1+h) - f(-1)}{h} = \lim_{h \to 0}\left(h^2 - 3h + 3\right) = 3$$

$$y - y_1 = m(x - x_1)$$

$$y - 1 = 3\left[x - (-1)\right]$$

$$y - 1 = 3x + 3$$

$$y = 3x + 4$$

35. a. & c.

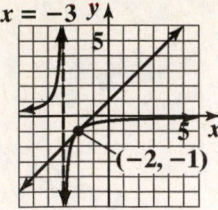

$x = -3$

$(-2, -1)$

b. $f(-2) = -1$, so the tangent line passes through $(-2, -1)$.

$$m_{\tan} = \lim_{h \to 0} \frac{f(-2+h) - f(-2)}{h} = \lim_{h \to 0} \frac{1}{h+1} = 1$$

$$y - y_1 = m(x - x_1)$$

$$y - (-1) = 1(x - (-2))$$

$$y + 1 = x + 2$$

$$y = x + 1$$

37. $f(x) = x^2$

a.

$$\frac{f(a+h) - f(a)}{h} = \frac{f(6+0.1) - f(6)}{0.1}$$

$$= \frac{f(6.1) - f(6)}{0.1}$$

$$= \frac{6.1^2 - 6^2}{0.1}$$

$$= 12.1$$

The average rate of change of the area is 12.1 square inches per inch as x changes from 6 to 6.1 inches.

$$\frac{f(a+h) - f(a)}{h} = \frac{f(6+0.01) - f(6)}{0.01}$$

$$= \frac{f(6.01) - f(6)}{0.01}$$

$$= \frac{6.01^2 - 6^2}{0.01}$$

$$= 12.01$$

The average rate of change of the area is 12.01 square inches per inch as x changes from 6 to 6.01 inches.

b.

$$f'(x) = \lim_{h \to 0} \frac{f(x+h) - f(x)}{h}$$

$$= \lim_{h \to 0} \frac{(x+h)^2 - x^2}{h}$$

$$= \lim_{h \to 0} \frac{x^2 + 2xh + h^2 - x^2}{h}$$

$$= \lim_{h \to 0} \frac{2xh + h^2}{h}$$

$$= \lim_{h \to 0} (2x + h)$$

$$= 2x + 0$$

$$= 2x$$

$$f'(6) = 2 \cdot 6 = 12$$

The instantaneous rate of change of the area with respect to x at the moment when $x = 6$ inches is 12 square inches per inch.

39. $f(x) = \pi x^2$

a. $\dfrac{f(a+h)-f(a)}{h} = \dfrac{f(2+0.1)-f(2)}{0.1}$

$\qquad\qquad\qquad\quad = \dfrac{f(2.1)-f(2)}{0.1}$

$\qquad\qquad\qquad\quad = \dfrac{\pi(2.1)^2 - \pi(2)^2}{0.1}$

$\qquad\qquad\qquad\quad = 4.1\pi$

The average rate of change of the area is 4.1π square inches per inch as x changes from 2 to 2.1 inches.

$\dfrac{f(a+h)-f(a)}{h} = \dfrac{f(2+0.01)-f(2)}{0.01}$

$\qquad\qquad\qquad\quad = \dfrac{f(2.01)-f(2)}{0.01}$

$\qquad\qquad\qquad\quad = \dfrac{\pi(2.01)^2 - \pi(2)^2}{0.01}$

$\qquad\qquad\qquad\quad = 4.01\pi$

The average rate of change of the area is 4.01π square inches per inch as x changes from 2 to 2.01 inches.

b. $f'(x) = \displaystyle\lim_{h \to 0} \dfrac{f(x+h)-f(x)}{h}$

$\qquad\quad = \displaystyle\lim_{h \to 0} \dfrac{\pi(x+h)^2 - \pi x^2}{h}$

$\qquad\quad = \displaystyle\lim_{h \to 0} \dfrac{\pi(x^2 + 2xh + h^2) - \pi x^2}{h}$

$\qquad\quad = \displaystyle\lim_{h \to 0} \dfrac{\pi x^2 + 2\pi xh + \pi h^2 + \pi x^2}{h}$

$\qquad\quad = \displaystyle\lim_{h \to 0} \dfrac{2\pi xh + \pi h^2}{h}$

$\qquad\quad = \displaystyle\lim_{h \to 0} (2\pi x + \pi h)$

$\qquad\quad = 2\pi x + \pi \cdot 0$

$\qquad\quad = 2\pi x$

$f'(2) = 2\pi \cdot 2 = 4\pi$

The instantaneous rate of change of the area with respect to x at the moment when $x = 2$ inches is 4π square inches per inch.

41. $f(x) = 4\pi x^2$

$$f'(x) = \lim_{h \to 0} \frac{f(x+h) - f(x)}{h}$$

$$= \lim_{h \to 0} \frac{4\pi(x+h)^2 - 4\pi x^2}{h}$$

$$= \lim_{h \to 0} \frac{4\pi(x^2 + 2xh + h^2) - 4\pi x^2}{h}$$

$$= \lim_{h \to 0} \frac{4\pi(x^2 + 2hx + h^2 - x^2)}{h}$$

$$= \lim_{h \to 0} \frac{4\pi(2hx + h^2)}{h}$$

$$= \lim_{h \to 0} 4\pi(2x + h)$$

$$= 4\pi(2x + 0)$$

$$= 8\pi x$$

$$f'(6) = 8\pi \cdot 6 = 48\pi$$

The instantaneous rate of change of the surface area with respect to x at the moment $x = 6$ inches is 48π square inches per inch.

43. $s(t) = -16t^2 + 64t$

a.
$$s'(t) = \lim_{h \to 0} \frac{s(t+h) - s(t)}{h}$$

$$= \lim_{h \to 0} \frac{-16(t+h)^2 + 64(t+h) - (-16t^2 + 64t)}{h}$$

$$= \lim_{h \to 0} \frac{-16(t^2 + 2ht + h^2) + 64t + 64b + 16t^2 - 64t}{h}$$

$$= \lim_{h \to 0} \frac{-16t^2 - 32ht - 16h^2 + 64t + 64h + 16t^2 - 64t}{h}$$

$$= \lim_{h \to 0} \frac{-32ht - 16h^2 + 64h}{h}$$

$$= \lim_{h \to 0} (-32t - 16h + 64)$$

$$= -32t - 16 \cdot 0 + 64$$

$$= -32t + 64$$

$$s'(1) = -32(1) + 64 = -32 + 64 = 32$$

The instantaneous velocity of the debris at 1 second is 32 feet per second.

$$s'(3) = -32(3) + 64 = -96 + 64 = -32$$

The instantaneous velocity of the debris at 3 seconds is –32 feet per second.

b. Set $s(t) = 0$.

$$-16t^2 + 64t = 0$$

$$t^2 - 4t = 0$$

$$t(t-4) = 0$$

$$t = 0 \quad t - 4 = 0$$

$$t = 4$$

$$s'(4) = -32(4) + 64 = -128 + 64 = -64$$

The instantaneous velocity of the debris just before it hits the ground is –64 feet per second.

45. $s(t) = -16t^2 + 96t + 4$

    a. $s'(t) = \lim_{h \to 0} \dfrac{s(t+h) - s(t)}{h}$

$$= \lim_{h \to 0} \frac{-16(t+h)^2 + 96(t+h) + 4 - (-16t^2 + 96t + 4)}{h}$$

$$= \lim_{h \to 0} \frac{-16(t^2 + 2ht + h^2) + 96t + 96h + 4 + 16t^2 - 96t - 4}{h}$$

$$= \lim_{h \to 0} \frac{-16t^2 - 32ht - 16h^2 + 96t + 96h + 4 + 16t^2 - 96t - 4}{h}$$

$$= \lim_{h \to 0} \frac{-32ht - 16h^2 + 96h}{h}$$

$$= \lim_{h \to 0} (-32t - 16h + 96)$$

$$= -32t - 16 \cdot 0 + 96$$

$$= -32t + 96$$

        $s'(2) = -32(2) + 96 = -64 + 96 = 32$

    The instantaneous velocity of the ball at 2 seconds is 32 feet per second.

        $s'(4) = -32(4) + 96 = -128 + 96 = -32$

    The instantaneous velocity of the ball at 4 seconds is –32 feet per second.

    b.   Set $s'(t) = 0$.

$$-32t + 96 = 0$$

$$32t = 96$$

$$t = 3$$

    The ball reaches its maximum height 3 seconds after it was hit.

        $s(3) = -16(3)^2 + 96(3) + 4 = -144 + 288 + 4 = 148$

    The maximum height obtained by the ball is 148 feet.

47. – 61.   Answers may vary.

63. $f(x) = \dfrac{x}{x-3}$

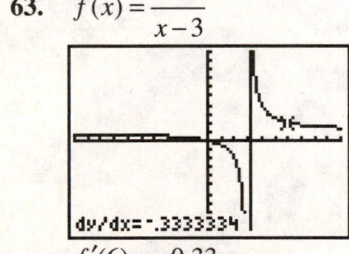

    $f'(6) \approx -0.33$

65. $f(x) = e^x \sin x$

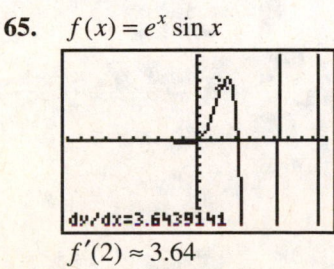

    $f'(2) \approx 3.64$

67. makes sense

69. makes sense

71. The slope of original graph is a negative constant. The derivative is represented by graph e.

73. The slope of original graph is a constantly increasing. The derivative is represented by graph d.

75. The slope of original graph is increasing then decreasing. The derivative is represented by graph b.

77. $A(r) = \pi r^2$

$$A'(r) = \lim_{h \to 0} \frac{\pi (r+h)^2 - \pi r^2}{h}$$

$$= \lim_{h \to o} \frac{\pi r^2 + 2\pi rh + \pi h^2 - \pi r^2}{h}$$

$$= \lim_{h \to 0} \frac{2\pi rh + \pi h^2}{h}$$

$$= \lim_{h \to 0} 2\pi r + \pi h$$

$$= 2\pi r$$

$2\pi r$ is the diameter of the circle

78. $f'(x) = \lim_{h \to 0} \dfrac{a(x+h)^2 + b(x+h) + c - (ax^2 + bx + c)}{h}$

$$= \lim_{h \to 0} \frac{ax^2 + 2axh + ah^2 + bx + bh + c - ax^2 - bx - c}{h}$$

$$= \lim_{h \to 0} \frac{2axh + ah^2 + bh}{h}$$

$$= \lim_{h \to 0} (2ax + ah^2 + bh) = 2ax + b$$

$$0 = 2ax + b$$

$$-b = 2ax$$

$$\frac{-b}{2a} = x$$

The derivative of the function at zero is $-\dfrac{b}{2a}$, which is the x-coordinate of the vertex.

79. Answers may vary.

Chapter 11 Review Exercises

1. $f(x) = \dfrac{x^3 - 1}{x - 1}$

x	0.99	0.999	0.9999 →←1.0001	1.001	1.01
f(x)	2.9701	2.9970	2.9997 →← 3.0003	3.0030	3.0301

$\lim\limits_{x \to 1} f(x) = 3$

2. $f(x) = \dfrac{\sqrt{x+1} - 1}{x}$

x	–0.01	–0.001	–0.0001 →← 0.0001	0.001	0.01
f(x)	0.50126	0.50013	0.50001 →← 0.49999	0.49988	0.49876

$\lim\limits_{x \to 0} f(x) = 0.5$

3. $f(x) = \dfrac{\sin 2x}{x}$

x	–0.1	–0.01	–0.001 →← 0.001	0.01	0.1
f(x)	1.9867	1.9999	2.0000 →← 2.0000	1.9999	1.9867

$\lim\limits_{x \to 0} f(x) = 2$

4. The graph shows that as x approaches –4, $f(x)$ approaches 0. Thus, $\lim\limits_{x \to -4} f(x) = 0$.

5. The graph shows that as x approaches –1, $f(x)$ approaches 1. Thus, $\lim\limits_{x \to -1} f(x) = 1$.

6. The graph shows that as x approaches 3, $f(x)$ approaches 3. Thus, $\lim\limits_{x \to 3} f(x) = 3$.

7. The graph of $f(x)$ at $x = -4$ is the point with coordinates (–4, 2). Thus, $f(-4) = 2$.

8. The graph of $f(x)$ at $x = 3$ is the point with coordinates (3, 3). Thus, $f(3) = 3$.

9. As x approaches –6 from the right-hand side $f(x)$ approaches 3, therefore, $\lim\limits_{x \to -6^+} f(x) = 3$.

10. As x approaches –4 from the left-hand side $f(x)$ approaches 5, therefore, $\lim\limits_{x \to -4^-} f(x) = 5$.

11. As x approaches –4 from the right-hand side $f(x)$ approaches 3, therefore, $\lim\limits_{x \to -4^+} f(x) = -3$.

12. $\lim\limits_{x \to -4^-} f(x) \neq \lim\limits_{x \to -4^+} f(x)$, therefore, $\lim\limits_{x \to -4} f(x)$ does not exist.

13. The graph of $f(x)$ at $x = -4$ is the point with coordinates (–4, 1). Thus, $f(-4) = 1$.

14. As x approaches –1 from the left-hand side $f(x)$ approaches 3, therefore, $\lim\limits_{x \to -1^+} f(x) = 3$.

15. As x approaches -1 from the left-hand side $f(x)$ approaches -3, therefore, $\lim\limits_{x \to -1^-} f(x) = -3$.

16. $\lim\limits_{x \to -1^-} f(x) \neq \lim\limits_{x \to -1^+} f(x)$, therefore, $\lim\limits_{x \to -1} f(x)$ does not exist.

17. The graph of $f(x)$ at $x = -1$ is the point with coordinates $(-1, 3)$. Thus, $f(-1) = 3$.

18. The graph has an open dot corresponding to $x = 2$, therefore, $f(2)$ does not exist.

19. As x approaches 2 from the left-hand side $f(x)$ approaches 5, therefore, $\lim\limits_{x \to 2^-} f(x) = 5$.

20. As x approaches 2 from the right-hand side $f(x)$ approaches 5, therefore, $\lim\limits_{x \to 2^+} f(x) = 5$.

21. $\lim\limits_{x \to 2^-} f(x) = \lim\limits_{x \to 2^+} f(x)$, therefore, $\lim\limits_{x \to 2} f(x)$ exists, and is 5.

22. As x gets closer to 5 the values of $f(x)$ get closer to 0. We conclude that $\lim\limits_{x \to 5} f(x) = 0$.

23. The graph of $f(x)$ at $x = 5$ is the point with coordinates $(5, 0)$. Thus, $f(5) = 0$.

24. $f(x) = \dfrac{x^2 - 9}{x - 3}$

Graph by first dividing out the factor $x - 3$. Then graph the equivalent function, $f(x) = x + 3$ where $x \neq 3$. Use the y-intercept 3, and the slope, 1. Because $x = 3$ is not included, show an open dot on the line corresponding to $x = 3$.

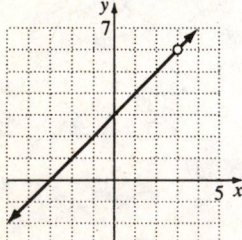

As x gets closer to 3, the values of $f(x)$ get closer to 6. We conclude that $\lim\limits_{x \to 3} f(x) = 6$.

25. $f(x) = \sin x$

Graph using the period $= 2\pi$, amplitude $= 1$, and no phase shift. Use the key points, $(0, 0)$, $\left(\dfrac{\pi}{2}, 1\right)$, $(\pi, 0)$, $\left(\dfrac{3\pi}{2}, -1\right)$, and $(2\pi, 0)$.

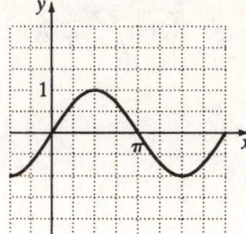

As x gets closer to $\dfrac{3\pi}{2}$, the values of $f(x)$ get closer to -1. We conclude that $\lim\limits_{x \to \frac{3\pi}{2}} f(x) = -1$.

26. $f(x) = \begin{cases} 1 - x & \text{if } x < 0 \\ \cos x & \text{if } x \geq 0 \end{cases}$

Graph $f(x) = 1 - x$ on the left side of $x = 0$ by using the y-intercept, 1, and the slope, -1. Graph $(x) = \cos x$ on the right side of $x = 0$ by using the key points $(0, 1)$, $\left(\dfrac{\pi}{2}, 0\right)$, $(\pi, -1)$, $\left(\dfrac{3\pi}{2}, 0\right)$, and $(2\pi, 1)$.

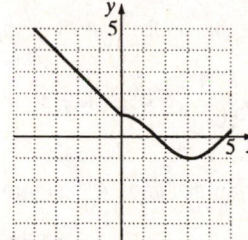

As x gets closer to 0, the values of $f(x)$ get closer to 1. We conclude that $\lim\limits_{x \to 0} f(x) = 1$.

27. $\lim\limits_{x \to 4} (2x^2 - 5x + 3) = 2(4)^2 - 5(4) + 3 = 15$

28. $\lim\limits_{x \to -1} (-2x^3 - x + 5) = -2(-1)^3 - (-1) + 5 = 8$

29. $\lim\limits_{x \to -3} (x^2 + 1)^3 = \left[\lim\limits_{x \to -3} (x^2 + 1) \right]^3$

$\qquad = \left[(-3)^2 + 1 \right]^3$

$\qquad = 10^3$

$\qquad = 1000$

30. $\lim\limits_{x\to 4}\sqrt{x^2+9}=\sqrt{\lim\limits_{x\to 4}(x^2+9)}$

$\qquad\qquad = \sqrt{4^2+9}$

$\qquad\qquad = \sqrt{25}$

$\qquad\qquad = 5$

31. $\lim\limits_{x\to 5}\dfrac{11x-3}{x^2+1}=\dfrac{\lim\limits_{x\to 5}(11x-3)}{\lim\limits_{x\to 5}(x^2+1)}$

$\qquad\qquad = \dfrac{11(5)-3}{5^2+1}$

$\qquad\qquad = \dfrac{52}{26}$

$\qquad\qquad = 2$

32. $\lim\limits_{x\to -4}\dfrac{x^2-16}{x+4}=\lim\limits_{x\to -4}\dfrac{(x+4)(x-4)}{x+4}$

$\qquad\qquad = \lim\limits_{x\to -4}(x-4)$

$\qquad\qquad = -4-4$

$\qquad\qquad = -8$

33. $\lim\limits_{x\to 7}\dfrac{5x-35}{x-7}=\lim\limits_{x\to 7}\dfrac{5(x-7)}{x-7}$

$\qquad\qquad = \lim\limits_{x\to 7}5$

$\qquad\qquad = 5$

34. $\lim\limits_{x\to 0}\dfrac{\sqrt{x+100}-10}{x}$

$\qquad = \lim\limits_{x\to 0}\dfrac{\sqrt{x+100}-10}{x}\cdot\dfrac{\sqrt{x+100}+10}{\sqrt{x+100}+10}$

$\qquad = \lim\limits_{x\to 0}\dfrac{x+100-100}{x\left(\sqrt{x+100}+10\right)}$

$\qquad = \lim\limits_{x\to 0}\dfrac{1}{\sqrt{x+100}+10}$

$\qquad = \dfrac{1}{\sqrt{0+100}+10}$

$\qquad = \dfrac{1}{20}$

35. $\lim\limits_{x\to -1}\dfrac{x^2-1}{x^2+x}=\lim\limits_{x\to -1}\dfrac{(x+1)(x-1)}{x(x+1)}$

$\qquad\qquad = \lim\limits_{x\to -1}\dfrac{x-1}{x}$

$\qquad\qquad = \dfrac{-1-1}{-1}$

$\qquad\qquad = 2$

36. $\lim\limits_{x\to 100}\dfrac{\sqrt{x}-10}{x-100}=\lim\limits_{x\to 100}\dfrac{\sqrt{x}-10}{x-100}\cdot\dfrac{\sqrt{x}+10}{\sqrt{x}+10}$

$\qquad\qquad = \lim\limits_{x\to 100}\dfrac{x-100}{(x-100)\left(\sqrt{x}+10\right)}$

$\qquad\qquad = \lim\limits_{x\to 100}\dfrac{1}{\sqrt{x}+10}$

$\qquad\qquad = \dfrac{1}{\sqrt{100}+10}$

$\qquad\qquad = \dfrac{1}{20}$

37. $\lim\limits_{x\to 0}\dfrac{\frac{1}{x+5}-\frac{1}{5}}{x}=\lim\limits_{x\to 0}\dfrac{\frac{5-(x+5)}{5(x+5)}}{x}$

$\qquad\qquad = \lim\limits_{x\to 0}\dfrac{5-x-5}{5x(x+5)}$

$\qquad\qquad = \lim\limits_{x\to 0}\dfrac{-x}{5x(x+5)}$

$\qquad\qquad = \lim\limits_{x\to 0}-\dfrac{1}{5(x+5)}$

$\qquad\qquad = -\dfrac{1}{5(0+5)}$

$\qquad\qquad = -\dfrac{1}{25}$

38. $f(x)=\begin{cases} x^2+1 \text{ if } x<2 \\ 3x+1 \text{ if } x\ge 2\end{cases}$

a. $\lim\limits_{x\to 2^-}f(x)=\lim\limits_{x\to 2^-}(x^2+1)=2^2+1=5$

b. $\lim\limits_{x\to 2^+}f(x)=\lim\limits_{x\to 2^+}(3x+1)=3(2)+1=7$

c. $\lim\limits_{x\to 2^-}f(x)\ne\lim\limits_{x\to 2^+}f(x)$, therefore,

$\lim\limits_{x\to 2}f(x)$ does not exist.

39. $f(x) = \begin{cases} \sqrt[3]{x^2+7} & \text{if } x < 1 \\ 4x & \text{if } x \geq 1 \end{cases}$

    a. $\lim\limits_{x \to 1^-} f(x) = \lim\limits_{x \to 1^-} \sqrt[3]{x^2+7}$

               $= \sqrt[3]{1^2+7}$

               $= \sqrt[3]{8}$

               $= 2$

    b. $\lim\limits_{x \to 1^+} f(x) = \lim\limits_{x \to 1^+} 4x = 4 \cdot 1 = 4$

    c. $\lim\limits_{x \to 1^-} f(x) \neq \lim\limits_{x \to 1^+} f(x)$, therefore,

         $\lim\limits_{x \to 1} f(x)$ does not exist.

40. $f(x) = \begin{cases} \frac{x^2-25}{x+5} & \text{if } x \neq -5 \\ 13 & \text{if } x = -5 \end{cases}$

    a. $\lim\limits_{x \to -5^-} f(x) = \lim\limits_{x \to -5^-} \frac{x^2-25}{x+5}$

            $= \lim\limits_{x \to -5^-} \frac{(x+5)(x-5)}{x+5}$

            $= \lim\limits_{x \to -5^-} (x-5)$

            $= -5-5$

            $= -10$

    b. $\lim\limits_{x \to -5^+} f(x) = \lim\limits_{x \to -5^+} \frac{x^2-25}{x+5}$

         $= \lim\limits_{x \to -5^+} \frac{(x+5)(x-5)}{x+5}$

         $= \lim\limits_{x \to -5^+} (x-5)$

         $= -5-5$

         $= -10$

    c. $\lim\limits_{x \to -5^-} f(x) = \lim\limits_{x \to -5^+} f(x)$, therefore,

         $\lim\limits_{x \to -5} f(x)$ exists, and is -10.

41. $f(x) = 3x^2 - 2x + 1$

     $f(4) = 3(4)^2 - 2(4) + 1 = 41$

     $\lim\limits_{x \to 4}(3x^2 - 2x + 1) = 3 \cdot 4^2 - 2 \cdot 4 + 1 = 41$

     $\lim\limits_{x \to 4} f(x) = f(4)$, therefore, f is continuous at 4.

42. $f(x) = \frac{x^2-9}{x+3}$

     $f(x)$ is undefined at $x = -3$, therefore,
     f is discontinuous at -3.

43. $f(x) = \begin{cases} \frac{x^2+5x}{x^2-5x} & \text{if } x \neq 0 \\ -2 & \text{if } x = 0 \end{cases}$

     $f(0) = -2$

     f is defined at 0.

     $\lim\limits_{x \to 0} f(x) = \lim\limits_{x \to 0} \frac{x^2+5x}{x^2-5x}$

            $= \lim\limits_{x \to 0} \frac{x(x+5)}{x(x-5)}$

            $= \lim\limits_{x \to 0} \frac{x+5}{x-5}$

            $= \frac{0+5}{0-5}$

            $= -1$

     $\lim\limits_{x \to 0} f(x) = -1 \neq f(0)$, therefore, f is discontinuous at 0.

44. $f(x) = \begin{cases} \frac{x^2+x}{x^2-3x-4} & \text{if } x \neq -1 \\ \frac{1}{5} & \text{if } x = -1 \end{cases}$

     $f(-1) = \frac{1}{5}$

     f is defined at -1.

     $\lim\limits_{x \to -1} f(x) = \lim\limits_{x \to -1} \frac{x^2+x}{x^2-3x-4}$

            $= \lim\limits_{x \to -1} \frac{x(x+1)}{(x+1)(x-4)}$

            $= \lim\limits_{x \to -1} \frac{x}{x-4}$

            $= \frac{-1}{-1-4}$

            $= \frac{1}{5}$

     $\lim\limits_{x \to -1} f(x) = \frac{1}{5} = f(-1)$, therefore, f is continuous at -1.

45. $f(x) = \begin{cases} 3x & \text{if } x < 2 \\ 5 & \text{if } x = 2 \\ x+4 & \text{if } x > 2 \end{cases}$

$f(2) = 5$

f is defined at 2.

$\lim\limits_{x \to 2^-} f(x) = \lim\limits_{x \to 2^-} 3x = 3 \cdot 2 = 6$

$\lim\limits_{x \to 2^+} f(x) = \lim\limits_{x \to 2^+} (x+4) = 2+4 = 6$

$\lim\limits_{x \to 2^-} f(x) = \lim\limits_{x \to 2^+} f(x)$, therefore, $\lim\limits_{x \to 2} f(x)$ exists and is 6.

$\lim\limits_{x \to 2} f(x) \neq f(2)$, therefore, f is discontinuous at 2.

46. $f(x) = x^3 + 5x^2 - 1$ is a polynomial function and, therefore, continuous for every number x.

47. $f(x) = \dfrac{x-1}{(x-1)(x+3)}$ is undefined at $x = 1$ and $x = -3$. So, f is discontinuous at 1 and –3.

48. $f(x) = \begin{cases} -1 & \text{if } x < 0 \\ 1 & \text{if } x \geq 0 \end{cases}$

–1 and 1 are continuous for every number x. We must investigate continuity at $x = 0$.

$f(0) = 1$

f is defined at 0.

$\lim\limits_{x \to 0^-} f(x) = \lim\limits_{x \to 0^-} (-1) = -1$

$\lim\limits_{x \to 0^+} f(x) = \lim\limits_{x \to 0^+} 1 = 1$

The left-and right-hand limits are not equal, so $\lim\limits_{x \to 0} f(x)$ does not exist. Therefore, f is discontinuous at 0.

49. $f(x) = \begin{cases} 4x & \text{if } x < 5 \\ x^2 - 5 & \text{if } x \geq 5 \end{cases}$

f is continuous for every number x except possibly where f changes its functional form at $x = 5$. We must investigate continuity at $x = 5$.

$f(5) = 5^2 - 5 = 20$

f is defined at 5.

$\lim\limits_{x \to 5^-} f(x) = \lim\limits_{x \to 5^-} 4x = 4 \cdot 5 = 20$

$\lim\limits_{x \to 5^+} f(x) = \lim\limits_{x \to 5^+} (x^2 - 5) = 5^2 - 5 = 20$

$\lim\limits_{x \to 5^-} f(x) = \lim\limits_{x \to 5^+} f(x)$, therefore, $\lim\limits_{x \to 5} f(x)$ exists and is 20.

$\lim\limits_{x \to 5} f(x) = f(5)$, therefore, f is continuous for every number x.

50. $f(x) = \begin{cases} \frac{x^2-4}{x+2} & \text{if } x \neq -2 \\ 4 & \text{if } x = -2 \end{cases}$

f is continuous for every number x except possibly at $x = -2$. We must investigate continuity at $x = -2$.

$f(-2) = 4$

f is defined at -2.

$\lim\limits_{x \to -2} f(x) = \lim\limits_{x \to -2} \frac{x^2-4}{x+2} = \lim\limits_{x \to -2} \frac{(x+2)(x-2)}{x+2} = \lim\limits_{x \to -2} (x-2) = -2-2 = -4$

$\lim\limits_{x \to -2} f(x) = -4 \neq f(-2)$, therefore, f is discontinuous at -2.

51. $f(x) = \begin{cases} \dfrac{x^2-121}{x-11} & \text{if } x \neq 11 \\ 20 & \text{if } x = 11 \end{cases}$

f is continuous for every number x except possibly at $x = 11$. We must investigate continuity at $x = 11$.

$f(11) = 20$

f is defined at 11.

$\lim\limits_{x \to 11} f(x) = \lim\limits_{x \to 11} \frac{x^2-121}{x-11} = \lim\limits_{x \to 11} \frac{(x+11)(x-11)}{x-11} = \lim\limits_{x \to 11} (x+11) = 11+11 = 22$

$\lim\limits_{x \to 11} f(x) = 22 \neq f(11)$, therefore, f is discontinuous at 11.

52. $f(x) = 2x^2 + 5x$ at $(1, 7)$

a. $m_{\tan} = \lim\limits_{h \to 0} \dfrac{f(a+h)-f(a)}{h} = \lim\limits_{h \to 0} \dfrac{f(1+h)-f(1)}{h} = \lim\limits_{h \to 0} \dfrac{2(1+h)^2 + 5(1+h) - [2(1)^2 + 5(1)]}{h}$

$= \lim\limits_{h \to 0} \dfrac{2 + 4h + 2h^2 + 5 + 5h - 2 - 5}{h} = \lim\limits_{h \to 0} \dfrac{9h + 2h^2}{h} = \lim\limits_{h \to 0} (9 + 2h) = 9 + 2 \cdot 0 = 9$

b. $y - y_1 = m(x - x_1)$

$y - 7 = 9(x - 1)$

$y - 7 = 9x - 9$

$y = 9x - 2$

53. $f(x) = x^2 - 7x - 4$ at $(-1, 4)$

a. $m_{\tan} = \lim\limits_{h \to 0} \dfrac{f(a+h)-f(a)}{h} = \lim\limits_{h \to 0} \dfrac{f(-1+h)-f(-1)}{h} = \lim\limits_{h \to 0} \dfrac{(-1+h)^2 - 7(-1+h) - 4 - \left[(-1)^2 - 7(-1) - 4\right]}{h}$

$= \lim\limits_{h \to 0} \dfrac{1 - 2h + h^2 + 7 - 7h - 4 - 1 - 7 + 4}{h} = \lim\limits_{h \to 0} \dfrac{-9h + h^2}{h} = \lim\limits_{h \to 0} (-9 + h) = -9 + 0 = -9$

b. $y - y_1 = m(x - x_1)$

$y - 4 = -9(x + 1)$

$y - 4 = -9x - 9$

$y = -9x - 5$

54. $f(x) = 3x^2 + 12x - 1; \ x = -2, x = 1$

    a. $f'(x) = \lim\limits_{h \to 0} \dfrac{f(x+h) - f(x)}{h} = \lim\limits_{h \to 0} \dfrac{3(x+h)^2 + 12(x+h) - 1 - (3x^2 + 12x - 1)}{h}$

        $= \lim\limits_{h \to 0} \dfrac{3x^2 + 6hx + 3h^2 + 12x + 12h - 1 - 3x^2 - 12x + 1}{h} = \lim\limits_{h \to 0} \dfrac{6hx + 3h^2 + 12h}{h}$

        $= \lim\limits_{h \to 0} (6x + 3h + 12) = 6x + 3 \cdot 0 + 12 = 6x + 12$

    b. $f'(-2) = 6(-2) + 12 = 0$

        $f'(1) = 6(1) + 12 = 18$

55. $f(x) = 2x^3 - x; \ x = -1, x = 1$

    a. $f'(x) = \lim\limits_{h \to 0} \dfrac{f(x+h) - f(x)}{h} = \lim\limits_{h \to 0} \dfrac{2(x+h)^3 - (x+h) - (2x^3 - x)}{h}$

        $= \lim\limits_{h \to 0} \dfrac{2x^3 + 6hx^2 + 6h^2 x + 2h^3 - x - h - 2x^3 + x}{h} = \lim\limits_{h \to 0} \dfrac{6hx^2 + 6h^2 x + 2h^3 - h}{h}$

        $= \lim\limits_{h \to 0} (6x^2 + 6hx + 2h^2 - 1) = 6x^2 + 6 \cdot 0 \cdot x + 2 \cdot 0^2 - 1 = 6x^2 - 1$

    b. $f'(-1) = 6(-1)^2 - 1 = 5$

        $f'(1) = 6(1)^2 - 1 = 5$

56. $f(x) = \dfrac{1}{x}; \ x = -2, x = 2$

    a. $f'(x) = \lim\limits_{h \to 0} \dfrac{f(x+h) - f(x)}{h} = \lim\limits_{h \to 0} \dfrac{\frac{1}{x+h} - \frac{1}{x}}{h} = \lim\limits_{h \to 0} \dfrac{\frac{x-(x+h)}{x(x+h)}}{h} = \lim\limits_{h \to 0} \dfrac{x - x - h}{xh(x+h)} = \lim\limits_{h \to 0} \dfrac{-1}{x(x+h)}$

        $= \dfrac{-1}{x(x+0)} = -\dfrac{1}{x^2}$

    b. $f'(-2) = -\dfrac{1}{(-2)^2} = -\dfrac{1}{4}$

        $f'(2) = -\dfrac{1}{2^2} = -\dfrac{1}{4}$

57. $f(x) = \sqrt{x}; \ x = 36, x = 81$

    a. $f'(x) = \lim\limits_{h \to 0} \dfrac{f(x+h) - f(x)}{h} = \lim\limits_{h \to 0} \dfrac{\sqrt{x+h} - \sqrt{x}}{h} = \lim\limits_{h \to 0} \dfrac{\sqrt{x+h} - \sqrt{x}}{h} \cdot \dfrac{\sqrt{x+h} + \sqrt{x}}{\sqrt{x+h} + \sqrt{x}}$

        $= \lim\limits_{h \to 0} \dfrac{x + h - x}{h\left(\sqrt{x+h} + \sqrt{x}\right)} = \lim\limits_{h \to 0} \dfrac{1}{\sqrt{x+h} + \sqrt{x}} = \dfrac{1}{\sqrt{x+0} + \sqrt{x}} = \dfrac{1}{2\sqrt{x}}$

    b. $f'(36) = \dfrac{1}{2\sqrt{36}} = \dfrac{1}{12}$

        $f'(81) = \dfrac{1}{2\sqrt{81}} = \dfrac{1}{18}$

58. $f(x) = 5x^2$

a. $\dfrac{f(a+h)-f(a)}{h} = \dfrac{f(2+0.1)-f(2)}{0.1} = \dfrac{f(2.1)-f(2)}{0.1} = \dfrac{5(2.1)^2 - 5(2)^2}{0.1} = 20.5$

The average rate of change of the volume with respect to x as x changes from 2 to 2.1 inches is 20.5 cubic inches per inch.

$\dfrac{f(a+h)-f(a)}{h} = \dfrac{f(2+0.01)-f(2)}{0.01} = \dfrac{f(2.01)-f(2)}{0.01} = \dfrac{5(2.01)^2 - 5(2)^2}{0.01} = 20.05$

The average rate of change of the volume with respect to x as x changes from 2 to 2.01 inches is 20.05 cubic inches per inch.

b. $f'(x) = \lim\limits_{h \to 0} \dfrac{f(x+h)-f(x)}{h} = \lim\limits_{h \to 0} \dfrac{5(x+h)^2 - 5x^2}{h} = \lim\limits_{h \to 0} \dfrac{5x^2 + 10hx + 5h^2 - 5x^2}{h}$

$= \lim\limits_{h \to 0} \dfrac{10hx + 5h^2}{h} = \lim\limits_{h \to 0}(10x + 5h) = 10x + 5 \cdot 0 = 10x$

$f'(2) = 10(2) = 20$

The instantaneous rate of change of the volume with respect to x at $x = 2$ inches is 20 cubic inches per inch.

59. $f(x) = \dfrac{4}{3}\pi x^3$

$f'(x) = \lim\limits_{h \to 0} \dfrac{f(x+h)-f(x)}{h} = \lim\limits_{h \to 0} \dfrac{\frac{4}{3}\pi(x+h)^3 - \frac{4}{3}\pi x^3}{h} = \lim\limits_{h \to 0} \dfrac{\frac{4}{3}\pi(x^3 + 3hx^2 + 3h^2 x + h^3 - x^3)}{h}$

$= \lim\limits_{h \to 0} \dfrac{\frac{4}{3}\pi(3hx^2 + 3h^2 x + h^3)}{h} = \lim\limits_{h \to 0} \dfrac{4}{3}\pi(3x^2 + 3hx + h^2) = \dfrac{4}{3}\pi(3x^2 + 3 \cdot 0 \cdot x + 0^2) = 4\pi x^2$

$f'(5) = 4\pi(5)^2 = 100\pi$

The instantaneous rate of change of the volume with respect to x at $x = 5$ inches is 100π cubic inches per inch.

60. $s(t) = -16t^2 + 80t + 5$

a. $s'(t) = \lim\limits_{h \to 0} \dfrac{s(t+h)-s(t)}{h} = \lim\limits_{h \to 0} \dfrac{-16(t+h)^2 + 80(t+h) + 5 - (-16t^2 + 80t + 5)}{h}$

$= \lim\limits_{h \to 0} \dfrac{-16t^2 - 32ht - 16h^2 + 80t + 80h + 5 + 16t^2 - 80t - 5}{h} = \lim\limits_{h \to 0} \dfrac{-32ht - 16h^2 + 80h}{h}$

$= \lim\limits_{h \to 0}(-32t - 16h + 80) = -32t - 16 \cdot 0 + 80 = -32t + 80$

$s'(2) = -32(2) + 80 = 16$

The instantaneous velocity of the ball 2 seconds after it is thrown is 16 feet per second.

$s'(4) = -32(4) + 80 = -48$

The instantaneous velocity of the ball 4 seconds after it is thrown is –48 feet per second.

b. Set $s'(t) = 0$.

$-32t + 80 = 0$

$32t = 80$

$t = 2.5$

The ball reaches its maximum height 2.5 seconds after it is thrown.

$s(2.5) = -16(2.5)^2 + 80(2.5) + 5 = 105$

The maximum height of the ball is 105 feet.

Chapter 11 Test

1. $f(x) = \dfrac{9-x}{3-\sqrt{x}}$

x	8.9	8.99	8.999 $\rightarrow \leftarrow$ 9.001	9.01	9.1
$f(x)$	5.9833	5.9983	5.9998 $\rightarrow \leftarrow$ 6.0002	6.0017	6.0166

$\lim\limits_{x \to 9} f(x) = 6$

2. The graph shows that as x approaches -2, $f(x)$ approaches -3. Thus, $\lim\limits_{x \to -2} f(x) = -3$.

3. The graph of $f(x)$ at $x = -2$ is the point with coordinates $(-2, -5)$. Thus, $f(-2) = -5$.

4. As x approaches 2 from the left-hand side $f(x)$ approaches 4, therefore, $\lim\limits_{x \to 2^-} f(x) = 4$.

5. As x approaches 2 from the right-hand side $f(x)$ approaches 6, therefore, $\lim\limits_{x \to 2^+} f(x) = 6$.

6. $\lim\limits_{x \to 2^-} f(x) \neq \lim\limits_{x \to 2^+} f(x)$, therefore, $\lim\limits_{x \to 2} f(x)$ does not exist.

7. The graph shows that as x approaches 4, $f(x)$ approaches 4. Thus, $\lim\limits_{x \to 4} f(x) = 4$.

8. $\lim\limits_{x \to -2} (x^2 + x + 1)^4 = \left[\lim\limits_{x \to -2} (x^2 + x + 1) \right]^4 = \left[(-2)^2 + (-2) + 1 \right]^4 = 81$

9. $\lim\limits_{x \to -1} \dfrac{x^2 - x - 2}{x + 1} = \lim\limits_{x \to -1} \dfrac{(x+1)(x-2)}{x+1} = \lim\limits_{x \to -1} (x - 2) = -1 - 2 = -3$

10. $\lim\limits_{x \to 9} \dfrac{\sqrt{x} - 3}{x - 9} = \lim\limits_{x \to 9} \dfrac{\sqrt{x} - 3}{x - 9} \cdot \dfrac{\sqrt{x} + 3}{\sqrt{x} + 3} = \lim\limits_{x \to 9} \dfrac{x - 9}{(x-9)\left(\sqrt{x} + 3\right)} = \lim\limits_{x \to 9} \dfrac{1}{\sqrt{x} + 3} = \dfrac{1}{\sqrt{9} + 3} = \dfrac{1}{6}$

11. $f(x) = \begin{cases} \dfrac{x^2 - 1}{x + 1} & \text{if } x \neq -1 \\ 6 & \text{if } x = -1 \end{cases}$

    $f(-1) = 6$

    f is defined at -1.

    $\lim\limits_{x \to -1} f(x) = \lim\limits_{x \to -1} \dfrac{x^2 - 1}{x + 1} = \lim\limits_{x \to -1} \dfrac{(x+1)(x-1)}{x+1} = \lim\limits_{x \to -1} (x - 1) = -1 - 1 = -2$

    $\lim\limits_{x \to -1} f(x) = -2 \neq f(-1)$,

    therefore, f is discontinuous at -1.

12. $f(x) = \begin{cases} 2-x & \text{if } x \le 2 \\ x^2 - 2x & \text{if } x > 2 \end{cases}$

$f(2) = 2 - 2 = 0$

f is defined at 2.

$\lim\limits_{x \to 2^-} f(x) = \lim\limits_{x \to 2^-} (2-x) = 2 - 2 = 0$

$\lim\limits_{x \to 2^+} f(x) = \lim\limits_{x \to 2^+} (x^2 - 2x) = 2^2 - 2(2) = 0$

$\lim\limits_{x \to 2^-} f(x) = \lim\limits_{x \to 2^+} f(x)$, therefore, $\lim\limits_{x \to 2} f(x)$ exists and is 0.

$\lim\limits_{x \to 2} f(x) = f(2)$, therefore, f is continuous at 2.

13. $f(x) + x^2 - 5x + 1$

$f'(x) = \lim\limits_{h \to 0} \dfrac{f(x+h) - f(x)}{h} = \lim\limits_{h \to 0} \dfrac{(x+h)^2 - 5(x+h) + 1 - (x^2 - 5x + 1)}{h}$

$= \lim\limits_{h \to 0} \dfrac{x^2 + 2hx + h^2 - 5x - 5h + 1 - x^2 + 5x - 1}{h} = \lim\limits_{h \to 0} \dfrac{2hx + h^2 - 5h}{h}$

$= \lim\limits_{h \to 0} (2x + h - 5) = 2x + 0 - 5 = 2x - 5$

14. $f(x) = \dfrac{10}{x}$

$f'(x) = \lim\limits_{h \to 0} \dfrac{f(x+h) - f(x)}{h} = \lim\limits_{h \to 0} \dfrac{\frac{10}{x+h} - \frac{10}{x}}{h} = \lim\limits_{h \to 0} \dfrac{\frac{10x - 10(x+h)}{x(x+h)}}{h} = \lim\limits_{h \to 0} \dfrac{10x - 10x - 10h}{xh(x+h)} = \lim\limits_{h \to 0} \dfrac{-10}{x(x+h)}$

$= \dfrac{-10}{x(x+0)} = -\dfrac{10}{x^2}$

15. $f(x) = x^2$ at $(-3, 9)$

$m_{\tan} = \lim\limits_{h \to 0} \dfrac{f(x+h) - f(x)}{h} = \lim\limits_{h \to 0} \dfrac{(x+h)^2 - x^2}{h} = \lim\limits_{h \to 0} \dfrac{x^2 + 2hx + h^2 - x^2}{h} = \lim\limits_{h \to 0} \dfrac{2hx + h^2}{h}$

$= \lim\limits_{h \to 0} (2x + h) = 2x + 0 = 2x$

$y - y_1 = m(x - x_1)$

$y - 9 = 2(-3)(x + 3)$

$y - 9 = -6x - 18$

$y = -6x - 9$

16. $s(t) = -16t^2 + 72t$

$s'(t) = \lim\limits_{h \to 0} \dfrac{s(t+h) - s(t)}{h} = \lim\limits_{h \to 0} \dfrac{-16(t+h)^2 + 72(t+h) - (-16t^2 + 72t)}{h}$

$= \lim\limits_{h \to 0} \dfrac{-16t^2 - 32ht - 16h^2 + 72t + 72h + 16t^2 - 72t}{h} = \lim\limits_{h \to 0} \dfrac{-32ht - 16h^2 + 72h}{h}$

$= \lim\limits_{h \to 0} (-32t - 16h + 72) = -32t - 16 \cdot 0 + 72 = -32t + 72$

$s'(3) = -32(3) + 72 = -24$

The instantaneous velocity of the ball 3 seconds after it was thrown is –24 feet per second.

Cumulative Review Exercises (Chapters P–11)

1.
$$\frac{1}{x+2} > \frac{3}{x+1}$$

$$\frac{1}{x+2} - \frac{3}{x+1} > 0$$

$$\frac{x+1-3(x+2)}{(x+2)(x+1)} > 0$$

$$\frac{-2x-5}{(x+2)(x+1)} > 0$$

Find the boundary points.

$-2x-5=0 \qquad x+2=0$

$\qquad x = -\frac{5}{2} \qquad x = -2$

$x+1=0$

$\qquad x = -1$

Interval	Number	Substitute	Conclusion
$\left(-\infty, -\frac{5}{2}\right]$	-3	$\frac{-2(-3)-5}{(-3+2)(-3+1)} > 0$	true
$\left[-\frac{5}{2}, -2\right)$	$-\frac{9}{4}$	$\frac{-2\left(-\frac{9}{4}\right)-5}{\left(-\frac{9}{4}+2\right)\left(-\frac{9}{4}+1\right)} > 0$	false
$(-2, -1)$	$-\frac{3}{2}$	$\frac{-2\left(-\frac{3}{2}\right)-5}{\left(-\frac{3}{2}+2\right)\left(-\frac{3}{2}+1\right)} > 0$	true
$(-1, \infty)$	0	$\frac{-2(0)-5}{(0+2)(0+1)} > 0$	false

The solution set is $\left\{ x \,\middle|\, x \le -\frac{5}{2} \text{ or } -2 < x < -1 \right\}$.

2. $2x^3 + 11x^2 - 7x - 6 = 0$

Use the Rational Zero Theorem.

The constant term is –6, which has factors: ± 1, ± 2, ± 3, and ± 6. The leading coefficient is 2, which has factors: ± 1 and ± 2.

$$\text{Possible rational zeros} = \frac{\text{Factors of the constant term}}{\text{Factors of the leading coefficient}} = \frac{\pm 1, \pm 2, \pm 3, \pm 6}{\pm 1, \pm 2} = \pm 1, \pm 2, \pm 3, \pm 6, \pm \frac{1}{2}, \pm \frac{3}{2}$$

Use trial-and-error and synthetic division to find the first rational zero. Try 1.

$$
\begin{array}{r|rrrr}
1 & 2 & 11 & -7 & -6 \\
  &   & 2 & 13 & 6 \\
\hline
  & 2 & 13 & 6 & 0 \\
\end{array}
$$

1 is a rational zero, so

$$2x^3 + 11x^2 - 7x - 6 = (x-1)(2x^2 + 13x + 6)$$
$$= (x-1)(2x+1)(x+6)$$

$$x - 1 = 0 \quad 2x + 1 = 0 \quad x + 6 = 0$$

$$x = 1 \qquad x = -\frac{1}{2} \qquad x = -6$$

The solution set is $\left\{1, -\frac{1}{2}, -6\right\}$.

3. $|2x+4| > 3$

$$2x + 4 < -3 \quad \text{or} \quad 2x + 4 > 3$$

$$2x < -7 \quad \text{or} \qquad 2x > -1$$

$$x < -\frac{7}{2} \quad \text{or} \qquad x > -\frac{1}{2}$$

The solution set is $\left\{x \mid x < -\frac{7}{2} \text{ or } x > -\frac{1}{2}\right\}$.

4. $\cos^2 x + \sin x + 1 = 0, \, 0 \le x < 2\pi$

$$0 = (1 - \sin^2 x) + \sin x + 1$$

$$0 = -\sin^2 x + \sin x + 2$$

$$0 = \sin^2 x - \sin x - 2$$

$$0 = (\sin x - 2)(\sin x + 1)$$

$$\sin x - 2 = 0 \quad \text{or} \quad \sin x + 1 = 0$$

$$\sin x = 2 \quad \text{or} \qquad \sin x = -1$$

$$\text{undefined} \quad \text{or} \qquad x = \sin^{-1}(-1)$$

$$x = \frac{3\pi}{2}$$

The solution set is $\left\{\frac{3\pi}{2}\right\}$.

4. $\log_4(x^2 - 9) - \log_4(x + 3) = 3$

$$\log_4 \frac{x^2 - 9}{x + 3} = 3$$

$$\frac{x^2 - 9}{x + 3} = 4^3$$

$$x^2 - 9 = 64(x + 3)$$

$$x^2 - 9 = 64x + 192$$

$$x^2 - 64x - 201 = 0$$

$$(x - 67)(x + 3) = 0$$

$$x - 67 = 0 \quad x + 3 = 0$$

$$x = 67 \qquad x = 3$$

Because $\log_4((-3)+3) = \log_4 0$ and $\log_4 0$ is undefined, $x = 3$ is an extraneous solution.

The solution set is $\{67\}$.

6. $f(x) = x^3 + x^2 - 12x$

$= x(x^2 + x - 12)$

$= x(x+4)(x-3)$

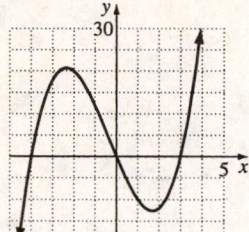

7. $f(x) = \dfrac{2x^2 - 5x + 2}{x^2 - 4}$

$= \dfrac{(2x-1)(x-2)}{(x+2)(x-2)}$

$= \dfrac{2x-1}{x+2}$, where $x \neq 2$

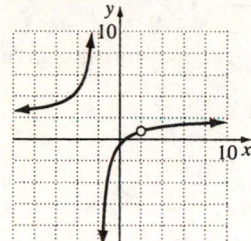

8. $f(x) = \begin{cases} -x+1 & \text{if } -1 \leq x < 1 \\ 2 & \text{if } x = 1 \\ x^2 & \text{if } x > 1 \end{cases}$

Graph the line $f(x) = -x + 1$ from $x = -1$ to $x = 1$ with
a closed dot at $(-1, 2)$ and an open dot at $(1, 0)$.
Graph a closed dot at $(1, 2)$, and graph the
parabola $f(x) = x^2$ on the right side of $x = 1$.

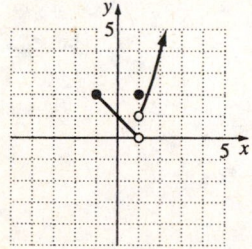

9. $y = 2\sin\left(2x + \dfrac{\pi}{2}\right)$

period $= \dfrac{2\pi}{B} = \dfrac{2\pi}{2} = \pi$

amplitude $= |A| = |2| = 2$

phase shift $= \dfrac{c}{B} = \dfrac{-\frac{\pi}{2}}{2} = -\dfrac{\pi}{2} \cdot \dfrac{1}{2} = -\dfrac{\pi}{4}$

Graph using the key points

$(0,\ 2)$, $\left(\dfrac{\pi}{4},\ 0\right)$, $\left(\dfrac{\pi}{2},\ -2\right)$, $\left(\dfrac{3\pi}{4},\ 0\right)$, and $(\pi,\ 2)$.

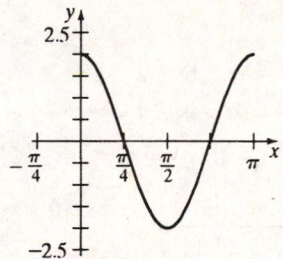

10. $y = \dfrac{1}{2}\sec 2\pi x, \ 0 \leq x \leq 2$

Use the reciprocal function $y = \dfrac{1}{2}\cos 2\pi x$, with key

points

$\left(0, \dfrac{1}{2}\right), \left(\dfrac{1}{4}, 0\right), \left(\dfrac{1}{2}, -\dfrac{1}{2}\right), \left(\dfrac{3}{4}, 0\right)$, and $\left(1, \dfrac{1}{2}\right)$.

Extend the graph one cycle to the right, and draw
vertical asymptotes
at the x-intercepts for guides.

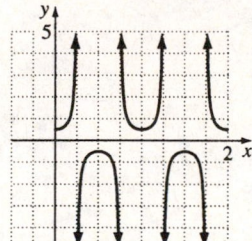

11. $x - 2y \leq 4$

$-2y \leq -x + 4$

$y \geq \dfrac{1}{2}x - 2, x \geq 2$

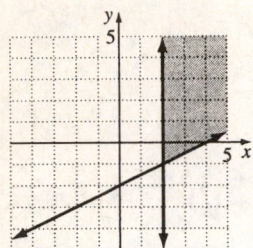

12. $x^2 - 4y^2 - 4x - 24y - 48 = 0$

$x^2 - 4x + 4 - 4(y^2 + 6y + 9) - 48 = 4 - 36$

$(x-2)^2 - 4(y+3)^2 = 16$

$\dfrac{(x-2)^2}{16} - \dfrac{(y+3)^2}{4} = 1$

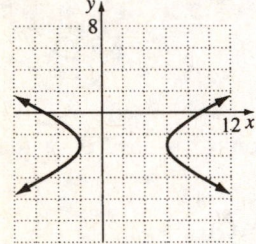

13. $f(x) = \sqrt{x}, g(x) = \sqrt{x-2} + 1$

$g(x)$ is the graph of $f(x)$ shifted 2 units to the right and 1 unit up.

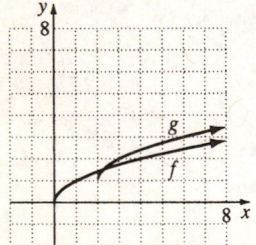

14. $x = 3 \sin t, y = 4 \cos t + 2; 0 \leq t \leq 2\pi$

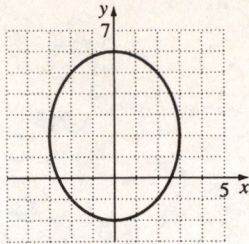

15. $2x^2 + 5xy + 2y^2 - \dfrac{9}{2} = 0$

Use the Amount of Rotation formula.

$\cos 2\theta = \dfrac{A - C}{B} = \dfrac{2 - 2}{5} = 0$

$\theta = \dfrac{1}{2}\cos^{-1} 0 = 45°$

With $\theta = 45°$, the rotation formulas for x and y are

$x = \dfrac{\sqrt{2}}{2}(x' - y')$ and $y = \dfrac{\sqrt{2}}{2}(x' + y')$.

Substituting these into the original equation gives the

hyperbola $\dfrac{x'^2}{1} - \dfrac{y'^2}{9} = 1$.

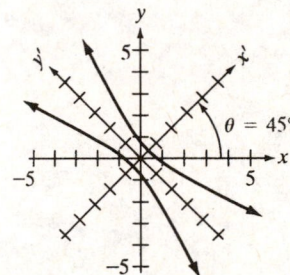

16. $f(x) = -2x^2 + 7x - 1$

$$f'(x) = \lim_{h \to 0} \frac{f(x+h) - f(x)}{h} = \lim_{h \to 0} \frac{-2(x+h)^2 + 7(x+h) - 1 - (-2x^2 + 7x - 1)}{h}$$

$$= \lim_{h \to 0} \frac{-2x^2 - 4hx - 2h^2 + 7x + 7h - 1 + 2x^2 - 7x + 1}{h} = \lim_{h \to 0} \frac{-4hx - 2h^2 + 7h}{h}$$

$$= \lim_{h \to 0} (-4x - 2h + 7) = -4x - 2 \cdot 0 + 7 = -4x + 7$$

17. $f(x) = 7x - 1$

$$x = 7y - 1$$
$$x + 1 = 7y$$
$$y = \frac{1}{7}x + \frac{1}{7}$$
$$f^{-1}(x) = \frac{1}{7}x + \frac{1}{7}$$

18. $\lim_{x \to -3} \dfrac{x^2 + x - 6}{x^2 + 2x - 3} = \lim_{x \to -3} \dfrac{(x+3)(x-2)}{(x+3)(x-1)}$

$$= \lim_{x \to -3} \frac{x-2}{x-1}$$
$$= \frac{-3-2}{-3-1}$$
$$= \frac{5}{4}$$

19. $(x^2 - 3y)^4 = \dfrac{4!}{0!(4-0)!}(x^2)^4 + \dfrac{4!}{1!(4-1)!}(x^2)^3(-3y) + \dfrac{4!}{2!(4-2)!}(x^2)^2(-3y)^2 + \dfrac{4!}{3!(4-3)!}x^2(-3y)^3$

$$+ \frac{4!}{4!(4-4)!}(-3y)^4$$

$$= 1 \cdot x^8 + 4x^6(-3y) + 6x^4 \cdot 9y^2 + 4x^2(-27y^3) + 1 \cdot 81y^4$$

$$= x^8 - 12x^6 y + 54x^4 y^2 - 108x^2 y^3 + 81y^4$$

20. $2x + y - 6 = 0$

$$y = -2x + 6$$

The slope is –2.

$$y - y_1 = m(x - x_1)$$
$$y + 3 = -2(x - 2)$$
$$y + 3 = -2x + 4$$
$$y = -2x + 1$$

21. $\mathbf{v} = -2\mathbf{i} + \mathbf{j}, \mathbf{w} = 4\mathbf{i} - 3\mathbf{j}$

$\mathbf{v} \cdot \mathbf{w} = (-2\mathbf{i} + \mathbf{j}) \cdot (4\mathbf{i} - 3\mathbf{j})$

$\qquad = -2(4) + 1(-3)$

$\qquad = -8 - 3$

$\qquad = -11$

$\cos\theta = \dfrac{\mathbf{v} \cdot \mathbf{w}}{|\mathbf{v}||\mathbf{w}|}$

$\theta = \cos^{-1}\left(\dfrac{\mathbf{v} \cdot \mathbf{w}}{|\mathbf{v}||\mathbf{w}|}\right)$

$\theta = \cos^{-1}\left(\dfrac{-11}{\sqrt{(-2)^2 + 1^2}\sqrt{4^2 + (-3)^2}}\right)$

$\theta = \cos^{-1}\left(\dfrac{-11}{\sqrt{5}\sqrt{25}}\right)$

$\theta = 170°$

22. $\dfrac{1}{x(x^2 + x + 1)} = \dfrac{A}{x} + \dfrac{Bx + C}{x^2 + x + 1}$

$\qquad 1 = A(x^2 + x + 1) + Bx^2 + Cx$

$\qquad 1 = (A + B)x^2 + (A + C)x + A$

$A = 1 \quad A + B = 0 \qquad A + C = 0$

$\qquad\quad B = -A \qquad C = -A$

$\qquad\quad B = -1 \qquad C = -1$

$\dfrac{1}{x(x^2 + x + 1)} = \dfrac{1}{x} - \dfrac{x + 1}{x^2 + x + 1}$

23. $\tan\theta + \cot\theta = \dfrac{\sin\theta}{\cos\theta} + \dfrac{\cos\theta}{\sin\theta}$

$\qquad\qquad = \dfrac{\sin^2\theta + \cos^2\theta}{\cos\theta\sin\theta}$

$\qquad\qquad = \dfrac{1}{\cos\theta\sin\theta}$

$\qquad\qquad = \sec\theta\csc\theta$

24. $\tan(\theta + \pi) = \dfrac{\tan\theta + \tan\pi}{1 - \tan\theta\tan\pi}$

$\qquad\qquad = \dfrac{\tan\theta + 0}{1 - \tan\theta(0)}$

$\qquad\qquad = \tan\theta$

25. $BA = \begin{bmatrix} 1 & 0 \\ 3 & 2 \\ 2 & 1 \end{bmatrix}\begin{bmatrix} 2 & 1 & 3 \\ 1 & -1 & 0 \end{bmatrix}$

$= \begin{bmatrix} 1(2) + 0(1) & 1(1) + 0(-1) & 1(3) + 0(0) \\ 3(2) + 2(1) & 3(1) + 2(-1) & 3(3) + 2(0) \\ 2(2) + 1(1) & 2(1) + 1(-1) & 2(3) + 1(0) \end{bmatrix}$

$= \begin{bmatrix} 2 & 1 & 3 \\ 8 & 1 & 9 \\ 5 & 1 & 6 \end{bmatrix}$

26. $r = 4\sin\theta$

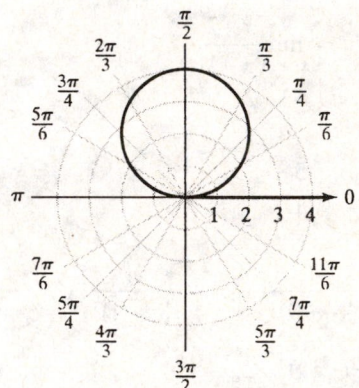

27. $h(x) = (x^2 - 3x + 7)^9$

$h(x) = (f \circ g)(x)$

$f(x) = x^9$

$g(x) = x^2 - 3x + 7$

28. $2x - y - 2z = -1$
    $x - 2y - z = 1$
     $x + y + z = 4$

$$\begin{bmatrix} 2 & -1 & -2 & | & -1 \\ 1 & -2 & -1 & | & 1 \\ 1 & 1 & 1 & | & 4 \end{bmatrix} \frac{1}{2}R_1$$

$$= \begin{bmatrix} 1 & -\frac{1}{2} & -1 & | & -\frac{1}{2} \\ 1 & -2 & -1 & | & 1 \\ 1 & 1 & 1 & | & 4 \end{bmatrix} -R_1 + R_2$$

$$= \begin{bmatrix} 1 & -\frac{1}{2} & -1 & | & -\frac{1}{2} \\ 0 & -\frac{3}{2} & 0 & | & \frac{3}{2} \\ 1 & 1 & 1 & | & 4 \end{bmatrix} -R_1 + R_3$$

$$= \begin{bmatrix} 1 & -\frac{1}{2} & -1 & | & -\frac{1}{2} \\ 0 & -\frac{3}{2} & 0 & | & \frac{3}{2} \\ 0 & \frac{3}{2} & 2 & | & \frac{9}{2} \end{bmatrix} R_2 + R_3$$

$$= \begin{bmatrix} 1 & -\frac{1}{2} & -1 & | & -\frac{1}{2} \\ 0 & -\frac{3}{2} & 0 & | & \frac{3}{2} \\ 0 & 0 & 2 & | & 6 \end{bmatrix} -\frac{2}{3}R_2$$

$$= \begin{bmatrix} 1 & -\frac{1}{2} & -1 & | & -\frac{1}{2} \\ 0 & 1 & 0 & | & -1 \\ 0 & 0 & 2 & | & 6 \end{bmatrix} \frac{1}{2}R_3$$

$$= \begin{bmatrix} 1 & -\frac{1}{2} & -1 & | & -\frac{1}{2} \\ 0 & 1 & 0 & | & -1 \\ 0 & 0 & 1 & | & 3 \end{bmatrix} \frac{1}{2}R_2 + R_1$$

$$= \begin{bmatrix} 1 & 0 & -1 & | & -1 \\ 0 & 1 & 0 & | & -1 \\ 0 & 0 & 1 & | & 3 \end{bmatrix} R_3 + R_1$$

$$= \begin{bmatrix} 1 & 0 & 0 & | & 2 \\ 0 & 1 & 0 & | & -1 \\ 0 & 0 & 1 & | & 3 \end{bmatrix}$$

The solution set is $\{(2, -1, 3)\}$.

29. $\displaystyle\sum_{i=1}^{6} 4(-2)^i$

$$S_6 = \frac{a_1(1 - r^6)}{1 - r} = \frac{4(-2)^1 \left[1 - (-2)^6\right]}{1 - (-2)} = 168$$

30. $\left[\sqrt{2}(\cos 15° + i\sin 15°)\right]^4 = (\sqrt{2})^4 \left[\cos 4(15°) + i\sin 4(15°)\right]$

$$= 4(\cos 60° + i\sin 60°)$$

$$= 4\left(\frac{1}{2} + \frac{\sqrt{3}}{2}i\right)$$

$$= 2 + 2i\sqrt{3}$$

31. $0.08(120,000 - x) + 0.18x = 10,000$

$$9600 - 0.08x + 0.18x = 10,000$$

$$0.1x = 400$$

$$x = 4000$$

$116,000 was loaned at 8% and $4000 was loaned at 18%.

32. $V = 9x^2 = 225$

$$x^2 = 25$$

$$x = 5$$

$x + 2(9) = 5 + 18 = 23$

The dimensions of the sheet metal should be $23\,\text{cm} \times 23\,\text{cm}$.

33. $200 = 2x + y, \quad 200 - 2x = y$

$$A(x) = x(200 - 2x)$$

$$= -2x^2 + 200x$$

$$x = \frac{-200}{2(-2)} = 50$$

$$200 - 2x = 200 - 2(50) = 100$$

$$A(100) = 50(200 - 2(50)) = 5000$$

The dimensions are 100 feet by 50 feet and the maximum area is 5000 square feet.

34. $T = C + (T_0 - C)e^{kt}$

a. $75°\text{F} = 72°\text{F} + (375°\text{F} - 72°\text{F})e^{k(60 \text{ min})}$

$$3°\text{F} = (303°\text{F})e^{k(60 \text{ min})}$$

$$\ln\left(\frac{3°\text{F}}{303°\text{F}}\right) = \ln e^{k(60 \text{ min})}$$

$$k = \frac{1}{60 \text{ min}}\ln\left(\frac{1}{101}\right)$$

$$\approx -0.0769 \text{ (min)}^{-1}$$

b. $250\,°\text{F}=72\,°\text{F}+(303\,°\text{F})e^{[-0.0769\ (\text{min})^{-1}]t}$

$178\,°\text{F}=(303\,°\text{F})e^{[-0.0769\ (\text{min})^{-1}]t}$

$\ln\left(\dfrac{178\,°\text{F}}{303\,°\text{F}}\right)=\ln e^{[-0.0769\ (\text{min})^{-1}]t}$

$t=\dfrac{\text{min}}{-0.0769}\ln\left(\dfrac{178}{303}\right)$

$\approx 6.9\ \text{min}$

The temperature of the pie will reach 250° F in approximately 6.9 minutes.

35. $6000=xy$

$\dfrac{6000}{x}=y$

$C=25x+5(2y)$

$=25x+10\left(\dfrac{6000}{x}\right)$

$=25x+\dfrac{60,000}{x}$

36.   Set the origin at the harbor with the y-axis pointing north.
Final position of the first ship:

$x_1=r\cos\theta=(23)\cos(90°-42°)=23\cos 48°\approx 15.4$

$y_1=r\sin\theta$

$\quad=23\sin 48°$

$\quad\approx 17.1$

Final position of the second ship:

$x_2=r\cos\theta$

$\quad=(72)\cos(90°+38°)$

$\quad=72\cos 128°$

$\quad\approx -44.3$

$y_2=r\sin\theta$

$\quad=72\sin 128°$

$\quad\approx 56.7$

$d=\sqrt{(x_1-x_2)^2+(y_1-y_2)^2}$

$\quad=\sqrt{(15.4+44.3)^2+(17.1-56.7)^2}$

$\quad\approx 71.6$

The two ships are approximately 71.6 miles apart.

37. $V = \dfrac{k}{P}$

$40 = \dfrac{k}{22}$

$k = 22(40)$

$k = 880$

$V = \dfrac{880}{30} = 29\dfrac{1}{3}$

The volume of the gas is $29\dfrac{1}{3}$ cubic inches.

38. $s(t) = -16t^2 + 40t$

$\begin{aligned}
s'(t) &= \lim_{h \to 0} \frac{s(t+h) - s(t)}{h} \\
&= \lim_{h \to 0} \frac{-16(t+h)^2 + 40(t+h) - (-16t^2 + 40t)}{h} \\
&= \lim_{h \to 0} \frac{-16t^2 - 32ht - 16h^2 + 40t + 40h + 16t^2 - 40t}{h} \\
&= \lim_{h \to 0} \frac{-32ht - 16h^2 + 40h}{h} \\
&= \lim_{h \to 0} (-32t - 16h + 40) \\
&= -32t - 16 \cdot 0 + 40 \\
&= -32t + 40
\end{aligned}$

$s'(2) = -32(2) + 40 = -24$

The instantaneous velocity of the ball 2 seconds after it was thrown is –24 feet per second.

39. $V = lwh$

$4 = x \cdot x \cdot y$

$\dfrac{4}{x^2} = y$

$A(x) = x^2 + 4xy$

$A(x) = x^2 + 4x\left(\dfrac{4}{x^2}\right)$

$A(x) = x^2 + \dfrac{16}{x}$

40. $f(x) = -2.32x^2 + 76.58x - 559.87, 12 \le x \le 17$

Set $f(x) = 70$.

$70 = -2.32x^2 + 76.58x - 559.87$

$0 = -2.32x^2 + 76.58x - 629.87$

Use the quadratic formula to solve for x.

$$x = \frac{-B \pm \sqrt{B^2 - 4AC}}{2A}$$

$$= \frac{-76.58 \pm \sqrt{76.58^2 - 4(-2.32)(-629.87)}}{2(-2.32)}$$

≈ 15.6 or 17.5 (17.5 is outside the domain.)

Seventy percent of U.S. students 15.6 years old say that their school is not drug free.